Contents

3 The Classical Superconductors

4 Thermodynamic Properties

5 Ginzburg–Landau Theory

6 *BCS Theory*

7 *Perovskite and Cuprate Crystallographic Structures*

8 Hubbard Models and Band Structure

9 Type II Superconductivity

10 *Magnetic Properties*

11 *Intermediate and Mixed States*

12 Critical States

13 Tunneling

14 *Transport Properties*

15 *Spectroscopic Properties*

Preface

When we wrote our 1988 book, *Cooper Oxide Superconductors*, our aim was to present an early survey of the experimental aspects of the field of high temperature superconductivity as an aid to researchers who were then involved in the worldwide effort to (a) understand the phenomenon of cuprate superconductivity and (b) search for ways to raise the critical temperature and produce materials suitable for the fabrication of magnets and other devices. A great deal of experimental data are now available on the cuprates, and their superconducting properties have been well characterized using high quality untwinned monocrystals and epitaxial thin films. Despite this enormous research effort, the underlying mechanisms responsible for the superconducting properties of the cuprates are still open to question. Nevertheless, we believe that the overall picture is now clear enough to warrant the writing of a textbook that presents our present-day understanding of the nature of the phenomenon of superconductivity, surveys the properties of various known superconductors, and shows how these properties fit into various theoretical frameworks. The aim is to present this material in a format suitable for use in a graduate-level course.

An introduction to superconductivity must be based on a background of fundamental principles found in standard solid state physics texts, and a brief introductory chapter provides this background. This initial chapter on the properties of normal conductors is limited to topics that are often referred to throughout the remainder of the text: electrical conductivity, magnetism, specific heat, etc. Other background material specific to particular topics is provided in the appropriate chapters. The presence of the initial normal state chapter makes the remainder of the book more coherent.

The second chapter presents the essential features of the superconducting state—the phenomena of zero resistance and perfect diamagnetism. Super current flow, the accompanying magnetic fields, and the transition to this ordered state that occurs at the transition temperature T_c are described. The third chapter surveys the properties of the various classes of superconductors, including the organics, the buckministerfullerenes, and the precursors to the cuprates, but not the high temperature superconductors themselves. Numerous tables and figures summarize the properties of these materials.

Having acquired a qualitative understanding of the nature of superconductivity, we now proceed, in five subsequent chapters, to describe various theoretical frameworks which aid in understanding the facts about superconductors. Chapter 4 discusses superconductivity from the viewpoint of thermodynamics and provides expressions for the free energy—the thermodynamic function that constitutes the starting point for the formulations of both the Ginzburg–Landau (GL) and the BCS theories. The GL theory is developed in Chapter 5 and the BCS theory in Chapter 6. GL is a readily understandable phenomenological theory that provides results that are widely used in the interpretation of experimental data, and BCS is a more fundamental, and mathematically challenging, theory that makes predictions that are often checked against experimental results. Most of Chapter 5 is essential reading, whereas much of the formalism of Chapter 6 can be skimmed during a first reading.

The theoretical treatment is interrupted by Chapter 7, which presents the details of the structures of the high temperature superconductors. This constitutes important background material for the band theory sections of Chapter 8, which also presents the Hubbard and related models, such as RVB and t–J. In addition, Chapter 8 covers other theoretical approaches involving, for example, spinons, holons, slave bosons, anyons, semions, Fermi liquids, charge and spin density waves, spin bags, and the Anderson interlayer tunneling scheme. This completes the theoretical aspects of the field, except for the additional description of critical state models such as the Bean model in Chapter 12. The Bean model is widely used for the interpretation of experimental results.

The remainder of the text covers the magnetic, transport, and other properties of superconductors. Most of the examples in these chapters are from the literature on the cuprates. Chapter 9 introduces Type II superconductivity and describes magnetic properties, Chapter 10 continues the discussion of magnetic properties, Chapter 11 covers the intermediate and mixed states, and Chapter 12, on critical state models, completes the treatment of magnetic properties. The next two chapters are devoted to transport properties. Chapter 13 covers various types of tunneling and the Josephson effect, and Chapter 14 presents the remaining transport properties involving the Peltier, Seebeck, Hall, and other effects.

When the literature was surveyed in preparation for writing this text, it became apparent that a very significant percentage of current research on superconductivity is being carried out by spectroscopists, and to accommodate this, Chapter 15 on spectroscopy was added. This chapter lets the reader know what the individual branches of spectroscopy can reveal about the properties of superconductors, and in addition, it provides an entrée to the vast literature on the subject.

This book contains extensive tabulations of experimental data on various superconductors, classical as well as high T_c types. Figures from research articles were generally chosen because they exemplify principles described in the text. Some other figures, particularly those in Chapter 3, provide correlations of extensive data on many samples. There are many cross-references between the chapters to show how the different topics fit together as one unified subject.

Most chapters end with sets of problems that exemplify the material presented and sets of references for additional reading on the subject. Other literature citations are scattered throughout the body of each chapter. Occasional reference is made to our earlier work, *Copper Oxide Superconductors*, for supplementary material.

One of us (C.P.P.) taught a graduate-level superconductivity course three times using lecture notes which eventually evolved into the present text. It was exciting to learn with the students while teaching the course and simultaneously doing research on the subject.

We thank the following individuals for their helpful discussions and comments on the manuscript: C. Almasan, S. Aktas, D. Castellanos, T. Datta, N. Fazyleev, J. B. Goodenough, K. E. Gray, D. U. Gubser, D. R. Harshman, A. M. Herman, Z. Iqbal, E. R. Jones, A. B. Kaiser, D. Kirvin, O. Lopez, M. B. Maple, A. P. Mills, Jr., S. Misra, F. J. Owens, M. Pencarinha, A. Petrile, W. E. Pickett, S. J. Poon, A. W. Sleight, O. F. Schuette, C. Sisson, David B. Tanner, H. Testardi, C. Uher, T. Usher, and S. A. Wolf. We also thank the graduate students of the superconductivity classes for their input, which improved the book's presentation. We appreciate the assistance given by the University of South Carolina (USC) Physics Department; our chairman, F. T. Avignone; the secretaries, Lynn Waters and Cheryl Stocker; and especially by Gloria Phillips, who is thanked for her typing and multiple emendations of the BCS chapter and the long list of references. Eddie Josie of the USC Instructional Services Department ably prepared many of the figures.

1

Properties of the Normal State

I. INTRODUCTION

This text is concerned with the phenomenon of superconductivity, a phenomenon characterized by certain electrical, magnetic, and other properties, many of which will be introduced in the following chapter. A material becomes superconducting below a characteristic temperature, called the *superconducting transition temperature* T_c, which varies from very small values (millidegrees or microdegrees) to values above 100 K. The material is called normal above T_c, which merely means that it is not superconducting. Elements and compounds that become superconductors are conductors—but not good conductors—in their normal state. The good conductors, such as copper, silver, and gold, do not superconduct.

It will be helpful to survey some properties of normal conductors before discussing the superconductors. This will permit us to review some background material and to define some of the terms that will be used throughout the text. Many of the normal state properties that will be discussed here are modified in the superconducting state. Much of the material in this introductory chapter will be referred to later in the text.

II. CONDUCTION ELECTRON TRANSPORT

The electrical conductivity of a metal may be described most simply in terms of the constituent atoms of the metal. The atoms, in this representation, lose their valence electrons, causing a background lattice of positive ions, called *cations*, to

Table 1.1 Characteristics of Selected Metallic Elements[a]

Z	Element	Valence	Radius (Å)	Xtal type	a (Å)	n_e $\left(\frac{10^{22}}{\text{cm}^3}\right)$	r_s (Å)	ρ, 77 K (μΩ cm)	ρ, 273 K (μΩ cm)	τ, 77 K (fs)	τ, 273 K (fs)	K_{th} $\left(\frac{\text{W}}{\text{cm K}}\right)$
11	Na	1	0.97	bcc	4.23	2.65	2.08	0.8	4.2	170	32	1.38
19	K	1	1.33	bcc	5.23	1.40	2.57	1.38	6.1	180	41	1.0
29	Cu	1	0.96	fcc	3.61	8.47	1.41	0.2	1.56	210	27	4.01
47	Ag	1	1.26	fcc	4.09	5.86	1.60	0.3	1.51	200	40	4.28
41	Nb	1	1.0	bcc	3.30	5.56	1.63	3.0	15.2	21	4.2	0.52
20	Ca	2	0.99	fcc	5.58	4.61	1.73		3.43		22	2.06
38	Sr	2	1.12	fcc	6.08	3.55	1.89	7	23	14	4.4	≈ 0.36
56	Ba	2	1.34	bcc	5.02	3.15	1.96	17	60	6.6	1.9	≈ 0.19
13	Al	3	0.51	fcc	4.05	18.1	1.10	0.3	2.45	65	8.0	2.36
81	Tl	3	0.95	bcc	3.88	10.5	1.31	3.7	15	9.1	2.2	0.5
50	Sn(W)	4	0.71	tetrg	$a = 5.82$ $c = 3.17$	14.8	1.17	2.1	10.6	11	2.3	0.64
82	Pb	4	0.84	fcc	4.95	13.2	1.22	4.7	19.0	5.7	1.4	0.38
51	Sb	5	0.62	rhomb	4.51	16.5	1.19	8	39	2.7	0.55	0.18
83	Bi	5	0.74	rhomb	4.75	14.1	1.13	35	107	0.72	0.23	0.09

[a] Notation: a, lattice constant; n_e, conduction electron density; $r_s = (3/4\pi n_e)^{1/3}$; ρ, resistivity; τ, Drude relaxation time; K_{th}, thermal conductivity; $L = \rho K_{th}/T$ is the Lorentz number; γ, electronic specific heat parameter; m^*, effective mass; R_H, Hall constant; Θ_D, Debye temperature; ω_p, plasma frequency in radians per femtosecond (10^{-15} s); IP, first ionization potential; WF, work function; E_F, Fermi energy; T_F, Fermi temperature in kilokelvins; k_F, Fermi wavenumber in mega reciprocal centimeters; and v_F, Fermi velocity in centimeters per microsecond.

form, and the now delocalized conduction electrons move between these ions. The number density n (electrons/cm^3) of conduction electrons in a metallic element of density ρ_m (g/cm^3), atomic mass number A (g/mole), and valence Z is given by

$$n = \frac{N_A Z \rho_m}{A}, \tag{1.1}$$

where N_A is Avogadro's number. The typical values listed in Table 1.1 are a thousand times greater than those of a gas at room temperature and atmospheric pressure.

The simplest approximation that we can adopt as a way of explaining conductivity is the Drude model. In the Drude model it is assumed that the conduction electrons

1. do not interact with the cations ("free-electron approximation") except when one of them collides elastically with a cation which happens, on average, $1/\tau$ times per second, with the result that the velocity v of the electron abruptly and randomly changes its direction ("relaxation-time approximation");
2. maintain thermal equilibrium through collisions, in accordance with Maxwell–Boltzmann statistics ("classical-statistics approximation");
3. do not interact with each other ("independent-electron approximation").

This model predicts many of the general features of electrical conduction phenomena, as we shall see later in the chapter, but it fails to account for many others, such as tunneling, band gaps, and the Bloch T^5 law. More satisfactory explanations of electron transport relax or discard one or more of these approximations.

Ordinarily, one abandons the free-electron approximation by having the electrons move in a periodic potential arising from the background lattice of positive ions. Figure 1.1 gives an example of a simple potential that is negative near the positive ions and zero between them. An electron moving through the lattice interacts with the surrounding positive ions, which are oscillating about their equilib-

L $\left(\frac{\mu\Omega\ \mathrm{W}}{\mathrm{K}^2}\right)$	γ $\left(\frac{\mathrm{mJ}}{\mathrm{mole\ K}^2}\right)$	$\frac{m^*}{m_e}$	$\frac{1}{R_H ne}$	Θ_D (K)	ω_p $\left(\frac{\mathrm{rad}}{\mathrm{fs}}\right)$	IP (eV)	WF (eV)	E_F (eV)	T_F (kK)	k_F (M cm^{-1})	v_F $\left(\frac{\mathrm{cm}}{\mu\mathrm{s}}\right)$
0.021	1.5	1.3	−1.1	150	8.98	5.14	2.75	3.24	37.7	92	107
0.022	2.0	1.2	−1.1	100	5.98	4.34	2.3	2.12	24.6	75	86
0.023	0.67	1.3	−1.4	310	3.85	7.72	4.6	7.0	81.6	136	157
0.023	0.67	1.1	−1.2	220		7.57	4.3	5.49	63.8	120	139
0.029	8.4	12		265		6.87	4.3	5.32	61.8	118	137
0.026	2.7	1.8	−0.76	230		6.11	2.9	4.69	54.4	111	128
0.030	3.6	2.0		150		5.69	2.6	3.93	45.7	102	118
0.042	2.7	1.4		110		5.21	2.7	3.64	42.3	98	113
0.021	1.26	1.4	+1.0	394	14.5	5.99	4.3	11.7	136	175	203
0.028	1.5	1.1		96		6.11	3.8	8.15	94.6	146	169
0.025	1.8	1.3		170		7.34	4.4	10.2	118	164	190
0.026	2.9	1.9		88		7.41	4.3	9.47	110	158	183
0.026	0.63	0.38		200		8.64	4.6	10.9	127	170	196
0.035	0.084	0.047		120		7.29	4.2	9.90	115	161	187

rium positions, and the charge distortions resulting from this interaction propagate along the lattice, causing distortions in the periodic potential. These distortions can influence the motion of yet another electron some distance away that is also interacting with the oscillating lattice. Propagating lattice vibrations are called *phonons*, so that this interaction is called the *electron–phonon interaction*. We will see later that two electrons interacting with each other through the intermediary phonon can form bound states and that the resulting bound electrons, called *Cooper pairs*, become the carriers of the super current.

The classical statistics assumption is generally replaced by the Sommerfeld approach. In this approach the electrons are

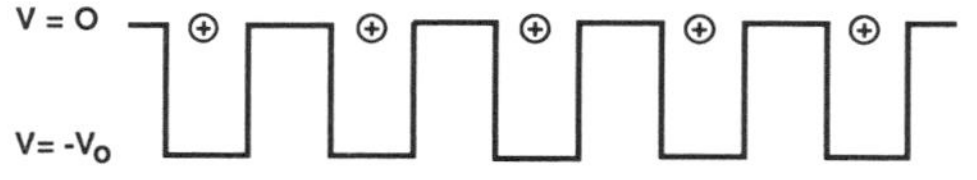

Figure 1.1 Muffin tin potential has a constant negative value $-V_0$ near each positive ion and is zero in the region between the ions.

assumed to obey Fermi–Dirac statistics with distribution function

$$f_0(\mathbf{v}) = \frac{1}{\exp\left[(mv^2/2 - \mu)/k_B T\right] + 1}, \tag{1.2}$$

(see the discussion in Section IX, where k_B is Boltzmann's constant, and the constant μ is called the *chemical potential*). In Fermi–Dirac statistics, noninteracting conduction electrons are said to constitute a Fermi gas. The chemical potential is the energy required to remove one electron from this gas under conditions of constant volume and constant entropy.

The relaxation time approximation assumes that the distribution function $f(\mathbf{v}, t)$ is time dependent and that when $f(\mathbf{v}, t)$ is disturbed to a nonequilibration configuration f^{col}, collisions return it back to its equilibrium state f^0 with time constant τ in accordance with the expression

$$\frac{df}{dt} = -\frac{f^{\mathrm{col}} - f^0}{\tau}. \tag{1.3}$$

Ordinarily, the relaxation time τ is assumed to be independent of the velocity, resulting in a simple exponential return to equilibrium:

$$f(\mathbf{v},t) = f^0(\mathbf{v}) + [f^{\mathrm{col}}(\mathbf{v}) - f^0(\mathbf{v})]e^{-t/\tau}. \quad (1.4)$$

In systems of interest $f(\mathbf{v},t)$ always remains close to its equilibrium configuration (1.2). A more sophisticated approach to collision dynamics makes use of the Boltzmann equation, and this is discussed in texts in solid state physics (e.g., Ashcroft and Mermin, 1976; Burns, 1985; Kittel, 1976) and statistical mechanics (e.g., Reif, 1965).

It is more realistic to waive the independent-electron approximation by recognizing that there is Coulomb repulsion between the electrons. In the following section, we will show that electron screening makes electron–electron interaction negligibly small in good conductors. The use of the Hartree–Fock method to calculate the effects of this interaction is too complex to describe here; it will be briefly discussed in Chapter 8, Section XVII.

When a method developed by Landau (1957a, b) is employed to take into account electron–electron interactions so as to ensure a one-to-one correspondence between the states of the free electron gas and those of the interacting electron system, the conduction electrons are said to form a Fermi liquid. Due to the Pauli exclusion principle, momentum-changing collisions occur only in the case of electrons at the Fermi surface. In what are called *marginal Fermi liquids* the one-to-one correspondence condition breaks down at the Fermi surface. Chapter 8, Section XVII provides a brief discussion of the Fermi liquid and the marginal Fermi liquid approaches to superconductivity.

III. CHEMICAL POTENTIAL AND SCREENING

Ordinarily, the chemical potential μ is close to the Fermi energy E_F and the conduction electrons move at speeds v_F corresponding to kinetic energies $\frac{1}{2}mv_F^2$ close to $E_F = k_B T_F$. Typically, $v_F \approx 10^6$ m/s for good conductors, which is 1/300 the speed of light; perhaps one-tenth as great in the case of high-temperature superconductors and $A15$ compounds in their normal state. If we take τ as the time between collisions, the mean free path l, or average distance traveled between collisions, is

$$l = v_F \tau. \quad (1.5)$$

For aluminum the mean free path is 1.5×10^{-8} m at 300 K, 1.3×10^{-7} m at 77 K, and 6.7×10^{-4} m at 4.2 K.

To see that the interactions between conduction electrons can be negligible in a good conductor, consider the situation of a point charge Q embedded in a free electron gas with unperturbed density n_0. This negative charge is compensated for by a rigid background of positive charge, and the delocalized electrons rearrange themselves until a static situation is reached in which the total force density vanishes everywhere. In the presence of this weak electrostatic interaction the electrons constitute a Fermi liquid.

The free energy F in the presence of an external potential is a function of the local density $n(r)$ of the form

$$F[n] = F_0[n] - e\int n(r)\Phi(r)\, d^3r, \quad (1.6)$$

where $\Phi(r)$ is the electric potential due to both the charge Q and the induced screening charge and $F_0[n]$ is the free energy of a noninteracting electron gas with local density n. Taking the functional derivative of $F[n]$ we have

$$\frac{\delta F[n]}{\delta n(r)} = \mu_0(r) - e\Phi(r) \quad (1.7)$$

$$= \mu, \quad (1.8)$$

where $\mu_0(r)$ is the local chemical potential of the free electron gas in the absence of charge Q and μ is a constant. At zero temperature, which is a good approximation because $T \ll T_F$, the local chemical potential is

$$\mu_0 = \frac{\hbar^2}{2m}(3\pi^2 n)^{2/3}. \tag{1.9}$$

Solving this for the density of the electron gas, we have

$$n(r) = \frac{1}{3\pi^2}\left\{\frac{2m}{\hbar^2}[\mu + e\Phi(r)]\right\}^{3/2}. \tag{1.10}$$

Typically the Fermi energy is much greater than the electrostatic energy so Eq. (1.10) can be expanded about $\Phi = 0$ to give

$$n(r) = n_0\left(1 + \frac{3}{2}\cdot\frac{e\Phi}{\mu}\right), \tag{1.11}$$

where $n_0 = [2m\mu/\hbar^2]^{3/2}/3\pi^2$. The total induced charge density is then

$$\begin{aligned}\rho_i(r) &= e[n_0 - n(r)]\\ &= -\frac{3}{2}\cdot\frac{n_0 e^2\Phi(r)}{\mu}.\end{aligned} \tag{1.12}$$

Poisson's equation for the electric potential can be written as

$$\nabla^2\Phi(r) - \lambda_{sc}^{-2}\Phi(r) = -4\pi Q\delta(r), \tag{1.13}$$

where the characteristic distance λ_{sc}, called the *screening length*, is given by

$$\lambda_{sc}^2 = \frac{1}{6\pi}\cdot\frac{\mu}{n_0 e^2}. \tag{1.14}$$

Equation (1.13) has the well-known Yukawa solution

$$\rho_i(r) = -\frac{Q}{r}e^{-r/\lambda_{sc}}. \tag{1.15}$$

Note that at large distances the potential of the charge falls off exponentially, and that the characteristic distance λ_{sc} over which the potential is appreciable decreases with the electron density. In good conductors the screening length can be quite short, and this helps to explain why electron–electron interaction is negligible. Screening causes the Fermi liquid of conduction electrons to act like a Fermi gas.

IV. ELECTRICAL CONDUCTIVITY

When a potential difference exists between two points along a conducting wire, a uniform electric field E is established along the axis of the wire. This field exerts a force $F = -eE$ that accelerates the electrons:

$$-eE = m\left(\frac{dv}{dt}\right), \tag{1.16}$$

and during a time t that is on the order of the collision time τ the electrons attain a velocity

$$v = -\left(\frac{eE}{m}\right)\tau. \tag{1.17}$$

The electron motion consists of successive periods of acceleration interrupted by collisions, and, on average, each collision reduces the electron velocity to zero before the start of the next acceleration.

To obtain an expression for the current density $\mathbf{J}$,

$$\mathbf{J} = ne\mathbf{v}_{av}, \tag{1.18}$$

we assume that the average velocity $\mathbf{v}_{av}$ of the electrons is given by Eq. (1.17), so we obtain

$$\mathbf{J} = \left(\frac{ne^2\tau}{m}\right)\mathbf{E}. \tag{1.19}$$

The dc electrical conductivity σ_0 is defined by Ohm's law,

$$\mathbf{J} = \sigma_0\mathbf{E} \tag{1.20}$$

$$= \frac{\mathbf{E}}{\rho_0}, \tag{1.21}$$

where $\rho_0 = 1/\sigma_0$ is the resistivity, so from Eq. (1.19) we have

$$\sigma_0 = \frac{ne^2\tau}{m}. \tag{1.22}$$

We infer from the data in Table 1.1 that metals typically have room temperature resistivities between 1 and 100 $\mu\Omega$ cm. Semiconductor resistivities have values from 10^4 to 10^{15} $\mu\Omega$ cm, and for insulators the resistivities are in the range from 10^{20} to 10^{28} $\mu\Omega$ cm.

Collisions can arise in a number of ways, for example, from the motion of atoms away from their regular lattice positions due to thermal vibrational motion—the dominant process in pure metals at high temperatures (e.g., 300 K), or from the presence of impurities or lattice imperfections, which is the dominant scattering process at low temperatures (e.g., 4 K). We see from a comparison of the data in columns 11 and 12 of Table 1.1 that for metallic elements the collision time decreases with temperature so that the electrical conductivity also decreases with temperature, the latter in an approximately linear fashion. The relaxation time τ has the limiting temperature dependences

$$\tau \approx \begin{cases} T^{-3} & T \ll \Theta_D \\ T^{-1} & T \gg \Theta_D, \end{cases} \tag{1.23}$$

as shown in Figure 1.2; here Θ_D is the Debye temperature. We will see in Section VI that, for $T \ll \Theta_D$, an additional phonon scattering correction factor must be taken into account in the temperature dependence of σ_0.

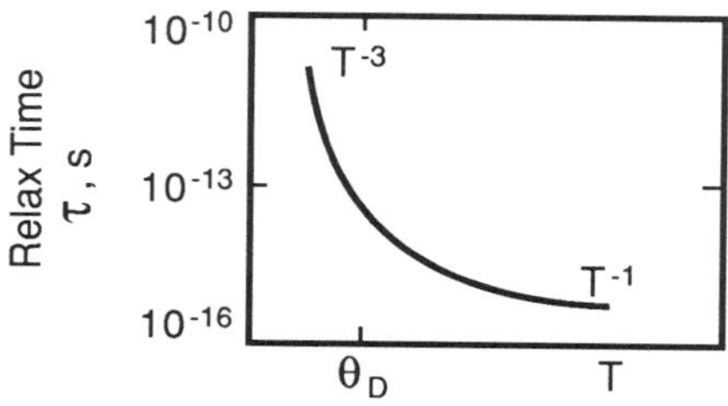

Figure 1.2 Typical temperature dependence of the conduction electron relaxation time τ.

V. FREQUENCY DEPENDENT ELECTRICAL CONDUCTIVITY

When a harmonically varying electric field $E = E_0 e^{-i\omega t}$ acts on the conduction electrons, they are periodically accelerated in the forward and backward directions as E reverses sign every cycle. The conduction electrons also undergo random collisions with an average time τ between the collisions. The collisions, which interrupt the regular oscillations of the electrons, may be taken into account by adding a frictional damping term $\mathbf{p}/\tau$ to Eq. (1.16),

$$\frac{d\mathbf{p}}{dt} + \frac{\mathbf{p}}{\tau} = -e\mathbf{E}, \tag{1.24}$$

where $\mathbf{p} = m\mathbf{v}$ is the momentum. The momentum has the same harmonic time variation, $\mathbf{p} = m\mathbf{v}_0 e^{-i\omega t}$. If we substitute this into Eq. (1.24) and solve for the velocity $\mathbf{v}_0$, we obtain

$$\mathbf{v}_0 = \frac{-e\mathbf{E}_0}{m} \cdot \frac{\tau}{1 - i\omega\tau}. \tag{1.25}$$

Comparing this with Eqs. (1.18) and (1.22) with $\mathbf{v}_0$ playing the role of $\mathbf{v}_{av}$ gives us the ac frequency dependent conductivity:

$$\sigma = \frac{\sigma_0}{1 - i\omega\tau}. \tag{1.26}$$

This reduces to the dc case of Eq. (1.22) when the frequency is zero.

When $\omega\tau \ll 1$ many collisions occur during each cycle of the E field, and the average electron motion follows the oscillations. When $\omega\tau \gg 1$, E oscillates more rapidly than the collision frequency, Eq. (1.24) no longer applies, and the electrical conductivity becomes predominately imaginary, corresponding to a reactive impedance. For very high frequencies, the collision rate becomes unimportant and the electron gas behaves like a plasma, an electrically neutral ionized gas in which the negative charges are mobile electrons and the positive charges are fixed in position. Electromagnetic wave phenomena can be described in terms of the frequency-dependent dielectric constant $\epsilon(\omega)$,

$$\epsilon(\omega) = \epsilon_0\left(1 - \frac{\omega_p^2}{\omega^2}\right), \qquad (1.27)$$

where ω_p is the plasma frequency,

$$\omega_p = \left(\frac{ne^2}{\epsilon_0 m}\right)^{1/2}. \qquad (1.28)$$

Thus ω_p is the characteristic frequency of the conduction electron plasma below which the dielectric constant is negative—so electromagnetic waves cannot propagate—and above which ϵ is positive and propagation is possible. As a result metals are opaque when $\omega < \omega_p$ and transparent when $\omega > \omega_p$. Some typical plasma frequencies $\omega_p/2\pi$ are listed in Table 1.1. The plasma wavelength can also be defined by setting $\lambda_p = 2\pi c/\omega_p$.

VI. ELECTRON–PHONON INTERACTION

We will see later in the text that for most superconductors the mechanism responsible for the formation of Cooper pairs of electrons, which carry the supercurrent, is electron–phonon interaction. In the case of normal metals, thermal vibrations disturb the periodicity of the lattice and produce phonons, and the interactions of these phonons with the conduction electrons cause the latter to scatter. In the high-temperature region ($T \gg \Theta_D$), the number of phonons in the normal mode is proportional to the temperature (cf. Problem 6). Because of the disturbance of the conduction electron flow caused by the phonons being scattered, the electrical conductivity is inversely proportional to the temperature, as was mentioned in Section IV.

At absolute zero the electrical conductivity of metals is due to the presence of impurities, defects, and deviations of the background lattice of positive ions from the condition of perfect periodicity. At finite but low temperatures, $T \ll \Theta_D$, we know from Eq. (1.23) that the scattering rate $1/\tau$ is proportional to T^3. The lower the temperature, the more scattering in the forward direction tends to dominate, and this introduces another T^2 factor, giving the Bloch T^5 law,

$$\sigma \approx T^{-5} \qquad T \ll \Theta_D, \qquad (1.29)$$

which has been observed experimentally for many metals.

Standard solid-state physics texts discuss Umklapp processes, phonon drag, and other factors that cause deviations from the Bloch T^5 law, but these will not concern us here. The texts mentioned at the end of the chapter should be consulted for further details.

VII. RESISTIVITY

Electrons moving through a metallic conductor are scattered not only by phonons but also by lattice defects, impurity atoms, and other imperfections in an otherwise perfect lattice. These impurities produce a temperature-independent contribution that places an upper limit on the overall electrical conductivity of the metal.

According to Matthiessen's rule, the conductivities arising from the impurity and phonon contributions add as reciprocals; that is, their respective individual resistivities, ρ_0 and ρ_{ph}, add to give the total resistivity

$$\rho(T) = \rho_0 + \rho_{ph}(T). \qquad (1.30)$$

We noted earlier that the phonon term $\rho_{ph}(T)$ is proportional to the temperature T at high temperatures and to T^5 via the Bloch law (1.29) at low temperatures. This means that, above room temperature, the impurity contribution is negligible, so that

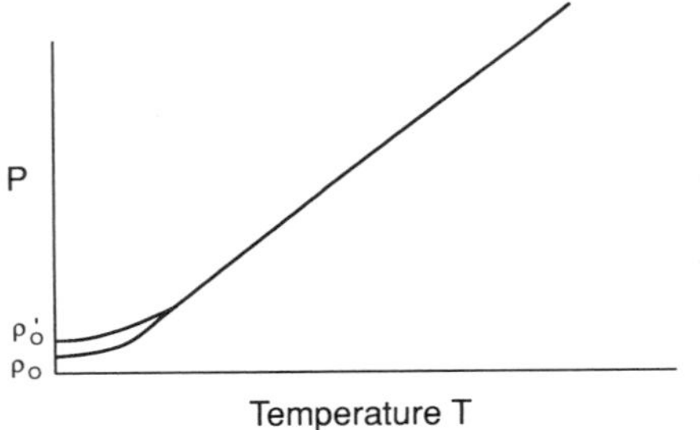

Figure 1.3 Temperature dependence of the resistivity ρ of a pure (ρ_0) and a less pure conductor. Impurities limit the zero temperature resistivity (ρ_0') in the latter case.

the resistivity of metallic elements is roughly proportional to the temperature:

$$\rho(T) \approx \rho(300\ \mathrm{K})\left[\frac{T}{300}\right] \qquad 300\ \mathrm{K} < T. \tag{1.31}$$

At low temperatures far below the Debye temperature, the Bloch T^5 law applies to give

$$\rho(T) = \rho_0 + AT^5 \qquad T \ll \Theta_D. \tag{1.32}$$

Figure 1.3 shows the temperature dependence of the resistivity of a high-purity (low ρ_0) and a lower-purity (larger ρ_0') good conductor.

Typical resistivities at room temperature are 1.5 to 2 $\mu\Omega$ cm for very good conductors (e.g., Cu), 10 to 100 for poor conductors, 300 to 10,000 for high-temperature superconducting materials, 10^4 to 10^{15} for semiconductors, and 10^{20} to 10^{28} for insulators. We see from Eqs. (1.31) and (1.32) that metals have a positive temperature coefficient of resistivity, which is why metals become better conductors at low temperature. In contrast, the resistivity of a semiconductor has a negative temperature coefficient, so that it increases with decreasing temperature. This occurs because of the decrease in the number of mobile charge carriers that results from the return of thermally excited conduction electrons to their ground states on donor atoms or in the valence band.

VIII. THERMAL CONDUCTIVITY

When a temperature gradient exists in a metal, the motion of the conduction electrons provides the transport of heat (in the form of kinetic energy) from hotter to cooler regions. In good conductors such as copper and silver this transport involves the same phonon collision processes that are responsible for the transport of electric charge. Hence these metals tend to have the same thermal and electrical relaxation times at room temperature. The ratio $K_{th}/\sigma T$, in which both thermal (K_{th}, J cm^{-1} s^{-1} K^{-1}) and electrical (σ, Ω^{-1} cm^{-1}) conductivities occur (see Table 1.1 for various metallic elements), has a value which is about twice that predicted by the law of Wiedermann and Franz,

$$\frac{K_{th}}{\sigma T} = \frac{3}{2}\left(\frac{k_B}{e}\right)^2 \tag{1.33}$$

$$= 1.11 \times 10^{-8}\ \mathrm{W\Omega/K^2}, \tag{1.34}$$

where the universal constant $\frac{3}{2}(k_B/e)^2$ is called the Lorenz number.

IX. FERMI SURFACE

Conduction electrons obey Fermi–Dirac statistics. The corresponding F–D distribution function (1.2), written in terms of the energy E,

$$f(E) = \frac{1}{\exp[(E-\mu)/k_BT]+1}, \tag{1.35}$$

is plotted in Fig. 1.4a for $T = 0$ and in Fig. 1.4b for $T > 0$. The chemical potential μ corresponds, by virtue of the expression

$$\mu \approx E_F = k_BT_F, \tag{1.36}$$

to the Fermi temperature T_F, which is typically in the neighborhood of 10^5 K. This means that the distribution function $f(\mathbf{v})$ is 1 for energies below E_F and zero above E_F, and assumes intermediate values only in a region k_BT wide near E_F, as shown in Fig. 1.4b.

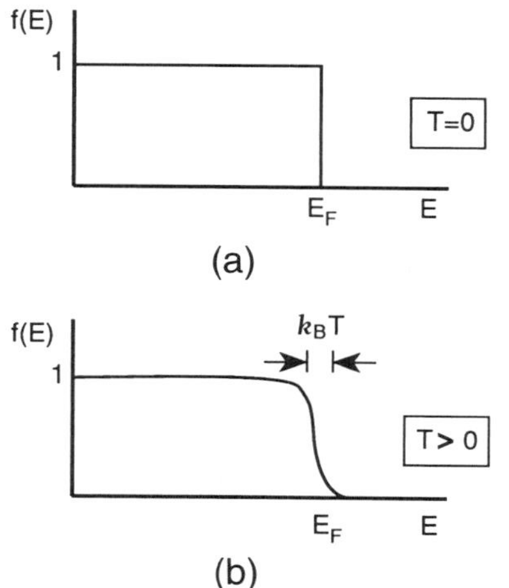

Figure 1.4 Fermi–Dirac distribution function $f(E)$ for electrons (a) at $T = 0$ K, and (b) above 0 K.

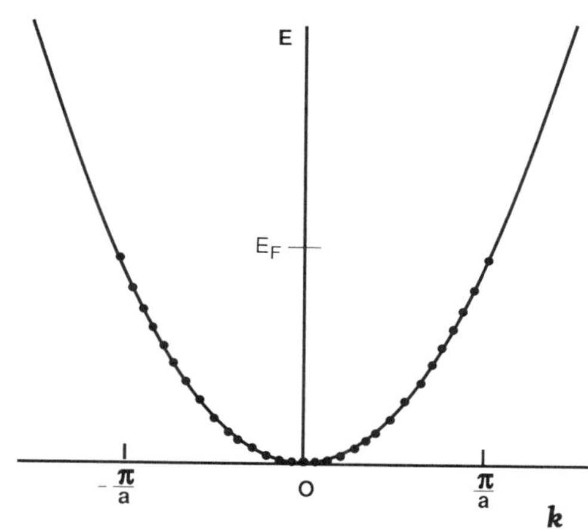

Figure 1.5 One-dimensional free electron energy band shown occupied out to the first Brillouin zone boundaries at $k = \pm\pi/a$.

The electron kinetic energy can be written in several ways, for example,

$$E_k = \tfrac{1}{2}mv^2 = \frac{p^2}{2m} = \frac{\hbar^2 k^2}{2m} = \frac{\hbar^2}{2m}\left(k_x^2 + k_y^2 + k_z^2\right), \qquad (1.37)$$

where $p = \hbar k$, and the quantization in k-space, sometimes called reciprocal space, means that each Cartesian component of **k** can assume discrete values, namely $2\pi n_x/L_x$ in the x direction of length L_x, and likewise for the y and z directions of length L_y and L_z, respectively. Here n_x is an integer between 1 and L_x/a, where a is the lattice constant; n_y and n_z are defined analogously. The one-dimensional case is sketched in Fig. 1.5. At absolute zero these k-space levels are doubly occupied by electrons of opposite spin up to the Fermi energy E_F,

$$E_F = \frac{\hbar^2 k_F^2}{2m}, \qquad (1.38)$$

as indicated in the figure. Partial occupancy occurs in a narrow region of width $k_B T$ at E_F, as shown in Fig. 1.4b. For simplicity we will assume a cubic shape, so that $L_x = L_y = L_z = L$. Hence the total number of electrons N is given as

$$N = 2\frac{\text{occupied } k\text{-space volume}}{k\text{-space volume per electron}} = 2\frac{4\pi k_F^3/3}{(2\pi/L)^3}. \qquad (1.39)$$

The electron density $n = N/V = N/L^3$ at the energy $E = E_F$ is

$$n = \frac{k_F^3}{3\pi^2} = \frac{1}{3\pi^2}\left(\frac{2mE_F}{\hbar^2}\right)^{3/2}, \qquad (1.40)$$

and the density of states $D(E)$ per unit volume, which is obtained from evaluating the derivative dn/dE of this expression (with E_F replaced by E), is

$$D(E) = \frac{d}{dE}n(E) = \frac{1}{2\pi^2}\left(\frac{2m}{\hbar^2}\right)^{3/2} E^{1/2} = D(E_F)(E/E_F)^{1/2}, \qquad (1.41)$$

and this is shown sketched in Fig. 1.6. Using Eqs. (1.36) and (1.38), respectively,

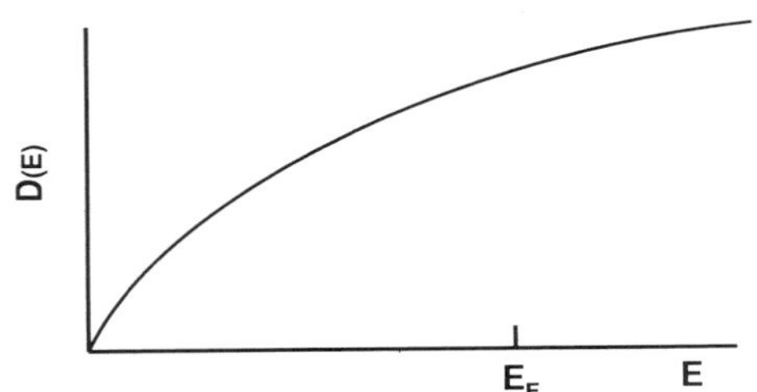

Figure 1.6 Density of states $D(E)$ of a free electron energy band $E = \hbar^2k^2/2m$.

the density of states at the Fermi level can be written in two equivalent ways,

$$D(E_F) = \begin{cases} \dfrac{3n}{2k_BT_F} \\ \dfrac{mk_F}{\hbar^2\pi^2} \end{cases}, \qquad (1.42)$$

for this isotropic case in which energy is independent of direction in k-space (so that the Fermi surface is spherical). In many actual conductors, including the high-temperature superconductors in their normal states above T_c, this is not the case, and $D(E)$ has a more complicated expression.

It is convenient to express the electron density n and the total electron energy E_T in terms of integrals over the density of states:

$$n = \int D(E)f(E)dE, \qquad (1.43)$$

$$E_T = \int D(E)f(E)EdE. \qquad (1.44)$$

The product $D(E)f(E)$ that appears in these integrands is shown plotted versus energy in Fig. 1.7a for $T = 0$ and in Fig. 1.7b for $T > 0$.

X. ENERGY GAP AND EFFECTIVE MASS

The free electron kinetic energy of Equation (1.37) is obtained from the

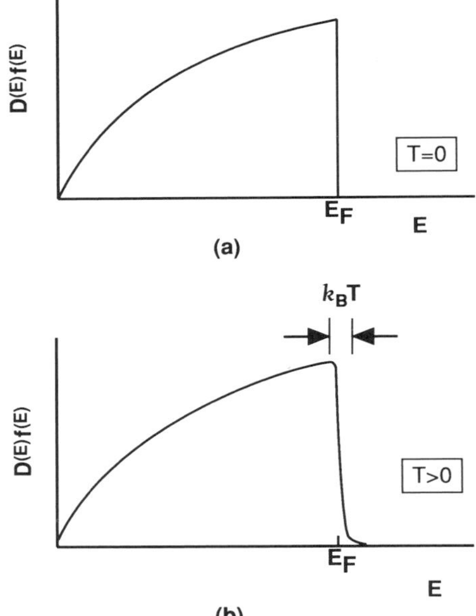

Figure 1.7 Energy dependence of occupation of a free electron energy band by electrons (a) at 0 K and (b) for $T > 0$ K. The products $D(E)f(E)$ are calculated from Figs. 1.4 and 1.6.

plane wave solution $\phi = e^{-i\mathbf{k}\cdot\mathbf{r}}$ of the Schrödinger equation,

$$-\left(\frac{\hbar^2}{2m}\right)\nabla^2\phi(\mathbf{r}) + V(\mathbf{r})\phi(\mathbf{r}) = E\phi(\mathbf{r}), \qquad (1.45)$$

with the potential $V(\mathbf{r})$ set equal to zero. When a potential, such as that shown in Fig. 1.1, is included in the Schrödinger equation, the free-electron energy parabola of Fig. 1.5 develops energy gaps, as shown in Fig. 1.8. These gaps appear at the boundaries $k = \pm n\pi/a$ of the unit cell in k-space, called the *first Brillouin zone*, and of successively higher Brillouin zones, as shown. The energies levels are closer near the gap, which means that the density of states $D(E)$ is larger there (see Figs. 1.9 and 1.10). For weak potentials, $|V| \ll E_F$, the density of states is close to its free-electron form away from the gap, as indicated in the figures. The number of points in k-space remains the same, that is, it is

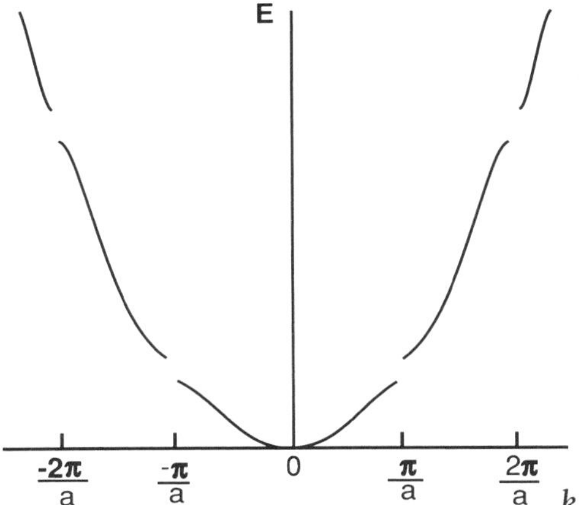

Figure 1.8 A one-dimensional free electron energy band shown perturbed by the presence of a weak periodic potential $V(x) \ll \hbar^2\pi^2/2ma^2$. The gaps open up at the zone boundaries $k = \pm n\pi/a$, where $n = 1, 2, 3, \ldots$.

conserved, when the gap forms; it is the density $D(E)$ that changes.

If the kinetic energy near an energy gap is written in free form,

$$E_{\mathrm{k}} = \frac{\hbar^2 k^2}{2m^*}, \tag{1.46}$$

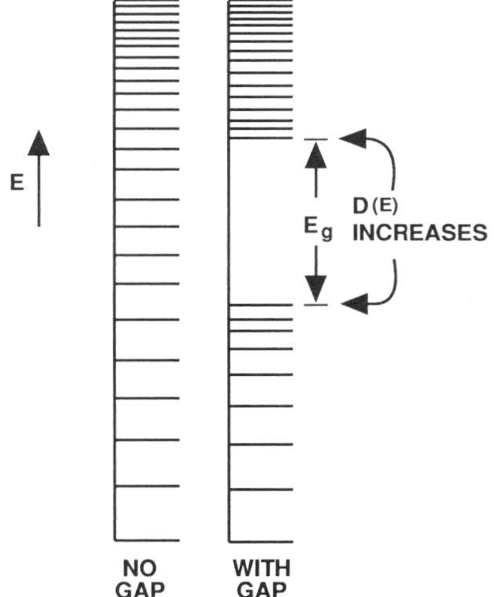

Figure 1.9 Spacing of free electron energy levels in the absence of a gap (left) and in the presence of a small gap (right) of the type shown in Fig. 1.8. The increase of $D(E)$ near the gap is indicated.

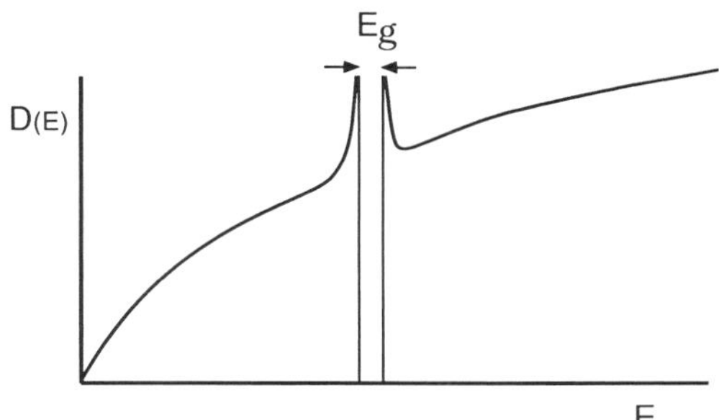

Figure 1.10 Energy dependence of the density of states $D(E)$ corresponding to the case of Fig. 1.9 in the presence of a gap.

the effective mass $m^*(k)$, which is different from the free-electron value m, becomes a function of k, which takes into account bending of the free-electron parabola near the gap. It can be evaluated from the second derivative of E_{k} with respect to k:

$$\frac{1}{m^*} = \frac{1}{\hbar^2}\left(\frac{d^2E_{\mathrm{k}}}{dk^2}\right)_{E_{\mathrm{F}}}. \tag{1.47}$$

This differentiation can be carried out if the shapes of the energy bands near the Fermi level are known. The density of states $D(E_{\mathrm{F}})$ also deviates from the free-electron value near the gap, being proportional to the effective mass m^*,

$$D(E_{\mathrm{F}}) = \frac{m^* k_{\mathrm{F}}}{\hbar^2\pi^2}, \tag{1.48}$$

as may be inferred from Eq. (1.42).

XI. ELECTRONIC SPECIFIC HEAT

The specific heat C of a material is defined as the change in internal energy U brought about by a change in temperature

$$C = \left(\frac{dU}{dT}\right)_V. \tag{1.49}$$

We will not make a distinction between the specific heat at constant volume and the specific heat at constant pressure because for solids these two properties are virtually indistinguishable. Ordinarily, the specific heat is measured by determining the heat input dQ needed to raise the temperature of the material by an amount dT,

$$dQ = CdT. \tag{1.50}$$

In this section, we will deduce the contribution of the conduction electrons to the specific heat, and in the next section we will provide the lattice vibration or phonon participation. The former is only appreciable at low temperatures while the latter dominates at room temperature.

The conduction electron contribution C_e to the specific heat is given by the derivative dE_T/dT. The integrand of Eq. (1.44) is somewhat complicated, so differentiation is not easily done. Solid-state physics texts carry out an approximate evaluation of this integral, to give

$$C_e = \gamma T, \tag{1.51}$$

where the normal-state electron specific heat constant γ, sometimes called the Sommerfeld constant, is given as

$$\gamma = \left(\frac{\pi^2}{3}\right) D(E_F) k_B^2. \tag{1.52}$$

This provides a way to experimentally evaluate the density of states at the Fermi level. To estimate the electronic specific heat per mole we set $n = N_A$ and make use of Eq. (1.42) to obtain the free-electron expression

$$\gamma_0 = \frac{\pi^2 R}{2T_F}, \tag{1.53}$$

where $R = N_A k_B$ is the gas constant. This result agrees (within a factor of 2) with experiment for many metallic elements.

A more general expression for γ is obtained by applying $D(E_F)$ from Eq. (1.48) instead of the free-electron value of (1.42). This gives

$$\gamma = \left(\frac{m^*}{m}\right) \gamma_0, \tag{1.54}$$

where γ_0 is the Sommerfeld factor (1.53) for a free electron mass. This expression will be discussed further in Chapter 3, Section XI, which treats heavy fermion compounds that have very large effective masses.

XII. PHONON SPECIFIC HEAT

The atoms in a solid are in a state of continuous vibration. These vibrations, called *phonon modes*, constitute the main contribution to the specific heat. In models of a vibrating solid nearby atoms are depicted as being bonded together by springs. For the one-dimensional diatomic case of alternating small and large atoms, of masses m_s and m_l, respectively, there are low-frequency modes called *acoustic* (A) *modes*, in which the two types of atoms vibrate in phase, and high-frequency modes, called *optical* (O) *modes*, in which they vibrate out of phase. The vibrations can also be longitudinal, i.e., along the line of atoms, or transverse, i.e., perpendicular to this line, as explained in typical solid-state physics texts. In practice, crystals are three-dimensional and the situation is more complicated, but these four types of modes are observed. Figure 1.11 presents a typical wave vector dependence of their frequencies.

It is convenient to describe these vibrations in k-space, with each vibrational mode having energy $E = \hbar\omega$. The Planck distribution function applies,

$$f(E) = \frac{1}{\exp(E/k_B T) - 1}, \tag{1.55}$$

where the minus one in the denominator indicates that only the ground vibrational level is occupied at absolute zero. There is no chemical potential because the number of phonons is not conserved. The total number of acoustic vibrational modes per unit volume N is calculated as in Eq.

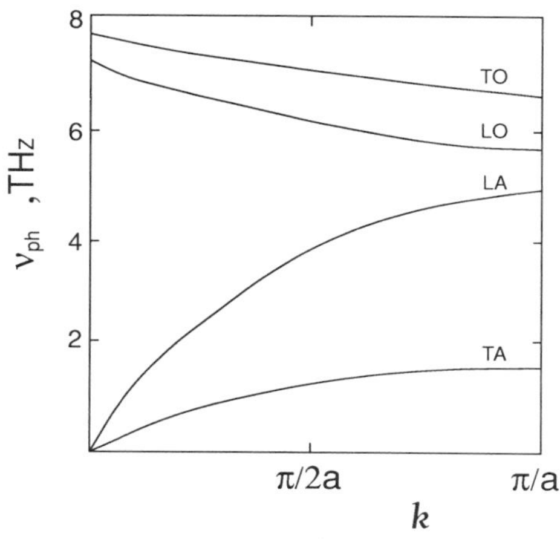

Figure 1.11 Typical dependence of energy E on the wave vector k for transverse (T), longitudinal (L), optical (O), and acoustic (A) vibrational modes of a crystal.

(1.39) with the factor 2 omitted since there is no spin,

$$N = \frac{\text{occupied } k\text{-space volume}}{k\text{-space volume per atom}}$$

$$= \frac{4\pi k_D^3/3}{(2\pi/L)^3}, \qquad (1.56)$$

where L^3 is the volume of the crystal and k_D is the maximum permissible value of k. In the Debye model, the sound velocity v is assumed to be isotropic ($v_x = v_y = v_z$) and independent of frequency,

$$v = \frac{\omega}{k}. \qquad (1.57)$$

Writing $\omega_D = v k_D$ and substituting this expression in Eq. (1.56) gives, for the density of modes $n = N/L^3$,

$$n = \frac{\omega_D^3}{6\pi^2 v^3}, \qquad (1.58)$$

where the maximum permissible frequency ω_D is called the *Debye frequency*.

The vibration density of states per unit volume $D_{ph}(\omega) = dn/d\omega$ is

$$D_{ph}(\omega) = \frac{\omega^2}{2\pi^2 v^3}, \qquad (1.59)$$

and the total vibrational energy E_{ph} is obtained by integrating the phonon mode energy $\hbar\omega$ times the density of states (1.59) over the distribution function (1.55) (cf. de Wette *et al.*, 1990)

$$E_{ph} = \int_0^{\omega_D} \left(\frac{\omega^2}{2\pi^2 v^3} \right) \frac{\hbar\omega \, d\omega}{e^{\hbar\omega/k_B T} - 1}. \qquad (1.60)$$

The vibrational or phonon specific heat $C_{ph} = dE_{ph}/dT$ is found by differentiating Eq. (1.60) with respect to the temperature,

$$C_{ph} = 3R \left(\frac{T}{\Theta_D} \right)^3 \int_0^{\Theta_D} \frac{x^4 e^x dx}{(e^x - 1)^2}, \qquad (1.61)$$

and Fig. 1.12 compares this temperature dependence with experimental data for Cu and Pb. The molar specific heat has the respective low- and high-temperature limits

$$C_{ph} = \left(\frac{12\pi^4}{5} \right) R \left(\frac{T}{\Theta_D} \right)^3 \qquad T \ll \Theta_D \qquad (1.62a)$$

$$C_{ph} = 3R \qquad T \gg \Theta_D \qquad (1.62b)$$

far below and far above the Debye temperature

$$\Theta_D = \frac{\hbar\omega_D}{k_B}, \qquad (1.63)$$

and the former limiting behavior is shown by the dashed curve in Fig. 1.12. We also see from the figure that at their superconducting transition temperatures T_c the element Pb and the compound LaSrCuO are in the T^3 region, while the compound YBaCuO is significantly above it.

Since at low temperatures a metal has an electronic specific heat term (1.51) that is linear in temperature and a phonon term (1.62a) that is cubic in T, the two can be experimentally distinguished by plotting C_{exp}/T versus T^2, where

$$\frac{C_{exp}}{T} = \gamma + AT^2, \qquad (1.64)$$

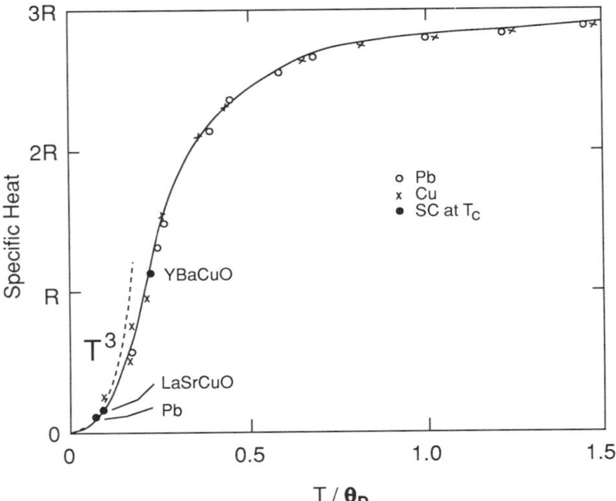

Figure 1.12 Temperature dependence of the phonon-specific heat in the Debye model compared with experimental data for Cu and Pb. The low-temperature T^3 approximation is indicated by a dashed curve. The locations of the three superconductors Pb, $(La_{0.925}Sr_{0.075})_2CuO_4$, and $YBa_2Cu_3O_{7-\delta}$ at their transition temperature T_c on the Debye curve are indicated (it is assumed that they satisfy Eq. (1.61)).

as shown in Fig. 1.13. The slope gives the phonon part A and the intercept at $T = 0$ gives the electronic coefficient γ.

Materials with a two-level system in which both the ground state and the excited state are degenerate can exhibit an extra contribution to the specific heat, called the *Schottky term*. This contribution depends on the energy spacing E_{Sch} between the ground and excited states. When $E_{Sch} \ll k_B T$, the Schottky term has the form aT^{-2} (Crow and Ong, 1990). The resulting upturn in the observed specific heat at low temperatures, sometimes called the *Schottky anomaly*, has been observed in some superconductors.

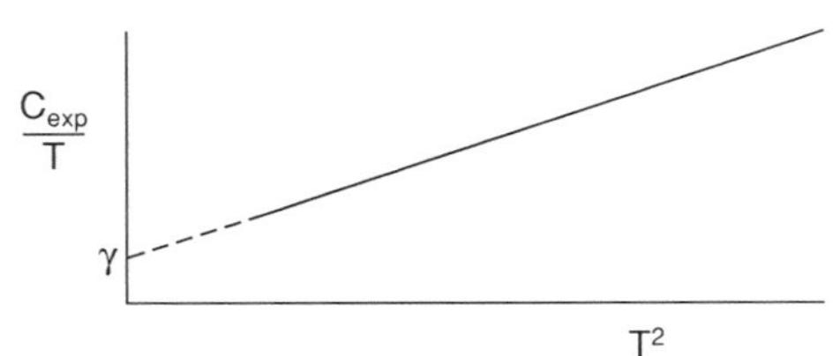

Figure 1.13 Typical plot of C_{exp}/T versus T^2 for a conductor. The phonon contribution is given by the slope of the line, and the free electron contribution γ is given by the intercept obtained by the extrapolation $T \to 0$.

XIII. ELECTROMAGNETIC FIELDS

Before discussing the magnetic properties of conductors it will be helpful to say a few words about electromagnetic fields, and to write down for later reference several of the basic equations of electromagnetism.

These equations include the two homogeneous Maxwell's equations

$$\nabla \cdot \mathbf{B} = 0, \tag{1.65}$$

$$\nabla \times \mathbf{E} + \frac{\partial \mathbf{B}}{\partial t} = 0, \tag{1.66}$$

and the two inhomogeneous equations

$$\nabla \cdot \mathbf{D} = \rho, \tag{1.67}$$

$$\nabla \times \mathbf{H} = \mathbf{J} + \frac{\partial \mathbf{D}}{\partial t}, \tag{1.68}$$

where ρ and $\mathbf{J}$ are referred to as the free charge density and the free current density, respectively. The two densities are said to be 'free' because neither of them arises from the reaction of the medium to the presence of externally applied fields, charges, or currents. The **B** and **H** fields and the **E** and **D** fields, respectively, are related through the expressions

$$\mathbf{B} = \mu\mathbf{H} = \mu_0(\mathbf{H} + \mathbf{M}), \qquad (1.69)$$

$$\mathbf{D} = \epsilon\mathbf{E} = \epsilon_0\mathbf{E} + \mathbf{P}, \qquad (1.70)$$

where the medium is characterized by its permeability μ and its permittivity ϵ, and μ_0 and ϵ_0 are the corresponding free space values. These, of course, are SI formulae. When cgs units are used, $\mu_0 = \epsilon_0 = 1$ and the factor 4π must be inserted in front of **M** and **P**.

The fundamental electric (**E**) and magnetic (**B**) fields are the fields that enter into the Lorentz force law

$$\mathbf{F} = q(\mathbf{E} + \mathbf{v} \times \mathbf{B}) \qquad (1.71)$$

for the force **F** acting on a charge q moving at velocity **v** in a region containing the fields **E** and **B**. Thus **B** and **E** are the macroscopically measured magnetic and electric fields, respectively. Sometimes **B** is called the magnetic induction or the magnetic flux density.

It is convenient to write Eq. (1.68) in terms of the fundamental field **B** using Eq. (1.69)

$$\nabla \times \mathbf{B} = \mu_0(\mathbf{J} + \nabla \times \mathbf{M}) + \mu_0 \frac{\partial \mathbf{D}}{\partial t}, \qquad (1.72)$$

where the displacement current term $\partial\mathbf{D}/\partial t$ is ordinarily negligible for conductors and superconductors and so is often omitted. The reaction of the medium to an applied magnetic field produces the magnetization current density $\nabla \times \mathbf{M}$ which can be quite large in superconductors.

XIV. BOUNDARY CONDITIONS

We have been discussing the relationship between the **B** and **H** fields within a medium or sample of permeability μ. If the medium is homogeneous, both μ and **M** can be constant throughout, and Eq. (1.69), with $\mathbf{B} = \mu\mathbf{H}$, applies. But what happens to the fields when two media of respective permeabilities μ' and μ'' are in contact? At the interface between the media the $\mathbf{B}'$ and $\mathbf{H}'$ fields in one medium will be related to the $\mathbf{B}''$ and $\mathbf{H}''$ fields in the other medium through the two boundary conditions illustrated in Fig. 1.14, namely:

1. The components of **B** normal to the interface are continuous across the boundary:

$$\mathbf{B}'_\perp = \mathbf{B}''_\perp . \qquad (1.73)$$

2. The components of **H** tangential to the interface are continuous across the boundary:

$$\mathbf{H}'_\parallel = \mathbf{H}''_\parallel . \qquad (1.74)$$

If there is a surface current density $\mathbf{J}_{\text{surf}}$ present at the interface, the second condition must be modified to take this into account,

$$\hat{\mathbf{n}} \times (\mathbf{H}' - \mathbf{H}'') = \mathbf{J}_{\text{surf}}, \qquad (1.75)$$

where **n** is a unit vector pointing from the double primed (″) to the primed region, as indicated in Fig. 1.14, and the surface current density $\mathbf{J}_{\text{surf}}$, which has the units ampere per meter, is perpendicular to the field direction. When $\mathbf{H}'$ and $\mathbf{H}''$ are measured along the surface parallel to each

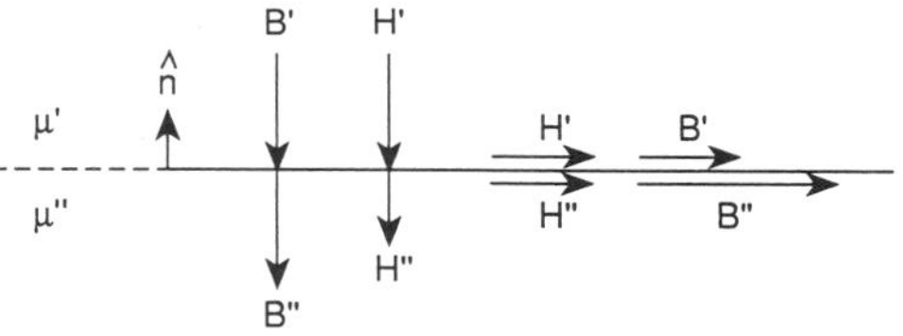

Figure 1.14 Boundary conditions for the components of the **B** and **H** magnetic field vectors perpendicular to and parallel to the interface between regions with different permeabilities. The figure is drawn for the case $\mu'' = 2\mu'$.

other, Eq. (1.75) can be written in scalar form:

$$H'_{\parallel} - H''_{\parallel} = J_{\text{surf}}. \tag{1.76}$$

In like manner, for the electric field case the normal components of D and the tangential components of E are continuous across an interface, and the condition on D must be modified when surface charges are present.

XV. MAGNETIC SUSCEPTIBILITY

It is convenient to express Eq. (1.69) in terms of the dimensionless magnetic susceptibility χ,

$$\chi = \frac{\mathbf{M}}{\mathbf{H}}, \tag{1.77}$$

to give

$$\mathbf{B} = \mu_0 \mathbf{H}(1 + \chi_{\text{SI}}) \quad \text{SI units} \tag{1.78a}$$

$$\mathbf{B} = \mathbf{H}(1 + 4\pi\chi_{\text{cgs}}) \quad \text{cgs units.} \tag{1.78b}$$

The susceptibility χ is slightly negative for diamagnets, slightly positive for paramagnets, and strongly positive for ferromagnets. Elements that are good conductors have small susceptibilities, sometimes slightly negative (e.g., Cu) and sometimes slightly positive (e.g., Na), as may be seen from Table 1.2. Nonmagnetic inorganic compounds are weakly diamagnetic (e.g., NaCl), while magnetic compounds containing transition ions can be much more strongly paramagnetic (e.g., $CuCl_2$).

The magnetization in Eq. (1.77) is the magnetic moment per unit volume, and the susceptibility defined by this expression is dimensionless. The susceptibility of a material doped with magnetic ions is proportional to the concentration of the ions in the material. In general, researchers who study the properties of these materials are more interested in the properties of the ions themselves than in the properties of the material containing the ions. To take this into account it is customary to use molar susceptibilities χ^{M}, which in the SI system have the units m^3 per mole.

It is shown in solid-state physics texts (e.g., Ashcroft and Mermin, 1976; Burns, 1985; Kittel, 1976) that a material containing paramagnetic ions with magnetic moments μ that become magnetically ordered at low temperatures has a high-temperature magnetic susceptibility that obeys the Curie–Weiss Law:

$$\chi^{\text{M}} = \frac{n\mu^2}{3k_{\text{B}}(T - \Theta)}, \tag{1.79a}$$

$$= \frac{C}{(T - \Theta)}, \tag{1.79b}$$

where n is the concentration of paramagnetic ions and C is the Curie constant. The Curie–Weiss temperature Θ has a positive sign when the low-temperature alignment is ferromagnetic and a negative sign when it is antiferromagnetic. Figure 1.15 shows the temperature dependence of χ^{M} for

Table 1.2 cgs Molar Susceptibility (χ_{cgs}) and Dimensionless SI Volume Susceptibility (χ) of Several Materials

Material	MW g/mole	Density g/cm³	χ_{cgs} cm³/mole	χ —
Free space	—	0.0	0	0
Na	22.99	0.97	1.6×10^{-5}	8.48×10^{-6}
NaCl	58.52	2.165	-3.03×10^{-5}	-1.41×10^{-5}
Cu	63.54	8.92	-5.46×10^{-6}	-9.63×10^{-6}
$CuCl_2$	134.6	3.386	1.08×10^{-3}	3.41×10^{-4}
Fe alloy	≈ 60	7–8	10^3–10^4	10^3–10^4
Perfect SC	—	—	—	−1

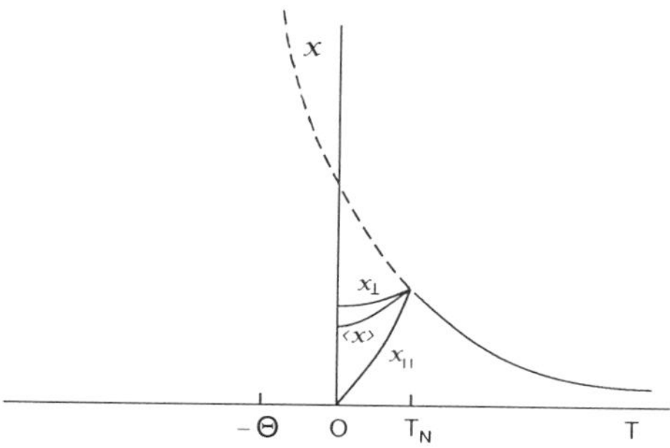

Figure 1.15 Magnetic susceptibility of a material that is paramagnetic above the Néel transition temperature T_N and antiferromagnetic with axial symmetry below the transition. The extrapolation of the paramagnetic curve below $T = 0$ provides the Curie–Weiss temperature Θ.

the latter case, in which the denominator becomes $T + |\Theta|$. The temperature T_N at which antiferromagnetic alignment occurs is referred to as the *Néel temperature*, and typically $T_N \neq \Theta$. When $\Theta = 0$, Eq. (1.79) is called the Curie law.

For a rare earth ion with angular momentum $J\hbar$ we can write

$$\mu^2 = g^2 \mu_B^2 J(J+1), \tag{1.80}$$

where $J = L + S$ is the sum of the orbital (L) and spin (S) contributions, $\mu_B = e\hbar/2m$ is the Bohr magneton, and the dimensionless Landé g factor is

$$g = \frac{3}{2} + \frac{S(S+1) - L(L+1)}{2J(J+1)}. \tag{1.81}$$

For a first transition series ion, the orbital angular momentum $L\hbar$ is quenched, which means that it is uncoupled from the spin angular momentum and becomes quantized along the crystalline electric field direction. Only the spin part of the angular momentum contributes appreciably to the susceptibility, to give the so-called spin-only result

$$\mu^2 = g^2 \mu_B^2 S(S+1), \tag{1.82}$$

where for most of these ions $g \approx 2$.

For conduction electrons the only contribution to the susceptibility comes from the electrons at the Fermi surface. Using an argument similar to that which we employed for the electronic specific heat in Section XI we can obtain the temperature-independent expression for the susceptibility in terms of the electronic density of states,

$$\chi = \mu_B^2 D(E_F), \tag{1.83}$$

which is known as the *Pauli susceptibility*. For a free electron gas of density n we substitute the first expression for $D(E_F)$ from Eq. (1.42) in Eq. (1.83) to obtain, for a mole,

$$\chi^M = \frac{3n\mu_B^2}{2k_B T_F}. \tag{1.84}$$

For alkali metals the measured Pauli susceptibility decreases with increasing atomic number from Li to Cs with a typical value $\approx 1 \times 10^{-6}$. The corresponding free-electron values from Eq. (1.84) are about twice as high as their experimental counterparts, and come much closer to experiment when electron–electron interactions are taken into account. For very low temperatures, high magnetic fields, and very pure materials there is an additional diamagnetic correction term χ_{Landau}, called *Landau diamagnetism*, which arises from the orbital electronic interaction with the magnetic field. For the free-electron model this correction has the value

$$\chi_{\text{Landau}} = -\tfrac{1}{3}\chi_{\text{Pauli}}. \tag{1.85}$$

In preparing Table 1.2 the dimensionless SI values of χ listed in column 5 were calculated from known values of the molar cgs susceptibility χ^M_{cgs}, which has the units cm^3 per mole, using the expression

$$\chi = 4\pi\left(\frac{\rho_m}{\text{MW}}\right)\chi^M_{\text{cgs}}, \tag{1.86}$$

where ρ_m is the density in g per cm^3 and MW is the molecular mass in g per mole. Some authors report per unit mass susceptibility data in emu/g, which we are calling χ^g_{cgs}. The latter is related to the dimensionless χ through the expression

$$\chi = 4\pi\rho_m \text{X}^g_{\text{cgs}}. \tag{1.87}$$

The ratio of Eq. (1.52) to Eq. (1.83) gives the free-electron expression

$$\frac{\gamma}{\chi^{\mathrm{M}}} = \frac{1}{3}\left(\frac{\pi k_{\mathrm{B}}}{\mu_{\mathrm{B}}}\right)^2, \tag{1.88}$$

where χ^{M} is the susceptibility arising from the conduction electrons. An experimental determination of this ratio provides a test of the applicability of the free-electron approximation.

This section has been concerned with dc susceptibility. Important information can also be obtained by using an ac applied field $B_0 \cos \omega t$ to determine $\chi_{\mathrm{ac}} = \chi' + i\chi''$, which has real part χ', called dispersion in phase with the applied field, and an imaginary lossy part χ'', called absorption, which is out of phase with the field (Khode and Couach, 1992). D. C. Johnston (1991) reviewed normal state magnetization of the cuprates.

XVI. HALL EFFECT

The Hall effect employs crossed electric and magnetic fields to obtain information on the sign and mobility of the charge carriers. The experimental arrangement illustrated in Fig. 1.16 shows a magnetic field B_0 applied in the z direction perpendicular to a slab and a battery that establishes an electric field E_y in the y direction that causes a current $I = JA$ to flow, where $J = nev$ is the current density. The Lorentz force

$$\mathbf{F} = q\mathbf{v} \times \mathbf{B}_0 \tag{1.89}$$

of the magnetic field on each moving charge q is in the positive x direction for both positive and negative charge carriers, as shown in Figs. 1.17a and 1.17b, respectively. This causes a charge separation to build up on the sides of the plate, which produces an electric field E_x perpendicular to the directions of the current (y) and

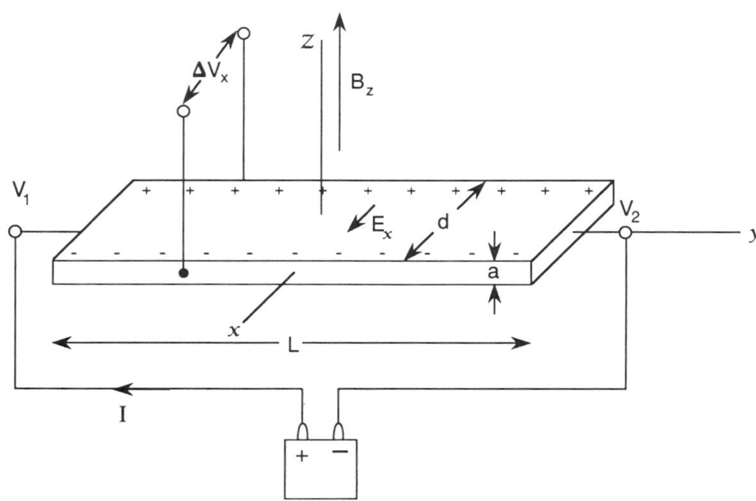

Figure 1.16 Experimental arrangement for Hall effect measurements showing an electrical current I passing through a flat plate of width d and thickness a in a uniform transverse magnetic field B_z. The voltage drop $V_2 - V_1$ along the plate, the voltage difference ΔV_x across the plate, and the electric field E_x across the plate are indicated. The figure is drawn for negative charge carriers (electrons).

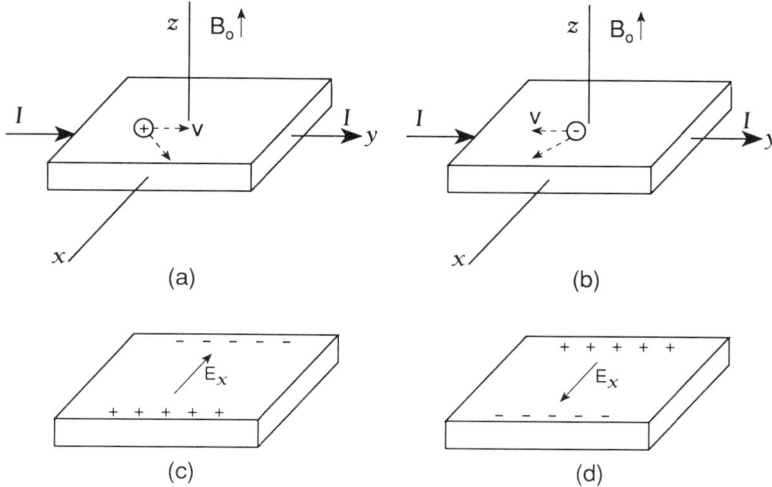

Figure 1.17 Charge carrier motion and transverse electric field direction for the Hall effect experimental arrangement of Fig. 1.16. Positive charge carriers deflect as indicated in (a) and produce the transverse electric field E_x shown in (c). The corresponding deflection and resulting electric field for negative charge carriers are sketched in (b) and (d), respectively.

magnetic (z) fields. The induced electric field is in the negative x direction for positive q, and in the positive x direction for negative q, as shown in Figs. 1.17c and 1.17d, respectively. After the charge separation has built up, the electric force $q\mathbf{E}_x$ balances the magnetic force $q\mathbf{v} \times \mathbf{B}_0$,

$$q\mathbf{E}_x = q\mathbf{v} \times \mathbf{B}_0, \tag{1.90}$$

and the charge carriers q proceed along the wire undeflected.

The Hall coefficient $R_{\rm H}$ is defined as a ratio,

$$R_{\rm H} = \frac{E_x}{J_y B_z}. \tag{1.91}$$

Substituting the expressions for $\mathbf{J}$ and $\mathbf{E}_x$ from Eqs. (1.18) and (1.90) in Eq. (1.91) we obtain for holes ($q = e$) and electrons ($q = -e$), respectively,

$$R_{\rm H} = \frac{1}{ne} \quad \text{(holes)} \tag{1.92a}$$

$$R_{\rm H} = -\frac{1}{ne} \quad \text{(electrons)}, \tag{1.92b}$$

where the sign of $R_{\rm H}$ is determined by the sign of the charges. The Hall angle $\Theta_{\rm H}$ is defined by

$$\tan \Theta_{\rm H} = \frac{E_x}{E_y}. \tag{1.93}$$

Sometimes the dimensionless Hall number is reported,

$$\text{Hall } \# = \frac{V_0}{R_{\rm H} e}, \tag{1.94}$$

where V_0 is the volume per chemical formula unit. Thus the Hall effect distinguishes electrons from holes, and when all of the charge carriers are the same this experiment provides the charge density n. When both positive and negative charge carriers are present, partial (or total) cancellation of their Hall effects occurs.

The mobility μ is the charge carrier drift velocity per unit electric field,

$$\mu = \frac{|v_{\rm av}|}{E}, \tag{1.95}$$

and with the aid of Eqs. (1.18), (1.21), and (1.92) we can write

$$\mu_{\mathrm{H}} = \frac{R_{\mathrm{H}}}{\rho}, \tag{1.96}$$

where the Hall mobility μ_{H} is the mobility determined by a Hall effect measurement. It is a valid measure of the mobility (1.95) if only one type of charge carrier is present.

By Ohm's law (1.21) the resistivity is the ratio of the applied electric field in the direction of current flow to the current density,

$$\rho = \frac{E_y}{J}. \tag{1.97}$$

In the presence of a magnetic field, this expression is written

$$\rho_{\mathrm{m}} = \frac{E_y}{J}, \tag{1.98}$$

where ρ_{m} is called the *transverse magnetoresistivity*. There is also a longitudinal magnetoresistivity defined when E and B_0 are parallel. For the present case the resistivity does not depend on the applied field, so $\rho_{\mathrm{m}} = \rho$. For very high magnetic fields ρ_{m} and ρ can be different. In the superconducting state ρ_{m} arises from the movement of quantized magnetic flux lines, called vortices, so that it can be called the flux flow resistivity ρ_{ff}. Finally, the Hall effect resistivity ρ_{xy} (Ong, 1991) is defined by

$$\rho_{xy} = \frac{E_x}{J_y}. \tag{1.99}$$

FURTHER READING

Most of the material in this chapter may be found in standard textbooks on solid state physics (e.g., Ashcroft and Mermin, 1976; Burns, 1985; Kittel, 1976).

PROBLEMS

1. Show that Eq. (1.61) for the phonon specific heat has the low- and high-temperature limits (1.62a) and (1.62b), respectively.
2. Aluminum has a magnetic susceptibility $+16.5 \times 10^{-6}$ cgs, and niobium, 195×10^{-6} cgs. Express these in dimensionless SI units. From these values estimate the density of states and the electronic specific heat constant γ for each element.
3. Copper at room temperature has 8.47×10^{22} conduction electrons/cm^3, a Fermi energy of 7.0 eV, and $\tau = 2.7 \times 10^{-14}$ s. Calculate its Hall coefficient, average conduction electron velocity in an electric field of 200 V/cm, electrical resistivity, and mean free path.
4. Calculate the London penetration depth, resistivity, plasma frequency, and density of states of copper at room temperature.
5. It was mentioned in Section 1.II that the chemical potential μ is the energy required to remove one electron from a Fermi gas under the conditions of constant volume and constant entropy. Use a thermodynamic argument to prove this assertion, and also show that μ equals the change in the Gibbs free energy when one electron is removed from the Fermi gas under the conditions of constant temperature and constant pressure.
6. Show that well above the Debye temperature the number of phonons in a normal mode of vibration is proportional to the temperature.
7. For the two-dimensional square lattice draw the third Brillouin zone in (a) the extended zone scheme and (b) the reduced zone scheme in which the third zone is mapped into the first zone. Show where each segment in the extended scheme goes in the first zone. Draw constant energy lines for $\epsilon = 2\epsilon_0$, $3\epsilon_0$, $4\epsilon_0$, $5\epsilon_0$. Sketch the Fermi surface for $\epsilon_{\mathrm{F}} = 4.5\epsilon_0$. Indicate the electron-like and hole-like regions.

2

The Phenomenon of Superconductivity

I. INTRODUCTION

A perfect superconductor is a material that exhibits two characteristic properties, namely zero electrical resistance and perfect diamagnetism, when it is cooled below a particular temperature T_c, called the *critical temperature*. At higher temperatures it is a normal metal, and ordinarily is not a very good conductor. For example, lead, tantalum, and tin become superconductors, while copper, silver, and gold, which are much better conductors, do not superconduct. In the normal state some superconducting metals are weakly diamagnetic and some are paramagnetic. Below T_c they exhibit perfect electrical conductivity and also perfect or quite pronounced diamagnetism.

Perfect diamagnetism, the second characteristic property, means that a superconducting material does not permit an externally applied magnetic field to penetrate into its interior. Those superconductors that totally exclude an applied magnetic flux are known as Type I superconductors, and they constitute the subject matter of this chapter. Other superconductors, called Type II superconductors, are also perfect conductors of electricity, but their magnetic properties are more complex. They totally exclude magnetic flux when the applied magnetic field is low, but only partially exclude it when the applied field is higher. In the region of higher magnetic fields their diamagnetism is not perfect, but rather of a mixed type. The basic properties of these mixed magnetism superconductors are described in Chapters 9 and 10.

II. A BRIEF HISTORY

In 1908, H. Kamerlingh Onnes initiated the field of low-temperature physics by liquifying helium in his laboratory at Leiden. Three years later he found that below 4.15 K of the dc resistance of mercury dropped to zero (Onnes, 1911). With that finding the field of superconductivity was born. The next year Onnes discovered that application of a sufficiently strong axial magnetic field restored the resistance to its normal value. One year later, in 1913, the element lead was found to be superconducting at 7.2 K (Onnes, 1913). Another 17 years were to pass before this record was surpassed, by the element niobium ($T_c = 9.2$ K) (vide Ginzburg and Kitzhnits, 1977, p. 2).

A considerable amount of time went by before physicists became aware of the second distinguishing characteristic of a superconductor—namely, its perfect diamagnetism. In 1933, Meissner and Ochsenfeld found that when a sphere is cooled below its transition temperature in a magnetic field, it excludes the magnetic flux.

The report of the Meissner effect led the London brothers, Fritz and Heinz, to propose equations that explain this effect and predict how far a static external magnetic field can penetrate into a superconductor. The next theoretical advance came in 1950 with the theory of Ginzburg and Landau, which described superconductity in terms of an order parameter and provided a derivation for the London equations. Both of these theories are macroscopic in character and will be described in Chapter 5.

In the same year it was predicted theoretically by H. Fröhlich (1950) that the transition temperature would decrease as the average isotopic mass increased. This effect, called the *isotope effect*, was observed experimentally the same year (Maxwell, 1950; Reynolds *et al.*, 1950). The isotope effect provided support for the electron–phonon interaction mechanism of superconductivity.

Our present theoretical understanding of the nature of superconductivity is based on the BCS microscopic theory proposed by J. Bardeen, L. Cooper, and J. R. Schrieffer in 1957 (we will describe it in Chapter 6). In this theory it is assumed that bound electron pairs that carry the super current are formed and that an energy gap between the normal and superconductive states is created. The Ginzburg–Landau (1950) and London (1950) results fit well into the BCS formalism. Much of the present theoretical debate centers around how well the BCS theory explains the properties of the new high-temperature superconductors.

Alloys and compounds have been extensively studied, especially the so-called *A*15 compounds, such as Nb_3Sn, Nb_3Ga, and Nb_3Ge, which held the record for the highest transition temperatures from 1954 to 1986, as shown in Table 2.1. Many other types of compounds have been studied in recent years, particularly the so-called heavy fermion systems in which the superconducting electrons have high effective masses of $100m_e$ or more. Organic superconductors have shown a dramatic

Table 2.1 Superconducting Transition Temperature Records through the Years[a]

Material	T_c (K)	Year
Hg	4.1	1911
Pb	7.2	1913
Nb	9.2	1930
$NbN_{0.96}$	15.2	1950
Nb_3Sn	18.1	1954
$Nb_3(Al_{\frac{3}{4}}Ge_{\frac{1}{4}})$	20–21	1966
Nb_3Ga	20.3	1971
Nb_3Ge	23.2	1973
$Ba_xLa_{5-x}Cu_5O_y$	30–35	1986
$(La_{0.9}Ba_{0.1})_2Cu_{4-\delta}$ at 1 GPa	52	1986
$YBa_2Cu_3O_{7-\delta}$	95	1987
$Bi_2Sr_2Ca_2Cu_3O_{10}$	110	1988
$Tl_2Ba_2Ca_2Cu_3O_{10}$	125	1988
$Tl_2Ba_2Ca_2Cu_3O_{10}$ at 7 GPa	131	1993
$HgBa_2Ca_2Cu_3O_{8+\delta}$	133	1993
$HgBa_2Ca_2Cu_3O_{8+\delta}$ at 25 GPa	155	1993
$Hg_{0.8}Pb_{0.2}Ba_2Ca_2Cu_3O_x$	133	1994
$HgBa_2Ca_2Cu_3O_{8+\delta}$ at 30 GPa	164	1994

[a] cf. Ginzburg and Kirzhnits, 1977.

rise in transition temperatures during the past decade.

On April 17, 1986, a brief article, entitled "Possible High T_c Superconductivity in the Ba–La–Cu–O System," written by J. G. Bednorz and K. A. Müller was received by the *Zeitschrift für Physik*, initiating the era of high-temperature superconductivity. When the article appeared in print later that year, it met with initial skepticism. Sharp drops in resistance attributed to "high-T_c" superconductivity had appeared from time to time over the years, but when examined they had always failed to show the required diamagnetic response or were otherwise unsubstantiated. It was only when a Japanese group (Uchida *et al.*, 1987) and Chu's group in the United States (Chu *et al.*, 1987b) reproduced the original results that the results found by Bednorz and Müller began to be taken seriously. Soon many other researchers became active, and the recorded transition temperature began a rapid rise.

By the beginning of 1987, scientists had fabricated the lanthanum compound, which went superconducting at close to 40 K at atmospheric pressure (Cava *et al.* 1987; Tarascon *et al.*, 1987c) and at up to 52 K under high pressure (Chu *et al.*, 1987a). Soon thereafter, the yttrium–barium system, which went superconducting in the low 90s (Chu *et al.*, 1988a; Zhao *et al.*, 1987), was discovered. Early in 1988, superconductivity reached 110 K with the discovery of BiSrCaCuO (Chu *et al.*, 1988b; Maeda *et al.*, 1988; Michel *et al.*, 1987), and then the 120–125 K range with TlBaCaCuO (Hazen *et al.*, 1988; Sheng and Herman, 1988; Sheng *et al.*, 1988). More recently, Berkley *et al.* (1993) reported

$$T_c = 131.8\ \text{K for } Tl_2Ba_2Ca_2Cu_3O_{10-x}$$

at a pressure of 7 GPa. Several researchers have reported T_c above 130 K for the Hg series of compounds $HgBa_2Ca_nCu_{n+1}O_{2n+4}$ with $n = 1, 2,$ sometimes with Pb doping for Hg (Chu *et al.*, 1993a; Iqbal *et al.*, 1994; Schilling *et al.*, 1993, 1994a). The transition temperature of the Hg compounds increases with pressure (Chu, 1994; Klehe *et al.*, 1992, 1994; Rabinowitz and McMullen, 1994) in the manner shown in Fig. 2.1a (Gao *et al.*, 1994) and onset T_c values in the 150 K range are found for pressures above 10 GPa (Chu *et al.*, 1993b, Ihara *et al.*, 1993). We see from Fig. 2.1b that the transitions are broad, with midpoint T_c located 7 or 8 K below the onset, and the zero resistivity point comes much lower still (Gao *et al.*, 1994).

The recent pace of change and improvement in superconductors exceeds that of earlier decades, as the data listed in Table 2.1 and plotted in Fig. 2.2 demonstrate. For 56 years the element niobium and its compounds had dominated the field of superconductivity. In addition to providing the highest T_c values, niobium compounds such as NbTi and Nb_3Sn are also optimal magnet materials: for NbTi, $B_{c2} = 10$ T and for Nb_3Sn, $B_{c2} = 22$ T at 4.2 K, where B_{c2} is the upper-critical field of a Type II superconductor, in the sense that it sets a limit on the magnetic field attainable by a magnet; thus application of an applied magnetic field in excess of B_{c2} drives a superconductor normal. The period from 1930 to 1986 can be called the Niobium Era of superconductivity. The new period that began in 1986 might become the Copper Oxide Era because, thus far, the presence of copper and oxygen has, with rare exceptions, been found essential for T_c above 40 K. It is also interesting to observe that Hg was the first known superconductor, and now 82 years later mercury compounds have become the best!

III. RESISTIVITY

Before beginning the discussion of super currents, we will examine the resistivity of superconducting materials in their normal state above the transition temperature T_c; we will then make some comments on the drop to zero resistance at T_c; finally, we will describe the measurements that have set upper limits on resistivity below T_c.

A. Resistivity above T_c

In Chapter 1, Section VII, we explained, and now we illustrate in Figs. 2.3–2.6, how the resistivity of a typical conductor depends linearly on temperature at high temperatures and obeys the T^5 Bloch law at low temperatures. Classical or low-temperature superconductors are in the Bloch law region if the transition temperature is low enough, as illustrated in Fig. 2.3a. High-temperature superconductors have transition temperatures that are in the linear region, corresponding to the resistivity plot of Fig. 2.3b. However, the situation is actually more complicated because the resistivity of single crystals of high-temperature superconductors is strongly anisotropic, as we will show later. Several theoretical treatments of the resistivity of cuprates have appeared recently (e.g., Griessen, 1990; Micnas *et al.*, 1987; Song and Gaines, 1991; Wu *et al.*, 1989; Zeyhe, 1991).

Good conductors such as copper and silver have room temperature resistivities of about 1.5 $\mu\Omega$ cm, whereas at liquid nitrogen temperatures the resistivity typically decreases by a factor in the range 3–8, as shown by the data in Table 1.1. The elemental superconductors, such as Nd, Pb, and Sn, have room temperature resistivities a factor of 10 greater than good conductors. The metallic elements Ba, Bi, La, Sr, Tl, and Y, which are also present in oxide superconductors, have room temperature resistivities 10 to 70 times that of Cu.

The copper-oxide superconductors have even higher room temperature resistivities, more than three orders of magnitude greater than that of metallic copper, which puts them within a factor of 3 or 4 of the semiconductor range, as shown by the data in Table 2.2. The resistivity of these materials above T_c decreases more or less linearly with decreasing temperature down to the neighborhood of T_c, with a drop by a factor of 2 or 3 from room temperature to this point, as shown in Fig. 2.4 and by the data in Table 2.2. Figure 2.5 shows that the linearity extends far above room temperature, especially for the lanthanum compound (Gurvitch and Fiory, 1987a, b, c; Gurvitch *et al.*, 1988). It has been linked to the two-dimensional character of electron transport (Micnas *et al.*, 1990).

We see from the figure that

$$YBa_2Cu_3O_{7-\delta}$$

begins to deviate from linearity at about 600–700 K, near the orthorhombic-to-tetragonal phase transformation (cf. Chapter 7, Section VI.D) where it changes from a metallic material below the transition to a semiconductor above. Heating causes a loss of oxygen, as shown in Fig. 2.6 which presents the dependence of resistivity on the oxygen partial pressure (Grader *et al.*, 1987). The temperature dependence of resistivity has been related to the loss of oxygen [cf. Eq. (X-1) from Poole *et al.*, 1988; cf. Chaki and Rubinstein, 1987; Fiory *et al.*, 1987].

The resistivity of poor metals at high temperatures tends to saturate to a temperature-independent value when the mean free path l approaches the wavelength $\lambda_F = 2\pi/k_F$ associated with the Fermi level, where k_F is the Fermi wave vector. The Ioffe–Regel criterion for the onset of this saturation is $k_F l \approx 1$. The quantity $k_F l$ for $YBa_2Cu_3O_{7-\delta}$ has been estimated to have a value of 30 for $T = 100$ K (Hagen *et al.*, 1988) and 3 for $T = 1000$ K (Crow and Ong, 1990). These considerations, together with the curves in Fig. 2.5, indicate that, in practice, the Ioffe–Regel criterion does not cause the resistivity to saturate in high-temperature superconductors. The *A*15 compound V_3Si, whose crys-

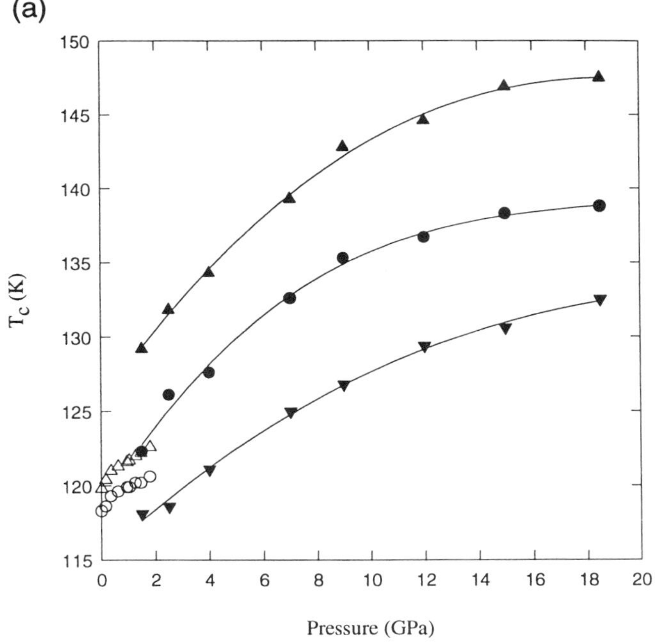

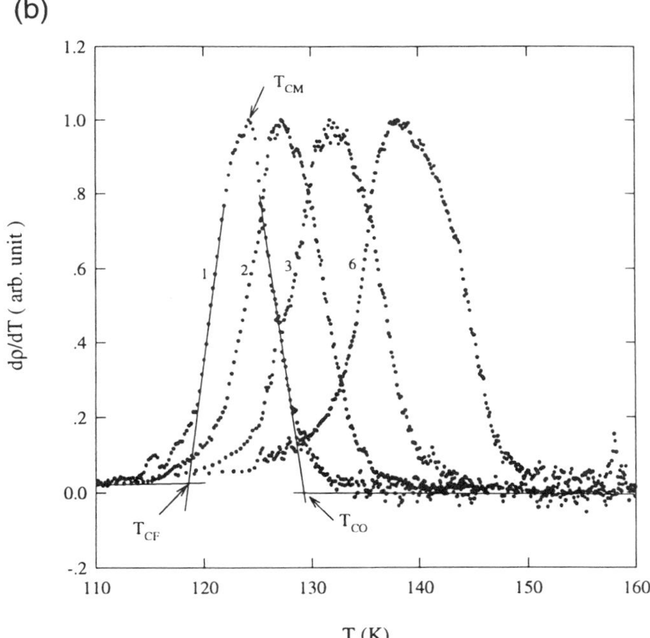

Figure 2.1 Effect of pressure on the transition temperature of the superconductor $HgBa_2Ca_2Cu_3O_{8+\delta}$ showing (a) pressure dependence of the onset (upper curve), midpoint (middle curve), and final off-set (lower curve) values of T_c, and (b) temperature dependence of the resistivity derivative $d\rho/dT$ at 1.5 (1), 4 (2), 7 (3), and 18.5 GPa (6). Definitions of T_{co} onset, T_{cm} midpoint, and T_{cf} final offset that are plotted in (a) are given in (b) (Gao *et al.*, 1993).

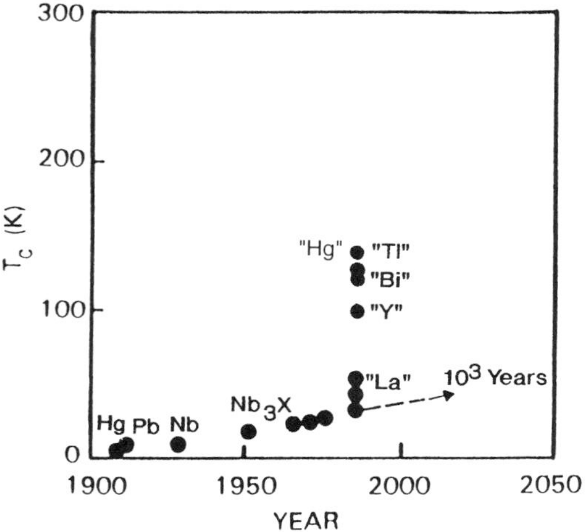

Figure 2.2 Increase in the superconducting transition temperature with time. A linear extrapolation of the data before 1986 predicts that room temperature would be reached in about 1000 years. From left to right $X = Sn$, $Al_{0.75}Ge_{0.25}$, Ga, and Ge for the data points of the *A*15 compound Nb_3X (Adapted from Fig. I-1, Poole *et al.*, 1988).

tal structure is stable up to 1950°C, does exhibit saturation in its resistivity-versus-temperature plot.

B. Resistivity Anisotropy

The resistivity of $YBa_2Cu_3O_{7-\delta}$ is around two orders of magnitude greater along the *c*-axis than parallel to the *a*, *b*-plane; thus $\rho_c/\rho_{ab} \approx 100$ and for $Bi_{2+x}Sr_{2-y}CuO_{6+\delta}$, $\rho_c/\rho_{ab} \approx 10^5$ (Fiory *et al.*, 1989). The temperature dependence of these resistivities, measured by the method described in the following section, exhibits a peak near T_c in the case of ρ_c, and this is shown in Fig. 2.7. When the data are fitted to the expressions (Anderson and Zou, 1988)

$$\rho_{ab} = \frac{A_{ab}}{T} + B_{ab}T, \tag{2.1}$$

$$\rho_c = \frac{A_c}{T} + B_cT, \tag{2.2}$$

by plotting $\rho_{ab}T$ and ρ_cT from the data of Fig. 2.7 versus T^2, a good fit is obtained, as shown in Fig. 2.8. The angular dependence of the resistivity is found to obey the expression (Wu *et al.*, 1991b)

$$\rho(\Theta) = \rho_{ab}\sin^2\Theta + \rho_c\cos^2\Theta \tag{2.3}$$

where Θ is the angle of the current direction relative to the *c* axis.

Typical measured resistivities of polycrystalline samples are much closer to the in-plane values. The anisotropy ratio $\rho_c/\rho_{ab} \approx 100$ is so large that the current encounters less resistance when it follows a longer path in the planes than when it takes the shorter path perpendicular to the planes, so it tends to flow mainly along the crystallite planes. Each individual current

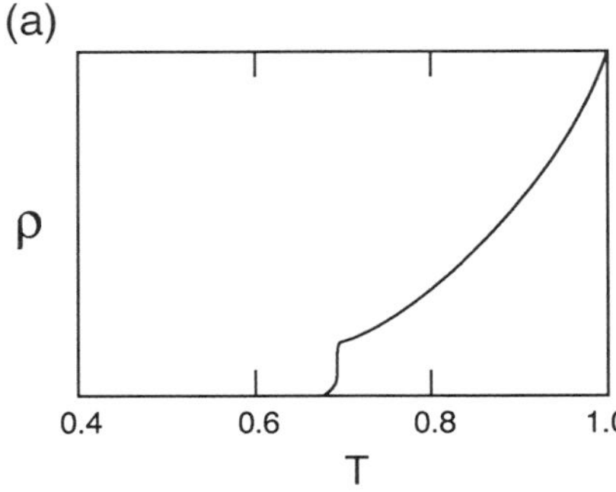

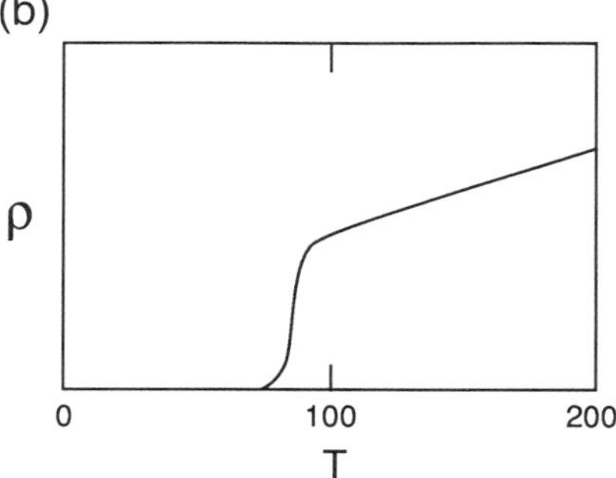

Figure 2.3 Abrupt drop of the resistivity to zero at the superconducting transition temperature T_c (a) for a low-temperature superconductor in the Bloch T^5 region and (b) for a high-temperature superconductor in the linear region.

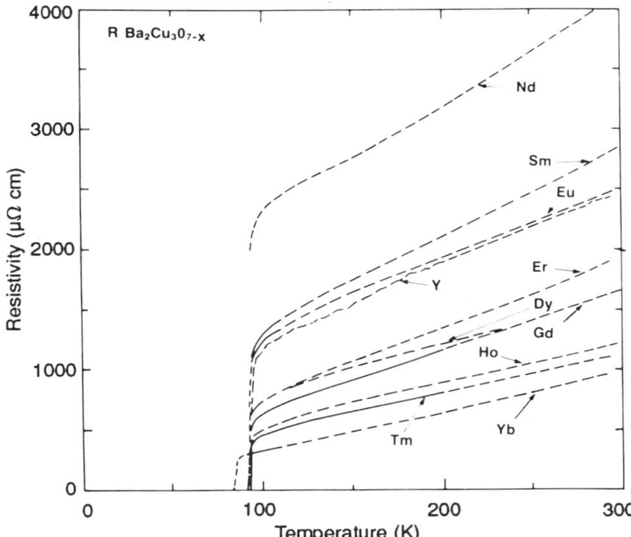

Figure 2.4 Temperature dependence of resistivity for various rare earth substituted $RBa_2Cu_3O_7$ compounds. For these compounds T_c is in the linear region (Tarascon *et al.*, 1987b).

zigzags from one crystallite to the next, so that its total path is longer than it would be if all of the crystallites were aligned with their planes parallel to the direction of the current. The increase in the resistivity of a polycrystalline sample beyond ρ_{ab} can be a measure of how much the average path length increases.

Polycrystalline samples should be compacted or pressed into pellets before resistivity measurements are made, in order to reduce the number of voids in the sample and minimize intergrain contact problems.

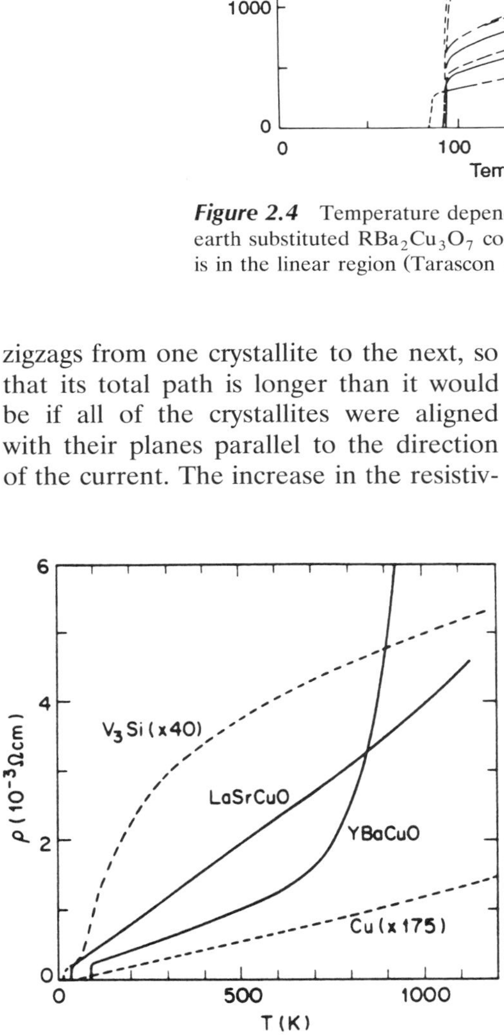

Figure 2.5 Comparison of the resistivities of $(La_{0.9125}Sr_{0.0825})_2CuO_4$ and $YBa_2Cu_3O_{7-\delta}$ with those of the *A*15 compound V_3Si ($T_c = 17.1$ K), and with nonsuperconducting copper (Gurvitch and Fiory, 1987a, b, c).

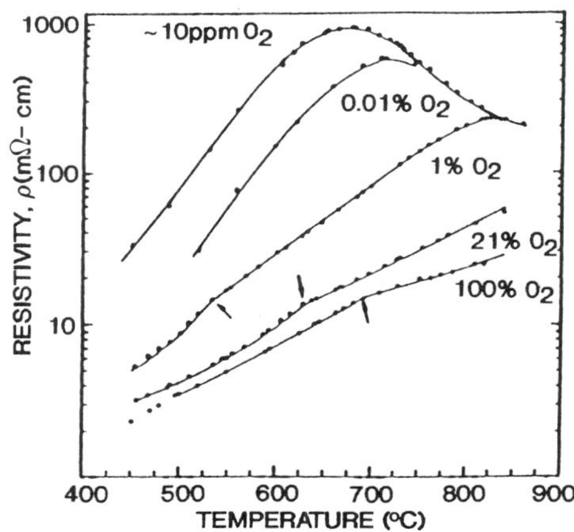

Figure 2.6 Temperature dependence of the resistivity of $YBa_2Cu_3O_{7-\delta}$ for various oxygen partial pressures (Grader *et al.*, 1987).

Table 2.2 Resistivity Data on Superconducting Single Crystals Slightly above T_c and near Room Temperature. The Slopes $\Delta\rho/\Delta T$ Are Averages for the Typical Range from 150 K to 290 K. (Earlier data on mostly polycrystalline samples are given in Table X-1 of Poole *et al.* (1988)[a])

Material	T K	ρ_{ab} $\mu\Omega$ cm	ρ_c mΩ cm	ρ_c/ρ_{ab}	$\Delta\rho_{ab}/\Delta T$ $\mu\Omega$ cm/K	$\Delta\rho_c/\Delta T$ mΩ cm/K	Reference
Re	50	0.8	0.0005	0.6			Volkenshteyn *et al.* (1978)
					0.075	0.055	
Re	275	17.5	0.013	0.7			Volkenshteyn *et al.* (1978)
TaS_2				450			Wattamaniuk *et al.* (1975); Hambourger and Di Salvo (1980)
2H–$TaSe_2$	4			1200			Martin *et al.* (1990)
2H–NbS_2				7000			Pfalzgraf and Spreckels (1987)
K_3C_{60} thin film[e]	290	[2.5 mΩ cm]					Palstra *et al.* (1992)
$(La_{0.925}Sr_{0.075})_2CuO_4$	40	950	19				Hidaka *et al.* (1989)
$(La_{0.925}Sr_{0.075})_2CuO_4$	50	2500					Preyer *et al.* (1991)
					13		
$(La_{0.925}Sr_{0.075})_2CuO_4$	290	5000					Preyer *et al.* (1991)
$(Nd_{0.925}Ce_{0.075})_2CuO_4$	30	1700	500	300			Crusellas *et al.* (1991)
					19	3.7	
$(Nd_{0.925}Ce_{0.075})_2CuO_4$	273	4800	1300	270			Crusellas *et al.* (1991)
$(Nd_{0.925}Ce_{0.075})_2CuO_4$	30	150					Suzuki and Hikita (1990)
					2.0		
$(Nd_{0.925}Ce_{0.075})_2CuO_4$	290	500					Suzuki and Hikita (1990)
$YBa_2Cu_3O_{7-\delta}$	100	200	18	90			Tozer *et al.* (1987)
$YBa_2Cu_3O_{7-\delta}$	300	450	13	30			Tozer *et al.* (1987)
$YBa_2Cu_3O_{7-\delta}$	100	250	9	36			Iye *et al.* (1988)
					1.6	0.06	

$YBa_2Cu_3O_{7-\delta}$	290	550	21	38			Iye *et al.* (1988)
$YBa_2Cu_3O_{7-\delta}$	100	150	16	115			Penney *et al.* (1988)
					1.0		
$YBa_2Cu_3O_{7-\delta}$	290	330	13	40			Penney *et al.* (1988)
$YBa_2Cu_3O_{7-\delta}$	100	70	10.5	150			Hagen *et al.* (1988)
					0.5	0.017	
$YBa_2Cu_3O_{7-\delta}$	290	180	11	60			Hagen *et al.* (1988)
$YBa_2Cu_3O_{7-\delta}$	110	220	19	86			Wu *et al.* (1991b)
					0.7	0.008	
$YBa_2Cu_3O_{7-\delta}$	290	380	19	50			Wu *et al.* (1991b)
$YBa_2Cu_3O_{7-\delta}$	100	30^b	2.5	85			Friedmann *et al.* (1990)
					0.3	0.01	
$YBa_2Cu_3O_{7-\delta}$	275	105^b	4.6	45			Friedmann *et al.* (1990)
$Bi_2Sr_2CuO_{6\pm\delta}$	25	90	14,000	1.6×10^5			Martin *et al.* (1990)
					0.9	−6	
$Bi_2Sr_2CuO_{6\pm\delta}$	290	275	6000	2.2×10^4			Martin *et al.* (1990)
$Bi_2Sr_{2.2}Cu_2O_8$	100	55^c	5200	9.5×10^4			Martin *et al.* (1988)
					0.46	15	
$Bi_2Sr_{2.2}Cu_2O_8$	300	150^c	8880	5.9×10^4			Martin *et al.* (1988)
$Bi_2Sr_2CaCu_2O_8$	300	(d)		10^5			Fiory *et al.* (1988)
$Bi_2Sr_2CaCu_2O_8$				10^5			Busch *et al.* (1992)
$Tl_2Ba_2CuO_6$	110	900					Mukaida *et al.* (1990)
$Tl_2Ba_2CaCu_2O_8$	110	3500					Mukaida *et al.* (1990)

[a] Typical semiconductors range from 10^4 to 10^{15} $\mu\Omega$ cm and insulators from 10^{20} to 10^{28} $\mu\Omega$ cm.
[b] Averages of $\rho_a = 40$, $\rho_b = 20$ $\mu\Omega$ cm at 100 K, and $\rho_a = 145$, $\rho_b = 65$ $\mu\Omega$ cm at 275 K. See Welp *et al.* (1990): $\rho_a/\rho_b \approx 1.2$ to 1.85 near room temperature.
[c] Averages of $\rho_a = 60$, $\rho_b = 50$ $\mu\Omega$ cm at 100 K, and $\rho_a = 180$, $\rho_b = 120$ $\mu\Omega$ cm at 300 K.
[d] $\rho_a/\rho_b \approx 2.5$ from 120 K to 540 K; see also Honma *et al.* (1991).
[e] Not a single crystal.

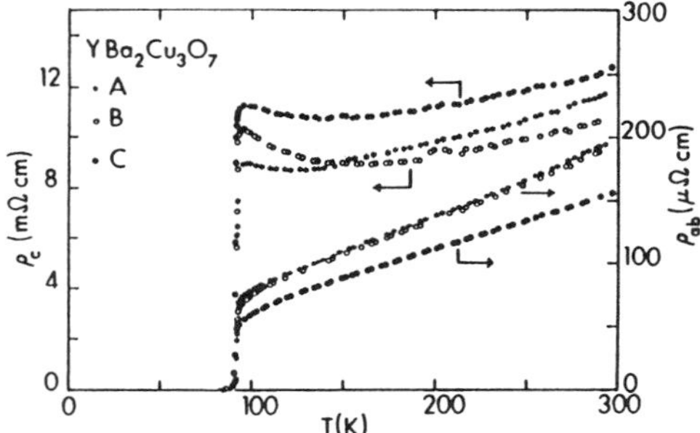

Figure 2.7 Resistivity for current flow parallel (ρ_{ab}) and perpendicular (ρ_c) to the CuO planes of $YBa_2Cu_3O_7$. Data are given for three samples A, B, and C. Note from the change in scale that $\rho_{ab} \ll \rho_c$ (Hagen *et al.*, 1988).

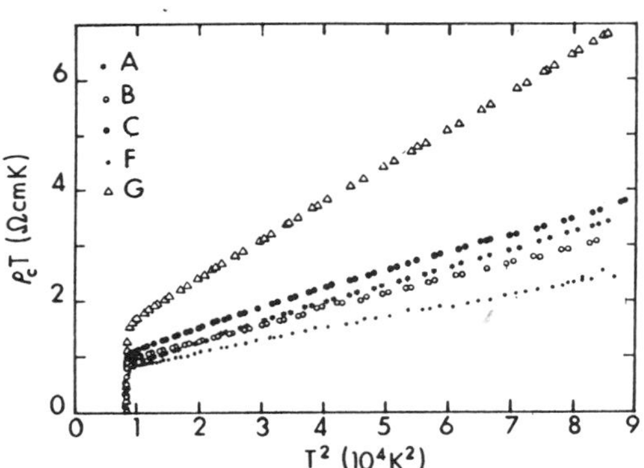

Figure 2.8 Plot of $\rho_c T$ versus T^2 to test the validity of Eq. (2.2) for five single crystals of $YBa_2Cu_3O_{7-\delta}$ (Hagen *et al.*, 1988).

Such compacted samples require appropriate heat treatments to maintain the proper oxygen content. Uniaxial compression tends to align the grains with their c-axes parallel so that the resulting compressed pellets have different resistivities when measured parallel to the compression direction compared to when they are measured perpendicular to this direction.

Hysteresis effects have been seen in the resistance-versus-temperature curves, as illustrated in Fig. X-1 of the monograph by Poole *et al.* (1988), for

$$(Y_{0.875}Ba_{0.125})_2CuO_{4-\delta}$$

(Tarascon *et al.*, 1987a). These hysteresis effects occur in the presence of both magnetic fields and transport currents, with the latter illustrated in Figure X-1.

C. Anisotropy Determination

The most common way of measuring the resistivity of a sample is the four-probe method sketched in Fig. 2.9. Two leads or probes carry a known current into and out of the ends of the sample, and two other leads separated by a distance L measure the voltage drop at points nearer the center where the current approximates uniform, steady-state flow. The resistance R between measurement points 3 and 4 is given by the ratio V/I of the measured

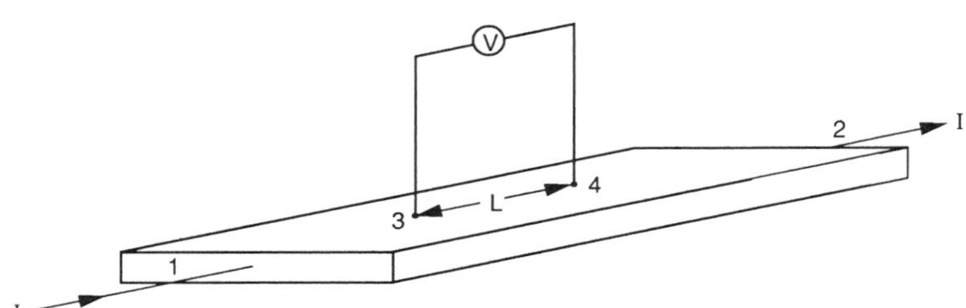

Figure 2.9 Experimental arrangement for the four-probe resistivity determination.

voltage to the input current, and the resistivity ρ is calculated from the expression

$$R = \frac{\rho L}{A}, \tag{2.4}$$

where A is the cross-sectional area. This four-probe technique is superior to a two-probe method in which uniform, steady-state current flow is not assured, and errors from lead and contact resistance are greater.

The four-probe method is satisfactory for use with an anisotropic sample if the sample is cut with one of its principal directions along the direction of current flow and if the condition $L \gg \sqrt{A}$ is satisfied. For a high-temperature superconductor, this requires two samples for the resistivity determination, one with the c-axis along the current flow direction and one with the c-axis perpendicular to this direction.

Transverse and longitudinal resistance determinations, R_t and R_l respectively, can both be made on a sample cut in the shape of a rectangular solid with $a = b$, with the shorter c-axis along the current direction, as shown in Fig. 2.10 (Hagen *et al.*, 1988). These resistances R_t and R_l are used to calculate the resistivity ρ_{ab} in the a,b-plane and the resistivity and ρ_c perpendicular to this plane, i.e., along c. The expressions that relate the resistances depend on the parameter x,

$$x = \frac{c}{a}\left(\frac{\rho_c}{\rho_{ab}}\right)^{1/2}, \tag{2.5}$$

where $\rho_c/\rho_{ab} \approx 100$ for $YBa_2Cu_3O_{7-\delta}$. For the limiting case $x \ll 1$, the measured resistances are given by (Montgomery, 1971):

$$R_t = \frac{a}{bc}\rho_{ab}\left[1 - \frac{4\ln 2}{\pi}\right] \qquad x \ll 1, \tag{2.6}$$

$$R_l = \frac{c}{ab}\rho_c\left[\frac{16\exp(-\pi/x)}{\pi x}\right] \qquad x \ll 1. \tag{2.7}$$

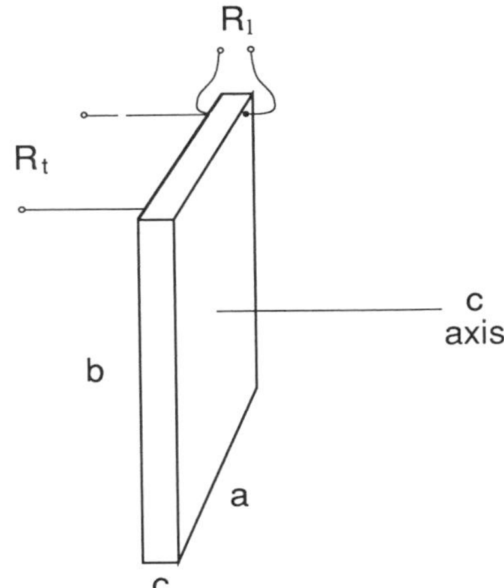

Figure 2.10 Experimental arrangement for measuring anisotropic resistivities (see explanation in text) (Hagen *et al.*, 1988).

and for the opposite limit $x \gg 1$ we have

$$R_t = \frac{a}{bc}\rho_{ab}\left[\frac{16x\exp(-\pi x)}{\pi}\right] \qquad x \gg 1, \tag{2.8}$$

$$R_l = \frac{c}{ab}\rho_c\left[1 - \frac{4\ln 2}{\pi x}\right] \qquad x \gg 1. \tag{2.9}$$

Both the resistance with the exponential factor and the correction term containing the factor $4\ln 2/\pi$ are small.

Contributions to the electrical conductivity in the normal state near T_c arising from fluctuations of regions of the sample into the superconducting state, sometimes called paraconductivity, have been observed and discussed theoretically (X. F. Chem *et al.*, 1993; Friedman *et al.*, 1989; Lawrence and Doniach, 1971; Shier and Ginsberg, 1966). Several recent theoretical articles treating resistivity have appeared (e.g., Gijs *et al.*, 1990a; Hopfengärtner *et al.*, 1991; Kumar and Jayannavar, 1992; Sanborn *et al.*, 1989; Yel *et al.*, 1991).

D. Sheet Resistance of Films: Resistance Quantum

When a current flows along a film of thickness d through a region of surface

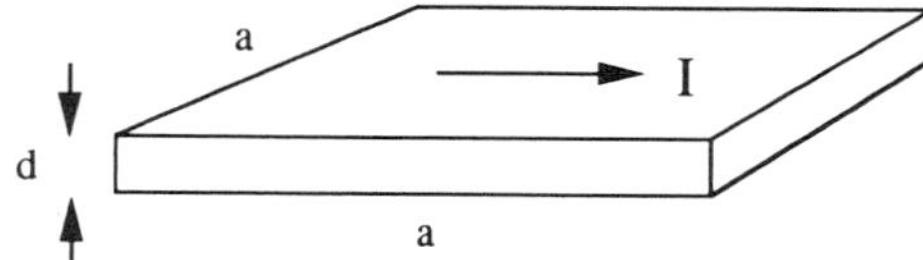

Figure 2.11 Geometrical arrangement and current flow direction for sheet resistance determination.

with dimensions $a \times a$, as shown in Fig. 2.11, it encounters the resistance R_s which, from Eq. (2.4), is given by

$$R_s = \frac{\rho a}{ad} = \frac{\rho}{d}. \quad (2.10)$$

The resistance ρ/d is called the *sheet resistance*, or the *resistance per square*, because it applies to a square section of film, as shown in Fig. 2.11, and is independent of the length of the side a. It is analogous to the surface resistance $R_s = \rho/\delta$ of a metallic surface interacting with an incident high-frequency electromagnetic wave, where δ is the skin depth of the material at the frequency of the wave.

There is a quantum of resistance $h/4e^2$ with the value

$$\frac{h}{4e^2} = 6.45 \text{ k}\Omega, \quad (2.11)$$

where the charge is $2e$ per pair. When the films are thin enough so that their sheet resistance in the normal state just above T_c exceeds this value, they no longer become superconducting (Hebard and Paalanen, 1990; Jaeger *et al.*, 1989; Lee and Ketterson, 1990; Li *et al.*, 1990; Pyun and Lemberger, 1991; Seidler *et al.*, 1992; Tanda *et al.*, 1991; Valles *et al.*, 1989; T. Wang *et al.*, 1991). It has been found experimentally (Haviland *et al.*, 1989) that bismuth and lead films deposited on germanium substrates become superconducting only when they have thicknesses greater than 0.673 nm and 0.328 nm, respectively. The variation in T_c with the sheet resistance for these two thin films is shown in Fig. 2.12. Figure 2.13 shows the sharp drop in resistivity at T_c for bismuth films with a range

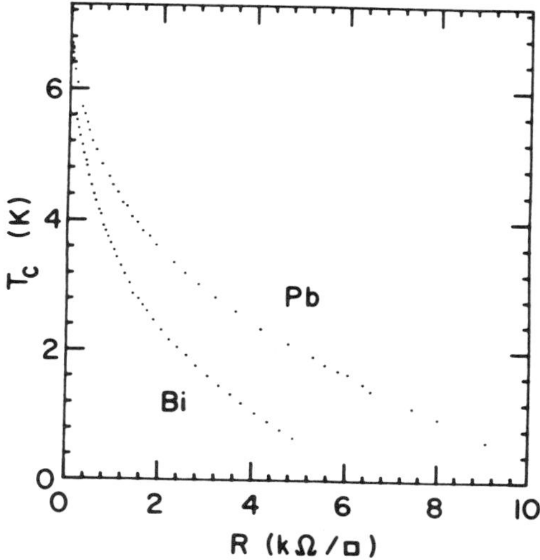

Figure 2.12 Dependence of the transition temperature T_c of Bi and Pb films on the sheet resistance (Haviland *et al.*, 1989).

of thickness greater than 0.673 nm. Thinner films exhibit resistivity increases down to the lowest measured temperatures, as shown in the figure. The ordinary transition temperatures, which occur for the limit $\rho/d \ll h/4e^2$, are 6.1 K for Bi films and 7.2 K for Pb.

Copper-oxide planes in high-temperature superconductors can be considered thin conducting layers, with thickness c for $YBa_2Cu_3O_{7-\delta}$ corresponding to a sheet resistance $\rho_{ab}/\frac{1}{2}c$. Using this layer approximation, the Ioffe–Regel parameter $k_F l$ mentioned in Section A can be estimated from the expression

$$k_F l = \frac{\text{conductance per square}}{\text{conductance quantum}} \quad (2.12)$$

$$= \frac{h/4e^2}{2\rho_{ab}/\frac{1}{2}c}, \quad (2.13)$$

where the conductances are the reciprocals of the resistances. Note that the two reported $k_F l$ values for $YBa_2Cu_3O_{7-\delta}$ calculated by this method and referred to earlier assumed $k_F = 4.6 \times 10^7 \text{ cm}^{-1}$.

It is of interest that metallic contacts of atomic size exhibit conduction jumps at

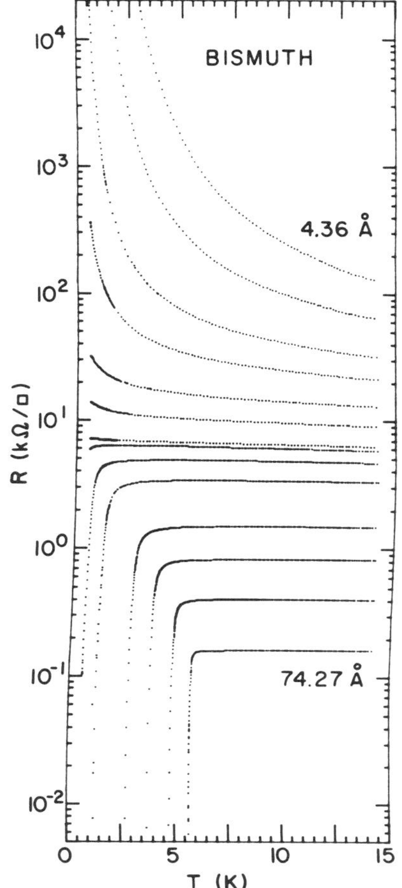

Figure 2.13 Temperature dependence of the sheet resistance of films of Bi deposited on Ge as a function of film thickness in the range from 4.36 Å to 74.27 Å (Haviland *et al.*, 1989).

integral multiples of $2e^2/h$ (Agraït *et al.*, 1993), and that the Hall effect resistance in one-dimensional objects, so-called *quantum wires*, is quantized to $h/2Ne^2$, where $N = 1, 2, 3, \ldots$ (Akera and Andu, 1989).

IV. ZERO RESISTANCE

In 1911, when Onnes was measuring the electrical resistance of mercury, he expected to find a temperature dependence of the type given by Eq. (1.30). Instead, to his surprise, he found that below 4.2 K the electrical resistance dropped to zero, as shown in Fig. 2.14. He had discovered superconductivity! At this temperature mercury transforms from the normal metallic state to that of a superconductor. Figure 2.3a shows the abrupt change to zero resistance for the case of an old superconductor, where T_c is in the low-temperature Bloch T^5 region, while Fig. 2.3b shows what happens in the case of a high-temperature superconductor where T_c is in the linear region.

A. Resistivity Drop at T_c

Figures 2.3a, 2.3b, 2.7, and 2.8 show the sharp drop in resistance that occurs at T_c. We will see later in the chapter that there is an analogous drop in susceptibility at T_c.

A susceptibility measurement is a more typical thermodynamic indicator of the superconducting state because magnetization is a thermodynamic state variable. Resis-

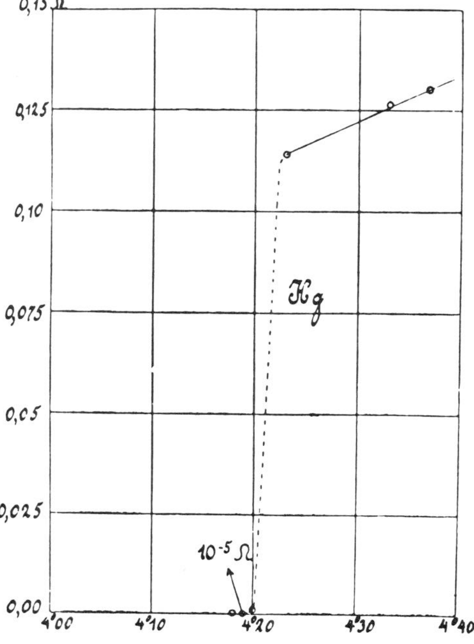

Figure 2.14 Resistivity-versus-temperature plot obtained by Kamerlingh Onnes when he discovered superconductivity in Leiden in 1911.

tivity, on the other hand, is easier to measure, and can be a better guide for applications. Generally, the T_c value determined from the resisitivity drop to zero occurs at a somewhat higher temperature than its susceptibility counterpart. This is because any tiny part of the material going superconductive loses its resistance, and $R = 0$ when one or more continuous superconducting paths are in place between the measuring electrodes. In contrast, diamagnetism measurements depend on macroscopic current loops to shield the **B** field from an appreciable fraction of the sample material, and this happens when full superconducting current paths become available. Therefore, filamentary paths can produce sharp drops in resistivity at temperatures higher than the temperatures at which there are pronounced drops in diamagnetism, which also require extensive regions of superconductivity. Such filamentary behavior can be described in terms of percolation thresholds (Gingold and Lobb 1990; Lin, 1991; Phillips, 1989b; Tolédano *et al.*, 1990; Zeng *et al.*, 1991).

B. Persistent Currents below T_c

To establish a transport current in a loop of superconducting wire, the ends of the wire may be connected to a battery in series with a resistor, thus limiting the current, as shown in Fig. 2.15. When switch S_2 is closed, current commences to flow in the loop. When switch S_1 is closed in order to bypass the battery and S_2 opened in order to disconnect the battery, the loop resistance drops to zero and the current flow enters the persistent mode. The zero resistance property implies that the current will continue flowing indefinitely.

Many investigators have established currents in loops of superconducting wire and have monitored the strength of the associated magnetic field through the loop over prolonged periods of time using, for example, a magnetometer with a pickup coil, as shown in Fig. 2.15. In experiments it was found that there is no detectable decay of the current for periods of time on the order of several years. The experiments established lower limits on the lifetime of the current and upper limits on the possible resistivity of superconducting materials. Currents in copper oxide superconductors persist for many months or in excess of a year, and resistivity limits have been reported as low as 10^{-18} (Yeh *et al.*, 1987) and 10^{-22} Ω cm (Kedve *et al.*, 1987). Super current lifetimes of low-temperature superconductors are also greater than a

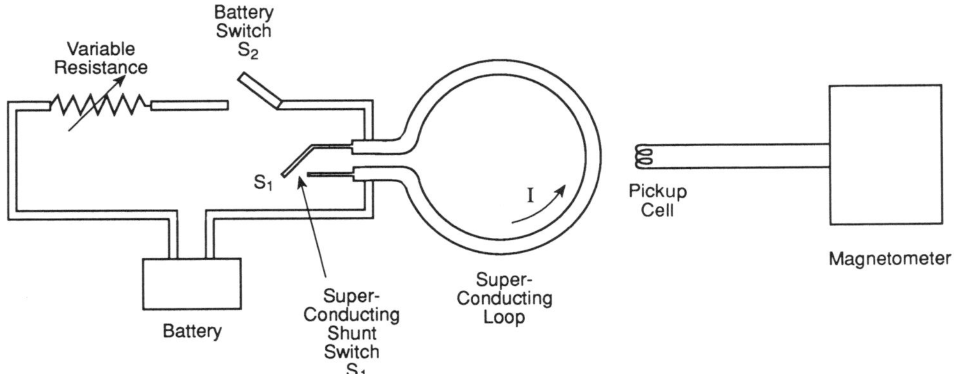

Figure 2.15 Experimental arrangement for establishing and measuring a persistent current. Switch S_2 is closed to send current through the loop and S_1 is closed to confine the current flow to the loop. The magnetometer measures the magnetic field through the loop and thereby determines the current.

year for $\rho < 10^{-23}$ Ω cm (Chandrasekhar, 1969). Persistent current flow has also been treated theoretically (e.g., Ambegaokar and Eckern, 1991; Cheun *et al.*, 1988; Kopietz, 1993; Riedel *et al.*, 1989; von Oppen and Riedel, 1991).

It will be instructive to estimate the minimum resistivity of a simple loop of superconducting wire of loop radius r and wire radius a. The inductance L of the loop is given by

$$L \approx \mu_0 r[\ln(8r/a) - 2]. \qquad (2.14)$$

The loop has 'length' $2\pi r$ and cross-sectional area πa^2, so that its resistance R is

$$R = \frac{2r\rho}{a^2}. \qquad (2.15)$$

This gives it a time constant $\tau = L/R$. Combining Eqs. (2.14) and (2.15) gives the product

$$\rho\tau \approx \tfrac{1}{2}\mu_0 a^2[0.0794 + \ln(r/a)]. \qquad (2.16)$$

Using $\mu_0 = 4\pi \times 10^{-7}$ H/m and typical loop dimensions of $a = 1.5$ mm and $r = 15$ cm gives

$$\rho\tau \approx 6.6 \times 10^{-10} \ \Omega \text{ cm s} \qquad (2.17)$$

for the product $\rho\tau$.

A super current I_s can be made to flow in the loop by subjecting it to a changing magnetic field below T_c, in accordance with Faraday's and Lenz' laws. The magnitude of the current that is flowing can be determined by measuring the induced magnetic field. At a point P along the axis a distance z above the loop, as shown in Fig. 2.16, this magnetic field has the following value, as given in standard general physics texts:

$$B(z) = \frac{\mu_0 I_s r^2}{2(r^2 + z^2)^{3/2}}, \qquad (2.18)$$

and once $B(z)$ is measured, I_s can be calculated.

If the super current persists unchanged for over a year ($\tau > 3.16 \times 10^7$ s) without any appreciable decrease (we are assuming that a 1% decrease is easily detectable), Eq. (2.17) can be used to place an upper limit on the resistivity:

$$\rho < 2.1 \times 10^{-17} \ \Omega \text{ cm}, \qquad (2.19)$$

which is in the range mentioned earlier, and is 11 orders of magnitude less than the resistivity of copper ($\rho = 1.56$ μ Ω cm). A similar loop of copper wire at room temperature has $\tau \approx 0.42$ ms, so that the current will be gone after several milliseconds.

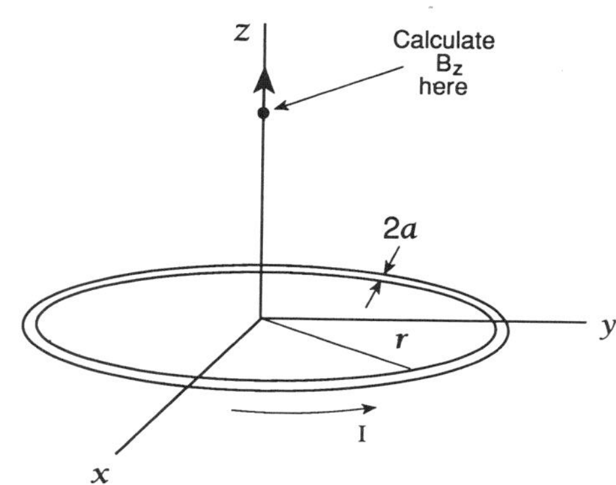

Figure 2.16 Magnetic field B_z along the axis of a circular loop of wire of radius r carrying the current I. The wire itself has radius a.

We will see in Section XIV that the drop to zero resistance can be explained in terms of a two–fluid model in which some of the normal electrons turn into super electrons which move through the material without resistance. The current carried by the flow of super electrons is then assumed to short circuit the current arising from the flow of normal electrons, causing the measured resistance to vanish.

V. TRANSITION TEMPERATURE

Before proceeding to the discussion of magnetic and transport properties of superconductors, it will be helpful to say a few words about the transition temperature. We will discuss it from the viewpoint of the resistivity change even though the onset of the energy gap and pronounced diamagnetism are more fundamental indices of T_c. Pechan and Horvath (1990) described a fast and inexpensive method for accurate determination of transition temperatures above 77 K.

Although the theoretical transition from the normal to the superconducting state is very sharp, experimentally it sometimes occurs gradually and sometimes abruptly. Figure 2.17 shows the gradual decrease in resistivity near T_c that was reported by Bednorz and Müller (1986) in the first published article on the new superconductors. We see from this figure that the range of temperatures over which the resistivity changes from its normal-state value to zero is comparable with the transition temperature itself. An example of a narrow transition centered at 90 K with width of ≈ 0.3 K is shown in Fig. 2.18. These two cases correspond to $\Delta T/T_c \approx 1/2$ and $\Delta T/T_c \approx 0.003$, respectively.

The sharpness of the drop to zero resistance is a measure of the goodness or purity of the sample. Figure 2.19 shows how the drop to zero in pure tin becomes broader and shifts to a higher temperature in an impure specimen. In a sense impure tin is a better superconductor because it has a higher T_c but worse because it has a broader transition. When high-temperature superconductors are doped with paramagnetic ions at copper sites, the

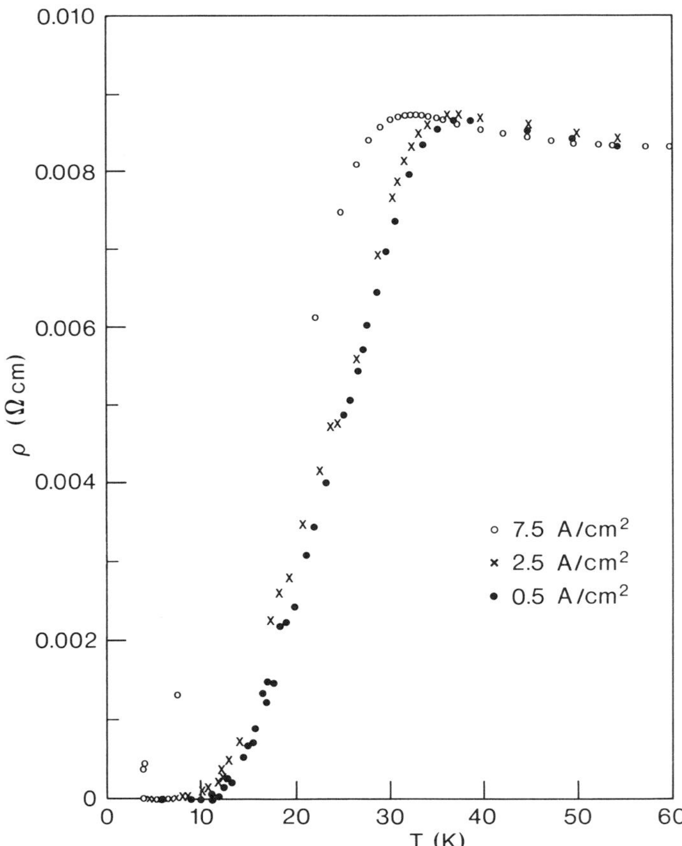

Figure 2.17 First reported drop to zero resistance for a high-temperature superconductor (Bednorz and Müller, 1986).

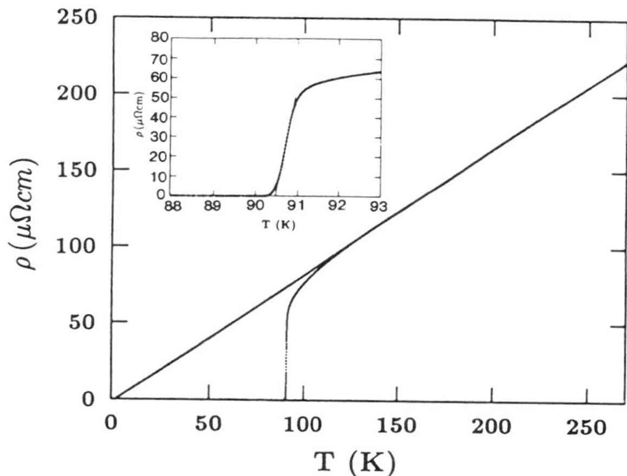

Figure 2.18 Sharp drop to zero resistance of a $YBa_2Cu_3O_7$ epitaxial film (Hopfengärtner *et al.*, 1991).

transition temperature both shifts to lower values and broadens, whereas doping at the yttrium sites of YBaCuO has very little effect on T_c, as may be seen by comparing the data plotted in Figs. 2.20 and 2.4, respectively. This can be explained in terms of delocalization of the super electrons on the copper oxide planes.

There are various ways of defining the position and sharpness, or width, of the superconducting transition temperature, and the literature is far from consistent on this point. Authors talk in terms of the onset, 5%, 10%, midpoint, 90%, 95%, and zero resistance points, and Fig. 2.21 shows some of these on an experimental resistivity curve. The onset, or 0% point, is where the experimental curve begins to drop below the extrapolated high-temperature linear behavior of Eq. (1.30), indicated by the dashed line in the figure. The T_c values that we cite or list in the tables are ordinarily midpoint values at which $\rho(T)$ has decreased by 50% below the onset. Many of the published reports of unusually high transition temperatures actually cited onset values, which can make them suspect. The current density can influence the resistive transition (Goldschmidt, 1989).

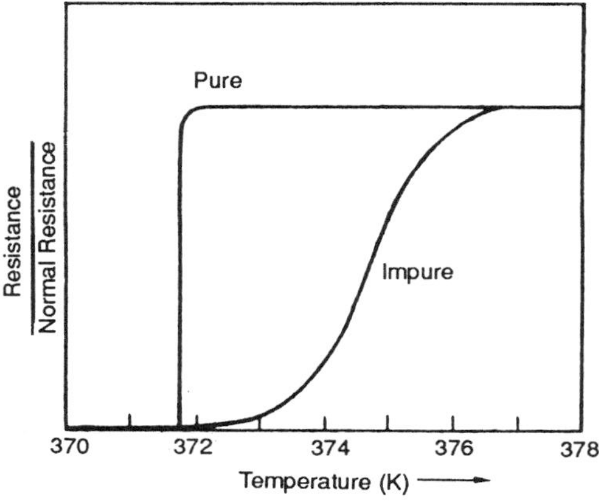

Figure 2.19 Narrow and broad superconducting resistivity drop in pure and impure tin, respectively. Reprinted from Rose Innes and Rhoderick (1978), p. 7, with kind permission of Pergamon Press, Headington Hill Hall, Oxford OX3 0BW, UK.

The point at which the first derivative of the resistivity curve, shown in Fig. 2.22b, reaches its maximum value could be selected as defining T_c, since it is the inflection point on the original curve (Azoulay, 1991; Datta *et al.*, 1988; Nkum and Datars, 1992; Poole and Farach, 1988). The width ΔT between the half-amplitude points of the first derivative curve, or the peak-to-peak width ΔT_{pp} of the second derivative curve sketched in Fig. 2.22c, are both good quantitative measures of the width of the transition. An asymmetry parameter, equal

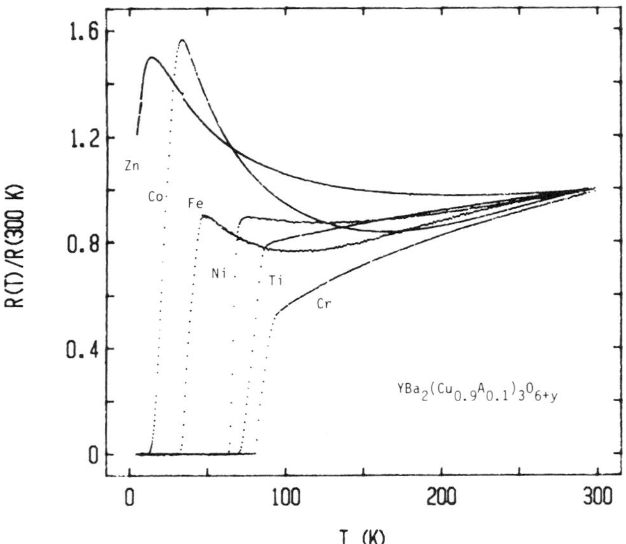

Figure 2.20 Influence of doping $YBa_2(Cu_{0.9}M_{0.1})_3O_{6+y}$ with the first transition series ions M = Ti, Cr, Fe, Co, Ni, and Zn on the resistivity transition near T_c (Xiao *et al.*, 1987a).

to $[(A-B)/(A+B)]$, may also be evaluated from Fig. 2.22c.

There appear to be enough data points near the midpoint of Fig. 2.22a to accurately define the transition, but the first and second derivative curves of Figs. 2.22b and 2.22c, respectively, show that this is not the case. This need for additional data points demonstrates the greater precision of the derivative method.

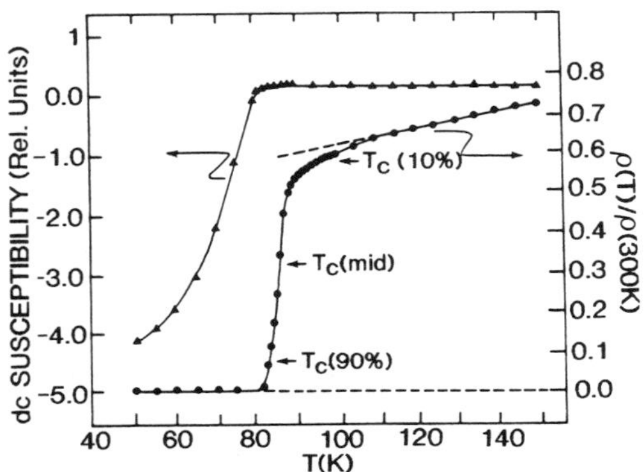

Figure 2.21 Temperature dependence of the resistivity and zero-field-cooled magnetization of $HoBa_2Cu_3O_7$. The 10%-drop, midpoint, and 90%-drop points are indicated on the resistivity curve (Ku *et al.*, 1987).

Phase transitions in general have finite widths, and a typical approach is to define T_c in terms of the point of most rapid change from the old to the new phase. Critical exponents are evaluated in this region near T_c. Ordinarily, less account is taken of the more gradual changes that take place at the onset or during the final approach to the new equilibrium state. The onset of superconductivity is important from a physics viewpoint because it suggests that superconducting regions are being formed, whereas the zero point is important from an engineering viewpoint because it is where the material can finally carry a super current.

VI. PERFECT DIAMAGNETISM

The property of perfect diamagnetism, which means that the susceptibility $\chi = -1$ in Eq. (1.78a),

$$\mathbf{B} = \mu_0 \mathbf{H}(1+\chi), \quad (2.20)$$

$$= \mu_0(\mathbf{H}+\mathbf{M}), \quad (2.21)$$

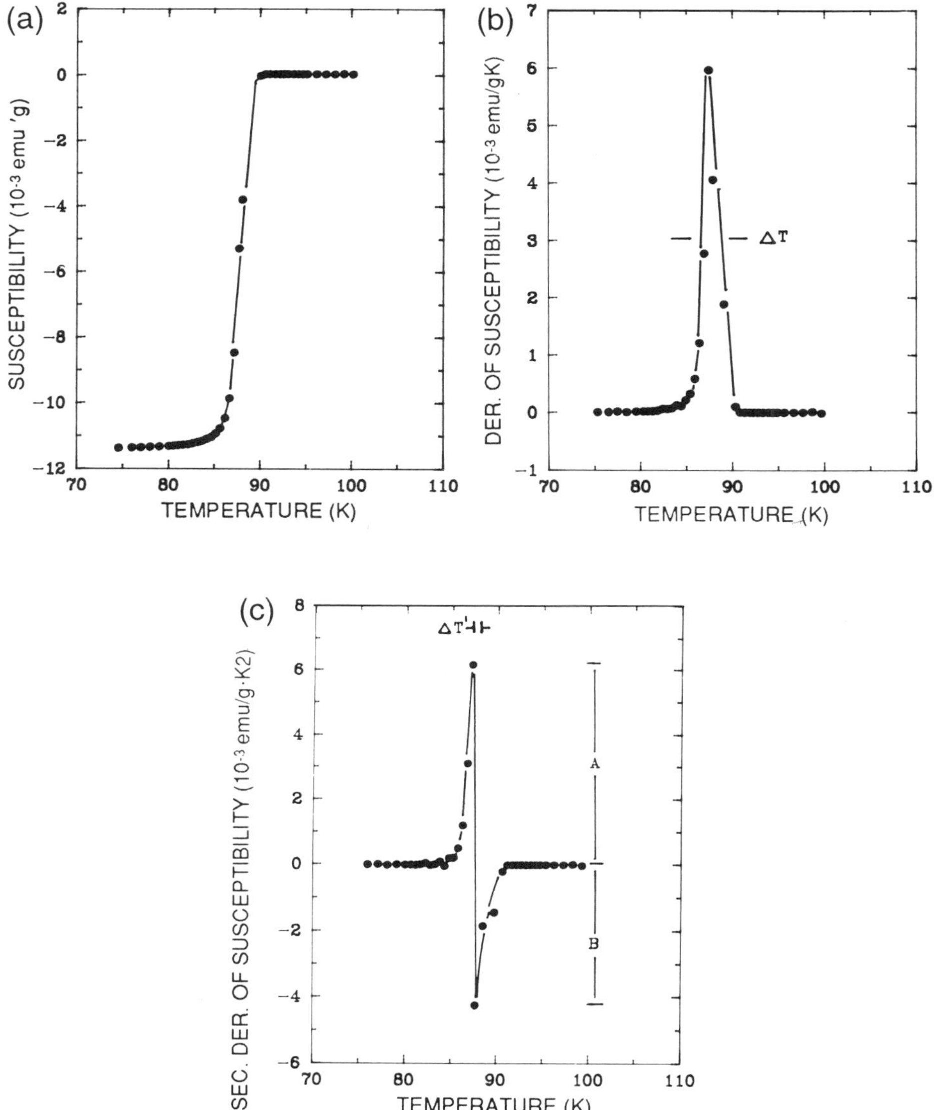

Figure 2.22 Zero-field-susceptibility of $YBa_2Cu_3O_7$ as a function of temperature in a magnetic field of 0.1 mT: (a) usual susceptibility plot χ; (b) first derivative plot $d\chi/dT$; and (c) second derivative plot $d^2\chi/d^2T$ (Almasan *et al.*, 1988).

is equivalent to the assertion that there can be no **B** field inside a perfect diamagnet because the magnetization **M** is directed opposite to the **H** field and thereby cancels it:

$$\mathbf{M} = -\mathbf{H}. \tag{2.22}$$

When a superconductor is placed between the pole pieces of a magnet, the **B** field lines from the magnet go around it instead of entering, and its own internal field remains zero, as shown in Fig. 2.23. This field distribution is the result of the superposition of the uniform applied field and a dipole field from the reversely magnetized superconducting sphere, as illustrated in Fig. 2.24 (Jackson, 1975; cf. Section 5-10).

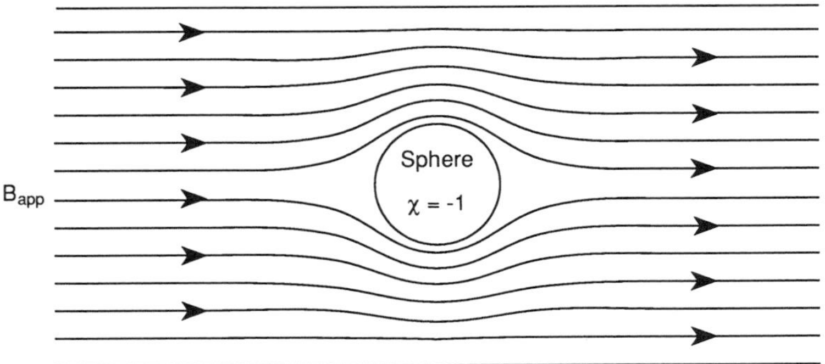

Figure 2.23 Curvature of magnetic field lines around a superconducting sphere in a constant applied field.

There are two aspects to perfect diamagnetism in superconductors. The first is flux exclusion: If a material in the normal state is zero field cooled (ZFC), that is, cooled below T_c to the superconducting state without any magnetic field present, and is then placed in an external magnetic field, the field will be excluded from the superconductor. The second aspect is flux expulsion: If the same material in its normal state is placed in a magnetic field, the field will penetrate and have almost the same value inside and outside because the permeability μ is so close to the free-space value μ_0. If this material is then field cooled (FC), that is, cooled below T_c in the presence of this field, the field will be expelled from the material, a phenomenon called the *Meissner effect*. These two processes are sketched on the left side of Fig. 2.25. Although ZFC and FC lead to the same result (absence of magnetic flux inside the sample below T_c), nevertheless the two processes are not equivalent, as we will see in Section IX.

Thompson *et al.* (1991) found that for a "defect-free" high-purity niobium sphere the ZFC and FC susceptibilities are almost identical. A second high-purity sphere of similar composition that exhibited strong pinning was also examined and the same ZFC results were obtained, except that no Meissner flux expulsion following field cooling was observed. The pinning was so strong that the vortices could not move out of the sample. Figure 2.25 is drawn for the

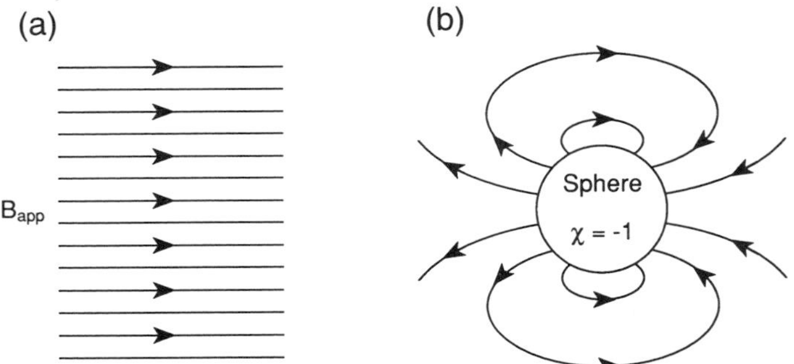

Figure 2.24 Sketch of constant applied magnetic field (a) and dipole field (b) that superimpose to provide the magnetic field lines shown in Fig. 2.23 (Jackson, 1975).

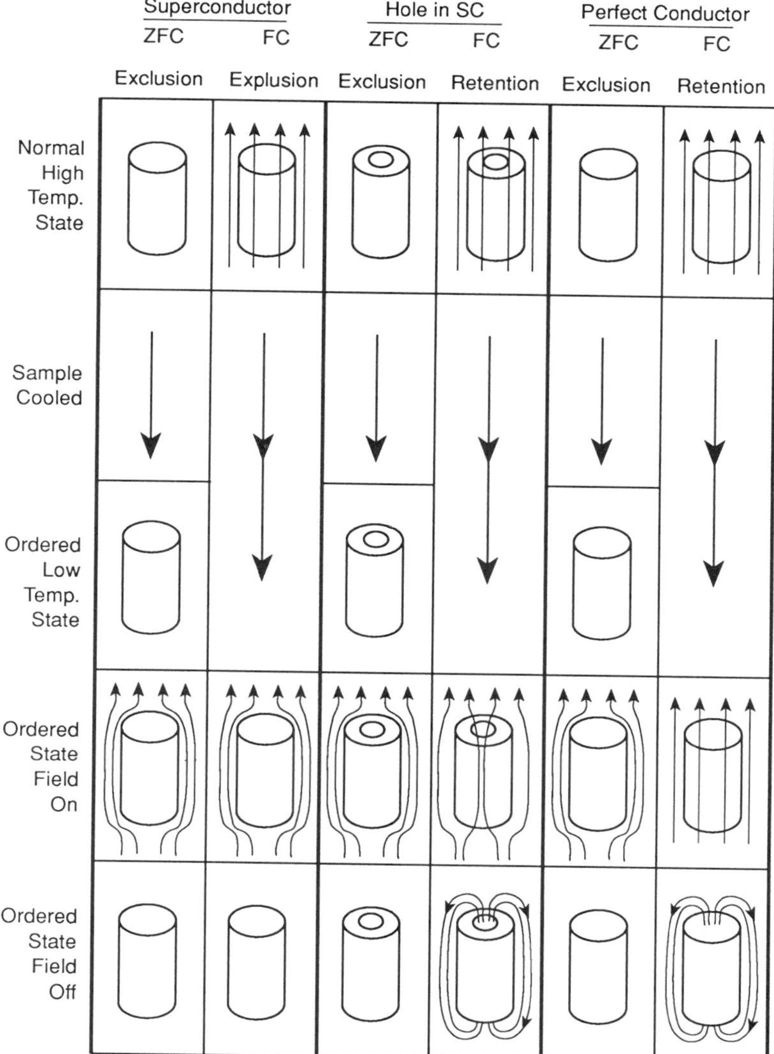

Figure 2.25 Effect of zero field cooling (ZFC) and field cooling (FC) of a solid superconducting cylinder (left), a superconducting cylinder with an axial hole (center), and a perfect conductor (right).

case of very weak pinning, in which virtually all of the flux is expelled from the superconducting material following field cooling.

VII. FIELDS INSIDE A SUPERCONDUCTOR

To further clarify the magnetic field configurations inside a superconductor, consider a long cylindrical sample placed in a uniform applied magnetic field with its axis in the field direction, as indicated in Fig. 2.26. Since there are no applied currents, the boundary condition at the surface given in Chapter 1, Section XIV,

$$H'_{\parallel} = H''_{\parallel}, \tag{2.23}$$

shows that the H field is uniform inside with the same value as the applied field:

$$H_{\text{app}} = H_{\text{in}} \tag{2.24}$$

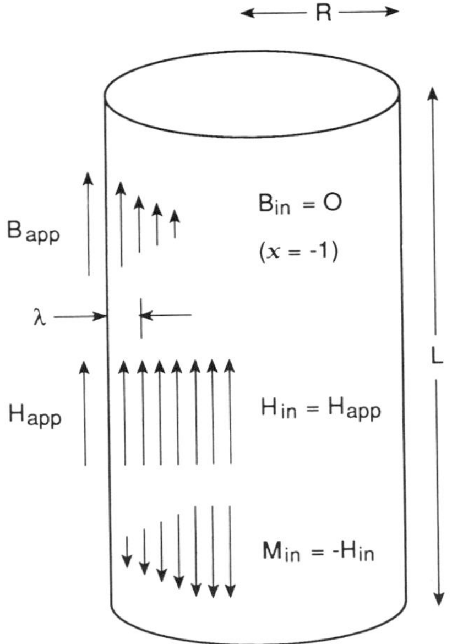

Figure 2.26 Boundary region and internal fields for a superconducting cylinder in an axial external magnetic field B_{app}.

The B field has only a z component with value $B_{\text{app}} = \mu_0 H_{\text{app}}$ outside and zero inside, $B_{\text{in}} = 0$. There is, however, a transition layer of thickness λ, called the *penetration depth*, at the surface of the superconductor where the B field drops exponentially from its value B_{app} on the outside to zero inside, in accordance with the expression

$$B(r) \approx B_0 \exp[-(R-r)/\lambda], \quad (2.25)$$

as shown in Fig. 2.27. Thus the B field exists only in the surface layer, and not in the bulk. Since

$$B_{\text{in}}(r) = \mu[H_{\text{in}} + M(r)] \quad (2.26)$$

with $H_{\text{in}} = H_{\text{app}}$, we have for $M(r)$

$$M(r) \approx -H_{\text{app}}\left\{1 - \exp\left[-\frac{(R-r)}{\lambda}\right]\right\}, \quad (2.27)$$

again subject to the assumption that $\lambda \ll R$, and this is also sketched in Fig. 2.27.

We will show later in Chapter 5, Sections VII and VIII, that this exponential

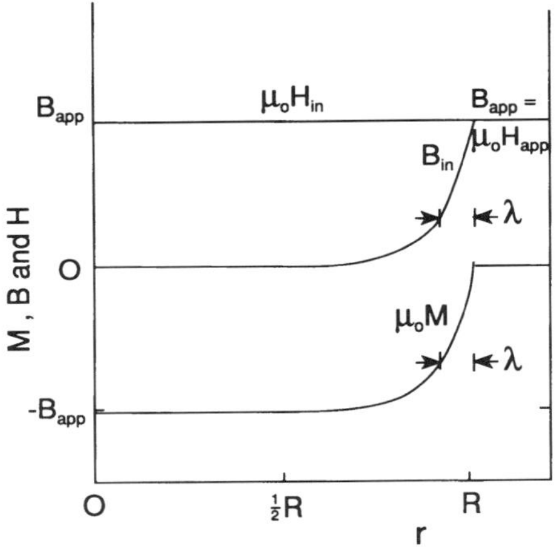

Figure 2.27 Plot of the fields B and $\mu_0 H$ and of the magnetization $\mu_0 M$ outside $(r > R)$ and inside $(r < R)$ a superconducting cylinder of radius R in an axial applied field B_{app}.

decay process arises naturally in the Ginzburg–Landau and London theories, and that these theories provide an explicit formula for what is called the *London penetration depth* λ_L, namely

$$\lambda_L = \left(\frac{m}{\mu_0 n_s e^2}\right)^{1/2}. \quad (2.28)$$

VIII. SHIELDING CURRENT

In the absence of any applied transport current we set $\mathbf{J} = 0$ (also $\partial D/\partial t = 0$) in Maxwell's equation, Eq. (1.72), to obtain

$$\nabla \times \mathbf{B}_{\text{in}} = \mu_0 \nabla \times \mathbf{M} \quad (2.29a)$$

$$= \mu_0 \mathbf{J}_{\text{sh}} \quad (2.29b)$$

where $\mathbf{J}_{\text{sh}}$ is called the *shielding* or *magnetization current density*:

$$\mathbf{J}_{\text{sh}} = \nabla \times \mathbf{M}. \quad (2.30)$$

Since $\mathbf{B}_{\text{in}}$ has only a z or axial component, the curl, expressed in terms of cylindrical coordinates, gives the following shielding current density flowing around the cylinder in the negative ϕ direction:

$$\mathbf{J}_{\text{sh}}(r) = -\frac{1}{\mu_0} \cdot \frac{d\mathbf{B}}{dr} \tag{2.31}$$

$$\approx -\left(\frac{\mathbf{B}_{\text{app}}}{\mu_0 \lambda}\right) \exp\left[-\frac{(R-r)}{\lambda}\right] \tag{2.32}$$

$$\approx -\mathbf{J}_0 \exp\left[-\frac{(R-r)}{\lambda}\right], \tag{2.33}$$

where

$$\mathbf{B}_{\text{app}} = \mu_0 \lambda \mathbf{J}_0, \tag{2.34}$$

again with $\lambda \ll R$, and this circular current flow is sketched in Fig. 2.28 and graphed in Fig. 2.29. In other words, the vectors $\mathbf{B}$ and $\mathbf{J}_{\text{sh}}$ do not exist in the bulk of the superconductor but only in the surface layer where they are perpendicular to each other, with $\mathbf{B}$ oriented vertically and $\mathbf{J}_{\text{sh}}$ flowing around the cylinder in horizontal circles. It may be looked upon as a circulating demagnetizing current that shields or screens the interior of the superconductor by producing a negative $\mathbf{B}$ field that cancels $\mathbf{B}_{\text{app}}$ so that $\mathbf{B}_{\text{in}} = 0$ inside.

Thus we see that the superconducting medium reacts to the presence of the applied field by generating shielding currents that cancel the interior $\mathbf{B}$ field. The reaction of the medium may also be looked upon as generating a magnetization $\mathbf{M}$ that cancels the interior $\mathbf{B}$ field, as was explained above. These are two views of the same phenomenon, since the shielding current density $\mathbf{J}_{\text{sh}}$ and the compensating magnetization $\mathbf{M}$ are directly related through Eq. (2.30). The negative $\mathbf{B}$ field that cancels $\mathbf{B}_{\text{app}}$ is really a magnetization in the negative z direction.

It is instructive to see how Eq. (2.34) is equivalent to the well-known formula

$$B_0 = \frac{\mu_0 NI}{L} \tag{2.35}$$

for the magnetic field B_0 of an N-turn solenoid of length L. Since each turn carries the current I, the total current is NI.

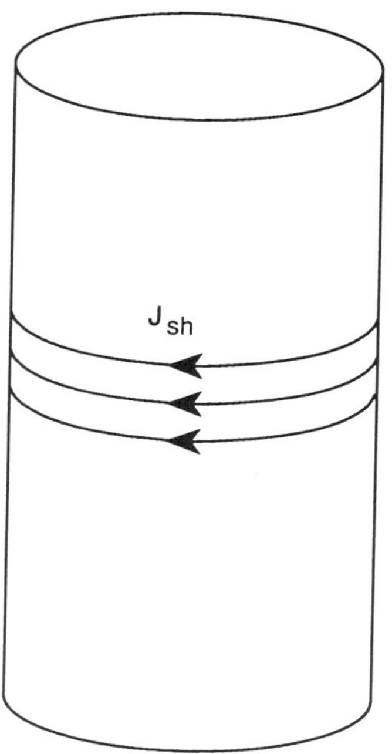

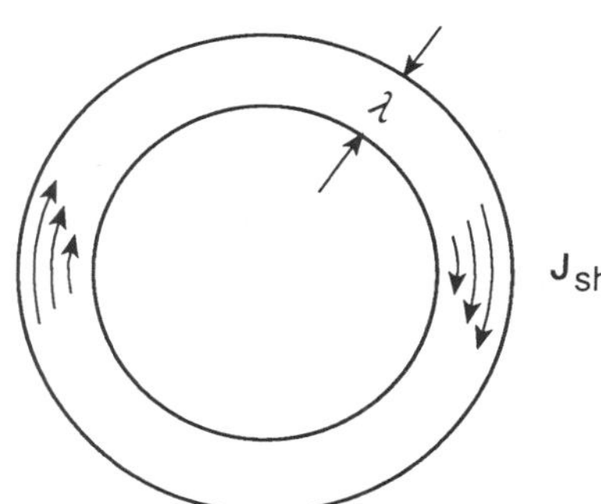

Figure 2.28 Shielding current flow J_{sh} in a surface layer of thickness λ around a superconducting cylinder in an axial applied magnetic field B_{app}.

This total current also equals the current density $\mathbf{J}_0$ times the area λL, corresponding to

$$NI = \lambda L J_0. \tag{2.36}$$

Substituting NI from this expression in Eq. (2.35) gives Eq. (2.34). Thus the circulating shielding current is equivalent to the effect of a solenoid that cancels the applied $\mathbf{B}$ field inside the superconductor.

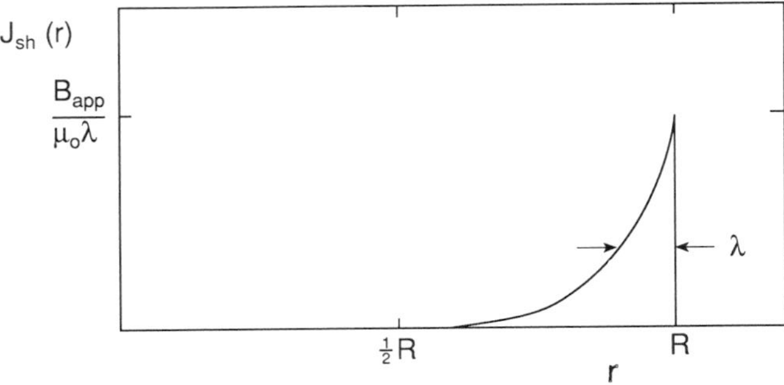

Figure 2.29 Dependence of the shielding current density J_{sh} on the position inside a superconducting cylinder of radius R in an applied axial field B_{app}. Note that J_{sh} has the value H_{app}/λ at the surface.

The dipole field of the superconducting sphere sketched in Fig. 2.24 may be considered as arising from demagnetizing currents circulating in its surface layers, as shown in Fig. 2.30. These demagnetizing currents provide the reverse magnetization that cancels the applied field to make $\mathbf{B} = 0$ inside, just as in the case of a cylinder.

IX. HOLE IN SUPERCONDUCTOR

As an example of how ZFC and FC can lead to two different final states of magnetism let us examine the case of a hole inside a superconductor.

Consider a cylindrical superconducting sample of length L and radius R with a concentric axial hole through it of radius r, as shown in Fig. 2.31. This will be referred to as an "open hole" because it is open to the outside at both ends. If this sample is zero-field-cooled in the manner described in Section VI, an axial magnetic field applied after cooling below T_c will be excluded from the superconductor and also from the open axial hole. Surface currents

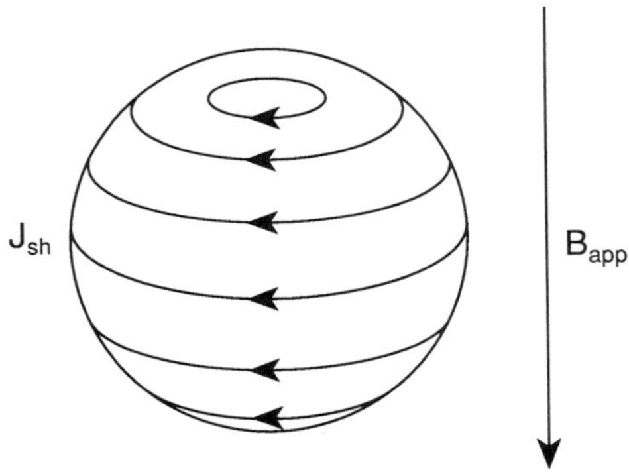

Figure 2.30 Shielding current flow around the surface of a superconducting sphere in an applied magnetic field B_{app}.

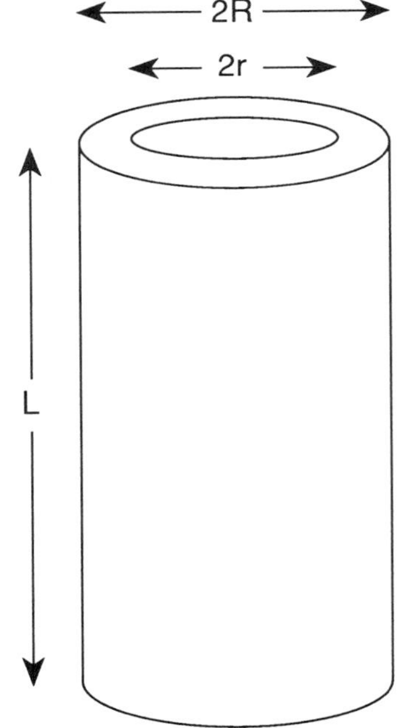

Figure 2.31 Superconducting tube of radius R with an axial hole of radius r.

shield the superconducting regions from the external field and bring about the flux exclusion shown in Fig. 2.25. These same surface currents also shield the hole from the applied field. This means that the superconductor plus the hole act like a perfect diamagnet under zero-field cooling. The entire volume $\pi R^2 L$, including the open hole volume $\pi r^2 L$, has an effective susceptibility of -1,

$$\chi_{\text{eff}} = -1. \tag{2.37}$$

If this same sample, still with an open hole, is field cooled, once it attains the superconducting state the magnetic flux will be expelled from the superconducting material, but will remain in the hole. The same outer surface currents flow to shield the superconductor from the applied field, but the surface currents flowing in the reverse direction around the inside surface of the cylinder, i.e., around the hole periphery as indicated in Fig. 2.32, cancel the effect of the outside surface currents and sustain the original magnetic flux in the hole. The volume of the superconducting material, $(\pi R^2 - \pi r^2)L$, has a susceptibility of -1, but the space in the open hole, $\pi r^2 L$, does not exhibit diamagnetism, so that for the hole $\chi = 0$. The effective susceptibility of the cylinder with the hole is the average of -1 for the superconducting material and 0 for the hole, corresponding to

$$\chi_{\text{eff}} = -\left[1 - \left(\frac{r}{R}\right)^2\right], \tag{2.38}$$

which reduces to -1 for no hole $(r = 0)$ and to 0 for $r = R$. This experimentally measurable result is different from the ZFC open hole case (2.37). Experimentally, it is found that the magnetic susceptibility is less negative for field-cooled samples than for zero-field-cooled samples, as shown by the data in Fig. 2.33. Mohamed *et al.* (1990) give plots of the ZFC and FC magnetic field distributions of a 16-mm diameter, 2-mm thick superconducting disk with a 3-mm diameter axial hole.

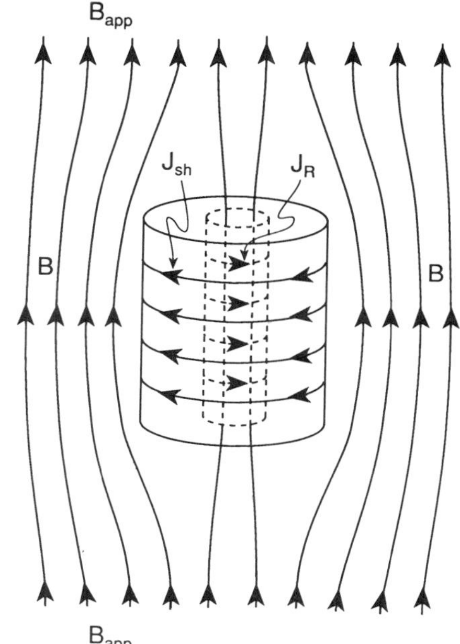

Figure 2.32 Magnetic field lines, shielding current flow J_{sh} on the outside surface, and reverse-direction shielding current flow J_R on the inside surface of a superconducting tube in an applied axial magnetic field. The magnetic field lines pass through the hole because the cylinder has been field cooled.

Another important case to consider is that of a totally enclosed hole of the type shown in Fig. 2.34, which we call a closed hole or cavity. It is clear that for ZFC the closed hole behaves the same as the open hole, that is, flux is excluded from it, with $\chi_{\text{eff}} = -1$, as shown in the fifth column of Fig. 2.25. Flux is also excluded for field cooling. To see this, we recall that the **B** field lines must be continuous and can only begin or end at the poles of a magnet. In the open hole case, the **B** field lines in the hole either join to the externally applied field lines or form loops that close outside the sample, as shown at the bottom of column 4 of Fig. 2.25. The **B** field lines have no way of leaving a closed hole to connect with the external field or to form closed loops outside, so such lines cannot exist inside a cavity completely surrounded by a superconducting material. Therefore, flux is expelled during field cooling, so

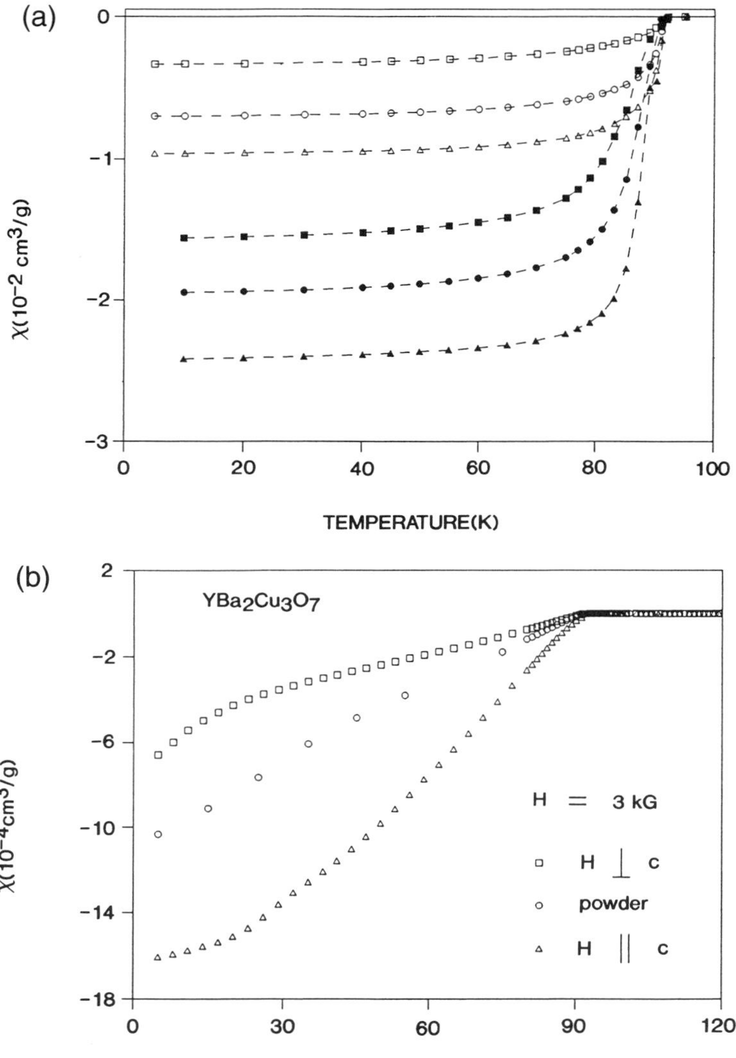

Figure 2.33 Zero-field-cooled (closed symbols) and field-cooled (open symbols) magnetic susceptibility of $YBa_2Cu_3O_7$ nonaligned powder (circles) and grain-aligned samples with the applied field parallel to the *c*-axis (triangles) and perpendicular to the *c*-axis (squares). Results are shown in an applied field of (a) 5 mT and (b) 0.3 T. Note the change in ordinate scale between the two figures (Lee and Johnston, 1990).

again $\chi_{\text{eff}} = -1$. Thus a superconductor with a cavity behaves like a solid superconductor with the difference that magnetization can exist only in the superconductor, not in the cavity.

In this section we have discussed the cases of open and closed holes in superconductors. We showed in Table 1.2 that the susceptibility of typical diamagnetic and paramagnetic samples is quite close to zero, so that the empty hole results also apply to holes filled with typical nonsuperconducting materials. Experimentally, we deal with samples with a known overall or external volume, but with an unknown fraction of this volume taken up by holes,

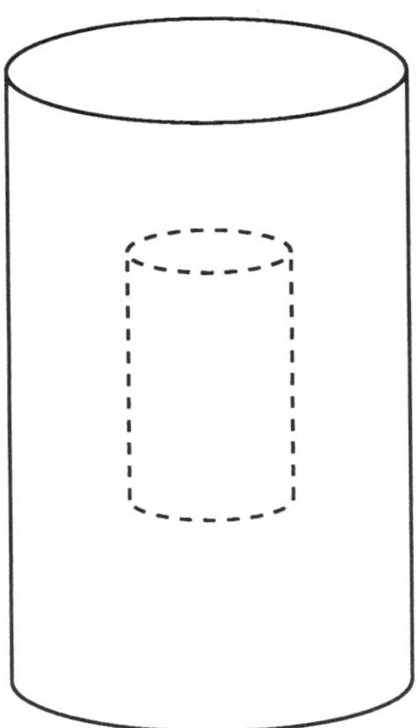

Figure 2.34 Superconducting cylinder with a totally enclosed hole.

intergranular spaces, and nonsuperconducting material that could respond to ZFC and FC preconditioning the same way as a hole.

If a sample is a mixture of a superconducting material and a non-superconducting material with the nonsuperconducting part on the outside so that the applied magnetic field can penetrate it under both ZFC and FC conditions, the average sample susceptibility will be the average of $\chi = 0$ for the normal material and $\chi = -1$ for the superconducting part. Thus both the ZFC and the FC measurement will give values of χ_{eff} that are less negative than -1. A granular superconducting sample can have an admixture of normal material on the outside or inside and space between the grains that produce ZFC and FC susceptibilities of the type shown in Fig. 2.33, where, typically, the measured susceptibilities are $\chi_{\text{zfc}} \approx -0.7$ and $\chi_{\text{fc}} \approx -0.3$.

X. PERFECT CONDUCTIVITY

We started this chapter by describing the perfect conductivity property of a superconductor—namely, the fact that it has zero resistance. Then we proceeded to explain the property of perfect diamagnetism exhibited by a superconductor. In this section we will treat the case of a perfect conductor, i.e., a conductor that has zero resistivity but the susceptibility of a normal conductor, i.e., $\chi \approx 0$. We will examine its response to an applied magnetic field and see that it excludes magnetic flux, but does not expel flux, as does a superconductor. We will start with a good conductor and then take the limit, i.e., letting its resistance fall to zero so that it becomes a hypothetical perfect conductor.

A static magnetic field penetrates a good conductor undisturbed because its magnetic permeability μ is quite close to the magnetic permeability of free space μ_0, as the susceptibility data of Table 1.2 indicate. Therefore, a good conductor placed in a magnetic field leaves the field unchanged, except perhaps for current transients that arise while the field is turned on and die out rapidly. In Section IV we estimated the decay time constant for a loop of copper wire 15 cm in diameter to be 0.42 ms.

Consider a closed current path within the conductor. When the magnetic field $\mathbf{B}_{\text{app}}$ is applied, the magnetic flux through this circuit $\Phi = \mathbf{A} \cdot \mathbf{B}_{\text{app}}$ changes, so that by Lenz' law a voltage $-\mathbf{A} \cdot d\mathbf{B}_{\text{app}}/dt$ is induced in the circuit and a current I flows, as indicated in Fig. 2.35, in accordance

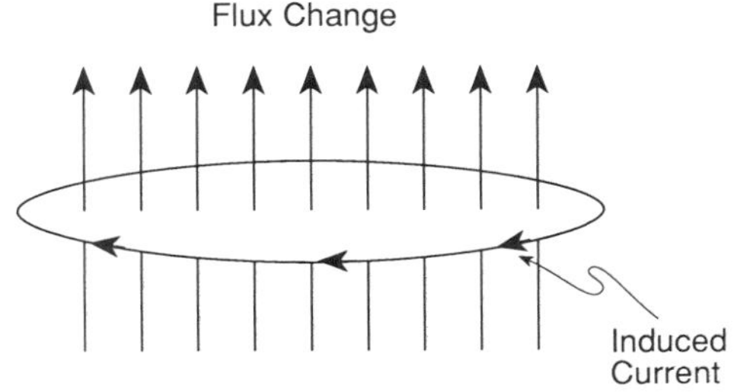

Figure 2.35 Magnetic field B rapidly established through a loop of wire and induced current I which, by Lenz' law, flows in a direction to oppose the establishment of this field.

with the expression

$$-\mathbf{A}\cdot\frac{d\mathbf{B}_{\text{app}}}{dt}=RI+L\frac{dI}{dt}. \quad (2.39)$$

The current rapidly dies out with time constant L/R. For a perfect conductor the resistance term in Eq. (2.39) vanishes. Solving the resultant equation,

$$-\mathbf{A}\cdot\frac{d\mathbf{B}_{\text{app}}}{dt}=L\frac{dI}{dt}, \quad (2.40)$$

gives

$$LI+\mathbf{A}\cdot\mathbf{B}_{\text{app}}=\Phi_{\text{Total}}, \quad (2.41)$$

which means that the total flux $LI+\mathbf{A}\cdot\mathbf{B}_{\text{app}}$ remains constant when the field is applied. If no fields or currents are present and the field $\mathbf{B}_{\text{app}}$ is applied, the flux LI will be induced to cancel that from the applied field and maintain the $\mathbf{B}=0$ state inside the perfect conductor. In real conductors the induced currents die out so rapidly that the internal $\mathbf{B}$ field builds up immediately to the applied field value. Hence a perfect conductor exhibits flux exclusion since a magnetic field turned on in its presence does not penetrate it. It will, however, not expel flux already present because flux that is already there will remain forever. In other words, an FC-perfect conductor retains magnetic flux.

Thus we find that a ZFC-perfect conductor excludes magnetic flux just like a ZFC superconductor. The two, however, differ in their field-cooled properties, the perfect conductor retaining flux and a superconductor excluding flux after FC. A perfect conductor acts like an open hole in a superconductor!

We do not know of any examples of perfect conductors in nature. The phenomenon has been discussed because it provides some insight into the nature of superconductivity.

XI. TRANSPORT CURRENT

In the previous section we discussed the shielding currents induced by the presence of applied magnetic fields. We saw how a field applied along the cylinder axis gives rise to currents circulating around this axis. When a current is applied from the outside and made to flow through a superconductor, it induces magnetic fields near it. An applied current is called a transport current, and the applied current density constitutes the so-called "free" current density term on the right side of Maxwell's inhomogeneous equation (1.68).

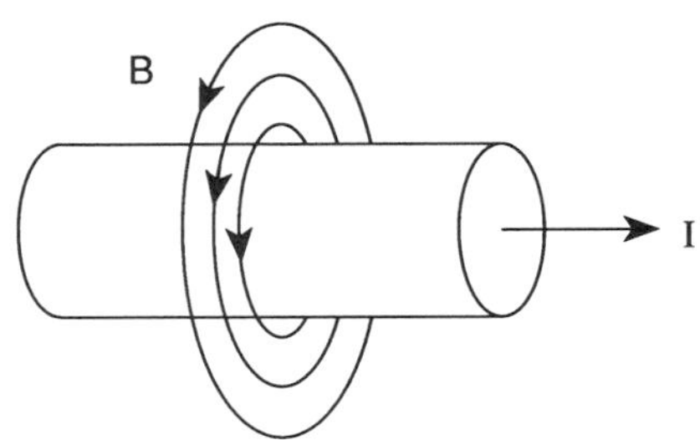

Figure 2.36 Magnetic field lines B around a wire carrying a current I.

Suppose that an external current source causes current I to flow in the direction of the axis of a superconducting cylinder of radius R, in the manner sketched in Fig. 2.36. We know from general physics that the wire has a circular B field around it, as indicated in the figure, and that this field decreases with distance r from the wire in accordance with the expression

$$B=\frac{\mu_0 I}{2\pi r} \qquad r\geq R, \quad (2.42)$$

as shown sketched in Fig. 2.37, with the following value on the surface:

$$B_{\text{surf}}=\frac{\mu_0 I}{2\pi R}. \quad (2.43)$$

We also know that if the current density were uniform across the cross section of the wire, the B field inside would be proportional to the distance from the axis, $B=B_{\text{surf}}(r/R)$, as shown in Fig. 2.37.

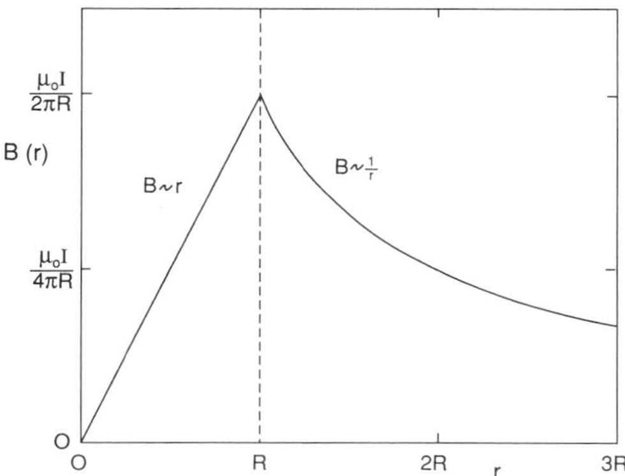

Figure 2.37 Dependence of the internal ($r < R$) and external ($r > R$) magnetic field on distance from the center of a normal conductor wire carrying a current of uniform density.

Since magnetic flux is excluded from inside a superconducting wire, the current density cannot be uniform, and instead the transport current must flow in a surface layer of thickness λ, as shown in Fig. 2.38, to maintain the B field equal to zero inside. This current density $J(r)$ must have the same exponential dependence on distance as given by Eq. (2.31) for the case of the shielding current:

$$J(r) = \frac{B_{\text{surf}}}{\mu_0 \lambda} \exp\left[-\frac{(R-r)}{\lambda}\right] \quad (2.44)$$

$$= \frac{I}{2\pi R\lambda} \exp\left[-\frac{(R-r)}{\lambda}\right]. \quad (2.45)$$

Figure 2.39 shows how the current distribution changes at the junction between a normal wire and a superconducting wire from uniform density flow in the normal conductor to surface flow in the superconductor. The total current I is the integral of the current density $J(r)$ from Eq. (2.45) over the cross section of the superconducting wire, with value

$$I = 2\pi R\lambda J, \quad (2.46)$$

where $J = J(R)$ is the maximum value of $J(r)$, which is attained at the surface, and the quantity $2\pi R\lambda$ is the effective cross-sectional area of the surface layer. Substituting the expression for I from Eq. (2.46) in Eq. (2.43) gives

$$B_{\text{surf}} = \mu_0 \lambda J, \quad (2.47)$$

which is the same form as Eq. (2.34) for the shielding current.

Comparing Eqs. (2.32) and (2.44) we obtain for the magnetic field inside the wire

$$B(r) = B_{\text{surf}} \exp\left[-\frac{(R-r)}{\lambda}\right] \qquad r \le R, \quad (2.48)$$

as shown sketched in Fig 2.40. In Chapter 5, Sections VII and IX, we show how to derive these various exponential decay expressions from the Ginzburg–Landau and London theories. Outside the wire the magnetic field exhibits the same decline with distance in both the normal and superconducting cases, as can be seen by comparing Figs. 2.37 and 2.40.

There is really no fundamental difference between the demagnetizing current and the transport current, except that in the present case of a wire their directions are orthogonal to each other. When a current is impressed into a superconductor it is called a transport current, and it induces a magnetic field. When a superconductor is placed in an external magnetic field, the current induced by this field is called *demagnetization current*. The current–field relationship is the same in both cases. This is why Eqs. (2.25) and (2.48) are the same.

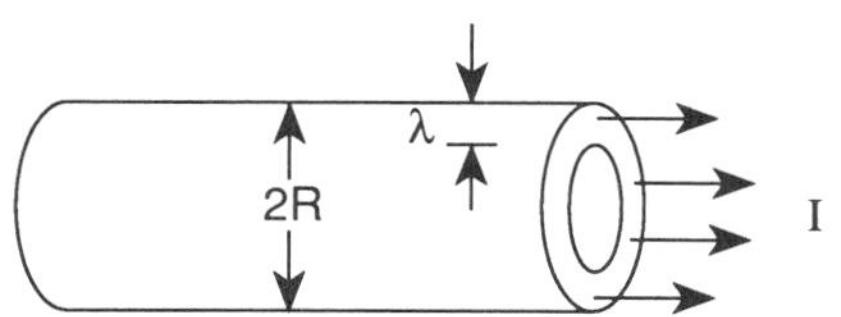

Figure 2.38 Transport current flow in a surface layer of thickness λ of a Type I superconducting wire of radius R.

XII. CRITICAL FIELD AND CURRENT

We noted in Section II that application of a sufficiently strong magnetic field

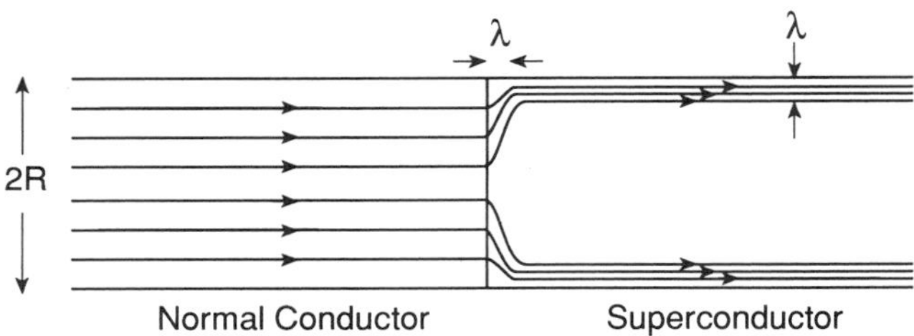

Figure 2.39 Current flow through a wire that is normal on the left and Type I superconducting on the right. Note that the penetration depth λ determines the thickness of both the transition region at the interface and that of the surface layer.

to a superconductor causes its resistance to return to the normal state value, and each superconductor has a critical magnetic field B_c above which it returns to normal. There is also a critical transport current density J_c that will induce this critical field at the surface and drive the superconductor normal. Comparing Eqs. (2.34) and (2.47), respectively, we have for both the demagnetizing and transport cases

$$B_c(T) = \mu_0 \lambda(T) J_c(T), \qquad (2.49)$$

where all three quantities are temperature dependent in a way that will be described in the following section. Either an applied field or an applied current can destroy the superconductivity if either exceeds its respective critical value. At absolute zero, we have

$$B_c(0) = \mu_0 \lambda(0) J_c(0), \qquad (2.50)$$

and this is often written

$$B_c = \mu_0 \lambda J_c, \qquad (2.51)$$

where $T = 0$ is understood.

A particular superconducting wire of radius R has a maximum current, called the critical current I_c, which, by Eq. (2.46), has the value

$$I_c = 2\pi R \lambda J_c. \qquad (2.52)$$

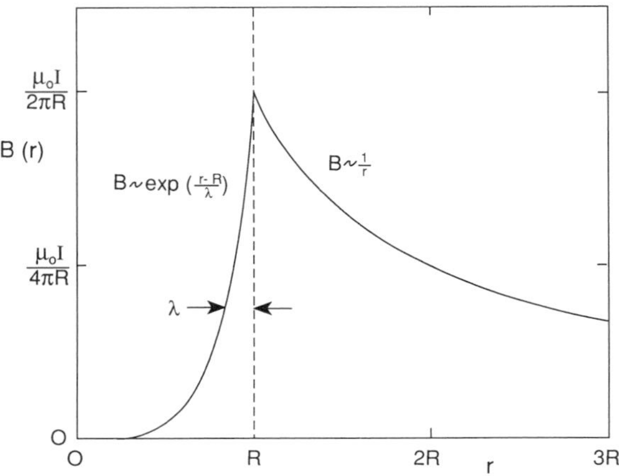

Figure 2.40 Dependence of the internal ($r < R$) and external ($r > R$) magnetic field on distance from the center of a superconducting wire carrying a current that is confined to the surface layer. This figure should be compared with Fig. 2.37.

Using Eq. (2.51), the value of the critical current may be written as

$$I_c = \frac{2\pi R B_c}{\mu_0} \quad (2.53a)$$

$$= 5 \times 10^6\, R H_c. \quad (2.53b)$$

The existence of a critical current in a superconducting wire above which the wire goes normal is called the *Silsbee effect*.

In Type I superconductors with thicknesses much greater than the penetration depth λ, internal magnetic fields, shielding currents, and transport currents are able to exist only in a surface layer of thickness λ. The average current carried by a superconducting wire is not very high when most of the wire carries zero current. To achieve high average super current densities, the wire must have a diameter less than the penetration depth, which is typically about 50 nm for Type I superconductors. The fabrication of such filamentary wires is not practical, and Type II superconductors are used for this application.

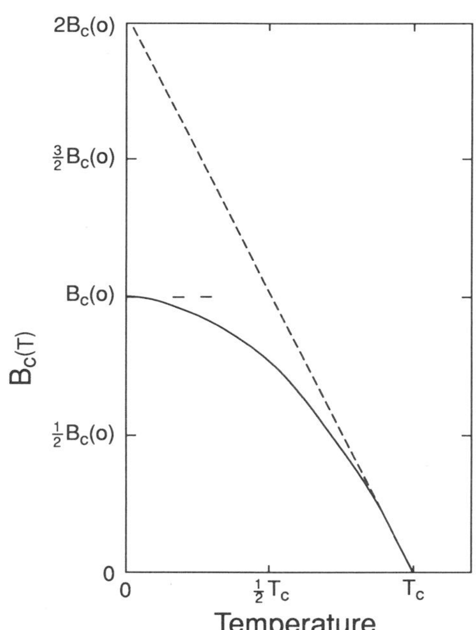

Figure 2.41 Temperature dependence of the critical field $B_c(T)$ corresponding to the behavior expressed by Eq. (2.56). The asymptotic behaviors near $T = 0$ and $T = T_c$ are indicated by dashed lines.

XIII. TEMPERATURE DEPENDENCES

In the normal region above the transition temperature there is no critical field ($B_c = 0$) and there is total magnetic field penetration ($\lambda = \infty$). As a superconductor is cooled down through the transition temperature T_c, the critical field gradually increases to its maximum value $B_c(0)$ at absolute zero ($T = 0$), while the penetration depth decreases from infinity to its minimum value $\lambda(0)$ at absolute zero. The explicit temperature dependences of $B_c(T)$ and $\lambda(T)$ are given by the Ginzburg–Landau theory that will be presented in Chapter 5, where $\lambda(0) = \lambda_L$ as given by Eq. (2.28),

$$\lambda(0) = \left(\frac{m}{\mu_0 n_s e^2}\right)^{1/2}, \quad (2.54)$$

which assumes that all of the conduction electrons are super electrons at $T = 0$. The critical current density may be written as the ratio

$$J_c(T) = \frac{B_c(T)}{\mu_0 \lambda(T)} \quad (2.55)$$

given in Eq. (2.49) in order to obtain the temperature dependence of $J_c(T)$. These temperature dependences have the form

$$B_c = B_c(0)\left[1 - \left(\frac{T}{T_c}\right)^2\right], \quad (2.56)$$

$$\lambda = \lambda(0)\left[1 - \left(\frac{T}{T_c}\right)^4\right]^{-1/2}, \quad (2.57)$$

$$J_c = J_c(0)\left[1 - \left(\frac{T}{T_c}\right)^2\right]\left[1 - \left(\frac{T}{T_c}\right)^4\right]^{1/2}, \quad (2.58)$$

and are sketched in Figs. 2.41, 2.42, and 2.43. Also shown by dashed lines in the

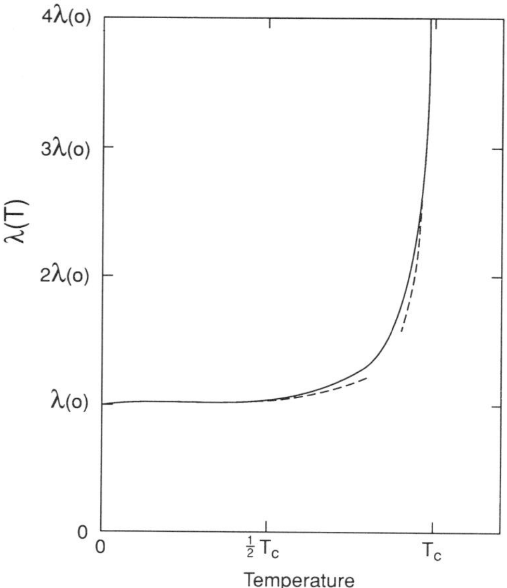

Figure 2.42 Temperature dependence of the penetration depth $\lambda(T)$ corresponding to Eq. (2.57). The asymptotic behaviors near $T = 0$ and $T = T_c$ are indicated by dashed lines.

figures are the asymptotic behaviors near the transition temperature $T \approx T_c$ (Nicol and Carbotte, 1991):

$$B_c \approx 2B_c(0)\left[1 - \frac{T}{T_c}\right], \tag{2.59}$$

$$\lambda \approx \tfrac{1}{2}\lambda(0)\left[1 - \frac{T}{T_c}\right]^{-1/2}, \tag{2.60}$$

$$J_c \approx 4J_c(0)\left[1 - \frac{T}{T_c}\right]^{3/2}. \tag{2.61}$$

Jiang and Carbotte (1992) give plots of $\lambda(0)/\lambda(T)$ for various theoretical models and anisotropies. The asymptotic behaviors near absolute zero, $T \to 0$, are as follows:

$$B_c = B_c(0)\left[1 - \left(\frac{T}{T_c}\right)^2\right], \tag{2.62}$$

$$\lambda \approx \lambda(0)\left[1 + \frac{1}{2}\left(\frac{T}{T_c}\right)^4\right], \tag{2.63}$$

$$J_c \approx J_c(0)\left[1 - \left(\frac{T}{T_c}\right)^2\right], \tag{2.64}$$

which are proven in Problems 5 and 6, respectively. Note that Eq. (2.62) is identical to Eq. (2.56). Some authors report other values of the exponents or expressions related to Eqs. (2.56)–(2.64) for B_c (Miu, 1992; Miu *et al.*, 1990), λ (Däumling and Chandrashekhar, 1992; Hebard *et al.*, 1989; Kanoda *et al.*, 1990; Kogan *et al.*, 1988), and J_c (Askew *et al.*, 1991; Freltoft *et al.*, 1991).

For later reference we give here the temperature dependence of the superconducting energy gap E_g in the neighborhood of T_c:

$$E_g \approx 1.74E_g(0)\left[1 - \frac{T}{T_c}\right]^{1/2} \tag{2.65}$$

(cf. Section VI, Chapter 6 for an explanation of the energy gap and a plot (Fig. 6.7) of this expression). Another length parameter that is characteristic of the superconducting state is the coherence length ξ; this parameter will be introduced in Chapter 5 and referred to frequently throughout the remainder of the text. It is reported to have a $[1 - T/T_c]^{-n}$ dependence, with $n = 1/2$ expected; the penetration depth also depends on

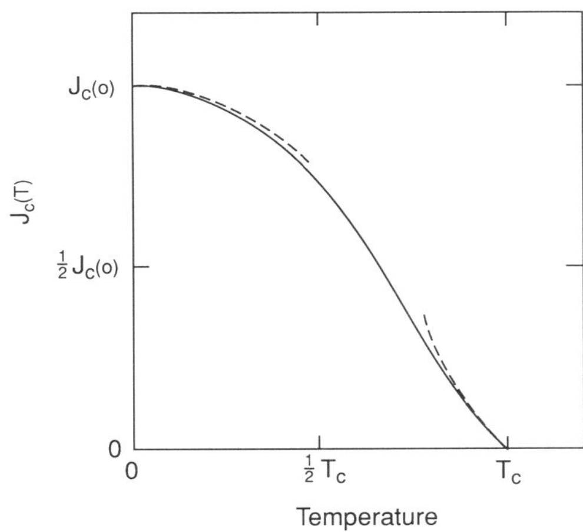

Figure 2.43 Temperature dependence of the critical current density $J_c(T)$ in accordance with Eq. (2.58). The asymptotic behavior near $T = 0$ and $T = T_c$ is indicated by dashed lines.

$[1-(T/T_c)]^{-n}$ near T_c (Chakravarty *et al.*, 1990; Duran *et al.*, 1991; Schneider, 1992).

XIV. CONCENTRATION OF SUPER ELECTRONS

Many properties of superconductors can be explained in terms of a two-fluid model that postulates a fluid of normal electrons mixed with a fluid of superconducting electrons. The two fluids interpenetrate but do not interact. A similar model of interpenetrating fluids consisting of normal and superfluid atoms is used to explain the properties of He^4 below its lambda point. When a superconductor is cooled below T_c, normal electrons begin to transform to the super electron state. The densities of the normal and the super electrons, n_n and n_s, respectively, are temperature dependent, and sum to the total density n of the conduction electrons,

$$n_n(T) + n_s(T) = n, \qquad (2.66)$$

where at $T=0$ we have $n_n(0)=0$ and $n_s(0)=n$.

If we assume that Eq. (2.54) is valid for any temperature below T_c,

$$\lambda(T) = \left(\frac{m}{\mu_0 n_s(T) e^2}\right)^{1/2}, \qquad (2.67)$$

then $\lambda(0) = (m/\mu_0 n e^2)^{1/2}$, and we can write

$$\frac{n_s}{n} = \left[\frac{\lambda(0)}{\lambda(T)}\right]^2, \qquad (2.68)$$

which becomes, with the aid of Eq. (2.57),

$$n_s \approx n\left[1 - \left(\frac{T}{T_c}\right)^4\right]. \qquad (2.69)$$

Figure 2.44 shows a sketch of n_s versus temperature. Substituting the latter in Eq. (2.61) gives for the normal electron density

$$n_n \approx n\left(\frac{T}{T_c}\right)^4. \qquad (2.70)$$

Equation (2.68) is useful for estimating super electron densities.

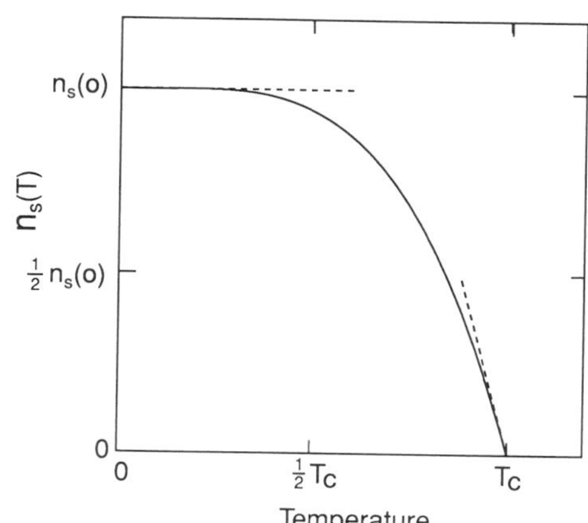

Figure 2.44 Temperature dependence of the density of superconducting electrons n_s as given by Eq. (2.69). The dashed lines show the slope $dn_s/dT = 0$ at $T=0$ and -4 at T_c.

XV. CRITICAL MAGNETIC FIELD SLOPE

We showed in the previous section that the critical magnetic field has the parabolic dependence on temperature given by Eq. (2.56), and this is plotted in Fig. 2.41. The slope of the curve near T_c is given by Eq. (2.59) and may also be written

$$\frac{dB_c(T)}{dT} = -\frac{2B_c(0)}{T_c}. \qquad (2.71)$$

For most Type I superconductors this ratio varies between -15 and -50 mT/K; for example, it has a value of -22.3 mT/K for lead.

A Type II superconductor has two critical fields, a lower-critical field B_{c1} and an upper-critical field B_{c2}, where $B_{c1} < B_{c2}$, as we will see in Chapter 9. These critical fields have temperature dependences similar to that of Eq. (2.71). Typical values of these two slopes for a high-temperature superconductor are

$$\frac{dB_{c1}}{dT} = -\frac{2B_{c1}(0)}{T_c} \approx -1\ \text{mT/K}, \qquad (2.72)$$

$$\frac{dB_{c2}}{dT} = -\frac{2B_{c2}(0)}{T_c} \approx -1.83\ \text{T/K}. \qquad (2.73)$$

For high-temperature superconductors the slopes of Eqs. (2.72) and (2.73) near T_c can be quite anisotropic.

XVI. CRITICAL SURFACE

The critical behavior of a superconductor may be described in terms of a critical surface in three-dimensional space formed by the applied magnetic field B_{app}, applied transport current J_{tr}, and temperature T, and this is shown in Fig. 2.45. The surface is bounded on the left by the $B_c(T)$ versus T curve (d–c–b–a) drawn for $J_{tr} = 0$; this curve also appears in Figs. 2.41 and 2.46. The surface is bounded on the right by the $J_c(T)$ versus T curve (g–h–i–a) drawn for $B_{app} = 0$, which also appears in Figs. 2.43 and 2.47. Figure 2.46 shows three $B_c(T)$ versus T curves projected onto the $J_{tr} = 0$ plane, while Fig. 2.47 presents three $J_c(T)$ versus T curves projected onto the $B_{app} = 0$ plane. Finally, Fig. 2.48 gives projections of three $J_c(T)$ versus $B_c(T)$ curves onto the $T = 0$ plane. The points a, b, ..., l in the various figures are meant to clarify how the projections are made. The notation $B_c(0) = B_c$ and $J_c(0) = J_c$ is used in these figures.

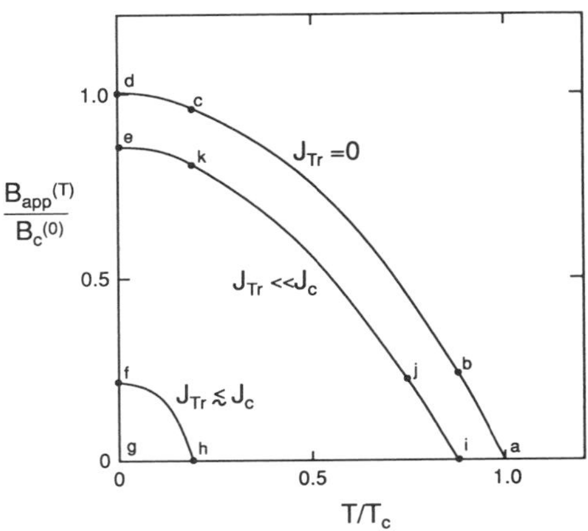

Figure 2.46 Projection of constant current curves of the critical surface of Fig. 2.45 on the B_{app}, T-plane. Projections are shown for $J_{tr} = 0$, $J_{tr} \ll J_c$, and $J_{tr} \approx J_c$. The $J_{tr} = 0$ curve is calculated from Eq. (2.56). The other two curves are drawn so as to have the same shape as the curve for $J_{tr} = 0$.

The x- and y-coordinates of this surface are, respectively, the applied magnetic field B_{app} and the applied transport current J_{tr}. The former does not include the

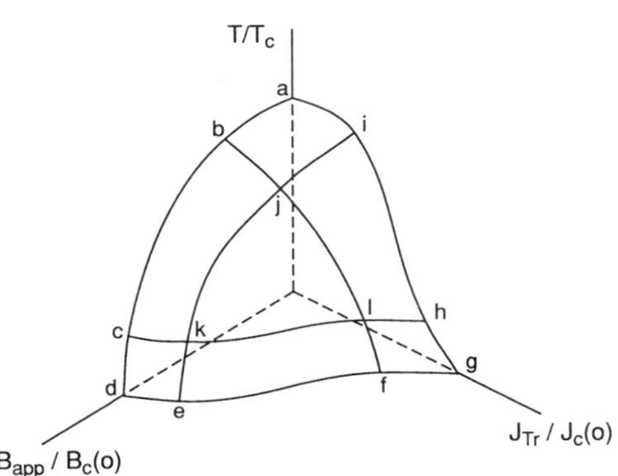

Figure 2.45 Critical surface of a superconductor. Values of applied field B_{app}, transport current J_{tr}, and temperature T corresponding to points below the critical surface, which are in the superconducting region, and points above the critical surface, which are in the normal region. The points on the surface labeled A, B, ..., L also appear in Figs. 2.46–2.49.

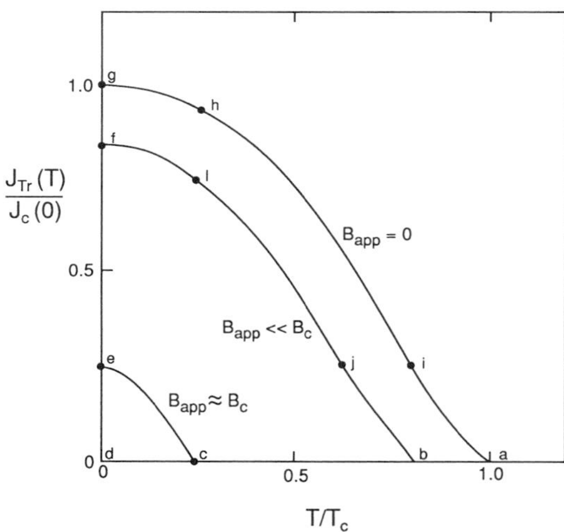

Figure 2.47 Projection of constant applied field curves of the critical surface of Fig. 2.45 onto the J_{tr}, T plane. Projections are shown for $B_{app} = 0$, $B_{app} \ll B_c$, and $B_{app} \approx B_c$. The $B_{app} = 0$ curve is calculated from Eq. (2.58). The other two curves are drawn to have the same shape as the curve for $B_{app} = 0$.

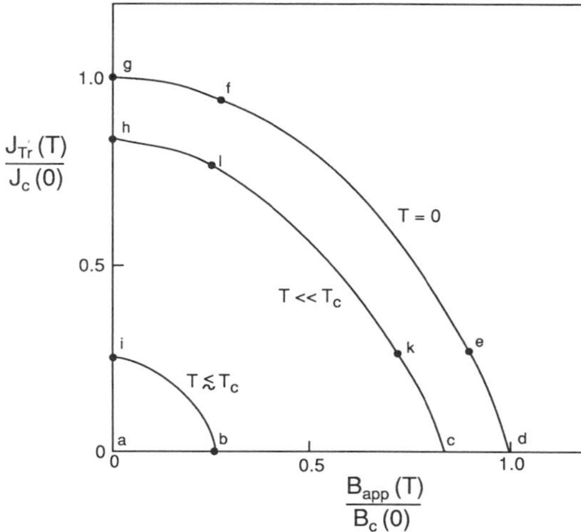

Figure 2.48 Projection of constant-temperature curves of the critical surface of Fig. 2.45 onto the J_{tr}, B_{app} plane. Projection isotherms are shown for $T = 0$, $T \ll T_c$, and $T \approx T_c$. The shapes given for these curves are guesses.

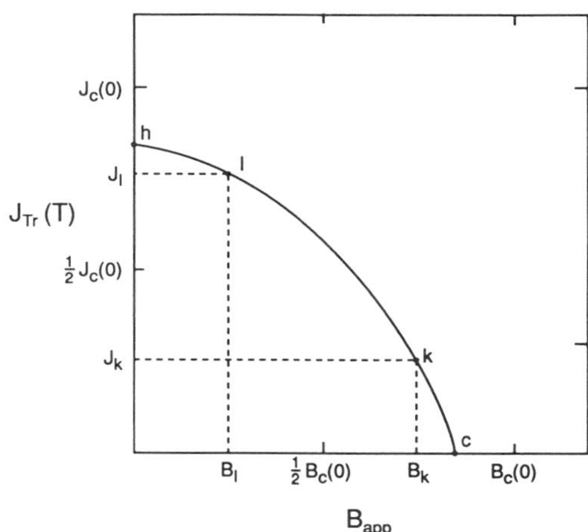

Figure 2.49 Projection of the h-l-k-c curve of Fig. 2.45 onto the $J_c(T)$ versus B_{app} plane showing the critical fields and current densities at the points k and l.

magnetic fields that are induced by the presence of transport currents, while the latter does not include shielding currents arising from the applied fields. What the critical surface means is that at a particular temperature T there is a characteristic critical field $B_c(T)$ that will drive the superconductor normal if applied in the absence of a transport current. Similarly there is a critical current density $J_c(T)$ that will drive the superconductor normal if it is applied in zero field. In the presence of an applied field a smaller transport current will drive the superconductor normal, and if a transport current is already passing through a superconductor, a smaller applied magnetic field will drive it normal. This is evident from the three constant temperature $B_c(T)$ versus $J_c(T)$ curves shown in Fig. 2.48. One of these (h–l–k–c) is redrawn in Fig. 2.49.

It will be instructive to illustrate the significance of Figs. 2.48 and 2.49 by an example. Consider the case of a long, cylindrical superconductor of radius $R \ll L$ with an applied transport current I_{tr} flowing along its axis and located in a magnetic field B_{app} along its axis, as indicated in Fig. 2.50. This situation is analyzed by taking into account the magnetic field produced at the surface by the transport current, assuming that $J_c(T)/J_c(0) = B_c(T)/B_c(0)$ and that the normalized $J_c(T)$-versus-$B_c(T)$ curve of Fig. 2.49 is an arc of a circle.

We can see from Eq. (2.43) that the transport current produces the magnetic field B_{tr},

$$B_{tr} = \frac{\mu_0 I_{tr}}{2\pi R} \tag{2.74}$$

at the surface of the cylinder. This magnetic field is at right angles to B_{app} at the surface, as shown in Fig. 2.51, so that the net field B_{net} at the surface is the square root of the sum of the squares of B_{app} and B_{tr}:

$$B_{net} = \left(B_{app}^2 + B_{tr}^2\right)^{1/2}. \tag{2.75}$$

Using Eq. (2.74) this equation can be written explicitly in terms of the transport current:

$$B_{net} = \left[B_{app}^2 + \left(\frac{\mu_0 I_{tr}}{2\pi R}\right)^2\right]^{1/2}. \tag{2.76}$$

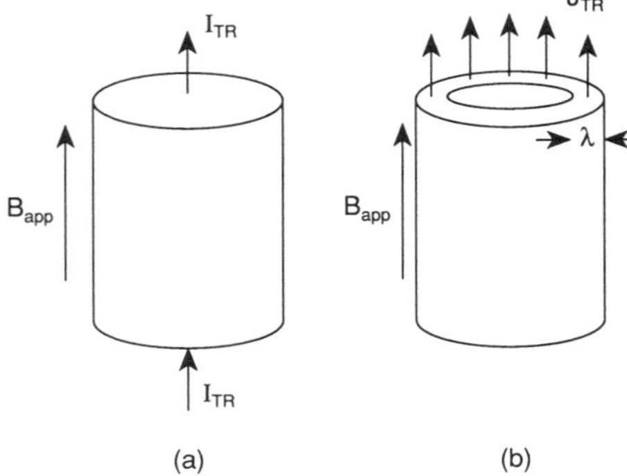

Figure 2.50 Type I superconducting cylinder (a) carrying a transport current I_{tr} of density J_{tr} in an applied magnetic field B_{app}, and (b) flow of this transport current in a surface layer of thickness λ.

The superconductor will go normal when the combination of B_{app} and I_{tr} is high enough to make B_{net} equal $B_c(T)$, the critical field for this temperature in the absence of transport currents:

$$B_c(T) = \left[B_{app}^2 + \left(\frac{\mu_0 I_{tr}}{2\pi R} \right)^2 \right]^{1/2}. \quad (2.77)$$

If we consider the case of the superconductor going normal at the point k of Fig. 2.49 then, in the notation of that figure, we have at this point,

$$B_{app} = B_k, \quad (2.78)$$

$$I_{tr} = 2\pi R \lambda J_k, \quad (2.79)$$

where in a typical experimental situation the applied quantities B_{app} and I_{tr} are often known.

This analysis was carried out by equating the vector sum of the applied field and the field arising from the transport current to the critical field $B_c(T)$. An alternate way of analyzing this situation is to equate the vector sum of the transport current density and the shielding current density to the critical current density $J_c(T)$ at the same temperature. This can be done with the aid of Fig. 2.52 to give the expression

$$J_c(T) = \left[\left(\frac{B_{app}}{\mu_0 \lambda} \right)^2 + \left(\frac{I_{tr}}{2\pi R \lambda} \right)^2 \right]^{1/2}, \quad (2.80)$$

which is the counterpart of Eq. (2.77).

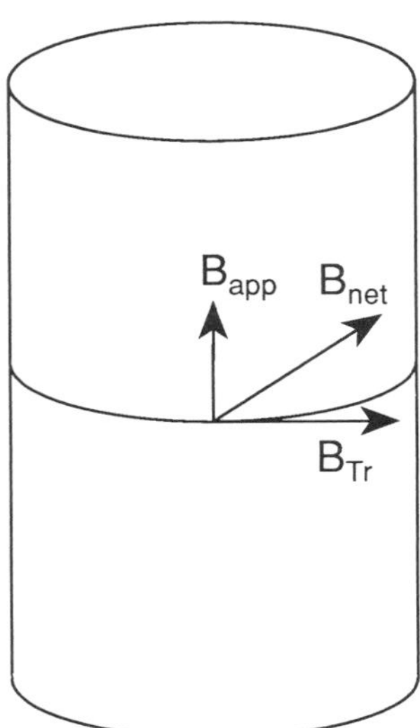

Figure 2.51 Net magnetic field B_{net} on the surface of a superconducting cylinder resulting from vector addition of the applied field B_{app} and the field B_{tr} produced by the transport current.

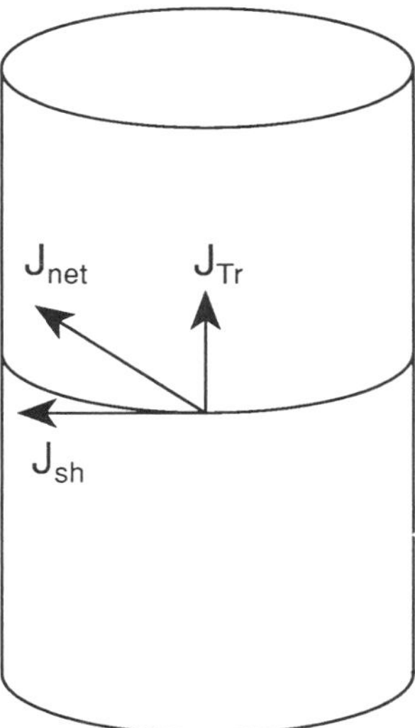

Figure 2.52 Net current density J_{net} flowing on the surface of a superconducting cylinder resulting from vector addition of the transport current density J_{tr} and the shielding current density J_{sh}.

In this section we assumed axially applied fields and currents and neglected demagnetizing effects that depend on the shape of the sample. More general cases are far more difficult to analyze.

FURTHER READING

Several superconductivity texts cover the material found in this chapter. Five of them may be cited: Kresin and Wolf, 1990; Orlando and Delin, 1991; Rose-Innes and Rhoderick, 1994; Tilley and Tilley, 1986; Tinkham, 1985. Ott (1993) surveyed the progress in superconductivity from 1980 to 1990 and provided a collection of reprinted articles. Other sources of introductory material are Hettinger and Steel (1994), Sheahan (1994), and Shi (1994). There are recent Landolt–Börnstein data tabulations on the classic superconductors by Flükiger and Klose (1993) and on the cuprates and related compounds by Kazei and Krynetskii (1993). The book by Hermann and Yakhmi (1993) is devoted to the thallium compounds.

Since most of the material in this chapter involves what might be called the basic properties of classical superconductors, we would like to direct the reader's attention to some representative older works devoted to superconductivity (Lynton, 1962) or containing introductory chapters on the subject (Lane, 1962, Chapter 9; London, 1961; Mendelssohn, 1960, Chapter 6). Probably the most important source book for classical superconductivity is the two-volume work by Parks (1969).

Many of the topics introduced in this chapter will be discussed at greater length later, and at that time more detailed references will be given.

PROBLEMS

1. A wire with a radius of 1 cm is produced from a superconductor with a transition temperature of 120 K. It is in a longitudinal magnetic field of 40 T at 60 K, and it is found that increasing the applied current to 10^3 A drives it normal. What are the values of the upper critical field, the critical current, and the critical current density for the wire at 60 K and in the limit $T \to 0$ K? Assume that B_c and B_{c2} exhibit the same temperature behavior.
2. A cylindrical superconductor of radius 200 cm with an axial hole in the center of radius 100 cm is located in a parallel magnetic field of 2 T at 300 K. It has a penetration depth of 2000 Å. What amount of flux is stored in the superconducting material and in the hole if the sample is cooled to 40 K, well below $T_c = 90$ K. If the applied field is reduced to 0.5 T, how will these stored fluxes change? What is the value of the current density on the outside surface and on the inside surface for these two cases?
3. What is the resistance of a 50-cm length of niobium wire of diameter 3 mm at 300 K? How much longer would a wire made of copper have to be in order to have the same resistance?
4. A superconducting wire 4 mm in diameter is formed into a loop of radius 7 cm. If a super current persists unchanged in this wire for 12 years, what is the approximate upper limit on the resistivity?
5. Show that Eqs. (2.59)–(2.61) provide the limiting behaviors of Eqs. (2.56)–(2.58), respectively, in the limit $T \to T_c$.
6. Show that Eqs. (2.62)–(2.64) provide the limiting behaviors of Eqs. (2.56)–(2.58), respectively, as the temperature approaches absolute zero.
7. Explain how the analysis of Fig. 2.49 that is given in Section XVI is based on the assumptions that were made concerning $J_c(T)$, $B_c(T)$, and the shape of the curve in the figure.
8. Derive Eq. (2.80).
9. Give the location of point l in Fig. 2.46, of point k in Fig. 2.47, and of point j in Fig. 2.48.
10. What is the concentration of super electrons at $T = 0$ K, $T = \frac{1}{4}T_c$, $T = \frac{1}{2}T_c$, and $T = 1.1T_c$ in a superconductor with a penetration depth of 150 nm? What is the concentration of normal conduction electrons at these temperatures?
11. A Type I superconductor has a critical field $B_c = 0.3$ T and a critical current

density $J_c = 2 \times 10^4$ A/cm^2 at 0 K. Find B_c, J_c, λ, and n_s at $T = \frac{1}{2}T_c$.

12. If a transport current density of 9000 A/cm^2 is flowing through the superconductor of Problem 11 at 0 K, what magnetic field will drive it normal?
13. A Type I superconducting wire 3 mm in diameter has a critical field $B_c = 0.4$ T and a critical current density $J_c = 3 \times 10^4$ A/cm^2 at 0 K. What is the maximum transport current that can flow through it at 0 K in an applied field of 0.35 T?
14. A Type I superconductor with $T_c = 7$ K has slope $dB_c/dT = 25$ mT/K at T_c. Estimate its critical field at $T = 6$ K.
15. Show that for a particular temperature T a plot of the critical surface B_{app} versus $\mu_0 \lambda J_{tr}$ is an arc of a circle a distance $B_c(T)$ from the origin.
16. If it is assumed that the a direction electrical conductivity arises from the planes and that the b direction conductivity is the sum of the contributions from the planes and chains (as explained in Section 7.VI), find σ_{plane} and σ_{chain} for $YBa_2Cu_3O_7$ at 100 K and 275 K.

3

The Classical Superconductors

I. INTRODUCTION

In this chapter we will survey the properties of various classes of elements and compounds that are superconductors below about 25 K. We will begin with the simplest group, namely the elements, and will proceed to discuss the binary, ternary, and larger compounds. Then we will treat the *A*15 compounds Nb_3Sn that, until the recent discovery of the high T_c types, produced the highest transition temperatures of all. Following this discussion we will review the Laves phases, the Chevrel phases, and the heavy fermions which were a focal point for research prior to the discovery of the cuprates. The section on charge-transfer organics is important because this class has experienced a rapid rise in transition temperatures in recent years. The oxides, especially

$$BaPb_{1-x}Bi_xO_3,$$

served as predecessors and prototypes for the high-temperature superconductors, while the cubic perovskite oxide $Ba_{1-x}K_xBiO_{1-y}$ superconducts at a higher temperature than the best of the *A*15 compounds. (For a discussion of the high-temperature superconductors, see Chapter 7.) Finally we introduce the reader to the more recently discovered K_3C_{60}-type fullerene compounds, sometimes called *buckyballs*, which have the symmetry of a soccer ball. The Villars–Phillips (1988) approach of Chapter 7, Section XIII introduces three metallic coordinates that can help to correlate some of the data presented in this chapter.

The new superconductors are oxides that contain the crucial transition element copper; most of them incorporate other transition elements such as yttrium, lanthanum, or some other rare earth. We will examine here the older type superconduc-

Table 3.1 Properties of the Superconducting Elements[a]

Z	Element[b]	N_e	Crystal Structure[c]	T_c (K)	Θ_D (K)	B_c (mT)	$2B_c/T_c$ (mT/K)	γ $\left(\frac{\text{mJ}}{\text{mole K}^2}\right)$	$\chi \times 10^6$ (cm^3/mole)	λ	μ_c^*
4	Be	2	hcp	0.026	940			0.21			
13	Al	3	fcc	1.18	420	10.5	18	1.4			
21	Sc	3	hcc	0.01	470			10.9			
22	Ti	4	hcp	0.40	415	5.6	28	3.3	155	0.38	0.17
23	V	5	bcc	5.40	383	141.0	52	9.82	300	0.60	0.17
30	Zn	12	hcp	0.85	316	5.4	12	0.66			
31	Ga	3	orthr	1.08	325	5.83	11	0.60			
40	Zr	4	hcp	0.61	290	4.7	15	2.77	129	0.41	0.15
41	Nb	5	bcc	9.25	276	206.0	45	7.80	212	0.82	0.15
42	Mo	6	bcc	0.92	460	9.6	21	1.83	89	0.41	0.10
43	Tc	7	hcp	7.8	411	141.0	36	6.28	270		
44	Ru	8	hcp	0.49	580	6.9	28	2.8	39	0.38	0.14
48	Cd	12	hcp	0.517	210	2.8	11	0.69			
49	In	3	tetrg	3.41	108	28.2	17	1.67			
50	Sn(w)	4	tetrg	3.72	195	30.5	16	1.78			
57	La(α)	3	hcp	4.88	152	80.0	33	9.8			
57	La(β)	3	fcc	6.3	140	110.0	37	11.3			
71	Lu	3	hcp	0.1		<40.0					
72	Hf	4	hcp	0.13	252	1.27	20	2.2	70	0.14	
73	Ta	5	bcc	4.47	258	82.9	37	6.15	162	0.75	
74	W	6	bcc	0.015	383	0.12	16	0.90	53	0.25	
75	Re	7	hcp	1.70	415	20.0	24	2.35	68	0.37	0.10
76	Os	8	hcp	0.66	500	7.0	21	2.35	13	0.44	0.12
77	Ir	9	fcc	0.11	425	1.6	29	3.2	24	0.35	
80	Hg(α)	12	trig	4.15	88	41.1	20	1.81			
80	Hg(β)	12	tetrg	3.9	93	33.9	17	1.37			
81	Tl	3	hcp	2.38	79	17.8	15	1.47		0.80	
82	Pb	4	fcc	7.20	96	80.3	22	3.1		1.55	
90	Th	4	fcc	1.38	165	16.0	23	4.32			
91	Pa	5		1.4							
95	Am	9	fcc	1.0							

[a] N_e is as defined in Fig. 3.1; Θ_D, Debye temperature; B_c, critical field; γ, electronic specific heat parameter; χ susceptibility; λ, electron–phonon coupling constant; μ_c^*, Coulomb pseudopotential; P, pressure; WF, work function; E_q, energy gap; and $D(E_F)$, density of states at the Fermi level.
Most of the data in the table come from Roberts (1976), Vonsovsky *et al.* (1982), and *Handbook of Chemistry and Physics*, 70th edition (1989–1990).

[b] Sn is the gray diamond structure α form below 13.2°C, and the white tetragonal β form above; La is the fcc β form above 310°C, and the hcp α form at lower temperatures.

[c] In Pearson notation fcc = A − 1, bcc = A − 2, hcp = A − 3, while Sn is A − 5, In is A − 6, Sb, Bi are A − 7.

tors, which contain transition elements, oxygen, or both. Many of these superconductors produced some of the higher transition temperatures in the past.

II. ELEMENTS

Superconductivity was first observed in 1911 in the element mercury with $T_c = 4.1$ K, as shown in Fig. 2.2. Two years later lead surpassed mercury with $T_c = 7.2$ K. Niobium with $T_c = 9.25$ K held the record for highest T_c for the longest period of time, from 1930 to 1954 when the *A*15 compounds came to prominence. Other relatively high-T_c elements are Tl (2.4 K), In (3.4 K), Sn (3.7 K), Ta (4.5 K), V (5.4 K), La (6.3 K), and Tc (7.8 K), as shown in Table 3.1. Figure 3.1 shows how the super-

dT_c/dP (K/GPa)	P (GPa)	α	WF (eV)	$E_g = 2\Delta$ (meV)	E_g/kT_c	$D(E_F)$ (states, atom eV)	Z
			5.0				4
			4.3	0.35	3.4		13
			5.9				21
0.6	0–1.4		4.33			≈ 1.4	22
6.3	0–2.5		4.3	1.6	3.4	≈ 2.1	23
		0.45	4.3	0.23	3.2		30
				0.33	3.5		31
15.0	0–2.0	0	4.05			≈ 0.8	40
−2.0	0–2.5		4.3	3.0	3.8	≈ 2.1	41
−1.4	0–2.5	0.37	4.6	0.26	3.4	0.65	42
−12.5	0–1.5		5.0	2.4	3.6		43
−2.3	0–1.8	0	4.7	0.15	3.5	0.91	44
		0.5	4.2	0.14	3.2		48
			3.8	1.05	3.6		49
		0.47	4.38	1.4	4.4		50
190	0–2.3			1.5	3.5		57
110							57
				0.028	3.3		71
−2.6	0–1.0			0.044	3.9	0.83	72
−2.6				≈ 1.7	≈ 3.5	≈ 1.7	73
			4.5	≈ 0.006	≈ 4.5	≈ 0.5	74
−2.3	0–1.8	0.23		0.78	3.4	0.76	75
−1.8		0.20		0.29	4.8	0.70	76
				0.048	5.6		77
		0.50	4.52	1.7	4.6		80
							80
		0.50	3.7	0.79	3.8		81
		0.48	4.3	2.7	4.3		82
				0.41	3.4		90
							91
							95

conducting elements are clustered in two regions of the periodic table, with the transition metals on the left and the nontransition metals on the right. Some elements become superconducting only as thin films, only under pressure, or only after irradiation, as indicated in the figure.

We see from Table 3.1 that the great majority of the superconducting elements have crystallographic structures of very high symmetry, either face-centered cubic (10, fcc), hexagonal close-packed (15, hcp), or body-centered cubic (11, bcc), with the unit cells sketched in Fig. 3.2. The fcc and hcp structures provide the densest possible crystallographic packing, with each atom surrounded by 12 equidistant nearest neighbors. Other cases include trigonal Hg, tetragonal In, tetragonal (white) Sn, and orthorhombic Ga.

Slightly more than half of the elements that are superconducting are members of different transition series, for example, the first transition series from scandium to zinc (5), which has an incomplete $3d^n$ electron shell; the second transition series from yttrium to cadmium (8), with $4d^n$ electrons; the third such series from lutecium to mercury (8), with $5d^n$ electrons; the rare earths from lanthanum to ytterbium (3), which have an incomplete $4f^n$ electron shell; and the actinides from

SYMBOL T_C — KELVINS
DEBYE TEMP — KELVINS
ELECT. SP HEAT — mJ/MOLE K
E - PH COUPL — DIMENSIONLESS
DEN OF ST N(E_F) — STATES/ATOM eV

1	2	3	4	5	6	7	8	9	10	11	12	3	4	5	6	7	8
Li FILM	Be 0.03																
												Al 1.2 423 1.4	Si FILM PRES	P PRES			
		Sc 0.01 470 10.9	Ti 0.4 415 3.3 0.54 1.4	V 5.4 383 9.8 1.0 2.1	Cr FILM						Zn 0.9 316 0.7	Ga 1.1 317 0.60	Ge FILM PRES	As PRES	Se PRES		
		Y PRES	Zr 0.6 290 2.8 0.22 0.8	Nb 9.3 276 7.8 0.85 2.0	Mo 0.9 460 1.8 0.35 0.6	Tc 7.8 411 6.3	Ru 0.5 580 2.8 0.47 0.9		Pd IRRAD		Cd 0.5 210 0.67	IN 3.4 108 1.7	Sn 3.7 196 1.8 (W)	Sb PRES	Te PRES		
Cs FILM PRES	Ba PRES	La 4.9 6.3 (α) (β)	Hf 0.1 252 2.2 0.14 0.8	Ta 4.4 258 6.2 0.75 1.7	W 0.02 383 0.9 0.25 0.5	Re 1.7 415 2.4 0.37 0.74	Os 0.7 500 2.4 0.44 0.68	Ir 0.1 425 3.2 0.4 0.35			Hg 4.2 75 1.8	Tl 2.4 88 1.5 0.8	Pb 7.2 102 3.1 1.55	Bi FILM PRES			
			Ce PRES					Eu FILM								Lu 0.1	
			Th 1.4 165 4.3	Pa 1.4	U PRES			Am 1.0 (β)									

Figure 3.1 Periodic table showing the superconducting elements together with their transition temperatures T_c and some of their properties (Poole *et al.*, 1988).

actinium to lawrencium (4), with $5f^n$ electrons (the number of superconductors in the class is given in parenthesis).

Among the elements niobium has the highest transition temperature, and perhaps not coincidentally it is also a constituent of higher T_c compounds such as Nb_3Ge. Niobium has not appeared prominently in the newer oxide superconductors.

Of the transition elements most commonly found in the newer ceramic-type superconductors, lanthanum is superconducting with a moderately high T_c (4.88 K for the α or hcp form and 6.3 K for the β or fcc form), yttrium becomes superconducting only under pressure ($T_c \approx 2$ K for pressure P in the range $110 \leq P \leq 160$ kbar), and copper is not known to superconduct. Studies of the transition temperature of copper alloys as a function of copper content have provided an extrapolated value of $T_c = 6 \times 10^{-10}$ K for Cu which, while nonzero, cannot be achieved experimentally. The nontransition elements oxy-

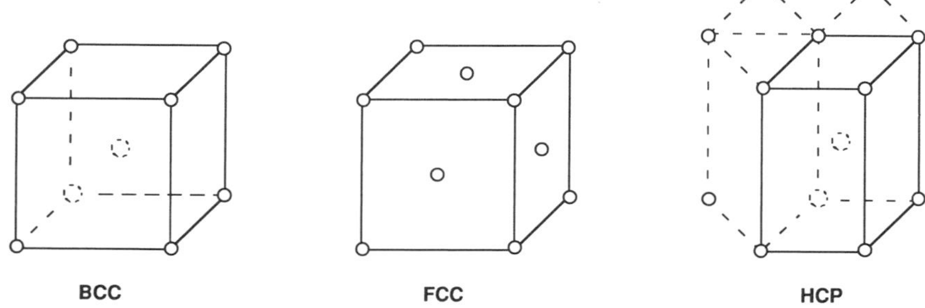

Figure 3.2 Body-centered cubic, face-centered cubic, and hexagonal close-packed unit cells.

gen and strontium present in these compounds do not superconduct, barium does so only under pressure ($T_c = 1$ K to 5.4 K under pressures from 55 to 190 kbar), bismuth likewise superconducts under pressure, and thallium is a superconductor with $T_c = 2.4$ K. Lead, added in low concentrations to stabilize the bismuth and thallium high-T_c compounds, is also a well-known elemental superconductor. Thus the superconducting properties of the elements are not always indicative of the properties of their compounds, although niobium seems to be an exception.

III. PHYSICAL PROPERTIES OF SUPERCONDUCTING ELEMENTS

Figure 3.1 gives the transition temperature T_c, Debye temperature Θ_D, Sommerfeld constant, or normal-state electronic specific heat constant γ from Eq. (1.51), $C_e = \gamma T$, dimensionless electron–phonon coupling constant λ (cf. Chapter 6), and density of states $D(E_F)$ at the Fermi level (1.42) for different elemental superconductors. The columns of the periodic table are labeled with the number of (valence) electrons N_e outside the closed shells. Table 3.1 lists various properties of some of these elements.

When an element has more than one isotope, the transition temperature often decreases with increasing isotopic mass M in accordance with the relation

$$M^{\alpha} T_c = \text{constant}, \qquad (3.1)$$

where $\alpha = 1/2$ for the simplified BCS model described in Chapter 6, Section V. This is to be expected for a simple metal because the phonon frequency is proportional to the square root of the atom's mass. However electron–phonon coupling can also be mass dependent, and deviations from Eq. (3.1) are not unusual.

Some elemental superconductors have isotope effect coefficients α close to $1/2$, such as Hg (0.50), Pb (0.48), Sn (0.47), and Zn (0.45). Most values of α listed in Table 1 for the transition metal superconductors are less than this BCS-theory estimate. For the two metals zirconium and ruthenium, both with $T_c < 0.1$ K, α is zero to within experimental error.

The BCS theory predicts that twice the energy gap 2Δ of a superconductor is 3.52 times $k_B T_c$ (cf. Chapter 6, Section VI) and from the data in Table 3.1 it is clear that this prediction is fairly well satisfied for the elements. In rhenium the energy gap is anisotropic, varying between 2.9 and $3.9 k_B T_c$, depending upon the direction. The anisotropies found for molybdenum and vanadium are half as large and almost within the experimental accuracy.

It has been found that some of the properties of an element correlate with the number N_e of its valence electrons in the same manner as the transition temperature (Vonsovsky *et al.*, 1982). Here N_e is the number of electrons outside the filled shells corresponding to the configuration of the next lower noble gas. Figure 3.3 shows that T_c is a maximum for transition metals with five and seven valence electrons; Figs. 3.4, 3.5, 3.6, and 3.7 show that the Sommerfeld factor γ of the conduction-electron heat capacity $C_e = \gamma T$, the

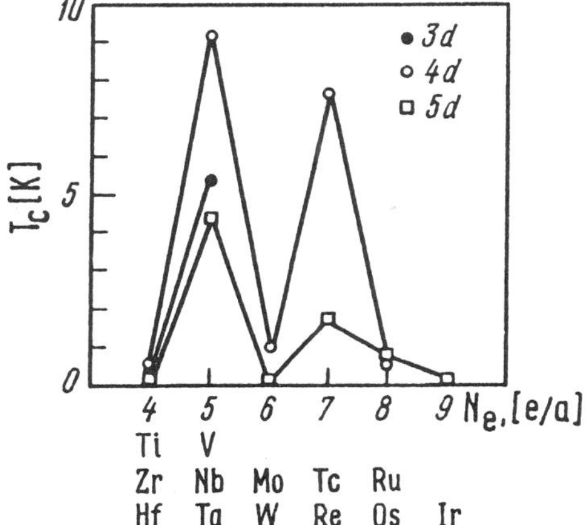

Figure 3.3 Dependence of transition temperature on the number of valence electrons N_e in the superconducting transition elements (Vonsovsky *et al.*, 1982, p. 184).

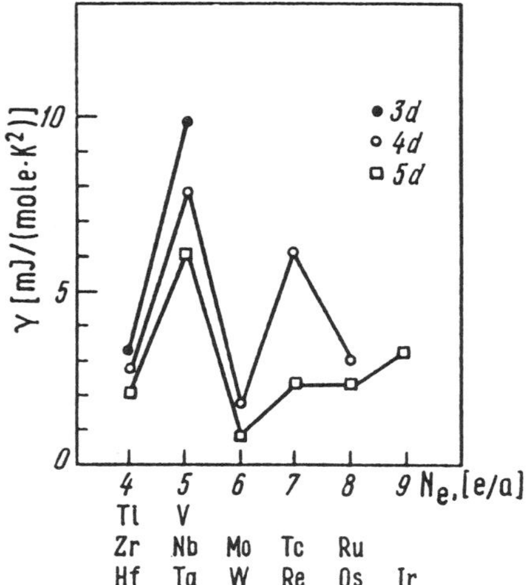

Figure 3.4 Dependence of electronic specific heat γ on N_e, as in Fig. 3.3 (Vonsovsky *et al.*, 1982, p. 184).

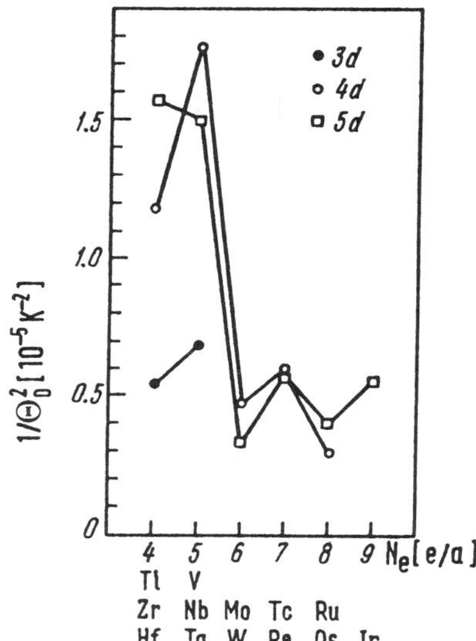

Figure 3.6 Dependence of inverse Debye temperature squared $1/\Theta_D^2$ on N_e, as in Fig. 3.3 (Vonsovsky *et al.*, 1982, p. 185).

magnetic susceptibility $\chi = M/\mu_0 H$, the square of the inverse of the Debye temperature Θ_D, and the electron–phonon coupling constant λ defined by Eq. (6.160) all exhibit similar behavior. These quantities, together with the dimensionless screened Coulomb interaction parameter μ_C^* (cf. Chapter 6), are tabulated in Table 3.1 for the superconducting elements. The correlation of the melting points of the transition metals, as plotted in Fig. 3.8, with the number of valence electrons N_e tends to be opposite to the correlation of T_c with

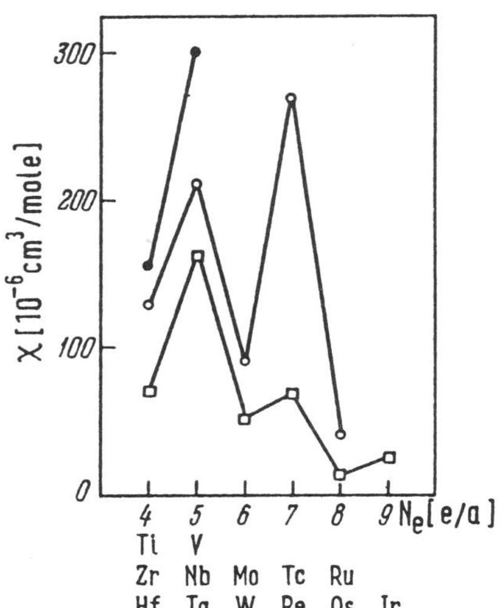

Figure 3.5 Dependence of magnetic susceptibility χ on N_e, as in Fig. 3.3 (Vonsovsky *et al.*, 1982, p. 185).

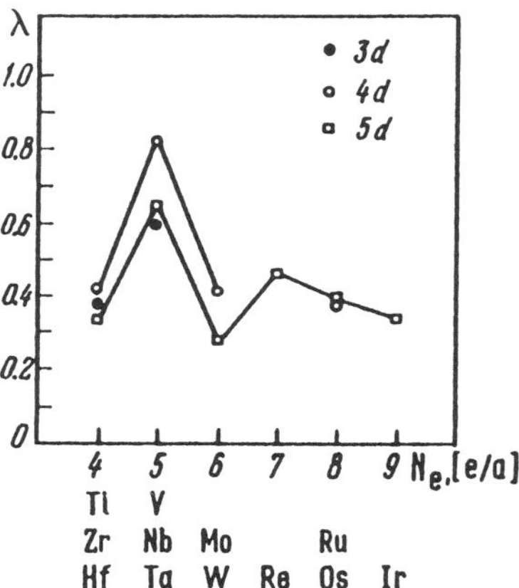

Figure 3.7 Dependence of electron–phonon coupling constant λ on N_e, as in Fig. 3.3 (Vonsovsky *et al.*, 1982, p. 211).

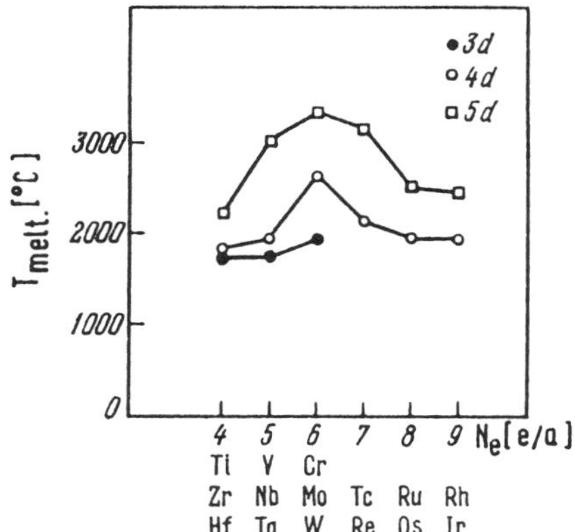

Figure 3.8 Dependence of melting temperature on N_e, as in Fig. 3.3 (Vonsovsky *et al.*, 1982, p. 186).

N_e—thus the highest melting points occur for six valence electrons for which T_c is the lowest in each series.

The chemical bonding of the transition metals is mainly ionic, but there can also be contributions of a covalent type. The amount of covalency is particularly strong in the two metals molybdenum and tungsten, each of which has five valence electrons. This fact has been used to account for the low-transition temperatures of these two elements.

Another important electronic parameter of a metal is its density of states $D(E_F)$ at the Fermi level, and Table 1 lists values of $D(E_F)$ for different elements. In several cases the value in the table is an average of several determinations with a large amount of scatter. For example, four reported values for niobium of 1.6, 1.8, 2.1, and 2.7 are given by Vonsovsky *et al.* (1982, p. 202) along with a rounded-off average of 2.1 states/atom eV. These large scatters lead one to suspect the accuracy of cases in which only one determination is available. The d electrons dominate this density of states, with small contributions from the remaining valence electrons. For example, in vanadium the percentage contributions to $D(E_F)$ from the s, p, d, and f electrons are 1%, 14%, 84%, and 1%, respectively, while for niobium the corresponding percentages are 3%, 14%, 81%, and 2%.

When a metal is subjected to high pressure, the density of states at the Fermi level changes. This change may be detected by the change in the conduction-electron heat capacity factor γ, since from Eq. (1.52) γ is proportional to $D(E_F)$. Sometimes the derivative dT_c/dP is positive, as in the case of vanadium (see curve for vanadium plotted in Fig. 3.9), so that here T_c increases with increasing pressure, and sometimes it is negative, as with tantalum (cf. Fig. 3.9), where high pressures lead to lower values of T_c. With some elements the situation is more complicated. For example, when niobium is subjected to high pressure T_c decreases until about 40 kbar is reached, then it begins to increase with increasing pressure and, eventually, above 150 kbar, surpasses its atmospheric value, as indicated in Fig. 3.9. Finally some elements, such as P, As, Se, Y, Sb, Te, Ba, Ce, and U, become superconducting only when subjected to high pressure.

T_c of some transition metals rises dramatically when the metal is in the form of thin films made by ion sputtering on various substrates. For example, the transition temperature of tungsten (bulk value 0.015 K) rises to 5.5 K in a film, that for molybdenum (bulk value 0.915 K) rises to 7.2 K,

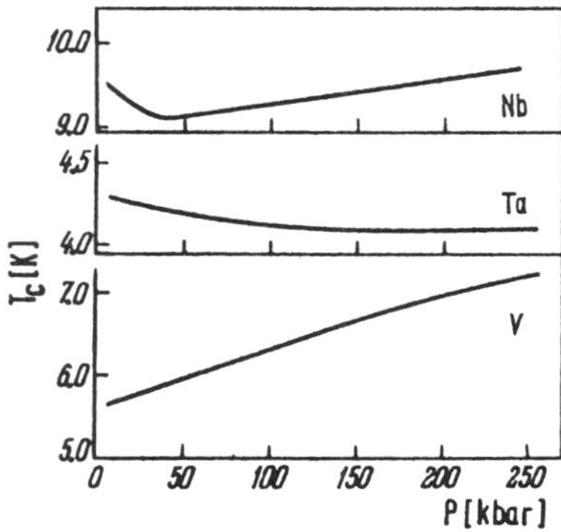

Figure 3.9 Dependence of transition temperature T_c on pressure for the elements Nb, Ta, and V (Vonsovsky *et al.*, 1982, p. 188).

and that for titanium (bulk value 0.40 K) rises to 2.52 K. Chromium and lithium only superconduct in the thin-film state, while other nonsuperconductors such as Bi, Cs, Ge, and Si can be made to superconduct either by applying pressure or by preparing them as thin films. Figure 3.1 summarizes this information.

IV. COMPOUNDS

Superconductivity researchers often employ the old Strukturbericht notation which uses the letter A to denote elements, B for AB compounds, C for AB_2 compounds, and D for A_mB_n binary compounds, with additional letters assigned to compounds containing three or more dissimilar atoms. Superconductors of the class Nb_3Sn were originally assigned to the β-W structure (Wyckoff, 1963, p. 42), which has two types of tungsten atoms, one type in the center and the other six on the faces of the cubic unit cell; these A_3B compounds came to be called the $A15$ compounds. The notation has endured despite the fact that other compounds such as UH_3 and Cr_3Si had been assigned to this structure (Wyckoff, 1964, p. 119) before Pearson described the notation in the 1958 Handbook. This notation is no longer widely used outside the superconducting community.

Several structures contain a large number of superconductors. Table 3.2 presents data on the principal superconductors, together with the transition temperature of a representative compound from each group. The table provides the Strukturbericht symbol, T_c of a prototype compound, and the number of superconducting compounds that are found in the listings of Phillips (1989a) and Vonsovsky *et al.* (1982) for each group.

All of the known nonradioactive elements are constituents of at least one superconducting compound, as Table 3.3 shows. The table catalogs the structure types with binary superconducting compounds, and gives the value of T_c for a representative superconductor of each element. It is clear that transition temperatures above 10 K are widely distributed among the elements and compounds.

On the whole, there is a tendency for the superconducting materials to be stoichiometric, i.e., with ratios of the constituent elements generally integral. Even some of the solid solutions, such as $Nb_{0.75}Zr_{0.25}$ and $Nb_{0.75}Ti_{0.25}$, have atom ratios that are easily expressed in terms of integers (Nb_3Zr and Nb_3Ti), though others, such as $Mo_{0.38}Re_{0.62}$, do not fit this format. Indeed, T_c is often sensitive to stoichiometry, and experiments in which Nb_3Ge gradually approached stoichiometry raised its measured T_c from 6 K to 17 K and, finally, to the previous record value of 23.2 K. Other materials have undergone the same evolution with the approach to ideal stoichiometry, such as Nb_3Ga (T_c going from 14.9 to 20.3 K), V_3Sn (T_c from 3.8 to 17.9 K), and V_3Ge (T_c increasing from 6.0 to 11.2 K). In contrast, there are compounds, such as Cr_3Os, Mo_3Ir, Mo_3Pt, and V_3Ir, in which the highest T_c does not occur at the ideal stoichiometric composition and in which T_c is generally less dependent on composition. For example $T_c = 0.16$ K in stoichiometric Cr_3Ir but $T_c = 0.75$ K in $Cr_{0.82}Ir_{0.18}$. Although less prevalent among the older superconducting types, this phenomenon is not unusual among the newer superconductors (cf. Vonsovsky *et al.*, 1982, for more details).

V. ALLOYS

An alloy is a solid solution or mixture in which the constituent atoms are randomly distributed on the lattice sites. An intermetallic compound, on the other hand, contains definite ratios of atoms that are crystallographically ordered in the sense that there is a unit cell that replicates itself throughout space to generate the lattice. Some alloys become ordered for particular

Table 3.2 Structure Types and Transition Temperatures of Representative Compounds of each Type[a]

Structure and Type	Example	T_c (K)	Nbr	Type	Reference[c]
$B1$, NaCl, fc cubic	MoC	14.3	26	a	Ph, 336, 369; Vo, 393
$B2$, CsCl, bc cubic	VRu	5.0	10	b	Ph, 362; Vo, 385
$B13$, MnP, *ortho*	GeIr	4.7	10	c	Ph, 341
$A12$, α-Mn, bc cubic	$Nb_{0.18}Re_{0.82}$	10	15	d	Ph, 368; Vo, 388
$B8_1$, NiAs, hex	$Pd_{1.1}Te$	4.1	18	e	Ph, 354
$D10_2$, Fe_3Th_7, hex, 3-7 compound	B_3Ru_7	2.6	12	f	Ph, 359
$D8_b$, CrFe, tetrag, σ-phase	$Mo_{0.3}Tc_{0.7}$	12.0	27	g	Ph, 347, Vo, 388
$C15$, $MgCu_2$, fc cubic, Laves	HfV_2	9.4	40	h	Ph, 370; Vo, 375
$C14$, $MgZn_2$, hex, Laves	$ZrRe_2$	6.8	19	i	Ph, 357; Vo, 375
$C16$, Al_2Cu, bc tetrag	$RhZr_2$	11.3	16	j	Ph, 350
$A15$, UH_3[b], cubic	Nb_3Sn	18	60	k	Ph, 336, 363; Vo, 259
$L1_2$, $AuCu_3$, cubic	La_3Tl	8.9	24	l	Ph, 362
Binary heavy fermions	UBe_{13}	0.9	9	m	Table 3.10
Miscellaneous binary compounds	MoN	14.8	170	n	Ph, Appendix C
$C22$, Fe_2P, Trig	HfPRu	9.9	11	o	Ph, 357
$E2_1$, $CaTiO_3$, cubic, perovskite	$SrTiO_3$	0.3		p	
$H1_1$, $MgAl_2O_4$, cubic, spinel	$LiTi_2O_4$	13.7	3	q	Ph, 339; Vo, 431
B_4CeCo_4, tetrag, ternary boride	YRh_4B_4	11.9	10	r	Ph, 347; Vo, 415
$PbMo_6S_8$, trig, Chevrel	$LaMo_6Se_8$	11.4	88	s	Ph, 361; Vo, 418
$Co_4Sc_5Si_{10}$, tetrag	$Ge_{10}As_4Y_5$	9.1	11	t	Ph, 348
fcc, buckminsterfullerene	$C_{60}Rb_2Cs$	31	12	u	

[a] The Pearson (1958) symbols (e.g., $A15$) are given for most of the structures. The numbers of compounds listed in column 4 were deduced from data given in the references of column 6.
[b] $A15$ is sometimes called the β-Mn or the Cr_3Si structure.
[c] Ph and Vo followed by page numbers denote the references Phillips (1989a) and Vonsovsky *et al.* (1982), respectively. Additional data may be found in Roberts (1976).

ratios of atoms. Both random and ordered materials can become superconducting.

First we will consider the random binary alloys. In these types of alloys two transition elements are mixed in all proportions. There are several possibilities for the transition temperature of such an alloy: it can be higher than that of both elements, between the T_c values of the constituents, or lower than either constituent taken by itself. The curve of T_c versus binary alloy concentration can be close to a straight line, concave downwards with a minimum, or concave upwards with an intermediate maximum value. These three alternatives are illustrated in Fig. 3.10. The figure shows how T_c varies with Nb content when niobium is alloyed with any one of the transition elements V, Zr, Mo, Ta, or W, with the plots arranged in the order in which the five transition elements are distributed around niobium in the periodic table. We see from Fig. 3.10 that the transition temperature reaches a maximum with an alloy containing 25 at-% zirconium. The figure also shows that it can be very small for alloys with Zr, Mo, or W.

To gain some understanding of the shapes of these curves systematic studies involving sequences of transition metals that are adjacent to each other in the periodic table have been carried out. (See Vonsovsky *et al.*, 1982.) The results indi-

Table 3.3 Number of Superconducting Binary Compounds $A_m B_n$ of the Elements Discussed in the Literature, and T_c of a Representative Compound of each Element among the Classical Superconductors (see column 5 of Table 3.2 for key to the compound types)

Element			Binary Compounds		Representative Compound		
Z	Symbol	T_c (K)[a]	Number	Types	Compound	T_c (K)	Type
1	H	—	1	n	HNb_2	7.3	n
2	Li	—	2	n	$LiTi_2O_4$	13.7	q
11	Na	—	3	n u	$Na_2Mo_6S_8$	8.6	s
19	K	—	2	h u	Bi_2K	3.58	h
37	Rb	—	2	h u	$C_{60}Rb_2K$	24	u
55	Cs	1.5^P	2	h u	Bi_2Cs	4.8	h
4	Be	8.6^F	1	n	BeTc	5.2	n
12	Mg	—	7	b n	HgMg	1.4	b
20	Ca	0.52	6	h l n	Au_5Ca	0.38	n
38	Sr	—	2	h	Rh_2Sr	6.2	h
56	Ba	5.0^P	3	n	$BaBi_{0.2}O_3Pb_{0.8}$	4.5	p
5	B	—	12	a f j r n	B_4LuRh_4	11.7	r
13	Al	1.18	17	d g h j k l m n	Nb_3Al	19	k
31	Ga	1.08	13	c j k l n	Nb_3Ga	21	k
49	In	3.41	9	a c k l n	V_3In	13.9	k
81	Tl	2.38	6	l n	$TlMo_6Se_8$	12.2	s
6	C	—	14	a n u	$C_{60}Rb_2Cs$	32	u
14	Si	6.7	14	c h k n	Nb_3Si	19	k
32	Ge	5.3	18	a c k m n	Nb_3Ge	23.2	k
50	Sn	3.7	16	a c k l m n	Nb_3S_n	18	k
82	Pb	7.20	13	h j k l n	Ta_3Pb	17	k
7	N	—	11	a n	NbN	17.3	a
5	P	5.8^P	6	n	PbP	7.8	n
33	As	0.5^P	9	c k n	$Ge_{10}As_4Y_5$	9.1	t
51	Sb	2.7^P	8	k n	Ti_3Sb	5.8	k
83	Bi	6.1^F	25	c e h l n	$BaBi_{0.2}O_3Pb_{0.8}$	4.5	p
8	O	—	4	a n	$LiTi_2O_4$	13.7	q
16	S	—	7	a n	$Sn_{0.6}Mo_6S_4$	14.2	s
34	Se	6.9^P	7	a n	$TlMo_6Se_8$	12.2	s
52	Te	3.9^P	12	a e i n	$Mo_6S_{4.8}Te_{3.2}$	2.5	s
9	F	—	0	—	$F_{0.12}K_{0.1}Li_{0.02}O_{2.88}W$	1.1	n
17	Cl	—	0	—	$Cl_3Mo_6Se_8$	9.1	s
35	Br	—	0	—	$Br_2Mo_6S_6$	13.8	s
53	I	—	0	—	$I_2Mo_6S_6$	14.0	s
21	Sc	0.01	10	b d h i n	$ScMo_6S_8$	3.6	s
22	Ti	0.40	10	a b d k n	$LiTi_2O_4$	13.7	q
23	V	5.40	21	a b g h k	V_3Si	17.2	k
24	Cr	—	7	g h k	Cr_3Os	4.68	k
25	Mn	—	9	b c j k m n	Mn_3Si	12.5	k
26	Fe	—	5	f g n	Fe_3Re_2	6.6	g
27	Co	—	12	d f g j n	$CoZr_2$	6.3	j
28	Ni	—	5	f j k n	Ni_3Th_7	1.98	f
29	Cu	—	8	b j m n	$Cu_{1.8}Mo_6S_8$	10.8	s
30	Zn	0.85	5	n	$Mo_{6.6}S_8Zn_{11}$	3.6	s
39	Y	2.5^P	20	a b f h i l n	B_4YRh_4	11.3	r
40	Zr	0.61	25	a d g h i j k l n	ZrN	10.7	a
41	Nb	9.25	41	a c d g h j k n	Nb_3Ge	23.2	k

(continues)

Table 3.3 (*continued*)

Element			Binary Compounds		Representative Compound		
Z	**Symbol**	**T_c (K)[a]**	**Number**	**Types**	**Compound**	**T_c (K)**	**Type**
42	Mo	0.92	24	a d g h k m n	Tc_3Mo	15.8	g
43	Tc	7.8	6	d g k n	Tc_3Mo	15.8	g
44	Ru	0.49	18	b f g h i k n	HfPRu	9.9	o
45	Rh	—	32	c f g h j k n	B_4LuRh_4	11.7	r
46	Pd	—	28	c d e g j k n	Bi_2Pd_3	4.0	e
47	Ag	—	6	n	$Ag_{1.6}Mo_{6.4}S_8$	9.1	s
48	Cd	0.517	1	n	CdHg	1.77	n
71	Lu	1.0	6	b h i l n	B_4LuRh_4	11.7	r
72	Hf	0.13	10	a d h i n	HfV_2	9.4	h
73	Ta	4.47	18	a c d g j k n	Ta_3Pb	17	k
74	W	0.015	12	a d g k n	W_3Re	11.4	k
75	Re	1.7	24	a d f g h i k n	Mo_3Re	15	k
76	Os	0.66	20	b d f g h i j k	$LaOs_2$	6.5	i
77	Ir	0.11	30	d f g h i k m n	$Ir_{0.4}Nb_{0.6}$	10	d
78	Pt	—	23	b c e f g h k m n	Nb_3Au	11.5	k
79	Au	—	17	b c h j k n	Ta_3Au	16	k
80	Hg	4.15	10	n	Hg_2Mg	4.0	n
57	La	4.9	21	a h n	$LaMo_6S_8$	7.1	s
58	Ce	1.7^P	4	h m n	B_4LuRh_4	11.7	r
59	Pr	—	0		$PrMo_6S_8$	4.0	s
60	Nd	—	0		$NdMo_6Se_8$	8.2	s
61	Pm	—	0				
62	Sm	—	0		B_4SmRh_4	2.7	r
63	Eu	—	1	n	$Eu_{0.012}La_{0.988}$	0.2	n
64	Gd	—	0		$GdMo_6Se_8$	5.6	s
65	Tb	—	0		$TbMo_6Se_8$	5.7	s
66	Dy	—	0		$Dy_{1.2}Mo_6Se_8$	8.2	s
67	Ho	—	0		$Ho_{1.2}Mo_6Se_8$	6.1	s
68	Er	—	0		B_4ErRh_4	8.7	r
69	Tm	—	0		B_4TmRh_4	9.8	r
70	Yb	—	0		$Yb_{1.2}Mo_6Se_8$	6.2	s
90	Th	1.38	27	a f h i j l n	Pb_3Th	5.55	l
91	Pa	1.4	0				
92	U	1.0	9	l m n	UPt_3	0.43	m

[a] P denotes T_c measured under pressure, and F indicates measurement on thin film.

cate that T_c varies with the number of valence electrons N_e in the manner illustrated in Fig. 3.11. The curves have two maxima of T_c, one near $N_e = 4.7$ and one near $N_e = 6.5$. The peak in the Nb versus Zr plot of Fig. 3.10 occurs close to $N_e = 4.7$. Amorphous alloys exhibit only one maximum for each series, as indicated in Fig. 3.12. Other properties, such as the electronic specific-heat factor γ that was defined in Eq. (1.51), magnetic susceptibility χ, and pressure derivative dT_c/dP have the dependences on electron concentration that are illustrated in Figs. 3.13, 3.14, and 3.15, respectively. The specific heat and susceptibility plots are similar to the T_c versus N_e graph of Fig. 3.11.

In addition to the correlation of the transition temperature with the valence electron concentration, there is also a correlation with the lattice properties. The body-centered cubic structure is the stable one for Ne in the range from 4.5–6.5, with hcp the stable structure outside this range,

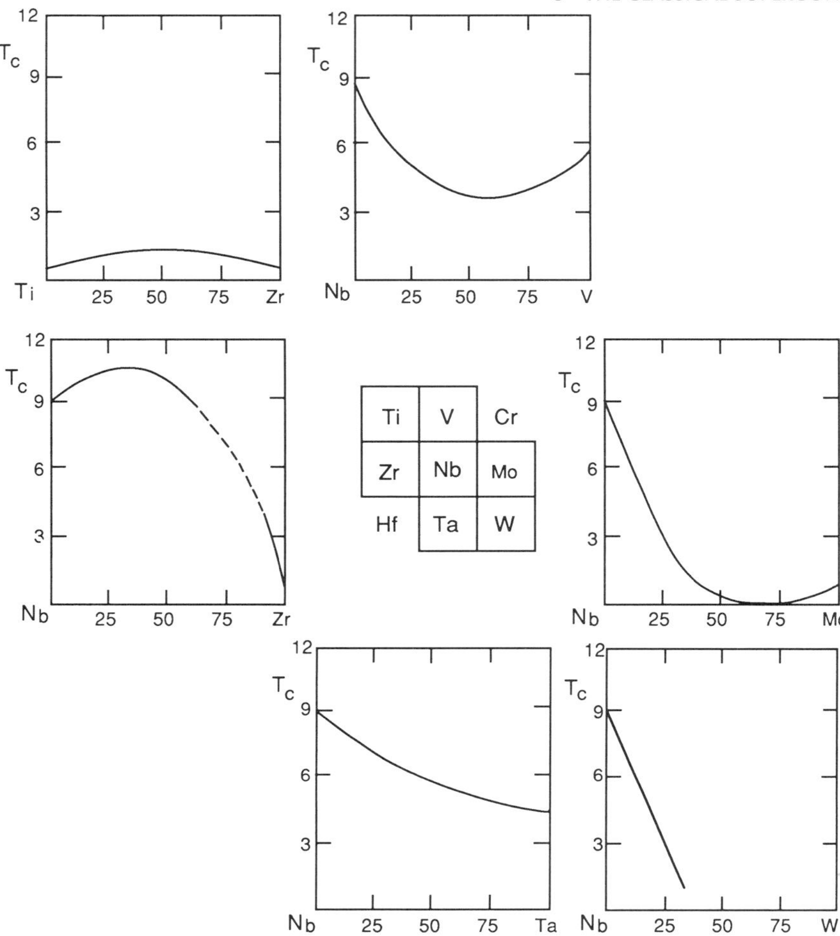

Figure 3.10 Dependence of transition temperature T_c on concentration for binary alloys of Nb with adjacent transition elements in the periodic table. The abscissae are expressed in percentages (adapted from Vonsovsky *et al.*, 1982, pp. 234, 235).

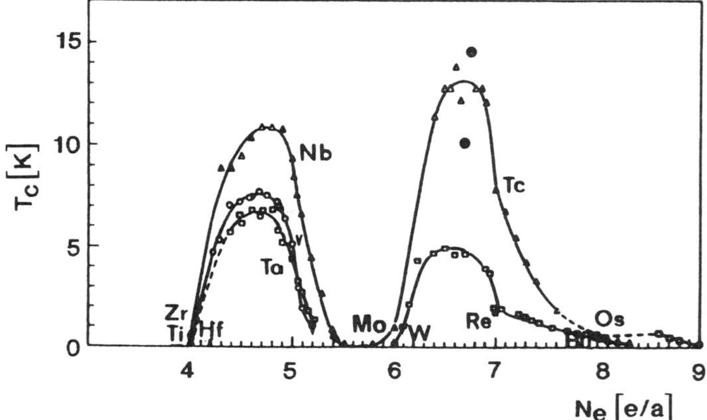

Figure 3.11 Dependence of transition temperature T_c on the number of valence electrons N_e in solid solutions of adjacent 3*d* (○), 4*d* (△), and 5*d* (□) elements in the periodic table. Dark symbols are for pure elements (Vonsovsky *et al.*, 1982, p. 239).

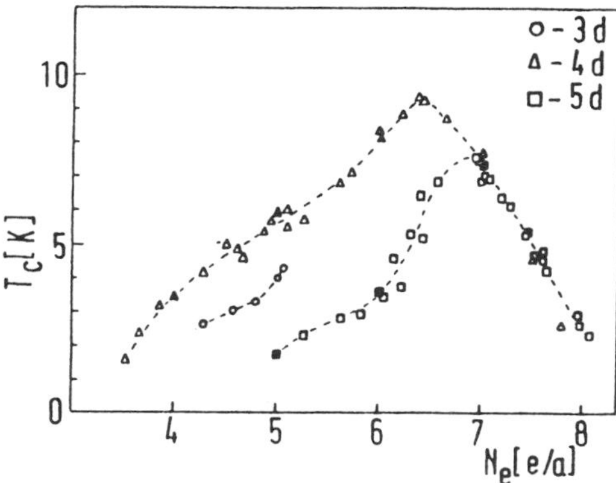

Figure 3.12 Dependence of transition temperature T_c on the number of valence electrons N_e in amorphous alloys of 3*d*, 4*d*, and 5*d* elements, using the notation of Fig. 3.11 (Vonsovsky *et al.*, 1982, p. 241).

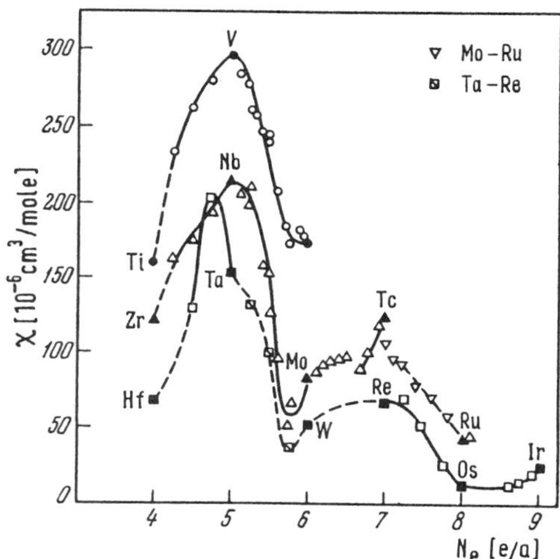

Figure 3.14 Dependence of magnetic susceptibility χ on N_e, as in Fig. 3.11. Points are plotted for Mo–Ru instead of Tc–Ru, and for Ta–Re instead of Ta–W for the same N_e, since data were not available for the preferred adjacent elements (Vonsovsky *et al.*, 1982, p. 242).

as shown in Fig. 3.16. The peaks in the plot of T_c versus N_e occur at the boundaries of instability—i.e., where a lattice rearrangement transition can occur between the two structure types. The lowest T_c occurs (Fig. 3.11) for $N_e \approx 5.5$–6, which is also where the bcc structure is most stable.

The alloy types listed in Table 3.2 are binary; most of them having their component elements in an atom ratio of 1:1, 1:2, or 1:3. In most cases at least one constituent is an elemental superconductor, and sometimes (e.g., $NbTc_3$, VRu) both elements superconduct. Occasionally, more than one intermetallic stoichiometry (e.g., $RhZr_2$ and $RhZr_3$, Nb_3Ge and $NbGe_2$) is superconducting. The binary superconductors in Table 3.2 have T_c values higher than the highest T_c of their constituents, although even here VRu is an exception. The high-T_c semiconducting and layered compounds tend to be binary also. Some of the compounds in Table 3.2 are ternary types and even $Lu_{0.75}Th_{0.25}Rh_4B_4$ is really the ternary compound MRh_4B_4 with Lu occupying three-quarters and Th one-quarter of the M sites.

Recent work on compounds and alloys has focused on K_xMoS_2 (Gupta *et al.*, 1991), $Mo_{79}Ge_{21}$ (Missert and Beasley, 1989), δ-MoN (Bezinge *et al.*, 1987), Mo–Ta superlattices (Maritato *et al.*, 1988), Ni_xZr_{1-x} (Mahini *et al.*, 1989; Schultz *et al.*, 1987), and V–Si (Kanoda *et al.*, 1989).

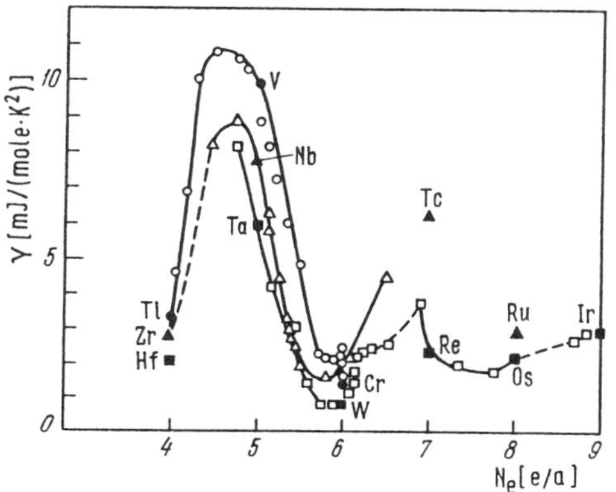

Figure 3.13 Dependence of electronic specific heat γ on N_e, as in Fig. 3.11 (Vonsovsky *et al.*, 1982, p. 237).

VI. MIEDEMA'S EMPIRICAL RULES FOR ALLOYS

Matthias (1953, 1955) interpreted the shape of the curve of T_c versus N_e as

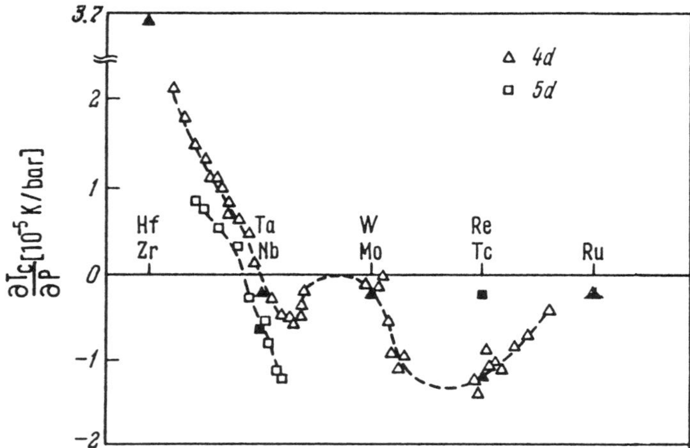

Figure 3.15 Dependence of pressure derivative dT_c/dP on N_e in alloys of adjacent $4d$ and $5d$ elements, following Fig. 3.11 (Vonsovsky *et al.*, 1982, p. 238).

indicating the presence of favorable and unfavorable regions of N_e and suggested rules for explaining the T_c-versus-concentration curves. One rule, for example, explains the increase of T_c in terms of the shift of N_e toward more favorable values; thus $N_e = 5$ (V, $T_c = 5.4$ K; Nb, $T_c = 9.3$ K; Ta, $T_c = 4.5$ K) and $N_e = 7$ (Tc, $T_c = 7.8$ K, Re, $T_c = 1.7$ K).

Miedema (1973, 1974) proposed an empirical method of correlating the concentration dependence of the transition temperature and other physical characteristics of alloys. The method assumes that the density of states $D(E_F)_{AB} = D_{AB}$ at the Fermi level of an alloy AB is an additive function of its constituents,

$$D_{AB} = f_A D_A(N_A) + f_B D_B(N_B), \quad (3.2)$$

where f_A and f_B are the mole fractions of the components A and B, and the densities of states D_A and D_B depend on the number of valence electrons N_A and N_B of atoms A and B, respectively (cf. Vonsovsky *et al.* (1982, Section 5.4) for evaluation of the N_i dependences of the functions D_i and γ_i in Eq. (3.2)). Equation (1.52) permits us to write a similar expression for the electronic heat capacity factor γ:

$$\gamma_{AB} = f_A \gamma_A(N_A) + f_B \gamma_B(N_B). \quad (3.3)$$

The density of states and γ depend upon the number of valence electrons per atom N_i in a similar manner for the $3d$, $4d$, and $5d$ transition ion series. Some electron concentration is transferred between the atoms during alloying, so that the number of valence electrons of the ith component N_i differs from its free atom value N_{ei} by the factor

$$N_i = N_{ei} + K\Delta\phi(1 - f_i), \quad (3.4)$$

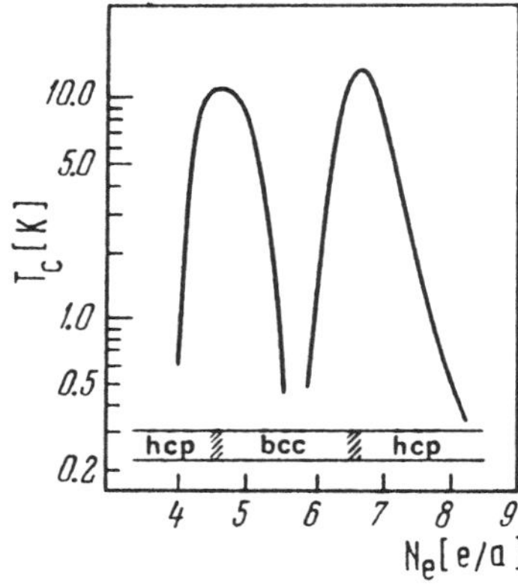

Figure 3.16 Sketch of the structure dependence of the transition temperature for alloys of adjacent transition elements, showing how T_c peaks at the boundaries between structure types (Vonsovsky *et al.*, 1982, p. 245).

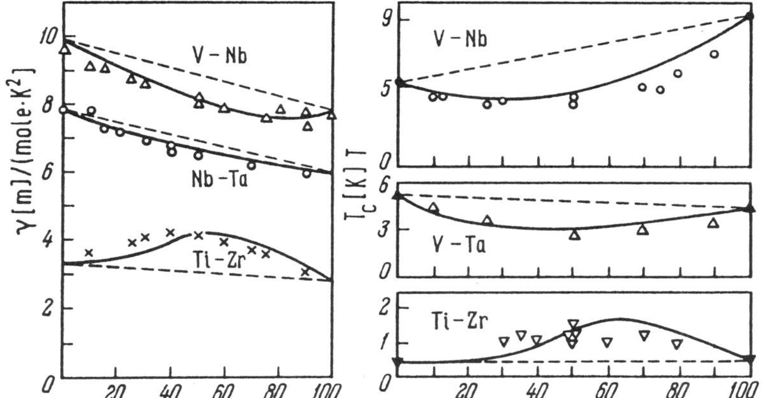

Figure 3.17 Comparison of calculated (——) and experimental (symbols) values of the specific heat coefficient γ and the transition temperature T_c for four isoelectronic binary alloys. The linear approximation (– – – –), which is obtained by setting $N_i = N_{ei}$ in Eqs. (3.3) and (3.6), does not fit the data very well (Vonsovsky *et al.*, 1982, p. 250).

where $\Delta\phi$ is the difference between the work functions (cf. Chapter 13, Section II, A) of the two pure metals involved in the alloy. Table 3.1 lists the work functions of the superconducting elements. Empirical expressions similar to Eqs. (3.2) and (3.3) have been written for the electron–phonon coupling constant λ_{AB},

$$\lambda_{AB} = f_A \lambda_A(N_A) + f_B \lambda_B(N_B), \quad (3.5)$$

and for the quantity $[\ln(T_c/\Theta_D)]^{-1}$ for the binary alloys,

$$\frac{1}{[\ln(T_c/\Theta_D)]_{AB}} = \frac{f_A}{[\ln(T_c/\Theta_D)]_A} + \frac{f_B}{[\ln(T_c/\Theta_D)]_B}. \quad (3.6)$$

Figure 3.17 shows that the agreement of Eqs. (3.3) and (3.6) with experiment is good for four isoelectronic alloy systems over the entire solid solution range. We see from this figure that the linear approximation obtained by setting $K = 0$ in Eq. (3.4), hence replacing N_i by N_{ei} in Eqs. (3.2)–(3.6), does not agree with experiment.

VII. COMPOUNDS WITH THE NaCl STRUCTURE

The $B1$ class of AB superconductors has the metallic atoms A and nonmetallic atoms B arranged on a sodium chloride-type lattice that consists of two interpenetrating fcc lattices with each atom of one type in the center of an octahedron whose vertices are occupied by atoms of the other type, as indicated in Fig. 3.18. As of 1981,

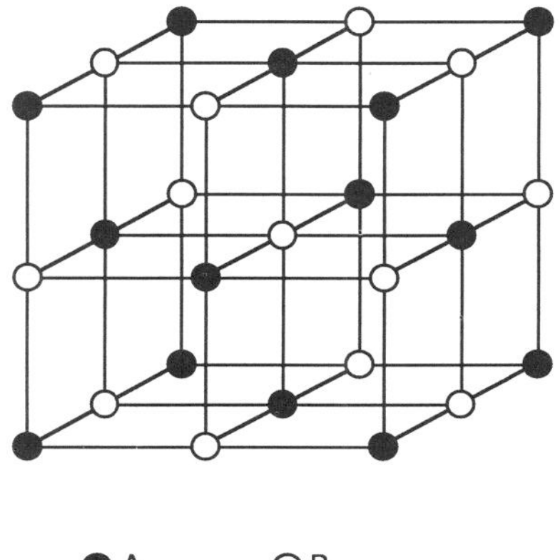

Figure 3.18 Structure of NaCl in which each atom of one type (○ or ●) is in the center of an octahedron whose vertices are occupied by atoms of the other type (● or ○).

26 $B1$ compounds (out of the 40 $B1$ compounds that had been tested) had been found to be superconducting. The carbides AC and nitrides AN, such as NbN with $T_c = 17$ K (Kim and Riseborough, 1990), had the 12 highest transition temperatures, while the metallic A atoms with T_c values above 10 K were Nb, Mo, Ta, W, and Zr. Niobium always seems to be the best! Three examples of superconducting NaCl-type compounds are given in Tables 3.2 and 3.3.

The NaCl-type superconductors are compositionally stoichiometric but not structurally so. In other words, these compounds have a small to moderate concentration of vacancies in the lattice, as indicated in Table 3.4. We see from the table that YS has 10% vacancies, which means that its chemical formulae should properly be written $Y_{0.9}S_{0.9}$. Nonstoichiometric NaCl-type compounds, such as $Ta_{1.0}C_{0.76}$, also exist.

Ordinarily the vacancies are random, but sometimes they are ordered on the metalloid (e.g., $Nb_{1.0}C_{0.75}$) or on the metallic (e.g., $V_{0.763}O_{1.0}$) sublattice, and can also produce a larger unit cell in, for example, $Ti_{1.0}O_{0.7}$. The vacancies can also be ordered on both sublattices in stoichiometric compounds, such as $Nb_{0.75}O_{0.75}$. Table 3.5 lists several NaCl-type compounds with ordering of the vacancies. It has been found that the metallic and nonmetallic atoms can be absent over broad composition ranges.

Table 3.4 Percentage of Vacancy Concentration $100(1 - x)$ in Stoichiometric Compounds A_xB_x with NaCl Structure[a]

A_x \ B_x	C	N	O	S	Se
Ti	2	4.0	15		
V	8.5	1.0	11–15		
Y				10	
Zr	3.5	3.5		20	16
Nb	0.5–3.0	1.3	25		
Hf	4				
Ta	0.5	2.0			

[a] After Vonsovsky *et al.*, 1982.

Table 3.5 Nonstoichiometric Compounds A_xB_y with NaCl Structure with Vacancy Ordering on One Sublattice and Stoichiometric Compounds A_xB_x with this Structure (shown in square brackets) with Vacancy Ordering on Two Sublattices[a]

A_x \ B_x	C	N	O
Ti			$Ti_{1.0}O_{0.7}$ [$Ti_{0.85}O_{0.85}$]
V	$V_{1.0}C_{0.84}$	$V_{1.0}N_{0.75}$	$V_{0.763}O_{1.0}$
Nb	$Nb_{1.0}C_{0.75}$		[$Nb_{0.75}O_{0.75}$]
Ta	$Ta_{1.0}C_{0.76}$		

[a] After Vonsovsky *et al.*, 1982, p. 394.

VIII. TYPE *A*15 COMPOUNDS

The highest transition temperatures for the older superconductors were obtained with the $A15$ intermetallic compounds A_3B, and extensive data are available on these compounds. Nb_3Sn can be considered the prototype of this class. These compounds have the (simple) cubic structure (Pm3n, O_h^3) sketched in Fig. 3.19 with the two B atoms in the unit cell at the body center $(\frac{1}{2}, \frac{1}{2}, \frac{1}{2})$ and apical $(0, 0, 0)$ positions, and the six A atoms paired on each face at the sites $(0, \frac{1}{2}, \frac{1}{4})$; $(0, \frac{1}{2}, \frac{3}{4})$; $(\frac{1}{2}, \frac{1}{4}, 0)$; $(\frac{1}{2}, \frac{3}{4}, 0)$; $(\frac{1}{4}, 0, \frac{1}{2})$; $(\frac{3}{4}, 0, \frac{1}{2})$, a configuration that amounts to the presence of chains of A atoms with spacing of one-half the lattice constant a. The A atom is any one of the transition elements (but not Hf) shown in the center of Fig. 3.10. The B element is either a transition element or is in row III (Al, Ga, In, Tl), row IV (Si, Ge, Sn, Pb), row V (P, As, Sb, Bi), or row VI (Te) of the periodic table. High transition temperatures occur when B is either a metal (Al, Ga, Sn) or a nonmetal (Si, Ge), but not a transition element. Table 3.6 shows seven A elements and 20 B elements for a total of 140 possibilities, 60 of

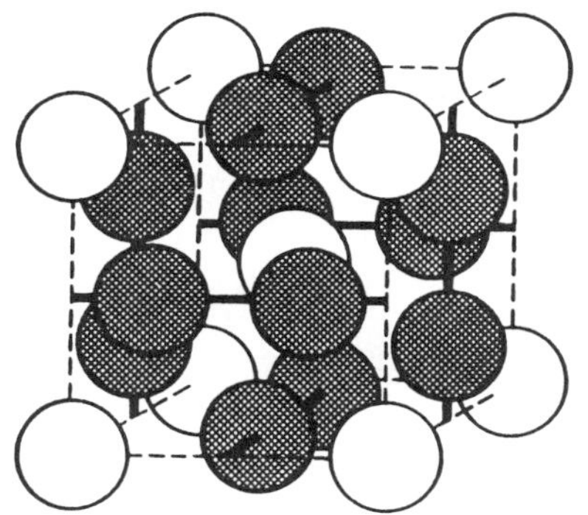

Figure 3.19 The unit cell of the *A*15 compound A_3B showing *B* atoms at the apical and body center positions and *A* atoms in pairs on the faces of the cube (Vonsovsky *et al.*, 1982, p. 260; see also Wyckoff, 1964, p. 119).

which superconduct. Two additional *A*15 superconductors are V_3Ni with $T_c = 0.3$ K and W_3Os with $T_c = 11.4$ K.

Stoichiometry is important, and paying attention to it has produced higher transition temperatures, as was explained in Section IV. Typical *A*15 compounds $A_{3+x}B_{1-x}$ have narrow ranges of homogeneity. They are homogeneous toward *A*, with deviations from stoichiometry, $x > 0$, that tend to maintain the chains intact. Some atypical compounds, such as $A_{3-x}B_{1+x}$, are homogeneous toward *B* so that the chains are affected, and can have their highest T_c values when they deviate from ideal stoichiometry. Figure 3.20 shows that there is a close correlation between the transition temperature and the valence electron concentration N_e. We see that high values of T_c occur for $N_e = 4.5$ (Nb_3Ga, $T_c = 20.3$ K), $N_e = 4.75$ (Nb_3Ge, $T_c = 23.2$ K), $N_e = 6.25$ (Nb_3Pt, $T_c = 10.9$ K), and $N_e = 6.5$ (TaAu, $T_c = 13$ K). The specific-heat factor γ plotted in Fig. 3.21 and the magnetic susceptibility χ (Vonsovsky *et al.*, 1982) show the same correlation (cf. Hellman and Geballe, 1987). The Villars–Phillips approach described in Chap-

Table 3.6 Superconducting Transition Temperatures T_c of some *A*15 Compounds A_3B^a

$B \backslash A_3$	Ti	V	Cr	Zr	Nb	Mo	Ta
Al		11.8			18.8	0.6	
Ga		16.8			20.3	0.8	
In		13.9			9.2		
Si		17.1			19	1.7	
Ge		11.2	1.2		23.2	1.8	8.0
Sn	5.8	7.0		0.9	18.0		8.4
Pb				0.8	8.0		17
As		0.2					
Sb	5.8	0.8			2.2		0.7
Bi				3.4	4.5		
Tc						15	
Ru			3.4			10.6	
Rh		1.0	0.3		2.6		10.0
Pd		0.08					
Re						15.0	
Os		5.7	4.7		1.1	12.7	
Ir	5.4	1.7	0.8		3.2	9.6	6.6
Pt	0.5	3.7			10.9	8.8	0.4
Au		3.2		0.9	11.5		16.0
Tl					9		

[a] After Phillips (1989a) and Vonsovsky *et al.* (1982).

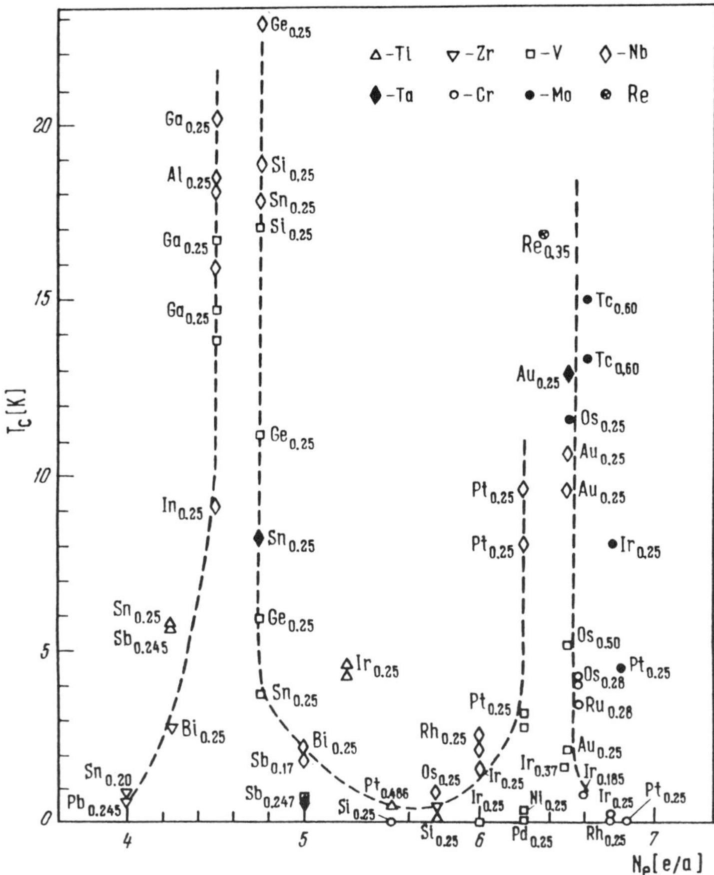

Figure 3.20 Dependence of transition temperature T_c on the number of valence electrons N_e in $A_{0.75}B_{0.25}$ compounds with the $A15$ structure. The A element is specified by the symbol at the top right, and the B element (see Fig. 3.19) is indicated at the experimental points (Vonsovsky *et al.*, 1982, p. 269).

ter 7, Section XIII adds two additional parameters for high T_c values.

The superconducting energy gap data vary over a wide range, with $2\Delta/k_BT_c$ in the range 0.2–4.8, low values probably representing poor junctions. The $A15$ group has some weak-coupled, BCS-like compounds, such as V_3Si with $2\Delta/k_BT_c \approx 3.5$, and some strong coupled compounds, such as Nb_3Sn with $2\Delta/k_BT_c \approx 4.3$ and Nb_3Ge with $2\Delta/k_BT_c \approx 4.3$. The electron–phonon coupling constant λ has been reported to vary between the weak coupling value of 0.1 and the strong coupling value of 2.0 (see Table 6.2).

Some $A15$ compounds undergo a reversible structural phase transformation above T_c from a high-temperature cubic phase to a low-temperature tetragonal phase that deviates very little from cubic ($|c-a|/a \approx 3 \times 10^{-3}$). At the transformation each atom remains close to its original site and the volume of the unit cell remains the same. Table 3.7 lists some transformation temperatures and $(c-a)/a$ ratios.

There is no isotope effect in this class of compounds, meaning that $\alpha = 0$ in Eq. (3.1). In addition there is a large scatter in the data on the change of T_c with pres-

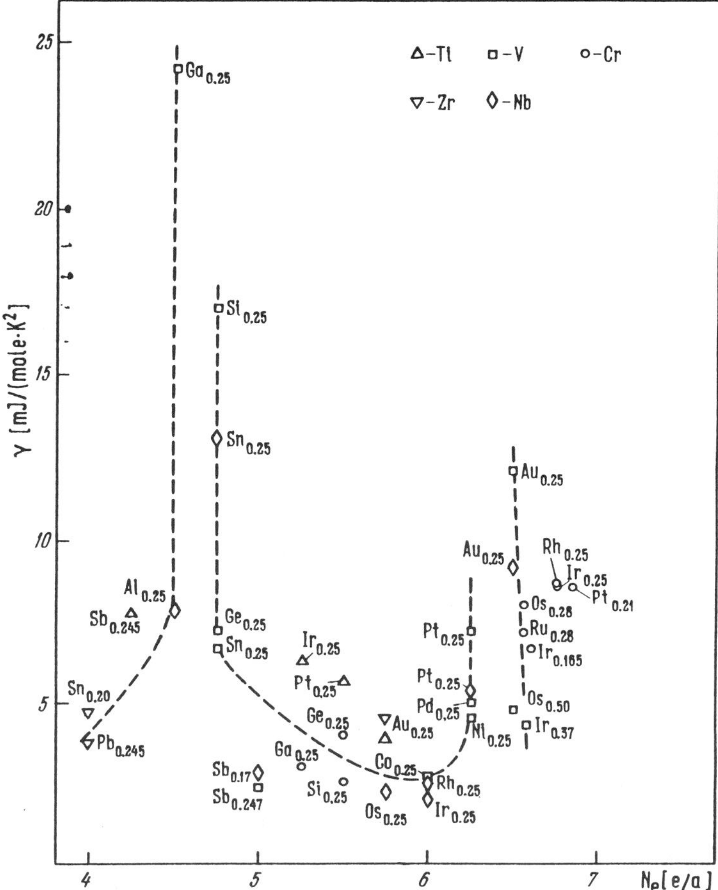

Figure 3.21 Dependence of electronic specific heat γ on N_e in $A15$ compounds, using the notation of Fig. 3.20 (Vonsovsky *et al.*, 1982, p. 271). The dependence of the magnetic susceptibility χ on N_e produces a similar plot (Vonsovsky *et al.*, 1982, p. 271).

sure, dT_c/dP, as Fig. 3.22 indicates (cf. Ota, 1987).

IX. LAVES PHASES

There are several dozen metallic AB_2 compounds called Laves phases which are superconducting; the transition temperatures of some of these compounds are listed in Table 3.8. The $C15$ Laves phases have the cubic (Fd3m, O_h^7) structure sketched in Fig. 3.23, and the $C14$ phases are hexagonal, as noted in Table 3.2. One additional Laves superconductor $HfMo_2$ has the $C36$ hexagonal structure with a larger unit cell. Some have critical temperatures above 10 K and high critical fields. For example, $Zr_{\frac{1}{2}}Hf_{\frac{1}{2}}V_2$ has $T_c = 10.1$ K, $B_{c2} = 24$ T, and a compound with a different Zr/Hf ratio has similar T_c and B_{c2} values with $J_c \approx 4 \times 10^5$ A/cm^2. These materials also have the advantage of not being as hard and brittle as some other intermetallics and alloys with comparable transition temperatures.

Table 3.7 Structural Transformation Temperature T_{str} and Anisotropy $(c-a)/a$ in the Low-Temperature Tetragonal Phase of Several *A*15 type Superconductors[a]

Compound	T_{str} (K)	T_c (K)	Anisotropy $(c-a)/a$	Reference
V_3Si	21	17	0.0024	Batterman and Barrett (1964)
Nb_3Sn	43	18	−0.0061	Mailfert *et al.* (1967)
V_3Ga	> 50	14.5	—	Nembach *et al.* (1970)
Nb_3Al	80	17.9	—	Kodess (1973, 1982)
$Nb_3(Al_{0.75}Ge_{0.25})$	105	18.5	−0.003	Kodess (1973, 1982)
$Nb_{3.1}(Al_{0.7}Ge_{0.3})$	130	17.4	—	Kodess (1973, 1982)

[a] cf. Vonsovsky *et al.*, 1982, p. 278.

X. CHEVREL PHASES

The Chevrel phases $A_xMo_6X_8$ are mostly ternary transition metal chalcogenides, where X is S, Se, or Te and A can be almost any element. These compounds have relatively high transition temperatures and critical fields B_{c2} of several teslas. However, the critical currents, typically 2 to 500 A/cm^2, are rather low. Substituting oxygen for sulphur in $Cu_{1.8}Mo_6S_8$ raises T_c (Wright *et al.*, 1987). Table 3.9 lists several dozens of these superconductors and their transition temperatures. Figure 3.24 compares the critical currents (see Fig. 10.26 for a comparison of the critical fields for several superconductors).

The trigonal structure sketched in Fig. 3.25, with space group R3, C_{3i}^2, is a simple cubic arrangement slightly distorted along the (111) axis of the Mo_6X_8-group building blocks, each consisting of a deformed cube with large X atoms at the vertices and small Mo atoms at the centers of the faces. The Mo_6X_8 group may be looked on as an Mo_6 octahedron inscribed in an X_8 cube. Mo_6X_{12}-group building blocks are also found. The distortions are not

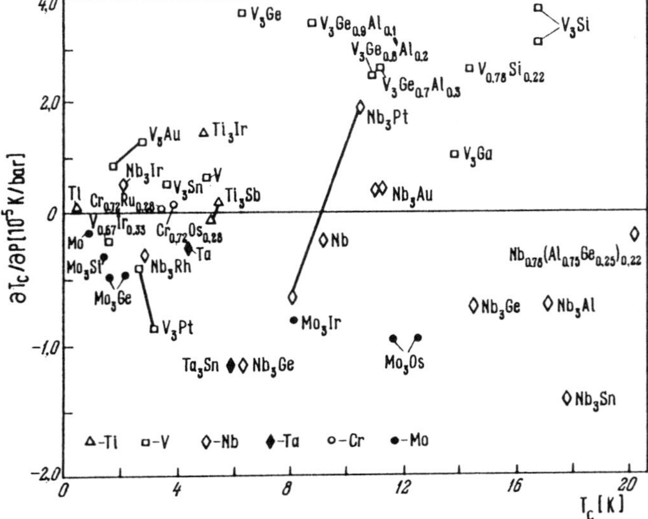

Figure 3.22 Dependence of pressure derivative dT_c/dP on N_e in *A*15 compounds, following Fig. 3.20 (Vonsovsky *et al.*, 1982, p. 288).

Table 3.8 Superconducting Transition Temperatures T_c of Selected Laves Phase (AB_2) Compounds[a]

A \ B_2	V	Mo	Re	Ru	Os	Rh	Ir	Pt	Te
Ca						6.4	6.2		
Sr						6.2	5.7		
Sc			4.2[b]	2.3[b]	4.6[b]	6.2	2.5	0.7	
Y			1.8[b]	2.4[b]	4.7[b]		2.1	0.5	
La				4.4	8.9		0.5		
Zr	9.6	0.13	6.8[b]	1.8[b]	3.0[b]		4.1		7.6[b]
Hf	9.4	0.07[b]	5.6[b]		2.7[b]				5.6[b]
Lu				0.9[b]	3.5[b]	1.3	2.9		
Th			5.0[b]	3.5			6.5		

[a] After Phillips (1989a) and Vonsovsky *et al.*, 1982.
[b] Hexagonal $C14$ structure. ($HfMo_2$ has hexagonal $C36$ structure).
Note. Unlabeled compounds have the cubic $C15$ structure.

shown in the figure. The parameter x in the formula $A_xMo_6X_8$ assumes various values such as $x = 1$ (e.g., YMo_6S_8, $LaMo_6S_8$), $x = 1.2$ (e.g., $V_{1.2}Mo_6Se_8$), $x = 1.6$ (e.g., $Pb_{1.6}Mo_6S_8$), and $x = 2$ (e.g., $Cu_2Mo_6Se_8$). This parameter can vary because of the large number of available sites between the cubes for the A cations. Most of the space is occupied by the large chalcogenide anions, which have radii of 0.184 nm (S), 0.191 nm (Se), and 0.211 nm (Te).

The electronic and superconducting properties depend mainly on the Mo_6X_8 group. No correlations are evident between the type of A ion and the superconducting properties. Magnetic order and superconductivity are known to coexist in Chevrel phase compounds. When A is a rare earth its magnetic state does not influence the superconducting properties, but when A is a transition metal ion the magnetic properties suppress the superconductivity. This may be explained on structural grounds by pointing out that the large rare earths occupy sites between the Mo_6X_8 groups, as shown in Fig. 3.25, where they are remote from the molybdenums with only X as nearest neighbors. The smaller transition ions, on the other hand, can fit into octahedral sites with six Mo as their nearest neighbors (Ø. Fischer, 1990).

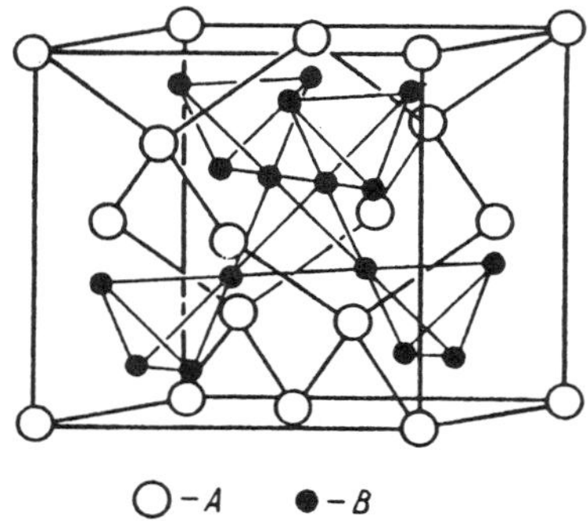

Figure 3.23 Crystal structure of the Laves phase (Vonsovsky *et al.*, 1982, p. 376).

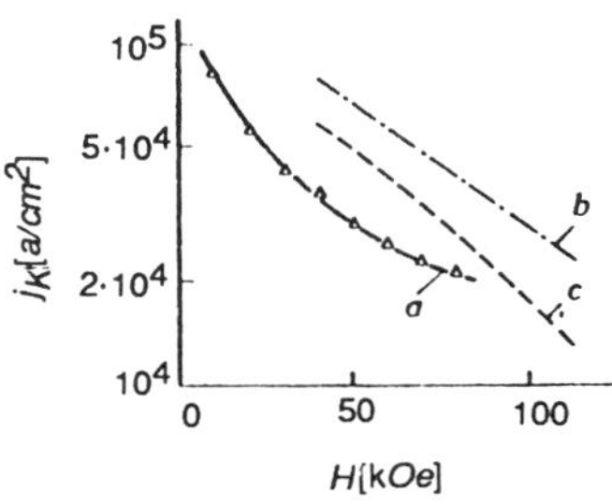

Figure 3.24 Comparison of the critical current densities on the applied field for a, $SnGa_{0.25}Mo_6S_8$, b, V_3Ga, and c, Nb_3Sn (Alekseevskii *et al.*, 1977).

Table 3.9 Superconducting Transition Temperatures of some Chevrel compounds[a]

$A_xMo_6S_8$	T_c (K)	$A_xMo_6Se_8$	T_c (K)	Misc. Compounds	T_c (K)
Mo_6S_8	1.6	Mo_6Se_8	6.4	$Pb_{0.9}Mo_6S_{7.5}$	15.2
$Cu_2Mo_6S_8$	10.7	$Cu_2Mo_6Se_8$	5.9	$PbGd_{0.2}Mo_6S_8$	14.3
$LaMo_6S_8$	6.6	$La_2Mo_6Se_8$	11.7	$PbMo_6S_8$	12.6
$PrMo_6S_8$	4.0	$PrMo_6Se_8$	9.2	$Sn_{1.2}Mo_6S_8$	14.2
$NdMo_6S_8$	3.5	$NdMo_6Se_8$	8.4	$SnMo_6S_8$	11.8
$Sm_{1.2}Mo_6S_8$	2.9	$Sm_{1.2}Mo_6Se_8$	6.8	$LiMo_6S_8$	4.0
$Tb_{1.2}Mo_6S_8$	1.7	$Tb_{1.2}Mo_6Se_8$	5.7	$NaMo_6S_8$	8.6
$Dy_{1.2}Mo_6S_8$	2.1	$Dy_{1.2}Mo_6Se_8$	5.8	KMo_6S_8	2.9
$Ho_{1.2}Mo_6S_8$	2.0	$Ho_{1.2}Mo_6Se_8$	6.1	$Br_2Mo_6S_6$	13.8
$Er_{1.2}Mo_6S_8$	2.0	$Er_{1.2}Mo_6Se_8$	6.2	$I_2Mo_6S_6$	14.0
$Tm_{1.2}Mo_6S_8$	2.1	$Tm_{1.2}Mo_6Se_8$	6.3	$BrMo_6Se_7$	7.1
$Yb_{1.2}Mo_6S_8$	≈ 8.7	$Yb_{1.2}Mo_6Se_8$	5.8	IMo_6Se_7	7.6
$Lu_{1.2}Mo_6S_8$	2.0	$Lu_{1.2}Mo_6Se_8$	6.2	$I_2Mo_6Te_6$	2.6

[a] See Phillips (1989a, pp. 339, 361) and Vonsovsky *et al.* (1982, p. 419) for more complete listings.

Recent Chevrel phase studies have focused on electron–phonon interactions (Furuyama *et al.*, 1989), the irreversibility line (Rossel *et al.*, 1991), the magnetooptical Kerr effect (Fumagalli and Schoenes, 1991), oxygen doping (Wright *et al.*, 1987), and phonon density of states (Brusetti *et al.*, 1990).

XI. HEAVY ELECTRON SYSTEMS

For several years prior to 1987 there was a great deal of interest in the study of heavy-electron superconductors, i.e., superconductors whose effective conduction-electron mass m^* is typically more than 100 electron masses. Physicists have given these materials a somewhat more pretentious name "heavy fermion superconductors." The first such superconductor, $CeCu_2Si_2$, was discovered in 1979 (Steglich *et al.*, 1979), and some time passed before the phenomenon was confirmed by the discovery of other examples, such as UBe_{13} (Ott *et al.*, 1983) and UPt_3 (Stewart

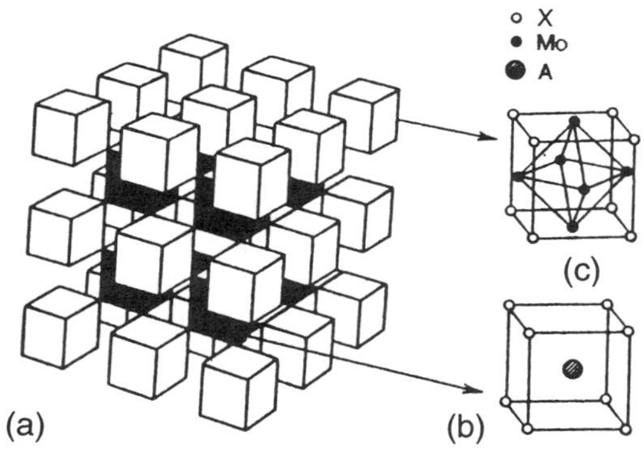

Figure 3.25 Structure of the Chevrel phase $A_xMo_6X_8$ (Vonsovsky *et al.*, 1982, p. 431).

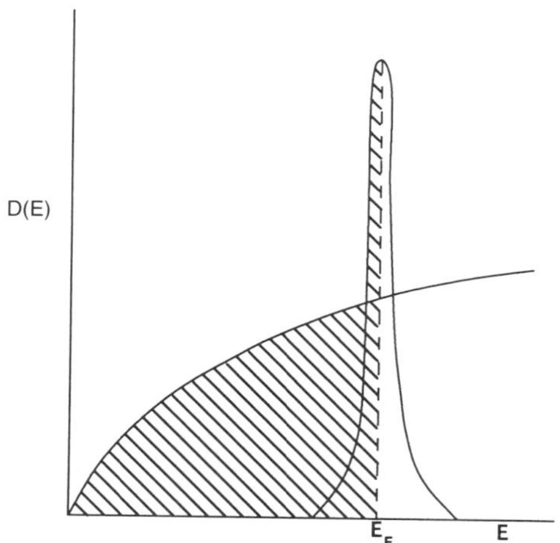

Figure 3.26 Density of states of a heavy fermion compound showing a peaked, narrow, half-occupied band at the Fermi level.

et al., 1984). Since then many additional cases have been found. Many of the investigators who are now active in the field of oxide superconductivity obtained their experience with the heavy-electron types. Work on heavy-electron superconductors has been reviewed in a number of studies (Coles, 1987; Ott, 1987; Stewart, 1984).

Since the superconducting charge carriers are Cooper pairs formed from heavy electrons and since these pairs are bosons, it would be more appropriate to call these compounds "heavy boson superconductors." This, however, is never done, so we will conform to the conventional usage. Ordinary superconductors are not usually called boson superconductors either.

The large effective mass has a pronounced effect on several properties of superconducting materials since it enters into the expression (Eq. (1.41)) for the electron density of states at the Fermi level:

$$D(E_F) = \frac{1}{2\pi^2}\left(\frac{2m^*}{\hbar^2}\right)^{3/2} E_F^{1/2}. \quad (3.7)$$

Heavy-electron compounds have densities of states that correspond to values of $m^* \approx 200m_e$ in Eq. (3.7), as shown by the data in Table 3.10.

The rare-earth element Ce has two $4f$ electrons and the actinide element U has three $5f$ electrons. In compounds each of these elements has f electron configurations that can mix as linear combinations or hybridize with the conduction electrons and together produce sharp energy bands near the Fermi level (Chapter 8, Section VI). The narrow width of such a band gives it a high density of states, and hence, by Eq. (3.7), a large effective mass. A situation of this type is sketched in Fig. 3.26 in which a narrow hybridization band superimposed on the usual conduction electron expression $D(E)\ \alpha\ \sqrt{E}$ from Fig. 1.7 is shown. The figure is drawn for the situation where the Fermi level is at the center of the hybridization band. Figure 3.27 shows the density of states at the Fermi level calculated for three values of the effective hybridization between a conduction electron and an f electron wave function (Hofmann and Keller, 1989). We see that the sharpness of the peak in $D(E_F)$ depends on the strength of this interaction. The electrons in these f shells of Ce and U are also responsible for the formation of the superconducting state. Other rare earths and actinides do not form these types of hybridization bands at the Fermi level.

Table 3.10 Properties of Several Heavy-Electron Superconductors[a]

Compound	T_c (K)	T_N (K)	Θ (K)	μ_{eff}/μ_B	m^*/m_e
$CeAl_3$			−43	2.62	
$CeCu_6$			−88	2.68	
$NpBe_{13}$		3.4	−42	2.76	
UBe_{13}	0.85	8.8	−70	3.1	192
UCd_{11}		5.0	−23	3.45	
UPt_3	0.43	5.0	−200	2.9	187
U_2Zn_{17}		9.7	−250	4.5	
$CeCu_2Si_2$	0.6	0.7	−140	2.6	220
UNi_2Al_3	1.0	4.6			
UPd_2Al_3	≈ 2.0	14.0			
URu_2Si_2	1.3	17.5			

[a] T_c is the superconducting transition temperature; T_N is the Néel temperature; Θ is the Curie–Weiss temperature; μ_{eff} is the effective magnetic moment and m^*/m_e is the ratio of the effective mass to the free electron mass. Some of the data are borrowed from Stewart (1984). Additional data are from Geibel *et al.* (1991a) on UNi_2Al_3, from Geibel *et al.* (1991b and 1991c) on UPd_2Al_2, and Issacs *et al.* (1990) on URh_2Si_2.

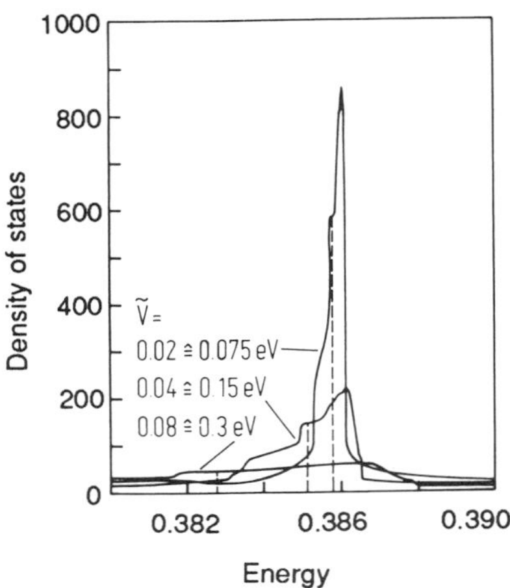

Figure 3.27 Density of states of a narrow hybridization band calculated for three values of the hybridization between a conduction electron and an f electron wave function (Hofmann and Keller, 1989).

Much of the evidence for the high effective mass comes from experimental observations in the normal state. For example, the conduction-electron contribution to the specific heat from Eq. (1.52),

$$\gamma = \tfrac{1}{3}\pi^2 D(E_F) k_B^2, \tag{3.8}$$

is proportional to the density of states (3.7), and hence is unusually large for heavy-electron compounds. The electronic specific-heat coefficients γ for heavy-electron superconductors are, on average, more than 10 times larger than those of other superconducting compounds, as may be seen by consulting Table 4.1. The discontinuity $(C_e - \gamma T_c)$ in the specific heat at the transition temperature is also correspondingly large for heavy-electron compounds, so the ratio $(C_s - \gamma T_c)/\gamma T_c$ is close to the usual BCS value of 1.43, as may be seen from the same table. Figure 3.28 shows a recent measurement of this discontinuity in the compound UPt_3.

Heavy-electron systems often exhibit two ordering transitions, a superconducting transition at T_c and an antiferromagnetic ordering transition at the Néel temperature T_N, with typical values given in Table 3.10. The superconducting transition is illustrated by the drop in magnetic susceptibility at T_c (shown in Fig. 3.29 for URu_2Si_2). When the magnetic moments of the ions couple antiferromagnetically, the magnetic susceptibility often exhibits Curie–Weiss behavior (cf. Eq. (1.79)). Many of these compounds have effective magnetic moments exceeding the Bohr magneton, as shown in Table 3.10. The table lists several other properties of these materials.

The heavy-electron superconductors have anisotropic properties that become apparent in such measurements as those for the critical fields, electrical resistivity, ultrasonic attenuation, thermal conductivity, and NMR relaxation. Figures 3.30 and 3.31, respectively, show typical anisotropies in the resistivity (Coles, 1987; Mydosh, 1987; Palst *et al.*, 1987) and upper-critical field (Assmus *et al.*, 1984; Stewart, 1984). The presence of a peak in the resistivity shown in Fig. 3.30 is characteristic of the heavy fermions.

The lower-critical fields B_{c1} are several mT at 0 K, while the upper-critical fields B_{c2} at 0 K approach 1 or 2 tesla, as shown by the example in Fig. 3.31. The critical field derivatives dB_{c2}/dT are high in absolute magnitude, such as -10 T/K for $CeCu_2Si_2$ and -44 T/K for UBe_{13}.

The London penetration depth λ_L is several thousand angstroms, consistent with the large effective mass m^*, which enters as a square root factor in the classical expression in (5.45)

$$\lambda_L = \left(\frac{m^*}{\mu_0 n_s e^2}\right)^{1/2}. \tag{3.9}$$

In ordinary superconductors magnetic impurities suppress the superconducting state due to the pair-breaking effect of the magnetic moments. In anisotropic superconductors, on the other hand, any kind of impurity is pair breaking. In particular, small amounts of nonmagnetic impurities replacing, for example, uranium, beryllium, or platinum, bring about a pronounced lowering of T_c.

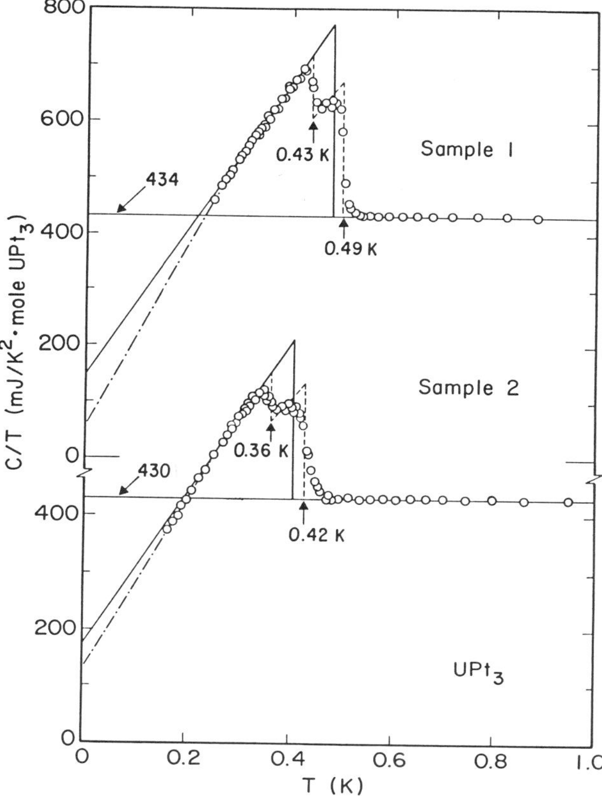

Figure 3.28 Specific heat of the heavy fermion compound UPt_3 in the neighborhood of the transition temperature. The solid lines correspond to a single ideally sharp transition while the dashed lines are fits to the data with two adjacent sharp transitions (Fisher *et al.*, 1989).

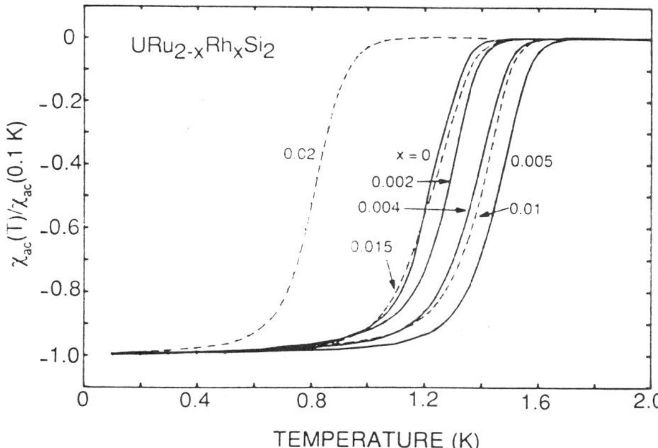

Figure 3.29 Dependence of magnetic susceptibility χ on temperature for the heavy fermion compounds $URu_{2-x}Rh_xSi_2$, with x in the range $0 \leq x \leq 0.02$ (Dalichaouch *et al.*, 1990).

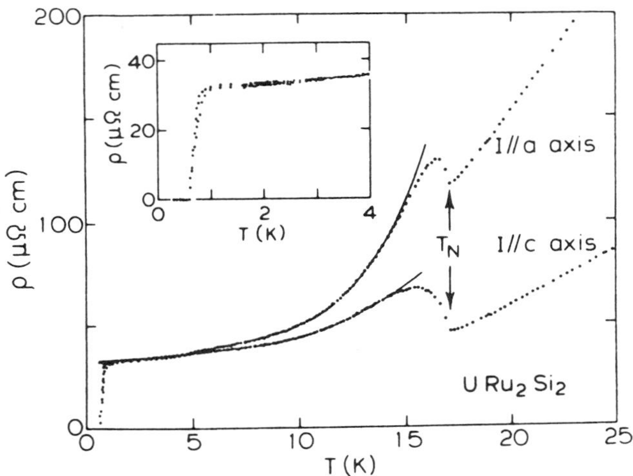

Figure 3.30 Peak near 16 K in the resistivity-versus-temperature plot of the heavy fermion compound URu_2Si. The inset shows the superconducting transition below 1 K in an expanded scale (Mydosh, 1987; see also Palst *et al.*, 1987).

Some researchers believe that the heavy-electron materials exhibit an unconventional type of superconductivity that does not involve the ordinary electron–phonon interaction (Bishop *et al.*, 1984; Coles, 1987; Gumhalter and Zlatic, 1990; D. W. Hess *et al.*, 1989; Ott, 1987; Ozaki and Machida, 1989; Rodriguez, 1987; Stewart, 1984). The same claim has been made for the newer oxide superconductors.

There have also been reports of high effective masses in the oxide superconductors, with, for example, $m^*/m \approx 12$ (Matsura and Miyake, 1987) in LaSrCuO, and $m^*/m \approx 5$ (Gottwick *et al.*, 1987), $m^*/m = 9$ (Salamon and Bardeen, 1987), and $m^*/m \approx 10^2$ in YBaCuO (Kresin and Wolf, 1987). Some recently discovered heavy-electron superconductors are $Ce_3Bi_4Pt_3$ (Riseborough, 1992), UNi_2Al_3 (Geibel *et al.*, 1991a; Krimmel *et al.*, 1992), UPd_2Al_3 (Geibel *et al.*, 1991b, c; Krimmel *et al.*, 1992; Sato *et al.*, 1992), and YBiPt (Fisk *et al.*, 1991).

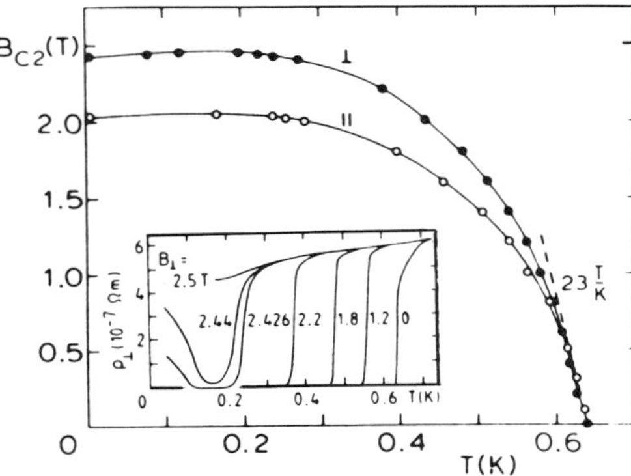

Figure 3.31 Upper crystal-field anisotropy in tetragonal heavy fermion superconductor $CeCu_2Si_2$ parallel to, respectively perpendicular to, the Ce planes. Note the lack of anisotropy at T_c where both orientations have the same slope, -23 T/K, as given by the dashed line. The inset shows the temperature dependence of the resistivity under various applied fields (Assmus *et al.*, 1984).

XII. CHARGE–TRANSFER ORGANICS

Organic compounds and polymers are ordinarily considered as electrical insulators, but it is now known that some of them are also good electrical conductors. For example, the organic compound 7, 7, 7, 8,-tetracyano-*p*-quinodimethane,

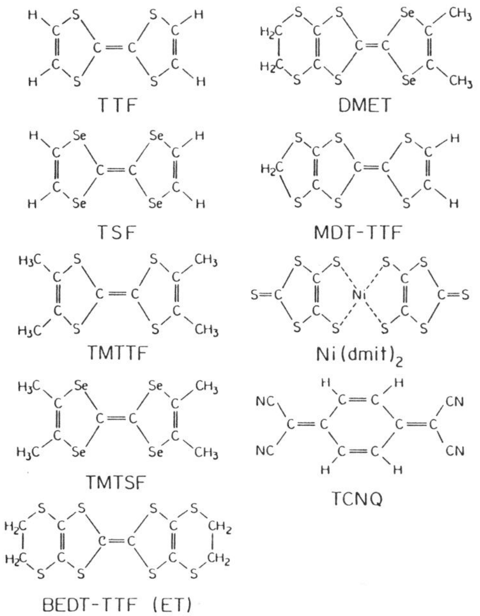

TTF DMET

TSF MDT-TTF

TMTTF Ni(dmit)$_2$

TMTSF TCNQ

BEDT-TTF (ET)

Figure 3.32 Structures of the principal molecules that form organic conductors and superconductors (Ishiguro and Yamaji, 1990, p. 2).

called TCNQ for short, forms highly conducting salts with a number of compounds, and the properties of these and other organic conductors were widely studied in the 1970s. Several of the $(TMTSF)_2X$ charge–transfer salts superconduct under pressure, where TMTSF is an electron donor and the monovalent counter ion X^- is, for example, AsF_6^-, ClO_4^-, FSO_3^-, PF_6^-, ReO_4^-, SbF_6^-, or TaF_6^-. Figure 3.32 gives the structural formulae of some of the principal organic molecules that play the role of electron donors in conducting and superconducting organics. Figure 3.33 shows the number and range of transition temperatures associated with each.

The electrical properties of organic conductors are often highly anisotropic. TCNQ salts behave as quasi-one-dimensional conductors, and salts of other organics, such as bis(ethylenedithia)tetrathiafulvalene, called BEDT-TTF for short, exhibit low-dimensional behavior (Brooks *et al.*, 1992; Fortune *et al.*, 1992), with T_c reported in excess of 13 K (Schirber *et al.*, 1991). In addition, the superconducting properties of the organics, such as the critical fields and the coherence length, are often anisotropic. For example, the triclinic compound β-$(ET)_2I_3$ has lower-critical fields $B_{c1} = 5$, 9, and 36 μT along the *a*, *b*, and *c* crystallographic directions, respectively and corresponding upper-critical fields $B_{c2} = 1.78$, 1.70, and 0.08 T (Schwenk *et al.*, 1985; Tokumoto *et al.*, 1985). The coherence lengths are $(\xi)_{\text{plane}} = 350$ Å in the conducting plane and $(\xi)_\perp = 23$ Å perpendicular to this plane, the latter being close to the longest lattice constant c = 15.3 Å (Ishiguro and Yamaji, 1990, p. 260). Other anisotropic properties are the plasma frequency (0.89 and 0.48 eV) and the effective mass $m^*/m_e = 2.0$ and $m^*/m_e = 7.0$) parallel to and perpendicular to the (−110) direction, respectively (Kuroda *et al.*, 1988).

At the present time the transition temperature T_c of the organics are in the range of typical classical superconductors. However, it is clear from Fig. 3.34 that these temperatures have been rising rapidly during the past decade, though the best of them is still far below the higher copper-oxide values. Some of the organics, such as

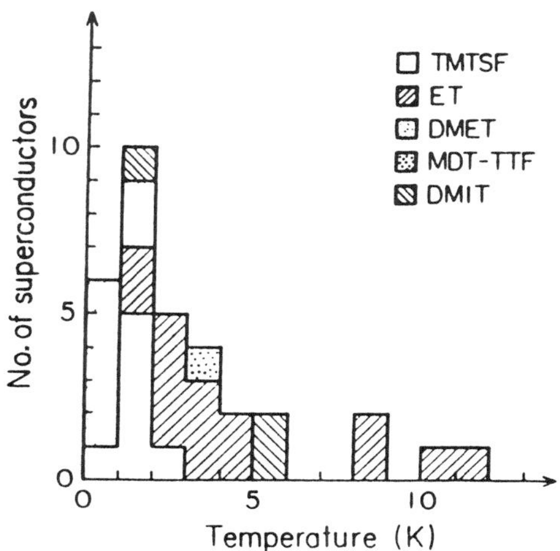

Figure 3.33 Number of known organic superconductors classified by their molecular type as a function of T_c (Ishiguro and Yamaji, 1990, p. 263).

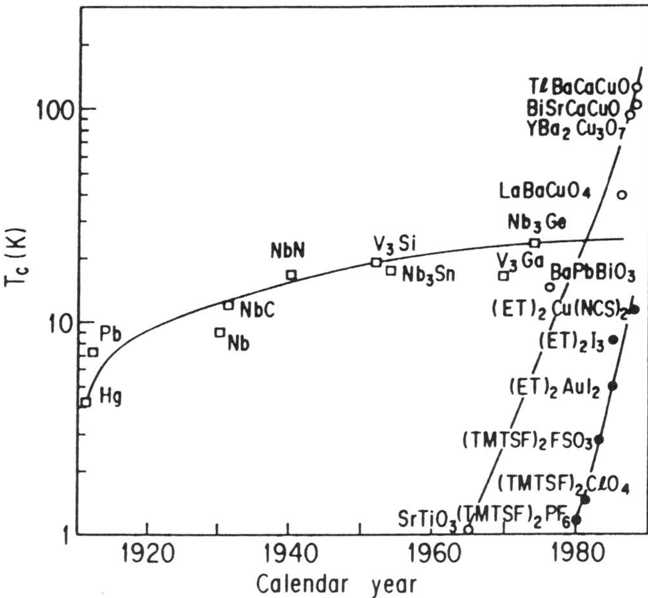

Figure 3.34 Increase in the transition temperature T_c with time. The classic types have leveled off, while the organics on the right are still rising (Ishiguro and Yamaji, 1990, p. 259).

BEDT-TTF, exhibit interesting similarities with the cuprates because of their layered structures (Farrell *et al.*, 1990b).

XIII. CHALCOGENIDES AND OXIDES

Many of the classical superconductors (for example, the Chevrel phases discussed in Section X) contain an element of row VI in the periodic table, namely O, S, Se, or Te, with oxygen by far the least represented among the group. The newer superconductors in contrast, are oxides. Since the presence of lighter atoms tends to raise the Debye temperature, oxides are expected to have higher Debye temperatures than the other chalcogenides (Gallo *et al.*, 1987, 1988). Thus the presence of group VI elements is a commonality that links the older and the newer superconductors.

The two oxide compounds listed in Table 3.2 are cubic and ternary. One is the well-known ferroelectric perovskite $SrTiO_3$, which has a very low transition temperature (0.03–0.35 K). Nb-doped $SrTiO_3$, with its small carrier concentration $N_e \approx 2 \times 10^{20}$ and high electron–phonon coupling, has $T_c = 0.7$ K (Baratoff and Binnig, 1981; Binnig *et al.*, 1980). The other cubic ternary oxide is the spinel $LiTi_2O_4$ with moderately high $T_c = 13.7$ K (Johnston *et al.*, 1973). The system $Li_xTi_{3-x}O_4$ is superconducting in the range $0.8 \leq x \leq 1.33$ with T_c in the range 7–13 K. It is interesting to note that the stoichiometric compound with $x = 1$ is near the composition where the metal-to-insulator transition occurs. A recent band structure calculation of this Li–Ti spinel (Satpathy and Martin, 1987) is consistent with resonance valence bond superconductivity (Chapter 8, Section VII, F) and a large electron–phonon coupling constant ($\lambda \approx 1.8$). Only three more of the 200 known spinels superconduct—namely, $CuRh_2Se_4$ with $T_c = 3.5$ K, CuV_2S_4 with $T_c = 4.5$, and $CuRh_2S_4$ with $T_c = 4.8$—so $LiTi_2O_4$ turns out to be the only spinel oxide superconductor.

XIV. BARIUM LEAD–BISMUTH OXIDE PEROVSKITE

In their pioneering article Bednorz and Müller (1986) called attention to the discovery of superconductivity in the mixed-valence compound $BaPb_{1-x}Bi_xO_3$ by Sleight and other researchers. (Bansil and Kaprzyk, 1991; Batlogg *et al.*, 1988; Gilbert *et al.*, 1978; Prassides *et al.*, 1992; Sleight *et al.*, 1975; Sleight, 1987; Suzuki *et al.*, 1981a, b; Thorn, 1987). It was pointed out in these studies that the stoichiometric form of this compound presumably has the composition $Ba_2Bi^{3+}Bi^{5+}O_6$; structurally, it is distorted perovskite, as explained in Chapter 7, Section IV. It is clear that an admixture with $BaPb^{4+}O_3$, together with some oxygen deficiency, could change the Bi^{5+}/Bi^{3+} ratio. The highest T_c for homogeneous oxygen-deficient mixed crystals of this system was 13 K, and this came with the comparatively low carrier concentration of 2×10^{21}–4×10^{21} (Than *et al.*, 1980). The intensity of the strong vibrational breathing mode near 100 cm^{-1} was found to be proportional to T_c (Bednorz and Müller, 1986; Masaki *et al.*, 1987). These results led Bednorz and Müller to reason that, "Within the BSC system, one may find still higher T_c's in the perovskite type or related metallic oxides, if the electron–phonon interactions and the carrier densities at the Fermi level can be enhanced further." It was their determination to prove the validity of this conjecture that led to the biggest breakthrough in physics of the present decade. Their choice of materials to examine was influenced by the 1984 article of Michel and Raveau (1984) on mixed-valent Cu^{2+}, Cu^{3+} lanthanum–copper oxides containing alkaline earths.

XV. BARIUM-POTASSIUM BISMUTH-OXIDE CUBIC PEROVSKITE

The compound $Ba_{1-x}K_xBiO_{3-y}$ with $T_c \approx 30$ K for $x \approx 0.4$ is structurally related to the Ba–Pb–Bi system. It was not discovered until after the advent of high-T_c (Cava *et al.*, 1988; Mattheiss *et al.*, 1988), and is of particular significance for several reasons. It is the first oxide superconductor without copper that has a transition temperature above that of all the *A*15 compounds. This high T_c occurs without the presence of a two-dimensional metal–oxygen lattice. It has the cubic perovskite structure (Kwei *et al.*, 1989; Pei *et al.*, 1990; Schneemeyer *et al.*, 1988) described in Chapter 7, Section II, which is also the prototype of the HTSC structures. This makes it similar to $BaPb_{1-x}Bi_xO_3$, which is not a cubic, but rather a distorted type of perovskite. The cubic potassium compound has attracted a great deal of attention with the hope that it might elucidate the mechanism of high-temperature superconductivity. Many experimental measurements have been made on this system, such as magnetization (Huang *et al.*, 1991b; Kwok *et al.*, 1989), photoemission (Hamada *et al.*, 1989; Jeon *et al.*, 1990; Nagoshi *et al.*, 1991), x-ray absorption (Salem-Sugui *et al.*, 1991), energy gap (Schlesinger *et al.*, 1989), and the irreversibility line (Shi *et al.*, 1991); a value for the isotope effect exponent $\alpha \approx 0.38$ has been reported (Hinks *et al.*, 1988b, 1989; W. Jin *et al.*, 1991).

Uemura *et al.* (1991) point out that $Ba_{1-x}K_xBiO_{3-y}$ shares with the cuprates, Chevrel phase compounds, heavy fermions, and organic superconductors a transition temperature T_c which is high relative to its N_s/m^* (carrier density-to-effective mass) ratio, and suggest that this feature distinguishes these superconductors from the ordinary BSC types.

XVI. BUCKMINSTERFULLERENES

The C_{60} molecule consists of 60 carbon atoms at the apices or vertices of a regular triacontaduohedron (32-sided figure) 7.1 Å in diameter, as shown in Fig. 3.35. This structure contains 12 regular pentagons and 20 hexagons; some details

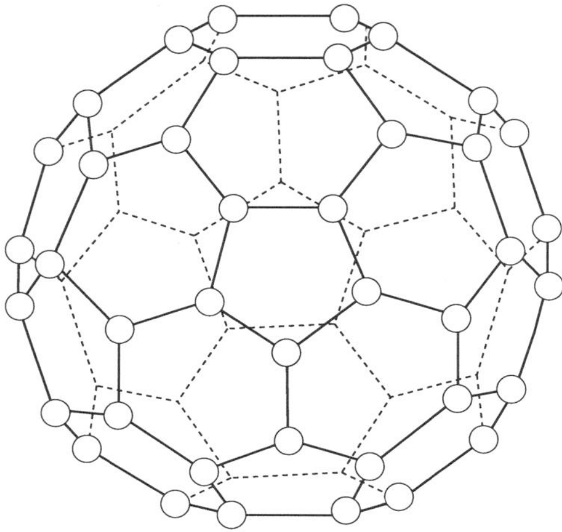

Figure 3.35 Structure of the buckminsterfullerene molecule C_{60}.

of its bond lengths and symmetry operations are given in Chapter 7, Section XI. One important fact to note is that all of the carbon atoms are equivalent to each other. Thus the C_{60} molecule might be called the world's smallest soccer ball! Because of its resemblance to the geodesic dome of architect R. Buckminster Fuller it has been referred to as buckminsterfullerene, or fullerene for short. The dual resemblances have prompted the sobriquet buckyball.

Table 3.11 Transition Temperatures T_c of some Alkali Metal-Doped C_{60} Compounds[a]

Compound	T_c, K
K_3C_{60}	19
K_2RbC_{60}	22
Rb_2KC_{60}	25
Rb_3C_{60}	29
Rb_2CsC_{60}	31
Cs_2RbC_{60}	33
Cs_3C_{60}	47[b]
$Rb_{2.7}Tl_{2.2}C_{60}$	45[c]

[a] Haddon, 1992, Fig. 7.
[b] Iqbal *et al.*, 1991.
[c] Novikov *et al.*, 1992.

The compound C_{60} is not itself a superconductor, but when alkali metals are added it becomes superconducting. The doped compound forms a face-centered cubic lattice with a lattice constant of 10.04 Å. The structure has two tetrahedral holes and one octahedral hole per C_{60} molecule. If all of these holes are occupied by alkali metals A, the resulting compound is A_3C_{60}. An example of such a compound is K_2RbC_{60} with the potassiums in the smaller tetrahedral holes and the rubidiums in the larger octahedral holes. The transition temperatures of several of these doped fullerenes are given in Table 3.11. The compound Rb_3C_{60} has been found to have an isotope effect exponent $\alpha = 0.37$ (Ramirez *et al.*, 1992a).

XVII. BOROCARBIDES

During 1994, there was considerable interest in the borocarbide class of superconductor with the general formula RM_2B_2C where M is usually Ni and R is a rare earth. Table 3.12 lists some of the superconducting compounds of this series with their T_C values. Transition temperatures in the low 20's have been reported for less well characterized members of the series. A variety of studies have been made of these compounds. Some are found to exhibit antiferromagnetism coexisting with the superconducting state, and reentrant

Table 3.12 Transition Temperatures T_C of some borocarbide RM_2B_2C and boronitride $R_3M_2B_2N_3$ compounds

Compound	T_C, K
$ErNi_2B_2C$	11
$HoNi_2B_2C$	8
$LuNi_2B_2C$	16
$ThNi_2B_2C$	8
$TmNi_2B_2C$	9
YNi_2B_2C	16
$ThPt_2B_2C$	6
$ThPd_2B_2C$	14
$La_3Ni_2B_2N_3$	12

superconductivity has been observed with an intermediate temperature range of normal state behavior. Related compounds such as the boronitride $La_3Ni_2B_2N_3$ are also found to be superconducting. Hilscher *et al.* (1994) provide a status report on these rare earth transition metal borocarbides.

FURTHER READING

The status of superconductivity research through 1969 is presented in the two-volume work edited by Parks. The properties of the superconducting transition elements and their compounds are surveyed in the book by Vonsovsky, Izyumov, and Kurmaev (1982). Several of the figures and some of the data in the tables of the present chapter were borrowed from this work. Roberts (1976) tabulated various properties of a large number of classical superconductors. The Villars–Phillips (1988) approach of Chapter 7, Section XIII, correlates many of the data on transition metal compounds. A review of the Chevrel compounds may be found in Fischer (1978).

The first examples of heavy-electron superconductors were reported by Steglich *et al.* (1979), Ott *et al.* (1983), and Stewart *et al.* (1984). Several reviews of the field have appeared over the years (e.g., Coles, 1987; Ott, 1987; Sauls, 1994; Stewart, 1984; Taillfer, 1995).

Organic superconductors have recently been surveyed in monographs by Ishiguro and Yamaji (1990) and Kresin and Little (1990), and compared with the cuprates in a brief review (Greene, 1990).

The *Accounts of Chemical Research* devoted the special issue of March, 1992 (Vol. 25, No. 3) to reviews of various properties of the buckminsterfullerenes. Other reviews of fullerenes are by Hebard (1994), Kadish and Ruoff (1994), and Yakhmi (1994).

The Strukturbericht notation for the crystal structures ($A15$, $C14$, etc.) used in Table 3.2 and elsewhere in this text is explained by Pearson (1958).

PROBLEMS

1. Identify the 120 symmetry operations of a buckyball. Hint: there is a 72° improper rotation about each fivefold axis of symmetry.
2. Show why the alloys of Fig. 3.17 contain isoelectronic elements.
3. Consider the following expression as an alternate to Eq. (3.3) for describing the electronic specific heat of alloys:

$$\gamma_{AB} = f_A \gamma_A(N_{eA}) + f_B \gamma_B(N_{eB}) + \alpha f_A f_B.$$

Evaluate the constant α for the three cases of Fig. 3.17, and compare the goodness of fit to the data with the results obtained from Eq. (3.3), as plotted in Fig. 3.17.

4

Thermodynamic Properties

I. INTRODUCTION

The first three chapters surveyed normal state conductivity, properties characteristic of superconductivity, and the principal types of superconducting materials. But none of the theoretical ideas that have been proposed to account for these phenomena were developed. In the present chapter we will refer to certain principles of thermodynamics as a way of providing some coherence to our understanding of the material that has been covered so far. In three following chapters we will deepen our understanding by examining in succession the London approach, the Ginzburg–Landau phenomenological theory, the microscopic theory of Bardeen, Cooper, and Schrieffer (BCS), the Hubbard model, and the band structure. Then, after having acquired some understanding of the theory, we will proceed to examine other aspects of superconductivity from the perspective of the theoretical background, with an emphasis on the high-transition temperature cuprates.

The overall behavior of the heat absorption process that will be examined in this chapter can be understood by deriving the thermodynamic functions of the normal state from the known specific heat–temperature dependence. The corresponding superconducting-state thermodynamic functions can then be deduced from the critical field dependence of the Gibbs free energy. We will begin by presenting experimental results on specific heat, following that with a derivation of the different thermodynamic functions associated with specific heat in the normal and superconducting states.

A specific heat determination is of interest because it provides a good measure of the range of applicability of the photon-mediated BCS theory (cf. Chapter 6, Section VI). This theory predicts characteristics of the discontinuity in specific heat at T_c.

II. SPECIFIC HEAT ABOVE T_C

One of the most extensively studied properties of superconductors is the specific heat. It represents a "bulk" measurement that sees the entire sample since all of the sample responds. Many other measurements are sensitive to only part of the sample, for example, microscopy in which only the surface is observed.

Above the transition temperature T_c the specific heat C_n of high-temperature superconductors tends to follow the Debye theory described in Chapter 1, Section XII (cf. Fig. 1.12, which shows the positions of the lanthanum and yttrium compounds on the Debye plot at their transition temperatures). We know from Eq. (1.64) that C_n of a normal metal far below the Debye temperature Θ_D is the sum of a linear term $C_e = \gamma T$ arising from the conduction electrons, a lattice vibration or phonon term $C_{ph} = AT^3$, and sometimes an additional Schottky contribution aT^{-2} (Crow and Ong, 1990) (cf. Chapter 1, Section XII).

$$C_n = aT^{-2} + \gamma T + AT^3. \qquad (4.1)$$

For the present we will ignore the Schottky term aT^{-2}. The C_{exp}/T versus T^2 plot of Fig. 4.1 shows how the yttrium compound obeys Eq. (4.1) at low temperatures and then deviates from it at higher temperatures, as expected for the Debye approximation. The normalized specific heat plots of Fig. 4.2 compare for the case of several metals the electronic and photon contributions to the specific heat at low temperatures.

In the free-electron approximation the electronic contribution to the specific heat per mole of conduction electrons is given by Eqs. (1.51) and (1.53), which we combine as follows:

$$C_e = \gamma T = \tfrac{1}{2}\pi^2 R\left(\frac{T}{T_F}\right)$$
$$= 4.93R\left(\frac{T}{T_F}\right). \qquad (4.2)$$

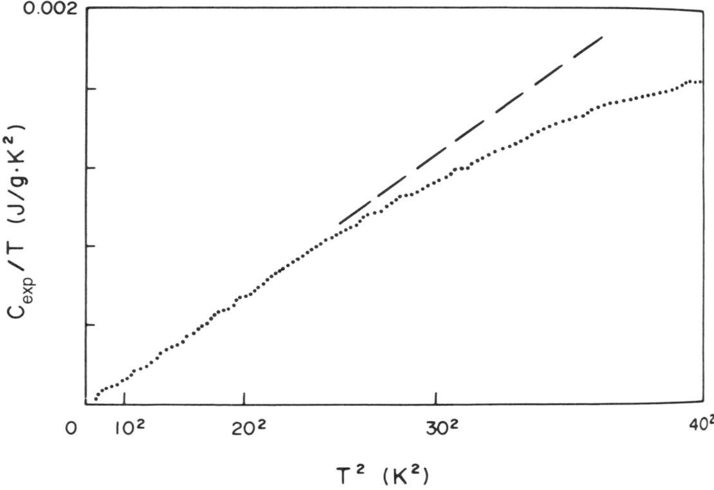

Figure 4.1 Plot of C_{exp}/T versus T^2 for $YBa_2Cu_3O_{7-\delta}$ showing how the deviation from linearity begins far below the transition temperature $T_c = 90\,K$ (Zhaojia *et al.*, 1987).

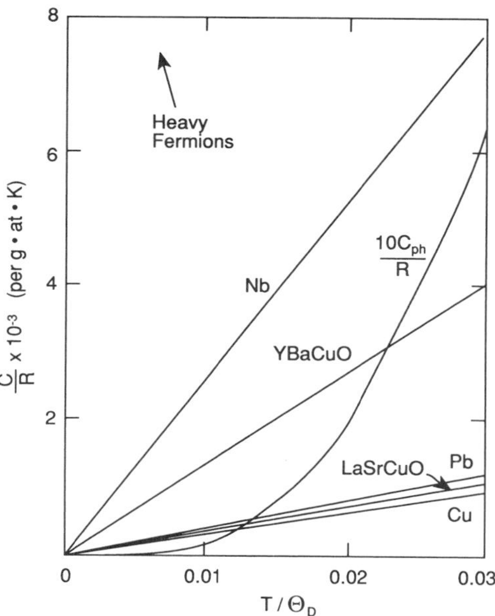

Figure 4.2 Comparison of the electronic specific heat $C_e = \gamma T$ of several conductors and superconductors at low temperature. The low-temperature Debye approximation $C_{ph} \approx AT^3$, multiplied by 10, is shown for comparison. The specific heats are normalized relative to the gas constant R and are expressed in terms of gram atoms. The heavy fermions are off scale on the upper left.

In the Debye approximation the phonon contribution to the specific heat per gram atom is given by Eq. (1.62a),

$$C_{ph} = AT^3 = \left(\frac{12\pi^4}{5}\right) R \left(\frac{T}{\Theta_D}\right)^3$$

$$= 234R\left(\frac{T}{\Theta_D}\right)^3, \qquad (4.3)$$

where $R = k_B N_A$ is the gas constant and T_F the Fermi temperature. For a typical high-temperature superconductor we see from Table 4.1 that $\gamma \approx 10$ mJ/mole Cu K^2, and Eq. (4.2) gives $T_F \approx 4.0 \times 10^3$ K. This is much smaller than typical good conductor values, such as 8.2×10^4 K for Cu, as listed in Table 1.1. This discrepancy can be accounted for in terms of effective masses, as will be explained subsequently.

The vibrational and electronic contributions to the specific heat at $T = T_c$ may be compared with the aid of Eqs. (4.2) and (4.3),

$$\frac{C_{ph}}{C_e} = \frac{AT_c^2}{\beta\gamma} \qquad (4.4)$$

$$= \left(\frac{47.5}{\beta} \cdot \frac{T_F}{\Theta_D^3}\right) T_c^2, \qquad (4.5)$$

where the factor β, which is the ratio of the number of conduction electrons to the number of atoms in the compound, is needed when C_e is expressed in terms of moles of conduction electrons and C_{ph} in terms of moles of atoms. When both specific heats are in the same units, β is set equal to 1. Typical values of the Fermi and Debye temperatures are 10^5 K and 350 K, respectively.

For most low-temperature superconductors the transition temperature T_c is sufficiently below Θ_D so that the electronic term in the specific heat is appreciable in magnitude, and sometimes dominates. This is not the case for high-temperature superconductors, however. Using measured values of γ and A we have shown in our earlier work (Poole *et al.*, 1988), that $AT_c^2 \gg \gamma$ for

$$(La_{0.9}Sr_{0.1})_2CuO_{4-\delta} \text{ and } YBa_2Cu_3O_{7-\delta},$$

so for oxide superconductors the vibrational term dominates at T_c, in agreement with the data plotted in Figs. 4.1 and 4.3.

If the conduction electrons have effective masses m^* that differ from the free-electron mass m, the conduction-electron specific heat coefficient γ is given by Eq. (1.54),

$$\gamma = \left(\frac{m^*}{m}\right)\gamma_0, \qquad (4.6)$$

where γ_0 is the ordinary electron counterpart of γ from Eq. (1.51). In the free-electron approximation we have from Eq. (4.2)

$$\gamma_0 = \frac{\frac{1}{2}\pi^2 R}{T_F}, \qquad (4.7)$$

Table 4.1 Debye temperature Θ_D, Density of States $D(E_F)$, and Specific Heat Data[a]

Material	T_c (K)	Θ_D (K)	γ_n^*	γ_n $\left(\frac{mJ}{mole\,K^2}\right)$	$(C_s - C_n)/T_c$ (mJ / mole K^2)	$(C_s - C_n)/\gamma T_c$	A (mJ / mole K^4)	$D(E_F)$ (states / eV)	Reference
Cd	0.55	252		0.67	0.91	1.36			
Al	1.2	423		1.36	1.97	1.45			
Sn, white	3.72	196		1.78	2.85	1.60			
Pb	7.19	102		3.14	8.51	2.71		1.55	
Nb	9.26	277		7.66	14.8	1.93		2.0	
$Zr_{0.7}Ni_{0.3}$	2.3	203		4.04	≈ 6.7	≈ 1.65	0.23		Sürgers *et al.* (1989)
Zr_3Pb (*A*15)	0.76			4					Vonsovsky *et al.* (1982, pp. 269ff.)
V_3Ge (*A*15)	11.2			7					Vonsovsky *et al.* (1982, pp. 269ff.)
V_3Si (*A*15)	17.1			17					Vonsovsky *et al.* (1982, pp. 269ff.)
Nb_3Sn (*A*15)	18.0			13					Vonsovsky *et al.* (1982, pp. 269ff.)
HfV_2 (laves)	9.2	187		21.7				2.30	Vonsovsky *et al.* (1982, p. 379)
$(Hf_{0.5}Zr_{0.5})V_2$ (laves)	10.1	197		28.3				2.97	Vonsovsky *et al.* (1982, p. 379)
ZrV_2 (laves)	8.5	219		16.5				1.86	Vonsovsky *et al.* (1982, p. 379)
$PbMo_6S_8$ (chevrel)	12.6			79					Vonsovsky *et al.* (1982, p. 420)
$PbMo_6Se_8$ (chevrel)	3.8			28					Vonsovsky *et al.* (1982, p. 420)
$SnMo_6S_8$ (chevrel)	11.8			105					Vonsovsky *et al.* (1982, p. 420)
YMo_6S_7 (chevrel)	6.3			34					Vonsovsky *et al.* (1982, p. 420)
UPt_3 (heavy fermion)	0.46		110	460	≈ 380	≈ 0.9	1525		Ellman *et al.* (1990); Fisher *et al.* (1989); Schuberth *et al.* (1992)
UCd_{11} (heavy fermion)	5	200		290.			115		de Andrade *et al.* (1991)
URu_2Si_2 (heavy fermion)	1.1			31	13	0.42			Ramirez *et al.* (1991)
$CeRu_2Si_2$ (heavy fermion)	≈ 0.8			340	1190	3.5			van de Meulen *et al.* (1991)
β-$(ET)_2I_3$ (organic)	1.1			24			19		Stewart *et al.* (1986)
$(TMTSF)_2ClO_4$ (organic)	1.2	213		10.5	17.5	1.67	11.4		Garoche *et al.* (1982)
K-$(ET)_2Cu(NCS)_2$ (organic)	9.3			34	1.5				Graebner *et al.* (1990)
K_3C_{60} (buckyball)	19	70			68			9.3	Ramir *et al.* (1992b); Novikov *et al.* (1992)
Rb_3C_{60} (buckyball)	30.5							10.9	Novikov *et al.* (1992)
Cs_3C_{60} (buckyball)	47.4							12.7	Novikov *et al.* (1992)

$BaPb_{1-x}Bi_xO_3$ (perovskite)	10			0.6	5	8		0.24	Junod (1990)
$Ba_{1-x}K_xBiO_3$ (perovskite)	24	210		< 0.06	0.4				Collocott *et al.* (1990b); Junod (1990); Stupp *et al.* (1989)
La_2CuO_4 (nonsuper-conducting)	—	385	≈ 0.7	3.4				1.5	Junod (1990)
$(La_{0.925}Ca_{0.075})_2CuO_4$	20	390	3		≈ 17	5.7			Junod (1990)
$(La_{0.925}Sr_{0.075})_2CuO_4$	37	360	≈ 2.2	4.5	≈ 9	2.0		1.9	Junod (1990); Sun *et al.* (1991)
$(La_{0.925}Ba_{0.075})_2CuO_4$	27	370	≈ 3.7		≈ 4	1.1			Junod (1990)
$(Nd_{0.925}Ce_{0.075})_2CuO_{4-\delta}$	18		53		0				Junod (1990); see Collocott *et al.* (1992)
$YBa_2Cu_3O_7$ (orthorhombic)	92	410	≈ 7	14	50	3.6	0.035	2.0	Collocott *et al.* (1990a); Junod (1990); Stupp *et al.* (1991)
$YBa_2Cu_3O_{6.5}$ (tetragonal)	—	350	≈ 8	5.5			0.44	0.8	Junod (1990); von Molnár *et al.* (1988)
$YBa_2Cu_4O_{8.5}$	80	350	≈ 12	4.9	14		2.1	2.1	Junod (1990); Junod *et al.* (1991)
$Bi_2Sr_2CaCu_2O_8$	95	250	≈ 3.4	≈ 8					Junod (1990); Fisher and Huse (1988); Urbach *et al.* (1989)
$Bi_2Sr_2Ca_2Cu_3O_{10}$	110	260			20				Junod (1990)
$(Bi_{1-x}Pb_x)_2Sr_2Ca_2Cu_3O_{10}$	105				25				Junod (1990)
$Tl_2Ba_2CuO_6$	80	240	≈ 0						Junod (1990)
$Tl_2Ba_2CaCu_2O_8$	110	260	< 6.8		19	> 2.8			Junod (1990)
$Tl_2Ba_2Ca_2Cu_3O_{10}$	125	280					2.0		Junod (1990); Urbach *et al.* (1989)
$HgBa_2Ca_2Cu_3O_8$	133				45				Schilling *et al.* (1994a, b)

[a] Some of the high-temperature superconductor values are averages from Junod (1990), in many cases with a wide scatter of the data. The density of states is expressed per atom for the elements and per copper atom for the high-temperature superconductors. For the latter γ_n is the electronic specific heat factor determined from normal state measurements, and γ_n^* is the value obtained from the limit $T \to 0$, as explained by Junod. The BCS theory predicts $(C_s - C_n)/\gamma_n T_c = 1.43$.

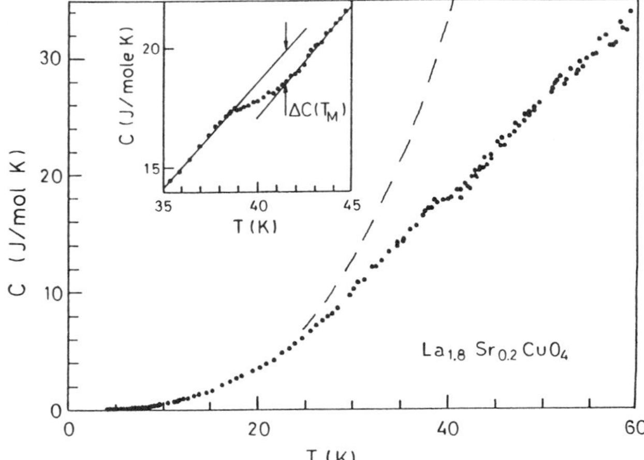

Figure 4.3 Discontinuity in the specific heat of $(La_{0.9}Sr_{0.1})_2CuO_4$ near 40 K. The inset shows the magnitude of the jump. The AT^3 behavior indicated by the dashed curve shows that the transition occurs beyond the region where the T^3 approximation is valid (Nieva *et al.*, 1987).

which gives for the effective mass ratio

$$\frac{m^*}{m} = \frac{\gamma T_F}{\frac{1}{2}\pi^2 R}. \qquad (4.8)$$

Table 1.1 lists effective mass ratios for the elemental superconductors calculated from this expression. The unusually low estimate of T_F given following Eq. (4.3) for a high-temperature superconductor can be explained in terms of a large effective mass. We see from Table 3.10 and Fig. 4.2 that large effective masses make the electronic term γ very large for the heavy fermions. The plot of T_c versus γ in Fig. 4.4 shows that the points for BCS phonon-mediated superconductors cluster in a region delimited by the dashed lines. The heavy fermions lie far to the right, as expected, while the oxide and cuprate compounds lie somewhat above those in the main group. A diagram similar to Fig. 4.4 may be found in Batlogg *et al.* (1987).

III. DISCONTINUITY AT T_c

The transition from the normal to the superconducting state in the absence of an applied magnetic field is a second-order phase transition, as we will show in Section XIV. This means that there is no latent heat, but nevertheless a discontinuity in the specific heat. The BCS theory, which will be explained in Chapter 6, predicts that the electronic specific heat jumps abruptly at T_c from the normal state value γT_c to the superconducting state value C_s with ratio

$$\frac{C_s - \gamma T_c}{\gamma T_c} = 1.43. \qquad (4.9)$$

Figure 4.5, as well as Fig. IX-12 of our earlier work (Poole *et al.*, 1988) show details of this jump for an element and for a high-temperature superconductor, respectively. For the latter case the magnitude of the jump is small compared to the magni-

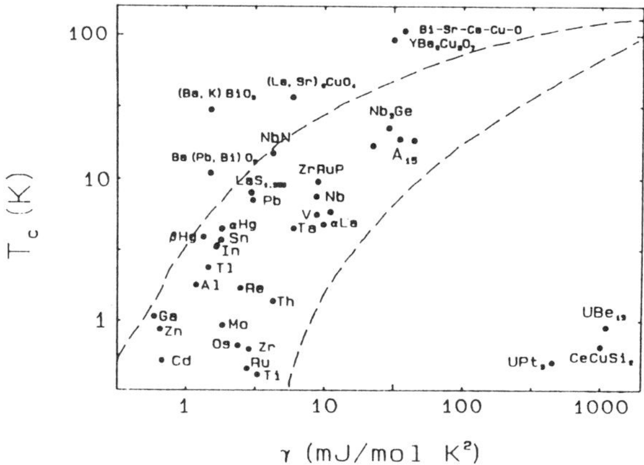

Figure 4.4 Comparison of the electronic specific-heat factor γ for a selection of superconductors and superconducting types over a wide range of T_c values. The dashed lines delimit the region of phonon-mediated superconductivity (Crow and Ong, 1990, p. 239).

tude of the total specific heat because it is superimposed on the much larger AT^3 vibrational term, as indicated in Fig. 4.3. This is seen if Eq. (4.9) is used to express Eq. (4.5) in the form

$$\frac{AT_c^3}{C_e - \gamma T_c} = \left(\frac{67.9}{\beta} \cdot \frac{T_F}{\Theta_D^3}\right) T_c^2. \quad (4.10)$$

Figure 4.6 illustrates how the small change at T_c is resolved by superimposing curves of C/T versus T^2 obtained in zero field and in a magnetic field large enough ($B_{app} > B_{c2}$) to destroy the superconductivity. It is clear from the figure that the data in the superconducting state extrapolate to zero, and that the normal state data extrapolate to γ at 0 K.

Many researchers have observed the jump in the specific heat at T_c (cf. Table 4.1 for results from a number of studies). Table 1 also lists experimental values of T_c, Θ_D, and γ, together with the ratios $(C_s - C_n)/T_c$ and $(C_s - C_n)/\gamma T_c$, for several elements and a number of copper-oxide superconductors. Some of the elements are close to the BCS value of 1.43, but the strongly coupled ones, Pb and Nb, which have large electron–phonon coupling constants λ, are higher. Several experimental results for YBaCuO are close to 1.43, as indicated in the table. Some researchers have failed to observe a specific heat discontinuity, however.

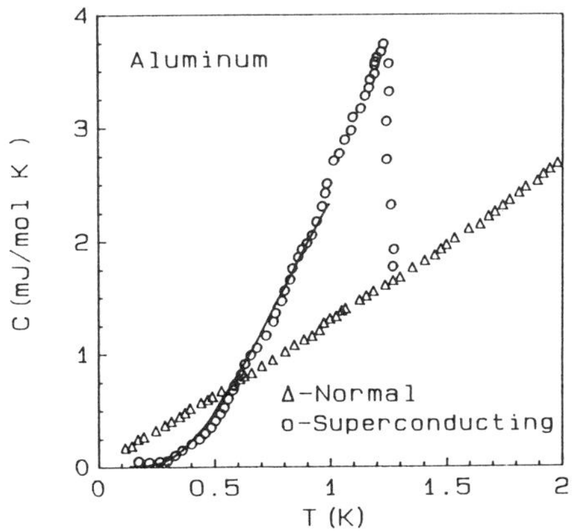

Figure 4.5 Specific heat jump in superconducting Al compared with the normal-state specific heat (Phillips, 1959; see Crow and Ong, 1990, p. 225).

IV. SPECIFIC HEAT BELOW T_c

For $T \ll T_c$ BCS theory predicts that the electronic contribution to the specific heat C_e will depend exponentially on temperature,

$$C_s \approx a \exp\left(-\frac{\Delta}{k_B T}\right), \qquad (4.11)$$

where 2Δ is the energy gap in the superconducting density of states. We see from Fig. 4.5 that the fit of this equation to the data for aluminum is good, with the specific heat falling rapidly to zero far below T_c, as predicted. The vibrational term AT^3 also becomes negligible as 0 K is approached, and other mechanisms become important, for example, antiferromagnetic ordering and nuclear hyperfine effects, two mechanisms that are utilized in cryogenic experiments to obtain temperatures down to the microdegree region.

V. DENSITY OF STATES AND DEBYE TEMPERATURE

The density of states at the Fermi level $D(E_F)$ can be estimated from Eq. (1.52):

$$D(E_F) = \frac{3}{\pi^2} \cdot \frac{\gamma}{R} \cdot \frac{1}{k_B}. \qquad (4.12)$$

For a typical high-temperature superconductor with $\gamma \approx 0.01$ J/mole Cu K we obtain

$$D(E_F) \approx \frac{4.5 \text{ states}}{\text{eV Cu atom}}. \qquad (4.13)$$

The Debye temperature may be estimated from the slope of the normal state C_n/T-versus-T^2 curve sketched in Fig. 1.13, since with the aid of Eq. (1.62a) we can write

$$\Theta_D^3 = \frac{12\pi^4 R}{\{5[\text{slope}]\}}. \qquad (4.14)$$

Typical values for Θ_D are from 200 to 350 K.

We have been using formulae that involve the free-electron approximation. To estimate the validity of this approximation we can make use of Eq. (1.88), which gives the ratio of γ to the magnetic susceptibility χ arising from the conduction electrons,

$$\frac{\gamma}{\chi} = \frac{1}{3}\left(\frac{\pi k_B}{\mu_B}\right)^2, \qquad (4.15)$$

in terms of well-known physical constants.

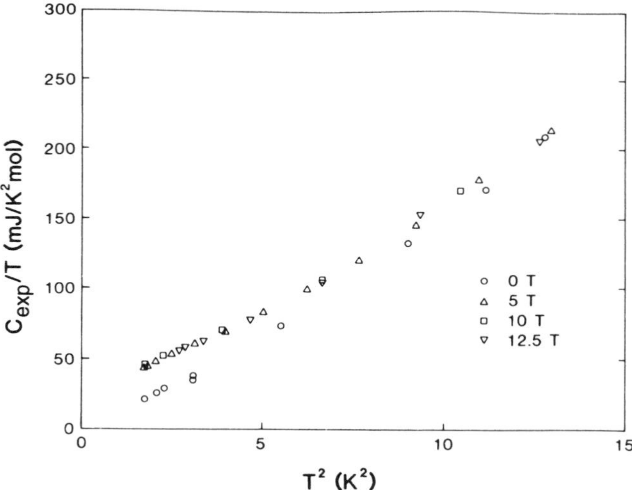

Figure 4.6 Plot of C_{exp}/T versus T^2 for monocrystals of the organic superconductor K–$(ET)_2Cu(NCS)_2$ in the superconducting state with no applied magnetic field, and also in the presence of applied fields that destroy the superconductivity. The superconducting state data extrapolate to a value $\gamma = 0$, while the normal-state extrapolation indicates $\gamma \approx 25$ mJ/mole K^2 (Andraka *et al.*, 1989).

VI. THERMODYNAMIC VARIABLES

We have been discussing the specific heat of a superconductor in its normal and superconducting states in the absence of an applied magnetic field. When a magnetic field is present the situation is more complicated, and we must be more careful in describing the specific heat as the result of a thermodynamic process. In this section we will develop some of the necessary background material required for such a description and in the following sections we will apply the description to several cases.

Later on, in Chapter 10, Section X, we will learn why the magnetic energy of a superconducting sample in a magnetic field depends on its shape and orientation. In the present chapter we will not be concerned with these demagnetization effects and will instead assume that the sample is in the shape of a cylinder and that the internal magnetization $\mathbf{M}$ is directed along the axis of the cylinder, as illustrated in Fig. 2.26. If an external field $\mathbf{B}_{\text{app}}$ is applied, it will also be directed along this axis. This means that the applied $\mathbf{B}$ field is related to the internal $\mathbf{H}$ field by means of the expression

$$\mathbf{B}_{\text{app}} = \mu_0 \mathbf{H}_{\text{in}}. \tag{4.16}$$

This geometry simplifies the mathematical expressions for the free energy, enthalpy, and other properties of a superconductor in the presence of a magnetic field. In the next few sections we will simplify the notation by using the symbol $\mathbf{B}$ instead of $\mathbf{B}_{\text{app}}$ for the applied magnetic field, but throughout the remainder of the text the symbol $\mathbf{B}_{\text{app}}$ will be used.

In treating the superconducting state it is convenient to make use of the free energy because (1) the superconductivity state is always the state of lowest free energy at a particular temperature, and (2) the free energies of the normal and superconducting states are equal at the transition temperature. We will use the Gibbs free energy $G(T, P, B) = G(T, B)$ rather than the Helmholtz free energy

$$F(T, V, M) = F(T, M),$$

where the variables P and V are omitted because pressure–volume effects are negligible for superconductors. The Gibbs free energy $G(T, B)$ is selected because the experimenter has control over the applied magnetic field B, whereas the magnetization $M(T, B)$ is produced by the presence of the field. The remaining thermodynamic functions will be expressed in terms of the two independent variables T and B.

In the treatment that follows we will be dealing with thermodynamic quantities on a per-unit-volume basis, so that G will denote the Gibbs free energy density and S the entropy density. For simplicity, we will generally omit the term density by, for example, calling G the Gibbs free energy.

The first law of thermodynamics for a reversible process expresses the conservation of energy. For a magnetic material the differential of the internal energy dU may be written in terms of the temperature T, the entropy S, the applied magnetic field $\mathbf{B}$, and the magnetization $\mathbf{M}$ of the material as

$$dU = TdS + \mathbf{B} \cdot d\mathbf{M}, \tag{4.17}$$

where the usual $-PdV$ term for the mechanical work is negligible and hence omitted, while the $+\mathbf{B} \cdot d\mathbf{M}$ term for the magnetic work is included. (Work is done when an applied pressure P decreases the volume of a sample or an applied magnetic field $\mathbf{B}$ increases its magnetization.) So these two work terms are opposite in sign. The work term $\mu_0 \mathbf{H} \cdot d\mathbf{M}$ that appears in Eq. (4.17) is positive. This equation does not include the term $d(\mathbf{B}^2/2\mu_0) = \mu_0^{-1} \mathbf{B} \cdot d\mathbf{B}$ for the work involved in building up the energy density of the applied field itself since we are only interested in the work associated with the superconductor.

We will be concerned with a constant applied field rather than a constant magnetization, so it is convenient to work with the enthalpy H' rather than the internal energy U,

$$H' = U - \mathbf{B} \cdot \mathbf{M}, \qquad (4.18)$$

with differential form

$$dH' = TdS - \mathbf{M} \cdot d\mathbf{B}. \qquad (4.19)$$

The second law of thermodynamics permits us to replace TdS by CdT for a reversible process, where C is specific heat,

$$CdT = TdS, \qquad (4.20)$$

which gives for the differential enthalpy

$$dH' = CdT - \mathbf{M} \cdot d\mathbf{B}. \qquad (4.21)$$

Finally we will be making use of the Gibbs free energy

$$G = H' - TS, \qquad (4.22)$$

and its differential form

$$dG = -SdT - \mathbf{M} \cdot d\mathbf{B}. \qquad (4.23)$$

Note the prime in the symbol H' for enthalpy to distinguish it from the symbol H for the magnetic field. For the balance of the chapter we will also be assuming that the vectors **B**, **H**, and **M** are parallel and write, for example, MdB instead of $\mathbf{M} \cdot d\mathbf{B}$.

The fundamental thermodynamic expressions (4.17)–(4.23) provide a starting point for discussing the thermodynamics of the superconducting state. Two procedures will be followed in applying these expressions to superconductors. For the normal state we will assume a known specific heat (4.1) and then determine the enthalpy by integrating Eq. (4.21), determine the entropy by integrating Eq. (4.20), and finally find the Gibbs free energy from Eq. (4.22). For the superconducting case we will assume a known magnetization and critical field, and determine the Gibbs free energy by integrating Eq. (4.23), the entropy by differentiating Eq. (4.23), the enthalpy from Eq. (4.22), and finally the specific heat by differentiating Eq. (4.21). The first procedure, called the specific heat-to-free energy procedure, goes in the direction C → H′ → S → G and the second, called the free energy-to-specific heat procedure, goes in the opposite direction G → S → H′ → C. The former procedure will be presented in the following section and the latter in the succeeding three sections. We will assume specific expressions for C and M, respectively, to obtain closed-form expressions for the temperature dependences of the difference thermodynamic variables. This will give us considerable physical insight into the thermodynamics of the superconducting state. These assumptions also happen to approximate the behavior of many real superconductors.

VII. THERMODYNAMICS OF A NORMAL CONDUCTOR

In this section we will use the specific heat-to-free energy procedure. We deduce in succession the enthalpy, entropy, and Gibbs free energy of a normal conductor by assuming that its low-temperature specific heat C_n is given by Eq. (4.1) with the Schottky term omitted:

$$C_n = \gamma T + AT^3. \qquad (4.24)$$

The enthalpy at zero magnetic field is obtained by setting $MdB = 0$,

$$MdB = \mu_0^{-1} \chi_n B dB \approx 0, \qquad (4.25)$$

where $\chi_n = \mu_0 M/B$. Integrating Eq. (4.21), we find that

$$\int dH'_n = \int_0^T [\gamma T + AT^3] dT, \qquad (4.26)$$

$$H'_n(T) = \tfrac{1}{2}\gamma T^2 + \tfrac{1}{4}AT^4, \qquad (4.27)$$

where it is assumed that γ and A are independent of temperature, and that $H'_n(0) = 0$. A similar calculation for the entropy involves integrating Eq. (4.20),

$$\int dS_n = \int_0^T [\gamma T + AT^3] \frac{dT}{T}, \qquad (4.28)$$

$$S_n(T) = \gamma T + \tfrac{1}{3}AT^3, \qquad (4.29)$$

where $S_n(0) = 0$. The normal-state Gibbs free energy at zero field may be determined either from Eq. (4.22) or by integrating Eq. (4.23) with MdB set equal to zero. It has the following temperature dependence:

$$G_n(T) = -\tfrac{1}{2}\gamma T^2 - \tfrac{1}{12}AT^4. \quad (4.30)$$

The normal-state specific heat, entropy, enthalpy, and Gibbs free energy from Eqs. (4.24), (4.29), (4.27), and (4.30) are plotted in Figs. 4.7, 4.8, 4.9, and 4.10, respectively, for the very-low-temperature region. Here the AT^3 term is negligible, and only the γT term is appreciable in magnitude.

In this section we have derived several thermodynamic expressions for a normal conductor in the absence of a magnetic field. The permeability μ of such a conductor is so close to that of free space μ_0 (cf. Chapter 1, Section XV, and Table 1.2), that the magnetic susceptibility χ_n is negligibly small and $M \approx 0$. Therefore, the thermodynamic quantities (4.24), (4.27), (4.29), and (4.30) are not appreciably influenced by a magnetic field, and we will assume that they are valid even when there is a magnetic field present. For example, we assume that

$$G_n(T, B) \approx G_n(T, 0) \quad (4.31)$$

and it is convenient to simplify the notation by writing

$$G_n(T, B) \approx G_n(T). \quad (4.32)$$

Thus all of the equations derived in this section are applicable when a magnetic field is also present.

VIII. THERMODYNAMICS OF A SUPERCONDUCTOR

If we had a well-established expression for the specific heat C_s of a superconduc-

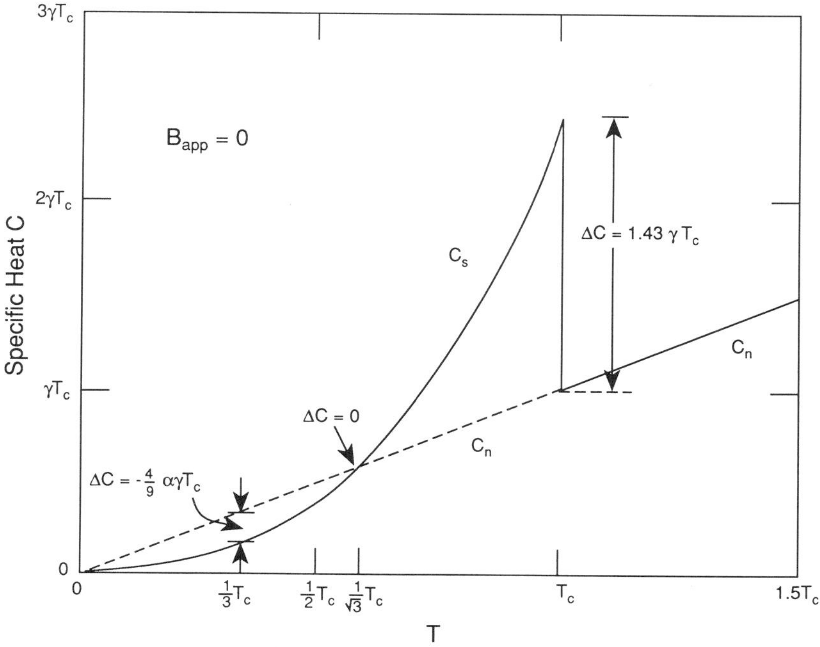

Figure 4.7 Temperature dependence of the normal-state C_n (– – –) and superconducting-state C_s (—) specific heats. The figure shows the specific heat jump $1.43\gamma T_c$ of Eq. (4.9) that is predicted by the BCS theory, the crossover point at $T = T_c/\sqrt{3}$, and the maximum negative jump $0.44\alpha\gamma T_c$ at $T = T_c/3$. In this and the following 12 figures it is assumed that only the linear electronic term γT exists in the normal state (i.e, $AT^3 = 0$).

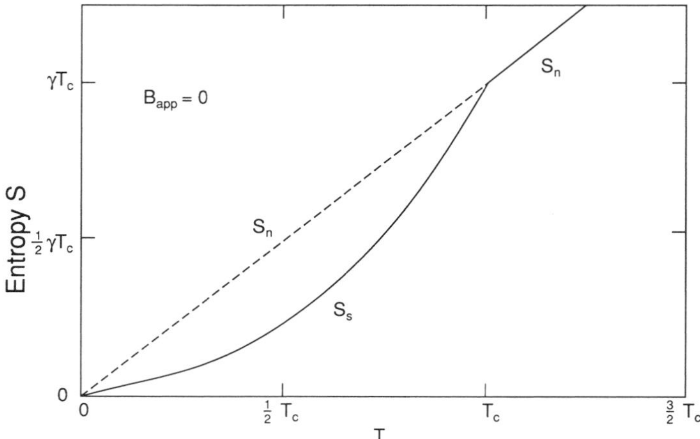

Figure 4.8 Temperature dependence of the normal-state S_n (---) and superconducting-state S_s (——) entropies. The transition is second order so there is no discontinuity in entropy at T_c.

tor below T_c it would be easy to follow the same C → H′ → S → G procedure to obtain the quantities H_s', S_s, and G_s, as in the case of a normal conductor. Unfortunately, there is no such expression, although many experimental data far below T_c have been found to follow the BCS expression (4.11). Equation (4.11) does not cover the entire temperature range of the superconducting region, however, and, in addition it does not integrate in closed form. Another complication is that the thermodynamic properties of the superconducting state are intimately related to its magnetic properties, as we will demonstrate below, and the specific heat relation (4.11) does not take magnetism into account.

Because of the close relationship between superconductivity and magnetism we will adopt the free energy-to-specific heat procedure and examine the Gibbs free energy of a superconductor in the presence of an applied magnetic field B. We will not

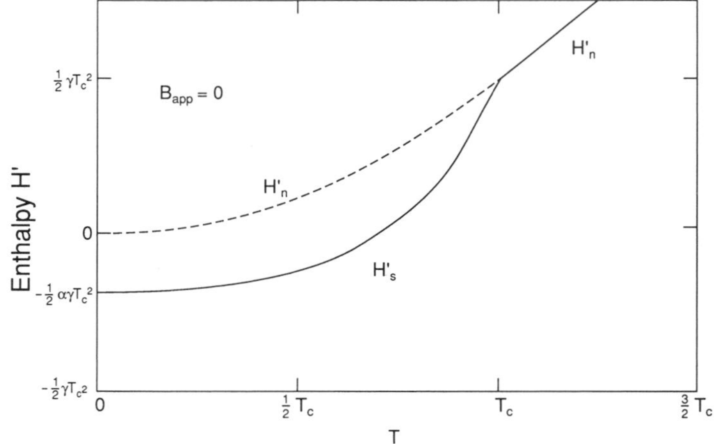

Figure 4.9 Temperature dependence of the normal state H_n (- - -) and superconducting state H_s (——) enthalpies. The transition is second order so there is no discontinuity in enthalpy, and hence no latent heat at T_c.

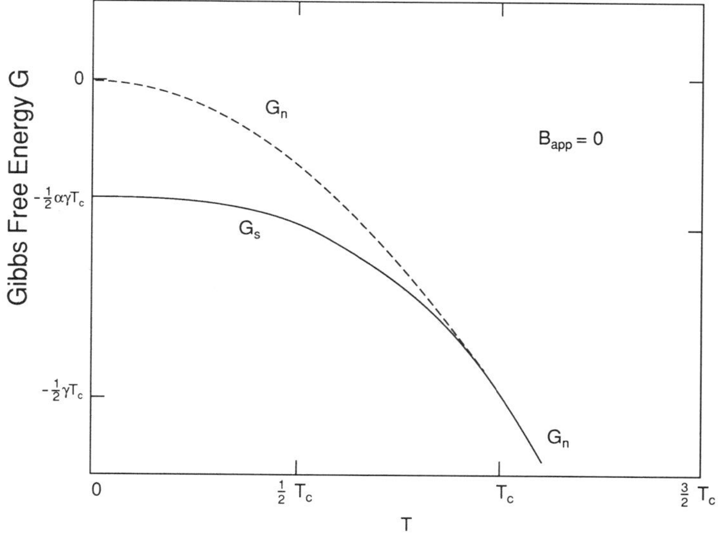

Figure 4.10 Temperature dependence of the normal-state G_n (− − −) and superconducting-state G_s (——) Gibbs free energies. Since the transition is second order, both G and its first derivative are continuous at T_c.

resort to any model for the temperature dependence of the specific heat or that of the critical field, so the results that will be obtained will be general. Then in the following two sections we will return to the specific model based on Eq. (4.24) to obtain more practical results.

We begin by seeking an expression for the free-energy difference $G_s(T, B) - G_n(T, B)$ between the superconducting and normal states to allow us to deduce S_s and H_s' by differentiation. To accomplish this we write down the differential of the Gibbs free energy from Eq. (4.23) assuming isothermal conditions ($dT = 0$),

$$dG = -MdB, \qquad (4.33)$$

and examine its magnetic field dependence in the superconducting and normal states.

We treat the case of a Type I superconductor that has the magnetization given by Eq. (2.22), $M = -H = -B/\mu_0$, and assume that surface effects involving the penetration depth are negligible. Demagnetization effects are also inconsequential, as explained in Section VI, so Eq. (4.33) becomes

$$dG_s = \mu_0^{-1} BdB. \qquad (4.34)$$

If this expression is integrated from $B = 0$ to a field B we obtain

$$G_s(T, B) = G_s(T, 0) + \tfrac{1}{2}\mu_0^{-1}B^2, \quad (4.35)$$

where, of course, the magnetic energy density $B^2/2\mu_0$ is independent of temperature. When the applied field B equals the critical field $B_c(T)$ for a particular temperature $T < T_c$, the free energy becomes

$$G_s(T, B_c(T)) = G_s(T, 0) + \tfrac{1}{2}\mu_0^{-1}B_c(T)^2 \qquad T = T_c(B), \quad (4.36)$$

and recalling that this is a phase transition for which $G_s = G_n$, we have

$$G_n(T) = G_s(T, 0) + \tfrac{1}{2}\mu_0^{-1}B_c(T)^2 \qquad T = T_c(B), \quad (4.37)$$

where $\frac{1}{2}\mu_0^{-1}B_c(T)^2$ is the magnetic-energy density associated with the critical field, and, from Eq. (4.32), $G_n(T)$ does not depend on the field. Subtracting Eq. (4.35) from Eq. (4.37) gives

$$G_s(T, B) = G_n(T) - \tfrac{1}{2}\mu_0^{-1}\left(B_c(T)^2 - B^2\right), \quad (4.38)$$

where, of course, $B < B_c(T)$. In the absence of an applied field Eq. (4.38) becomes

$$G_s(T,0) = G_n(T) - \tfrac{1}{2}\mu_0^{-1}B_c(T)^2 \quad (B=0), \tag{4.39}$$

so the Gibbs free energy in the superconducting state depends on the value of the critical field at that temperature. This confirms that there is indeed a close relationship between superconductivity and magnetism. Figure 4.11 shows that the curves for $G_s(T,0)$ and $G_n(T)$ intersect at the temperature T_c, while those for $G_s(T,B)$ and $G_n(T)$ intersect at the temperature $T_c(B)$. The figure also shows that $\frac{1}{2}\mu_0^{-1}B_c^2$ is the spacing between the curves of $G_s(T,0)$ and $G_n(T)$, and that $\frac{1}{2}\mu_0^{-1}B^2$ is the spacing between the curves of $G_s(T,B)$ and $G_s(T,0)$. The figure is drawn for a particular value of the applied field corresponding to $T_c(B) = \frac{1}{2}T_c$.

Since we know that the Gibbs free energy of the superconducting state depends only on the applied magnetic field and the temperature, we can proceed to write down general expressions for the other thermodynamic functions that can be obtained through differentiation of $G_s(T)$ with respect to the temperature when the applied field H is kept constant. The value of H, of course, does not depend on T.

For the entropy we have, using Eq. (4.23),

$$S_s - S_n = -\frac{d}{dT}[G_s - G_n], \tag{4.40}$$

and for the free energy, from Eq. (4.38),

$$S_s(T) = S_n(T) + \frac{B_c(T)}{\mu_0}\frac{d}{dT}B_c(T). \tag{4.41}$$

The entropy $S_s(T,B)$ does not depend explicitly on the applied field, so it is denoted $S_s(T)$. From this expression, together with Eqs. (4.22) and (4.39), we can write down

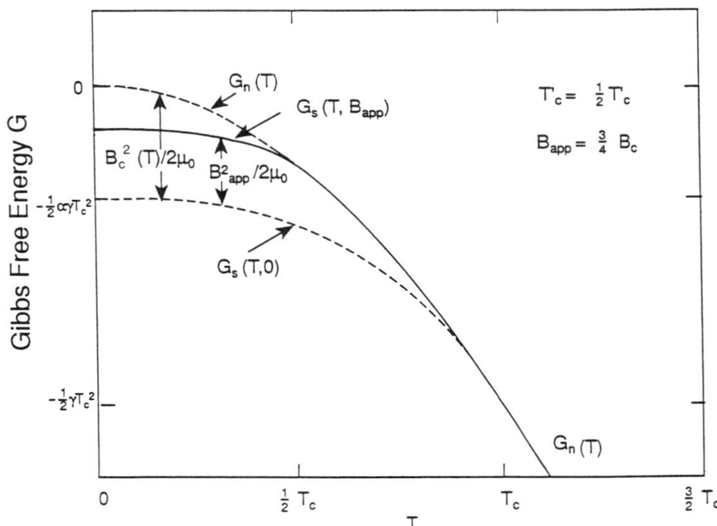

Figure 4.11 Effect of an applied magnetic field $B_{app} = 0.75B_c$ on the Gibbs free energy $G_s(T,B)$ in the superconducting state. In this and the succeeding figures dashed curves are used to indicate both the normal-state extrapolation below T_c' and the zero-field superconducting state behavior, where T_c' denotes the transition temperature when there is a field present.

the enthalpy at constant field,

$$H_s'(T, B) = H_n'(T) - \tfrac{1}{2}\mu_0^{-1}\left[B_c(T)^2 - B^2\right] + \frac{TB_c(T)}{\mu_0}\frac{d}{dT}B_c(T), \quad (4.42)$$

which shows that the enthalpy does depend on the magnetic energy $B^2/2\mu_0$. We see from Eq. (4.21) that the enthalpy can be differentiated to provide the specific heat at constant field:

$$C_s(T) = C_n(T) + \mu_0^{-1}TB_c(T)\frac{d^2}{dT^2}B_c(T) + \frac{T}{\mu_0}\left|\frac{d}{dT}B_c(T)\right|^2. \quad (4.43)$$

The specific heat does not depend explicitly on the applied field, so we write $C_s(T)$ instead of $C_s(T, B)$. We will see below that the terms in this expression that depend on $B_c(T)$ become negative at the lowest temperatures, making $C_s(T)$ less than $C_n(T)$. At zero field ($B = 0$) the transition temperature is T_c itself, and we know (e.g., from Eq. (4.45) in Section IX) that $B_c(T_c) = 0$, so that only the second term on the right exists under this condition:

$$C_s(T) = C_n(T) + \frac{T}{\mu_0}\left|\frac{d}{dT}B_c(T)\right|^2 \quad T = T_c. \quad (4.44)$$

This is known as Rutger's formula. It provides the jump in the specific heat at T_c that is observed experimentally, as shown in Figs. 4.3, 4.5, and 4.6. We will show in Section XIII how this expression can be used to evaluate the electronic specific-heat factor γ.

IX. SUPERCONDUCTOR IN ZERO FIELD

We will develop the thermodynamics of a Type I superconductor in the absence of a magnetic field using the free energy-to-specific heat procedure G → S → H′ → C. We apply the general expression (4.39) to the particular case in which the free energy of the normal state is given by Eq. (4.30) and the critical magnetic field $B_c(T)$ has a parabolic dependence on temperature,

$$B_c(T) = B_c(0)\left[1 - \left(\frac{T}{T_c}\right)^2\right], \quad (4.45)$$

given by Eq. (2.56). Substituting these expressions in Eq. (4.39) gives

$$G_s(T, 0) = -\tfrac{1}{2}\gamma T^2 - \tfrac{1}{12}AT^4 - \tfrac{1}{2}\mu_0^{-1}B_c(0)^2 \times\left[1 - \left(\frac{T}{T_c}\right)^2\right]^2, \quad (4.46)$$

which is plotted in Fig. 4.10 with A set equal to zero.

The difference between the entropies in the normal and superconducting states is obtained by substituting the expressions from Eqs. (4.29) and (4.45) in Eq. (4.41) and carrying out the differentiation:

$$S_s(T) = \gamma T + \tfrac{1}{3}AT^3 - 2\mu_0^{-1}B_c(0)^2 \times\left(\frac{T}{T_c^2}\right)\left[1 - \frac{T^2}{T_c^2}\right]. \quad (4.47)$$

The last term on the right is zero for both $T = 0$ and $T = T_c$, so $S_s = S_n$ for both limits. The former result is expected from the third law of thermodynamics. Differentiation shows that the last term on the right is a maximum when $T = T_c/\sqrt{3}$, so that the difference $S_s - S_n$ is a maximum for this temperature. The entropy S_s with $A = 0$ is plotted in Fig. 4.8.

The enthalpy $H_s'(T, 0)$ of the superconducting state in zero field is obtained from Eq. (4.22),

$$H_s'(T, 0) = \tfrac{1}{2}\gamma T^2 + \tfrac{1}{4}AT^4 - \tfrac{1}{2}\mu_0^{-1}B_c(0)^2 \times\left[1 - \frac{T^2}{T_c^2}\right]\left[1 + 3\frac{T^2}{T_c^2}\right], \quad (4.48)$$

and the specific heat of the superconducting state from Eq. (4.20),

$$C_s = T\left(\frac{dS_s}{dT}\right)_H, \quad (4.49)$$

by differentiating Eq. (4.47) at constant field, to give

$$C_s(T) = \gamma T + AT^3 + 2\mu_0^{-1}B_c(0)^2 \times \left(\frac{T}{T_c^2}\right)\left[3\frac{T^2}{T_c^2} - 1\right]. \quad (4.50)$$

The last term on the right changes sign at $T = T_c/\sqrt{3}$. Expressions (4.48) and (4.50) are plotted in Figs. 4.9 and 4.7, respectively, with the AT^3 term set equal to zero.

The results given in this section are for a Type I superconductor in zero field with electronic specific heat given by Eq. (4.24) and a critical field with the temperature dependence of Eq. (4.45). Figures 4.7, 4.8, 4.9, and 4.10 show plots of the temperature dependence of the thermodynamic functions C_s, S_s, H_s', and G_s under the additional assumption $A = 0$.

X. SUPERCONDUCTOR IN A MAGNETIC FIELD

In the previous section we derived Eq. (4.38) for the Gibbs free energy $G_s(T, B)$ of the superconducting state in the absence of an applied magnetic field B. With the aid of Eqs. (4.30), (4.38), and (4.45) this can be written in the following form for the case of an applied field:

$$G_s(T, B) = -\tfrac{1}{2}\gamma T^2 - \tfrac{1}{12}AT^4 - \tfrac{1}{2}\mu_0^{-1} \times \left[B_c(0)^2\left(1 - \frac{T^2}{T_c^2}\right)^2 - B^2\right]. \quad (4.51)$$

Since the applied field B does not depend on the temperature, the entropy obtained from Eq. (4.40) by differentiating the Gibbs free energy (4.51) assuming the presence of a field is the same as in the case where there is no magnetic field present,

$$S_s(T) = \gamma T + \tfrac{1}{3}AT^3 - 2\mu_0^{-1}B_c(0)^2 \times \frac{T}{T_c^2}\left[1 - \frac{T^2}{T_c^2}\right]. \quad (4.52)$$

The enthalpy obtained from Eq. (4.22) does depend explicitly on this field,

$$H_s'(T, B) = \tfrac{1}{2}\gamma T^2 + \tfrac{1}{4}AT^4 - \tfrac{1}{2}\mu_0^{-1}B_c(0)^2\left(1 - \frac{T^2}{T_c^2}\right) \times \left(1 + 4\frac{T^2}{T_c^2}\right) + \tfrac{1}{2}\mu_0^{-1}B^2, \quad (4.53)$$

but the specific heat from Eq. (4.20) does not,

$$C_s(T) = \gamma T + AT^3 + 2\mu_0^{-1}B_c(0)^2 \times \frac{T}{T_c^2}\left(3\frac{T^2}{T_c^2} - 1\right), \quad (4.54)$$

where Eqs. (4.52) and (4.54) are the same as their zero-field counterparts (4.47) and (4.50), respectively. The field-dependent G_s and H_s' terms of Eqs. (4.51) and (4.53), on the other hand, differ from their zero-field counterparts (4.46) and (4.48) by the addition of the magnetic-energy density $B^2/2\mu_0$.

In a magnetic field the sample goes normal at a lower temperature than in zero field. We denote this magnetic-field transition temperature by $T_c(B) = T_c'$, where, of course, $T_c(0) = T_c$ and $T_c' < T_c$. This transition from the superconducting to the normal state occurs when the applied field H equals the critical field $B_c(T)$ given by Eq. (4.45) at that temperature. Equation (4.45) may be rewritten in the form

$$T_c' = T_c\left[1 - \frac{B}{B_c(0)}\right]^{1/2} \quad (4.55)$$

to provide an explicit expression for the transition temperature T_c' in an applied field B. We show in Problem 7 that this same expression is obtained by equating the Gibbs free energies $G_s(T, B)$ and $G_n(T)$

for the superconducting and normal states at the transition point,

$$G_s(T, B) = G_n(T) \quad T = T_c', \quad (4.56)$$

At the transition temperature $T_c' = T_c(B)$ the superconducting and normal state entropies (4.52) and (4.29), respectively, differ. Their difference gives the latent heat L of the transition by means of the standard thermodynamic expression

$$L = (S_n - S_s)T_c(B) \quad (4.57)$$

$$= 2\mu_0^{-1}B_c^2\left[\frac{T_c(B)}{T_c}\right]^2 \times \left\{1 - \left[\frac{T_c(B)}{T_c}\right]^2\right\}. \quad (4.58)$$

We show in Problem 9 that this same result can be obtained from the enthalpy difference $L = H_n' - H_s'$. The latent heat is a maximum at the particular transition temperature $T_c(B) = T_c/\sqrt{2}$, as may be shown by setting the derivative of Eq. (4.58) with respect to temperature equal to zero. We see from this equation that there is no latent heat when the transition occurs in zero field, i.e., when $T = T_c$, or at absolute zero, $T = 0$. In addition to the latent heat, there is also a jump in the specific heat at $T_c(B)$ which will be discussed in the following section.

Figures 4.12, 4.13, 4.14, and 4.15 show the temperature dependences of the thermodynamic functions C, S, H', and G, respectively, for high applied fields in which $T_c(B)$ is far below T_c. Figures 4.16, 4.17, 4.18, and 4.19 show these same plots for low applied fields in which $T_c(B)$ is slightly below T_c. All of these plots are for the case $A = 0$. We see from Figs. 4.12, 4.16, 4.13, and 4.17, respectively, that the specific heat C_s and entropy S_s curves (assuming the presence of a magnetic field) coincide with their zero-field counterparts below $T_c(B)$ and with their normal-state counterparts above $T_c(B)$. In contrast, from Figs. 4.14, 4.18, 4.15, and 4.19 it is clear that the enthalpy H_s' and Gibbs free en-

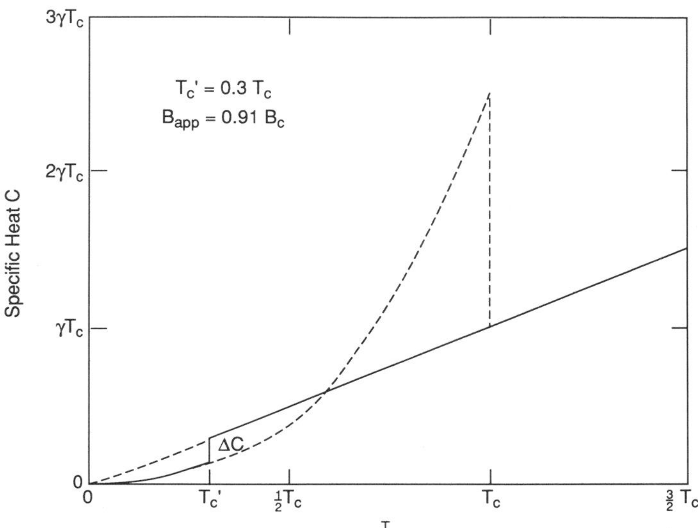

Figure 4.12 Temperature dependence of the specific heat in the normal and superconducting states in the presence of a strong applied magnetic field. The downward jump in specific heat ΔC at T_c' is indicated.

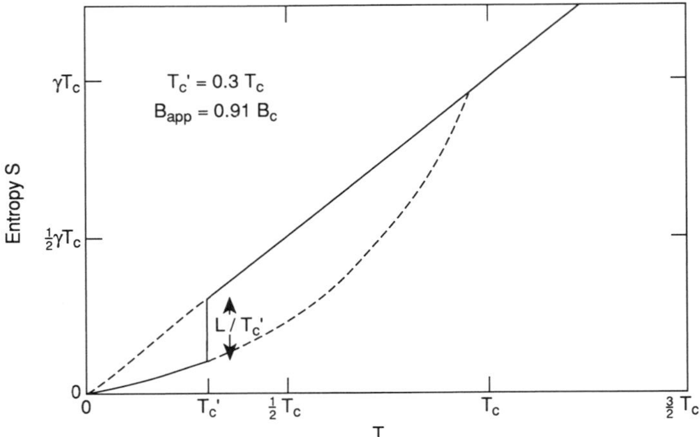

Figure 4.13 Temperature dependence of the entropy in the normal and superconducting states in the presence of a strong applied magnetic field. The latent heat factor L/T_c' of the jump in entropy at T_c' is indicated.

ergy G_s curves in a magnetic field lie between their normal-state and zero-field superconducting state counterparts below the transition point $T_c(B)$, and coincide with the normal-state curves above the transition. These plots also show the jumps associated with the specific heat and the latent heat as well as the continuity of the Gibbs free energy at the transition.

Figure 4.20 shows the experimentally determined Gibbs free-energy surface of $YBa_2Cu_3O_7$ obtained by plotting $G(T, B) - G(T, 0)$ versus temperature and the applied field close to the superconduct-

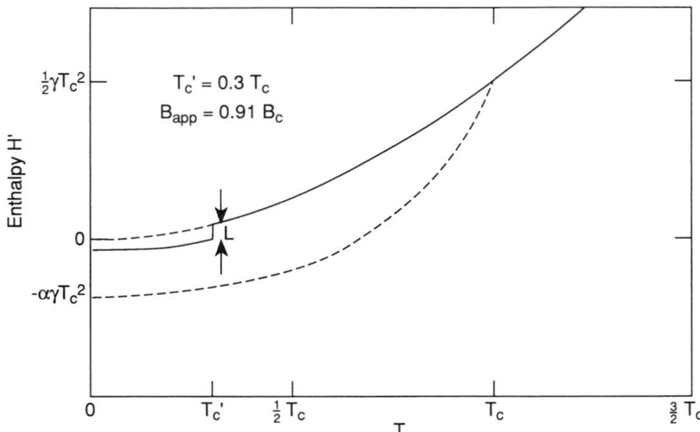

Figure 4.14 Temperature dependence of the enthalpy in the normal and superconducting states in the presence of a strong applied magnetic field. The jump in entropy at the transition temperature T_c' is equal to the latent heat L, as indicated.

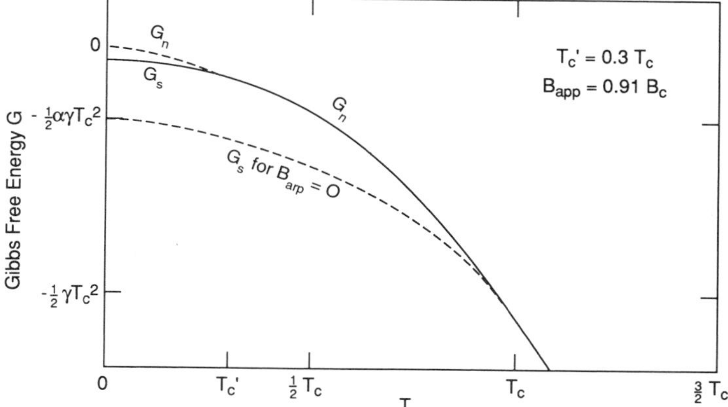

Figure 4.15 Temperature dependence of Gibbs free energy in the normal and superconducting states in the presence of a strong applied magnetic field. The transition is first order so that there is no discontinuity in free energy at the transition temperature T_c', but there is a discontinuity in the derivative. The normal (G_n) and superconducting (G_s) branches of the upper curve are indicated. The lower dashed (– – –) curve shows the Gibbs free energy in the superconducting state at zero field ($B_{app} = 0$) for comparison.

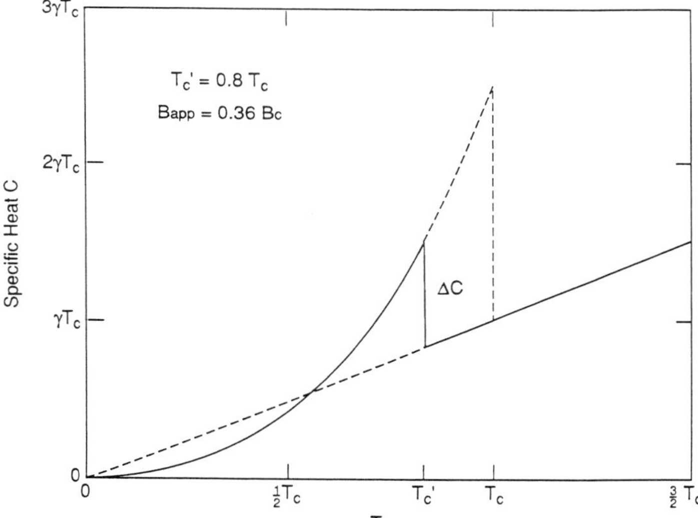

Figure 4.16 Temperature dependence of specific heat in the normal and superconducting states in the presence of a weak applied magnetic field. The jump in specific heat ΔC at T_c' is upward, in contrast to the downward jump shown in Fig. 4.12 for the high-field case.

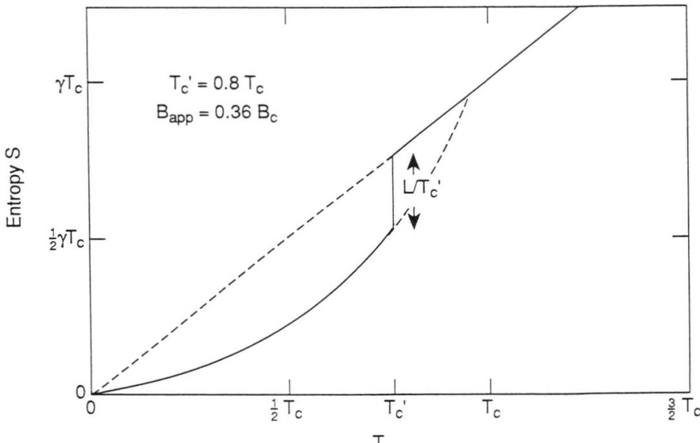

Figure 4.17 Temperature dependence of entropy in the normal and superconducting states in the presence of a weak applied magnetic field showing the jump in entropy L/T_c' at T_c', as expected for a first-order transition.

ing transition temperature (Athreya *et al.*, 1988). The free-energy differences are obtained by integrating Eq. (4.33) using measured magnetization data for $M(T, B)$:

$$G(T,0) - G(T,B) = MdB'. \quad (4.59)$$

This procedure is possible because close to the transition temperature magnetic flux moves easily and reversibly into and out of the material, which makes the magnetization a thermodynamic variable. Magnetization is linear in $(T_c - T)^2$ near T_c. The free-energy surface varies with the magnetic field all the way up to 92 K. Fang *et al.* (1989) determined free-energy surfaces for thallium-based superconductors.

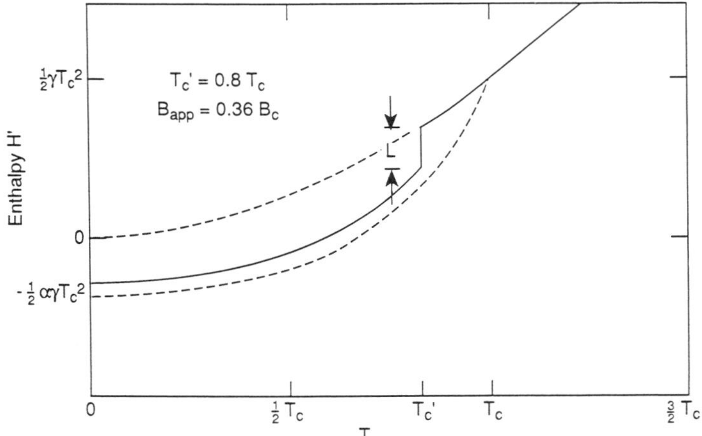

Figure 4.18 Temperature dependence of enthalpy in the normal and superconducting states in the presence of a weak applied magnetic field, showing the presence of a latent heat jump L at the transition temperature T_c', indicating a first-order transition.

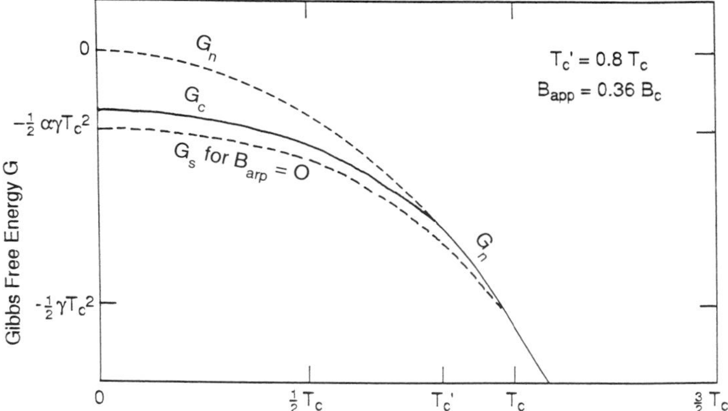

Figure 4.19 Temperature dependence of Gibbs free energy in the normal and superconducting states in the presence of a weak applied magnetic field using the notation of Fig. 4.15. The transition is first order so that the change in G at T_c' is continuous, but the change in its derivative is discontinuous (cf. Athreya *et al.*, 1988).

XI. NORMALIZED THERMODYNAMIC EQUATIONS

The equations for $G_s(T, B)$, $S_s(T)$, and $C_s(T)$ given in the previous section, together with $H_s'(T, B)$ of Problem 9, can be written in normalized form by defining two dimensionless independent variables,

$$t = \frac{T}{T_c} \qquad b = \frac{B}{B_c} \tag{4.60}$$

and two dimensionless parameters,

$$a = \frac{AT_c^2}{\gamma} \qquad \alpha = \frac{B_c^2}{\mu_0 \gamma T_c^2}. \tag{4.61}$$

The resulting normalized expressions for g_s, s_s, and h_s' are given in Table 4.2. Also given in the table are the normalized specific heat jump $\Delta C/\gamma T_c$ and the normalized latent heat $L/\gamma T_c^2$. These expressions are valid under the condition

$$t^2 + b < 1. \tag{4.62}$$

The sample becomes normal when either t or b are increased to the point where $t^2 + b = 1$, and the value of t that satisfies this expression is called t':

$$t'^2 + b = 1. \tag{4.63}$$

This is the normalized equivalent of Eq. (4.55), where $t' = T_c'/T_c$ is the normalized transition temperature in a magnetic field.

The normalized specific heat jump has the following special values:

$$\frac{\Delta C}{\gamma T_c} = 2\alpha t'(3t'^2 - 1)$$
$$= \begin{cases} 0 & t' = 0 \\ -\dfrac{4\alpha}{9} & t' = \dfrac{1}{3} \quad \text{(max)} \\ 0 & t' = \dfrac{1}{\sqrt{3}} \\ 4\alpha & t' = 1 \end{cases}, \tag{4.64}$$

where $4\alpha/9$ is its maximum magnitude of $\Delta C/\gamma T_c$ for reduced temperatures in the range $0 < t' < 1/\sqrt{3}$, as indicated in Fig. 4.12. The normalized latent heat has the special values

$$\frac{L}{\gamma T_c^2} = 2\alpha t'^2(1 - t'^2)$$
$$= \begin{cases} 0 & t' = 0 \\ \frac{1}{2}\alpha & t' = \dfrac{1}{\sqrt{2}} \quad \text{(max)}, \\ 0 & t' = 1 \end{cases} \tag{4.65}$$

where its maximum $\frac{1}{2}\alpha$ is at $t' = 1/\sqrt{2}$.

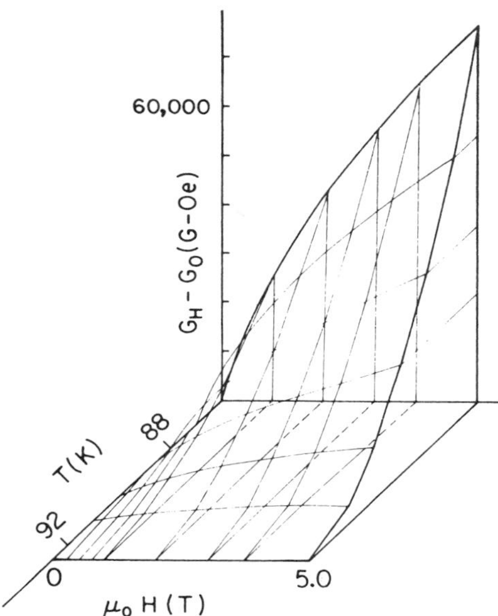

Figure 4.20 Free-energy surface for $YBa_2Cu_3O_7$ close to the transition temperature (Athreya *et al.*, 1988).

XII. SPECIFIC HEAT IN A MAGNETIC FIELD

A number of authors have measured or calculated the specific heat of high-temperature superconductors in a magnetic field (Hikami and Fujita, 1990a, b; Riecke *et al.*, 1989; Quade and Abrahams, 1988; Watson *et al.*, 1989). Reeves *et al.*, (1989) found that the quantity C/T of $YBa_2Cu_3O_{7-\delta}$ in an applied magnetic field is linear in T^2 in the range $4\ \text{K} < T < 6\ \text{K}$ in accordance with the expression

$$C = [\gamma + \gamma'(B)]T + [A - A'(B)]T^3, \tag{4.66}$$

which is compared in Fig. 4.21 with experimental data for applied fields up to 3 T. It was also found that $\gamma = 4.38$ mJ/mole K^2 and $A = 0.478$ mJ/mole K^4, with the coefficients $\gamma'(B)$ and $A'(B)$ increasing as the applied magnetic field was increased.

Table 4.2 Normalized Equations for the Thermodynamic Functions of a Superconductor in an Applied Magnetic Field B[a]

Gibbs Free Energy	$g_s = \dfrac{G_s}{\gamma T_c^2} = -\frac{1}{2}t^2 - \frac{1}{12}at^4 - \frac{1}{2}\alpha[(1-t^2)^2 - b^2]$
Entropy	$s_s = \dfrac{S_s}{\gamma T_c} = t + \frac{1}{3}at^3 - 2\alpha t(1-t^2)$
Specific Heat	$c_s = \dfrac{C_s}{\gamma T_c} = t + at^3 + 2\alpha t(3t^2 - 1)$
Enthalpy	$h'_s = \dfrac{H'_s}{\gamma T_c^2} = \frac{1}{2}t^2 + \frac{1}{4}at^4 - \frac{1}{2}\alpha[(1-t^2)(1+3t^3-b^2)]$
Specific Heat Jump	$\dfrac{\Delta C}{\gamma T_c} = 2\alpha t'(3t'^2 - 1)$
Latent Heat	$\dfrac{L}{\gamma T_c^2} = 2\alpha t'^2(1 - t'^2)$

Definitions of normalized variables (t, b) and parameters:

$t = \dfrac{T}{T_c}$	$b = \dfrac{B}{B_c(0)}$	$a = \dfrac{AT_c^2}{\gamma}$
$t' = \dfrac{T'}{T_c}$	$b' = \dfrac{B_c(T')}{B_c(0)}$	$\alpha = \dfrac{[B_c(0)]^2}{\mu_0 \gamma T_c^2}$

[a] The first four expressions are valid under the condition $t^2 + b < 1$ of Eq. (4.62), and the last two are valid at the transition point given by $t'^2 + b^2 = 1$ from Eq. (4.63).

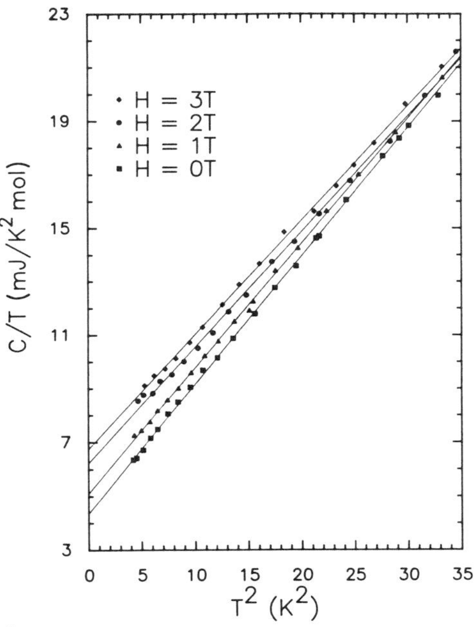

Figure 4.21 Low-temperature specific heat of $YBa_2Cu_3O_{7-\delta}$ in a magnetic field. The straight lines are fits of Eq. (4.66) to the data for each field value (Reeves *et al.*, 1989).

At the highest measured field of 3 T, it turned out that

$$\frac{\gamma}{\gamma'} = 0.54, \qquad \frac{A}{A'} = 0.11. \tag{4.67}$$

Reeves *et al.*, also mention that other workers have obtained results that differ from those described by Eq. (4.66).

Bonjour *et al.* (1991), Inderhees *et al.* (1991), and Ota *et al.* (1991) measured the magnetic-field dependence of the anisotropies in the specific heat near T_c. The results obtained by Inderhees *et al.* for untwinned $YBa_2Cu_3O_{7-\delta}$, which are presented in Fig. 4.22, turned out to be similar to those obtained by the other two groups. We see that increasing the magnetic field shifts the specific-heat jump to lower temperatures and broadens it, especially for an applied field parallel to the *c*-axis.

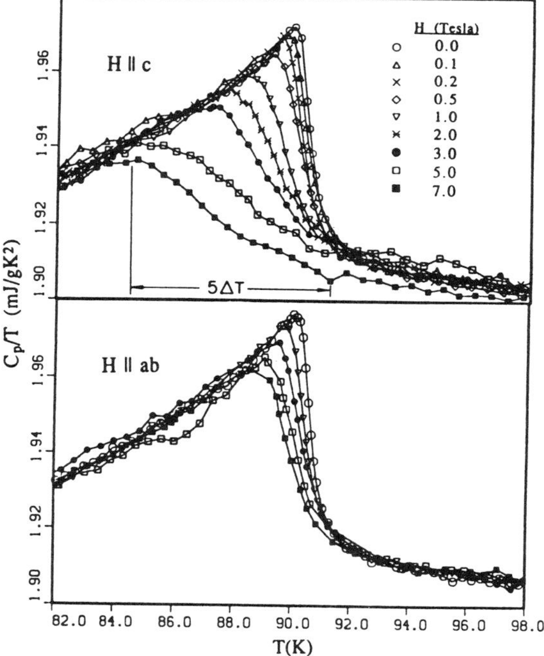

Figure 4.22 Specific heat jump of untwinned $YBa_2Cu_3O_{7-\delta}$ near T_c for different applied magnetic fields aligned parallel (top) and perpendicular (bottom) to the *c*-axis (Inderhees *et al.*, 1991).

Ebner and Stroud (1989) obtained a good approximation to the specific heat curves of Fig. 4.22 with $B \| c$ by including fluctuations in the Ginzburg–Landau free energy (cf. Chapter 5, Section III) and carrying out Monte-Carlo simulations. Figure 4.23 shows how the difference between the specific heat measured at zero field C_0 and that measured in the field C_H depends on the value of the applied field at a temperature of 88 K, which is close to T_c. The difference is about five times larger in the parallel field orientation than in the perpendicular field orientation.

Bonjour *et al.* (1991) used their own specific heat data to determine the dependence of the entropy difference $S_0 - S_H$ on the applied field, where $S_0(T)$ is the entropy in the absence of the field and $S_H(T, B)$ the entropy assuming the presence of a field; their results are given in Fig. 4.24. They were aided by recent magnetic data of Welp *et al.* (1989; cf. Hake, 1968) in deducing the experimental entropy. Bonjour *et al.* compared their measured entropies with the following generalization of Eq. (4.41) to the mixed state of a Type II superconductor:

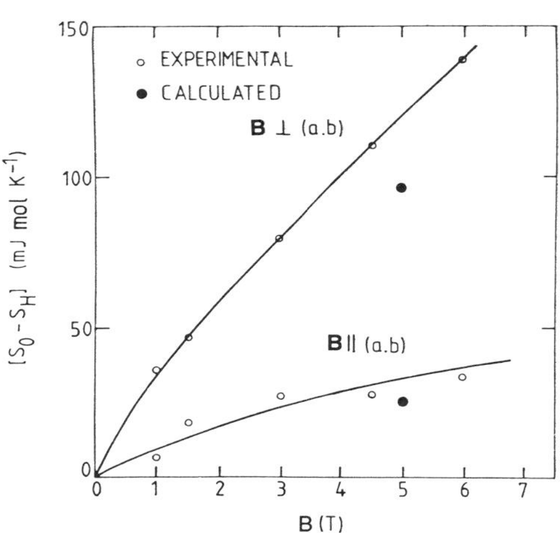

Figure 4.24 Magnetic field dependence of the entropy difference of parallel and perpendicular fields showing measured (○) and calculated (●) values for $YBa_2Cu_3O_7$ (Bonjour *et al.*, 1991).

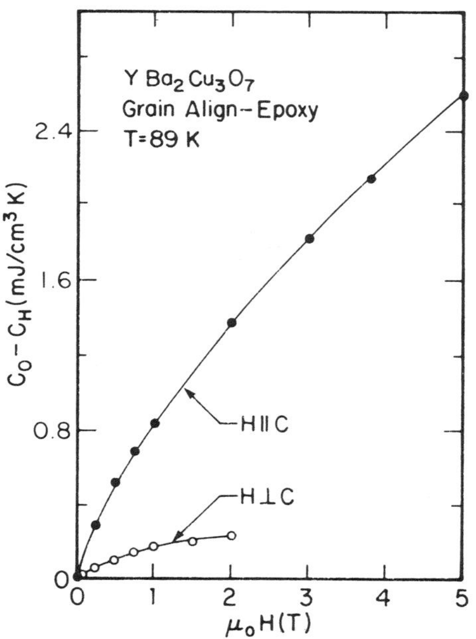

Figure 4.23 Magnetic field dependence of the specific heat difference for parallel and perpendicular fields (Athreya *et al.*, 1988).

$$S_H(T, B) = S_n(T) + \chi'(T) \times \frac{B_{c2}(T) - B}{\mu_0} \cdot \frac{d}{dT} B_{c2}(T), \quad (4.68)$$

where $\chi' = \mu_0 dM/dB$ is called the 'differential susceptibility.' This gives

$$S_0(T_i) - S_H(T_i, B) = \chi'(T_i) \times \frac{B}{\mu_0} \cdot \frac{d}{dT} B_{c2}(T_i), \quad (4.69)$$

where $T_i = 80$ K is the temperature at which all the specific heat curves are still superimposed. The values for $B = 5$ T calculated from Eq. (4.68) using the data of Welp *et al.* are reasonably close to the measured values, as indicated in the figure.

XIII. EVALUATING THE SPECIFIC HEAT

Earlier in the chapter we mentioned the jump in the specific heat in zero field (4.44), in a magnetic field (4.43), and as predicted by the BCS theory (4.9). We also gave expressions for the temperature dependence of the specific heat in the superconducting state, one of which (Eq. (4.11)) appeared to be incompatible with the other two expressions (Eqs. (4.50) and (4.54)). In this section we compare these results and use them to evaluate the electronic specific-heat coefficient γ for zero field, after which we will write down an expression for the jump in the specific heat in a magnetic field.

At the transition temperature $T = T_c$ in zero field, Eq. (4.50), with $A = 0$, simplifies to

$$C_s(T_c) - C_n(T_c) = \frac{4[B_c(0)]^2}{\mu_0 T_c}, \quad (4.70)$$

where $C_n(T_c) = \gamma T_c$. If the BCS prediction (4.9) is substituted in Eq. (4.70), we obtain for the normalized specific heat factor α of Eq. (4.61)

$$\alpha = 0.357. \quad (4.71)$$

The curves of Figs. 4.7–4.9 were drawn for this value. Since $B_c^2/2\mu_0$ is an energy density expressed in units J/m^3 and γ is given in units mJ/mole K, it is necessary to multiply γ by the density ρ and divide it by the molecular weight (MW) in Eq. (4.71), giving us the BCS expression

$$\frac{[B_c(0)]^2(\text{MW})}{T_c^2 \rho\gamma} = 449, \quad (4.72)$$

where B_c is expressed in units mT, γ in mJ/mole K^2, ρ in g/cm^3, and T_c in degrees Kelvin. It is reasonable to assume that this expression will be a good approximation for Type I superconductors, and we see from the last column of Table 4.3 that this is indeed the case for the elemental superconductors. Equation (4.72) was derived for materials in which the number density of the conduction electrons is the same as the number density of the atoms. For materials in which this is not the case, the effective electron density $\beta\rho$ can be used, where β is the factor introduced in

Table 4.3 Variation of the Dimensionless Ratio $B_c^2(\text{MW})/T_c^2\rho\gamma$ of Several Elemental Superconductors[a]

Element	T_c K	$\frac{B_c^2(\text{MW})}{T_c^2\rho\gamma}$ $\frac{\text{mT}^2\ \text{cm}^3}{\text{mJ}}$	$\frac{\text{Ratio}}{\text{BCS Ratio}}$
W	0.015	676	1.51
Ir	0.11	569	1.27
Ru	0.49	577	1.29
Zr	0.61	300	0.67
Os	0.66	403	0.90
Re	1.7	522	1.16
Sn	3.72	615	1.37
V	5.4	571	1.27
Pb	7.20	733	1.63
Tc	7.80	443	0.99
Nb	9.25	697	1.55
BCS theory	—	449	1.00

[a] $B_c = B_c(0)$, MW is molecular weight, ρ density, and γ electronic specific heat.

Eq. (4.4), to give

$$\frac{[B_c(0)]^2(\text{MW})}{T_c^2 \beta\rho\gamma} = 449. \quad (4.73)$$

Equations (4.54) and (4.11) constitute entirely different dependences of $C_s(T)$ on temperature, and it is of interest to compare them. In normalized form, with A set equal to zero, they are

$$\frac{C_s(T)}{\gamma T_c} = \frac{T}{T_c}\left[0.285 + 2.145\left(\frac{T}{T_c}\right)^2\right], \quad (4.74)$$

$$\frac{C_s(T)}{\gamma T_c} = 14\exp\left(-1.76\frac{T_c}{T}\right), \quad (4.75)$$

where α has the BCS value 0.357 and the coefficient 1.76 in the exponential expression is chosen because the BCS theory predicts $\Delta = 1.76kT_c$ in Eq. (4.11). The coefficient 14 is selected to normalize Eq. (4.75) to the BCS value (4.11); i.e., $C_s(T_c) = 2.43\gamma T_c$ at the transition point. Figure 4.25 compares the temperature dependence of (4.74) and (4.75), and shows that they are close at all but the lowest

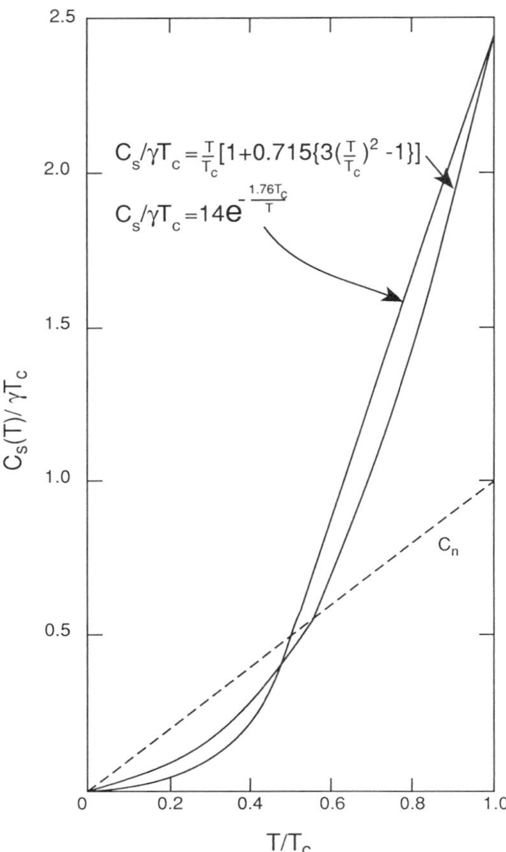

Figure 4.25 Comparison of the thermodynamic and BCS expressions (4.74) and (4.75), respectively, for the specific heat ratio $C_s/\gamma T_c$ normalized to the same value at $T = T_c$.

temperatures. Equation (4.74) is slightly lower for T near T_c and Eq. (4.75) is significantly lower for $T \ll T_c$.

The first of these expressions for $C_s(T)$, i.e., Eq. (4.74), is based on Eq. (4.45), which is a good approximation to the temperature dependence of the critical field $B_c(T)$ near the critical temperature. However, the temperature derivatives of Eq. (4.45) that enter into the $C_s(T)$ expression (4.43) are not expected to be valid quantitatively far below T_c. The second expression, Eq. (4.75), on the other hand, is based on excitation of quasiparticles to energies above the superconducting ground state, and is valid at temperatures far below T_c where most of the electrons that contribute to the superconductivity are condensed as Cooper pairs in the ground-energy state. Therefore, we might expect an experimental $C_s(T)$-versus-T curve to approximate Eq. (4.75) far below T_c, as in the case of the superconducting Al data shown in Fig. 4.5.

Now that we have found explicit expressions for the specific heat in the superconducting state in the absence of a magnetic field, let us examine the case when there is a field present. We will continue to assume that $\gamma T = C_n(T)$ and $A = 0$, and that the BCS expression (4.10) is valid in zero field at T_c. Thus, in a magnetic field Eq. (4.74) can be written

$$C_s(T) - C_n(T) = 0.715\gamma T\left[3\left(\frac{T}{T_c}\right)^2 - 1\right], \tag{4.76}$$

and the jump in specific heat at the transition temperature T_c' in a magnetic field is given by

$$C_s(T_c') - C_n(T_c') = 0.715\gamma T\left[3\left(\frac{T_c'}{T_c}\right)^2 - 1\right]. \tag{4.77}$$

This change in specific heat $C_s - C_n$ is negative for $T_c' < T_c/\sqrt{3}$ and positive for $T_c' > T_c/\sqrt{3}$. This means that with increasing temperature there is an upward jump in the specific heat for $T_c' < T_c/\sqrt{3}$ and a downward jump for $T_c' > T_c/\sqrt{3}$, as shown in Figs. 4.12, and 4.16, respectively. We also see that no jump at all occurs at the crossover point of the normal-state and superconducting curves, where $T_c' = T_c/\sqrt{3}$. In addition to the jump in specific heat, there is also latent heat present, Eq. (4.57), in the presence of a magnetic field.

XIV. ORDER OF THE TRANSITION

We mentioned in Section III that the transition from the normal to the superconducting state in the absence of a magnetic field is a second-order phase transition, which means that the Gibbs free energy and its temperature derivative are continuous at the transition:

$$G_s(T_c) = G_n(T_c), \tag{4.78}$$

$$\frac{dG_s}{dT} = \frac{dG_n}{dT}. \tag{4.79}$$

This can be seen from Eq. (4.39), using the condition $B_c(T_c) = 0$ from Eq. (4.45). Therefore, there is no latent heat, but there is a discontinuity in the specific heat given, for example, by Eqs. (4.44) and (4.70).

We showed in Section X that the transition from the superconducting to the normal state in the presence of a magnetic field does have a latent heat given by Eq. (4.58) and, therefore, is a first-order phase transition.

XV. THERMODYNAMIC CONVENTIONS

There are several conventions in vogue for formulating the thermodynamic approach to superconductivity. Some of these conventions make use of the total internal energy U_{tot}, which includes the energy of the magnetic field $B^2/2\mu_0 = \frac{1}{2}\mu_0 H^2$ that would be present in the absence of the superconductor, whereas others, including the one adopted in the present work, use the internal energy U, which excludes this field energy. The total internal energy and internal energy are related through the expressions

$$U_{tot} = U + \frac{B^2}{2\mu_0}, \tag{4.80}$$

$$dU_{tot} = TdS + \mathbf{H}\cdot d\mathbf{B}, \tag{4.81}$$

$$= dU + d\left(\frac{B^2}{2\mu_0}\right). \tag{4.82}$$

Some authors, including ourselves, deduce the properties of superconductors with the aid of the Gibbs free energy G defined in Eq. (4.22), while others resort to G_{tot}, where

$$G_{tot} = H'_{tot} - TS. \tag{4.83}$$

Still other authors instead employ the Helmholtz free energy F or F_{tot}, where

$$F = U - TS, \tag{4.84}$$

$$F_{tot} = U_{tot} - TS \tag{4.85}$$

$$= F + \frac{B^2}{2\mu_0}. \tag{4.86}$$

An added complication in making comparison between results arrived at by different authors arises because some authors use the cgs system instead of SI units.

XVI. CONCLUDING REMARKS

In the beginning of this chapter we discussed the experimental results of specific heat measurements, and then proceeded to develop the thermodynamic approach to superconductivity, an approach in which the specific heat plays a major role. Some of the expressions that were derived are fairly general. Others, however, are for the particular model in which the specific heat (4.24) in the normal state obeys the linear low-temperature relation γT and the critical field (4.45) has a simple parabolic dependence $[1-(T/T_c)^2]$ on temperature. Some expressions make use of the additional assumption that the BCS expression $C_s(T_c) = 2.43\gamma T_c$ of Eq. (4.9) is also valid. It is believed that these models provide a good physical picture of the thermodynamics of the superconducting state. A more appropriate description for the high-temperature superconductors would include the AT^3 term in the specific heat. It is, of course, also true that real superconductors have more complex temperature dependences than is implied by these simple models. The theoretical approaches presented in the following two chapters are needed to achieve a more basic understanding of the nature of superconductivity.

FURTHER READING

Several superconductivity texts discuss thermodynamics and thermal properties (Crow and Ong 1990; Orlando and Delin, 1991; Rose Innes and Rhoderick, 1994; Tinkham, 1985).

Some thermodynamics and statistical mechanics texts also have sections on superconductivity (e.g., Isihara, 1971; Reif, 1965), and some solid-state physics texts provide brief introductions to the specific heat and thermodynamics of superconductors (Ashcroft and Mermin, 1976; Burns, 1985; Kittel, 1976).

Studies have appeared that review the specific heats of superconductors (Junod, 1990); (Nevitt *et al.*, 1987; Phillips *et al.*, 1991); Salamon (1989) discusses the thermodynamics of superconductors, including specific heat.

PROBLEMS

1. Consider a metallic element such as copper that contributes one electron per atom to the conduction band. Show that in the free-electron approximation the electronic and phonon contributions to the specific heat will be equal at the temperature

$$T = \Theta_D(5/24\pi^2)^{1/2}\left(\frac{\Theta_D}{T_F}\right)^{1/2}$$

2. Show that the factor β in Eq. (4.5) has the value 1 for an element, $1/7$ for the LaSrCuO compound, and $3/13$ for the YBaCuO compound.
3. A superconductor has a Fermi energy of 3 eV. What is the density of states at the Fermi level and the electronic specific-heat factor γ. If this superconductor has an effective mass m^* of 81, what will be the value of these quantities? What other measurable quantities depend on the effective mass?
4. Consider a BCS-type superconductor with transition temperature $T_c = 20$ K and a critical field $B_c(0) = 0.2$ T. What is its electronic specific-heat factor γ? What are the values of its specific heat, entropy, Gibbs free energy, and enthalpy in the superconducting state at 10 K, both in zero field and in an applied magnetic field of 0.1 T? (Ignore the vibrational contribution to the specific heat.)
5. With the initial conditions of the previous problem, what applied magnetic field will drive the superconductor normal at 10 K? What will be the latent heat? What will be the change in the specific heat at the transition? (Ignore the vibrational contribution to the specific heat.)
6. Show that the following expressions for the enthalpy are valid:

$$\begin{aligned} H'_{\text{tot}} &= U_{\text{tot}} - H\cdot B \\ &= H' - \tfrac{1}{2}\mu_0 H^2, \\ dH'_{\text{tot}} &= TdS - B\cdot dH \\ &= dH' - d\left(\tfrac{1}{2}\mu_0 H^2\right). \end{aligned}$$

7. Show that equating the superconducting- and normal-state Gibbs free energies $G_s(T, H) = G_n(T)$ at the critical temperature leads to Eq. (4.55):

$$T'_c = T_c\left[1 - \frac{B}{B_c(0)}\right]^{1/2}.$$

8. Calculate the transition temperature T'_c, jump in specific heat, jump in entropy, jump in enthalpy, and the values of the Gibbs and Helmholtz free energies at the temperature $T = T'_c$ of a Type I superconductor in an applied magnetic field $B_{\text{app}} = \frac{1}{2}B_c$. Express your answers in terms of γ and T_c, assuming that $\alpha = 4.0$ and $A = \gamma/3T_c^2$.
9. We know from thermodynamics that at the transition temperature $T = T_c(B)$ in an applied magnetic field B, the latent heat equals the difference in enthalpy, $L = H'_n - H'_s$. Show that this difference gives Eq. (4.57).
10. Derive the expression for the enthalpy of a superconductor in a magnetic field, and show that in its normalized form it agrees with the expression $H'_s/\gamma T_c^2$ in Table 4.2.
11. Show that the specific heat jump in a magnetic field has the maximum

$4\alpha\gamma T_c/9$ in the range $0 < T_c < 1/\sqrt{3}$, and that the latent heat has the maximum $\frac{1}{2}\alpha\gamma T_c^2$.

12. Show that the following normalized thermodynamic expressions are valid,

$$du = tds + b \cdot dm,$$
$$cdt = tds,$$
$$h' = u - b \cdot m,$$
$$g = h' - ts,$$

and write down expressions for the normalized internal energy u and magnetization m.

13. Derive Eq. (4.70) from Rutger's formula.

14. Sketch a three-dimensional Gibbs free energy surface analogous to the surface presented in Fig. 4.20 using the equations in Section X.

5

Ginzburg–Landau Theory

I. INTRODUCTION

The previous chapter presented the thermodynamic approach to the phenomenon of superconductivity. We used the Gibbs free energy since in the absence of a magnetic field the Gibbs free energy is continuous across the superconducting-to-normal-state transition. The situation below the transition temperature T_c was handled by assuming a known magnetization and a known critical field, which were then used to calculate the various thermodynamic functions. This approach cannot really be called a theory because it simply incorporates known properties of superconductors into a standard treatment of thermodynamics in the presence of an applied magnetic field.

To gain more understanding of the phenomenon of superconductivity let us examine some simple but powerful theories that have been developed in efforts to explain it. In the present chapter we will consider the phenomenological approach proposed by Ginzburg and Landau (GL) in 1950. This approach begins by adopting certain simple assumptions that are later justified by their successful prediction of many properties of superconducting materials. The assumptions describe superconductivity in terms of a complex order parameter ϕ the physical significance of which is that $|\phi|^2$ is proportional to the density of super electrons. The order parameter is minimally coupled to the electromagnetic field, and in the presence of a magnetic field $\mathbf{B} = \nabla \times \mathbf{A}$ the momentum operator $-i\hbar\nabla\phi$ becomes $[-i\hbar\nabla\phi + e^*\mathbf{A}]$, where e^* is the charge associated with the "super electrons." The free energy is a minimum with respect to variations of both ϕ and $\mathbf{A}$. The London equations, dating from 1935, follow as a natural consequence of the GL theory, as we show in Section IX (London and London, 1935).

In the next chapter we will examine the more fundamental Bardeen–Cooper–Schrieffer (BCS) microscopic theory that first appeared in 1957. Soon after this theory was published, its correct prediction of many observable properties of superconductors was recognized. The earlier GL theory, on the other hand, was not widely accepted outside the Soviet Union until Gor'kov showed in 1959 that it is derivable from the BCS theory.

This chapter will concentrate on the case of isotropic superconductors. Formulations of the GL theory and of the London Model are also available for the anisotropic case (e.g., Coffey, 1993; Doria *et al.*, 1990; Du *et al.*, 1992; Klemm, 1993, 1994; Wang and Hu, 1991), and more specifically for the cuprates (Horbach *et al.*, 1994; Schneider, *et al.*, 1991; Wilkin and Moore, 1993). Time dependent processes have also been treated (Malomed and Weber, 1991; Stoof, 1993).

II. ORDER PARAMETER

Many phenomena in nature, such as the boiling of liquids and ferromagnetism, involve a transition from an ordered to a disordered phase. Each of these transitions can be characterized by an appropriate order parameter that has one value in the high-temperature disordered state and another in the low-temperature ordered state. The order parameter may be thought of as characterizing the extent to which the system is "aligned."

In the case of boiling, the order parameter might be the density, which is high in the liquid state and low in the gaseous state. The magnetic order parameter is often taken as the magnetization; it is zero in the high-temperature paramagnetic region, where the spins are randomly oriented, and nonzero at low temperatures, where the spins are ferromagnetically aligned.

In the normal conduction state the electric current is carried by a Fermi gas of conduction electrons, as was explained in Chapter 1. The GL theory assumes that in the superconducting state the current is carried by super electrons of mass m^*, charge e^*, and density n^* which are connected by the relationships

$$m^* = 2m, \tag{5.1a}$$

$$e^* = \pm 2e, \tag{5.1b}$$

$$n_s^* = \tfrac{1}{2}n_s \tag{5.1c}$$

with their electron counterparts m, e, and n_s, respectively. The actual "mass" here is the *effective* mass, and it need not be twice the mass of a free electron. The charge is negative for electron-type charge carriers, as is the case with many classical superconductors, and positive for hole conduction, as with most of the high-temperature superconductors. The super electrons begin to form at the transition temperature and become more numerous as the temperature falls. Therefore, their density n_s^* is a measure of the order that exists in the superconducting state. This order disappears above T_c, where $n_s^* = 0$, although fluctuations in n_s^* can occur above T_c. More generally $n_s^* \leq \frac{1}{2}n_s$, and Eq. (5.1c) gives us the limiting value of n_s^* for $T = 0$.

The Ginzburg–Landau theory, to be described in the following section, is formulated in terms of the complex order parameter $\phi(\mathbf{r})$, which may be written in the form of a product involving a phase factor Θ and a modulus $|\phi(\mathbf{r})|$,

$$\phi(\mathbf{r}) = |\phi(\mathbf{r})|e^{i\Theta} \tag{5.2}$$

whose square, $|\phi|^2$, is the super electron density,

$$n_s^* = |\phi|^2. \tag{5.3}$$

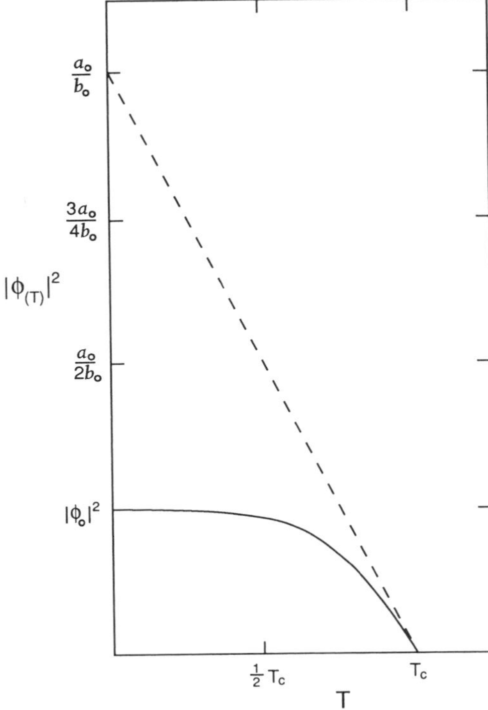

Figure 5.1 Temperature dependence of the order parameter $|\phi|^2$ showing its value $|\phi_0|^2$ at $T = 0$, and the linear behavior (– – –) near T_c, which extrapolates to the ordinate value a_0/b_0. This figure is drawn under the assumption $|\phi_0|^2 = \frac{1}{4}a_0/b_0$ to agree with Fig. 2.44.

The parameter ϕ is zero above T_c and increases continuously as the temperature falls below T_c, as shown in Fig. 5.1.

III. GINZBURG–LANDAU EQUATIONS

We saw in the previous chapter that the thermodynamic properties of the superconducting state can be described in terms of the Gibbs free energy density G. Ginzburg and Landau assumed that, close to the transition temperature below T_c, the Gibbs free energy per unit volume $G_s[\phi]$ may be expanded as a local functional of the order parameter,

$$G_s[\phi] = G_n + \frac{1}{V}\int d^3\mathbf{r}\left[\frac{1}{2m^*}\times(-i\hbar\nabla + e^*\mathbf{A})\phi^*\cdot(i\hbar\nabla + e^*\mathbf{A})\phi + \left(\frac{1}{2\mu_0}\right)B^2(\mathbf{r}) - \mu_0\mathbf{H}(\mathbf{r})\cdot\mathbf{M}(\mathbf{r}) + a\phi\phi^* + \tfrac{1}{2}b\phi\phi^*\phi\phi^* + \cdots\right], \tag{5.4}$$

where G_n is the free-energy density of the normal state, $\mathbf{A}$ is the magnetic vector potential, and a and b are functions of the temperature only. If the material is normal, $\mathbf{B} = \mu_0\mathbf{H}$, $\mathbf{M} = 0$, and the magnetic contribution is $\frac{1}{2}\mu_0 H^2$. In regions of perfect superconductivity $\mathbf{B} = 0$ and $\mathbf{M} = -\mathbf{H}$, and the magnetic contribution is $\mu_0 H^2$. In equilibrium the superconductor distributes currents in such a way as to minimize the total free energy.

The assumption is made that over a small range of temperatures near T_c the parameters a and b have the approximate values

$$a(T) \approx a_0\left[\frac{T}{T_c} - 1\right], \tag{5.5a}$$

$$b(T) \approx b_0, \tag{5.5b}$$

where a_0 and b_0 are both defined as positive, so that $a(T)$ vanishes at T_c and is negative below T_c.

To determine $\phi(\mathbf{r})$ we require that the free energy be a minimum with respect to variations in the order parameter. Taking the variational derivative (Arfken, 1985, Chapter 17) of the integrand in (5.4) with respect to ϕ^* with ϕ held constant gives the first GL equation:

$$\frac{1}{2m^*}(i\hbar\nabla + e^*\mathbf{A})^2\phi + a\phi + b|\phi|^2\phi = 0. \tag{5.6}$$

In the London–Landau gauge (sometimes called the Coulomb or radiation gauge)

$$\nabla \cdot \mathbf{A} = 0, \tag{5.7}$$

the first GL equation can be expanded into the form

$$\frac{1}{2m^*}(\hbar^2\nabla^2\phi - 2i\hbar e^*\mathbf{A}\cdot\nabla\phi - e^{*2}\mathbf{A}^2\phi) - a\phi - b|\phi|^2\phi = 0. \tag{5.8}$$

The free energy is also a minimum with respect to variations in the vector potential **A**, where

$$\mathbf{B} = \nabla \times \mathbf{A}. \tag{5.9}$$

Taking the variational derivative of G with respect to **A** we obtain the second GL equation:

$$\nabla \times (\nabla \times \mathbf{A}) + \frac{i\hbar e^*}{2m^*}(\phi^*\nabla\phi - \phi\nabla\phi^*) + \frac{e^{*2}}{m^*}\mathbf{A}|\phi|^2 = \mathbf{0}. \tag{5.10}$$

In Cartesian coordinates this equation, expressed in terms of the London–Landau gauge (5.7), can be simplified by writing $-\nabla^2\mathbf{A}$ in place of $\nabla \times (\nabla \times \mathbf{A})$ (see Problem 7). If we substitute the expression for **B** from Eq. (5.9) into the Maxwell expression (Ampère's law),

$$\nabla \times \mathbf{B} = \mu_0 \mathbf{J}, \tag{5.11}$$

and compare the result with Eq. (5.10), we find the following proper gauge-invariant expression for the current density:

$$\mu_0\mathbf{J} = -\frac{i\hbar e^*}{2m^*}(\phi^*\nabla\phi - \nabla\phi^*\phi) - \frac{e^{*2}}{m^*}\mathbf{A}|\phi|^2. \tag{5.12}$$

Thus the Ginzburg–Landau theory gives us two coupled differential equations, (5.8) and (5.10), involving the order parameter and vector potential, which can be solved to determine the properties of the superconducting state. For most applications the equations must be solved numerically. However, there are some simple cases in which exact closed-form solutions can be found, and others in which useful approximate solutions can be obtained. We will examine some of these cases, and then transform the GL equations to a normalized form and discuss the solution for more complex cases. When these equations are written in a normalized form, the coherence length, penetration depth, and quantum of magnetic flux, called the *fluxoid*, appear as natural parameters in the theory.

IV. ZERO-FIELD CASE DEEP INSIDE SUPERCONDUCTOR

To get a feeling for the behavior of ϕ, let us first consider the zero-field case ($\mathbf{A} = \mathbf{0}$) with homogeneous boundary conditions (zero gradients, $\nabla^2\phi = 0$). The absence of gradients corresponds to a region deep inside a superconductor where the super electron density does not vary with position. Integration of Eq. (5.4) can be carried out directly for this zero field–zero gradient case, to give for the Gibbs free energy density G_s of the superconductor

$$G_s = G_n + a|\phi|^2 + \tfrac{1}{2}b|\phi|^4, \tag{5.13}$$

where from Eqs. (5.5) b is positive and a negative below T_c. The GL equation (5.8) provides the minimum for this free energy,

$$a\phi + b|\phi|^2\phi = 0, \tag{5.14}$$

and all of the terms of the second GL equation (5.10) vanish. The phase of ϕ is arbitrary, so we can take ϕ to be real. Equation (5.14) has one solution, $\phi = 0$, corresponding to the normal state and one solution for $a < 0$ at $T < T_c$, with lower free energy:

$$|\phi|^2 = -\frac{a}{b} = \frac{|a|}{b}. \tag{5.15}$$

Using the approximations (5.5a) and (5.5b) for a and b, respectively, we have

$$|\phi|^2 = \frac{a_0}{b_0}\left(1 - \frac{T}{T_c}\right), \tag{5.16}$$

and this linear temperature dependence is shown in Fig. 5.1 for the region near $T \approx T_c$. For lower temperatures $|\phi|^2$ is expected to deviate from linearity on its approach to its 0 K value, $|\phi_0|^2 < a_0/b_0$, as shown in the figure. From Eq. (5.3) we have for the super electron density

$$n_s^* = \frac{a_0}{b_0}\left[1 - \frac{T}{T_c}\right], \tag{5.17}$$

which agrees with Eq. (2.69) in the superconducting region near T_c.

When the expressions for ϕ from Eqs. (5.15) and (5.16) are substituted into Eq. (5.13), we obtain for the minimum Gibbs free energy density

$$\begin{aligned} G_s &= G_n - \frac{1}{2}\left(\frac{a^2}{b}\right) \\ &= G_n - \frac{1}{2}\left(\frac{a_0^2}{b_0}\right)\left[1 - \frac{T}{T_c}\right]^2, \end{aligned} \tag{5.18}$$

where $\frac{1}{2}(a^2/b)$, called the condensation energy per unit volume of the super electrons, is the energy released by transformation of normal electrons to the super electron state. The condensation energy can be expressed in terms of the thermodynamic critical field $\mathbf{B}_c$ as follows:

$$\frac{1}{2}\left(\frac{a^2}{b}\right) = \frac{B_c^2}{2\mu_0}. \tag{5.19}$$

Figure 5.2 presents a plot of $G_s - G_n$ from Eq. (5.13) versus ϕ for the three ratios of temperatures $T/T_c = 1$, 0.9, and

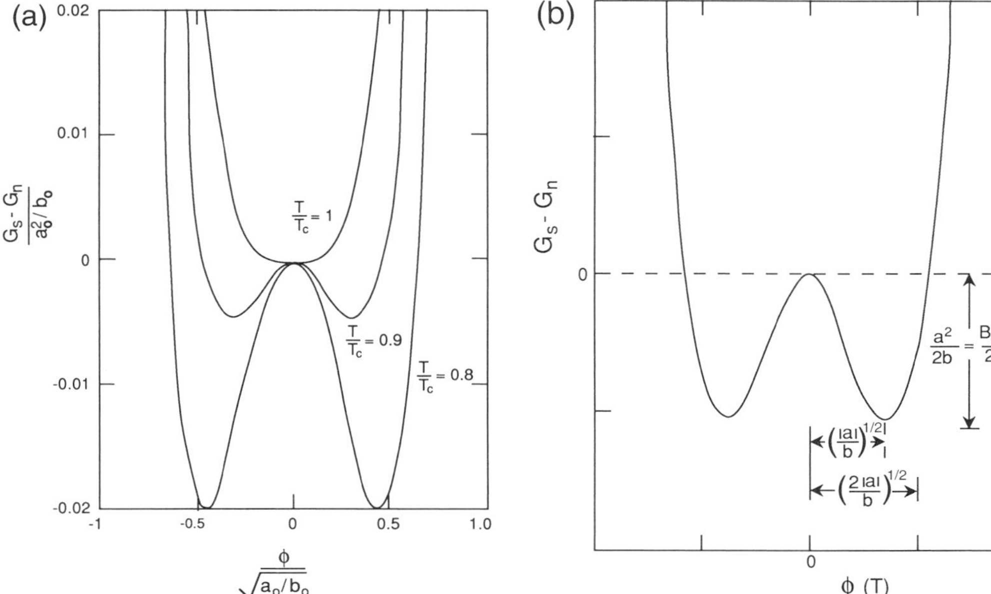

Figure 5.2 Dependence of the difference $G_s - G_n$ between the Gibbs free energy in the normal and superconducting states on the order parameter ϕ. (a) Normalized plots for $T/T_c = 1$, 0.9, and 0.8, and (b) Plot for $T/T_c = 0.85$ showing the minimum free-energy difference $G_s - G_n = a^2/2b$, which occurs for $\phi = (|a|/b)^{1/2}$, and the zero, $G_s - G_n = 0$, at $\phi = (2|a|/b)^{1/2}$.

0.8. The minimum for each curve occurs at $\phi = (|a|/b)^{1/2}$ given by Eq. (5.15), and $G_s - G_n = 0$ at $\phi = (2|a|/b)^{1/2}$. These coordinates for the minimum and crossover points of the $T/T_c = 0.8$ curve are indicated in the figure. The equilibrium superconducting state exists at the minimum of each curve. The minimum gets deeper, and the order parameter ϕ for the minimum increases, as the temperature is lowered, as shown. The magnitude of the free-energy minimum at 0 K cannot be written down because the temperature dependence of Eq. (5.16) can only be a good approximation near T_c.

V. ZERO-FIELD CASE NEAR SUPERCONDUCTOR BOUNDARY

Next we consider the case of zero field with inhomogeneous boundary conditions, which means that gradients can exist. Setting $\mathbf{A} = \mathbf{0}$ in the second GL equation (5.10) gives

$$\phi^* \nabla \phi = \phi \nabla \phi^*, \tag{5.20}$$

which means, from Eq. (5.2), that the phase Θ of the order parameter is independent of position. The first GL equation, Eq. (5.8), with $\mathbf{A}$ set equal to zero, provides us with a differential equation for the order parameter:

$$-\frac{\hbar^2}{2m^*}\nabla^2\phi + a\phi + b|\phi|^2\phi = 0. \tag{5.21}$$

Since the phase of the order parameter is constant we select ϕ to be real.

We assume that the right half-space, $x > 0$, is filled with a superconductor and that the left half-space, $x < 0$, is a vacuum or normal material, as shown in Fig. 5.3. Therefore, ϕ is a function of x, the gradient operator ∇ only has an x component, and we can write Eq. (5.21) in one-dimensional form:

$$-\frac{\hbar^2}{2m^*}\frac{d^2\phi}{dx^2} + a\phi + b|\phi|^2\phi = 0. \tag{5.22}$$

When we change variables by letting

$$\phi = \left(\frac{|a|}{b}\right)^{1/2} f, \tag{5.23}$$

the normalized order parameter f satisfies the "nonlinear Schrödinger equation"

$$\frac{\hbar^2}{2m^*|a|}\cdot\frac{d^2 f}{dx^2} + f(1 - f^2) = 0. \tag{5.24}$$

If we define the dimensionless variable η as

$$\eta = \frac{x}{\xi}, \tag{5.25}$$

where

$$\xi^2 = \frac{\hbar^2}{2m^*|a|}, \tag{5.26}$$

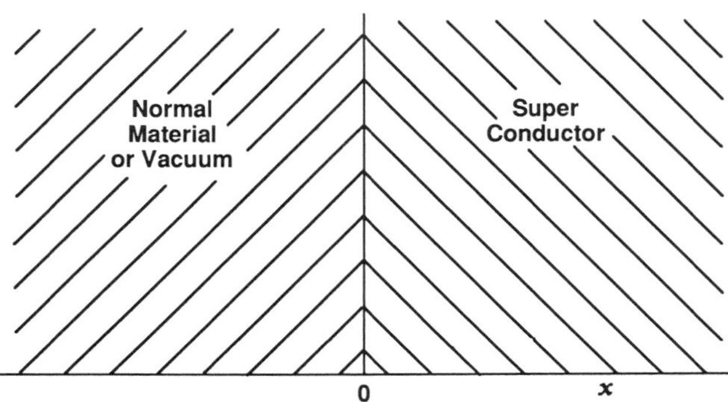

Figure 5.3 Interface between a normal material on the left ($x < 0$) and a superconductor on the right ($x > 0$).

Eq. (5.24) assumes the simplified dimensionless form

$$\frac{d^2f}{d\eta^2} + f(1 - f^2) = 0. \tag{5.27}$$

It may be easily verified by direct substitution that Eq. (5.27) has the solution

$$f = \tanh\frac{\eta}{\sqrt{2}}. \tag{5.28}$$

This can be written in terms of the original variable ϕ,

$$\phi = \phi_\infty \tanh\frac{x}{\sqrt{2}\,\xi}, \tag{5.29}$$

where

$$\phi_\infty = \left(\frac{|a|}{b}\right)^{1/2}, \tag{5.30}$$

with $\phi \to 0$ as $x \to 0$ and $\phi \to \phi_\infty$ as $x \to \infty$. Therefore, ξ is the characteristic length over which ϕ can vary appreciably. The parameter ξ, called the *coherence length*, is one of the two fundamental length scales associated with superconductivity. Its significance is shown graphically in Fig. 5.4, in which we see that ϕ is close to ϕ_∞ far inside the superconductor, is zero at the interface with the normal material, and has intermediate values in a transition layer near the interface with a width on the order of ξ.

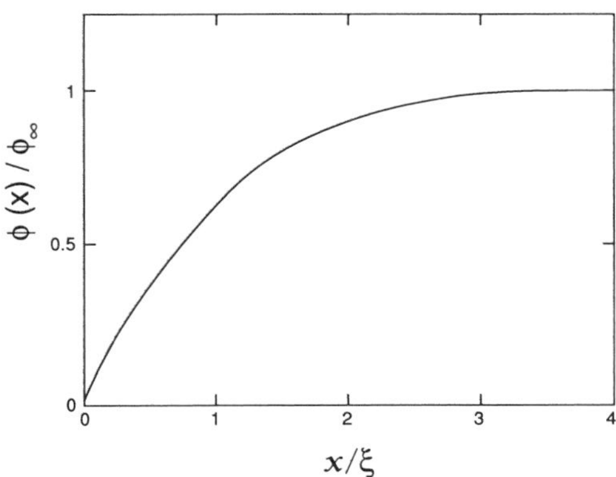

Figure 5.4 Dependence of the order parameter $\phi(x)$ on distance x inside a superconductor. The order parameter is large for $x > \xi$, where ξ is the coherence length.

Substituting Eq. (5.20) in Eq. (5.12) shows that for $\mathbf{A} = \mathbf{0}$ the current density $\mathbf{J}$ vanishes. This is to be expected since from the Maxwell expression (5.11) we know that electric currents cannot exist if there are no associated magnetic fields present.

VI. FLUXOID QUANTIZATION

Now that we have determined the order parameter ϕ for the case $B = 0$, we will proceed to investigate the situation when there is an applied magnetic field. In the presence of such a field an interesting result follows from Eq. (5.12). If we write $\phi(\mathbf{r})$ as the product of a modulus and a phase factor, as in Eq. (5.2), the gradient of ϕ will have the form

$$\nabla\phi = i\phi\nabla\Theta + e^{i\Theta}\nabla|\phi(\mathbf{r})|, \tag{5.31}$$

and the total current from Eq. (5.12) will be given by

$$\mu_0\mathbf{J} = \frac{\hbar e^*}{m^*}|\phi|^2\nabla\Theta - \frac{e^{*2}}{m^*}|\phi|^2\mathbf{A}. \tag{5.32}$$

Dividing Eq. (5.32) by $\hbar e^*|\phi|^2/m^*$ and taking the line integral around a closed contour gives

$$\frac{m^*}{e^{*2}}\oint\frac{\mu_0\mathbf{J}}{|\phi|^2}\cdot d\mathbf{l} = \frac{\hbar}{e^*}\oint\nabla\Theta\cdot d\mathbf{l} - \oint\mathbf{A}\cdot d\mathbf{l}. \tag{5.33}$$

For the order parameter to be single valued the line integral over the phase Θ around a closed path must be a multiple of 2π,

$$\oint\nabla\Theta\cdot d\mathbf{l} = 2\pi n, \tag{5.34}$$

where n is an integer. Equation (5.33) can now be written

$$\frac{m^*}{e^{*2}}\oint\frac{\mu_0\mathbf{J}}{|\phi|^2}\cdot d\mathbf{l} + \oint\mathbf{A}\cdot d\mathbf{l} = n\Phi_0, \tag{5.35}$$

where the quantum of flux Φ_0 has the value

$$\Phi_0 = \frac{h}{e^*}, \tag{5.36}$$

in agreement with experiment (e.g., Cabrera *et al.*, 1989; Gough *et al.*, 1987; S. Hasegawa *et al.*, 1992).

It is convenient to express the line integral of **A** in Eq. (5.35) in terms of the magnetic flux Φ through the closed contour. Applying Stokes' theorem we find

$$\oint \mathbf{A}\cdot d\mathbf{l} = \int \mathbf{B}\cdot d\mathbf{S} \tag{5.37}$$

$$= \Phi, \tag{5.38}$$

and Eq. (5.35) becomes

$$\frac{m^*}{e^{*2}} \oint \frac{\mu_0 \mathbf{J}}{|\phi|^2}\cdot d\mathbf{l} + \Phi = n\Phi_0. \tag{5.39}$$

This expression is valid for all superconductors, and can be applied to the intermediate and mixed states described in Chapter 11. Equation (5.39) expresses the condition whereby the sum of the enclosed flux Φ and the line integral involving the current density **J** is quantized.

We will see later that for Type II superconductors quantized flux occurs in vortices, which have a core region of very high field, and a field outside which decreases with distance in an approximately exponential manner far from the core. Figure 5.5 sketches two such vortices. When a contour is taken in a region of space that contains vortices, the integer n in Eq. (5.39) corresponds to the number of cores included within the path of integration. Figure 5.5 shows contours enclosing $n = 0$, 1, and 2 core regions. Here we are assuming that all the vortices have the same polarity, i.e., the magnetic field points in the same direction in all the vortices. Equation (5.39) is easily generalized to include the presence of positively and negatively directed vortices.

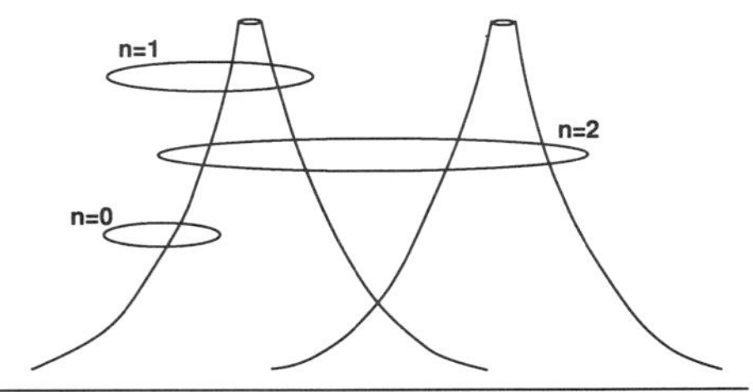

Figure 5.5 Integration paths for Eq. (5.35) encircling no cores ($n = 0$), encircling one core ($n = 1$), and encircling two cores ($n = 2$).

The Little–Parks (1962, 1964) experiment demonstrated this flux quantization by measuring the magnetic field dependence of the shift in T_c of a thin-walled superconducting cylinder in an axial applied field.

VII. PENETRATION DEPTH

In Section V we found how the order parameter changes with distance in the neighborhood of the boundary of a superconductor, and this provided us with the first fundamental length scale—the coherence length ξ. In this section we will investigate the behavior of the internal magnetic field in the neighborhood of a boundary when there is an applied field outside. This will give us the penetration depth λ_L, the second of the two fundamental length scales of superconductivity.

We begin by returning to the semi-infinite geometry of Fig. 5.3 with a uniform magnetic field oriented in the z direction. In the London–Landau gauge (5.7) the vector potential for a constant magnetic field B_0 outside the superconductor ($x < 0$) is

$$\mathbf{A} = A_y(x)\hat{\mathbf{j}} \tag{5.40}$$

with

$$A_y(x) = xB_0 + A_0 \qquad x < 0, \tag{5.41}$$

where the constant A_0 is selected for continuity with the solution $A_y(x)$ inside the superconductor, as shown in Fig. 5.6. This constant does not affect the field $\mathbf{B}(x)$.

In order to determine how the phase of the order parameter varies throughout the interior of the superconductor, let us evaluate the line integrals of Eq. (5.35) along a rectangular contour in the x, y-plane that is closed at $x = x_0$ and $x_1 \to \infty$,

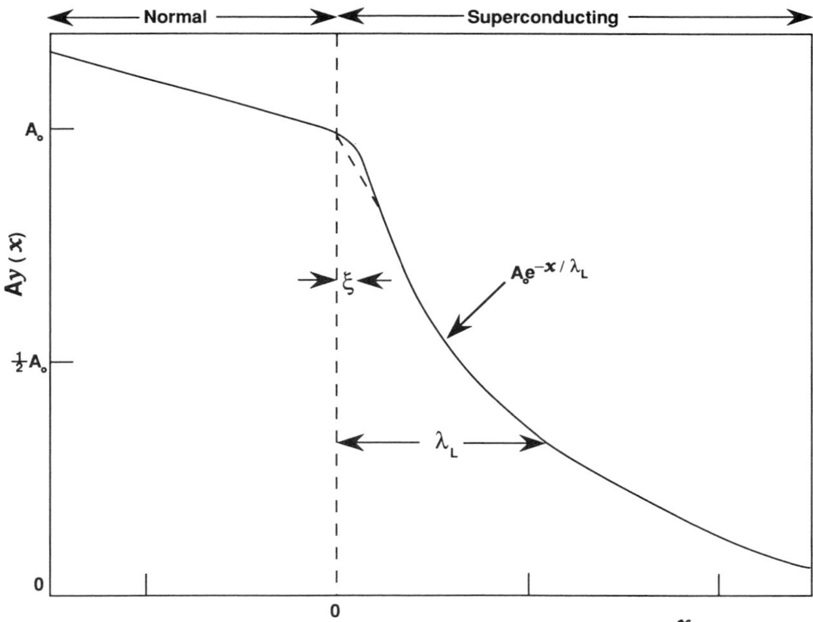

Figure 5.6 Dependence of the vector potential $A(x)$ on distance x for the case of Fig. 5.3. $A(x)$ depends linearly on x outside the superconductor (left), where there is a constant applied magnetic field, and decays exponentially inside the superconductor (right), becoming very small for $x \gg \lambda_L$, where λ_L is the London penetration depth.

as indicated in Fig. 5.7. This is done for a contour of arbitrary width L, as shown. Since **A** is a vector in the y direction, it is perpendicular to the upper and lower horizontal parts of the contour, which are along x, so that the integral of $\mathbf{A}\cdot d\mathbf{l}$ vanishes along these paths. We also observe that no current flows into the superconductor, so that $J_x = 0$ and the line integrals of $\mathbf{J}\cdot d\mathbf{l}$ along these same upper and lower horizontal paths vanish. When we take the limit $x_1 \to \infty$ the two line integrals along this vertical x_1 path vanish because **A** and **J** are zero far inside the superconductor. As a result only the line integrals along the x_0 vertical path contribute, and they may be written down immediately because there is no y dependence for the fields and currents:

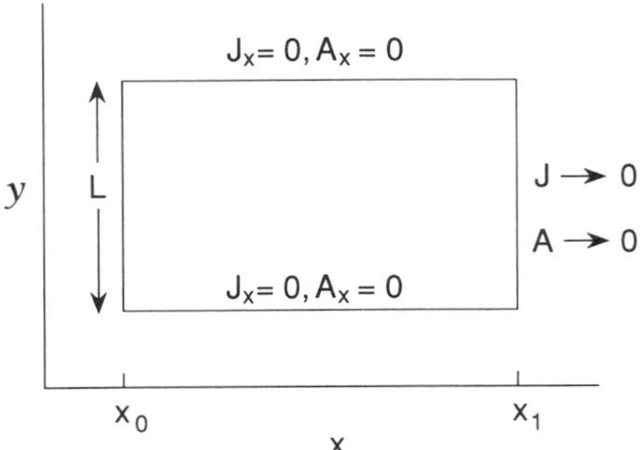

Figure 5.7 Integration path inside a superconductor for determining the phase Θ of the order parameter ϕ.

$$L\cdot\left[\frac{m^*J_y(x_0)}{e^{*2}|\phi(x_0)|^2}+A_y(x_0)\right]=n\Phi_0. \tag{5.42}$$

Since the width L is arbitrary and n is quantized, it follows that $n = 0$. Then, from the quantization condition (5.34) and the arbitrariness of the path, we conclude that the phase Θ of the order parameter (5.2) is constant everywhere throughout the superconductor, and we set it equal to 0.

Furthermore, since x_0 in Eq. (5.42) is arbitrary, it follows that

$$J_y(x) = -\frac{e^{*2}|\phi(x)|^2}{m^*}A_y(x), \quad (5.43)$$

which tells us how **J** is related to **A**.

To determine the x dependence of A_y, note that Eq. (5.20) is valid for a constant phase and that the second GL equation (5.10) reduces to the expression

$$\frac{d^2}{dx^2}A_y(x) = \frac{\mu_0 e^{*2}|\phi(x)|^2}{m^*}A_y(x). \quad (5.44)$$

We seek to solve this equation far enough inside the superconductor, $x \gg \xi$, so that the order parameter attains its asymptotic value, $\phi \to \phi_\infty$, independent of x. It is convenient to define the London penetration depth λ_L, the second of the two fundamental length scales of a superconductor:

$$\lambda_L^2 = \frac{m^*}{\mu_0 e^{*2}|\phi_\infty|^2}. \quad (5.45)$$

This permits us to write Eq. (5.44) in the form

$$\frac{d^2}{dx^2}A_y(x) = \frac{A_y(x)}{\lambda_L^2}, \quad (5.46)$$

which has a simple exponential solution inside the superconductor,

$$A_y(x) = A_0 \exp(-x/\lambda_L) \qquad x > 0, \quad (5.47)$$

for the case $\xi \ll \lambda_L$ which is plotted in Fig. 5.6. The preexponential factor A_0 makes $A_y(x)$ from Eqs. (5.41) and (5.47) match continuously across the boundary at $x = 0$.

In writing Eq. (5.46) we implicitly assumed that the London penetration depth λ_L is greater than the coherence length ξ. For distances from the surface x in the range $0 < x \ll \xi$ we know from Eq. (5.29) and the power series expansion of $\tanh(x/\sqrt{2}\,\xi)$ for small values of the argument that

$$\phi(x) \approx \phi_\infty \frac{x}{\sqrt{2}\,\xi} \qquad 0 \ll x \ll \xi, \quad (5.48)$$

so that $\phi(x)$ is much less than ϕ_∞. In this range the effective penetration depth exceeds the London value (5.45), so $A_y(x)$ decays more gradually there, as indicated in Fig. 5.6.

To obtain the fields from the potentials we apply the curl operation $\mathbf{B} = \nabla \times \mathbf{A}$. Only the z component exists, as assumed initially,

$$B_z(x) = B_0 x < 0, \quad (5.49)$$

$$B_z(x) = \frac{-A_0}{\lambda_L}\exp(-x/\lambda_L) \qquad \xi < x < \infty,$$
$$= B_0 \exp(-x/\lambda_L) \quad (5.50)$$

where $A_0 = -\lambda_L B_0$ from the boundary condition at the surface ($x = 0$). The distance dependences of Eqs. (5.48) and (5.49), together with the more gradual decay in the range $0 < x < \xi$, are shown in Fig. 5.8. We conclude that for this case the applied field has the constant value $\mathbf{B}_0$ outside the superconductor, decays exponentially with distance inside, and becomes negligibly small beyond several penetration depths within, as shown.

From Eq. (5.43) we find that far inside the superconductor

$$\mu_0 \lambda_L^2 J_y(x) = -A_y(x) \qquad \xi \ll x < \infty, \quad (5.51)$$

as $\phi(x) \to \phi_\infty$, and hence that $J_y(x)$ also satisfies Eq. (5.46) with the distance dependence

$$J_y(x) = \frac{A_0}{\mu_0 \lambda_L^2}\exp(-x/\lambda_L)$$
$$= J_0 \exp(-x/\lambda_L) \qquad \xi < x < \infty, \quad (5.52)$$

which is the same as $A_y(x)$ of Eq. (5.47). In the range $0 < x < \xi$, which is near the

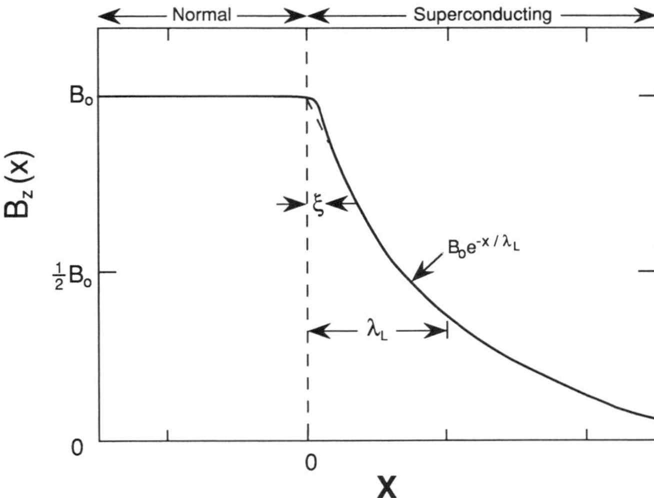

Figure 5.8 Exponential decay of a constant applied magnetic field $B_z(x)$ inside a superconductor for the case $\lambda_L > \xi$. Note the small deviation from exponential behavior within a coherence length ξ of the surface.

surface, we see from Eq. (5.43) that $J_y(x)$ is less than this value, as indicated in Fig. 5.9. Thus, we see that $B_z(x)$ decays less and that the current density $J_y(x)$ has a magnitude less than its value beyond the coherence length. In the remainder of the chapter we will ignore these surface effects for $x < \xi$ and only take into account the exponential decay in terms of the penetration-depth distance parameter.

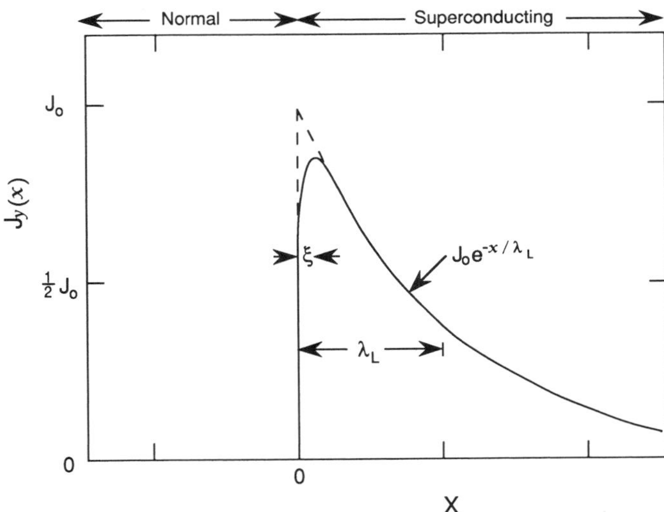

Figure 5.9 Dependence of the current density $J_y(x)$ on distance x inside a superconductor for the case $\lambda_L > \xi$.

VIII. CRITICAL CURRENT DENSITY

An electric current is accompanied by a magnetic field. To obtain an expression for the current density independent of the magnetic field, the vector potential can be eliminated between the current density equation and the second GL equation. We will do this deep inside the superconductor where the order parameter $\phi(\mathbf{r})$ depends on position only through the phase $\Theta(r)$.

For this situation the order parameter, written in the form

$$\phi(\mathbf{r}) = \phi_0 e^{i\Theta(r)}, \qquad (5.53)$$

has the gradient

$$\nabla\phi(\mathbf{r}) = i\phi\nabla\Theta(r), \qquad (5.54)$$

and the current density (5.12) is

$$\mathbf{J} = \frac{\hbar e^*}{m^*}\phi_0^2\left(\nabla\Theta - \frac{2\pi}{\Phi_0}\mathbf{A}\right), \qquad (5.55)$$

where Φ_0 is given by Eq. (5.36). Substituting the expression for the order parameter from (5.53) in Eq. (5.8) and multiplying on the left by $e^{-i\Theta}$ gives

$$\frac{\hbar^2}{2m^*|a|}\phi_0 e^{-i\Theta}\left(i\nabla + \frac{2\pi}{\Phi_0}\mathbf{A}\right)^2 e^{i\Theta} + \phi_0 - \left(\frac{b}{|a|}\right)\phi_0^3 = 0. \qquad (5.56)$$

If the Laplacian $\nabla^2\phi$ is negligible, this becomes

$$\frac{\hbar^2}{2m^*|a|}\left(\nabla\Theta - \frac{2\pi}{\Phi_0}\mathbf{A}\right)^2 + 1 - \frac{\phi_0^2}{(|a|/b)} = 0. \qquad (5.57)$$

The factor $[\nabla\Theta - (2\pi/\Phi_0)\mathbf{A}]$ can be eliminated between Eqs. (5.55) and (5.57) to give for the current density

$$J_s = \frac{\Phi_0}{2\pi\mu_0\lambda_L^2\xi}f^2(1-f^2)^{1/2}, \qquad (5.58)$$

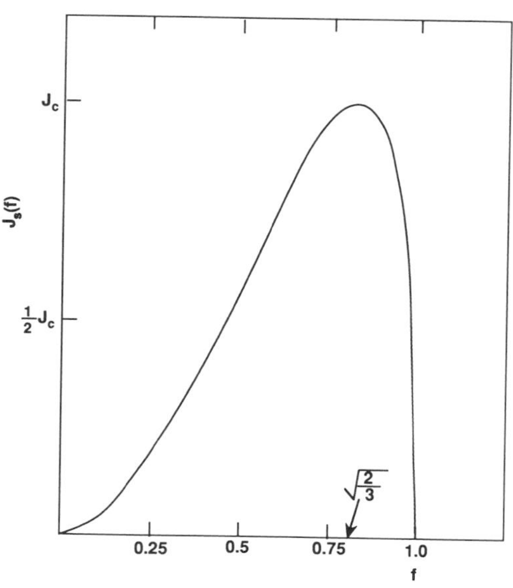

Figure 5.10 Dependence of the super current density $\mathbf{J}_s$ on the normalized order parameter f. $\mathbf{J}_s(f)$ reaches a maximum at $f = (2/3)^{1/2}$.

where f is given by Eq. (5.23) (we have used Eqs. (5.26) and (5.45) here). Figure 5.10 shows how J_s depends on f. The largest possible current density shown in the figure, called the *critical current density* J_c, is obtained by maximizing Eq. (5.58) through differentiation with respect to f^2. This gives $f^2 = \frac{2}{3}$, and we obtain what is sometimes called the Ginzburg–Landau critical current density,

$$J_c = \frac{\Phi_0}{3\sqrt{3}\,\pi\mu_0\lambda_L^2\xi}. \qquad (5.59)$$

This can also be written in terms of the thermodynamic critical field (9.10):

$$J_c = \frac{2\sqrt{2}\,B_c}{3\sqrt{3}\,\mu_0\lambda_L}. \qquad (5.60)$$

From Eq. (2.61) this has the following temperature dependence near T_c:

$$J_c = \frac{8\sqrt{2}\,B_c(0)}{3\sqrt{3}\,\mu_o\lambda_L(0)}\left(1 - \frac{T}{T_c}\right)^{3/2}. \qquad (5.61)$$

Thus J_c becomes zero at the critical temperature, and we know from Eq. (2.58) that it is a maximum at $T = 0$.

IX. LONDON EQUATIONS

In 1935 the London brothers, Fritz and Heinz, proposed a simple theory to explain the Meissner effect, which had been discovered two years earlier. They assumed that the penetration depth λ_L is a constant independent of position. The equations which they derived, now called the first London equation,

$$\mathbf{E} = \mu_0 \lambda_L^2 \frac{d}{dt} \mathbf{J}, \tag{5.62}$$

and the second London equation,

$$\mathbf{B} = -\mu_0 \lambda_L^2 \nabla \times \mathbf{J}, \tag{5.63}$$

were used to explain the properties of superconductors.

These two equations are easily obtained from the GL theory with the aid of Eq. (5.51) expressed in vector form:

$$\mu_0 \lambda_L^2 \mathbf{J} = -\mathbf{A}. \tag{5.64}$$

If the vector potential expression (5.9) is substituted in Maxwell's equation (1.66), we obtain

$$\nabla \times \left(\mathbf{E} + \frac{d\mathbf{A}}{dt} \right) = 0, \tag{5.65}$$

and with the aid of Eq. (5.51) we can then write down the first London equation (5.62). The second London equation (5.63) is obtained by substituting the expression for $\mathbf{A}$ from Eq. (5.64) in Eq. (5.9). It should be compared with Eq. (1.72), which, in the absence of magnetization and displacement currents, becomes Ampère's law:

$$\nabla \times \mathbf{B} = \mu_0 \mathbf{J}. \tag{5.66}$$

Thus we see that Maxwell's and London's equations link the magnetic field $\mathbf{B}$ and the current density $\mathbf{J}$ in such a way that if one is present in the surface layer so is the other.

If the expression for the current density $\mathbf{J}$ from Eq. (5.66) is substituted in Eq. (5.63) we obtain

$$\nabla^2 \mathbf{B} = \frac{\mathbf{B}}{\lambda_L^2}, \tag{5.67}$$

and eliminating $\mathbf{B}$ between these same two expressions gives

$$\nabla^2 \mathbf{J} = \frac{\mathbf{J}}{\lambda_L^2}. \tag{5.68}$$

Thus, recalling (5.46), we see that $\mathbf{A}$, $\mathbf{B}$, and $\mathbf{J}$ all obey the same differential equation. In Cartesian coordinates, Eqs. (5.67) and (5.68) correspond to the Helmholtz equation well known from mathematical physics (Arfken, 1985). In the following section we will provide applications of these equations to the phenomena of magnetic field penetration and surface current flow. There has been some recent literature on the London equations (Daemen and Gubernatis, 1991b, 1992b; Hao and Clem, 1991; Ivanchenko, 1993).

X. EXPONENTIAL PENETRATION

In Section VII we deduced the exponential decay of the magnetic field $\mathbf{B}$ and the current density $\mathbf{J}$, Eqs. (5.50) and (5.52), respectively, inside a superconductor in the presence of an external magnetic field $\mathbf{B}_0$, and in the previous section we wrote down the Helmholtz equations (5.67) and (5.68), respectively, for these same two cases. In the present section we will apply these equations to several practical situations involving magnetic field penetration and surface current flow in superconductors with rectangular and cylindrical shapes. Both shielding and transport currents will be discussed.

Consider a flat superconducting slab oriented in the y, z-plane in the presence of an applied magnetic field $\mathbf{B}_0$ in the z direction, as illustrated in Fig. 5.11. The

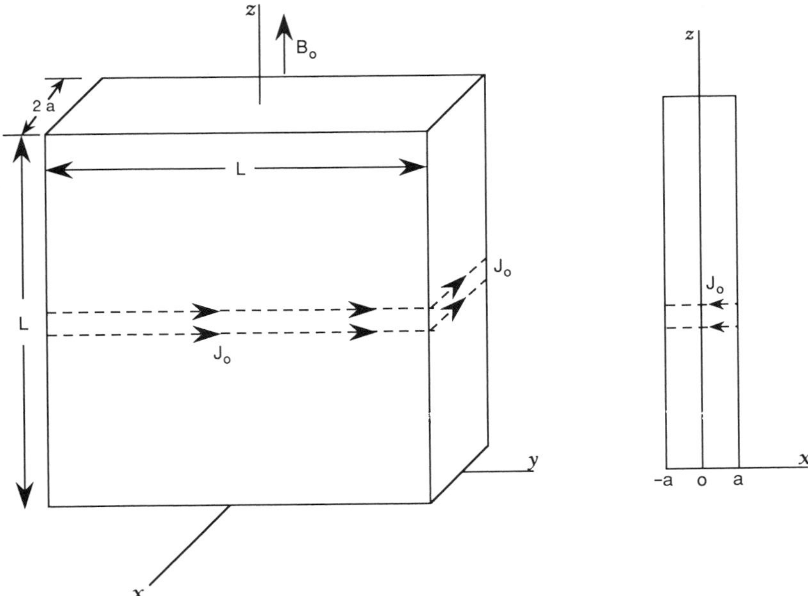

Figure 5.11 Flat superconducting slab with thickness $2a$ much less than the two broad dimensions L. The applied magnetic field $\mathbf{B}_0$ and super current flow $\mathbf{J}_0$ have the indicated directions. The thickness parameter a should not be confused with the GL parameter a of Eq. 5.4.

slab is of length L, width L, and thickness $2a$, as indicated in the figure; we assume that $a \ll L$. The solution to Helmholtz equation (5.67) which satisfies the boundary conditions $B_z(-a) = B_z(a) = \mathbf{B}_0$ at the edges is

$$B_z(x) = \frac{B_0 \cosh\left(\dfrac{x}{\lambda_L}\right)}{\cosh\left(\dfrac{a}{\lambda_L}\right)} \qquad -a < x < a. \tag{5.69}$$

This is sketched in Fig. 5.12 for $\lambda_L \ll a$ and in Fig. 5.13 for $\lambda_L \gg a$. For the former case we have

$$B_z(x) \approx B_0 \exp\left(\frac{-(a - |x|)}{\lambda_L}\right) \qquad \lambda_L \ll a. \tag{5.70}$$

We show in Problem 5 that in the latter case the penetration is linear near each boundary,

$$B_z(x) \approx B_0\left[1 - \frac{a(a-x)}{\lambda_L^2}\right] \qquad \lambda_L \gg a, \quad 0 \ll x < a, \tag{5.71a}$$

and

$$B_z(x) \approx B_0\left[1 - \frac{a(a+x)}{\lambda_L^2}\right] \qquad \lambda_L \gg a, \quad -a < x \ll 0, \tag{5.71b}$$

with the value in the center

$$B_z(0) \approx B_0\left[1 - \frac{a^2}{2\lambda_L^2}\right] \qquad x = 0, \tag{5.72}$$

as indicated in Fig. 5.13.

To derive the corresponding expressions for the current density we find from Eq. (5.11) that for this case $\mathbf{J}$ and $\mathbf{B}$ are related through the expression

$$\mu_0 J_y = \frac{dB_z}{dx}, \tag{5.73}$$

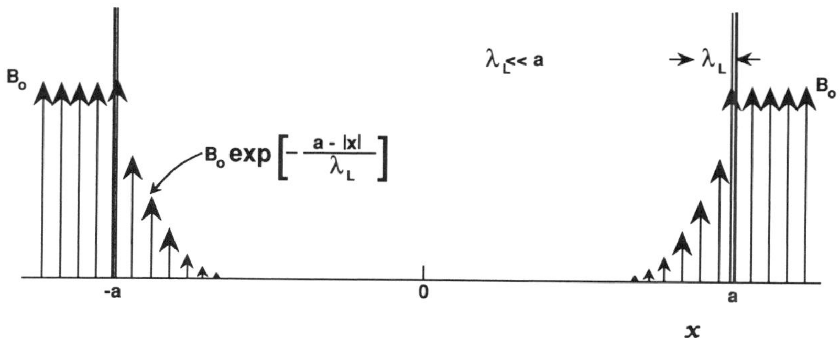

Figure 5.12 Exponential decay of a magnetic field inside a superconductor for the case $\lambda_L \ll a$. Both this figure and Fig. 5.13 are symmetric about the midpoint $x = 0$.

and differentiating $B_z(x)$ in Eq. (5.69) gives

$$\mu_0 J_y(x) = \frac{B_0}{\lambda_L} \cdot \frac{\sinh \dfrac{x}{\lambda_L}}{\cosh \dfrac{a}{\lambda_L}} \qquad -a < x < a. \tag{5.74}$$

Thus the magnetic field **B** and the current density **J** are mutually perpendicular, as indicated in Fig. 5.11. The current density flows around the slab in the manner shown in Fig. 5.14, and is positive on one side and negative on the other. It has the maximum magnitude $J_y(0) = J_0$ on the surface, $x = \pm a$, where

$$J_0 = \frac{B_0}{\mu_0 \lambda_L} \tanh \frac{a}{\lambda_L}, \tag{5.75}$$

and this gives for $J_y(x)$

$$J_y(x) = J_0 \frac{\sinh\left(\dfrac{x}{\lambda_L}\right)}{\sinh\left(\dfrac{a}{\lambda_L}\right)} \qquad -a < x < a. \tag{5.76}$$

This expression for the current density satisfies Helmholtz equation (5.68), as expected. For $a \gg \lambda_L$ the current density flows in a surface layer of thickness λ_L, while for the opposite limit, $a \ll \lambda_L$, it flows through the entire cross section, in accordance with Figs. 5.15 and 5.16, respectively. In the latter case the distance

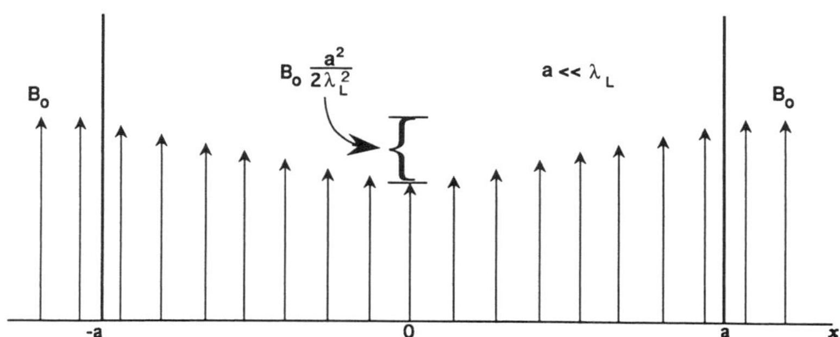

Figure 5.13 Decrease of the magnitude of a magnetic field inside a superconductor for the case $\lambda_L \gg a$. The field in the center is $[1 - \frac{1}{2}(a/\lambda_L)^2]$ times the field $\mathbf{B}_0$ outside.

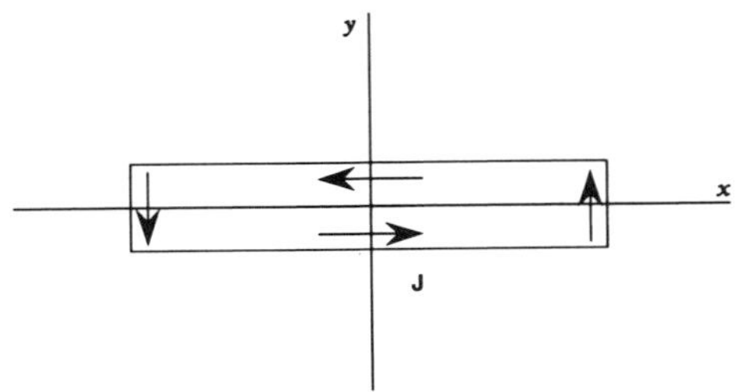

Figure 5.14 Cross section in the x, y-plane of the slab of Fig. 5.11 showing the shielding super current flow for an applied magnetic field $\mathbf{B}_0$ in the z direction.

dependence is linear such that

$$J(x) \approx \frac{J_0 x}{a} \qquad -a < x < a, \quad (5.77)$$

as shown in Fig. 5.16.

The super current which flows in the surface layer may be looked upon as generating a magnetic field in the interior that cancels the applied field there. Thus the encircling currents are called *shielding currents* in that they shield the interior from the applied field.

The case of the long superconducting cylinder shown in Fig. 5.17 in an external axial magnetic field B_0 is best treated in cylindrical coordinates, and, as we show in Chapter 9, Section III.B, the solutions are modified Bessel functions. In the limit $\lambda_L \ll R$, the surface layer approximates a planar layer, and the penetration is approximately exponential,

$$B_z(r) \approx B_0 \exp\left[\frac{-(R-r)}{\lambda_L}\right]$$

$$\lambda_L \ll R \qquad 0 < r < R, \quad (5.78)$$

as illustrated in Fig. 5.18, and expected on intuitive grounds. Figure 5.17 presents three-dimensional sketches of the fields and currents.

Another example to consider is a flow of transport current moving in a surface layer in the axial direction, as shown in Fig. 5.19. Note the magnetic field lines encircling the wire outside and decaying into the surface layer. Figures 5.17 and

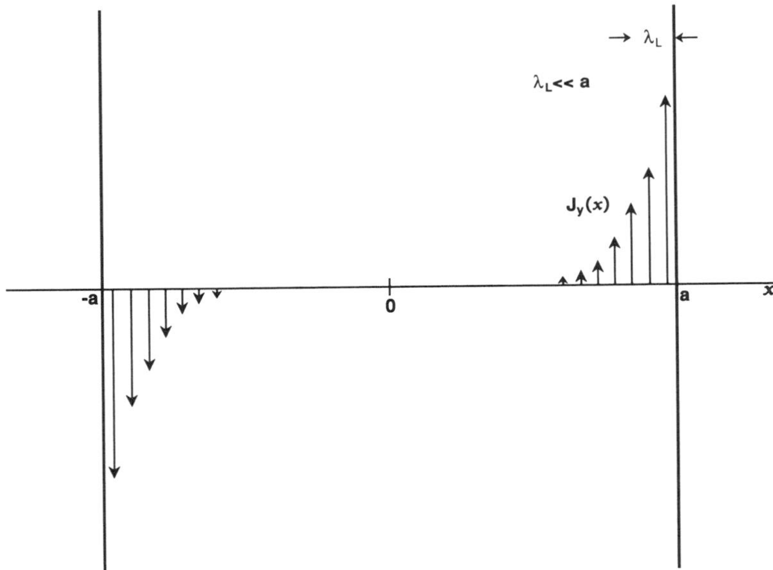

Figure 5.15 Current density $J_y(x)$ inside the superconducting slab for the case $a \gg \lambda_L$. This figure and Fig. 5.16 are antisymmetric about the origin $x = 0$.

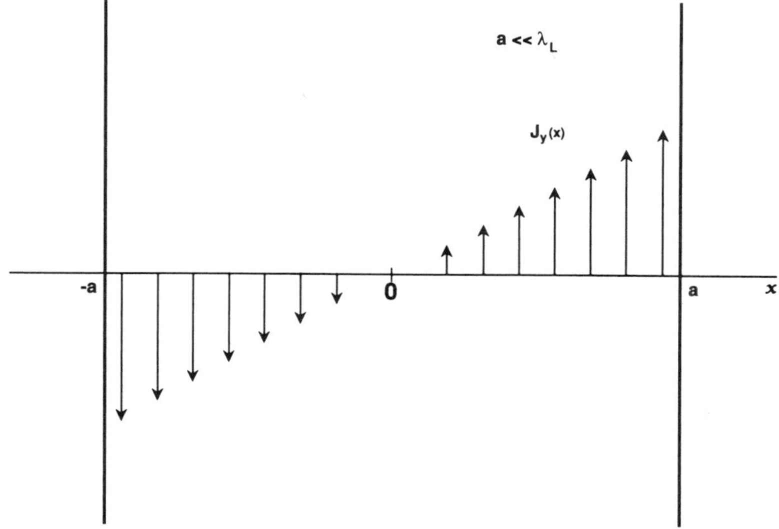

Figure 5.16 Current density $J_y(x)$ inside the superconducting slab for the case $a \ll \lambda_L$. Note that the magnitude of $\mathbf{J}(x)$ decreases linearly with distance x.

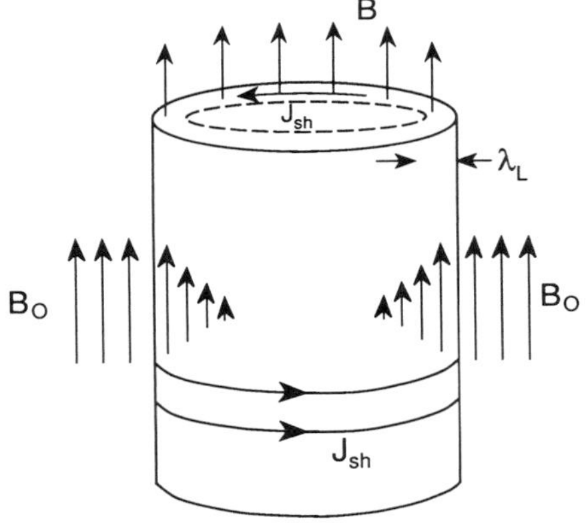

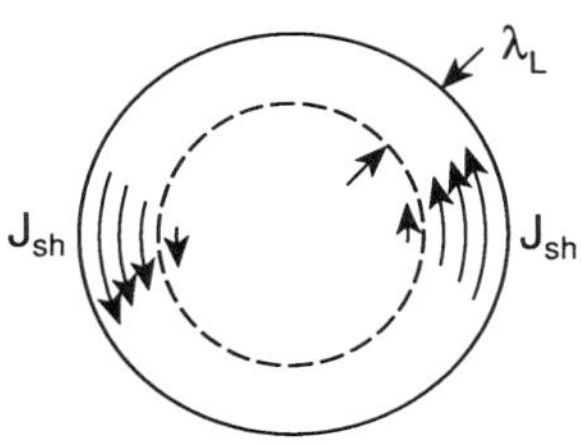

Figure 5.17 Sketch of a Type I superconducting cylinder in an external magnetic field $\mathbf{B}_{\text{app}} = \mathbf{B}_0$ directed along its axis, an arrangement referred to as parallel geometry. The penetration of the magnetic field **B** into the superconductor and the current flow $\mathbf{J}_{\text{sh}}$ near the surface are shown. The London penetration depth λ_L is also indicated.

5.19 compare these shielding and transport current cases. The figures are drawn for the limit $a \gg \lambda_L$ and apply to Type I superconductors that exclude the **B** field and current flow from the interior. They also apply to Type II superconductors in low applied fields below but near the transition temperature, $T < T_c$, since in this case the internal field $\mathbf{B}_{\text{in}}$ is small and the behavior approximates Type I.

XI. NORMALIZED GINZBURG–LANDAU EQUATIONS

In Section IV we wrote down the one-dimensional zero field GL equation normalized in terms of a dimensionless coordinate (5.25) and a dimensionless order parameter (5.23), and this simplified the process of finding a solution. Before proceeding to more complex cases it will be helpful to write down the general GL equations (Eqs. (5.6) and (5.10)) in fully normalized form in terms of the coherence length (5.26), London penetration depth (5.45), and flux quantum (5.36).

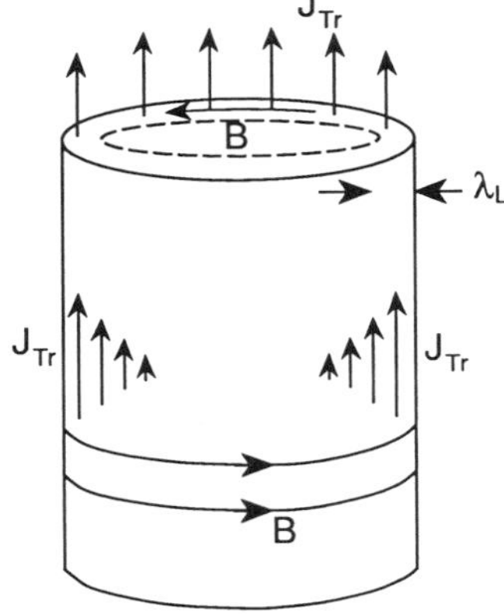

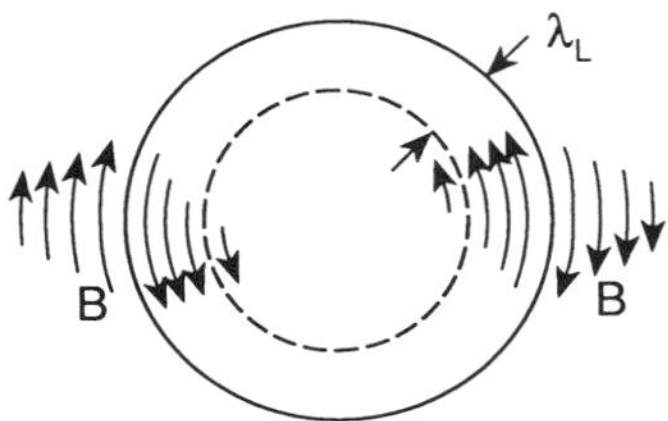

Figure 5.19 Sketch of the current density $\mathbf{J}_{\text{tr}}$ and magnetic field **B** near the surface of a Type I superconducting cylinder carrying a transport current.

To accomplish this we express the coordinates as dimensionless variables divided by the coherence length ξ. Thus we have, for example, ρ/ξ, ϕ, z/ξ in cylindrical coordinates, and use the differential operator symbols $\boldsymbol{\nabla}$ and ∇^2,

$$\boldsymbol{\nabla} \to \xi\, \boldsymbol{\nabla}, \tag{5.79}$$

$$\nabla^2 \to \xi^2 \nabla^2, \tag{5.80}$$

to designate differentiation with respect to these normalized coordinates. The order parameter ϕ is normalized as in Eq. (5.23),

$$\phi = \left(\frac{|a|}{b}\right)^{1/2} f, \tag{5.81}$$

the vector potential **A** is normalized in terms of the flux quantum Φ_0,

$$\mathbf{A} = \left(\frac{\Phi_0}{2\pi\xi}\right)\mathscr{A}, \tag{5.82}$$

and we make use of the Ginzburg–Landau parameter κ, which is defined as the ratio of the penetration depth to the coherence length,

$$\kappa = \frac{\lambda_{\text{L}}}{\xi}. \tag{5.83}$$

Using this notation the GL equations (5.6) and (5.10), respectively, expressed in the London–Landau gauge (5.7), $\boldsymbol{\nabla}\cdot\mathscr{A} = 0$, assume the normalized forms

$$-(i\boldsymbol{\nabla} - \mathscr{A})^2 f + f(1 - f^2) = 0, \tag{5.84a}$$

$$\kappa^2 \boldsymbol{\nabla} \times (\boldsymbol{\nabla} \times \mathscr{A}) + \tfrac{1}{2} i (f^* \boldsymbol{\nabla} f - f \boldsymbol{\nabla} f^*) + \mathscr{A} f^2 = 0. \tag{5.84b}$$

We can also define a dimensionless current density **j** from

$$\mathbf{J} = \frac{\Phi_0}{2\pi\lambda_{\text{L}}^2\, \xi\mu_0}\, \mathbf{j}, \tag{5.85}$$

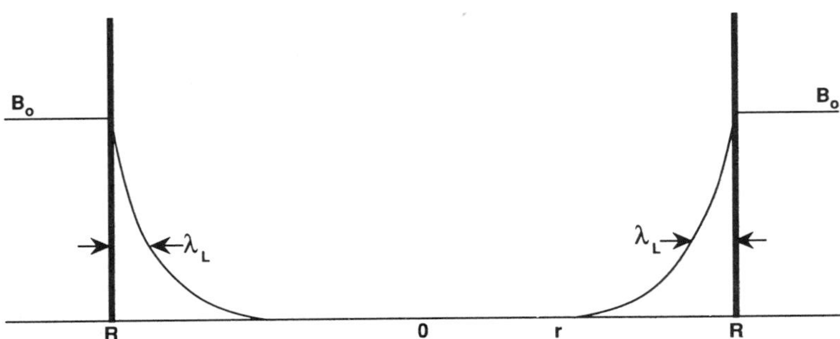

Figure 5.18 Magnetic field $\mathbf{B}(r)$ inside the Type I superconducting cylinder of Fig. 5.17 for the case $\lambda_{\text{L}} \ll R$.

which gives us

$$\mathbf{j} = \kappa^2 \boldsymbol{\nabla} \times (\boldsymbol{\nabla} \times \mathscr{A}). \qquad (5.86)$$

Equation (5.84a) can be expanded and Eq. (5.84b) written as follows:

$$\boldsymbol{\nabla}^2 f - 2i\mathscr{A}\cdot\boldsymbol{\nabla} f - \mathscr{A}^2 f + f(1 - f^2) = 0, \qquad (5.87a)$$

$$\mathbf{j} = -\tfrac{1}{2}i(f^*\boldsymbol{\nabla} f - f\boldsymbol{\nabla} f^*) - \mathscr{A}f^2. \qquad (5.87b)$$

Thus the coherence length, penetration depth, and flux quantum are the natural normalization parameters for transforming the GL equations into dimensionless form. In the following section we will use these normalized equations to elucidate various properties of superconductors.

XII. TYPE I AND TYPE II SUPERCONDUCTIVITY

In Chapter 11 we will discuss how bulk normal and superconducting phases coexist in equilibrium in an external magnetic field $\mathbf{B}_{\text{app}}$. We now wish to investigate this "mixed state" by considering a plane interface between a normal phase filling the left half-space $z < 0$ and a superconducting phase in the right half-space $z > 0$, as indicated in Fig. 5.3.

We expect the superconducting order parameter to vanish at the interface and, as we have seen, begin approaching its bulk equilibrium value within a characteristic length ξ. On the other hand, surface currents flow in a surface layer of width $\approx \lambda_{\text{L}}$, and full exclusion of magnetic flux occurs only deep inside the superconductor. Here we are interested in calculating the effect of the interface on the free energy of the state. This, in turn, leads naturally to the idea of a "surface tension" between the superconducting and normal phases.

Deep within either of the homogeneous phases the free-energy density at the critical field $B_{\text{app}} = \mu_0 H_{\text{c}}$ is equal to $G_{\text{n}0} + \frac{1}{2}\mu_0 H_{\text{c}}^2$. The free-energy density of the associated mixed state, including the interface, is

$$G(z) = \begin{cases} G_{\text{n}0} + \frac{1}{2}\mu_0 H_{\text{c}}^2 & z < 0 \\ G_{\text{n}0} - \frac{1}{2}b|\phi|^4 + \dfrac{1}{2\mu_0} \\ \qquad \times (B^2 - 2\mu_0^2 \mathbf{H}_{\text{c}}\cdot\mathbf{M}) & z > 0 \end{cases}, \qquad (5.88)$$

where we have used Eq. (5.4) subject to the minimization restriction (5.6) for the half-space $z > 0$. The surface tension σ_{ns} is defined as the difference in free energy per unit area between a homogeneous phase (either all normal or all superconducting) and a mixed phase. Therefore, we can write

$$\sigma_{\text{ns}} = \int dz \Big[-\tfrac{1}{2}b|\phi|^4 + \frac{1}{2\mu_0} \times (B^2 - 2\mu_0^2 \mathbf{H}_{\text{c}}\cdot\mathbf{M}) - \tfrac{1}{2}\mu_0 \mathbf{H}_{\text{c}}^2 \Big], \qquad (5.89)$$

since the integrand vanishes for $z < 0$. With the aid of the expression $B = \mu_0(H + M)$ this becomes

$$\sigma_{\text{ns}} = \int dz \big[-\tfrac{1}{2}b|\phi|^4 + \tfrac{1}{2}\mu_0 M^2 \big]. \qquad (5.90)$$

Note that as $z \to \infty$, $M \to -H_{\text{c}}$, and by Eq. (5.15), $|\phi|^2 \to |a|/b$, so from Eq. (5.19) the integrand vanishes far inside the superconductor where $z > \lambda_{\text{L}}$, and the principal contribution to the surface tension comes from the region near the boundary.

If $\sigma_{\text{ns}} > 0$, the homogeneous phase has a lower free energy than the mixed phase, and therefore the system will remain superconducting until the external field exceeds B_{c}, at which point it will turn completely normal. Superconductors of this variety are called Type I. However, if $\sigma_{\text{ns}} < 0$, the superconductor can lower its free energy by spontaneously developing normal regions that include some magnetic flux. Since the greatest saving in free energy is achieved by maximizing the surface area : flux ratio, these normal regions will be as small as possible consistent with the quantization of fluxoid. Thus the flux enters in discrete flux quanta.

Returning to Eq. (5.90), the first term represents the free energy gained by condensation into the superconducting state, while the second is the cost of excluding flux from the boundary layer. Roughly speaking, the order parameter attains its bulk value over a characteristic length ξ, while the super currents and magnetic flux are confined to a distance on the order of λ_L from the surface. If we define the dimensionless magnetization m by

$$M^2 = \left(\frac{a^2}{\mu_0 b}\right) m^2 \tag{5.91}$$

and make use of the dimensionless order parameter (5.81), Eq. (5.90) becomes

$$\sigma_{ns} = \frac{a^2}{2b} \int dz(-f^4 + m^2), \tag{5.92}$$

which can be written

$$\sigma_{ns} = \frac{a^2}{2b} \int dz[(1-f^4) - (1-m^2)]. \tag{5.93}$$

Equation (5.28) gives $f = \tanh(z/\sqrt{2}\,\xi)$ (see Problem 10 for an expression for the distance dependence of m). We can estimate σ_{ns} by observing that $f^4 = m^2 = 1$ in the bulk, that f^4 is small only over a distance on the order of ξ, and that m^2 is small only over a distance on the order of λ_L. This gives the approximate result

$$\sigma_{ns} \approx \frac{B_c^2}{2\mu_0}(\xi - \lambda_L), \tag{5.94}$$

where we have used Eq. (5.19). The value of the integral is the difference between the area under the two terms of the integrand, as shown plotted in Fig. 5.20. If $\xi > \lambda_L$, the surface tension is positive and we have Type I behavior. On the other hand, for $\xi < \lambda_L$, σ_{ns} is negative, and the superconductor is unstable with respect to the formation of a normal–superconducting interface, i.e., vortices form and Type II behavior appears.

We could also argue that λ_L is basically the width of an included vortex, i.e., the radius within which most of the flux is confined, and ξ is the distance over which the super electron density rises from $n_s = 0$ at the center of the vortex to its full bulk value, i.e., the distance needed to "heal the wound." A long coherence length ξ prevents the superconductor's n_s from rising quickly enough to provide the shielding current required to contain the flux, so no vortex can form.

Ginzburg and Landau (1950) showed that σ_{ns} vanishes for $\kappa = \lambda_L/\xi = 1/\sqrt{2}$, so as a convention we adopt the following criterion:

$$\begin{aligned} \kappa &< \frac{1}{\sqrt{2}} \qquad \text{Type I} \\ \kappa &> \frac{1}{\sqrt{2}} \qquad \text{Type II.} \end{aligned} \tag{5.95}$$

For Type II superconductors in very weak applied fields, $B_{app} \ll B_c$, the Meissner effect will be complete, but as B_{app} is increased above the lower critical field B_{c1}, where $B_{c1} < B_c$, vortices will begin to penetrate the sample. The magnetization of the sample then increases until the upper critical field B_{c2} is reached, at which point the vortex cores almost overlap and the bulk superconductivity is extinguished. Superconductivity may persist in a thin sheath up to an even higher critical field B_{c3}, where the sample goes completely normal.

XIII. UPPER CRITICAL FIELD B_{c2}

To calculate the upper critical field B_{c2} of a Type II superconductor we will examine the behavior of the normalized GL equation (5.87a) in the neighborhood of this field. For this case the order parameter is small and we can assume $B_{in} \approx B_{app}$. This suggests neglecting the nonlinear term

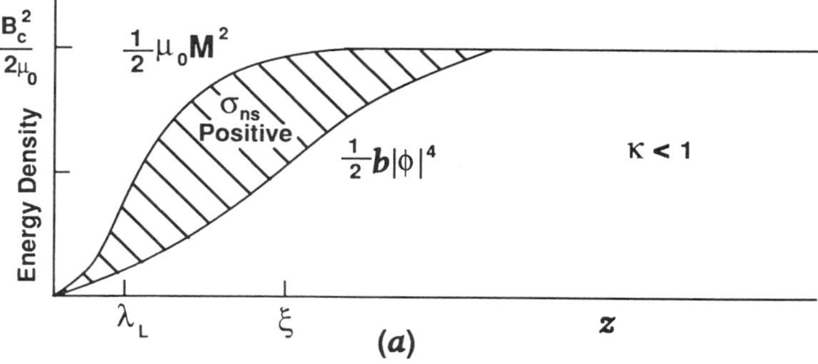

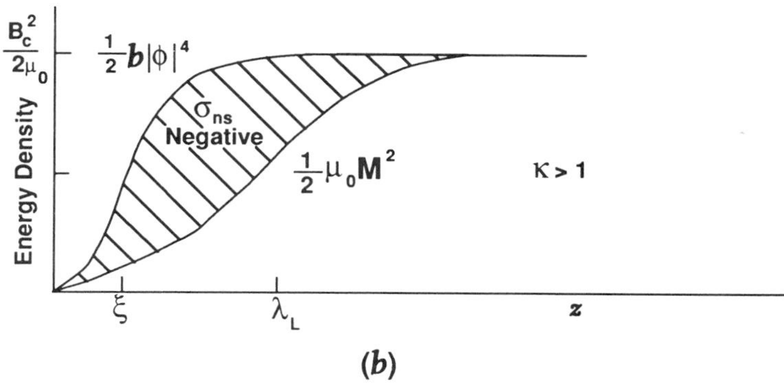

Figure 5.20 Order parameter $|\phi|$ and magnetization **M** inside a superconductor which is Type I ($\kappa < 1$, a) and inside a superconductor which is Type II ($\kappa > 1$, b), where $\kappa = \lambda_L/\xi$. The surface energy σ_{ns} is positive for the Type I case and negative for Type II.

f^3 in Eq. (5.87a) and following Eq. (5.41), taking for the normalized vector potential

$$\mathscr{A} = b_0 u, \tag{5.96}$$

where u is a dimensionless Cartesian coordinate perpendicular to the directions of both the applied field and the vector potential. From Eq. (5.82) we have for the magnitude of b_0,

$$b_0 = \left(\frac{2\pi\xi^2}{\Phi_0}\right) B_{\text{app}}. \tag{5.97}$$

Deep inside the superconductor the normalized order parameter f is independent of position so that the term $\mathscr{A} \cdot \nabla f$ in the GL equation (5.87a) is zero.

The linearized GL equation now has the form

$$\nabla^2 f - \mathscr{A}^2 f + f = 0. \tag{5.98}$$

This equation has bounded solutions only for special values of b_0. By analogy with the harmonic-oscillator Schrödinger equation from quantum mechanics, we can take $f \approx e^{\alpha u/2}$, which on substitution in (5.98) gives $\alpha = 2 = b_0$. Solutions can be found for larger values of b_0, but these are not of physical interest. Identifying the upper critical field with the applied field of Eq. (5.97) for this solution, we have

$$B_{c2} = \frac{\Phi_0}{\pi\xi^2}. \tag{5.99}$$

This expression has an appealing physical explanation. If we assume that in the upper critical field the cores of the vortices are nearly touching and that the flux contained in each core is $\approx \Phi_0$, the average magnetic field is $B_{c2} \approx \Phi_0/\pi\xi^2$.

Obviously, the existence of an upper critical field requires that $B_{c2} > B_c$, the thermodynamic critical field. By Eq. (9.12) the ratio of B_{c2} to B_c is

$$\frac{B_{c2}}{B_c} = \sqrt{2}\,\kappa. \qquad (5.100)$$

Therefore, for $\kappa < 1/\sqrt{2}$, $B_{c2} < B_c$ and no vortex state exists. In this way we can see that the condition for a superconductor to be Type II is $\kappa > 1/\sqrt{2}$.

XIV. QUANTUM VORTEX

For the case of a semi-infinite superconductor in a magnetic field it was found, by the arguments of Section VII, that the phase Θ of the order parameter remains fixed throughout the superconductor. Here we will consider a different geometry in which the phase of the order parameter is nontrivial.

In Type II superconductors it is observed that magnetic flux is completely excluded only for external fields $B < B_{c1}$. Above the lower-critical field, B_{c1}, flux penetrates in discrete flux quanta in the form of flux tubes, or vortices. In this section we will obtain approximate expressions for the fields associated with such a vortex, both in the core region and far outside the core. We assume that the external magnetic field B_{app} is applied along the z direction, parallel to the surface, and that currents flow at the surface, canceling the field inside. We are concerned with a vortex that is far enough inside the superconductor so that exponential decay of the external fields, as given by Eq. (5.50), drops essentially to zero.

States with more than one quantum of flux are also possible (Sachdev, 1992), but the energy scales as n^2, so single-flux quanta are energetically favored. This is because, according to Eq. (5.15), the parameter a scales as n, from Eq. (5.12) J scales as n, and from Eq. (5.11) B scales as n. Therefore n noninteracting vortices have n times the energy of a single vortex, but one multiquantum vortex has a magnetic energy $(nB)^2$, which scales as n^2.

A. Differential Equations

To treat this case we assume that there is no flux far inside the superconductor. If the applied field $B_{app} \approx B_{c1}$ a single quantum of flux Φ_0 enters in the form of a vortex with axis parallel to the applied field. The simplest assumption we could make about the shape of the vortex is to assume that it is cylindrically symmetric, so that in its vicinity the order parameter (5.81) has the form of Eq. (5.2), corresponding to

$$f(x,\Theta) = f(x)e^{i\Theta}, \qquad (5.101)$$

where $(x,\Theta) = (\rho/\xi, \Theta)$ are normalized polar coordinates. The vector potential has the form $\mathbf{A} = A(x)\hat{\mathbf{\Theta}}$, so that we can write for its normalized counterpart (5.82)

$$\mathscr{A}(x) = \mathscr{A}(x)\hat{\mathbf{\Theta}}. \qquad (5.102)$$

This is a two-dimensional problem since neither $f(x)$ nor $\mathscr{A}(x)$ have a z dependence. It is easy to show that $\nabla \times \mathscr{A}$ has only a z component (this we do by working out the curl operation in cylindrical coordinates), which is to be expected, since the magnetic field $\mathbf{B} = \nabla \times \mathbf{A}$ is known to be parallel to z.

If we substitute these functions in the two GL equations (5.84) and perform the Laplacian and double curl operations in cylindrical coordinates, we obtain

$$\left[\frac{1}{x}\frac{d}{dx}\left(x\frac{df}{dx}\right) - \frac{f}{x^2} + \left(\frac{2\mathscr{A}}{x}\right)f - \mathscr{A}^2 f\right] + f(1-f^2) = 0, \qquad (5.103)$$

$$\frac{d}{dx}\left[\frac{1}{x}\cdot\frac{d}{dx}(x\mathscr{A})\right] + \frac{1}{\kappa^2}f^2\left(\frac{1}{x} - \mathscr{A}\right) = 0, \qquad (5.104)$$

where $x = \rho/\xi$ and the current density equation (5.87b) becomes

$$j = f^2\left(\frac{1}{x} - \mathscr{A}\right) = 0. \qquad (5.105)$$

In constructing a solution to Eqs. (5.103) and (5.104) we must be guided by two requirements, first that the magnetic field and current density must be finite everywhere and, second, that the solution must have a finite free energy per unit length along the z-axis. If the free energy per unit length were infinite, the total free energy would diverge and render the solution unphysical. Further, we anticipate from the Meissner effect and Eqs. (5.50) and (5.52), that the magnetic field and the current density will decay exponentially far from the axis of the vortex.

B. Solutions for Small Distances

We seek to solve Eqs. (5.103) and (5.105) for the short-distance limit, namely in the core where $x < 1$. Since the first term in Eq. (5.105) has the factor $1/x$, it is necessary for the order parameter f to vanish as $x \to 0$ in order for the current density to remain finite in the core. By symmetry and continuity, the current density must vanish on the axis of the vortex, and it is expected to be small everywhere in the core. Maxwell's equation, Eq. (5.11), tells us that in this situation the magnetic field **B** is approximately constant in the core and we can write

$$\mathbf{A} = \tfrac{1}{2}B_0\,\rho\,\hat{\mathbf{\Theta}}, \qquad (5.106)$$

or, in dimensionless units, with $x = \rho/\xi$

$$\mathscr{A} = \tfrac{1}{2}xb_0\hat{\mathbf{\Theta}}, \qquad (5.107)$$

recalling Eq. 5.82, and from Eq. (5.97),

$$b_0 = \frac{2\pi\xi^2}{\Phi_0}B_0. \qquad (5.108)$$

If we now use this approximate solution (5.107) for the vector potential in Eq. (5.103) and neglect the f^3 term because we expect $f \ll 1$ in the core, we will have

$$\frac{1}{x}\frac{d}{dx}\left(x\frac{df}{dx}\right) + \left[(b_0+1) - \tfrac{1}{4}b_0^2x^2 - \frac{1}{x^2}\right]f = 0. \qquad (5.109)$$

This equation has exactly the form of Schrödinger's equation for the two-dimensional harmonic oscillator. We know from quantum mechanics texts (e.g., Pauling and Wilson, 1935, p. 105) that the constant term in the square brackets $(b_0 + 1)$ is the eigenvalue, the coefficient of the x^{-2} term is the z component of the angular momentum, i.e., $m = 1$, and, for the lowest eigenvalue, the coefficient of the x^2 term is related to the other two terms by the expression

$$(b_0 + 1) = 2(m+1)\left(\tfrac{1}{4}b_0^2\right)^{1/2}. \qquad (5.110)$$

Solving this for b_0 gives

$$b_0 = 1. \qquad (5.111)$$

Substituting Eq. (5.111) in Eq. (5.108) gives the magnetic field on the axis of the vortex:

$$B_0 = \frac{\Phi_0}{2\pi\xi^2}. \qquad (5.112)$$

The solution to the 'Schrödinger' equation, Eq. (5.109), is

$$f = Cxe^{-x^2/4}, \qquad (5.113)$$

where C is a constant. This function reaches its maximum at $x = \sqrt{2}$, which is outside the core, so, to a first approximation, f continuously increases in magnitude with increasing radial distance throughout the core region. This behavior is shown by the dashed curve in and near the core region of Fig. 5.21.

We can use the results of Problem 8 to obtain a better approximation to the vec-

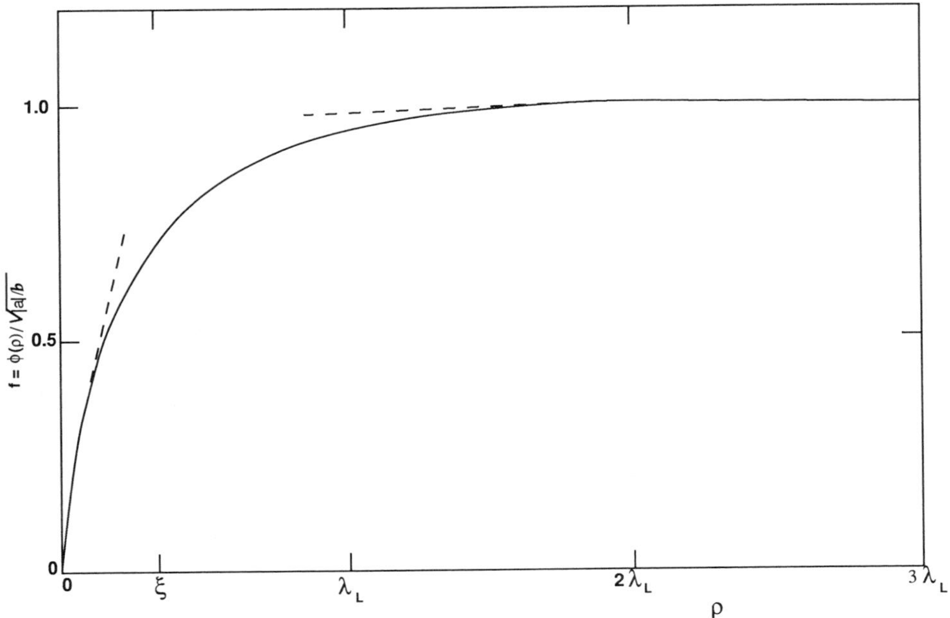

Figure 5.21 Dependence of the order parameter $|\phi|$ on distance ρ from the core of a vortex. The asymptotic behaviors near the core and far from the core are indicated by dashed lines.

tor potential and magnetic field in the core region,

$$\mathbf{A}(\rho) = \tfrac{1}{2}B_0\left[\rho - \alpha\left(\frac{\rho^3}{\xi^2}\right)\right]\hat{\boldsymbol{\Theta}}, \quad (5.114)$$

$$\mathbf{B}(\rho) = B_0\left[1 - 3\alpha\left(\frac{\rho}{\xi}\right)^2\right]\hat{\mathbf{k}}, \quad (5.115)$$

where $\alpha \ll 1$. These expressions are plotted as dashed curves in the core regions of Figs. 5.22 and 5.23, respectively.

C. Solutions for Large Distances

To obtain a solution far from the vortex core, $x \gg 1$, it is convenient to simplify Eqs. (5.103) to (5.105) by means of a change of variable,

$$\mathscr{A}' = \mathscr{A} - \frac{1}{x}, \quad (5.116)$$

which gives

$$\frac{d^2f}{dx^2} + \frac{1}{x}\cdot\frac{df}{dx} - \mathscr{A}'^2 f + f(1-f^2) = 0. \quad (5.117)$$

$$\frac{d^2\mathscr{A}'}{dx^2} + \frac{1}{x}\cdot\frac{d\mathscr{A}'}{dx} - \frac{\mathscr{A}'}{x^2} - \frac{f^2\mathscr{A}'}{\kappa^2} = 0, \quad (5.118)$$

$$j = -f^2\mathscr{A}', \quad (5.119)$$

where the derivatives have been multiplied out. It should be pointed out that the curl of $(1/x)\hat{\boldsymbol{\Theta}}$ vanishes in the region under consideration, so that $\nabla\times\mathscr{A}' = \nabla\times\mathscr{A}$, and hence the $1/x$ term of Eq. (5.116) does not contribute to the magnetic field (see, however, Problem 9).

For the approximation $f \approx 1$, the change of variable $x = \kappa y$ puts Eq. (5.118) into the form of a first-order ($n = 1$) modi-

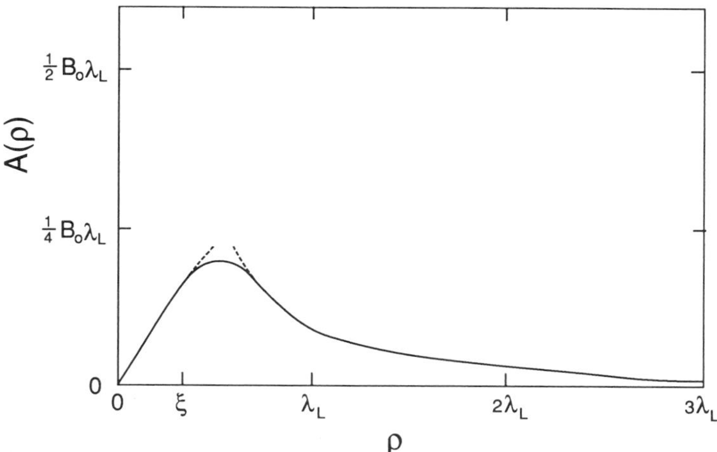

Figure 5.22 Distance dependence of the vector potential $\mathbf{A}(\rho)$ associated with a vortex in the notation of Fig. 5.21.

fied Bessel equation:

$$y^2 \frac{d^2\mathscr{A}'}{dy^2} + y\frac{d\mathscr{A}'}{dy} - (y^2+1)\mathscr{A}' = 0. \tag{5.120}$$

The solution to this equation which satisfies the boundary conditions $\mathscr{A}'(y) \to 0$ as $y \to \infty$ is

$$\mathscr{A}'(y) = A'_\infty K_1(y), \tag{5.121}$$

where $K_1(y)$ is a modified first-order Bessel function. For large distances, $x \gg \kappa$, it has the asymptotic form

$$\mathscr{A}'(x) = A_\infty \frac{e^{-x/\kappa}}{\sqrt{x}}, \tag{5.122}$$

where $A'_\infty = (2/\pi\kappa)^{1/2}A_\infty$. Figure 5.22 shows the asymptotic long-distance behavior of $A(\rho)$.

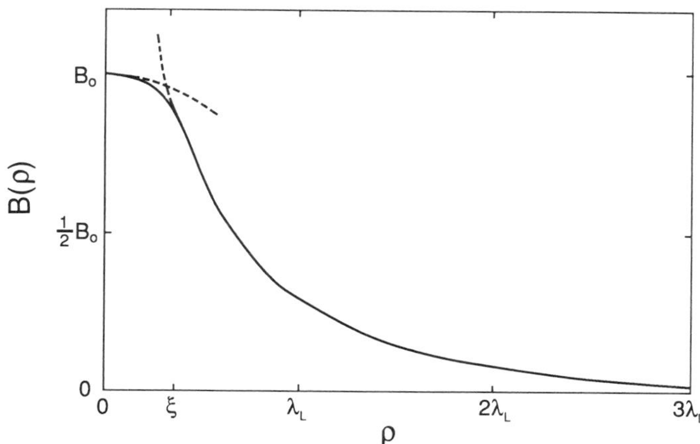

Figure 5.23 Distance dependence of the magnetic field $\mathbf{B}(\rho)$ encircling a vortex in the notation of Fig. 5.21.

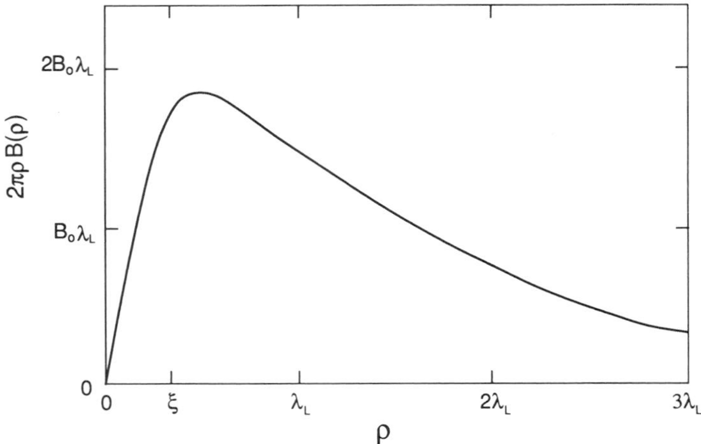

Figure 5.24 Distance dependence of the function $2\pi\rho\mathbf{B}(\rho)$ in the notation of Fig. 5.21. This function is proportional to the amount of magnetic flux at a distance ρ from the origin, and the integrated area under the curve is one fluxoid, $h/2e$.

Taking the curl $\mathbf{B} = \nabla \times \mathbf{A}$ in cylindrical coordinates (cf. Eq. (5.86)) provides the corresponding magnetic field for $x \gg \kappa$,

$$B_z(\rho) \approx \frac{e^{-\rho/\lambda_L}}{(\rho/\xi)^{1/2}} \qquad \rho \gg \lambda_L, \quad (5.123)$$

where we have restored the original coordinate $\rho = x\xi = y\lambda_L$. Figure 5.23 shows a plot of $B_z(\rho)$ versus ρ for large ρ and indicates the $\rho = 0$ value of Eq. (5.115).

To find the radial dependence of the order parameter far from the core, where the material is in the superconducting state, we have $f \approx 1$, so we can write

$$f(x) = 1 - g(x), \qquad (5.124)$$

where $g(x) \ll 1$, and hence $f(1 - f^2) \approx 2g$. As a result Eq. (5.117) assumes the form

$$\frac{d^2}{dx^2} g(x) + \frac{1}{x} \cdot \frac{d}{dx} g(x) + \mathscr{A}'^2 - 2g(x) = 0. \qquad (5.125)$$

Far from the core, $x \gg \kappa$, the behavior of $g(x)$ for $\kappa \gg 1$ is determined by that of $\mathscr{A}'(x)$, and we have

$$g(x) \approx g_\infty \frac{e^{-2x/\kappa}}{x}, \qquad (5.126)$$

where g_∞ is positive.

Comparing Figs. 5.22 and 5.23 we see that A increases in the core region, reaches a maximum near the inflection point of the B curve, and decreases outside the core. The quantity $2\pi\rho\mathbf{B}(\rho)$ is proportional to the amount of magnetic flux at a particular distance from the vortex axis; it is shown plotted against ρ in Fig. 5.24. The integrated area under this curve equals one fluxoid, $h/2e$.

FURTHER READING

The GL theory was first proposed by Ginzburg and Landau in 1950. Its value became more apparent after Gor'kov (1959) showed that it is a limiting case of the BCS theory. The theory was extended to the limit of high κ by Abrikosov (1957) in the same year that the BCS theory was proposed. The London and London (1935), London (1950), and related Pippard (1953) equations follow from the GL theory.

Several recent texts have included discussions of the Ginzburg–Landau theory, such as those by Lynn (1990a, Chapter 2), Belitz (1990), Orlando and Delin

(1991, Chapter 10), Tilley and Tilley (1986, Chapter 8), Tinkham (1985, Chapter 4), and Van Duzer and Turner (1981, Chapters 7 and 8). More advanced treatments are given by de Gennes (1966) and Fetter and Walecka (1971). Rochlin (1975) presented a simplified version of approximate solutions to the GL equations.

The Ginzburg–Landau theory appeared over 40 years ago, and the extent of its literature is enormous. We conclude this chapter by mentioning some of the recent articles in which the theory has been applied: (a) high-temperature superconductors (Bulaevskii, 1991; Bulaevskii and Vagner, 1991; Busiello and Uzunov, 1990; Chela-Flores *et al.*, 1988; Das, 1989; De Cesare, 1991; Doniach, 1990; Eab and Tang, 1989; Hao and Clem, 1992; Hikami and Fujita, 1990a, b; Kadowaki *et al.* 1995; Kapitulnik *et al.*, 1988; Marsigeio, 1992; Tewordt *et al.*, 1989); (b) anisotropic superconductors with unconventional electron pairing state or unconventional order parameter (Blagoeva *et al.*, 1990; Blatter *et al.*, 1992; Busiello *et al.*, 1991; Choi and Muzikar, 1989; Choi and José, 1989; Palumbo *et al.*, 1990b); Suwa *et al.*, 1989; Yip *et al.*, 1990; B. R. Zhao *et al.*, 1991; Z. Zhou *et al.*, 1991); (c) magnetic properties (Akera *et al.*, 1991; Arias and Joannopoulos, 1989; Busiello *et al.*, 1991; De Cesare, 1991; Hikami *et al.*, 1991; Klein and Pöttinger, 1991; Palumbo *et al.*, 1990a; Parmar and Bhattacharjee, 1992; Shrivastava and Sinha, 1984; Sigrist *et al.*, 1989); (d) fluctuations (Blagoeva *et al.*, 1989, 1990; Kapitulnik *et al.*, 1988; Lobb, 1987; Ullah and Dorsey, 1991); and (e) tunneling (Yip *et al.*, 1990).

PROBLEMS

1. Derive the first GL equation, Eq. (5.6), from the Gibbs free energy integral (5.4).
2. Show that minimizing the term $B^2/2\mu_0$ with respect to the vector potential **A** in Eq. (5.4) gives the expression $\nabla^2\mathbf{A}/\mu_0$ that is found in Eq. (5.10). Hint: write

$$\frac{\partial}{\partial A_n(x)}\left[\epsilon_{ijk}\,\epsilon_{ilm}\int \partial_j A_k(y)\,\partial_l A_m(y)\,d^2y\right],$$

bring the partial differentiation inside the integral, and integrate by parts.
3. Show that $f(\eta)=\coth(\eta/\sqrt{2})$ is also a solution to Eq. (5.27), and explain why it is not used.
4. Show that Eq. (5.17) is consistent with Eq. (2.69) in the superconducting region near T_c, and express the ratio a_0/b_0 in terms of the density n of conduction electrons.
5. Derive Eqs. (5.71a) and (5.71b).
6. Justify Eq. (5.97): $b_0=(2\pi\xi^2/\Phi_0)B_{\text{app}}$.
7. Show that $\nabla\times(\nabla\times\mathbf{A})=-\nabla^2\mathbf{A}$ in Cartesian coordinates, assuming the London–Landau gauge. Why is this not true when the coordinate system is non-Cartesian?
8. Assume the following power series solutions to Eqs. (5.103) and (5.104) in the region of the core:

$$f(x)\approx\Sigma a_n x^n \qquad x\ll 1,$$

$$f(x)\approx\Sigma f_n x^n \qquad x\ll 1.$$

(a) Show that the lowest-order terms that exist are f_1 and a_1, that the even order terms vanish, and that

$$a_3=-\frac{f_1^2}{8\kappa^2},$$

$$f_3=-\tfrac{1}{4}(a_1+\tfrac{1}{2})f_1.$$

(b) Show that the distance dependence of the order parameter, vector potential, magnetic field, and current density in the neighborhood of the origin are given by

$$|\phi(\rho)|\approx\left(\frac{|a|}{b}\right)^{1/2}\left[f_1\left(\frac{\rho}{\xi}\right)-|f_3|\left(\frac{\rho}{\xi}\right)^3\right] \qquad \rho\ll\xi,$$

$$A_\phi(\rho)\approx\left(\frac{\Phi_0}{2\pi\xi}\right)\left[a_1\left(\frac{\rho}{\xi}\right)-|a_3|\left(\frac{\rho}{\xi}\right)^3\right] \qquad \rho\ll\xi,$$

$$B_z(\rho)\approx\left(\frac{\Phi_0}{\pi\xi^2}\right)\left[a_1-2|a_3|\left(\frac{\rho}{\xi}\right)^2\right] \qquad \rho\ll\xi.$$

(c) Show that the expression for $|\phi(\rho)|$ agrees with Eq. (5.113).

9. Show that $\oint \mathscr{A}' \cdot d\mathbf{l} = 0$, whereas $\oint \mathscr{A} \cdot d\mathbf{l} = 2\pi$ for contours at infinity. What is the significance of the $1/x$ term in Eq. (5.116)?
10. Show that the dimensionless magnetization m defined by Eq. (5.125) can be written

$$m = \frac{M}{H_c},$$

and has the distance dependence

$$m = -(1 - e^{-z/\lambda_L}).$$

in Eq. (5.93) (assume zero demagnetization factor).
11. Derive Eq. (5.123) for the magnetic field far from a vortex. Find the first higher-order term that is neglected in writing out this expression.

6

BCS Theory

I. INTRODUCTION

Chapter 5 presents the Ginzburg–Landau (GL) theory, which originated in 1950. Despite the fact that it is a phenomenological theory, it has had surprising success in explaining many of the principal properties of superconductors. Nevertheless, it has limitations because it does not explain the microscopic origins of superconductivity. In 1957 Bardeen, Cooper, and Schrieffer (BCS) proposed a microscopic theory of superconductivity that predicts quantitatively many of the properties of elemental superconductors. In addition, the Landau–Ginzburg theory can be derived from the BCS theory, with the added bonus that the charge and mass of the "particle" involved in the superconducting state emerge naturally as $2e$ and $2m_e$, respectively.

With the discovery of the heavy fermion and copper-oxide superconductors it is no longer clear whether the BCS theory is satisfactory for all classes of superconductors. The question remains open, although there is no doubt that many of the properties of high-temperature superconductors are consistent with the BCS formalism.

To derive the BCS theory it is necessary to use mathematics that is more advanced than that which is employed elsewhere in this book, and the reader is referred to standard quantum mechanics texts for the details of the associated derivations. If the chapter is given a cursory initial reading without working out the intermediate steps in the development, an overall picture of BCS can be obtained. For didactic purposes we will end the chapter by describing the simplified case of a square well electron–electron interaction potential, which is also the case treated in the original formulation of the theory.

II. COOPER PAIRS

One year before publication of the BCS theory, Cooper (1956) demonstrated that the normal ground state of an electron gas is unstable with respect to the formation of "bound" electron pairs. We have used quotation marks here because these electron pairs are not bound in the ordinary sense, and the presence of the filled Fermi sea is essential for this state to exist. Therefore this is properly a many-electron state.

In the normal ground state all one-electron orbitals with momenta $k < k_{\mathrm{F}}$ are occupied, and all the rest are empty. Now, following Cooper, let us suppose that a weak attractive interaction exists between the electrons. The effect of the interaction will be to scatter electrons from states with initial momenta (k_1, k_2) to states with momenta (k_1', k_2'). Since all states below the Fermi surface are occupied, the final momenta (k_1', k_2') must be above k_{F}. Clearly, these scattering processes tend to increase the kinetic energy of the system. However, as we shall now see, the increase in kinetic energy is more than compensated by a decrease in the potential energy if we allow states above k_{F} to be occupied in the many-electron ground state.

We begin by considering the Schrödinger equation for two electrons interacting via the potential V,

$$\left[-\frac{\hbar^2}{2m}(\nabla_1^2+\nabla_2^2)+V(r_1-r_2)\right]\Psi(r_1,r_2) = (E+2E_{\mathrm{F}})\Psi(r_1,r_2). \quad (6.1)$$

In (6.1) the spin part of the wavefunction has been factored out and the energy eigenvalue E is defined relative to the Fermi level ($2E_{\mathrm{F}}$). Most superconductors are spin-singlet, so the orbital part to the wavefunction, $\Psi(r_1, r_2)$, must be symmetric.

As with any two-body problem, we begin by defining the center of mass coordinate,

$$R = \tfrac{1}{2}(r_1 + r_2), \quad (6.2)$$

and the relative coordinate,

$$r = r_1 - r_2. \quad (6.3)$$

In terms of these coordinates (6.1) becomes

$$\left[-\frac{\hbar^2}{4m}\nabla_R^2 - 2\frac{\hbar^2}{2m}\nabla_r^2\right]\Psi(R,r) + V(r)\Psi(R,r) = (E+2E_{\mathrm{F}})\Psi(R,r). \quad (6.4)$$

The center of mass and relative coordinates now separate and we can write

$$\Psi(R,r) = \Phi(R)\Psi(r). \quad (6.5)$$

$\Phi(R)$ is simply a plane wave,

$$\Phi(R) = e^{iK\cdot R}, \quad (6.6)$$

while for the relative coordinate wavefunction $\Psi(r)$ we have

$$\left[-2\frac{\hbar^2}{2m}\nabla_r^2 + V(r)\right]\Psi(r) = \left(E + 2E_{\mathrm{F}} - \frac{\hbar^2K^2}{4m}\right)\Psi(r). \quad (6.7)$$

Since we are interested in the ground state, we can set $K = 0$. There are solutions for $K \neq 0$ that lie close to the $K = 0$ states and are needed to describe states in which a persistent current flows.

We now express $\Psi(r)$ as a sum over states with momenta $p > k_{\mathrm{F}}$,

$$\Psi(r) = \frac{1}{\sqrt{V}}\sum_p{}' a(p)e^{ip\cdot r}. \quad (6.8)$$

In (6.8) Σ_p' denotes a summation over all $|p| > k_{\mathrm{F}}$. Substitution of the expression for

$\Psi(r)$ from (6.8) in (6.7) then gives the Schrödinger equation in momentum space,

$$\left[2(\mathscr{E}_p - E_\mathrm{F}) - E\right]a(p) + \sum_{p'} V(p,p')a(p') = 0, \quad (6.9)$$

where

$$V(p,p') = \frac{1}{V}\int d^3r e^{-i(p-p')\cdot r}V(r). \quad (6.10)$$

In order to simplify the solution of (6.9) we assume that

$$V(p,p') = \begin{cases} -V_0 & 0 \le \mathscr{E}_p - E_\mathrm{F} \le \hbar\omega_\mathrm{D} \\ & \text{and } 0 \le \mathscr{E}_{p'} - E_\mathrm{F} \le \hbar\omega_\mathrm{D} \\ 0 & \text{otherwise.} \end{cases} \quad (6.11)$$

In (6.11), $\hbar\omega_\mathrm{D}$ is a typical phonon energy, which reflects the idea that attraction between electrons arises via exchange of virtual phonons.

With the potential (6.11) the interaction term in (6.9) becomes

$$\sum_{p'}{}' V(p,p')a(p') = -V_0 K \Theta(\hbar\omega_\mathrm{D} - E_\mathrm{F} - \mathscr{E}_p), \quad (6.12)$$

where $\Theta(x)$ is the ordinary step function and

$$K = \sum_p{}' a(p) \quad (6.13)$$

is a constant. Solving (6.9) for $a(p)$, we have

$$a(p) = \frac{V_0 K}{2(\mathscr{E}_p - E_\mathrm{F}) - E} \times \Theta(\hbar\omega_\mathrm{D} - E_\mathrm{F} - \mathscr{E}_p). \quad (6.14)$$

Note that our Cooper pair involves momenta only in the narrow region $\mathscr{E}_p - E_\mathrm{F} \le \hbar\omega_\mathrm{D}$ just above the Fermi surface. We can now self-consistently evaluate the constant K in (6.13),

$$K = V_0 K \sum_p{}' \frac{1}{2(\mathscr{E}_p - E_\mathrm{F}) - E} \times \Theta(\hbar\omega_\mathrm{D} - E_\mathrm{F} - \mathscr{E}_p). \quad (6.15)$$

If we assume that $K \ne 0$, this leads to an implicit equation for the eigenvalue E,

$$1 = V_0 \sum_p{}' \frac{1}{2(\mathscr{E}_p - E_\mathrm{F}) - E} \times \Theta(\hbar\omega_\mathrm{D} - E_\mathrm{F} - \mathscr{E}_p). \quad (6.16)$$

The sum over the momenta can be expressed as an integral over the energies in terms of the density of states $D(\mathscr{E})$. Since typically $\hbar\omega_\mathrm{D} \ll E_\mathrm{F}$, $D(\mathscr{E})$ is well approximated inside the integral by its value at the Fermi surface, $D(E_\mathrm{F})$. Thus we have

$$1 = V_0 D(E_\mathrm{F}) \int_{E_\mathrm{F}}^{E_\mathrm{F} + \hbar\omega_p} \frac{1}{2(\mathscr{E} - E_\mathrm{F}) - E}\, d\mathscr{E} = \tfrac{1}{2} V_0 D(E_\mathrm{F}) \ln\left(\frac{E - 2\hbar\omega_\mathrm{D}}{E}\right). \quad (6.17)$$

Solving for E we have

$$E = -\frac{2\hbar\omega_\mathrm{D}}{\exp[2/V_0 D(E_\mathrm{F})] - 1}. \quad (6.18)$$

In the weak-coupling limit $V_0 D(E_\mathrm{F}) \ll 1$ and the exponential dominates the denominator in (6.18) so that

$$E \approx -2\hbar\omega_\mathrm{D} \exp\left(-\frac{2}{V_0 D(E_\mathrm{F})}\right). \quad (6.19)$$

This result is remarkable in several ways. First, it tells us that the pair state we have constructed will always have a lower energy than the normal ground state no matter how small the interaction V_0. This is why we say the normal ground state is unstable with respect to the formation of Cooper pairs. Second, we see in (6.18) a hierarchy of very different energy scales,

$$E_\mathrm{F} \gg \hbar\omega_\mathrm{D} \gg |E|, \quad (6.20)$$

which, if we assume that $k_\mathrm{B} T_c \cong |E|$, explains why the superconducting transition temperature is so small compared with the Debye temperature,

$$\theta_D = \frac{\hbar \omega_D}{k_B}.$$

If one Cooper pair lowers the ground state by $-|E|$, then, clearly many pairs will lower the energy even further, and one might be tempted to conclude that all the electrons should pair up in this fashion. Such a state would then resemble a Bose–Einstein condensate of Cooper pairs. However, we must keep in mind that if we do away entirely with the normal Fermi sea the state we have constructed collapses. We can use these intuitive ideas to guide our thinking, but to arrive at the true BCS ground state we must go beyond simple one- and even two-electron pictures and realize that the superconducting state is a highly correlated many-electron state.

III. BCS ORDER PARAMETER

In this section and those that follow we will present the formal details of the BCS theory. The most natural mathematical language to use in this case is "second quantization," where all the observables are expressed as functions of the fundamental fermion creation and annihilation operators, which act on a special Hilbert space called Fock space.

The fermion creation operator $\psi_\sigma^\dagger(x)$ "creates" a fermion (i.e., electron) with z component of spin $\sigma = +\frac{1}{2}$ or $-\frac{1}{2}$, localized at the point x. Similarly, the Hermitian conjugate, or annihilation, operator $\psi_\sigma(x)$ removes an electron with spin σ at point x.

Two advantages are gained with the use of the language of second quantization. First, from the point of view of quantum statistical mechanics, the natural ensemble to employ in a many-body theory is the grand canonical ensemble, where the number of particles is not fixed but controlled by the chemical potential μ. Second, the complications that arise from the exclusion principle embodied in the antisymmetry of the many-fermion wavefunction under particle exchange are naturally and economically taken into account algebraically through the canonical anticommutation relations

$$\begin{aligned}\{\psi_\alpha(x), \psi_\beta^\dagger(y)\} &= \psi_\alpha(x)\psi_\beta^+(y) \\ &\quad + \psi_\beta^+(y)\psi_\alpha(x) \\ &= \delta_{\alpha\beta}\,\delta(x-y) \quad (6.21a)\end{aligned}$$

and

$$\begin{aligned}\{\psi_\alpha(x), \psi_\beta(y)\} &= \psi_\alpha(x)\psi_\beta(y) \\ &\quad + \psi_\beta(y)\psi_\alpha(x) \\ &= 0. \quad (6.21b)\end{aligned}$$

As in any theory of a phase transition the first task is to identify the "order parameter." An order parameter is an operator whose thermal average vanishes in the high-temperature, disordered (or symmetric) phase and assumes a nonzero value in the low-temperature, ordered, or broken-symmetry phase. From the demonstration by Cooper that the ground state of the normal Fermi gas is unstable to the formation of "pairs" and the idea that a macroscopic number of such pairs could undergo Bose condensation into a single orbital level, we take as the superconducting order parameter the "pair function"

$$\Phi_{\alpha\beta}(x, y) = \left\langle \psi_\alpha^\dagger(x)\psi_\beta^\dagger(y)\right\rangle. \quad (6.22)$$

Note that the complex conjugate of the pair function is

$$\Phi_{\alpha\beta}^*(x, y) = \left\langle \psi_\beta(y)\psi_\alpha(x)\right\rangle. \quad (6.23)$$

The angular brackets in (6.22) and (6.23) indicate a thermal average taken over all states weighted by their Boltzmann factors. The antisymmetry of the order parameter under change of coordinates and spin labels follows from the anticommutation relation (6.21b):

$$\Phi_{\alpha\beta}(x, y) = -\Phi_{\beta\alpha}(y, x). \quad (6.24)$$

The Hamiltonian for the electron system is

$$H = \int dx \left[\frac{\hbar^2}{2m} \nabla\psi_\sigma^\dagger \cdot \nabla\psi_\sigma - \mu\psi_\sigma^\dagger \psi_\sigma \right] + \tfrac{1}{2}\int dx \int dy \psi_\sigma^\dagger(x)\psi_{\sigma'}^\dagger(y) \times V(x-y)\psi_{\sigma'}(y)\psi_\sigma(x), \quad (6.25)$$

where the first term is the kinetic energy, μ is the chemical potential, and $V(x-y) = V(y-x)$ is the interaction potential between electrons. The spin label subscripts σ, σ' are summed over wherever they are repeated (summation convention).

The potential in (6.25) is taken to be a function of $x-y$ only, which makes H translationally invariant. In real solids, of course, the potential is not so simple, but the theory is much easier to work out if we assume translational symmetry, and none of the essential physics is lost.

The BCS theory is based on the reasonable assumption that the fluctuation in the order parameter is much smaller than the order parameter itself for $T < T_c$. We first write the trivial identity

$$\psi_\alpha^\dagger(x)\psi_\beta^\dagger(y) = \left\langle \psi_\alpha^\dagger(x)\psi_\beta^\dagger(y)\right\rangle + \left[\psi_\alpha^\dagger(x)\psi_\beta^\dagger(y) - \left\langle \psi_\alpha^\dagger(x)\psi_\beta^\dagger(y)\right\rangle \right]. \quad (6.26)$$

The operator in square brackets in (6.26) is the fluctuation in the order parameter. This is a mean field approach in which the first term on the right side plays the role of the mean field, and the second term is the first-order deviation from the mean field. Inserting (6.26) and its Hermitian conjugate in the interaction term in the Hamiltonian, we have

$$\begin{aligned}
&\tfrac{1}{2}\int dx \int dy \psi_\sigma^\dagger \psi_{\sigma'}^\dagger(y) V(x-y)\psi_{\sigma'}(y)\psi_\sigma(x) \\
&= \tfrac{1}{2}\int dx \int dy \langle \psi_\sigma^\dagger(x)\psi_{\sigma'}^\dagger(y)\rangle \times V(x-y)\langle \psi_{\sigma'}(y)\psi_\sigma(x)\rangle \\
&\quad + \tfrac{1}{2}\int dx \int dy \langle \psi_\sigma^\dagger(x)\psi_{\sigma'}^\dagger(y)\rangle V(x-y) \times [\psi_{\sigma'}(y)\psi_\sigma(x) - \langle \psi_{\sigma'}(y)\psi_\sigma(x)\rangle] \\
&\quad + \tfrac{1}{2}\int dx \int dy \, [\psi_\sigma^\dagger(x)\psi_{\sigma'}^\dagger(y) - \langle \psi_\sigma^\dagger(x)\psi_{\sigma'}^\dagger(y)\rangle] \times V(x-y)\langle \psi_{\sigma'}(y)\psi_\sigma(x)\rangle \\
&\quad + \tfrac{1}{2}\int dx \int dy \times [\psi_\sigma^\dagger(x)\psi_{\sigma'}^\dagger(y) - \langle \psi_\sigma^\dagger(x)\psi_{\sigma'}^\dagger(y)\rangle] \times V(x-y) \times [\psi_{\sigma'}(y)\psi_\sigma(x) - \langle \psi_{\sigma'}(y)\psi_\sigma(x)\rangle].
\end{aligned} \quad (6.27)$$

In (6.27) we have ordered the right side in ascending powers of the fluctuation. If we keep all four terms we will be back where we started—an intractable problem. However, if the last term is dropped because it is second order in the fluctuations, and hence, by assumption, very small, we will have

$$H \cong H_1 + H_{\mathrm{BCS}}, \quad (6.28)$$

where H_1, given by

$$H_1 = -\tfrac{1}{2}\int dx \int dy \langle \psi_\sigma^\dagger(x)\psi_{\sigma'}^\dagger(y)\rangle \times V(x-y)\langle \psi_{\sigma'}(y)\psi_\sigma(x)\rangle, \quad (6.29)$$

is just a constant and

$$\begin{aligned}
H_{\mathrm{BCS}} = &\int dx \left[\frac{\hbar^2}{2m} \nabla\psi_\sigma^\dagger \cdot \nabla\psi_\sigma(x) - \mu\psi_\sigma^\dagger(x)\psi_\sigma(x) \right] \\
&+ \tfrac{1}{2}\int dx \int dy \big[\Phi_{\alpha\beta}(x,y) \times V(x-y)\psi_\beta(y)\psi_\alpha(x) \\
&+ \psi_\alpha^\dagger(x)\psi_\beta^\dagger(y) V(x-y)\Phi_{\alpha\beta}^*(x,y)\big]
\end{aligned} \quad (6.30)$$

is the Hamiltonian that describes the superconducting system.

We know from statistical mechanics that the Helmholtz free energy F of a system is given by

$$F = -\frac{1}{\beta}\ln \operatorname{Tr} \exp(-\beta H), \quad (6.31)$$

where $\beta = 1/k_B T$, Tr denotes a sum over a complete set of states in the Hilbert space, and H is the Hamiltonian of the system. Since H_1 is a c-number (i.e. a complex number), it factors out of the trace, to give

$$F = -\tfrac{1}{2}\int dx \int dy \Phi_{\alpha\beta}(x,y) \times V(x-y)\Phi^*_{\alpha\beta}(x,y) - \frac{1}{\beta}\ln[\operatorname{Tr}(\exp(-\beta H_{\text{BCS}}))]. \quad (6.32)$$

The free energy must be a minimum with respect to variations of the order parameter, so we have

$$\frac{\delta F}{\delta \Phi^*_{\alpha\beta}(x,y)} = -\tfrac{1}{2}\Phi_{\alpha\beta}(x,y)V(x-y) + \frac{\operatorname{Tr}[\exp(-\beta H_{\text{BCS}})\delta H_{\text{BCS}}/\delta\Phi^*_{\alpha\beta}(x,y)]}{\operatorname{Tr}\exp(-\beta H_{\text{BCS}})} \quad (6.33)$$

$$= 0. \quad (6.34)$$

The variational derivative of H_{BCS} with respect to Φ is easily calculated, and we find that (6.12) is satisfied if

$$\Phi_{\alpha\beta}(x,y) = \frac{\operatorname{Tr}\left[\exp(-\beta H_{\text{BCS}})\psi^\dagger_\alpha(x)\psi^\dagger_\beta(y)\right]}{\operatorname{Tr}\exp(-\beta H_{\text{BCS}})}. \quad (6.35)$$

Comparing Eq. (6.21) with (6.35) we see that within an approximation that neglects terms that are quadratic in fluctuations of the order parameter, Φ is given self-consistently by (6.30) and (6.35).

IV. GENERALIZED BCS THEORY

We introduce the gap function

$$\Delta_{\alpha\beta}(x,y) \equiv \left\langle \psi^\dagger_\alpha(x)\psi^\dagger_\beta(y)\right\rangle V(x-y) \quad (6.36)$$

with the property

$$\Delta^*_{\alpha\beta}(x,y) = \left\langle \psi_\beta(y)\psi_\alpha(x)\right\rangle V(x-y), \quad (6.37)$$

so that

$$H_{\text{BCS}} = \int dx\left(\frac{\hbar^2}{2m}\nabla\psi^\dagger_\sigma\cdot\nabla\psi_\sigma - \mu\psi^\dagger_\sigma\psi_\sigma\right) + \tfrac{1}{2}\int dx\int dy \times\left[\Delta_{\alpha\beta}(x,y)\psi_\beta(y)\psi_\alpha(x) + \Delta^*_{\alpha\beta}(x,y)\psi^\dagger_\alpha(x)\psi^\dagger_\beta(y)\right]. \quad (6.38)$$

Note that H_{BCS} in (6.38) is bilinear in the field variables. Such a Hamiltonian can be diagonalized by a suitable linear transformation in much the same way as the simple harmonic oscillator can be solved by introducing the well-known raising and lowering operators (which are linear combinations of the operators p and x). In the present context the transformation is called the Bogoliubov transformation, and while the algebra may appear formidable at times, conceptually it is no more complex than the algebraic solution of a simple harmonic oscillator. Ultimately we will reduce (6.38) to a collection of noninteracting fermions, or quasiparticles, and determine their excitation spectrum $\lambda(k)$.

To write this Hamiltonian in a more compact form we define a new 2-component vector of fields

$$\Psi_\alpha = \begin{bmatrix}\psi_\sigma(x)\\ \psi^\dagger_\sigma(x)\end{bmatrix} \qquad \alpha = +,- \quad (6.39)$$

and the BCS Hamiltonian becomes

$$H_{\text{BCS}} = \tfrac{1}{2}\Psi^\dagger_\alpha H_{\alpha\beta}\Psi_\beta + H_0, \quad (6.40)$$

$$H_0 = \tfrac{1}{2}\operatorname{Tr}\left[-\frac{\hbar^2}{2m}\nabla^2 - \mu\right], \quad (6.41)$$

where the subscript $\alpha = (\alpha, \sigma, x)$ plays a threefold role, standing for (a) the new label introduced in (6.38) (note that $+$ indicates the upper and $-$ the lower component); (b) the spin label (which assumes the values $+$ and $-$); and (c) the coordinate. The summation convention is extended so that when such an index is repeated, sums over the two discrete labels and integration over the coordinate are implied. The constant term H_0 in (6.41) follows from the anticommutation relation (6.21a) when the fields are reordered. The trace of the differential operator can be evaluated by taking its expectation value in any complete basis set of states. Although the constant is infinite, we will encounter a second constant later on and their sum will actually be well-behaved.

Since the vector of fields (6.39) has two components, it follows that the Hamiltonian kernel $H_{\alpha\beta}$ in (6.40) can be written as a 2×2 matrix:

$$H_{\alpha\beta} = \begin{bmatrix} \left(\frac{\hbar^2}{2m} \nabla_x \cdot \nabla_y - \mu\right) \delta_{\alpha\beta}(x-y) & \Delta^*_{\alpha\beta}(x,y) \\ -\Delta_{\alpha\beta}(x,y) & -\left(\frac{\hbar^2}{2m} \nabla_x \cdot \nabla_y - \mu\right) \delta_{\alpha\beta}(x-y) \end{bmatrix}. \quad (6.42)$$

Note that $H_{\alpha\beta}$ is Hermitian:

$$(H^\dagger)_{\alpha\beta} = \begin{bmatrix} \left(\frac{\hbar^2}{2m} \nabla_x \cdot \nabla_y - \mu\right) \delta_{\alpha\beta}(x-y) & -\Delta^*_{\beta\alpha}(y,x) \\ \Delta_{\beta\alpha}(y,x) & -\left(\frac{\hbar^2}{2m} \nabla_x \cdot \nabla_y - \mu\right) \delta_{\alpha\beta}(x-y) \end{bmatrix}. \quad (6.43)$$

Using the antisymmetry of the gap function,

$$\Delta_{\alpha\beta}(y,x) = -\Delta_{\alpha\beta}(x,y), \quad (6.44)$$

we have

$$H^\dagger_{\alpha\beta} = H_{\alpha\beta}. \quad (6.45)$$

As with any bilinear form, the transformation that diagonalizes the matrix $H_{\alpha\beta}$ can be constructed from its eigenvectors, which we write in the form

$$\varphi^{(n)}_\alpha = \begin{pmatrix} u^{(n)}_\alpha \\ v^{(n)}_\alpha \end{pmatrix}. \quad (6.46)$$

These eigenvectors satisfy the eigenvalue equation

$$H_{\alpha\beta}\, \varphi^{(n)}_\beta = \lambda_n \varphi^{(n)}_\alpha. \quad (6.47)$$

Just as the indices α and β play a threefold role in equations such as (6.42), so does the label n of the eigenvector and eigenvalue in (6.47). Since $H_{\alpha\beta}$ is a 2×2 matrix of operators, the eigenvectors will carry a new index $n = +, -$ labelling the two linearly independent eigenvectors. In addition, the eigenvectors will carry their own spin index, σ, and finally a third index, which in a sense is conjugate to the coordinate. For translationally invariant cases this index is simply the wavevector k, and we shall use this notation when making things explicit. By introducing a vector of fields, Ψ_α, we have expanded the state space by a factor of 2. We shall see, however, that there is a symmetry between the two parts so that all the physical information is contained in just one part.

We now show that if λ is an eigenvalue of $H_{\alpha\beta}$, there exists a second eigenvalue $-\lambda$. That this is so can be seen by considering the case $\Delta = 0$. Then $H_{\alpha\beta}$ has the diagonal form the diagonal form

$$H_{\alpha\beta} = \begin{bmatrix} \left(\frac{\hbar^2}{2m}\nabla_x\cdot\nabla_y - \mu\right)\delta_{\alpha\beta}(x-y) & 0 \\ 0 & -\left(\frac{\hbar^2}{2m}\nabla_x\cdot\nabla_y - \mu\right)\delta_{\alpha\beta}(x-y) \end{bmatrix}. \quad (6.48)$$

The eigenvectors are plane waves so that Eq. (6.47) separates,

$$\varphi_\sigma^{+,\mu}(x;k) = \begin{bmatrix} u_\sigma^\mu(k) \\ 0 \end{bmatrix} e^{ikx},$$
$$\varphi_\sigma^{-,\mu}(x;k) = \begin{bmatrix} 0 \\ v_\sigma^\mu(k) \end{bmatrix} e^{ikx}, \quad (6.49)$$

and the eigenvalues in the two cases are, respectively,

$$\lambda_+(k) = \frac{\hbar^2k^2}{2m} - \mu = \xi(k), \quad (6.50a)$$

$$\lambda_-(k) = -\left(\frac{\hbar^2k^2}{2m} - \mu\right) = -\xi(k). \quad (6.50b)$$

Another way to see this is to note that the trace of the kernel matrix is zero, and since the trace of a Hermitian matrix is invariant under diagonalization, the eigenvalues of (6.47) must be negative inverses. In general, if we write the eigenvector $\varphi = (u, v)$ and $H\varphi = \lambda\varphi$, we find that $\varphi^* = (v^*, u^*)$ is also an eigenvector with eigenvalue $-\lambda$. To see this we first note that (6.47) can be cast in the form

$$H_0u + \Delta^*v = \lambda u,$$
$$-H_0v - \Delta u = \lambda v, \quad (6.51)$$

where u and v are the components of φ. If we now multiply both sides of (6.51) by (-1), take the complex conjugate, and rearrange we get

$$H_0v^* + \Delta^*u^* = -\lambda v^*,$$
$$-H_0u^* - \Delta v^* = -\lambda u^*, \quad (6.52)$$

from which it follows that (v^*, u^*) is also an eigenstate with eigenvalue $-\lambda$. We will use the convention that the eigenvector written as (u, v) has the positive eigenvalue, and in the future write out the sums all over the $n = +, -$ index explicitly in terms of the u and v.

At this point the introduction of a second branch $\lambda_-(k)$ of the energy spectrum that simply mirrors the "normal" branch, $\lambda_+(k)$, is a purely mathematical device. However, as we see in Fig. 6.1, we can assign some physical meaning to the new energy band by looking at the normal ground state.

In a system of noninteracting fermions, the ground state is constructed by occupying every state below the Fermi level. Since we are measuring all energies relative to the Fermi level (recall that $E_F = \mu$ $(T = 0)$), this means that in the ground state all levels with negative energy are occupied.

If we look at Fig. 6.1 we see that part of both branches lie below the Fermi surface. The "+" branch is interpreted in terms of ordinary electron states, so that in the ground state all electron states below the Fermi surface are assumed to be occupied. The new "−" branch may be interpreted in terms of "holes." This interpretation is justified by noting that the creation operator for a state on the "−" branch is simply the electron annihilation operator. Since a hole represents the absence of an electron, all hole states above the Fermi surface are "occupied" (or, equivalently, all electron states are "unoccupied").

If we turn our attention to the positive energy states shown in Fig. 6.1, we see that there are, of course, ordinary electron states above the Fermi surface, but we can

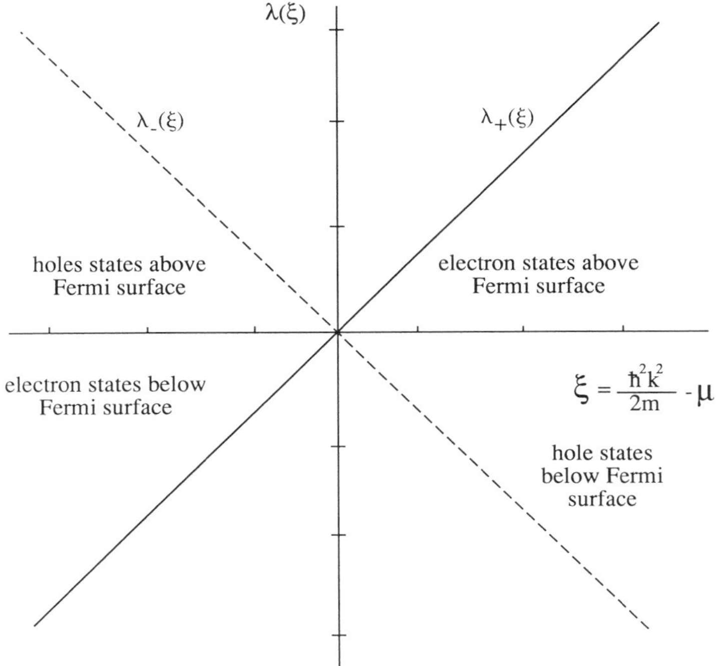

Figure 6.1 Energy spectrum in the normal state.

also create an excited state by removing an electron from below the Fermi surface; that is, create a hole with positive energy. These states correspond to the upper part of the "$-$" branch.

One final observation is in order at this point. The two branches of the energy spectrum in Fig. 6.1 are degenerate right at the Fermi surface ($\xi = 0$), and, as is generally the case in quantum mechanics, introduction of a potential lifts this degeneracy and strongly modifies the nature of the states close to the point of degeneracy. A gap opens up that separates the positive and negative branches of the energy spectrum. This gap explains many of the unusual properties of the superconducting state.

The eigenvectors of $H_{\alpha\beta}$ are, as usual, both orthonormal and complete. The orthogonality condition

$$\sum_{\alpha} \varphi_{\alpha}^{n*} \varphi_{\alpha}^{n'} = \delta_{nn'}, \tag{6.53}$$

written out explicitly in terms of u and v, is

$$\sum_{\alpha} \int dx \big[u_{\alpha}^{\mu*}(x;k) u_{\alpha}^{\nu}(x,k') + v_{\alpha}^{\mu*}(x;k) v_{\alpha}^{\nu}(x;k') \big] = \delta_{\mu\nu}\, \delta_{k,k'}, \tag{6.54a}$$

$$\sum_{\sigma} \int dx \big[u_{\sigma}^{\mu*}(x;k) v_{\sigma}^{\nu*}(x,k') + v_{\sigma}^{\mu*}(x;k) u_{\sigma}^{\nu*}(x;k') \big] = 0, \tag{6.54b}$$

$$\sum_{\sigma} \int dx [v_{\sigma}^{\mu}(x;k) u_{\sigma}^{\nu}(x,k') + u_{\sigma}^{\mu}(x;k) v_{\sigma}^{\nu}(x;k')] = 0, \tag{6.54c}$$

$$\sum_{\sigma} \int dx \big[v_{\sigma}^{\mu}(x;k) v_{\sigma}^{\nu*}(x,k') + u_{\sigma}^{\mu}(x;k) u_{\sigma}^{\nu*}(x;k') \big] = \delta_{\mu\nu}\, \delta_{k,k'}. \tag{6.54d}$$

Similarly, the completeness relation

$$\sum_{n} \varphi_{\alpha}^{n*} \varphi_{\beta}^{n} = \delta_{\alpha\beta}, \tag{6.55}$$

when written out in terms of u and v, is

$$\sum_{\mu k}\left[u_\alpha^{\mu *}(x;k)u_\beta^\mu(y;k)+v_\alpha^\mu(x;k)v_\beta^{\mu *}(y;k)\right]=\delta_{\alpha\beta}\,\delta(x-y), \tag{6.56a}$$

$$\sum_{\mu k}\left[v_\alpha^{\mu *}(x;k)u_\beta^\mu(y;k)+u_\alpha^\mu(x;k)v_\beta^{\mu *}(y;k)\right]=0, \tag{6.56b}$$

$$\sum_{\mu k}\left[v_\alpha^{\mu *}(x;k)u_\beta^\mu(y;k)+u_\alpha^\mu(x;k)v_\beta^{\mu *}(y;k)\right]=0, \tag{6.56c}$$

$$\sum_{\mu k}\left[v_\alpha^{\mu *}(x;k)v_\beta^\mu(y;k)+u_\alpha^\mu(x;k)u_\beta^{\mu *}(y;k)\right]=\delta_{\alpha\beta}\,\delta(x-y). \tag{6.56d}$$

We now express the original fermion operators Ψ_α as linear superpositions of new quasiparticle operators A_n by means of the Bogoliubov transformation:

$$\Psi_\alpha=\sum_n \varphi_\alpha^n A_n. \tag{6.57}$$

The Bogoliubov transformation has an inverse; multiplying (6.57) by $\varphi_\alpha^{m^*}$ and summing on α we obtain

$$\sum_\alpha \varphi_\alpha^{m^*}\Psi_\alpha=\sum_{n,\alpha}\varphi_\alpha^{m^*}\varphi_\alpha^n A_n = A_m, \tag{6.58}$$

where we have used the orthogonality relation (6.53). Expanding (6.58), we have

$$A_\mu^+(k)=\int dx\sum_\sigma\left[v_\alpha^{\mu *}(x;k)\psi_\sigma(x)+v_\sigma^{\mu *}(x;k)\psi_\sigma^\dagger(x)\right], \tag{6.59a}$$

$$A_\mu^-(k)=\int dx\sum_\sigma\left[v_\alpha^{\mu}(x;k)\psi_\sigma(x)+u_\sigma^{\mu}(x;k)\psi_\sigma^\dagger(x)\right]. \tag{6.59b}$$

Taking the Hermitian conjugate of $A_\mu^-(k)$ we have

$$\left[A_\mu^-(k)\right]^\dagger=\int dx\sum_\sigma\left[v_\alpha^{\mu}(x;k)\psi_\sigma(x)+u_\sigma^{\mu}(x;k)\psi_\sigma^\dagger(x)\right], \tag{6.60}$$

from which it is clear that $A_\mu^-(k)$ is the Hermitian conjugate of $A_\mu^+(k)$. We can simplify our notation somewhat and write for the quasiparticle operators

$$a_\mu(k)=A_\mu^+(k), \tag{6.61a}$$

$$a_\mu^\dagger(k)=A_\mu^-(k). \tag{6.61b}$$

We can also show that the quasiparticle operators obey the canonical anticommutation relations

$$\left\{a_\mu(k),a_\nu^\dagger(k')\right\}=\delta_{\mu\nu}\,\delta_{kk'}. \tag{6.62}$$

Using (6.59a, b), (6.62) becomes

$$\begin{aligned}\left\{a_\mu(k),a_\nu^\dagger(k')\right\} &=\int dx\int dy\sum_{\alpha,\beta}\left[u_\alpha^\mu(x;k)\psi_\alpha(x,k)\right.\\ &\quad+v_\alpha^\mu(x,k)\psi_\alpha^\dagger(x),v_\beta^\nu(y;k')\psi_\beta(y)\\ &\quad\left.+u_\beta^\nu(y;k')\psi_\beta^\dagger(y)\right].\end{aligned} \tag{6.63}$$

Applying the canonical commutation relations for the original operators ψ and $\psi^\dagger$, we find that (6.63) becomes

$$\begin{aligned}\left[a_\mu(k),a_\nu^\dagger(k')\right] &=\int dx\int dy\sum_{\alpha,\beta}\left[u_\alpha^{\mu *}(x;k')u_\beta^\mu(y,k)\right.\\ &\quad\left.+v_\alpha^{\mu *}(x;k)v_\beta^\nu(y;k')\right]\delta_{\alpha\beta}\,\delta(x-y)\\ &=\int dx\sum_\alpha u_\alpha^{\mu *}(x;k')u_\alpha^\nu(x,k')\\ &\quad+v_\alpha^{\mu *}(x;k)v_\alpha^\nu(x;k').\end{aligned} \tag{6.64}$$

If we now apply the orthogonality condition (6.54a) we have what we wished to show, namely that the quasiparticle operators also obey the canonical anticommutation relation,

$$\left[a_\mu(k),a_\nu^\dagger(k')\right]=\delta_{\mu\nu}\,\delta_{k,k'}. \tag{6.65}$$

We can now write out the Bogoliubov transformation (6.57) explicitly in terms of the u and v components:

$$\psi_\sigma(x) = \sum_{k\mu} \left[u_\sigma^\mu(x;k) a_\mu(k) + v_\sigma^{\mu*}(x;k) a_\mu^\dagger(k) \right], \quad (6.66a)$$

$$\psi_\sigma^\dagger(x) = \sum_{k\mu} \left[u_\sigma^{\mu*}(x;k) a_\mu^\dagger(k) + v_\sigma^\mu(x;k) a_\mu(k) \right]. \quad (6.66b)$$

This provides us with the two components of the vector of fields (6.39) expressed as linear combinations of the quasiparticle operators.

Now we are in a position to rewrite the BCS Hamiltonian (6.40) in terms of the quasiparticle operators. Substituting (6.57) and its Hermitian conjugate in (6.40), we have

$$\begin{aligned} H_{\text{BCS}} &= \tfrac{1}{2} \sum_{n,m,\alpha,\beta} \varphi_\alpha^{n*} A_n^\dagger H_{\alpha\beta} \varphi_\beta^m A_m \\ &= \tfrac{1}{2} \sum_{n,m} \varphi_\alpha^{n*} A_n^\dagger \lambda_{n'} \varphi_\alpha^{n'} A_m \\ &= \tfrac{1}{2} \sum_n \lambda_n A_n^\dagger A_n. \end{aligned} \quad (6.67)$$

Expanding the latter summation, using Eqs. (6.61) and (6.60), we obtain an expression for the generalized BCS Hamiltonian in terms of quasiparticle operators,

$$\begin{aligned} H_{\text{BCS}} &= \tfrac{1}{2} \sum_{\mu k} \lambda_\mu(k) \times \left[a_\mu^\dagger(k) a_\mu(k) - a_\mu(k) a_\mu^\dagger(k) \right] \\ &= \sum_{\mu k} \lambda_\mu(k) a_\mu^\dagger(k) a_\mu(k) + \tfrac{1}{2} \mathrm{Tr}\left(-\frac{\hbar^2}{2m} \nabla^2 - \mu \right) - \tfrac{1}{2} \sum_{\mu k} \lambda_\mu(k), \end{aligned} \quad (6.68)$$

where the second term is unchanged from Eq. (6.41).

In (6.68) we have finally reduced the BCS Hamiltonian to a collection of noninteracting fermions. The number operator

$$n_\mu(k) = a_\mu^\dagger(k) a_\mu(k) \quad (6.69)$$

has the usual spectrum (0, 1) for unoccupied and occupied states, respectively, and the partition function is

$$Z = Z_0 \prod_{\mu,k} [1 + e^{-\beta \lambda_\mu(k)}], \quad (6.70)$$

where

$$Z_0 = \exp\left\{ -\frac{\beta}{2} \left[\mathrm{Tr}\left(-\frac{\hbar^2 \nabla^2}{2m} - \mu \right) - \sum_{\mu,k} \lambda_\mu(k) \right] \right\}. \quad (6.71)$$

Equation (6.68) is in a form corresponding to noninteracting fermions, but the energies $\lambda_\mu(k)$ of the system depend on the gap Δ through the functions u and v.

The components (6.66) of the vector of fields (6.39) can be used to evaluate the order parameter (6.21). Inserting Eq. (6.66b) in Eq. (6.21),

$$\begin{aligned} \Phi_{\alpha\beta}(x,y) = \sum_{k\mu} \sum_{k'v} \Big\langle & \left[u_\alpha^{\mu*}(x;k) a_\mu^\dagger(k) + v_\alpha^\mu(x;k) a_\mu(k) \right] \\ & \times \left[u_\alpha^{\nu*}(x;k) a_v^\dagger(k') + v_\beta^\nu(x;k') a_v(k') \right] \Big\rangle, \end{aligned} \quad (6.72)$$

and carrying out thermal averaging gives

$$\begin{aligned} \Phi_{\alpha\beta}(x,y) &= \sum_{k,\mu} u_\alpha^{\mu*}(x;k) v_\beta^\mu(y,k) n(\mu,k) \\ &\quad + v_\alpha^\mu(x;k) u_\beta^\mu(y,k) [1 - n(\mu,k)], \end{aligned} \quad (6.73)$$

where $n(\mu, k)$ is the Fermi–Dirac distribution function,

$$n(\mu,k) = [\exp[\beta \lambda(\mu,k)] + 1]^{-1}. \quad (6.74)$$

Equation (6.72) is a self-consistent equation for Φ, since u, v, and λ all depend on Φ in terms of Δ in Eq. (6.51).

V. SINGLET PAIRING IN A HOMOGENEOUS SUPERCONDUCTOR

The fundamental electron operators are spin $-\frac{1}{2}$, and the order parameter (6.22) can therefore be either a spin singlet ($S=0$) or triplet ($S=1$). By far the most common situation in superconductors is the spin–singlet case, while in the superfluid phase of ^{3}He, the order parameter is in the triplet state. In this section we will consider the spin–singlet case in detail. We will also assume that the superconductor is homogeneous and isotropic, which, by extensive use of the Fourier transform, simplifies the algebra enormously. This will provide us with the particular results of the BCS theory that are ordinarily compared with experiment.

Let us separate the spin and coordinate dependences of the order parameter:

$$\Phi_{\alpha\beta}(x,y)=\chi_{\alpha\beta}\Phi(x,y). \quad (6.75)$$

For a spin–singlet state $\chi_{\alpha\beta}=-\chi_{\beta\alpha}$, and since the order parameter is overall antisymmetric it follows that the space part is symmetric: $\Phi(x,y)=\Phi(y,x)$. Let $T(a)$ be a translation operator, which, acting on the field operator $\psi_\alpha(x)$, gives

$$T(a)\psi_\sigma(x)T^{-1}(a)=\psi_\sigma(x+a). \quad (6.76)$$

By hypothesis the translations commute with the Hamiltonian, so that

$$\langle \psi_\alpha^\dagger(x)\psi_\beta^\dagger(y)\rangle$$

$$=\langle T(a)\psi_\alpha^\dagger(x)T^{-1}(a)T(a)\psi_\beta^\dagger(y)T^{-1}(\boldsymbol{a})\rangle \quad (6.77a)$$

$$=\langle \psi_\alpha^\dagger(x+a)\psi_\beta^\dagger(y+a)\rangle \quad (6.77b)$$

or

$$\Phi_{\alpha\beta}(x,y)=\Phi_{\alpha\beta}(x+a,y+a). \quad (6.78)$$

Since a is an arbitrary translation, (6.78) implies that

$$\Phi_{\alpha\beta}(x,y)=\Phi_{\alpha\beta}(x-y). \quad (6.79)$$

The antisymmetry of the spin part of the order parameter defines it uniquely, and we can write

$$\chi_{\alpha\beta}=\epsilon_{\alpha\beta}, \quad (6.80)$$

where

$$\epsilon_{\alpha\beta}=\begin{cases}0 & \text{if } \alpha=\beta\\ 1 & \text{if } \alpha=+,\beta=-\\ -1 & \text{if } \alpha=-,\beta=+.\end{cases} \quad (6.81)$$

The orbital part of the order parameter is conveniently expressed in terms of its Fourier transform,

$$\Phi(x-y)=\frac{1}{\Omega}\sum_k e^{ik(x-y)}\Phi(k), \quad (6.82)$$

where Ω is the volume of the system. The electron–electron interaction is also only a function of $x-y$, and we may write

$$V(x-y)=\frac{1}{\Omega}\sum_k e^{ik(x-y)}V(k). \quad (6.83)$$

The gap function, $\Delta_{\alpha\beta}(x-y)$, can also be Fourier transformed:

$$\Delta_{\alpha\beta}(x-y)=\epsilon_{\alpha\beta}\frac{1}{\Omega}\sum_k e^{ik(x-y)}\Delta(k). \quad (6.84)$$

The relation between $\Delta(k)$, $\Phi(k)$, and $V(k)$ follows by convolution:

$$\Delta(k)=\frac{1}{\Omega}\sum_p V(p)\Phi(k-p). \quad (6.85)$$

Since the kernel operator $H_{\alpha\beta}$ is translationally invariant, its eigenvectors will be plane waves, so that we will have

$$u_\alpha^\mu(x;k)=\frac{1}{\sqrt{\Omega}}u_\alpha^\mu(k)e^{ik\cdot x}, \quad (6.86a)$$

$$v_\alpha^\mu(x;k)=\frac{1}{\sqrt{\Omega}}v_\alpha^\mu(k)e^{ik\cdot x}. \quad (6.86b)$$

Substitution of (6.86a, b) in (6.49) then gives the two coupled equations

$$\left(\frac{\hbar^2k^2}{2m}-\mu\right)u_\alpha^\mu(k)+\Delta_{\alpha\beta}^*(k)v_\beta^\mu(k)$$

$$=\lambda_\mu(k)u_\alpha^\mu(k) \quad (6.87a)$$

and

$$\left(\frac{\hbar^2 k^2}{2m} - \mu\right) v_\alpha^\mu(k) + \Delta_{\alpha\beta}(k) u_\beta^\mu(k) = -\lambda_\mu(k) v_\alpha^\mu(k). \quad (6.87b)$$

By (6.87b),

$$\left[\frac{\hbar^2 k^2}{2m} - \mu + \lambda_\mu(k)\right] v_\alpha^\mu(k) = -\Delta_{\alpha\beta}(k) u_\beta^\mu(k). \quad (6.88)$$

Multiplying (6.87b) by $[(\hbar^2 k^2/2m) - \mu + \lambda_\mu(k)]$ then gives

$$\left[\frac{\hbar^2 k^2}{2m} - \mu + \lambda_\mu(k)\right] \cdot \left[\frac{\hbar^2 k^2}{2m} - \mu - \lambda_\mu(k)\right] \times u_\alpha^\mu(k) - \Delta^*_{\alpha\beta}(k)\Delta_{\beta\gamma}(k) u_\gamma^\mu(k) = \left[\frac{\hbar^2 k^2}{2m} - \mu + \lambda_\mu(k)\right] \lambda_\mu(k) u_\alpha^\mu(k). \quad (6.89)$$

The sum over spin labels in the second term on the left is

$$\Delta^*_{\alpha\beta}(k)\Delta_{\beta\gamma}(k) = \epsilon_{\alpha\beta}\epsilon_{\beta\gamma}|\Delta(k)|^2 = -\delta_{\alpha\gamma}|\Delta(k)|^2, \quad (6.90)$$

so that (6.89) now becomes

$$\xi(k)[\xi(k) + \lambda(k)] + |\Delta(k)|^2 = [\xi(k) + \lambda(k)]\lambda(k), \quad (6.91)$$

which gives us the well-known dependence of the energy $\lambda(k)$ on the gap $\Delta(k)$,

$$\lambda(k) = \pm\sqrt{\xi^2(k) + |\Delta(k)|^2}, \quad (6.92)$$

where the quantity $\xi(k)$,

$$\xi(k) = \frac{\hbar^2 k^2}{2m} - \mu, \quad (6.93)$$

is the energy in the absence of a gap $[\Delta(k) = 0]$ measured relative to the chemical potential.

The quasiparticle spectrum (6.92) for the case of a gap $\Delta(k) = \Delta$ that is independent of k is shown in Fig. 6.2. In the superconducting ground state all negative energy states are occupied (− branch) and

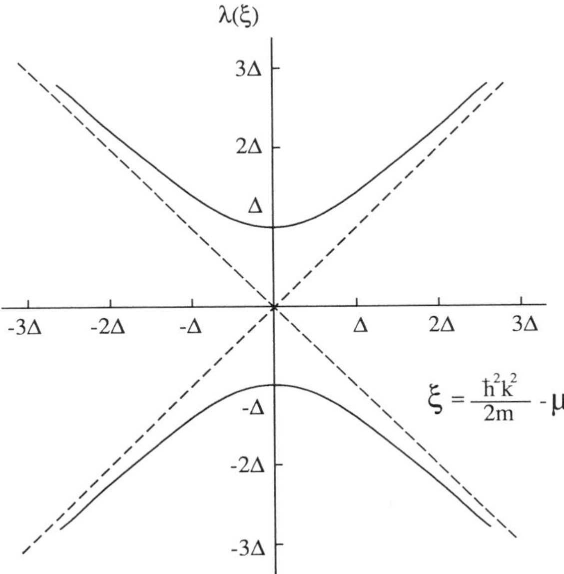

Figure 6.2 Quasiparticle energy spectrum in the superconducting state. The dashed lines refer to the pure electron–hole branches in the normal state shown in Fig. 6.1.

all positive energy states are empty. Near the Fermi surface the quasiparticle states are strong superpositions of hole and electron states. Far from the Fermi surface the quasiparticles retain their electron- or hole-like character.

We can derive explicit expressions for u and v from (6.87a) and (6.87b) if we first define

$$\bar{v}_\alpha^\mu(k) = -\epsilon_{\alpha\beta} v_\beta^\mu(k). \qquad (6.94)$$

We then find for u and $\bar{v}$

$$[\xi(k) - \lambda_\mu(k)] u_\alpha^\mu(k) - \Delta^*(k)\bar{v}_\alpha^\mu(k) = 0, \qquad (6.95a)$$

$$[\xi(k) + \lambda_\mu(k)] \bar{v}_\alpha^\mu(k) - \Delta(k)\bar{u}_\alpha^\mu(k) = 0. \qquad (6.95b)$$

In the absence of spin-dependent forces we expect that the Bogoliubov transformation should reduce to the identity transformation above the critical temperature ($\Delta = 0$). Therefore, for $\lambda(k) > 0$ and $\Delta = 0$ we should have

$$\left.\begin{aligned} u_\alpha^\mu(k) &= \delta_\alpha^\mu \\ \bar{v}_\alpha^\mu(k) &= 0 \end{aligned}\right\} \quad \xi(k) > 0 \qquad (6.96a)$$

and

$$\left.\begin{aligned} u_\alpha^\mu(k) &= 0 \\ \bar{v}_\alpha^\mu(k) &= \delta_\alpha^\mu \end{aligned}\right\} \quad \xi(k) < 0. \qquad (6.96b)$$

We can then argue by continuity that for all temperatures,

$$u_\alpha^\mu(k) = \delta_\alpha^\mu u(k) \qquad (6.97a)$$

and

$$\bar{v}_\alpha^\mu(k) = \delta_\alpha^\mu v(k). \qquad (6.97b)$$

From (6.97b) we see that

$$v_\alpha^\mu(k) = \epsilon_\alpha^\mu v(k). \qquad (6.98)$$

We can now solve for $|u|^2(k)$ and $|v|^2(k)$, finding

$$|u|^2(k) = \tfrac{1}{2}\left[1 + \frac{\xi(k)}{\lambda(k)}\right], \qquad (6.99a)$$

$$|v|^2(k) = \tfrac{1}{2}\left[1 - \frac{\xi(k)}{\lambda(k)}\right], \qquad (6.99b)$$

which are shown in Fig. 6.3.

In terms of the original one-electron picture, a positive-energy quasiparticle state is a superposition of an electron with kinetic energy above the Fermi surface ($\xi > 0$) and a hole below the Fermi surface.

As noted previously, there are also negative-energy quasiparticle states. These are found from the positive-energy solutions by the substitution $u \leftrightarrow v^*$. The concepts of negative- and positive-energy quasiparticle states become clear if we consider the case $\Delta = 0$. The positive-energy quasiparticles are then identical to ordinary electrons above the Fermi surface, while negative-energy quasiparticles are just ordinary electrons below the Fermi surface. For $\Delta \neq 0$ this picture is significantly modified only in a region of width Δ near the Fermi surface. Both above and below T_c the negative-energy states are occupied in the ground state. In the superconducting state the positive-energy states are separated in energy by a gap 2Δ, and in this respect the state resembles a semiconductor more than a metal. The unique electromagnetic properties of the super-

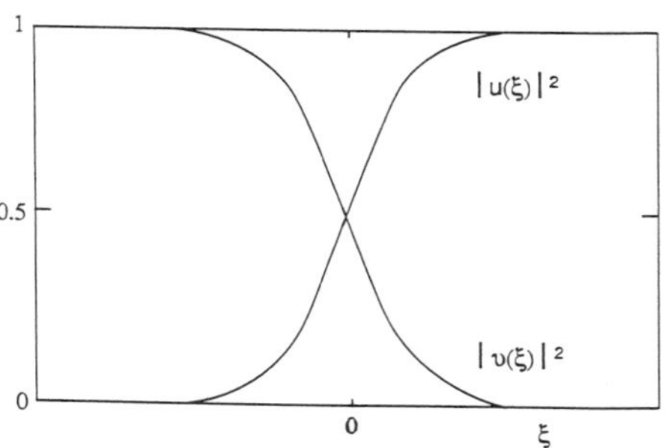

Figure 6.3 Bogoliubov amplitudes for a positive energy quasiparticle. For negative energy quasiparticles the roles of u and v are reversed.

conducting state do not arise from excitation of quasiparticles but rather from the response of the ground state itself to external fields. However, the gap manifests itself in the form of the low-temperature specific heat and in the properties of normal–superconductor tunnel junctions that involve excitation of quasiparticle states.

It will be instructive to deduce a closed-form expression for the superconducting density of states $D_s(\lambda)$ for the case plotted in Fig. 6.2 (energy gap Δ independent of k). From Eq. (6.92) the energies of the normal and superconducting states are

$$\lambda(\xi) = \begin{cases} \xi & \text{normal state} \\ \sqrt{\xi^2 + \Delta^2} & \text{superconducting state.} \end{cases} \tag{6.100}$$

There is a one-to-one correspondence between the points in k-space for the normal and superconducting states. From Eq. (6.93) we see that the one-to-one correspondence also exists for the values of ξ. This permits us to write for the two densities of states

$$D_s(\lambda)d\lambda = D_n(0)d\xi, \tag{6.101}$$

where we are assuming that the normal density of states $D_n(\xi)$ has constant value $D_n(0)$ in the neighborhood of the gap.

From Eq. (6.92) we have for a constant gap

$$\frac{d\xi}{d\lambda} = \frac{\lambda}{\sqrt{\lambda^2 - \Delta^2}}, \tag{6.102}$$

which gives

$$D_s(\lambda) = \begin{cases} \dfrac{D_n(0)\lambda}{\sqrt{\lambda^2 - \Delta^2}} & \lambda > \Delta \\ 0 & -\Delta < \lambda < \Delta, \\ \dfrac{-D_n(0)\lambda}{\sqrt{\lambda^2 - \Delta^2}} & \lambda < -\Delta \end{cases} \tag{6.103}$$

where no states exist in the gap. Figure 6.4 shows the energy dependence of $D_s(\lambda)$,

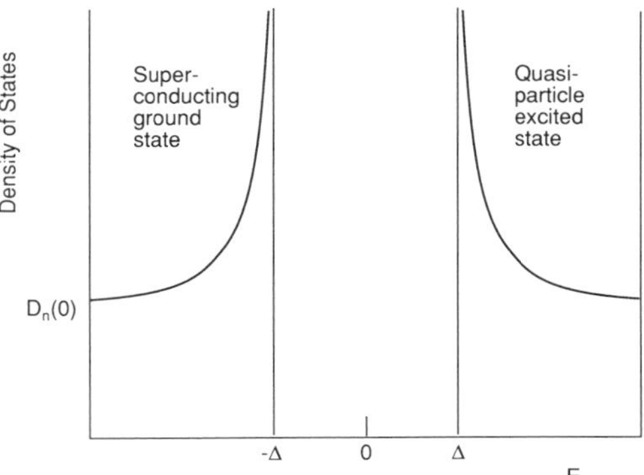

Figure 6.4 Superconductivity density of states $D_s(E)$ superimposed on the normal state DOS, assuming that the latter has a constant value $D_n(0)$ in the neighborhood of the superconducting energy gap Δ.

indicating the behavior near the gap. Figure 6.5 shows how $D_s(\lambda)$ approaches the normal state density of states $D_n(E)$ far from the gap.

VI. SELF-CONSISTENT EQUATION FOR THE ENERGY GAP

We have, at least formally, solved for the spectrum of the BCS Hamiltonian. This solution is incomplete, however, because our answer for the quasiparticle spectrum, (6.92), depends on the gap function, $\Delta(k)$. The gap function in turn is related to the order parameter Φ through the definition given in (6.36). Since the order parameter is the thermal average of an operator, it depends implicitly on the energy spectrum and therefore on Δ. Thus to complete the solution of the BCS theory we must show that this tangled system of self-consistent equations has a nontrivial solution, i.e., a solution for which $\Delta \neq 0$.

We jump into this self-consistent circle of dependencies with Eq. (6.73). Substituting our previous results for the Bogoliubov amplitudes u and v and applying a Fourier transformation, we find that

$$\Phi(k) = -\frac{\Delta(k)}{2\lambda(k)}[1 - 2n(\lambda)]. \tag{6.104}$$

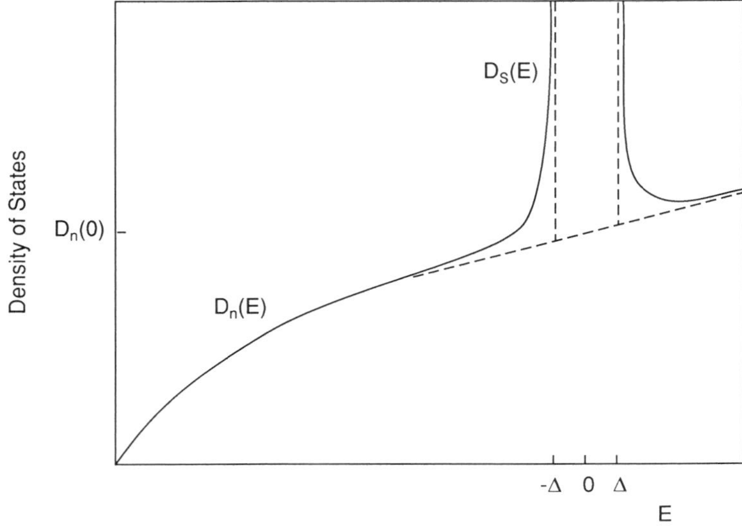

Figure 6.5 Dependence of the superconducting and normal densities of states, $D_s(E)$ and $D_n(E)$, respectively, over a broad energy range. Figure 6.4 shows the details in the neighborhood of the gap.

Substituting this result in (6.85) then gives

$$\Delta(k) = -\frac{1}{\Omega}\sum_p V(p)\frac{\Delta(k-p)}{2\lambda(k-p)} \times\{1-2n[\lambda(k-p)]\}. \quad (6.105)$$

Note that Δ appears both explicitly and implicitly (through λ) on the right side of (6.105). This equation is a nonlinear integral equation, and to have any hope of extracting useful results from it we will be forced to make some approximations.

The first approximation is to assume that the Fourier transformation of the effective electron–electron potential, $V(p)$, is independent of p. In real-space terms this means the interaction is very short range. We also assume, following Cooper, that the interaction is attractive, so we take

$$V(p) = -V_0. \quad (6.106)$$

The second approximation is to assume that the gap function, $\Delta(k)$, is independent of the direction of the wave vector k and nonzero only in the neighborhood of the Fermi surface,

$$\Delta(k) = \begin{cases} \Delta_0 & |\xi(k)| \le \hbar\omega_D \\ 0 & \text{otherwise}. \end{cases} \quad (6.107)$$

Note that in (6.107) we have introduced a new energy scale, the Debye energy $\hbar\omega_D$ under the assumption that the underlying mechanism responsible for the effective attractive interaction between electrons is due to phonons. This assumption, however, is not essential and one can well imagine the exchange of other types of collective excitations contributing to the electron–electron interaction.

Substituting expressions (6.106) and (6.107) in the gap equation (6.105), we have

$$\Delta_0 = V_0\Delta_0\frac{1}{\Omega}\sum_p{}' \frac{1}{2\lambda(k-p)} \times\{1-2n[\lambda(k-p)]\}. \quad (6.108)$$

The prime on the summation sign in (6.108) indicates that only momenta satisfying the constraint

$$|\xi(p)| \le \hbar\omega_D \quad (6.109)$$

are included in the sum. The sum over momenta can be replaced by an integral over energies within this thin energy shell, which is centered on the Fermi surface, by introducing the density of states,

$$\frac{1}{\Omega}\sum_{p}\frac{1}{2\lambda(k-p)}[1-2n\lambda(k-p)]$$
$$=\int_{-\hbar\omega_D}^{\hbar\omega_D} d\xi D_n(\xi)\frac{1}{2\lambda(\xi)}$$
$$\times\{1-2n[\lambda(\xi)]\}. \quad (6.110)$$

We can further approximate (6.110) by observing that over a narrow range of energies, $-\hbar\omega_D \leq \xi \leq \hbar\omega_D$, the density of states varies very little and can be replaced by its value at the Fermi surface, $D(0)$, so that we now have for (6.108)

$$1=\frac{V_0 D_n(0)}{2}\int_{-\hbar\omega_D}^{\hbar\omega_D} d\xi$$
$$\times\frac{1}{\lambda(\xi)}[1-2n(\xi)]. \quad (6.111)$$

This equation is still too difficult to solve exactly for arbitrary temperature $T \leq T_c$. However, we can solve it for the two special cases $T=0$ and $T=T_c$.

At $T=0$ there are no quasiparticles and $n(\xi)=0$. We then have

$$1=\frac{V_0 D_n(0)}{2}\int_{-\hbar\omega_D}^{\hbar\omega_D} d\xi\frac{1}{\sqrt{\xi^2+\Delta_0^2}}. \quad (6.112)$$

This is an elementary integral and we find

$$1=V_0 D_n(0)\sinh^{-1}\left(\frac{\hbar\omega_D}{\Delta_0}\right). \quad (6.113)$$

Solving for Δ_0 gives

$$\Delta_0=\frac{\hbar\omega_0}{\sinh[1/V_0 D_n(0)]}. \quad (6.114)$$

In the weak-coupling limit, $V_0 D(0) \ll 1$, this becomes

$$\Delta_0 \cong \frac{\hbar\omega_D}{2}\exp\left[-\frac{1}{V_0 D_n(0)}\right], \quad (6.115)$$

which is very close to the result we found for Cooper pairs in (6.19).

At the transition temperature, $\Delta(T_c)=0$ so that $\lambda(\xi)=\xi$ and (6.111) becomes

$$1=v_0 D(0)\int_0^{\hbar\omega_D}\left[\frac{\tanh(\xi/k_B T_0)}{\xi}\right]d\xi. \quad (6.116)$$

We first integrate by parts:

$$\int_0^{\hbar\omega_D} dx\,\frac{1}{x}\tanh x=\ln x\cdot\tanh x\bigg|_0^{\hbar\omega_D/kT_c}$$
$$-\int_0^{\hbar\omega_D/kT_c} dx\,\ln(x)\text{sech}(x). \quad (6.117)$$

We anticipate that $\hbar\omega_D \gg k_B T_c$ in the weak-coupling limit, so that the upper limit of integration in (6.117) is large. In the remaining integral on the right-hand side the integrand is exponentially small for large x, and we can safely set the upper limit of integration to infinity. This definite integral has the value ($\gamma=0.5772$)

$$\int_0^{\infty} dx\,\ln x\cdot\text{sech}^2 x=-\ln\left(\frac{4e^{\gamma}}{\pi}\right). \quad (6.118)$$

Reinserting these results in (6.116) we find

$$k_B T_c=\frac{2e^{\gamma}}{\pi}\hbar\omega_D\exp[-1/V_0 D_n(0)]. \quad (6.119)$$

Figure 6.6 shows that there is a very wide scatter in the way T_c varies with $D(0)$ in various superconductors. Note that the expressions for the zero-temperature gap Δ_0 and the critical temperature T_c depend in the same way on the strength of the electron–electron interaction and the density of states. Taking their ratio, we find

$$\frac{\Delta_0}{k_B T_c}=\frac{\pi}{e^{\gamma}}=1.76. \quad (6.120)$$

In Table 6.1 we show the experimental ratio $2\Delta_0 : k_B T_c$ which in the BCS theory has the value 3.52. It is clear that many superconductors follow the simple relation in (6.120) (cf. Fig. 1X-4 of our earlier work

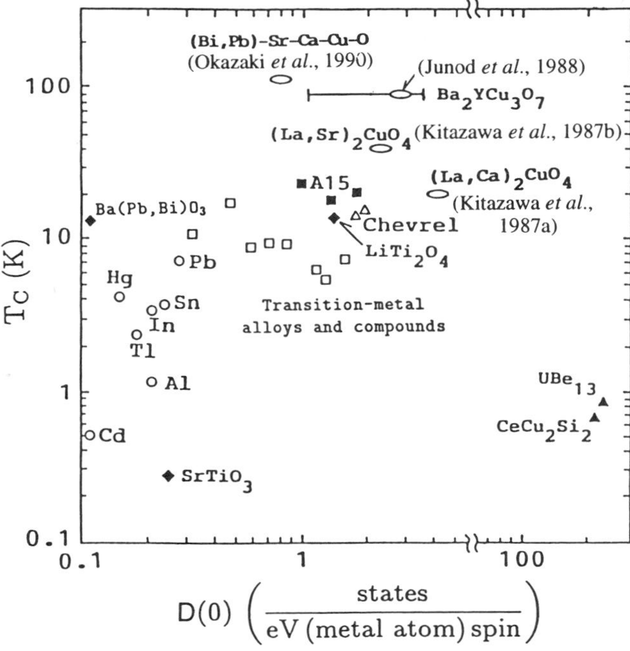

Figure 6.6 Dependence of superconducting transition temperature T_c on the density of states $D(0)$ for various superconductors (Okazaki *et al.*, 1990).

(Poole, 1988)). The full temperature-dependent function $\Delta(T)$ is shown in Fig. 6.7.

In the beginning of our discussion we introduced an effective electron–electron potential $V(x-y)$ without saying anything about its form or origin, other than noting that in the BCS theory exchange of virtual phonons gives rise to an attractive interaction. A detailed calculation of electron–phonon coupling is beyond the scope of this book so we will simply refer to the central result of Eliashberg. Note that the dimensionless constant $V_0 D(0)$ enters into our expression for the transition temperature (6.119). For the attractive phonon-mediated interaction between electrons Eliashberg defines the dimensionless electron-phonon coupling constant λ

$$\lambda = 2\int_0^\infty \frac{\alpha^2(\omega)D_{\text{ph}}(\omega)}{\omega}\,d\omega, \quad (6.121)$$

where $D_{\text{ph}}(\omega)$ is the phonon density of states and $\alpha^2(\omega)$ is the electron–phonon coupling strength.

Superconductors are characterized according to the magnitude of λ,

$$\begin{aligned} &\lambda \ll 1 && \text{weak coupling} \\ &\lambda \sim 1 && \text{intermediate coupling} \\ &\lambda \gg 1 && \text{strong coupling.} \end{aligned} \quad (6.122)$$

In addition to the attractive phonon-mediated part of the electron–electron interaction, there is, of course, a residual repulsive screened Coulomb interaction, which is characterized by the dimensionless parameter μ_c^*. The net electron–electron interaction is the sum of these two parts, so that in (6.119) we make the substitution

$$V_0 D_n(0) = \lambda - \mu_c^*, \quad (6.123)$$

Table 6.1 Comparison of Energy Gaps for Various Superconductors[a]

Material	T_c, K	$2\Delta_0$, meV	$2\Delta_0/k_BT_c$
Hf	0.13	0.044	3.9
Cd	0.52	0.14	3.2
Zn	0.85	0.23	3.2
Al	1.2	0.35	3.4
In	3.4	1.05	3.6
Hg	4.2	1.7	4.6
Pb	7.2	2.7	4.3
Nb	9.3	3.0	3.8
V_3Ge (*A*15)	11.2	3.1	3.2
V_3Si (*A*15)	17.1	5.4	3.7
Nb_3Sn (*A*15)	18.1	4.7	3.0
K_3C_{60}	19	5.9	3.6
Rb_3C_{60}	29	7.5	3.0
$Ba_{0.6}K_{0.4}BiO_3$	18.5	5.9	3.7
$(Nd_{0.925}Ce_{0.075})_2CuO_4$	21	7.4	4.4
$(La_{0.925}Sr_{0.075})_2CuO_4$	36	13	4.3
$YBa_2Cu_3O_{7-\delta}$	87	30	4.0
$Bi_2Sr_2Ca_2Cu_3O_{10}$	108	53	5.7
$Tl_2Ba_2CaCu_2O_8$	112	44	4.5
$Tl_2Ba_2Ca_2Cu_3O_{10}$	105	28	3.1
$HgBa_2Ca_2Cu_3O_8$	131	48	4.3

[a] Data on elements from Meservey and Schwartz (1969); data on the *A*15 compounds from Vonsovsky *et al.* (1982); and data on high-temperature superconductors from T. Hasegawa *et al.* (1991). K_3C_{60} and Rb_3C_{60} values are from Degiorgi *et al.* (1992), and $HgBa_2Ca_2Cu_3O_8$ data are from Schilling *et al.* (1994b). Many of the $2\Delta_0/k_BT_c$ ratios are averages of several determinations, sometimes with considerable scatter; the $2\Delta_0$ values are calculated from columns 2 and 4. The BCS value of $2\Delta_0/k_BT_c$ is 3.52. Table 3.1 provides energy gap data for many additional elements.

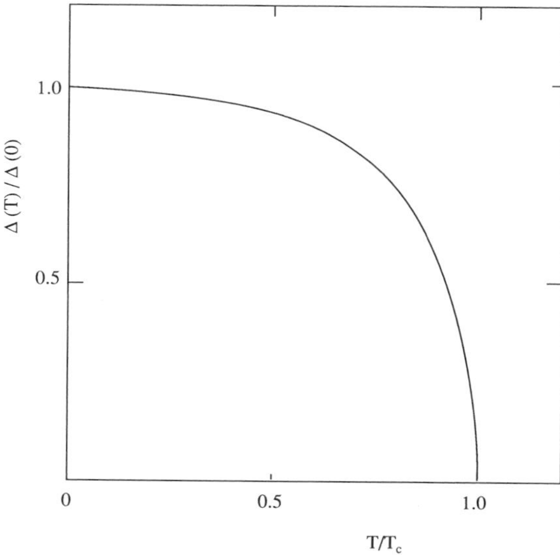

Figure 6.7 Temperature dependence of the BCS gap function Δ.

Table 6.2 Electron–Phonon Coupling Constants λ and Coulomb Interaction Parameters μ_c^{*} [a]

Material	T_c (K)	λ	μ_c^*	References
Ru	0.49	0.47	0.15	Table 3.1
Zr	0.61	0.22	0.17	Table 3.1
Os	0.66	0.44	0.12	Table 3.1
Mo	0.92	0.35	0.09	Table 3.1
Re	1.7	0.37	0.1	Table 3.1
Pb	7.2	1.55		Ginzburg and Kirzhnits (1977, p. 171)
Nb	9.3	0.85		Ginzburg and Kirzhnits (1977, p. 171)
NbC	11.1	0.61		Ginzburg and Kirzhnits (1977, p. 171)
TaC	11.4	0.62		Ginzburg and Kirzhnits (1977, p. 171)
V_3Ge	6.1	0.7		Vonsovsky *et al.* (1982, p. 303)
V_3Si	17.1	1.12		Vonsovsky *et al.* (1982, p. 303)
Nb_3Sn	18.1	1.67		Ginzburg and Kirzhnits (1977, p. 171)
Nb_3Ge	23.2	1.80		Vonsovsky *et al.* (1982, p. 303)
K_3C_{60}	16.3	0.51		Novikov *et al.* (1992)
Rb_3C_{60}	30.5	0.61		Novikov *et al.* (1992)
Cs_3C_{60}	47.4	0.72		Novikov *et al.* (1992)
$Ba(Pb, Bi)O_3$	12	1.3		Schlesinger *et al.* (1989)
$(La_{0.913}Sr_{0.087})CuO_4$		0.1		Gurvitch and Fiory (1987a, b, c)
$(La_{0.913}Sr_{0.087})CuO_4$	35	2.0	0.18	Rammer (1987)
$YBa_2Cu_3O_7$		0.2		Gurvitch and Fiory (1987a, b, c)
$YBa_2Cu_3O_7$		0.3		Tanner and Timusk (1992, p. 416)
$YBa_2Cu_3O_7$	90	2.5	0.1	Kirtley *et al.* (1987)
$Bi_2Sr_2CuO_6$		0.2		Tanner and Timusk (1992, p. 416)
$Bi_2Sr_2CaCu_2O_8$		0.3		Tanner and Timusk (1992, p. 416)
$Tl_2Ba_2CaCu_2O_8$		0.3		Foster *et al.* (1990)

[a] High and low estimates are given for high-temperature superconductors, some of which are averages of several investigators. Skriver and Mertig (1990) give coupling constants from rare earths.

and the transition temperature is then given by

$$T_c = 1.13\theta_0 \exp\left(\frac{-1}{\lambda - \mu_c^*}\right). \quad (6.124)$$

Values of λ and μ_c^* reported in the literature for various superconductors are listed in Table 6.2.

A number of other expressions for T_c have appeared in the literature. McMillan (1968) gives the following semiempirical formula:

$$T_c = \frac{\Theta_D}{1.45} \exp\left[-\frac{1.04(1+\lambda)}{\lambda - \mu_c^*(1+0.62\lambda)}\right]. \quad (6.125)$$

We will not attempt to derive or justify (6.121) to (6.125) but merely state them as they are often cited in the literature.

VII. RESPONSE OF A SUPERCONDUCTOR TO A MAGNETIC FIELD

In this section we consider the BCS Hamiltonian in the presence of a static magnetic field given by the vector potential **A**,

$$\mathbf{B} = \nabla \times \mathbf{A}. \quad (6.126)$$

The BCS Hamiltonian becomes

$$H_{\text{BCS}} = \int dx \left[\frac{\hbar^2}{2m} \left(\nabla - \frac{ie}{\hbar} A \right) \psi_\sigma^\dagger \times \left(\nabla + \frac{ie}{\hbar} A \right) \psi_\sigma - \mu \psi_\sigma^\dagger(x) \psi_\sigma(x) \right] + \tfrac{1}{2} \int dx \int dy\, \Delta_{\alpha\beta}(x,y) \psi_\beta(y) \psi_\alpha(x) + \psi_\alpha^\dagger(x) \psi_\beta^\dagger(y) \Delta_{\alpha\beta}^*(x,y), \quad (6.127)$$

where $-e$ is the charge of the electron and m is its mass.

As in (6.33), we require the free energy to be a minimum with respect to variations in the order parameter, and in addition, in the presence of a magnetic field, a minimum with respect to variation of the vector potential. Adding the free energy of the field itself to (6.32), the total free energy is

$$F = -\tfrac{1}{2} \int dx \int dy \Phi_{\alpha\beta}(x,y) V(x-y) \times \Phi_{\alpha\beta}^*(x,y) - \frac{1}{\beta} \ln \operatorname{Tr} \exp(-\beta H_{\text{BCS}}) + \frac{1}{2\mu_0} \int dx B^2(x). \quad (6.128)$$

Taking the variational derivative of (6.128) with respect to the order parameter reproduces (6.33), while the variational derivative with respect to the vector potential gives

$$\frac{\delta F}{\delta \mathbf{A}_k(z)} = -\tfrac{1}{2} \int dx \int dy \left[\frac{\delta \Phi_{\alpha\beta}(x,y)}{\delta \mathbf{A}_k(z)} V(x-y) \Phi_{\alpha\beta}^*(x,y) + \Phi_{\alpha\beta}(x,y) V(x-y) \frac{\delta \Phi_{\alpha\beta}^*(x,y)}{\delta \mathbf{A}_k(z)} \right] + \left\langle \frac{+iq\hbar}{2m} (\psi_\sigma^\dagger \nabla_k \psi_\sigma - \nabla_k \psi_\sigma^\dagger \psi_\sigma) + \frac{q^2}{m} \mathbf{A}_k \psi_\sigma^\dagger \psi_\sigma \right\rangle + \tfrac{1}{2} \int dx \int dy \langle \Delta_{\alpha\beta}(x,y) \psi_\beta(y) \psi_\alpha(x) + \psi_\alpha^\dagger(x) \psi_\beta^\dagger(y) \Delta_{\alpha\beta}^*(x,y) \rangle + \frac{1}{\mu_0} \nabla \times \mathbf{B} = 0. \quad (6.129)$$

It can be shown that by (6.33) the first and third terms cancel, while the second and fourth terms take the form of Ampère's law,

$$\nabla \times \mathbf{B} = \mu_0 \mathbf{J}, \quad (6.130)$$

where the gauge-invariant current density is

$$\mathbf{J}(x) = +\frac{ie\hbar}{2m} \langle \psi_\sigma^\dagger(x) \nabla \psi_\sigma(x) - \nabla \psi_\sigma^\dagger(x) \psi_\sigma(x) \rangle - \frac{e^2}{m} \mathbf{A} \langle \psi_\sigma^\dagger(x) \psi_\sigma(x) \rangle. \quad (6.131)$$

Expressing the fields $\psi_\sigma(x)$ in terms of the quasiparticle operators (6.66) then gives

$$\begin{aligned} \mathbf{J}(x) = {} & +\frac{ie\hbar}{2m} \sum_{k\mu} \{ [u_\sigma^{\mu *}(x;k) \nabla u_\sigma^\mu(x;k) - \nabla u_\sigma^{\mu *}(x;k) u_\sigma^\mu(x;k)] n(\mu,k) \\ & + [v_\sigma^\mu(x;k) \nabla v_\sigma^{\mu *}(x;k) - \nabla v_\sigma^\mu(x;k) v_\sigma^{\mu *}(x;k)] \times [1 - n(\mu,k)] \} \\ & - \frac{e^2}{m} \mathbf{A}(x) \sum_{k\mu} \{ u_\sigma^{\mu *}(x;k) \times u_\sigma^\mu(x;k) n(\mu,k) + v_\sigma^\mu(x;k) v_\sigma^{\mu *}(x;k) [1 - n(\mu,k)] \}. \end{aligned} \quad (6.132)$$

We now wish to calculate the current induced by a weak, slowly varying magnetic field. In the presence of a vector potential the fundamental equations determining u and v become

$$\frac{-\hbar^2}{2m}\left(\nabla + \frac{ie}{\hbar}\mathbf{A}\right)^2 u_\alpha^\mu(x;k) - \mu u_\alpha^\mu(x;k)$$
$$+\int dy\, \Delta^*_{\alpha\sigma}(x,y) v_\sigma^\mu(y;k)$$
$$= \lambda(\mu,k) u_\alpha^\mu(x,k) \qquad (6.133a)$$

and

$$\frac{-\hbar^2}{2m}\left(\nabla - \frac{ie}{\hbar}\mathbf{A}\right)^2 v_\alpha^\mu(x;k) - \mu v_\alpha^\mu(x;k)$$
$$+\int dy\, \Delta_{\alpha\sigma}(x,k) u_\sigma^\mu(y;k)$$
$$= \lambda(\mu,k) v_\sigma^\mu(x,k). \qquad (6.133b)$$

If we expand the kinetic energy and drop the term in $\mathbf{A}^2$ we have, in the gauge $\nabla\cdot\mathbf{A} = 0$,

$$\frac{-\hbar^2}{2m}\left(\nabla^2 + \frac{2ie}{\hbar}\mathbf{A}\cdot\nabla\right) u_\alpha^\mu(x;k)$$
$$+\int dy\, \Delta^*_{\alpha\sigma}(x,y) v_{\sigma'}^\mu(y;k)$$
$$= \lambda(\mu,k) u_\alpha^\mu(x,k) \qquad (6.134a)$$

and

$$\frac{-\hbar^2}{2m}\left(\nabla^2 + \frac{2ie}{\hbar}\mathbf{A}\cdot\nabla\right) v_\alpha^\mu(x;k)$$
$$+\int dy\, \Delta_{\alpha\sigma}(x,y) u_\sigma^\mu(y;k)$$
$$= \lambda(\mu,k) v_\alpha^\mu(x,k). \qquad (6.134b)$$

If there is no magnetic field present, the solutions to (6.133–6.134) will be plane waves,

$$u_\sigma^\mu(x;k) = u_\sigma^\mu(k) e^{ik\cdot x},$$
$$v_\sigma^\mu(x;k) = v_\sigma^\mu(k) e^{ik\cdot x}. \qquad (6.135)$$

Substituting these in (6.134) then gives

$$\left[\frac{\hbar^2k^2}{2m} - \mu + \frac{\hbar e}{m}\mathbf{k}\cdot\mathbf{A}(x)\right] u_\sigma^\mu(x;k)$$
$$+\int dy\, \Delta^*_{\sigma\alpha}(x,y) v_\alpha^\mu(y;k)$$
$$= \lambda(\mu,k) u_\sigma^\mu(x;k), \qquad (6.136a)$$

$$\left[\frac{\hbar^2k^2}{2m} - \mu - \frac{\hbar e}{m}\mathbf{k}\cdot\mathbf{A}(x)\right] v_\sigma^\mu(x;k)$$
$$+\int dy\, \Delta_{\sigma\alpha}(x,y) u_\alpha^\mu(y;k)$$
$$= \lambda(\mu,k) v_\sigma^\mu(x;k'), \qquad (6.136b)$$

or, on rearranging terms,

$$\left(\frac{\hbar^2k^2}{2m} - \mu\right) u_\sigma^\mu(x;k)$$
$$+\int dy\, \Delta^*_{\sigma\alpha}(x,y) v_\alpha^\mu(y;k)$$
$$= \left[\lambda - \frac{\hbar e}{m}\mathbf{k}\cdot\mathbf{A}(x)\right] u_\sigma^\mu(x;k), \qquad (6.137a)$$

$$\left(\frac{\hbar^2k^2}{2m} - \mu\right) v_\sigma^\mu(x;k)$$
$$+\int dy\, \Delta_{\sigma\alpha}(x,y) u_\alpha^\mu(y;k)$$
$$= \left[\lambda - \frac{\hbar e}{m} k\cdot\mathbf{A}(x)\right] v_\sigma^\mu(x;k). \qquad (6.137b)$$

If we assume that $\mathbf{A}(x)$ is slowly varying, an approximate solution to (6.137a, b) is to let the eigenvalue λ follow the vector potential adiabatically, i.e.,

$$\lambda(k) \cong \lambda_0(k) + \frac{\hbar e}{m}\mathbf{k}\cdot\mathbf{A}(x). \qquad (6.138)$$

Under this assumption u and v are given by their solutions in the absence of a magnetic field.

The current density now becomes, to first order in **A**,

$$J_i(x) = -\frac{e^2\hbar^2}{m^2}\sum_{k,\mu} \times\left[u_\sigma^{\mu*}(k)u_\sigma^\mu(k) + v_\sigma^{\mu*}(k)v_\sigma^\mu(k)\right] \times k_i\mathbf{k}\cdot\mathbf{A}(x)\frac{\partial n}{\partial\lambda} - \frac{e^2}{m}nA_i(x), \quad (6.139)$$

where we have used the fact that J vanishes for $\mathbf{A}=\mathbf{0}$ and noted that

$$\sum_{k,\mu} u_\sigma^{\mu*}(x,k)v_\sigma^\mu(x;k)n(\mu,k) + v_\sigma^\mu(x,k)v_\sigma^{\mu*}[1-n(\mu,k)] = n, \quad (6.140)$$

where n is the total electron density.

For singlet pairing, in which the spins of the electrons of a Cooper pair are antiparallel, appropriate choices for $u_\sigma^\mu(k)$ and $v_\sigma^\mu(k)$ are

$$u_\sigma^\mu(k) = \tfrac{1}{2}\delta_\sigma^\mu\left(1+\frac{\xi(k)}{\lambda(k)}\right), \quad (6.141a)$$

$$v_\sigma^\mu(k) = \tfrac{1}{2}\epsilon_\sigma^\mu\left(1-\frac{\xi(k)}{\lambda(k)}\right), \quad (6.141b)$$

which gives

$$J_i(x) = \frac{2e^2\hbar^2}{m^2\Omega}\sum_k k_i[k\cdot\mathbf{A}(x)] \times\frac{\partial n}{\partial\lambda} - \frac{e^2}{m}nA_i(x). \quad (6.142)$$

Taking limits so that the sum becomes an integral in the usual way, we have

$$\mathbf{J}(x) = \frac{e^2}{m}n_s\mathbf{A}(x), \quad (6.143)$$

where the density of "super electrons" is given by

$$n_s = n - \frac{\hbar^2}{3\pi^2 m}\int_0^\infty k^4\left[-\frac{\partial n}{\partial\lambda}(k)\right]dk. \quad (6.144)$$

Note that we have used the spherical symmetry of the integral in (6.144) to write

$$\int k_ik_j\frac{\partial n}{\partial\lambda}(k)d^3k = \frac{\delta_{ij}}{3}\int k^2\frac{\partial n}{\partial\lambda}(k)d^3k. \quad (6.145)$$

Equation (6.145) expresses the Meissner effect since, by Maxwell's equation,

$$\nabla\times\mathbf{B} = \mu_0\mathbf{J}, \quad (6.146)$$

the vector potential (again $\nabla\cdot\mathbf{A}=0$) satisfies

$$\nabla^2\mathbf{A} = \lambda_L^{-2}\mathbf{A}, \quad (6.147)$$

where

$$\lambda_L^2 = \frac{m}{\mu_0 e^2 n_s} \quad (6.148)$$

is the London penetration depth.

At low temperatures the average quasiparticle occupation number

$$n(\lambda) = \frac{1}{e^{\beta\lambda}+1} \quad (6.149)$$

vanishes and the number of super electrons equals the total number of electrons in the metal.

FURTHER READING

Some of the classical articles on the BCS theory have already been mentioned at the beginning of the chapter. The article by Cooper (1956), predicting the formation of the "pairs" that bear his name, provided the setting for the BCS theory formulated by Bardeen, Cooper, and Schrieffer in 1957, and elaborated upon in the books by de Gennes (1966), Fetter and Walecka (1971), and Schrieffer (1964).

The textbook by Tinkham (1985) provides a good introduction to the BCS theory, and Tilley and Tilley (1986) give a briefer introduction.

Gorkov (1959) showed that the Ginzburg–Landau theory, which was discussed in the previous chapter, follows from the BCS theory. This provided a solid theoretical foundation for the GL theory.

Chapters 4–14 of the book, Theories of High Temperature Superconductivity (Halley, 1988), discuss applications of the BCS theory. Allen (1990) reviewed the BCS approach to electron pairing. The pairing state in $YBa_2Cu_3O_{7-\delta}$ is discussed by Annett *et al.* (1990). We will cite some representative articles.

The weak and strong limits of BCS have been discussed (Cohen, 1987; Cohen and Penn, 1990; Entin-Wohlman and Imry, 1989; Nasu, 1990). There is a crossover between a BCS proper regime of weakly coupled, real space-overlapping Cooper pairs and a Bose–Einstein regime involving a low density boson gas of tightly bound fermion pairs (Pistolesi and Strinati, 1994; Quick *et al.*, 1993; Tokumitu *et al.*, 1993). The BCS theory has been applied to high temperature superconductors (Berlinsky *et al.*, 1993; Ihm and Yu, 1989; Japiassu *et al.*, 1992; Jarrell *et al.*, 1988; Kitazawa and Tajima, 1990; Lal and Joshi, 1992; Lu *et al.*, 1989; Marsiglio, 1991; Marsiglio and Hirsch, 1991; Penn and Cohen, 1992; Pint and Schachinger, 1991; Sachdev and Wang, 1991).

The present chapter, although based in part on the electron–phonon coupling mechanism (Jiang and Carbotte, 1992b; Kirkpatrick and Belitz, 1992; Kresin *et al.*, 1993; Marsiglio and Hirsch, 1994; Nicol and Carbotte, 1993; Zheng *et al.*, 1994), is nevertheless much more general in its formalism. Unconventional phonon or nonphonon coupling can also occur (Bussmmann-Holder and Bishop, 1991; Cox and Maple, 1995; Dobroliubov and Khlebnikov, 1991; Keller, 1991; Klein and Aharony, 1992; Krüger, 1989; Spathis *et al.*, 1992; Tsay *et al.*, 1991; Van Der Marel, 1990), involving, for example, excitons (Bala and Olés, 1993; Gutfreund and Little, 1979; Takada, 1989), plasmons (quantized plasma oscillations; Côte and Griflin, 1993; Cui and Tsai, 1991; Ishii and Ruvalds, 1993), polaritons (Lue and Sheng, 1993), polarons (electron plus induced lattice polarization; Kabanov and Mashtakov, 1993; Konior, 1993; Nettel and MacCrone, 1993; Wood and Cooke, 1992) and bipolarons (de Jongh, 1992; Emin, 1994; Khalfin and Shapiro, 1992). Both *s*-wave and *d*-wave pairings have been considered (Anlage *et al.*, 1994; Carbotte and Jiang, 1993; Côte and Griflin, 1993; Lenck and Carbotte, 1994; Li *et al.*, 1993; Scalapino, 1995; Wengner and Östlund, 1993; Won and Maki, 1994).

Some authors question the applicability of BCS to high temperature superconductors (Collins *et al.*, 1991; Kurihara, 1989). Tesanovic and Rasolt (1989) suggested a new type of superconductivity in very high magnetic fields in which there is no upper critical field. The BCS theory has been examined in terms of the Hubbard (Falicov and Proetto, 1993; Micnas *et al.*, 1990; Sofo *et al.*, 1992) and Fermi liquid (Horbach *et al.*, 1993, Ramakumar, 1993) approaches, which are discussed in Chapter 8.

Carbotte (1990) reviewed Eliashberg (1960a, b) theory and its relationship with BCS. Representative articles concern (a) high temperature superconductors (Jin *et al.*, 1992; Lu *et al.*, 1989; Marsiglio, 1991; Monthoux and Pines, 1994; Sulewski *et al.*, 1987; Wermbter and Tewordt, 1991a; Williams and Carbotte, 1991), (b) anisotropies (Combescot, 1991; Lenck *et al.*, 1990; Radtke *et al.*, 1993; Zhao and Callaway, 1994), (c) transport properties (Kulic and Zeyher, 1994; Ullah and Dorsey, 1991), (d) weak coupling limits (Combescot, 1990; Crisan, 1887), and (e) strong coupling limits (Bulaevskii *et al.*, 1988; Heid, 1992 (Pb); Rammer, 1991).

7

Perovskite and Cuprate Crystallographic Structures

I. INTRODUCTION

Chapter 3 shows that the majority of single-element crystals have highly symmetrical structures, generally fcc or bcc, in which their physical properties are the same along the three crystallographic directions *x*, *y*, and *z*. The NaCl-type and *A*15 compounds are also cubic. Some compounds do have lower symmetries, showing that superconductivity is compatible with many different types of crystallographic structure, but higher symmetries are certainly more common. In this chapter we will describe the structures of the high-temperature superconductors, almost all of which are tetragonal or orthorhombic.

In Chapter 3, we also gave some examples of the role played by structure in determining the properties of superconductors. The highest transition temperatures in alloys of transition metals are at the boundaries of instability between the bcc and hcp forms. The NaCl-type compounds have ordered vacancies on one or another lattice site. The magnetic and superconducting properties of the Chevrel phases depend on whether the large magnetic cations (i.e., positive ions) occupy eightfold sites surrounded by chalcogenide ions or whether the small magnetic ions occupy octahedral sites surrounded by Mo ions.

The structures described here are held together by electrons that form ionic or covalent bonds between the atoms. No account is taken of the conduction electrons, which are delocalized over the copper oxide planes and form Cooper pairs responsible for the superconducting properties below T_c. The following chapter will be devoted to explaining the role of these conduction electrons within the frameworks of the Hubbard model and band theory. Whereas the present chapter describes atom positions in coordinate space, the following chapter relies on a reciprocal lattice elucidation of these same materials.

We begin with a description of perovskite and explain some reasons that perovskite undergoes various types of distortions. This prototype exhibits a number of characteristics that are common to the high-temperature superconducting cuprates (see Section V). We will emphasize the structural commonalities of these materials and make frequent comparisons between them. Our earlier work (Poole *et al.*, 1988) and the comprehensive review by Yvon and François (1989) may be consulted for more structural detail on the atom positions, interatomic spacings, site symmetries, etc., of these compounds. There have been reports of superconductivity in certain other cuprate structures (e.g., Murphy *et al.*, 1987), but these will not be reported on in this chapter.

There is a related series of layered compounds $Bi_2O_2(M_{m-1}R_mO_{3m+1})$ called Aurivillius (1950, 1951, 1952) phases, with the 12-coordinated $M =$ Ca, Sr, Ba, Bi, Pb, Cd, La, Sm, Sc, etc., and the 6-coordinated transition metal $R =$ Nb, Ti, Ta, W, Fe, etc. The $m = 1$ compound Bi_2NbO_6 belongs to the same tetragonal space group $I4/mmm$, D_{4h}^{17} as the lanthanum, bismuth, and thallium high temperature superconductors (Medvedeva *et al.*, 1993).

We assume that all samples are well made and safely stored. Humidity can affect composition, and Garland (1988) found that storage of $YBa_2Cu_3O_{7-\delta}$ in 98% humidity exponentially decreased the diamagnetic susceptibility with a time constant of 22 days.

II. PEROVSKITES

Much has been written about the high-temperature superconductors being perovskite types, so we will begin by describing the structure of perovskites. The prototype compound barium titanate, $BaTiO_3$, exists in three crystallographic forms with the following lattice constants and unit cell volumes (Wyckoff, 1964):

$$\begin{array}{lll} \text{cubic:} & a=b=c=4.0118\ \text{Å} & V=64.57\ \text{Å}^3 \\ \text{tetragonal:} & a=b=3.9947,\ c=4.0336 & V=64.37\ \text{Å}^3 \\ \textit{ortho}\text{rhombic:} & a=4.009\sqrt{2}\ \text{Å},\ b=4.018\sqrt{2}\ \text{Å},\ c=3.990\ \text{Å} & V=2(64.26)\ \text{Å}^3 \end{array} \quad (7.1)$$

For all three cases the crystallographic axes are mutually perpendicular. We will comment on each case in turn.

A. Cubic Form

Above 201°C barium titanate is cubic and the unit cell contains one formula unit $BaTiO_3$ with a titanium atom on each apex, a barium atom in the body center, and an oxygen atom on the center of each edge of the cube, as illustrated in Fig. 7.1. This corresponds to the barium atom, titanium atom, and three oxygen atoms being placed in positions with the following x, y, and z coordinates:

$$\begin{array}{lll} \text{E site: Ti} & (0,0,0) & \text{Ti on apex} \\ \text{F site: O} & (0,0,\tfrac{1}{2});\ (0,\tfrac{1}{2},0);\ (\tfrac{1}{2},0,0) & \text{three oxygens centered on edges} \\ \text{C site: Ba} & (\tfrac{1}{2},\tfrac{1}{2},\tfrac{1}{2}) & \text{Ba in center.} \end{array} \quad (7.2)$$

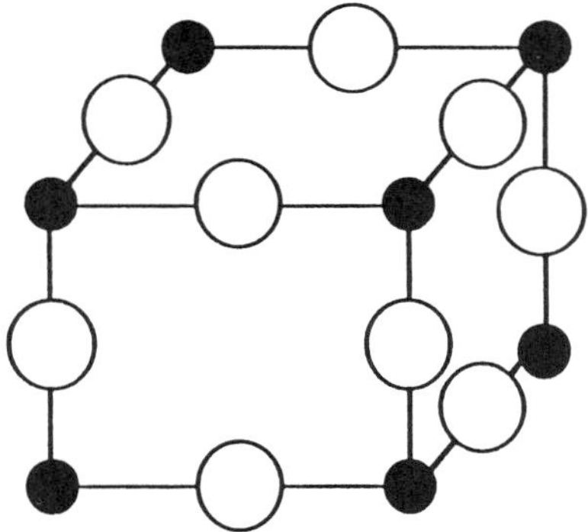

Figure 7.1 Barium titanate ($BaTiO_3$) perovskite cubic unit cell showing titanium (small black circles) at the vertices and oxygen (large white circles) at the edge-centered positions. Ba, not shown, is at the body center position (Poole *et al.*, 1988, p. 73).

The barium in the center has 12 nearest-neighbor oxygens, so we say that it is 12-fold coordinated, while the titanium on each apex has 6-fold (octahedral) coordination with the oxygens, as may be seen from the figure. (The notation *E* for edge, *F* for face, and *C* for center is adopted for reasons that will become clear in the discussion which follows.) Throughout this chapter we will assume that the *z*-axis is oriented vertically, so that the *x* and *y* axes lie in the horizontal plane.

Ordinarily, solid-state physics texts place the origin (0, 0, 0) of the perovskite unit cell at the barium site, with titanium in the center and the oxygens at the centers of the cube faces. Our choice of origin facilitates comparison with the structures of the oxide superconductors.

This structure is best understood in terms of the sizes of the atoms involved. The ionic radii of O^{2-} (1.32 Å) and Ba^{2+} (1.34 Å) are almost the same, as indicated in Table 7.1, and together they form a perfect fcc lattice with the smaller Ti^{4+} ions (0.68 Å) located in octahedral holes surrounded entirely by oxygens. The octahedral holes of a close-packed oxygen lattice have a radius of 0.545 Å; if these holes were empty the lattice constant would be $a = 3.73$ Å, as noted in Fig. 7.2a. Each titanium pushes the surrounding oxygens outward, as shown in Fig. 7.2b, thereby increasing the lattice constant. When the titanium is replaced by a larger atom, the lattice constant expands further, as indicated by the data in the last column of Table 7.2. When Ba is replaced by the smaller Ca (0.99 Å) and Sr (1.12 Å) ions, by contrast, there is a corresponding decrease in the lattice constant, as indicated by the data in columns 3 and 4, respectively, of Table 7.2. All three alkaline earths, Ca, Sr, and Ba, appear prominently in the structures of 3 high-temperature superconductors.

B. Tetragonal Form

At room temperature barium titanate is tetragonal and the deviation from cubic, $(c - a)/\frac{1}{2}(c + a)$, is about 1%. All of the atoms have the same x, y coordinates as in the cubic case, but are shifted along the *z*-axis relative to each other by ≈ 0.1 Å, producing the puckered arrangement shown in Fig. 7.3. The distortions from the ideal structure are exaggerated in this sketch. The puckering bends the Ti–O–Ti group so that the Ti–O distance increases while the Ti–Ti distance remains almost

Table 7.1 Ionic Radii for Selected Elements[a]

Small	Cu^{2+}	0.72 Å	Bi^{5+}	0.74 Å
Small–Medium	Cu^{+}	0.96 Å	Y^{3+}	0.94 Å
	Bi^{3+}	0.96 Å	Tl^{3+}	0.95 Å
	Ca^{2+}	0.99 Å	Bi^{3+}	0.96 Å
	Nd^{3+}	0.995 Å		
Medium–Large	Hg^{2+}	1.10 Å		
	Sr^{2+}	1.12 Å	La^{3+}	1.14 Å
	Pb^{2+}	1.20 Å	Ag^{+}	1.26 Å
Large	K^{+}	1.33 Å	O^{2-}	1.32 Å
	Ba^{2+}	1.34 Å	F^{-}	1.33 Å

[a] See Table VI-2 of Poole *et al.* (1988) for a more extensive list.

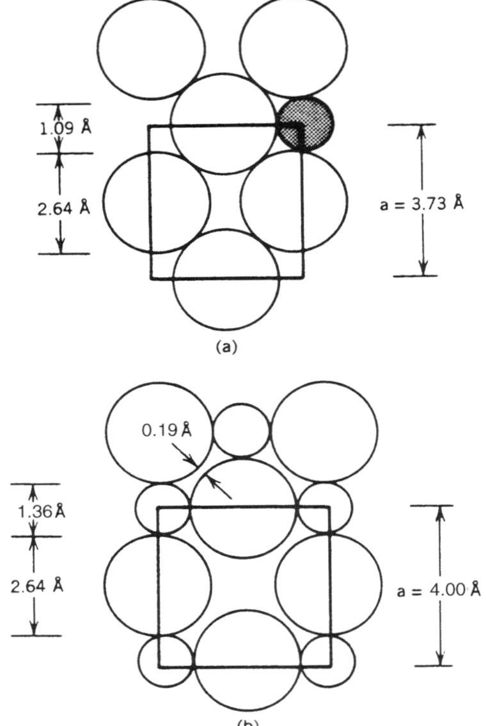

Figure 7.2 Cross section of the perovskite unit cell in the $z = 0$ plane showing (a) the size of the octahedral hole (shaded) between oxygens (large circles), and (b) oxygens pushed apart by the transition ions (small circles) in the hole sites. For each case the lattice constant is indicated on the right and the oxygen and hole sizes on the left (Poole *et al.*, 1988, p. 77).

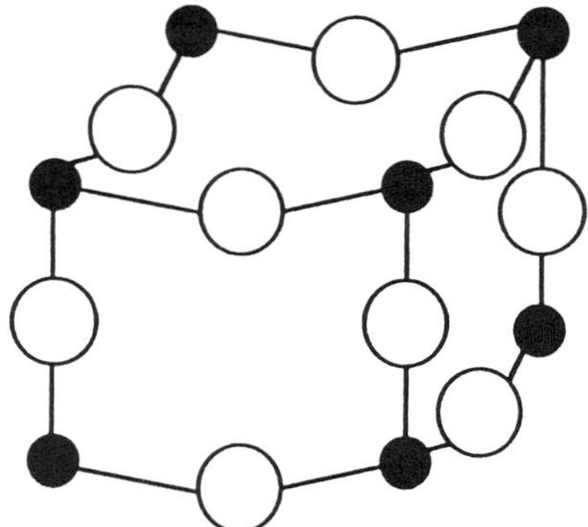

Figure 7.3 Perovskite tetragonal unit cell showing puckering of Ti–O layers that are perfectly flat in the cubic cell of Fig. 7.1. The notation of Fig. 7.1 is used (Poole *et al.*, 1988, p. 75).

the same. This has the effect of providing more room for the titanium atoms to fit in their lattice sites. We will see later that a similar puckering distortion occurs in the high-temperature superconductors as a way of providing space for the Cu atoms in the planes.

C. Orthorhombic Form

There are two principal ways in which a tetragonal structure distorts to form an orthorhombic phase. The first, shown at

Table 7.2 Dependence of Lattice Constants a of Selected Perovskites AMO_3 on Alkaline Earth A and Ionic Radius of Transition Metal Ion M^{+4}; the Alkaline Earth Ionic Radii are 0.99 Å (Ca), 1.12 Å (Sr), and 1.34 Å (Ba)[a]

Transitional metal	Transitional metal radius, Å	Lattice constant a, Å		
		Ca	Sr	Ba
Ti	0.68	3.84	3.91	4.01
Fe	—	—	3.87	4.01
Mo	0.70	—	3.98	4.04
Sn	0.71	3.92	4.03	4.12
Zr	0.79	4.02	4.10	4.19
Pb	0.84	—	—	4.27
Ce	0.94	3.85	4.27	4.40
Th	1.02	4.37	4.42	4.80

[a] Data from Wyckoff (1964, pp. 391ff).

the top of Fig. 7.4, is for the b-axis to stretch relative to the a-axis, resulting in the formation of a rectangle. The second, shown at the bottom of the figure, is for one diagonal of the ab square to stretch and the other diagonal to compress, resulting in the formation of a rhombus. The two diagonals are perpendicular, rotated by 45° relative to the original axes, and become the a', b' dimensions of the new orthorhombic unit cell, as shown in Fig. 7.5. These a', b' lattice constants are $\approx \sqrt{2}$ times longer than the original constants, so that the volume of the unit cell roughly doubles; thus, it contains exactly twice as many atoms. (The same $\sqrt{2}$ factor appears in Eq. 7.1 in our discussion of the lattice constants for the orthorhombic form of barium titanate.)

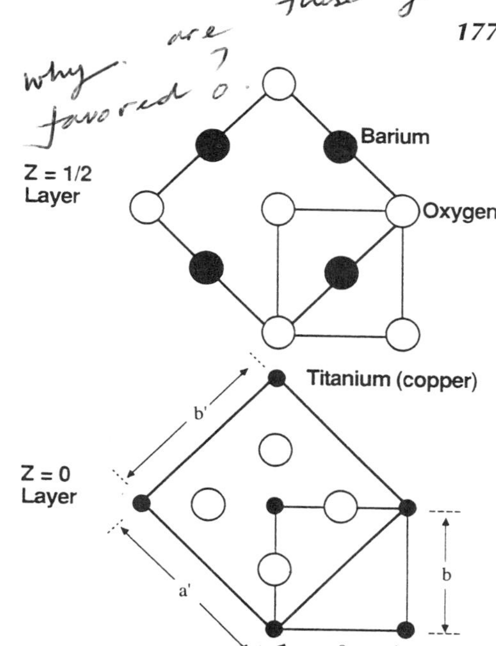

Figure 7.5 Rhombal expansion of monomolecular tetragonal unit cell (small squares, lower right) to bimolecular orthorhombic unit cell (large squares) with new axes 45° relative to the old axes. The atom positions are shown for the $z = 0$ and $z = \frac{1}{2}$ layers (Poole *et al.*, 1988, p. 76).

When barium titanate is cooled below 5°C it undergoes a diagonal- or rhombal-type distortion. The atoms have the same z coordinates ($z = 0$ or $\frac{1}{2}$) as in the cubic phase, so the distortion occurs entirely in the x, y-plane, with no puckering of the atoms. The deviation from tetragonality, as given by the percentage of anisotropy,

$$\% \text{ ANIS} = \frac{100|b - a|}{\frac{1}{2}(b + a)} = 0.22\%, \quad (7.3)$$

is less than that of most orthorhombic copper oxide superconductors. We see from Fig. 7.5 that in the cubic phase the oxygen atoms in the $z = 0$ plane are separated by 0.19 Å. The rhombal distortion increases this O–O separation in one direction and decreases it in the other, in the manner indicated in Fig. 7.6a, to produce the Ti nearest-neighbor configuration shown in Fig. 7.6b. This arrangement helps to fit the titanium into its lattice site.

The transformation from tetragonal to orthorhombic is generally of the rhombal type for $(La_{1-x}Sr_x)_2CuO_4$ and of the rectilinear type for $YBa_2Cu_3O_{7-\delta}$.

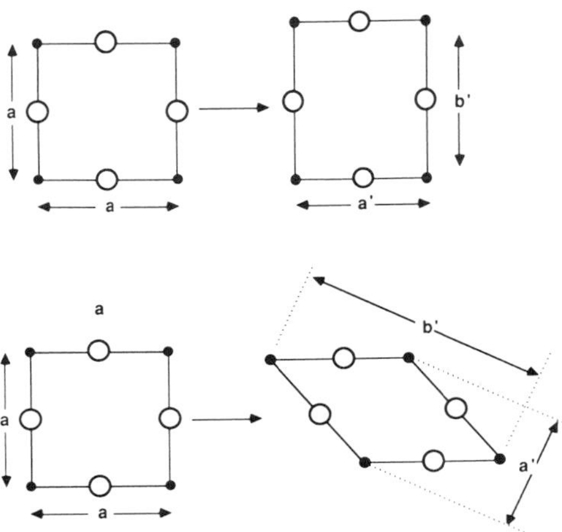

Figure 7.4 Rectangular- (top) and rhombal- (bottom) type distortions of a two-dimensional square unit cell of width a (Poole *et al.*, 1989).

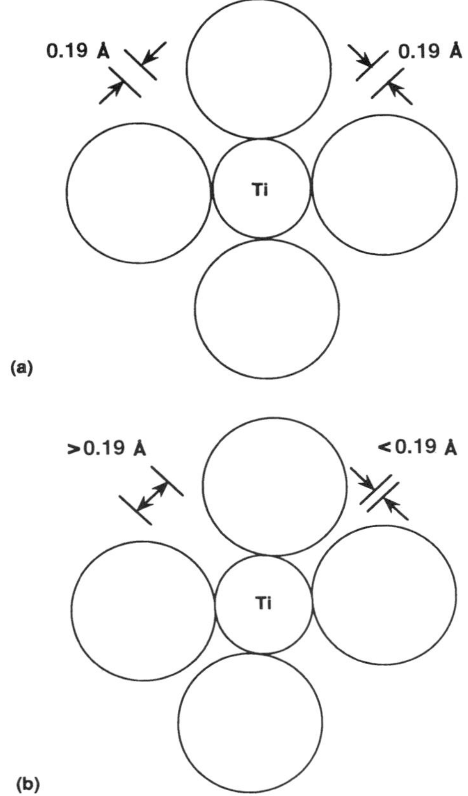

Figure 7.6 Shift of the oxygens in the *a*, *b*-plane around the titanium atom of perovskite from the room-temperature tetragonal (and cubic) configuration (a) to the rhombal configuration (b) of its low-temperature orthorhombic structure.

D. Planar Representation

Another way of picturing the structure of perovskite is to think of the atoms as forming horizontal planes. If we adopt the notation [E F C] to designate the occupation of the E, F, and C sites, the sketches of perovskite presented in Figs. 7.1 and 7.3 follow the scheme

$$
\begin{array}{lll}
z = 1 & [\mathrm{TiO_2}-] & \text{Ti at E, O at two F sites} \\
z = \frac{1}{2} & [\mathrm{O{-}Ba}] & \text{O at E, Ba at C} \\
z = 0 & [\mathrm{TiO_2}-] & \text{Ti at E, O at two F sites.}
\end{array}
\tag{7.4}
$$

The planes at the heights $z = 0$, $\frac{1}{2}$, and 1 can be labeled using this notation. The usefulness of this labeling scheme will be clarified in Section V.

This completes our treatment of the structure of perovskite. We encountered many features that we will meet again in the analogous superconductor cases, and established notation that will be useful in describing the structure of the cuprates. However, before proceeding we present details about a cubic and a close-to-cubic perovskite superconductor in the following two sections.

III. CUBIC BARIUM POTASSIUM BISMUTH OXIDE

The compound

$$\mathrm{Ba}_{1-x}\mathrm{K}_x\mathrm{BiO}_{3-y},$$

which forms for $x > 0.25$, crystallizes in the cubic pervoskite structure with $a = 4.29$ Å (Cava *et al.*, 1988; Jin *et al.*, 1992; Mattheiss *et al.*, 1988). K^+ ions replace some of the Ba^{2+} ions in the C site, and Bi ions occupy the E sites of Eq. (7.2) (Hinks *et al.*, 1988b; Kwei *et al.*, 1989; Pei *et al.*, 1990; Salem-Sugui *et al.*, 1991; Schneemeyer *et al.*, 1988). Some oxygen sites are vacant, as indicated by *y*. Hinks *et al.* (1989) and Pei *et al.* (1990) determined the structural phase diagram (cf. Kuentzler *et al.*, 1991; Zubkus *et al.*, 1991). We should note from Table 7.1 that the potassium (1.33 Å) and barium (1.32 Å) ions are almost the same size, and that Bi^{5+} (0.74 Å) is close to Ti^{4+} (0.68 Å). Bismuth represents a mixture of the valence states Bi^{3+} and Bi^{5+} which share the Ti^{4+} site in a proportion that depends on *x* and *y*. The larger size (0.96 Å) of the Bi^{3+} ion causes the lattice constant *a* to expand 7% beyond its cubic $BaTiO_3$ value. Oxygen vacancies help to compensate for the larger size of Bi^{3+}.

It is noteworthy that $\mathrm{Ba}_{1-x}\mathrm{K}_x\mathrm{BiO}_{3-y}$ becomes superconducting at a temperature (≈ 40 K for $x \approx 0.4$) that is higher than the T_c of all of the *A*15 compounds. This compound, which has no copper, has been widely studied in the quest for clues that would elucidate the mechanism of high-temperature superconductivity. Features of $Ba_{1-x}K_xBiO_{3-y}$, such as the fact that it contains a variable valence state ion and utilizes oxygen vacancies to achieve charge compensation, reappear in the high-temperature superconducting compounds.

IV. BARIUM LEAD BISMUTH OXIDE

In 1983 Mattheiss and Hamann referred to the 1975 "discovery by Sleight *et al.* of high-temperature superconductivity" in the compound $BaPb_{1-x}Bi_xO_3$ in the composition range $0.05 \leq x \leq 0.3$ with T_c up to 13 K. Many consider this system, which disproportionates $2Bi^{4+} \rightarrow Bi^{3+} + Bi^{5+}$ in going from the metallic to the semiconducting state, as a predecessor to the LaSrCuO system.

The metallic compound $BaPbO_3$ is a cubic perovskite with the relatively large lattice constant (Wyckoff, 1964; cf. Nitta *et al.*, 1965; Shannon and Bierstedt, 1970) listed in Table 7.3. At room temperature semiconducting $BaBiO_3$ is monoclinic ($a \approx b \approx c/\sqrt{2}$, $\beta = 90.17°$), but close to orthorhombic (Chaillout *et al.*, 1985; Cox and Sleight, 1976, 1979; cf. Federici *et al.*, 1990; Jeon *et al.*, 1990; Shen *et al.*, 1989). These two compounds form a solid solution series $BaPb_{1-x}Bi_xO_3$ involving cubic, tetragonal, orthorhombic, and monoclinic modifications. Superconductivity appears in the tetragonal phase, and the metal-to-insulator transition occurs at the tetragonal-to-orthorhombic phase boundary $x \approx 0.35$ (Gilbert *et al.*, 1978; Koyama and Ishimaru, 1992; Mattheiss, 1990; Mattheiss and Hamann, 1983; Sleight, 1987; cf. Bansil *et al.*, 1991; Ekino and Akimitsu, 1989a, b; Papaconstantopoulous *et al.*, 1989).

The compound resembles

$$Ba_{1-x}K_xBiO_{3-y}$$

with its variable Bi valence states, but it differs in not exhibiting superconductivity in the cubic phase.

V. PEROVSKITE-TYPE SUPERCONDUCTING STRUCTURES

In their first report on high-temperature superconductors Bednorz and Müller (1986) referred to their samples as "metallic, oxygen-deficient ... perovskite-like mixed-valence copper compounds." Subsequent work has confirmed that the new superconductors do indeed possess these characteristics.

In the oxide superconductors Cu^{2+} replaces the Ti^{4+} of perovskite, and in most cases the TiO_2-perovskite layering is retained as a CuO_2 layering with two oxygens per copper. Because of this feature of CuO_2 layers, which is common to all of the high-temperature superconductors, such superconductors exhibit a uniform lattice size in the a, b-plane, as the data in Table 7.3 demonstrate. The compound $BaCuO_3$ does not occur because the Cu^{4+} ion does not form, but this valence constraint is overcome by replacement of Ba^{2+} by a trivalent ion, such as La^{3+} or Y^{3+}, by a reduction in the oxygen content, or by both. The result is a set of "layers" containing only one oxygen per cation located between each pair of CuO_2 layers, or none at all. Each high-temperature superconductor has a unique sequence of layers.

We saw from Eq. (7.2) that each atom in perovskite is located in one of three types of sites. In like manner, each atom at the height z in a high-temperature superconductor occupies either an Edge (E) site on the edge $(0, 0, z)$, a Face (F) site on the midline of a face ($(0, \frac{1}{2}, z)$ or $(\frac{1}{2}, 0, z)$ or both), or a Centered (C) site centered within the unit cell on the z-axis $(\frac{1}{2}, \frac{1}{2}, z)$. The site occupancy notation [E F C] is used because many cuprates contain a succession of [Cu O_2 −] and [− O_2 Cu] layers in which the Cu atom switches between edge and centered sites, with the oxygens

Table 7.3 Crystallographic Characteristics of Oxide Superconducting and Related Compounds[a]

Compound	Symbol	Symm	Type	Enlarg.	Form. units	a_0(Å)	c_0(Å)	c_0/Cu	%Anis	T_c (K)	Comments
$BaTiO_3$	—	C	A	1	1	4.012	4.012	—	0	—	$T > 200°C$
$BaTiO_3$	—	T	A	1	1	3.995	4.03	—	0	—	20°C
$BaTiO_3$	—	O	A	$\sqrt{2}$	2	$4.013\sqrt{2}$	3.990	—	0.23	—	T < 5°C
$BaPbO_3$	—	C	A	1	1	4.273	4.273	—	0	0.4	
$BaPb_{0.7}Bi_{0.3}O_3$	—	T	S	$\sqrt{2}$	4	$4.286\sqrt{2}$	4.304	—	0	12	
$BaBiO_3$	—	M	A	$\sqrt{2}$	2	$4.355\sqrt{2}$	4.335	—	0.13	—	$\beta = 90.17°$
$Ba_{0.6}K_{0.4}BiO_3$	—	C	A	1	1	4.293	4.293	—	0	30	
La_2CuO_4	0201	T	S	1	2	3.81	13.18	6.59	0	35	Sr, doped
La_2CuO_4	0201	O	S	$\sqrt{2}$	4	$3.960\sqrt{2}$	13.18	6.59	6.85	35	Sr, doped
$YBa_2Cu_3O_8$	0213	T	A	1	1	3.902	11.94	3.98	0	—	
$YBa_2Cu_3O_7$	0213	O	A	1	1	3.855	11.68	3.89	1.43	92	
$Bi_2Sr_2CaCu_2O_8$	2212	T	S	$5\sqrt{2}$	20	$3.81\sqrt{2}$	30.6	7.65	0	84	
$Bi_2Sr_2Ca_2Cu_3O_{10}$	2223	O	S	$5\sqrt{2}$	20	$3.83\sqrt{2}$	37	6.17	0.57	110	
$Tl_2Ba_2CuO_6$	2201	T	S	1	2	3.83	23.24	11.6	0	90	
$Tl_2Ba_2CaCu_2O_8$	2212	T	S	1	2	3.85	29.4	7.35	0	110	
$Tl_2Ba_2Ca_2Cu_3O_{10}$	2223	T	S	1	2	3.85	35.88	5.98		125	
$TlBa_2CuO_5$	1201		A	1	1		9.5	9.5		< 17	
$TlBa_2CaCu_2O_7$	1212		A	1	1		12.7	6.35		91	
$TlBa_2Ca_2Cu_3O_9$	1223		A	1	1		15.9	5.3		116	
$TlBa_2Ca_3Cu_4O_{11}$	1234		A	1	1		19.1	4.78		122	
$TlBa_2Ca_4Cu_5O_{13}$	1245		A	1	1		22.3	4.46		< 120	
$HgBa_2CuO_4$	1201	T	A	1	1	3.86	9.5	9.5		95	
$HgBa_2CaCu_2O_6$	1212	T	A	1	1	3.86	12.6	6.3		122	
$HgBa_2Ca_2Cu_3O_8$	1223	T	A	1	1	3.86	17.7	5.2		133	

[a] Symbol, symmetry (cubic C, tetragonal T, orthorhombic O, monoclinic M); type (aligned A, staggered S); enlargement in a, b-plane (diagonal distortion $\sqrt{2}$, superlattice 5); formula units per unit cell; lattice parameters (a_0, c_0, c_0, and c_0 per Cu ion); % anisotropy; and transition temperature T_c. For the orthorhombic compounds tabulated values of a_0 are averages of a_0 and b_0.

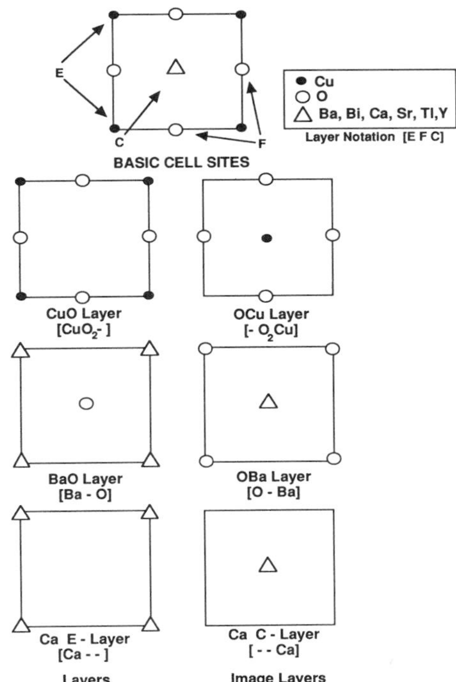

Figure 7.7 Types of atom positions in the layers of a high-temperature superconductor structure, using the edge, face, center notation [*E F C*]. Typical site occupancies are given in the upper right (Poole *et al.*, 1989).

remaining at their face positions. Similar alternations in position take place with Ba, O, and Ca layers, as illustrated in Fig. 7.7.

Hauck *et al.* (1991) proposed a classification of superconducting oxide structures in terms of the sequence (1) superconducting layers [Cu O_2 –] and [– O_2 Cu], (2) insulating layers, such as [Y – –] or [– – Ca], and (3) hole-donating layers, such as [Cu O^b –] or [Bi – O].

The high-temperature superconductor compounds have a horizontal reflection plane ($\perp$ to z) called σ_h at the center of the unit cell and another σ_h reflection plane at the top (and bottom). This means that every plane of atoms in the lower half of the cell at the height z is duplicated in the upper half at the height $1-z$. Such atoms, of course, appear twice in the unit cell, while atoms right on the symmetry planes only occur once since they cannot be reflected. Figure 7.8 shows a [Cu O_2 –] plane at a height z reflected to the height $1-z$. Note how the puckering preserves the reflection symmetry operation. Superconductors that have this reflection plane, but lack end-centering and body-centering op-

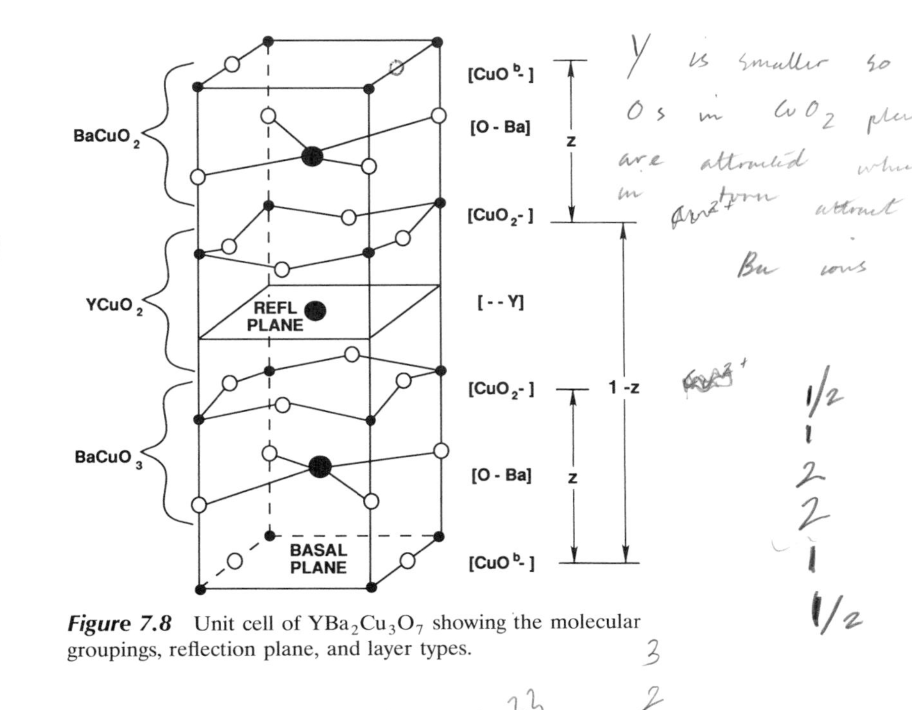

Figure 7.8 Unit cell of $YBa_2Cu_3O_7$ showing the molecular groupings, reflection plane, and layer types.

erations (see Section VII), are called *aligned* because all of their copper atoms are of one type; either all on the edge $(0,0,z)$ in E positions or all centered $(\frac{1}{2},\frac{1}{2},z)$ at C sites. In other words, they all lie one above the other on the same vertical lines, as do the Cu ions in Fig. 7.8.

VI. ALIGNED $YBa_2Cu_3O_7$

The compound $YBa_2Cu_3O_7$, sometimes called YBaCuO or the 123 compound, in its orthorhombic form is a superconductor below the transition temperature $T_c \approx 92$ K. Figure 7.8 sketches the locations of the atoms, Fig. 7.9 shows the arrangement of the copper oxide planes, Fig. 7.10 provides more details on the unit cell, and Table 7.4 lists the atom positions and unit cell dimensions (Beno *et al.*, 1987; Capponi *et al.*, 1987; Hazen *et al.*, 1987; Jorgensen *et al.*, 1987; Le Page *et al.*, 1987; Siegrist *et al.*, 1987; Yan and Blanchin, 1991; see also Schuller *et al.*, 1987). Considered as a perovskite derivative, it can be looked upon as a stacking of three perovskite units $BaCuO_3$, $YCuO_2$, and $BaCuO_2$, two of them with a missing oxygen, and this explains why $c \approx 3a$. It is, however, more useful to discuss the compound from the viewpoint of its planar structure.

A. Copper Oxide Planes

We see from Fig. 7.9 that three planes containing Cu and O are sandwiched between two planes containing Ba and O and one plane containing Y. The layering scheme is given on the right side of Fig. 7.8, where the superscript b on O indicates that the oxygen lies along the b-axis, as shown. The atoms are puckered in the two [Cu O_2 −] planes that have the [− − Y] plane between them. The third copper oxide plane [Cu O^b −], often referred to as

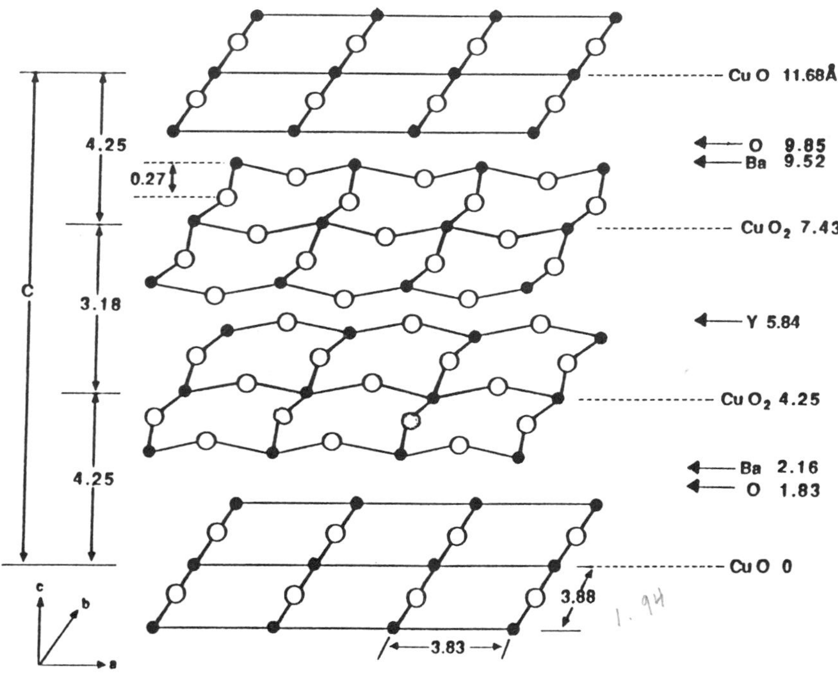

Figure 7.9 Layering scheme of orthorhombic $YBa_2Cu_3O_7$ with the puckering indicated. The layers are perpendicular to the c-axis (Poole *et al.*, 1988, p. 101).

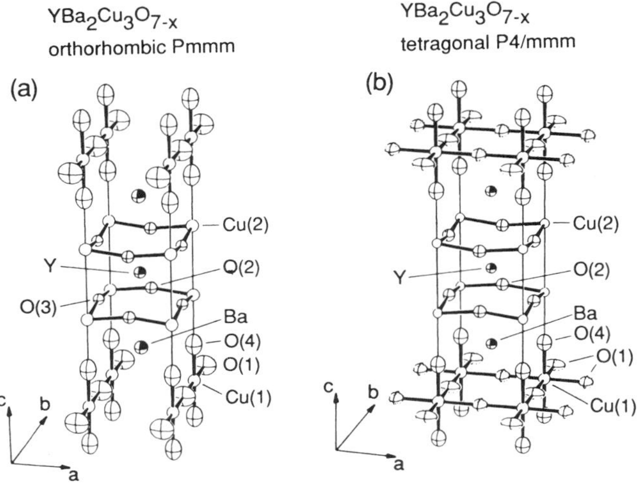

Figure 7.10 Sketches of the superconducting orthorhombic (left) and nonsuperconducting tetragonal (right) YBaCuO unit cells. Thermal vibration ellipsoids are shown for the atoms. In the tetragonal form the oxygen atoms are randomly dispersed over the basal plane sites (Jorgensen *et al.*, 1987a, b; also see Schuller *et al.*, 1987).

Table 7.4 Normalized Atom Positions in the $YBa_2Cu_3O_7$ Orthorhombic Unit Cell (dimensions $a = 3.83$ Å, $b = 3.88$ Å, and $c = 11.68$ Å)

Layer	Atom	x	y	z
[Cu O –]	Cu(1)	0	0	1
	O(1)	0	$\frac{1}{2}$	1
[O – Ba]	O(4)	0	0	0.8432
	Ba	$\frac{1}{2}$	$\frac{1}{2}$	0.8146
[Cu O_2 –]	Cu(2)	0	0	0.6445
	O(3)	0	$\frac{1}{2}$	0.6219
	O(2)	$\frac{1}{2}$	0	0.6210
[– – Y]	Y	$\frac{1}{2}$	$\frac{1}{2}$	$\frac{1}{2}$
[Cu O_2 –]	O(2)	$\frac{1}{2}$	0	0.3790
	O(3)	0	$\frac{1}{2}$	0.3781
	Cu(2)	0	0	0.3555
[O – Ba]	Ba	$\frac{1}{2}$	$\frac{1}{2}$	0.1854
	O(4)	0	0	0.1568
[Cu O –]	O(1)	0	$\frac{1}{2}$	0
	Cu(1)	0	0	0

"the chains," consists of –Cu–O–Cu–O– chains along the *b* axis in lines that are perfectly straight because they are in a horizontal reflection plane σ_h; where no puckering can occur. Note that, according to the figures, the copper atoms are all stacked one above the other on edge (E) sites, as expected for an aligned-type superconductor. Both the copper oxide planes and the chains contribute to the superconducting properties.

B. Copper Coordination

Now that we have described the planar structure of YBaCuO it will be instructive to examine the local environment of each copper ion. The chain copper ion Cu(1) is square planar-coordinated and the two coppers Cu(2) and Cu(3) in the plane exhibit fivefold pyramidal coordination, as indicated in Fig. 7.11. The ellipsoids at the atom positions of Fig. 7.10 provide a measure of the thermal vibrational motion which the atoms experience, since the amplitudes of the atomic vibrations are indicated by the relative size of each of the ellipsoids.

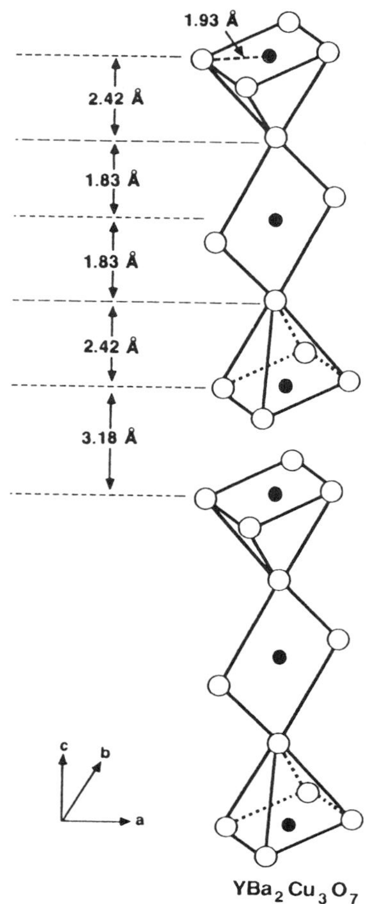

Figure 7.11 Stacking of pyramid, square–planar, and inverted pyramid groups along the *c*-axis of orthorhombic $YBa_2Cu_3O_7$ (adapted from Poole *et al.*, 1988, p. 100).

C. Stacking Rules

The atoms arrange themselves in the various planes in such a way as to enable them to stack one above the other in an efficient manner, with very little interference from neighboring atoms. Steric effects prevent large atoms such as Ba (1.34 Å) and O (1.32 Å) from overcrowding a layer or from aligning directly on top of each other in adjacent layers. In many cuprates stacking occurs in accordance with the following two empirical rules:

1. Metal ions occupy either edge or centered sites, and in adjacent layers alternate between E and C sites.
2. Oxygens are found in any type of site, but they occupy only one type in a particular layer, and in adjacent layers they are on different types of sites.

Minor adjustments to make more room can be brought about by puckering or by distorting from tetragonal to orthorhombic.

D. Crystallographic Phases

The $YBa_2Cu_3O_{7-\delta}$ compound comes in tetragonal and orthorhombic varieties, as shown in Fig. 7.10, and it is the latter phase which is ordinarily superconducting. In the tetragonal phase the oxygen sites in the chain layer are about half occupied in a random or disordered manner, and in the orthorhombic phase are ordered into –Cu–O– chains along the *b* direction. The oxygen vacancy along the *a* direction causes the unit cell to compress slightly so that $a < b$, and the resulting distortion is of the rectangular type shown in Fig. 7.4a. Increasing the oxygen content so that $\delta < 0$ causes oxygens to begin occupying the vacant sites along *a*. Superlattice ordering of the chains is responsible for the phase that goes superconducting at 60 K.

YBaCuO is prepared by heating in the 750–900°C range in the presence of various concentrations of oxygen. The compound is tetragonal at the highest temperatures, increases its oxygen content through oxygen uptake and diffusion (Rothman *et al.*, 1991) as the temperature is lowered, and undergoes a second-order phase transition of the order–disorder type at about 700°C to the low-temperature orthorhombic phase, as indicated in Fig. 7.12

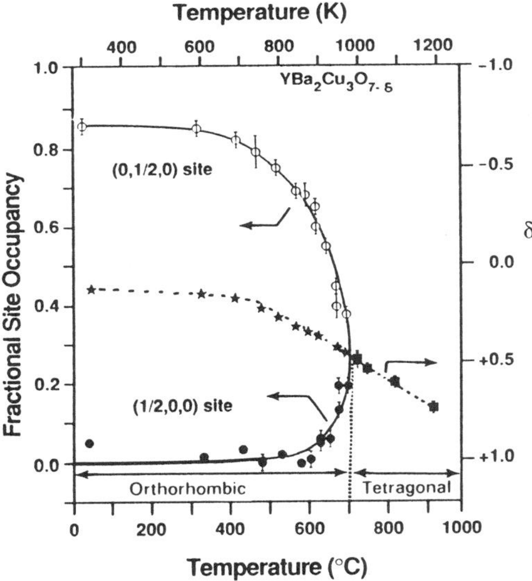

Figure 7.12 Fractional occupancies of the $(\frac{1}{2},0,0)$ (bottom) and $(0,\frac{1}{2},0)$ (top) sites (scale on left), and the oxygen content parameter δ (center, scale on right) for quench temperatures of YBaCuO in the range 0–1000°C. The δ parameter curve is the average of the two site-occupancy curves (adapted from Jorgensen *et al.*, 1987a; also see Schuller *et al.*, 1987; see also Poole *et al.*, 1988).

(Jorgensen *et al.*, 1987, 1990; Schuller *et al.*, 1987; cf. Beyers and Ahn, 1991; Metzger *et al.*, 1993; Fig. 8). Quenching by rapid cooling from a high temperature can produce at room temperature the tetragonal phase sketched on the right side of Fig. 7.10, and slow annealing favors the orthorhombic phase on the left. Figure 7.12 shows the fractional site occupancy of the oxygens in the chain site $(0,\frac{1}{2},0)$ as a function of the temperature in an oxygen atmosphere. A sample stored under sealed conditions exhibited no degradation in structure or change in T_c four years later (Sequeira *et al.*, 1992). Ultra-thin films tend to be tetragonal (Streiffer *et al.*, 1991).

E. Charge Distribution

Information on the charge distributions around atoms in conductors can be obtained from knowledge of their energy bands (see description in Chapter 8). This is most easily accomplished by carrying out a Fourier-type mathematical transformation between the reciprocal k_x, k_y, k_z-space (cf. Chapter 8, Section II) in which the energy bands are plotted and the coordinate x, y, z-space, where the charge is distributed. We will present the results obtained for $YBa_2Cu_3O_7$ in the three vertical symmetry planes (x,z, y,z, and diagonal), all containing the z-axis through the origin, shown shaded in the unit cell of Fig. 7.13.

Contour plots of the charge density of the valence electrons in these planes are sketched in Fig. 7.14. The high density at the Y^{3+} and Ba^{2+} sites and the lack of contours around these sites together indicate that these atoms are almost completely ionized, with charges of +3 and +2, respectively. It also shows that these ions are decoupled from the planes above and below. This accounts for the magnetic isolation of the Y site whereby magnetic ions substituted for yttrium do not interfere with the superconducting properties. In contrast, the contours surrounding the Cu and O ions are not characteristic of an

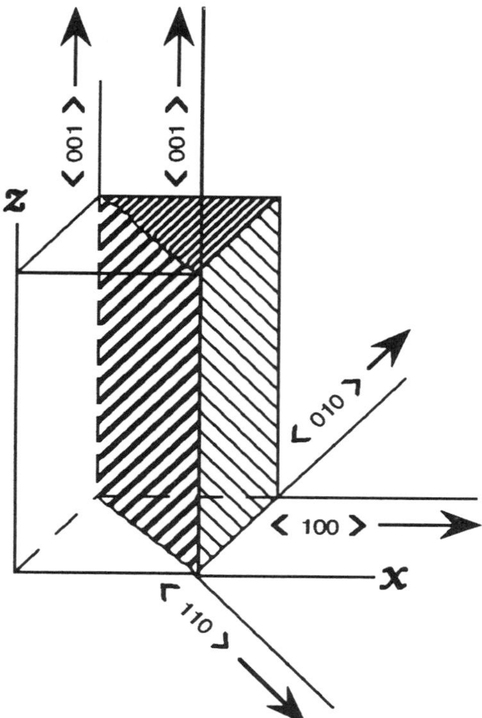

Figure 7.13 Three vertical crystallographic planes (x, z-, y, z-, and diagonal) of a tetragonal unit cell of $YBa_2Cu_3O_7$, and standard notation for the four crystallographic directions.

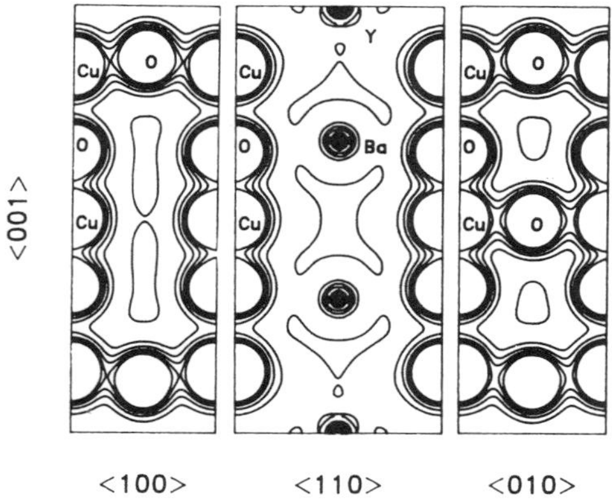

Figure 7.14 Charge density in the three symmetry planes of YBaCuO shown shaded in Fig. 7.13. The x, z, diagonal and the y, z planes are shown from left to right, labeled ⟨100⟩, ⟨110⟩, and ⟨010⟩, respectively. These results are obtained from band structure calculations, as will be explained in the following chapter (Krakauer and Pickett, 1988).

ordinary ionic compound. The short Cu–O bonds in the planes and chains (1.93–1.96 Å) increase the charge overlap. The least overlap appears in the Cu(2)–O(4) vertical bridging bond, which is also fairly long (2.29 Å). The Cu, O charge contours can be represented by a model that assigns charges of +1.62 and −1.69 to Cu and O, respectively, rather than the values of +2.33 and −2.00 expected for a standard ionic model, where the charge +2.33 is an average of +2, +2, and +3 for the three copper ions. Thus the Cu–O bonds are not completely ionic, but partly covalent.

F. YBaCuO Formula

In early work the formula

$$YBa_2Cu_3O_{9-\delta}$$

was used for YBaCuO because the prototype triple pervoskite $(YCuO_3)(BaCuO_3)_2$ has nine oxygens. Then crystallographers showed that there are eight oxygen sites in the 14-atom YBaCuO unit cell, and the formula $YBa_2Cu_3O_{8-\delta}$ came into widespread use. Finally, structure refinements demonstrated that one of the oxygen sites is systematically vacant in the chain layers, so the more appropriate expression $YBa_2Cu_3O_{7-\delta}$ was introduced. It would be preferable to make one more change and use the formula $Ba_2YCu_3O_{7-\delta}$ to emphasize that Y is analogous to Ca in the bismuth and thallium compounds, but very few workers in the field do this, so we reluctantly adopt the usual "final" notation. In the Bi–Tl compound notation of Section IX, B, $Ba_2YCu_3O_{7-\delta}$ would be called a 0213 compound. We will follow the usual practice of referring to $YBa_2Cu_3O_{7-\delta}$ as the 123 compound.

G. $YBa_2Cu_4O_8$ and $Y_2Ba_4Cu_7O_{15}$

These two superconductors are sometimes referred to as the 124 compound and the 247 compound, respectively. They have the property that for each atom at position (x, y, z) there is another identical atom at

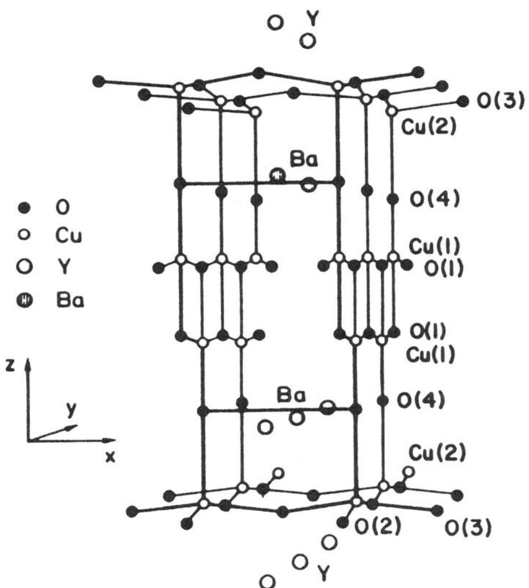

Figure 7.15 Crystal structure of $YBa_2Cu_4O_8$ showing how, as a result of the side-centering symmetry operation, the atoms in adjacent Cu–O chains are staggered along the y direction, with Cu above O and O above Cu (Heyen *et al.*, 1991; modified from Campuzano *et al.*, 1990).

position $(x, y + \frac{1}{2}, z + \frac{1}{2})$. In other words, the structure is side centered. This property prevents the stacking rules of Section C from applying.

The chain layer of $YBa_2Cu_3O_7$ becomes two adjacent chain layers in $YBa_2Cu_4O_8$, with the Cu atoms of one chain located directly above or below the O atoms of the other, as shown in Fig. 7.15 (Campuzano *et al.*, 1990; Heyen *et al.*, 1990a, 1991; Iqbal, 1992; Kaldis *et al.*, 1989; Marsh *et al.*, 1988; Morris *et al.*, 1989a). The transition temperature remains in the range from 40 K to 80 K when Y is replaced by various rare earths (Morris *et al.*, 1989). The double chains do not exhibit the variable oxygen stoichiometry of the single ones.

The other side-centered compound, $Y_2Ba_4Cu_7O_{15}$, may be considered according to Torardi, "as an ordered 1 : 1 intergrowth of the 123 and 124 compounds

$$(YBa_2Cu_3O_7 + YBa_2Cu_4O_8 = Y_2Ba_4Cu_7O_{15})\text{"}$$

(Bordel *et al.*, 1988, Gupta and Gupta, 1993). The 123 single chains can vary in their oxygen content, and superconductivity onsets up to 90 K have been observed. This compound has been synthesized with several rare earths substituted for Y (Morris *et al.*, 1989b).

VII. BODY CENTERING

In Section V we discussed aligned-type superconductor structures that possess a horizontal plane of symmetry. Most high-temperature superconductor structures have, besides this σ_h plane, an additional symmetry operation called body centering whereby for every atom with coordinates (x, y, z) there is an identical atom with coordinates as determined from the following operation:

$$x \rightarrow x \pm \tfrac{1}{2}, \quad x \rightarrow y \pm \tfrac{1}{2}, \quad z \rightarrow z \pm \tfrac{1}{2} \qquad (7.5)$$

Starting with a plane at the height z this operation forms what is called an image plane at the height $z + \frac{1}{2}$ in which the edge atoms become centered, the centered atoms become edge types, and each face atom moves to another face site. In other words, the body-centering operation acting on a plane at the height z forms a body centered plane, also called an image plane, at the height $z \pm \frac{1}{2}$. The signs in these operations are selected so that the generated points and planes remain within the unit cell. Thus if the initial value of z is greater than $\frac{1}{2}$, the minus sign must be selected, viz., $z \rightarrow z - \frac{1}{2}$. Body centering causes half of the Cu–O planes to be [Cu O_2 –], with the copper atoms at edge sites, and the other half to be [– O_2 Cu], with the copper atoms at centered sites.

Let us illustrate the symmetry features of a body-centered superconductor by considering the example of $Tl_2Ba_2CaCu_2O_8$. This compound has an initial plane [Cu O_2 –] with the copper and oxygen atoms at the vertical positions $z = 0.0540$ and 0.0531, respectively, as shown in Fig.

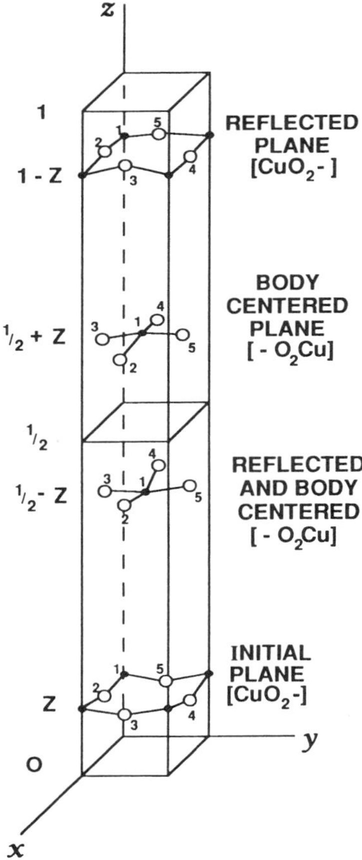

Figure 7.16 Body-centered tetragonal unit cell containing four puckered CuO_2 groups showing how the initial group (bottom) is replicated by reflection in the horizontal reflection plane ($z = \frac{1}{2}$), by the body centering operation, and by both.

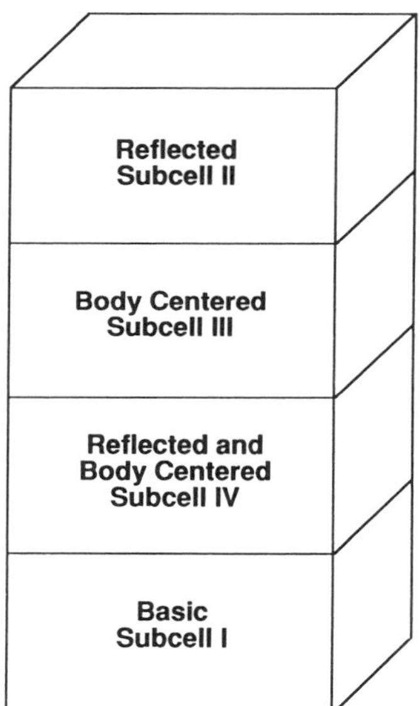

Figure 7.17 Body-centered unit cell divided into four regions by the reflection and body centering operations.

7.16. For illustrative purposes the figure is drawn for values of z closer to 0.1. We see from the figure that there is a reflected plane [Cu O_2 −] at the height $1 - z$, an image (i.e., body centered) plane [− O_2 Cu] of the original plane at the height $\frac{1}{2} + z$, and an image plane [− O_2 Cu] of the reflected plane (i.e., a reflected and body centered plane) at the height $\frac{1}{2} - z$. Figure 7.16 illustrates this situation and indicates how the atoms of the initial plane can be transformed into particular atoms in other planes (see Problem 5). Figure 7.17 shows how the configurations of the atoms in one-quarter of the unit cell, called the *basic subcell*, or subcell I, determine their configurations in the other three subcells II, III, and IV through the symmetry operations of reflection and body centering.

VIII. BODY-CENTERED La_2CuO_4 AND Nd_2CuO_4

The body-centered compound

$$M_2CuO_4$$

has three structural variations in the same crystallographic space group, namely the $M = La$ and $M = Nd$ types, and a third mixed variety (Xiao *et al.*, 1989). Table 7.5 lists the atom positions of the first two types, and Fig. 7.18 presents sketches of the structures of all three. Each will be discussed in turn.

A. Unit Cell Generation of La_2CuO_4 (T Phase)

The structure of the more common La_2CuO_4 variety, often called the T phase,

Table 7.5 Atom Positions in the La_2CuO_4 and Nd_2CuO_4 Structures

La_2CuO_4 structure					Nd_2CuO_4 structure				
Layer	**Atom**	x	y	z	**Layer**	**Atom**	x	y	z
	O(1)	$\frac{1}{2}$	0	1		O(1)	$\frac{1}{2}$	0	1
[Cu O_2 –]	Cu	0	0	1	[Cu O_2 –]	Cu	0	0	1
	O(1)	0	$\frac{1}{2}$	1		O(1)	0	$\frac{1}{2}$	1
	La	$\frac{1}{2}$	$\frac{1}{2}$	0.862	[– – Nd]	Nd	$\frac{1}{2}$	$\frac{1}{2}$	0.862
[O – La]									
	O(2)	0	0	0.818		O(3)	0	$\frac{1}{2}$	$\frac{3}{4}$
					[– O_2 –]				
	O(2)	$\frac{1}{2}$	$\frac{1}{2}$	0.682		O(3)	$\frac{1}{2}$	0	$\frac{3}{4}$
[La – O]									
	La	0	0	0.638	[Nd – –]	Nd	0	0	0.638
	O(1)	$\frac{1}{2}$	0	$\frac{1}{2}$		O(1)	$\frac{1}{2}$	0	$\frac{1}{2}$
[– O_2 Cu]	Cu	$\frac{1}{2}$	$\frac{1}{2}$	$\frac{1}{2}$	[– O_2 Cu]	Cu	$\frac{1}{2}$	$\frac{1}{2}$	$\frac{1}{2}$
	O(1)	0	$\frac{1}{2}$	$\frac{1}{2}$		O(1)	0	$\frac{1}{2}$	$\frac{1}{2}$
	La	0	0	0.362	[Nd – –]	Nd	0	0	0.362
[La – O]									
	O(2)	$\frac{1}{2}$	$\frac{1}{2}$	0.318		O(3)	$\frac{1}{2}$	0	$\frac{1}{4}$
					[– O_2 –]				
	O(2)	0	0	0.182		O(3)	0	$\frac{1}{2}$	$\frac{1}{4}$
[O – La]									
	La	$\frac{1}{2}$	$\frac{1}{2}$	0.138	[– – Nd]	Nd	$\frac{1}{2}$	$\frac{1}{2}$	0.138
	O(1)	0	$\frac{1}{2}$	0		O(1)	0	$\frac{1}{2}$	0
[Cu O_2 –]	Cu	0	0	0	[Cu O_2 –]	Cu	0	0	0
	O(1)	$\frac{1}{2}$	0	0		O(1)	$\frac{1}{2}$	0	0

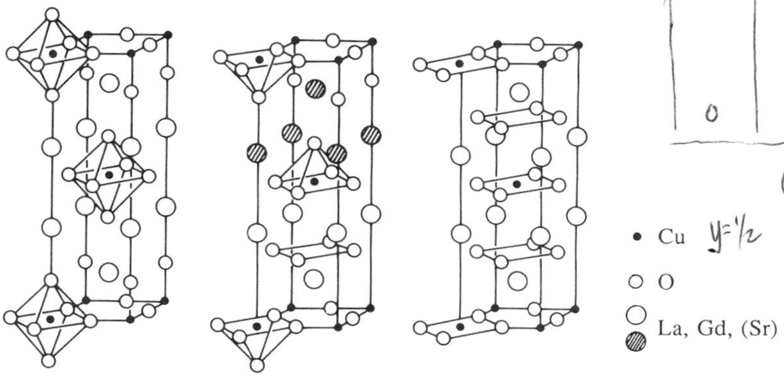

• Cu
○ O
◯ La, Gd, (Sr)

(a) T phase (b) T* phase (c) T′ phase

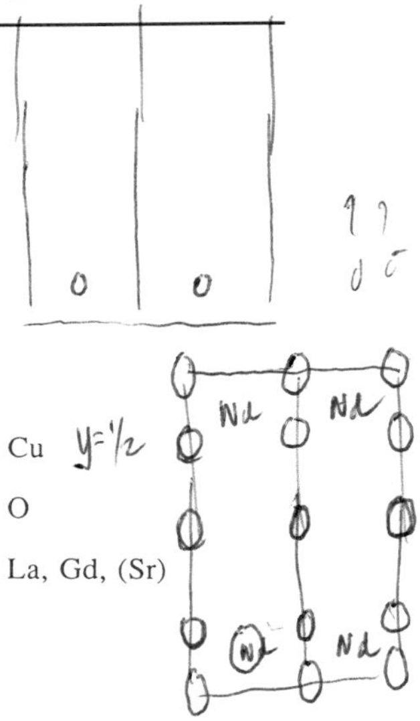

Figure 7.18 (a) Regular unit cell (T phase) associated with hole-type $(La_{1-x}Sr_x)_2CuO_4$ superconductors, (b) hybrid unit cell (T* phase) of the hole-type $La_{2-x-y}R_ySr_xCuO_4$ superconductors, and (c) alternate unit cell (T′ phase) associated with electron-type $(Nd_{1-x}Ce_x)_2CuO_4$ superconductors. The La atoms in the left structure become Nd atoms in the right structure. The upper part of the hybrid cell is T type, and the bottom is T′. The crystallographic space group is the same for all three unit cells (Xiao *et al.*, 1989; see also Oguchi, 1987; Ohbayashi *et al.*, 1987; Poole *et al.*, 1988, p. 83; Tan *et al.*, 1990).

can be pictured as a stacking of CuO_4La_2 groups alternately with image (i.e., body centered) La_2O_4Cu groups along the c direction, as indicated on the left side of Fig. 7.19 (Cavaet *et al.*, 1987; Kinoshita *et al.*, 1992; Longo and Raccah, 1973; Ohbayashi *et al.*, 1987; Onoda *et al.*, 1987; Zolliker *et al.*, 1990). Another way of visualizing the structure is by generating it from the group $Cu_{\frac{1}{2}}O_2La$, comprising the layers [O–La] and $\frac{1}{2}$[Cu O_2 –] in subcell I shown on the right side of Fig. 7.19 and also on the left side of Fig. 7.20. (The factor $\frac{1}{2}$ appears because the [Cu O_2 –] layer is shared by two subcells.) Subcell II is formed by reflection from subcell I, and subcells III and IV are formed from I and II via the body-centering operation in the manner of Figs. 7.16 and 7.17. Therefore, subcells I and II together contain the group CuO_4La_2, and subcells III and IV together contain its image (body centered) counterpart group La_2O_4Cu. The BiSrCaCuO and TlBaCaCuO structures to be discussed in Section IX can be generated in the same manner, but with much larger repeat units along the c direction.

B. Layering Scheme

The La_2CuO_4 layering scheme consists of equally-spaced, flat CuO_2 layers with their oxygens stacked one above the other, the copper ions alternating between the (0, 0, 0) and $(\frac{1}{2}, \frac{1}{2}, \frac{1}{2})$ sites in adjacent layers, as shown in Fig. 7.21. These planes are body-centered images of each other, and are perfectly flat because they are

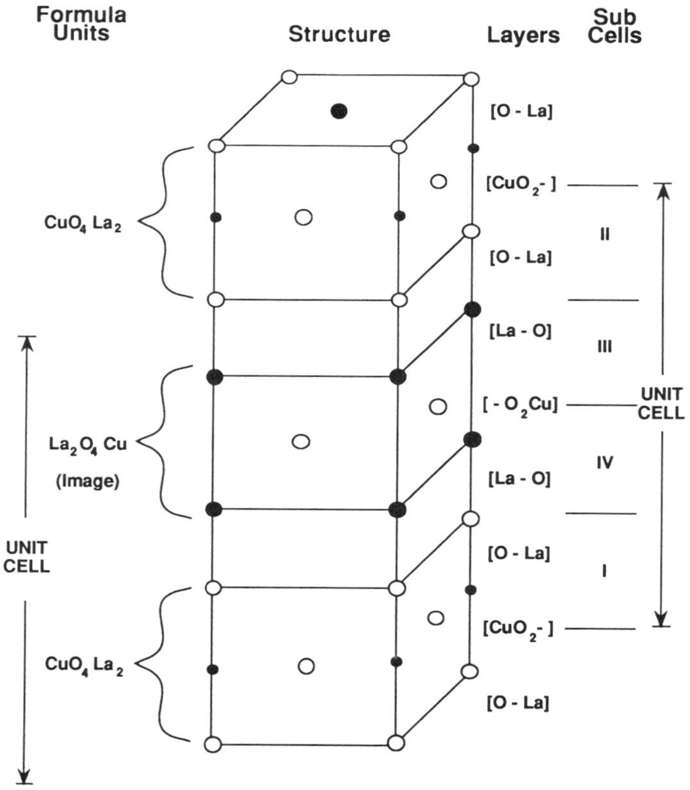

Figure 7.19 Structure of La_2CuO_4 (center), showing the formula units (left) and the level labels and subcell types (right). Two choices of unit cell are indicated, the left-side type unit cell based on formula units, and the more common right-side type unit cell based on copper-oxide layers.

La_2CuO_4	Sub Cell	Nd_2CuO_4
[CuO_2-]	——	[CuO_2 -]
[O - La]	II	[- - Nd]
	——	[- O_2 -]
[La - O]	III	[Nd - -]
[- O_2Cu]	——	[- O_2Cu]
[La - O]	IV	[Nd - -]
	——	[- O_2 -]
[O - La]	I	[- - Nd]
[CuO_2 -]	——	[CuO_2 -]

Figure 7.20 Layering schemes of the La_2CuO_4 (T, left) and Nd_2CuO_4 (T′, right) structures. The locations of the four subcells of the unit cell are indicated in the center column.

reflection planes. Half of the oxygens, O(1), are in the planes, and the other half, O(2), between the planes. The copper is octahedrally coordinated with oxygen, but the distance 1.9 Å from Cu to O(1) in the CuO_2 planes is much less than the vertical distance of 2.4 Å from Cu to the apical oxygen O(2), as indicated in Fig. 7.22. The La is ninefold coordinated to four O(1) oxygens, to four O(2) at $(\frac{1}{2}, \frac{1}{2}, z)$ sites, and to one O(2) at a $(0, 0, z)$ site.

C. Charge Distribution

Figure 7.23 shows contours of constant-valence charge density on a logarithmic scale drawn on the back x, z-plane and on the diagonal plane of the unit cell sketched in Fig. 7.13. These contour plots are obtained from the band structure calculations described in Chapter 8, Section XIV. The high-charge density at the lanthanum site and the low charge density around this site indicate an ionic state

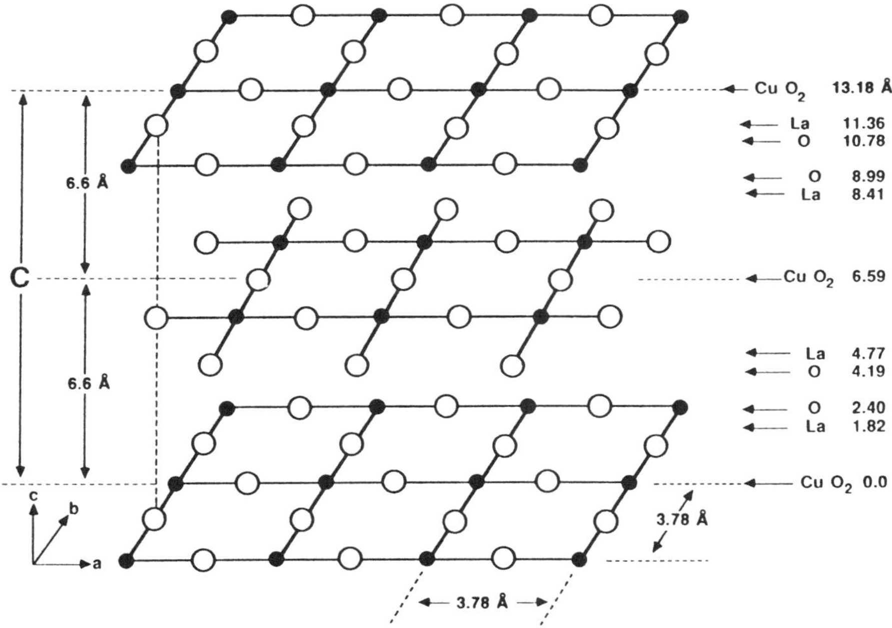

Figure 7.21 CuO_2 layers of the La_2CuO_4 structure showing horizontal displacement of Cu atoms in alternate layers. The layers are perpendicular to the c-axis (Poole *et al.*, 1988, p. 87).

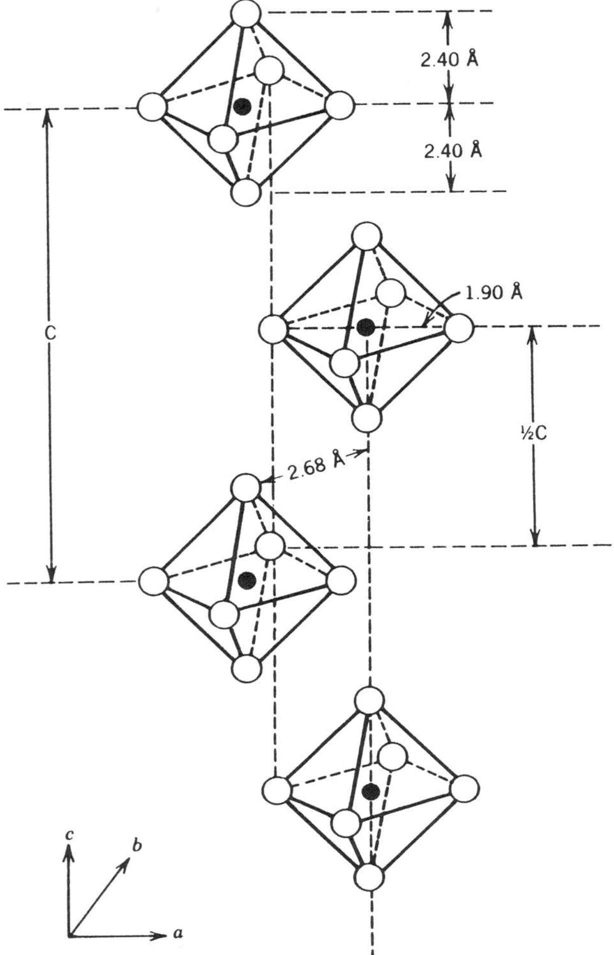

Figure 7.22 Ordering of axially distorted CuO_6 octahedra in La_2CuO_4 (Poole *et al.*, 1988, p. 88).

La^{3+}. The charge density changes in a fairly regular manner around the copper and oxygen atoms, both within the CuO_2 planes and perpendicular to these planes, suggestive of covalency in the Cu–O bonding, as is the case with the $YBa_2Cu_3O_7$ compound.

D. Superconducting Structures

The compound La_2CuO_4 is itself an antiferromagnetic insulator and must be doped, generally with an alkaline earth, to exhibit pronounced superconducting properties. The compounds $(La_{1-x}M_x)_2CuO_4$, with 3% to 15% of $M =$ Sr or Ba replacing La, are orthorhombic at low temperatures and low M contents and are tetragonal otherwise; superconductivity has been found on both sides of this transition. The orthorhombic distortion can be of the rectangular or of the rhombal type, both of which are sketched in Fig. 7.4. The phase diagram of Fig. 7.24 shows the tetragonal, orthorhombic, superconducting, and antiferromagnetically ordered regions for the lanthanum compound (Weber *et al.*, 1989; cf. Goodenough *et al.*, 1993). We see that

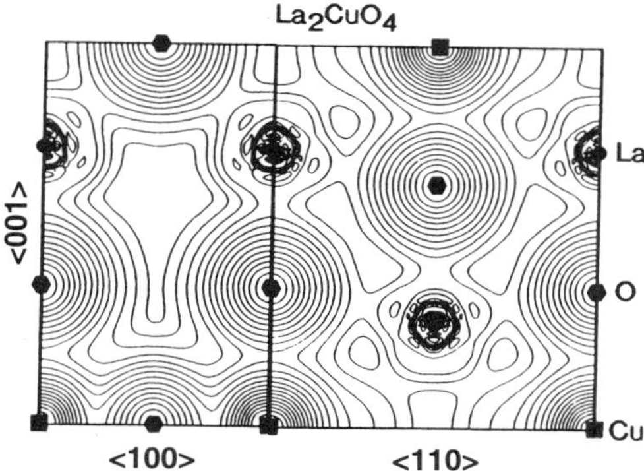

Figure 7.23 Contour plots of the charge density of La_2CuO_4 obtained from band structure calculations. The x, z-crystallographic plane labeled ⟨100⟩ is shown on the left and the diagonal plane labeled ⟨110⟩ on the right. The contour spacing is on a logarithmic scale (Pickett, 1989).

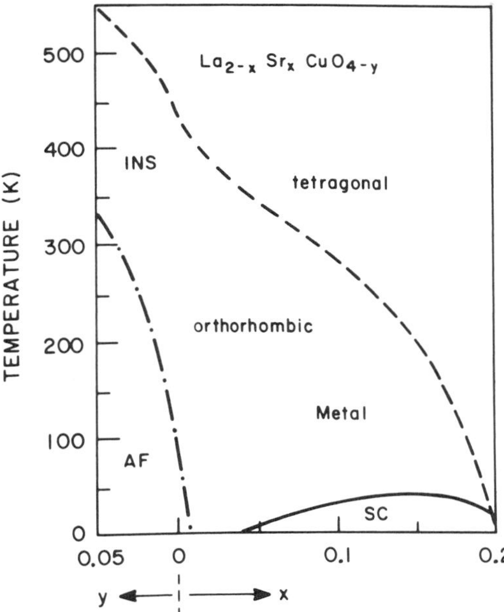

Figure 7.24 Phase diagram for hole-type $La_{2-x}Sr_xCuO_{4-y}$ indicating insulating (INS), antiferromagnetic (AF), and superconducting (SC) regions. Figure VI-6 of Poole *et al.* (1988) shows experimental data along the orthorhombic-to-tetragonal transition line. Spin-density waves (SDW) are found in the AF region (Weber *et al.*, 1989).

the orthorhombic phase is insulating at high temperatures, metallic at low temperatures, and superconducting at very low temperatures. Spin-density waves, to be discussed in Chapter 8, Section XIX, occur in the antiferromagnetic region.

E. Nd_2CuO_4 Compound (T′ Phase)

The rarer Nd_2CuO_4 structure (Skantakumar *et al.*, 1989; Sulewski *et al.*, 1990; Tan *et al.*, 1990) given on the right side of Fig. 7.18 and Table 7.5 has all of its atoms in the same positions as the standard La_2CuO_4 structure, except for the apical O(2) oxygens in the [O–La] and [La–O] layers, which move to form a [– O_2 –] layer between [– – La] and [La – –]. These oxygens, now called O(3), have the same x, y coordinate positions as the O(1) oxygens, and are located exactly between the CuO_2 planes with $z = \frac{1}{4}$ or $\frac{3}{4}$. We see from Fig. 7.18 that the CuO_6 octahedra have now lost their apical oxygens, causing Cu to become square planar-coordinated CuO_4 groups. The Nd is eightfold coordinated to four O(1) and four O(3) atoms, but with slightly different Nd–O distances. The CuO_2 planes, however, are identical in the two structures. Superconductors with this Nd_2CuO_4 structure are of the electron type, in contrast to other high-temperature superconductors, in which the current carriers are holes. In particular, the electron superconductor $Nd_{1.85}Ce_{0.15}CuO_{4-\delta}$ with $T_c = 24$ K has been widely studied (Fontcuberta and Fàbrega, 1995, a review chapter; Allen 1990; Alp *et al.*, 1989b; Barlingay *et al.*, 1990; Ekino and Akimitsu, 1989a, b; Lederman *et al.*, 1991; Luke *et al.*, 1990; Lynn *et al.*, 1990; Sugiyama *et al.*, 1991; Tarason *et al.*, 1989a). Other rare earths, such as Pr (Lee *et al.*, 1990) and Sm (Almasan *et al.*, 1992) have replaced Nd.

The difference of structures associated with different signs attached to the current carriers may be understood in terms of the doping process that converts undoped material into a superconductor. Lanthanum and neodymium are both trivalent, and in the undoped compounds they each contribute three electrons to the nearby oxygens,

$$\begin{aligned} La &\rightarrow La^{3+} + 3e^-, \\ Nd &\rightarrow Nd^{3+} + 3e^-, \end{aligned} \tag{7.6}$$

to produce O^{2-}. To form the superconductors a small amount of La in La_2CuO_4 can be replaced with divalent Sr, and some Nd in Nd_2CuO_4 can be replaced with tetravalent Ce, corresponding to

$$\begin{aligned} Sr &\rightarrow Sr^{2+} + 2e^- \quad (\text{in } La_2CuO_4) \\ Ce &\rightarrow Ce^{4+} + 4e^- \quad (\text{in } Nd_2CuO_4). \end{aligned} \tag{7.7}$$

Thus, Sr doping decreases the number of electrons to produce hole-type carriers, while Ce doping increases the electron concentration and the conductivity is electron type.

There are also copper-oxide electron superconductors with different structures, such as $Sr_{1-x}Nd_xCuO_2$ (Smith *et al.*, 1991) and $TlCa_{1-x}R_xSr_2Cu_2O_{7-\delta}$, where R is a rare earth (Vijayaraghavan *et al.*, 1989). Electron- and hole-type superconductivity in the cuprates has been compared (Katti and Risbud, 1992; Medina and Regueiro, 1990).

F. $La_{2-x-y}R_xSr_yCuO_4$ Compounds (T* Phase)

We have described the T structure of La_2CuO_4 and the T′ structure of Nd_2CuO_4. The former has O(2) oxygens and the latter O(3) oxygens, which changes the coordinations of the Cu atoms and that of the La and Nd atoms as well. There is a hybrid structure of hole-type superconducting lanthanum cuprates called the T* structure, illustrated in Fig. 7.18b, in which the upper half of the unit cell is the T type with O(2) oxygens and lower half the T′ type with O(3) oxygens. These two varieties of halfcells are stacked alternately along the tetragonal c-axis (Akimitsu *et al.*, 1988; Cheong *et al.*, 1989b; Kwei *et al.*, 1990; Tan *et al.*, 1990). Copper, located in the base of an oxygen pyramid, is fivefold-coordinated CuO_5. There are two inequivalent rare earth sites; the ninefold-coordinated site in the T-type halfcell is preferentially occupied by the larger La and Sr ions, while the smaller rare earths R (i.e., Sm, Eu, Gd, or Tb) prefer the eightfold-coordinated site in the T′ half-cell. Tan *et al.* (1991) give a phase diagram for the concentration ranges over which the T and T* phases are predominant.

IX. BODY-CENTERED BiSrCaCuO AND TlBaCaCuO

Early in 1988 two new superconducting systems with transition temperatures considerably above those attainable with YBaCuO, namely the bismuth- and thallium-based materials, were discovered. These compounds have about the same a and b lattice constants as the yttrium and lanthanum compounds, but with much larger unit cell dimensions along c. We will describe their body-centered structures in terms of their layering schemes. In the late 1940s some related compounds were synthesized by the Swedish chemist Bengt Aurivillius (1950, 1951, 1952).

A. Layering Scheme

The $Bi_2Sr_2Ca_nCu_{n+1}O_{6+2n}$ and $Tl_2Ba_2Ca_nCu_{n+1}O_{6+2n}$ compounds, where n is an integer, have essentially the same structure and the same layering arrangement (Barry *et al.*, 1989; Siegrist *et al.*, 1988; Torardi *et al.*, 1988a; Yvon and François, 1989), although there are some differences in the detailed atom positions. Here there are groupings of CuO_2 layers, each separated from the next by Ca layers with no oxygen. The CuO_2 groupings are bound together by intervening layers of BiO and SrO for the bismuth compound, and by intervening layers of TlO and BaO for the thallium compound. Figure 7.25 compares the layering scheme of the $Tl_2Ba_2Ca_nCu_{n+1}O_{6+2n}$ compounds with $n = 0, 1, 2$ with those of the lanthanum and yttrium compounds. We also see from the figure that the groupings of [Cu O_2 −] planes and [− O_2 Cu] image

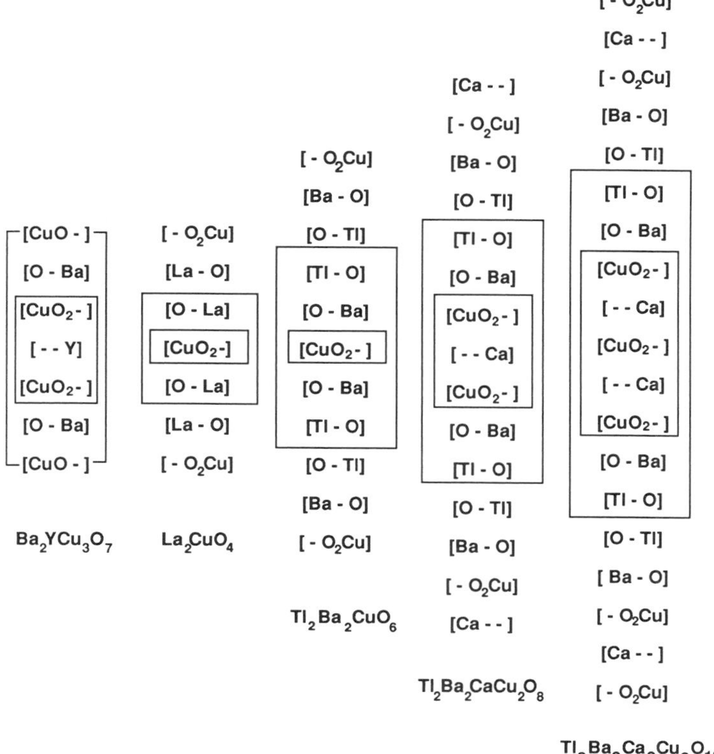

Figure 7.25 Layering schemes of various high-temperature superconductors. The CuO_2 plane layers are enclosed in small inner boxes, and the layers that make up a formula unit are enclosed in larger boxes. The Bi–Sr compounds $Bi_2Sr_2Ca_nCu_{n+1}O_{6+2n}$ have the same layering schemes as their Tl–Ba counterparts shown in this figure.

(i.e., body centered) planes repeat along the *c*-axis. It is these copper-oxide layers that are responsible for the superconducting properties.

A close examination of this figure shows that the general stacking rules mentioned in Section VI.C for the layering scheme are satisfied, namely metal ions in adjacent layers alternate between edge (E) and centered (C) sites, and adjacent layers never have oxygens on the same types of sites. The horizontal reflection symmetry at the central point of the cell is evident. It is also clear that $YBa_2Cu_3O_7$ is aligned and that the other four compounds are staggered.

Figure 7.26 (Torardi *et al.*, 1988a) presents a more graphical representation of the information in Fig. 7.25 by showing the positions of the atoms in their layers. The symmetry and body centering rules are also evident on this figure. Rao (1991) provided sketches for the six compounds $Tl_mBa_2Ca_nCu_{n+1}O_x$ similar to those in Fig. 7.26 with the compound containing one ($m = 1$) or two thallium layers ($m = 2$), where $n = 0, 1, 2$, as in the Torardi *et al.* figure.

B. Nomenclature

There are always two thalliums and two bariums in the basic formula for $Tl_2Ba_2Ca_nCu_{n+1}O_{6+2n}$, together with n calciums and $n + 1$ coppers. The first three members of this series for $n = 0$, 1, and 2 are called the 2201, 2212, and 2223 compounds, respectively, and similarly for their

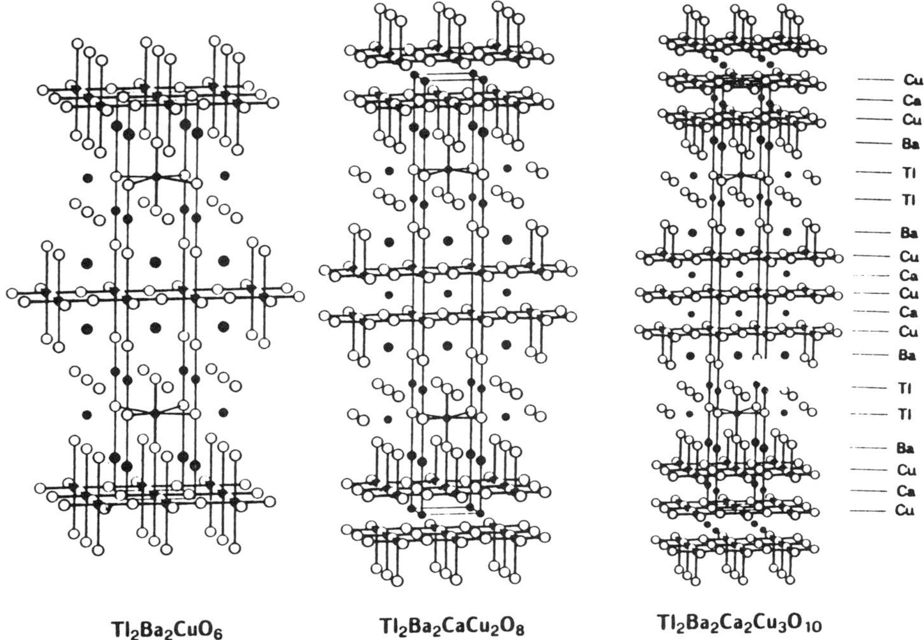

Figure 7.26 Crystal structures of $Tl_2Ba_2Ca_nCu_{n+1}O_{6+2n}$ superconducting compounds with $n = 0, 1, 2$ arranged to display the layering schemes. The $Bi_2Sr_2Ca_nCu_{n+1}O_{6+2n}$ compounds have the same respective structures (Torardi *et al.*, 1988a).

BiSr analogues $Bi_2Sr_2Ca_nCu_{n+1}O_{6+2n}$. Since Y in $YBa_2Cu_3O_7$ is structurally analogous to Ca in the Tl and Bi compounds, it would be more consistent to write $Ba_2YCu_3O_7$ for its formula, as noted in Section VI.F. In this spirit $Ba_2YCu_3O_{7-\delta}$ might be called the 0213 compound, and $(La_{1-x}M_x)_2CuO_{4-\delta}$ could be called 2001.

C. Bi–Sr Compounds

Now that the overall structures and interrelationships of the BiSr and TlBa high-temperature superconductors have been made clear in Figs. 7.25 and 7.26 we will comment briefly about each compound. Table 7.3 summarizes the characteristics of these and related compounds.

The first member of the BiSr series, the 2201 compound with $n = 0$, has octahedrally coordinated Cu and $T_c \approx 9$ K (Torardi *et al.*, 1988b). The second member, $Bi_2(Sr, Ca)_3Cu_2O_{8+\delta}$, is a superconductor with $T_c \approx 90$ K (Subramanian *et al.*, 1988a; Tarascon *et al.*, 1988b). There are two [Cu O_2 –] layers separated from each other by the [– – Ca] layer. The spacing from [Cu O_2 –] to [– – Ca] is 1.66 Å, which is less than the corresponding spacing of 1.99 Å between the levels [Cu O_2 –] and [– – Y] of YBaCuO. In both cases the copper ions have a pyramidal oxygen coordination of the type shown in Fig. 7.11. Superlattice structures have been reported along *a* and *b*, which means that minor modifications of the unit cells repeat approximately every five lattice spacings, as explained in Sect. IX.E. The third member of the series, $Bi_2Sr_2Ca_2Cu_3O_{10}$, has three CuO_2 layers separated from each other by [– – Ca] planes and a higher transition temperature, 110 K, when doped with Pb. The two Cu ions have pyramidal coordination, while the third is square planar.

Charge-density plots of

$$Bi_2Sr_2CaCu_2O_8$$

indicate the same type of covalency in the Cu–O bonding as with the $YBa_2Cu_3O_7$ and La_2CuO_4 compounds. They also indicate very little bonding between the adjacent [Bi – O] and [O – Bi] layers.

D. Tl–Ba Compounds

The TlBa compounds

$$Tl_2Ba_2Ca_nCu_{n+1}O_{6+2n}$$

have higher transition temperatures than their bismuth counterparts (Iqbal *et al.*, 1989; Subramanian *et al.*, 1988b; Torardi *et al.*, 1988a). The first member of the series, namely $Tl_2Ba_2CuO_6$ with $n = 0$, has no [– – Ca] layer and a relatively low transition temperature of ≈ 85 K. The second member ($n = 1$), $Tl_2Ba_2CaCu_2O_8$, called the 2212 compound, with $T_c = 110$ K has the same layering scheme as its Bi counterpart, detailed in Figs. 7.25 and 7.26. The [Cu O_2 –] layers are thicker and closer together than the corresponding layers of the bismuth compound (Toby *et al.*, 1990). The third member of the series, $Tl_2Ba_2Ca_2Cu_3O_{10}$, has three [Cu O_2 –] layers separated from each other by [– – Ca] planes, and the highest transition temperature, 125 K, of this series of thallium compounds. It has the same copper coordination as its BiSr counterpart. The 2212 and 2223 compounds are tetragonal and belong to the same crystallographic space group as La_2CuO_4.

We see from the charge-density plot of $Tl_2Ba_2CuO_6$ shown in Fig. 7.27 that Ba^{2+} is ionic, Cu exhibits strong covalency, especially in the Cu–O plane, and Tl also appears to have a pronounced covalency. The bonding between the [Tl – O] and [O – Tl] planes is stronger than that between the [Bi – O] and [O – Bi] planes of Bi–Sr.

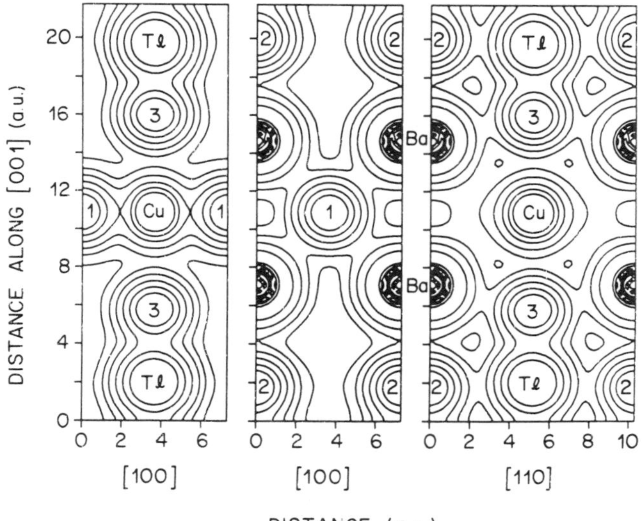

Figure 7.27 Contours of constant charge density on a logarithmic scale in two high-symmetry crystallographic planes of $Tl_2Ba_2CuO_6$. Oxygen atoms O(1), O(2), and O(3) are denoted 1, 2, and 3, respectively. The planar Cu–O1 binding is strongest (Hamann and Mattheiss, 1988; see Pickett, 1989).

E. Modulated Structures

The x-ray and neutron-diffraction patterns obtained during crystal structure determinations of the bismuth cuprates $Bi_2Sr_2Ca_nCu_{n+1}O_{6+2n}$ exhibit weak satellite lines with spacings that do not arise from an integral multiple of the unit cell dimensions. These satellites have modulation periods of 21 Å, 19.6 Å, and 20.8 Å, respectively, for the $n = 0$, 1, and 2 compounds (Li *et al.*, 1989). Since the lattice constant $a = 5.41$ Å ($b = 5.43$ Å) for all three compounds, this corresponds to a superlattice with unit cell of dimensions $\approx 3.8a$, b, c, with the repeat unit along the a direction equal to $\approx 3.8a$ for all three compounds. A modulation of $4.7b$ has also been reported (Kulik *et al.*, 1990). This structural modulation is called incommensurate because the repeat unit is not an integral multiple of a.

Substitutions dramatically change this modulation. The compound

$$Bi_2Sr_2Ca_{1-x}Y_xCu_2O_y$$

has a period that decreases from about $4.8b$ for $x = 0$ to the commensurate value $4.0b$ for $x = 1$ (Inoue *et al.*, 1989; Tamegai *et al.*, 1989). Replacing Cu by a transition metal (Fe, Mn, or Co) produces nonsuperconducting compounds with a structural modulation that is commensurate with the lattice spacing (Tarascon *et al.*, 1989b). A modulation-free bismuth–lead cuprate superconductor has been prepared (Manivannan *et al.*, 1991). Kistenmacher (1989) examined substitution-induced superstructures in $YBa_2(Cu_{1-x}M_x)_3O_7$. Superlattices with modulation wavelengths as short as 24 Å have been prepared by employing ultra-thin deposition techniques to interpose insulating planes of $PrBa_2Cu_3O_7$ between superconducting Cu–O layers of $YBa_2Cu_3O_7$ (Jakob *et al.*, 1991; Lowndes *et al.*, 1990; Pennycook *et al.*, 1991; Rajagopal and Mahanti, 1991; Triscone *et al.* 1990). Tanaka and Tsukada (1991) used the Kronig–Penney model (Tanaka and Tsukada, 1989a, b) to calculate the quasiparticle spectrum of superlattices.

F. Aligned Tl–Ba Compounds

A series of aligned thallium-based superconducting compounds that have the general formula $TlBa_2Ca_nCu_{n+1}O_{5+2n}$ with n varying from 0 to 5 has been reported (Ihara *et al.*, 1988; Rona, 1990). These constitute a series from 1201 to 1245. They have superconducting transition temperatures almost as high as the $Tl_2Ba_2Ca_nCu_{n+1}O_{6+2n}$ compounds. Data on these compounds are listed in Table 7.3.

G. Lead Doping

In recent years a great deal of effort has been expended in synthesizing lead-doped superconducting cuprate structures (Itoh and Uchikawa, 1989). Examples involve substituting Pb for Bi (Dou *et al.*, 1989; Zhengping *et al.*, 1990), for Tl (Barry *et al.*, 1989; Mingzhu *et al.*, 1990), or for both Bi and Tl (Iqbal *et al.*, 1990). Different kinds of Pb, Y-containing superconductors have also been prepared (cf. Mattheiss and Hamann, 1989; Ohta and Maekawa, 1990; Tang *et al.*, 1991; Tokiwa *et al.*, 1990, 1991).

X. ALIGNED HgBaCaCuO

The series of compounds

$$HgBa_2Ca_nCu_{n+1}O_{2n+4},$$

where n is an integer, are prototypes for the Hg family of superconductors. The first three members of the family, with $n = 0, 1, 2$, are often referred to as Hg-1201, Hg-1212, and Hg-1223, respectively. They have the structures sketched in Fig. 7.28 (Tokiwa-Yamamoto *et al.*, 1993; see also Martin *et al.*, 1994; Putilin *et al.*, 1991). The lattice constants are $a = 3.86$ Å for all of them, and $c = 9.5$, 12.6, and 15.7 Å for $n = 0, 1, 2$, respectively. The atom positions of the $n = 1$ compound are listed in Table 7.6 (Hur *et al.*, 1994). The figure is drawn

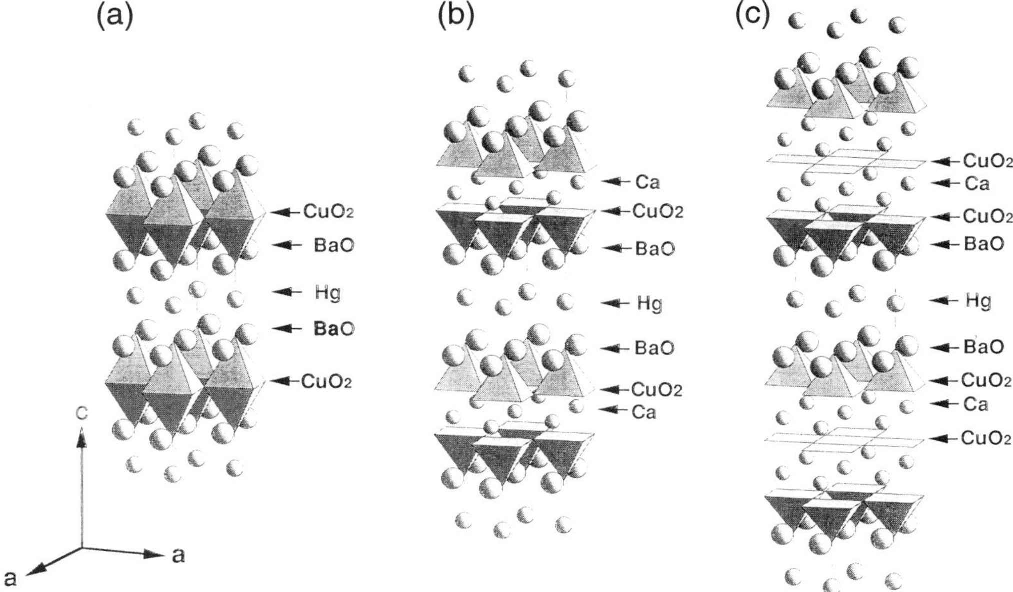

Figure 7.28 Structural models for the series $HgBa_2Ca_nCu_{n+1}O_{2n+4}$. The first three members with $n = 0, 1, 2$ are shown (parts a, b, and c, respectively) (Tokiwa-Yamamoto *et al.*, 1993).

Table 7.6 Normalized Atom Positions in the Tetragonal Unit Cell of $HgBa_2Ca_{0.86}Sr_{0.14}Cu_2O_{6+\delta}$[a]

Layer	Atom	x	y	z
	Hg	0	0	1
[Hg – –]				
	O(3)	$\frac{1}{2}$	$\frac{1}{2}$	1
	O(2)	0	0	0.843
[O – Ba]				
	Ba	$\frac{1}{2}$	$\frac{1}{2}$	0.778
	Cu	0	0	0.621
[Cu O_2 –]	O(1)	0	$\frac{1}{2}$	0.627
	O(1)	$\frac{1}{2}$	0	0.627
[– – Ca]	Ca, Sr	$\frac{1}{2}$	$\frac{1}{2}$	$\frac{1}{2}$
	O(1)	$\frac{1}{2}$	0	0.373
[Cu O_2 –]	O(1)	0	$\frac{1}{2}$	0.373
	Cu	0	0	0.379
	Ba	$\frac{1}{2}$	$\frac{1}{2}$	0.222
[O – Ba]				
	O(2)	0	0	0.157
	O(3)	$\frac{1}{2}$	$\frac{1}{2}$	0
[Hg – –]				
	Hg	0	0	0

[a] Unit cell dimensions $a = 3.8584$ Å and $c = 12.6646$ Å, space group is $P4/mmm$, D_{4h}^1. The Hg site is 91% occupied and the O(3) site is 11% occupied ($\delta = 0.11$). The data are from Hur *et al.* (1994).

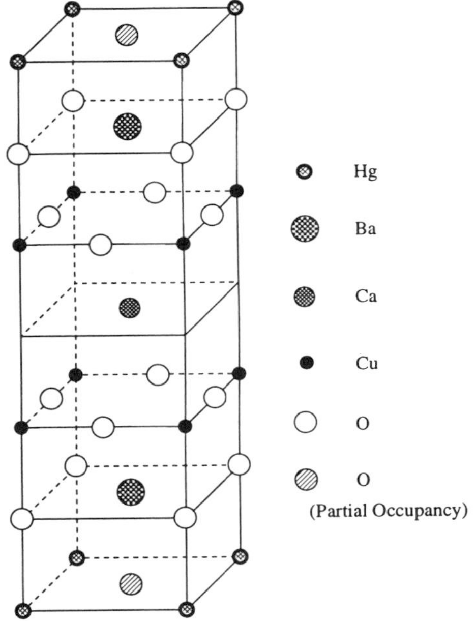

Figure 7.29 Schematic structure of the $HgBa_2CaCu_2O_{6+\delta}$ compound which is also called Hg-1212 (Meng *et al.*, 1993a).

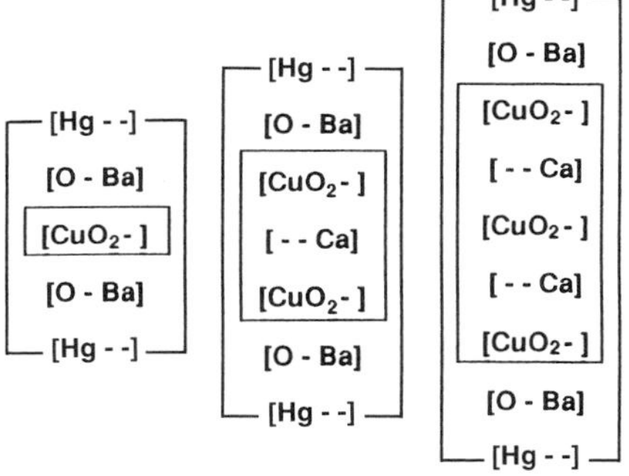

Figure 7.30 Layering schemes of three $HgBa_2Ca_nCu_{n+1}O_{2n+4}$ compounds, using the notation of Fig. 7.25.

with mercury located in the middle layer of the unit cell, while the table puts Hg at the origin (000) and Ca in the middle ($\frac{1}{2}\frac{1}{2}\frac{1}{2}$). Figure 7.29 presents the unit cell for the $n = 1$ compound $HgBa_2CaCu_2O_{6+\delta}$ drawn with Ca in the middle (Meng *et al.*, 1993a). The symbol δ represents a small excess of oxygen located in the center of the top and bottom layers, at positions $\frac{1}{2}\frac{1}{2}0$ and $\frac{1}{2}\frac{1}{2}1$ which are labeled "partial occupancy" in the figure. If this oxygen were included the level symbol would be [Hg − O] instead of [Hg − −]. These Hg compound structures are similar to those of the series $TlBa_2Ca_nCu_{n+1}O_{2n+4}$ mentioned above in Section IX.F.

We see from Fig. 7.28 that the copper atom of Hg-1201 is in the center of a stretched octahedron with the planar oxygens O(1) at a distance of 1.94 Å, and the apical oxygens O(2) of the [O − Ba] layer much further away (2.78 Å). For $n = 1$ each copper atom is in the center of the base of a tetragonal pyramid, and for $n = 2$ the additional CuO_2 layer has Cu atoms which are square planar coordinated. The layering scheme stacking rules of Section VI.C are obeyed by the Hg series of compounds, with metal ions in adjacent layers alternating between edge (E) and centered (C) sites, and oxygen in adjacent layers always at different sites. We see from Table 7.6 that the [O − Ba] layer is strongly puckered and the [Cu O_2 −] layer is only slightly puckered.

The relationships between the layering scheme of the $HgBa_2Ca_nCu_{n+1}O_{2n+4}$ series of compounds and those of the other cuprates may be seen by comparing the sketch of Fig. 7.30 with that of Fig. 7.25. We see that the $n = 1$ compound $HgBa_2CaCu_2O_6$ is quite similar in structure to $YBa_2Cu_3O_7$ with Ca replacing Y in the center and Hg replacing the chains [Cu O −]. More surprising is the similarity between the arrangement of the atoms in the unit cell of each

$$HgBa_2Ca_nCu_{n+1}O_{2n+4}$$

compound and the arrangement of the atoms in the semi-unit cell of the corresponding

$$Tl_2Ba_2Ca_nCu_{n+1}O_{2n+6}$$

compound. They are the same except for the replacement of the [Tl − O] layer by [Hg − −], and the fact that the thallium compounds are body centered and the Hg ones are aligned.

Supercells involving polytypes with ordered stacking sequences of different phases, such as Hg-1212 and Hg-1223, along the *c* direction have been reported. The stoichiometry is often

$$Hg_2Ba_4Ca_3Cu_5O_x$$

corresponding to equal numbers of the Hg-1212 and Hg-1223 phases (Phillips, 1993; Schilling *et al.*, 1993, 1994).

Detailed structural data have already been reported on various Hg family compounds such as $HgBa_2CuO_{4+\delta}$ (Putlin *et al.*, 1993) and the $n = 1$ compound with partial Eu substitution for Ca (Putlin *et al.*, 1991). The compound

$$Pb_{0.7}Hg_{0.3}Sr_2Nd_{0.3}Ca_{0.7}Cu_3O_7$$

has Hg in the position (0.065, 0, 0), slightly displaced from the origin of the unit cell (Martin *et al.*, 1994). Several researchers have reported synthesis and pretreatment procedures (Adachi *et al.*, 1993; Itoh *et al.*, 1993; Isawa 1994a; Meng, 1993b; Paranthaman, 1994; Paranthaman *et al.*, 1993). Lead doping for Hg has been used to improve the superconducting properties (Iqbal *et al.*, 1994; Isawa *et al.*, 1993; Martin *et al.*, 1994).

XI. BUCKMINSTERFULLERENES

The compound C_{60}, called buckminsterfullerene, or fullerene for short, consists of 60 carbon atoms at the vertices of the dotriacontohedron (32-sided figure) that is sketched in Fig. 3.35 and discussed in Chapter 3, Section XVI. The term fullerene is used here for a wider class of compounds C_n with n carbon atoms, each of whose carbon atoms is bonded to three other carbons to form a closed surface, with the system conjugated such that for every resonant structure each carbon has two single bonds and one double bond. The smallest possible compound of this type is tetrahedral C_4, which has the three resonant structures shown in Fig. 7.31. Cubic C_8 is a fullerene, and we show in Problem 17 that it has nine resonant structures. Icosahedral C_{12} is also a fullerene, but octahedral C_6 and dodecahedral C_{20} are not because their carbons are bonded to more than three neighbors. These hypothetical smaller C_n compounds have never been synthesized, but the larger ones, such as C_{60}, C_{70}, C_{76}, C_{78}, and C_{82}, have been made and characterized. Some of them have several forms, with different arrangements of polygons. Clusters of buckminsterfullerenes, such as icosahedral $(C_{60})_{13}$, have also been studied (T. P. Martin *et al.*, 1993).

There are several interesting geometrical characteristics of fullerenes (Chung and Sternberg, 1993). Since each carbon (vertex) joins three bonds (edges) and each edge has two vertices, the number of edges E in a structure C_n is 50% greater than the number of vertices V. There is a general theorem in topology, called Euler's Theorem, that the number of faces F of a

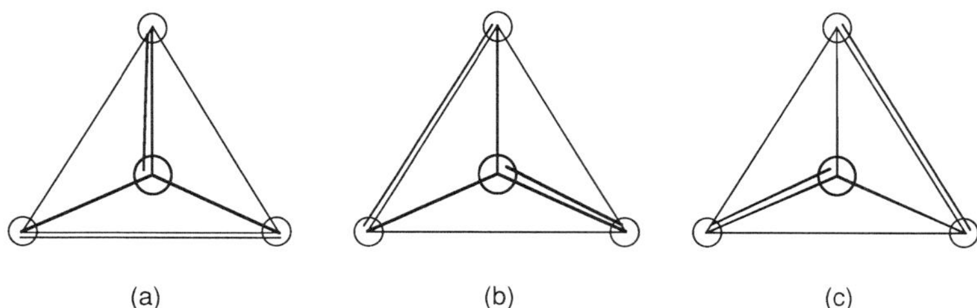

Figure 7.31 The three resonant structures of the (hypothetical) tetrahedral compound C_4.

polyhedron is given by the formula

$$F = E - V + 2. \tag{7.8}$$

In a fullerene C_n where $n = V$ three edges meet at each vertex, so we have

$$E = 3V/2, \tag{7.9}$$

$$F = \frac{V}{2} + 2. \tag{7.10}$$

It is shown in Problem 16 that

$$E = \tfrac{1}{2} \sum_s sF_s \tag{7.11a}$$

$$V = \tfrac{1}{3} \sum_s sF_s, \tag{7.11b}$$

where F_s is the number of faces with s sides, and of course,

$$F = \sum_s F_s. \tag{7.12}$$

Combining Eqs. (7.10)–(7.12) gives the fullerene face formula

$$\sum_s (6 - s)F_s = 12. \tag{7.13}$$

This expression does not place any restrictions on the number of hexagons (F_s), but it does severely limit the number of other polyhedra. The two smallest hypothetical fullerenes, the tetrahedron and the cube, have no hexagons, and the larger ones consist of 12 pentagons (F_s), from Eq. (7.13), and numerous hexagons. For example, the molecule C_{60} with $V = 60$ has 12 pentagons and 20 hexagons. Table 7.7 gives the geometric characteristics of the five Platonic solids, the solids generated by truncating all of their vertices, and several other regular polygons, most of which are fullerenes. The fullerenes of current interest are C_{60} and larger molecules consisting of 12 pentagons and numerous hexagons, such as C_{70}, C_{76}, C_{78}, and C_{82}. Some have several varieties, such as the isomers of C_{78} with the symmetries C_{2v}, D_3, and D_{3h} (Diederich and Whetten, 1992).

The outer diameter of the C_{60} molecule is 7.10 Å and its van der Waals separation is 2.9 Å, so that the nearest-neighbor distance (effective diameter) in a

Table 7.7 Characteristics of Several Regular Solids[a]

Figure	Vertices	Edges	Faces	Face (polygon) type
Tetrahedron	4	6	4	all equilateral triangles
Octahedron[b]	6	12	8	all equilateral triangles
Cube	8	12	6	all squares
Icosahedron[b]	12	30	20	all equilateral triangles
Dodecahedron (pentagonal)	20	30	12	all regular pentagons
Hexadecahedron	28	42	16	12 pentagons, 4 hexagons
Truncated tetrahedron	12	18	8	4 equilateral triangles, 4 hexagons
Truncated octahedron	24	36	14	6 squares, 8 hexagons
Truncated cube	24	36	14	8 equilateral triangles, 6 octagons
Dotriacontohedron (truncated icosahedron)	60	90	32	12 regular pentagons, 20 hexagons
Truncated dodecahedron	60	90	32	20 equilateral triangles, 12 decagons
Heptatriacontohedron	70	105	37	12 pentagons (2 regular), 25 hexagons
Tetracontahedron	76	114	40	12 pentagons, 28 hexagons
Hentetracontohedron	78	116	41	12 pentagons, 29 hexagons
Dotetracontohedron	84	126	44	12 pentagons, 32 hexagons
Large Fullerene	n	$\frac{3}{2}n$	$\frac{1}{2}n + 2$	12 pentagons, $\frac{1}{2}n - 10$ hexagons

[a] The first five solids are the Platonic solids, and the seventh to eleventh are truncations of the Platonic solids. When carbons occupy the vertices all correspond to fullerenes except the octahedron and the icosahedron for which $3V \neq 2E$. The smallest compounds in this table have never been synthesized.

[b] Not a fullerene because the vertices have more than three edges.

solid is 10.0 Å. The bonds shared by a five-membered and a six-membered ring are 1.45 Å long, while those between two adjacent six-membered rings are 1.40 Å long. Above 260 K these molecules form a face centered cubic lattice with lattice constant 14.2 Å; below 260 K it is simple cubic with $a = 7.10$ Å (Fischer *et al.*, 1991; Kasatani *et al.*, 1993; Troullier and Martins, 1992). When C_{60} is doped with alkali metals to form a superconductor it crystallizes into a face centered cubic lattice with larger octahedral and smaller tetrahedral holes for the alkalis. The C_{60} ions are orientationally disordered in the lattice (Gupta and Gupta, 1993).

XII. SYMMETRIES

Earlier in this chapter we mentioned the significance of the horizontal reflection plane σ_h characteristic of the high-temperature superconductors, and noted that most of these superconductors are body centered. In this section we will point out additional symmetries that are present. Table VI-14 of our earlier work (Poole *et al.*, 1988) lists the point symmetries at the sites of the atoms in a number of these compounds.

In the notation of group theory the tetragonal structure belongs to the point group $4/mmm$ (this is the newer international notation for what in the older Schönflies notation was written D_{4h}). The unit cell possesses the inversion operation at the center, so when there is an atom at position (x, y, x), there will be another identical atom at position $(-x, -y, -z)$. The international symbol $4/mmm$ indicates the presence of a fourfold axis of symmetry C_4 and three mutually perpendicular mirror planes m. The Schönflies notation D_{4h} also specifies the fourfold axis, h signifying a horizontal mirror plane σ_h and D indicating a dihedral group with vertical mirror planes.

We see from Fig. 7.32 that the z-axis is a fourfold (90°) symmetry axis called C_4, and that perpendicular to it are twofold (180°) symmetry axes along the x and y directions, called C_2, and also along the diagonal directions (C_2') in the midplane. There are two vertical mirror planes σ_v, two diagonal mirror planes σ_d which are also vertical, and a horizontal mirror plane σ_h. Additional symmetry operations that are not shown are a 180° rotation C_2^z around the z axis,

$$C_2^z = C_4^z C_4^z, \tag{7.14}$$

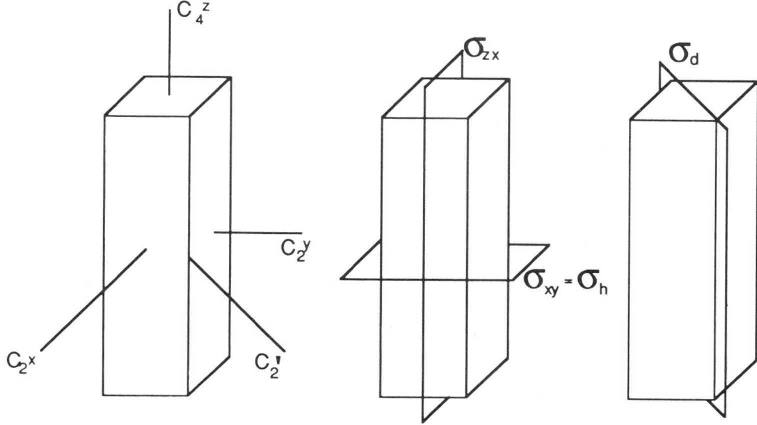

Figure 7.32 Symmetry operations of the tetragonal unit cell showing a fourfold rotation axis C_4, three twofold axes C_2, and reflection planes of the vertical $\sigma_{zx} = \sigma_v$, horizontal $\sigma_{xy} = \sigma_h$; and diagonal σ_d types.

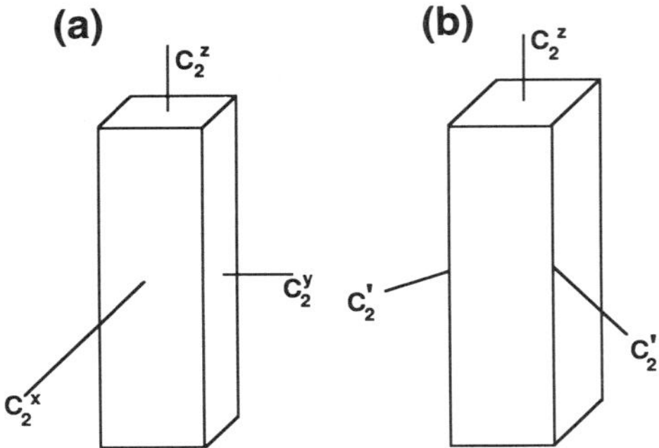

Figure 7.33 Rotational symmetry operations of an orthorhombic unit cell (a) with rectangular distortion, and (b) with rhombal distortion from an originally tetragonal cell.

and the improper fourfold rotation S_4^z around z that corresponds to C_4^z followed by, or preceded by, σ_h,

$$S_4^z = C_4^z \sigma_h = \sigma_h C_4^z, \tag{7.15}$$

where C_4^z and σ_h commute.

The orthorhombic structure has *mmm*, D_{2h} symmetry. We see from Fig. 7.33 that both the rectangular and rhombal unit cells, which correspond to Figs. 7.4a and 7.4b, respectively, have three mutually perpendicular twofold axes, and that they also have three mutually perpendicular mirror planes σ, which are not shown. The two cases differ in having their horizontal axes and vertical planes oriented at 45° to each other.

Cubic structures, being much higher in symmetry, have additional symmetry operations, such as fourfold axes C_4^x, C_4^y, and C_4^z along each coordinate direction, threefold axes C_3 along each body diagonal, and numerous other mirror planes. These can be easily seen from an examination of Fig. 7.1. Buckyballs belong to the icosohedral group, which has twofold (C_2), fivefold (C_5), and sixfold (C_6) rotation axes, horizontal reflection planes, inversion symmetry, and sixfold (S_6) and tenfold (S_{10}) improper rotations, for a total of 120 individual symmetry operations in all (Cotton, 1963).

XIII. CRYSTAL CHEMISTRY

In Chapter 3 we briefly described the structures of some classical superconductors, and in this chapter we provided a more detailed discussion of the structures of the cuprate superconductors. The question arises of how structure is related to the presence of metallic and superconducting properties.

Villars and Phillips (1988; Phillips, 1989a) proposed to explain the combinations of elements in compounds that are favorable for superconductivity at relatively high temperatures by assigning three metallic coordinates to each atom, namely an electron number N_e, a size r, and an electronegativity X. The electron numbers are given in Table 3.1 for most of the elements, with $N_e = 3$ for all of the rare earths and actinides; several correlations of N_e with T_c have already been given in Chapter 3. The sizes and electronegativities were determined empirically from a study of some 3,000 binary intermetallic compounds of types AB, AB_2, AB_3, and A_2B_5. The resulting values for each atom are listed in Fig. 7.34 together with their electron numbers. These values, although arrived at empirically on the basis of the constraint of self-consistency, do have a spectroscopic basis, and thus are called, respectively, *spectroscopic radii* and *spectroscopic electronegativities*.

The metallic coordinates of the atoms can be employed to calculate the three Villars–Phillips (VP) coordinates for each compound, namely (a) average number of valence electrons $N_v = \langle N_e \rangle_{av}$, (b) spectroscopic electronegativity difference ΔX, and (c) spectroscopic radius difference ΔR, where we are using the VP notation. For example, for the compound NbN, with $T_c = 17.3$ K, we have, using the data from Fig. 7.34,

$$\begin{aligned} N_v &= \tfrac{1}{2}(4+5) = 4.5, \\ \Delta R &= 2.76 - 0.54 = 2.22, \\ \Delta X &= 2.03 - 2.85 = -0.82. \end{aligned} \tag{7.16}$$

H 1 2.10[a] 1.25[b]																
Li 1 0.90 1.61	Be 2 1.45 1.08											B 3 1.90 0.795	C 4 2.37 0.64	N 5 2.85 0.54	O 6 3.32 0.465	F 7 3.78 0.405
Na 1 0.89 2.65	Mg 2 1.31 2.03											Al 3 1.64 1.675	Si 4 1.98 1.42	P 5 2.32 1.24	S 6 2.65 1.10	Cl 7 2.98 1.01
K 1 0.80 3.69	Ca 2 1.17 3.00	Sc 3 1.50 2.75	Ti 4 1.86 2.58	V 5 2.22 2.43	Cr 6 2.00 2.44	Mn 7 2.04 2.22	Fe 8 1.67 2.11	Co 9 1.72 2.02	Ni 10 1.76 2.18	Cu 11 1.08 2.04	Zn 12 1.44 1.88	Ga 3 1.70 1.695	Ge 4 1.99 1.56	As 5 2.27 1.415	Se 6 2.54 1.285	Br 7 2.83 1.20
Rb 1 0.80 4.10	Sr 2 1.13 3.21	Y 3 1.41 2.94	Zr 4 1.70 2.825	Nb 5 2.03 2.76	Mo 6 1.94 2.72	Tc 7 2.18 2.65	Ru 8 1.97 2.605	Rh 9 1.99 2.52	Pd 10 2.08 2.45	Ag 11 1.07 2.375	Cd 12 1.40 2.215	In 3 1.63 2.05	Sn 4 1.88 1.88	Sb 5 2.14 1.765	Te 6 2.38 1.67	I 7 2.76 1.585
Cs 1 0.77 4.31	Ba 2 1.08 3.402	La 3 1.35 3.08	Hf 4 1.73 2.91	Ta 5 1.94 2.79	W 6 1.79 2.735	Re 7 2.06 2.68	Os 8 1.85 2.65	Ir 9 1.87 2.628	Pt 10 1.91 2.70	Au 11 1.19 2.66	Hg 12 1.49 2.41	Tl 3 1.69 2.235	Pb 4 1.92 2.09	Bi 5 2.14 1.997	Po 6 2.40 1.90	At 7 2.64 1.83
Fr 1 0.70[a] 4.37[b]	Ra 2 0.90[a] 3.53[b]	Ac 3 1.10[a] 3.12[b]														

Ce 3 1.1[a] 4.50[b]	Pr 3 1.1[a] 4.48[b]	Nd 3 1.2[a] 3.99[b]	Pm 3 1.15[a] 3.99[b]	Sm 3 1.2[a] 4.14[b]	Eu 3 1.15[a] 3.94[b]	Gd 3 1.1[a] 3.91[b]	Tb 3 1.2[a] 3.89[b]	Dy 3 1.15[a] 3.67[b]	Ho 3 1.2[a] 3.65[b]	Er 3 1.2[a] 3.63[b]	Tm 3 1.2[a] 3.60[b]	Yb 3 1.1[a] 3.59[b]	Lu 3 1.2[a] 3.37[b]
Th 3 1.3[a] 4.98[b]	Pa 3 1.5[a] 4.96[b]	U 3 1.7[a] 4.72[b]	Np 3 1.3[a] 4.93[b]	Pu 3 1.3[a] 4.91[b]	Am 3 1.3[a] 4.89[b]								

Figure 7.34 Periodic table listing metallic valences (upper right), sizes (center), and electronegativities (bottom) in the box of each element, according to the Villars–Phillips model (Phillips, 1989a, p. 321).

The VP coordinates for the *A*15 compound Ge_3Nb with $T_c = 23.2$ K are calculated as follows:

$$\begin{aligned} N_v &= \tfrac{1}{4}(4 + 3 \times 5) = 4.75, \\ \Delta R &= \tfrac{1}{2}(1.56 - 2.76) = -0.60, \quad (7.17) \\ \Delta X &= \tfrac{1}{2}(1.99 - 2.03) = -0.02. \end{aligned}$$

The text by Phillips (1989a) tabulates the VP coordinates for more than 60 superconductors with $T_c > 10$ K and for about 600 additional superconductors with transition temperatures in the range $1 < T_c < 10$ K.

When the points for the 600 compounds with lower transition temperatures are plotted on a three-dimensional coordinate system with axes N_v, ΔX, and ΔR, they scatter over a large range of values, but when the points for compounds with $T_c > 10$ K are plotted, they are found to cluster in three regions, called islands, as shown in Fig. 7.35. Island A contains the *A*15 compounds plus some complex intermetallics, island B consists mainly of the NbN family plus some borides and carbides, and island C has closely clustered Chevrel phases, with the high-T_c cuprates on the left. When ternary ferroelectric oxides with Curie temperatures that exceed 500°C are plotted in the same diagram as the superconductors they cluster between the Chevrel group and the cuprates. These ferroelectric oxides are not superconductors, though Phillips (1989a) suggested that doping them with Cu and alkaline earths could produce superconductors with high transition temperatures.

Thus we see that the high transition temperatures of classical superconductors are favored by particular structures and by particular combinations of metallic coordinates for each of these structures. The Villars–Phillips approach provides both structural and atomic criteria for the presence of high T_c.

We have discussed the Phillips approach to a crystal chemistry explanation of the superconductivity of the cuprates. Other researchers have offered alternate, in some cases somewhat related, approaches to understanding the commonali-

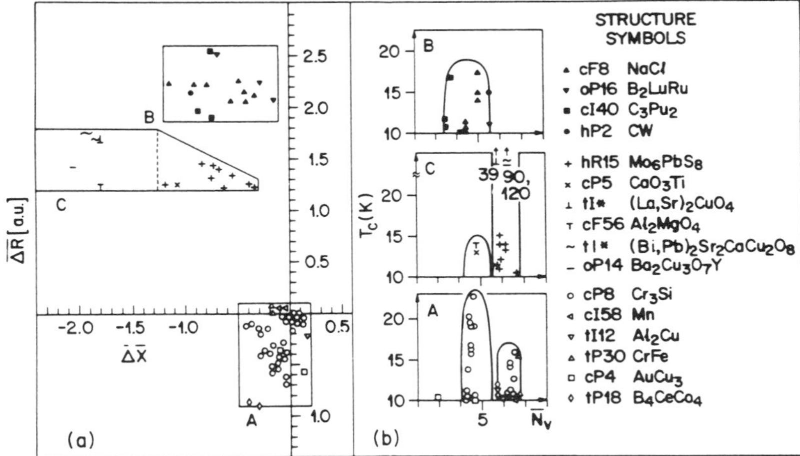

Figure 7.35 Regions in the Villars–Phillips configuration space where superconductivity occurs at relatively high temperatures (Phillips, 1989a, p. 324; Villars and Phillips, 1988).

ties of the various high-temperature and classical superconductors (Adrian, 1992; Schneider, 1992; Tajima and Kitazawa, 1990; Whangbo and Torardi, 1991; Torrace, 1992; Yakhmi and Iyer, 1992; Zhang and Sato, 1993).

XIV. COMPARISON WITH CLASSICAL SUPERCONDUCTOR STRUCTURES

Many elements such as copper and lead are face centered cubic, while many other elements, such as niobium, are body centered cubic, with $a = 3.30$ Å for Nb. The *A*15 compounds, such as Nb_3Se, are (simple) cubic with lattice constant $a \approx 3.63\sqrt{2}$ and have parallel chains of Nb atoms 5.14 Å apart. Other types of classical superconductors, such as the Laves and Chevrel phases, are cubic or close to cubic. The new oxide superconductors are tetragonal or orthorhombic close to tetragonal, and they all have $a \approx b \approx 3.85$ Å, which is somewhat greater than the value for the *A*15 compounds. The third lattice constant c varies with the compound, with the values 13.2 Å for LaSrCuO, 11.7 Å for YBaCuO, and ≈ 23 to 36 Å for the BiSrCaCuO and TlBaCaCuO compounds. These differences occur because the number of copper–oxygen and other planes per unit cell, as well as the spacings between them, vary from compound to compound due to the diverse arrangements of atoms between the layers. Thus relatively high-symmetry crystal structures are characteristic of many superconductors.

XV. CONCLUSIONS

Almost all the high-temperature oxide superconductors have point symmetry D_{4h} ($a = b$) or symmetry close to D_{4h} ($a \approx b$). These superconductors consist of horizontal layers, each of which contains one positive ion and either zero, one, or two oxygens. The copper ions may be coordinated square planar, pyramidal, or octahedral, with some additional distortion. Copper oxide layers are never adjacent to each other, and equivalent layers are never adjacent. The cations alternate sites vertically, as do the oxygens. The copper oxide layers are either flat or slightly puckered, in contrast to the other metal oxide layers, which are generally far from planar. The highest T_c compounds have metal layers (e.g., Ca) with no oxygens between the copper oxide planes.

FURTHER READING

The Wyckoff series, *Crystal Structures* (1963, Vol. 1; 1964, Vol. 2; 1965, Vol. 3; 1968, Vol. 4) provides a comprehensive tabulation of crystal structures, but many important classical superconductors such as the *A*15 compounds are not included. The *International Tables for X-Ray Crystallography* (Henry and Lonsdale, 1965, Vol. 1) provide the atom positions and symmetries for all of the crystallographic space groups. The Strukturbericht notation, e.g., *A*15 for Nb_3Ge, is explained in Pearson's compilation (1958).

Details of cuprate crystallographic structures are given by Beyers and Shaw (1989; $YBa_2Cu_3O_7$), Burns and Glazer (1990), Hazen (1990), Poole *et al.* (1988, Chapter 6), Santoro (1990), and Yvon and François (1989). Phillips (1989a) provides an extensive discussion of the crystal chemistry of the cuprates. Our earlier work (Poole *et al.*, 1988, p. 107) lists the site symmetries in perovskite and cuprate structures. Billinge *et al.* (1994) reviewed lattice effects in high temperature superconductors, and Zhu (1994) reviewed structural defects in $YBa_2Cu_3O_{7-\delta}$.

The microstructure of high temperature superconductors studied by electron microscopy are reviewed by Chen (1990), Gai and Thomas (1992), Gross and Koelle (1994), and Shekhtman (1993). Oxygen stoichiometry in HTSC's is reviewed by Chandrashekhar *et al.*, (1994), Green and Bagley (1990) and by Routbert and Rothman (1995). Electron-doped superconductors are reviewed by Almasan and Maple (1991) and by Fontcuberta and Fàbrega (1995).

The March 1992 special issue of *Accounts of Chemical Research* (Vol. 25, No. 3) is devoted to reviews of buckminsterfullerenes. Two recent books are edited by Billups and Ciofolini (1993) and by Kroto and Walton (1993), and the review by Dresselhaus *et al.* (1994) are devoted to fullerenes. The thallium compounds were reviewed by Hermann and Yakhimi (1993) and the mercury superconductors by Chu (1995).

PROBLEMS

1. Show that the radius of the octahedral hole in an fcc close-packed lattice of atoms of radius r_0 is equal to $[\sqrt{2} - 1]r_0$. What is the radius of the hole if the lattice is formed from oxygen ions?
2. Show that the radius of the tetrahedral hole in an fcc close-packed lattice of atoms of radius r_0 is equal to $[(3/2)^{1/2} - 1)r_0$. What is the radius of the hole if the lattice is formed from oxygen ions?
3. The "image perovskite" unit cell is generated from the unit cell of Fig. 7.1 by shifting the origin from the point $(0,0,0)$ to the point $(\frac{1}{2},\frac{1}{2},\frac{1}{2})$. Sketch this "image" cell. Show that the planes of atoms in this cell are the image planes related by the body centering operation to those of the original perovskite. This image cell is the one that usually appears to represent perovskite in solid-state physics texts.
4. Calculate the distance between the yttrium atom and its nearest-neighbor Ba, Cu, and O atoms in the superconductor $YBa_2Cu_3O_7$.
5. Write down the x, y, z coordinates for the five numbered atoms in the initial plane of Fig. 7.16. Give the explicit symmetry operations, with the proper choice of sign in Eq. (7.5) for each case, that transform these five atoms to their indicated new positions on the other three planes.
6. Explain how the international and Schönflies symbols, *mmm* and D_{2h} respectively, are appropriate for designating the point group for the orthorhombic superconductors.
7. What are the symmetry operations of the *A*15 unit cell of Fig. 3.19?
8. The D_{2h} point group consists of eight symmetry operations that leave an orthorhombic cell unchanged, namely an identity operation E that produces no change, three twofold rotations C_2^i along $i = x,\ y,\ z$, three mirror reflection planes σ_{ij}, and an inversion i. Examples of these symmetry operations are

$$\begin{array}{llll} E & x \to x & y \to y & z \to z \\ C_2^x & x \to x & y \to -y & z \to -z \\ \sigma_{xy} & x \to x & y \to y & z \to -z \\ i & x \to -x & y \to -y & z \to -z. \end{array}$$

A group has the property that successive application of two symmetry operations produces a third. Thus, we have, for example,

$$C_2^x \sigma_{xy} = \sigma_{zx}$$
$$C_2^y C_2^x = C_2^z$$
$$i C_2^y = \sigma_{zx}$$
$$\sigma_{zx} \sigma_{yz} = C_2^z.$$

These results have been entered into the following multiplication table for the D_{2h} group. Fill in the remainder of the table. Hint: each element of a group appears in each row and each column of the multiplication table once and only once.

	E	C_2^x	C_2^y	C_2^z	i	σ_{xy}	σ_{yz}	σ_{zx}
E								
C_2^x						σ_{zx}		
C_2^y		C_2^z						
C_2^z								
i			σ_{zx}					
σ_{xy}								
σ_{yz}								
σ_{zx}							C_2^z	

9. Construct the multiplication table for the D_{4h} point group which contains the 16 symmetry elements that leave a tetragonal unit cell unchanged. Which pairs of symmetry elements A and B do not commute, i.e., such that $AB \neq BA$? Hint: follow the procedures used in Problem 8.
10. Draw diagrams analogous to those in Fig. 7.25 for the first two members of the aligned series $TlBa_2Ca_nCu_{n+1}O_{5+2n}$, where $n = 0, 1$.
11. Draw the analogue of Fig. 7.19 for the Nd_2CuO_4 compound, showing the location of all of the Cu and O atoms. How do Figs. 7.21 and 7.22 differ for Nd_2CuO_4?
12. Calculate the Villars–Phillips coordinates for the three superconductors MoP_3, V_3Sn, and NbTi.
13. Select one of the compounds ($Tl_2Ba_2CuO_6$, $Bi_2Sr_2CaCu_2O_8$, $Bi_2Sr_2Ca_2Cu_3O_{10}$, $Tl_2Ba_2Ca_2Cu_3O_6$) and construct a table for it patterned after Tables 7.5 or 7.6.
14. Locate a twofold (C_2), fivefold (C_5), and sixfold (C_6) rotation axis, and also a reflection plane σ_h in the buckyball sketch of Fig. 3.35. How many of each type of operation are there?
15. We can see by examining Fig. 3.35 that a buckyball has inversion symmetry. Identify a sixfold (S_6) and tenfold (S_{10}) improper rotation axis, where an improper rotation is understood to involve a sequential inversion and a proper rotation. How many S_6 and how many S_{10} axes are there?
16. Show that the total number of edges E in a fullerene is given by
$$E = \tfrac{1}{2} \sum_s sF_s,$$
and the number of vortices is
$$V = \tfrac{1}{3} \sum_s sF_s,$$
where F_s is the number of faces with s sides.
17. Show that the cubic fullerene compound C_8 has nine resonant structures.

8

Hubbard Models and Band Structure

I. INTRODUCTION

In the previous chapter we discussed the crystallographic structures of high-temperature superconducting compounds. In these compounds positively charged atoms occupy sites where they are surrounded by negatively charged oxygen nearest neighbors. Some cations, such as Ca^{2+} and La^{3+}, are ionically bound with their valence electrons transferred to oxygen. Other metal ions, such as Cu^{2+} and Tl^{3+}, have a strong admixture of covalency, with an appreciable amount of their electron density distributed in chemical bonds with the neighboring oxygens. We did not take into account electrons that are delocalized among many atoms, and hence the structure studies did not explain why these materials are metallic conductors.

Before proceeding to the superconductors themselves, some background on energy bands and bonding orbitals will be reviewed, based in part on material found in Chapter 1, Sections II, IV, V, IX, and X. We will begin by discussing the energy bands that arise from free electrons, after which we will add a periodic potential and take into account the atomic orbitals of the valence electrons and their overlap between adjacent atoms. Next, we will introduce the simplified Hubbard model, which takes into account correlations between electrons, and comment on two extensions of the model, called the *t-J* and resonant-valence-bond (RVB) models. Finally, we will examine the results of more complex calculations in which a separate Schrödinger equation is written for each such electron in the unit cell, with the atoms and their inner electronic shells providing the periodic potential.

The results show that the electrical and superconducting properties arise from a single band or very small number of bands and the densities of states of the band or bands near the Fermi level. The particular atoms associated with these bands will be identified, and their role will be discussed. For example, the band structure results confirm the dominance of the copper-oxide planes in determining these properties. The calculations clarify the crucial importance of oxygen content and doping for understanding the superconductivity of the cuprates. The presence of Fermi surface nesting suggests the possible onset of charge density or spin-density waves. Electronic charge distributions in coordinate space are also determined. These results as well as other useful information are provided by the band structure.

The accuracy of band structure calculations can be checked by comparing them with experiment. The most straightforward comparison is with photoemission and x-ray spectroscopy, which will be discussed in Chapter 15. The electron-energy spectra exhibit peaks in the same places as the density-of-states (DOS) calculations, but often are less well resolved. Other experimental results, such as optical reflectivity and magnetic susceptibility, also provide a check on the bands. At the end of the chapter we will present figures that compare theory with experiment.

The band structure results do indeed provide an explanation for many properties of superconductors, but they also have their limitation. For example, the calculated bands and densities of states sometimes fit the experimental data well, but sometimes do not. The antiferromagnetism of La_2CuO_4 is not explained. More fundamentally, the bands do not accurately foretell which compounds will superconduct, and at what temperature. The results are more descriptive of superconductivity, rather than predictive of it. Nevertheless, they do provide insights that could not be obtained otherwise, as is clear from Warren E. Pickett's (1989) review.

II. RECIPROCAL SPACE AND BRILLOUIN ZONE

In the previous chapter we gave the atomic positions of the atoms of superconducting compounds in the direct lattice of the x, y, z-coordinate space. To describe the energy bands in conductors and superconductors it is more convenient to employ a so-called reciprocal space. This reciprocal space is sometimes called k-space or momentum space because its coordinates k_x, k_y, and k_z arise from the quantum mechanical expression for the momentum $\mathbf{p} = \hbar\mathbf{k}$. A free electron has a kinetic energy (Eq. 1.37) but no potential energy, and we will find it convenient to write this kinetic energy E_k in the form

$$E_k = E_0\left(\frac{a}{\pi}\right)^2 k^2 \tag{8.1}$$

with

$$E_0 = \frac{\pi^2\hbar^2}{2ma^2}, \tag{8.2}$$

where $k = 2\pi/\lambda$ and a is the lattice constant. The basis vectors **A**, **B**, and **C** in reciprocal space are related to their respective counterparts **a**, **b**, and **c** in coordinate space as follows,

$$\begin{aligned} \mathbf{A} &= \frac{2\pi\mathbf{b}\times\mathbf{c}}{\mathbf{a}\cdot(\mathbf{b}\times\mathbf{c})}, \\ \mathbf{B} &= \frac{2\pi\mathbf{c}\times\mathbf{a}}{\mathbf{a}\cdot(\mathbf{b}\times\mathbf{c})}, \\ \mathbf{C} &= \frac{2\pi\mathbf{a}\times\mathbf{b}}{\mathbf{a}\cdot(\mathbf{b}\times\mathbf{c})}, \end{aligned} \tag{8.3}$$

as is shown in standard solid-state physics texts. The particular direct lattice basis vectors **a**, **b**, and **c** that were used in Chapter 7 to describe the crystal structures are mutually perpendicular to each other, and from Eqs. (8.3) the corresponding reciprocal lattice basis vectors are also mutually perpendicular, with the respective magnitudes

$$A = 2\pi/a, \qquad B = 2\pi/b, \qquad C = 2\pi/c. \tag{8.4}$$

For orthorhombic structures ($a < b < c$), $A > B > C$ and all are of different lengths; for tetragonal crystals ($a = b \neq c$), we have $A = B \neq C$, while for the cubic case ($a = b = c$), the reciprocal space basis vectors are all equal in length $\mathbf{A} = \mathbf{B} = \mathbf{C}$. We will use the term, *reciprocal lattice vector* to denote a vector $\mathbf{G}$ in k-space that has the form

$$\mathbf{G} = n_x\mathbf{A} + n_y\mathbf{B} + n_z\mathbf{C}, \tag{8.5}$$

where n_x, n_y, and n_z are integers. A vector of this type connects points on the reciprocal lattice.

The unit cell in the reciprocal space is called the Brillouin zone (BZ), or more precisely the first Brillouin zone, with volume $\mathbf{A}\cdot(\mathbf{B}\times\mathbf{C}) = (2\pi)^3/abc$. Figure 8.1 shows a sketch of a two-dimensional Brillouin zone for a square lattice ($a = b$) with its corners at the points $(\pm\pi/a, \pm\pi/a)$ and with the area $4\pi^2/a^2$. The spacing between points in k-space is $2\pi/L$, as shown in Fig. 8.2, where L is the overall length and width of the crystal. The number of k-space points in the BZ of Fig. 8.1 is $(L/a)^2$.

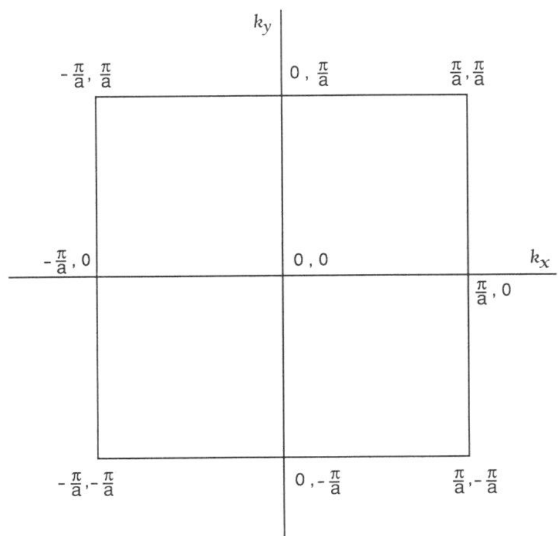

Figure 8.1 Two-dimensional Brillouin zone for a square lattice in k-space. The corners are at the points $(\pm\pi/a, \pm\pi/a)$, and the area of the zone is $4\pi^2/a^2$. The figure gives the coordinates (k_x, k_y) of several special points.

The BZ for the square lattice has three special symmetry points, the origin Γ, the k-axis boundary point X, and the corner point M, with the respective coordinates $(0, 0)$, $(\pi/a, 0)$ and $(\pi/a, \pi/a)$. These points define three special directions in k-space, $\Gamma \to X$, $\Gamma \to M$, and $X \to M$, as indicated in Fig. 8.3. We will be interested in determining how the electron energies change as we move along these directions in k-space.

III. FREE-ELECTRON BANDS IN TWO DIMENSIONS

In this section we will write down expressions for the energy of free-conduction electrons on a square lattice and deduce the energy bands and Fermi surface. (The study of Fermi surfaces is sometimes called Fermiology.) Energy bands are plots of energy versus coordinate k along particular directions in k-space, and the Fermi surface is the boundary between regions of k-space that are full of electrons and regions where there are no electrons at $T = 0$. It will be easy to generalize these results to more complex lattices and to three dimensions.

At the beginning of the previous section we mentioned that all of the energy (8.1) of a free electron is kinetic. For the two-dimensional case this may be written

$$E_k = E_0(a/\pi)^2(k_x^2 + k_y^2), \tag{8.6}$$

where we are making use of the unit of energy E_0 defined by Eq. (8.2). In order to express all of the energies in terms of k_x and k_y values within the (first) Brillouin zone, corresponding to $-\pi/a < k_x$, $k_y <$

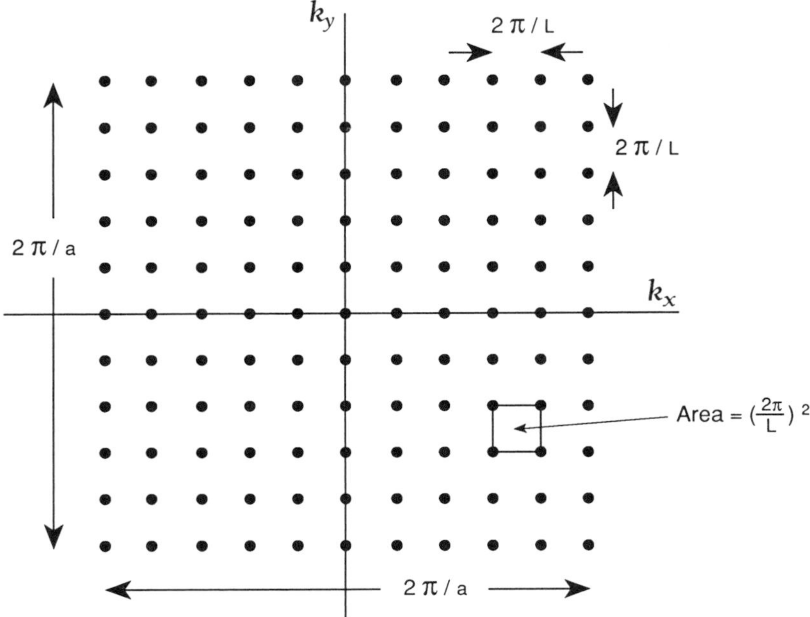

Figure 8.2 Points of rectangular lattice in two-dimensional k-space with spacing $2\pi/L$. The figure is drawn for the case $L = 10a$.

$+\pi/a$, reciprocal lattice vectors (8.5) can be applied to bring outside points back into the zone. In other words, each $\mathbf{k}'$ that extends outside the BZ can be written as the vector sum $\mathbf{G} + \mathbf{k}$, where $\mathbf{k}$ is within the BZ and $\mathbf{G}$ is a reciprocal lattice vector as in (8.5). If we do this, the free electron energy, (8.6), becomes

$$E_{\mathrm{k}} = E_0\left[(2n_x + k_x a/\pi)^2 + (2n_y + k_y a/\pi)^2\right]. \quad (8.7)$$

The k_x, k_y dependence of the energy E_{k} shown plotted in Fig. 8.4 consists of a set of what are called *energy bands*, each band labeled with its n_x, n_y values. For example, the lowest band, labelled (0,0) in the figure, has energy

$$E_0\left[(k_x a/\pi)^2 + (k_y a/\pi)^2\right],$$

which is obtained by setting $n_x = n_y = 0$ in Eq. (8.7).

It will be instructive to write down expressions for the energies along the three special directions in the Brillouin zone, namely Γ to X, Γ to M, and X to M, as shown in Fig. 8.3. Along the path from Γ

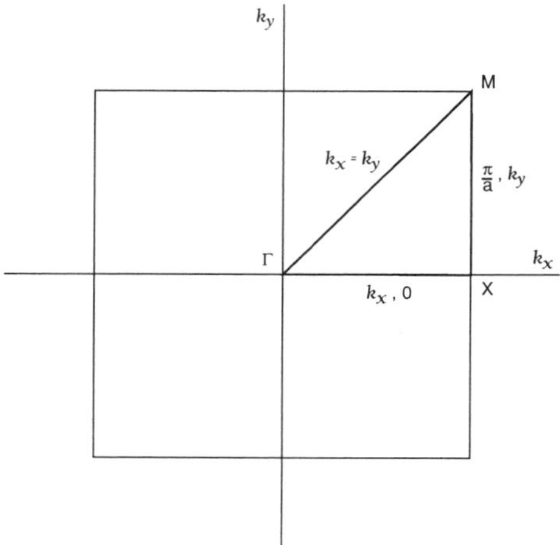

Figure 8.3 Special symmetry points of Brillouin zone of Fig. 8.1. Γ the center (0,0), X the midpoint of the side $(\pi/a, 0)$, and M the corner $(\pi/a, \pi/a)$ of the zone.

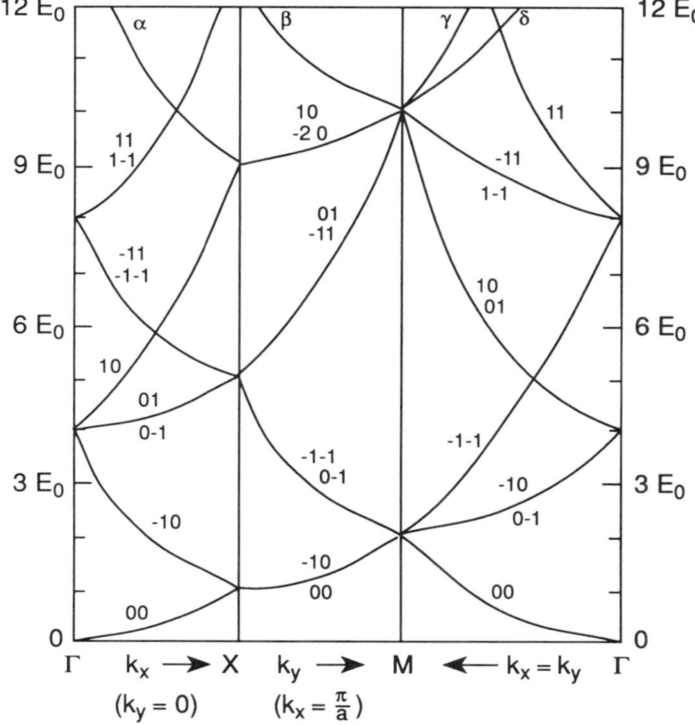

Figure 8.4 Energy bands for the free-electron approximation on a square lattice. The energies are plotted along the three principal directions $\Gamma \to X$, $X \to M$ and $M \to \Gamma$ defined in Fig. 8.3, and the levels are labeled with their (n_x, n_y) values from Eq. (8.7). We see from the figure that the lowest energy band (0,0) is doubly degenerate from X to M. Problem 5 asks for the identity of the four bands labeled α, β, γ, and δ at the top of the figure.

to X, where $k_y = 0$, Eq. (8.7) becomes

$$E_k = E_0\left[\left(2n_x + \frac{k_x a}{\pi}\right)^2 + (2n_y)^2\right] \qquad \Gamma \text{ to } X, \quad (8.8)$$

and the (0,0) band is lowest, as indicated on the left side of Fig. 8.4. Along the path from Γ to M, where $k_x = k_y$, we have

$$E_k = E_0\left[\left(2n_x + \frac{k_x a}{\pi}\right)^2 + \left(2n_y + \frac{k_x a}{\pi}\right)^2\right] \qquad \Gamma \text{ to } M \quad (8.9)$$

and the lowest energy band is still (0,0). The path X to M is along the π/a, k_y line with $k_x = \pi/a$, so that Eq. (8.7) has the form

$$E_k = E_0\left[(2n_x + 1)^2 + \left(2n_y + \frac{k_y a}{\pi}\right)^2\right] \qquad X \text{ to } M \quad (8.10)$$

and the lowest, or X to M band is doubly degenerate, with the (n_x, n_y) values (0,0) and $(-1,0)$, as shown in the center panel of Fig. 8.4.

We can see from an examination of Fig. 8.4 that at particular points on the Γ, X, and M axes energy bands from the left and right come together. In each case the bands that converge from both sides have the same (n_x, n_y) labels. For example, at the energy E_0 on the x-axis the two energy bands (0,0) and $(-1,0)$ coming from point

Γ on the left converge to meet the doubly degenerate $(0,0)$, $(-1,0)$ band coming to X from point M on the right. We also see from this figure that each particular band can be traced in the cyclic path $\Gamma \to X \to M \to \Gamma$. Thus the band $(-1,0)$ starts with energy $4E_0$ at point Γ, this energy decreases to E_0 at X, rises to $2E_0$ at M, and then returns to its starting value of $4E_0$ at Γ.

IV. NEARLY FREE ELECTRON BANDS

The discussion in the previous section was for the free-electron case. We will now examine what happens when a weak periodic potential $V(r)$ is taken into account. Perturbation theory will be used to solve the Schrödinger equation

$$H\Psi = \epsilon(k)\Psi \tag{8.11}$$

for the one-electron Hamiltonian

$$H = \frac{p^2}{2m} + V(\mathbf{r}), \tag{8.12}$$

and since the potential is periodic it can be expressed as a Fourier sum,

$$V(\mathbf{r}) = \sum_{\mathbf{G}} e^{i\mathbf{G}\cdot\mathbf{r}} V_{\mathbf{G}}, \tag{8.13}$$

over reciprocal lattice vectors (8.5). We know from Bloch's theorem, which is proven in standard solid-state physics texts, that the one-electron Hamiltonian (8.12) has eigenfunctions $\Psi_k(\mathbf{r})$ of the form

$$\Psi_k(\mathbf{r}) = \left(\frac{1}{\sqrt{N}}\right) \sum_{\mathbf{G}} e^{i(\mathbf{k}+\mathbf{G})\cdot\mathbf{r}} a_{k+G}, \tag{8.14}$$

where again $\mathbf{k}$ lies in the first BZ and $\mathbf{G}$ is a reciprocal lattice vector. Substituting the expressions in Eqs. (8.12)–(8.14) in Eq. (8.11) gives

$$\left[\frac{\hbar^2(k+G)^2}{2m} - \epsilon(k)\right] a_{k+G} + \sum_{G'} V_{G-G'} a_{k+G'} = 0, \tag{8.15}$$

which is nothing else than the Schrödinger equation written down in reciprocal space, without any approximations.

Since the potential energy (8.13) is small compared to the kinetic energy of the Hamiltonian (8.12), the eigenstates are nearly pure plane waves. If we assume that one of the Fourier coefficients, a_{k+G}, is large compared to the other coefficients, the Hamiltonian (8.15) can be solved by perturbation theory to give

$$\epsilon(k) = \frac{\hbar^2(k+G_0)^2}{2m} - \sum_{\mathbf{G}} \frac{|V_{G-G_0}|^2}{\epsilon_0(k+G) - \epsilon_0(k+G_0)}, \tag{8.16}$$

where the reciprocal lattice vector $\mathbf{G}_0$ has been excluded from the summation. This expression is only valid for values of k for which the denominators of the summand are all much greater than $|V_{G-G_0}|$. Note that for each reciprocal lattice vector $\mathbf{G}_0$ there is a branch or band of the excitation spectrum, so that there are N bands in all.

When degeneracies occur in the free-electron energies, one of more of the denominators in Eq. (8.16) vanishes, so the equation is no longer valid and we have to apply degenerate perturbation theory. For the case of double degeneracy, this leads to the quadratic equation

$$\epsilon^2(k) - \epsilon(k)[\epsilon_0(k+G_1) + \epsilon_0(k+G_2)] - \left[|V_{G_1-G_2}|^2 - \epsilon_0(k+G_1)\epsilon_0(k+G_2)\right] = 0, \tag{8.17}$$

which is easily solved to give the eigenvalues $\epsilon(k)$,

$$\epsilon(k) = \tfrac{1}{2}[\epsilon_0(k+G_1) + \epsilon_0(k+G_2)] \pm \tfrac{1}{2}\left\{4|V_{G_1-G_2}|^2 + [\epsilon_0(k+G_1) - \epsilon_0(k+G_2)]^2\right\}^{1/2}. \tag{8.18}$$

Any value of $\mathbf{k}$ for which there exists a pair of reciprocal lattice vectors $\mathbf{G}_1$ and $\mathbf{G}_2$ that make $\epsilon_0(k+\mathbf{G}_1) = \epsilon_0(k+\mathbf{G}_2) = \epsilon_0$ gives $\epsilon(k) = \epsilon_0 + V_{\mathbf{G}_1-\mathbf{G}_2}$ in Eq. (8.18). There is an energy gap E_g given by

$$E_g = 2|V_{\mathbf{G}_1-\mathbf{G}_2}|, \tag{8.19}$$

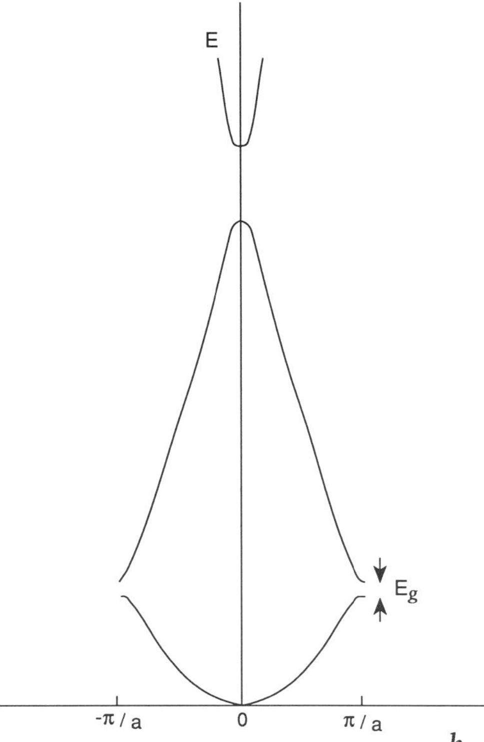

Figure 8.5 One-dimensional energy bands in the reduced zone representation for a weak periodic potential showing the gaps at the zone boundaries.

as shown for $k = \pi/a$ in the one-dimensional case of Fig. 8.5. The reciprocal lattice vectors $\mathbf{G}_1 = 0$ and $\mathbf{G}_2 = -2\pi/a$ make $|k + \mathbf{G}_1| = |k + \mathbf{G}_2|$ for this point and produce the gap. Energy gaps occur at points of high symmetry in reciprocal space, such as the zone center Γ and the boundary points X and M in the BZ of Fig. 8.3. Problem 6 asks the reader to find appropriate reciprocal lattice vectors $\mathbf{G}_1$ and $\mathbf{G}_2$ for these three points.

In two and three dimensions it is possible to have three, four, or more degenerate free-electron bands, as illustrated in Fig. 8.4, and this leads to cubic, quartic, and higher-order equations to solve instead of the rather simple quadratic equation (8.17). Serious band structure calculations can involve diagonalization of the Hamiltonian over a set of hundreds of basis states.

V. FERMI SURFACE IN TWO DIMENSIONS

A. Fermi Surface

Because electrons obey the Fermi–Dirac distribution (Eq. 1.2), at absolute zero the N conduction electrons fill the points in reciprocal space up to the highest occupied level, called the *Fermi level*, with higher levels empty. For the free-electron approximation in two dimensions, the occupied region will be bounded by a circle $k_x^2 + k_y^2 = k_F^2$ of radius k_F centered at the point Γ in reciprocal space, where k_F is related to the Fermi energy by

$$E_F = \hbar^2 k_F^2/2m = E_0(a/\pi)^2 k_F^2. \quad (8.20)$$

This circle constitutes the two-dimensional "Fermi surface." Points in reciprocal space are separated by a distance $2\pi/L$, as shown in Fig. 8.2, so that the area per point is $(2\pi/L)^2$. Since two electrons of opposite spin direction can occupy each point in reciprocal space, the total number of electrons N is twice the number of k_x, k_y points that are inside the Fermi surface. This, in turn, is simply twice the area πk_F^2 of the circle divided by the area per point $(L/2\pi)^2$,

$$N = \frac{L^2 k_F^2}{2\pi}. \quad (8.21)$$

The electron density $n = N/A = N/L^2$, or number of electrons per unit area, is

$$n = k_F^2/2\pi = mE_F/\pi\hbar^2. \quad (8.22)$$

The density of states $D(E)$ per unit area that is obtained by evaluating the derivative of this expression, dn/dE, with E_F replaced by E,

$$D(E) = m/\pi\hbar^2, \quad (8.23)$$

is a constant, independent of the energy. This result is for free electrons in the absence of a periodic potential; the situa-

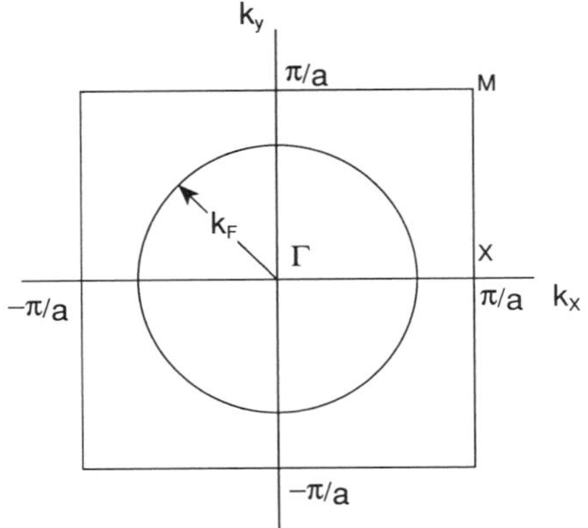

Figure 8.6 Circular Fermi surface centered at the zone center Γ in the reduced zone representation with $k_F < \pi/a$. Electrons occupy the central region out to k_F.

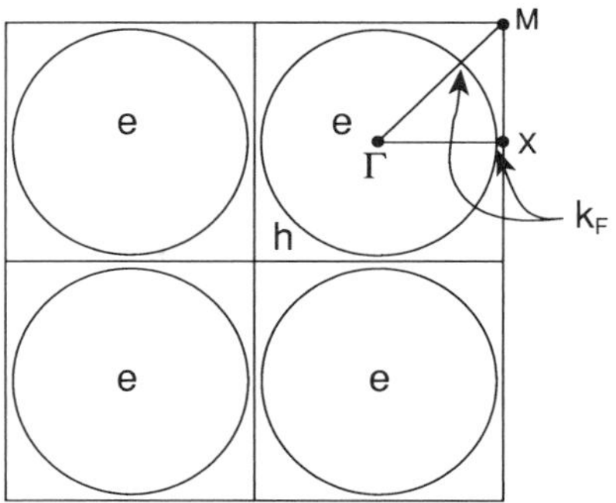

Figure 8.7 Extended zone representation of the Fermi surface of Fig. 8.6 with $k_F < \pi/a$. The electron (e) and hole (h) regions are indicated.

tion is more complicated in the case of actual superconductors.

Equations (8.20)–(8.23) may be compared with their three-dimensional counterparts, which are given in Chapter 1. Problem 8 asks the reader to determine the analogous one-dimensional expressions.

B. Closed Fermi Surface

If, for illustrative purposes, we assume in the two-dimensional case that k_F is less than the zone boundary π/a, the occupied part of the first Brillouin zone turns out to be the area inside the circle drawn in Fig. 8.6 with the radius $k_F < \pi/a$. We will be interested in representing the Fermi surface in what is called the *periodic zone scheme*. In such a scheme several Brillouin zones are presented adjacent to each other in the same diagram, as in Fig. 8.7. The areas inside the circles are labeled e because they are occupied by electrons, while the area outside is labeled h to indicate the absence of electrons, which is equivalent to occupancy by holes. This type of Fermi surface is said to be *closed* since electron regions of adjacent zones in the periodic scheme are not connected.

The energy-level diagram in Fig. 8.8 shows how the lowest band, labeled (0, 0) in Fig. 8.4, is occupied up to the Fermi energy E_F. We see from a comparison of Figs. 8.7 and 8.8 that when the Fermi level lies completely within the first BZ it crosses the two paths Γ to X and Γ to M, with the path X to M entirely above the Fermi surface, and hence devoid of electrons. Unless the Fermi surface approaches or crosses the zone boundary the presence of a periodic lattice potential has very little effect.

C. Open Fermi Surface

Now let us consider the case when k_F is greater than the zone boundary π/a but less than $\sqrt{2}\,\pi/a$. We see from Fig. 8.9 that the Fermi surface now lies beyond the first Brillouin zone at the point X and is inside the first BZ at M. For free electrons there is no energy gap and the Fermi sur-

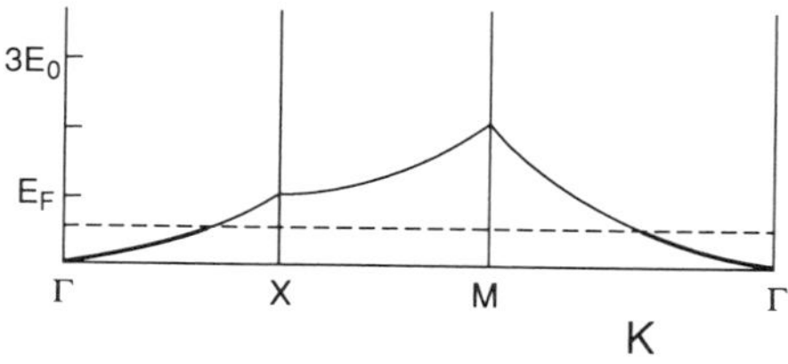

Figure 8.8 Electron occupancy around the center point Γ (darkened portion) of the lowest (0, 0) band along the principal directions of Fig. 8.3 for $k_F < \pi/a$.

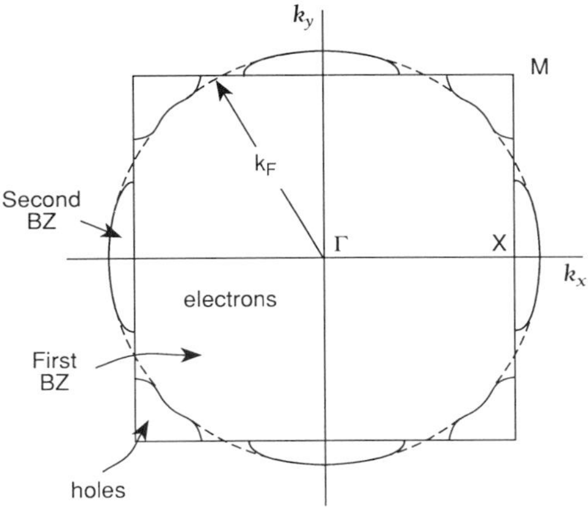

Figure 8.9 Circular Fermi "surface" centered at the zone center Γ in the reduced zone representation with $\pi/a < k_F < \sqrt{2}\,\pi/a$. The surface is drawn for the two-dimensional analogue of Fig. 8.5 in which a periodic potential causes gaps to appear at the zone boundaries. The dashed curves at these gaps are for the free-electron approximation (circular Fermi surface).

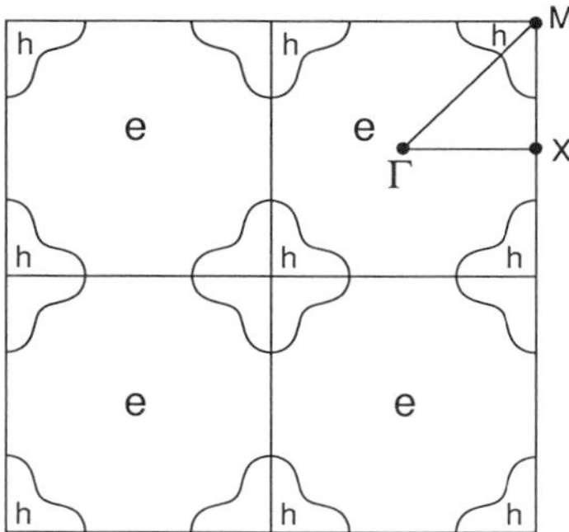

Figure 8.10 *Extended* zone representation of the Fermi surface of Fig. 8.9 with $\pi/a < k_F < \sqrt{2}\,\pi/a$. The electron (e) and hole (h) regions are indicated.

face is a circle, shown dashed in the figure. More realistically, the presence of a lattice potential produces an energy gap, and the Fermi surface becomes distorted near the boundary, bending toward the boundary and intersecting it perpendicular to the interface, as indicated in the figure. This is the two-dimensional analogue of the way the one-dimensional energy bands of Fig. 1.8 bend toward the zone boundary when the gap opens up. We will examine the electron distribution in the presence of a gap.

The Fermi surface encloses electrons, so most of the BZ is electron-like, but there are closed regions around point M occupied by holes, as indicated in Fig. 8.9. This is shown more clearly in the extended zone scheme of Fig. 8.10. Since the electron regions of adjacent zones connect across the boundaries, this type of Fermi surface is called open. The energy-level diagram in Fig. 8.11 shows how the lowest band, (0, 0) of Fig. 8.4, is occupied up to the Fermi energy E_F, which now lies above point X but below point M. We see from Figs. 8.10 and 8.11 that the Fermi level extends beyond the first BZ in the k_x direction and crosses the two paths X to M and Γ to M, with the path Γ to X lying entirely below the Fermi surface, and hence full of electrons.

VI. ELECTRON CONFIGURATIONS

Thus far we have been discussing the free-electron energy bands in two dimensions. For actual compounds there are additional factors to be taken into account. The unit cell contains several atoms with each atom contributing one or more electrons, as listed in Table 8.1. A separate Schrödinger equation is written down for each electron, and these equations are coupled together and must be solved self-consistently. An initial guess for the wave-

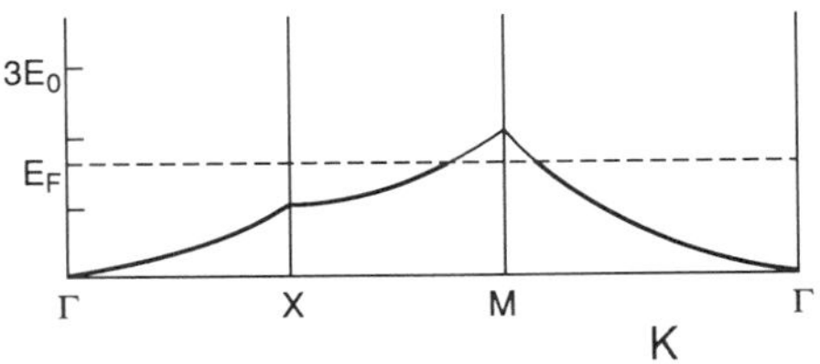

Figure 8.11 Electron occupancy (darkened portion) of the lowest (0, 0) band along the three principal directions of Fig. 8.3 for $\pi/a < k_F < \sqrt{2}\,\pi/a$. The band is full everywhere except near the corner point M.

Table 8.1 Electron Configurations of Selected Atoms Commonly Used for Band Structure Calculations of Superconductors[a]

Atom number	Symbol	Core[b, c]	Atom configuration	No. valence electrons	Ion	Ion configuration	No. electrons
8	O	Be 4	$[2s^2]2p^4$	4	O^{1-}	$2p^5$	5
					O^{2-}	$2p^6$	6
14	Si	Ne 10	$3s^2 3p^2$	4	Si^{4+}	—	0
19	K	Ar 18	$[3p^6]4s$	1	K^+	—	0
20	Ca	Ar 18	$4s^2$	2	Ca^{2+}	—	0
23	V	Ar 18	$3d^3 4s^1 4p^1$	5	V^{3+}	$3d^2$	2
29	Cu	Ar 18	$3d^{10} 4s^1$	11	Cu^{1+}	$3d^{10}$	10
					Cu^{2+}	$3d^9$	9
					Cu^{3+}	$3d^8$	8
38	Sr	Kr 36	$5s^2$	2	Sr^{2+}	—	0
39	Y	Kr 36	$4d^1 5s^2$	3	Y^{3+}	—	0
41	Nb	Kr 36	$4d^3 5s^1 5p^1$	5	Nb^{4+}	$4d^1$	1
50	Sn	--46	$5s^2 5p^2$	4	Sn^{4+}	—	0
56	Ba	Xe 54	$[5p^6]6s^2$	2	Ba^{2+}	—	0
57	La	Xe 54	$5d^1 6s^2$	3	La^{3+}	—	0
80	Hg	--78	$[5d^{10}]6s^2$	2	Hg^{2+}	$[5d^{10}]$	0
81	Tl	-- 78	$[5d^{10}]6s^2 6p^1$	3	Tl^{3+}	$[5d^{10}]$	0
82	Pb	--78	$[5d^{10}]6s^2 6p^2$	4	Pb^{4+}	$[5d^{10}]$	0
83	Bi	--78	$[5d^{10}]6s^2 6p^3$	5	Bi^{3+}	$[5d^{10}]6s^2$	2
					Bi^{4+}	$[5d^{10}]6s^1$	1
					Bi^{5+}	$[5d^{10}]$	0

[a] Core electrons listed in square brackets are sometimes included in the basis set.
[b] The core of Sn is Kr plus the fourth transition series ($4d^{10}$) closed shell.
[c] The core of Tl, Pb, and Bi is Xe plus the rare earth ($4f^{14}$) and fifth transition series ($5d^{10}$) closed shells.

functions of these electrons is used to calculate the potential, the Schrödinger equations are solved with this potential to obtain new wavefunctions, and these new wavefunctions are then used to produce an improved potential. The process is repeated until the difference between the new potential and the previous potential is less than some predetermined limit.

A. Electronic Configurations and Orbitals

Table 8.1 gives the electronic configurations of several atoms that occur commonly in superconductors. For each atom the table gives the total number of electrons, the number of electrons in the core that do not directly enter the calculations, the configuration of the outer electrons, and the configuration of an ion that may be present if a simple ionic picture is adopted. The notation used is nl^N, where n is the principal quantum number corresponding to the level number, the orbital quantum number l is 0 for an s state, 1 for a p state and 2 for a d state, and N is the number of electrons in each l state. A full l state contains $2(2l+1)$ electrons, corresponding to 2, 6, and 10 for s, p, and d states, respectively. The wavefunctions of these outer electrons are called *orbitals*.

The various s, p, and d orbitals have the unnormalized analytical forms given in Table 8.2; the electronic charge distribution in space of the d orbitals is sketched in Fig. 8.12. Each orbital represents the charge of one electron; the sign on each lobe is the sign of the wavefunction. For example, we see from the table that p_z is given by $r\cos\Theta$, which is positive along the positive z-axis ($\Theta = 0$), negative along the negative z-axis ($\Theta = \pi$), and zero in

Table 8.2 Unnormalized Analytical Expressions in Cartesian and Polar Coordinates for the *s*, *p*, and *d* Orbitals[a]

Orbital	**Cartesian form**	**Polar form**
s	1	1
p_x	$\frac{x}{r}$	$\sin\Theta\cos\phi$
p_y	$\frac{y}{r}$	$\sin\Theta\sin\phi$
p_z	$\frac{z}{r}$	$\cos\Theta$
d_{xy}	$\frac{xy}{r^2}$	$\sin^2\Theta\sin\phi\cos\phi$
d_{yz}	$\frac{yz}{r^2}$	$\sin\Theta\cos\Theta\sin\phi$
d_{zx}	$\frac{zx}{r^2}$	$\sin\Theta\cos\Theta\cos\phi$
$d_{x^2-y^2}$	$\frac{x^2-y^2}{r^2}$	$\sin^2\Theta(\cos^2\phi-\sin^2\phi)$
d_z	$\frac{3z^2-r^2}{r^2}$	$3\cos^2\Theta-1$

[a] $l = 0$, 1, and 2, respectively.

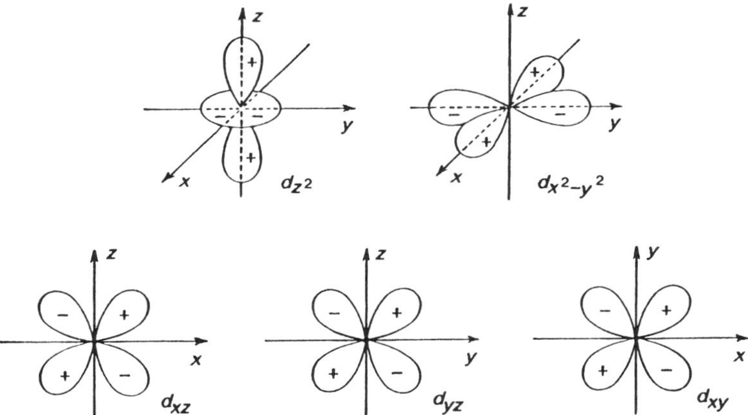

Figure 8.12 Spacial distribution of electron density for the five *d* orbitals. The signs (±) on the lobes are for the wavefunction; the sign of the electric charge is the same for each lobe of a particular orbital (Ballhausen, 1962).

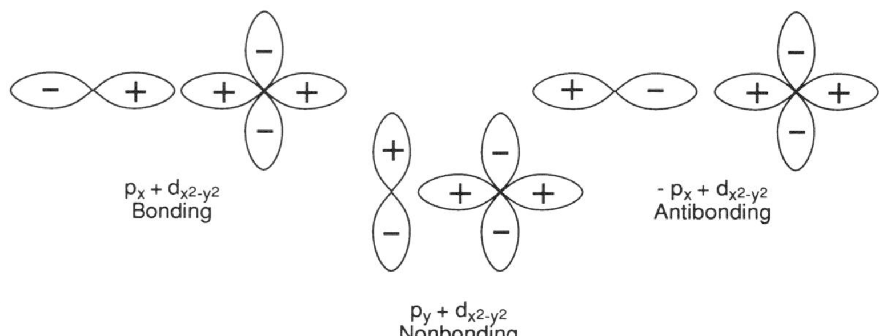

Figure 8.13 Examples of bonding (left), nonbonding (center), and antibonding (right) configurations involving the $d_{x^2-y^2}$ orbital with p_x and p_y orbitals.

the x, y-plane ($\Theta = \frac{1}{2}\pi$). Linear combinations of atomic orbitals, called *hybrid orbitals*, are used to form covalent bonds that hold the atoms together, as illustrated in Fig. 8.13 for bonding between an oxygen p orbital (p_x) and a copper d orbital ($d_{x^2-y^2}$). The figure shows, first, a bonding case in which the signs of the two orbitals that form the hybrid are the same in the region of overlap, second, an antibonding case in which the signs are opposite where overlap occurs, and, third, a nonbonding case where there is no appreciable overlap.

Each orbital can accommodate two electrons of opposite spin. In the usual case the two electrons enter the low-lying bonding level to form a chemical bond that holds the atoms together while the antibonding level remains empty, as illustrated in Fig. 8.14. The "bonding overlap" case is called a sigma (σ) bond, and a Cu–O bond of this type can be called a $3d_{x^2-y^2}$–$2p\sigma$ bond. We will see later that the band structures of superconductors generally consist of many fully occupied bonding levels, called *valence bands*, which lie below the Fermi level, many unoccupied antibonding levels well above it, and one or more partly occupied hybrid orbitals that pass through the Fermi level.

The same approach may be used to treat holes as well as electrons. For example, the copper ion ($3d^9$) may be looked upon as a filled d shell ($3d^{10}$) plus one $3d$ hole, while the oxygen mononegative ion ($2p^5$) may be treated as a filled p shell ($2p^6$) plus one $2p$ hole.

B. Tight-Binding Approximation

In Section IV we talked about the nearly free electron case in which the potential energy is small and the eigenfunctions approximate plane waves. Here the energy bands are broad and overlapping. At the beginning of the present section we discussed atomic orbitals, and noted that when the atomic potential energy is dominant the eigenfunctions approximate atomic orbitals centered on individual atoms. This is the case for core electrons whose energy levels lie deep within the atom and do not depend on k.

In this section we will discuss the intermediate case in which valence-electron orbitals centered on adjacent atoms over-

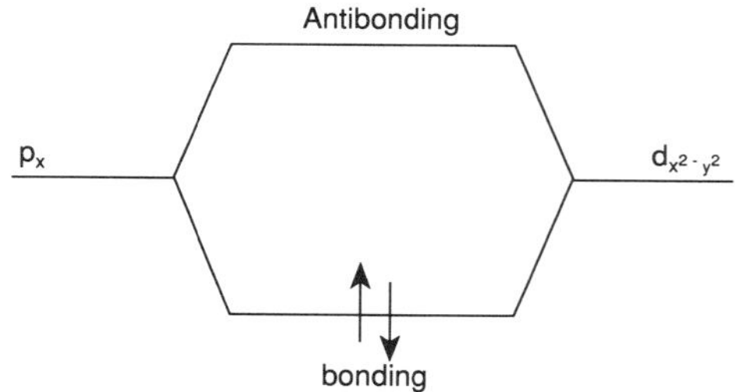

Figure 8.14 Hybridization of a copper $d_{x^2-y^2}$ orbital with an oxygen p_x orbital to form a low-energy bonding configuration and a high-energy antibonding configuration. Two antiparallel electrons that form a chemical bond are shown in the bonding level.

lap, as shown in Fig. 8.15. Here the isolated atom picture is no longer valid, but the overlap is not sufficiently great to obscure the identity of the individual atomic contribution. This limit, which corresponds to narrow bands with appropriate atomic quantum numbers assigned to each band, is referred to as the tight-binding approximation. This approach is employed in the Hubbard model as well as in the full band structure calculations to be discussed in Sections VIII – XVI.

To clarify the nature of this approach we will examine the case of one atomic state—for example, an s state—which is well isolated in energy from nearby states. A possible basis set for a crystal made up of N such atoms includes states in which the electron is localized on one atom, $\phi(r-R)$, where $\phi(r)$ is an atomic wave function and R is a direct lattice vector. These states overlap and are not orthogonal, and the overlap integral defined by

$$\gamma(R-R') = \int d^3r\phi^*(r-R)\phi(r-R') \tag{8.24}$$

is a measure of nonorthogonality. When the overlap integral for nearest-neighbor atoms is small, the atomic states are approximately orthogonal.

These states, however, do not behave under lattice transformations, $\mathbf{r}' \to \mathbf{r} + \mathbf{R}$, according to Bloch's theorem

$$\Psi_k(r+R) = e^{ik\cdot R}\Psi_k(r). \tag{8.25}$$

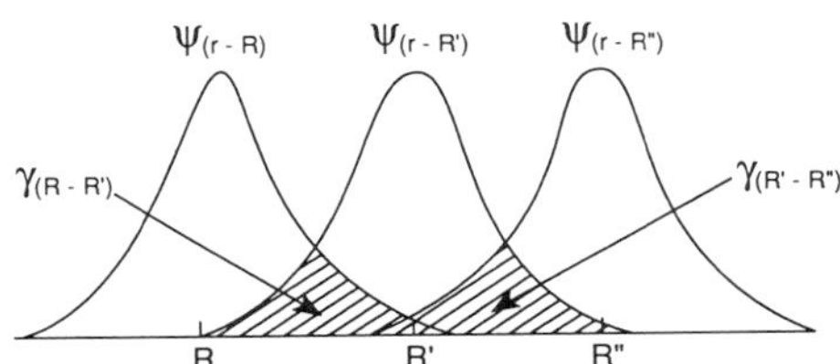

Figure 8.15 Overlap $\gamma(R_i - R_j)$ of atomic wavefunctions for nearby atoms.

We can remedy this situation by constructing Bloch states, i.e., linear combinations of localized states of the form

$$\Psi_k(r) = (N)^{-1/2}\sum_R e^{ik\cdot r}\phi(r-R), \tag{8.26}$$

which are orthogonal but not normalized,

$$\int d^3r\Psi^*_{k'}(r)\Psi_k(r) = e^{-i(k-k')\cdot R}\int d^3r\Psi^*_{k'}(r)\Psi_k(r) \tag{8.27}$$

$$= \begin{cases} 0 & k' \leq k \\ \Gamma(k) & k' = k. \end{cases} \tag{8.28}$$

We show in Problem 9 that $\Gamma(k)$ is the Fourier transform of the overlap integral

$$\Gamma(k) = \sum_R e^{-ik\cdot R}\gamma(R). \tag{8.29}$$

Since the full Hamiltonian H is symmetric under lattice transformations and the Bloch states $\Psi_k(r)$ for different values of k are orthogonal, it follows that in the simple one-band approximation $\Psi_k(r)$ is an eigenstate of H. The eigenvalue $\epsilon(k)$ can be evaluated by calculating the expectation value $\int d^3r\Psi^*_{k'}(r)H\Psi_k(r)$. We show in Problem 10 that this gives

$$\epsilon(k) = \epsilon_a + \frac{B(k)}{\Gamma(k)}, \tag{8.30}$$

where the energy of an isolated atom ϵ_a includes a kinetic-energy part $p^2/2m$ and a Coulomb part $u(R)$,

$$u(R) = \int d^3r\phi^*(r-R) \times [V_a(r-R)]\phi(r-R). \tag{8.31}$$

$\Gamma(k)$ is given by Eq. (8.29), and $B(k)$ is the Fourier transform of the exchange integral $\beta(R)$,

$$B(k) = \sum_R e^{-ik\cdot R}\beta(R), \tag{8.32}$$

$$\beta(R-R') = \int d^3r\phi^*(r-R) \times \left[\sum_{R''} V_a(r-R'')\right]\phi(r-R'), \tag{8.33}$$

where $V_a(r - R'')$ is the potential due to the nucleus at the lattice position R'' and the summation in Eq. (8.32) excludes the case $R = R'$.

We can see from Fig. 8.15 that overlap integrals fall off very rapidly with distance. The same is true of the exchange integrals. Therefore, we are justified in retaining in (8.29) and (8.32) only terms in which R' is either R or a nearest neighbor to R, which gives

$$\Gamma(k) \approx \gamma_0 + \sum_R e^{-ik \cdot R} \gamma(R)$$
$$\approx \gamma_0 + 2\gamma_1(\cos k_x a + \cos k_y a) \quad (8.34)$$

and

$$B(k) \approx 2\beta(\cos k_x a + \cos k_y a), \quad (8.35)$$

so that

$$\epsilon(k) \approx \epsilon_a + \frac{2\beta}{\gamma_0}(\cos k_x a + \cos k_y a), \quad (8.36)$$

where we have assumed that $\beta \ll \gamma_0$ and $\gamma_1 \ll \gamma_0$. The width of the band measured along the line Γ–X, for example, is $\Delta\epsilon = \epsilon(0,0) - \epsilon(\pi/a, 0) = 4\beta/\gamma_0$. The exchange integral β determines the degree of dispersion in the band. If β is very small, as is the case for well-separated atoms, the bands are nearly flat and centered about the atomic energy level ϵ_a. Since $\epsilon(k)$ is a slowly varying function of k, it is possible to form a superposition of states with different k but essentially the same energy. If we wish, we can construct a wave packet that is centered on one atom, and in this way return to our original set of atomic orbitals.

Typical band structure calculations take into account several atomic orbitals $\phi_n(r - R)$ for each atom, so we can define the Bloch states

$$\Psi_{nk}(r) = (N)^{-1/2} \sum_R e^{ik \cdot r} \phi_n(r - R), \quad (8.26a)$$

where n now labels each of the atomic states in the basis. The formalism leading to Eq. (8.30) must now be generalized somewhat to include a greater number of overlap integrals,

$$\gamma_{nm}(R - R') = \int d^3r \phi_n^*(r - R) \phi_m(r - R'), \quad (8.24a)$$

and exchange integrals,

$$\beta_{nm}(R - R') = \int d^3r \phi_n^*(r - R) \times \left[\sum_{R''} V_a(r - R'') \right] \phi_m(r - R'), \quad (8.37)$$

and their respective Fourier transform counterparts $\Gamma_{nm}(k)$ and $B_{nm}(k)$. The eigenstates are linear combinations of atomic orbitals, and the associated method is often referred to as LCAO.

Having covered some background on energy bands and chemical bonding, we will proceed to describe the Hubbard model which, despite its simplicity, exhibits many properties that are characteristic of superconductors (Anderson 1987a; cf. Hirsch 1985a, b, 1987). This will prepare the way for understanding the more complex band structures of the elements, of the *A*15 compounds, and, finally, of the high-temperature superconductors.

VII. HUBBARD MODELS

Many of the most interesting properties of materials, such as magnetic ordering and superconductivity, require theories that go beyond the independent-electron approximation. In order to understand the phenomena it is necessary to take into account electron correlations. The simplest model of correlated electrons is the one-state Hubbard (1963, 1964) model, and so we will begin with it. Since this model is expressed in terms of basic functions, called *Wannier functions* (Anderson, 1959), we will describe this basis before presenting the Hubbard model itself. The Hubbard model literature is far too extensive for us to attempt a survey.

A. Wannier Functions and Electron Operators

In the tight-binding approximation we made use of the Bloch states $\Psi_k(r)$ as given by Eq. (8.26), which are superpositions of atomic functions $\phi(r-R)$ over lattice positions. These states are orthogonal, and $[\Gamma(k)]^{-1/2}$ from Eq. (8.29) constitutes their normalization constant. A summation of normalized Bloch states over all of the k-states of a band,

$$W(r-R) = [N]^{1/2} \sum_k [\Gamma(k)]^{-1/2} e^{-ik\cdot R} \Psi_k, \quad (8.38)$$

provides a new wavefunction $W(r-R)$ associated with the atom at lattice position R. This wavefunction represents a Wannier state.

The Wannier states are useful because the overlap of two Wannier states at different lattice sites R and R' is zero. For two different bands with indices n and n', the associated Wannier states $W_n(r-R)$ and $W_{n'}(r-R)$ are orthogonal, corresponding to the general orthonormality condition

$$\int d^3r W_n^*(r-R) W_{n'}(r-R') = \delta_{nn'} \delta_{RR'}. \quad (8.39)$$

The Wannier states are also complete:

$$\sum_{r'} \sum_R W_n^*(r-R) W_n(r'-R) = \delta(r-r'). \quad (8.40)$$

If the band is narrow each $W(r-R)$ is localized about one lattice point R. Using Wannier states as a basis, the electron operator for a single band Ψ_σ can be written as

$$\Psi_\sigma(r) = [N]^{-1/2} \sum_R a_\sigma(R) W(r-R). \quad (8.41)$$

The operators $a_\sigma^\dagger(\mathrm{R})$ and $a_\sigma(R)$ are, respectively, the electron creation operator and the electron annihilation operator, and together they form the number operator $n_\sigma(R)$ for electrons of spin σ in the Wannier state $W(r-R)$,

$$n_\sigma(R) = a_\sigma^\dagger(R) a_\sigma(R), \quad (8.42)$$

where no summation is intended. These operators satisfy the anticommutation rule (cf. (6.62))

$$\alpha_\sigma^\dagger(R) a_{\sigma'}(R') + a_{\sigma'}(R') a_\sigma^\dagger(R) = \delta_{\sigma\sigma'} \delta_{RR'}, \quad (8.43)$$

and have the properties

$$\begin{aligned} a_-|0\rangle &= 0 \\ a_-|-\rangle &= |0\rangle \\ a_-|+\rangle &= 0 \\ a_-|\pm\rangle &= |+\rangle \\ a_-^\dagger|0\rangle &= |-\rangle \\ a_-^\dagger|-\rangle &= 0 \\ a_-^\dagger|+\rangle &= |\pm\rangle \\ a_-^\dagger|\pm\rangle &= 0 \end{aligned} \quad (8.44)$$

Similar expressions can be written for the spin-up operators, $a_+^\dagger$ and a_+. The wavefunctions $|j\rangle$ correspond to sites occupied by a spin-up electron $|+\rangle$, by a spin-down electron $|-\rangle$, or by two electrons of opposite spin $|\pm\rangle$, while $|0\rangle$ denotes a vacant site. We will use these expressions and extensions of them to the two-electron case $|ij\rangle$ to evaluate the energy of some Hubbard Hamiltonians.

B. One-State Model

In a one-state Hubbard model there is one electron orbital per unit cell. To construct this model we begin, as in the tight-binding approximation, with electrons localized in atomic-like states at the positions R of the atoms. We assume that there is only one valence orbital per atom and that each atom can accommodate 0, 1, or 2 electrons.

The Hamiltonian consists of a kinetic energy term proportional to a "hopping amplitude" $t > 0$ that represents the electron correlation; a term $-\mu\hat{N}$, where μ is the chemical potential and $\hat{N}$ is the total number of electrons; and an on-site Coulomb repulsion term (8.31), $U > 0$. We thus have

$$H = -t \sum_{R,R',\sigma} \left[a_\sigma^\dagger(R) a_\sigma(R') + a_\sigma^\dagger(R') a_\sigma(R) \right] - \mu \sum_{R,\sigma} a_\sigma^\dagger(R) a_\sigma(R) + U \sum_R n_+(R) n_-(R), \quad (8.45)$$

where $\mu = 0$ for the undoped case when the orbitals are half filled with electrons. The kinetic energy is the sum of two hermitian conjugates. The "hopping amplitude" t given by

$$t = -\frac{\hbar^2}{2m} \int d^3r \nabla W^*(r-R) \cdot \nabla W(r-R') \quad (8.46)$$

is a measure of the contribution from an electron hopping from one site to another neighboring site. It is assumed that the overlap of Wannier functions separated by more than one lattice spacing is negligible, and that t is the same for all nearest-neighbor pairs R, R'. The chemical potential term is included because we are interested in the change in the properties of the model as the number of electrons is varied. The Coulomb repulsion is assumed to be the same for all sites. The Hamiltonian takes into account only nearest-neighbor correlation and on-site Coulomb repulsion.

The Hamiltonian (8.45), simple as it appears, embodies a great deal of physics. This Hamiltonian, or a generalization of it, is the starting point for a number of theories of high-T_c superconductivity, the Mott insulator transition, and other phenomena related to highly correlated many-electron systems. Allen (1990) went so far as to state what he called the Hubbard hypothesis: "The fundamental physics of the oxide superconductors is contained in the Hamiltonian (8.45) on a two-dimensional square lattice for small numbers of holes."

In Sections VII.D, VII.E, and VII.F we will discuss examples of the limiting case $U \gg t$. Typical values have been given of $t \sim 0.25$–0.5 eV and $U \sim 3$–4 eV (Ruckenstein *et al.*, 1988). The opposite limit, $U \ll t$, has also been discussed (e.g., see Varma et al., 1988).

C. Electron–Hole Symmetry

As we shall see, the Hubbard Hamiltonian exhibits an electron–hole symmetry. Such symmetry is of considerable importance, in that most high-temperature superconductors are hole types with a close to a half-full conduction band. To demonstrate this symmetry it is convenient to begin by writing out the generalized Hamiltonian (8.45) in the more symmetric form

$$H = -t \sum_{R,R',\sigma} \left[a_\sigma^\dagger(R) a_\sigma(R') + a_\sigma^\dagger(R') a_\sigma(R) \right] - \mu \sum_{R,\sigma} a_\sigma^\dagger(R) a_\sigma(R) + U \sum_R [n_+(R) - \tfrac{1}{2}][n_-(R) - \tfrac{1}{2}]. \quad (8.47)$$

This can be done because the extra terms simply shift the chemical potential and the zero point of energy.

The next step is to consider a transformation to "hole" operators by recalling that a site in k-space that is missing an electron is occupied by a hole, so destruction of a hole is equivalent to creation of an electron. Accordingly, we can define the "hole" operators

$$b_\sigma^\dagger(R) = a_{-\sigma}(R), \qquad b_\sigma(R) = a_{-\sigma}^\dagger(R), \quad (8.48)$$

where σ and $-\sigma$ denote opposite spin directions. We then have for the number operator, from Eq. (8.42),

$$n_\sigma(R) = a_\sigma^\dagger(R)a_\sigma(R) \tag{8.49}$$
$$= b_{-\sigma}(R)b_{-\sigma}^\dagger(R)$$
$$= 1 - b_{-\sigma}^\dagger(R)b_{-\sigma}(R)$$
$$= 1 - \tilde{n}_{-\sigma}(R), \tag{8.50}$$

where the anticommutation rule (8.43) is applied to obtain the number operator $\tilde{n}_\sigma(R) = b_\sigma^\dagger(R)b_\sigma(R)$ for holes. Using these expressions, the Hamiltonian can be written in terms of hole operators,

$$\begin{aligned} H = t \sum_{R,R',\sigma} & \left[b_\sigma^\dagger(R)b_\sigma(R') + b_\sigma^\dagger(R')b_\sigma(R)\right] \\ & + \mu N_s + \mu \sum_{R,\sigma} b_\sigma^\dagger(R)b_\sigma(R) \\ & + U\sum_R [\tilde{n}_+(R) - \tfrac{1}{2}][\tilde{n}_-(R) - \tfrac{1}{2}], \end{aligned} \tag{8.51}$$

where N_s is the number of sites on the lattice.

The thermodynamic potential Ω is given by

$$\Omega(\mu, T; t, U)$$
$$= -k_B T \log\{\mathrm{Tr}[e^{-H/kT}]\} \tag{8.52}$$
$$= -2\mu N_s + \Omega(-\mu, T: -t, U). \tag{8.53}$$

If the lattice is bipartite, that is, if it can be decomposed into two sublattices, A and B, with the property that an atom of one sublattice has atoms of the other sublattice as its nearest neighbors, the electron–hole transformation can be modified to

$$b_\sigma(R) = \begin{cases} a_{-\sigma}^\dagger(R) & R \in A \\ -a_{-\sigma}^\dagger(R) & R \in B \end{cases}. \tag{8.54}$$

This has the effect $t \to -t$, so that Eq. (8.53) becomes

$$\Omega(\mu, T; t, U)$$
$$= -2\mu N_s + \Omega(-\mu, T; t, U). \tag{8.55}$$

The average number of particles $N(\mu, T)$ is

$$N(\mu, T) = -d\Omega/d\mu \tag{8.56}$$
$$= 2N_s - N(-\mu, T), \tag{8.57}$$

and for $\mu = 0$ we obtain

$$N(0, T) = N_s, \tag{8.58}$$

that is, one electron per site. Because of the simple relationship (8.57) we need only consider $\mu \geq 0$; the properties of the system for $\mu < 0$ are then easily found by means of the electron–hole transformation (8.54).

D. Half-Filling and Antiferromagnetic Correlations

The high-temperature superconductors and related compounds typically have a set of full low-lying bands, an almost half-full hybrid band near the Fermi level, and a set of empty bands at higher energy. The half-full, one-state Hubbard model stimulates this hybrid band, and thereby provides a simple approximation to the materials. Later, we will see how a three-state Hubbard model more closely approximates the behavior of high-temperature superconductors.

We take $\mu = 0$ so that $\Sigma\langle n_\sigma(\mathrm{R})\rangle = 1$. Thus there is on average one electron per lattice site. If the on-site repulsion is large, $U \gg t$, states with double occupancy will be suppressed. Therefore, let us consider as a subset of all states of the half-filled band those states with exactly one electron at each site. The Coulomb repulsion term $U\Sigma n_+(R)n_-(R)$ only exists for a site that is doubly occupied, so it vanishes in this case. Were it not for the hopping term, these states would all be degenerate with zero energy. Here we show that hopping removes the degeneracy, and, in addition, show by simple perturbation theory that the effect of the hopping term is to lower the energy of an antiferromagnetic pair relative to a ferromagnetic pair of nearest-neighbor electrons.

Consider a nearest-neighbor pair of electrons that occupies one of the two antiferromagnetic states, with antiparallel spins

$$|+\,-\rangle \text{ and } |-\,+\rangle, \tag{8.59}$$

or one of the two ferromagnetic states, with parallel spins

$$|+\,+\rangle \text{ and } |-\,-\rangle, \tag{8.60}$$

where + denotes spin up, and − signifies spin down. In addition to these four states, there are two higher energy states

$$|0\,\pm\rangle \text{ and } |\pm\,0\rangle, \tag{8.61}$$

in which both electrons are localized on the same atom.

We wish to apply perturbation theory up to second order for the case $U \gg t$ corresponding to the expression

$$E_i = E_i^{(0)} + E_i^{(1)} + E_i^{(2)} \tag{8.62}$$

for the ith-order energy, where

$$E_i = \langle \Psi_i | H_{\text{Coul}} | \Psi_i \rangle + \langle \Psi_i | H_{\text{hop}} | \Psi_i \rangle + \sum_{j \neq i} \frac{|\langle \Psi_i | H_{\text{hop}} | \Psi_j \rangle|^2}{E_j - E_i}. \tag{8.63}$$

As noted earlier, there is no Coulomb contribution for the four states (8.59) and (8.60), so that the zero-order term vanishes, $E_i^{(0)} = 0$. The first-order energy must be evaluated from the expression

$$E_i^{(1)} = \langle \Psi_i | H_{\text{hop}} | \Psi_i \rangle, \tag{8.64}$$

where the hopping operator H_{hopp} of Eq. (8.45) has the explicit form

$$H_{\text{hop}} = -t[a_+^{\dagger}(1)a_+(2) + a_+^{\dagger}(2)a_+(1) + a_-^{\dagger}(1)a_-(2) + a_-^{\dagger}(2)a_-(1)]. \tag{8.65}$$

Since each term of this operator changes the spin state, the first-order energy (8.64) also vanishes.

Second-order perturbation theory entails evaluating the matrix elements $\langle \Psi_i | H_{\text{hop}} | \Psi_j \rangle$ for $i \neq j$. The only nonvanishing terms are

$$\begin{aligned}\langle +\,- | H_{\text{hop}} | 0\,\pm \rangle &= \langle +\,- | H_{\text{hop}} | \pm\,0 \rangle \\ &= \langle -\,+ | H_{\text{hop}} | 0\,\pm \rangle \\ &= \langle -\,+ | H_{\text{hop}} | \pm\,0 \rangle \\ &= -t.\end{aligned} \tag{8.66}$$

The denominator $E_j - E_i$ of the second-order term of Eq. (8.63) is $-U$, which gives for the energy of the two antiferromagnetic states

$$E_{+-} = E_{-+} = -\frac{2t^2}{U}. \tag{8.67}$$

Since the energy of the two ferromagnetic states is zero (Problem 17),

$$E_{++} = E_{--} = 0, \tag{8.68}$$

the hopping term has the effect of lowering the energy of an antiferromagnetic pair relative to that of a ferromagnetic pair.

E. *t-J* Model

In one variant of the Hubbard model, called the t-J model, the Coulomb repulsion U term in the Hamiltonian of Eq. (8.45) is replaced by the Heisenberg term $J\Sigma \mathbf{S}_i \cdot \mathbf{S}_j$ (Anderson, 1978a, b; Rodriguez and Douçot, 1990; Zhang and Rice, 1988; Halley, 1988, Chapters 15–20). This model may be obtained as the large-U limit ($U \gg t$) of the Hubbard model (Harris and Lange, 1967; Marde *et al.*, 1990). We will justify it in terms of the discussion in the previous section.

If we restrict our attention to states with one electron per site, we can describe the states completely in terms of the spin-$\frac{1}{2}$ operators for the electrons at each site, $S_k(R)$. Although we have employed a basis in which one component of the spin, conventionally the z-component, is diagonal, the Hubbard Hamiltonian is invariant under rotations in spin space. To second order in perturbation theory, the effective Hamiltonian for the subset of states can therefore be written in Heisenberg form,

$$H_{\text{Heis}} = -J\Sigma \mathbf{S}(R) \cdot \mathbf{S}(R'), \tag{8.69}$$

where the exchange coupling is

$$J = 4t^2/U. \tag{8.70}$$

Typical values of the parameters, $t \approx 0.4$ eV and $U \approx 3.5$ eV (Ruckenstein *et al.*, 1987), give J close to the experimentally determined value $J \approx 0.14-0.16$ eV (Gagliano and Bacci, 1990; Singh *et al.*, 1989). We conclude that at half filling, in the limit $U/t \to \infty$, the ground state of the Hubbard model is equivalent to that of a Heisenberg antiferromagnet.

The Heisenberg term used in the $t-J$ model corresponds to the Hamiltonian

$$H = -t\Sigma(a_{i\sigma}^+ a_{j\sigma} + a_{j\sigma}^+ a_{i\sigma}) + J\Sigma \mathbf{S}_i \cdot \mathbf{S}_j. \tag{8.71}$$

Undoped materials can be understood in terms of the two-dimensional Heisenberg term (Chakravarty *et al.*, 1988; Dagotto *et al.*, 1990), and in the presence of empty sites the hopping matrix element t can represent holes propagating in a fluctuating antiferromagnetic background arising from this $J\Sigma \mathbf{S}_i \cdot \mathbf{S}_j$ term (Poilblanc and Dagotto, 1990; Zotos *et al.*, 1990). Many believe that the t-J model embraces the essential features of high-temperature superconductivity (Ohkawa, 1990), and there is an extensive literature on the subject.

The symbol t-J is used to designate the two terms in the Hamiltonian (8.71). In the same spirit, the ordinary Hubbard model represented by the Hamiltonian (8.45) could be called the t-U model, but this notation is never employed. The t-J and Hubbard models have been compared (e.g., Bhattacharya and Wang, 1992; Dagotto *et al.*, 1992).

F. Resonant-Valence Bonds

The Hubbard model (8.51) is deceptively simple, and while some exact results for the ground state are known in the limit $U/t \to \infty$ and for the case of half filling, the nature of the ground states for arbitrary U and t are not known. An alternate choice of basis for elucidating the nature of the low-lying levels of the model makes use of "resonant-valence-bond" (RVB) states (Anderson, 1987a, b; Emery and Reiter, 1988; Feine, 1993; Kivelson, 1989; Wittmann and Stolze, 1993; and Zhang and Rice, 1988). Electrons at nearest-neighbor sites are paired into "bonds" and linear combinations of the basis states are used to construct eigenstates of the total electronic spin (Allen, 1990).

For two electrons the total spin S can be 1 or 0, and the corresponding triplet ($S = 1$, $M = 0, \pm 1$) and singlet ($S = M = 0$) wavefunctions are, respectively,

$$\left.\begin{aligned} |11\rangle &= |++\rangle \\ |10\rangle &= \frac{1}{\sqrt{2}}(|+-\rangle + |-+\rangle) \\ |1-1\rangle &= |--\rangle \end{aligned}\right\} \quad \text{triplet}, E = 0 \tag{8.72}$$

$$|VB\rangle = \frac{1}{\sqrt{2}}(|+-\rangle - |-+\rangle) \qquad \text{singlet}, E = -4t^2/U, \tag{8.73}$$

where we have used the notation $|SM\rangle$. There are two additional ionized states $|A\rangle$ and $|B\rangle$ which are both singlets,

$$|A\rangle = \frac{1}{\sqrt{2}}(|0\pm\rangle - |\pm 0\rangle) \qquad E = U \tag{8.74a}$$

$$|B\rangle = \frac{1}{\sqrt{2}}(|0\pm\rangle + |\pm 0\rangle)$$

$$E = U + \frac{4t^2}{U}, \tag{8.74b}$$

and, of course, a state in which there are no electrons,

$$|\text{vacuum}\rangle = |00\rangle.$$

The singlet state with two electrons labeled $|VB\rangle$ is the lowest in energy, and it constitutes the "valence bond" state. The energies of these two electron states as given in (8.72)–(8.74b) are sketched in Fig. 8.16.

The RVB theory starts by assuming that the ground state of insulating La_2CuO_4 is a linear combination of $|VBij\rangle$ states for various pairs of electrons i, j,

$$|\text{RVB}\rangle = c_{12}|VB12\rangle + c_{34}|VB34\rangle + c_{56}|VB56\rangle + \ldots, \quad (8.75a)$$

and in the presence of a single hole with spin down (h −) paired with a spin-up electron (e +) this becomes

$$|\text{RVB, 1 hole}\rangle = c_{12}|h-\rangle|e+\rangle + c_{34}|VB34\rangle + c_{56}|VB56\rangle + \ldots, \quad (8.75b)$$

where the hole can hop between sites. The RVB approach has been used to explain high-temperature superconductivity (e.g., Bose, 1991; Jain and Ray, 1989; Klein *et al.*, 1991; Miyazaki *et al.*, 1989; Ye and Sachdev, 1991; Yndurain and Martinez, 1991), but has not found wide application (but see Johnson, 1990; Kallio *et al.*, 1989; Nagaoa and Lee, 1991; Wijngaarden *et al.*, 1990). Kurihara (1989) proposed a hole–spin model of high temperature superconductivity.

G. Spinons, Holons, Slave Bosons, Anyons, and Semions

In the RVB formulation of the two-dimensional large-U Hubbard model, the spin and charge degrees of freedom are separated. The elementary excitations are either neutral spin-$\frac{1}{2}$ solitons (fermion solitary excitations, Dodd *et al.*, 1982; Doi *et al.*, 1992; Drazi and Johnson, 1989; Kivshar, 1991; Rokhsar, 1990; Tighe *et al.*, 1993), called *spinons*, or charged boson particles, called *holons* (Allen, 1990; Anderson and Zou, 1988; Kivelon, 1989; Nori *et al.*, 1990; Schmeltzer, 1994; Schofield and Wheatley, 1993; Sinha, 1992; Wang, 1989; Xing and Liu, 1991; Zou and Anderson, 1988).

There is a version of the *t-J* model, called the *slave boson representation*, in which the Hamiltonian contains spinon and holon creation and annihilation operators (Schönhammer, 1990; Sheng *et al.*, 1990). The so-called slave boson operators are introduced to specify and keep track of empty and singly occupied sites (Arrigoni *et al.*, 1990; Jolicoeur and LeGuillou, 1991; Rodriguez and Douçot, 1992). From a more general viewpoint, the slave boson formulation maps a purely fermionic model onto an effective bosonic one (Lilly *et al.*, 1990), and the slave boson fields provide a larger quantum mechanical (Fock) space that is subject to constraints involving the

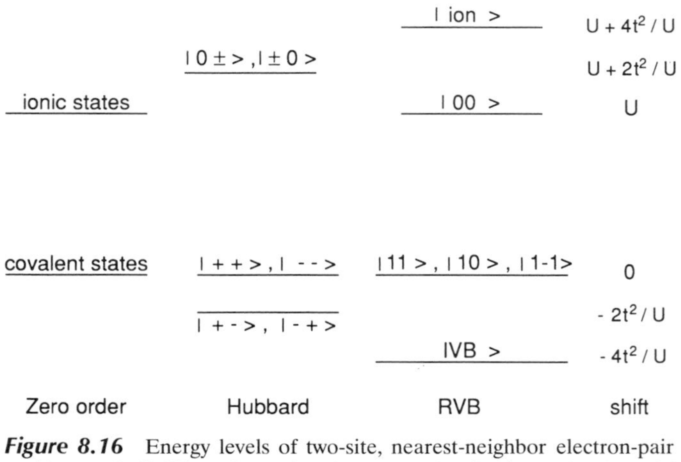

Figure 8.16 Energy levels of two-site, nearest-neighbor electron-pair problem at half filling, shown for the zero-order (left), Hubbard model (center), and resonant valence-bond (right) approximations.

slave boson operators (Jolicoeur and LeGuillou, 1991; Rodriguez and Douçot, 1992; Zhang *et al.*, 1993; Zou and Anderson, 1988). In addition, the slave bosons carry a conserved quantum number, namely charge (Zou and Anderson, 1988).

In quantum mechanics we learn that the wavefunction $U(r_2, r_1)$ for the exchange of two particles is given by

$$U(r_2, r_1) = e^{i\Theta} U(r_2, r_1) \qquad (8.76)$$

where $\Theta = 0$ for bosons which are symmetric under interchange, and $\Theta = \pi$ for fermions which are antisymmetric. Wilczek (1982a, b) coined the word anyon for a particle with some other value Θ in the range $-\pi < \Theta < \pi$, and such a particle is said to have fractional statistics. It violates the usual time reversal (T) and parity (P) conservation laws. Anyons have been used to explain the fractional quantum Hall effect in semiconductor heterojunctions (Arovas *et al.*, 1984; Halperin, 1984; Prange and Girvin, 1987). Kalmeyer and Laughlin (1987; Laughlin, 1988a, b) suggested that semions, anyons with $\Theta = \pm \pi/2$,

$$U(r_2, r_1) = iU(r_2, r_1), \qquad (8.77)$$

might be the charge carriers responsible for high temperature superconductivity. In principle this can be tested by experiments involving T, P symmetry, but the results have not been encouraging (Shen and Lu, 1993; Zhou and Chen, 1993). Some studies of an anyon gas suggest that its ground state is superconducting, that a collective excitation is a phonon mode, and that a single particle excitation can be identified as a vortex (Choi *et al.*, 1992; Gelfand and Halperin, 1992; Mori, 1991; Zhang *et al.*, 1990).

H. Three-State Model

The one-state Hubbard model that we discussed in Section VII.B involved one orbital per unit cell, with two parameters, a hopping amplitude t and an on-site Coulomb repulsion U. We will now extend the model to a unit cell, shown outlined in Fig. 8.17, containing a Cu atom that contributes a $d_{x^2-y^2}$ orbital, and two oxygens one of which contributes a p_x orbital, the other a p_y orbital (Jefferson *et al.*, 1992). This corresponds to the configuration of the CuO_2 plane shown in the figure, which is believed to be responsible for the superconducting properties of the cuprates. In this plane each copper has four oxygen nearest neighbors and each oxygen has two copper nearest neighbors. The figure is drawn with the signs on the orbital lobes selected so that all of the overlaps are of the bonding type. The convention chosen for assigning positive (e.g., $+p_x$) and negative (e.g., $-p_y$) signs to the orbitals is given in the figure.

The Hamiltonian is formulated in terms of operators $d^+_{i\sigma}$ and $p^+_{\alpha i\sigma}$ which create $Cu(3d_{x^2-y^2})$ and $O(2p_x, 2p_y)$ holes, respectively, at Cu and O sites, by generalizing the procedure used to write down the one-band expression (8.45). The subscript α assumes the values x and y for p_x and p_y orbitals, respectively. The presence of three types of orbitals increases the number of terms in the Hamiltonian (Entel and Zielinski, 1990; see also Dopf *et al.*, 1992; Feiner *et al.*, 1992; Grilli *et al.*, 1991; Hybertsen *et al.*, 1992; Ramasesha and Rao, 1991; Ruckenstein *et al.*, 1987; Zhang and Rice, 1988):

$$\begin{aligned} H = {} & \sum_{i,\sigma} t_d d^\dagger_{i\sigma} d_{i\sigma} + \sum_{i,\sigma} t_p p^\dagger_{\alpha i\sigma} p_{\alpha i\sigma} \\ & + 2i\gamma \sum_{\alpha,i,\sigma} (-1)^\alpha S_{\alpha i} \\ & \times (d^\dagger_{i\sigma} p_{\alpha i\sigma} - p^\dagger_{\alpha i\sigma} d_{i\sigma}) \\ & + 4\beta \sum_{\substack{i,\sigma \\ \alpha\neq\beta}} S_{\alpha i} S_{\beta i} p^\dagger_{\alpha i\sigma} p_{\beta i\sigma} + U_{dd} \\ & \times \sum_{i,\sigma} d^\dagger_{i\sigma+} d_{i\sigma+} d^\dagger_{i\sigma-} d_{i\sigma-} \\ & + U_{pp} \sum_{\alpha,i,\sigma} p^\dagger_{\alpha i\sigma+} p_{\alpha i\sigma+} p^\dagger_{\alpha i\sigma-} p^\dagger_{\alpha i\sigma-} V_{dp} \\ & \times \sum_{i,\sigma} d^\dagger_{i\sigma} d_{i\sigma} p^\dagger_{\alpha i\sigma} p_{\alpha i\sigma}, \end{aligned} \qquad (8.78)$$

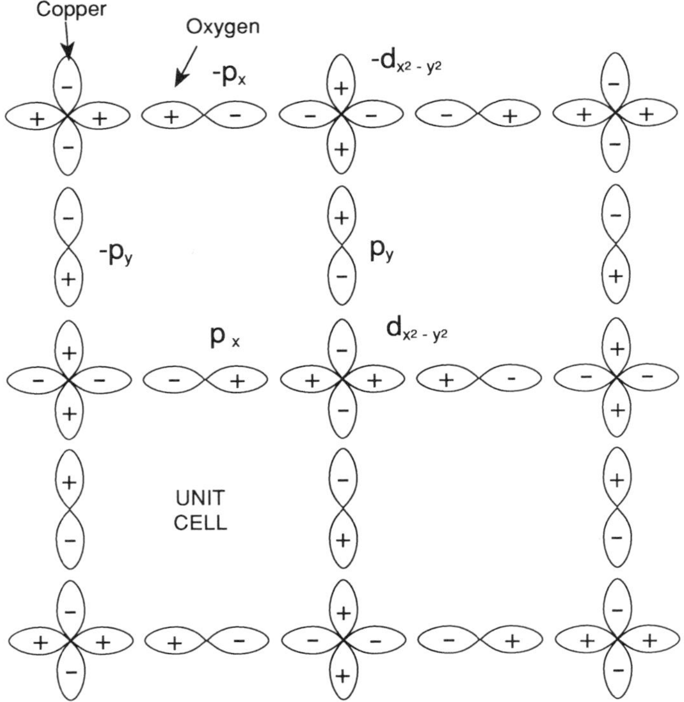

Figure 8.17 Orbitals used for the three-state model of Cu–O planes. Each copper contributes a $d_{x^2-y^2}$ orbital and each oxygen contributes either a p_x or a p_y orbital, as shown. The unit cell contains one of each type of ion, and hence one of each type of orbital. The figure shows four unit cells.

where

$$S_{\alpha i} = \sin(\tfrac{1}{2}k_\alpha a) \qquad \alpha = 1 \text{ for } x,\ \alpha = 2 \text{ for } y. \tag{8.79}$$

The first two terms represent the kinetic energies of the d and p electrons, which have hopping amplitudes t_d and t_p, respectively. The third and fourth mixed terms which are of type $d^\dagger_{i\sigma}$, $p_{i\sigma}$ and $p^\dagger_{xi\sigma}$, $p_{yi\sigma}$ contain the respective hybridization constants γ and β. The fifth and sixth terms describe intra-atomic (on-site) Coulomb repulsions (8.31) involving d and p orbitals, respectively, and the seventh term accounts for interatomic Coulomb repulsion between d and p holes on neighboring Cu and O atoms. The undoped reference state consists of Cu^+ with the $3d^{10}$ configuration and O^{2-} with the $2p^6$ configuration, so that all the p and d orbitals are occupied. Charge transfer mechanisms have been examined within this model (Jarrell *et al.*, 1988; Weber, 1988).

I. Energy Bands

The energy bands calculated by Entel and Zielinksi (1990) using the Hamiltonian (8.78) are plotted in Fig. 8.18a, and the density of states, together with the partial contributions from the p electrons (n_p) and d electrons (n_d), is plotted in Fig. 8.18b. The conditions for the calculation are clear from the labels in the energy scale of Fig. 8.18a: $t_p = \epsilon_p = 0$, $t_d = \epsilon_d = 3.6$ eV, $U_{pp} = 4$ eV, $U_{dd} = 10.5$ eV, and $V_{dp} = 0$. We see that a predominately p-type band rises above the Fermi level at the corner point M, and that the Fermi surface con-

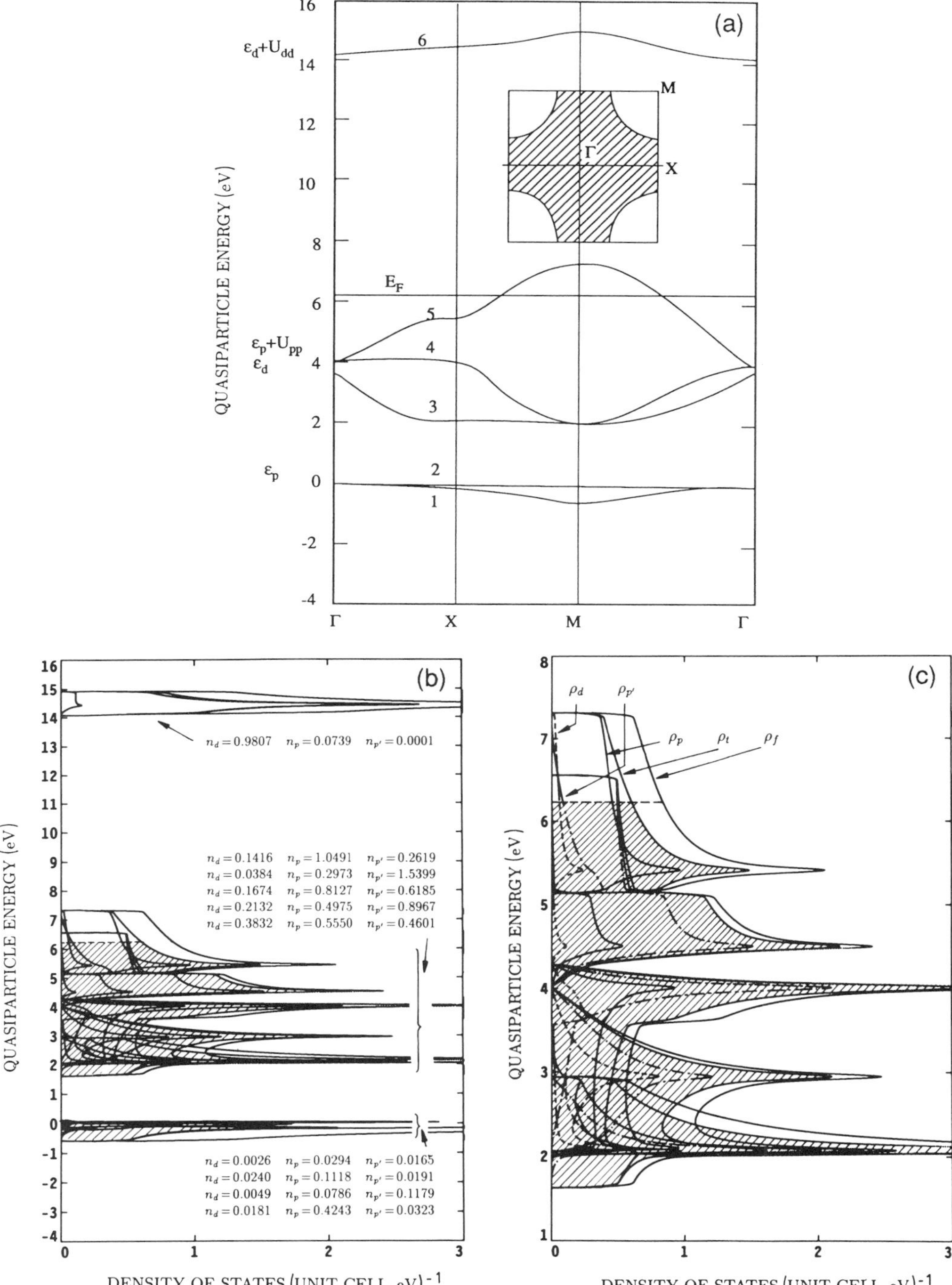

Figure 8.18 Band structure for the three-state Hubbard model for 4.5 electrons per unit cell. The model assumes one d state and two bonding p states per unit cell: (a) Energy bands with the Fermi energy below the gap, (b) density of states, and (c) density of states near E_F on an expanded scale. Energies involving $\epsilon_p = 0$, $\epsilon_d = 3.6$ eV, U_{pp} 4.0 eV, and $U_{dd} = 10.5$ eV indicated in the ordinate scale. Total density of states ρ_t together with the partial p and d contributions ρ_p and ρ_d and their occupation numbers n_p and n_d, respectively, are shown. The free-electron result ρ_f is included for comparison (from Entel and Zielinski, (1990), with modifications in (a) and (c)).

sists of the p-like pocket of holes indicated in the Brillouin zone shown as an inset to Fig. 8.18a. This is similar to the case illustrated in Figs. 8.10 and 8.11. The low-lying, mainly p-type pair of bonding bands, and the higher, predominately d-type antibonding band, are separated from the three bands near the Fermi level by insulating gaps.

Since there are three orbitals in the set chosen for the calculation, and since each can contain two electrons of opposite spin, it takes six electrons to fill all of the bands. As electrons are added to the system, the Fermi level rises in the manner shown in Fig. 8.19b. Note the two sharp jumps in E_F where the two band gaps are located. The increase in the occupation numbers n_p and n_d with the increase in the number of electrons is given in Fig. 8.19c. The amount of dispersion, or range of energy, over which a band is spread is called the *width* of the band. The width depends on the number of electrons in the unit cell in the manner shown in Fig. 8.19a. The predominately p-type band (No. 5) that passes through the Fermi level in Fig. 8.18a for the $n = 4.5$ electron case has the greatest dispersion.

We see from Figs. 8.18a and 8.18b that there are regions where the energy bands are flat, and at these energies the density of states becomes large, corresponding to what are called van Hove singularities (cf. XIII. A). Singularities of this type will be encountered later in the chapter when we present density of states plots of the $A15$ compounds, $Ba_{1-x}K_xBiO_3$, and the high-temperature superconductors. A DOS peak associated with a CuO_2 plane van Hove singularity can cause a peak in the transition temperature T_c as a function of hole doping (Mahan, 1993; Markiewicz, 1991a, 1991b; Pattnaik *et al.*, 1992; Penn and Cohen, 1992). Time dependent effects have been treated by Malomed and Weber (1991) and Stoof (1993).

The ability to calculate occupation numbers n_i for individual electrons associated with the various bands permits us to determine the contribution of the electrons from each atom in the unit cell of a superconductor to the density of states. In the following sections we will present sev-

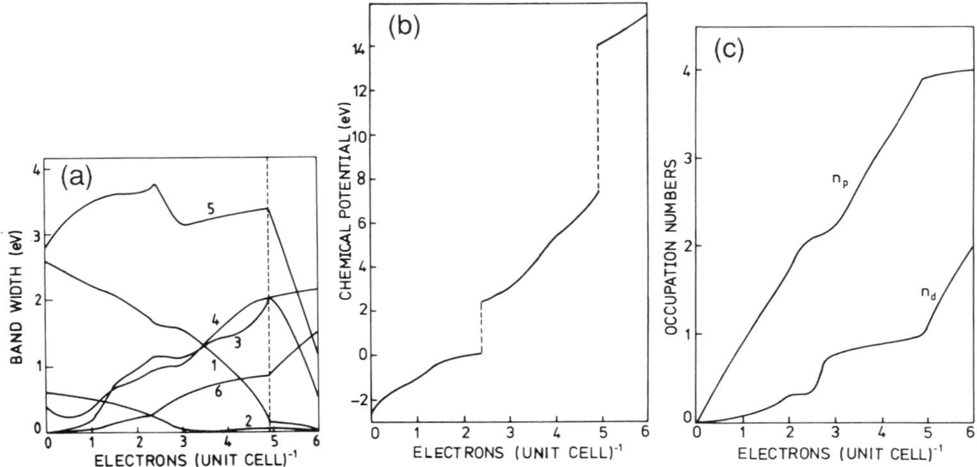

Figure 8.19 Three-state Hubbard model showing the variation with band filling of (a) the quasiparticle band widths, (b) the Fermi energy (chemical potential) at $T = 100$ K, and (c) the occupation numbers n_p and n_d. The metal-to-insulator transition occurs at the discontinuity in the Fermi energy near $n = 5$ (Entel and Zielinski, 1990).

eral such plots of individual atom density of states versus energy.

J. Metal–Insulator Transition

When the fifth band of Fig. 8.18a is almost full, the Fermi level lies near the top of the band. Adding more electrons fills this band and additional electrons enter the sixth band far above it. This causes the Fermi level to rise above the energy gap to the level of the top band. When this occurs the system undergoes a transition from a metal to an insulator. Evidence for this transition is seen in the discontinuity in slope for some of the curves in Fig. 8.19a at the position of the upper energy gap where there are five electrons per cell. Figure 8.20 presents the metal-to-insulator phase diagram, showing the metallic and insulating regions as a function of the p electron Coulomb repulsion energy U_{pp} and the energy difference between the p and d hopping integrals, $\Delta = \frac{1}{2}(t_d - t_p)$, called the *charge-transfer energy* (Entel and Zielinski, 1990). We see from the figure that a charge-transfer insulator arising from the presence of a charge-density wave forms for $U_{dd} \gg \Delta$, and that a Mott insulator (MI) forms for $U_{dd} \ll \Delta$. The present calculation is for the ratio $U_{dd}/\Delta = 5.8$, so that it corresponds to the metallic region above the ordinate value $U_{dd}/\gamma > 5$, which could be close to the transition to a charge-transfer insulator. The p orbital Coulomb repulsion energy U_{pp} has much less effect on the phase diagram than its d orbital counterpart U_{dd}. The phase diagram of an actual superconductor can be more complicated, as Fig. 7.24 suggests. Hebard (1994b) discusses the nature of the superconductor–insulator transition.

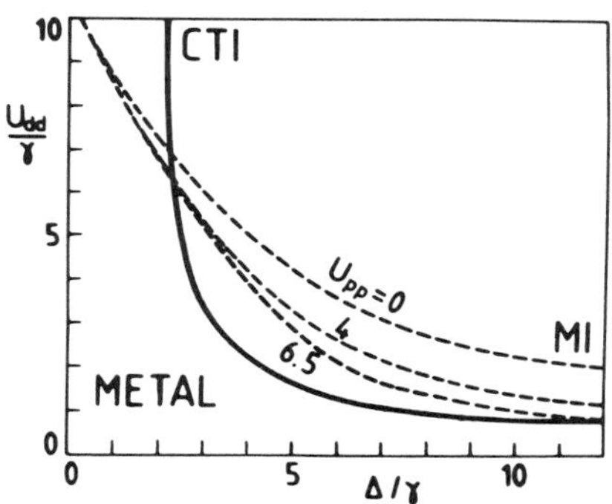

Figure 8.20 Metal-to-insulator phase diagram for the three-state Hubbard model with $V_{dp} = 0$, where the charge transfer energy is given by $\Delta = \frac{1}{2}(t_d - t_p)$. The regions where the charge-transfer insulator (CTI) and the Mott insulator (MI) occur are indicated. For the charge-transfer insulator, which arises from the presence of a charge-density wave, $n_p > n_d$ for the band beneath the gap (Entel and Zielinski, 1990).

The orbital character of the bands depends on U_{pp}. This is shown in Fig. 8.21 for the fifth band when the number of electrons in the unit cell is maintained at the value 4.9. There is a strong dependence of the ratio n_d/n_p on U_{pp} since decreasing U_{pp} from its chosen value of 4 eV to below 3.6 eV raises band 3 above bands 4 and 5, as is clear from the proximity of these bands at the point Γ of Fig. 8.18a. This causes the "new fifth band" to become more d-like so that we have $n_d > n_p$ for low values of U_{pp}, as indicated in Fig. 8.21, instead of the opposite ratio $n_d < n_p$ for the large U_{pp} case of Fig. 8.18.

Now that we have surveyed several varieties of Hubbard models we will proceed to describe the band structure results obtained for various types of superconductors. Many of the features characteristic of the three-state Hubbard model will be

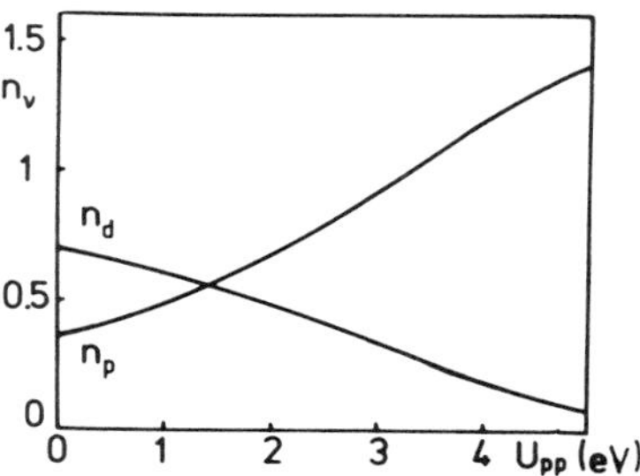

Figure 8.21 Dependence of the occupation numbers n_p and n_d in the band beneath the gap on the intra-atomic Coulomb repulsion energy U_{pp} for $n =$ 4.9 electrons per unit cell in the three-state Hubbard model (Entel and Zielinski, 1990).

found duplicated in the more sophisticated band structures. We will begin with the simplest system, namely the elements, after which we will examine the *A*15 compounds, and finally we will present the band structures of the high-temperature superconductors.

VIII. TRANSITION METAL ELEMENTS

We know from Fig. 3.1 that many of the transition elements, which have the configuration d^N, where $0 < N < 10$, are superconductors. If the Fermi surface is well within the BZ, there will be electron regions around the center point Γ and the Fermi surface will be closed, in analogy to the case of Fig. 8.6. Higher electron concentrations will move the Fermi surface out to the boundaries of the BZ zone and convert it to an open type. An example is the bcc element vanadium, which has the electron configuration $3d^3 4s^2$ with $T_c = 5.4$ K. Its Fermi surface, shown in Fig. 8.22b, is analogous to the two-dimensional case of Fig. 8.9. Electron regions cross the boundary, producing the open surface illustrated in the figure. Unoccupied or hole regions exist between these boundary crossings, as shown. The three bcc nd^3 hole-type superconductors V ($n = 3$), Nb ($n = 4$), and Ta ($n = 5$) in group V of the periodic table all have this type of open Fermi surface. The superconducting bcc nd^4 elements Mo ($n = 4$) and W ($n = 5$) have an extra d electron that fills these unoccupied regions and produces pockets of electrons in another Brillouin zone, thereby forming closed Fermi surfaces, as shown in Fig. 8.23. The principal quantum number n has very little effect on the energy bands.

The fact that the three series of d^N transition elements have a wide range of Fermi surface characteristics, some being electron types and some hole types, and the additional fact that most of them superconduct, demonstrates that superconductivity is not confined to particular configurations of energy bands. However, some configurations can be more favorable for high-transition temperatures (cf. Hong and Hirsch, 1992).

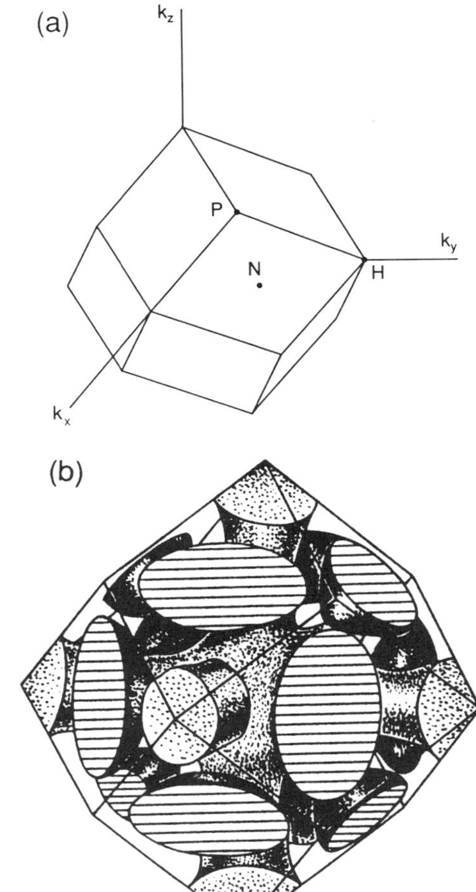

Figure 8.22 Brillouin zone of a bcc metal (top), and of an open Fermi surface (bottom) of the nd^3 elements V, Nb, and Ta. The electron regions cross over into the next unit cell along the k_x-, k_y-, and k_z-axes (point H) and at the centers of the faces (around point P, shown cross-hatched) (from Vonsovsky *et al.*, 1982, p. 199).

IX. *A*15 COMPOUNDS

In an *A*15 compound A_3B the B atoms lie on a bcc structure while the A atoms are on the faces of a unit cube aligned along three mutually perpendicu-

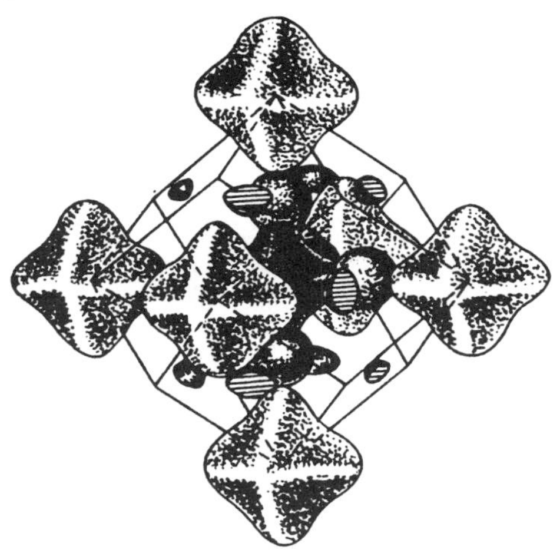

Figure 8.23 Closed Fermi surface of the nd^4 elements Mo and W (Vonsovsky *et al.*, 1982, p. 199).

lar families of nonintersecting linear chains, as was explained in Chapter 3, Section VIII (cf. Fig. 3.19). (The origin of the notation *A*15 is given in Chapter 3, Section IV.) The *A* element is always one of the nine transition elements with configuration nd^N where $n, N = 2, 3, 4$. Sometimes *B* is also a transition element, and relatively high transition temperatures occur when *B* is either a metal (Al, Ga, Sn) or a semi-metal (Si, Ge), which indicates that T_c does not correlate with ordinary chemical bonding parameters, such as the degree of covalency or the difference in electronegativity. We saw in Chapter 3, Section VIII, that T_c does, however, correlate with the electron concentration N_e and with the closeness to stoichiometry.

Figure 8.24 shows the energy bands and density of states calculated for Nb_3Sn ($T_c = 18$ K) using the electronic configurations $4d^3 5s^1 5p^1$ for Nb and $5s^2 5p^2$ for Sn, as listed in Table 8.1, for a total of $3 \times 5 + 4 = 19$ orbitals, and hence 38 bands (Klein *et al.*, 1978, 1979; Vonsovsky *et al.*, 1982, pp. 299ff.). The principal quantum number n has very little effect, whereas the number of d orbitals has a pronounced effect on the results, as in the case of the transition metal elements.

The upper part of the figure shows that there is a flat band that stays close to the Fermi level and is doubly degenerate most of the time. This provides a high density of states $D(E_F)$ at the Fermi level, which is conducive to a high transition temperature through the BCS relation (6.119)

$$T_c = 1.134\Theta_D \exp\left[-\frac{1}{VD(E_F)}\right], \quad (8.80)$$

where Θ is the Debye temperature and V the Cooper pair interaction potential. The lower part of the figure indicates that the large value of $D(E_F)$ arises principally from the p and d niobium electrons. We see from Fig. 8.25 that *A*15 compounds with T_c above 4 K have comparable $D(E_F)$, with a large variation between the calculated values and those evaluated from specific heat and susceptibility measurements. The results do not correlate well with the BCS plot of Eq. (8.80) which is included as a dashed curve in the figure (Tüttö *et al.*, 1979; Vonsovsky *et al.*, 1982, p. 342).

Several theoretical models had been proposed to explain superconductivity in the *A*15 compounds before the first principles band structure calculations were carried out. These models involved either postulating a wide conduction band from the 4s and 4p states of *B* atoms overlapping a narrow d band from the *A* atoms (Clogston and Jaccarino, 1961), emphasizing the role of the mutually perpendicular chains of transition ions (Labbé, 1967; Labbé *et al.*, 1967; Weger, 1964), assuming a step function for the density of states (Cohen *et al.*, 1971), attributing anomalous properties to the Fermi surface passing through the symmetry point X on the Brillouin zone boundary (Gor'kov, 1973, 1974), taking into account space group symmetry (Lee and Birman, 1978; Lee *et al.*, 1977a, b), or using orbitals from the six *A* chain atoms to approximate the bands near the Fermi surface (Bhatt, 1977, 1978). The first principles band structure calculations disagree with all of these models except Bhatt's even though the parameters of

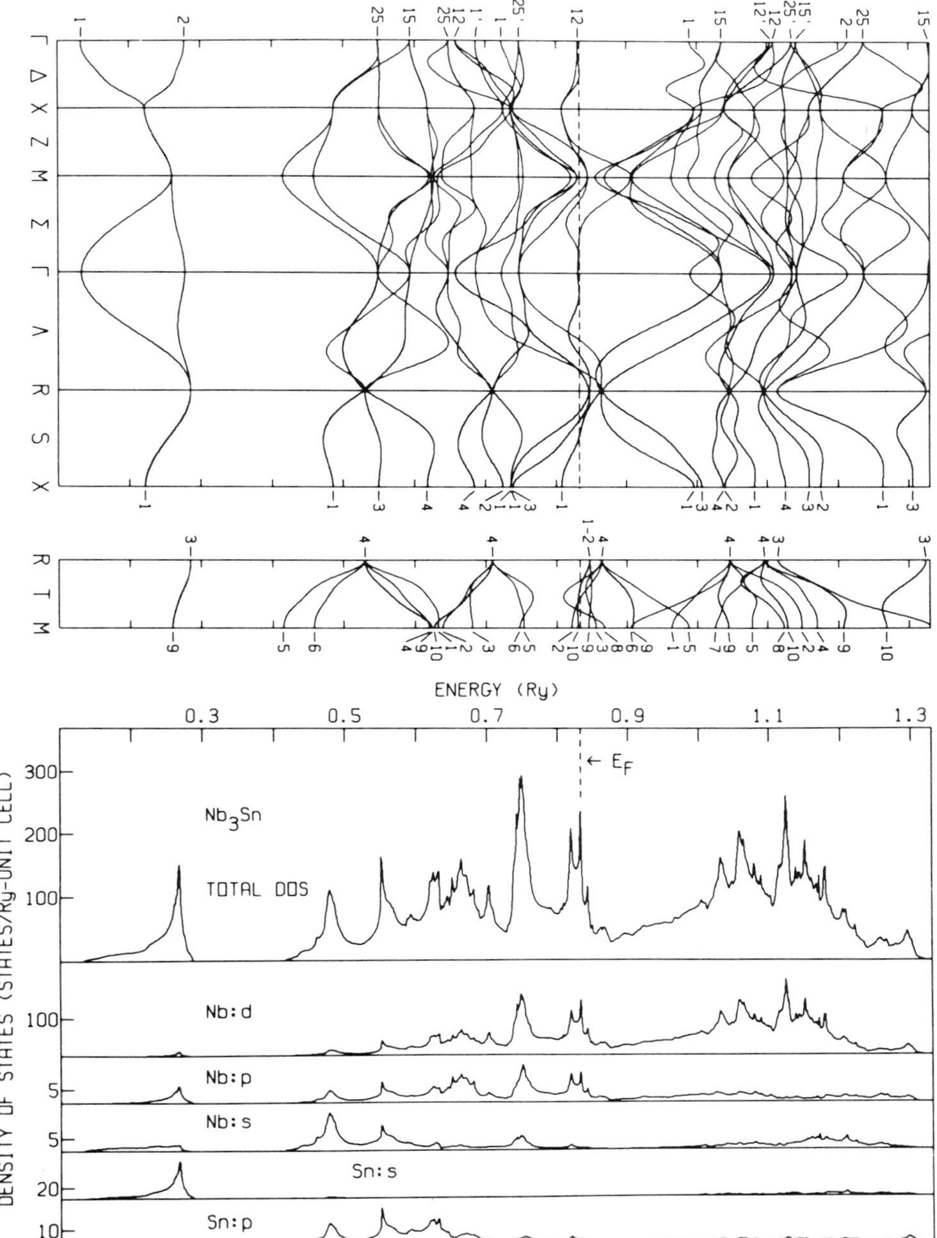

Figure 8.24 Energy bands and density of states of Nb_3Sn with $T_c = 18.0$ K obtained from a self-consistent calculation (Klein *et al.*, 1978).

Bhatt's model were selected to provide agreement with the calculated bands. As already noted, these results do not correlate well with Eq. (8.80) either (see Chapter 6 of Vonsovsky *et al.* (1982) for further details).

X. BUCKMINSTERFULLERENES

A number of band structure and other theoretical calculations of the electronic properties of the undoped ($x = 0$) and doped M_xC_{60} fullerenes, where $M = K$,

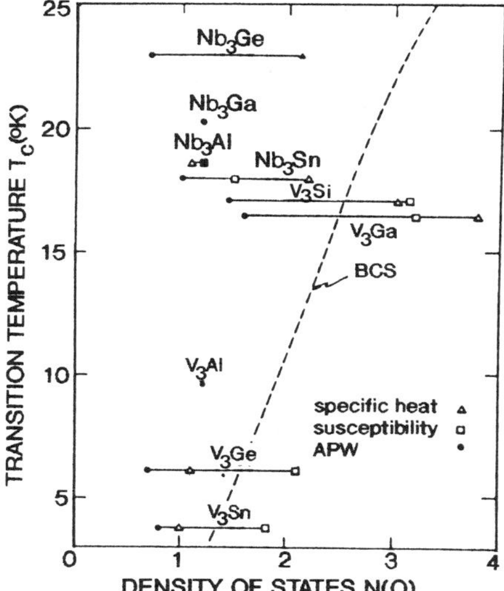

Figure 8.25 Superconducting transition temperature T_c as a function of the density of states at the Fermi level for several *A*15 compounds. The figure shows values of $D(E_F)$ evaluated from augmented plane wave (APW) band-structure calculations as well as values determined from specific heat and magnetic susceptibility measurements (Tüttö *et al.*, 1979).

Rb, Cs, $x = 0, 1, 2, \ldots, 6$, and $M_3 = RbK_2$, Rb_2K, or Rb_2Cs have been reported in the literature (Huang *et al.*, 1992). We will now comment on a few of these calculations.

Y.-N. Xu *et al.* (1991) compared the band structures of semiconducting fcc C_{60}, conducting fcc K_3C_{60}, and semiconducting bcc K_6C_{60}, and found good agreement with electron energy loss and NMR measurements. The calculations included the energy bands along various directions in the Brillouin zone, the density of states $D(E_F)$, Fermi surface, dielectric function $\epsilon(w)$, etc. Contour plots of the valence electron density of fcc C_{60} and K_3C_{60} crystals have been published (Martins *et al.*, 1992; Saito and Nishino, 1991). Novikov *et al.* (1992) showed that the electron–phonon coupling constant λ, T_c, and $D(E_F)$ all increase in the order of substitution $M =$ K, Rb, Cs, as indicated by the data in Tables 6.2, 3.11, and 4.1, respectively.

XI. $BaPb_{1-x}Bi_xO_3$ SYSTEM

We mentioned in Chapter 7, Section IV, that the solid solution series

$$BaPb_{1-x}Bi_xO_3$$

undergoes several crystallographic modifications from cubic $BaPbO_3$ to monoclinic $BaBiO_3$ as x varies from 0 to 1. Superconductivity is observed in the range $0.05 \leq x \leq 0.3$ and the highest transition temperature, $T_c \approx 13$ K, occurs for $x \approx 1/4$. Mattheiss and Hamann (1983) carried out a complete band structure calculation. A simplified version (Mattheiss, 1985) is presented in Fig. 8.26 for $x = 0$, $\frac{1}{2}$, and 1. The metallic properties are influenced mainly by the broad antibonding band arising from the Pb, Bi $6s$–O $2p\sigma$ hybrid orbitals that lie above the gap and pass through the Fermi level.

The significance of this hybrid band is best explained in terms of an ionic description $Ba^{2+}Pb^{4+}_{1-x}Bi^{4+}_xO^{2-}_3$ of the compound that assumes that the Bi ions are tetravalent. We see from Table 8.1 that the three ions Ba^{2+}, Pb^{4+}, and O^{2-} have closed shells of electrons which, together with the closed shell electrons of Bi^{4+}, fill all of the low-lying bands. Each Bi^{4+} ion has a single $6s$ electron outside the closed shells which is part of the hybrid band, so that x represents the fraction of electrons in this band. The highest T_c occurs for $x \approx \frac{1}{4}$, and for this small value there are so few electrons that the Fermi surface is nearly spherical, with the electrons confined to a region around the central point Γ of the BZ, analogous to the case shown in Figs. 8.6 and 8.7. Such a small spherical Fermi surface is not conducive to a high superconducting transition temperature. Phillips (1989a, Section IV-2) provides more details on this system. We will see in the following section how the transition temperature in the perovskite system is raised by bringing the hybrid band closer to half-full.

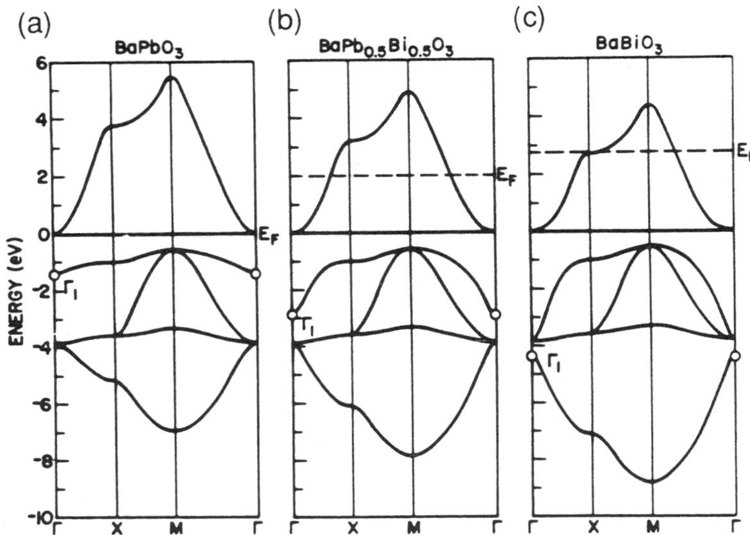

Figure 8.26 Energy bands of the perovskite compounds $BaPb_{1-x}Bi_xO_3$ compared for (a) $x = 0$, (b) $x = \frac{1}{2}$, and (c) $x = 1$ (Mattheiss, 1985, p. 120).

XII. $Ba_{1-x}K_xBiO_3$ SYSTEM

The superconducting cubic perovskite $Ba_{1-x}K_xBiO_3$ has the structure of $BaBiO_3$; the highest transition temperatures $T_c \approx 30$ K are obtained for $x \approx 0.4$. Mattheiss and Hamann (1988) calculated the band structure of the $x = \frac{1}{2}$ compound $Ba_{\frac{1}{2}}K_{\frac{1}{2}}BiO_3$, making use of the atomic configurations $Ba(5p^6 6s^2)$, $K(3p^6 4s)$, $Bi(6s^2 6p^3)$, and $O(2s^2 2p^4)$. The energy bands are presented in Fig. 8.27, and the corresponding density of states is compared with that of $BaBiO_3$ in Fig. 8.28. The bands of $Ba_{\frac{1}{2}}K_{\frac{1}{2}}BiO_3$ in the direction Γ to X to M to Γ of the BZ are similar to those in Fig. 8.26 for the $BaPb_{1-x}Bi_xO_3$ system, the principal difference being the presence of extra bands arising from the use of a much larger basis set of one-electron states for the calculations.

The occupancy of the Bi $6s$–O $2p\sigma$ hybrid band near E_F determines the electronic properties. This occupancy can be described by the ionic formula $Ba^{2+}_{1-x}K^{1+}_xBi^{4+}_{1-x}Bi^{5+}_xO^{2-}_3$ in which the Bi ions change valence to accommodate the substitution of monovalent K for divalent Ba. We see from Table 8.1 that all of the ions in this compound, except for Bi^{4+}, have closed shells, and that Bi^{4+} contributes a $6s$ electron to the hybrid band. As a result, the hybrid band contains $(1-x)$ electrons, whereas the $x = 0.4$ compound $Ba_{0.6}K_{0.4}BiO_3$ has 0.6 electron in this band. The Fermi surface now extends further out toward the point X, which tends to distort it away from spherical. Figure 8.29 compares the number of electrons in the hybrid band for the $BaPb_{1-x}Bi_xO_3$ and $Ba_{1-x}K_xBiO_3$ compounds that we have been discussing. The highest T_c compound in each group is shown. We see that the two series of perovskites have the same end-point compound $BaBiO_3$ for which the band is half-full, but they differ in their "band-empty" compound. (Note the unfortunate choice of notation, whereby x denotes the band occupancy for $BaPb_{1-x}Bi_xO_3$ whereas $(1-x)$ gives the band occupancy for $Ba_{1-x}K_xBiO_3$.)

Calculations were carried out for three ordered $BaK(BiO_3)_2$ compounds in which planes of K atoms and planes of Ba atoms alternated in the (0, 0, 1), (1, 1, 0), and (1, 1, 1) directions, respectively. The results

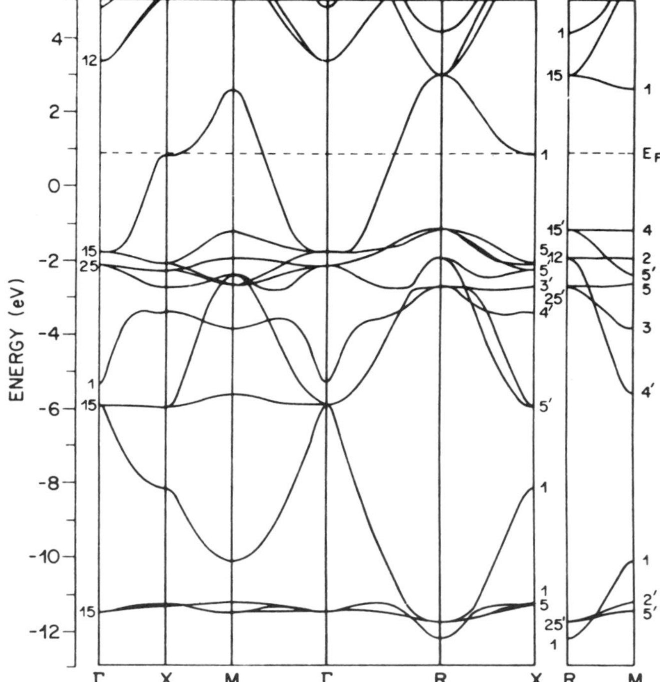

Figure 8.27 Energy bands of the cubic perovskite superconductor $Ba_{\frac{1}{2}}K_{\frac{1}{2}}BiO_3$ obtained from self-consistent linear augmented plane wave (LAPW) calculations (Mattheiss and Hamann, 1988).

show that the ordering of K and Ba has only a slight effect on the electronic states near E_F, and that these two cations have comparable densities of states. The important factor in raising T_c seems to involve increasing the electron count of the $6s$–$2p\sigma$ hybrid band to bring it closer to half-full. Further energy band and DOS information on $Ba_{1-x}K_xBiO_3$ may be found in the studies of Hamada *et al.* (1989), W. Jin *et al.*, (1991, 1992), and Liechtenstein *et al.*, (1991).

XIII. BAND STRUCTURE OF $YBa_2Cu_3O_7$

We will begin our discussion of the energy bands of the high-temperature superconductors with the compound $YBa_2Cu_3O_7$ since its structure is of the aligned type, as explained in Chapter 7, Section VI, and hence its Brillounin zone is simpler than those of the cuprates which are body centered. This section will describe the band structure reported by Pickett *et al.* (1990) and Pickett (1989; cf. Costa-Quintana *et al.*, 1989; Curtiss and Tam, 1990; Krakaer *et al.*, 1988; Singh *et al.*, 1990; J. Yu *et al.*, 1991). The bands reflect the principal structural features of the compound—the presence of two CuO_2 planes containing the Cu(2), O(2), and O(3) atoms, and a third plane containing chains of Cu(1)–O(1) atoms along the b direction, as shown in Fig. 8.30 and described in Chapter 7, Section VI. The *ortho*-rhombic Brillouin zone, also shown in the figure, has a height-to-width ratio

$$(2\pi/c)/(2\pi/a) = a/c \approx 0.33,$$

which is the reciprocal of the height-to-width ratio of the unit cell in coordinate space.

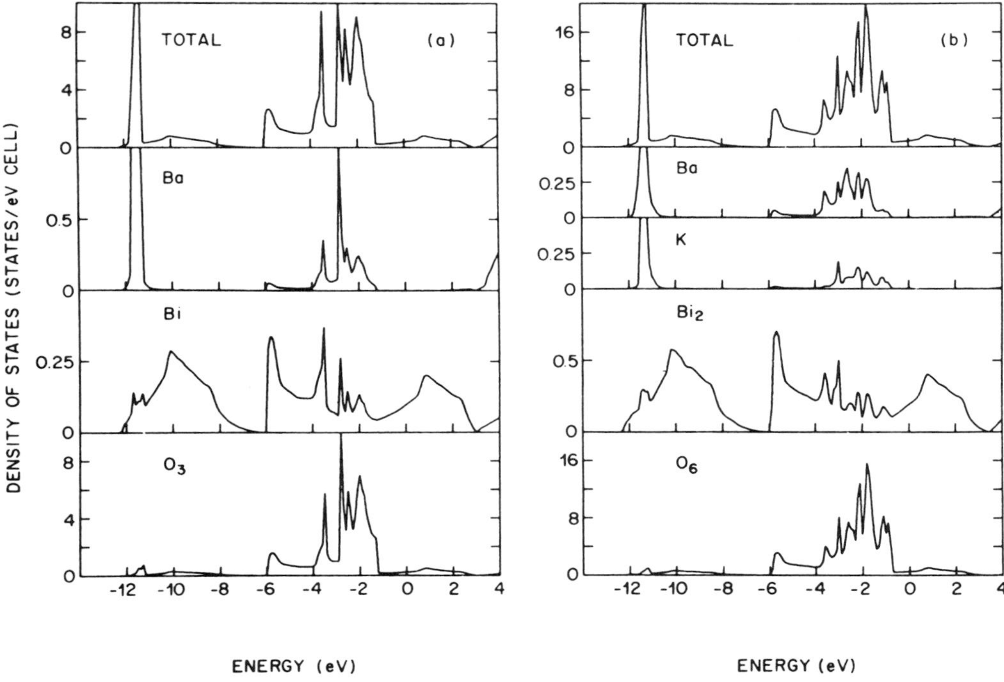

Figure 8.28 Calculated density of states of (a) cubic $BaBiO_3$, and (b) tetragonal ordered alloy $BaKBi_2O_6$ with planes of Ba and K alternating along the *c*-axis. The units are scaled to permit direct comparison of the results (Mattheiss and Hamann, 1988).

A. Energy Bands and Density of States

The energy bands of $YBa_2Cu_3O_7$ near the Fermi surface are presented in Fig. 8.31 along the principal directions connecting the symmetry points Γ, X, Y, and S in the central ($k_z = 0$) horizontal plane of the Brillouin zone. The bands change very little at the corresponding symmetry points Z, U, T, and R of the top ($k_z = \frac{1}{2}$) plane. The highest transition temperature is found in the case of a small amount of oxygen deficiency, $\delta \approx 0.1$, in the formula $YBa_2Cu_3O_{7-\delta}$. This slightly less than half-full condition, where $\delta = 0$ for half-full, means that there are missing electrons near E_F, and that the conductivity is of the hole type. This is confirmed by Hall effect measurements, as will be seen in Chapter 14, Section VI.

The two narrow CuO plane-related bands are shown in Fig. 8.31 strongly dispersed, i.e., rising far above the Fermi surface at the corner points S and R of the Brillouin zone of Fig. 8.30. These two bands are almost identical in shape, resembling the Hubbard band that is shown rising above E_F at the corner point M of the Brillouin zone in the inset of Fig. 8.18a. The much broader chain band that is shown strongly dispersed in both the S and the Y (also R and T) or chain directions far above E_F arises from the Cu–O sigma bonds along the chains formed from the oxygen p_y and copper $d_{x^2-y^2}$ orbitals, as illustrated in Fig. 8.32. There is another chain band which undergoes very little dispersion, staying close to the Fermi surface, but it rises slightly above E_F at S and R,

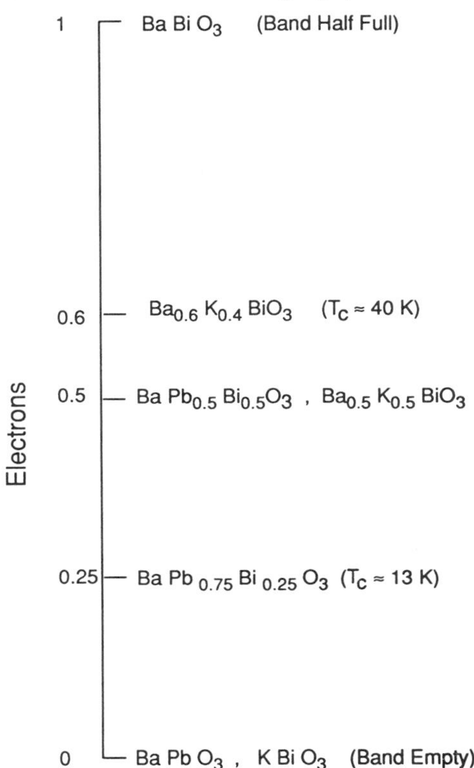

Figure 8.29 Number of electrons outside the closed shells in several perovskite compounds. Transition temperatures are given for two superconducting members of the series.

as shown in the figure. Oxygen deficiency depopulates the chains, and this is reflected in the absence of chain bands in the semiconducting compound $YBa_2Cu_3O_6$ (Yu *et al.*, 1987).

The lack of appreciable dispersion in the bands along k_z demonstrated by the similarity of Figs. 8.31a and 8.31b means that the effective mass m^* defined by Eq. (1.47),

$$\frac{1}{m^*} = \frac{1}{\hbar^2}\left(\frac{d^2E_{\mathrm{k}}}{dk_z^2}\right)_{E_{\mathrm{F}}}, \tag{8.81}$$

is very large, and since the electrical conductivity σ from Eq. (1.22) is inversely proportional to the effective mass, this means that the conductivity is low along the z direction. In contrast, the plane and chain bands appear parabolic in shape along k_x and k_y, corresponding to a much lower effective mass. This explains the observed anisotropy in the normal-state electrical conductivity, $\sigma_c/\sigma_{ab} \approx 20$.

The total density of states and the contribution of each copper and oxygen to $D(E)$ are presented in Fig. 8.33. There are peaks (called *van Hove singularities*, VII, I)

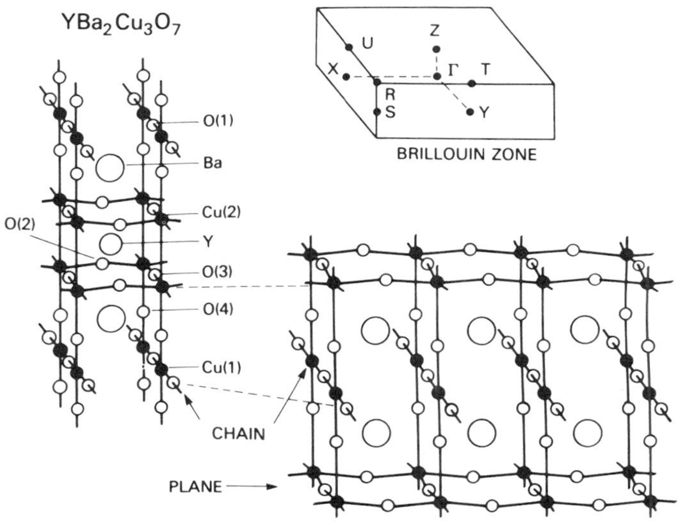

Figure 8.30 Two views of the $YBa_2Cu_3O_7$ unit cell and sketch of the corresponding Brillouin zone (Krakauer *et al.*, 1988).

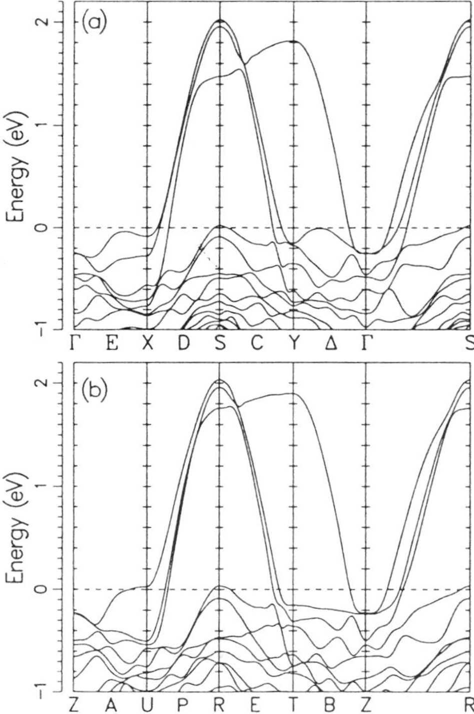

Figure 8.31 Energy bands of $YBa_2Cu_3O_7$ along the principal directions in the Brillouin zone sketched in Fig. 8.30. The two plane bands rise sharply above the Fermi level at points S and R, and the broader chain band is high in energy between S and Y, and between R and T (Pickett *et al.*, 1990).

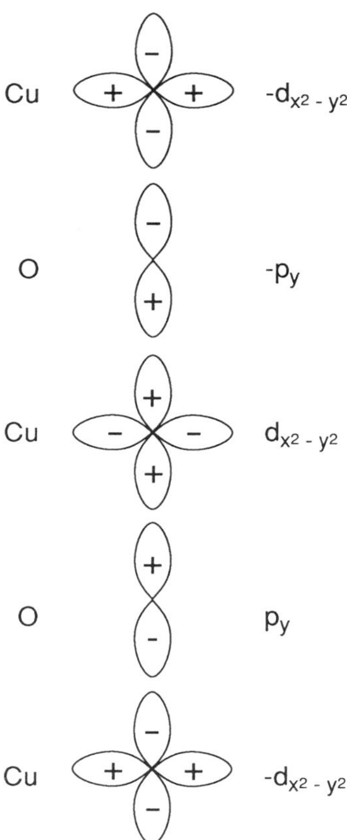

Figure 8.32 Sigma bonding between Cu(1) and O(1) orbitals along chains in the crystallographic b direction of $YBa_2Cu_3O_7$.

in the partial DOS of the planar oxygens O(2) and O(3) at energies at which the planar Cu(2) atoms also have peaks, whereas chain Cu(1) have peaks that match those of the O(1) and O(4) oxygens bonded to it. This suggests that the planes and chains are partially decoupled from each other, as we might have expected, since the bridging oxygen O(4) is 1.8 Å from Cu(1) compared to the much longer distance 2.3 Å from Cu(2). Super current can flow along both the planes and chains.

B. Fermi Surface: Plane and Chain Bands

The Fermi surface of $YBa_2Cu_3O_7$ is plotted in Fig. 8.34 for the $k_z = 0$ and $k_z = \pi/a$ horizontal planes of the Brillouin zone, with the symmetry points S and R, respectively, selected as the origin, in accordance with Fig. 8.35. The close similarity between the respective left ($k_z = 0$) and right ($k_z = \pi/a$) panels of the figure confirms that there is very little dispersion in the vertical (k_z) direction. The shapes of these Fermi surfaces can be deduced from where the corresponding bands cross the Fermi surface in Fig. 8.31.

The two plane–band Fermi surfaces presented in the four center panels of Fig. 8.34 consist of large regions of holes in the center, with narrow bands of electrons around the periphery. The surface of one plane band makes contact with the zone boundary between points Y and Γ due to the small maximum that rises above E_F between these points in Fig. 8.31. When

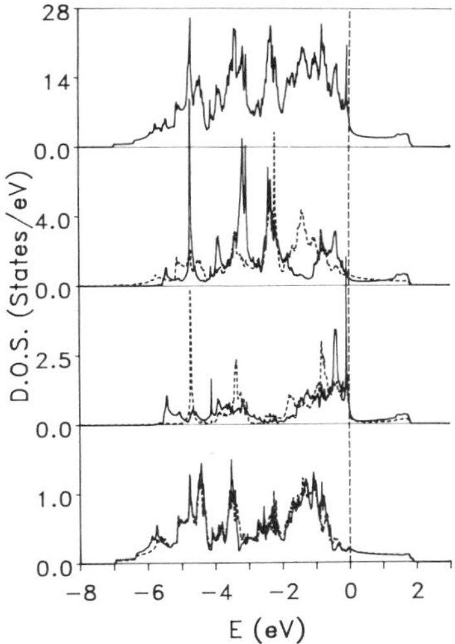

Figure 8.33 Density of states of $YBa_2Cu_3O_7$. The top panel shows the total density of states, the second panel gives the partial DOS of Cu(1) [——] and Cu(2) [- - - -], the third panel presents the same for O(1) [——] and O(4) [- - - -], and the bottom one gives O(2) [——] and O(3) [- - - -] (Krakauer *et al.*, 1988).

0.2 fewer electrons are present, the Fermi surface is lowered and the contact with the zone boundary extends over a larger range between Y and Γ. Adding 0.2 more electron raises the Fermi surface above the small relative maximum near Y, and the zone boundary is no longer reached. Thus the oxygen content, which determines the electron concentration, influences details of the shape of the Fermi surface. As already noted in Section A, the highest T_c occurs for $\delta \approx 0.1$. The regions of holes around the central points R and S (not shown) resemble their counterpart around point M of the three-state Hubbard model.

The two chain–band Fermi surfaces sketched in the upper and lower pairs of panels of Fig. 8.34 differ considerably. The upper panels show the highly dispersed chain band that lies above E_F everywhere except from Γ to X, where it is below E_F, as shown in Fig. 8.31. This corresponds to a narrow one-dimensional-like slab containing electrons that extends continuously from cell to cell along the direction Γ–X–Γ..., and parallel to it there is a wide hole-type slab along Y–S–Y..., as shown at the top of Fig. 8.34. The edges of this wide slab are mostly parallel, a condition needed for nesting, but are not close enough to the boundary to create strong nesting, as will be explained in Section XVIII. The other chain band lies entirely below the Fermi level except for a narrow region near point S in Fig. 8.31. Figure 8.34 shows this small hole region in the zone centers S and R of the lower two panels.

C. Charge Distribution

We have been describing the details of the band structure in reciprocal space. A Fourier transformation from reciprocal space to coordinate space provides the charge distributions at and between the atoms in the unit cell. Contour plots of the overall valence charge density in coordinate space that were calculated by this method are sketched in Fig. 7.14 for three high-symmetry planes of the unit cell.

XIV. BAND STRUCTURE OF $(La_{1-x}Sr_x)_2CuO_4$

The second high-temperature superconductor that we will discuss,

$$(La_{1-x}Sr_x)_2CuO_4,$$

is made by doping the semiconducting prototype compound La_2CuO_4, which is not itself a superconductor. Substituting Sr for La has the effect of changing the electron content of the unit cell through the addition of holes; this lowers the Fermi level and strongly affects the superconducting properties. Conductivity is of the hole type. The geometry of the Brillouin zone is somewhat complicated because the crystallographic structure is body centered. We wish to describe the band structure (Chen and Callaway, 1989; DeWeert *et al.*, 1989; Krakauer *et al.*, 1987; Kulkarni *et al.*, 1991; Pickett, 1989; Pickett *et al.*, 1987).

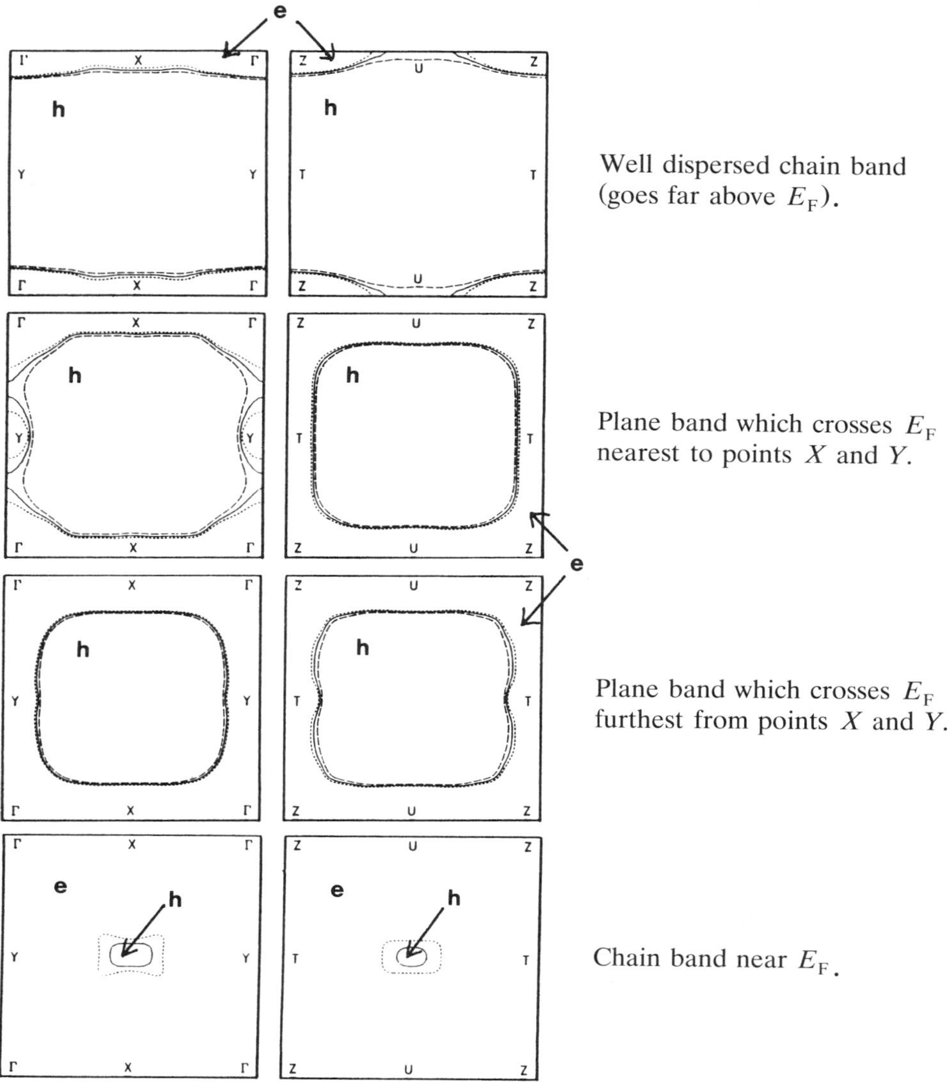
e
Γ X Γ
Z U Z
h
Y Y
T T
Well dispersed chain band (goes far above E_F).
Plane band which crosses E_F nearest to points X and Y.
Plane band which crosses E_F furthest from points X and Y.
Chain band near E_F.

Figure 8.34 Fermi surfaces of $YBa_2Cu_3O_{7-\delta}$ calculated from the band structure with the electron (e) and hole (h) regions indicated. The left panels are for the midplane of the Brillouin zone ($k_z = 0$) with point S at the center and those on the right are for the top plane ($k_z = \pi/a$) with point R at the center. Points S and R are not labeled. The two top panels show a well dispersed chain band that rises far above the Fermi level. The second pair of panels shows the first plane band; it crosses the Fermi level closest to points X and Y in Fig. 8.31. The third pair of panels presents the second plane band with the crossing further from these points. The bottom panels display the chain band, which is mostly below the Fermi surface, rising above it only near point S (and point R) in Fig. 8.31. The solid lines are for the calculated Fermi surface with $\delta = 0$, the short dashed curve is for $\delta = 0.2$ containing 0.2 fewer electron, and the long-dashed curve is for $\delta = -0.2$ with 0.2 more electron (Krakauer *et al.*, 1988).

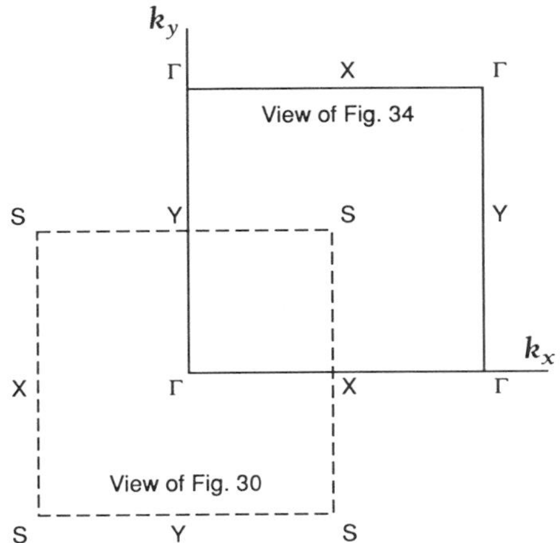

Figure 8.35 Relationship between Brillouin zone boundary in Fig. 8.30 (dashed lines) when point Γ is at the center and Fig. 8.34 (solid lines) when S is chosen as the center.

A. Orbital States

For the undoped compound La_2CuO_4, the copper contributes five d states while each of the four oxygens provides three p states for a total of 17 orbitals. These orbitals are occupied by 33 electrons, which may be counted in one of two ways. Counting by ions, there are nine ($3d^9$) from Cu^{2+} and six ($2p^6$) from each of the four O^{2-} ions. The ionic state La^{3+} has no valence electrons. Counting by atoms, there are three from each La, 11 from Cu, and four from each oxygen. These numbers are listed in Table 8.1. Contour plots of the valence charge density around these atoms in coordinate space that were calculated from the band structure are sketched in Fig. 7.23, and discussed in Chapter 7, Section VIII.C.

If 10% of the La is replaced by Sr to give the superconducting compound

$$(La_{0.9}Sr_{0.1})_2CuO_4,$$

charge neutrality can be maintained by a change in copper valence to $Cu^{2+}_{0.8}Cu^{3+}_{0.2}$, with an average of $Cu^{2.2+}$. From Table 8.1 we see that Cu^{2+} has the configuration $3d^9$ and Cu^{3+} the configuration $3d^8$, which means that Cu contributes 8.8 electrons, instead of 9, and the total number of electrons is 32.8, not 33. If, instead, all of the copper remains in the Cu^{2+} state and charge neutrality is preserved by virtue of a deficit of oxygen, we obtain the compound $(La_{0.9}Sr_{0.1})_2CuO_{3.9}$, with Cu contributing nine electrons and oxygen 23.4, for a total of 32.4. Intermediate cases, such as $(La_{0.9}Sr_{0.1})_2CuO_{4-\delta}$, also occur in which there is some change in Cu valence, plus some deficit δ of oxygen. In practice, the Sr content x is usually known, but δ is ordinarily not determined.

B. Energy Bands and Density of States

The calculated energy bands are presented in Fig. 8.36 with the Fermi energy selected as the zero of energy. The bcc Brillouin zone sketched in Fig. 8.37 shows the symmetry points and directions along which the bands in Fig. 8.36 are drawn. Figure 8.38 shows how adjacent zones fit together in reciprocal space.

There are 16 low-lying "bonding" bands below the Fermi energy fully occupied by 32 of the 33 valence electrons, one bonding "hybrid" band near E_F and 17 empty higher-energy "antibonding" bands. The hybrid band contains exactly one electron for La_2CuO_4 and 0.8 electron for the doped compound $(La_{0.9}Sr_{0.1})_2CuO_4$, as already noted. Its principal excursion above E_F in the X direction is analogous to that of its counterpart at point M in the three-state Hubbard model of Fig. 8.18a. That the bands are in fact two-dimensional is seen by observing in Fig. 8.36 how flat they are along the vertical paths from Γ to Z ((0, 0, 0) to (0, 0, 1)), from (1, 0, 0) to (1, 0, 1), and from Δ to U ((h, h, 0) to (h, h, 1), where h is arbitrary). We can again argue from Eq. (8.81) that the effective mass m^* is large in the k_z direction, and hence that the electrical conductivity is low for current flow perpendicular to the planes.

The total density of states (DOS) shown in Fig. 8.39 is highest at 3 eV above E_F where many bands are close together,

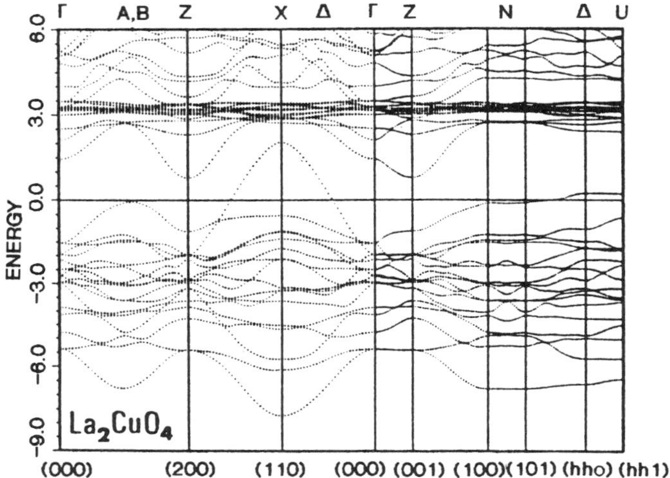

Figure 8.36 Energy bands of La_2CuO_4 along the symmetry directions of the Brillouin zone defined in Fig. 8.37, with E_F taken as the zero of energy. These bands also apply to the doped compound $(La_{0.9}Sr_{0.1})_2Cu_{4-\delta}$. This compound has fewer electrons so its Fermi energy is below (adapted from Pickett *et al.*, 1987).

and is smallest from 1 eV below E_F to 2 eV above E_F; at E_F there is a pronounced sparseness of bands (Aligia, 1989; Gold and Ghazali, 1987). We see from the figure that the DOS near E_F arises from the copper ions and oxygens in the plane, which indicates that the hybrid band is occupied by electrons in the CuO_2 planes. The value of DOS at the Fermi surface is smaller than for the *A*15 compounds, as may be seen by comparison with Fig. 8.24.

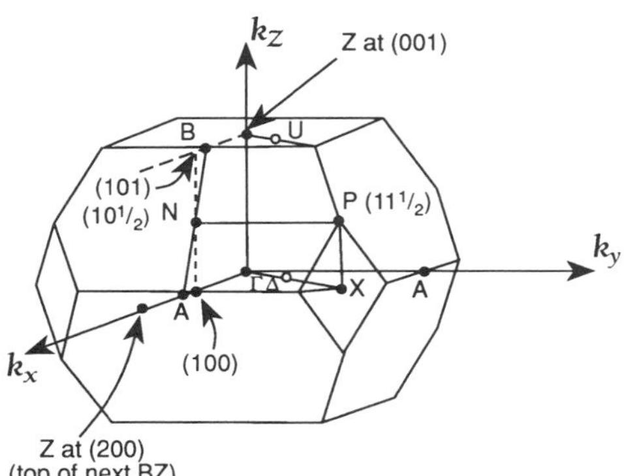

Figure 8.37 Brillouin zone of body-centered La_2CuO_4 with the symmetry points indicated. The symbols Δ and *U* denote general points along the [110] directions (adapted from Pickett, 1989).

C. Brillouin Zone

Perhaps a few words should be said to clarify the geometry of the Brillouin zone. This is important because the bismuth and thallium compounds which we will be discussing belong to the same crystallographic space group and have the same Brillouin zone.

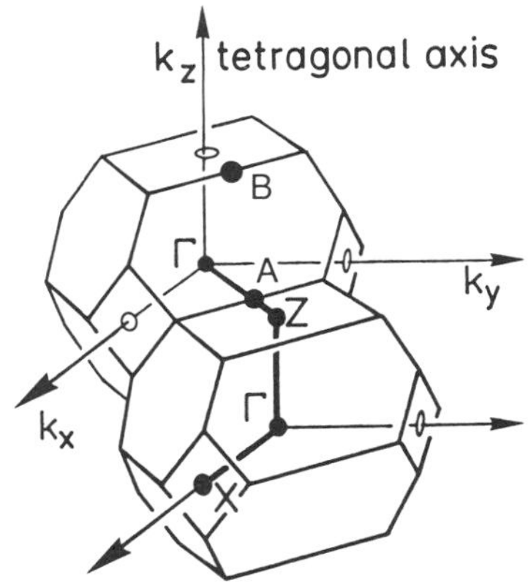

Figure 8.38 Geometrical relationship between adjacent La_2CuO_4 Brillouin zones (Kulkarni *et al.*, 1991). Note that the k_x, k_y axes in this figure are rotated by 90° relative to the axes of Fig. 8.37. The description in the text is in terms of the axes of Fig. 8.37.

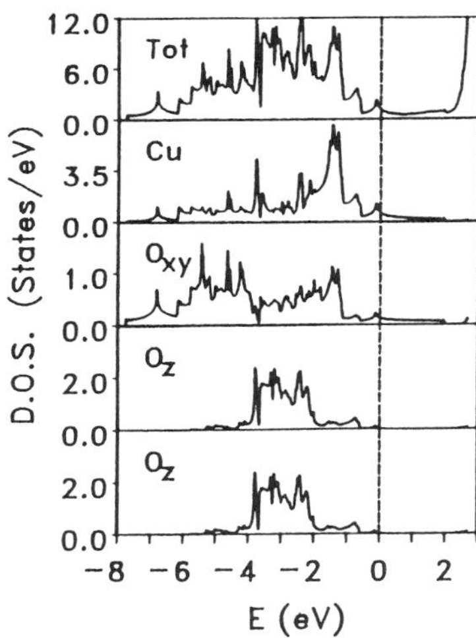

Figure 8.39 Total density of states (top) and partial densities of states for the individual atoms of La_2CuO_4 determined from the band structure. The O_{xy} oxygens are in the CuO_2 planes (Pickett *et al.*, 1987).

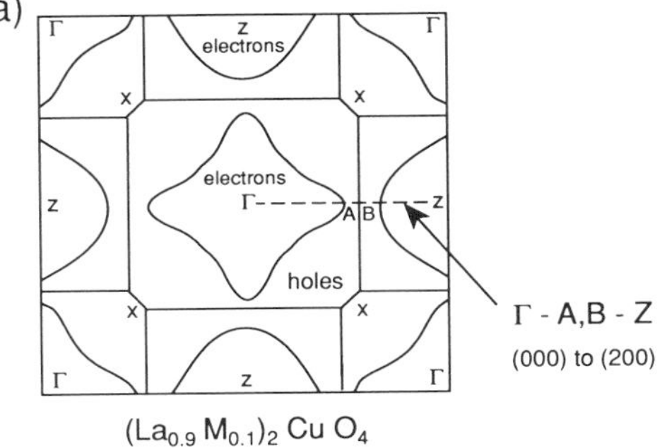

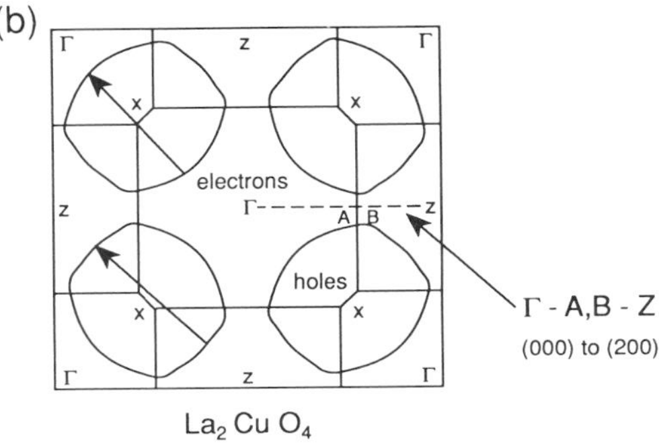

Figure 8.40 Fermi surfaces (a) of La_2CuO_4 with enclosed holes, and (b) of $(La_{0.9}M_{0.1})_2CuO_4$ with enclosed electrons. See Figs. 8.37 and 8.38 for a clarification of the path Γ–*A*–*B*–*Z*. The bold arrows on (b) approximate nesting wave vectors drawn somewhat longer than the Γ–*X* distance. The changeover from enclosed holes to enclosed electrons occurs at a van Hove singularity (Pickett, 1989).

Consider the path along the k_x direction of Fig. 8.37 from the center point Γ of one Brillouin zone to the point *Z* at the top of the next Brillouin zone. Details of this horizontal path from Γ to *Z* are shown in Fig. 8.38. It runs horizontally from Γ at the origin (0, 0, 0) of one unit cell to the position (2, 0, 0) corresponding to the point *Z* of the next unit cell. The figure also shows a vertical Γ to *Z* path that starts at the origin (0, 0, 0) and proceeds upwards along k_z to (0, 0, 1). Therefore, both points (0, 0, 1) and (2, 0, 0) are justifiably called *Z*.

Figure 8.40 sketches the Brillouin zone in the k_x,k_y-plane located at $k_z = 0$. The zone boundary around the origin Γ is an octagon of four long sides with center points *A* and four short sides with center points *X*. The boundary around point *Z* at the top of the next zone where $k_z = \pi/a$ is a square with point *B* at the center of its edge, as indicated. Half of this square appears four times in Fig. 8.40, with the points *Z* and *B* indicated. The horizontal path Γ–*A*, *B*–*Z* of Fig. 8.40 is the same path as the (0, 0, 0)–(1, 0, 0)–(2, 0, 0) of Figs. 8.37 and 8.38 which passes from Γ to *A* through the center plane of one Brillouin zone and proceeds from *B* to *Z* along the top of the next zone.

D. Fermi Surface

We see from Fig. 8.36 that the Fermi level is higher than all of the low-lying bands at points Γ and *Z* and is below the hybrid band at *X*. As a result, there are electron regions around points Γ and *Z* and holes around point *X*, as shown in Fig. 8.40. The Fermi level of the compound La_2CuO_4 is above the low-lying bands along the horizontal path from Γ through *A*, *B* to *Z*, as indicated in Fig. 8.36, so the Fermi surface consists of the interconnected regions of electrons, with isolated patches of holes around point *X*, as depicted in Fig. 8.40b. The doped compound $(La_{0.9}Sr_{0.1})_2CuO_4$ has 0.2 fewer electrons and its Fermi surface is lowered, falling below the top of the band maximum at point A, B, so the Fermi surface consists of interconnected regions of holes with isolated islands of electrons at points Γ and Z, as indicated in Fig. 8.40a. Figure 8.40a is similar to nearly free electron Fig. 8.7 with the electrons confined to isolated regions in k space, and Fig. 8.40b resembles nearly free electron Fig. 8.10 with its isolated holes. Extended zone schemes of the lanthanum compound Fermi surface have appeared (Ong *et al.*, 1987; Pickett, 1989; Trugman, 1990). The Fermi surface of the related electron superconductor

$$(Nd_{1-x}Ce_x)_2CuO_{4-\delta}$$

has also been determined (King *et al.*, 1993).

Since the point Z is a distance π/c above Γ, as shown in Fig. 8.37, the electron regions of the superconductor $(La_{0.9}Sr_{0.1})_2CuO_{4-\delta}$ extend vertically, from Γ to Z to Γ, and so on, forming a cylinder of irregular cross-section (Klemm and Liu, 1991) containing electrons surrounded by holes. The shapes of this cross section around the point Γ and around the point Z are shown in Fig. 8.40a. In like manner, the compound La_2CuO_4 has vertical cylinders of holes centered at the point X and surrounded by electrons. The change from cylinders of holes to cylinders of electrons is reminiscent of the electron–hole symmetry that was discussed in Section VII.C.

The sketch of the "Fermi surface" of La_2CuO_4 in Fig. 8.40 is not in accord with experiment because La_2CuO_4 is an antiferromagnetic insulator with a 2eV energy gap, and hence no Fermi surface may be defined for it. This shows the limitations of band structure calculations.

E. Orthorhombic Structure

Both cubic and orthorhombic body-centered crystal structures have Brillouin zones that are face centered in reciprocal space, as depicted in Fig. 8.22a for the cubic case. However, it is a peculiarity of the body-centered tetragonal structure that its Brillouin zone is body-centered. This occurs because a conventional tetragonal unit cell that is face centered reduces to a conventional tetragonal body-centered cell of half the volume, as illustrated in Fig. 8.41. This construction is not valid in the cubic case because the resulting smaller body-centered cell would be elongated relative to its base, and hence would have tetragonal rather than cubic symmetry. It is not valid for the orthorhombic case either because the base would no longer be a square, so that the basis vectors would no longer be orthogonal, corresponding to monoclinic symmetry. The construction is valid only for the tetragonal case.

The abrupt change in the conventional unit cell illustrated in Fig. 8.41 that takes place at the tetragonal-to-orthorhombic transition is associated with a smooth, continuous change in the Brillouin zone. This is illustrated in Fig. 8.42, which sketches the smooth transformation of the body-centered tetragonal Brillouin zone to both

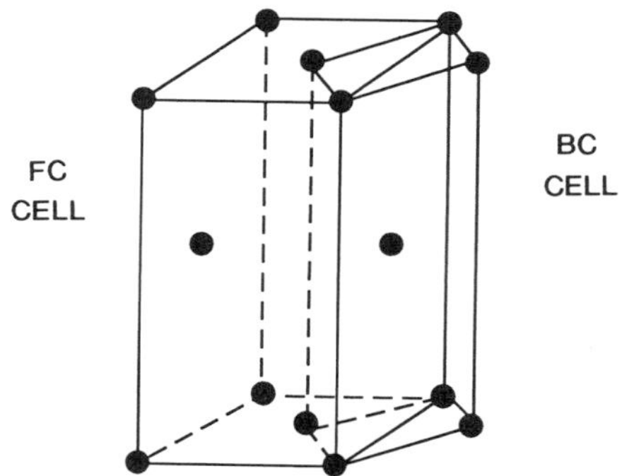

Figure 8.41 Relationship between larger face-centered and smaller body-centered tetragonal conventional unit cell. Only the latter is recognized as a valid unit cell in the tetragonal system. *Ortho*-rhombic distortion produces a body-centered or a face-centered *ortho*-rhombic cell, depending on whether the stretching is along the rectilinear or the diagonal direction, respectively, of the smaller body-centered tetragonal cell (see Fig. 7.4).

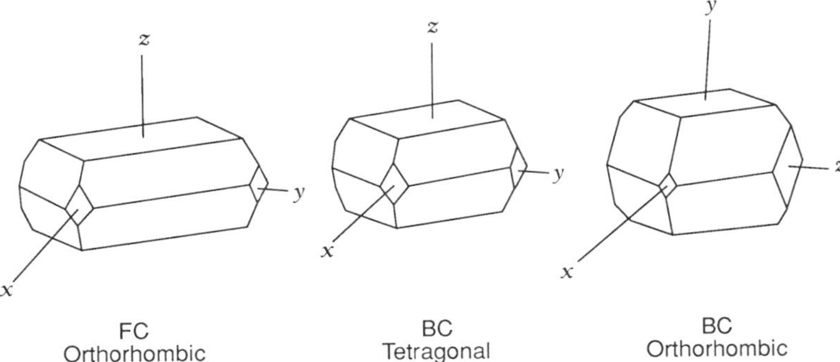

Figure 8.42 Distortion of body-centered tetragonal Brillouin zone (center) to a face-centered (left) and to a body-centered (right) orthorhombic Brillouin zone. The former corresponds to the superconductor case of a body-centered crystal structure and the latter to a face-centered crystal structure.

body-centered and face-centered orthorhombic. The change is also smooth and continuous for the corresponding Wigner–Seitz cell in coordinate space at this transition.

XV. BISMUTH AND THALLIUM COMPOUNDS

We have devoted considerable space to the $YBa_2Cu_3O_7$ and $(La_{1-x}Sr_x)_2CuO_4$ compounds, but will discuss the bismuth and thallium compounds only very briefly. Band-structure calculations have been carried out for several members of these series. Let us consider the results that have been found for one member of each series. The compounds belong to the same crystallographic space group as La_2CuO_4, as noted in Chapter 7, Section I, so that their Brillouin zones are similar in shape to that shown in Fig. 8.37, except for being compressed in the k_z direction. The compression occurs because of the size of the ratio c/a for $Bi_2Sr_2CaCu_2O_8$ and its thallium analogue; for these compounds $c/a = 8.0$, whereas it is much less, $c/a = 3.5$, for La_2CuO_4. The structure of $Bi_2Sr_2CaCu_2O_8$ sketched in Fig. 8.43 is similar to that of $Tl_2Ba_2CuO_6$, except for the fact that the former has a Ca layer plus a second CuO_2 plane. The energy bands of the thallium compound and the DOS plots of both compounds are presented in Figs. 8.44, 8.45, and 8.46, respectively (Kulkarni *et al.*, 1991; Pickett, 1989). Superlattice structures were not taken into account in the calculations. The real space charge distributions of these compounds, as obtained from band structure calculations, were discussed in Chapter 7. The Bi and Tl series of compounds are hole-type superconductors, like their La and Y counterparts.

In many ways the band structures of the two compounds are very similar, and also resemble those of the La and Y compounds. The CuO_2 planes contribute nearly half-filled $3d$, $2p$ bands that cross E_F and peak sharply in the X direction, as in the Hubbard model. This is shown in Fig. 8.44 for $Tl_2Ba_2CuO_6$. This figure also shows that the bands are flat in the vertical Γ to Z direction, and that there is less c direction dispersion than in the La and Y compounds, making $Bi_2Sr_2CaCu_2O_8$ and $Tl_2Ba_2CuO_6$ more two-dimensional. The large effective mass in the k_z direction, as calculated from Eq. (8.81), causes the electrical conductivity to be relatively small in the z direction. Figures 8.45 and 8.46 show that the density of states at and slightly below E_F arises mainly from the Cu d orbital and the Ol p orbital of the CuO_2 planes. Thus these planes are responsible

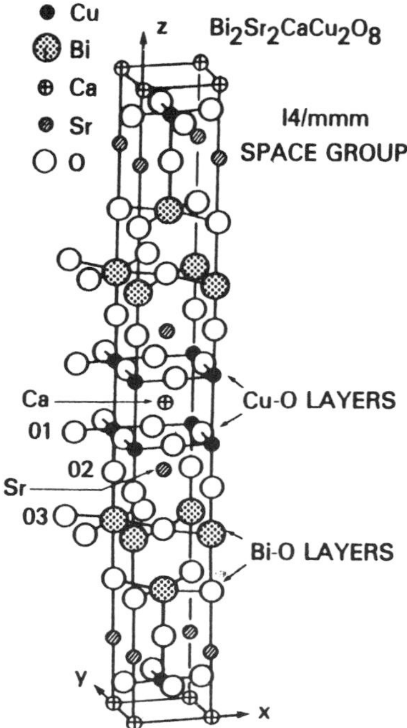

Figure 8.43 Atom positions in unit cell of $Bi_2Sr_2CaCu_2O_8$. The compound $Tl_2Ba_2CaCu_2O_8$ has the same structure with Tl replacing Bi and Ba replacing Sr. The compounds $Bi_2Sr_2CuO_6$ and $Tl_2Ba_2CuO_6$ have the same structure type with the Ca atom and an adjacent CuO_2 plane removed (Pickett, 1989).

for the common features of all of the cuprates.

In addition to these shared properties, the bismuth and thallium atoms contribute features that are specific to each compound. $Bi_2Sr_2CaCu_2O_8$ has a pair of slightly filled Bi $6p$ bands high above E_F that drop below the Fermi level along the horizontal Γ to Z direction and hybridize strongly with the CuO bands just below E_F (Krakauer and Pickett, 1988; Pickett, 1989; Szpunar and Smith, 1992), thereby providing additional carriers in the Bi–O planes. Thus the Bi and Cu layers jointly contribute to the metallic properties. One Tl $6s$–O $2p$ antibonding band lying mostly above E_F drops slightly below the Fermi level at the Γ point, as indicated in Fig. 8.44, and forms a pocket of electrons at Γ. The Tl s–O p bands indicate the presence of metallic TlO layers between the CuO_2 planes that help to carry the current. The density of states of the Tl and associated O(2) atoms, shown in Fig. 8.45, remains high over a somewhat wider range of energies than is the case for the other cuprates, as may be seen by comparing Figs. 8.33, 8.39, and 8.46. Chan *et al.* (1991) and Tatarskii *et al.* (1993) sketch the Fermi surface.

XVI. MERCURY COMPOUNDS

Band structure calculations have been carried out for the Hg-1201 and the Hg-1221 members of the

$$HgBa_2Ca_nCu_{n+1}O_{2n+4+\delta}$$

series of compounds, and we will summarize some of the results that were obtained. Section X of Chapter 7 pointed out the structural similarities of these mercury compounds to the yttrium and bismuth, thallium cuprates, and these should be kept in mind while reading the present section.

The band structure calculations were carried out by adding to the usual orbital set the following upper core level orbitals: Cu $3s$ and $3p$, O $2s$, Ba $5s$ and $5p$, and Hg $5p$ and $5d$, with $\delta = 0$. Ca $3s$ and $3p$ orbitals were added for Hg-1223. Figure 8.47 and 8.48 show the energy bands and density of states for the $n = 0$ compound (Singh, 1993), and Figs. 8.49 and 8.50 present their counterparts for the $n = 2$ compound (Rodriguez, 1994; Singh, 1994). In addition the articles provide Fermi surface across sections for the two compounds.

The bands of $HgBa_2CuO_4$ presented in Fig. 8.47 are highly two dimensional, with very little dispersion in the z direction shown from Γ to Z in the figure. Only one band crosses the Fermi surface, and that is the $pd\sigma^*$ band derived from the CuO_2 planes, which is characteristic of all the layered cuprates. An unoccupied band

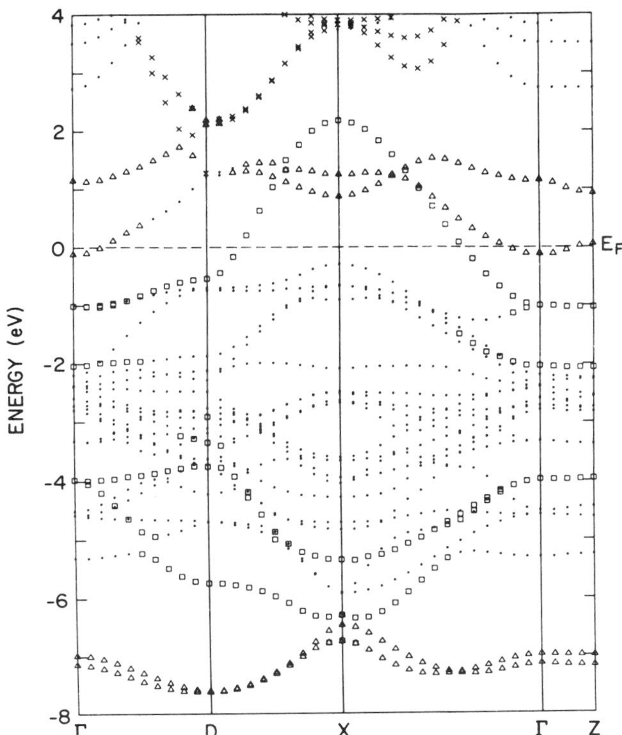

Figure 8.44 Energy bands of tetragonal $Tl_2Ba_2CuO_6$. Particular symbols are employed to denote the bands of predominantly Tl *s* and O(2) *p* character (triangles), those with strong Cu–O pd σ character (squares), and the low-lying Tl p_x, p_y bands (crosses). One Tl–O band crosses the Fermi level to form an electron pocket at Γ (Hamann and Mattheiss, 1988).

arising from hybridization of Hg $6p_z$ and O(2) $2p_z$ orbitals lies above the Fermi level, and approaches it without ever crossing it. As a result the $pd\sigma^*$ band is exactly half full, and the stoichiometric compound ($\delta = 0$) is expected to be a Mott insulator. This is in contrast to the case of $Ba_2Tl_2CuO_6$, where we saw in Fig. 8.44 that the hybridization band does dip below the Fermi level.

We see from Fig. 8.48 that the density of states of $HgBa_2CuO_4$ at the Fermi level is quite small because only the $pd\sigma^*$ band passes through this level, and it is quite steep there. Near but below the Fermi level the DOS arises mainly from the Cu and O(1) atoms of the CuO_2 planes. The hybrid band is largely responsible for the DOS shown above E_F in Fig. 8.48.

The $HgBa_2CuO_{4+\delta}$ calculations were carried out for stoichiometric materials, i.e., $\delta = 0$, so no account was taken of the possibility of the presence of oxygen in the Hg plane at the $\frac{1}{2}\frac{1}{2}0$, $\frac{1}{2}\frac{1}{2}1$ site of O(3) listed in Table 7.6. The table mentions 11% oxygen occupancy of this site, and this would result in the presence of Cu^{3+} ions as observed, corresponding to hole doping in the CuO_2 planes and metallic behavior.

In addition calculations were carried out of the positron charge density that is

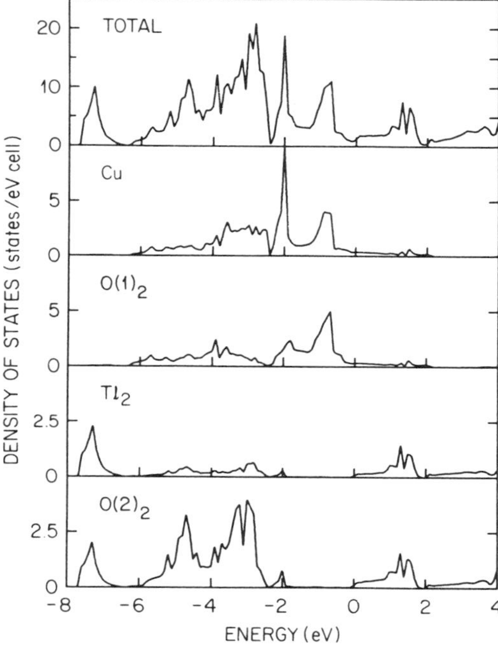

Figure 8.45 Total (top) and partial densities of states from the band structure of $Tl_2Ba_2CuO_6$. The Cu–O densities are similar to those from the other cuprates. The Tl_2 and O(2) spectral density extends just to E_F from above (Hamann and Mattheiss, 1988).

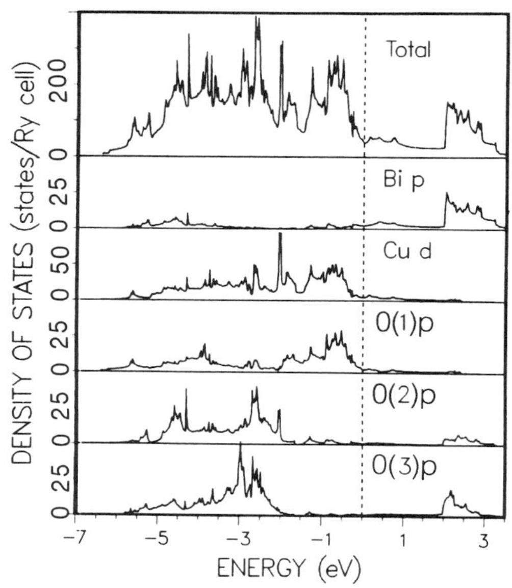

Figure 8.46 Total (top) and partial densities of states from the band structure of $Bi_2Sr_2CaCu_2O_8$. The distributions of Cu and O are similar to those from the other cuprates. Most of the Bi density lies above E_F, but two Bi derived bands do drop to or below E_F (Krakauer and Pickett, 1988).

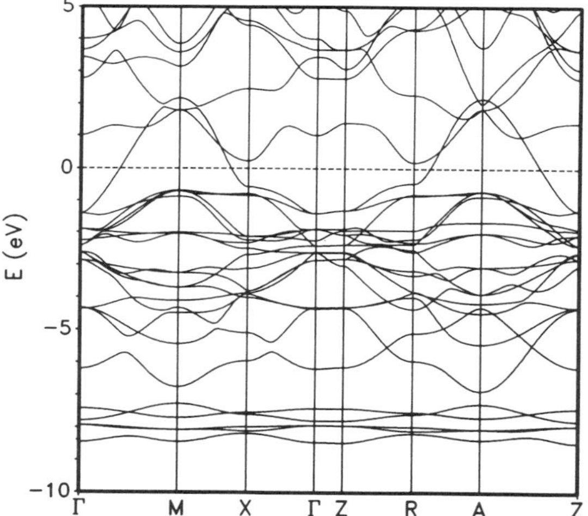

Figure 8.47 Energy bands of stoichiometric $HgBa_2CuO_4$ ($\delta = 0$) along the Γ–M–X–Γ directions defined by Fig. 8.3 for the $k_z = 0$ plane, and along the corresponding Z–A–R–Z directions of the $k_z = \frac{1}{2}$ plane in reciprocal space. The horizontal dashed line at $E = 0$ denotes the Fermi level (Singh *et al.*, 1993a).

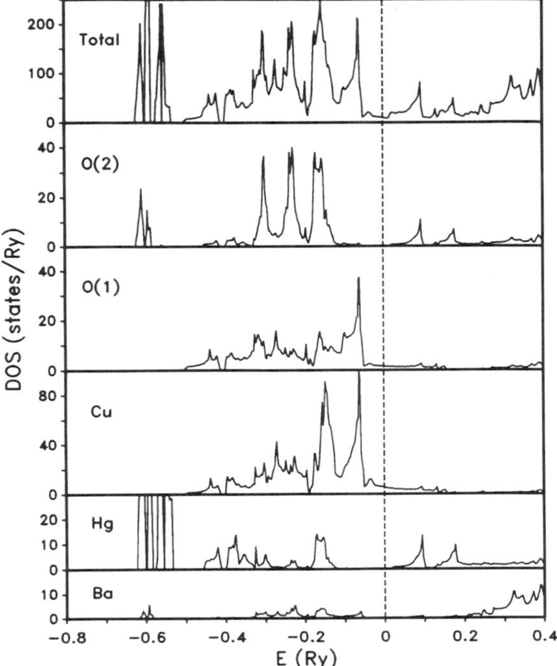

Figure 8.48 Total (top panel) and local site densities of states derived from the band structure of $HgBa_2CuO_4$ shown in Fig. 8.47, where 1 Ry = 13.6 eV, O(1) is in the CuO_2 layer, and O(2) is in the [O–Ba] layer (Singh *et al.*, 1993a).

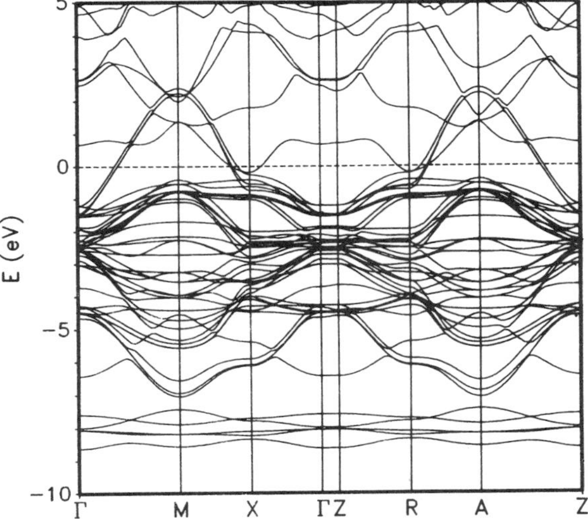

Figure 8.49 Energy bands of stoichiometric compound $HgBa_2Ca_2Cu_3O_8$ ($\delta = 0$) using the notation of Fig. 8.47. The five lowest energy bands shown are the Hg 5*d* manifold, and because of their flatness they produce the large Hg DOS shown in Fig. 8.50 at $E = 0.6$ Ry (Singh *et al.*, 1993b).

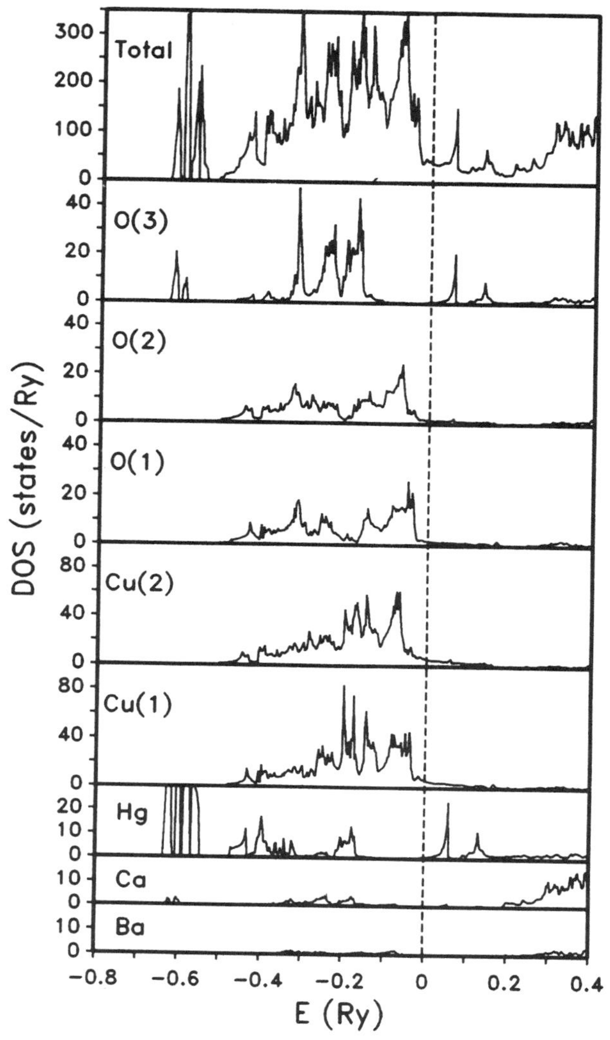

Figure 8.50 Total (top panel) and local site densities of states derived from the bands of $HgBa_2Ca_2Cu_3O_8$ shown in Fig. 8.49, where 1 Ry = 13.6 eV. O(1) is in the central Cu(1) layer, O(2) is in the Cu(2) layer, and apical O(3) is in the [O–Ba] layer (Singh *et al.*, 1993b).

predicted for angular correlation of annihilation radiation (ACAR) experiments carried out with positron irradiation, and the resulting contour plots are shown in Fig. 8.51 for two planes of the crystallographic unit cell. The charge density is smallest at the atom positions because the positrons avoid the positively charged atomic cores. It is largest at the empty O(3) site along the $\langle 110 \rangle$ direction.

Table 8.3 shows the calculated electric field gradients V_{ii} at all of the $HgBa_2CuO_4$ atom positions, where $V_{xx} = dE_x/dx$, etc. For each site Laplace's equation is obeyed, $V_{xx} + V_{yy} + V_{zz} = 0$, and at sites of tetragonal symmetry $V_{xx} = V_{yy}$. The gradients are smallest in magnitude for Cu in the planar direction, and largest for Hg.

The energy bands of the stoichiometric $HgBa_2Ca_2Cu_3O_{8+\delta}$ compound ($\delta = 0$) are shown in Fig. 8.49 and the density of states plots are in Fig. 8.50. There are some very strong similarities with $HgBa_2CuO_4$, but there are also some

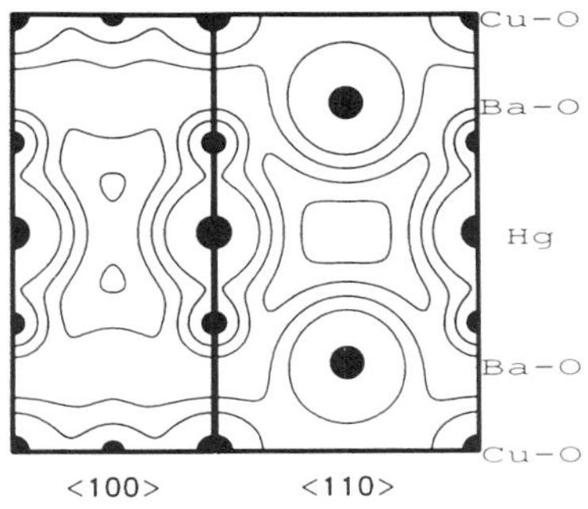

Figure 8.51 Contour plot of positron charge density normalized to one positron per $HgBa_2CuO_4$ unit cell determined by ACAR. Adjacent contours are separated by 0.0005 e^+/a.u. The density is largest in the voids (O(3) site) of the Hg layer. A uniformly distributed positron would have a density 0.00104 e^+/a.u. (Singh *et al.*, 1993a).

marked differences. There are three $pd\sigma^*$ well dispersed antibonding bands that cross the Fermi surface, one for each CuO_2 layer, and they are almost superimposed. The hybrid band dips below the Fermi surface at the X and R points and takes electrons from the $pd\pi^*$, band so there are less electrons in the CuO_2 planes, corresponding to the bands of these planes being less than half full. Thus this three-layer Hg compound is self-doped: the Hg layer produces hole doping of the CuO_2 layers. This is analogous to the $Tl_2Ba_2CuO_6$ case mentioned above.

Table 8.3 Electric Field Gradients at the Atom Sites of $HgBa_2CuO_4$ in Units of 10^{22} V/m^2. The x, y, z Directions Are along the Crystallographic a, b, c Directions, Respectively, but for the O(1) Site the x Direction is Toward the Neighboring Cu Atoms. Four of the Sites have Tetragonal Symmetry: $V_{xx} = V_{yy}$

Site	V_{xx}	V_{yy}	V_{zz}
Cu	0.096	0.096	−0.192
O(1)	1.31	−0.87	−0.44
Ba	−0.33	−0.33	0.66
O(2)	−0.70	−0.70	1.39
Hg	3.85	3.85	−7.70

We see from Fig. 8.50 that the density of states near but below the Fermi level arises from the three CuO_2 planes, while that slightly above E_F is due to Hg and the oxygen atom O(3) in the Hg plane. Subsequent band structure calculations carried out by Singh and Pickett (1994) showed that oxygen occupation of the $\frac{1}{2}\frac{1}{2}0$, $\frac{1}{2}\frac{1}{2}1$ site produces additional hole doping of the CuO_2 planes. They mentioned that measured values of δ are 0.06, 0.22, and 0.4 for the $n = 0$, $n = 1$, and $n = 2$ mercury compounds, corresponding to 0.12, 0.22, and 0.27 holes per Cu atom, respectively, for these three compounds.

XVII. FERMI LIQUIDS

In Section III we treated conduction electrons as noninteracting particles obeying Fermi–Dirac statistics, i.e., as constituting a Fermi gas. When the electrons continue to obey FD statistics and interact with each other in such a way that their properties remain close to those of a Fermi gas, they constitute what may be called a Fermi liquid.

Landau (1957) developed a method of taking account of electron–electron interactions in such a manner as to maintain a one-to-one correspondence between the states of a free-electron gas and those of the interacting electron system. The existence of such a one-to-one correspondence constitutes the usual definition of a Fermi liquid. In such a liquid the Pauli exclusion principle permits electrons at the Fermi surface to experience only momentum-changing collisions. Elementary excitations of quasiparticles and quasiholes correspond to those of a Fermi gas. In what are called marginal Fermi liquids, the one-to-one correspondence condition breaks down at the Fermi surface, but many of the properties continue to resemble those of a Fermi liquid (Baym and Pethick, 1991; Levine, 1993; Williams and Carbotte, 1991; Zimanyi and Bedell, 1993).

Transport properties of the cuprates, such as resistivity and the Hall effect, are often described by considering the conduction electrons as forming a normal Fermi liquid (Crow and Ong, 1990; Tsuei *et al.*, 1989). Varma *et al.* (1989) ascribed anomalies in these normal-state properties to marginal Fermi liquid behavior. Anderson (1987c) attributes the superconductivity of the cuprates to the breakdown of Fermi liquid theory and suggested the applicability of what are called Luttinger liquids (Anderson, 1990a, b). The Hubbard model (Si and Kotliar, 1993, Yarlagadda and Kurihara, 1993), and the *t-J* model, introduced in Section VII.E, has also been applied to Fermi liquids (Blok and Monien, 1993; Castellani *et al.*, 1991; Sá de Melo *et al.*, 1992; Morse and Lubensky, 1991; Z. Wang *et al.*, 1991; Pickett *et al.*, 1992) provide some perspective on Fermi liquids applied to the cuprates.

XVIII. FERMI SURFACE NESTING

The phenomenon of *Fermi surface nesting* occurs when the two sheets of the Fermi surface are parallel and separated by a common reciprocal lattice wave vector $\mathbf{G} = \mathbf{k}_{\text{nest}}$. We see from the bottom of Fig. 8.40 that the two sheets of the Fermi surface of $(La_{1-x}Sr_x)_2CuO_4$ on either side of the hole region are roughly parallel, and that the reciprocal lattice spacing across the region is close to the wave vector $(110)\pi/a$, of length $\sqrt{2}\,\pi/a$, which is also the distance from Γ to X (Crow and Ong, 1990; Emery, 1987a; Pickett, 1989; Virosztek and Ruvalds, 1991; Wang *et al.*, 1990). Two such nesting wave vectors are shown as bold arrows in the figure. The lanthanum compound approaches near-perfect nesting with $\mathbf{k}_{\text{nest}}$ equal to a reciprocal lattice vector.

If more electrons are added to the bands, the hole region contracts and the nesting vector turns out to be less than **G**. Decreasing the number of electrons enlarges the hole region, but this also leads to an instability in which the Fermi surface switches from the configuration of isolated hole regions shown at the bottom of the figure to the isolated electron regions in the top drawing. Further doping decreases the size of these electron regions until the spanning vector **k** is no longer a reciprocal lattice vector **G**, and the Fermi surface is no longer nested.

More generally, nesting may lead to instabilities in the Fermi surface even in the absence of this "switchover" instability. This can be seen from the viewpoint of perturbation theory by considering the fact that the energy denominator in the perturbation expression $1/(E_k - E_{k+\mathbf{G}})$ of Eq. (8.63) can approach zero over a wide range of k values for a nesting wave vector $\mathbf{k}_{\text{nest}}$. The result might be the generation of either a charge-density wave (CDW), a spin-density wave (SDW), or both.

The yttrium compound also exhibits a nesting feature, as may be seen from the parallel Fermi surfaces of Fig. 8.34. However, the spanning k-vectors are not close to the reciprocal lattice vectors, the energy denominators $(E_k - E_{k+\mathbf{G}})$ do not become small as in the lanthanum case, and no instability develops.

XIX. CHARGE-DENSITY WAVES, SPIN-DENSITY WAVES, AND SPIN BAGS

We noted in Sections XIV.C and XV.A contour plots of the charge density around the atoms of a superconductor can be determined from a band-structure calculation. Examples of such plots were given in Figs. 7.14 and 7.23 for the yttrium and lanthanum compounds, respectively. In this section we will be concerned with the charge density of an independent electron gas that is calculated self-consistently (cf. Section VI) using the Hartree–Fock approximation referred to in Chapter 1, Section II.

If this method is applied to an independent electron gas using a potential that is periodic in space and independent of spin, self-consistency can be obtained with solutions involving charge density which is periodic in space. In other words, the solution is a charge-density wave. The CDW can have periodicities that are incommensurate with the lattice spacings, and as noted above in Section XVIII, CDW is favored by Fermi surface nesting. A crystal structure distortion and the opening up of a gap at the Fermi surface accompany the formation of a CDW and stabilize it by lowering the energy. The presence of the gap can cause the material to be an insulator. Sufficiently high doping of La_2CuO_4 with Sr or Ba disables the nesting and destabilizes the CDW, thereby converting the material to a metal and making it superconducting.

In quasi-one-dimensional metals the CDW instability that leads to a structure distortion is called a *Peierls instability* (Burns, 1985), a phenomenon which has been discussed in the superconductor literature (e.g., Crow and Ong, 1990; Fesser *et al.*, 1991; Gammel *et al.*, 1990; Nathanson *et al.*, 1992; Ugawa *et al.*, 1991; Wang *et al.*, 1990). This CDW represents a nonmagnetic solution to the Hartree–Fock equations for an independent electron gas. Solutions with uniform charge densities that are fully magnetized can also be obtained. Overhauser (1960, 1962) assumed an exchange field that oscillates periodically in space and obtained a lower energy solution involving a nonuniform spin density whose magnitude or direction varies periodically in space. This solution corresponds to a spin density wave. Generally, the SDW is incommensurate because its periodicities are not multiples of the crystallographic lattice parameters, and, as noted, a nesting Fermi surface is favorable for the excitation of a SDW state (Burns, 1985). The antiferromagnetic insulator state of La_2CuO_4 mentioned at the end of Section XIV.D involves antiparallel nearest-neighbor spins, so that it is the SDW of maximal amplitude and shortest wavelength (Phillips, 1989a). The antiferromagnetism of itinerant electrons, as in a conductor, can be described as a spin-density wave (Kampf and Schrieffer, 1990; Wang *et al.*, 1990).

One attempt to understand the mechanism of high-temperature superconductivity has involved the use of what are called *spin bags* (Allen, 1990; Anisimov *et al.*, 1992; Goodenough and Zhou, 1990; Schrieffer *et al.*, 1988; Weng *et al.*, 1990). Consider the case of a half-filled band and an appropriately nested Fermi surface, so that the gap Δ_{SDW} extends over the Fermi surface and the system is an antiferromagnetic insulator. The presence of a quasiparticle alters the nearby sublattice magnetization and forms a region of reduced antiferromagnetic order. Such a region is called a spin bag because it is a metallic domain of depleted spin immersed in a surrounding SDW phase. This spin bag, which moves together with the quasiparticle, has a radius r_{bag} equal to the SDW coherence length, namely $r_{bag} = \hbar v_F/\Delta_{SDW}$ (Kampf and Schrieffer, 1990). Two spin bags attract each other to form a Cooper pair and, as a result, two holes tend to lower their energy by sharing one common bag.

XX. MOTT-INSULATOR TRANSITION

If the lattice constant of a conductor is continuously increased, the overlap between the orbitals on neighboring atoms will decrease, and the broad conduction bands will begin to separate into narrow atomic levels. Beyond a certain nearest-neighbor distance what is called a *Mott transition* occurs, and the electrical conductivity of the metal drops abruptly to a very small value. The metal has thus been transformed into what is called a Mott insulator.

One of the chronic failures of band-structure calculations has been an inability to obtain the observed insulating gaps in oxides such as FeO and CoO. In the prototype Mott insulator NiO the calculated gap has been observed to be an order of magnitude too small (Wang *et al.*, 1990). The

Mott insulator problem involves learning how to accurately predict the electronic properties of materials of this type.

For the large U case of Eq. (8.45) the ground state of the undoped ($x = 0$) Hubbard model system is a Mott insulator, sometimes called a Mott–Hubbard insulator. In the cuprates almost all of the Cu ions are in the $3d^9$ state, and there is one hole on each site of the system. There exists a large gap for excitations to levels where two antiparallel holes can occupy the same site. This gap of several eV, called the Mott–Hubbard gap, is too large to permit significant thermal excitation to occur. The system can, however, lower its energy by having individual holes make virtual hops to and from antiparallel neighbors. This virtual hopping process can be maximized by having the spin system assume an antiferromagnetically ordered configuration. Such ordering has been found experimentally in insulating members of the yttrium and lanthanum families of compounds (Birgeneau *et al.*, 1987; Crow and Ong, 1990). Some of the cuprate superconductors order antiferromagnetically at very low temperatures. The Mott transition in high-temperature superconductors has been widely discussed (Aitchison and Mavromators, 1989; Arrigoni and Strinati, 1991; Brandt and Sudbø, 1991; Cha *et al.*, 1991; Dai *et al.*, 1991; Hallberg *et al.*, 1991; Hellman *et al.*, 1991; Ioffe and Kalmeyer, 1991; Ioffe and Kotliar, 1990; Kaveh and Mott, 1992; Khurana, 1989; Mila, 1989; Millis and Coopersmith, 1991; Reedyk *et al.*, 1992a; Schulz, 1990; Spalek and Wojcik, 1992; Torrance *et al.*, 1992).

XXI. ANDERSON INTERLAYER TUNNELING SCHEME

P. W. Anderson (1958), as well as N. N. Bogoliubov (1958), provided alternate formulations of the basic ideas of the Bardeen *et al.* (1957) theory of superconductivity (vide Chapter 6) a year after its publication. More recently Anderson and his co-workers developed an interlayer tunneling model (Anderson, 1987a, 1994a; Anderson and Zou, 1988; Chakravarty *et al.*, 1993; Wheatley *et al.*, 1988; Zou and Anderson, 1988) which, he claims, provides the only mechanism that can "plausibly account for the high transition temperatures" of the cuprates (Anderson, 1994). In his "last paper of this character," Anderson (1991) lists many of the workers who contributed to the development of the ideas behind the theory. We will give a brief description of his approach and then present some supporting evidence for the model.

The theory is based on the two-dimensional character of the CuO_2 planes above T_c, and the three-dimensional behavior that sets in below T_c due to the condensation of hole bosons and the coherent Josephson-type tunneling of quasiparticle holon pairs between the layers. Strong electronic correlation effects suppress coherent interlayer single particle tunneling. Assuming an electron–phonon mechanism, an energy gap equation was derived. The gap is anisotropic with an s-state character. It is not a BCS type because it contains a phonon enhancement term involving the gap parameter $T_J(\mathbf{k})$, and T_c is predicted to be proportional to T_J. The theory has affinities with the Hubbard one-band model and a Luttinger Fermi liquid.

Anderson believes that his model is internally consistent and also compatible with the various experimental constraints. These constraints arise from the properties of the CuO_2 planes, despite the differences in the structure of several cuprates outside the planes. The anisotropy is large for some quantities such as the thermal conductivity (6:1) and the in-plane resistivity ($\rho_{ab}/\rho_c \approx 10^2 - 10^3$:1) together with its temperature dependence ($\rho_{ab} \approx T$, $\rho_c \approx 1/T$), and ρ_c exceeds the Mott limit. Other parameters such as the penetration depth, the coherence length, and the critical field have a less dramatic anisotropy ($\approx \sqrt{10:1}$).

Various measurements indicate a crossover from two- to three-dimensional behavior as the material is cooled below T_c. There is a lack of strong infrared absorption of c-axis polarized photons, and at low doping all cuprates become antiferromagnetic. These properties are compatible with the theory (Anderson, 1992; Hsu and Anderson, 1989).

A dialogue between Philip Anderson and Robert Schrieffer on the theory of high T_c was published in the June 1991 issue of *Physics Today*, and we conclude this section with a quotation from the introduction to the dialogue: Anderson's views have focused, in his own words, "on a non-Fermi-liquid normal state with separate spin and charge excitations, and deconfinement by interlayer Josephson tunneling as the driving force for the superconductivity," and Schrieffer, for his part, has pursued "the interplay between antiferromagnetism and superconductivity, extending the pairing theory beyond the Fermi-liquid regime in terms of spin polarons or 'bags'."

Anderson and Schrieffer agreed that understanding the physics of the normal state is a prerequisite for understanding high T_c, they both favor a one-band Hubbard model, and they both accept the importance of a single Fermi surface arising from a strongly hybridized admixture of copper 3d wavefunctions of $x^2 - y^2$ symmetry and oxygen $p\sigma$ orbitals. A major point of contention is Schrieffer's emphasis on spin bags which attract primarily within the two-dimensional CuO_2 plane and Anderson's stress on the role of hole pair hopping between neighboring planes.

XXII. COMPARISON WITH EXPERIMENT

A number of researchers have compared their experimental data with band-structure calculations. In this section we will present some of their results. Because of these comparisons we are able to estimate the accuracy of the calculations and to evaluate their usefulness for making predictions concerning new compounds. X-ray and photoemission (PE) spectroscopy provide the most straightforward comparison, so they will be discussed first, after which other experimental results that also give checks on the bands will be reported. An additional energy-band comparison is given by Takahashi *et al.* (1989).

X-ray and PE spectroscopy provide direct probes of the energy-band structure of a solid, as is explained in Chapter 15. The detected photons or electrons have energies that reflect the locations of the deep-lying energy levels characteristic of the core of the atom as well as the bands near the Fermi level that arise from the valence states of particular ions.

Figure 8.52 compares the experimentally determined x-ray emission bands (a and b) and optical density of states (c) of elemental niobium with the density of states calculated by three investigators (Anderson *et al.*, 1973; Geguzin *et al.*, 1973; Mattheiss, 1970). The four calculated spectra vary somewhat. Due to the lack of resolution the experimental spectra show the calculated bands A, B, and C as shoulders. Figure 8.53 shows that the x-ray K-emission data of the two $A15$ compounds V_3Si and V_3Ga (Kurmaev *et al.*, 1974) are close to the curves obtained from band-structure calculations (Klein *et al.*, 1978).

Figure 8.54 compares calculated and observed photoemission data of

$$YBa_2Cu_3O_{7-\delta}$$

that were obtained for three photon energies. The Cu d and O p valence electron contributions are shown separate from the total calculated spectrum. The agreement with experiment is qualitatively but not quantitatively good.

Figure 8.55 provides further comparison with experiment for $YBa_2Cu_3O_{6.9}$ and $(La_{0.925}Sr_{0.075})_2CuO_4$, respectively. The experimental results for direct photoemission are shown on the lower left of each figure ($E < 0$), and results for inverse photoemission are on the lower right ($E > 0$).

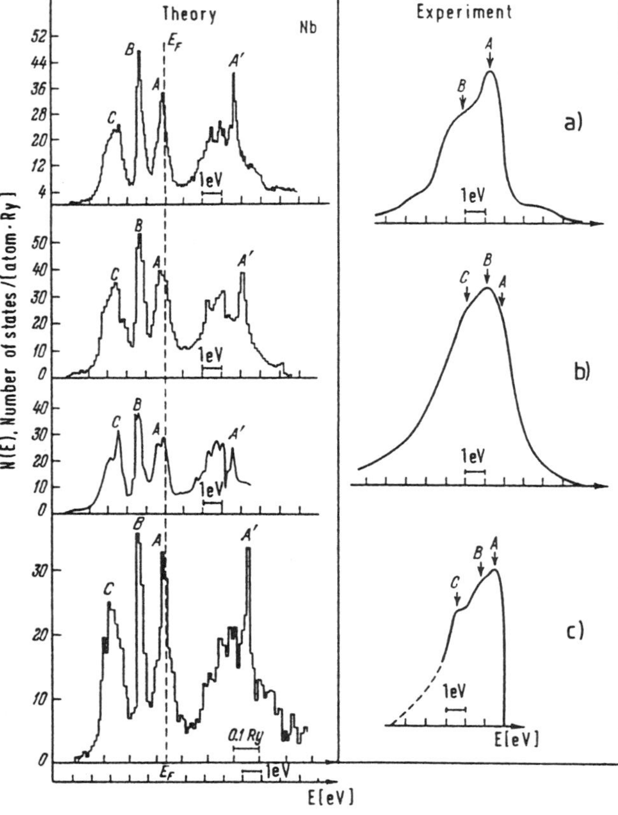

Figure 8.52 Four calculations of the density of states of elemental niobium are shown on the left. The experimental results on the right show (a) $M_{IV,V}$ x-ray emission band (Nemnovov *et al.*, 1969), (b) L_{III} x-ray emission band (Belash *et al.*, 1974), and (c) optical density of states determined by photoemission (Eastman 1969). (Figure from Vonsovsky *et al.*, 1982, p. 194.)

The curves may be compared with the densities of states calculated by several groups, shown above the experimental results. The calculations provide much more resolution of structure, and the agreement with experiment is qualitative. The band-structure calculations vary, as do experiments carried out using different energies, as shown. The shapes of the direct PE spectra of $YBa_2Cu_3O_{6.9}$ and

$$(La_{0.925}Sr_{0.075})_2CuO_4,$$

which are shifted slightly in E_F in the inset to the upper figure, are quite close.

Figure 8.56 shows that the measured magnetic susceptibility (Rehwald *et al.*, 1972) of the *A*15 compound Nb_3Sn is close to that calculated from the band structure (Klein *et al.*, 1979). We see from Fig. 8.57 that the calculated density of states of the $BaPb_{1-x}Bi_xO_3$ system does not compare favorably with DOS determinations from thermopower and specific heat measurements, but the calculated optical reflectivity of this system does agree well with experiment.

The 2D-ACAR variant of positron-annihilation spectroscopy that is dis-

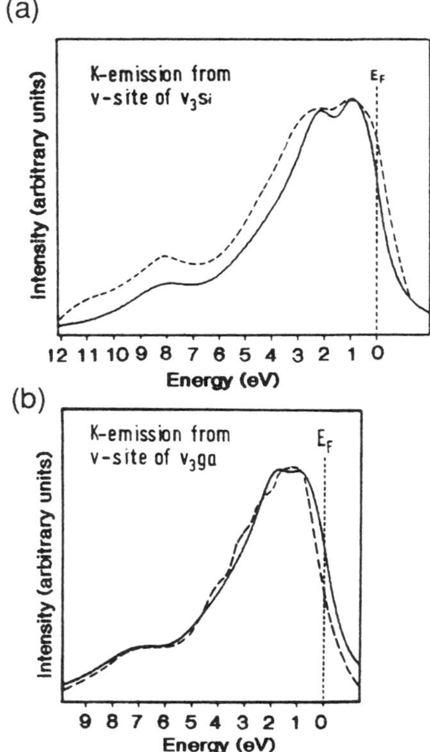

Figure 8.53 X-ray K emission bands of (a) V_3Si, and (b) V_3Ga deduced from a band-structure calculation (solid curve) and compared with experimental data (dashed curve) (Vonsovsky *et al.*, 1982, p. 305).

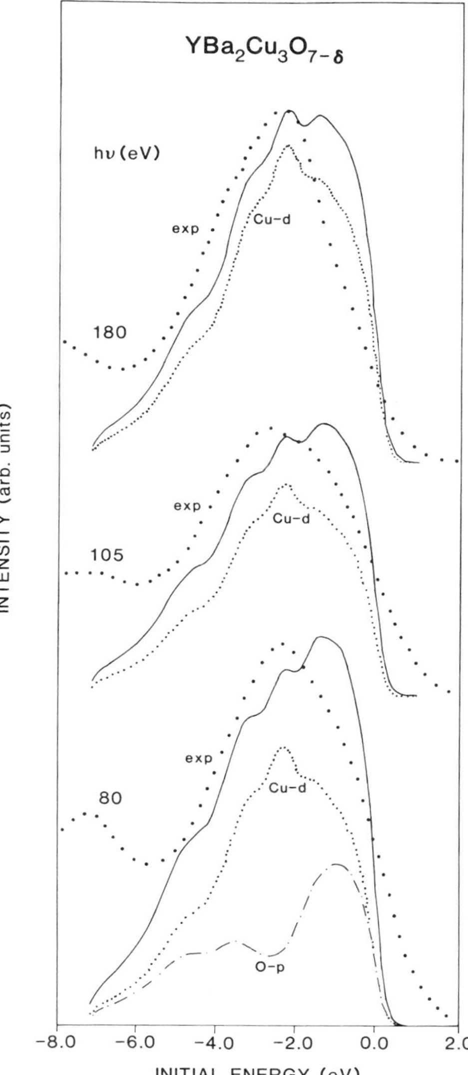

Figure 8.54 Comparison of experimental ultraviolet photoemission data (···) with the intensities calculated from the total band structure (——) and from the Cu–*d* band (– – –) alone. The O*p* contribution (–·–·–) is shown for the case of 80 eV. The different photon energies are labeled (Redinger *et al.*, 1987).

cussed in Chapter 15, Section VII, provides information on the topology of the Fermi surface. Tanigawa *et al.* (1988; cf. Klemm and Liu, 1991) provide a sketch of the electron and hole surfaces of $La_2CuO_{4-\delta}$ determined by this technique, as well as the momentum-space density plots that were employed for reconstructing the Fermi surface.

XXIII. DISCUSSION

In this chapter we presented the Hubbard model approach and then described band-structure calculations that have been carried out for several types of superconductors. The calculated bands, densities of states, and Fermi surface plots together provide a good explanation of the normal-state properties of the various materials. For example, they describe well the

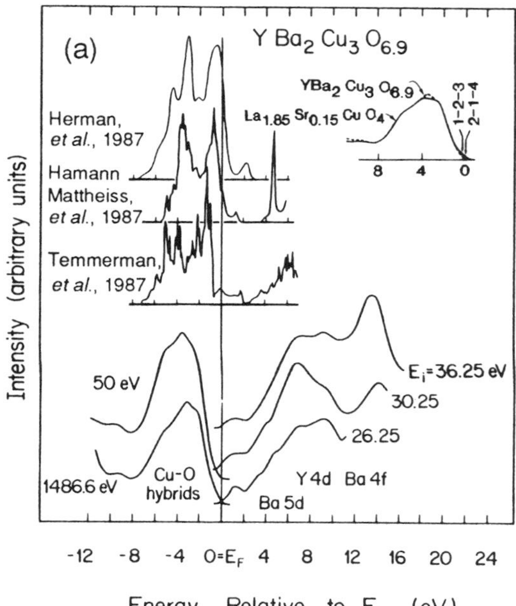

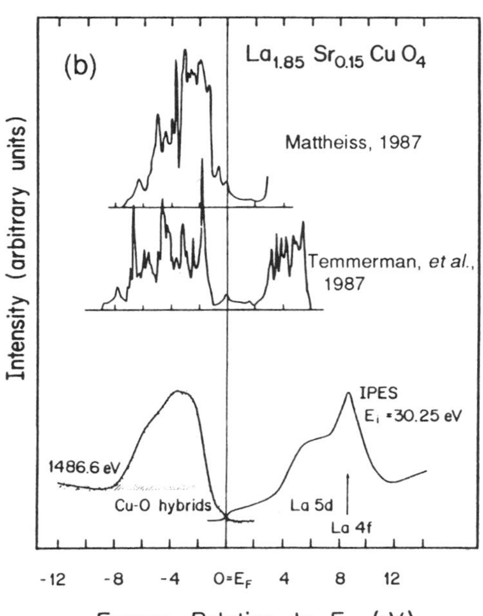

Figure 8.55 Comparison of experimental photoemission ($E < 0$) and inverse photoemission (IPES, $E > 0$) spectra for $YBa_2Cu_3O_{6.9}$ and $(La_{0.925}Sr_{0.075})_2CuO_4$ with calculated density-of-states plots. The inset at the upper right demonstrates the similarity of the valence band spectra from $YBa_2Cu_3O_{6.9}$ and $(La_{0.925}Sr_{0.075})_2CuO_4$ (Meyer *et al.*, 1988).

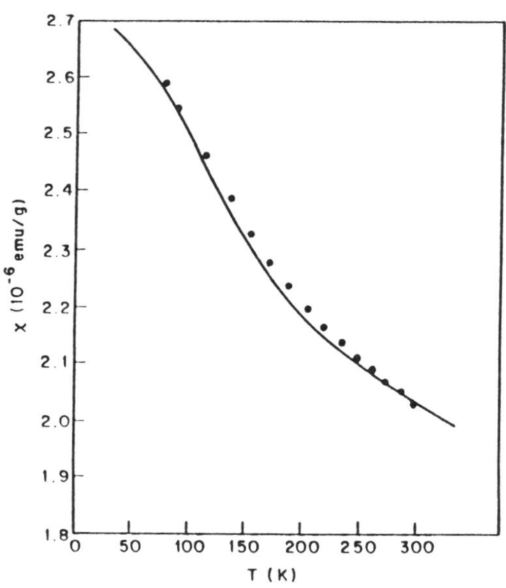

Figure 8.56 Temperature dependence of the magnetic susceptibility χ of Nb_3Sn deduced from a band-structure calculation (solid curve) compared with experimental data (Klein *et al.*, 1979).

planar nature of the conductivity in the high-temperature superconducting compounds and yield plots of electron density that elucidate the chemical bonding of the atoms. The calculated density of states at the Fermi level does not correlate well with known T_c values in various compounds, however.

In the previous section we saw that some of the results of band-structure calculations agree well with experiment, whereas others exhibit rather poor agreement. We also found differences in the calculated results published by different investigators. This would appear to indicate that these calculations are unreliable for estimating properties that have not yet been measured, and thus are more descriptive of superconductivity than predictive of it. In particular, investigators have not been successful in anticipating whether or not new compounds will be superconductors, and if so, with how high a T_c. Nevertheless, it is our belief that band structures do provide insight into the na-

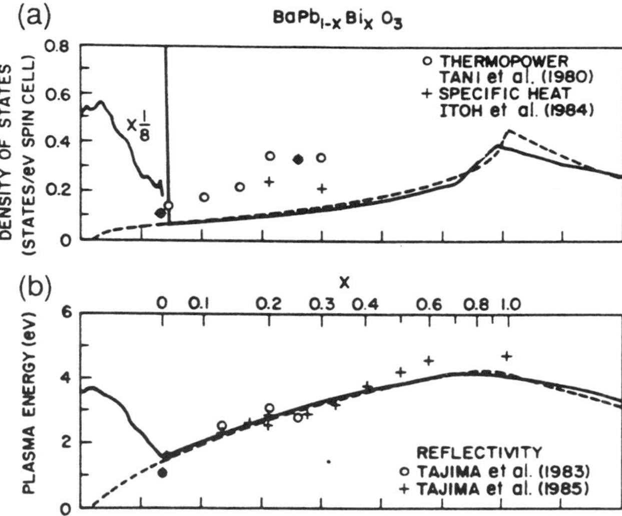

Figure 8.57 Comparison of $BaPb_{1-x}Bi_xO_3$ density of states with thermopower and specific heat data (a), and comparison of calculated plasma energy with optical reflectivity data (b) (Mattheiss, 1985, p. 121).

ture of superconductors that could not be obtained otherwise, in addition to elucidating their normal-state properties.

FURTHER READING

Some textbooks on solid-state physics contain good introductions to band theory (Aschroft and Mermin, 1976; Burns, 1985; Kittel, 1976). Calloway (1964) provides a more advanced, but still readable treatment.

There is a very extensive literature on Hubbard models (Auerbach, 1994; Dagotto, 1994; Konior, 1994; Kuzemsky, 1994), but the model has not had much impact on experimentalists. Micnas *et al*. (1990) examine the relationships between the BCS and Hubbard approaches. Hass (1989) surveyed the conventional and Hubbard band approaches, and compared them with available HTSC experimental data.

The band theory of transition metal elements and *A*15 compounds has been reviewed by Vonsovsky *et al*. (1982). The bands of high-temperature superconductors have been reviewed by Pickett (1989) and Wang (1990).

Fermi liquid theory is discussed in the books by Baym and Pethick (1991) and by Tilley and Tilley (1986, Chapter 9), and reviewed by Muzikar (1994) and Rainer and Sauls (1994).

Mott-insulator transitions have been reviewed by Spalek and Honig (1991). Electrons, phonons, and their interactions are surveyed by Cohen (1994).

PROBLEMS

1. Show how to derive Eq. (8.7) with the aid of the reciprocal lattice vector (8.5).
2. Draw the first, second, third, and fourth Brillouin zones for the case of Fig. 8.9.
3. Show that the energies for the Γ, *X*, and *M* points in the Brillouin zone of Fig. 8.3 are given by

$$E_\Gamma = 4E_0\left[n_x^2 + n_y^2\right] \qquad \text{Point } \Gamma$$

$$E_X = E_0\left[(2n_x+1)^2 + (2n_y)^2\right] \qquad \text{Point } X$$

$$E_M = E_0\left[(2n_x+1)^2 + (2n_y+1)^2\right] \qquad \text{Point } M.$$

4. Write down the energies $E_j(n_x, n_y)$ for each of the eight special points Γ, *X*, and *M* that appear in Fig. 8.4, giving every (n_x, n_y) combination for each energy. For example, for the lowest energy *M* point five bands come together, but only three of them (0,0), (−1,0), and (−1,−1) have distinct labels, so we have

$$E_M(0,0) = E_M(-1,0) = E_M(-1,-1) = 2E_0.$$

Do the other seven cases.

5. Identify one (n_x, n_y) label for each of the four bands α, β, γ, and δ at the top of Fig. 8.4 and find the energies of the bands α, γ, and δ at the next higher Γ point along with the energy of band β at the next higher point (point X).
6. Write down a pair of reciprocal lattice vectors $\mathbf{G}_1$ and $\mathbf{G}_2$ for which there exists a degeneracy in Eq. (8.18) that produces the energy gap (8.19) at the zone center point Γ of Fig. 8.3. Do the same for the points X and M
7. Draw Fig. 8.4 for the case of a periodic potential that produces small energy gaps.
8. Find the quantities E_F, N, n, and $D(E_F)$ for a one-dimensional free electron gas. What is the largest value that $D(E)$ can have, and for what energy? Compare these four one-dimensional results with their two- and three-dimensional counterparts.
9. Show that $\Gamma(\mathbf{k})$ is the Fourier transform of the overlap integral $\gamma(\mathbf{R})$ (Eq. (8.29)), where

$$\Gamma(\mathbf{k}) = \sum_{\mathbf{R}} e^{-i\mathbf{k}\cdot\mathbf{R}} \gamma(\mathbf{R}).$$

10. Calculate the expectation value

$$\int d^3 r \Psi_{k'}^*(r) H \Psi_k(r),$$

and show that the energy in the one-band case is given by Eq. (8.30):

$$\epsilon(k) = \epsilon_a + \frac{B(k)}{\Gamma(k)}.$$

11. Write down expressions similar to Eqs. (8.34) and (8.35) for a diatomic lattice of the NaCl or ZnS type. Why are equations of this type not valid for diatomic lattices in general?
12. Show that the only nonvanishing operations involving one-electron number operators are

$$n_+|+\rangle = |+\rangle \qquad n_-|-\rangle = |-\rangle$$
$$n_+|\pm\rangle = |\pm\rangle \qquad n_-|\pm\rangle = |\pm\rangle$$

13. Why does the first-order energy shift of the hopping term $\mathrm{H}_{\mathrm{hop}}$ (8.62) vanish?
14. What is the total number of electrons outside the closed shells in the compound $(La_{0.9}Sr_{0.15})_2CuO_{3.9}$? How many are contributed by each atom and how many by each ion? What is the average valence of copper?
15. Show that the denominator $E_j - E_i$ of the second-order term in Eq. (8.63) is equal to the Coulomb repulsion term U for the states $|j\rangle = |0\pm\rangle$ and $|\pm 0\rangle$.
16. Prove that the only nonvanishing terms of $\langle \Psi_i | \mathrm{H}_{\mathrm{hop}} | \Psi_j \rangle$ for $i \neq j$ are given by Eq. (8.66).
17. Show that the energy of the ferromagnetic states (8.60) is zero, as given by Eq. (8.68).
18. Prove Eq. (8.70), namely that $J = 4t^4/U$.
19. Show that in the Hubbard model the ionic states $|0\pm\rangle$ and $|\pm 0\rangle$ have the energy $U + 2t^2/U$, as shown in Fig. 8.16.
20. Show that in the RVB model the two ionic states $|00\rangle$ and $|\mathrm{ion}\rangle$ have the respective energies U and $U + 4t^2/U$, as shown in Fig. 8.16.
21. Show that in the RVB model the states $|\mathrm{VB}\rangle$ and $|\mathrm{ion}\rangle$ mix, and have the eigenenergies $E_i = \frac{1}{2}U \pm (\frac{1}{4}U^2 + 4t^2)^{1/2}$. Show that these reduce to those shown in Fig. 8.16 for the limit $U \gg t$.
22. Find the length of the distances Γ to Z (vertical), Γ to Z (horizontal), Γ to X, Γ to N, Γ to P, and X to P in the Brillouin zone of La_2CuO_4 (Fig. 8.37). Express the answers in $(\mathrm{nm})^{-1}$.
23. Explain why $S = 0.06$, 0.22, and 0.40 for the $n = 0$, 1, 2 Hg compounds $HgBa_2Ca_nCu_{n+1}O_{2n+4}$ give 0.12, 0.22, and 0.27, respectively, holes per Cu atom.

9

Type II Superconductivity

I. INTRODUCTION

In Chapter 2 we discussed Type I superconductors, which are superconductors that exhibit zero resistance and perfect diamagnetism. They are also perfect diamagnets for applied magnetic fields below the critical field B_c, and become normal in higher applied fields. Their coherence length exceeds their penetration depth so it is not energetically favorable for boundaries to form between their normal and superconducting phases. The superconducting elements, with the exception of niobium, are all Type I.

We showed in Chapter 5, Section XII, that when the penetration depth λ is larger than the coherence length ξ, it becomes energetically favorable for domain walls to form between the superconducting and normal regions. When such a superconductor, called Type II, is in a magnetic field, the free energy can be lowered by causing domains of normal material containing trapped flux to form with low-energy boundaries created between the normal core and the surrounding superconducting material. When the applied magnetic field exceeds a value referred to as the lower critical field, B_{c1}, magnetic flux is able to penetrate in quantized units by forming cylindrically symmetric domains called *vortices*. For applied fields slightly above B_{c1}, the magnetic field inside a Type II superconductor is strong in the normal cores of the vortices, decreases with distance from the cores, and becomes very small far away. For much higher applied fields the vortices overlap and the field inside the superconductor becomes strong everywhere. Eventually, when the applied field reaches a value called the upper critical field B_{c2}, the material becomes normal. Alloys and compounds exhibit Type II superconductivity, with mixed-type magnetic behavior and partial flux penetration above B_{c1}. Type II superconductors also have zero resistance, but their perfect diamagnetism occurs only below the lower critical field B_{c1}. The superconductors used in practical applications, which have relatively high transition temperatures, carry large currents and often operate in large magnetic fields, are all of Type II. Their properties will be described in this chapter. In the latter part of the chapter we will examine the properties of the vortices, discussing how they confine flux, how they interact and how they move about.

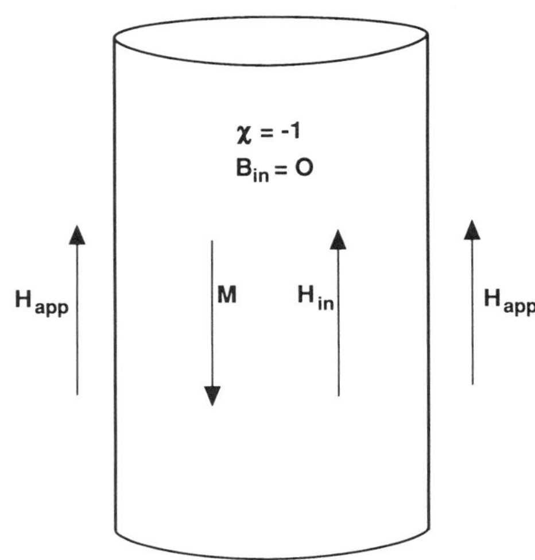

Figure 9.1 Internal fields produced inside a perfectly superconducting cylinder ($\chi = -1$) in an external magnetic field $\mathbf{B}_{app} = \mu_0 \mathbf{H}_{app}$ applied parallel to its axis. This arrangement is referred to as parallel geometry.

II. INTERNAL AND CRITICAL FIELDS

A. Magnetic Field Penetration

The general expression (1.69)

$$\mathbf{B} = \mu_0(\mathbf{H} + \mathbf{M}) \tag{9.1}$$

is valid both inside and outside a superconducting sample in an applied field. For simplicity we will examine the case of an elongated cylindrical superconductor with its axis in the direction of the applied magnetic field, as shown in Fig. 9.1. For this "parallel" geometry the boundary condition (1.74) requires the H fields outside ($H_{app} = B_{app}/\mu_0$) and inside (H_{in}) to be equal at the surface of the sample,

$$H_{app} = H_{in}. \tag{9.2}$$

If we apply Eq. (9.1) to the fields inside a Type I superconductivity and recall that $B_{in} = 0$, we obtain for the magnetization in the sample, with the aid of Eq. (9.2),

$$\mu_0 M = -B_{app}. \tag{9.3}$$

Above the critical field B_c the material becomes normal, the magnetization M becomes negligibly small, and $B_{in} \approx B_{app}$. This situation is indicated in Fig. 2.27 and plotted in Fig. 9.2, with the field $\mu_0 H_{in}$ below B_c indicated by a dashed line in the latter figure.

The corresponding diagram for a Type II superconductor has two critical fields, B_{c1}, the field where flux begins to penetrate, and B_{c2}, the field where the material becomes normal. For this case, again applying the boundary condition (9.2), the internal field and magnetization given by

$$\left.\begin{aligned} \mu_0 M &= -B_{app} \\ B_{in} &= 0 \end{aligned}\right\} \quad 0 \le B_{app} \le B_{c1}, \tag{9.4a}$$

$$\mu_0 M = -(B_{app} - B_{in}) \quad B_{c1} \le B_{app} \le B_{c2}, \tag{9.4b}$$

are shown plotted in Figs. 9.3 and 9.4, respectively. The dashed line ($\mu_0 H_{in}$) in Fig. 9.3 represents asymptote of B_{in} as it approaches B_{c2}. Also shown in these two figures is the thermodynamic critical field B_c defined by the expression

$$\int_{B_{c1}}^{B_c} (B_{app} + \mu_0 M) dB_{app} = \mu_0 \int_{B_c}^{B_{c2}} (-M) dB_{app}, \tag{9.5}$$

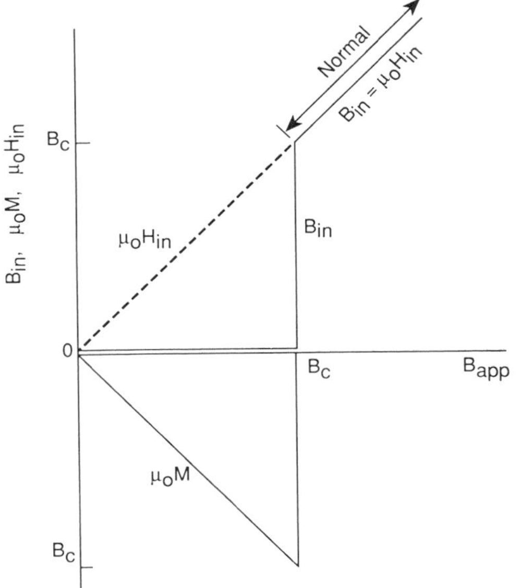

Figure 9.2 Internal fields B_{in} and H_{in} and magnetization M for an ideal Type I superconductor. Use is made of the permeability μ_0 of free space in this and the following two figures so that B_{in}, $\mu_0 H_{\text{in}}$, and $\mu_0 M$ have the same units, in accordance with Eq. (9.1).

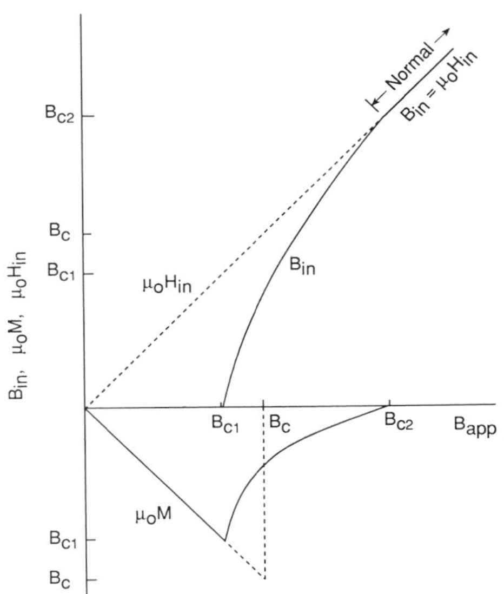

Figure 9.3 Internal fields B_{in} and H_{in} and magnetization M for an ideal Type II superconductor, using the notation of Fig. 9.2.

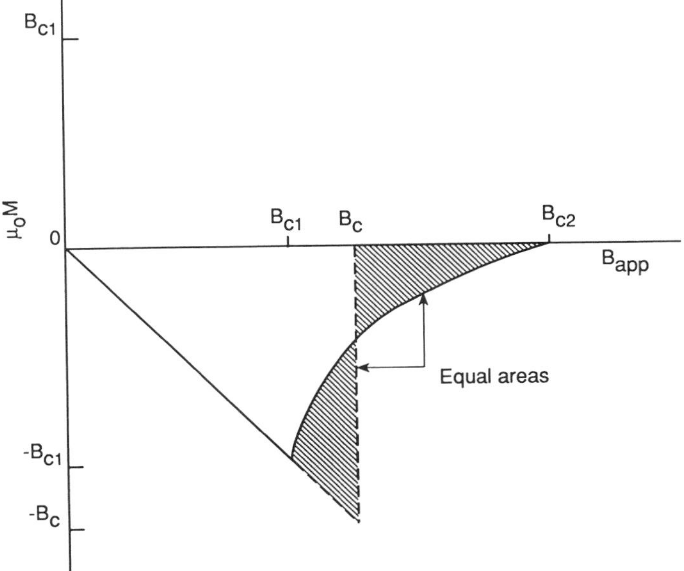

Figure 9.4 Dependence of magnetization M on the applied field for an ideal Type II superconductor. The equality of the areas separated by the thermodynamic critical field B_c is indicated.

which makes the two areas shown shaded in Fig. 9.4 equal. Figures 9.2, 9.3, and 9.4 are idealized cases; in practice, the actual magnetization and internal field curves are rounded, as indicated in Fig. 9.5.

We used the parallel geometry arrangement because it avoids the complications of demagnetization effects; these will be discussed later in Chapter 10, Sections X and XI. For this geometry the demagnetization factor N, which is a measure of these complications, is zero.

B. Ginzburg–Landau Parameter

In Chapter 5, Sections V and VII, respectively, we introduced two characteristic length parameters of a superconductor—the coherence length ξ and the penetration depth λ. Their ratio is the Ginzburg–Landau parameter κ,

$$\kappa = \frac{\lambda}{\xi}. \tag{9.6}$$

The density of super electrons n_s, which characterizes the superconducting state, increases from zero at the interface with a normal material to a constant value far inside, and the length scale for this to occur is the coherence length ξ. An external magnetic field **B** decays exponentially to zero inside a superconductor, with length scale λ. Figure 9.6 plots these distance dependences of n_s and **B** near the

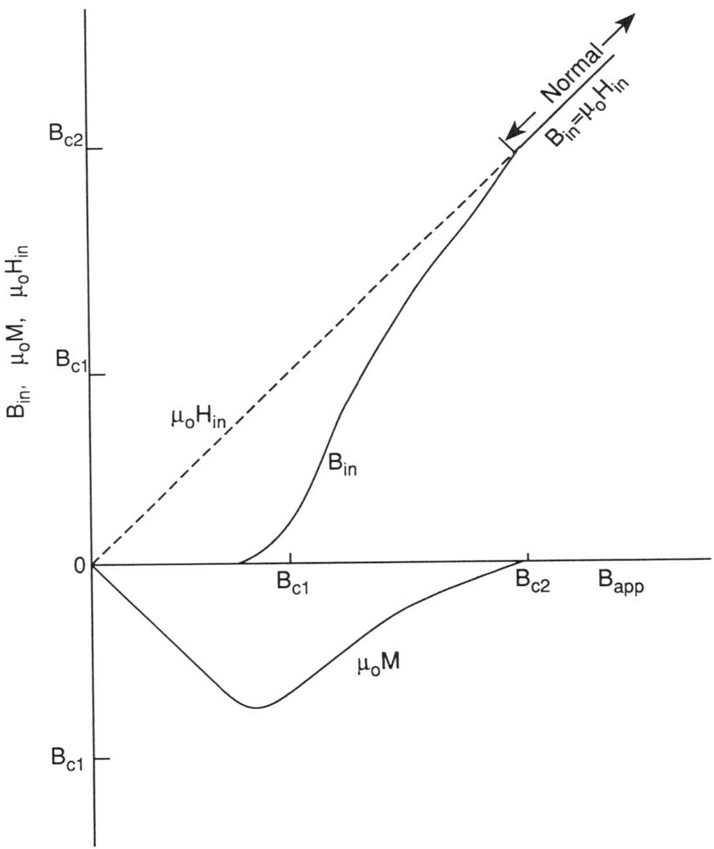

Figure 9.5 Dependence of internal magnetic field B_{in} and magnetization M on an applied field B_{app} for a nonidealized Type II superconductor in which the curves near the lower-critical field are rounded. This is in contrast to the idealized cases of Figs. 9.3 and 9.4 which exhibit abrupt changes in B_{in} and M when the applied field passes through the value B_{c1}.

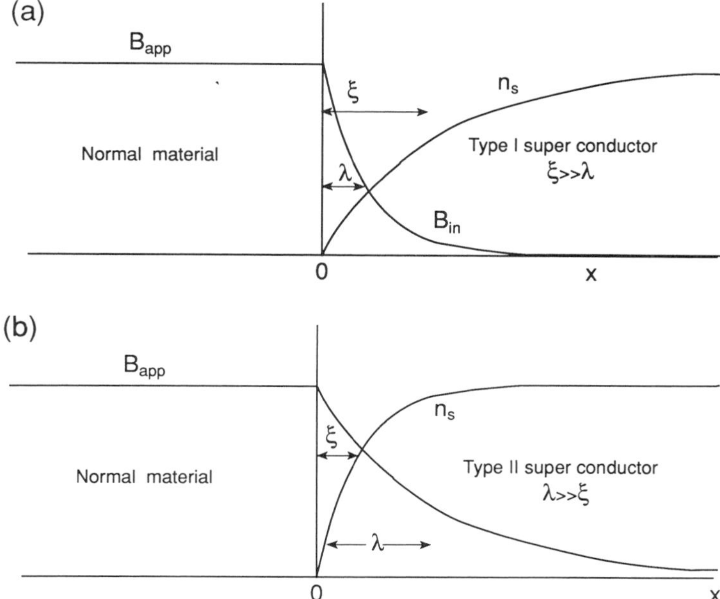

Figure 9.6 Increase in the number of superconducting electrons n_s and decay of the magnetic field B_{in} with distance x from the surface of the superconductor. The coherence length ξ and penetration depth λ associated with the change in n_s and B_{in}, respectively, are shown. (a) Type I superconductor, with $\xi > \lambda$, and (b) Type II superconductor, with $\lambda > \xi$.

boundary of a superconductor with a normal material for the two cases $\kappa < 1$ and $\kappa > 1$.

For a Type I superconductor the coherence length is the larger of the two length scales, so superconducting coherence is maintained over relatively large distances within the sample. This overall coherence of the superconducting electrons is not disturbed by the presence of external magnetic fields.

When, on the other hand, the material is Type II and the penetration depth λ is the larger of the two length parameters, external magnetic fields can penetrate to a distance of several or more coherence lengths into the sample, as shown in Fig. 9.6b. Thus, near the interface relatively large magnetic field strengths coexist with high concentrations of superconducting electrons. In addition, inside the superconductor we find tubular regions of confined magnetic flux (the vortices) as already noted. These have an effective radius of a penetration depth beyond which the magnetic field decays approximately exponentially to zero, in the manner illustrated in Fig. 9.7. As the applied field increases, more and more vortices form and their magnetic fields overlap, as indicated in Fig. 9.7b. Type II material is said to be in a mixed state over the range $B_{c1} < B_{app} < B_{c2}$ of applied fields.

Values of ξ, λ, and κ for a number of superconducting materials are given in Table 9.1. Superconductors are classified as Type I or Type II depending on whether the parameter κ is less than or greater than $1/\sqrt{2}$, respectively. We see from the table that all the elements (except for Nb) are Type I and that all the compounds are Type II, with the copper-oxide superconductors having the highest κ values, on the order of 100. Many of the data in the table are averages from several earlier compilations that do not agree very closely. The scatter in the values of ξ and λ listed in Table 9.2 for five of the elements is

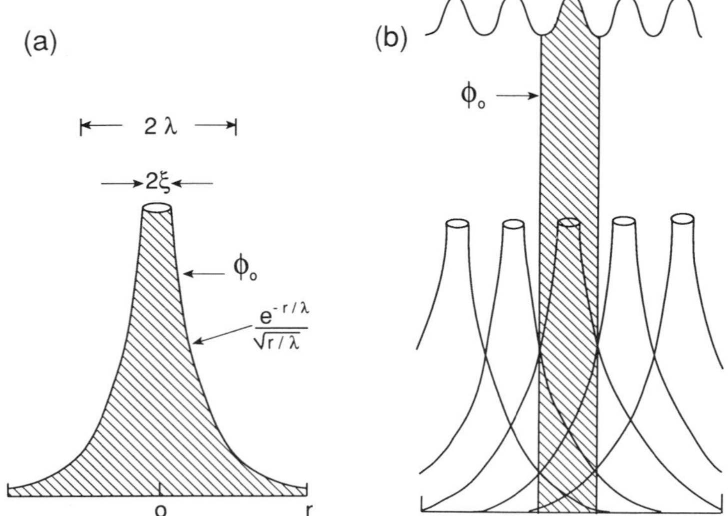

Figure 9.7 Sketch of (a) the magnetic field around an individual vortex, and (b) the field (top) from a group of nearby overlapping vortices (bottom). The coherence-length radius of the core, penetration-depth radius of the field outside the core, and decay of B at large distances are indicated for an individual vortex. The region of fluxoid quantization is cross-hatched for the individual vortex and for one of the overlapping vortices.

comparable to that of the high-temperature superconductors in Table III-1 of our earlier work (Poole *et al.*, 1988).

C. Critical Fields

In Chapter 4, Section VIII, and Chapter 5, Section IV, we saw that a Type I superconductor has a critical field B_c, and in Eq. (4.37) we equated the difference $G_n - G_s$ in the Gibbs free energy between the normal and the superconducting states to the magnetic energy $B_c^2/2\mu_0$ of this critical field,

$$G_n - G_s = \frac{B_c^2}{2\mu_0}. \tag{9.7}$$

Since this is a thermodynamic expression, B_c is called the *thermodynamic critical field*. Both Type I and Type II superconductors have thermodynamic critical fields. In addition, a Type II superconductor has lower- and upper-critical fields, B_{c1} and B_{c2}, respectively, given by

$$B_{c1} = \frac{\Phi_0 \ln \kappa}{4\pi\lambda^2}, \tag{9.8}$$

$$B_{c2} = \frac{\Phi_0}{2\pi\xi^2}. \tag{9.9}$$

These can be expressed in terms of the thermodynamic critical field B_c,

$$B_c = \frac{\Phi_0}{2\sqrt{2}\,\pi\lambda\xi}, \tag{9.10}$$

as follows:

$$B_{c1} = \frac{B_c \ln \kappa}{\sqrt{2}\,\kappa}, \tag{9.11}$$

$$B_{c2} = \sqrt{2}\,\kappa B_c. \tag{9.12}$$

It is also of interest to write down the ratio and the product of the two critical fields:

$$B_{c2}/B_{c1} = 2\kappa^2/\ln \kappa, \tag{9.13a}$$

$$(B_{c1}B_{c2})^{1/2} = B_c(\ln \kappa)^{1/2}. \tag{9.13b}$$

Figure 9.3 shows the position of the lower and upper critical fields as well as the thermodynamic critical field on the magne-

Table 9.1 Coherence Length ξ, Penetration Depth λ, and Ginzburg–Landau Parameter κ of Various Superconductors[a]

Material	T_c (K)	ξ (nm)	λ (nm)	κ (λ/ξ)	Source
Cd	0.56	760	110	0.14	Meservey and Schwartz (1969)
Al[a]	1.18	550	40	0.03	Table 9.2
In[a]	3.41	360	40	0.11	Table 9.2
Sn[a]	3.72	180	42	0.23	Table 9.2
Ta	4.4	93	35	0.38	Buckel (1991)
Pb[a]	7.20	82	39	0.48	Table 9.2
Nb[a]	9.25	39	50	1.28	Table 9.2
Pb–In	7.0	30	150	5.0	Orlando and Delin (1991)
Pb–Bi	8.3	20	200	10	Orlando and Delin (1991)
Nb–Ti	9.5	4	300	75	Orlando and Delin (1991)
Nb–N	16	5	200	40	Orlando and Delin (1991)
$PbMo_6S_8$ (Chevrel)	15	2	200	100	Orlando and Delin (1991)
V_3Ga (*A*15)	15	≈ 2.5	90	≈ 35	Orlando and Delin (1991)
V_3Si (*A*15)	16	3	60	20	Orlando and Delin (1991)
Nb_3Sn (*A*15)	18	3	65	22	Orlando and Delin (1991)
Nb_3Ge (*A*15)	23.2	3	90	30	Orlando and Delin (1991)
K_3C_{60}	19	2.6	240	92	Holczer *et al.* (1991)
Rb_3C_{60}	29.6	2.0	247	124	Sparn *et al.* (1992)
$(La_{0.925}Sr_{0.075})_2CuO_4$[b]	37	2.0	200	100	Poole *et al.* (1988)
$YBa_2Cu_3O_7$[b]	89	1.8	170	95	Poole *et al.* (1988)
HgBaCaCuO	126	2.3			Gao *et al.* (1993)
$HgBa_2Ca_2Cu_3O_{8+\delta}$	131			100	Schilling *et al.* (1994b)

[a] Figures are rounded averages from Table 9.2.
[b] Averages of the polycrystalline data from our earlier Table III-1 (1988).

tization curve, and Table 9.3 lists the critical fields of a few Type II superconductors. Section IV gives expressions similar to Eqs. (9.8)–(9.12) for anisotropic cases.

When the applied magnetic field is perpendicular to the surface of the superconductor, the upper critical field is truly B_{c2}. When it is parallel to the surface, however, it turns out that the superconducting state can persist in a thin surface sheath for applied surface fields up to the higher value $B_{c3} = 1.69\ B_{c2}$ (Saint-James and de Gennes, 1963; Saint-James *et al.*, 1969; Van Duzer and Turner, 1981, p. 319;

Table 9.2 Coherence Length ξ and Penetration Depth λ of Five Superconducting Elements from Several Reports[a]

Parameter	Al	In	Sn	Pb	Nb	Reference
Coherence length ξ, nm		360	175	510	39	Buckel (1991)
	1360	275	94	74		Huebener (1979)
	1600	360	230	90	40	Orlando and Delin (1991)
	1600	440	230	83	38	Van Duzer and Turner (1981)
Penetration depth λ, nm		24	31	32	32	Buckel (1991)
	51	47	52	47		Huebener (1979)
	50	65	50	40	85	Orlando and Delin (1991)
	16	21	36	37	39	Van Duzer and Turner (1981)

[a] Some of the data are reported averages from earlier primary sources. Table 9.1 lists rounded averages calculated from these values, with the entry $\xi = 510$ nm for Pb excluded.

Table 9.3 Critical Fields of Selected Type II Superconductors[a]

Material	T_c (K)	B_{c1} (mT)	B_c (mT)	B_{c2} (T)	Reference
Nb wire, RRR = 750	9.3	181.0	0.37	2.0	Roberts (1976)
Nb wire, cold-drawn	9.3	248.0		≈ 10.0	Roberts (1976)
$In_{0.95}Pb_{0.05}$ (alloy)	3.7	31.8	37.5	0.049	Roberts (1976)
$Mo_{\approx 0.1}Nb_{\approx 0.9}$ (alloy)	6.4	29.0	78.5	0.414	Roberts (1976)
$Mo_{0.66}Re_{0.34}$ (alloy)	11.8	38.1	0.113		Roberts (1976)
$Nb_{0.99}Ta_{0.01}$ (alloy)	8.8	173.0	20.4	0.445	Roberts (1976)
Nb–Ti	9.5			13.0	Orlando and Delin (1991); Van Duzer and Turner (1981)
CTa (NaCl Structure)	≈ 10.0	22.0	81.0	0.46	Roberts (1976)
Nb–N (NaCl Structure)	16.0	9.3		15.0	Orlando and Delin (1991); Roberts (1976)
Cr_3Ir (*A*15)	0.75	16.8		1.05	Roberts (1976)
V_3Ge (*A*15)	6.8			≈ 5.0	Roberts (1976)
V_3Ga (*A*15)	15.0			23.0	Orlando and Delin (1991); Van Duzer and Turner (1981)
V_3Si (*A*15)	16.0	55.0	670.0	23.0	Roberts (1976)
Nb_3Sn (*A*15)	18.2	35.0	440.0	23.0	Roberts (1976)
Nb_3Ge (*A*15)	23.1			37.0	Orlando and Delin (1991); Van Duzer and Turner (1981)
HfV_2 (Laves)	9.2	187.0		21.7	Vonsovsky *et al.* (1982, p. 376)
$(Hf_{0.5}Zr_{0.5})V_2$ (Laves)	10.1	197.0		28.3	Vonsovsky *et al.* (1982, p. 376)
ZrV_2 (Laves)	8.5	219.0		16.5	Vonsovsky *et al.* (1982, p. 376)
$NbSe_2$	7.2	7.2	204.0	17.4	Roberts (1976)
$PbMo_6Se_8$ (Chevrel)	3.8			3.8	Vonsovsky *et al.* (1982, p. 420)
$LaMo_6S_8$ (Chevrel)	≈ 6.5			5.4	Vonsovsky *et al.* (1982, p. 420)
$LaMo_6Se_8$ (Chevrel)	11.0			44.5	Vonsovsky *et al.* (1982, p. 420)
$SnMo_6S_8$ (Chevrel)	11.8			34.0	Vonsovsky *et al.* (1982, p. 420)
$PbMo_6S_8$ (Chevrel)	15.0			60.0	Orlando and Delin (1991)
$U_{0.97}Th_{0.03}Be_{13}$ (heavy fermion)	0.35	4.0			Rauchschwalbe *et al.* (1987)
UPt_3 (heavy fermion)	0.46			1.9	Schenström *et al.* (1989)
$U_{0.985}La_{0.015}Be_{13}$ (heavy fermion)	0.57			3.8	Dalichaouch *et al.* (1991)
UBe_{13} (heavy fermion)	0.9			6.0	Maple *et al.* (1984)
K_3C_{60} (buckyball)	19.0	13.0	0.38	32.0	Boebinger *et al.* (1992); Foner *et al.* (1992); Holczer *et al.* (1991); C. E. Johnson *et al.* (1992); Z. H. Wang *et al.* (1993)
Rb_3C_{60} (buckyball)	29.6	12.0	0.44	57.0	Foner *et al.* (1992); C. E. Johnson *et al.* (1992); Sparn *et al.* (1992)
$HgBa_2CuO_{4+\delta}$	99		10^3	> 35	Thompson *et al.* (1993)

[a] Some of the data are averages from more than one source.

Walton *et al.*, 1974; Yuan and Whitehead, 1991).

The temperature dependence of the thermodynamic critical field B_c is given in Chapter 2, Section XIII. The lower and upper critical fields of Type II superconductors have a similar temperature dependence. The fields B_{c2} needed to extinguish Type II superconductivity are much larger than those B_c that are sufficient for extinguishing the Type I variety. These large upper-critical fields make Type II superconductors suitable for magnet applications.

Quoted upper-critical fields are usually given for 4.2 K or for extrapolations to 0 K. Values of technological interest are the 4.2 K fields for the low-temperature superconductors and the 77 K fields for the high-temperature superconductors. For example, B_{c2} for the standard magnetic material NbTi is 10 T at 4 K and can be 30 T or more at 77 K for high-temperature superconductors, as shown in Fig. 9.8 (Fischer, 1978; Newhouse, 1969, p. 1268; Vonsovsky *et al.*, 1982, p. 431). Theoretical articles have appeared that discuss upper-critical fields (e.g., Brézin *et al.*, 1990; Estrera and Arnold, 1989; Norman, 1990; Pérez-González and Carbotte, 1992; Pérez-González *et al.*, 1992; Santhanam and Chi, 1988; Theodorakis and Tesanovic, 1989).

III. VORTICES

We have seen that an applied magnetic field B_{app} penetrates a superconductor in the mixed-state, $B_{c1} < B_{app} < B_{c2}$. Penetration occurs in the form of tubes, called vortices (see Fig. 9.9), which serve to confine the flux (Abrikosov 1957; Belitz 1990). The highest field is in the core,

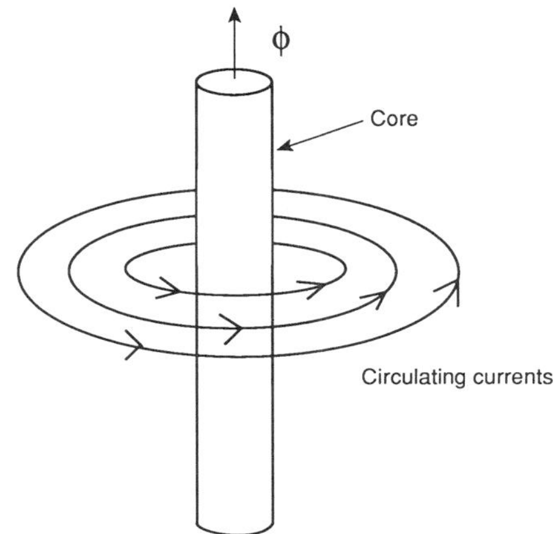

Figure 9.9 Sketch of shielding currents circulating around a vortex core.

which has a radius ξ. The core is surrounded by a region of larger radius λ within which magnetic flux and screening currents flowing around the core are present together, as shown in Fig. 9.9. The current density J_s of these shielding currents decays with distance from the core in an approximately exponential manner. Analytical expressions for the distance dependence of B and J are derived in the fol-

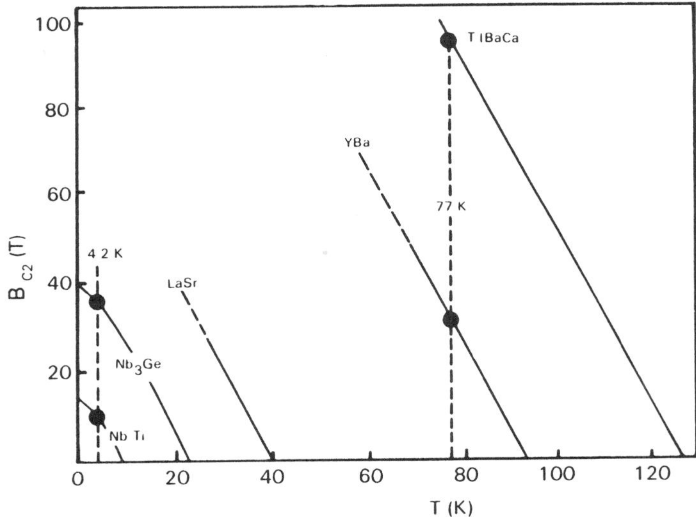

Figure 9.8 Comparison of the temperature dependence of the upper-critical fields B_{c2} of Nb–Ti, Nb_3Ge, LaSrCuO, YBaCuO, and TlBaCaO. The slopes are close to the Pauli limit 1.83 T/K (Poole *et al.* 1988, p. 8).

lowing section for the high-kappa ($\kappa \gg 1$) approximation.

A. Magnetic Fields

Equation (5.39) provides us with an expression for the magnetic flux passing through a region,

$$\int \mathbf{B} \cdot d\mathbf{S} + \frac{\mu_0 m^*}{e^{*2}} \int \frac{\mathbf{J} \cdot d\mathbf{l}}{|\phi|^2} = n\Phi_0, \quad (9.14)$$

where n is the number of vortex cores enclosed by the integrals. For isolated vortex $n = 1$ because it is energetically more favorable for two or more quanta to form separate vortices rather than to coexist together in the same vortex. Integration of **B** over the cross-sectional area of an isolated vortex can be taken from $r = 0$ to $r = \infty$, so the surface integral is numerically equal to the flux quantum Φ_0,

$$\int \mathbf{B} \cdot d\mathbf{S} = \Phi_0, \quad (9.15)$$

and the line integral vanishes because **J** becomes negligibly small at large distances. The quantum condition (9.15) fixes the total magnetic flux in an isolated vortex at one fluxoid, including flux in the core and in the surrounding layer. The possibility of vortices containing two or more quanta has been discussed (Buzdin, 1993; Sachdev, 1992; Tokuyasu *et al.*, 1990).

As the applied magnetic field increases, the density of vortices increases and they begin to overlap, making the vortex–vortex nearest-neighbor distance less than the penetration depth. The high-density case can be treated by assuming that the magnetic field at any point is a linear superposition of the fields from all of the overlapping vortices. At high densities B_{in} becomes very large and the variation of the field in the space between the cores becomes very small, as indicated in Fig. 9.10. Nevertheless, the quantization condition still applies and each vortex has, on average, one quantum of flux Φ_0 associated with it, as indicated in Fig. 9.7b. For a regular two-dimensional lattice arrangement of vortices, Eq. (9.15) holds as long as integration is carried out over the vortex unit cell; the line integral (9.14) of the current density vanishes when it is taken around the periphery of this cell.

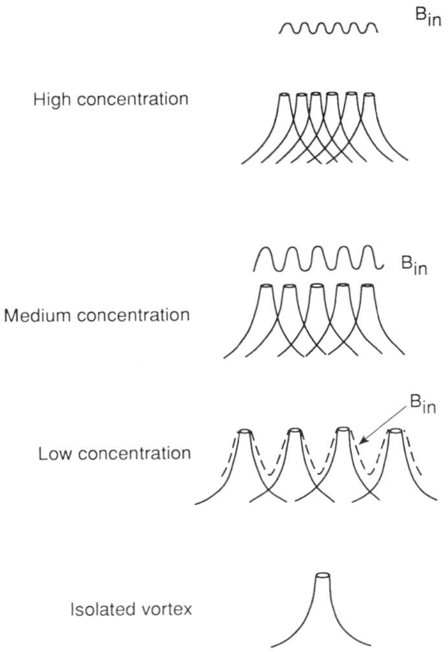

Figure 9.10 Sketch showing how the magnetic field B_{in} inside a superconductor increases as the concentration of vortices increases and their fields increasingly overlap.

When λ is much larger than ξ, as is the case with the high-temperature superconductors, there is considerable overlap of vortices throughout most of the mixed-state range, and the magnetic flux is present mainly in the surrounding region, rather than in the actual cores.

There is no limit to the length of a vortex. Along the axis, which is also the applied field direction, the magnetic field lines are continuous. Thus the flux does not begin and end inside the superconductor, but instead enters and leaves at the superconductor surface, which is also where the vortices begin and end. This is illustrated in Fig. 9.11.

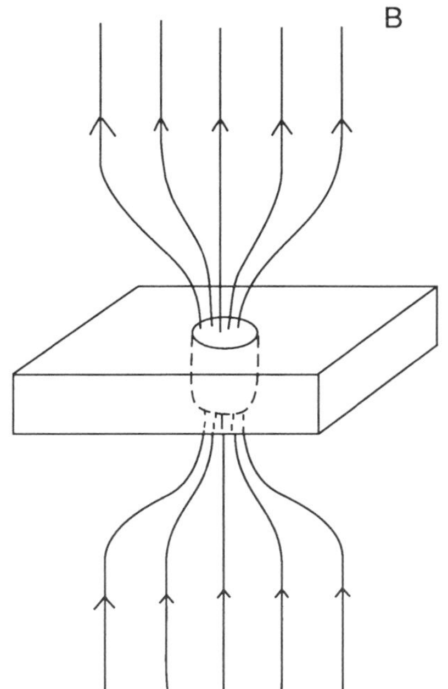

Figure 9.11 Passage of external magnetic field lines through a flat-plate superconductor in the region of a vortex.

We have seen that a vortex has a core radius equal to the coherence length ξ and a surrounding outer region with radius equal to the penetration depth λ. Such an entity can only exist in a Type II superconductor, where λ is greater than ξ. A vortex does not exist, and is not even a meaningful concept under Type I conditions ($\xi > \lambda$).

Scanning tunneling microscope studies of superconducting surfaces (H. F. Hess *et al.*, 1989, 1990, 1991; Karrai *et al.*, 1992; Renner *et al.*, 1991) reveal an enhancement of the differential tunneling conductance (see Chapter 13, Section V) in the vortex core. This has been attributed to the presence of bound states of quasiparticles in the core. As the magnetic field increases, additional vortices form accompanied by the breakup of Cooper pairs, and more and more quasiparticles or normal electrons become localized in the vortex cores (Daemen and Overhauser, 1989; Gygi and Schlüter, 1990a, b, 1991; Klein, 1989, 1990; Overhauser and Daemen, 1989; Shore *et al.*, 1989; Ullah *et al.*, 1990).

B. High-Kappa Approximation

To obtain a description of vortices that is more quantitative in nature as opposed to the rather qualitative description presented in the previous section, it will be helpful to have a closed-form expression for the distance dependence of the confined magnetic fields. For the high-κ limit, $\lambda \gg \xi$, which is valid for the copper-oxide superconductors that typically have $\kappa \approx 100$, we can make use of the Helmholtz equations that were derived from the London formalism in Chapter 5, Section IX. The vortex is assumed to be infinitely long and axially symmetric so that there are no z or angular dependences of its field distribution. The problem is thus equivalent to the two-dimensional problem of determining the radial dependences.

The magnetic field of the vortex is in the z direction, and its radial dependence outside the vortex core is obtained from the Helmholtz Eq. (5.67). Here we will write the Helmholtz equation in cylindrical coordinates for the two-dimensional case of axial symmetry without assuming any angular dependence,

$$\frac{\lambda^2}{r}\frac{d}{dr}\left(r\frac{d}{dr}\right)\mathbf{B} - \mathbf{B} = 0, \qquad (9.16)$$

This equation has an exact solution,

$$B(r) = \frac{\Phi_0}{2\pi\lambda^2}K_0(r/\lambda), \qquad (9.17)$$

where $K_0(r/\lambda)$ is a zeroth-order modified Bessel function. With the aid of Eq. (9.8) this can be written

$$B(r) = B_{c1}\frac{K_0(r/\lambda)}{\frac{1}{2}\ln(\kappa)}. \qquad (9.18)$$

To obtain the current density we substitute Eq. (9.17) in the Maxwell equation for $\mathbf{B}_{\text{in}}$,

$$\nabla \times \mathbf{B}_{\text{in}} = \mu_0 \mathbf{J}_s, \qquad (9.19)$$

to obtain

$$J_s(r) = \frac{\Phi_0}{2\pi\mu_0\lambda^3} K_1(r/\lambda) \quad (9.20)$$

$$= J_{01} \frac{K_1(r/\lambda)}{\frac{1}{2}\ln(\kappa)}, \quad (9.21)$$

where $K_1(r/\lambda)$ is a first-order modified Bessel function, and the characteristic current density J_{c1} is defined in analogy with Eq. (2.51),

$$J_{c1} = B_{c1}/\mu_0\lambda. \quad (9.22)$$

The function $K_1(r/\lambda)$ results from differentiation of Eq. (9.19), as expected from the modified Bessel function recursion relation $K_1(x) = -dK_0(x)/dx$ (Arfken, 1985, p. 614). The current density also satisfies the Helmholtz equation, Eq. (5.68), expressed in cylindrical coordinates (Eq. (9.16)) as

$$\frac{\lambda^2}{r}\frac{d}{dr}\left(r\frac{d}{dr}\right)\mathbf{J}_s + \mathbf{J}_s = 0. \quad (9.23)$$

Figure 9.12 compares the distance dependences of the modified Bessel functions $K_0(r/\lambda)$ and $K_1(r/\lambda)$ associated with $B(r)$ and $J_s(r)$, respectively.

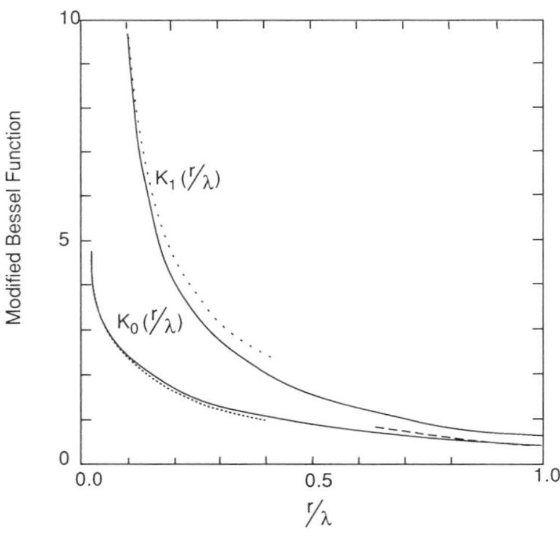

Figure 9.12 Comparison of the distance dependence of the zero-order (K_0) and first-order (K_1) modified Bessel functions associated with the magnetic field and current density, respectively, of a vortex. Asymptotic behavior at short distances (·······) is indicated. Both modified Bessel functions have the same large-distance asymptotic behavior (– – – – – – –), as shown (Aktas, 1993).

These modified Bessel functions have asymptotic behaviors at small radial distances,

$$K_0\left(\frac{r}{\lambda}\right) \approx \ln\left(\frac{2\lambda}{r}\right) - \gamma \qquad r \ll \lambda, \quad (9.24)$$

$$\approx \ln\left(\frac{1.123\lambda}{r}\right) \qquad r \ll \lambda, \quad (9.25)$$

$$K_1\left(\frac{r}{\lambda}\right) \approx \frac{\lambda}{r} \qquad r \ll \lambda, \quad (9.26)$$

where $\gamma = 0.57721566\ldots$ is the Euler–Mascheroni constant (Arfken, 1985, p. 284) and the factor $2e^{-\gamma} = 1.123$. These expressions show that $K_1(r) \gg K_0(r)$ near the core, where $r \ll \lambda$, as indicated in Fig. 9.12. At large distances the corresponding expressions are

$$K_0\left(\frac{r}{\lambda}\right) \approx \frac{\exp(-r/\lambda)}{(2r/\pi\lambda)^{1/2}} \qquad r \gg \lambda \quad (9.27)$$

$$K_1\left(\frac{r}{\lambda}\right) \approx \frac{\exp(-r/\lambda)}{(2r/\pi\lambda)^{1/2}} \qquad r \gg \lambda. \quad (9.28)$$

Figure 9.12 compares the asymptotic behaviors with the actual functions $K_0(r/\lambda)$ and $K_1(r/\lambda)$. These large-distance expressions permit us to express the magnetic field and current density far from the core in the form

$$B \approx B_{c1}\frac{(2\pi)^{1/2}}{\ln(\kappa)}\frac{\exp(-r/\lambda)}{(r/\lambda)^{1/2}} \qquad r \gg \lambda, \quad (9.29)$$

$$J_s \approx J_c\frac{(2\pi)^{1/2}}{\ln(\kappa)}\frac{\exp(-r/\lambda)}{(r/\lambda)^{1/2}} \qquad r \gg \lambda. \quad (9.30)$$

We see from Eqs. (9.24) and (9.26) that both B and J_s are singular at $r = 0$. Since the core is so small in the high-kappa approximation, it is appropriate to remove the singularity by assuming that the magnetic field in the core is constant with the value B(0) given by Eq. (9.17) for $r = \xi$. Even if the mathematical singularity were not removed, the total flux would still remain finite as $r \to 0$, as is proven in Problem 5. In Problem 3 we derive the following expression for the fraction of the total flux of the vortex that is present in the core:

$$\Phi_{\text{core}} \approx (\Phi_0/2\kappa^2)(\ln 2\kappa + \tfrac{1}{2} - \gamma). \quad (9.31)$$

Figure 9.13 sketches the dependence of $\Phi_{\text{core}}/\Phi_0$ on κ.

Since the magnetic field in the sample is confined to vortices, the total flux is Φ_0 times the number of vortices, and the average internal field B_{in}, given by

$$B_{\text{in}} = N_A \Phi_0, \quad (9.32)$$

is proportional to N_A, the number of vortices per unit area. For high applied fields much larger than B_{c1} but, of course, less than B_{c2}, the internal field is approximately proportional to the applied field (see Fig. 9.3), and therefore the density of vortices becomes approximately proportional to the applied field.

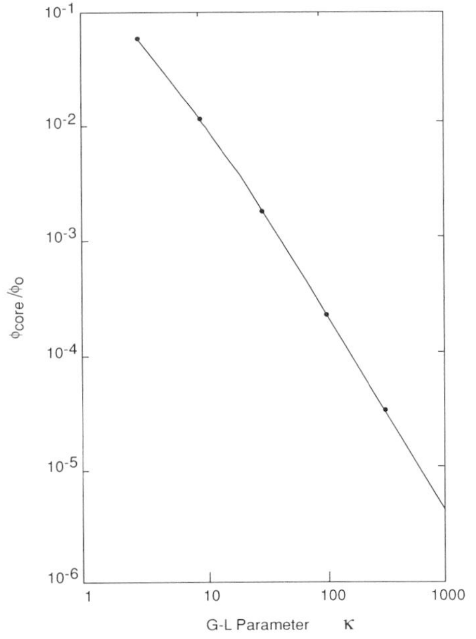

Figure 9.13 Fraction of the total flux quantum, $\Phi_{\text{core}}/\Phi_0$, present in the core of an isolated vortex as a function of the GL parameter κ.

C. Average Internal Field and Vortex Separation

Since interaction between the vortices is repulsive, as we will show in Section V.A, the vortices assume the arrangement that will keep them furthest apart—namely, the two-dimensional hexagonal lattice structure illustrated in Fig. 9.14. To observe this structure using what is called the Bitter (1931) technique, the surface is decorated by exposing it to a gas containing tiny suspended magnetic particles that adhere to the vortex cores and show up well on a photographic plate (Dolan *et al.*, 1989; Gammel *et al.*, 1987; Grier *et al.*, 1991; Vinnikov and Grigor'eva, 1988). The imaging can also be done with a scanning-tunneling (H. F. Hess *et al.*, 1989) or scanning-electron microscope with Lorentz electron microscopy, or with electron holography (Bonevich *et al.*, 1993). Individual vortices have been studied by magnetic force microscopy (Hug *et al.*, 1995; Moser *et al.*, 1995a, b).

The vortices arrange themselves in a hexadic pattern when their density is so high as to make the repulsive interactions between them appreciable in magnitude. Each vortex will then occupy the area $\frac{1}{2}\sqrt{3}\,d^2$ of the unit cell sketched in Fig. 9.15, where d is the average separation of the vortices. Such a structure has been observed on the surfaces of classical as well as high-temperature superconductors. The average field B_{in} inside the supercon-

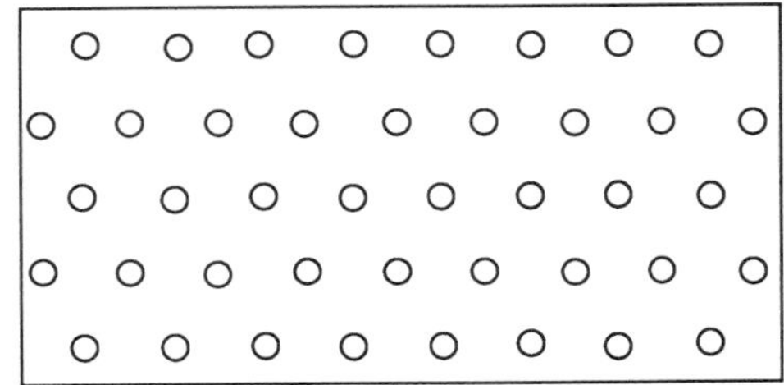

Figure 9.14 Two-dimensional hexagonal lattice of vortex cores.

ductor is given by

$$B_{in} = \frac{\Phi_0}{\frac{1}{2}\sqrt{3}\,d^2}, \tag{9.33}$$

and the number of vortices N equals the total cross-sectional area divided by the area per vortex $\frac{1}{2}\sqrt{3}\,d^2$,

$$N = \frac{A_T}{\frac{1}{2}\sqrt{3}\,d^2}. \tag{9.34}$$

The vortex lattice structure is not always the hexagonal type depicted in Fig. 9.14 for it can also depend on the magnetic field direction. In low-κ Type II alloys of, for example, Nb, Pb, Tc, or V, the vortices form a square lattice when the magnetic field is parallel to fourfold crystallographic symmetry axis and a hexagonal lattice when $\mathbf{B}_{app}$ is along a threefold axis. When $\mathbf{B}_{app}$ is along a twofold axis direction, a distorted hexagonal lattice is observed (Huebener, 1979, pp. 75ff; Obst, 1971). For $YBa_2Cu_3O_{7-\delta}$ in tilted applied fields, an SEM micrograph shows "a pinstripe array of vortex chains" lying in the $\mathbf{B}_{app}$, c plane (Gammel and Bishop, 1992), and for the applied field perpendicular to the c direction, chains of oval-shaped vortices are observed (Dolan *et al.*, 1989b).

D. Vortices near Lower Critical Field

When the flux first penetrates the superconductor at $\mathbf{B}_{app} = \mathbf{B}_{c1}$ the vortices are near the surface and isolated. As the applied field increases more vortices enter, and their mutual repulsion and tendency to diffuse causes them to migrate inward. Eventually they become sufficiently dense and close enough to experience each others' mutual repulsive forces, so they begin to arrange themselves into a more or less regular pattern resembling that in Fig. 9.14. For this case we can find an expression for the average internal field $\langle B_{in} \rangle$ in terms of the average separation d by eliminating Φ_0 from Eqs. (9.8) and (9.33):

$$\langle B_{in} \rangle = B_{c1} \frac{8\pi}{\sqrt{3}\,\ln\kappa} \frac{\lambda^2}{d^2}. \tag{9.35}$$

Therefore, the separation of vortices when the average internal field equals the lower-critical field is given by

$$d = 2\left(\frac{\pi}{\sqrt{3}\cdot\frac{1}{2}\ln\kappa}\right)^{1/2}\lambda \tag{9.36}$$

$$= \frac{3.81\lambda}{\sqrt{\ln\kappa}} \qquad \langle B_{in} \rangle = B_{c1}. \tag{9.37}$$

Since $\sqrt{\ln 10} = 1.52$ and $\sqrt{\ln 100} = 2.15$, the value of κ does not have much effect on the separation of the vortices. Figure 9.16 shows the dependence of the average internal field on their separation $\frac{1}{2}d/\lambda$.

The process of vortex entry into, and exit from, a superconductor is actually more complicated than this. It can occur in a surface sheath similar to the one that remains superconducting for applied fields in the range $B_{c2} < B_{app} < B_{c3}$, as mentioned in Section II.C. Walton *et al.* (1974) assumed the presence of a surface layer with "nascent" vortices that turn into nucleation sites for the formation of vortices. As the applied field increases, the interface between the surface region containing the vortices and the field-free bulk is able to move inward by diffusion at a velocity

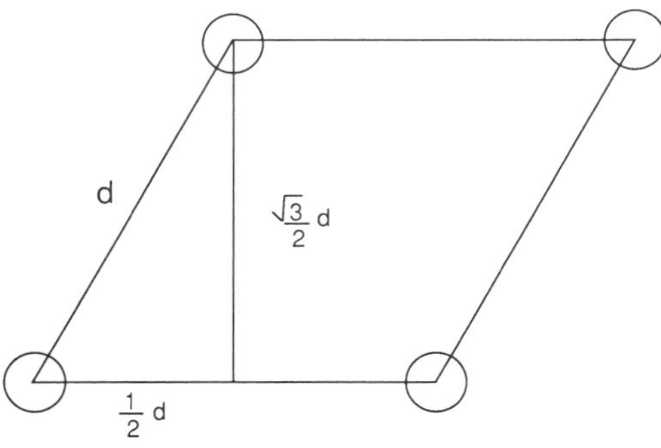

Figure 9.15 Vortex unit cell for the hexagonal lattice of Fig. 9.14. The area of the cell is $\frac{1}{2}\sqrt{3}\,d^2$.

proportional to the field gradient (Frahm *et al.*, 1991).

Since it is the applied field rather than the internal field which is known experimentally, it is of interest to determine how the separation of vortices depends on the ratio B_{app}/B_{c1} between the applied field and the lower-critical field. It is assumed that the vortices distribute themselves in a regular manner to produce a uniform internal field, $B_{in} = \langle B_{in} \rangle$. Figure 9.17, an enlargement of the low-field part of Fig. 9.5, shows a typical example illustrating how an internal field increasing with increasing applied field in the neighborhood of the lower-critical field. From the slope of the curve of B_{in} versus B_{app}, we conclude that the internal field will reach the value B_{c1} when the applied field approaches the upper limit of the range $B_{c1} < B_{app} < 2B_{c1}$, and we estimate that this might occur for $B_{app} \approx 1.8 B_{c1}$. Using this figure to convert from B_{in} to B_{app} graphs of B_{app} versus $\frac{1}{2}d/\lambda$ are plotted in Fig. 9.16.

We see from the figure that when the applied field is slightly greater than the lower-critical field, the vortices are much further apart than just one penetration depth. At higher applied fields, the vortices move closer together until near an applied field $B_{app} \approx 2B_{c1}$ their separation is about two penetration depths. For higher applied fields appreciable overlap occurs ($\frac{1}{2}d < \lambda$). Thus, relatively low fields produce a concentration of vortices high enough for the vortices to be treated as a continuum rather than as isolated entities.

E. Vortices near Upper Critical Field

When the applied magnetic field approaches the upper-critical field B_{c2} given by Eq. (9.9) the vortices are very close together, with their separation d somewhat greater than the coherence length ξ. Equating $\langle B_{in} \rangle$ and B_{c2} in Eqs. (9.9) and (9.33), respectively, we obtain an expression for the vortex nearest-neighbor distance d in terms of the coherence length:

$$d = 2\xi(\pi/\sqrt{3})^{1/2} \tag{9.38}$$

$$\approx 2.69\xi. \tag{9.39}$$

Since $d > 2\xi$, the cores do not quite touch for this highest density case.

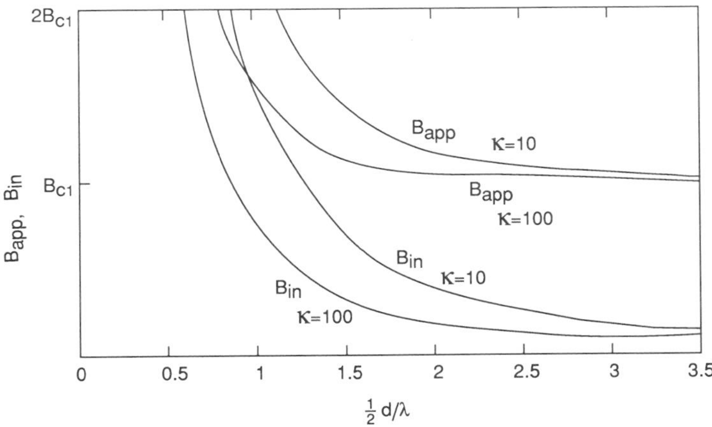

Figure 9.16 Relationship between applied and internal fields, B_{app} and B_{in}, and half the ratio $\frac{1}{2}d/\lambda$ between the vortex separation d and the penetration depth λ, for two values of κ. For much smaller separations $d \ll \lambda$, $\mathbf{B}_{in}$ approaches $\mathbf{B}_{app}$.

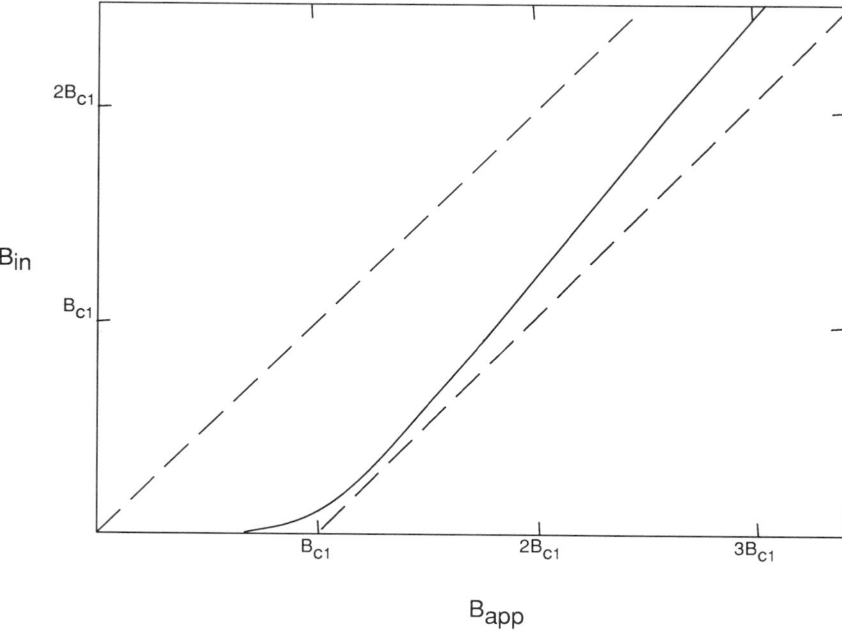

Figure 9.17 Magnitude of internal magnetic field B_{in} (solid line) in a Type II superconductor in the neighborhood of the lower-critical field. $\mathbf{B}_{in}$ approaches the dashed unit slope line through the origin for very high applied fields.

F. Contour Plots of Field and Current Density

In Section III.B we wrote down the closed-form expressions (9.17) and (9.20), respectively, for the magnetic field and current density of an isolated vortex, and in Fig. 9.12 we plotted the distance dependence of these quantities for an isolated vortex. In this section we will provide plots that were constructed from calculations of the position dependence of the field and the current density associated with densely packed vortices (carried out by Aktas, 1993; Aktas *et al.*, 1994). The computations involved adding the contributions of the many overlapping vortices in the neighborhood of a particular vortex, as indicated in Figs. 9.7b and 9.10. This meant taking into account hundreds of vortices, but it was sufficient to carry out the calculations in only one-twelfth of the unit cell because the smaller calculational cell defined by the triangle V–S–M–C–V of Fig. 9.18 replicates itself 12 times in the vortex unit cell of Fig. 9.15.

The magnetic field is a maximum at each vortex position V, of course, and a minimum at the midpoint M between three vortices. Figure 9.19 plots this calculated field change along the path $V \rightarrow S \rightarrow M \rightarrow C \rightarrow V$ for the case of vortices with "radius" $\lambda = 1000$ Å and separation $d = 400$ Å, which corresponds to considerable overlap. We see from the figure that the field increases along the two paths $M \rightarrow S$ and $M \rightarrow C$. There is a saddle point S midway between the two vortices, with the magnitude of the field decreasing slightly from S to M and increasing appreciably from S to V.

Figure 9.18 portrays the current density encircling the vortex cores V in one direction and flowing around the minimum points M in the opposite direction. Along the path from one vortex to the next the current density passes through zero and reverses direction at the saddle point S.

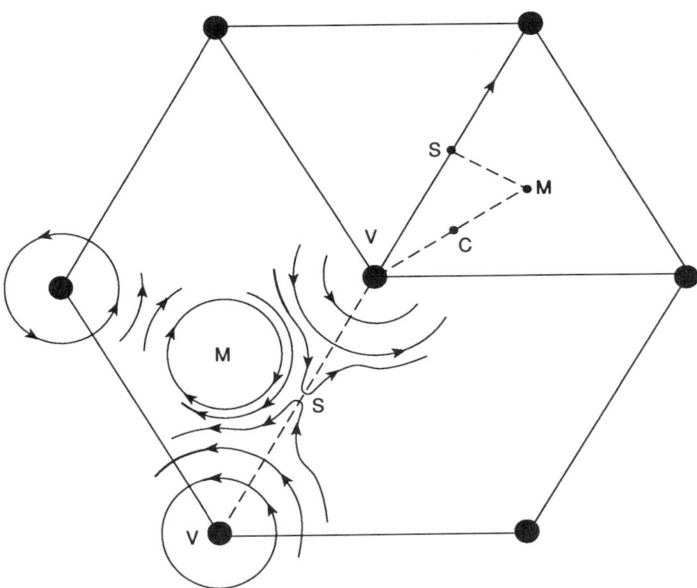

Figure 9.18 Geometrical relationships of a vortex unit cell showing the midpoint M between three vortices V at which the internal field B_{in} is a minimum and the saddle point S midway between two vortices. The triangle V–S–M is the calculational cell, which is one-twelfth the vortex unit cell sketched in Fig. 9.15 and contains all of the information on the fields and currents. Also shown are the current circulation around the vortices V and the midpoint M (Aktas *et al.*, 1994).

Along the path from the minimum point M to a vortex V there is a curvature change point C at which the current flow switches between clockwise and counterclockwise circulation. Points V, M, and S are well defined geometrically; the position of point C along the line from V to M has to be calculated. Figure 9.20 plots the current density calculated along the path V → S → M → C → V around the periphery of the calculational cell. A comparison of Figs. 9.19 and 9.20 indicates that the current density tends to be fairly constant over more of the unit cell than is the case with the magnetic field.

The previous few paragraphs describe the internal magnetic field on a mesoscopic scale, with resolution over distances comparable with the penetration depth. Ordinarily, we are interested in the value of the macroscopic internal field, which is an average over these mesoscopic field variations. Forkl *et al.* (1991) used a magneto-optical Faraday effect technique to determine the distribution of the macroscopic internal field inside a disk-shaped sample of $YBa_2Cu_3O_{7-\delta}$; the results are given in Figs. 9.21 and 9.22. Other investigators have published similar internal-field profiles (Flippen, 1991; Glatzer *et al.*, 1992; Mohamed *et al.*, 1989, 1990) and surface-field profiles (Brüll *et al.*, 1991; H. Muller *et al.*, 1991).

G. Closed Vortices

The vortices that we have been discussing are of the open type, in the sense that they begin and end at the surface of the superconductor. Here the flux is continuous with flux entering and leaving from the outside, as indicated in Fig. 9.11. We recall from Figs. 2.36 and 5.19 that a transport current flowing in a superconductor

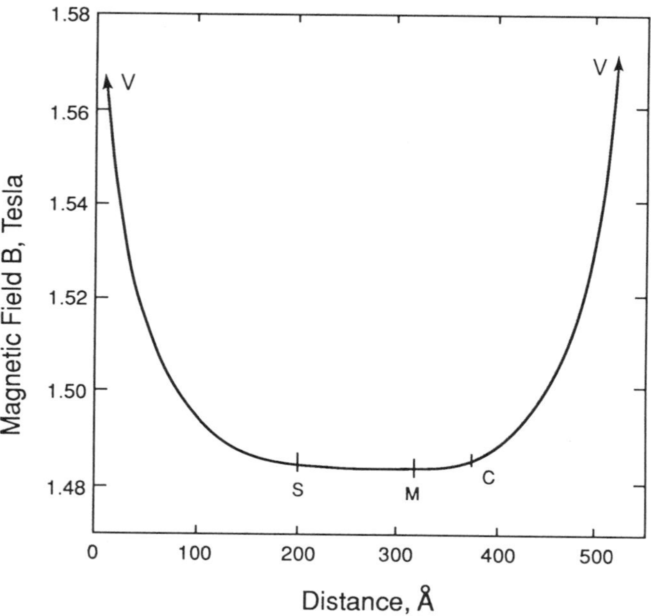

Figure 9.19 Magnitude of internal magnetic field along the three principal directions V → S, S → M, and M → V from the origin at the vortex V in the calculational cell of Fig. 9.18 for $d = 400$ Å and $\lambda = 1000$ Å (Aktas, 1993).

has encircling magnetic field lines, and that the portion of this encircling magnetic flux inside the superconducting material will be in the form of vortices that close in on themselves, basically vortices with loops of totally confined flux. The encircling flux in the region outside the superconductor is not quantized. When both transport and screening current are present, some of the vortices close in on themselves, and some do not.

IV. VORTEX ANISOTROPIES

Section III.B gave expressions for the magnetic field and current densities associated with a vortex in a high-κ isotropic superconductor. Many superconductors, such as those of the high-T_c type, are not isotropic, however, and in this section we will examine the configurations of the resulting vortices. These configurations depend on the coherence length and the penetration depth, which in turn depend on the anisotropies of the carrier effective mass, so we will say a few words about these parameters first. We will emphasize the case of axial symmetry; for this case the *a* and *b* directions are equivalent to each other. Such a geometry is exact for tetragonal, and a good approximation for orthorhombic high-temperature superconductors. The reader is referred to the text by Orlando and Delin (1991) for derivations of the various expressions in this section.

An alternative approach to that presented here considers an isolated vortex as having elastic properties that are analogous to those of a string under tension (Hanaguri *et al.*, 1994; Toner, 1991a; Widom *et al.*, 1992). The vortex is assumed to be held in place at its end points by the coupling to the external magnetic field, and when distorted it tends to return to a linear configuration. The flux-line lattice in an anisotropic superconductor is more

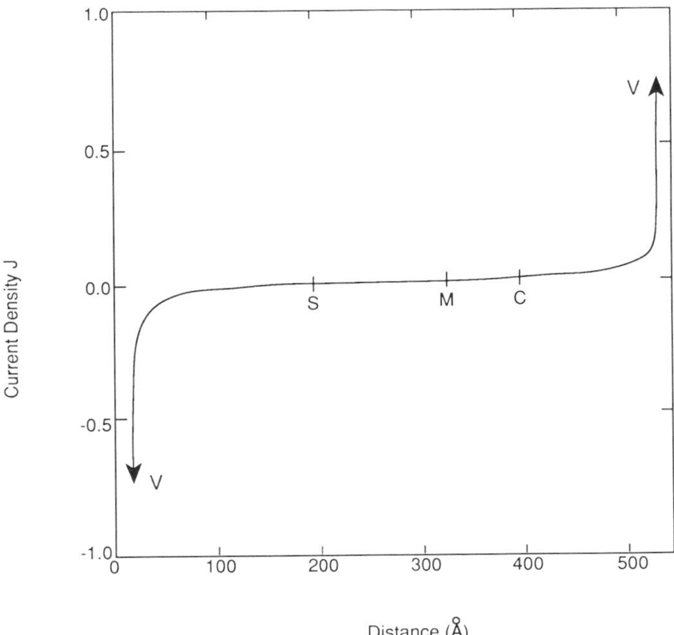

Figure 9.20 Current density along the three principal directions V → S, S → M, and M → V measured from the origin at the vortex V in the calculational cell of Fig. 9.18 for $d = 400$ Å and $\lambda = 1000$ Å. The curvature change point C along the path M → V at which the current flow direction shifts from clockwise around V to counterclockwise around M (cf. Fig. 9.18) is indicated. Comparison with Fig. 9.19 shows that the current density exhibits more abrupt changes than the fields (Aktas, 1993).

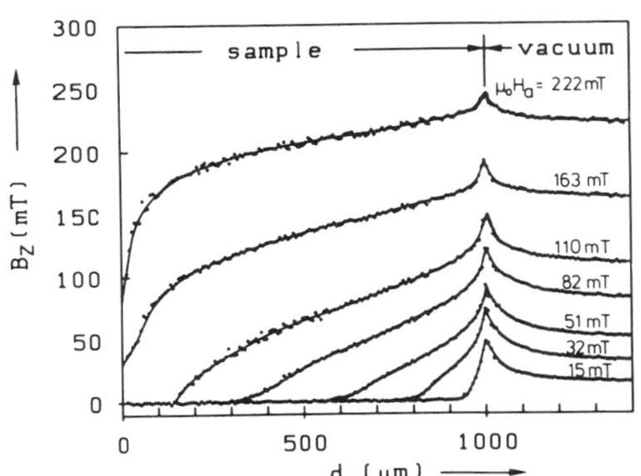

Figure 9.21 Radial distribution of internal magnetic field in a superconducting rod of 1 mm radius for applied fields from 15 mT to 222 mT. The penetration depth $\lambda \approx 20$ μm (Forkl *et al.*, 1991).

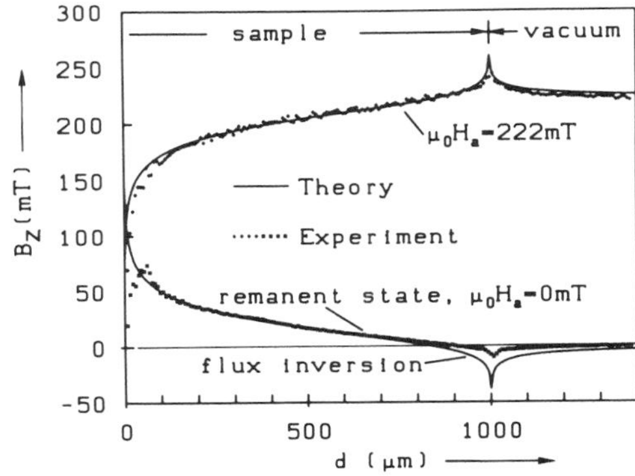

Figure 9.22 Radial distribution of internal magnetic field in the superconducting rod of Fig. 9.21 following application of a field of 222 mT and in the remanent state following removal of this field (Forkl *et al.*, 1991).

complicated, and has been treated using anisotropic elasticity theory (Sardella, 1992; Section 9-G-1).

A. Critical Fields and Characteristic Lengths

The electron or hole carriers of an anisotropic superconductor have effective masses m^* that depend on the direction, with the principal values m_a, m_b, and m_c along the three principal directions a, b, and c of the crystal. The Ginzburg–Landau theory tells us that, on the basis of Eqs. (5.26) and (5.45), respectively [cf. Eq. (6.148)],

$$\xi^2 = \hbar^2/2m^*|a|, \tag{9.40}$$

$$\lambda^2 = \frac{m^*}{\mu_0 e^{*2}|\phi_\infty|^2}, \tag{9.41}$$

the coherence length ξ is inversely proportional, and the penetration depth λ is directly proportional, to the square root of m^*, where the factors a and ϕ_∞ of the GL theory depend on the temperature. Since the anisotropy arises from the effective mass, to a first approximation we can write for the three principal directions

$$\xi_a\sqrt{m_a} = \xi_b\sqrt{m_b} = \xi_c\sqrt{m_c}, \tag{9.42a}$$

$$\lambda_a/\sqrt{m_a} = \lambda_b/\sqrt{m_b} = \lambda_c/\sqrt{m_c}. \tag{9.42b}$$

Multiplying these expressions together term by term gives

$$\xi_a\lambda_a = \xi_b\lambda_b = \xi_c\lambda_c, \tag{9.43}$$

which is the basic characteristic length relationship of anisotropic superconductors.

Many superconductors, such as the cuprates, are axially symmetric with in-plane ($m_a = m_b = m_{ab}$) and axial-direction (m_c) effective masses. We define the ratio Γ by

$$\Gamma = m_c/m_{ab}, \tag{9.44}$$

with reported values $\Gamma \geq 29$ for $YBa_2Cu_3O_{7-\delta}$ (Farrell *et al.*, 1988), $\Gamma \geq 3000$ for $Bi_2Sr_2CaCu_2O_8$ (Farrell *et al.*, 1989b), and $\Gamma \geq 10^5$ for $Tl_2Ba_2CaCuO_8$ (Farrell *et al.*, 1990a). Using Eqs. (9.42)–(9.44) we can show that the coherence length ξ_{ab} and penetration depth λ_{ab} in the a, b plane are related to their values ξ_c and λ_c along the c direction through the expression (Kes *et al.*, 1991).

$$\Gamma = (\xi_{ab}/\xi_c)^2 = (\lambda_c/\lambda_{ab})^2, \tag{9.45}$$

where for the cuprates (Hikita *et al.*, 1987; Worthington *et al.*, 1987) we have

$$\xi_c < \xi_{ab} \ll \lambda_{ab} < \lambda_c. \tag{9.46}$$

Some reported values of these quantities are listed in Table 9.4.

The GL parameter κ_i for the magnetic field in the ith principal direction is (Chakravarty *et al.*, 1990)

$$\kappa_i = \left|\frac{\lambda_j\lambda_k}{\xi_j\xi_k}\right|^{1/2}, \tag{9.47}$$

which gives for the cuprates with the applied field in the a, b-plane (κ_{ab}) and along the c direction (κ_c), respectively,

$$\kappa_{ab} = \left|\frac{\lambda_{ab}\lambda_c}{\xi_{ab}\xi_c}\right|^{1/2} \tag{9.48a}$$

$$\kappa_c = \lambda_{ab}/\xi_{ab}. \tag{9.48b}$$

In the next two sections we will employ these quantities to write down explicit expressions for the core perimeter, magnetic fields, and current densities of vortices in the presence of anisotropies. An average GL parameter $K_{av} = (\lambda_1\lambda_2\lambda_3/\xi_1\xi_2\xi_3)^{1/3}$ has also been defined (Clem and Coffey, 1990).

B. Core Region and Current Flow

The vortices described in Section III.B for $\kappa \gg 1$ were axially symmetric, with the shielding current flowing in circular paths around the axis, as illustrated in Fig. 9.9. Axial symmetry is also observed for the cuprates when the applied field is aligned along c. When, however, it is along the b (or a) direction, the core cross-section is an ellipse with semi-axes ξ_{ab} and ξ_c along the a and c directions, as indicated in Fig. 9.23. The current flows in an elliptical path with semi-axes $\alpha\lambda_c$ and $\alpha\lambda_{ab}$, as shown in Fig. 9.24, where α is a numerical factor that depends on the distance of the current from the core. We know from Eq. (9.43) that $\xi_{ab}/\xi_c = \lambda_c/\lambda_{ab}$, so the core and current flow ellipses have the same ratio $\sqrt{\Gamma}$ of semi-major to semi-minor axis, and hence the same eccentricity. The equation for the current flow ellipse with $\alpha = 1$ is

$$\frac{x^2}{\lambda_c^2} + \frac{z^2}{\lambda_{ab}^2} = 1 \qquad \mathbf{B}_{\text{app}} || b, \quad (9.49a)$$

where x and z are the Cartesian coordinates of points on the perimeter. The corresponding current flow equation for the applied field along the c direction is a circle,

$$\frac{x^2 + y^2}{\lambda_{ab}^2} = 1 \qquad \mathbf{B}_{\text{app}} || c. \quad (9.49b)$$

Equations (9.49a) and (9.49b) also correspond to loci of constant magnetic field around the vortex. They are plotted on the left and right sides, respectively, of Fig. 9.23 to indicate the relative size of the two vortices. Expressions analogous to Eqs. (9.49) can be written down for the perimeters of the cores.

Since the current paths are ellipses, each increment of current ΔI flows in a channel between a pair of ellipses, in the manner illustrated in Fig. 9.24. When the channel is narrow, as it is at the top and bottom of the figure, the flow is fast, so the current density is large, as indicated. Conversely, in the wider channels on the left and right, the flow is slow and J is small, as indicated. The increment of current ΔI through each part of the channel is the same, so the product J times the width must be constant, and we can write

$$J_x \, \Delta z = J_z \, \Delta x. \qquad (9.50)$$

This is a special case of the fluid mechanics expression $J_1 A_1 = J_2 A_2$ for current flow in a pipe of variable cross section A.

C. Critical Fields

The isotropic expressions for the critical fields given in Section III.B can be

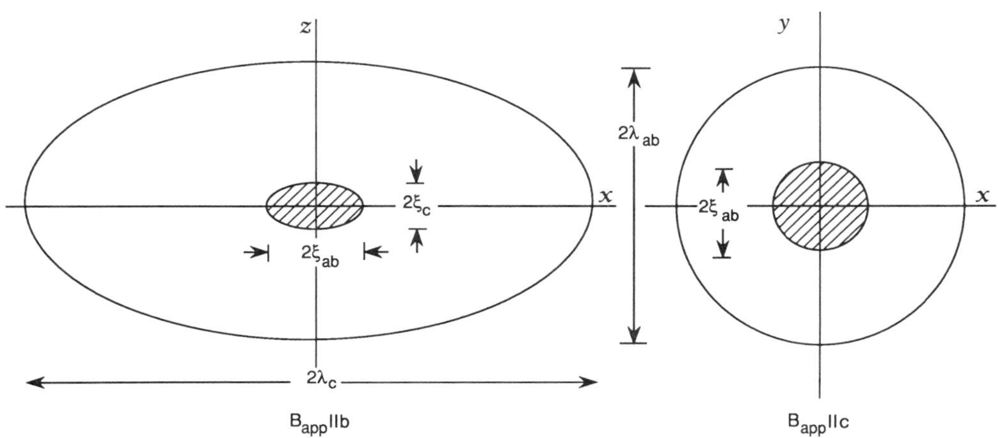

Figure 9.23 Shape of the core (shaded) and the perimeter one penetration length from the center of a vortex for an applied magnetic field along the b (left) and c (right) crystallographic directions, respectively. The magnetic field is constant along each ellipse and along each circle. The figure is drawn for the condition $\lambda_c = 2\lambda_{ab} = 6\xi_{ab} = 12\xi_c$.

Table 9.4 Coherence Lengths ξ_i and Penetration Depths λ_i of Various Superconductors in the Symmetry Plane (called the *a*, *b*-plane) and in the Axial Direction (called the *c*-axis), and Values of the Anisotropy Ratio Γ

Material	T_c (K)	ξ_{ab} (nm)	ξ_c (nm)	λ_{ab} (nm)	λ_c (nm)	Γ (m_c/m_{ab})	Reference
$NbSe_2$		7.7	2.3	69	230	11	Salamon (1989)
UPt_3 (heavy fermion)	0.46			782	707		Broholm *et al.* (1990)
K-$(ET)_2Cu[NCS]_2$	9			980			Harshman *et al.* (1990)
K-$(ET)_2Cu[NCS]_2$	10					$> 4 \times 10^4$	Farrell *et al.* (1990b)
K-$(ET)_2Cu[N(CN)_2]_2Br$	11.4			650			Lang *et al.* (1992a, b)
K-$(ET)_2Cu[N(CN)_2]_2Br$	11.6	3.7	0.4			86	Kwok *et al.* (1990b)
$(La_{0.925}Sr_{0.075})_2CuO_4$	34	2.9					Hase *et al.* (1991)
$(La_{0.91}Sr_{0.09})_2CuO_4$	30	3.3		283			Li *et al.* (1993)
$(Nd_{0.925}Ce_{0.075})_2CuO_4$[a]	21.5			80	100		Wu *et al.* (1993)
$(Nd_{0.9}Ce_{0.1})_2CuO_4$[a]						≈ 600	O and Markert (1993)
$(Sm_{0.925}Ce_{0.075})_2CuO_{4-\delta}$[a]	11.4	7.9	1.5				Dalichaouch *et al.* (1990b)
$(Sm_{0.925}Ce_{0.075})_2CuO_{4-\delta}$[a]	18	4.8					S. H. Han *et al.* (1992)
$YBa_2Cu_3O_{6.5}$	62	2.0	0.45			19	Vandervoort *et al.* (1991)
$YBa_2Cu_3O_{6.9}$	83			142	> 700		Harshman *et al.* (1989)
$YBa_2Cu_3O_{6.94}$	91.2	1.7		150			Ossandon *et al.* (1992a)
$YBa_2Cu_3O_{7-\delta}$	66			260			Lee and Ginsberg (1991)
$YBa_2Cu_3O_{7-\delta}$	90	2.5	0.8			10	Chaudhari *et al.* (1987)
$YBa_2Cu_3O_{7-\delta}$	89	3.4	0.7	26	125	25	Worthington *et al.* (1987)

$YBa_2Cu_3O_{7-\delta}$	92.4	4.3	0.7	27	180	41	Gallagher (1988)
$YBa_2Cu_3O_{7-\delta}$	92	1.2	0.3	89	550	≈ 27	Salamon (1989)
$YBa_2Cu_3O_{7-\delta}$	90	1.3	0.2	130	450	≈ 25	Krusin-Elbaum *et al.* (1989)
$YBa_2Cu_3O_{7-\delta}$	92	1.6	0.3				Welp *et al.* (1989)
$EuBa_2Cu_3O_{7-\delta}$	95	2.7	0.6			20	Hikita *et al.* (1987)
$EuBa_2Cu_3O_{7-\delta}$	94	3.5	0.38				Y. Tajima *et al.* (1988)
$TmBa_2Cu_3O_{7-\delta}$	86	7.4	0.9			68	Noel *et al.* (1987)
$Y_{0.8}Pr_{0.2}Ba_2Cu_3O_{7+\delta}$	73	2.4	0.78			9.5	Jia *et al.* (1992)
$Bi_2Sr_2CaCu_2O_8$	84	1.1					Johnston and Cho (1990)
$Bi_2Sr_2CaCu_2O_8$	109				500		Maeda *et al.* (1992)
$Bi_2Sr_2Ca_2Cu_3O_{10}$	109	2.9	0.09				Matsubara *et al.* (1992)
$Bi_2Sr_2Ca_2Cu_3O_{10}$	111	1.0	0.02				Q. Li *et al.* (1992)
$(Bi, Pb)_2Sr_2CaCu_2O_8$	91	2.0			178		H. Zhang *et al.* (1992)
$(Bi_{0.9}Pb_{0.1})_2Sr_2CaCu_2O_{8+\delta}$	91.1	2.04	0.037				W. C. Lee *et al.* (1991)
$(Bi_{0.9}Pb_{0.1})_2Sr_2Ca_2Cu_3O_{10+\delta}$	103	1.18					W. C. Lee *et al.* (1991)
$Pb_2Sr_2(Y, Ca)CaCu_3O_8$	76	1.5	0.3	258	643	$\approx 12^b$	Reedyk *et al.* (1992b)
$Tl_2Ba_2CaCu_2O_{8-\delta}$	100			182			Ning *et al.* (1992)
$Tl_2Ba_2Ca_2Cu_3O_{10}$	123			173	480	8	Thompson *et al.* (1990)
$Tl_2Ba_2Ca_2Cu_3O_{10}$	100					$\geq 10^5$	Farrell *et al.* (1990a)
$HgBa_2CuO_{4+\delta}$	93	2.1		117			Thompson *et al.* (1993)
$HgBa_2Ca_2Cu_3O_{8+\delta}$	133	1.3		130	3500	730	Schilling *et al.* (1994b)

Note: The axial direction is along the c-axis for high temperature superconductors, and along the a-axis for typical organic materials.
[a] This is an electron superconductor.
[b] An estimate, since the ratios ξ_{ab}/ξ_c and λ_c/λ_{ab} differ.

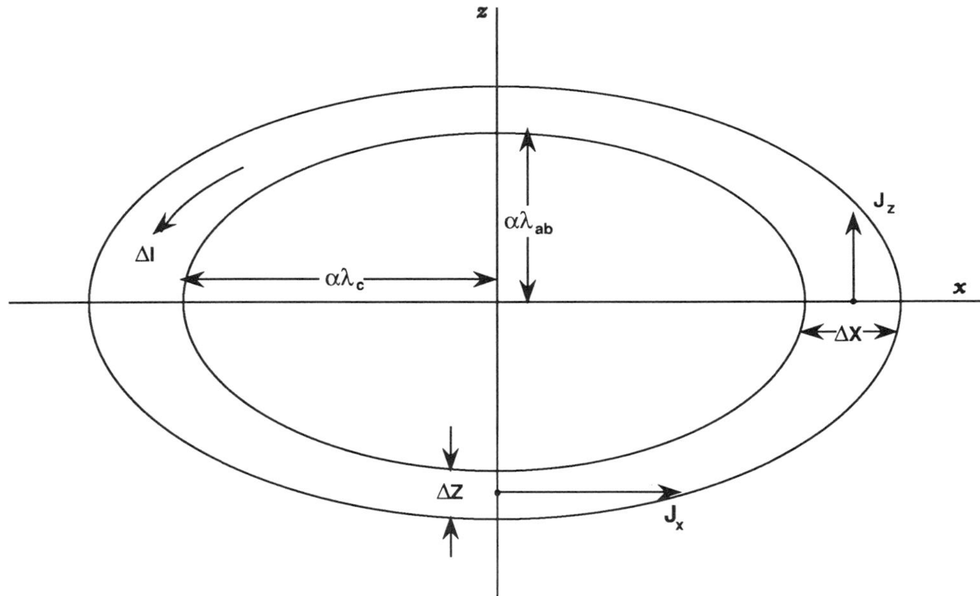

Figure 9.24 Flow of differential current $\Delta\mathbf{I}$ in an elliptical path around a vortex for an applied magnetic field along the b crystallographic axis. The current densities are $J_x = \Delta I/\Delta z\,\Delta y$ on the z-axis, and $J_z = \Delta I/\Delta x\,\Delta y$ on the x-axis.

modified for anisotropy by using the GL parameter κ_i (9.47) for the applied field in the i direction, and inserting the appropriate characteristic lengths for the directions perpendicular to i. When the applied field is along the ith principal direction, expressions (9.8) and (9.9) for the critical fields $B_{c1}(i)$ and $B_{c2}(i)$ become

$$B_{c1}(i) = \frac{\Phi_0 \ln \kappa_i}{4\pi\lambda_j\lambda_k}, \tag{9.51a}$$

$$B_{c2}(i) = \frac{\Phi_0}{2\pi\xi_j\xi_k}. \tag{9.51b}$$

These can be written in terms of the thermodynamic critical field $\mathbf{B}_c$,

$$B_{c1} = \frac{\ln \kappa_i}{\sqrt{2}\,\kappa_i} B_c, \tag{9.52a}$$

$$B_{c2} = \sqrt{2}\,\kappa_i B_c, \tag{9.52b}$$

where $\mathbf{B}_c$ itself (9.10),

$$B_c = \frac{\Phi_0}{2\sqrt{2}\,\pi\xi_i\lambda_i}, \tag{9.53}$$

is independent of the direction since, from Eq. (9.43), the product $\xi_i\lambda_i$ has the same value for $i = a, b, c$. The expressions for the ratio

$$B_{c2}(i)/B_{c1}(i) = 2\kappa_i^2/\ln \kappa_i, \tag{9.54}$$

and the product of these critical fields,

$$[B_{c1}(i)B_{c2}(i)]^{1/2} = B_c[\ln \kappa_i]^{1/2}, \tag{9.55}$$

are generalizations of Eqs. (9.13). For the particular case of axial symmetry we have for the critical fields in the a, b-plane,

$$B_{c1}(ab) = \frac{\Phi_0 \ln \kappa_{ab}}{4\pi\lambda_{ab}\lambda_c}, \tag{9.56a}$$

$$B_{c2}(ab) = \frac{\Phi_0}{2\pi\xi_{ab}\xi_c}, \tag{9.56b}$$

and along the c direction,

$$B_{c1}(c) = \frac{\Phi_0 \ln \kappa_c}{4\pi\lambda_{ab}^2}, \quad (9.57a)$$

$$B_{c2}(c) = \frac{\Phi_0}{2\pi\xi_{ab}^2}, \quad (9.57b)$$

where κ_{ab} and κ_c are defined by Eqs. (9.48). The ratio of the upper critical fields,

$$\frac{B_{c2}(c)}{B_{c2}(ab)} = \sqrt{\Gamma}, \quad (9.58)$$

is a particularly simple expression. Table 9.5 provides some experimentally determined values of these critical field anisotropies.

D. High-Kappa Approximation

In Section III.B we wrote down expressions for the radial dependence of the magnetic field and current density associated with a vortex in an isotropic superconductor. Now we will generalize these expressions to account for the presence of anisotropy.

When the applied magnetic field is in the z direction, along the c-axis, the vortex has axial symmetry and the magnetic field and current densities have the distance dependence,

$$B_z(x,y) = \frac{\Phi_0}{2\pi\lambda_{ab}^2} K_0\left[(x^2+y^2)^{1/2}/\lambda_{ab}\right], \quad (9.59)$$

$$\mathbf{J}_s(x,y) = \frac{\Phi_0}{2\pi\mu_0\lambda_{ab}^3} K_1\left[(x^2+y^2)^{1/2}/\lambda_{ab}\right] \times \left(\frac{y\mathbf{i} - x\mathbf{j}}{(x^2+y^2)^{1/2}}\right), \quad (9.60)$$

where K_0 and K_1 are zeroth- and first-order modified Bessel functions, respectively.

When the applied magnetic field is in the x direction, along the a-axis, the vortex no longer has axial symmetry, and the distance dependences are more complicated:

$$B_x(y,z) = \frac{\Phi_0}{2\pi\lambda_{ab}\lambda_c} K_0\left[\left(\frac{y^2}{\lambda_c^2} + \frac{z^2}{\lambda_{ab}^2}\right)^{1/2}\right], \quad (9.61)$$

$$\mathbf{J}_s(y,z) = \frac{\Phi_0}{2\pi\mu_0\lambda_{ab}\lambda_c} \times \frac{K_1\left[\left(\frac{y^2}{\lambda_c^2} + \frac{z^2}{\lambda_{ab}^2}\right)^{1/2}\right]}{\left(\frac{y^2}{\lambda_c^2} + \frac{z^2}{\lambda_{ab}^2}\right)^{1/2}} \times \left[\frac{y}{\lambda_c^2}\mathbf{k} - \frac{z}{\lambda_{ab}^2}\mathbf{j}\right]. \quad (9.62)$$

Equation (9.62) is obtained from (9.61) with the aid of relation $\nabla \times \mathbf{B} = \mu_0 \mathbf{J}_s$. Analogous expressions can be written down for B_{app} along y. The asymptotic equations (9.24)–(9.28) for the modified Bessel functions can also be applied to the anisotropic case.

When the applied magnetic field is aligned at an oblique angle relative to the c direction, the expressions for the magnetic field and current density in the neighborhood of a vortex become very complicated, and we will not try to specify them.

E. Pancake Vortices

In high-temperature superconductors the coherence length ξ_c along the c-axis is less than the average spacing between the copper-oxide planes, and hence the coupling between the planes tends to be weak. The Lawrence–Doniach model (1971; Bulaevskii, 1973; Bulaevskii *et al.*, 1992; Clem, 1989, 1991) assumes that the superconductor consists of parallel superconducting layers that are weakly Josephson coupled to each other. A vortex perpendicular to these layers, which conventionally would be considered a uniform cylinder of confined flux surrounded by circulating

Table 9.5 Critical Fields of Selected Anisotropic Type II Superconductors

Material	T_c (K)	B_{c1}^{ab} (mT)	B_{c1}^{c} (mT)	B_c (T)	B_{c2}^{ab} (T)	B_{c2}^{c} (T)	$-dB_{c2}^{ab}/dT$ (T/K)	$-dB_{c2}^{c}/dT$ (T/K)	Reference
$CeCu_2Si_2$ (heavy fermion)	0.63				2.0	2.4			Assmus *et al.* (1984)
β-$(ET)_2I_3$ (organic)	1.5	7^a	36		1.74^b	0.08			Ishiguro and Yamaji (1990)
β-$(ET)_2I_3$ 1.6 kbar	7.2				25	2.7			Ishiguro and Yamaji (1990)
β-$(ET)_2IBr_2$ (organic)	2.3	390	1600		3.48^c	1.5			Ishiguro and Yamaji (1990)
β-$(ET)_2AuI_2$ (organic)	4.2	400	2050		≈ 6.35	≈ 0.8			Ishiguro and Yamagi (1990)
K-$(ET)_2Cu[N(CN)_2]Br$	11.6						20	2.2	Kwok *et al.* (1990b)
$(Sm_{0.925}Ce_{0.05})_2CuO_4$ (electron type)	11.4				28.2	5.2	3.6	0.1	Dalichaouch *et al.* (1990b)
$(La_{0.95}Ca_{0.05})_2CuO_4$	≈ 14.0				> 20	> 13	4	0.3	Hidaka *et al.* (1987)
$(La_{0.9}Ca_{0.1})_2CuO_4$	30			0.2		32		1.5	Li *et al.* (1993)
$(La_{0.93}Ca_{0.07})_2CuO_4{}^d$	≈ 34.0	7	30						Naito *et al.* (1990)
$(Sm_{0.93}Ce_{0.07})_2CuO_4$	18.0					9.8	0.74		S. H. Han *et al.* (1992)
$YBa_2Cu_3O_{6.5}$	62	2.5	8.3	0.38	380	87	8.7	2.0	Vandervoort *et al.* (1991)
$YBa_2Cu_3O_{6.94}$	91.2		32			115		1.8	Ossandon *et al.* (1992a)
$YBa_2Cu_3O_{7-\delta}$		53	520						Dinger *et al.* (1987)
$YBa_2Cu_3O_{7-\delta}$		70	130				≈ 1.0	0.65	Song *et al.* (1987)
$YBa_2Cu_3O_{7-\delta}$	88.8	≤ 5	500	2.65	140	29	2.3	0.46	Worthington *et al.* (1987)
$YBa_2Cu_3O_{7-\delta}$	92.4	≤ 5	500	1.93	240	34	3.8	0.54	Gallagher (1988)
$YBa_2Cu_3O_{7-\delta}$	24	103	≈ 1.7				14		Salamon (1989)
$YBa_2Cu_3O_{7-\delta}$					110	40	3.4	1.0	Nakao *et al.* (1989)
$YBa_2Cu_3O_{7-\delta}$	90	18	53	≈ 1.8					Krusin-Elbaum *et al.* (1989)
$YBa_2Cu_3O_{7-\delta}$	92						10.5	1.9	Welp *et al.* (1989)
$EuBa_2Cu_3O_{7-\delta}$	95				190	45	3.0	0.7	Hikita *et al.* (1987)
$EuBa_2Cu_3O_{7-\delta}$	94.8				245	28	3.8	0.41	Y. Tajima *et al.* (1988)
$Y_{0.8}Pr_{0.2}Ba_2Cu_3O_{7-\delta}$	73				174	56	3.4	1.1	Jia *et al.* (1992)
$Bi_2Sr_2CaCu_2O_{8+\delta}$	90	85							Maeda *et al.* (1992)
$(Bi, Pb)_2Sr_2CaCu_2O_8$	91			0.65		≈ 89		1.4	L. Zhang *et al.* (1992)
$Bi_2Sr_2Ca_2Cu_3O_{10+\delta}$	109						16	0.5	Matsubara *et al.* (1992)
$Pb_2Sr_2(Y, Ca)Cu_3O_8$	76	9.5	50.5	0.80	590	96	11	1.75	Reedyk *et al.* (1992b)
$HgBa_2Ca_2Cu_3O_{8+\delta}$	131		45			190		2.0	Schilling *et al.* (1994b)

Note: Some of the thermodynamic critical fields B_c were calculated from Eq. (9.53) using data from Table 9.4.
[a] Average of $B_{c1}^{a} = 5$ mT, $B_{c1}^{b} = 9$ mT; [b] Average of $B_{c2}^{a} = 1.78$ T, $B_{c2}^{b} = 1.70$ T; [c] Average of $B_{c2}^{a} = 3.36$ T, $B_{c2}^{b} = 3.60$ T; [d] $-dB_{c1}^{ab}/dT = 1.8$ mT/K, $-dB_{c1}^{c}/dT = 5.5$ mT/K.

currents, is looked upon in this model as a stacking of two-dimensional (2D) pancake-shaped vortices, one pancake vortex per layer with surrounding, nearly circular current patterns confined to the layer. The stacked 2D Abrikosov vortices shown in Fig. 9.25 are coupled together by means of Josephson vortices whose axes thread through the Josephson junctions between the superconducting layers, stretching from the center of each pancake vortex to the center of the adjacent vortices above and below. The field and current distributions for the individual pancake vortices in this stack, aligned along *c*, as well as in a "leaning tower" or tilted stack of such vortices have been calculated (Clem, 1991).

Thermal agitation can shake the stack, decouple pancake vortices in adjacent layers, and even cause the stack to break up, as in a Kosterlitz–Thouless-type transition. Figure 9.26 shows a segment of a vortex displaced but still coupled. In this model melting might occur in the direction perpendicular to the layers, with the vortices within each layer forming 2D solids.

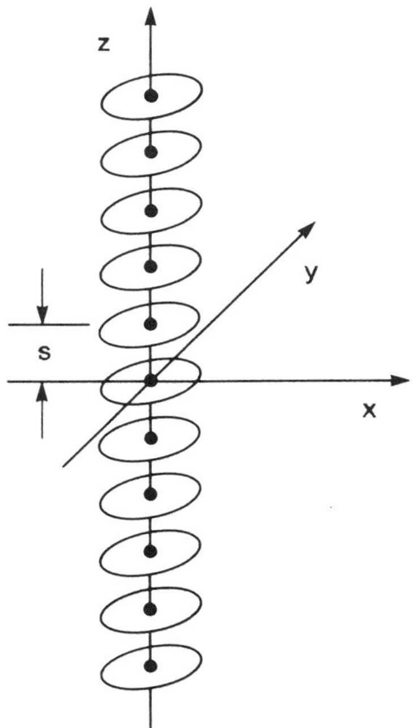

Figure 9.25 Stack of two-dimensional pancake vortices aligned along the *c* direction (Clem, 1991).

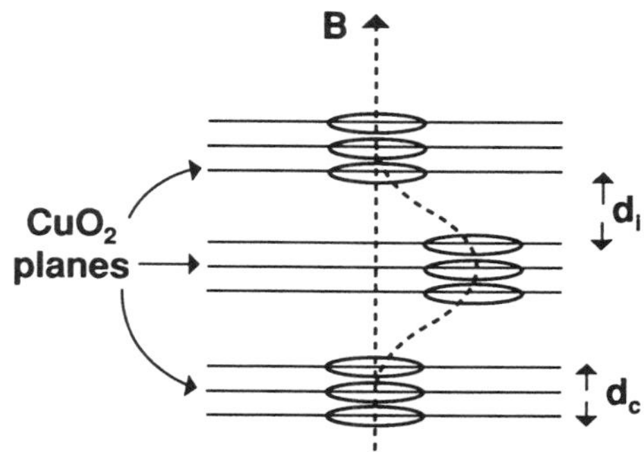

Figure 9.26 Pancake vortex model showing a segment of vortex displaced to the right (Maley, 1991).

F. Flux Trapping

When the applied magnetic field is inclined at an angle with respect to the *c*-axis of an anisotropic superconductor, neither the internal magnetic field nor the magnetization is oriented in the same direction as B_{app} (Felner *et al.*, 1989; D. H. Kim *et al.*, 1991b; Kolesnik *et al.*, 1992; L. Liu *et al.*, 1992; Tuominen *et al.*, 1990; K. Watanabe *et al.*, 1991; Welp *et al.*, 1989). This complicates the trapping of magnetic flux and alignment of the vortices. Elliptically shaped magnetic field contours around vortices have been published for the case of anisotropy (Thiemann *et al.*, 1989). In addition, the upper-critical field and critical current both depend on the orientation (K. Watanabe *et al.*, 1991). Transverse magnetization of an Abrikosov lattice, which is absent in an isotropic superconductor, has been determined for the anisotropic case using torque measurements (Farrell *et al.*, 1989b; Gray *et al.*, 1990).

V. INDIVIDUAL VORTEX MOTION

The mutual-repulsion Lorentz force $\mathbf{J}_i \times \mathbf{\Phi}_j$ between vortices arising from the interaction of the current density $\mathbf{J}_i$ of one vortex with the flux $\mathbf{\Phi}_j$ of another causes the vortices to become arranged in the hexagonal equilibrium configuration of Fig. 9.14. During the equilibration process each moving vortex experiences a frictional or damping force $\mathbf{f} = \beta \mathbf{v}$ that retards its motion, and a second force, called the *Magnus force*, which is given by $\alpha n_s e(\mathbf{v} \times \mathbf{\Phi}_0)$, where n_s is the density of the superconducting electrons. (The origin of the Magnus force will be explained in Section E.) Many vortices become trapped at pinning centers and hinder the motion of nearby vortices. When the pinning forces $\mathbf{F}_P$ are not sufficiently strong to prevent flux motion, the superconductor is called soft; otherwise it is called hard. When transport current $\mathbf{J}_{tr}$ is present, the Lorentz force $\mathbf{J}_{tr} \times \mathbf{\Phi}_0$ acts to unpin the vortices and induce a collective flux motion. When the pinning forces still dominate, this very slow motion is called *flux creep*, and when the Lorentz force dominates, the faster motion is called *flux flow*.

The forces acting on the vortex, such as $\mathbf{F}_P$ and $\mathbf{J} \times \mathbf{\Phi}_0$, are actually forces per unit length, but we have simplified the notation by referring to them as simply forces.

A. Vortex Repulsion

It is shown in electrodynamics texts that the Lorentz force density $\mathbf{f}$ for the interaction between an electric current density $\mathbf{J}$ and a magnetic field $\mathbf{B}$ is given by

$$\mathbf{f} = \mathbf{J} \times \mathbf{B}. \tag{9.63}$$

The force between two vortices may be considered as arising from the interaction between the magnetic field $\mathbf{B}$ of one vortex and the current density $\mathbf{J}$ present at the position of this field and arising from the other vortex, as shown in Fig. 9.27. It is assumed that both vortices are infinitely long and axially symmetric and that they are aligned parallel to each other a distance d apart. Since $\mathbf{f}$ is the force per unit volume, the total force $\mathbf{F}$ is obtained by integrating the current density over the volume containing the $\mathbf{B}$ field,

$$\mathbf{F} = \int \mathbf{J} \times \mathbf{B} r\, dr\, d\phi\, dz, \tag{9.64}$$

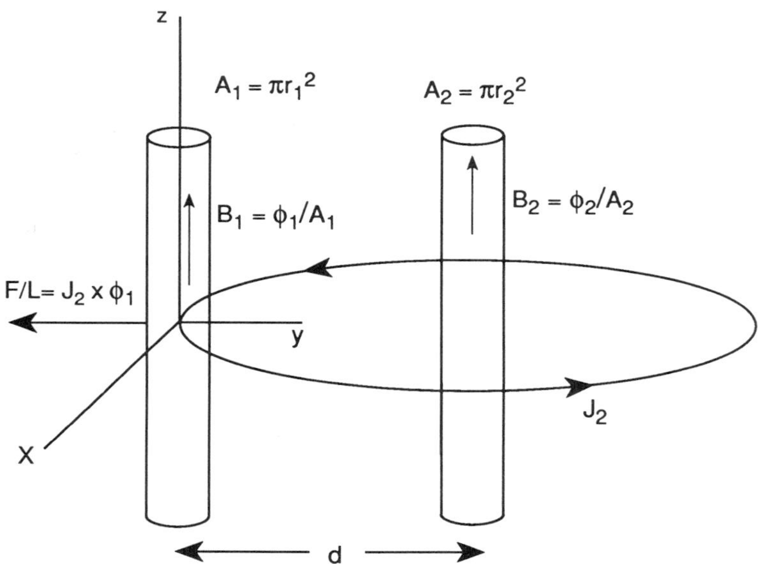

Figure 9.27 Repulsive interaction involving the magnetic field $\mathbf{B}_1$ in a vortex core with current density $\mathbf{J}_2$ from another parallel vortex $\mathbf{B}_2$. The repulsive force $F/L = \mathbf{J}_2 \times \mathbf{\Phi}_1$ is shown.

where cylindrical coordinates, r, ϕ, and z, have been used, and $r = (x^2 + y^2)^{1/2}$. Since there is no z dependence, it is more appropriate to calculate the force per unit length, which is given by

$$\mathbf{F}/L = \int \mathbf{J}(r') \times \mathbf{B}(r) r\,dr\,d\phi, \quad (9.65)$$

where from Fig. 9.28 the distance r' is given by

$$r' = (r^2 + d^2 - 2rd \cos \phi)^{1/2}. \quad (9.66)$$

In the high-κ approximation the expressions in Eqs. (9.17) and (9.20) for $\mathbf{B}(r)$ and $\mathbf{J}(r')$, respectively, can be substituted in the integral,

$$F/L = \frac{\Phi_0^2}{4\pi^2 \mu_0 \lambda^5} \int K_0\left(\frac{r}{\lambda}\right) K_1\left(\frac{r'}{\lambda}\right) r\,dr\,d\phi, \quad (9.67)$$

which can be evaluated to give the force per unit length. This force, which is indicated in Fig. 9.27, is repulsive and moves the vortices apart.

The current density and magnetic field strength vary over the region of integration in the manner illustrated in Fig. 9.29, and Eq. (9.67) cannot be integrated in closed form. If the vortices are far enough apart as to make the current density effectively constant throughout the region of integration, $\mathbf{J}(r')$ may be approximated by $\mathbf{J}(d)$ and taken outside the integral,

$$\mathbf{F}/L = \mathbf{J}(d) \times \int \mathbf{B}(r) r\,dr\,d\phi. \quad (9.68)$$

We know that the magnetic field $\mathbf{B}$ integrated over the cross-section of a vortex equals a fluxoid $\mathbf{\Phi}_0$ oriented along z, giving us

$$\mathbf{F}/L = \mathbf{J} \times \mathbf{\Phi}_0. \quad (9.69)$$

As before, the force $\mathbf{F}$ is along the negative y direction, so it is more convenient to write it as a scalar, F. Inserting Eq. (9.20) for $\mathbf{J}(d)$ and using the approximation (9.28) for $r \gg \lambda$, we obtain

$$\frac{F}{L} = \frac{\Phi_0^2}{2\mu_0(2\pi\lambda^5)^{1/2}} \frac{\exp(-d/\lambda)}{\sqrt{d}} \qquad d \gg \lambda. \quad (9.70)$$

Thus the repulsive interaction between vortices is very weak when the vortices are far apart. The forces between vortices are fairly short range with the penetration depth a measure of the range, and they must be sufficiently close together, compared to λ, for their interaction to be appreciable. We saw from Fig. 9.17 that an applied field $B_{app} \approx 1.8 B_{c1}$ can be strong enough to bring vortices sufficiently close together for this interaction to be effective.

An analogous case occurs when a wire carrying an electric current $\mathbf{I} = \mathbf{J}A$ interacts with a magnetic field $\mathbf{B}$. Equation (9.63) applies to this case also, and we write

$$\mathbf{F}/L = \mathbf{I} \times \mathbf{B}. \quad (9.71)$$

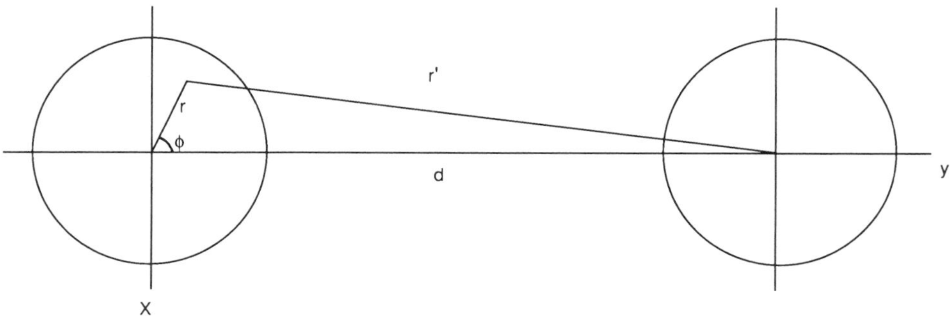

Figure 9.28 Coordinates for calculating the repulsive force between two vortices.

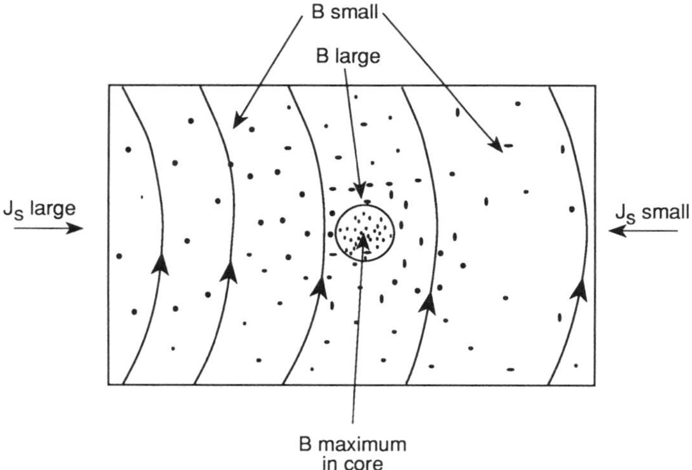

Figure 9.29 Cross section through a vortex core (center) with the strength of its magnitude field **B** directed upward from the page and proportional to the density of the dots. The current density lines $\mathbf{J}_s$ arising from another vortex located to the left become more widely separated toward the right, away from the other vortex. The Lorentz force density $\mathbf{J} \times \mathbf{B}$ is directed to the right and serves to move the two vortices apart.

In this case a wire carrying a current I_2 is encircled by magnetic field lines B_2 and a second parallel current I_1 a distance d away interacts with B_2 and experiences a force of attraction,

$$F/L = 2\mu_0 I_1 I_2/d, \qquad (9.72)$$

as shown in Fig. 9.30. This represents a much slower fall-off with distance than its vortex counterpart (9.70). Thus there is a close analogy between interaction of vortices with a current density "field" and interaction of current-carrying wires with a magnetic field. The vortices confine B and interact with J, while the wires confine J and interact with B. A crucial difference is the sign of the interaction; one repels and the other attracts.

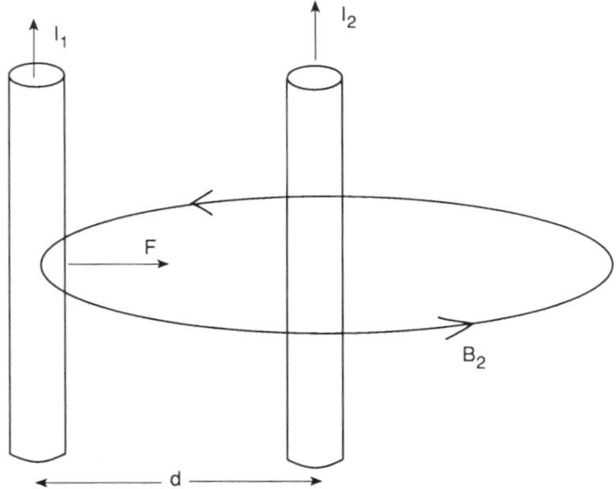

Figure 9.30 Attractive interaction involving a current $\mathbf{I}_1$ and the magnetic field $\mathbf{B}_2$ from a second parallel current $\mathbf{I}_2$ following in the same direction. The force of attraction **F** is shown.

In an anisotropic superconductor the B field of a vortex is not necessarily parallel to the vortex core, but depends on the location of this field relative to the vortex axis and the c direction. There even exist special applied field orientations that are tilted with respect to the principal axes for which adjacent vortices experience an attractive interaction (Bolle *et al.*, 1991; Daemen *et al.*, 1992; Kogan *et al.*, 1990).

B. Pinning

Pinning forces in general are not well understood, and it will be helpful to make a few qualitative observations. Various models and theories to explain pinning have been proposed (e.g., Brass *et al.*, 1989; Coffey, 1992; Daemen and Gubernatis, 1991; Glyde *et al.*, 1992; Levitovi, 1991; Wördenweber, 1992).

A pinning force F_P is a short-range force that holds the core of a vortex in place at, *inter alia*, a point defect (Giapintzakis *et al.*, 1992; Hylton and Beasley, 1990), columnar defect (Nelson and Vinokur, 1992; Prost *et al.*, 1993), screw dislocation (Ivlev and Thompson, 1992; S. Jin *et al.*, 1991), oxygen vacancy (Chudnovsky, 1990; Feenstra *et al.*, 1992), inclusion (Murakami *et al.*, 1991; Sagdahl *et al.*, 1991; Shi *et al.*, 1989, 1990a, b), grain boundary (Müller *et al.*, 1991), twin boundary (Kwok *et al.*, 1990a; Lairson *et al.*, 1990; J.-Z. Liu *et al.*, 1991; Svensmark and Falicov, 1990), intragranular or intergranular nonsuperconducting region (Davidov *et al.*, 1992; Jung *et al.*, 1990), or praseodymium (Pr) doping (Paulius *et al.*, 1993; Radousky, 1992). The density of pinning centers can be high, with average separations of 100 Å or less (Martin *et al.*, 1992; Tessler *et al.*, 1991). Pinning can be an activated process involving pinning barriers (Campbell *et al.*, 1990; Kopelvich *et al.*, 1991; McHenry *et al.*, 1991; Steel and Graybeal, 1992; Zhu *et al.*, 1992), with typical values between 1 and 12 eV for granular $YBa_2Cu_3O_{7-\delta}$ (Nikolo and Goldfarb, 1989). Vortices can undergo thermally activated hopping between pinning centers (Fisher *et al.*, 1991; Liu *et al.*, 1991; Martin and Hebard, 1991).

The Lorentz force needed to depin a single vortex equals the pinning force. The force per unit length needed to produce this depinning, F_p, has been found to have the temperature dependence

$$F_p = F_p(0)[1 - (T/T_c)]^n, \quad (9.73)$$

with $F_p(0)$ varying over a wide range from 10^{-12} to 4×10^{-4} N/m and n ranging from 1.5 to 3.5 (Fukami *et al.*, 1991a; Goldstein and Moulton, 1989; O. B. Hyun *et al.*, 1989; O. B. Hyun *et al.*, 1987; Job and Rosenberg, 1992; Park *et al.*, 1992; Shindé *et al.*, 1990; Wadas *et al.*, 1992; Wu and Sridhar, 1990). A distribution of pinning (activation) energies have been reported in the range of hundreds of meV (e.g., Civale *et al.*, 1990; Ferrari *et al.*, 1989; Fukami *et al.*, 1991b; J.-J. Kim *et al.*, 1991a, b; Mohamed and Jung, 1991; Nikolo *et al.*, 1992). Kato *et al.* (1991) suggested that for $T = 0.9T_c = 9$ K, a vortex that is 2λ from a pinning center moves toward it at ≈ 1000 m/sec to be trapped in $\approx 10^{-9}$ sec. Several workers have found the pinning force to be a maximum for applied fields of $\approx \frac{1}{4}\mathbf{B}_{c2}$, with $\mathbf{B}_{app}$ both along and perpendicular to the c direction (Cooley *et al.*, 1992; Fukami *et al.*, 1989; Satchell *et al.*, 1988).

In tetragonal high-temperature superconductors such as the bismuth and thallium types, the dominant pinning mechanism is generally relatively weak interactions between vortices and randomly distributed defects. In orthorhombic superconductors, such as the yttrium compound, twin boundaries provide stronger pinning to the flux lines. High-temperature superconductors tend to have lower pinning forces than classical superconductors (Ferrari *et al.*, 1991).

Some authors take into account a harmonic pinning force $F_p = -kx$ or $F_p = -k\sin(qx)$ or a stochastic force due to thermal fluctuations (Chen and Dong, 1991; Golosovsky *et al.*, 1991, 1992; Inui *et al.* 1989). Harmonic pinning can be important for oscillating applied fields and thermal fluctuations.

Ionizing radiation increases the concentration of pinning centers (Civale *et al.*, 1991b; Fleisher *et al.*, 1989; Gerhäuser *et al.*, 1992; Konczykowski *et al.*, 1991; Weaver *et al.*, 1991), which can have the effect of increasing the critical current density. For example, neutron irradiation of $HgBa_2CuO_{8+\delta}$ increased the area of the high field (± 1 T) hysteresis loop, and hence raised the value of J_c, by one or two orders of magnitude (Schwartz *et al.*, 1994). This enhancement of J_c is much greater than that obtained with other cuprates and suggests a scarcity of pinning centers before exposure to the neutron flux. Irradiation can be used to maximize J_c or to optimize the pinning force density for a given applied field B_{app} and superconductor type (Kahan, 1991; Vlcek *et al.*, 1992).

C. Equation of Motion

We will examine the case of an isolated vortex $\mathbf{\Phi}_0$ in a region of constant current density **J**, as shown in Fig. 9.31. When the pinning force exceeds the Lorentz force,

$$\mathbf{F}_P > \mathbf{J} \times \mathbf{\Phi}_0, \qquad (9.74)$$

the vortex is held in place and no motion can occur. When the Lorentz force exceeds the pinning force, motion begins. The vortex with an effective mass per unit length m_ϕ (Blatter *et al.*, 1991a; Coffey, 1994; Coffey and Clem, 1991; Gittleman and Rosenblum, 1968; van der Zant *et al.*, 1991) is set into motion and accelerated by the Lorentz force $\mathbf{J} \times \mathbf{\Phi}_0$, and the two velocity-dependent forces come into play. One possible equation of motion is

$$\mathbf{J} \times \mathbf{\Phi}_0 - \alpha n_s e(\mathbf{v} \times \mathbf{\Phi}_0) - \beta \mathbf{v} = m_\phi \frac{d\mathbf{v}}{dt}. \qquad (9.75)$$

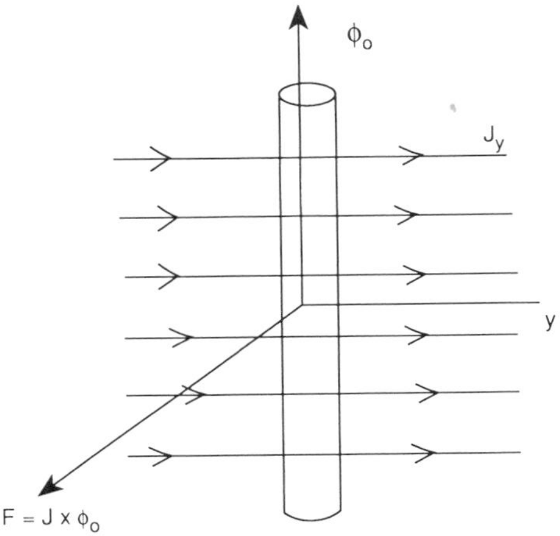

Figure 9.31 Lorentz force $\mathbf{F} = \mathbf{J} \times \mathbf{\Phi}$ exerted on a vortex by a perpendicular transport current **J**.

There is an initial period of acceleration that is too short to be observed, followed by steady-state motion at the terminal velocity $v(\infty)$. The steady-state motion is governed by the equation (Heubener, 1979)

$$\mathbf{J} \times \mathbf{\Phi}_0 - \alpha n_s e(\mathbf{v} \times \mathbf{\Phi}_0) - \beta \mathbf{v} = \mathbf{0}, \qquad (9.76)$$

with the Lorentz force $\mathbf{J} \times \mathbf{\Phi}_0$ balanced by the two velocity-dependent forces. A more general expression, written in terms of an unspecified dissipative force **f**,

$$\mathbf{J} \times \mathbf{\Phi}_0 - \alpha n_s e(\mathbf{v} \times \mathbf{\Phi}_0) - \mathbf{f} = \mathbf{0}, \qquad (9.77)$$

reduces to Eq. (9.76) for the case $\mathbf{f} = \beta \mathbf{v}$ that is suggested by intuition. We will comment on this more general expression at the end of the section.

D. Onset of Motion

At the onset of motion the velocity is very low and the two velocity-dependent terms in Eq. (9.75) can be neglected. This means that the initial velocity and acceleration are along the $\mathbf{J} \times \mathbf{\Phi}_0$ or x direction. As motion continues the velocity $\mathbf{v}(t)$ increases in magnitude toward a terminal value $\mathbf{v}(\infty) = \mathbf{v}_\phi$ with time constant τ_ϕ as well as shifting direction. We will show below that this terminal velocity vector lies in the x, y-plane in a direction between **J** and $\mathbf{J} \times \mathbf{\Phi}_0$.

To estimate the magnitude of τ_ϕ, we recall from hydrodynamics that the time constant for the approach of an object moving in a fluid to its terminal velocity is proportional to the effective mass, and we also know that the effective mass is proportional to the difference between the mass of the object and the mass of fluid which it displaces. In other words, it is proportional to the difference between the density of the object and the density of the medium. Vortex motion involves the movement of circulating super currents through a background medium comprised of super electrons of comparable density, and the closeness of these densities causes m_ϕ, and hence τ_ϕ, to be very small. The terminal velocity is reached so rapidly that only the final steady-state motion need be taken into account. Gurevich and Küpfer (1993) investigated the time scales involved in flux motion and found values ranging from 1 to 10^4 sec. Carretta and Corti (1992) reported an NMR measurement of partial flux melting with correlation times of tens of microseconds.

E. Magnus Force

The Magnus effect involves the force, sometimes called the lift force, that is exerted on a spinning object moving through a fluid medium. This force arises from the Bernoulli equation for streamline (non-turbulent) flow,

$$\tfrac{1}{2}\rho v^2 + P = \text{const}, \qquad (9.78)$$

where $\frac{1}{2}\rho v^2$ is the kinetic energy density, P is the pressure, and the gravity term ρgh is negligible and hence omitted.

When a vortex is moving through a medium at the speed v_ϕ, as shown in Fig. 9.32a, from the viewpoint of an observer on the vortex the medium is moving at the speed $-v_\phi$, as indicated in Fig. 9.32b. On one side of the vortex the velocity v_s of the circulating current adds to the velocity of the medium and on the other side it subtracts from it, in accordance with Fig. 9.32a, and for these two cases Eq. (9.78) assumes the scalar form

$$\tfrac{1}{2}\rho(v_\phi \pm v_s)^2 + P_\pm = \text{const}. \qquad (9.79)$$

Since the kinetic energy is greater for the positive sign, it follows from Eq. (9.79) that $P_+ < P_-$. This pressure difference causes a force F_{lift} to be exerted on the moving vortex in a direction at right angles to its velocity, toward the lower pressure side. The resulting deviated path is shown in Fig. 9.32c. The sideways acting force, called the Magnus force, is given by $-\alpha n_s e(\mathbf{v}_\phi \times \mathbf{\Phi}_0)$, as indicated in Eqs. (9.75)–(9.77). The Magnus coefficient α has different values in different models.

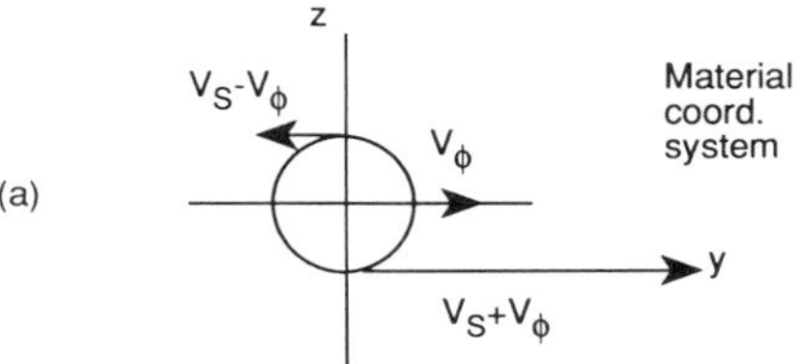

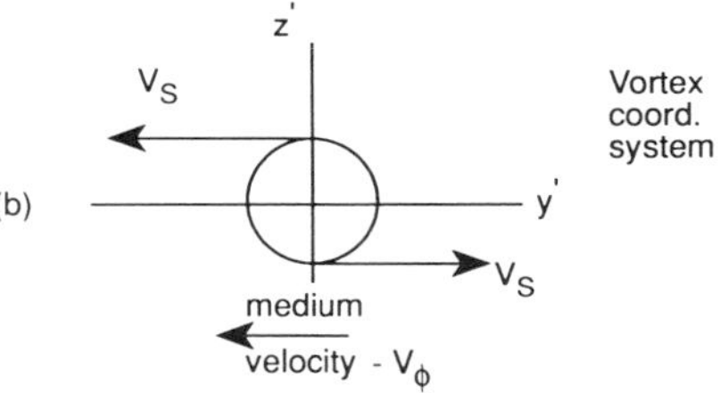

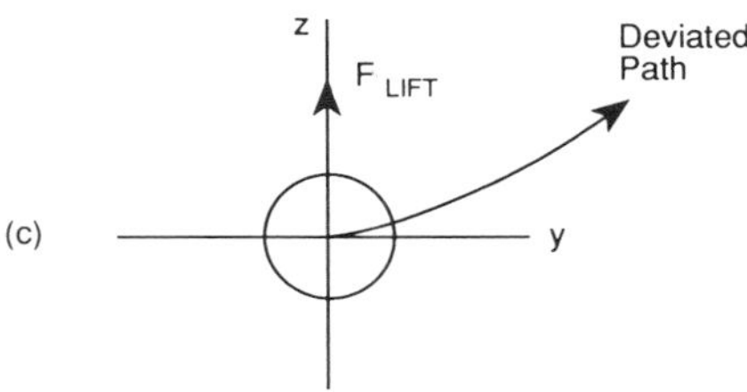

Figure 9.32 Vortex moving in a superconducting medium (a) relative to an x, y coordinate system fixed in the medium, (b) viewed from an x', y' coordinate system fixed on the vortex, and (c) resulting deviated path in the medium arising from the Magnus (lift) force.

F. Steady-State Vortex Motion

For the case of steady-state vortex motion with the viscous retarding force **f** given by $\beta\mathbf{v}$, as in Eq. (9.76), the vectors $\beta\mathbf{v}$ and $\alpha n_s e(\mathbf{v} \times \mathbf{\Phi}_0)$ are mutually perpendicular, as illustrated in Fig. 9.33a. The vortex velocity vector shown in Fig. 9.33b has the magnitude v_ϕ given by

$$v_\phi = \frac{J\Phi_0}{\left[\beta^2 + (\alpha n_s e\Phi_0)^2\right]^{1/2}}, \qquad (9.80)$$

and subtends the angle Θ_ϕ, where

$$\tan\Theta_\phi = \alpha n_s e\Phi_0/\beta, \qquad (9.81)$$

with the $\mathbf{J} \times \mathbf{\Phi}_0$ direction. Thus we see that the greater the viscous drag coeffi-

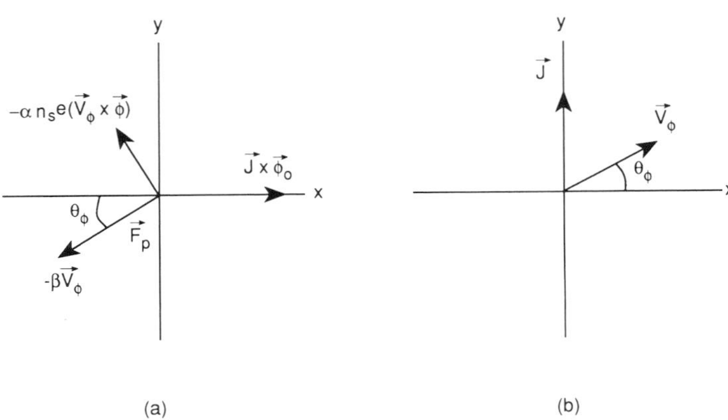

Figure 9.33 Vortex motion at the terminal velocity showing (a) the balance of forces, and (b) the directions of the current flow (**J**) and vortex motion ($\mathbf{v}_\phi$). These figures are drawn for the case $\mathbf{f} = \beta \mathbf{v}_\phi$ of Eq. (9.77).

cient β and the greater the Magnus force coefficient αn_s, the slower the velocity v_ϕ of the vortices.

Various theories have been proposed to explain steady-state vortex motion. It is common to assume that the core of the vortex is normal, that $T \ll T_c$ (so that normal electrons outside the core can be neglected), that $\alpha = 1$, and that the vortices move freely without any influence from pinning. Relaxation is so rapid that the distance a vortex moves in one relaxation period, τ_ϕ, is small compared to the core radius ξ. Bardeen and Stephen (1965), van Vijfeijken and Niessen (1965a, b), and Nozières and Vinen (1966) have proposed models for steady-state vortex motion based on assumptions of this type. Nozières and Vinen assumed that $\alpha = 1$, and wrote Eq. (9.77) in the form

$$n_s e(\mathbf{v}_s - \mathbf{v}_\phi) \times \Phi_0 - \mathbf{f} = 0, \quad (9.82)$$

asserting that the intuitive choice $\mathbf{f} = \beta \mathbf{v}_\phi$ does not agree with experiment. Another possible choice is $\mathbf{f} = \beta \mathbf{v}_s$. Different theories of vortex motion make predictions of the Hall resistivity and the Hall angle Θ_H of Eq. (1.93) and these predictions can be checked with experiment (Chien *et al.*, 1991; Jing and Ong, 1990; Wang and Ting, 1992b). The drag force coefficient β can be looked upon as proportional to a fluxon viscosity. Viscosity coefficients have been determined for the yttrium and bismuth compounds (Golosovsky *et al.*, 1992; Matsuda *et al.*, 1994).

G. Intrinsic Pinning

Pinning forces were introduced in Section B for the isotropic case. When the applied field is in the c direction there is very little anisotropy in the a and b directions, so that motion proceeds as in the isotropic case. When the field is applied in the a, b-plane, the vortex lines are parallel to the layers. They have their lowest energy when their cores are located between the layers (Carneiro, 1992). In the absence of pinning centers, vortex motion within the layers is uninhibited, but motion perpendicular to the layers is hindered by what is called *intrinsic pinning*. Intrinsic pinning is observable in untwinned $YBa_2Cu_3O_{7-\delta}$ when the applied field is aligned along the a, b-planes (Chakravarty *et al.*, 1990; Ivlev and Kopnin, 1990; Ivlev *et al.*, 1991a, b; Kwok *et al.*, 1991; Tachi and Takahashi, 1989; *vide* also Feinberg and Villard, 1990). The simultaneous presence of two species of vortices, with different orientations, has also been discussed (Daemen *et al.*, 1993).

H. Vortex Entanglement

A typical vortex is much longer than its core radius and can be pinned in several places (Niel and Evetts, 1992). For example, a straight vortex aligned along the c direction of a 1.2 μm-thick grain of $YBa_2Cu_3O_7$ has a length of about 1000 unit cells ($c = 1.194$ nm), 500 core radii ($\xi \approx 2.6$ nm), and five penetration depths ($\lambda \approx 0.26$ μm). Such a vortex can twist and turn in the material, and sections of it might move while others remain pinned; it could also undergo a perpendicular excursion along the a, b-plane of a high-temperature superconductor. Another scenario is for two vortices to become entangled, and if this happens they might undergo a reconnection interaction by interchanging segments. These configurations and motions are much harder to handle mathematically in the case of an array of vortices, but they are probably a more accurate representation of the situation than the more idealized case of straight, parallel vortices we have been discussing.

VI. FLUX MOTION

The previous section dealt with the motion of individual vortices. Now we wish to talk about the motion of vortices that are packed together sufficiently densely to be considered as a continuum, or to be thought of as moving about together in groups called *flux bundles*.

A. Flux Continuum

We saw in Section III.D that the vortices in a Type II superconductor are about two penetration depths apart when the applied magnetic field B_{app} reaches a value $\approx 1.8B_{c1}$. For $B_{app} > 2B_{c1}$ there is a strong overlap of the magnetic fields from neighboring vortices, and the internal field B_{in} exhibits spatial variations about an average value that are sketched in Figs. 9.7b and 9.10. The overlap is so great that it is reasonable to consider an array of vortices as constituting a continuum of magnetic flux.

When in motion such a continuum has some of the properties of a highly viscous fluid. Many vortices move as a unit or in large groups under the action of perturbing forces. There are two regimes of flux motion, both of which involve dissipation (Palstra *et al.*, 1988). The first is flux creep, when the pinning force dominates (Anderson, 1962), and the second flux flow, when the Lorentz force dominates (Kim *et al.*, 1964; Tinkham, 1964).

Flux motion is strongly dependent on vortex pinning. Strong pinning centers hold individual vortices in place independently of the presence of the weaker interaction forces from nearby vortices, while weak pinning centers compete with nearby vortices in their ability to hold an individual vortex in place. A large collection of weak pinning centers produces what is called collective pinning (Larkin and Ovchinnikov, 1974; Ovchinnikov and Ivler, 1991). In this case individual vortices cannot move about freely because of constraints from their neighbors so that a relatively small number of pinning centers can restrain the motion of many nearby vortices.

We know that the condensed phases of the liquid and solid state of a particular material are characterized by a fixed density, whereas the gas phase can have a wide range of density. An interesting feature of the condensed vortex phases is that they do not occur for a fixed density, but rather over a range of densities from ρ_{min} to ρ_{max}. This range may be approximated by the ratio between λ and ξ,

$$\frac{\rho_{max}}{\rho_{min}} \approx \kappa^2, \tag{9.83}$$

where the Ginzburg–Landau parameter $\kappa = \lambda/\xi$ enters as a square because the densities ρ are two-dimensional. For high-temperature superconductors with $\kappa \approx 100$ this range of densities is 10^4. Density variations and fluctuations can occur during flux motion.

B. Entry and Exit

The presence of the applied field at the surface of the superconductor induces vortices to form right inside the surface, and a relatively high local concentration can accumulate. An increase in the applied field causes more vortices to enter and move inward by diffusion and by virtue of mutual repulsion, with some of the vortices becoming pinned during migration. The Lorentz force density $\mathbf{J} \times \mathbf{B}$ associated with the interactions between vortices acts like a magnetomechanical pressure serving to push the flux inward. The vortices relax to a new equilibrium distribution consistent with the new screening currents associated with the increase in the applied field.

Foldeaki *et al.* (1989) found that in $YBa_2Cu_3O_{7-\delta}$ there is a large difference between the rate of flux flow in the case of expulsion of flux caused by removal of the applied field $\mathbf{B}_{app}$ following field cooling versus the rate at which flux penetrates the sample when $\mathbf{B}_{app}$ is turned on following zero field cooling. The activation energies in the former case, 14–28 meV, are significantly smaller than those, 34 to 67 meV, in the latter case.

C. Two-Dimensional Fluid

Ordinarily, we think of a gas as a collection of molecules that are so widely separated that the interactions between them are negligible, and that the molecules move independently of each other except when undergoing elastic collisions that change their directions and velocities. A liquid is a collection of molecules that are held closely together by short-range attractive forces, with thermal energy causing them to move around while remaining in contact, thereby preserving short-range order. In the solid state the nearest-neighbor attractive forces dominate over thermal effects, and the molecules become fixed in position in a regular lattice arrangement of the type shown on Fig. 9.14, with long-range order.

Molecules confined to a surface can exist in two-dimensional gas, liquid, and solid states. If some of the molecules in such a two-dimensional fluid become attached to the pinning centers, the motion of the fluid will be restricted by having to flow past the pinned molecules. An array of vortices in a superconductor has many of the properties of these two-dimensional states of matter, but there are some fundamental differences between the two cases. First the interactions between the vortices are repulsive rather than attractive. In addition, the forces have two characteristic lengths, a short coherence length that constitutes a closest-approach distance, and a (much) greater penetration depth which is a measure of the range of the repulsive interaction. The vortices are also much longer than their core diameters so they can become twisted and distorted, as was already mentioned in Section V.H.

When the average separation of vortices is much greater than the penetration depth, they will form a two-dimensional gas in which they are able to move independently of each other, assuming pinning is absent. We can deduce from Fig. 9.16 that applied fields only slightly above the lower-critical field—e.g., $B_{app} \approx 1.1 B_{c1}$ for $\kappa \approx 100$—can produce relatively closely spaced vortices (see Problem 6). In addition, the lower-critical field is often not a sharply defined quantity. As a result of these factors, the range of applied fields over which a vortex gas state might be able to exist is too small to be significant. It is the condensed phases which are mainly of interest.

D. Dimensionality

A flux fluid is three-dimensional because it occupies a volume of space, but it moves in a plane perpendicular to the internal field direction, so its motion is often treated as two-dimensional. When the applied field $\mathbf{B}_{app}$ is along the *c* direction of a cuprate superconductor, the vortices break up into pancake vortices that are confined to the CuO_2 layers. If the layers then become decoupled from each other, the resulting flux flow can be looked upon as a movement of pancake vortices in each layer that are independent of each other. If a strong pinning center exists in one layer, the pancake vortices in that layer will flow around it. Pancake vortices in the layers above and below will not experience that pinning center, however.

There is also a dimensionality in the super current flow. We see from Fig. 7.26 that the cuprates have groups of closely spaced and, hence, strongly coupled CuO_2 layers. These CuO_2 layers are also more widely separated from the next group above and below, with the coherence length in the c direction less than the distance between pairs of superconducting layers or groups of layers. For example, the $Tl_2Ba_2Ca_nCu_{n+1}O_{6+2n}$ compound has $n+1$ closely spaced CuO_2 planes with intervening Ca ions that are separated from the next such planar group by layers of BaO and TlO. Adjacent groups of planes are uncoupled from each other and conductivity is two-dimensional (2D) whenever the coherence length along the c direction is less than $s/\sqrt{2}$, where s is the spacing between successive planar groups (Bulaevskii, 1973). For example,

$$YBa_2Cu_3O_{7-\delta}$$

at 0 K has $\xi_c(0) \approx 3$ Å, and $s = 11.9$ Å is the lattice parameter along c, so $\xi_c \ll s/\sqrt{2}$ and the superconductivity is two-dimensional (2D).

The coherence length $\xi_c(T)$ has the temperature dependence

$$\xi_c(T) \approx \xi_c(0)(1-(T/T_c))^{-1/2}, \quad (9.84)$$

and when $\xi_c(T)$ is equal to $s/\sqrt{2}$ there is a crossover between 2D and 3D behavior, the latter occurring for $\xi_c > s/\sqrt{2}$. Marcon *et al.* (1992) report crossover temperatures of $0.99T_c$ for $Bi_2Sr_2CaCu_2O_{10}$ and $0.88T_c$ for

$$YBa_2Cu_3O_{7-\delta}.$$

The presence of an applied field lowers the critical temperature, $T_c(B) < T_c$, and this depresses the crossover temperature as well (Božovic, 1991; Farrell *et al.*, 1990c; Gray *et al.*, 1992; Koorevaar *et al.*, 1990; Weber and Jensen, 1991).

E. Solid and Glass Phases

When vortices become sufficiently numerous so that their charge and current densities overlap appreciably, they form a condensed phase, either a liquid phase if the temperature is relatively high or a solid near absolute zero. In the absence of pinning and anisotropy the vortices form the hexadic pattern of Fig. 9.14 with each vortex stretched between its two end points like an elastic string, as noted in Section IV. This configuration has long-range order, so the state is called a *flux lattice*. When random pinning centers are present, the spatial structure will reflect their distribution and the long-range order will be disturbed. The result is what is called a *vortex glass* (Chudnovsky, 1991). The portions of the flux lines between the pinning sites are held in place by repulsion from nearby vortices and form local hexadic arrangements, so short-range order is present. Bitter pattern decorations of $YBa_2Cu_3O_7$ crystals display this short-range order (Dolan *et al.*, 1989a; Gammel *et al.*, 1987).

A flux-glass phase can also form and exhibit short-range order (Chudnovsky, 1989; Fisher, 1989). The positional order of a vortex glass is analogous to the magnetic order of a spin glass (Binder and Young, 1986; Fisher and Huse, 1988). Both a flux lattice and a flux glass are solid phases, since in these phases the vortices remain fixed in place so long as the temperature is low enough, the applied field remains the same, and there is no transport current. Several researchers have reported evidence for a vortex-glass phase in epitaxial films (Koch *et al.*, 1989) and monocrystals (Rossel *et al.*, 1989a, b), other investigators have studied flux melting and the transition to the vortex-glass state (Charalambous *et al.*, 1992; Dekker *et al.*, 1992; Dorsey *et al.*, 1992; Koka and Shrivastava, 1990a, b; Safar *et al.*, 1992; Yeh, 1990; Yeh *et al.*, 1992a, b). Mechanical oscillation methods have been employed to study the melting transition (Gammel *et al.*, 1989; Gupta *et al.*, 1991; F. Kober *et al.*, 1991; Luzuriaga *et al.*, 1992; E. Rodriguez *et al.*, 1990). The subject has also been examined theoretically (Gingras, 1992; Toner, 1991b).

F. Moving Flux

The very slow flux motion at temperatures far below T_c is referred to as *flux creep*. When a magnetic field is applied to a superconducting sample for $T \ll T_c$, the field penetrates very slowly. Plots of the magnetization versus the logarithm of time tend to be linear, with flux continuing to enter the sample several hours later. Figure 9.34 shows the time dependence of the magnetization in a superconductor that was zero field cooled then exposed to a 5-T field that was subsequently decreased to 3 T. In this experiment (Kung *et al.*, 1992; see also Pencarinha *et al.*, 1994) the time dependence of the magnetization of $YBa_2Cu_3O_7$ containing Y_2BaCuO_5 particles (green phase) functioning as pinning centers to enhance $\mathbf{J}_c$ was monitored for several hours after the field had been reduced to 3 T. Experimental results are shown for temperatures between 5 K and 50 K. Similar results have been obtained with an electron spin resonance surface probe technique (Pencarinha *et al.*, 1994). We see from the figure that the magnetization is greater in magnitude and decays faster at the lower temperatures. The logarithmic time dependence of M, shown at low temperatures as in Fig. 9.34, often becomes nonlogarithmic at higher temperatures (Lairson *et al.*, 1990b; J. Z. Liu *et al.*, 1992; Safar *et al.*, 1989; Shi *et al.*, 1991).

For weak pinning the vortex lattice reacts elastically to an applied force, such as the Lorentz force from a transport current. In the case of strong pinning, untrapped vortices move past trapped vortices and flux flows along channels between regions of trapped flux (Brechet *et al.*, 1990). This latter flow can involve groups of vortices moving cooperatively as a unit, forming what are called *flux bun-*

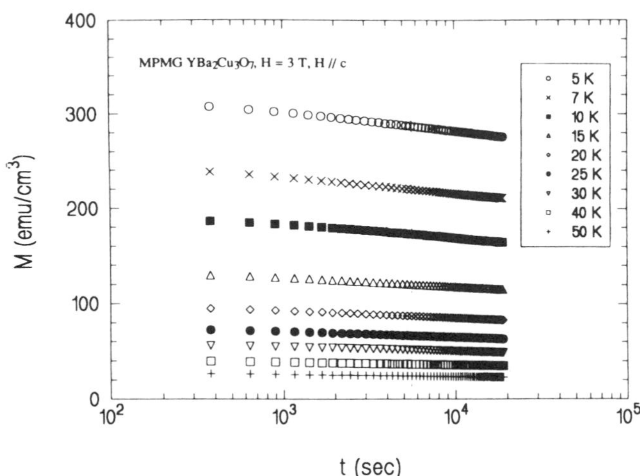

Figure 9.34 Magnetic relaxation of $YBa_2Cu_3O_7$ in a 3-T field showing the linear dependence of magnetization on the logarithm of time for temperatures between 5 K and 50 K (Kung *et al.*, 1992; see also Pencarinha *et al.*, 1994).

dles (Geim *et al.*, 1992; C. A. Wang *et al.*, 1992; Zeldov *et al.*, 1989) containing from 4 (Stoddart *et al.*, 1993) to 10^4 (Plaçais and Simon, 1989) vortices. Energy barriers can hinder flux creep (Anderson, 1962; Anderson and Kim, 1964) which involves thermally activated jumps of flux bundles (Cross and Goldfarb, 1991). In high-temperature superconductors the flux creep rate is much greater for flow parallel to the planes than for flow perpendicular to the planes (Biggs *et al.*, 1989). Some recent articles have discussed flux creep in thin films (Fischer *et al.*, 1989; Hettinger *et al.*, 1989), monocrystals (Xenikikos and Lemberger, 1990; Zuo *et al.*, 1991), and oriented powders (Ossandon *et al.*, 1992b; Y. Xu *et al.*, 1989, 1990) of $YBa_2Cu_3O_{7-\delta}$. Theoretical treatments have also been provided (Anderson, 1962; Beasley *et al.*, 1969).

G. Transport Current in a Magnetic Field

Suppose that a transport current I of uniform density J flows along a superconducting wire located in a transverse magnetic field. If the pinning forces are not sufficiently strong to prevent flux motion, i.e., the superconductor is soft, then: (1) The current exerts a force $\mathbf{J} \times \mathbf{\Phi}_0$ on the vortices, causing them to move from one side of the wire to the other. Viscous drag limits this motion to a constant velocity v_ϕ and the Magnus force causes it to occur at the angle Θ_ϕ shown in Fig. 9.33, as already noted; (2) through Maxwell's equation $\nabla \times \mathbf{B} = \mu_0 \mathbf{J}$ a constant magnetic field gradient is established across the sample. When the applied field is in the z direction and the current flows in the x direction, the gradient is given by

$$\frac{d}{dy} B_z(y) = \mu_0 J_x; \tag{9.85}$$

this situation is sketched in Fig. 9.35. (Wilson, 1983); (3) the flux flow corresponds to a magnetic field $\mathbf{B}$ moving across the sample at the constant speed v_ϕ. Such a moving magnetic field generates an electric field

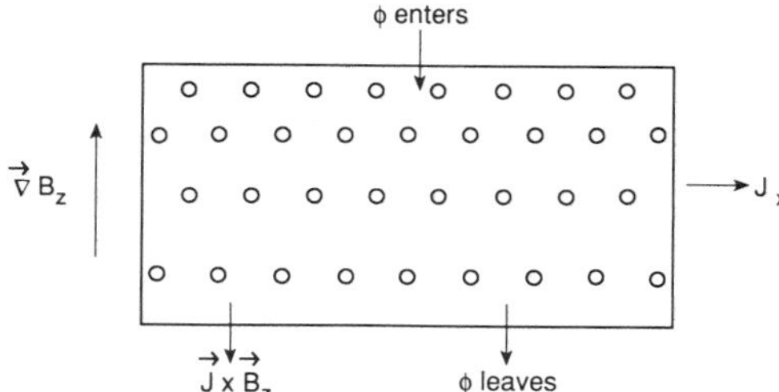

Figure 9.35 Triangular lattice of vortices with gradient $\nabla B_z = dB_z(y)/dy$ in the y direction due to application of a transport current density J_x in addition to the magnetic field B_z. The direction of the Lorentz force $\mathbf{J} \times \mathbf{B}$ is shown. In the absence of pinning forces the current density causes the vortices to move downward at a constant velocity, with new vortices entering the superconductor at the top and old vortices leaving at the bottom. Pinning forces can stop this motion and provide dissipationless current flow. Figure 9.14 shows this vortex lattice in the same applied field but without a transport current.

$$\mathbf{E} = \mathbf{v}_\phi \times \mathbf{B} \tag{9.86}$$

in the superconductor which is perpendicular to both $\mathbf{v}_\phi$ and $\mathbf{B}$. The electric field has a component with the same direction as $\mathbf{J}$, giving rise to the ohmic loss $\mathbf{J} \cdot \mathbf{E}$,

$$\mathbf{J} \cdot \mathbf{E} = \mathbf{J} \cdot (\mathbf{v}_\phi \times \mathbf{B}). \tag{9.87}$$

Another way of viewing the situation is to consider flux flow as a rate of change of flux which, by Faraday's law, produces a voltage drop in the superconductor along the direction of the current flow. The resistivity associated with this flow, according to Ohm's law, provides the mechanism for heat dissipation. Pinning of the vortices prevents the flux from flowing, and the result is no voltage drop and zero resistance.

The pinning forces must be weak enough to permit the initial vortex movement required for the establishment of the flux gradient of Eq. (9.85), and strong enough to prevent the continuous vortex motion that produces the heat dissipation of Eq. (9.87). Pinning forces are ordinarily quite weak. Intensive present-day research and development efforts are aimed at achieving sufficiently strong pinning forces for high current densities.

The mutual repulsion between vortices acts to set up a uniform vortex density, and, hence to oppose the establishment of the flux gradient sketched in Fig. 9.35. The pinning must be strong enough to maintain the gradient against these opposing forces. The stronger the pinning, the greater the magnitude of the gradient that can be maintained, and hence, from Eq. (9.85), the greater the current density that can flow without dissipation. This means that the highest current density that can flow in a superconductor, as given by the critical value $\mathbf{J}_c$, increases with an increase in the pinning strength.

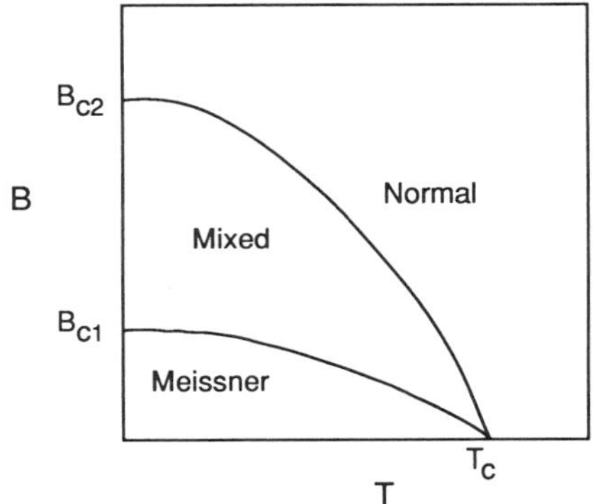

Figure 9.36 Simplified magnetic phase diagram showing the mixed and Meissner states of a Type II superconductor separated by the $B_{c1}(T)$ line.

H. Dissipation

Transport currents and thermal fluctuations can both induce flux motion and produce dissipation. This can involve, for example, the release and transportation of vortices to other pinning centers (Lairson *et al.*, 1991), and the pinning and depinning of flux bundles.

Heat produced by flux motion flows away from the region of generation toward the boundary of the material. A steady state is established in which the interior of the superconductor is at a somewhat higher temperature. We know from Chapter 2, Sections XII and XVI, that the critical current density J_c depends on the applied field and the temperature, so the heat that is generated will limit the value of J_c; the sample will go normal if the applied current density exceeds J_c. High critical currents can be attained by preparing samples with favorable distributions of pinning centers. This is especially important for magnet wire, which must carry high currents in the presence of strong magnetic fields.

I. Magnetic Phase Diagram

Figure 9.36 is a simplified phase diagram of the magnetic states of a Type II superconductor based on a Meissner phase of perfect diamagnetism (absence of vortices) at the lowest temperatures and a mixed (vortex-lattice) phase at higher temperatures. From the foregoing discussion it is reasonable to assume that the situation is actually much more complicated, however. Several more realistic phase diagrams have been suggested in the literature (Farrell *et al.*, 1991a; Fisher, 1990; Gammel *et al.*, 1991; Gerber *et al.*, 1992; Glazman and Koshelev, 1991a, b; Huang *et al.*, 1991a; Kes *et al.*, 1991; Gerber Marchetti and Nelson, 1990; Safar *et al.*, 1993; Schaf *et al.*, 1989; Zwerger, 1990). We will describe one such diagram (Yeh, 1989, 1991; Yeh and Tsuei, 1989).

Figure 9.37 depicts, in addition to the Meissner phase, a flux solid phase with vortices pinned or otherwise held in place, and a flux liquid phase with many vortices unpinned or free to move reversibly, but with dissipation. These two phases are separated by what is called the *irreversibility line* T_{irr}. In the narrow region, called the *plasma phase*, thermal fluctuations create positively and negatively oriented vortices, called *intrinsic vortices*. These latter vortices are more numerous than the field-induced (extrinsic) vortices we will be describing in the following section.

Some authors call the boundary between the flux liquid and the condensed flux phase the *melting line*; along this line depinning takes place (Hébard *et al.*, 1989), and flux creep becomes flux flow.

A number of theoretical treatments involving the structure and dynamics of the various phases of the flux state have appeared. The vortex configurations in these phases have been simulated by Monte Carlo (Hetzel *et al.*, 1992; Li and Teitel, 1991, 1992; Minnhagen and Olsson, 1991; Reger *et al.*, 1991; Ryu *et al.*, 1992) and other calculational methods (Aktas *et al.*, 1994; Jensen *et al.*, 1990; Kato *et al.*, 1993). Computer-generated contour drawings of vortices and vortex motion have been created (Brass and Jensen, 1989; Brass *et al.*, 1989; Kato *et al.*, 1991; Schenström *et al.*, 1989; Tokuyasu *et al.*, 1990; Xia and Leath, 1989).

VII. FLUCTUATIONS

Thermal fluctuations can have important effects on the properties of superconductors. In the present section we will give several examples of these effects.

A. Thermal Fluctuations

Thermal fluctuations increase with temperature, and as they do so they increase the extent to which vortices vibrate. Isolated flux lines acting as stretched strings can undergo longitudinal or transverse vibrations, but at higher concentrations the vibrations are more localized along the length of the core with numerous nearby vortices participating. Thermally induced fluctuations are better described as localized vibrations of a flux-line lattice with amplitudes and frequencies that depend on the wave vector dependent shear (c_{66}), bulk (c_B), and tilt (c_{44}) elastic constants (Brandt, 1989, 1990, 1992; Brandt and Sudbø, 1991; Houghten *et al.*, 1989; Kogan and Campbell, 1989; Shrivastava, 1990; Sudbø and Brandt, 1991a, b; Yeh *et al.*, 1990). When the vibrations become large enough they cause the solid-flux phase to disorder into a flux liquid consisting of mobile, pulsating vortices (Fisher *et al.*, 1991). According to the usual Lindemann criterion, melting occurs when the

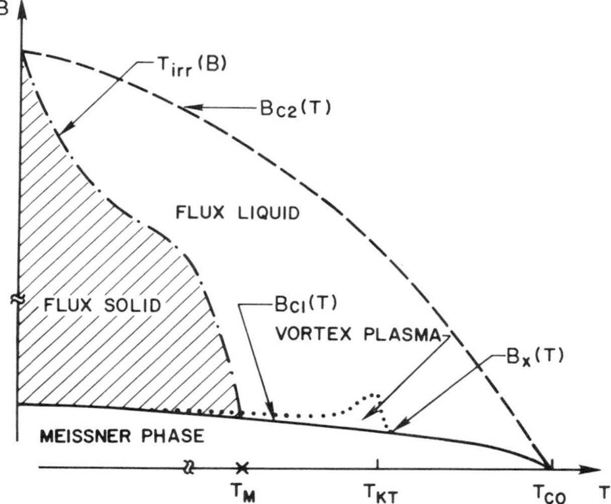

Figure 9.37 More complex magnetic phase diagram showing the Meissner phase, flux solid, and flux liquid regions separated by the irreversibility line (T_{irr}), plasma phase, lower- ($B_{c1}(T)$) and upper- ($B_{c2}(T)$) critical field curves, and melting (T_M) and Kosterlitz–Thouless (T_{KT}) temperatures (Yeh, 1989; Yeh and Tsuei, 1989).

root mean-square fluctuation amplitude $u_{\rm rms}$ exceeds the quantity $\approx 10^{-1}d$, where d is the average vortex separation introduced in Section III.C (Blatter and Ivler, 1993; Lindemann, 1910; Sengupta *et al.*, 1991). The thermal fluctuations can introduce noise and otherwise influence measurements of, for example, electrical conductivity (Jensen and Minnhagen, 1991; Song *et al.*, 1992), specific heat (Riecke *et al.*, 1989), and NMR relaxation (Bulut and Scalapino, 1992).

In granular samples the superconducting grains can couple together by means of Josephson weak links (cf. Chapter 13, Section VI.A) over a range of coupling energies. When the thermal energy k_BT exceeds the Josephson coupling energy of a pair of grains, the two grains can become uncoupled so that supercurrent no longer flows between them.

B. Characteristic Length

The flux quantum Φ_0 is associated with a characteristic length Λ_T which is determined by equating the quantized flux energy to the thermal energy. The energy U_M of a magnetic field in a region of volume V is given by

$$U_M = (B^2/2\mu_0)V \tag{9.88a}$$

$$= (\Phi^2/2\mu_0)(V/A^2), \tag{9.88b}$$

where $B = \Phi/A$. If this is equated to the thermal energy k_BT for a quantum of flux, and if we write $A^2/V = 2\pi/\Lambda_T$, we obtain for the characteristic length

$$\Lambda_T = \frac{\Phi_0^2}{4\pi\mu_0 k_B T} \tag{9.89}$$

$$= \frac{1.97}{T}\ \text{cm}, \tag{9.90}$$

where T is the temperature in degrees Kelvin. This is much larger than other characteristic lengths, such as ξ and λ, except in the case of temperatures extremely close to T_c, where Λ_T can become very large (cf. Eq. (2.57), Fig. 2.42). Therefore, fluctuation effects are expected to be weak in superconductors. In the high-temperature cuprates several factors combine to enhance the effects of thermal fluctuations: (1) higher transition temperature, (2) shorter coherence length ξ, (3) large magnetic penetration length λ, (4) quasi-two-dimensionality, and (5) high anisotropy (Fisher *et al.*, 1991; Nelson and Seung, 1989; Vinokur *et al.*, 1990; cf. discussion of Schnack and Griessen, 1992).

C. Entanglement of Flux Lines

At the lowest temperatures a hexagonal flux lattice is expected, perhaps with irregularities due to pinning. At higher temperatures thermal agitation becomes pronounced and can cause an individual vortex that is pinned in more than one place to undergo transverse motion between the pinning sites. This induces a wandering of vortex filaments and leads to an entangled flux liquid phase, as illustrated in Fig. 9.38. A pair of flux lines passing close to each other can be cut, interchanged, or reattached (LeBlanc *et al.*, 1991; Marchetti, 1991; Nelson and Le Doussal, 1990; Nelson and Seung, 1989; Obukhov and Rubinstein, 1990; Sudbø and Brandt, 1991a, b).

D. Irreversibility Line

Another characteristic of a glass state is irreversibility. This can manifest itself in

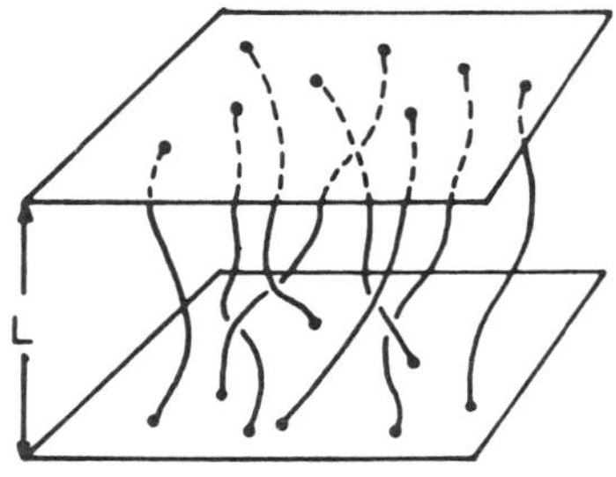

Figure 9.38 Sketch of vortex lines of an entangled flux lattice (Nelson and Seung, 1989).

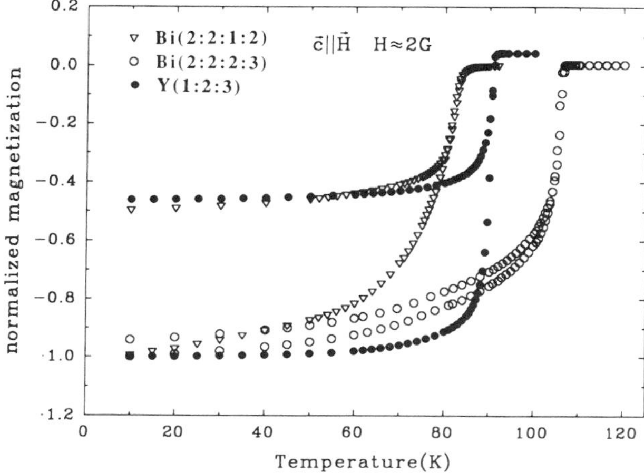

Figure 9.39 Temperature dependence of normalized magnetization of $YBa_2Cu_3O_7$, $Bi_2Sr_2CaCu_2O_8$, and $Bi_2Sr_2Ca_2Cu_3O_{10}$ for a 0.1-mT field applied parallel to the *c*-axis. Both field-cooled (upper curves) and zero-field-cooled (lower curves) data are shown for each superconductor (Y. Xu and Suenaga, 1991).

resistivity, susceptibility, and other measurable parameters (Ramakrishnan *et al.*, 1991). This history-dependent property was observed by Müller *et al.* (1987) at the beginning of the high-T_c era. The magnetization data plots presented in Fig. 9.39 for the three superconductors $YBa_2Cu_3O_7$, $Bi_2Sr_2CaCu_2O_8$, and $Bi_2Sr_2Ca_2Cu_3O_{10}$ all have a temperature T_{irr} above which the zero-field-cooled and field-cooled points superimpose, and below which the ZFC data are more negative than the FC ones (de Andrade *et al.*, 1991). The irreversibility temperature $T_{irr}(B)$ has been determined for a series of applied fields B, and results for the yttrium sample of Fig. 9.39 are plotted in Fig. 9.40. The linearity of these irreversibility line plots shows that

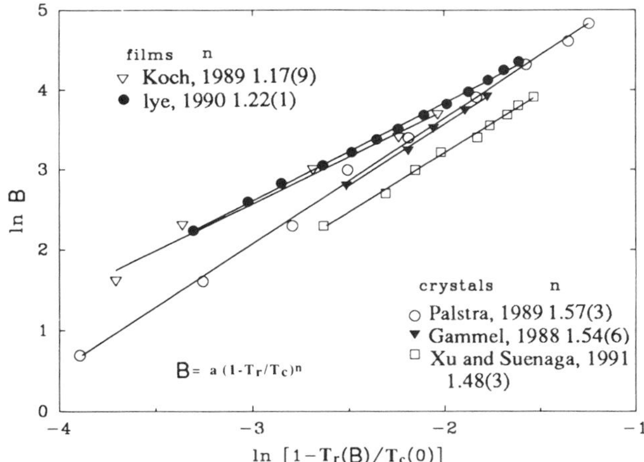

Figure 9.40 Log–Log plot of applied field versus temperature for $YBa_2Cu_3O_7$ films and monocrystals where T_r is the irreversibility temperature (Y. Xu and Suenaga, 1991).

there is a power-law relationship,

$$B \approx a\left[1 - \frac{T_{\mathrm{irr}}(B)}{T_{\mathrm{c}}(0)}\right]^n, \qquad (9.91)$$

where a is a proportionality constant (Lombardo *et al.*, 1992; Sagdahl *et al.*, 1990; Y. Xu and Suenaga, 1991). The slopes of the lines give $n \approx 1.2$ for thin films and $n \approx 1.5$ for crystals. Analogous plots are available for the flux-lattice melting temperature T_M(B), vortex glass-liquid transition temperature T_g (Koch *et al.*, 1989), and resistive transition temperature T_R(B). The resistivity plots exhibit a similar type of irreversibility as magnetization plots. Some recent experimental studies of the irreversibility line of cuprates have been carried out by means of resistivity (Jeanneret *et al.*, 1989; Yeh, 1989; Yeh and Tsuei, 1989) and susceptibility (Geshkenbein *et al.*, 1991; Khoder *et al.*, 1990; Perez *et al.*, 1991; Pureur and Schaf, 1991) measurements. Results are available for the new mercury superconductors (Chu, 1994; Iwasa *et al.*, 1994b, Huang *et al.*, 1994). Safar *et al.* (1989, 1991) mentioned that the melting and irreversibility lines tend to merge at higher fields, and this trend is clear from Fig. 9.40.

The irreversibility line is not very sensitive to the type and distribution of defects, although these defects have a pronounced effect on the critical current density (Cival *et al.*, 1990, 1991a, b).

Six samples of $YBa_2Cu_3O_{6.38}$ with transition temperatures in the range 7–17 K, adjusted by varying the quenching temperature, were studied, and the irreversibility temperature $T_{\mathrm{irr}}(B)$ was found to depend on T_c. Plotting T_{irr}/T_c versus $[1 - (T_{\mathrm{irr}}/T_c)]$ on a log–log scale, however, caused all of the data to follow the same universal curve (Seidler *et al.*, 1991). This type of scaling of irreversibility curves has been interpreted in terms of the Bean model (Wolfus *et al.*, 1988). The classical superconductors Nb_3Sn and Nb–Ti, when produced in the form of fine multifilamentary wires, exhibit magnetization reversibility with an irreversibility temperature $T_{\mathrm{irr}}(B)$ that may be close to the flux-lattice melting temperature (Suenaga *et al.*, 1991; cf. Drulis *et al.*, 1991). This is not the case for the cuprates, for which T_{irr} is interpreted as the depinning temperature.

E. Kosterlitz–Thouless Transition

We mentioned in Section VI.I that thermal fluctuations at low temperatures result in the production of vortex–antivortex pairs, called intrinsic vortices, where the flux and screening currents of an antivortex flow in a direction opposite to that of the vortex. A vortex and antivortex attract each other; at low temperatures they form bound pairs that dissociate at what is called the Kosterlitz–Thouless temperature T_{KT}, indicated in Fig. 9.37 (Berezinskiv, 1971; Creswick *et al.*, 1992; Kosterlitz and Thouless, 1972, 1973; Matlin *et al.*, 1989; Nelson, 1980; Pradhan *et al.*, 1993; Scheidl and Hackenbroich, 1992; Yeh, 1989; Yeh and Tsuei, 1989). In the cuprates the bound pairs tend to reside between planes (Dasgupta and Ramakrishnan, 1991). Experimental evidence for the KT transition in superconductors has been reported (Ando *et al.*, 1991a, b; Ban *et al.*, 1989; cf. Rogers *et al.*, 1992), and the subject has been discussed theoretically by a number of researchers (Blatker *et al.*, 1991b; Choi and Kim, 1991; Ding, 1992; Fazio and Schön, 1991; Gouvea *et al.*, 1989; Minnhagen and Olsson, 1991; Wallin, 1990).

By way of summary, some of the concepts and processes discussed in the previous sections are visualized in Fig. 9.41 (Brandt, 1990). The original article should be consulted for those parts of the figure that have not been discussed here.

VIII. QUANTIZED FLUX

In chapter 5, Section VI, we showed that the magnetic flux $\Phi = \mathbf{B} \cdot \mathbf{A}$ in a super-

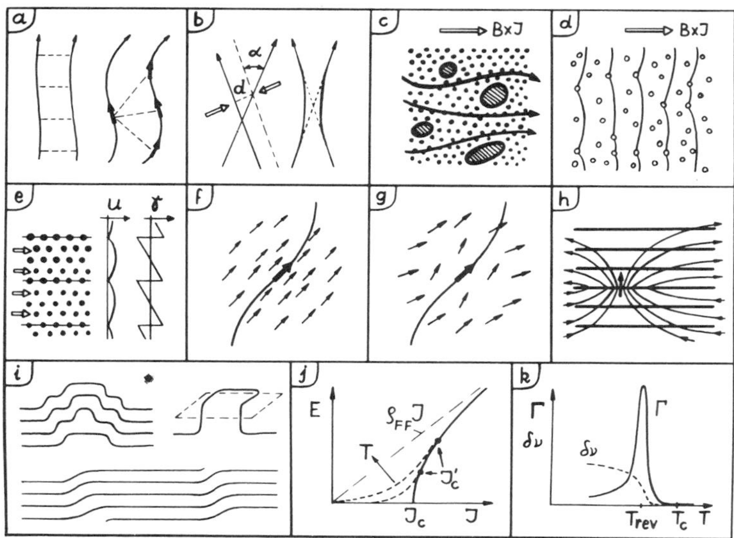

Figure 9.41 Schematic visualization of superconductor concepts and processes. (a) left: simplified 2D interaction between nearly parallel vortex lines, right: more realistic 3D interaction between vortex segments; (b) pair of vortices approaching, crossing, and reconnecting; (c) soft vortex liquid flowing between pinning regions; (d) vortex lattice or liquid, with each vortex pinned by many small pins, subjected to a Lorentz driving force acting toward the right; (e) vortex lattice pinned at equidistant lattice planes parallel to the Lorentz force, which presses it through these channels; plots showing the vortex displacements u and the zigzag shear strain v; (f) magnetic field (small arrows) arising from a vortex segment (dark arrow) in an isotropic superconductor; (g) the same in an anisotropic superconductor; (h) magnetic field lines associated with a point pancake vortex (vertical arrow) on a superconducting layer shown bent by the other layers (horizontal lines) and forced to become parallel to these layers; (i) vortex kinks (upper left), kink pairs (upper right), and a 3D kink structure (lower); (j) current–voltage curves for thermally activated vortex motion; and (k) damping Γ and frequency enhancement $\delta\nu$ of high-temperature superconducting vibrating reed in a longitudinal field. (See original article (Brandt, 1990) for details.)

conductor is quantized with the fluxoid

$$\Phi_0 = h/2e \tag{9.92}$$

as the unit of quantization, where Φ_0 is the flux present in one vortex. However, magnetic flux is not generally quantized as it is in superconductors.

Electric flux Φ_E defined by the expression

$$\Phi_E = \mathbf{E} \cdot \mathbf{A} \tag{9.93}$$

is also not generally quantized, but we know from Gauss' law that a fixed amount of electric flux equal to $\Phi_E = e/\epsilon_0$ is present around an isolated electron. From this viewpoint an electron is a unit electric monopole with e/ϵ_0 its associated quantum of electric flux.

Some physicists have speculated on the possibility of magnetic monopoles, a hypothetical magnetic analogue of electric charges, with the fundamental "magnetic charge" unit h/e. If an isolated magnetic monopole were to exist, the region of space around it could be expected to contain a quantum of magnetic flux, by analogy with the case of an electric monopole. Vortices are more like elongated magnetic dipoles, since a short vortex in isolation would have a dipole field around it.

A region of space containing magnetic monopoles would be described by adding magnetic charge density ρ_m and magnetic current density $\mathbf{J}_m$ terms to the right side of the two homogeneous Maxwell's equations (1.65) and (1.66), respectively. At present there is no firm experimental evidence for the existence of magnetic monopoles, although superconducting loops have been used in the search for them, such as in the experiment of Cabrera *et al.* (1989) at Stanford.

FURTHER READING

Other superconductivity texts cover some of the material presented in this chapter (e.g., Kresin and Wolf, 1990; Orlando and Delin, 1991; Rose-Innes and Rhoderick, 1994; Tilley and Tilley, 1986; Tinkham, 1985).

Flux lattice melting has also been reviewed (Brandt, 1991). Tinkham and Lobb (1989) survey flux creep and irreversibility. The dynamics of the vortex system in high temperature superconductors have been surveyed by Blatter *et al.* (1994), Farrell (1994), and Sengupta and Shi (1994). Anisotropy effects are reviewed by Cooper and Gray (1994), and Gray (1994).

PROBLEMS

1. Consider a Type II superconductor with a Ginzburg–Landau parameter $\kappa = 100$, transition temperature $T_c = 100$ K, and Debye temperature of 200 K. Use standard approximation formulae to estimate its upper critical field, lower critical field, thermodynamic field, energy gap, and electronic and vibrational specific heats.
2. Consider three vortices that form an equilateral triangle 3 μm on a side, with one of the vortices pinned. The temperature $T = \frac{1}{4}T_c$ and the penetration depth $\lambda(0) = 3000$ Å. What is the force per unit length on each vortex and in what direction will the two unpinned vortices move?
3. Show that for $\kappa \gg 1$, the quantity of magnetic flux in the core of an isolated vortex is given by

$$\Phi_{\text{core}} \approx \Phi_0/2\kappa^2(\ln 2\kappa + 1/2 - \gamma),$$

where γ is the dimensionless Euler–Mascheroni constant. What fraction of the total flux is in the core of an isolated vortex of Ti_2Nb, of Nb_3Sn, or of a typical high-temperature superconductor?
4. We have seen that parallel vortices inside a superconductor repel each other and become distributed throughout the interior. Show that charges of the same sign inside a normal conductor repel each other and become distributed over the surface. Hint: use Gauss' law and the continuity equation.
5. Show that the integrals of the asymptotic forms of the modified Bessel functions $K_0(r/\lambda)$ and $K_1(r/\lambda)$ of Eqs. (9.24)–(9.26) do not diverge as $r \to 0$, but rather provide finite fields and currents, respectively, in the neighborhood of the origin.
6. For a high-temperature superconductor ($\kappa \approx 100$), how high an applied field B_{app} is needed, relative to B_{c1}, to cause the average separation d between nearest-neighbor vortices to reach (a) 100λ, (b) 30λ, (c) 10λ, (d) 3λ, or (e) 1λ?
7. The small argument limit, $r \ll \lambda$, of the zero-order modified Bessel function is given by different authors in different forms:

$$K_0(r/\lambda) \approx -\ln\left(\frac{r}{\lambda}\right) - \gamma + \ln 2$$

Arfken, 1985, pp. 284, 612

$$-\ln\left(\frac{r}{\lambda}\right)$$

Abramowitz and Stegun, eds., 1970, p. 375

$$-\ln\left(\frac{r}{2\lambda}\right) - 0.5772\ldots$$

Jackson, 1975, p. 108

$$+\ln\left(\frac{\lambda}{r}\right)+0.12\ldots$$

Tinkham, 1985, p. 147

$$+\ln\left(\frac{1.123\lambda}{r}\right) \quad \text{Present work}$$

Which of these are equivalent?

8. At what points along the path $V \to S \to M \to C \to V$, starting and ending at the same vortex, V, does the current density pass through zero and at what points does it change sign? Explain these changes, and explain why J has opposite signs at the beginning and at the end of the plot in Fig. 9.20.
9. Derive Eqs. (9.57) and (9.58).
10. The eccentricity e of an ellipse with semi-major and semi-minor axes a and b, respectively, is defined by

$$e=(a^2-b^2)^{1/2}/a.$$

Show that for a high-temperature superconductor the two expressions

$$a/b=\sqrt{\Gamma},$$

$$e=(\Gamma-1)^{1/2},$$

are valid for both the core and the current flow ellipses of a vortex, where Γ is the effective mass ratio of Eq. (9.44).

11. Show that for an applied magnetic field aligned along the y direction of a high-temperature superconductor with $(x^2+z^2)^{1/2} \gg \lambda_c$, the current densities $J_s(x,z)$ of a vortex at points along the x- and z-axes, respectively, are given by

$$\mu_0\lambda_c J_s(x,0)=B_y(x,0)$$

$$\mu_0\lambda_{ab} J_s(0,z)=B_y(0,z).$$

10

Magnetic Properties

I. INTRODUCTION

Superconductivity can be defined as the state of perfect diamagnetism, and consequently researchers have always been interested in the magnetic properties of superconductors. The previous chapter provided an introduction to these magnetic properties. The emphasis there was on the role played by the quantized vortices, on the properties of magnetic flux considered as a continuum, and on flux in motion. The present chapter will extend the discourse to a number of additional magnetic properties.

We begin with a discussion of magnetization, zero field cooling, and field cooling, with comments on the granularity and porosity of high-temperature superconductors. Next we will explain how magnetization depends on the shape of the material and how this shape dependence affects the measured susceptibility. Both ac and dc susceptibilities will be treated. Finally, we will show how samples can be categorized in terms of traditional magnetic behavior, such as diamagnetism, paramagnetism, and antiferromagnetism. The chapter will conclude with remarks on ideal Type II superconductors and on magnets.

In the present chapter we do not always distinguish between Type I and Type II superconductors since many of the results that will be obtained here apply to both types of superconductors.

In the following chapter we will discuss some additional magnetic properties of superconductors, such as the intermediate and mixed states of Type I and Type II superconductors, respectively. In Chapter 12 we will present the Bean model, a model that provides a good description of some magnetic properties, especially hysteresis loops.

II. SUSCEPTIBILITY

A material in the mixed state of a Type II superconductor contains magnetic flux in vortices that are embedded in a superconducting matrix with $\chi = -1$. From a macroscopic perspective we average over this structure and consider the material to be homogeneous with uniform susceptibility in the range $-1 < \chi < 0$ and also constant throughout the volume. The internal fields B_{in}, H_{in}, and M are also averages that are uniform at this level of observation. In this chapter we will be working with these average quantities and ignore the underlying mesoscopic structure.

We saw in Chapter 1 how the B and H fields within a homogeneous medium are related to the magnetization M and the susceptibility χ through Eqs. (1.69), (1.77), and (1.78a),

$$B_{in} = \mu_0(H_{in} + M) \tag{10.1}$$

$$= \mu_0 H_{in}(1 + \chi) \tag{10.2}$$

$$\chi = \frac{M}{H_{in}} \tag{10.3}$$

where μ_0 is the permeability of free space and χ is an intrinsic property of the medium.

In the general case the susceptibility is a symmetric tensor with components χ_{ij} because of the off-diagonal components $(i \neq j)$ the vector fields $\mathbf{B}_{in}$, $\mathbf{H}_{in}$, and $\mathbf{M}$ are in different directions. In the principal coordinate system the susceptibility tensor is diagonal with components χ_x, χ_y, and χ_z along the three orthogonal principle directions. High-temperature superconductors are planar with values $\chi_a \approx \chi_b$ in the plane of the CuO_2 layers different from χ_c, which is measured along the c direction perpendicular to the layers. This axial anisotropy manifests itself in the large difference in the critical fields of single crystals when measured parallel to and perpendicular to the CuO_2 layers, as shown in Table 9.5. Several figures in the present chapter will illustrate this anisotropy. However, for the present we will restrict our attention to the isotropic case, for which $\chi = \chi_x = \chi_y = \chi_z$.

III. MAGNETIZATION AND MAGNETIC MOMENT

The magnetization $\mathbf{M}$ is the magnetic moment per unit volume. This means that the overall magnetic moment $\boldsymbol{\mu}$ of a sample is the volume integral of $\mathbf{M}$ throughout it,

$$\boldsymbol{\mu} = \int \mathbf{M}\, dV. \tag{10.4}$$

Many magnetic studies of superconductors are carried out using samples with shapes that can be approximated by ellipsoids. When the magnetic field $\mathbf{B}_{app} = \mu_0 \mathbf{H}_{app}$ is applied along or perpendicular to the symmetry axis of such a sample, the internal fields $\mathbf{B}_{in}$ and $\mathbf{H}_{in}$, and the magnetization $\mathbf{M}$ as well, are uniform and parallel to the applied field, with M given by

$$M = \frac{\mu}{V}, \tag{10.5}$$

where V is the sample volume. Long thin cylinders and thin films are limiting cases of this general ellipsoidal geometry.

We begin by analyzing the parallel geometry case of a superconductor in the shape of a long cylinder located in an applied field directed along its axis, as shown in Fig. 9.1. For this case the fields can be written as scalars. We wish to express the internal fields in terms of the known applied field B_{app}:

$$B_{app} = \mu_0 H_{app}. \tag{10.6}$$

For this particular geometry the boundary condition (1.74) shows that the internal field H_{in} equals H_{app}. From Eqs. (10.2) and (10.3) the internal fields are given by

$$B_{in} = B_{app}(1 + \chi), \qquad (10.7)$$

$$H_{in} = \frac{B_{app}}{\mu_0}, \qquad (10.8)$$

$$M = \frac{\chi B_{app}}{\mu_0}. \qquad (10.9)$$

Experimentally, it is the magnetic moment μ, given by

$$\mu = \frac{\chi V B_{app}}{\mu_0}, \qquad (10.10)$$

which is measured, for example, by a Superconducting Quantum Interference Device (SQUID) magnetometer. Since V and B_{app} are known, Eqs. (10.10) and (10.9) can be used to determine the susceptibility and magnetization, respectively. In these expressions χ is the volume susceptibility corresponding to the magnetic moment per unit field per unit volume. We assume that χ is independent of the applied field B_{app}, and that M is proportional to B_{app} through Eq. (10.9).

For an ideal superconductor the property of perfect diamagnetism means that $\chi = -1$, so that Eqs. (10.7)–(10.10) become, respectively,

$$B_{in} = 0, \qquad (10.11)$$

$$H_{in} = \frac{B_{app}}{\mu_0}, \qquad (10.12)$$

$$M = -\frac{B_{app}}{\mu_0}, \qquad (10.13)$$

$$\mu = -\frac{V B_{app}}{\mu_0}, \qquad (10.14)$$

and we see that the internal field B_{in} vanishes. This is the case illustrated in Fig. 9.1. The fact that B_{in} vanishes can also be explained in terms of the shielding currents (see Fig. 5.19) which flow on the surface and act like a solenoid to produce a field B_{in} which cancels B_{app}. This was discussed at length in Chapter 2, Section VIII.

Figure 10.1 shows the experimentally measured magnetization curve for

$$(La_{0.9}Sr_{0.1})_2CuO_4$$

plotted against the applied field. The applied field reaches a maximum at 30 mT,

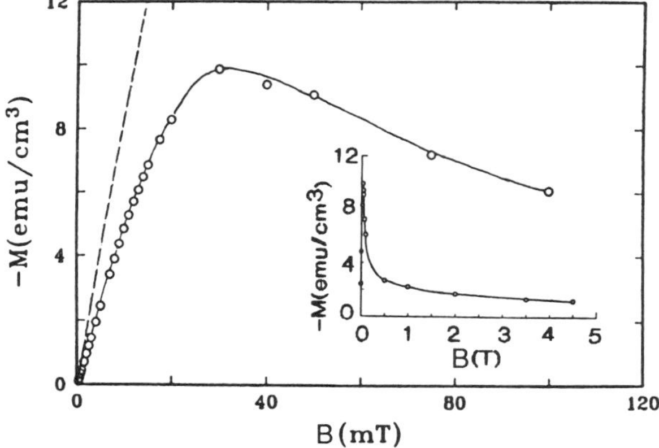

Figure 10.1 Zero-field magnetization of annealed $(La_{0.9}Sr_{0.1})_2CuO_4$ in applied magnetic fields up to 100 mT at a temperature of 5 K. The maximum of the curve occurs near the lower-critical field $B_{c1} \approx 30$ mT. The dashed line is the low-field asymptote for perfect diamagnetic shielding (Maletta *et al.*, 1987). The inset shows the magnetization in applied fields up to 4.5 T.

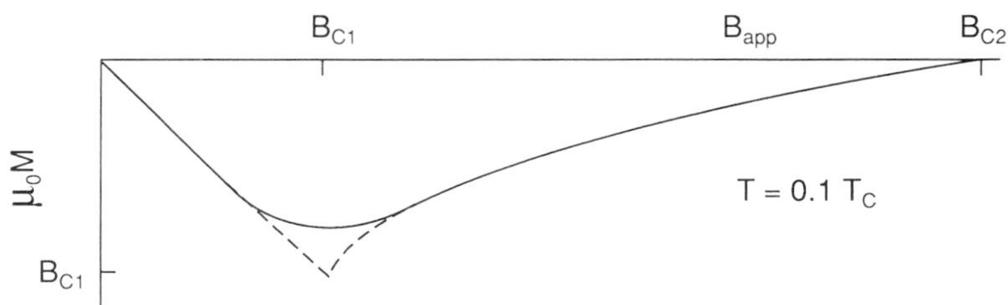

Figure 10.2 Typical magnetization curve for $T = 0.1T_c$ (cf. Fig. 10.3, which is drawn to the same scale).

which is approximately the lower critical field B_{c1} (Maletta *et al.*, 1987; see Müller *et al.*, 1987). The upper critical field is well beyond the highest field used, 4.5 T, as shown in the inset to the figure. Note that the abscissa scale is in terms of milliteslas for the main figure, and in terms of teslas for the inset.

We saw in Fig. 9.36 that the critical fields B_{c1} and B_{c2} are highest at 0 K and that they decrease continuously with increasing temperature until they become zero at the transition temperature T_c. Thus, a magnetization curve, such as that presented in Fig. 10.1, contracts as temperature increases. This situation is illustrated graphically by Figs. 10.2 and 10.3, which show sketches of magnetization curves at two temperatures $T = 0.1T_c$ and $T = 0.7T_c$.

IV. MAGNETIZATION HYSTERESIS

Many authors have reported hysteresis in the magnetization of superconductors, meaning that the magnetization depends on the previous history of how magnetic fields were applied. Hysteresis is observed when the magnetic field is increased from zero to a particular field, then scanned back through zero to the negative of this field, and finally brought back to zero again. Figure 10.4 sketches a low-field hysteresis loop showing the coercive field B_{coer}, or value of the applied field that reduces the magnetization to zero, and the remanent magnetization M_{rem}, or magnitude of the magnetization when the applied field passes through zero.

Figures 10.5 and 10.6, respectively, show how low-field hysteresis loops vary with changes in the scanning-field range and temperature. It is clear from these figures that the hysteresis loop is thin and close to linear when the scan range is much less than the lower-critical field and when the temperature is close to T_c. Decreasing the temperature broadens the loop. The larger the magnetic field excursion, the more the loop becomes elongated horizontally, which increases the ratio $B_{coer}/\mu_0 M_{rem}$ between the coercive field and the remanent magnetization.

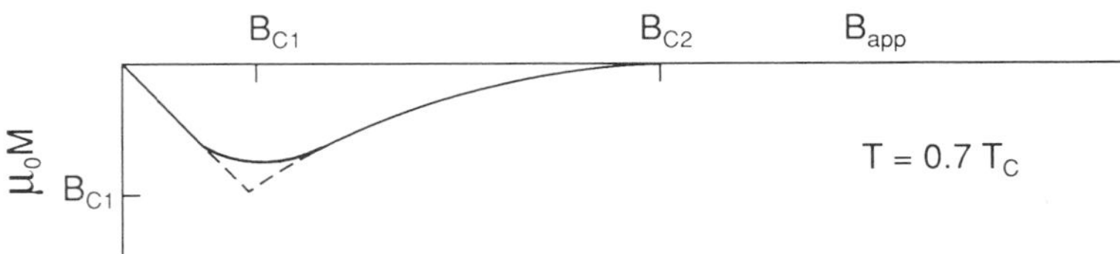

Figure 10.3 Typical magnetization curve for $T = 0.7T_c$ drawn to the same scale as Fig. 10.2.

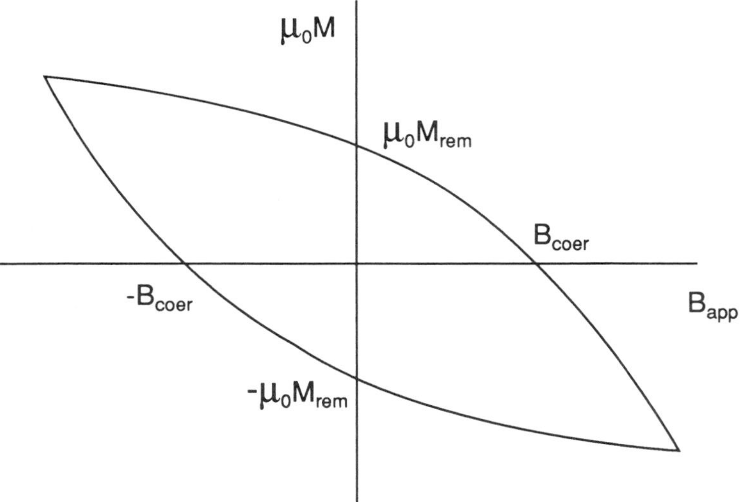

Figure 10.4 Typical low-field hysteresis loop showing the coercive field B_{coer}, where magnetization is zero, and the remanent magnetization M_{rem} which remains when the applied field is reduced to zero.

Figure 10.7 shows how hysteresis loops traversed over a broad field range vary with the temperature. Each loop has a peak near the lower-critical field B_{c1}. Beyond this point flux penetrates and the magnetization begins to decrease gradually. Ideally, no flux penetrates below B_{c1}, but in practice some of it does, as Fig. 10.1 suggests. The large hysteresis is indicative of flux pinning. It is observed that as the temperatures is lowered, the loop increases in area, as shown in the figures. Paranthaman *et al.* (1993) obtained similar results with the superconductor

$$\mathrm{HgBaCuO}_{4+\delta}.$$

Chapter 12 will present a model, called the critical-state model, which provides an explanation for the shapes of many hysteresis loops.

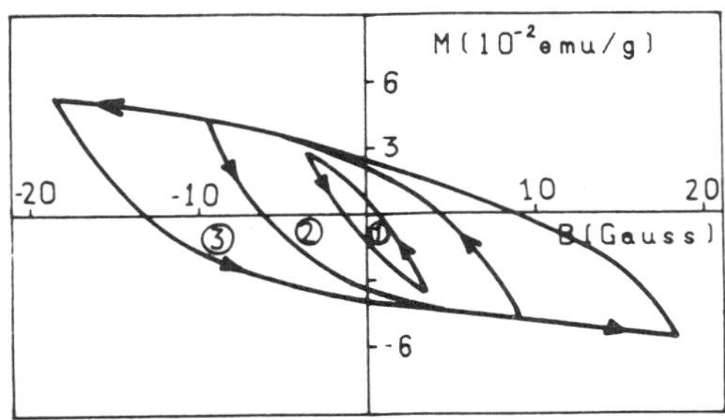

Figure 10.5 Low-field hysteresis loops of $(\mathrm{La}_{0.9}\mathrm{Sr}_{0.1})_2\mathrm{CuO}_4$ at 4.5 K cycled over different ranges of field up to 2 mT (Marcus *et al.*, 1987).

V. ZERO FIELD COOLING AND FIELD COOLING

In Chapter 2 we discussed the magnetic properties of a perfectly diamagnetic material with a hole that is either open or closed to the outside. We examined these two cases for the conditions of (a) zero field cooling (ZFC), a condition characterized by flux exclusion from both the open hole and the enclosed cavity, a phenomenon called *diamagnetic shielding*, and (b) field cooling (FC), a condition characterized by flux expulsion from the cavity but not from the hole, a phenomenon called the *Meissner effect*. For both cases, the flux is absent from the superconducting portion. Hence, the overall sample can

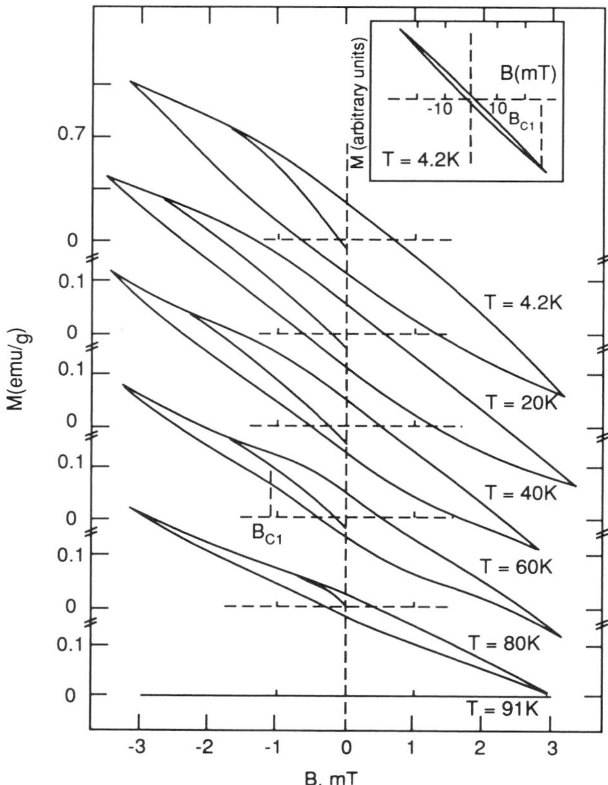

Figure 10.6 Low-field hysteresis loops of $YBa_2Cu_3O_7$ cycled over the same field scan, $-3\ \mathrm{mT} \le B_{\mathrm{app}} \le 3\ \mathrm{mT}$, over a range of temperatures. The loops gradually collapse as the temperature increases. The virgin curve for the initial rise in magnetization is given for each loop (Senoussi *et al.*, 1988).

exclude more flux when it is zero field cooled than it expels when it is is field cooled. The difference between the amount of excluded flux and the amount of expelled flux is the *trapped flux*.

To clarify some of the principles involved in ZFC and FC experiments, we will examine the rather idealized case of a cylindrical sample of total volume V_{T} that contains a volume V_{s} of perfectly superconducting material ($\chi = -1$), a cylindrical hole of volume V_{h} open at the top and bottom, and a totally enclosed cylindrical cavity of volume V_{c},

$$V_{\mathrm{T}} = V_{\mathrm{s}} + V_{\mathrm{h}} + V_{\mathrm{c}}, \qquad (10.15)$$

as shown in Fig. 10.8. The hole and cavity could either be empty or contain normal material; since the effect in the two cases is the same, we will consider them empty. The magnetic field B_{app} is applied parallel to the cylinder axis, as indicated in Fig. 10.9; demagnetizing effects arising from the lack of cylindrical symmetry will not be taken into account.

For this composite sample the measured or effective magnetic moment μ_{eff} can receive contributions from three individual components,

$$\mu_{\mathrm{eff}} = \mu_{\mathrm{s}} + \mu_{\mathrm{h}} + \mu_{\mathrm{c}}, \qquad (10.16)$$

with μ_{s} due to the superconducting material itself, μ_{h} resulting from the presence of the open hole, and μ_{c} due to the en-

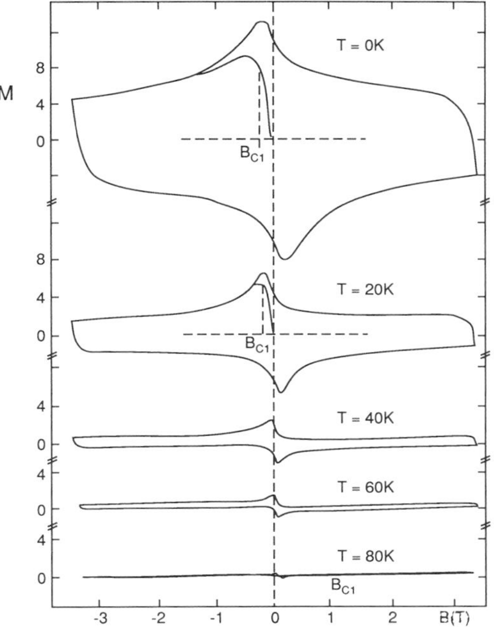

Figure 10.7 High-field hysteresis loops of $YBa_2Cu_3O_7$ cycled over the same field scan, $-3\ T \le B_{app} \le 3$ T, over a range of temperatures. The loops gradually collapse as the temperature increases. The deviation of the virgin curve from linearity occurs near the lower-critical field $\mathbf{B}_{c1}$, which increases as the temperature is lowered (Senoussi *et al.*, 1988).

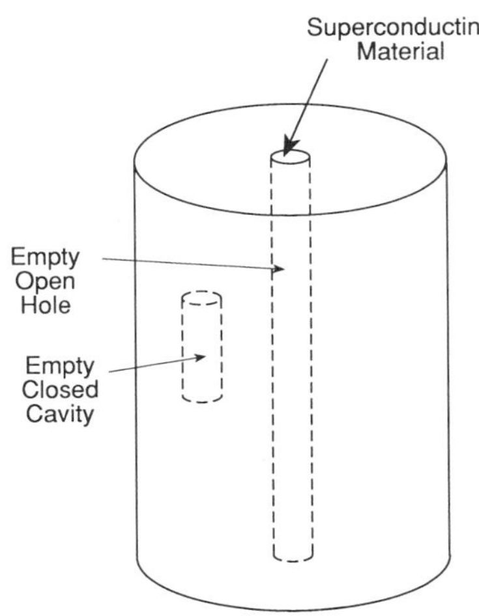

Figure 10.8 Cylindrical superconducting sample with hole of volume V_h open at the top and bottom, and a totally enclosed cavity of volume V_c.

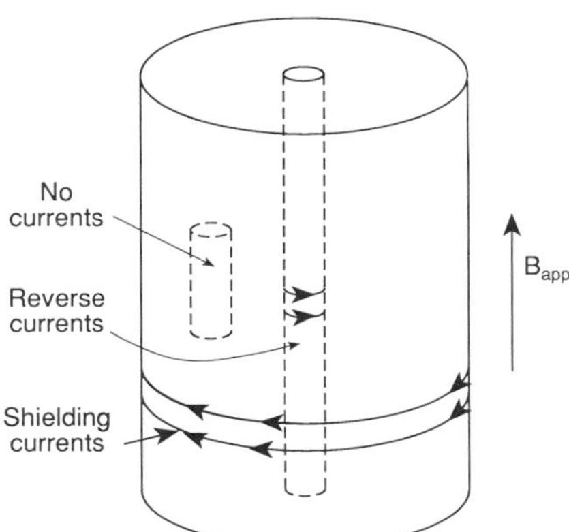

Figure 10.9 The superconducting cylinder sketched in Fig. 10.10 after field cooling in an axial applied field, showing the shielding currents flowing around the outside, the reverse-direction current flow around the walls of the open hole, and the absence of currents in the enclosed cavity.

closed cavity. In the case of zero field cooling, the circulating surface currents shield the superconductor, hole, and cavity, so Eq. (10.10), with $\chi = -1$, becomes

$$\mu_{zfc} = -(V_s + V_h + V_c)\frac{B_{app}}{\mu_0}. \quad (10.17)$$

For field cooling, the magnetic field is trapped in the open hole, while surface currents shield the superconductor itself and the enclosed cavity from this field, which gives for the magnetic moment

$$\mu_{fc} = -(V_s + V_c)\frac{B_{app}}{\mu_0}. \quad (10.18)$$

Associated with the effective magnetic moment (10.16) there is an effective magnetization M_{eff} defined by Eq. (10.5) in terms of the total volume (10.15),

$$M_{eff} = \frac{\mu_{eff}}{V_T} = \chi_{eff}\frac{B_{app}}{\mu_0}. \quad (10.19)$$

which can be employed to write down the ZFC and FC magnetization, respectively. The corresponding susceptibilities χ_{zfc} and χ_{fc} are determined in Problem 1. We know from Eqs. (10.9) and (10.10) and the above expressions that the ratios between the FC and ZFC moments, magnetizations, and susceptibilities all have the same value,

$$\frac{\mu_{fc}}{\mu_{zfc}} = \frac{M_{fc}}{M_{zfc}} = \frac{\chi_{fc}}{\chi_{zfc}} = \frac{V_s + V_c}{V_s + V_h + V_c}, \quad (10.20)$$

and that this value is independent of the units used.

If field cooling is carried out in an applied field $B_{fc} = \mu_0 H_{fc}$ that differs from the field B_{app} which is applied to measure the magnetic moment, we obtain, neglecting hysteresis (see Problem 3),

$$\mu_0 \mu_{fc} = -(V_s + V_c)B_{app} + (B_{fc} - B_{app})V_h, \quad (10.21)$$

which reduces to Eq. (10.18) when $B_{app} = B_{fc}$. Thus, the field trapped in the hole acts like a magnetization with the same magnitude and direction as the quantity $(B_{fc} - B_{app})$. Ordinarily, field cooling is carried out in the same field as the susceptibility measurements, so that $B_{app} = B_{fc}$ and Eq. (10.18) applies.

As long as the sample is kept below T_c the field B_{fc} remains in the open hole irrespective of whether the outside field is turned off or another applied field is turned on. B_{fc} is maintained in the hole by surface currents circulating in opposite directions around the inside of the superconducting tube, as shown in Fig. 10.9 and explained in Chapter 2, Sections VIII and IX. This trapped flux subtracts from the diamagnetic response to make the measured susceptibility and magnetization less negative for the Meissner effect (FC) than for diamagnetic shielding (ZFC). This is shown in Fig. 10.10 for the rubidium fullerene compound (C.-C. Chen *et al.*, 1991; Politis *et al.*, 1992), where the ZFC data points are far below the corresponding FC data, as expected from Eq. (10.20). The earliest HgBaCaCuO compound samples produced FC susceptibilities that were far above ZFC ones (Adachi *et al.*, 1993; Gao *et al.*, 1993; Meng *et al.*, 1993b, Schilling *et al.*, 1993).

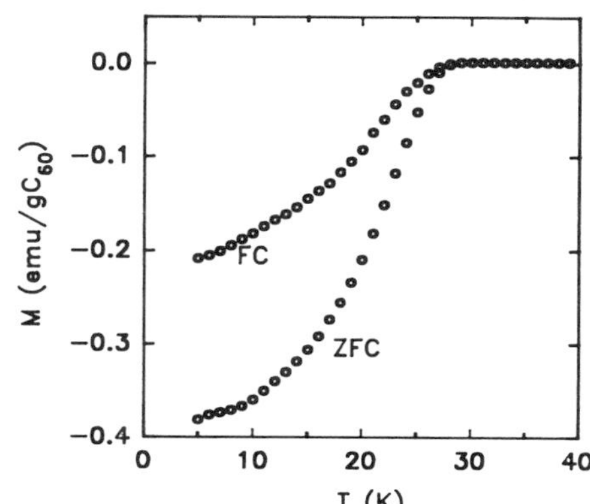

Figure 10.10 Rb_3C_{60} powder sample showing that the zero-field-cooled magnetic susceptibility is more negative than its field-cooled counterpart (C.-C. Chen *et al.*, 1991).

Clem and Hao (1993) examined the four cases of ZFC, FC with data collected on cooling (FCC), FC with data collected on warming (FCW), and remanence. In the fourth case the applied field is turned off after the specimen has been FC, and the remanent magnetization is measured as a function of increasing temperature.

VI. GRANULAR SAMPLES AND POROSITY

The analysis of the previous section can help us understand experimental susceptibility data on granular samples. The grains sometimes consist of a mixture of superconducting and normal material of about the same density, with empty space between and perhaps within the material. The two densities can be comparable when the sample preparation procedure does not completely transform the starting materials into the superconducting phase. A well-made granular superconductor does not contain any normal material, but it does have intergranular and perhaps intragranular spaces, either of which can trap flux. The field-cooled moment can be significantly less than the zero-field-cooled moment, as shown by the data in Table VIII.1 of previous work (Poole *et al.*, 1988).

A quantitative measure of the degree of granularity of a sample is its porosity P, which is defined by

$$P = (1 - \rho/\rho_{\text{x-ray}}), \qquad (10.22)$$

where the density ρ of the sample is

$$\rho = \frac{m}{V_T} \qquad (10.23)$$

and the x-ray density is calculated from the expression

$$\rho_{\text{x-ray}} = \frac{[\text{MW}]}{V_0 N_A}, \qquad (10.24)$$

where N_A is Avogadro's number and V_0 is the volume of the sample per formula unit, with the value

$$V_0 = abc \qquad (YBa_2Cu_3O_{7-\delta}), \qquad (10.25)$$

$$V_0 = \tfrac{1}{2}abc \qquad (\text{LaSrCuO, BiSrCaCuO, TlBaCaCuo}), \qquad (10.26)$$

where a, b, and c are the lattice constants and the La, Bi, and Tl compounds have assigned to them two formula units per unit cell, as explained in Chapter 7.

Porosity is a measure of the proportion of empty spaces or voids within and between the solid material or grains of a sample. Problem 4 shows how V_s, V_h, and V_c can be determined from measurements of ρ, χ_{zfc} and χ_{fc}. The x-ray density calculated from the unit cell dimensions of

$$YBa_2Cu_3O_7$$

is 6.383 g/cm^3. Typical densities of granular samples vary from 4.3 to 5.6 g/cm^3, corresponding to porosities between 33% and 12%, respectively (Blendell *et al.*, 1987; Mathias *et al.*, 1987).

Porosity can be reduced by applying pressure to the material. For example, a sample of $YBa_2Cu_3O_7$ with a 5:1 ratio between flux exclusion and flux expulsion was compressed at 20–30 kbar to a claimed 100% of theoretical density, $\rho = \rho_{\text{x-ray}}$, bringing the measure flux expulsion to within about 11% of the theoretical value (Venturini *et al.*, 1987). Researchers have also found 100% flux shielding and 95% flux expulsion in $YBa_2Cu_3O_7$ at 4.2 K (Larbalestier *et al.*, 1987a). Good single crystals, of course, have a porosity of zero.

VII. MAGNETIZATION ANISOTROPY

The magnetic properties of high-temperature superconductors are highly anisotropic, with magnetization and susceptibility depending on the angle which the applied field makes with the c-axis. We saw in Chapter 9, Section IV, that anisotropy here is a result of the difference in the values of the coherence length, penetration depth, and effective mass measured along the c direction as opposed to values obtained from measurements in the a, b-plane. Particles of anisotropic superconductors in a magnetic field experience a torque which tends to align them with the field (Kogan, 1988). Anisotropy effects can be determined by employing single crystals, epitaxial films, or grain-aligned powders. Epitaxial films are generally single-crystal films with the c-axis perpendicular to the plane. It is, of course, preferable to work with untwinned single crystals or epitaxial films. However, these are not always available, and much good research has been carried out with aligned granular samples.

Grain alignment is a technique that converts a collection of randomly oriented grains into a set of grains with their c-axes preferentially pointing in a particular direction. This alignment can be brought about by uniaxial compression, by application of a strong magnetic field to grains embedded in, for example, epoxy, or by melting a random powder sample and reforming it in the presence of a temperature gradient (Farrell *et al.*, 1987). It is much easier to fabricate grain-aligned samples than single crystals. Grain-aligned samples, however, cannot compete with single crystals in terms of degree of alignment. Untwinned monocrystals are needed for perfect alignment.

Another technique for preparing samples with monocrystal characteristics is melt-textured growth (L. Gao *et al.*, 1991; Jin *et al.*, 1988; Murakami *et al.*, 1991). In melt-textured growth a granular material is melted and then slowly cooled in a thermal gradient to produce a high degree of texturing. The effect is to reduce weak-link grain boundaries and increase critical currents.

Figure 10.11a shows that both the ZFC and FC susceptibilities of $YBa_2Cu_3O_7$ are greater in magnitude (i.e., more negative) for the applied field aligned parallel to the c-axis than they are for B_{app} aligned perpendicular to c (i.e., along the copper-oxide planes); these measurements were made with grain-aligned samples. The figure shows that the susceptibility data for a nonaligned power are between the results for $B_{app\|c}$ and $B_{app\perp c}$. Figure 10.11c shows that the susceptibility is much less for field cooling in the field $B_{fc} = 0.3$ T, again with the data for $B_{app\|c}$ lying below the data for $B_{app\perp c}$.

VIII. MEASUREMENT TECHNIQUES

Experimentally, susceptibility, a dimensionless quantity, is determined from

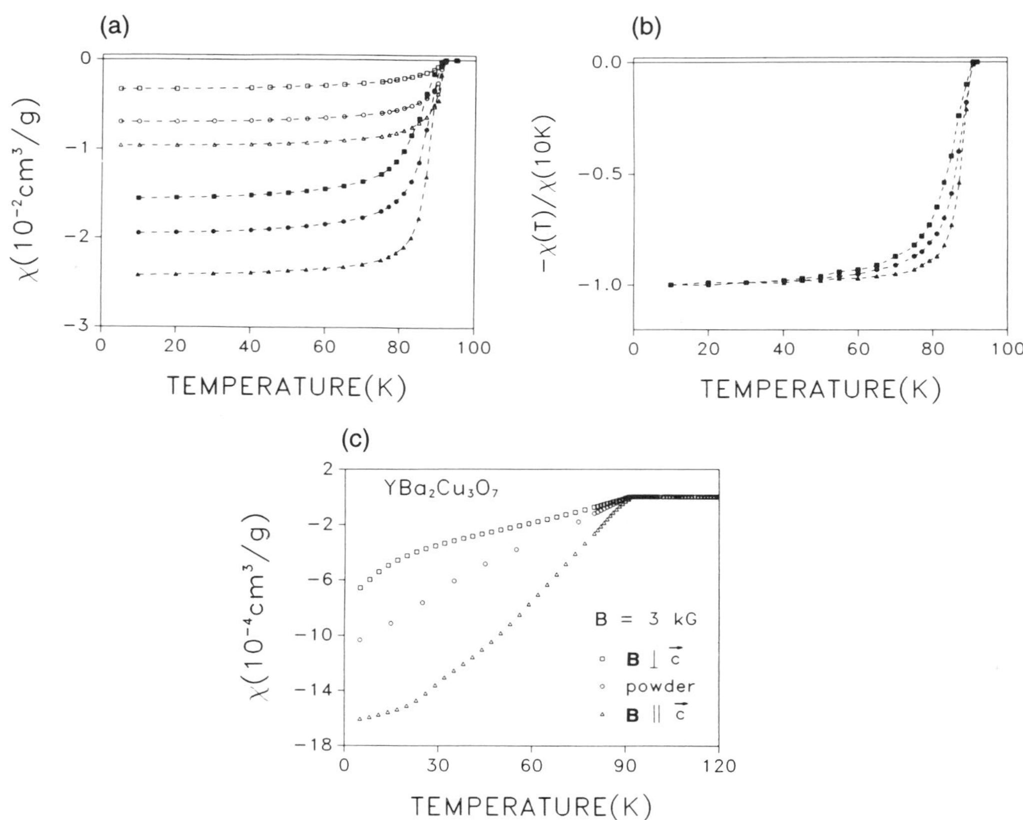

Figure 10.11 (a) Zero-field-cooled (closed symbols) and field-cooled (open symbols) susceptibility versus temperature for nonaligned powder (circles) and grain-aligned samples of $YBa_2Cu_3O_7$ in a field of 5 mT with $B_{app\|c}$ (triangles) and $B_{app\perp c}$ (squares), (b) normalized susceptibilities for the zero-field-cooled samples of (a), (c) field-cooled measurements in 0.3 T, plotted with the same symbol convention (Lee and Johnston, 1990).

the measured magnetic moment μ of the sample with the aid of Eq. (10.10),

$$\chi = \frac{\mu_0 \mu}{V_T B_{app}}. \quad (10.27)$$

For a small sample the overall volume V_T can be estimated by viewing it under a microscope. This is sometimes called the volume susceptibility, although in actuality the parameter is dimensionless. Many investigators determine sample size by weighting and report what is sometimes called the mass susceptibility χ_{mass}, defined by

$$\chi_{mass} = \frac{\chi}{\rho} = \frac{\mu_0 \mu}{(\rho V_T) B_{app}}, \quad (10.28)$$

where ρV_T is the mass of the sample. This quantity has the dimensions m^3/kg in the SI system and cm^3/g in the cgs system.

Many susceptibility and magnetism measurements are carried out with a SQUID, a dc measuring instrument (see Section III). In this device, which is sketched in Fig. 10.12, a magnetized sample that has been moved into a sensor coil causes the flux through the coil to change. The current produced by this flux change is passed to the multiturn coil on the left side of the figure where it is amplified by the increase in the number of turns. The SQUID ring with its weak links detects this flux change in a manner that will be discussed in Chapter 13, Section VIII.I. The change in flux provides the magnetic moment by the expression

$$\mu_0 \mu = \Delta\Phi, \quad (10.29)$$

and from Eq. (10.27) we have for the susceptibility,

$$\chi = \Delta\Phi / V B_{app}. \quad (10.30)$$

The data presented in Figs. 10.10 and 10.11 were obtained with a SQUID magnetometer. More classical techniques, such as the vibrating sample magnetometer or perhaps the Gouy or Faraday balance, are less frequently employed. One can make ac susceptibility measurements using a low-frequency mutual inductance bridge operating at, for example, 200 Hz.

IX. COMPARING SUSCEPTIBILITY AND RESISTIVITY RESULTS

We saw in Section V that the susceptibility of a composite sample is a linear combination of the contributions from its component parts. Thus, susceptibility measurements determine the magnetic state of an entire sample, and also give a better indication of the degree to which the sample has transformed to the super-

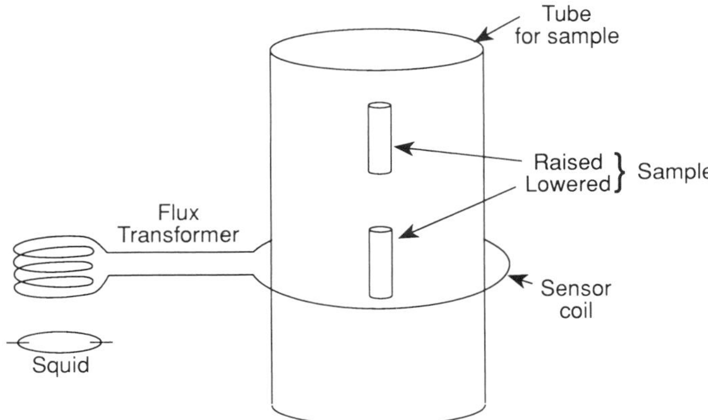

Figure 10.12 The change of magnetic flux in a sensor coil loop that has been produced by raising or lowering a sample induces a current which is transferred to a multiloop coil where it is measured by a Superconducting Quantum Interference Device (SQUID).

conducting state. Resistivity measurements, on the other hand, merely show whether or not continuous superconducting paths are in place. In addition, while a dc susceptibility measurement provides a better experimental indicator of the overall superconducting state, a resistivity measurement is a better practical guide for application purposes. We should also note that magnetization is a thermodynamic state variable (cf. Chapter 4, Section VI), whereas resistivity is not. The properties of zero resistance and perfect diamagnetism are the two classic ways of defining superconductivity. In an ideal homogeneous material both measurements should provide the same transition temperature.

The transition temperatures determined by magnetic susceptibility and resistivity measurements sometimes differ somewhat. When the transition is sharp, resistivity can drop sharply to zero at a temperature slightly above the onset of the susceptibility or magnetization transition, as shown in Fig. 2.21. When the transition is broad, the χ-versus-T and ρ-versus-T curves often overlap considerably. Many articles provide susceptibility and resistivity curves for the same sample. Figure III-5 from our previous work (Poole *et al.*, 1988) compares the resistivity, Meissner magnetization, ac susceptibility, and specific heat transitions for the same

$$YBa_2Cu_3O_7$$

sample (Junod *et al.*, 1988).

X. ELLIPSOIDS IN MAGNETIC FIELDS

In Section III we treated the case of a cylindrically shaped sample in a parallel magnetic field, noting that this geometry was chosen to avoid demagnetization effects that could complicate the calculation of the internal magnetic field and magnetization. Some commonly used superconductor arrangements in magnetic fields, such as thin films in perpendicular fields, have very pronounced demagnetization effects. In practice, these arrangements constitute limiting cases of ellipsoids, so that in the present section we will analyze the case of an ellipsoid in an applied field. Then we will show how some common geometries are good approximations to elongated and flattened ellipsoids. Many of the results of this and the following few sections are applicable to both Type I and Type II superconductors.

When an ellipsoid with permeability μ is placed in a uniform externally applied magnetic field $\mathbf{B}_{app}$ oriented along one of its principal directions, its internal fields $\mathbf{B}_{in}$ and $\mathbf{H}_{in}$ will be parallel to the applied fields, and hence all of the fields can be treated as scalars. Their values will be determined by applying Eqs. (10.1) and (10.2) to the internal fields

$$\begin{aligned} B_{in} &= \mu H_{in} = \mu_0 (H_{in} + M) \\ &= (1 + \chi)\mu_0 H_{in} \end{aligned} \tag{10.31}$$

and the applied fields

$$B_{app} = \mu_0 H_{app}, \quad M_{app} = 0 \tag{10.32}$$

and utilizing the demagnetization expression

$$\frac{NB_{in}}{B_{app}} + \frac{(1-N)H_{in}}{H_{app}} = 1, \tag{10.33}$$

where N is the demagnetization factor, to relate the internal and applied fields. The demagnetization factors along the three principal directions of the ellipsoid are geometrical coefficients that obey the normalization condition

$$N_x + N_y + N_z = 1, \tag{10.34}$$

with the largest value along the shortest principal axis and the smallest value along the longest principal axis. We will confine our attention to situations in which the external field is oriented along a principal direction since all the other orientations are much more complicated to analyze. In the following section we will give explicit expressions for the demagnetization factors associated with a sphere, a disk, and a rod.

Solving for B_{in}, H_{in}, and M in Eqs. (10.31) and (10.33) gives

$$B_{in} = B_{app}\frac{1+\chi}{1+\chi N}, \qquad (10.35)$$

$$H_{in} = \frac{B_{app}/\mu_0}{1+\chi N}, \qquad (10.36)$$

$$M = \frac{B_{app}}{\mu_0}\cdot\frac{\chi}{1+\chi N} \qquad (10.37)$$

for the internal fields and magnetization expressed in terms of the applied fields. We should bear in mind that the susceptibility χ is negative for a superconductor, so that the denominators in these expressions become small when χ approaches -1 and N approaches 1.

For an ideal superconducting material $\chi = -1$. Equations (10.35)–(10.37) now assume a simpler form:

$$B_{in} = 0, \qquad (10.38)$$

$$H_{in} = \frac{B_{app}/\mu_0}{1-N}, \qquad (10.39)$$

$$M = -\frac{B_{app}/\mu_0}{1-N}. \qquad (10.40)$$

These expressions are applicable to Type I superconductors subject to the condition $B_{app} < (1-N)B_c$, as will be explained in Chapter 11, Section IV. They apply to Type II superconductors when $B_{app} < (1 - N)B_{c1}$, but for higher applied fields Eqs. (10.35)–(10.37) must be used since $-1 < \chi < 0$. Sometimes the transition from the Meissner to the vortex state is not sharply defined and a precise value of B_{c1} cannot be determined.

XI. DEMAGNETIZATION FACTORS

It will be helpful to write down formulae for the demagnetization factors for sample shapes that are often encountered in practice. For a sphere all three factors are the same, $a = b = c$ and $N_x = N_y = N_z$, so that from the normalization condition (10.34) we obtain

$$N = \frac{1}{3} \qquad \text{(sphere)}. \qquad (10.41)$$

For an ellipsoid of revolution with the z direction selected as the symmetry axis, the semi-major axes $a = b \neq c$ along the x-, y-, and z-axes, and the demagnetization factors are $N_{\parallel} = N_z$ and $N_{\perp} = N_x = N_y$, subject to the normalization condition

$$N_{\parallel} + 2N_{\perp} = 1 \qquad (10.42)$$

of Eq. (10.34). An oblate ellipsoid, i.e., one flattened in the x,y-plane, has $c < a$ with $N_{\parallel} > N_{\perp}$, and von Hippel (1954) gives (cf. Osborn, 1945; Stone, 1945; Stratton, 1941),

$$N_{\parallel} = \frac{1}{\epsilon^2} - \frac{[1-\epsilon^2]^{1/2}}{\epsilon^3}\sin^{-1}\epsilon \qquad c < a, \qquad (10.43)$$

where the oblate eccentricity ϵ is

$$\epsilon = [1-(c^2/a^2)]^{1/2} \qquad c < a. \quad (10.44)$$

For a prolate ellipsoid, i.e., one elongated along its symmetry axis so that $c > a$ and $N_{\parallel} < N_{\perp}$, we have again from von Hippel (1954)

$$N_{\parallel} = \frac{1-\epsilon^2}{\epsilon^2}\left[\frac{1}{2\epsilon}\ln\left(\frac{1+\epsilon}{1-\epsilon}\right) - 1\right] \qquad c > a, \qquad (10.45)$$

where the prolate eccentricity ϵ is

$$\epsilon = [1-(a^2/c^2)]^{1/2} \qquad c > a. \quad (10.46)$$

Of especial interest are samples in the shape of a disk, which may be considered the limiting case of a very flattened oblate ellipsoid, $c \ll a$, with the demagnetization factors

$$N_{\parallel} \approx 1, \qquad N_{\perp} \approx 0, \qquad \text{(flat disk)}, \qquad (10.47)$$

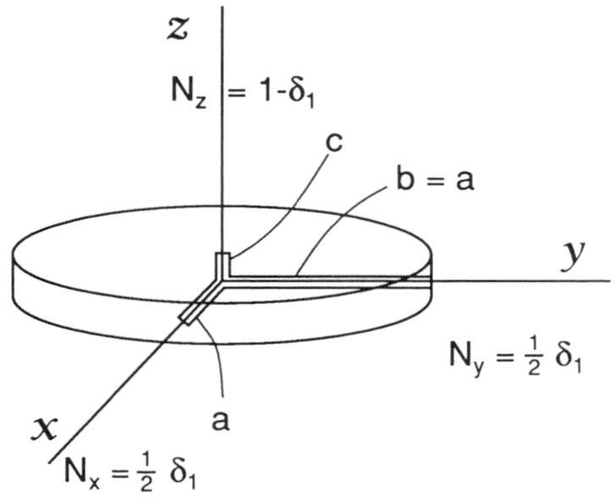

Figure 10.13 Demagnetization factors $N_x = N_y = \frac{1}{2}\delta_1 \ll 1$ and $N_z = 1 - \delta_1$ of an oblate ellipsoid with semi-major axes $a = b \gg c$ along the x, y, and z directions, respectively.

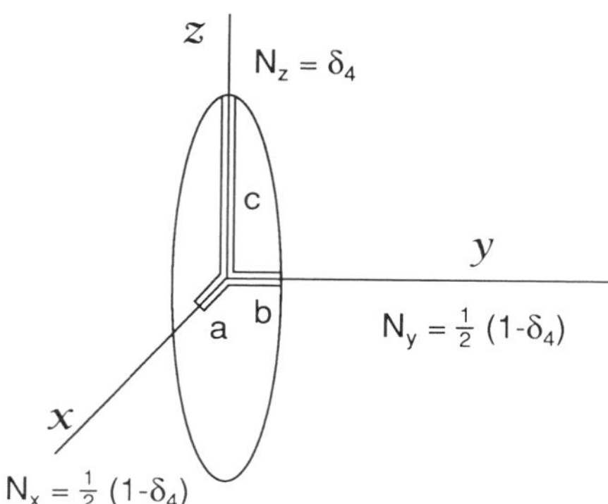

Figure 10.14 Demagnetization factors $N_x = N_y = \frac{1}{2}(1 - \delta_4)$ and $N_z = \delta_4 \ll 1$ of a prolate ellipsoid with $a = b \ll c$, using the notation of Fig. 10.13.

or in the shape of a rod, which is the limit of an elongated prolate ellipsoid, $c \gg a$, with the values

$$N_{\parallel} \approx 0, \qquad N_{\perp} \approx \tfrac{1}{2} \qquad \text{(long cylinder)}. \tag{10.48}$$

Correction factors δ_i to the limiting values of N_i given in Eqs. (10.47) and (10.48) are shown in Figs. 10.13 and 10.14, respectively, and listed in Table 10.1. Problems 6 and 7 give explicit expressions for these factors. Figure 10.15 shows how the parallel and perpendicular components of N depend on the length-to-diameter ratio of the ellipsoid. D.-X. Chen *et al.* (1991) reviewed demagnetization factors for cylinders; other pertinent articles are Bhagwat and Chaddah (1992), Kunchur and Poon (1991), and Trofimov *et al.* (1991). The electric case of depolarization factors is mathematically equivalent (Stratton, 1941).

XII. MEASURED SUSCEPTIBILITIES

From the theoretical viewpoint the magnetic susceptibility χ is a fundamental property of a material. It can be anisotropic, but for the present we will treat the isotropic case. It is defined by Eq. (10.3) as the ratio between the two quantities M and H_{in} in the interior of a superconductor,

$$\chi = M/H_{in}. \tag{10.49}$$

In practice, research workers often report an experimentally determined susceptibil-

Table 10.1 Demagnetization Factors for Ellipsoids of Revolution with Semi-axes $a=b$ and c for the Case of a Disk (oblate, $c < a$), Sphere ($c=a$), or Rod (prolate, $c > a$)[a]

Shape	Condition	$N_{\perp}$	$N_{\parallel}$
disk limit	$c \to 0$	0	1
flat disk	$c \ll a$	$\frac{1}{2}\delta_1$	$1 - \delta_1$
oblate	$c \approx a$	$\frac{1}{3} - \frac{1}{2}\delta_2$	$\frac{1}{3} + \delta_2$
sphere	$c = a$	$\frac{1}{3}$	$\frac{1}{3}$
prolate	$c \approx a$	$\frac{1}{3} + \frac{1}{2}\delta_3$	$\frac{1}{3} - \delta_3$
long rod	$c \gg a$	$\frac{1}{2} - \frac{1}{2}\delta_4$	δ_4
rod limit	$c \to \infty$	$\frac{1}{2}$	0

[a] Values of the correction factors δ_i are given in Problems 6 and 7.

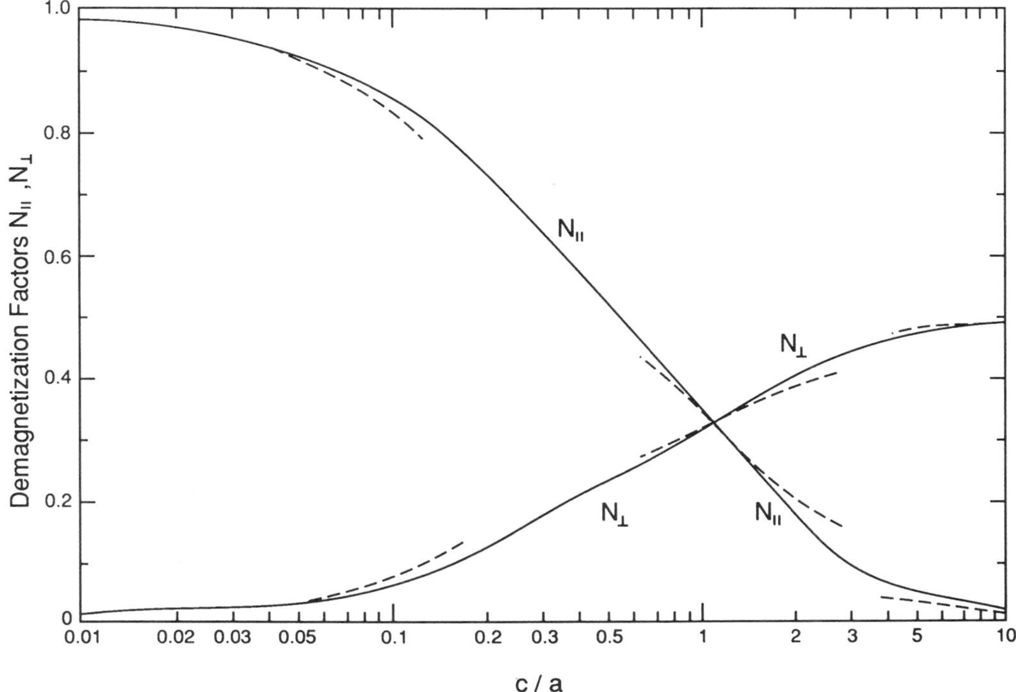

Figure 10.15 Dependence on the ratio c/a of the demagnetization factors $N_\perp = N_x = N_y$ perpendicular to the axis of an ellipsoid with semi-major axes $a = b \neq c$ and $N_\parallel = N_z$ along the axis. The solid lines were calculated using the exact expressions (10.42) to (10.46), and the dashed lines from the approximation formulae of Table 10.1 and Problems 6 and 7.

ity χ_{exp} that has been calculated from measured values of the magnetization and the applied field B_{app}, as follows:

$$\chi_{exp} = \mu_0 M / B_{app}. \qquad (10.50)$$

This is the definition of susceptibility that often appears in solid state physics books. Equations (10.49) and (10.50) are only equivalent for the case of "parallel geometry," in which the applied field is along the axis of a cylinder and the demagnetization factor N is zero: $H_{in} = B_{app}/\mu_0$.

When N is not zero, Eq. (10.49) is still valid because χ is a property of the material independent of its shape. Substituting Eq. (10.37) in Eq. (10.50) gives the expression

$$\chi_{exp} = \chi/(1 + N\chi), \qquad (10.51)$$

which may be solved for the intrinsic susceptibility in terms of the experimentally measured value

$$\chi = \chi_{exp}/(1 - N\chi_{exp}). \qquad (10.52)$$

The susceptibility must be in dimensionless units to apply this expression. Equation (10.52) shows that $|\chi| \leq |\chi_{exp}|$, where both χ and χ_{exp} are negative. Some authors set $N\chi = -N$ in Eq. (10.52) so as to write the approximate expression

$$\chi \approx \chi_{exp}(1 - N), \qquad (10.53)$$

which, however, underestimates the magnitude of χ, especially when N is appreciable and $|\chi|$ is small.

XIII. SPHERE IN A MAGNETIC FIELD

In this section we will examine a case that is commonly treated in electromagnetic theory and solid-state physics texts —that of a sphere in a magnetic field. This will provide us with closed-form expressions for the fields and the magnetization, both inside and outside the sphere as well as on its surface.

We mentioned in the previous section that for a sphere $N = 1/3$, so that using Eqs. (10.35)–(10.37) and (10.52) we have, respectively, for the two internal fields, magnetization, and susceptibility,

$$B_{\text{in}} = B_{\text{app}} \frac{3(1+\chi)}{3+\chi}, \qquad (10.54)$$

$$H_{\text{in}} = \frac{3B_{\text{app}}/\mu_0}{3+\chi}, \qquad (10.55)$$

$$M = \frac{B_{\text{app}}}{\mu_0} \cdot \frac{3\chi}{3+\chi}, \qquad (10.56)$$

$$\chi = \frac{3\chi_{\text{exp}}}{3-\chi_{\text{exp}}}. \qquad (10.57)$$

The B field immediately outside a superconducting sphere of radius a placed in a uniform external magnetic field B_{app} may be calculated from the standard formula for the magnetic scalar potential Φ_{out} given in electrodynamics texts (e.g., Jackson, 1975, p. 150),

$$\Phi_{\text{out}} = -\left[r - \frac{\chi}{\chi+3} \cdot \frac{a^3}{r^2}\right] B_{\text{app}} \cos\Theta, \qquad (10.58)$$

where Θ is the angle of the position vector $\mathbf{r}$ relative to the applied field direction. This is the solution to Laplace's equation

$$\nabla^2 \Phi = 0, \qquad (10.59)$$

which for the case of axial symmetry in spherical coordinates has the form

$$\frac{1}{r^2}\frac{d}{dr}\left(r^2 \frac{d\Phi}{dr}\right) + \frac{1}{r^2 \sin\Theta} \cdot \frac{d}{d\Theta}\left(\sin\Theta \frac{d\Phi}{d\Theta}\right) = 0, \qquad (10.60)$$

where the potential $\Phi_{\text{out}}(r, \Theta)$ depends on the polar angle Θ, but not on the azimuthal angle ϕ. This solution is subject to two boundary conditions, first, that B_r and H_Θ are continuous across the surface at $r = a$, and second, that $B = B_{\text{app}}$ far from the sphere where $r \gg a$.

The first term of Eq. (10.58),

$$rB_{\text{app}} \cos\Theta = zB_{\text{app}},$$

corresponds to the potential of the uniform applied field. The second term is known to be the magnetic field produced by a magnetic dipole of moment $\mu = a^3 H_{\text{app}}\chi/(\chi+3)$. The radial component B_r of the field outside,

$$B_r = -\frac{\partial \Phi_{\text{out}}}{\partial r} \qquad (10.61)$$

$$= \left[1 + \frac{2\chi}{\chi+3} \cdot \frac{a^3}{r^3}\right] B_{\text{app}} \cos\Theta, \qquad (10.62)$$

has a value at the surface $r = a$ of

$$B_r = \left[\frac{3(\chi+1)}{\chi+3}\right] B_{\text{app}} \cos\Theta. \qquad (10.63)$$

Setting $\chi = -1$ shows that this radial field vanishes at the surface for a perfect diamagnet. The polar angle component B_Θ outside,

$$B_\Theta = -\frac{1}{r} \cdot \frac{\partial \Phi_{\text{out}}}{\partial \Theta} \qquad (10.64)$$

$$= \left[1 - \frac{\chi}{\chi+3} \cdot \frac{a^3}{r^3}\right] B_{\text{app}} \sin\Theta, \qquad (10.65)$$

has a value at the surface of

$$B_\Theta = \left[\frac{3}{\chi+3}\right] B_{\text{app}} \sin\Theta. \qquad (10.66)$$

This field reaches a maximum along the equator, i.e., when $\Theta = \pi/2$. The magnetic field lines around the sphere, which are sketched in Fig. 2.23, are closest together at this maximum field position along the equator.

Equations (10.63) and (10.66) show that for the case $\chi = -1$ of perfect diamagnetism, the external field is parallel to the surface with no radial component. This field may be looked upon as inducing a current density in the surface of the sphere that circulates along circles of longitude that are oriented perpendicular to the z-axis, as illustrated in Fig. 2.30. These currents serve to cancel the B field that would otherwise be present inside the sphere. The presence of the factor $\sin\Theta$ in Eq. (10.66) means that the current density along a particular longitude circle at the latitude Θ is proportional to the radius ρ of the circle on which it flows, where $\rho = r\sin\Theta$, as indicated in Fig. 10.16. This causes each such current element to produce the same magnitude of magnetic field within the sphere, as expected.

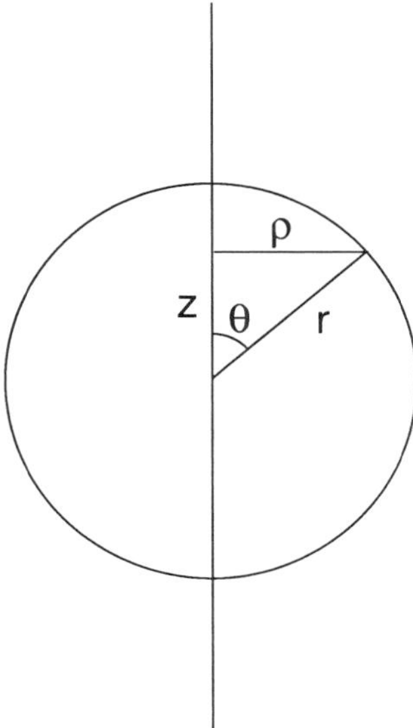

Figure 10.16 Coordinates for describing current flow along the surface of a sphere in a magnetic field applied along the z direction.

We will see in Chapter 11, Section XI, that the results of this section apply directly to a sphere in the mixed state of a Type II superconductor for applied fields in the range $\frac{2}{3}B_{c1} < B_{app} < B_{c2}$. For applied fields below $\frac{2}{3}B_{c1}$ the Meissner state exists with $\chi = -1$. For a Type I superconductor in the applied field range $\frac{2}{3}B_c < B_{app} < B_c$, the formalism applies with χ chosen so that $H_{in} = B_c/\mu_0$. For the condition $B_{app} < \frac{2}{3}B_c$ we have $\chi = -1$, as will be clear from the discussion in Chapter 11, Section IV.

XIV. CYLINDER IN A MAGNETIC FIELD

On several occasions we have discussed the case of a long diamagnetic cylinder in an axially applied magnetic field, as shown in Fig. 9.1. In this "parallel geometry" the demagnetization factor N is zero. Inserting $N = 0$ in Eqs. (10.35)–(10.37) gives Eqs. (10.7)–(10.9), which we have already obtained for this case. These reduce to Eqs. (10.11)–(10.13), respectively, for the ideal Type I superconductor with $\chi = -1$. Since $N = 0$, the boundary condition—i.e., that H is continuous across the interface—leaves the H field undisturbed by the presence of the superconductor. The fields outside the cylinder are then

$$B_{out} = B_{app} \tag{10.67}$$

$$H_{out} = B_{app}/\mu_0, \tag{10.68}$$

independent of position.

An alternate arrangement that sometimes occurs in practice is the perpendicular geometry sketched in Fig. 10.17, whereby the cylinder axis remains in the z direction but the magnetic field is applied along x. For this case we see from Table 10.1 that $N = \frac{1}{2}$ so that the fields inside are, from Eqs. (10.35)–(10.37),

$$B_{in} = B_{app}\frac{2(1+\chi)}{2+\chi}, \tag{10.69}$$

$$H_{in} = \frac{B_{app}}{\mu_0}\frac{2}{2+\chi}, \tag{10.70}$$

$$M = \frac{B_{app}}{\mu_0}\frac{2\chi}{2+\chi}. \tag{10.71}$$

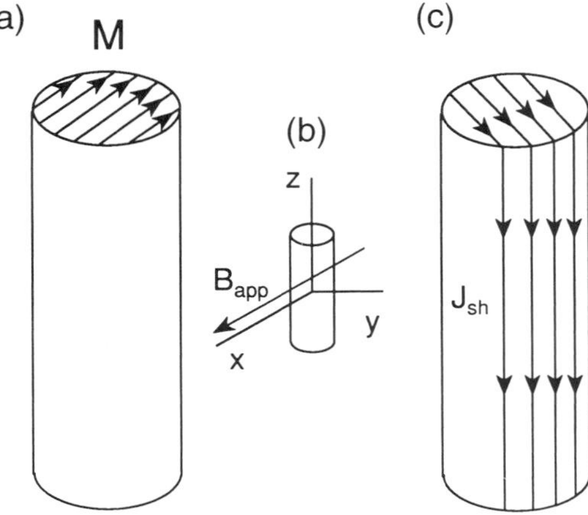

Figure 10.17 Magnetization M (a) and shielding current flow J (c) of a superconducting rod located in an applied magnetic field with the perpendicular geometry arrangement (b).

The calculation of the fields outside for this geometry is more complicated. There is no z dependence, so this is a two-dimensional problem in which Laplace's equation is solved in cylindrical coordinates with the $\partial^2/\partial z^2$ term omitted,

$$\frac{1}{\rho}\cdot\frac{d}{d\rho}\left(\rho\frac{\partial\Phi}{\partial\rho}\right)+\frac{1}{\rho^2}\cdot\frac{\partial^2\Phi}{\partial\phi^2}=0, \quad (10.72)$$

subject to two boundary conditions—first that H parallel to the surface and B perpendicular to the surface are both continuous; second, that far from the sample B is in the x direction with the magnitude B_{app}. The solution for the magnetic scalar potential $\Phi(\rho,\phi)$ is

$$\Phi_{out}=-\left[\rho-\frac{\chi}{\chi+2}\cdot\frac{a^2}{\rho}\right]B_{app}\cos\phi, \quad (10.73)$$

which is similar to the case of the sphere given in Eq. (10.58). The differences arise from the particular forms of the differential operators in Eqs. (10.60) and (10.72). The first term in this potential,

$$\rho B_{app}\cos\phi = xB_{app},$$

corresponds to the potential of the uniform applied field, where $x=\rho\cos\phi$.

The radial component B_ρ of the field outside,

$$B_\rho=-\frac{\partial\Phi_{out}}{\partial\rho} \quad (10.74)$$

$$=\left[1+\frac{\chi}{\chi+2}\cdot\frac{a^2}{\rho^2}\right]B_{app}\cos\phi, \quad (10.75)$$

has a value at the surface of

$$B_\rho=\left[\frac{2(\chi+1)}{\chi+2}\right]B_{app}\cos\phi, \quad (10.76)$$

which vanishes for the perfect superconductor case of $\chi=-1$. The azimuthal component B_ϕ outside,

$$B_\phi=-\frac{1}{\rho}\cdot\frac{\partial\Phi_{out}}{\partial\phi} \quad (10.77)$$

$$=\left[1-\frac{\chi}{\chi+2}\cdot\frac{a^2}{\rho^2}\right]B_{app}\sin\phi, \quad (10.78)$$

becomes at the surface

$$B_\phi=\left[\frac{2}{\chi+2}\right]B_{app}\sin\phi. \quad (10.79)$$

This surface field is zero along x and reaches a maximum value along the y direction, where $\phi=\pi/2$ and $B_\phi/\mu_0=H_{in}$ (cf. Eq. (10.70)). A sketch of the magnetic field lines around the cylinder would resemble that of Fig. 2.23. This case is equivalent to a two-dimensional problem with all of the field lines lying in the x,y-plane.

For perfect diamagnetism, $\chi=-1$, we see from Eqs. (10.76) and (10.79) that immediately outside the cylinder the external magnetic field is parallel to the surface with no radial component. Longitudinal surface currents J_z flow along the surface in the $+z$ direction on one side and in the $-z$ direction on the other, forming closed loops at the ends that sustain the magnetization inside, as indicated in Fig. 10.17. These currents serve to cancel the B field that would otherwise be present inside the cylinder. The factor $\sin\phi$ in Eq. (10.78) causes the surface-current density to produce the uniform magnetic field of Eq. (10.69) inside the cylinder.

XV. ac SUSCEPTIBILITY

Earlier in the chapter we discussed susceptibilities determined in constant magnetic fields. Now let us consider what happens when the external field varies harmonically in time (D.-X. Chen *et al.*, 1990c; vide Q. Y. Chen, 1992; Hein *et al.*, 1992; Khode and Couach, 1992). An ac field $B_0 \cos \omega t$ applied to the sample causes the magnetization $M(t)$ to trace out a magnetic hysteresis loop in the course of every cycle of the applied field. The initial loop for the first cycle will be different frɔm all the other cycles, as suggested by the initial curves starting from the middle of the loops of Figs. 10.6 and 10.7, but after several cycles a state of dynamic equilibrium is attained in which the magnetization $M(t)$ repeatedly traces out the same curve, perhaps of the types shown in Figs. 10.5 or 10.6, during every period of oscillation.

If the magnetization were to change linearly with the applied field, the response would be $M(t) = M_0 \cos \omega t$ in phase with the applied field, with $M_0 = \chi B_0/\mu_0$. The shape of the loop causes $M(t)$ to become distorted in shape and shift in phase relative to the applied field, causing it to acquire an out-of-phase component that varies as $\sin \omega t$. We can define the in-phase dispersion χ' and the out-of-phase (quadrature) absorption χ'' susceptibilities (Matsumoto *et al.*, 1991):

$$\chi' = \frac{\mu_0}{\pi B_0} \int M(t) \cos \omega t \, d\omega t, \quad (10.80a)$$

$$\chi'' = \frac{\mu_0}{\pi B_0} \int M(t) \sin \omega t \, d\omega t. \quad (10.80b)$$

Higher harmonic responses χ_n' and χ_n'' at the frequencies $n\omega t$ have also been studied (Ghatak *et al.*, 1992; Ishida and Goldfarb, 1990; Ishida *et al.*, 1991; Jeffries *et al.*, 1989; Ji *et al.*, 1989; Johnson *et al.*, 1991; Yamamoto *et al.*, 1992). Note that the absorption susceptibility is proportional to the energy dissipation. Unfortunately, in practice it is not practical to measure $M(t)$, so that a different approach must be followed.

The usual mutual inductance method for determining χ' and χ'' involves placing the sample in the coil of an LC tuned circuit to establish an alternating magnetic field $B_0 \cos \omega t$ in the superconductor and to detect the voltage induced in a detector pickup coil coupled to the coil of the LC circuit. The presence of the sample changes the effective inductance and resistance of the LC circuit, and this change is reflected in the form of the current induced in the detector coil. The component of the induced signal which is in phase with the applied field is proportional to the dispersion χ', while the out-of-phase component is proportional to the absorption χ''. These two responses can be separated instrumentally by a lock-in detector that compares the phase of the output signal with that of the reference signal $B_0 \cos \omega t$.

Figures 10.18 and 10.19 present the temperature dependence of the dispersion χ' and absorption χ'' components of the ac susceptibility determined for applied fields of the form

$$B_{app} = B_{dc} + B_0 \cos \omega t \quad (10.81)$$

at the frequency $\omega/2\pi = 73$ Hz. Figure 10.18 shows the results for three alternating field amplitudes B_0 with $B_{dc} = 0$, and Fig. 10.19 illustrates the effect of simultaneously applying a dc field. We see from the figures that for a particular applied field, χ' decreases continuously as the temperature is lowered, also that the drop in χ' is sharper and occurs closer to T_c for lower values of B_0 and B_{dc}. The peak in the χ''-versus-temperature curve is near the center of the sharp diamagnetic change in χ', as expected, inasmuch as magnetic susceptibilities, like dielectric constants, obey Kramers–Kronig relations (cf. Chapter 15, Section II.E; Poole and

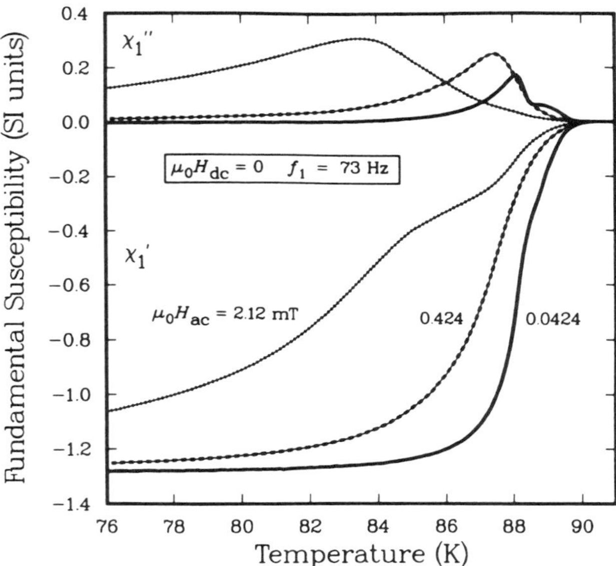

Figure 10.18 Real (χ') and imaginary (χ'') components of the susceptibility of $YBa_2Cu_3O_{7-\delta}$ measured in the applied ac magnetic fields with $\mu_0 H_{ac} = 0.0424$, 0.424, and 2.12 mT as a function of the temperature below T_c for a frequency of 73 Hz. For this experiment $\mu_0 H_{dc} = 0$; the data were not corrected for the demagnetization factor (Ishida and Goldfarb, 1990).

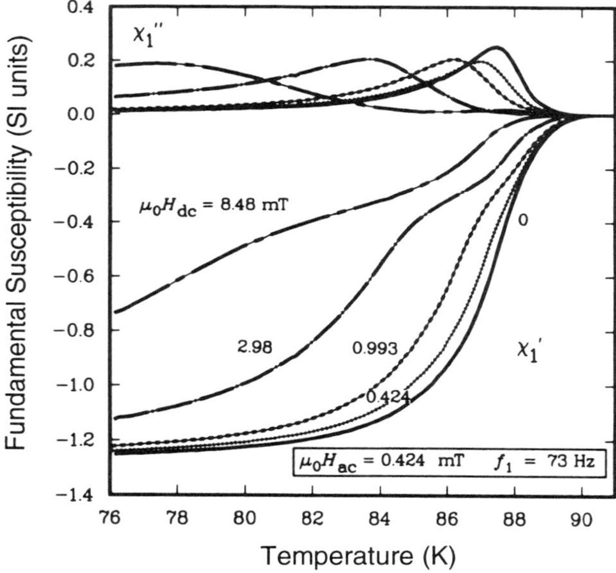

Figure 10.19 Real (χ') and imaginary (χ'') components of the susceptibility of $YBa_2Cu_3O_{7-\delta}$ measured in superimposed ac and dc magnetic fields as a function of the temperature below T_c for a frequency of 73 Hz. The conditions were $\mu_0 H_{ac} = 0.424$ mT, $\mu_0 H_{dc} = 0$, 0.424, 0.993, 2.98, and 8.48 mT; the data were not corrected for the demagnetization factor (Ishida and Goldfarb, 1990).

Farach, 1971, Chapter 20). Recent data on $HgBa_2CuO_{4+\delta}$ at high pressure exhibit this behavior (Klehe *et al.*, 1992). These K–K relations permit χ' to be calculated from knowledge of the frequency dependence of $\chi''(\omega)$, and vice versa. Increasing the applied field shifts the χ'' peak to lower temperatures and broadens it (D.-X. Chen *et al.*, 1988; Goldfarb *et al.*, 1987a, b; Ishida and Goldfarb, 1990; Puig *et al.*, 1990; K. V. Rao *et al.*, 1987). These χ_{ac} response curves depend only slightly on frequency below 1 kHz so that the magnetization is able to follow the variation in the applied field.

The ac susceptibility results can be thought of in terms of the temperature dependence of the lower-critical field $B_{c1}(T)$ (cf. Figs. 9.36 and 10.28). A low applied field at low temperature will be far below $B_{c1}(T)$, thus in Fig. 10.18 the χ' curve for $B_0 = 0.0424$ approaches total diamagnetic shielding. A high applied field near but still below T_c will exceed $B_{c1}(T)$ so that χ', will be smaller in magnitude and closer to its normal state value, as shown in Fig. 10.19.

It is more customary to interpret ac susceptibility data in terms of one of the critical-state models that will be introduced in Chapter 12 (Chen and Sanchez, 1991) with a temperature-dependent critical current (Ishida and Goldfarb, 1990; Johnson *et al.*, 1991; LeBlanc and LeBlanc, 1992). Ji *et al.* (1989) assumed the two-fluid model temperature dependence of Eq. (2.56). Here magnetic flux in the form of vortices alternately enters and leaves the sample as the magnetization cycles around the hysteresis loop. The maximum of χ'' can be interpreted as occurring near the applied field $B_{app} = B^*$, where the critical current and internal field just reach the center of the sample. Sample geometry (Forsthuber and Hilscher, 1992) and size effects (Skumryev *et al.*, 1991) have also been reported.

Clem (1992) suggested that there are three main mechanisms responsible for ac susceptibility losses: (a) flux flow losses, which can also be called eddy current losses or viscous losses, arising in the absence of pinning centers, when time-varying currents arising from the oscillating applied magnetic field induce fluxons to move, (b) hysteresis losses occuring near pinning centers that impede the flux motion, as well as wherever vortices of opposite sense annihilate each other, and (c) surface pinning losses arising from a surface barrier to vortex entry and exit (Hocquet *et al.*, 1992; Mathieu and Simon, 1988). An additional complication in granular superconductors is the presence of both intergranular and intragranular shielding currents.

In a granular superconductor the ac susceptibility is expected to receive contributions from intergranular current flow in loops through Josephson junctions at the boundaries between grains as well as from intragranular shielding current flow within the individual grains (J. H. Chen, 1990a, b; Lam *et al.*, 1990; Lera *et al.*, 1992; Müller and Pauza, 1989). The χ''-versus-T curves can exhibit both intergranular and intragranular peaks. Coreless Josephson vortices at the junctions and the more common Abrikosov vortices inside the grains alternately sweep in and out of the sample during each cycle around the hysteresis loop.

XVI. TEMPERATURE-DEPENDENT MAGNETISM

Diamagnetism is an intrinsic characteristic of a superconductor. Superconductors exhibit other types of magnetic behavior as well, due to, for example, the presence of paramagnetic rare earth and transition ions in their structure.

Susceptibility above T_c can have a temperature-independent contribution χ_0 arising from the conduction electrons along with a temperature-dependent Curie–Weiss term due to the presence of paramagnetic ions,

$$\chi = \chi_0 + \frac{K\mu^2}{3k_B(T-\Theta)} \quad (10.82)$$

$$= \chi_0 + \frac{C}{T-\Theta}, \quad (10.83)$$

where μ is the magnetic moment of the paramagnetic ions, K is a parameter that incorporates the concentration of paramagnetic ions and the conversion factor (1.86) for volume susceptibility, and C is the Curie constant. The Curie–Weiss temperature Θ is negative for ferromagnetic coupling between the magnetic ions and positive for antiferromagnetic coupling. Below T_c the large diamagnetism generally overwhelms the much smaller terms of Eq. (10.82), and they become difficult to detect.

A. Pauli Paramagnetism

The constant term χ_0 in Eq. (10.82) is often Pauli-like, arising from the conduction electrons (cf. Eq. (1.84)). We see from Eq. (1.83) that χ_{Pauli} provides an estimate of the density of states $D(E_F)$ at the Fermi level.

B. Paramagnetism

Most superconductors are paramagnetic above T_c. For example, it has been found (Tarascon *et al.*, 1987b) that the susceptibility of $YBa_2Cu_3O_{7-\delta}$ above T_c has a temperature dependence that obeys the Curie–Weiss law (1.79), with $\mu \approx 0.3\mu_B$/mole of Cu and $\Theta \approx -20$ to -30 K for oxygen contents δ in the range 0–0.6. Removing more oxygen increases μ and decreases Θ, but the samples no longer superconduct. These measured moments are less than the Cu^{2+} spin-only value of $1.9\mu_B$ given by Eq. (1.82),

$$\mu = g[S(S+1)]^{1/2}\mu_B = 1.9\mu_B, \quad (10.84)$$

where $S = \frac{1}{2}$ and $g \approx 2.2$.

Oxide materials in which magnetic rare earths replace lanthanum or yttrium provide linear plots of $1/\chi$ versus T above T_c as shown by the solid curves in Fig. 10.20, indicating paramagnetic behavior.

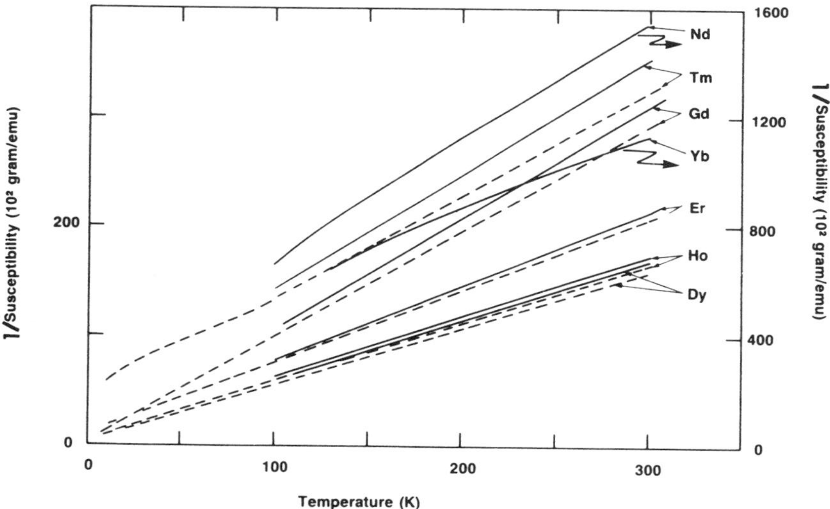

Figure 10.20 Temperature dependence of the reciprocal susceptibility $(1/\chi)$ for a series of rare earth (R) substituted $RBa_2Cu_3O_{7-\delta}$ superconductors over the temperature range 100–300 K in a field of 1 T (solid lines). Data for the corresponding nonsuperconducting vacuum-annealed compounds (dashed lines) are shown for comparison. The linear behavior is indicative of paramagnetism (Tarascon *et al.*, 1987b).

For some compounds the temperature-independent term χ_0 of Eq. (10.82) is zero. Vacuum annealing of the samples destroyed the superconductivity and gave linear Curie–Weiss plots below T_c, shown by the dashed curves in the figure, which provide Θ from the extrapolated intercept at $T = 0$. The magnetic moments μ were very close to the values $g(J(J+1))^{1/2}$ expected from Eq. (1.80) for rare earth ions. The positive sign for Θ indicates that these ions interact antiferromagnetically, with the susceptibility behavior above T_c corresponding to Fig. 1.15. The results suggest nearly complete decoupling of the Cu–O planes responsible for the superconducting properties from the planes containing the rare earth ions responsible for the magnetic properties. Such decoupling of the magnetic and superconducting properties was observed in Chapter 3, Section X, for the Chevrel phases; it also occurs with the heavy fermions (Jee *et al.*, 1990; Konno and Veda, 1989).

The paramagnetic contribution to χ arising from the Curie–Weiss law below T_c should appear as a rise in χ or M near $T = \Theta$. Such a rise is indeed noticeable at temperatures low enough for the diamagnetic contribution to have already come close to the asymptotic value $\chi(0)$ expected experimentally at absolute zero. In practice, this paramagnetism is often too weak to observe. However, we see from that data shown in Fig. 10.21 that it is enhanced at high applied fields. In fact, the highest fields used, $B_{app} > 1.5$ T, are strong enough to overwhelm the diamagnetic contribution and drive the magnetization positive. This rise in M is also partly due to the decrease in the diamagnetism as B_{app} is increased. The inset to this figure shows how the Meissner fraction, which is the value of χ_{fc} expressed as a percentage of its value (-1) for perfect diamagnetism, depends on the applied field.

The susceptibility above T_c of the series of compounds $YBa_2(Cu_{0.9}A_{0.1})_3O_{7-\delta}$, where A is a first transition series element, is an average of the contributions from the A and Cu ions. It has been found

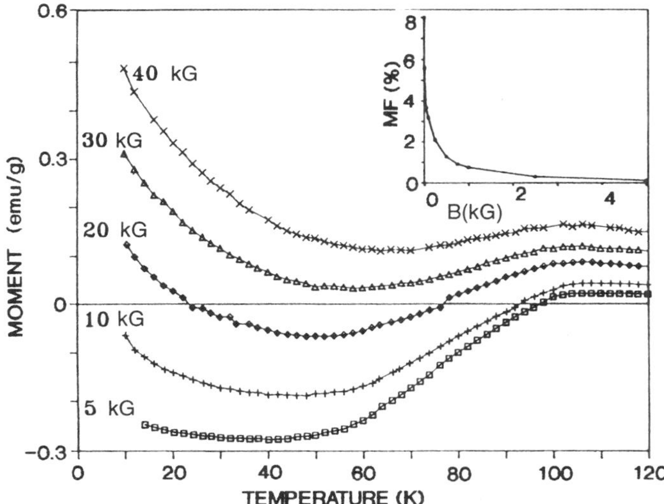

Figure 10.21 Appearance of a paramagnetic contribution at the low-temperature end of a field-cooled magnetization determination. The contribution becomes dominant as the field B_{fc} was increased from 0.5 to 4 T (i.e., from 5 to 40 kG), as shown. The inset gives the Meissner fraction (MF) as a function of the applied field from 0 to 0.5 T (Wolfus *et al.*, 1989).

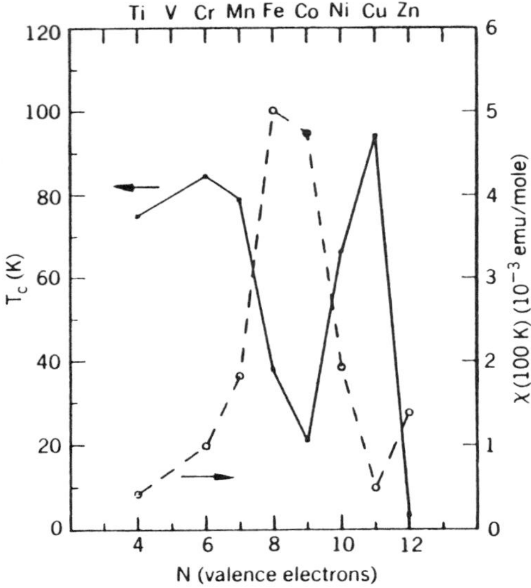

Figure 10.22 Dependence of the transition temperature T_c (——) and magnetic susceptibility at 100 K (– – –) on the number of valence electrons for the series of compounds $YBa_2(Cu_{0.9}A_{0.1})_3O_{7-\delta}$, where A is a $3d$ transition element, as shown (Xiao *et al.*, 1987a).

to obey Eq. (10.82) with an effective magnetic moment given by (Xiao *et al.*, 1987a, b),

$$\mu_{\text{eff}}^2 = 0.1\mu_A^2 + 0.9\mu_{\text{Cu}}^2, \quad (10.85)$$

where μ_A and μ_{Cu} are the moments of the A and Cu atoms, respectively. We see from Fig. 10.22 that the depression of T_c correlates with the size of the magnetic moment of the substituted transition ion—the larger the moment, the lower the T_c value. Others have reported similar results (e.g., Maeno *et al.*, 1987; Oseroff *et al.*, 1987).

C. Antiferromagnetism

Cuprate superconductors generally have a negative Curie–Weiss temperature Θ indicative of antiferromagnetic coupling (Chapter 1, Section XV). The undoped compound La_2CuO_4 is an antiferromagnet below the Néel temperature $T_N \approx 245$ K, which is considerably lower than the

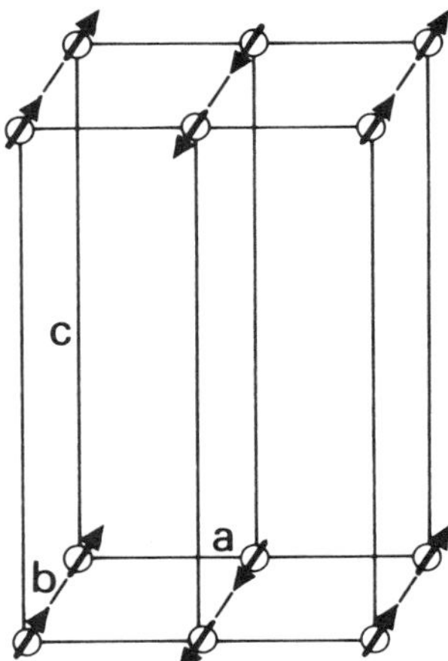

Figure 10.23 Magnetic spin structure of ordered Er ions in antiferromagnetic $ErBa_2Cu_3O_{7-\delta}$ determined by neutron diffraction (Chattopadhyay *et al.*, 1989).

tetragonal-to-orthorhombic transition temperature $T_{\text{t-o}} = 525$ K. The copper spins are ordered in the CuO_2 planes in the manner shown in Fig. VIII-18 of our earlier book (1988; cf. also Freltoft *et al.*, 1988; Kaplan *et al.*, 1989; Thio *et al.*, 1988; Yamada *et al.*, 1989). Antiferromagnetic spin fluctuations in these CuO_2 planes, called *antiparamagnons*, have also been discussed (Statt and Griffin, 1993).

Compounds formed by replacing the yttrium in $YBa_2Cu_3O_{7-\delta}$ by a rare-earth ion tend to align antiferromagnetically at low temperature (Lynn, 1992). For example, the Er moments $\mu = 4.8\mu_B$ in $ErBa_2Cu_3O_{7-\delta}$ order in the a, b-plane with antiferromagnetic coupling along a and ferromagnetic coupling along b and c, in the manner shown in Fig. 10.23. The neutron-magnetic reflection intensity plotted in Fig. 10.24 versus temperature provided the Néel temperature $T_N \approx 0.5$ K (Chattopadhyay *et al.*, 1989; Lynn *et al.*, 1989; Paul *et al.*, 1989). Below $T_N \approx 2.2$ K, the Gd moments in $GdBa_2Cu_3O_{7-\delta}$ align along the c-axis with antiferromagnetic coupling to all Gd nearest neighbors, as illustrated in Fig. VIII-19 of our earlier work (1988; cf. also Dunlap *et al.*, 1988;

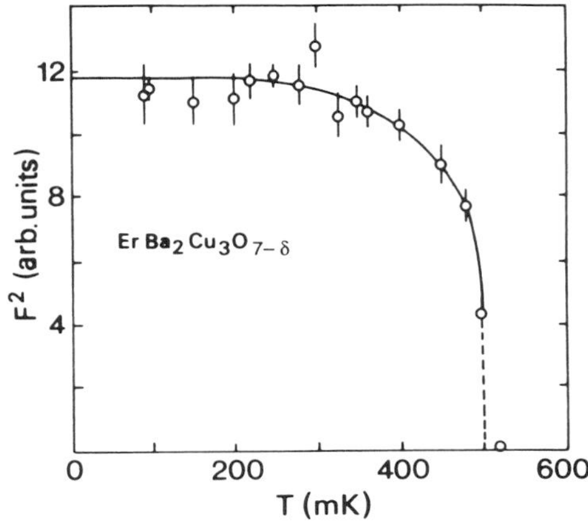

Figure 10.24 Temperature dependence of the intensity of reflected neutrons from the $ErBa_2Cu_3O_{7-\delta}$ sample of Fig. 10.23 showing the Néel temperature $T_N \approx 0.5$ K far below $T_c = 88$ K (Chattopadhyay *et al.*, 1989).

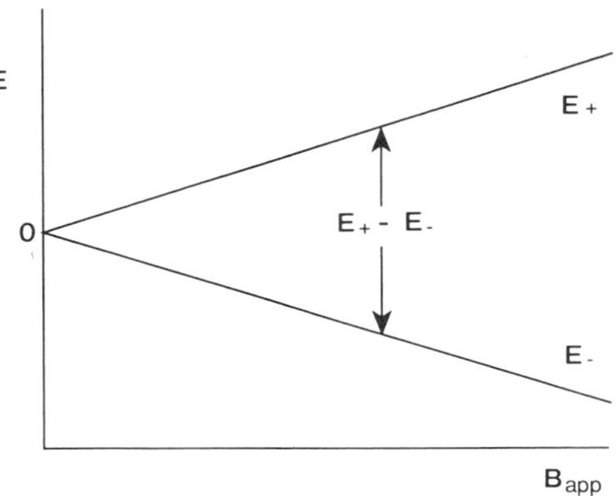

Figure 10.25 Zeeman energy level splitting $E_+ - E_-$ of electrons resulting in the breakup of Cooper pairs by becoming comparable to the energy gap 2Δ when B_{app} reaches the value B_{c2}.

Mook *et al.*, 1988; Niedermayer *et al.*, 1993; Paul *et al.*, 1988; Watson *et al.*, 1989). Other magnetic ions, such as Dy, Ho, Nd, Pr, and Sm, when substituted for Y, also produce antiferromagnetic ordering (Dy: Fischer *et al.*, 1988; Zhang *et al.*, 1992; Ho: Fischer *et al.*, 1988; Nd: Yang *et al.*, 1989; Pr: Kebede *et al.*, 1989; Sm: Yang *et al.*, 1989). For $x < 6.4$ the undoped compound $YBa_2Cu_3O_x$ is an antiferromagnetic nonsuperconductor with aligned Cu ions, and $T_N \approx 500$ K for $x \approx 6$ (Miceli *et al.*, 1988; Rossat-Mignod *et al.*, 1988; Tranquada, 1990; Tranquada *et al.*, 1992).

XVII. PAULI LIMIT AND UPPER CRITICAL FIELDS

An electron spin in a magnetic field has the Zeeman energy

$$E = g\mu_B \mathbf{B}_{app} \cdot S, \qquad (10.86)$$

$$E_\pm = \pm \tfrac{1}{2} g\mu_B B_{app} \qquad (10.87)$$

shown in Fig. 10.25, where $\mu = g\mu_B \mathbf{S}$ is the spin magnetic moment, $g = 2.0023$ for a free electron, and μ_B is the Bohr magneton. We will approximate the g-factor by 2, and, of course, $S = \frac{1}{2}$. If the Zeeman energy level splitting (Poole and Farach, 1987) indicated in the figure,

$$E_+ - E_- = 2\mu_B B_{app}, \qquad (10.88)$$

becomes comparable with the energy gap E_g, the field will be strong enough to break up the Cooper pairs and destroy the superconductivity. The magnetic field B_{Pauli} that brings this about is called the Pauli limiting field. It has the value

$$B_{Pauli} = \frac{E_g}{2\sqrt{2}\,\mu_B}, \qquad (10.89)$$

where the factor $\sqrt{2}$ comes from a more detailed calculation. Inserting the BCS gap ratio $E_g = 3.53 k_B T_c$, this becomes

$$B_{Pauli} = 1.83 T_c. \qquad (10.90)$$

The data in Table 10.2 demonstrate that this provides an approximation to experimentally determined upper-critical fields B_{c2}. This limiting field has also been called the paramagnetic limit or the Clogston–Chandrasekhar limit (Chandrasekhar, 1962; Clogston, 1962; Pérez-González and Carbotte, 1992).

For many Type II superconductors both the ratio $B_{c2}(0)/T_c$ and the slope dB_{c2}/dT at T_c are close to the Pauli value

Table 10.2 Comparison of Upper-Critical Field $B_{c2}(0)$, Slope dB_{c2}/dT at T_c, and Pauli Limiting Fields $B_{Pauli} = 1.83T_c$ of Selected Type II Superconductors

Material	T_c (K)	B_{c2} (T)	B_{c2}/T_c (T/K)	$-dB_{c2}/dT$ (T/K)	B_{Pauli} (T)
$CeCu_2Si_2$ (heavy fermion)	0.5	2.4	4.8	23	0.9
UBe_{13} (heavy fermion)	0.9	6	6.7	44	1.7
Nb(44%)–Ti (alloy)	9.3	15	1.6	2.4	17
$Gd_{0.2}PbMo_6S_8$ (Chevrel)	14	61	4.4	6.8	26
Nb_3Sn (*A*15)	18	20	1.1	2.5	33
$Nb_3(Al_{0.75}Ge_{0.25})$ (*A*15)	20.7	43.5	2.1	3.0	38
Nb_3Ge (*A*15)	23.1	38	1.7	2.3	42
$YBa_2Cu_3O_7$ (HTSC)	92	120–200	1.3–2.2	0.7–4.6	168

1.83 T/K, as shown by the data listed on Table 10.2 and plotted in Fig. 10.26. The zero-temperature upper-critical fields $B_{c2}(0)$ of high-temperature superconductors are generally too high to measure directly, but they can be estimated from the Pauli limit or from the empirical expression $B_{c2} \approx (2T_c/3)dB_{c2}/dT$, which can be deduced from the data in Table 10.2.

Upper critical fields $B_{c2}(T)$ and their temperature derivatives dB_{c2}/dT often depend on the orientation of the applied

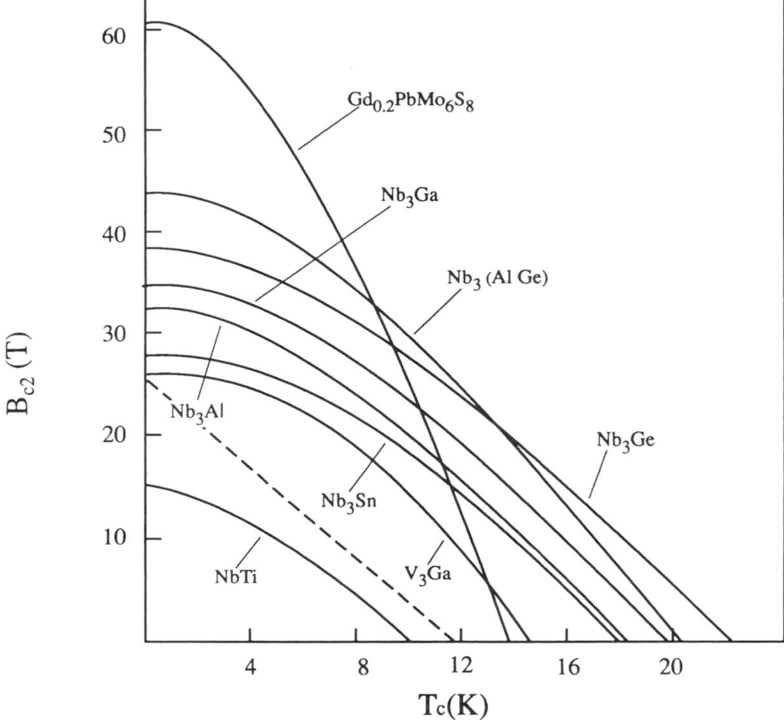

Figure 10.26 Relation between upper-critical field B_{c2} and temperature for the best classical superconductors, some of which are used for fabricating commercial magnets. In the high-temperature limit $(2/3\,T_c < T < T_c)$ many of the curves have a slope that is close to the value 1.83 T/K of the Pauli limit (dashed curve) (Wilson, 1983, p. 302).

magnetic field. This is especially true for the high-temperature superconductors because of their planar structures. These types of critical fields and their temperature derivatives at T_c are larger when the external field is applied perpendicular to the c-axis (i.e., parallel to the Cu–O planes) than when it is applied parallel to this axis, as shown in Fig. 10.27. This order is reversed for the lower critical field, as shown in Fig. 10.28; in other words, $B_{c1\perp c} < B_{c1\parallel c} \ll B_{c2\parallel c} < B_{c2\perp c}$.

This reversal is associated with the reversal in the order of sizes of the penetration depths and coherence lengths given by Eq. (9.46), $\xi_c < \xi_{ab} \ll \lambda_{ab} < \lambda_c$. Therefore, we have, from Eqs. (9.51) and (9.52), the lower critical field ratio

$$\frac{B_{c1\perp c}}{B_{c1\parallel c}} = \frac{\lambda_{ab}}{\lambda_c} \frac{\ln \kappa_{ab}}{\ln \kappa_c} < 1 \quad (10.91)$$

and its upper critical field counterpart

$$\frac{B_{c2\perp c}}{B_{c2\parallel c}} = \frac{\xi_{ab}}{\xi_c} > 1, \quad (10.92)$$

where κ_{ab} and κ_c are given by Eqs. (9.48) and $\kappa_{ab} > \kappa_c$. These inequalities may be verified from the data in Tables 9.4 and 9.5.

Tesanovic (1991), Tesanovic and Rasolt (1989), and Tesanovic *et al.* (1991) discussed the possibility of reentrant superconducting behavior in applied fields far exceeding B_{c2}.

XVIII. IDEAL TYPE II SUPERCONDUCTOR

A Type II superconductor has several characteristic parameters, such as its Ginzburg–Landau parameter κ, transition temperature T_c, energy gap E_g, coherence length ξ, penetration depth λ, upper-

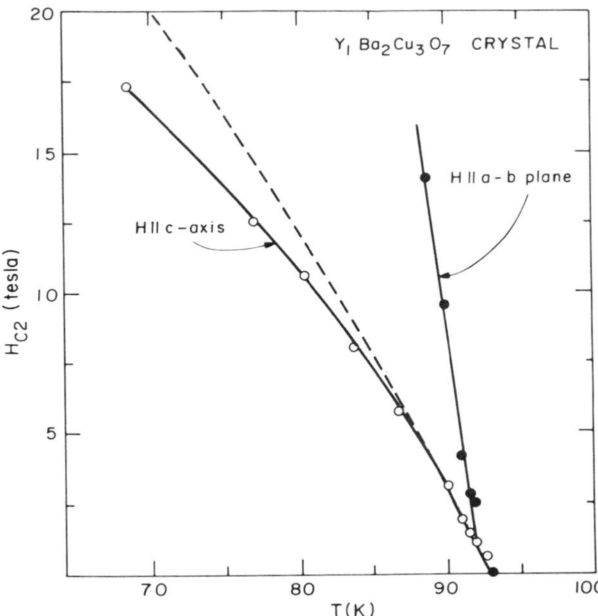

Figure 10.27 Anisotropy in the upper-critical fields of $YBa_2Cu_3O_7$. The initial slope $dB_{c2\parallel c}/dT$ at T_c (– – –) is -0.96 T/K, while its counterpart $dB_{c2\perp c}/dT$ at T_c is -4 T/K (Moodera *et al.*, 1988).

critical field B_{c2}, lower critical field B_{c1}, thermodynamic critical field B_c, and critical current density J_c. We have seen how these various parameters are related by simple theoretical expressions, so that if any two of them are specified, the others can be estimated. This suggests defining an ideal isotropic Type II superconductor as one whose parameters have "ideal" relationships with each other.

Consider such a Type II superconductor with $\kappa = 100$ and $T_c = 90$ K. Its energy gap is obtained from the BCS relation (6.120)

$$E_g = 3.528 k_B T_c = 27.5 \text{ meV}. \quad (10.93)$$

The Pauli limit (10.90) provides an estimate of the upper-critical field,

$$B_{c2} = 1.83 T_c = 165 \text{ T}. \quad (10.94)$$

Equation (9.9) gives the coherence length ξ,

$$\xi = \left(\frac{\Phi_0}{2\pi B_{c2}} \right)^{1/2} = 1.26 \text{ nm}, \quad (10.95)$$

and from the definition (9.6) of the Ginzburg–Landau parameter we obtain the penetration depth λ,

$$\lambda = \kappa \xi = 126 \text{ nm}. \quad (10.96)$$

Equations (9.10) and (9.11), respectively, give the thermodynamic and lower critical fields,

$$B_c = \frac{B_{c2}}{\sqrt{2}\,\kappa} = 1.16 \text{ T}, \quad (10.97)$$

$$B_{c1} = \frac{B_c \ln \kappa}{\sqrt{2}\,\kappa} = 37.9 \text{ mT}. \quad (10.98)$$

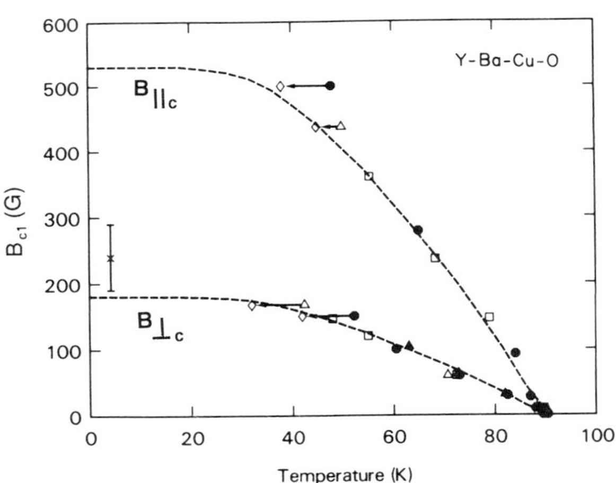

Figure 10.28 Anisotropy in the lower-critical fields of $YBa_2Cu_3O_7$. The initial slope $dB_{c1\|c}/dT$ at T_c is -1.4 mT/K, and that of $dB_{c1\perp c}/dT$ at T_c is -0.40 mT/K. The low-temperature extrapolations give 53 ± 5 mT for the applied field parallel to c and 18 ± 2 mT for B_{app} perpendicular to c. Yeshuran *et al.* (1988) obtained the 6 K values, $B_{c1\|c} = 90 \pm 10$ mT (not shown) and $B_{c1\perp c} = 25 \pm 5$ mT (shown as x with vertical error bar). The dashed curves are BCS fits to the data (Krusin-Elbaum *et al.*, 1989). Recall that 10G = 1 mT.

The critical current density J_c at 0 K is given by Eq. (2.51):

$$J_c = B_c/\mu_0 \lambda = 6.95 \times 10^8 \text{ A/cm}^2. \quad (10.99)$$

This approximates what has been called the depairing current density.

$$J_{\text{depair}} = 10B_c/4\pi\lambda\mu_0 \quad (10.100)$$

$$= 5.53 \times 10^8 \text{A/cm}^2 \quad (10.101)$$

where for YBaCuO the thermodynamic field $B_c \approx 1$ T, penetration depth $\lambda \approx 0.2$ μm, and $T_c \approx 92$ K. These "ideal" values are good approximations to the experimentally determined values for typical high-temperature superconductors.

A better approximation than Eq. (10.99) to the upper limit of the critical current density is given by the Ginzburg–Landau expression

$$J_{\text{cmax}} = \left[\frac{2}{3}\left(1 - \frac{T}{T_c}\right)\right]^{3/2} J_{\text{depair}}. \quad (10.102)$$

This gives $J_c \approx 3 \times 10^8$ A/cm^2 at 0 K and $J_c \approx 1.2 \times 10^7$ at 77 K, respectively. Jiang *et al.* (1991) reported $J_c \approx 1.3 \times 10^9$ A/cm^2 for microbridges of

$$YBa_2Cu_3O_{7-\delta}$$

films. Achievable critical currents are typically one-tenth the limiting values calculated from Eq. (10.102), as indicated by the data in Table I-2 of our earlier work (Poole *et al.*, 1988).

XIX. MAGNETS

Superconducting magnet design requires simultaneously achieving high critical fields, high critical currents, and suitably malleable wire. The slope $dB_{c2}/dT \approx -2$ T/K of YBaCuO is typical, and gives a critical field of 30 T at the temperature of liquid nitrogen, as shown in Fig. 9.8 and in Table I-2 of our previous work (Poole *et al.*, 1988). This high critical field is for the case of the externally applied field **B** aligned perpendicular to the *c*-axis, i.e., parallel to the crystallographic conducting planes. When $\mathbf{B}_{\text{app}}$ is parallel to the *c*-axis, the critical field is four or five times lower, as already noted in Section XVII.

The standard magnet materials Nb_3Sn and Nb–Ti have critical fields of about 24 T and 10 T, respectively, at 4.2 K, which are not much lower than that of YBaCuO at 77 K. Operating YBaCuO at temperatures much below 77 K will, of course, provide higher critical fields, and TlBaCaCuO, with its much higher T_c (125 K), is even better at 77 K. The problem is to obtain high-T_c superconductors that can carry large transport currents and in addition, have the proper ductility and possess the appropriate mechanical properties. This, however, has yet to be achieved. Vortex pinning must also be optimized to control flux creep.

FURTHER READING

Other superconductivity texts cover some of the material presented in this chapter (e.g., Buckel, 1991; Burns, 1992; Kresin and Wolf, 1990; Orlando and Delin, 1991; Rose Innes and Rhoderick, 1994; Tilley and Tilley, 1986; Tinkham, 1985).

Orlando and Delin (1991, Section 6.6) give a good account of anisotropy.

A number of workshops and conferences have been devoted to magnetic interactions in high-temperature superconductors (e.g., Bennett *et al.*, 1989). Portis (1993) discusses magnetic properties.

The monograph *Superconducting Magnets* (Wilson, 1983) discusses magnet technology before the high-T_c era and Osamura provides a recent coverage. A book by Moon (1994) on superconducting levitation has appeared.

The volume edited by Hein *et al.* (1991) provides much information on the magnetic susceptibilities of superconductors.

Malozemoff (1989) reviewed magnetic properties of high-temperature superconductors. Birgeneau and Shirane (1989) reviewed structural and magnetic excitations in the cuprates as studied by neutron scattering. Lynn (1990b) discussed Cu–O and rare earth magnetism. Markert, Dalichaouch, and Maple (1989) reviewed the effects of substituting magnetic ions, such as rare earths, in high-temperature superconductors.

PROBLEMS

1. Show that a superconductor containing a volume V_{ex} of voids which cannot store any flux has the following ZFC and FC susceptibilities and porosity:

$$\chi_{zfc} = -\frac{V_s + V_h + V_c}{V_s + V_h + V_c + V_{ex}},$$

$$\chi_{fc} = -\frac{V_s + V_c}{V_s + V_h + V_c + V_{ex}},$$

$$P = \frac{V_h + V_c + V_{ex}}{V_s + V_h + V_c + V_{ex}}.$$

2. A granular, 10-mg sample of $YBa_2Cu_3O_{7-\delta}$ has a density of 3.19 g/cm^3 and the susceptibilities $\chi_{zfc} = -0.8$ and $\chi_{fc} = -0.4$. Find the porosity and the volumes of the purely superconducting, normal material, open hole-like, enclosed cavity-like and non-flux storing portions of the sample. Assume that there is no normal material present.
3. Show that the measured magnetic moment is given by Eq. (10.21),

$$\mu_0 \mu_{fc} = -(V_s + V_c)B_{app} + (B_{fc} - B_{app})V_h,$$

when field cooling is carried out in a magnetic field B_{fc} that differs from the field B_{app} applied for the measurement.
4. Show that the sample of Problem 1 has the following superconducting, open-hole, closed-cavity, and non-flux storing volumes given by, respectively,

$$V_s = (1 - P)V_T,$$

$$V_h = -(\chi_{zfc} - \chi_{fc})V_T,$$

$$V_c = (P - 1 - \chi_{fc})V_T,$$

$$V_{ex} = (1 + \chi_{zfc})V_T,$$

where, of course, χ_{zfc} and χ_{fc} are both negative.
5. Show that (10.43) and (10.45) both have the limiting behavior $N_{\parallel} \to 1/3$ as $\epsilon \to 0$.
6. Show that δ_1 and δ_2 of Table 10.1 are given by

$$\delta_1 = \frac{1}{2}\pi\frac{c}{a},$$

$$\delta_2 = \frac{4}{15}\left(1 - \frac{c}{a}\right).$$

7. Show that δ_3 and δ_4 of Table 10.1 are equal to

$$\delta_3 = \frac{4}{15}\left(1 - \frac{a}{c}\right),$$

$$\delta_4 = \left[\frac{1}{2}\ln\left(2\frac{c^2}{a^2}\right) - 1\right]\frac{a^2}{c^2}.$$

8. Show that the expressions that were deduced in this chapter for the magnetic fields inside and outside a sphere obey the boundary conditions (1.73) and (1.74) at the surface $r = a$.
9. Show that the expressions that were deduced in this chapter for the magnetic fields inside and outside a cylinder in a perpendicular magnetic field obey the boundary conditions (1.73) and (1.74) at the surface.
10. Show that the Curie law, which is based on the assumption that $g\mu_B B_{app}/k_B T \ll 1$, is still applicable for the highest-field, lowest-temperature data of Fig. 10.26. What is the value of the ratio B_{app}/T for which $g\mu_B B_{app}/k_B T = 1$ for $g = 2.0$?
11. Show that for the condition $\chi_\infty = 0$, a plot of χ'' versus χ' over the frequency range $\omega_0 \le \omega \le \infty$ is a semicircle of radius $\frac{1}{2}\chi_0$. Identify the five points at which $(\omega - \omega_0)\tau$ is equal to 0, $\frac{1}{2}$, 1, 4, and ∞ on the semicircle. How would the plot change for $\chi_\infty = \frac{1}{4}\chi_0$?

11

Intermediate and Mixed States

I. INTRODUCTION

The previous two chapters discussed the magnetic properties of superconductors. In these chapters we showed that the superconducting state can only exist in a material when the external B field at the surface is less than the critical field B_c for a Type I superconductor and less than the upper critical field B_{c2} for a Type II superconductor. For a rod-shaped sample in a parallel external field, the demagnetization factor is zero, so the field H at the surface equals the externally applied field, $H_{in} = H_{app}$, and the situation is not complicated. For other field directions and other sample shapes, the fact that the demagnetization factor does not vanish complicates matters because it raises the question whether the external field can exceed the critical field over part of the surface but not over the remainder. When this occurs with a Type I superconductor, the sample lowers its free energy by going into an intermediate state involving partial penetration of the external field into the interior. In the present chapter we will examine how this happens. We will discuss thin films, the domains that form in thin films in the intermediate state and the magnetic field configurations associated with thin films. We will also treat the intermediate state induced by transport current in a wire. Most of the chapter will be devoted to these discussions of the intermediate state of a Type I superconductor.

A Type II superconductor exists in what is called the Meissner state of total flux exclusion, $B_{in} = 0$, for applied fields in the range $B_{app} < B_{c1}$ and in the mixed state of partial flux penetration when the applied field is in the range $B_{c1} < B_{app} < B_{c2}$

between the lower and upper critical fields. The way in which the demagnetization factor affects these Meissner and mixed states will be discussed at the end of the chapter. For now, bear in mind that the term, *intermediate state*, applies to Type I superconductors, and the term, *mixed state*, to Type II superconductors.

II. INTERMEDIATE STATE

We learned in Chapter 1 that the magnetic fields inside any material, including a superconductor, satisfy the general expression (1.69)

$$B_{\text{in}} = \mu_0(H_{\text{in}} + M). \tag{11.1}$$

We saw in Chapter 2 that an ideal Type I superconductor has the following internal magnetic fields:

$$\left.\begin{aligned} B_{\text{in}} &= 0 \\ M &= -H_{\text{in}} \\ H_{\text{in}} &< \frac{B_c}{\mu_0} \end{aligned}\right\} \begin{array}{c}\text{Type I}\\ \text{Superconducting}\\ \text{State}\end{array} \tag{11.2}$$

In this chapter we will see that in the intermediate state of an ideal Type I superconductor, the H_{in} field is pinned at the value B_c/μ_0. This provides us with the relationships

$$\left.\begin{aligned} H_{\text{in}} &= \frac{B_c}{\mu_0} \\ \mu_0 M &= B_{\text{in}} - B_c \end{aligned}\right\} \begin{array}{c}\text{Type I}\\ \text{Intermediate}\\ \text{State}\end{array} \tag{11.3}$$

for the fields inside. Finally, above T_c the normal state exists with the field configurations

$$\left.\begin{aligned} M &\approx 0 \\ B_{\text{in}} &\approx \mu_0 H_{\text{in}} \end{aligned}\right\} \begin{array}{c}\text{Normal}\\ \text{State}\\ \text{Above } T_c\end{array} \tag{11.4}$$

where we have written $M \approx 0$ since in the normal state $|\chi| \ll 1$, as we showed in Chapter 1, Section XV.

In the previous chapter we were concerned with the internal fields of a superconducting ellipsoid in an applied magnetic field, and we made use of expression (10.33),

$$NB_{\text{in}} + (1-N)\mu_0 H_{\text{in}} = B_{\text{app}}, \tag{11.5}$$

where N is the demagnetization factor. The magnetostatic properties of the intermediate state follow from Eqs. (11.3) combined with (11.5). Since it is not obvious why the internal field H_{in} is pinned at the value B_c/μ_0 in the intermediate state, we will provide some justification for this in the next section before applying Eq. (11.5) to elucidate the properties of this state.

III. SURFACE FIELDS AND INTERMEDIATE-STATE CONFIGURATION

We saw in Chapter 10, Section XIII, that the surface field $B_{\text{surf}}(\Theta)$ immediately outside a perfectly superconducting sphere has no radial component; it is parallel to the surface, with the value at the angle Θ given by Eq. (10.66) with $\chi = -1$,

$$B_{\text{surf}}(\Theta) = \tfrac{3}{2} B_{\text{app}} \sin\Theta. \tag{11.6}$$

If the applied field B_{app} is less than $2B_c/3$, the surface field will be less than B_c for all angles, and the perfectly superconducting state can exist. If, on the other hand, the applied field is greater than $2B_c/3$, we see from Eq. (11.6) that there will be a range of angles near $\Theta = \pi/2$,

$$\Theta_c < \Theta < \pi - \Theta_c, \tag{11.7}$$

where

$$\Theta_c = \sin^{-1}\left(\frac{2B_c}{3B_{\text{app}}}\right), \tag{11.8}$$

for which the surface field will exceed B_c. Thus the sphere is unable to remain perfectly superconducting. In the range of applied fields

$$\tfrac{2}{3} B_c < B_{\text{app}} < B_c, \tag{11.9}$$

the surface field must decompose into superconducting and normal regions that

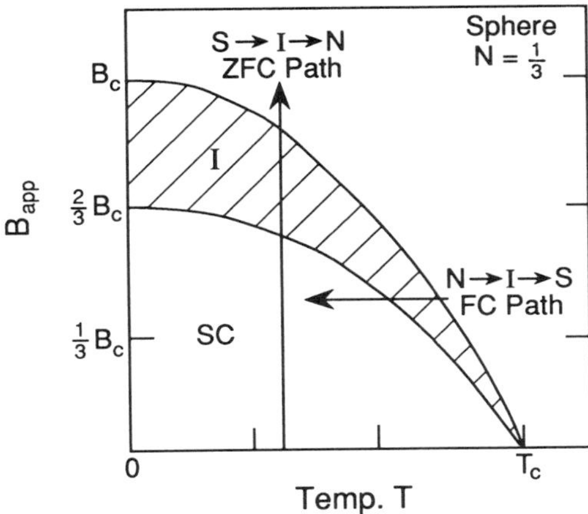

Figure 11.1 Dependence of the critical field $B_c(T)$ of a Type I superconductor on temperature (upper curve), where $B_c = B_c(0)$ is the critical field at 0 K. The figure also plots the curve $\frac{2}{3}B_c(T)$, which is the lower limit of the intermediate state (I) of a superconducting sphere. The vertical arrow indicates the path S → I → N traversed by a zero-field-cooled sphere as increasing applied fields bring it from the superconducting state (SC) at $B_{app} = 0$ through the intermediate state to the normal state (N) that exists for $B_{app} > B_c(T)$. The horizontal path $N \rightarrow$ I → S traversed by field cooling of the sphere is also shown.

prevent the average internal field H_{in} from exceeding the critical value H_c. In other words, the sphere must enter the intermediate state. For lower applied fields it is perfectly superconducting, whereas for higher applied fields it is in the normal state, as indicated in Fig. 11.1. The arrows in the figure show how the intermediate state can be traversed by varying either the applied magnetic field or the temperature.

One possibility for an intermediate state is for the sphere to go normal in a band around the equator delimited by Θ_c of Eq. (11.8). This, however, would not satisfy the boundary conditions. Another possibility would be for a normal outer layer to surround a superconducting inner region. But such states do not exist because it is energetically more favorable for a sphere to split into small regions of normal material adjacent to regions of superconducting material. We know that below T_c the superconducting state is energetically favored. Thus, the formation of the intermediate state is the way in which a material can continue to possess some of this favorable superconducting-state energy while still satisfying the boundary conditions on the surface field.

It is easier to picture how the intermediate state forms by considering the case of a Type I superconducting film in a perpendicular magnetic field. We assume a demagnetization factor $N = 0.9$ corresponding to the phase diagram of Fig. 11.2. When the applied field is raised to a value slightly above $0.1B_c(T)$, small regions of normal material appear embedded in a superconducting matrix, as shown in Fig. 11.3a. This is reminiscent of the vortex lattice that forms in Type II superconductors. We see from the sequence of structures in Figs. 11.3a–11.3f that as the applied field increases along the vertical path of Fig. 11.2, normal regions grow at the expense of the superconducting regions. For very high applied fields, as in Fig. 11.3f, when most of the material has be-

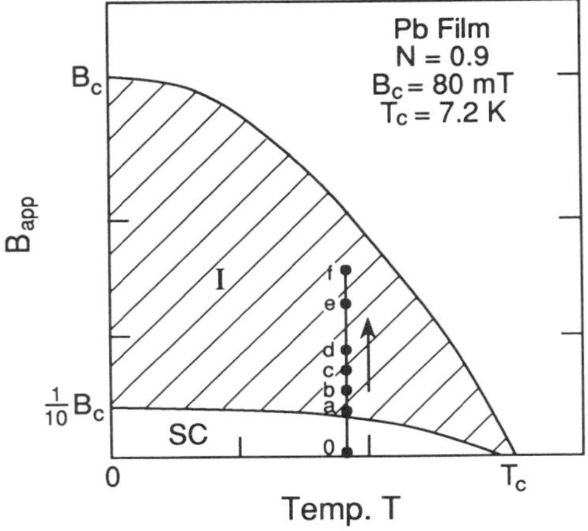

Figure 11.2 Magnetic phase diagram of a film with demagnetization factor $N = 0.9$ in a perpendicular magnetic field. This large demagnetization factor causes the intermediate state to be quite extensive. The vertical isothermal path S → I traversed by the film in passing through the succession of configurations depicted in Figs. 11.3a–11.3f is indicated.

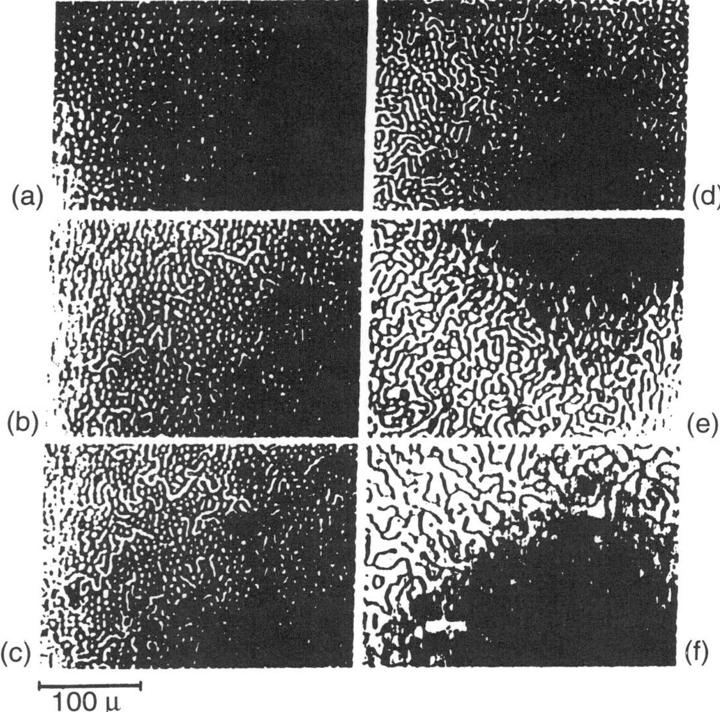

Figure 11.3 Intermediate state domain configurations of a 9.3-μm thick superconducting Pb film in a perpendicular magnetic field at 4.2 K. The structure is shown for the following field values: (a) 9.5 mT, (b) 13.2 mT, (c) 17.8 mT, (d) 21.8 mT, (e) 34.8 mT, and (f) 40.9 mT. The critical field $B_c = 80$ mT for Pb. The photographs were obtained using a magneto-optical method in which normal and superconducting regions are displayed as bright and dark, respectively. Figure 11.2 plots this sequence of increasing fields (Huebener, 1979, p. 22).

come normal, there is still a tendency for the extensive normal regions to be surrounded by what appear to be filaments of superconducting material. This has been referred to as a *closed topology*, that is, normal regions surrounded by a superconducting phase (Huebener, 1979).

We have just described the passage through the intermediate state for increasing values of the applied field experienced by a sample that has been precooled in zero field (ZFC) at a particular temperature below T_c. We showed that this S → I path proceeded along the vertical line in Fig. 11.2. Figure 11.1 shows a vertical S → I → N path for a ZFC sphere that starts in the superconducting state, passes through the intermediate state as the field increases at constant temperature, and finally reaches the normal state for $B_{app} > B_c(T)$. Another way of attaining the intermediate state is by field cooling the sphere along the N → I → S path in the same figure. As the sample gradually cools through the intermediate region, it expels flux by forming superconducting regions embedded in a normal matrix, corresponding to what is called an *open topology*, the opposite of what happens in the case S → I → N.

IV. TYPE I ELLIPSOID

Now that we have clarified the nature of the intermediate state, it will be instructive to write down the equations of the internal fields in a perfectly superconducting ellipsoid. For such an ellipsoid the factor $2B_c/3$ in Eq. (11.9) becomes $(1-N)B_c$, and for the purely superconducting state we have, from Eqs. (10.35)–(10.37),

$$B_{app} < (1-N)B_c, \tag{11.10}$$

$$B_{in} = 0, \tag{11.11}$$

$$H_{in} = B_{app}/(1-N)\mu_0, \tag{11.12}$$

$$\mu_0 M = -B_{app}/(1-N), \tag{11.13}$$

$$\chi = -1. \tag{11.14}$$

This states exists over the range given by Eq. (11.10).

The equations for the intermediate state obtained by combining Eqs. (11.3) and (11.5) are

$$(1-N)B_c < B_{app} < B_c, \tag{11.15}$$

$$B_{in} = \frac{1}{N}\left[B_{app} - (1-N)B_c\right], \tag{11.16}$$

$$H_{in} = B_c/\mu_0, \tag{11.17}$$

$$\mu_0 M = -\frac{1}{N}(B_c - B_{app}), \tag{11.18}$$

$$\chi = -\frac{1}{N}(1 - B_{app}/B_c). \tag{11.19}$$

The last equation (11.19), will be derived in the following section. The fields given by Eqs. (11.16)–(11.18) are, of course, averages over the normal and superconducting regions, as depicted in Figs. 11.3a–11.3f.

Setting $N = 1/3$ in Eqs. (11.15)–(11.19) recovers the expressions for a sphere. Figures 11.4, 11.5, 11.6, and 11.7 show how the various fields and the susceptibility of a sphere vary with the applied field in the perfectly superconducting and intermediate-state ranges. The internal field B_{in} is zero up to $\frac{2}{3}B_c$ and then increases linearly to its normal state value, as shown in Fig. 11.4. Figure 11.5 shows how H_{in} increases at first, and then remains pinned at H_c throughout the intermediate state. The magnitude of the magnetization, presented in Fig. 11.6, increases more rapidly than it would for a parallel cylinder, then drops linearly to zero. We see from Eqs. (11.4) and (11.19) and Fig. 11.7 that the susceptibility stays pinned at the value $\chi = -1$ until the intermediate state is reached and then drops linearly to zero.

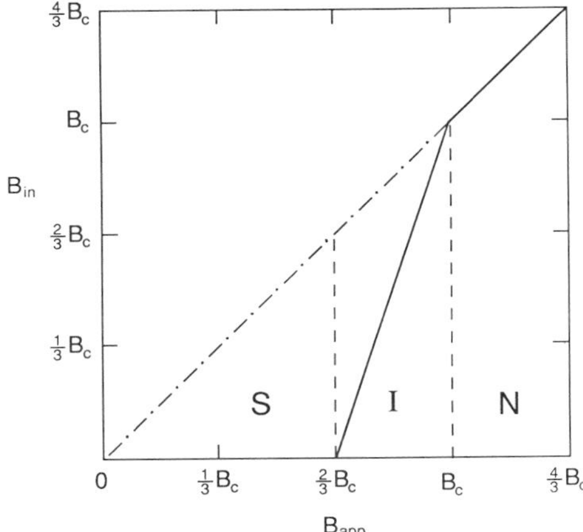

Figure 11.4 Internal magnetic field B_{in} in the Meissner (S) and intermediate (I) states of a Type I superconducting sphere ($N = 1/3$) as a function of the applied field B_{app} (Eqs. (11.11) and (11.16), respectively). In this and Figs. 11.5, 11.6, and 11.7, the solid lines represent the function that is being plotted, vertical dashed lines indicate the boundaries of the Meissner, intermediate, and normal regions, and the unit slope line (– · – · –) gives the behavior for zero demagnetization factor ($N = 0$).

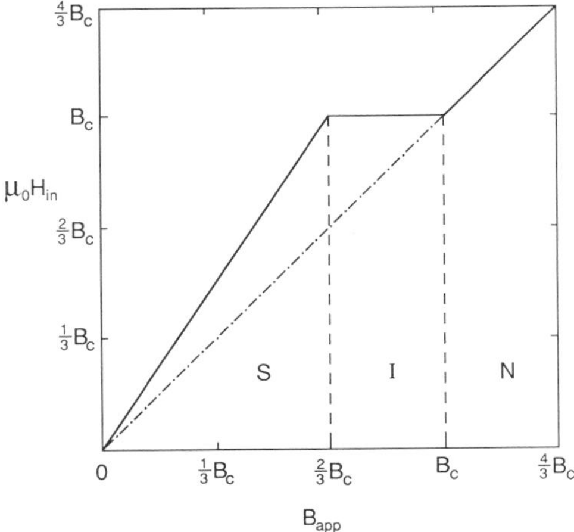

Figure 11.5 Internal field H_{in} (Eqs. (11.12) and (11.17)) for the case of Fig. 11.4.

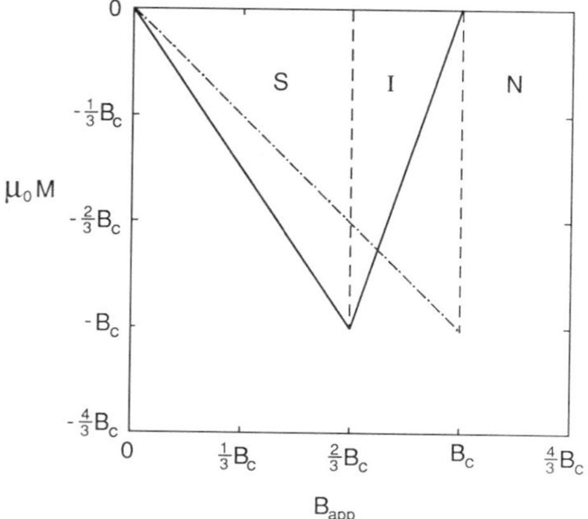

Figure 11.6 Magnetization M (Eqs. (11.13) and (11.18)) for the case of Fig. 11.4.

V. SUSCEPTIBILITY

We have seen that a material in the intermediate state is an admixture of normal and superconducting regions that coexist at the mesoscopic level. Viewed from a macroscopic perspective, we average over this structure, considering the material to be homogeneous with uniform susceptibility $\chi = M/H_{in}$ given by

$$\chi = \frac{M}{H_c}. \tag{11.20}$$

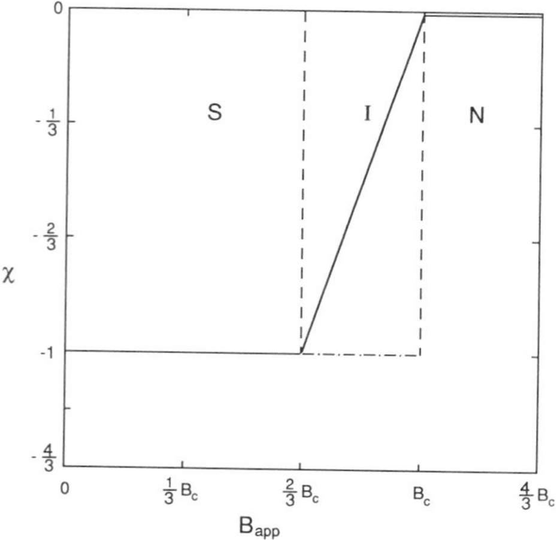

Figure 11.7 Magnetic susceptibility χ (Eqs. (11.14) and (11.19)) for the case of Fig. 11.4.

Substituting the expression from Eq. (11.18) in the latter expression gives Eq. (11.19). Since χ is an average of the value -1 in the superconducting regions and 0 in the normal regions, for the intermediate state it lies in the range

$$-1 < \chi < 0, \tag{11.21}$$

assuming the value $\chi = -1$ for $B_{app} \leq (1-N)B_c$ and $\chi = 0$ for $B_{app} \geq B_c$.

Susceptibility is an intrinsic property of a material defined by Eq. (1.77) in terms of the internal field H_{in},

$$\chi = \frac{M}{H_{in}}. \tag{11.22}$$

When the magnetization of a sample is measured in an applied magnetic field $B_{app} = \mu_0 H_{app}$, the experimentally determined susceptibility is often deduced from the expression

$$\chi_{exp} = M/H_{app}, \tag{11.23}$$

and the ratio of these gives for the intermediate state

$$\chi = \chi_{exp} B_{app}/B_c. \tag{11.24}$$

We can also use a more general expression (10.52),

$$\chi = \chi_{exp}/(1 - N\chi_{exp}), \tag{11.25}$$

to obtain χ from χ_{exp}.

VI. GIBBS FREE ENERGY FOR THE INTERMEDIATE STATE

The intermediate state can be described in thermodynamic terms using the formalism for the Gibbs free energy that was developed in Chapter 4, Sections VII–X. In this section we sometimes simplify the notation by writing B for the applied field B_{app} and B_c for the critical field $B_c(T)$ at a finite temperature.

Consider the case of a zero-field-cooled Type I superconductor in an applied magnetic field that is isothermally increased from $B_{app} = 0$ to $B_{app} = B_c(T)$

along a vertical S → I → N path of the type shown in Figs. 11.1 and 11.2. To calculate the Gibbs free energy density along this path, we begin by integrating Eq. (4.33),

$$G(B) = K - \int_0^B M dB, \quad (11.26)$$

where K is a constant to be evaluated. For the pure superconducting region, we have from Eq. (11.13),

$$M = -B/\mu_0(1-N). \quad (11.27)$$

For the intermediate state region M is given by Eq. (11.18),

$$M = -(B_c - B)/N\mu_0, \quad (11.28)$$

where B is the externally applied field. The free energy density is easily calculated for the superconducting region (11.10) by substituting the expression for M from Eq. (11.27) in (11.26). Carrying out the integration we obtain

$$G(B_{app}) = K + \frac{1}{1-N}\frac{B_{app}^2}{2\mu_0}, \quad (11.29)$$

which is an expression that is valid for $B_{app} < (1-N)B_c$. For higher applied fields, the interval of integration must be split into two parts,

$$G(B_{app}) = K - \int_0^{(1-N)B_c} M dB - \int_{(1-N)B_c}^{B} M dB. \quad (11.30)$$

Inserting the appropriate expressions (11.27) and (11.28) for the magnetization and carrying out the two integrations, we obtain after some cancellation of terms

$$G(B_{app}) = K - (1/2N\mu_0)(B_c - B_{app})^2 + B_c^2/2\mu_0, \quad (11.31)$$

an expression that is valid for $(1-N)B_c < B_{app} < B_c$. At the onset of the normal state ($B_{app} = B_c$), this reduces to the expression

$$G(B_c) = K + B_c^2/2\mu_0 \quad (11.32)$$

for the free energy density of what is now the normal state. Further integration beyond B_c does not yield anything more because in the normal state $\chi \approx 0$, and hence $M \approx 0$, as we showed in Table 1.2.

The constant K is selected as the negative of the condensation energy $B_c^2/2\mu_0$

$$K = -B_c^2/2\mu_0 \quad (11.33)$$

to make the free energy vanish at the upper critical field. This makes $G(B_{app})$ the free energy of the superconducting state relative to that of the normal state at $T = 0$. The quantity (11.33) is the condensation energy of Eq. (5.19) found from the Ginzburg–Landau theory and indicated in Fig. (5.2b).

If we define two normalized quantities

$$g(b) = G(B_{app})/[B_c^2/2\mu_0], \quad (11.34a)$$

$$b = B_{app}/B_c, \quad (11.34b)$$

Eqs. (11.29) and (11.31) become, respectively, for the superconducting and intermediate regions at $T = 0$,

$$g(b) = -1 + \frac{b^2}{1-N} \qquad 0 < b < 1-N, \quad (11.35a)$$

$$g(b) = \frac{-(1-b)^2}{N} \qquad 1-N < b < 1, \quad (11.35b)$$

with the special values

$$g(b) = \begin{cases} -1 & b = 0 \\ -N & b = 1-N \\ 0 & b = 1 \end{cases} \quad (11.36)$$

at the boundaries of the various regions.

Equations (11.35) are plotted in Fig. 11.8 for the cases of parallel geometry ($N = 0$), a cylinder in a parallel field ($N = 0.1$), a sphere ($N = \frac{1}{3}$), a long cylinder in a perpendicular field ($N = \frac{1}{2}$), and a disk in a perpendicular field ($N = 0.9$, $N \approx 1$). We see from this figure that a disk or flat plate is in the intermediate state for almost all applied fields below B_c. (This will be discussed in Section VIII.) The transformation from the superconducting to the intermediate state occurs when G crosses the

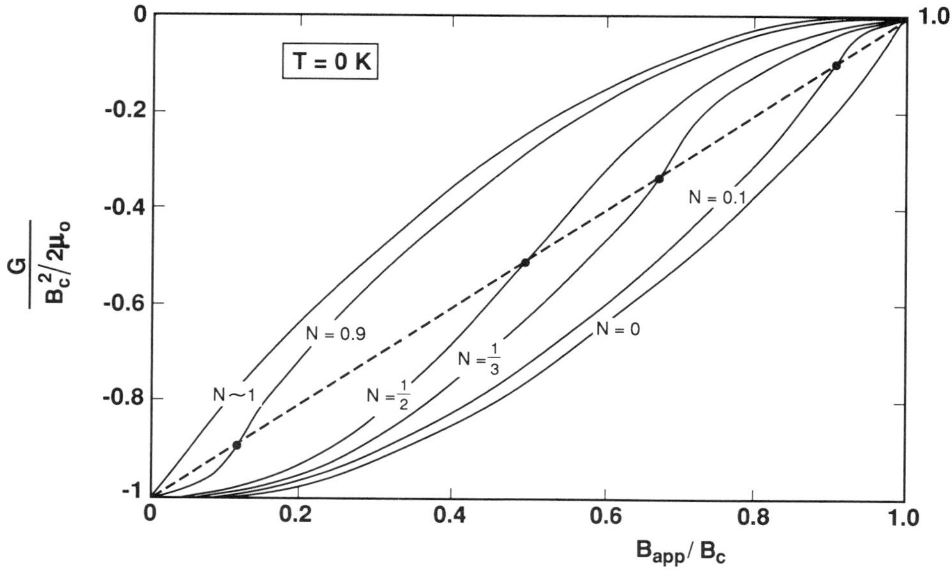

Figure 11.8 Dependence of the Gibbs free energy G on the applied magnetic field B_{app} for six values of the demagnetization factor, N(0, 0.1, 1/3, 1/2, 0.9, and ≈ 1), all at $T = 0$ K. Equations (11.35) were used to plot the curves.

dashed line of unit slope in Fig. 11.8. For each case the superconducting state exists at lower fields (the curve for G below the dashed line in the figure), while the intermediate state exists above the line.

Figure 11.8 presents plots of the Gibbs free energy from Eqs. (11.35) versus the applied field for several demagnetization factors at $T = 0$. Related figures in Chapter 4 (Figs. 4.10, 4.11, 4.15, and 4.19) present plots of the Gibbs free energy versus temperature for several applied fields with $N = 0$. Problem 8 shows how to combine the relevant expressions—(11.35) for $T = 0$, and (4.51) for $N = 0$—to obtain more general expressions that are valid in the superconducting and intermediate states when all three quantities T, B_{app}, and N have nonzero values. Using these results, the Gibbs free energy $G(B_{app})$ for $N = 0, \frac{1}{2}$, and 1 is plotted in Fig. 11.9 versus the applied field for the reduced temperatures $t = T/T_c = 0$, 0.55, 0.7, 0.8, 0.9, and 1.0, as indicated in Fig. 11.10. We see from these plots that when the temperature is increased, the range of applied fields over which the material superconducts decreases. However, the fraction of this range that is in the intermediate state remains the same since it depends only on the demagnetization factor.

VII. BOUNDARY-WALL ENERGY AND DOMAINS

We have been discussing the field configurations and the Gibbs free energy in superconductors without taking into account the details of the resulting domains. In this and the following two sections we will discuss these domains and their significance. We begin with the case of a Type I superconductor placed in an applied magnetic field which is in the intermediate range (11.15), and then comment on a Type II superconductor in the mixed state with $B_{c1} < B_{app} < B_{c2}$.

We adopt the model of a Type I superconductor in the intermediate state that splits into domains of normal material with $\chi \approx 0$ embedded in pure superconducting

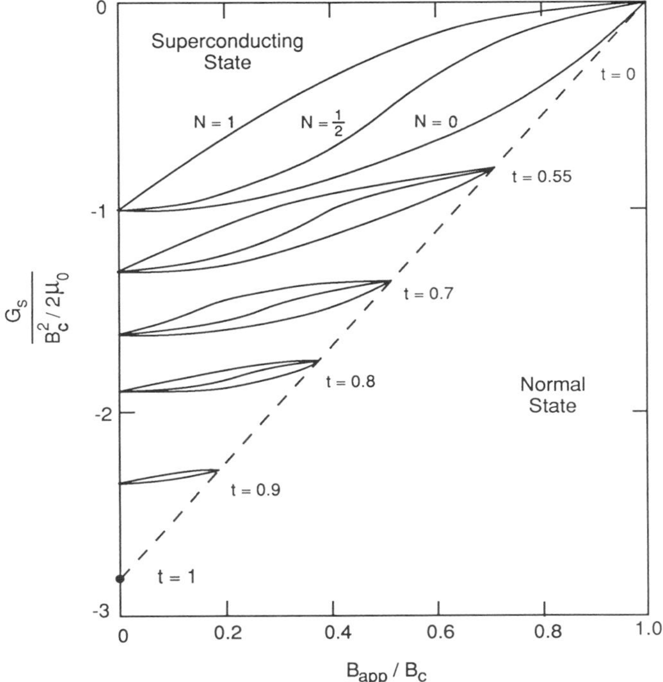

Figure 11.9 Dependence of the Gibbs free energy G on the applied magnetic field B_{app} for demagnetization factors $N = 0$, $\frac{1}{2}$, and 1 at several reduced temperatures $t = T/T_c$, as indicated in Fig. 11.10. The sample is in the superconducting or mixed state to the upper left of the dashed line and in the normal state to the lower right of the line. The figure is drawn for the condition $\mu_0 \gamma\, T_c^2/B_c^2 = 1/\alpha = 2.8$.

regions with $\chi = -1$. The boundary between these regions contains a density of magnetic energy $B_c^2/2\mu_0$. The superconducting regions exclude the magnetic field B_{in} and are lower in energy because of the Cooper pair condensation energy. The super electron density n_s extends into the normal region by a distance equal to the coherence length ξ. The magnetic field within the boundary, which is of thickness ξ, contributes a positive energy the magnitude of which is diminished by the effect of the penetration depth λ associated with the decay of the magnetic energy. As a result of this effect, the boundary is effectively shortened, and has thickness

$$d_{bound} \approx (\xi - \lambda). \qquad (11.37)$$

The overall energy density per unit area of the boundary layer has the value

$$E_{bound} = \left(\frac{B_c^2}{2\mu_0}\right) d_{bound} \qquad (11.38)$$

for a Type I superconductor in the intermediate state.

For Type II superconductors, λ exceeds ξ and the effective domain wall thickness from Eq. (11.37) is negative, so that the boundary energy is negative. In other words, the boundary wall energy is positive for small κ and negative for large κ, where $\kappa = \lambda/\xi$ is the Ginzburg–Landau parameter. For large κ it becomes energetically favorable for the superconducting material to split into domains of large magnetic field strength with positive energies and surrounding transition re-

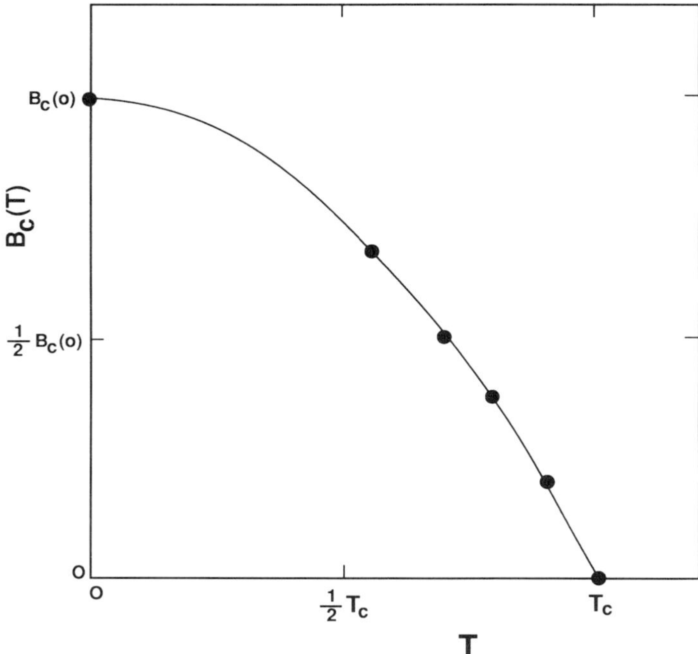

Figure 11.10 Dependence of the critical field $B_c(T)$ of a Type I superconductor on temperature. The points on the curve designate the values of the critical fields for the reduced temperature points indicated on the dashed line of Fig. 11.9.

gions of negative boundary energy, as shown in Fig. 11.11. The crossover between the positive and negative wall energies actually occurs at $\kappa = 1/\sqrt{2}$, so that domain formation becomes energetically favorable for $\kappa > 1/\sqrt{2}$. For the mixed state, which occurs for $\kappa > 1/\sqrt{2}$, the "domains" are, of course, vortices with cores of normal material, as sketched in Fig. 9.7. This energy argument makes it plausible to conclude that the mixed or vortex state forms in Type II superconductors, which have $\kappa > 1/\sqrt{2}$, but not in Type I superconductors.

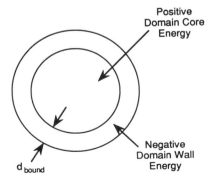

Figure 11.11 Domain core in the normal state with positive energy surrounded by a transition layer boundary of thickness d_{bound} with negative energy for the case $\kappa > 1/\sqrt{2}$.

VIII. THIN FILM IN APPLIED FIELD

Many studies have been carried out with films that are thin in comparison with their length and width. When such a material is placed in a magnetic field B_{app} that is oriented perpendicular to its surface, the field penetrates in the intermediate state, as shown in Fig. 11.12. We illustrate the case for $N = \frac{5}{6}$, so the factor $1 - N = \frac{1}{6}$ for this film. Figures 11.13–11.16 show plots of Eqs. (11.10)–(11.19) giving the depen-

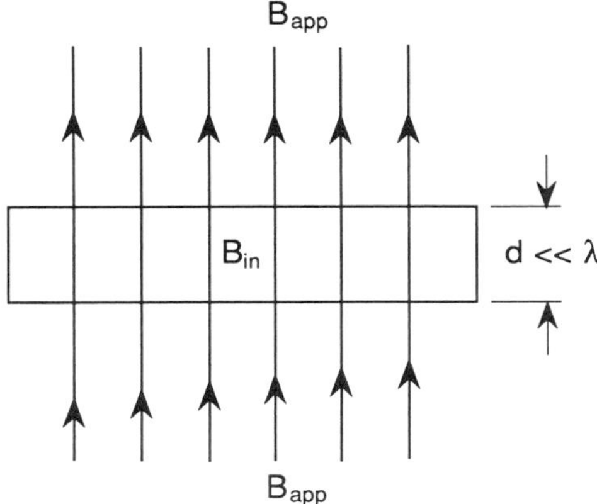

Figure 11.12 Magnetic field penetration through a superconducting thin film whose thickness d is small compared to the penetration depth λ.

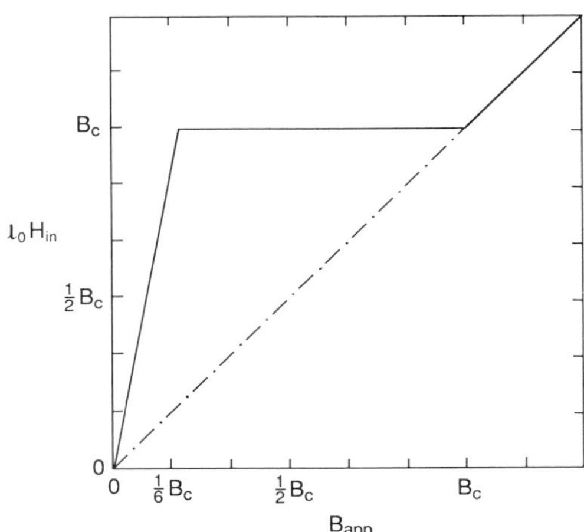

Figure 11.14 Internal field H_{in} (Eqs. (11.12) and (11.17)) in the film sketched in Fig. 11.12.

dences of B_{in}, H_{in}, M, and χ, respectively, on the applied field. The great extent of the intermediate state is evident from these plots.

Let us examine in more detail the case of a disk-shaped film of radius a and thickness $t \ll a$. We know from Table 10.1 and Problem 6 of Chapter 10 that this film has a demagnetization factor $N \approx 1 - \frac{1}{2}\pi t/a$, so that, from Eq. (11.15), the intermediate state extends over the range

$$\frac{\frac{1}{2}\pi t}{a} < \frac{B_{\text{app}}}{B_{\text{c}}} < 1. \tag{11.39}$$

This means that a superconducting film with a thickness-to-diameter ratio $t/2a = 10^{-3}$ and a lower critical field $B_{\text{c1}} = 50$ mT is driven to the intermediate state by the

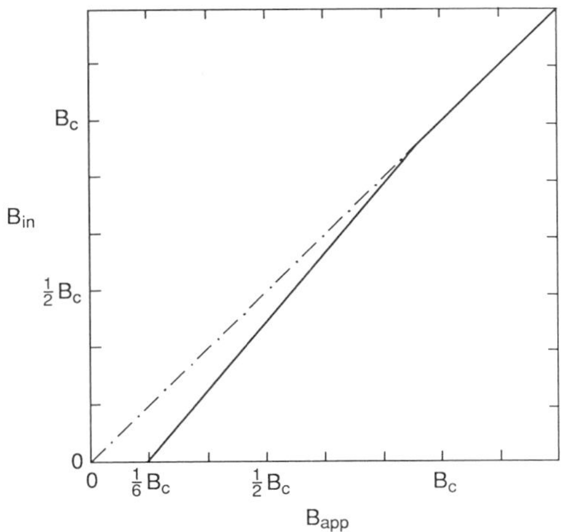

Figure 11.13 Internal magnetic field B_{in} in a superconducting film ($N = 5/6$) as a function of the applied field B_{app} (Eqs. (11.11) and (11.16)) (cf. case of a sphere, Fig. 11.4). Here and in Figs. 11.14 and 11.15, the unit slope line (– · – · –) represents the behavior for zero demagnetization factor ($N = 0$).

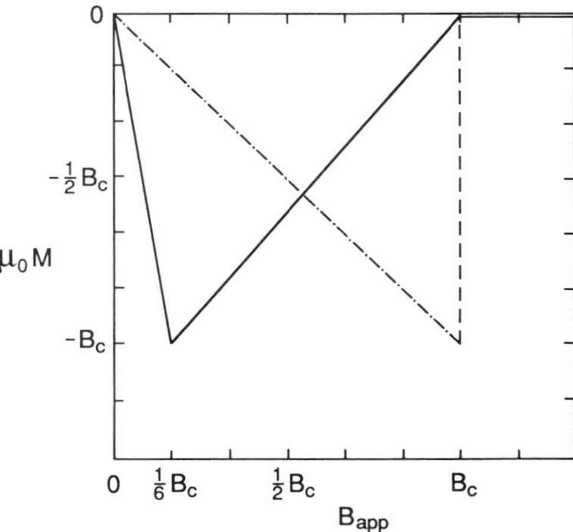

Figure 11.15 Magnetization M (Eqs. (11.13) and (11.18)) for the film sketched in Fig. 11.12.

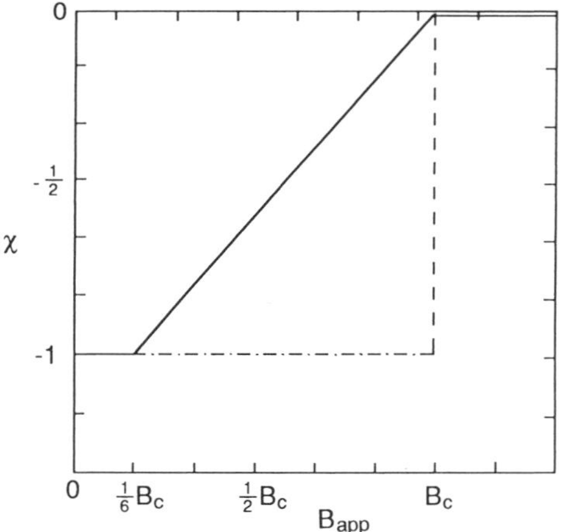

Figure 11.16 Magnetic susceptibility χ (Eqs. (11.14) and (11.19)) of the film sketched in Fig. 11.12.

Earth's typical field of 50 μT. The internal field from Eq. (11.16),

$$B_{\text{in}} = B_{\text{app}} - \frac{\frac{1}{2}\pi t}{N_a}(B_{\text{c}} - B_{\text{app}}), \quad (11.40)$$

is close to the applied field, $B_{\text{in}} \approx B_{\text{app}}$, over most of the intermediate state to a much greater extent than that shown in Fig. 11.13, since $\pi t/2N_a \ll \frac{1}{6}$. When the film thickness becomes comparable with or less than the penetration depth, the situation is more complicated, however.

IX. DOMAINS IN THIN FILMS

We just saw that a thin film in a perpendicular field below T_{c} is in the intermediate state for almost all applied fields below B_{c}. We will proceed to investigate the nature of this state. From the boundary conditions, the macroscopic B fields are continuous across the boundary, as shown in Fig. 11.12.

$$B_{\text{in}} = B_{\text{app}}. \quad (11.41)$$

If we take into account the domain structure, a fraction of the material f_{n} will be normal material with $B_{\text{in}} = B_{\text{c}}$ and a fraction f_{s} will be perfectly superconducting with $B_{\text{in}} = 0$, where

$$f_{\text{s}} + f_{\text{n}} = 1. \quad (11.42)$$

The magnetic field avoids the superconducting parts and passes through the normal regions, as shown in Fig. 11.17. This means that

$$B_{\text{in}} = f_{\text{n}} B_{\text{c}}, \quad (11.43)$$

so that

$$B_{\text{app}} = f_{\text{n}} B_{\text{c}}, \quad (11.44)$$

from Eq. (11.41). The boundaries or domain walls between the normal and superconducting regions also contain the field B_{c}, but it is assumed that the fraction of material f_{w} in the domain walls is small and can be neglected. The figure gives the energy density $u = B^2/2\mu_0$ at three positions in the field.

Figure 11.18 sketches the domain structure in a thin film of length L, width W, and thickness d that is oriented perpendicular to the direction of the applied field. The figure shows normal regions of width D_{n} and superconducting regions of width D_{s} separated by domain walls of thickness δ. In drawing this figure we have assumed that the domains are all the same size and equally spaced. This simplifies the mathematics without changing the physics of the situation. We can define a repetition length D for the domain pattern

$$D_{\text{s}} + D_{\text{n}} + 2\delta = D, \quad (11.45)$$

where each domain has two walls. If we neglect the thickness of the domain walls, as is done on Fig. 11.17, by assuming that $\delta \ll D_{\text{s}}, D_{\text{n}}$, we can write

$$D_{\text{s}} + D_{\text{n}} = D, \quad (11.46)$$

where

$$\begin{aligned} D_{\text{s}} &= f_{\text{s}} D, \\ D_{\text{n}} &= f_{\text{n}} D. \end{aligned} \quad (11.47)$$

The number of domains N_{d} is equal to the length L divided by the domain repetition length D:

$$N_{\text{d}} = L/D. \quad (11.48)$$

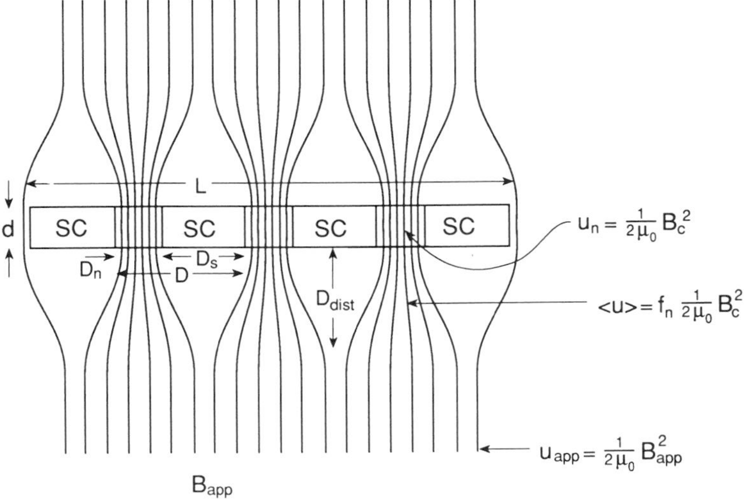

Figure 11.17 Passage of magnetic field lines B of a perpendicular applied field through normal domain regions with diameter D_n of a superconducting thin film. The distortion distance D_{dist}, normal regions in which $B_{in} = B_c$, and superconducting regions (SC) in which $B_{in} = 0$ are shown. The energy densities in the normal regions u_n, in regions far from the film u_{app}, and in regions near the film $\langle u \rangle$ are indicated.

The magnitude of the B field is appreciably distorted from its far value $\mathbf{B}_{app}$ only within a distance D_{dist} of the surface given by

$$1/D_{dist} = 1/D_s + 1/D_n, \quad (11.49)$$

and with the aid of Eq. (11.47) we obtain

$$D_{dist} = f_s f_n D \quad (11.50)$$

for the field distortion distance indicated in Fig. 11.17.

Let us try to justify Eq. (11.49) by a hydrodynamic analogy. Consider the B lines as the flow lines of a fluid passing through slots or holes in a barrier, where the density of flow lines represents the speed. The speed changes appreciably near the holes only if the diameter of the holes is small and near the material between the holes only when the holes take up most of the space, as shown in Fig. 11.19. Therefore $D_{dist} \approx D_n$ in the former case, when

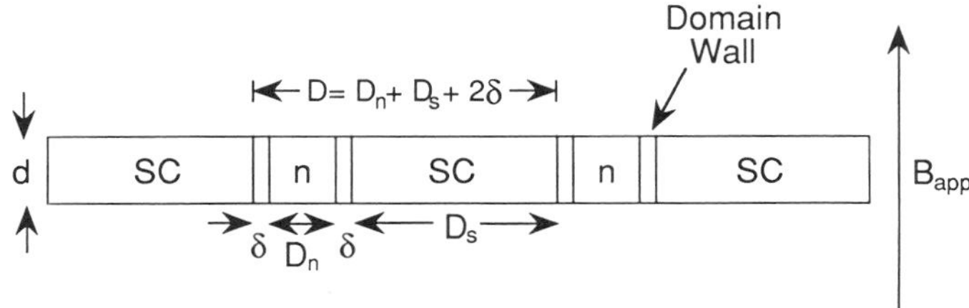

Figure 11.18 Thin film in a perpendicular applied magnetic field showing normal regions (n) of length D_n, superconducting regions (SC) of length D_s, and domain walls of thickness δ separating these regions. The repetition length $D = D_n + D_s + 2\delta$ is also shown.

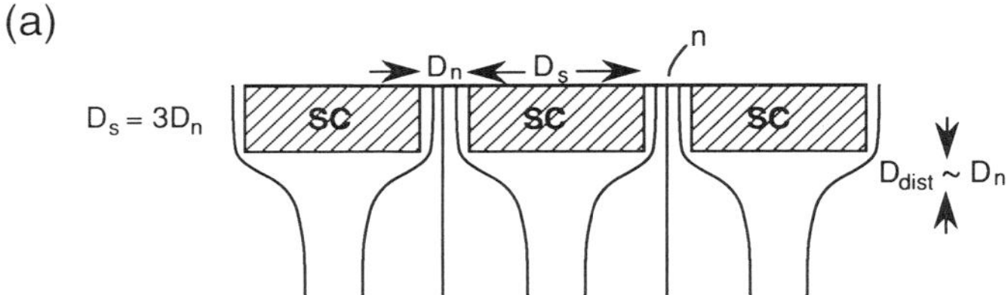

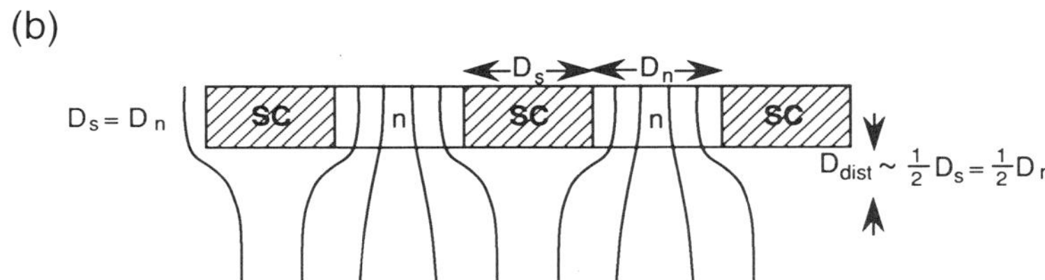

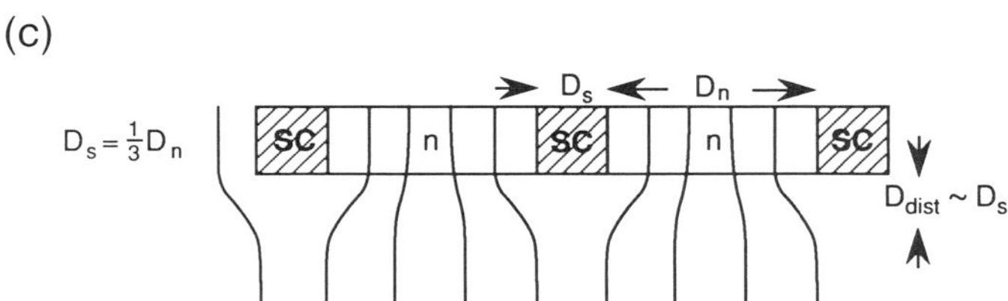

Figure 11.19 Illustration of the distortion distance D_{dist} for the cases $D_s = 3D_n$ with $D_{\text{dist}} \approx D_n$ (a), $\frac{1}{2}D_n = \frac{1}{2}D_s \approx D_{\text{dist}}$ (b), and $D_s = \frac{1}{3}D_n$ with $D_{\text{dist}} \approx D_s$ (c), where the wall thickness δ is negligible, i.e., $\delta \ll D_n$, D_s.

$D_n \ll D_s$, whereas $D_{\text{dist}} \approx D_s$ in the latter case, when $D_n \gg D_s$. Figure 11.19 illustrates the cases $D_s = 3D_n$, $D_s = D_n$, and $D_s = \frac{1}{3}D_n$.

The distortion of the applied magnetic field brought about by the presence of the film, as sketched in Figs. 11.17 and 11.19, serves to increase the Gibbs free energy G of the system. Figure 11.17 shows the magnetic energy densities of each region. The Gibbs free energy is also increased by the magnetic energy stored in the domain walls. A long repetition length reduces the number of domains and decreases the domain wall energy, whereas a short repetition length lessens the distortion of the field and decreases the field distortion energy at the expense of more domain wall energy. An intermediate domain repetition length is best, and we will derive an expression for the optimum value of D.

The magnetic energy stored in each pair of domain walls $2W\delta d(\frac{1}{2}B_c^2/\mu_0)$ gives for the total domain wall contribution to the Gibbs free energy

$$G_w = 2N_d(W\delta d)\left(\tfrac{1}{2}B_c^2/\mu_0\right) \tag{11.51}$$

$$= 2(WL\delta d/D)\left(\tfrac{1}{2}B_c^2/\mu_0\right), \tag{11.52}$$

where we have used Eq. (11.48). The change in field energy arises from the difference between energy density near the surface, $\frac{1}{2}f_n B_c^2/\mu_0$, and the energy density far from the film, $\frac{1}{2}B_{\text{app}}^2/\mu_0$, where the factor f_n in the former expression takes into account the fact that this field exists only near the normal regions, as shown in Figs. 11.17 and 11.19. This gives for the magnetic field contribution to the Gibbs free energy

$$G_f = 2WLD_{\text{dist}}\left[f_n\left(\tfrac{1}{2}B_c^2/\mu_0\right) - \left(\tfrac{1}{2}B_{\text{app}}^2/\mu_0\right)\right] \tag{11.53}$$

$$= 2(WLD)f_s^2 f_n^2\left(\tfrac{1}{2}B_c^2/\mu_0\right), \tag{11.54}$$

where we have used Eqs. (11.42), (11.44), and (11.50). The total Gibbs free energy G is the sum of these two, namely $G = G_w + G_f$. Per unit area of film we have

$$G/WL = 2\left(\tfrac{1}{2}B_c^2/\mu_0\right)(\delta d/D + f_s^2 f_n^2 D). \tag{11.55}$$

The domain repetition distance D obtained by minimizing the total free energy per unit area,

$$\frac{dG}{dD} = 0, \tag{11.56}$$

is equal to

$$D = (d\delta)^{1/2}/f_s f_n, \tag{11.57}$$

where the product $f_s f_n$ is easily deduced from Eqs. (11.42) and (11.44):

$$f_s f_n = B_{\text{app}}(B_c - B_{\text{app}})/B_c^2. \tag{11.58}$$

The thinner the film, the smaller the value of D and the greater the number of domains. D is large at both ends of the range, where either f_n or f_s is small, corresponding to $B_{\text{app}} \ll B_c$ and $B_{\text{app}} \approx B_c$, respectively. It is smallest in the middle, where $f_n = f_s = \frac{1}{2}$ and $B_{\text{app}} = \frac{1}{2}B_c$.

This derivation assumes that the domain wall thickness δ is much less than the domain sizes D_n and D_s. It does not take into account how the surface energy depresses the value of the critical field. When the film thickness is much greater than the domain wall, $d \gg \delta$, the critical field B_c is shifted to a lower value B'_c given by (Tinkham, 1985, p. 95)

$$B'_c = \left[1 - 2(\delta/d)^{1/2}\right]B_c \qquad d \gg \delta. \tag{11.59}$$

A similar depression of the lower critical field B_{c1} occurs with Type II superconductors.

X. CURRENT-INDUCED INTERMEDIATE STATE

The intermediate state that we have been discussing arose from the presence of a magnetic field that exceeded the critical field at the surface of the superconductor. We will now examine the intermediate state that forms when a transport current from a current generator produces a magnetic field at the surface of a superconducting wire that exceeds the critical field. In this section we simplify the notation by writing I instead of I_T for the transport current.

We know from Eq. (2.42) that the magnetic field at the surface of a wire of radius a carrying a current I is in the ϕ direction with magnitude

$$H = I/2\pi a. \tag{11.60}$$

This surface field is independent of the distribution of current density inside the wire. When the surface field reaches a particular value, called the critical field $B_c = \mu_0 H_c$, we say that the transport current has attained its critical value I_c, corresponding to

$$H_c = I_c/2\pi a. \tag{11.61}$$

If the current exceeds the critical value, the magnetic field at the surface will exceed H_c and the wire will no longer be able to remain a perfect superconductor. This destruction of pure superconductivity by a transport current is called the *Silsbee effect*. If the wire were to respond by going normal at the surface, the superconducting core would constitute a wire of even smaller radius $r < a$ carrying the same amount of current. From Eq. (11.60) it is clear that the field at the surface of the wire would then be even larger than the

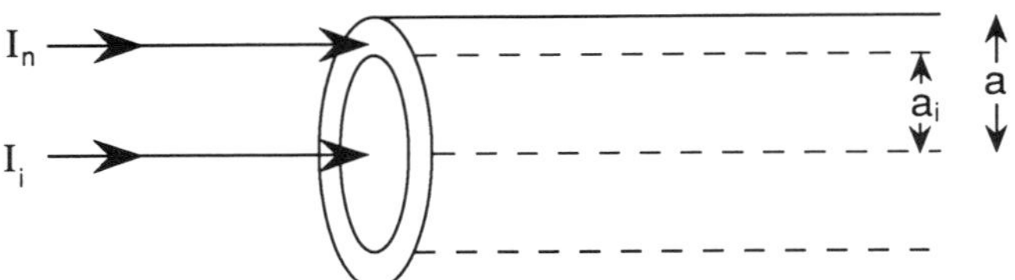

Figure 11.20 Division of applied transport current flowing through a superconducting wire of radius a into a portion I_n flowing in the outer normal region and a portion I_i in the intermediate state flowing in the core of radius a_i.

field that drove the surface normal. Following this line of reasoning, the entire wire would become normal. There is, however, a second solution to the problem, proposed by F. London (1937; see also F. London, 1950), whereby the wire is assumed to have a core of radius a_i that is in an intermediate state and an outer layer that is normal. We will show how this could occur.

When the total current I exceeds I_c, part of the current I_n flows in the outer normal region and the remainder I_i along a parallel path through the core, which is in the intermediate state,

$$I = I_n + I_i, \qquad (11.62)$$

as indicated in Fig. 11.20. The current density J_n in the outer layer is constant, with a value given by

$$J_n = \frac{I_n}{\pi(a^2 - a_i^2)} \qquad a_i < r < a. \qquad (11.63)$$

For this normal current I_n to flow there must be a potential difference between the two ends of the wire, which means that there is an axial electric field E in the normal layer given by Eq. (1.21),

$$E = J_n \rho_n, \qquad (11.64)$$

where both J_n and E are unknown.

The intermediate state has regions of superconducting material and regions of normal material, so there is also an electric field $E(r)$ present. All the magnetic fields that are present are independent of time, so, from the Maxwell curl relation $\nabla \times \mathbf{E} = -\partial \mathbf{B}/\partial t = 0$, E is also independent of the radial distance r. Therefore, $E(r)$ has the same value E in the normal and in the intermediate state region, and we can write an analogue of Eq. (11.64) for the intermediate state,

$$E = J_i(r)\rho_i(r) \qquad 0 < r < a_i, \qquad (11.65)$$

where $J_i(r)$ satisfies the boundary condition for continuity of the current density at $r = a_i$. Thus we obtain

$$J_i(a_i) = J_n, \qquad (11.66)$$

$$\rho_i(a_i) = \rho_n. \qquad (11.67)$$

We saw in Section II that the intermediate state is characterized by the pinning of the internal field H_{in} at the critical value,

$$H_{in} = H_c \qquad 0 < r < a_i. \qquad (11.68)$$

This permits us to write Eq. (11.61) for the intermediate state,

$$H_c = I_i(r)/2\pi r \qquad r < a_i, \qquad (11.69)$$

where

$$I_i(r) = \int_0^r J_i(r)(2\pi r)dr. \qquad (11.70)$$

We show in Problem 10 that the current density in the intermediate state is given by

$$J_i(r) = (a_i/r)J_n \qquad r \le a_i, \qquad (11.71)$$

as illustrated in Fig. 11.21, where

$$H_c = J_n a_i. \qquad (11.72)$$

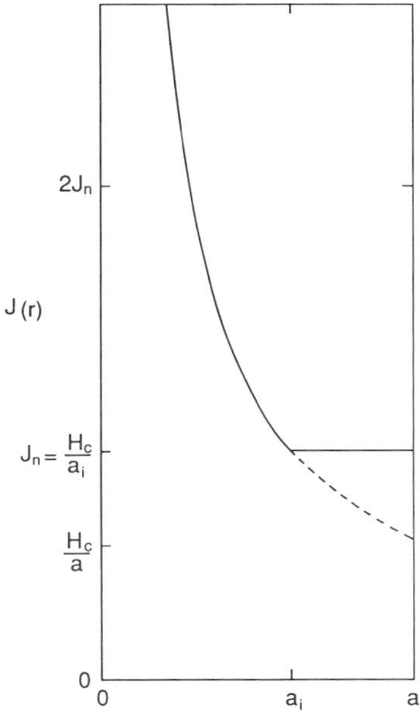

Figure 11.21 Dependence of the current density $J(r)$ on distance r from the center in a superconducting wire of radius a that is in the intermediate state. $J(r)$ decreases with r in the intermediate-state core region of radius a_i and has a constant value $J_n = H_c/a_i$ in the outer layer, where $a_i \le r \le a$, as shown. The extrapolation (– – –) of the core region behavior of $J(r)$ to the value H_c/a at the surface is indicated.

Combining Eqs. (11.64) and (11.72) gives for the radius of the intermediate state region

$$a_i = H_c \rho_n / E, \tag{11.73}$$

which depends on the applied current in the manner shown in Fig. 11.22. The resistivity in the intermediate state is proportional to the radius r,

$$\rho_i(r) = \left(\frac{r}{a_i}\right)\rho_n \qquad r \le a_i, \tag{11.74}$$

as shown in Fig. 11.23 (see Problem 11). This means that the longitudinal path length of the current through the normal region at the radius r is proportional to r. This fits the configuration illustrated in Fig. 11.24, which had been suggested by F. London. In this configuration the core has a sequence of conical regions of superconducting material arranged along the axis with interlaced regions of normal material.

It is shown in Problem 16 that the total current flowing through the wire is given by

$$I = \pi(a^2 + a_i^2)J_n. \tag{11.75}$$

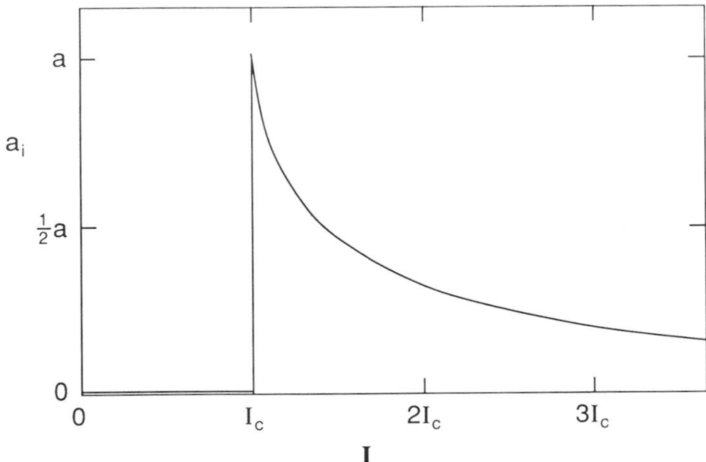

Figure 11.22 Dependence of the core radius a_i of a superconducting wire on the applied transport current I. The intermediate state occurs for transport currents exceeding the value $I_c = 2\pi a H_c$.

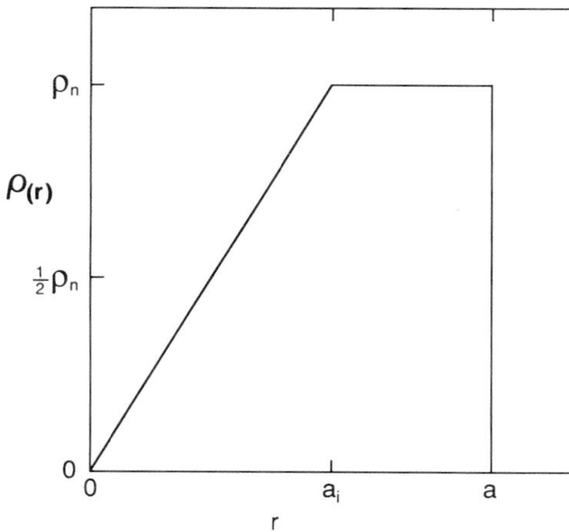

Figure 11.23 Dependence of the resistivity $\rho(r)$ of a superconducting wire in the intermediate state on the radial distance r from the center of the wire. Note that $\rho(r)$ increases linearly with r in the intermediate-state core region, where $r < a_i$, and has the normal state value ρ_n in the outer region, $r > a_i$, of the wire.

By Eqs. (11.64), (11.72), and (11.73), this may be written as a quadratic equation in terms of the electric field,

$$I(E) = \pi a^2 \frac{E}{\rho_n} + \pi H_c^2 \frac{\rho_n}{E}, \quad (11.76)$$

which has the solution

$$E = \tfrac{1}{2} I \frac{\rho_n}{\pi a^2} \left\{ 1 \pm \left[1 - (I_c/I)^2 \right]^{1/2} \right\}. \quad (11.77)$$

The average resistivity $\langle \rho \rangle$ is $E/\langle J \rangle$ and the average current density $\langle J \rangle$ through the wire is $I/\pi a^2$, so we can write

$$\langle \rho \rangle = E(\pi a^2/I), \quad (11.78)$$

and Eq. (11.77) gives us the London model expression

$$\frac{\langle \rho \rangle}{\rho_n} = \tfrac{1}{2} \left\{ 1 + \left[1 - \left(\frac{I_c}{I} \right)^2 \right]^{1/2} \right\}, \quad (11.79)$$

which is valid for $I > I_c$. The positive sign has been selected in Eq. (11.77) because it gives the proper asymptotic behavior of $\langle \rho \rangle \to \rho_n$ for $I \gg I_c$.

The various resistivity results in the London model can be grouped together as follows:

$$\frac{\langle \rho \rangle}{\rho_n} = \begin{cases} 0 & I < I_c \\ \frac{1}{2} & I = I_c \\ \frac{1}{2}\left\{1 + \left[1 - (I_c/I)^2\right]^{1/2}\right\} & I > I_c \\ 1 - \frac{1}{4}(I_c/I)^2 & I \gg I_c. \end{cases} \quad (11.80)$$

We see that at the point $I = I_c$ the resistivity jumps discontinuously from 0 to the value $\frac{1}{2}\rho_n$. It then slowly approaches ρ_n for higher applied currents in accordance with Eq. (11.80), as shown in Fig. 11.25. Experimental data, such as those plotted in the figure, ordinarily show a larger jump of $\frac{2}{3}$

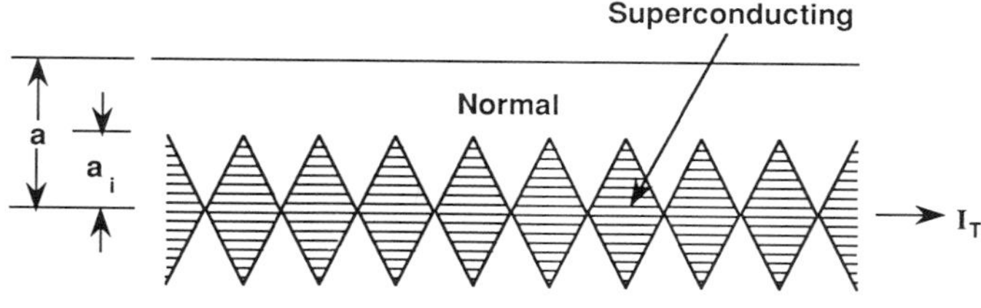

Figure 11.24 London model for the intermediate-state structure of a wire of radius a carrying a transport current in excess of the critical value $I_c = 2\pi a H_c$, where a_i is the radius of the core region (F. London, 1937, 1950).

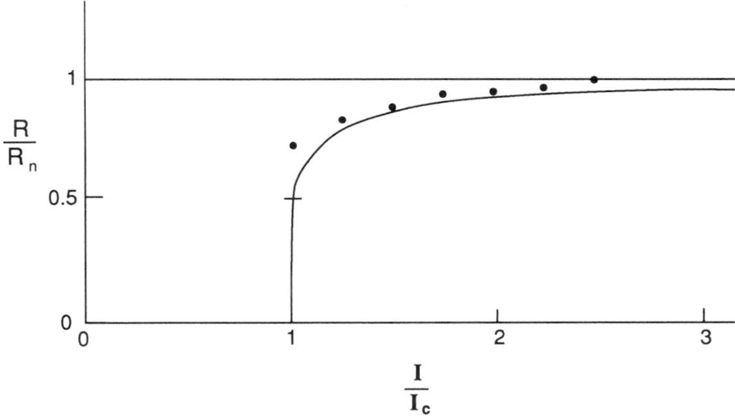

Figure 11.25 London model dependence of the resistance of the intermediate state of a superconducting wire on the value of the transport current I for a fixed temperature T. Experimental data on indium wires at 2.02 K (Watson and Huebener, 1974) are shown for comparison.

or more at $I = I_c$ instead of $\frac{1}{2}$. A refined version of the London model developed by Baird and Mukherjee (1968, 1971) exhibits a larger jump and fits the data better (Huebener, 1979, p. 213; Watson and Huebener, 1974). It is interesting that the intermediate state persists above I_c, and that some superconductivity is predicted to remain in the core region for all finite currents.

Tinkham (1985, p. 102) shows that there is also a dependence of the resistivity on temperature, $\Delta T = T_c - T$, in the neighborhood of the transition temperature, and this is shown in Fig. 11.26,

We have discussed the current-induced intermediate state for superconducting wire of circular cross-section. The same phenomenon can occur with samples of other shapes, such as tapes or thin films, but the mathematical analysis can be more complex.

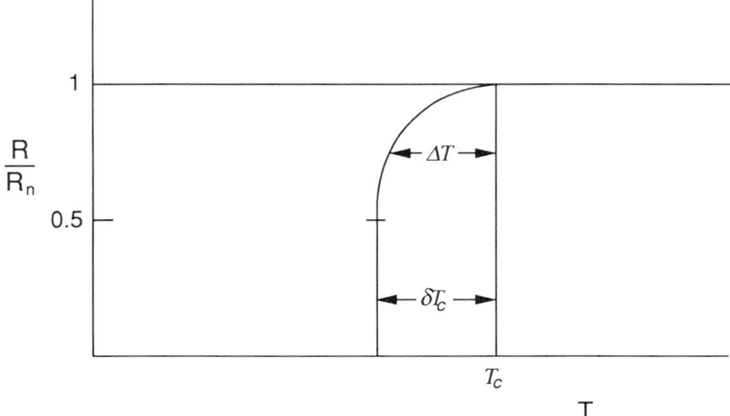

Figure 11.26 Temperature dependence of the resistance in the intermediate state of a current-carrying wire showing T_c decreasing by the variable amount ΔT for resistance in the range $\frac{1}{2}R_n < R < R_n$, and by a fixed amount δT in the range $R < \frac{1}{2}R_n$. The figure is drawn for the case of a constant applied transport current.

XI. MIXED STATE IN TYPE II SUPERCONDUCTORS

In this chapter we have been explaining how demagnetization effects in Type I materials produce the intermediate state in an effort to prevent surface fields from exceeding the critical field $B_{c'}$ and driving the material normal. In a Type II superconductor, when B_{app} becomes large enough to produce surface fields equal to the lower critical field, B_{c1}, vortices appear inside the material near the surface and the mixed state forms at a lower field $B_{c1'}$ than it would under the condition $N = 0$. Ideally, a Type II superconductor is in the $\chi = -1$ or Meissner state for $B_{app} < B_{c1'}$ and in the mixed state above $B_{c1'}$.

The mathematical description of the Meissner state of a Type II superconductor is similar to that of the Type I case that was presented above in Section IV, with B_{c1} replacing B_c at the upper limit of Eq. (11.10). This limit is the shifted lower critical field $B_{c1'}$ mentioned above,

$$B_{c1'} = (1 - N)B_{c1}. \qquad (11.81)$$

Thus we can rewrite Eqs. (11.10)–(11.14) for a Type II superconducting ellipsoid in the Meissner state as follows:

$$B_{app} < B_{c1'} \qquad (11.82)$$

$$B_{in} = 0, \qquad (11.83)$$

$$H_{in} = B_{app}/(1 - N)\mu_0, \qquad (11.84)$$

$$\mu_0 M = -B_{app}/(1 - N), \qquad (11.85)$$

$$\chi = -1. \qquad (11.86)$$

This state exists over the range given by Eq. (11.82).

When B_{app} exceeds $B_{c1'}$, vortices begin to form and the material enters the mixed state. Since this state can exist for H_{in} greater than $B_{c1'}/\mu_0$, the value of H_{in} does not become pinned at the critical field, as in the Type I case, but continues to increase as the applied field increases. This removes the constraint of Eq. (11.2), so the fields are given by the following more general expressions of Eqs. (10.35)–(10.37):

$$B_{c1'} < B_{app} < B_{c2}, \qquad (11.87)$$

$$B_{in} = B_{app}\frac{1 + \chi}{1 + \chi N}, \qquad (11.88)$$

$$H_{in} = \frac{B_{app}}{\mu_0(1 + \chi N)}, \qquad (11.89)$$

$$M = \frac{B_{app}}{\mu_0}\frac{\chi}{(1 + \chi N)}. \qquad (11.90)$$

These equations resemble those of the mixed state of a Type II ellipsoid, rather than those, (11.15)–(11.19), of the corresponding Type I intermediate state. The presence of the demagnetization factor N in the denominator of expressions (11.88)–(11.90) causes the internal fields to be larger than they would be for the case $N = 0$ of a cylinder in a parallel applied field.

Pakulis (1990) proposed a mixed state in zero field with normal regions called *thermons* that contain no magnetic flux.

FURTHER READING

Several superconductivity texts cover the material found in this chapter (e.g., Kresin and Wolf, 1990; Orlando and Delin, 1991; Rose Innes and Rhoderick, 1994; Tilley and Tilley, 1986; Tinkham, 1985).

The mixed and intermediate states are also discussed in several monographs (e.g., Huebener, 1979; Poole *et al.*, 1988; Van Duzer and Turner, 1981; Wilson, 1983). The book by Kovachev (1991) and the review by Kadowaki *et al.* (1995) discusses energy dissipation in the mixed state.

PROBLEMS

1. Find the range of angles for which the magnetic field at the surface of a perfectly superconducting sphere ($\chi = -1$) is (a) greater than, (b) equal to, or (c) less than the applied field B_{app}.
2. What are the smallest and the largest possible values of the angle Θ_c of Eq. (11.8) and at what applied fields do they occur?

3. Deduce the equations for B_{in}, H_{in}, M, and χ in the intermediate state of a Type I superconducting cylinder in a perpendicular magnetic field.
4. Show that the magnetic fields at the surface of a sphere in an applied magnetic field in the intermediate state have the following radial and azimuthal components:

$$B_r = B_{in} \cos \Theta,$$
$$B_\Theta = B_c \sin \Theta.$$

5. Show that the magnetic fields at the surface of a sphere in an applied magnetic field in the perfectly diamagnetic state have the following radial and azimuthal components:

$$B_r = 0,$$
$$B_\Theta = \tfrac{3}{2} B_{app} \sin \Theta.$$

6. Show that the expressions

$$\chi_{exp} = \chi/(1 + N\chi),$$
$$\chi = \chi_{exp}/(1 - N\chi_{exp}),$$
$$\chi_{exp} = \frac{1}{N}\left(1 - \frac{B_c}{B_{app}}\right),$$

are valid for the intermediate state of an ellipsoid.
7. Show that the boundary between the superconducting and intermediate state of an ellipsoid in a magnetic field lies along the dashed line of unit slope shown in Fig. 11.8. Show that along this boundary the Gibbs free energy is $-NB_c^2/2\mu_0$.
8. Show that the normalized Gibbs free energy $g(b,t)$ of an ellipsoid in an applied magnetic field is given by the expressions

$$g(b,t) = -t^2/\alpha - (1-t^2)^2 + b^2/(1-N)$$
$$0 < b < (1-N)(1-t^2)$$
$$g(b,t) = -t^2/\alpha - (1-t^2-b)^2/N$$
$$(1-N)(1-t^2) < b < (1-t^2).$$

These are plotted in Fig. 11.9 for several values of N and t using the BCS expression $1/\alpha = 2.8$.
9. Show that the dashed line of Fig. 11.9 corresponds to the expression $G_s = B_c^2/2\mu_0[(B_{app}/B_c) - 1]$.
10. Derive Eq. (11.71),

$$J_i(r) = J_n(a_i/r) \qquad r \le a_i,$$

and show that

$$J_i(r) = H_c/r \qquad r < a_i.$$

11. Derive Eq. (11.74),

$$\rho_i(r) = \rho_n(r/a_i) \qquad r \le a_i,$$

and show that

$$\rho_i(r) = rE/H_c \qquad r < a_i.$$

12. Show that the radius a_i of the intermediate-state region of a current-carrying superconducting wire is given by

$$a_i = \frac{aI_c}{I\left\{1 + [1 - (I_c/I)]^{1/2}\right\}}.$$

13. Show that the radius a_i of the intermediate-state region of a current-carrying superconducting wire has the limiting behavior $a_i = a$ for $I = I_c$ and that $a_i \approx aI_c/2I$ for $I \gg I_c$.
14. Show that Eq. (11.77) is the solution to Eq. (11.76).
15. Show that in the intermediate state the current density averaged over the whole wire has the value

$$\langle J \rangle = \frac{H_c}{a_i}\left(1 + \frac{a_i^2}{a^2}\right).$$

Since this is greater than J_n, we see that the formation of the intermediate state causes more current to flow in the core than would have happened if the wire had become normal.
16. Show that the total current flowing through the intermediate-state region of the wire in Fig. 11.24 is $2\pi a_i^2 J_n$, and that the total current flowing through the wire is given by Eq. (11.75), namely $I = \pi(a^2 + a_i^2)J_n$.

12

Critical States

I. INTRODUCTION

In Chapter 6 we described the Bardeen–Cooper–Schrieffer (1957) microscopic theory that had been devised to explain the nature of superconductivity, subsequently showing that many of the properties predicted by the BCS theory are satisfied by the classical superconductors, and perhaps by the cuprate superconductors as well. In Chapter 5 we delineated the Ginzburg–Landau (1950) phenomenological theory, a theory which is helpful for explaining many other properties of superconductors. Chapter 8 presented the Hubbard model and band theory viewpoints on superconductivity. There is yet another approach, introduced in two works by Bean (1962, 1964), which is too simplified to be called a theory and is instead referred to simply as a model. The Bean model, which has been employed by many experimentalists as an aid in the interpretation of their data, belongs to a class of critical-state models; we discuss it in detail and apply it to a number of cases. We will also compare the Fixed Pinning and Kim models with the Bean model to give some perspective on critical states. Other critical-state models will be reviewed at the end of the chapter.

These models postulate that for low applied fields or currents, the outer part of the sample is in a so-called "critical state" with special values of the current density and magnetic field, and that the interior is shielded from these fields and currents. The Bean model assumes that the super current density always has the magnitude J_c in the critical state, while the Fixed Pinning model assumes that the pinning force is constant in the critical state. In all the models the magnetic field **B** and the super current density **J** are coupled through the Maxwell relation $\nabla \times \mathbf{B} = \mu_0 \mathbf{J}$, so either one can be calculated from knowledge of the other. When the fields and currents are applied simultaneously and then reversed in direction, they produce modified critical states in the outer parts of the sample, consistent with the assumption of the particular model. High values of the applied fields or currents cause the critical state to penetrate to the innermost parts of the superconductor. The models do not take into account the existence of a lower-critical field B_{c1} or the difference between the Meissner and the mixed states. We do not claim that these models really explain the nature of superconductivity. Rather, they provide a convenient means of describing some experimentally observed phenomena.

In this chapter we will confine our attention to simple geometries, such as the parallel geometries illustrated in Figs. 2.26 and 9.1, in which the applied magnetic field is parallel to the surface and the demagnetization effects discussed in Chapter 10, Sections X and XI, do not have to be taken into account. The literature can be consulted for critical-state models involving ellipsoidal samples (Bhagwat and Chaddah, 1990, 1992; Chaddah and Bhagwat, 1992; Krasnov, 1992; Krasnov *et al.*, 1991; Navarro and Campbell, 1991).

II. CURRENT–FIELD RELATIONS

A. Transport and Shielding Current

Electric currents that flow through a superconductor owing to the action of an external current or voltage source are called *transport currents*. Those which arise in the presence of an externally applied magnetic field and cancel the magnetic flux inside the superconductor are called *screening currents*. Figure 12.1a shows the induced shielding current produced by an applied magnetic field and Fig. 12.1b, the induced magnetic field produced by an applied transport current. More complicated cases in which both a transport current and a magnetic field are applied to the superconductor will be discussed here. In such cases transport and screening currents are present simultaneously.

B. Maxwell Curl Equation and Pinning Force

We mentioned earlier that the magnetic field and current density that are

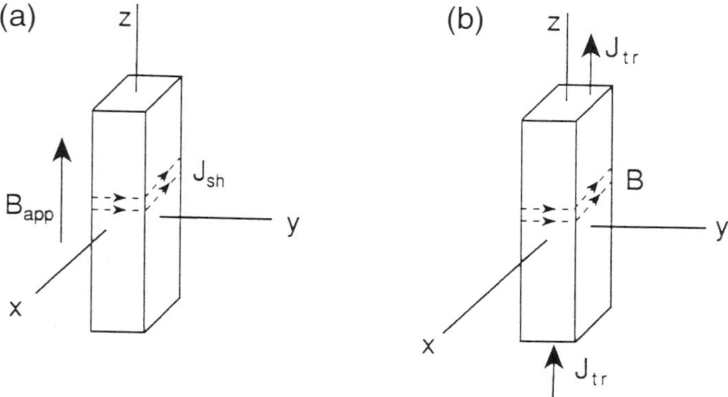

Figure 12.1 (a) Shielding currents J_{sh} induced in a superconducting rod of rectangular cross section by an external magnetic field B_{app} along its axis, and (b) magnetic field B induced in (and around) the same superconducting rod by a transport current J_{tr} flowing along the axis.

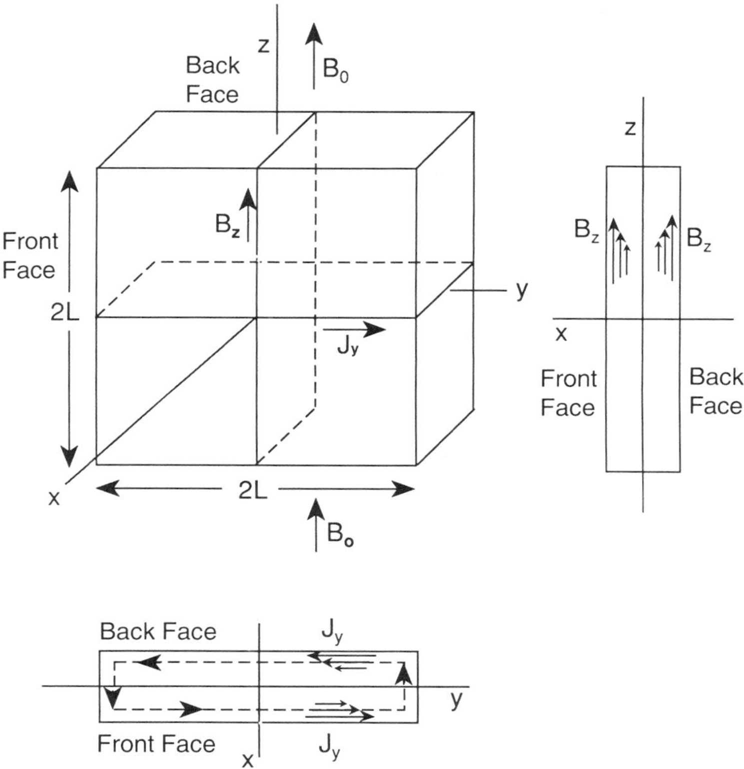

Figure 12.2 Superconducting slab of thickness $2a$ oriented in the y, z-plane with an externally applied magnetic field B_{app} directed along z. The induced shielding current density J_y flowing in the y direction inside the front and back faces is shown.

present in a superconductor are related through the Maxwell curl equation,

$$\nabla \times \mathbf{B} = \mu_0 \mathbf{J}. \tag{12.1}$$

This means that the **B** and **J** vectors are perpendicular at every point in space. We will examine the case of a rectangular slab oriented as shown in Fig. 12.2 in the presence of a magnetic field $\mathbf{B}_{\mathrm{app}}$ along the z direction. We assume that the magnetic field inside the slab, $\mathbf{B}_{\mathrm{in}} = B_z \mathbf{k}$, is along the z direction and that the current density $\mathbf{J} = J_y \mathbf{j}$ has a component only in the y direction; the current density component J_x at the ends of the loops is neglected. For this case we find from Eq. (12.1) that

$$\frac{d}{dx} B_z(x) = \mu_0 J_y(x), \tag{12.2}$$

which means that the field and current density depend only on x.

The internal magnetic field $B_z(x)$ is equal to the product $n(x)\Phi_0$ of the number of vortices per unit area $n(x)$ times the flux per vortex Φ_0. Equation (12.2) becomes

$$\Phi_0 \frac{d}{dx} n(x) = \mu_0 J_y(x). \tag{12.3}$$

We assume that the vortices in the material are in a static equilibrium configuration. The curl of the magnetic field strength (12.1) produces a gradient in the vortex density (12.3) in a direction perpendicular to the current flow direction, and this is sketched in Fig. 9.35. The pinning force density $\mathbf{F}_{\mathrm{p}}$ holds the vortices in place, while the Lorentz force density $\mathbf{J} \times \mathbf{B}$ acting on the vortices is balanced by $\mathbf{F}_p$,

$$\mathbf{F}_p = \mathbf{J} \times \mathbf{B} \quad (12.4)$$

$$= (\nabla \times \mathbf{B}) \times \mathbf{B}/\mu_0, \quad (12.5)$$

where we have again used Eq. (12.1). For the present case $\mathbf{F}_p$ only has an x component, with magnitude

$$F_{p_x} = J_y B_z \quad (12.6)$$

$$= \frac{1}{\mu_0} \cdot \frac{dB_z}{dx} B_z \quad (12.7a)$$

$$= \frac{1}{\mu_0} \cdot \frac{d}{dx} \cdot \tfrac{1}{2} B_z^2. \quad (12.7b)$$

C. Determination of Current–Field Relationships

Equations (12.1)–(12.7) must be satisfied when $\mathbf{B}$ is in the z direction and $\mathbf{J}$ in the y direction. There are many configurations of $B_z(x)$, $J_y(x)$, and $F_p(x)$ that meet this requirement. The Bean model assumes $J_y = \text{const}$, while the Fixed Pinning model assumes $F_p = \text{const}$; all the other models assume a more complex relationship between the internal field and the current density. For most models the relationship between $J_y(x)$ and $B_z(x)$ for the slab geometry is generally of the form

$$J_y(B_z) = \frac{J_K}{f(B_z)}, \quad (12.8)$$

where $f(B_z)$ is a function of the magnetic field and J_K is independent of the field, but can depend on the temperature. J_y is substituted into Eq. (12.2) and the resultant differential equation is solved to obtain $B_z(x)$, the position dependence of the internal field. Finally, this result is substituted back in Eq. (12.8) to give $J_y(x)$, and Eq. (12.6) immediately provides $F_p(x)$.

III. CRITICAL-STATE MODELS

A. Requirements of Critical-State Model

When a magnetic field is turned on, it enters a superconductor, its magnitude inside the superconductor decreasing with distance from the surface. If the applied field is weak enough, the internal field will be zero beyond a certain distance measured inward from the surface. Critical current flows where the field is present, in accordance with the Maxwell equation (12.1); this is called a *critical state*. As one moves inward, the critical current density generally increases as the field decreases, in accordance with Eq. (12.8). The current density is also zero beyond the point at which the internal field vanishes. When the applied field increases in magnitude, the internal field and current densities penetrate further and for sufficiently strong fields are present throughout the sample. Each critical-state model is based on a particular assumed relationship between the internal field and the critical-current density which satisfies these requirements.

B. Examples of Models

We now give current–field relationships for several well-known models:

$$J(B) = J_c \qquad \text{Bean (1962, 1964)} \quad (12.9)$$

$$J(B) = \frac{J_c}{|B(x)|/B_K} \qquad \text{Fixed Pinning} \quad (12.10)$$

(Ji *et al*., 1989; Le Blanc and LeBlanc, 1992)

$$J(B) = \frac{J_c}{|B(x)/B_K|^{1/2}} \qquad \text{Square Root (Le Blanc and Le Blanc, 1992)} \quad (12.11)$$

$$J(B) = \frac{J_c}{1 + |B(x)|/B_K} \qquad \text{Kim (Kim et al., 1962, 1963)} \quad (12.12)$$

$$J(B) = J_c \exp[-|B(x)|/B_K] \qquad \text{Exponential (Fietz et al., 1964)} \quad (12.13)$$

$$J(B) = J_c - J_c'|B(x)|/B_K \qquad \text{Linear (Watson, 1968)} \tag{12.14}$$

$$J(B) = \frac{J_c}{1 + [|B(x)|/B_K]^2} \qquad \text{Quadratic (Leta et al., 1992)} \tag{12.15}$$

$$J(B) = J_c(1 - |B(x)|/B_K)\Theta(B_K - |B(x)|) \qquad \text{Triangular Pulse (Dersch and Blatter, 1988)} \tag{12.16}$$

$$J(B) = \frac{J_c}{[1 + |B(x)|/B_K]^{\beta}} \qquad \text{Generalized (Lam et al., 1990; M. Xu et al., 1990)} \tag{12.17}$$

In these expressions the internal field $B = B(x)$, where x is the distance from the center toward the surface. In most of the models J_c is the critical current in the absence of an applied field.

C. Model Characteristics

Each of the critical-state models depends on a parameter B_K associated with the internal field and a parameter J_K associated with the critical-current density. Both of these parameters can depend on the temperature. The quantity

$$\Theta[B_K - |B(x)|]$$

is called the Heaviside step function. One can also write a more general power-law model $J(B) = A|B(x)|^{-n}$ (Askew *et al.*, 1991; Irie and Yamafuzi, 1967; Yeshurun *et al.*, 1988), which reduces to the Fixed Pinning and Square Root cases for $n = 1$ and $1/2$, respectively. The so-called generalized model reduces to the Bean model for $\beta = 0$ and to the Kim model for $\beta = 1$. The Kim model resembles the Fixed Pinning model for high applied fields

$$(|B(x)| \gg B_K),$$

while the exponential model, linear model with $J_K = J_K'$, and Kim model all reduce to the Bean model for low applied fields ($|B(x)| \ll B_K$). Later we will develop more precise definitions for the cases of low and high field.

Many of these models are old, such as the Bean, Kim, and linear varieties, as well as the exponential model (Chen and Goldfarb, 1989; Chen *et al.*, 1990a, b; Karasik *et al.*, 1971; Kumar and Chaddah, 1989; Sanchez and Chen, 1991; Sanchez *et al.*, 1991).

The explicit expressions for $B(x)$ and $J(x)$ that are obtained by solving the differential equation (12.2) for the functions (12.9)–(12.17) depend on boundary conditions, such as the strength of the applied field, the size, shape, and orientation of the sample, and the previous magnetic history. Examples of these solutions will be given for several commonly encountered cases. We will introduce the Fixed Pinning model for didactic purposes, but will emphasize the Bean model, by far the most widely employed of the critical-state models, as well as the Kim model, which has also been extensively applied. The different models have been compared by a number of authors (Chen and Goldfarb, 1989; Chen and Sanchez, 1991; Ghatak *et al.*, 1992; Ginzburg *et al.*, 1991; Ji *et al.*, 1989; Johnson *et al.*, 1991; Lera *et al.*, 1992; Sanchez *et al.*, 1991; Wahid and Jaggi, 1991).

IV. FIXED PINNING MODEL

We now wish to describe the Fixed Pinning model for the slab case of Fig. 12.2 with application of a magnetic field B_0 that is so weak it does not penetrate all the way to the center of the sample. The model is unphysical in the limit of vanishingly small applied fields because it predicts infinite critical current. Nevertheless, it can provide us with some insight into the nature of critical-state models.

The assumption of constant pinning gives, from Eq. (12.6),

$$J_y(B_z) = \frac{F_p}{B_z(x)}, \qquad (12.18)$$

where from the notation of Eq. (12.10),

$$F_p = J_c B_K. \qquad (12.19)$$

The boundary conditions are that the internal field at the surface, $x = \pm a$, equals the applied field B_0, and that there is a depth, $x = \pm a'$ inside the superconductor at which the internal field drops to zero,

$$B_z(\pm a) = B_0, \qquad (12.20a)$$

$$B_z(\pm a') = 0. \qquad (12.20b)$$

The differential equation obtained by substituting Eq. (12.18) in Eq. (12.2) has the solution

$$B_z(x) = B_0 \left(\frac{a' + x}{a' - a} \right)^{1/2} \qquad -a \le x \le -a', \qquad (12.21a)$$

$$B_z(x) = 0 \qquad -a' \le x \le a', \qquad (12.21b)$$

$$B_z(x) = B_0 \left(\frac{x - a'}{a - a'} \right)^{1/2} \qquad a' \le x \le a, \qquad (12.21c)$$

which is plotted in Fig. 12.3. In writing down these expressions we are limiting ourselves to the case of a field-free region, $B_z(x) = 0$, near the center $(-a' < x < a')$ of the slab, as shown in Fig. 12.3.

Substituting the expressions in (12.21) in Eq. (12.18) gives the current densities

$$J_y(x) = \tfrac{1}{2} J_c \left(\frac{a' + x}{a' - a} \right)^{-1/2} \qquad -a \le x \le -a', \qquad (12.22a)$$

$$J_y(x) = 0 \qquad -a' \le x \le a', \qquad (12.22b)$$

$$J_y(x) = -\tfrac{1}{2} J_c \left(\frac{x - a'}{a - a'} \right)^{-1/2} \qquad a' \le x \le a, \qquad (12.22c)$$

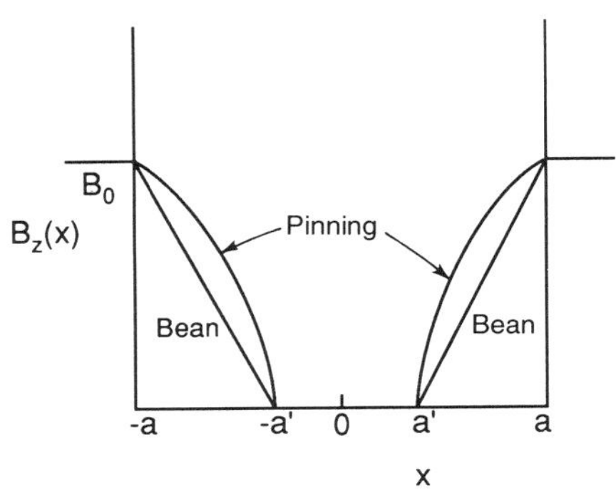

Figure 12.3 Comparison of the Bean and Fixed Pinning models for the penetration of an external magnetic field B_{out} into a superconductor with the slab geometry of Fig. 12.2. The low-field case is illustrated. Most of the chapter will be devoted to the Bean model.

which are plotted in Fig. 12.4, where J_c is given by

$$J_c = \frac{B_0}{\mu_0(a - a')}. \qquad (12.23)$$

Equation (12.18) provides the pinning forces,

$$F_p = \tfrac{1}{2} J_c B_0 \qquad -a \le x \le -a', \qquad (12.24a)$$

$$F_p = 0 \qquad -a' \le x \le a', \qquad (12.24b)$$

$$F_p = -\tfrac{1}{2} J_c B_0 \qquad a \le x \le a', \qquad (12.24c)$$

which are plotted in Fig. 12.5, these are independent of x in each region, as assumed. The factor $\frac{1}{2}$ is included in Eqs. (12.22) and (12.24) to give the Bean and Fixed Pinning models the same total current and the same average pinning force.

We see from Eqs. (12.21), (12.22), and (12.24) and Figs. 12.3, 12.4, and 12.5, respectively, that $B_z(x)$ is symmetric about the point $x = 0$, and that $J_y(x)$ and $F_p(x)$ are antisymmetric. The antisymmetry of $J_y(x)$ is also shown by the reverse directions of the J_y arrows on either side of the slab in Fig. 12.2. The choices of sign correspond to pinning forces that act to restrain

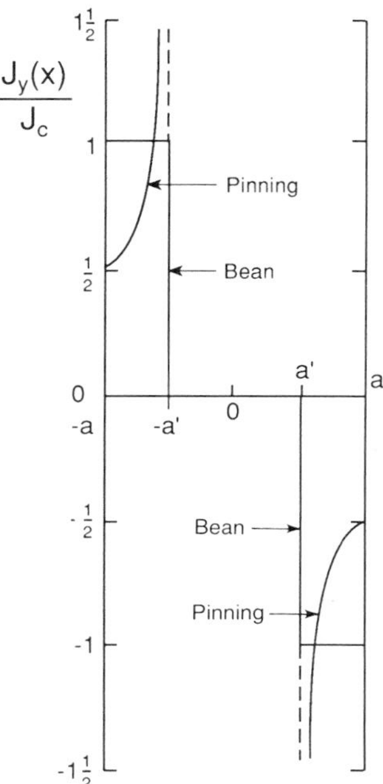

Figure 12.4 Comparison of the Bean and Fixed Pinning models for the dependence of the current density $J_y(x)$ on distance x inside a superconducting slab for the case of Fig. 12.3. The current density is produced by the presence of the external magnetic field B_0 at the surface of the slab and is normalized to give the same total current for each model.

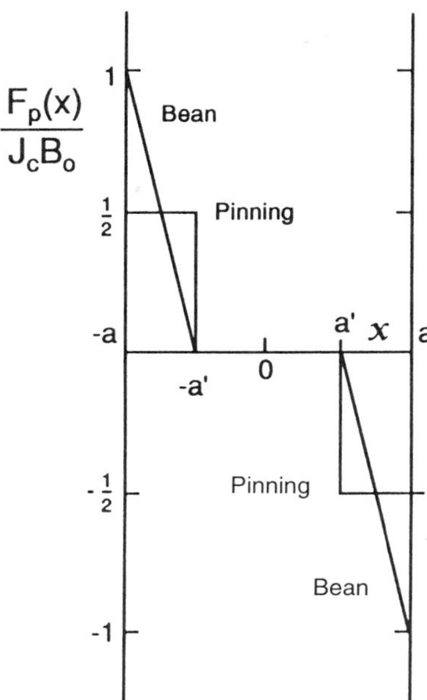

Figure 12.5 Comparison of the Bean and Fixed Pinning models for the dependence of the pinning force $F_p(x)$ on distance x inside a superconducting slab for the case of Fig. 12.3 in an external magnetic field B_0.

the Lorentz force density $\mathbf{J}_y \times \mathbf{B}_z$ from moving the vortices toward the center of the sample.

V. BEAN MODEL

The Bean model (1962, 1964) for super current flow is by far the most widely used of the critical-state models that have been proposed for describing the field and current distribution in a superconductor. The model assumes that wherever the current flows, it flows at the critical density J_c and that the internal magnetic field is given by Eq. (12.1).

A. Low-Field Case

We will first write down solutions for what is called the low-field case. In this case there is a field- and current-free region $(-a' < x < a')$ near the center. In the next section we will provide solutions for the high-field case, i.e., the case in which the fields and currents exist throughout the superconductor.

We again assume the geometry of Fig. 12.2 and follow the same procedures as with the Fixed Pinning model to write down the solutions for the low-field case. In the critical states near the surface, the current density equals $\pm J_c$. It vanishes in the center where $-a' < x < a'$, which corresponds to

$$J_y(x) = J_c \qquad -a \le x \le -a', \quad (12.25a)$$

$$J_y(x) = 0 \qquad -a' \le x \le a', \quad (12.25b)$$

$$J_y(x) = -J_c \qquad a' \le x \le a. \quad (12.25c)$$

Equation (12.2) requires that $B_z(x)$ depend linearly on x in regions where $J_y =$

$\pm J_c$, so that we have for the internal magnetic fields

$$B_z(x) = B_0\left(\frac{a'+x}{a'-a}\right) \qquad -a \le x \le -a', \tag{12.26a}$$

$$B_z(x) = 0 \qquad -a' \le x \le a', \tag{12.26b}$$

$$B_z(x) = B_0\left(\frac{x-a'}{a-a'}\right) \qquad a' \le x \le a. \tag{12.26c}$$

These expressions match the boundary condition $B_z(0) = B_0$ on the two surfaces $x = \pm a$. The quantities J_c and B_0 are related to each other by the expression

$$J_c = \frac{B_0}{\mu_0(a-a')}, \tag{12.27}$$

as in the Fixed Pinning model (12.23). With the aid of Eq. (12.7), we obtain the pinning forces

$$F_p(x) = J_c B_0\left(\frac{a'+x}{a'-a}\right) \qquad -a \le x \le -a', \tag{12.28a}$$

$$F_p = 0 \qquad -a' \le x \le a', \tag{12.28b}$$

$$F_p(x) = -J_c B_0\left(\frac{x-a'}{a-a'}\right) \qquad a' \le x \le a. \tag{12.28c}$$

These equations for $B_z(x)$, $J_y(x)$, and $F_p(x)$ are plotted in Figs. 12.6a for a finite value of a' and in Fig. 12.6b for $a' = 0$.

Figures 12.3, 12.4, and 12.5, respectively, compare Eqs. (12.26), (12.25), and (12.28) with their Fixed Pinning counterparts. As in the other models, $B_z(x)$ is symmetric about the point $x = 0$, while the other two functions $J_y(x)$ and $F_p(x)$ are antisymmetric about this point.

B. High-Field Case

Now that we have explained the low-field Bean model let us introduce its high-field counterpart. The two may be related in terms of a characteristic field B^* proportional to the radius a, as given by

$$B^* = \mu_0 J_c a. \tag{12.29}$$

B^* has the property that when $B_0 = B^*$ the fields and currents are able to reach the center of the slab, as shown in Fig. 12.6b. Thus there are two cases to consider, one for small applied fields,

$$B_0 < B^*, \tag{12.30a}$$

which was discussed in the previous section, and the other for high applied fields,

$$B_0 > B^*. \tag{12.30b}$$

It is easy to show that at high field the currents and fields, respectively, are given by the expressions

$$J_y(x) = J_c \qquad -a \le x \le 0, \tag{12.31a}$$

$$J_y(x) = -J_c \qquad 0 \le x \le a, \tag{12.31b}$$

$$B_z(x) = B_0 - B^*\left(\frac{a+x}{a}\right) \qquad -a \le x \le 0, \tag{12.32a}$$

$$B_z(x) = B_0 + B^*\left(\frac{x-a}{a}\right) \qquad 0 \le x \le a. \tag{12.32b}$$

It is left as an exercise (Problem 3) to write down the pinning forces at high field.

The magnitude of the critical-current density J_c is fixed by the characteristics of the particular superconductor, and depends on such factors as the superconducting material, granularity, twinning, concentration of defect centers, etc. The applied field can be varied, and Fig. 12.6 shows how the internal field, current density, and pinning force of Eqs. (12.26), (12.32), (12.25), (12.31), and (12.28) vary with the ratio B_0/B^* for the Bean model. Figure 12.6a is for the low-field case $B_0 < B^*$, Fig. 12.6b, with $B_0 = B^*$ the boundary between

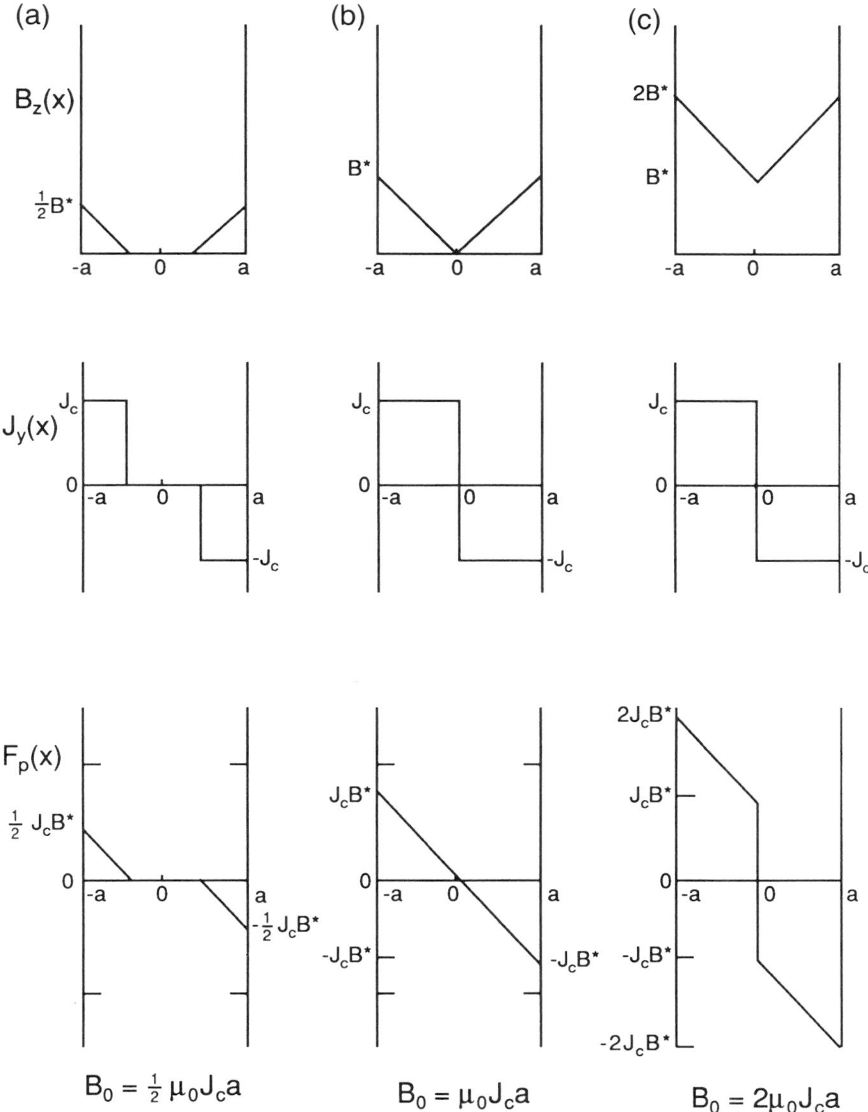

Figure 12.6 Dependence of the internal magnetic field $B_z(x)$, current density $J_y(x)$, and pinning force $F_p(x)$ on the strength of the applied magnetic field B_0 for normalized applied fields given by (a) $B_0/\mu_0 J_c a = \frac{1}{2}$, (b) $B_0/\mu_0 J_c a = 1$, and (c) $B_0/\mu_0 J_c a = 2$. This and subsequent figures are drawn for the Bean model. There is a field-free region in the center for case (a), while case (b) represents the boundary between the presence versus the absence of such a region.

the two cases, and Fig. 12.6c is for high fields, $B_0 > B^*$.

The figures that we have drawn are for zero-field-cooled samples in which the applied field had been increased from its initial value $B_{app} = 0$ to the value B_0, as shown. In particular, the three sets of curves drawn in Fig. 12.6 were obtained by increasing the applied field from 0 to $\frac{1}{2}B^*$, then to B^*, and finally to the value $2B^*$. We will see in Section VI that reversing the field leaves some flux trapped, which is reflected in the shape of the plots for $B_z(x)$ versus x.

C. Transport Current

We have discussed the Fixed Pinning and Bean models for a thin slab in an applied magnetic field. Analytic expressions were deduced for the magnetic field, current density, and pinning force for these cases. Let us now apply the same Bean model analysis to the case of a transport current in which a fixed amount of current passes along the slab or wire, and both external and internal magnetic fields are induced by this current. The extension of the results to the Kim model with transport currents will be left as an exercise.

We will again discuss the case of a slab oriented in the y, z-plane, but this time assuming an applied transport current of magnitude I flowing in the positive y direction, as shown in Fig. 12.7. The current distributes itself in the x, z-cross section in accordance with the Bean model. Thus the critical-current density J_c is adjacent to the outer boundary, between $x = -a$ and $x = -a'$, with zero current in the center, as shown in Fig. 12.8a. Since the cross-sectional area is $4(a - a')L$, the transport current is given by

$$I = 4(a - a')LJ_c. \tag{12.33}$$

This current flow produces internal magnetic fields with the orientations shown in the figure. The external fields that are induced outside the slab will not be of concern to us.

The equations for $B_z(x)$, $J_y(x)$, and $F_p(x)$ for the case of a transport current are the same as Eqs. (12.25)–(12.32) for the applied field case, except for several

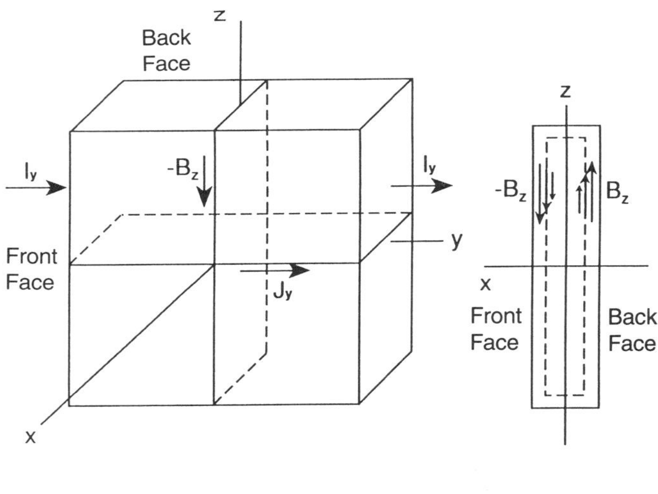

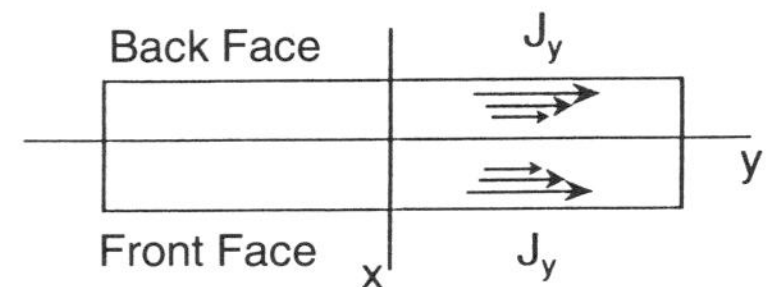

Figure 12.7 Superconducting slab of length $2L$, height $2L$ and thickness $2a$ oriented in the y, z-plane with an applied transport current I_y flowing in the positive y direction. The induced internal magnetic fields B_z are indicated. The induced external fields are not taken into account, and hence are not shown.

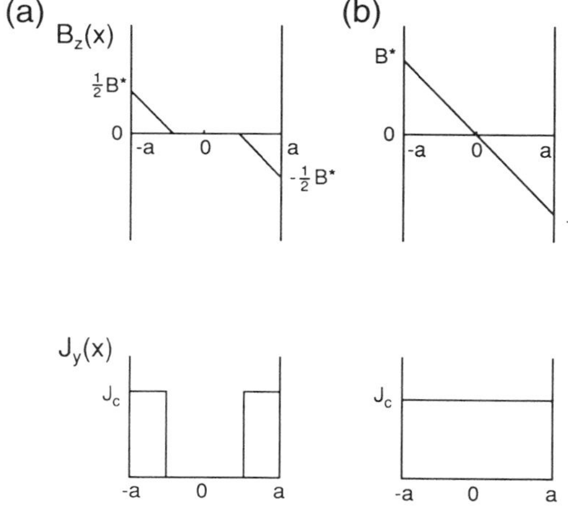

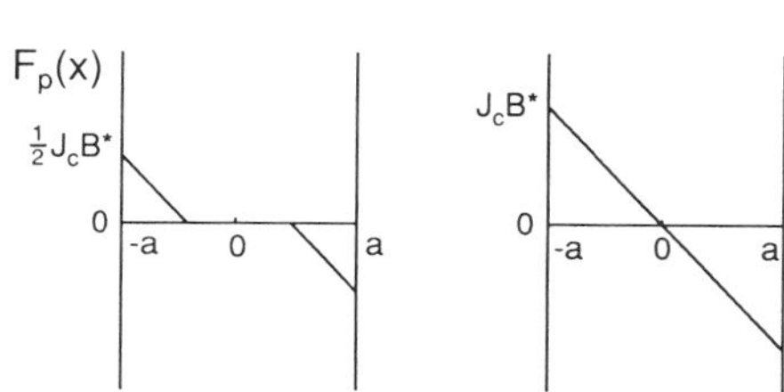

Figure 12.8 Dependence of the internal magnetic field $B_z(x)$, current density $J_y(x)$, and pinning force $F_p(x)$ on the strength of an applied transport current for the Bean model. Figures are drawn for applied currents I that are (a) less than the critical current $I_c = 4aLJ_c$, or (b) equal to the critical current I_c. Applied currents in excess of the critical current cause the wire to go normal.

reversals of sign, so we will not bother to write them down. Figures 12.8a provide plots of these three quantities for $I \approx \frac{1}{2}I_c$, where

$$I_c = 4aLJ_c. \tag{12.34}$$

We see from the figure that the induced magnetic field is in opposite directions on either side of the slab, while the internal pinning force has the same direction as in the screening current case. A comparison of Figs. 12.6 and 12.8 shows that B and J reverse their symmetries about the point $x = 0$, B being symmetric and J antisymmetric in the screening case, and B antisymmetric and J symmetric in the transport case; the pinning force F_p is antisymmetric in both cases. This is because in both cases the Lorentz force $\mathbf{J} \times \mathbf{B}$ acts to move the current–field configurations inward, while the pinning force opposes this motion and holds the fields (or vortices) in place. Figure 12.8b shows the field, current density, and pinning force when the wire is carrying its maximum possible transport current (12.34), namely, its critical current I_c.

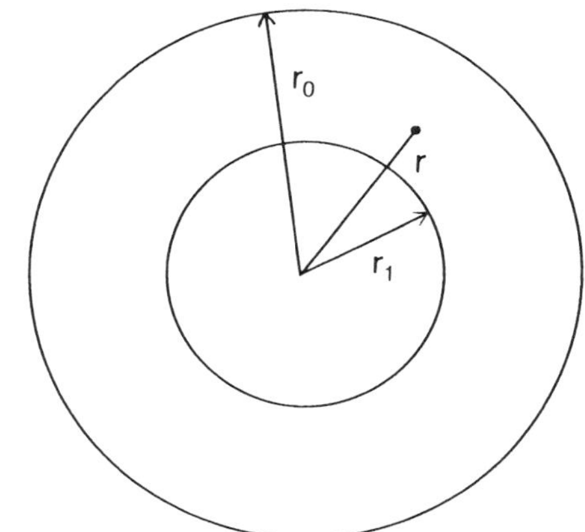

Figure 12.9 Circular cross section of a superconducting wire of radius r_0 carrying a transport current. In the Bean model magnetic fields $B_\phi(r)$ and current densities $J_z(r)$ are present in the outer annular region $r_1 \le r \le r_0$ and absent from the inner circular region $0 \le r \le r_1$. The radius r_1 depends on the magnitude of the transport current through Eq. (12.36).

D. Circular Cross Section

We have been discussing the Bean model in slabs of rectangular cross section. The other geometry of greatest practical importance is that of a superconducting wire with a circular cross section of radius r_0, as shown in Fig. 12.9. We will apply the Bean model to the case of transport current flowing through the wire.

This model assumes that the current flows in a layer at the surface with an inner radius r_1, so that it assigns the following values to the current density

$$J_z = 0 \qquad 0 < r < r_1, \quad (12.35a)$$

$$J_z = J_c \qquad r_1 < r < r_0, \quad (12.35b)$$

with the total current I given by

$$I = \pi J_c(r_0^2 - r_1^2). \quad (12.36)$$

Figure 12.10b shows the radial dependence of the current density corresponding to Eqs. (12.35).

The internal magnetic field B_ϕ will be in the form of loops encircling the origin. The analogue of Eq. (12.2) in cylindrical coordinates is

$$\frac{1}{r}\frac{d}{dr}\left[rB_\phi(r)\right] = \mu_0 J_z(r). \quad (12.37)$$

This gives the following values for the magnetic field induced by the transport current:

$$B_\phi = 0 \qquad 0 < r < r_1, \quad (12.38a)$$

$$B_\phi = \frac{\mu_0 J_c}{2}\left(\frac{r^2 - r_1^2}{r}\right) \qquad r_1 < r < r_0, \quad (12.38b)$$

$$B_\phi = \frac{\mu_0 I}{2\pi r} \qquad r_0 < r < \infty. \quad (12.38c)$$

The latter result is the standard expression found in electrodynamics texts for the magnetic field around a current-carrying wire. The pinning force given by $\mathbf{F}_p = \mathbf{J} \times \mathbf{B}$ is in the radial direction with the magnitude $F_p = J_z B_\phi$, given by

$$F_p = 0 \qquad 0 < r < r_1, \quad (12.39a)$$

$$F_p = \frac{\mu_0 J_c^2}{2}\left(\frac{r^2 - r_1^2}{r}\right) \qquad r_1 < r < r_0. \quad (12.39b)$$

Equations (12.35), (12.38), and (12.39) are plotted in Fig. 12.10.

If the wire is carrying its critical current I_c,

$$I_c = \pi r_0^2 J_c, \quad (12.40)$$

the current density, magnetic field, and pinning force inside the wire will be given by setting $r_1 = 0$

$$J_z = J_c \qquad 0 < r < r_0, \quad (12.41)$$

$$B_\phi = \tfrac{1}{2} J_c^2 r \qquad 0 < r < r_0, \quad (12.42)$$

$$F_p = \tfrac{1}{2} J_c^2 r \qquad 0 < r < r_0, \quad (12.43)$$

and the field outside is, from Eq. (12.38c),

$$B_\phi = \frac{\mu_0 I_c}{2\pi r} \qquad r_0 < r < \infty. \quad (12.44)$$

E. Combining Screening and Transport Current

When both an applied magnetic field and an applied transport current are pre-

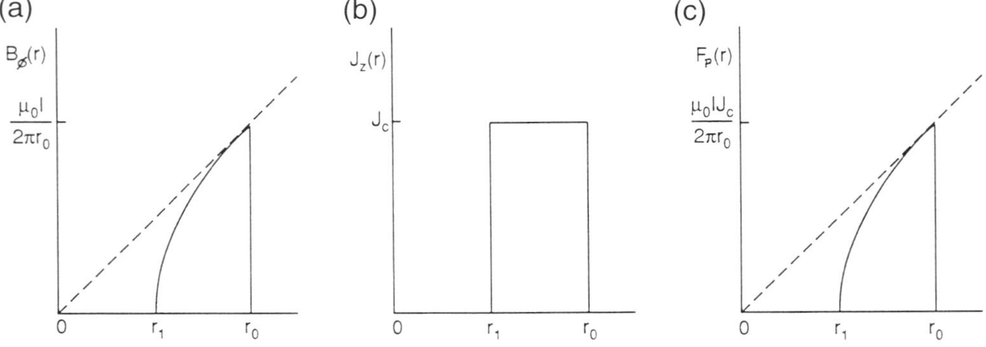

Figure 12.10 Bean model for transport current flow in a superconducting wire with the circular cross section of Fig. 12.9, showing the radial dependence of (a) the internal magnetic field $B_\phi(r)$, (b) current density $J_z(r)$, and (c) pinning force $F_p(r)$.

sent, the situation is more complicated. The Bean model conditions still apply—namely, wherever the current flows it flows at the critical density J_c. For the slab geometry, if there is an internal magnetic field $\mathbf{B}_{in}$, it will have a gradient (12.2) equal to $\mu_0 \mathbf{J}_c$. Figure 12.11 sketches the slab with an applied external field and a transport current present simultaneously. The induced currents and fields combine with the applied fields to produce the net current densities and magnetic fields, which have the x dependence plotted in Fig. 12.12. This figure is drawn for the high-field case, in which $B_{app} > \mu_0 J_c a$; such a case occurs in magnet wire wound as a solenoid to produce a strong magnetic field.

The difference ΔB between the magnetic fields on the left and right sides of the slab, where $x = -a$ and $x = a$, respectively, is related to the net average current density in the slab arising from the flow of transport current. It is left as an exercise (Problem 7) to show that

$$\Delta B = = \mu_0 J_c (x_R - x_L), \quad (12.45)$$

where the notation for this equation is given in the figure.

Section VI.G will examine more complicated cases involving the simultaneous presence of a transport current and an external field.

F. Pinning Strength

Pinning forces set limits on the amount of resistanceless current that can be carried by a superconductor. We know from

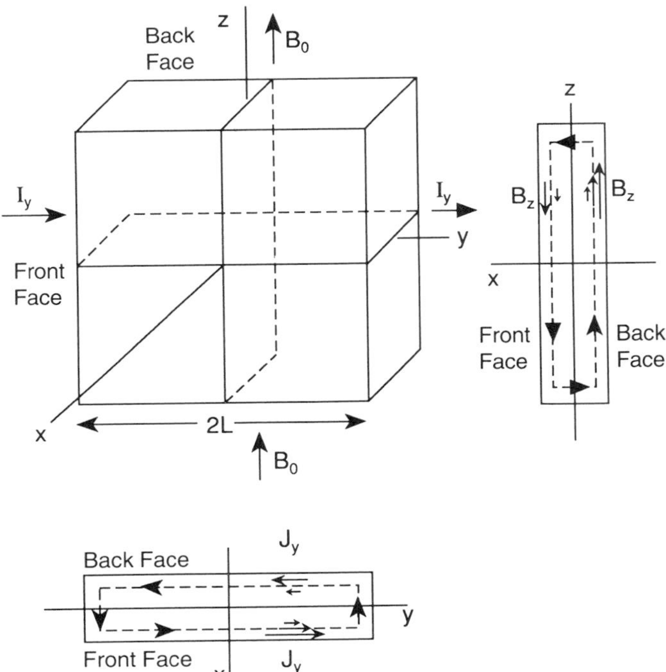

Figure 12.11 Superconducting slab of length $2L$, height $2L$ and thickness $2a$ oriented in the y, z-plane with an applied transport current I_y flowing in the y direction and an applied magnetic field B_0 oriented in the z direction. The internal magnetic fields $B_z(x)$ and current densities $J_y(x)$ shown in the figure are superpositions of those arising from the applied field and the transport current cases of Figs. 12.2 and 12.7, respectively.

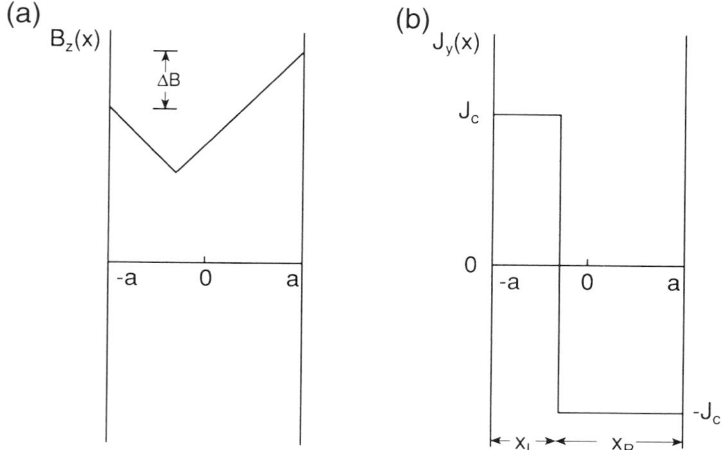

Figure 12.12 Dependence of the internal field $B_z(x)$ and the current density $J_y(x)$ on position x inside the slab of Fig. 12.11 when both a magnetic field and a transport current are applied. The figure is drawn for the high-field case.

Eqs. (12.28) that in the Bean model the pinning force has its maximum magnitude F_p at the edge of the sample, where $x = \pm a$,

$$F_p = J_c B_0. \qquad (12.46)$$

We will consider how the value of F_p affects the field and current distributions for a constant applied field B_0. If the slab is a soft superconductor, so that the pinning forces are too weak to hold the vortices in place, from Chapter 9, Section V.C it is clear that F_p is zero in the vortex equation of motion. Setting $F_p = 0$ in Eq. (12.46) gives $J_c = 0$, resistanceless current cannot flow, and, from Eq. (12.2), the magnetic field penetrates the entire cross section, as shown by the curves for "no pinning" in Fig. 12.13. For weak pinning, F_p is small, J_c is small, from Eq. (12.2) the slope dB_z/dx is small, and the magnetic field and current still penetrate the entire sample, but $B_z(x)$ is weaker in the center. If the slab is a hard superconductor, so that F_p is large, J_c is also large, the slope dB_z/dx is steep, and the field and current only exist near the surface of the sample, as shown by the curves for "strong pinning" in Fig. 12.13. The figure is drawn for the shielding current case, though it can also be concluded that increasing the pinning strength increases the transport supercurrent capacity of a wire (M. B. Cohn *et al.*, 1991).

VI. REVERSED CRITICAL STATES AND HYSTERESIS

Up until now we have made the implicit assumption that the sample was zero field cooled and then subjected to an applied field that only increased in value. As the surface field B_o is increased, it begins to proceed inward. Figure 12.14 shows the internal field configurations brought about by a series of six successive increases in the applied field. We know from Eqs. (12.32) that the applied field B^* causes the innermost internal field to just reach the center point $x = 0$, and that twice this surface field, $2B^*$, causes the field at the center point to be B^*, as illustrated in the figure.

The present section will examine what happens when the field decreases from a maximum value. We will not try to write down the equations for all the different cases because they are very complicated,

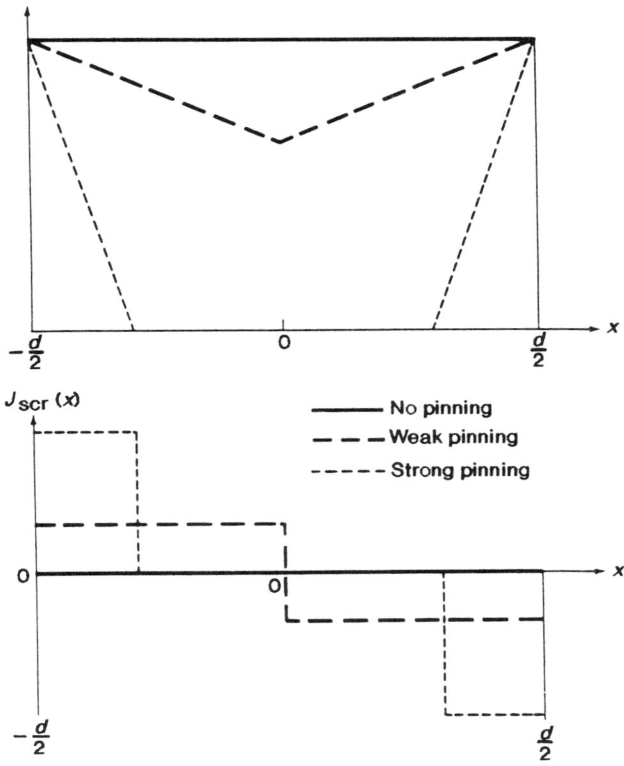

Figure 12.13 Internal magnetic field $B_z(x)$ and current density $J_y(x)$ for strong (– – – – –), weak (— — — —), and zero (———) pinning in a superconducting slab in an external magnetic field, as in Fig. 12.2 (von Duzer and Turner, 1981, p. 338).

as illustrated in Problem 9, without being very instructive. Instead, we will provide a qualitative discussion by sketching how the field and current configurations change as the field decreases.

A. Reversing Field

The first three panels of Fig. 2.15 show the field and current configurations as the applied field is increased from $0.5B^*$ to $2.5B^*$, while the next three panels show the configurations for a decrease in B_{app} from the maximum value $2.5B^*$ to the minimum $-2.5B^*$. This beings at the surface by a decrease in the internal field there. A B versus x line with the opposite slope moves inward, as shown in Figs. 12.15d and 12.15e. The result is that the flux is trapped inside during the field-lowering process as shown shaded in Fig. 16b. Thus the field inside exceeds that at the surface, and the average field inside is larger than the surface value. It will be clear from the discussion below that the amount of trapped flux reaches a maximum when the applied field has decreased through the range $\Delta B = 2B^*$ and that further decreases in the field maintain the amount of trapped flux constant. Finally, the applied field drops below zero and a negative applied field forms a critical state in the opposite direction, as shown. The figure also shows the current flow patterns for each step in the field-lowering process.

We see from Fig. 12.15 that increasing the applied field beyond B^* produces a critical state with the maximum amount of

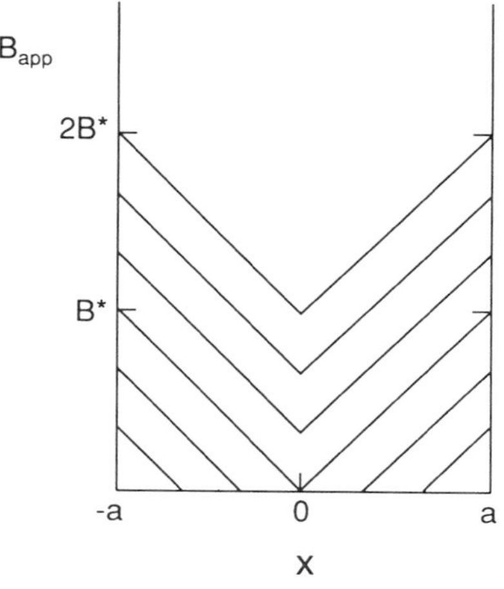

Figure 12.14 Internal field in a superconducting slab for increasing values of the applied field. When B_{app} reaches the value B^*, the internal field reaches the center; when $B_{app} = 2B^*$, the field at the center is B^*.

shielded flux. The subsequent decrease of the field by $2B^*$ or more produces a critical state with the maximum amount of trapped flux. Figure 12.16 shows these two cases and depicts the shielded and trapped flux as shaded regions. The area of the shaded region has the magnitude B^*a, which means that the maximum flux trapped per unit length Φ_{max}/L is

$$\frac{\Phi_{max}}{L} = B^*a. \tag{12.47}$$

Flux shielding occurs when the average field $\langle B \rangle$ inside the superconductor is lower in magnitude than the applied field, whereas flux trapping occurs when the average internal field exceeds the applied field. It is a simple matter to calculate

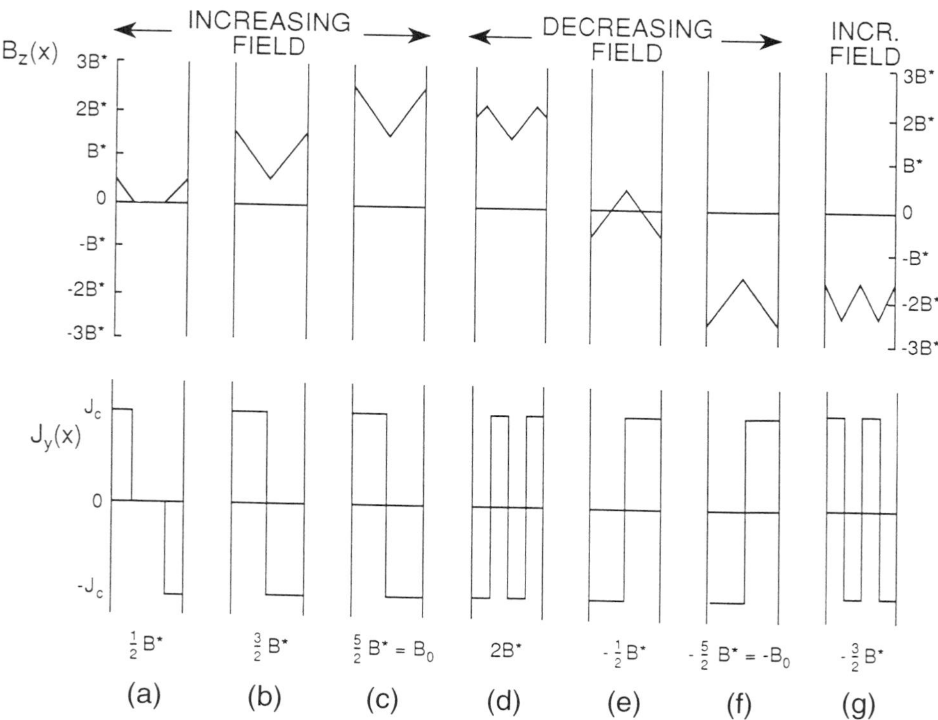

Figure 12.15 Applied field cycle with B_{app} starting at 0, increasing from $0.5B^*$ to $2.5B^*$, decreasing through zero to the negative value $-2.5B^*$, and then beginning to increase again. Plots are shown of the internal field $B_z(x)$ and the current density $J_y(x)$ for successive values of B_{app}.

these averages. Thus we have, for positive $B_0 > B^*$ and $B_{app} > 0$,

$$\langle B \rangle = B_{app} - \tfrac{1}{2}B^* \qquad \text{(shielding)} \tag{12.48a}$$

$$\langle B \rangle = B_{app} + \tfrac{1}{2}B^* \qquad \text{(trapping)}, \tag{12.48b}$$

corresponding to Figs. 12.16a and 12.16b, respectively. We see from Fig. 12.15 that these two cases are associated with current flow in opposite directions.

If the applied field is increased from zero to B_0 and then decreased back to zero, the amount of trapped flux will depend on whether the maximum field B_0 is less than B^*, between B^* and $2B^*$, or greater than $2B^*$, as shown in Fig. 12.17. Problem 11 involves calculating the average trapped fields and the amount of trapped flux per unit length for these cases.

B. Average Internal Field

In Eqs. (12.48a) and (12.48b) we gave the average internal fields for the conditions of maximum shielding and maximum trapping when $B_0 > B^*$. It is important to know how to determine the average internal field throughout a complete raising and lowering cycle.

In Fig. 12.18 we show how the average internal field $\langle B \rangle$ is calculated for the case in which this applied field is initially raised to the value $B_0 > B^*$ and then decreased down into the range $B_0 - B^* < B_{app} < B_0$. We begin by setting $\langle B \rangle$ equal to the applied field plus the average trapped field above B_{app} minus the average shielding field below B_{app},

$$\langle B \rangle = B_{app} + \tfrac{1}{4}(B_0 - B_{app})x - \tfrac{1}{2}(B_{app} - B_0 + B^*)(1 - x), \tag{12.49}$$

where x is defined in the figure. The trapped field exists over the distance xa and the shielding field over the distance $(1-x)a$. From the figure we see that

$$x = \frac{B_0 - B_{app}}{B^*}. \tag{12.50}$$

Substituting Eq. (12.50) in Eq. (12.49) gives

$$\langle B \rangle = B_0 - \tfrac{1}{2}B^* - \frac{(B_0 - B_{app})^2}{4B^*} \qquad B_0 - B^* < B_{app} < B_0. \tag{12.51}$$

Figure 12.19 presents plots of the average field $\langle B \rangle$ versus the applied field B_{app} when the latter is cycled over the

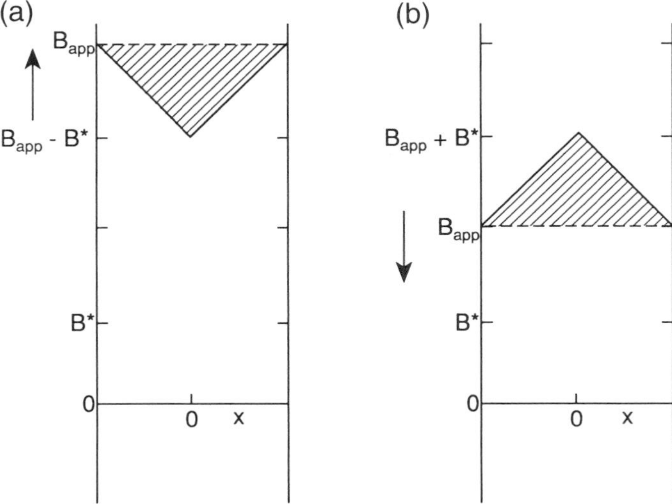

Figure 12.16 Examples of (a) shielded flux (shaded) for an increasing applied field, and (b) trapped flux (shaded) for a decreasing applied field.

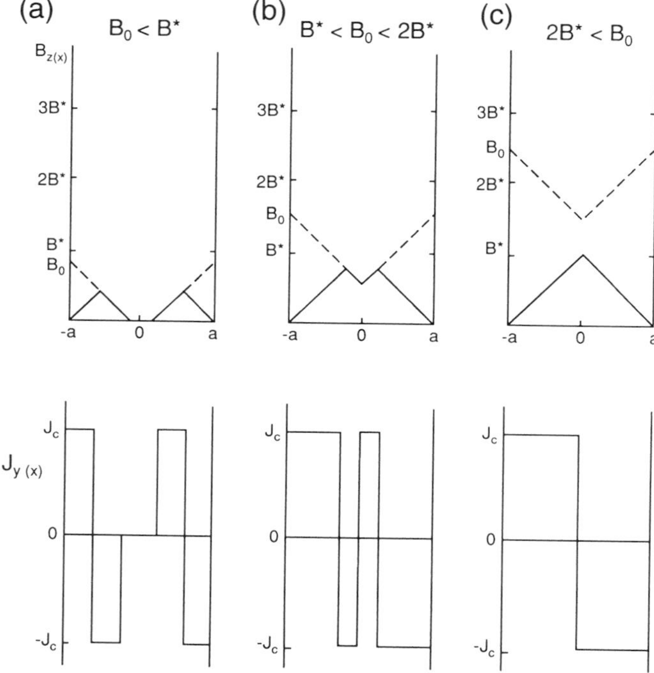

Figure 12.17 Flux trapped in a superconducting slab after the applied field B_{app} is raised to the value B_0 and then decreased back to zero. Cases are shown for (a) $B_0 = 0.5B^*$, (b) $B_0 = 1.5B^*$, and (c) $B_0 = 2.5B^*$. The internal field $B_z(x)$ and current density $J_y(x)$ are shown (solid lines) for each case after returning to zero field. The dotted lines show the maximum internal field configuration achieved for $B_{app} = B_0$.

range $-B_0 \le B_{app} \le B_0$. The figure shows three different shapes of loops for $\langle B \rangle$ versus B_{app} assuming three different values of B_0, as follows:

(1) We see from Fig. 12.19a that for a small maximum field $B_0 < B^*$ the average internal field is quite small over the entire range of applied fields. This is because the field does not penetrate far into the interior, as we see from Fig. 12.17a.

(2) Figure 12.19b illustrates the intermediate case for the maximum field B_0 in the range $B^* < B_0 < 2B^*$. The loop appears similar in shape to the low-field loop, but relatively broader and rotated toward the 45° line in the figure.

(3) For the limit of a large maximum field—namely, $B_0 > 2B^*$, which corresponds to Fig. 12.17c—we see that over much of the range of applied fields the average internal field $\langle B \rangle$ increases linearly with B_{app} on one side of the loop and decreases linearly with B_{app} on the other side.

A comparison of these three loops indicates that for each loop the highest average field $\langle B \rangle_{max}$ is at point A where the applied field equals B_0. Figure 12.20 shows that this maximum average field increases with B_0 and has the value

$$\langle B \rangle_{max} = B_0 - \tfrac{1}{2}B^*$$

for $B_0 > B^*$.

Each $\langle B \rangle$-versus-B_{app} loop has several special points. These are specially noted in Fig. 12.19:

(1) The maximum point a, where $B_{app} = B_0$; a′ denotes the minimum point

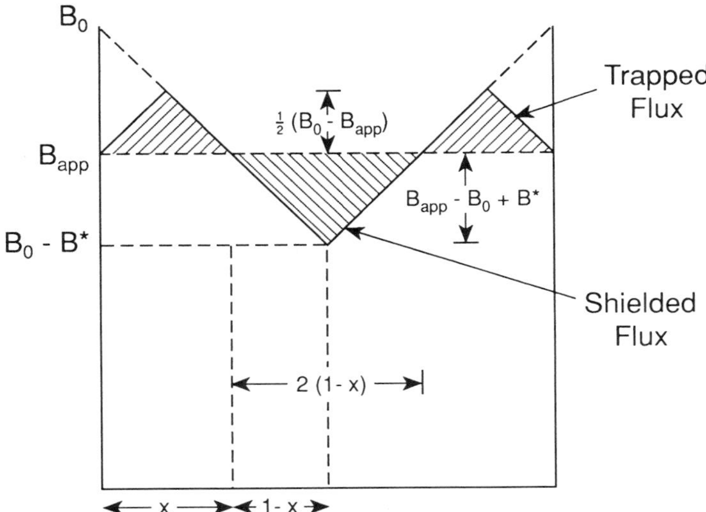

Figure 12.18 Internal field configuration when an applied field B_{app} equal to $B_0 > B^*$ is decreased down to a value in the range $B_0 - B^* < B_{app} < B_0$. The trapped and shielded fluxes are shown as well as the differences in field used to calculate the average field $\langle B \rangle$.

where $B_{app} = -B_0$. Both $\langle B \rangle$ and M are maxima at point A and minima at point a′.

(2) The zero magnetization point b, where $M = 0$ and $\langle B \rangle = B_{app}$. At this point the ⟨B⟩-versus-B_{app} loop intersects the dashed 45° line through the origin.

(3) The zero applied field point c, where $B_{app} = 0$.

(4) The zero average field point d, where $\langle B \rangle = 0$ and $\mu_0 M = B_{app}$. This point lies on a 45° line (not shown) through the origin of a $\mu_0 M$-versus-B_{app} loop. It is shown in Problem 13 and in Fig. 12.21, that for $B_0 = B^*$ this point is at the position $B_{app} = (1 - \sqrt{2})B^*$.

(5) The change-over point e between the quadratic behavior of Eq. (12.51) and the linear behavior of Eq. (12.48). This point does not exist for $B_0 < B^*$, it is the same as point a′ for $B_0 = B^*$, as shown in Fig. 12.21, between a′ and d for $B^* < B_0 < 2B^*$, as indicated in Fig. 12.19b, and between c and b for $2B^* < B_0$, as we see from Fig. 12.19c.

These definitions are summarized in Table 12.1. Table 12.2 gives explicit expressions for the average field $\langle B \rangle$ and the magnetization $\mu_0 M$ at these points. Low-field loops with $B_0 < B^*$ are missing point e. It is clear from symmetry that the upper and lower parts of a hysteresis loop have corresponding points.

C. Magnetization

The magnetization M is given by the relation (1.69),

$$M = \frac{B}{\mu_0} - H. \tag{12.52}$$

For the slab case depicted in Fig. 12.2 the boundary condition (1.76) shows that H is the same outside and inside the superconductor, with the value $\mu_0 H = B_{app}$. Ordinarily, we think of M as the average magnetization, $M = \langle M \rangle$. Bearing this in mind, Eq. (12.52), written for average quantities inside the superconductor, is as follows:

$$\mu_0 M = \langle B \rangle - B_{app}. \tag{12.53}$$

Thus we see that the magnetization determines how great is the difference between

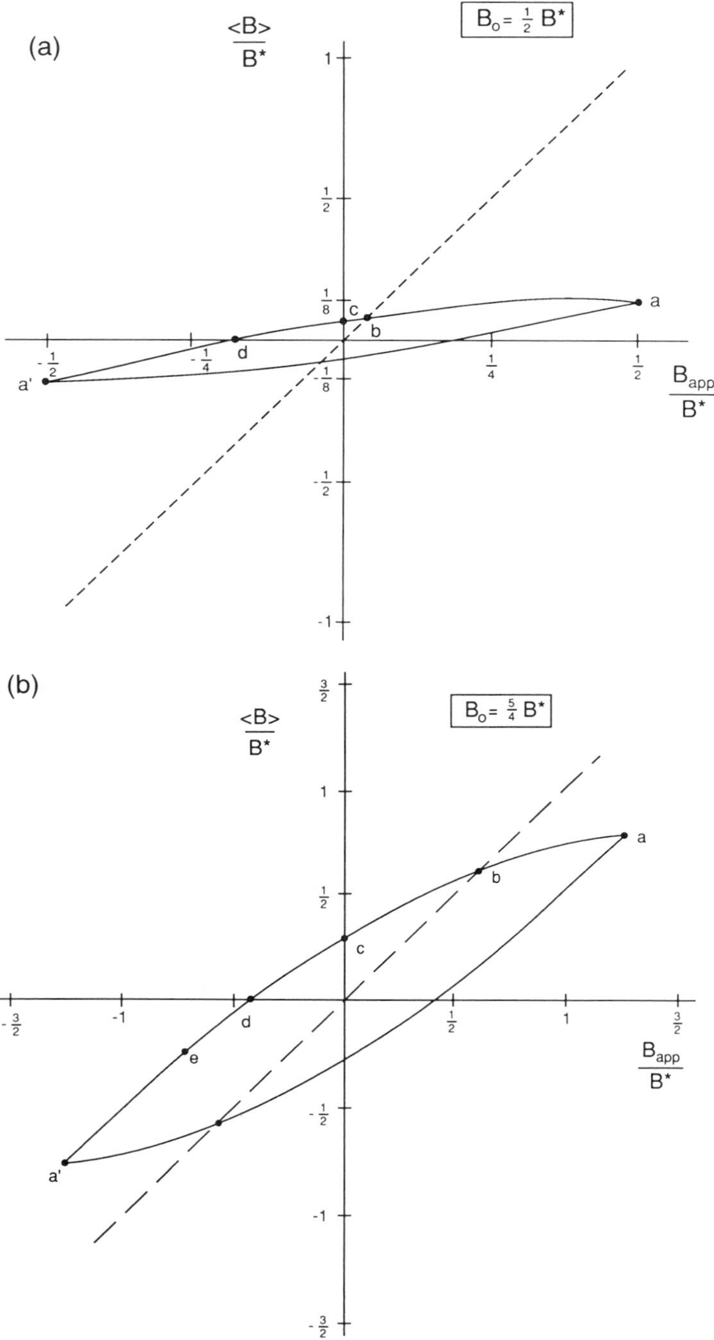

Figure 12.19 Dependence of the average field $\langle B \rangle$ on the applied field B_{app} when the latter is cycled over the range $-B_0 \leq B_{app} \leq B_0$ for three cases: (a) $B_0 = \frac{1}{2}B^*$, (b) $B_0 = \frac{5}{4}B^*$, and (c) $B_0 = 3B^*$. The special points a, b, c, d, and e traversed during the cycles are defined in the text and have the characteristics given in Tables 12.1 and 12.2. In this figure and Figs. 12.20, 12.21, 12.22, and 12.23, the abcissa and the ordinate are normalized relative to B^*. *(Continues)*

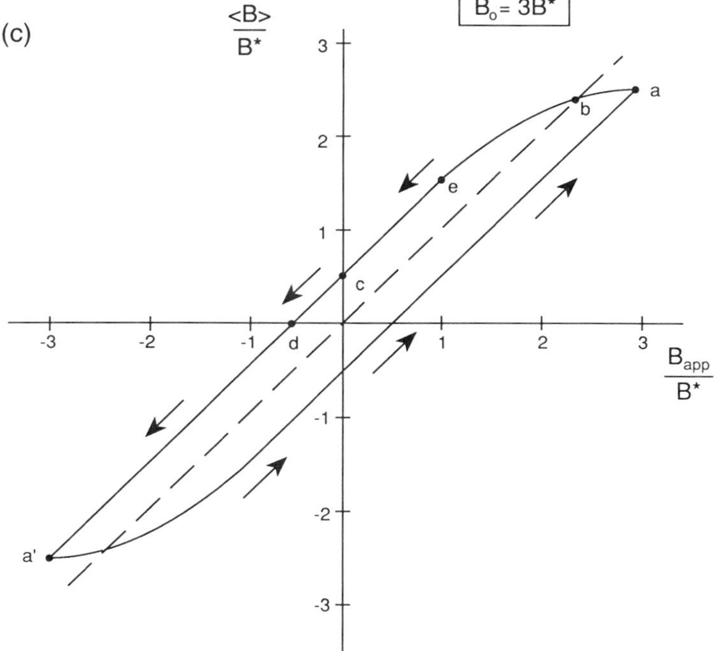

Figure 12.19 (*Continued*)

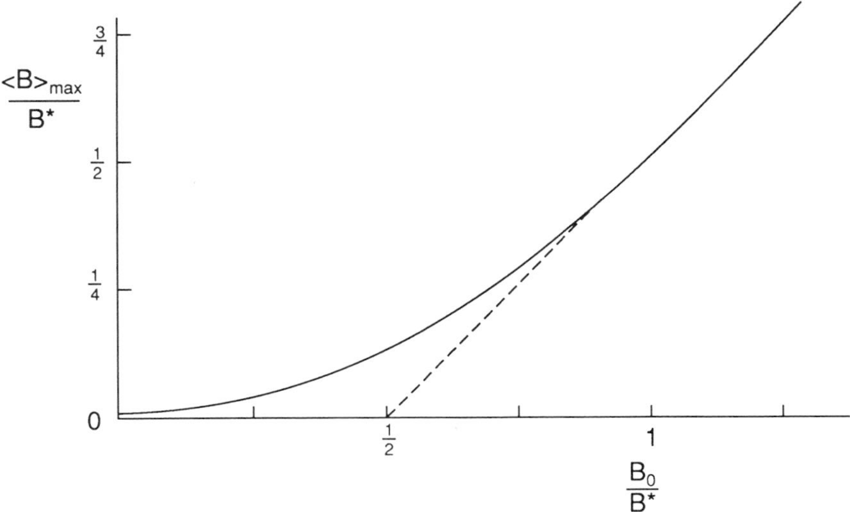

Figure 12.20 Dependence on the ratio B_0/B^* of the maximum average field $\langle B\rangle_{\max}$ found at point a on the loop. We see from the figure that the maximum average field is small for $B_0 < \frac{1}{2}B^*$ and has the value $\langle B\rangle_{\max} = B_0 - \frac{1}{2}B^*$ for $B_0 > B^*$.

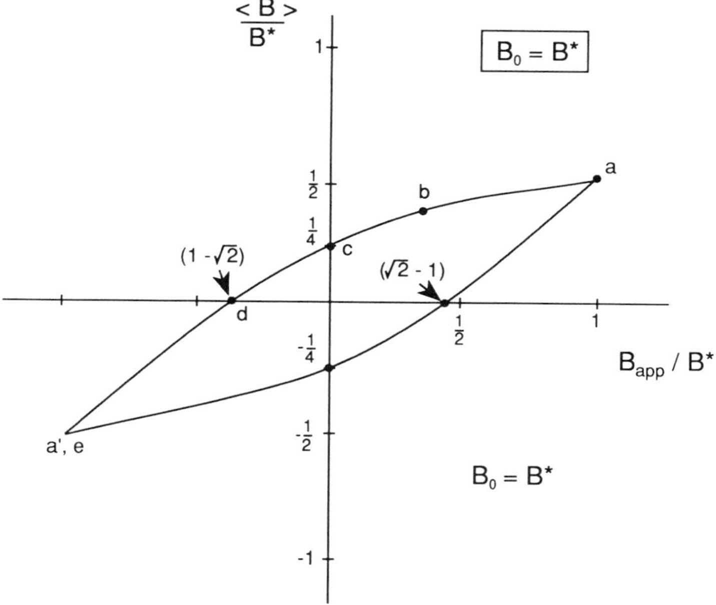

Figure 12.21 Average field $\langle B \rangle$ versus the applied magnetic field B_{app} cycled over the range $-B_0 \le B_{app} \le B_0$ for the case $B_0 = B^*$.

the average internal field and the applied field. Using this expression, we can write down the magnetization for the situation sketched in Fig. 12.18. Substituting Eq. (12.51) in Eq. (12.53) gives

$$\mu_0 M = B_0 - B_{app} - \tfrac{1}{2}B^* - \frac{(B_0 - B_{app})^2}{4B^*}$$

$$B_0 - B^* < B_{app} < B_0. \quad (12.54)$$

This gives $\mu_0 M = -\frac{1}{2}B^*$ for $B_{app} = B_0$, which corresponds to the maximum flux shielding of Eq. (12.48a) and to point a on the loops of Fig. 12.19. The magnetization is $+\frac{1}{2}B^*$ for the maximum flux trapping case of Eq. (12.48b) and for point a′ on the loops of Fig. 12.19.

Table 12.1 Definitions of the Special Points a, b, c, d, and e on the Hysteresis Loops of Figs. 12.19, 12.21, 12.22, and 12.23

Point	Characteristics	
a	$B_{app} = B_0$	end point of loop
b	$M = 0$	$\langle B \rangle = B_{app}$
c	$B_{app} = 0$	near midpoint of loop
d	$\langle B \rangle = 0$	$\mu_0 M = -B_{app}$
e	onset of linear portion of loop (exists for $B_0 > B^*$ and absent for $B_0 < B^*$)	

D. Hysteresis Loops

The loops for the average internal field that are presented in Figs. 12.19 and 12.21 could be called hysteresis loops, but the term is ordinarily reserved for loops involving a magnetization M. Figure 12.22 shows plots of $\mu_0 M$ versus B_{app} drawn to the same scale as the average field cases of Fig. 12.19. The magnetization loop is narrow and inclined at close to a 45° angle for $B \ll B^*$, in accordance with Fig. 12.22a. Figure 12.23 shows the intermediate field case for $B_0 = B^*$. The magnetization saturates at $M = B^*/\mu_0$ over most of the range for the case $B_0 \gg B^*$, as indicated in Fig. 12.22c. The hysteresis loops are labeled with the same points a, b, c, d, and e that were introduced in the previous section; values for the magnetization at these points are given in Table 12.2. These loops may

Table 12.2 Expressions for the Average Magnetic Field $\langle B \rangle$ and the Magnetization $\mu_0 M$ at Various Points on the Hysteresis Loops of Figs. 12.19, 12.21, 12.22, and 12.23 over a Range of Maximum Fields B_0 Relative to the Full Penetration Field B^*. Point e is absent for $B_0 < B^*$

Point	Range of B_0	Applied field B_{app}	Average field $\langle B \rangle$	Magnetization $\mu_0 M$
a	$0 \le B_0 \le B^*$	B_0	$\frac{1}{2}B_0^2/B^*$	$-B_0 + \dfrac{\frac{1}{2}B_0^2}{B^*}$
a	$B^* \le B_0$	B_0	$B_0 - \frac{1}{2}B^*$	$-\frac{1}{2}B^*$
b	$B_0 \ll B^*$	$\dfrac{\frac{1}{2}B_0^2}{B^*}$	$\frac{1}{2}B_0^2/B^*$	0
b	$B^* \le B_0$	$B_0 - (2-\sqrt{2})B^*$	$B_0 - (2-\sqrt{2})B^*$	0
c	$0 \le B_0 \le B^*$	0	$\frac{1}{4}B_0^2/B^*$	$\dfrac{\frac{1}{4}B_0^2}{B^*}$
c	$B^* \le B_0 \le 2B^*$	0	$B_0 - \frac{1}{2}B^* - \dfrac{\frac{1}{4}B_0^2}{B^*}$	$B_0 - \frac{1}{2}B^* - \dfrac{\frac{1}{4}B_0^2}{B^*}$
c	$2B^* \le B_0$	0	$\frac{1}{2}B^*$	$\frac{1}{2}B^*$
d	$0 \le B_0 \le \frac{3}{2}B^*$	$B_0 - [2B^*(2B_0 - B^*)]^{1/2}$	0	$-B_0 + [2B^*(2B_0 - B^*)]^{1/2}$
d	$\frac{3}{2}B^* \le B_0$	$-\frac{1}{2}B^*$	0	$\frac{1}{2}B^*$
e	$B^* \le B_0$	$B_0 - 2B^*$	$B_0 - \frac{3}{2}B^*$	$\frac{1}{2}B^*$

be compared with their experimentally determined low-field counterparts shown in Figs. 10.5 and 10.6; the former figure illustrates the onset of the saturation phenomenon. The high-field hysteresis loops of Fig. 10.7 are saturated over most of their range, but exhibit the additional feature of a discontinuity at the field B_{c1}, which is not taken into account in the Bean model.

If the magnetization is taken around a cycle of the loop, the net work done by the external field, expressed as the energy loss Q per unit volume, is equal to the area enclosed by the loop:

$$Q = \oint M dB. \tag{12.55}$$

Figure 12.24 shows that the area of the low-field hysteresis loop is proportional to its length $\approx \sqrt{2}B_0$ times its width $\approx B_0^2/2B^*$, using values estimated from Table 12.2. The area may be expressed as

$$Q = \frac{2B_0^2}{\mu_0}\Gamma(\beta), \tag{12.56}$$

where the loss factor $\Gamma(\beta)$ has the approximate value

$$\Gamma(\beta) \approx \frac{\beta}{3} \qquad \beta < 1 \tag{12.57}$$

where

$$\beta = \frac{B_0}{B^*}. \tag{12.58}$$

For high fields, $B_0 \gg B^*$, the area of the loop, to a first approximation, equals its height B^* times the horizontal distance $2B_0$ between the points a and a′, which gives $Q \approx 2B_0 B^* = 2B_0^2/\beta$, as may be seen from Fig. 12.22. The correction factor $\approx 4B^{*2}/3$ must be subtracted to account for the rounded part at the ends. Adding these two parts gives the energy loss (12.56), where the energy loss factor $\Gamma(\beta)$ is approximated as

$$\Gamma(\beta) \approx \frac{1}{\beta} - \frac{2}{3\beta^2} \qquad \beta > 1, \tag{12.59}$$

where β is given by Eq. (12.58). Figure 12.25 shows the dependence of the calculated loss factor $\Gamma(\beta)$ on β for the slab

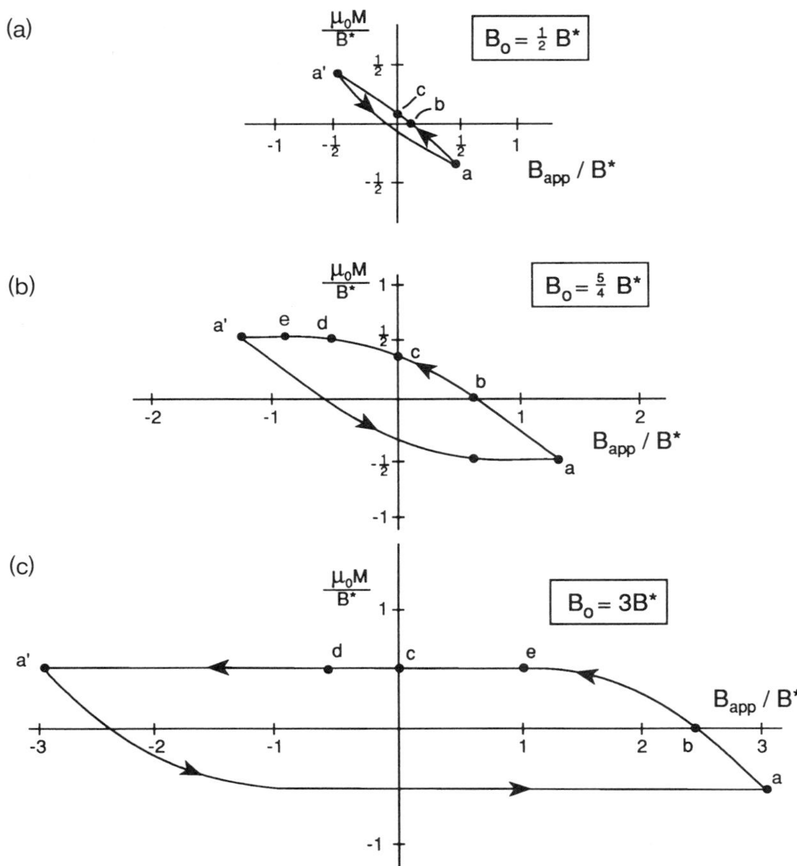

Figure 12.22 Hysteresis loops of magnetization $\mu_0 M$ versus applied magnetic field B_{app} cycled over the range $-B_0 \le B_{app} \le B_0$ for three cases: (a) $B_0 = \frac{1}{2}B^*$, (b) $B_0 = \frac{5}{4}B^*$, and (c) $B_0 = 3B^*$. The magnetization has the values listed in Table 12.2 for the special points a, b, c, d, and e indicated on the loops. These loops are drawn for the same ratios B_0/B^* as the average field loops of Fig. 12.19.

case, which has the limiting approximations (12.57) and (12.59), and also for a cylinder in a parallel and perpendicular field. The low- and high-field approximations for $\Gamma(\beta)$ of the cylinder are as follows:

$$\Gamma(\beta) \approx \frac{2\beta}{3} - \frac{\beta^2}{3} \qquad \beta < 1, \quad (12.60a)$$

$$\Gamma(\beta) \approx \frac{2}{3\beta} - \frac{1}{3\beta^2} \qquad \beta > 1. \quad (12.60b)$$

Figure 12.26 compares calculated values of $\Gamma(\beta)$ for a cylinder with experimental measurements.

E. Magnetization Current

We have seen that for high applied fields satisfying the condition $B_{app} \gg B^*$, the average internal field $\langle B \rangle$ varies between $B_{app} + \frac{1}{2}B^*$ and $B_{app} - \frac{1}{2}B^*$, so that the magnetization $\mu_0 M = \langle B \rangle - B_{app}$ varies over the range $\pm \frac{1}{2}B^*$. We know from Eq. (12.29) that

$$B^* = \mu_0 J_c a. \qquad (12.61)$$

A high-field hysteresis loop provides the difference,

$$M_+ - M_- = J_c a, \qquad (12.62)$$

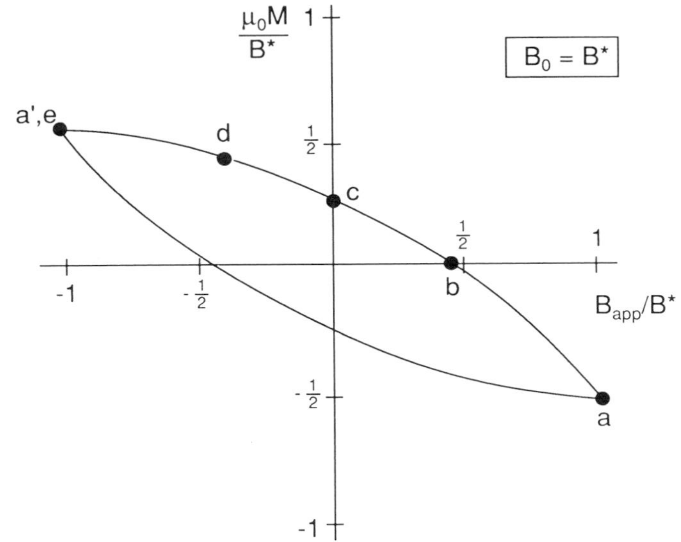

Figure 12.23 Magnetization hysteresis loop for the case $B_0 = B^*$. The average field $\langle B \rangle$ versus B_{app} is plotted in Fig. 12.21 for this same case.

between the upper and lower magnetization plateaux, where

$$\mu_0 M_+ = \tfrac{1}{2} B^*, \qquad (12.63a)$$

$$\mu_0 M_- = -\tfrac{1}{2} B^*, \qquad (12.63b)$$

as indicated in Fig. 12.22c. This gives us an expression for the critical current in terms of measured values of the magnetization through the Bean model formula,

$$J_c = \frac{2(M_+ - M_-)}{d} \qquad (12.64)$$

$$= 1.59 \times 10^6 \frac{\mu_0 \Delta M}{d} \quad (\text{A/m}^2), \qquad (12.65)$$

where $\mu_0 \Delta M = \mu_0(M_+ - M)$ is expressed in teslas while d is the diameter of the sample grains, measured in meters. More precisely, this represents a high-field Bean model formula. Such an indirect method of measuring J_c using hysteresis loops is widely employed (Biggs *et al.*, 1989; Crabtree *et al.*, 1987; Frucher and Campbell, 1989; Kohiki *et al.*, 1990; Kumakura *et al.*, 1987; Nojima and Fujita, 1991; Sun *et al.*, 1987; van den Berg *et al.*, 1989; Xiao *et al.*, 1987b), and constitutes one of the most important applications of the Bean model.

Many authors use electromagnetic units for the magnetization, which corresponds to the expression

$$J_c = \frac{30(M_+ - M_-)}{d} \quad \text{A/cm}^2, \qquad (12.66)$$

where d is now in centimeters.

This method has been widely applied for determining J_c. An example is the measurement by Shimizu and Ito (1989) of the critical currents of 16 $YBa_2Cu_3O_{7-\delta}$ samples with a range of particle diameters from 3 to 53 μm. Figure 12.27 shows some of their hysteresis loops. Shimizu and Ito determined the critical current from the plots of ΔM versus d shown in Figs. 12.28 and 12.29 for three applied field strengths

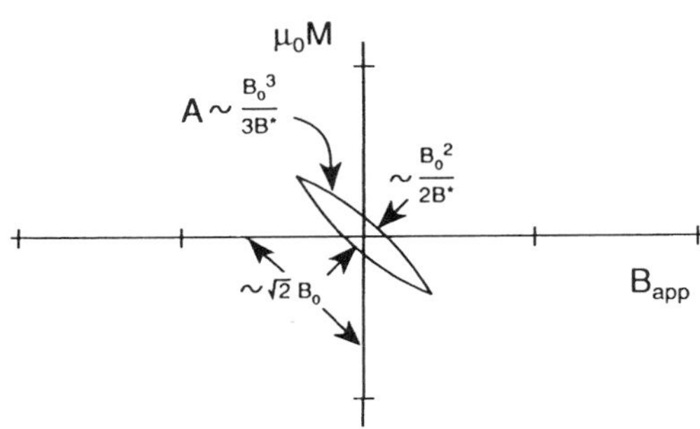

Figure 12.24 Estimating the area of a magnetization hysteresis loop at low field, $B_0 \ll B^*$.

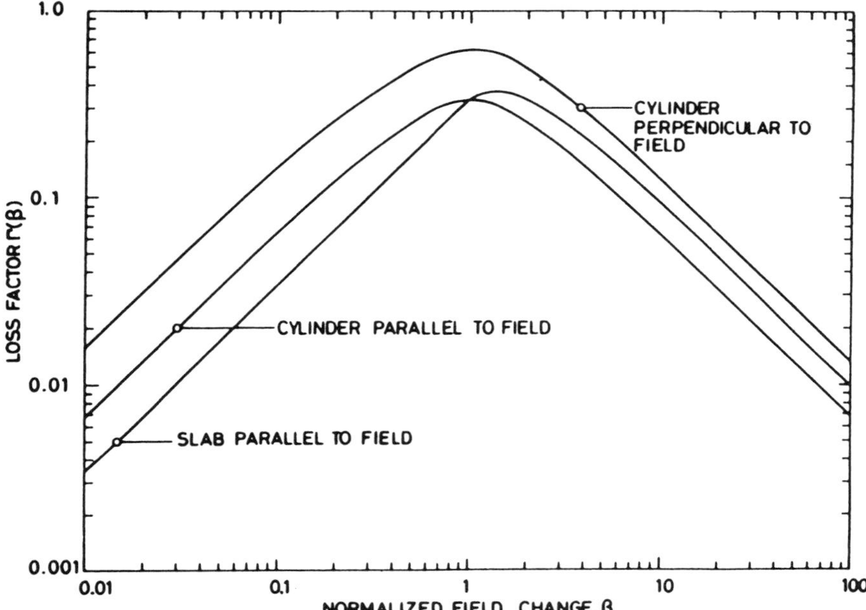

Figure 12.25 Energy loss factor $\Gamma(\beta)$ for different superconductor shapes and orientations in an applied magnetic field. The case of a slab parallel to the applied field is treated in the text, with the low-field side, $B_0 < B^*$ approximated by Eq. (12.57) and the high-field side, $B_0 > B^*$, by Eq. (12.59). Equation (12.60) provides analytical expressions for $\Gamma(\beta)$ in the case of a cylinder oriented parallel to the field (Wilson, 1983, p. 164).

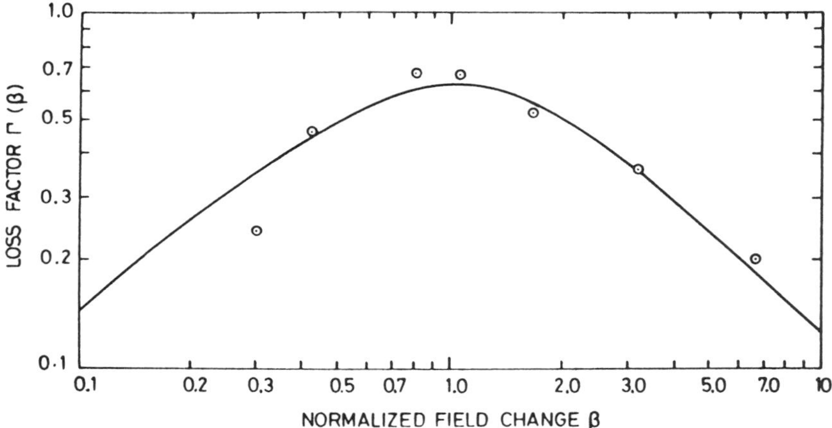

Figure 12.26 Comparison of the calculated loss factor $\Gamma(\beta)$ for a superconducting cylinder in a transverse field, plotted in Fig. 12.25, with experimental measurements (C. R. Walters, in Wilson, 1983, p. 169).

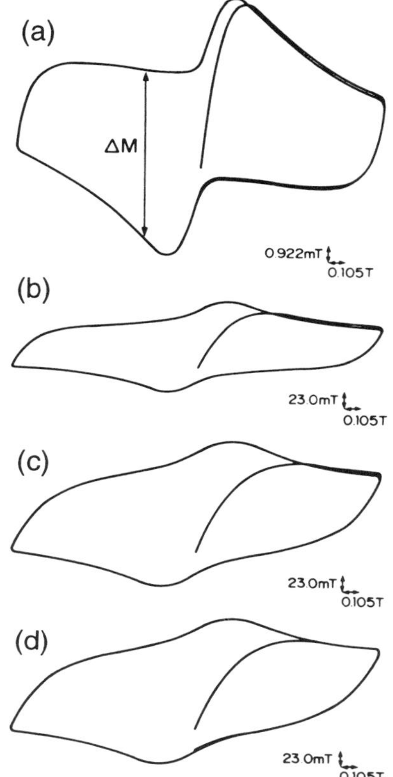

Figure 12.27 Experimental magnetization hysteresis loops of $YBa_2Cu_3O_{7-\delta}$ at 4.2 K used to determine the critical current J_c with powder diameters (a) 3 μm, (b) 15 μm, (c) 36 μm, and (d) 53 μm (Shimizu and Ito, 1989).

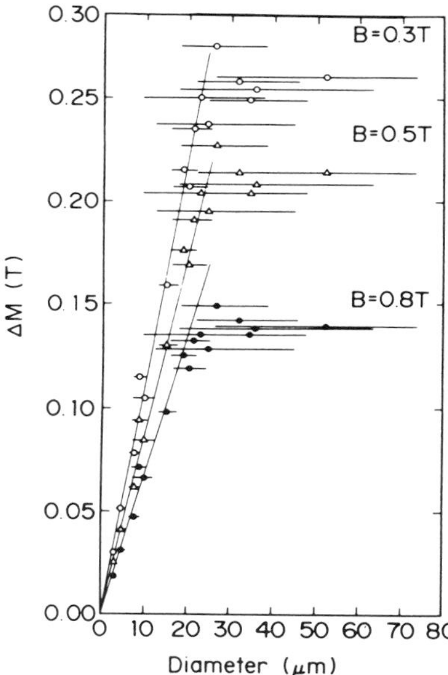

Figure 12.28 Dependence of the magnetization parameter ΔM on the particle diameter d for a series of $YBa_2Cu_3O_{7-\delta}$ samples at 4.2 K. The horizontal lines are a measure of the distribution in diameter, as explained in the text. The critical current is determined from the slope $\Delta M/d$ of the lines on the left using Eq. (12.65) (Shimizu and Ito, 1989).

at temperatures of 4.2 K and 77 K. The horizontal bar through each datum point gives a range of diameters between 25% and 75% of the diameter distribution. The results were $J_c = 2 \times 10^6$ A/cm^2 at 4.2 K in a field of 0.3 T and 7×10^4 A/cm^2 at 77 K in a field of 0.03 T. The experimental results presented in Figs. 12.28 and 12.29 show that, for small particle diameters, the magnetization is indeed proportional to the diameter (Shimizu and Ito, 1989; Tkaczyk *et al.*, 1992; cf. Babic *et al.*, 1992; Dersch and Blatter, 1988), which means, from Eq. (12.63), that the measured critical current is independent of the diameter. We also see from these figures that the magnetiza-

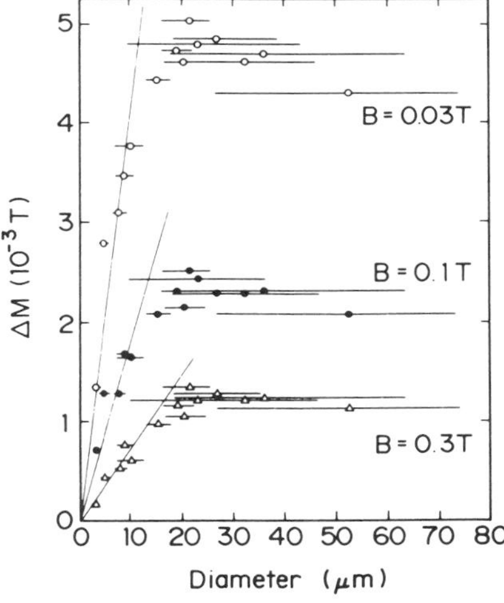

Figure 12.29 Dependence of the magnetization parameter ΔM on the particle diameter d for a series of $YBa_2Cu_3O_{7-\delta}$ samples at 77 K, using the same notation and J_c determination as in Fig. 12.28.

tion saturates for particle diameters greater than 20 μm, suggesting that appreciable magnetization current cannot flow through boundaries more than 20 μm wide. Using the Bean expression (12.64) in this saturation region provides values of J_c that are too low.

F. Critical Currents

Single crystal experiments show that critical currents J_c measured with the applied magnetic field oriented perpendicular to the planes, i.e., in the c direction, are smaller than critical currents J_c measured with the applied field parallel to the planes (Crabtree *et al.*, 1987; Dinger *et al.*, 1987; Worthington *et al.*, 1987). In both cases the current flows perpendicular to the field direction, but in the particular case of a field applied in the c direction the current flow is entirely in the a, b-, or "easy flow" plane. The highest values of J_c, over 10^9 A/cm^2 (Jiang *et al.*, 1991), have been reported for epitaxial thin films grown with the c-axis perpendicular to the plane and for single crystals with macroscopically aligned planes. When twinning is present it interchanges the a and b directions, but the planarity can persist across the twinning interfaces.

Single crystals are far superior current carriers than polycrystalline samples. Grinding, compressing, and heat treatment of polycrystalline samples improves intergrain contact and optimizes oxygen contact, and can also appreciably increase the critical-current density. For a particular sample J_c increases as the temperature is lowered in accordance with Fig. 2.43, so that the values listed for 4.2 K are larger than their 77 K counterparts. The data in

Table 12.3 Critical Current Densities of Monocrystals (MC) and Epitaxial Thin Films (EF), where 1 MA$=10^6$ A, together with Measurement Temperatures and Values of the Applied Magnetic Fields[a]

Material	Type	J_c (MA/cm^2)	T_{meas} (K)	B_{app} (T)	Comments	References
$YBa_2Cu_3O_{7-\delta}$	MC	1.4	5	0–1	$B_{app} \perp a, b$-plane	Crabtree *et al.* (1987)
$YBa_2Cu_3O_{7-\delta}$	MC	0.01	77	0.1	$B_{app} \perp a, b$-plane	Crabtree *et al.* (1987)
$YBa_2Cu_3O_{7-\delta}$	MC	0.0043	77	1.0	$B_{app} \perp a, b$-plane	Crabtree *et al.* (1987)
$YBa_2Cu_3O_{7-\delta}$		0.059	5	0.4	B_{app} in a, b-plane	Song *et al.* (1987)
$YBa_2Cu_3O_{7-\delta}$		0.044	5	0.4	$B_{app} \perp a, b$-plane	Song *et al.* (1987)
$YBa_2Cu_3O_{7-\delta}$	MC	0.16	4.5	~0	B_{app} in a,b-plane	Worthington *et al.* (1987)
$YBa_2Cu_3O_{7-\delta}$	MC	3.2	4.5	~0		Worthington *et al.* (1987)
$YBa_2Cu_3O_{7-\delta}$	EF	60	4.2	0		Roas *et al.* (1990)
$YBa_2Cu_3O_{7-\delta}$	EF	40	4.2	8	B_{app} in a, b-plane	Roas *et al.* (1990)
$YBa_2Cu_3O_{7-\delta}$	EF	6	4.2	8	$B_{app} \perp a, b$-plane	Roas *et al.* (1990)
$YBa_2Cu_3O_{7-\delta}$	EF	5	77	0		Roas *et al.* (1990)
$YBa_2Cu_3O_{7-\delta}$	EF	0.17	77	8	B_{app} in a, b-plane	Roas *et al.* (1990)
$YBa_2Cu_3O_{7-\delta}$	EF	0.04	77	5.5	$B_{app} \perp a, b$-plane	Roas *et al.* (1990)
$YBa_2Cu_3O_{7-\delta}$	MC	0.01	77	0	as-made xtal	vanDover *et al.* (1990)
$YBa_2Cu_3O_{7-\delta}$	MC	0.2	77	0	after irradiation	vanDover *et al.* (1990)
$YBa_2Cu_3O_{7-\delta}$	MC	20	4	0	after irradiation	vanDover *et al.* (1990)
$YBa_2Cu_3O_{7-\delta}$	EF	1300	4	0	500-nm thick	Jiang *et al.* (1991)
$Bi_2Sr_2CaCu_2O_8$	EF	9	4	0		Schmitt *et al.* (1991)
$Bi_2Sr_2CaCu_2O_8$	EF	10	30	0		Yamasaki *et al.* (1993)
$Bi_2Sr_2CaCu_2O_8$	EF	0.1	77	8		Yamasaki *et al.* (1993)
$HgBa_2CuO_{4+\delta}$		0.07	20	0.8	grain aligned	Lewis *et al.* (1993)

[a] See Table X-2 of Poole *et al.* (1988) for additional data.

Table 3 demonstrate that applied magnetic fields decrease the critical-current density.

Very high values of J_c are needed for magnet materials. A niobium–titanium filament has been reported with J_c as high as 3.7×10^5 A/cm^2 at 5 T (Chengren and Larbalestier, 1987). The Superconducting Super Collider and Relativistic Heavy Ion Collider accelerators require 5-μm wire filaments that support J_c of at least 2.8×10^5 A/cm^2 in a 5 T magnetic field (Gregory *et al.*, 1987). At present, J_c values of technologically suitable oxide materials at 77 K are too small for such high-field magnet applications. Although the oxide superconductors do no yet compete with the older superconductor in terms of critical currents, they are superior in terms of critical field (B_{c2}) capability, as shown in Fig. 9.8.

Some workers have reported that values of J_c determined directly from transport currents are smaller than those that have been determined indirectly from magnetization currents deduced from the hysteresis loops (Hampshire *et al.*, 1988; Larbalestier *et al.*, 1987a; Togano *et al.*, 1987). In high-temperature superconductors this could be caused by granularity and intergrain contact, where the transport current J_c is limited by intergranular effects and the shielding current J_c from the magnetization measurements limited by intragranular effects (Thompson *et al.*, 1989). Comparable values of magnetization and transport critical currents have been reported for epitaxial thin films.

G. Reversing Fields with Transport Currents

In Section V.E we discussed the simultaneous presence of magnetization and transport current in a sample. In this section we will discuss changes that occur in the internal field and current density configurations when changes are made in the applied field or in the transport current. Two examples will be presented.

Consider the case of a superconducting slab in a magnetic field carrying both transport current and flux-shielding magnetization current. We showed in Section V.E that $B_z(x)$ and $J(x)$ have the configurations that are sketched in Fig. 12.12 and repeated in Fig. 12.30a. If the external field is decreased, the internal field and current distributions change in the manner illustrated in Fig. 12.30b, which has regions of both trapped and shielded flux. Figure 12.30c shows the situation when the field has decreased sufficiently to have trapped all of the internal flux, and the current flow is now greater on the left side of the slab instead of on the right. Figure 12.30d shows that further decreases in that applied field have no effect on the current flow. The ratio of forward to backward flowing current remains the same throughout this field-decreasing process because the same amount of transport current is flowing. Physically, the change corresponds to a reversal in direction of the flow of the magnetization current, since it now produces flux trapping rather than flux shielding.

The second case to consider is that of a sample in a magnetic field that contains shielded flux. If a transport current is added, the field and current readjust to the distributions shown in Fig. 12.31. There are, of course, many other ways for the magnetization and transport currents to combine together besides the two cases which we have presented as representative examples.

VII. KIM MODEL

The two original critical-state models, which date back to 1962, are the Bean model, which we discussed at length in Section V and the Kim model (Anderson, 1962; Bean, 1962, 1964; Chaddah *et al.*, 1989; Chen and Goldfarb, 1989; Müller, 1989). The Kim model assumes that the critical-current density $J_c(B_z)$ depends on

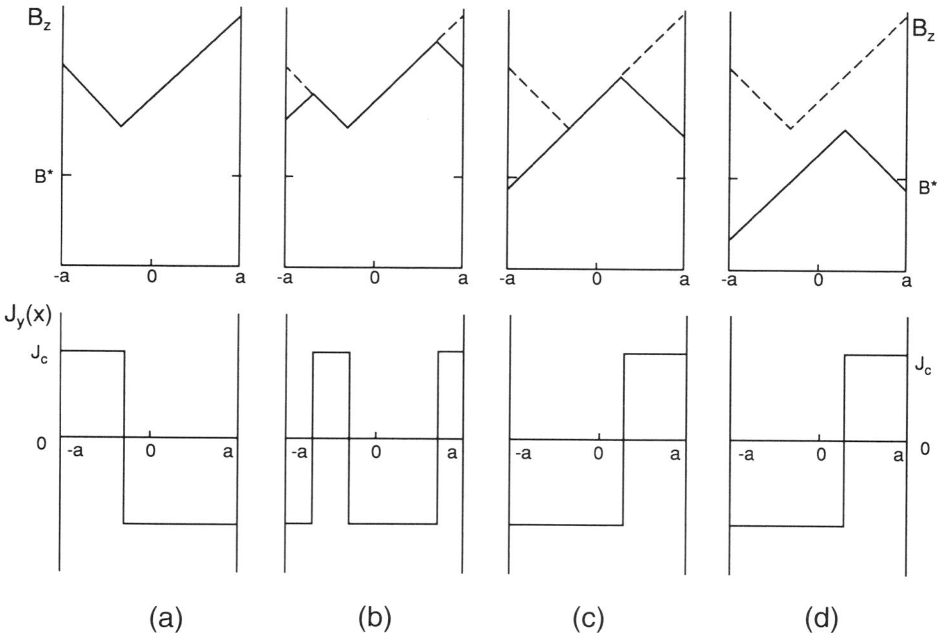

Figure 12.30 Changes in the combined transport current–applied field configuration shown in Fig. 12.12 when the applied field is gradually decreased in the sequence from left to right, as explained in the text. The overall effect on the current density is to reverse the locations of the forward and backward flowing current, while maintaining the ratio between them constant (adapted from Wilson, 1983, p. 171).

the internal magnetic field B_z through Eq. (12.12),

$$J_c(B_z) = \frac{J_c B_K}{B_K + B_z(x)}, \quad (12.67)$$

where, again, the slab geometry of Fig. 12.2 is assumed. This expression contains two parameters, the critical current J_c and a magnetic field parameter B_K. The other two models that we have examined each depended on only a single parameter —namely, J_c for the Bean model and F_p for the Fixed Pinning model. Most of the other critical-state models introduced in Section III.B in Eqs. (12.9)–(12.17) also require two parameters for their specification.

If we follow the usual procedure of substituting $J_c(B_z)$ from Eq. (12.67) in Eq. (12.2), solving the resultant differential equation, and then substituting the solution $B_z(x)$ in Eqs. (12.67) and (12.6), subject to the usual boundary conditions (12.20), we obtain the expressions

$$B_z(x) = B_K\left\{\left[1 + \beta\left(\frac{x + a'}{a}\right)\right]^{1/2} - 1\right\} \qquad -a \le x \le -a', \quad (12.68a)$$

$$B_z(x) = 0 \qquad -a' \le x \le a', \quad (12.68b)$$

$$B_z(x) = B_K\left\{\left[1 + \beta\left(\frac{x - a'}{a}\right)\right]^{1/2} - 1\right\} \qquad a' \le x \le a, \quad (12.68c)$$

$$J_y(x) = \frac{J_c}{[1 + \beta\{(x + a')/a\}]^{1/2}} \qquad -a \le x \le -a' \quad (12.69a)$$

$$J_y(x) = 0 \qquad -a' \le x \le a' \quad (12.69b)$$

$$J_y(x) = -\frac{J_c}{[1 + \beta\{(x - a')/a\}]^{1/2}} \qquad a' \le x \le a \quad (12.69c)$$

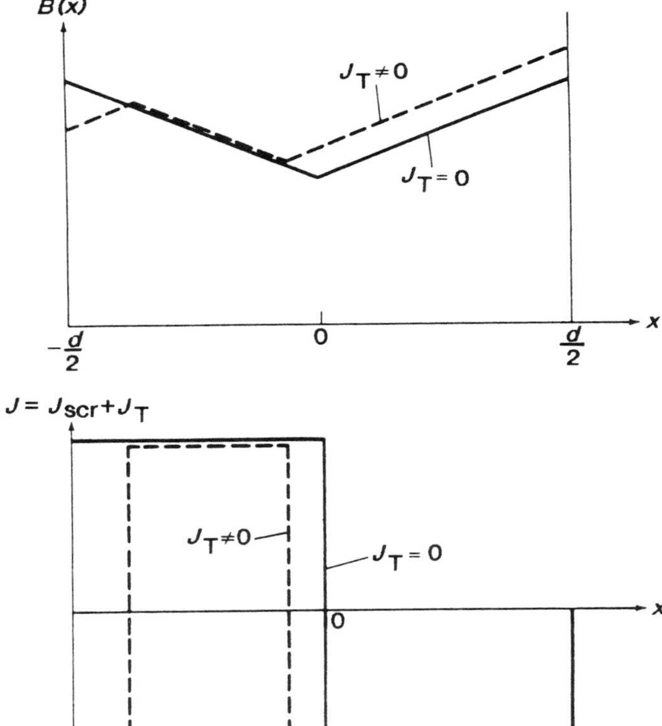

Figure 12.31 Effect of adding transport current to an already magnetized slab for the two conditions $B_{\text{app}} > 2B^*$ and $I_{\text{tr}} < 4aLJ_c$ (cf. Eq. (12.34), and the configuration of Fig. 12.11). The internal field $B_z(x)$ and current density $J_y(x)$ change from their initial $I_{\text{tr}} = 0$ distributions (———) to a new configuration (– – –), as shown (Van Duzer and Turner, 1981, p. 339).

Equation (12.19) provides the pinning forces

$$F_p(x) = B_K[J_c - J_y(x)] \qquad -a \le x \le -a' \tag{12.70a}$$

$$F_p = 0 \qquad -a' \le x \le a' \tag{12.70b}$$

$$F_p(x) = -B_K[J_c - J_y(x)] \qquad a \le x \le a', \tag{12.70c}$$

where β is given by

$$\beta = \frac{2\mu_0 J_c a}{B_K}. \tag{12.71}$$

Equations (12.68) and (12.69) are plotted in Figs. 12.32 and 12.33, respectively, for three values of β, with x in the range $-a \le x \le -a'$.

The expressions in (12.68)–(12.70) are valid for applied fields so weak that the internal field and currents do not reach the center of the slab. The boundary condition (12.20a) requires that $B_z(x) = B_0$ for $x = a$, so that from Eq. (12.68c) we have

$$B_0 = B_K\{[1 + \beta((a - a')/a)]^{1/2} - 1\}. \tag{12.72}$$

The applied field B^* at which the internal field reaches the center is determined by setting $a' = 0$ in this expression, to give

$$B^* = B_K\left[(1+\beta)^{1/2} - 1\right]. \quad (12.73)$$

Therefore, the low-field condition $B_{app} < B^*$ ensures that Eqs. (12.68)–(12.70) are valid, assuming there exists a field-free region at the center ($-a' < x < a'$). We will not bother to write down the fields and currents for the high-field case $B_{app} > B^*$.

If we compare the expressions for the Kim model, Eqs. (12.68)–(12.70), with their Bean model and Fixed Pinning model counterparts, Eqs. (12.25)–(12.28) and Eqs. (12.21)–(12.24), respectively, we see that the Kim model reduces to the Bean model in the limit $\beta \ll 1$ and to the Fixed Pinning model in the limit $\beta \gg 1$. This can be inferred from Figs. 12.32 and 12.33, which compare the Kim model for three different values of β with the Bean and Fixed Pinning models. Another parameter for comparing the models is the ratio of the current flow at the center $J_y(0)$ to the current flow at the surface $J_y(a)$ when $B_{app} = B^*$. For the Kim model it is given by

$$J_y(0)/J_y(a) = (1+\beta)^{1/2}. \quad (12.74)$$

This ratio has a value of 1 for the Bean model and infinity for the Fixed Pinning model.

The parameters defined by Eqs. (12.71), (12.73), and (12.74) can be used to compare the various models introduced in Section III.B. Table 12.4 enumerates these parameters for five of the models.

The advantages of the Kim model, particularly for the study of intergranular

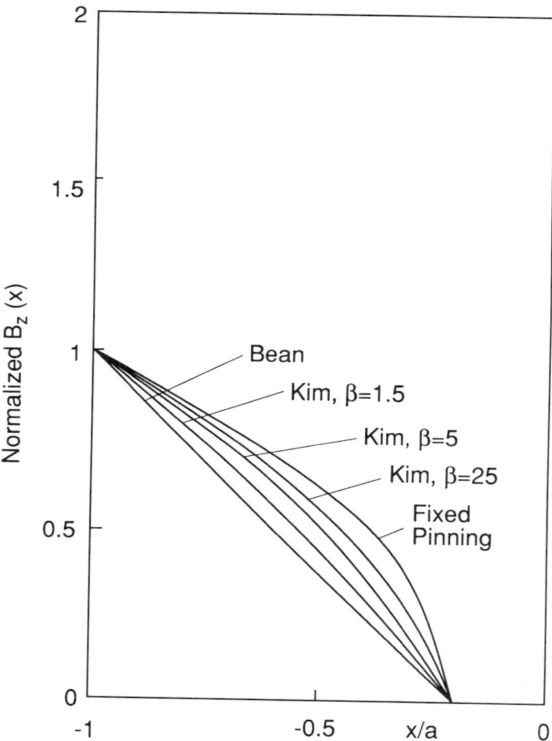

Figure 12.32 Dependence of the internal field $B_z(x)$ on distance x into a zero-field-cooled superconductor in the range $-a < x < 0$ for $a' = 0.2a$. Curves are shown for the Bean and Fixed Pinning models, as well as for the Kim model with $\beta = 1.5$, 5, and 25. Compare Fig. 12.3.

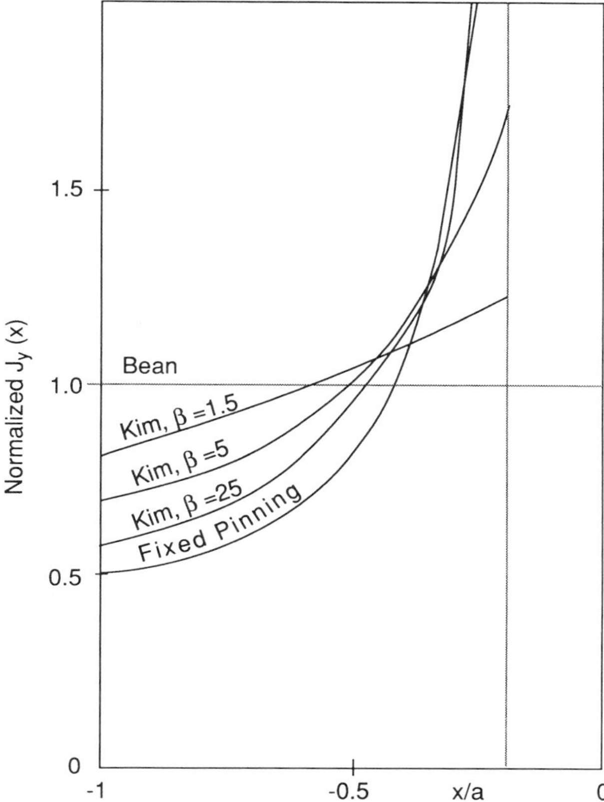

Figure 12.33 Dependence of the critical-current density $J_y(x)$ on distance x into a zero-field-cooled superconductor in the range $-a < x < 0$ for $a' = 0.2a$. Curves are shown for the Bean and Fixed Pinning models, as well as for the Kim model with $\beta = 1.5$, 5, and 25. Compare with Fig. 12.4.

current flow and ac susceptibility determinations, have been discussed (Ishida and Goldfarb, 1990; Ji *et al.*, 1989; Müller, 1989; Pan and Doniach, 1994; cf. Sanchez *et al.*, 1991).

VIII. COMPARISON OF CRITICAL-STATE MODELS WITH EXPERIMENT

The Bean model makes a simplifying assumption about the way in which the magnetic field penetrates a superconductor which is qualitatively in agreement with the experimental observations of magnetic field penetration that are presented in Figs. 9.21 and 9.22. We see from the first of these figures that the lowest applied field, 15 mT, only penetrates a short distance, $\approx 50\ \mu$m, and that higher fields penetrate further, until the center is attained. Even higher fields exist throughout the cross section, but are stronger closer to the surface. Figure 9.22 shows that removing the applied field leaves a remanent state in which there is more flux trapped in the center than near the outer surface, as predicted by the Bean model. There are also features of the measured fields of these figures that do not conform to the Bean model, such as the peak at the surface of the sample, the decrease in the slope as the applied field is increased, and the drop in the value of the internal field as the

Table 12.4 Characteristics of Several Critical-State Models[a]

Model	$J(B)$	β	B^*	$J(0)/J(a)$
Bean	J_c	—	$B_K = \mu_0 J_c a$	1
Fixed Pinning	$\dfrac{J_c B_K}{\lvert B\rvert}$	—	$B_K = \mu_0 J_c a$	∞
Kim	$\dfrac{J_c}{1 + \dfrac{\lvert B\rvert}{B_K}}$	$\dfrac{\mu_0 J_c a}{B_K}$	$B_K[(1+\beta)^{1/2} - 1]$	$(1+\beta)^{1/2}$
Exponential	$J_c \exp\left(-\dfrac{\lvert B\rvert}{B_K}\right)$	$\dfrac{\mu_0 J_c a}{B_K}$	$B_k \ln(1+\beta)$	$1+\beta$
Triangular Pulse	$J_c\left(1 - \dfrac{\lvert B\rvert}{B_K}\right)\Theta(B_K - \lvert B\rvert)$	$\dfrac{\mu_0 J_c a}{B_K}$	$B_K(1 - e^{-\beta})$	e^{β}

[a] $J(B)$, the local magnetic field dependence of the critical current; β, model parameter; B^*, applied magnetic field at which full penetration occurs; and $J(0)/J(a)$, ratio of critical current in the center to critical current at the surface when $B_{app} = B^*$.

center of the sample is approached. Nevertheless, despite this lack of agreement in details, the qualitative agreement with the experimental measurements is good.

Another test of the validity of critical-state models is provided by magnetization and susceptibility measurements. We have already discussed the widespread use of the Bean model magnetization expression (12.65) for the experimental determination of critical currents.

Lera *et al.* (1992) found that the Kim, exponential, and quadratic models fit the ac susceptibility data of $YBa_2Cu_3O_{7-\delta}$ well with the same value of J_c, but with different values of the characteristic fields. D.-X. Chen and Sanchez (1991) found that the exponential model gave a better fit to such data.

Figure 12.34 compares the hysteresis loops calculated from the Bean, Kim, and exponential models (Chen *et al.*, 1990a; Yamamoto *et al.*, 1993). The low-field loops of all three models are similar and also resemble some of the experimental loops of Figs. 10.5 and 10.6. The high-field Kim model loop looks like the 20-K experimental loop of Fig. 10.7. The highest field Bean model hysteresis loop is somewhat similar to the highest field loop of Fig. 10.5.

A Type II superconductor in an ac magnetic field generates harmonic components of magnetization. These were studied by Bean (1964) and led him to propose his famous model for explaining the production of odd harmonics by conventional superconductors. High-temperature superconductors in the presence of dc and ac fields also generate harmonics (Chen and Goldfarb, 1989; Ghatak *et al.*, 1992; Johnson *et al.*, 1991), including even harmonics that cannot be accounted for by the Bean model. Ji *et al.* (1989) extended the model to explain the presence of even harmonics. A crossover between Bean model intergranular behavior at low-ac fields to Kim model intragranular behavior for high fields has been discussed (Ghatak *et al.*, 1992; Ji *et al.*, 1989; Müller, 1991).

The Bean model has also been used to determine the anisotropic lower-critical field B_{c1} of $(La_{1-x}Sr_x)_2CuO_4$ (Krusin-

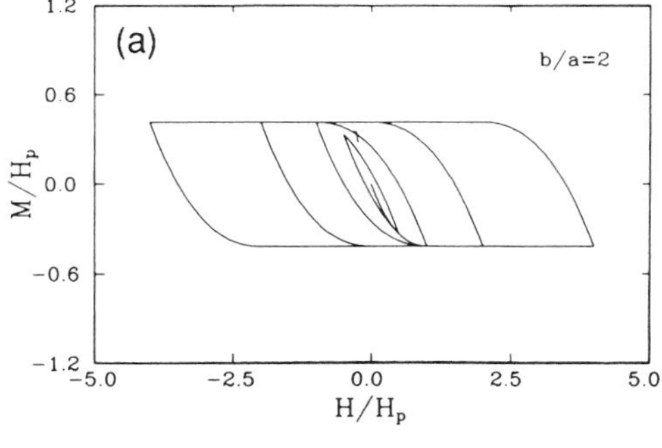

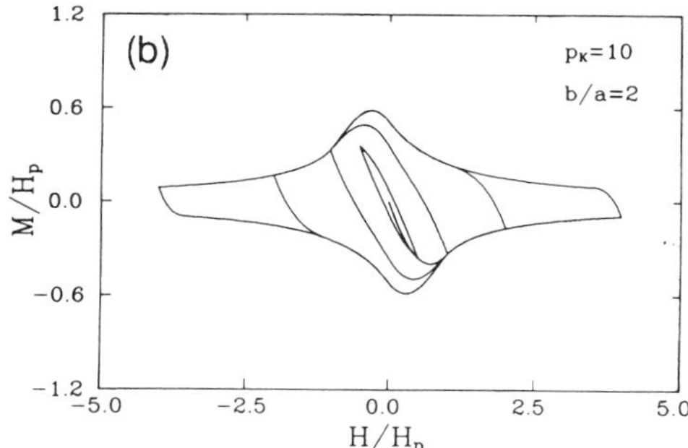

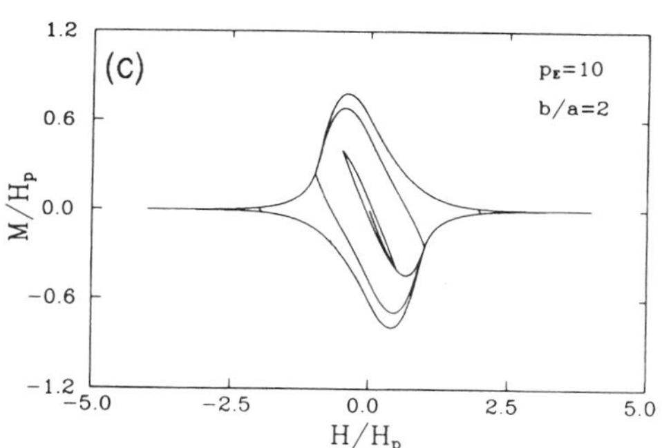

Figure 12.34 Comparison of the magnetization hysteresis loops calculated using the (a) Bean model, (b) Kim mode, and (c) exponential model (from Chen *et al.*, 1990a; vide also Chen and Goldfarb, 1989; Karasik *et al.*, 1971).

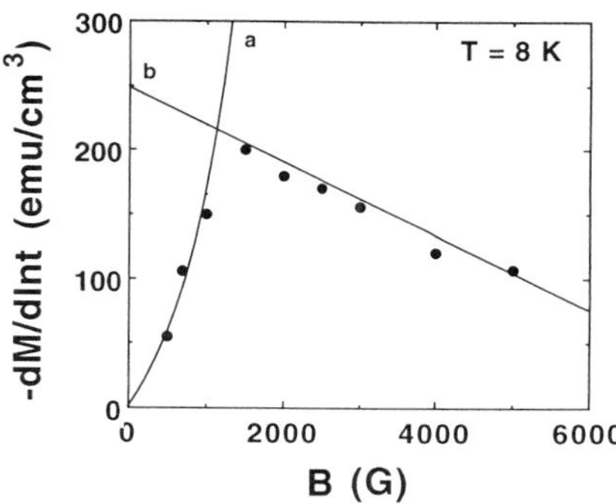

Figure 12.35 Plot of $dM/d(\ln t)$ versus the applied magnetic field for a $Bi_2Sr_2CaCu_2O_x$ monocrystal. The solid lines are fits of the Kim model for B_{app} below (a) and above (b) the full penetration field B^* to experimental data points (Shi *et al.*, 1990b).

Elbaum *et al.*, 1989; Naito *et al.*, 1990) and to explain microwave absorption in superconductors (Portis *et al.*, 1988).

Critical-state models have been employed to explain magnetic relaxation, in which plots of the magnetization tend to exhibit logarithmic dependences on time, as shown in Fig. 9.34 (Kung *et al.*, 1992; Schnack and Griessen, 1992; Shi *et al.*, 1990b; Vinokur *et al.*, 1991). The logarithmic time derivative of the magnetization $dM/d(\ln t)$ increases with increasing field for $B_{app} < B^*$, exhibiting a slower decrease for $B_{app} > B^*$, where B^* is the value of the applied field for the onset of complete flux penetration of the sample. Figure 12.35 shows that Kim model expressions for these two regions provide good fits to experimental magnetic relaxation data on a $Bi_2Sr_2CaCu_2O_x$ monocrystal (Shi and Xu, 1991; Shi *et al.*, 1990b).

IX. CONCLUDING REMARKS

The Bean model qualitatively reproduces many of the experimentally observed properties of Type II superconductors, such as the hysteresis loops discussed in Chapter 10, Section IV. It is also widely used in contemporary research for the understanding and interpretation of experimental data. One example is work on time-dependent effects, such as flux creep and magnetic relaxation phenonema. Perhaps the most important application is the determination of critical currents by magnetization measurements, as was explained in Section VI.E.

FURTHER READING

The original references (Bean, 1962, 1964, and Kim *et al.*, 1962, 1963) are recommended reading.

The Bean model is also described in several texts (Orlando and Delin, 1991; Rose Innes and Rhoderick, 1994; Tilley and Tilley, 1986; Tinkham, 1985) and monographs (Hueberer, 1979; Van Duzer and Turner, 1981; Wilson, 1983). References for other critical-state models are given in Section III.

PROBLEMS

1. Find solutions analogous to Eqs. (12.31) and (12.32) for the Kim model in the high-field case; in this case there is no current-field-free region in the center. Show that this occurs for the condition $B_0 > \mu_0 J_c a$.
2. Show how to formulate the Kim model in terms of a field B^* analogous to the field B^* of Eqs. (12.30), and derive expressions for $B_z(x)$, $J_y(x)$, and $F_p(x)$ for the two conditions $B_0 < B^*$ and $B_0 > B^*$.
3. Show that the expression [Eqs. (12.23) and (12.27)]
$$J_c = \frac{B_0}{\mu_0(a-a')}$$
is valid for both the low-field Kim model and the low-field Bean model. Write down the corresponding expressions for the pinning forces in the high-field Bean model.
4. Show that the condition
$$\frac{I}{L} = \int J_y(x)dx = J_c(a-a')$$
leads to the definition (12.23) of J_c.
5. Draw figures analogous to those shown in Figs. 12.6a, 12.6b, and 12.6c for the Kim model.
6. Draw figures analogous to all those shown in Figs. 12.8a and 12.8b for the Kim model.
7. Derive Eq. (12.45),
$$\Delta B = \mu_0 J_c(x_R - x_L),$$
where the notation is given in Fig. 12.12.
8. The applied field B_{app} is increased from 0 to the value B_0, where $0 < B^* < B_0$. Show that if it is then decreased down into the range
$$-B_0 < B_{app} < B_0 - 2B^*,$$
the average field will have the value
$$\langle B \rangle = B_{app} + \tfrac{1}{2}B^*.$$
What is the magnetization?
9. Write down analytic expressions for the magnetic field $B_x(x)$ of Fig. 12.18 over the full range of x values.
10. The applied field B_{app} is increased from 0 to the value B_0, where $0 < B_0 < B^*$. Show that if it is then decreased down into the range $0 < B_{app} < B_0$, the magnetization will be given by
$$\mu_0 M = B_{app} + \frac{(B_0 - B_{app})^2}{4B^*} - \frac{B_0^2}{2B^*}.$$
What is the average field $\langle B \rangle$?
11. The applied field B_{app} is increased to the value $3B^*$, then decreased to $2B^*$, and increased again to $2.5B^*$. Find the average field $\langle B \rangle$ and the magnetization M.
12. Calculate the magnetizations associated with the magnetic field configurations of Fig. 12.22a, 12.22b, and 12.22c.
13. Show that when $B_0 = B^*$, the point d at which $\langle B \rangle = 0$ occurs at the position $B_{app} = (1 - \sqrt{2})B^*$ (cf. Fig. 12.21).
14. Show that when $B_0 = B^*$, points d and e occur at the same spot on the hysteresis loop.
15. For what values of the ratio B_0/B^* will point e be found between points a′ and d in Fig. 12.19b, and when will it be found between d and c?
16. Write Eq. (12.54) for the applied field range
$$-B_0 < B_{app} < -B_0 + B^*.$$
17. Find the internal magnetic fields and pinning forces associated with the exponential model.
18. Justify the form of $\Gamma(\beta)$ in Eqs. (12.60a) and (12.60b).

13

Tunneling

I. INTRODUCTION

In the previous chapter we introduced critical-state models with an emphasis on the Bean model. This gave us a chance not only to provide simple explanations for some of the magnetic phenomena that had been discussed in Chapters 9–11, but also to discuss critical currents and, thereby, introduce the area of transport properties, the subject of the present and succeeding chapter. Transport properties are of importance because the principal applications of superconductors are based upon taking advantage of the ability of superconductors to carry electric current without any loss.

The chapter begins with a discussion of tunneling and super current flow in the absence of externally applied fields. After covering introductory material on tunneling, we will discuss the Josephson effect and macroscopic quantum phenomena. It will be shown how tunneling measurements provide energy gap values. The following chapter will examine several transport processes that involve applied fields and thermal effects.

II. THE PHENOMENON OF TUNNELING

Tunneling, or barrier penetration, is a process whereby an electron confined to a region of space by an energy barrier is nevertheless able to penetrate the barrier through a quantum mechanical process and emerge on the other side. The example shown in Fig. 13.1 involves electrons with kinetic energy $E_{\mathrm{KE}} = \frac{1}{2}mv^2$ confined to remain on the left side of a barrier by the potential V_{b}, where $\frac{1}{2}mv^2 < eV_{\mathrm{b}}$. We show

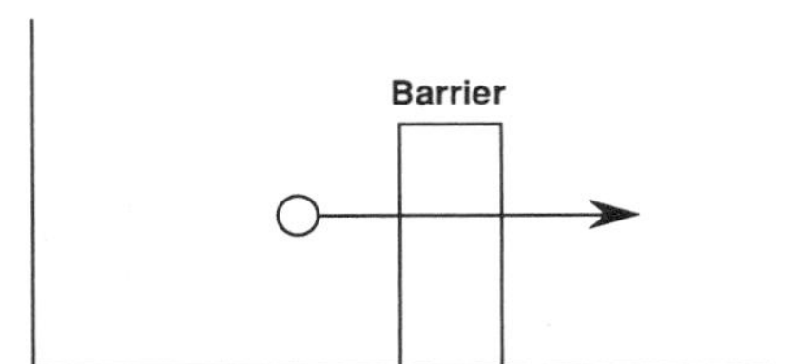

Figure 13.1 Tunneling of electrons through a barrier when the kinetic energy of the electrons is less than the barrier energy eV_b.

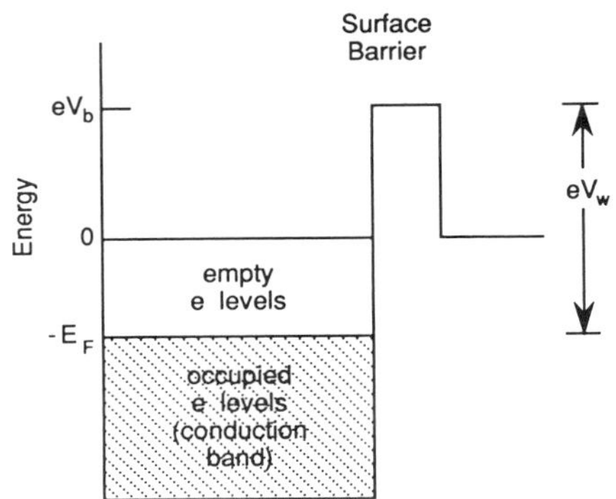

Figure 13.2 Energy-level diagram of a conductor showing the levels occupied below the Fermi energy E_F and the energy barrier eV_b at the surface. The minimum energy (work function) eV_w for extracting an electron is also indicated.

in this figure an electron tunneling through the barrier to the right side where it ends up with the same kinetic energy. Such a phenomenon can occur because there is a quantum mechanical probability per unit time that the electron will penetrate the barrier and escape, as explained in standard quantum mechanics texts. Tunneling phenomena are rather common in physics. For example, radioactive decay of nuclei is explained by a barrier penetration model, with half-lives varying from less than nanoseconds to many centuries.

A. Conduction-Electron Energies

The surface of a normal metal has a dipole charge density layer that produces a barrier potential V_b. The conduction electrons inside the metal move in a region where they experience an attractive potential, so that their energies are negative with the value $-|E_F|$, as indicated in Fig. 13.2. At absolute zero the conduction-band levels are filled up to the Fermi level E_F and are empty above, corresponding to Fig. 1.4a. In some of the figures dark shading is used to indicate occupied levels.

To remove an electron from the interior of a metal one must apply a potential equal to or greater than the work function potential V_w. The minimum energy eV_w that can extract an electron is

$$eV_w = eV_b + |E_F|. \tag{13.1}$$

Metals differ in their eV_w, eV_b, and E_F values, so that a proper treatment of how an electron is transferred between two metals in contact through an insulating barrier should take these factors into account. However, to simplify the mathematics we will ignore these surface potential effects and assume that two metals in contact at the same potential have the same Fermi energy.

Tunneling phenomena are sensitive to the degree of occupation of the relevant energy levels by electrons. Hence, in energy level diagrams it is helpful to include information about the occupation of the levels involved in the tunneling. Figure 1.4b plots the temperature dependence of the Fermi–Dirac distribution function, giving the fractional occupation of levels in the conduction band. We have replotted this function in Fig. 13.3 with energy as the ordinate, $f(E)$ as the abscissa, and electron occupation indicated by shading. Figure 13.4a presents a sketch of a conduction band that is filled at absolute zero and separated from an upper energy band by a gap. Figure 13.4b shows this same diagram at a finite temperature, combined with the distribution function plot of Fig. 13.3 to show the level populations.

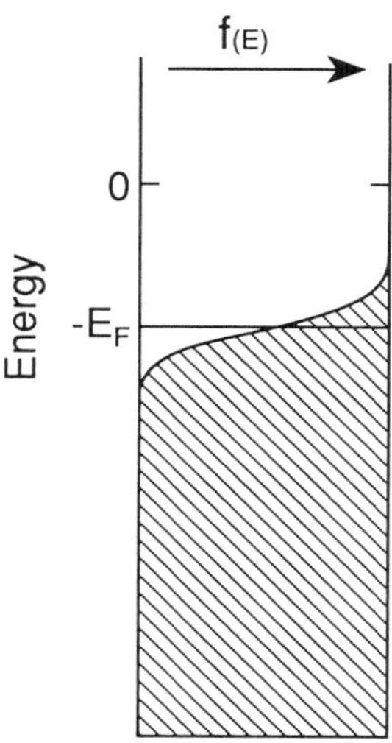

Figure 13.3 Energy-level diagram of a conductor with electron distribution $f(E)$ at a finite temperature near the Fermi level E_F plotted at the top. The shading in this and subsequent figures indicates electron occupancy.

B. Types of Tunneling

Tunneling can occur through an insulating layer, I, between two normal materials (N–I–N), such as semiconductors, between a normal metal and a superconductor (N–I–S), or between two superconductors (S–I–S). Proximity junctions (S–N–S), in which the Cooper pair and quasiparticle transfer across the junction via the proximity effect, are discussed in Section VI.F. Junctions involving semiconductors, such as the S-Semicond and S-Semicond-S types, will not be discussed here (Furusaki *et al.*, 1991, 1992; Kastalsky *et al.*, 1991; van Wees *et al.*, 1991). The dc and ac Josephson effects involve particular types of tunneling phenomena across a barrier between two superconductors. In the next several sections we will examine energy level diagrams, and then provide a qualitative picture of various tunneling processes, concluding with a more quantitative presentation.

III. ENERGY LEVEL SCHEMES

Before we discuss the N–I–N, N–I–S, and S–I–S types of tunneling it will be instructive to examine the energy level systems that are involved in each. Two conventions for representing the energy levels will be introduced, called, respectively, the *semiconductor representation* and the *Bose condensation representation.*

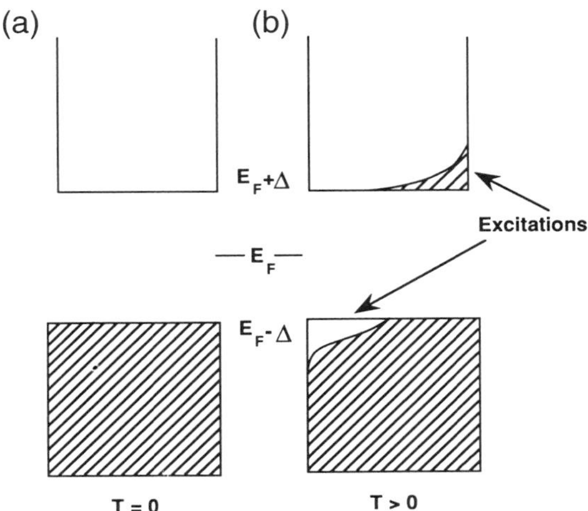

Figure 13.4 Semiconductor representation of the energy level occupancy of a superconductor (a) at $T = 0$, and (b) at $T > 0$. The band gap 2Δ and level populations are shown for each case, using the convention of Fig. 13.3.

A. Semiconductor Representation

A superconductor is considered to have an energy gap $E_g = 2\Delta$ between a lower energy band which is full of super electrons at absolute zero and an upper energy band which is empty at that temperature, as shown in Figs. 13.4a. At higher temperatures some of the electrons are raised from the lower band to the upper band as illustrated in Fig. 13.4b; these excited electrons are often referred to as *quasiparticles*. They act like normal conduction electrons, that is, they behave approximately like free electrons moving at the Fermi velocity v_F, as discussed in Chapter 1, Section II. When an electron jumps down from the bottom of the quasiparticle band to the top of the super electron band, its energy falls by the amount 2Δ. Following conventional semiconductor terminology, we assign the equivalent Fermi energy to the center of the gap. This single-electron picture, called the semiconductor representation of a superconductor, does not take into account the phenomenon of electron pairing.

B. Boson Condensation Representation

Another way of representing a superconductor at $T = 0$ is by a single level for the super electrons, as shown in Fig. 13.5a. This is justified by the argument that the Cooper pairs which occupy this level are paired electrons with zero spin, and hence are boson particles which obey Bose–Einstein statistics. For bosons there is no Pauli exclusion principle, so it is possible for all of them to have the same energy. Thus the transition to the superconducting state is an example of boson condensation, a phenomenon that is explained in quantum mechanics texts. The condensation takes place when the electrons drop into the single-superconducting level where they exist as Cooper pairs, as shown in Fig. 13.5a. This mode of presenting the energy level diagram is called the *boson condensation representation*.

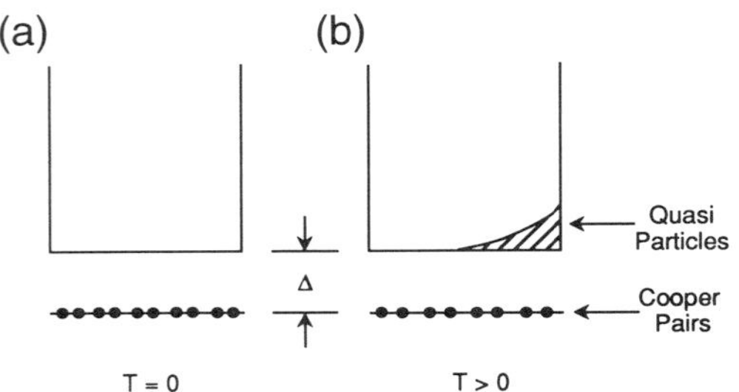

Figure 13.5 Boson condensation representation of the energy level occupancy of a superconductor (a) at $T = 0$, and (b) at $T > 0$, showing the level populations in the quasiparticle band for each case. Figure 13.4 presents the corresponding semiconductor representation.

The Cooper-pair binding energy E_g is shared by two electrons, so that $\Delta = \frac{1}{2}E_g$ is the binding energy per electron. In this representation the Cooper-pair level is located a distance Δ below the bottom of the conduction band, as shown in Fig. 13.5. At absolute zero all the conduction electrons are condensed in the Cooper-pair level. Above absolute zero some of the pairs break up and the individual electrons are excited to the bottom of the conduction band, as shown in Fig. 13.5b. These electrons which are produced by the breakup of Cooper pairs are the quasi-particles mentioned above.

We will find the boson condensation representation a little more convenient than the semiconductor representation for analysis of the tunneling of super electrons. Before proceeding, however, let us say a few words about tunneling in general, after which we will comment on normal electron tunneling.

IV. TUNNELING PROCESSES

We start with a brief qualitative description of the three types of tunneling processes, following it with a more detailed examination.

A. Conditions for Tunneling

Three conditions must be satisfied for tunneling to occur. First, there must be a barrier between the source and destination locations of the tunneling electrons preventing direct electron transport. Second, the total energy of the system must be conserved in the process, which is why single-electron tunneling occurs between levels that have the same energy on either side of the barrier. In two-electron tunneling, one electron gains as much energy as the other electron loses. Third, tunneling proceeds to energy states that are empty since otherwise the Pauli exclusion principle would be violated. A bias voltage that lowers the energy levels on the positive side relative to levels on the negative side is often applied. This can serve to align occupied energy levels on one side of the barrier with empty levels on the other so as to enable tunneling between the two sides.

There are sign and direction rules that apply to the description of tunneling processes. When a metal (or superconductor) is positively biased relative to another metal, so that its potential is +V, its energy levels are lowered, as shown in Fig. 13.6b. The electron tunneling direction is toward the metal with the positive bias, but the tunneling current flows in the opposite direction, as indicated in the figure. This is because, by convention, current flow is expressed in terms of positive charges so that negatively charged electrons must flow in the opposite direction. Thus current flows toward the negative bias, as shown in Fig. 13.6c.

In drawing energy level diagrams, one of the two metals is normally grounded so that its Fermi level does not change when the bias is applied. The Fermi level falls for a metal that is biased positive relative to ground, and raised for one biased negative.

B. Normal Metal Tunneling

Consider two normal metals grounded at absolute zero and separated by an insulating barrier. Their Fermi levels are aligned as shown in Fig. 13.6a, so no tunneling occurs. A positive-bias voltage is then applied to one of the metals, lowering its energy levels, as shown in Fig. 13.6b, so that the electrons are now able to tunnel from the top of the conduction band of the grounded metal to the empty continuum levels of the positively biased metal, as shown. The number of empty levels that can receive electrons is proportional to the bias, so that the current flow is also proportional to it, as shown in Fig. 13.6c. The magnitude of the tunneling current is, of course, small compared to the current that flows in the absence of the barrier. Such a process satisfies the three conditions for tunneling—namely, presence of a barrier, energy conservation, and empty target levels.

C. Normal Metal-to-Superconductor Tunneling

Next, we consider the case of an insulating barrier between a superconductor and a normal metal. N–I–S tunneling occurs through the processes outlined in Fig. 13.7, where the top three diagrams are the semiconductor representation and the three middle diagrams sketch the boson condensation representation. In the unbiased cases of Figs. 13.7b and 13.7e no tunneling occurs because there is no way for energy to be conserved by electrons tunneling to empty target levels. This is also true for the range $-\Delta/e < V < +\Delta/e$ of biases. For a positive bias, $V \geq \Delta/e$, electrons can tunnel from the conduction band of the normal metal to the empty states above the gap of the superconductor, as shown in Figs. 13.7c and 13.7f. The

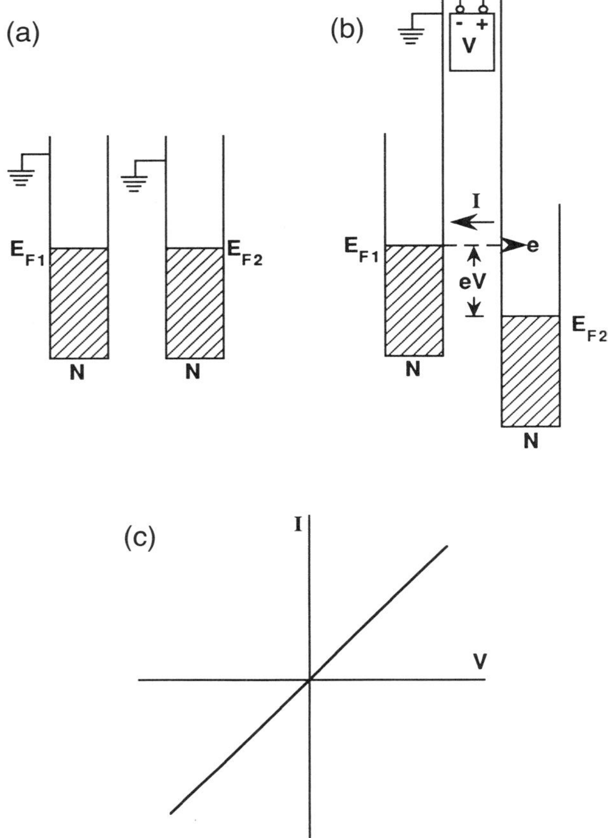

Figure 13.6 Normal metal tunneling showing (a) Fermi levels aligned for zero applied voltage, and hence zero tunneling current, (b) application of a positive bias voltage V to metal 2, lowering its Fermi level by eV relative to metal 1 and causing electrons to tunnel from left to right (metal 1 to metal 2), corresponding to current flow opposite in direction to the electron flow, as indicated by the arrows, and (c) linear dependence of the tunneling current on the applied voltage. Sketch (b) is drawn with metal 1 grounded, so that when the bias is applied, the Fermi level of metal 1 remains fixed while that of metal 2 falls.

figures appear similar in both representations because Cooper pairs do not participate.

For a negative bias, $V \leq -\Delta/e$, the process must be considered more carefully since the explanation is different in the two representations. In the boson condensation picture shown in Fig. 13.7d tunneling involves the breakup of a Cooper pair, with one electron of the pair tunneling down to the top of the normal-metal conduction band and the other jumping upward to the quasiparticle energy band of the superconductor. Thus the paired electrons separate to create a quasiparticle in the superconductor and transfer a conduction electron to the normal metal, with energy conserved in the process (Hu *et al.*, 1990; Rajam *et al.*, 1989; van den Brink *et al.*, 1991; Worsham *et al.*, 1991). Both elec-

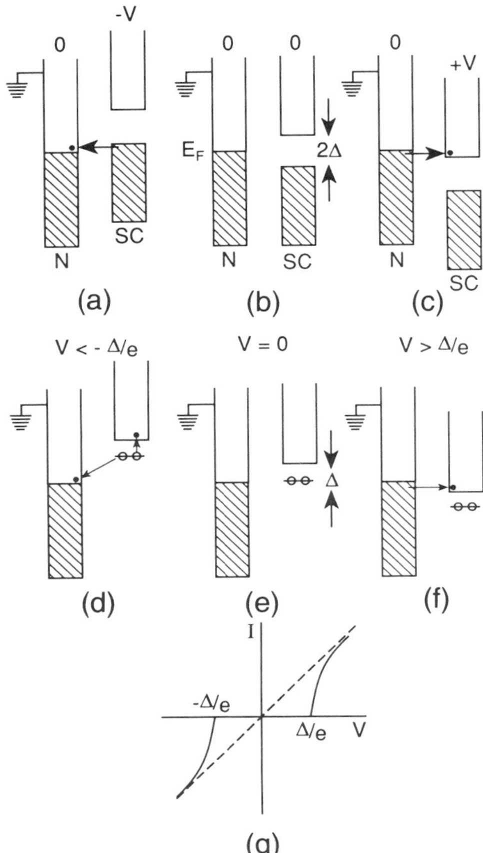

Figure 13.7 Normal metal–superconductor tunneling. The semiconductor representation shows (a) super electron tunneling (SC → N) for $V < -\Delta/e$, (b) zero tunneling current for the $V = 0$ case of the Fermi level in the gap ($-\Delta/e < V < \Delta/e$), and (c) normal electron tunneling (N → SC) for $V > \Delta/e$, which are also shown in the boson condensation representation (d), (e), and (f), respectively. The arrows show the electron tunneling directions, which are opposite to the current flow directions; the current–voltage characteristic is given in (g). The normal metal is grounded so that the superconductor bands shift downward when a positive bias is applied.

trons of the Cooper pair are accounted for. In the semiconductor representation, only the electron that is transferred to the normal metal is taken into account, as shown in Fig. 13.7a. This electron leaves behind it a hole in an otherwise filled band, it is this hole which constitutes the quasiparticle. Figure 13.7g shows how the experimentally measured current flow between the metal and the superconductor depends on the bias.

D. Superconductor-to-Superconductor Tunneling

Finally, let us consider the case of two identical superconductors. S–I–S tunneling occurs through the processes depicted in Fig. 13.8 for the two representations. Over the range of biases $-2\Delta/e < V < +2\Delta/e$ an electron in the semiconductor representation can tunnel from the superconducting state of one superconductor to become a quasiparticle in the normal state of the other, as shown in Figs. 13.8a and 13.8c. This has its counterpart explanation in Figs. 13.8d and 13.8f where we see how a Cooper pair in the higher of the two boson condensation levels can break up, with one electron jumping up to become a quasiparticle in its own excited level and the other electron jumping down to become a quasiparticle in the other superconductor. As the bias voltage increases beyond the range $-2\Delta/e < V < +2\Delta/e$, the current increases abruptly in magnitude and then approaches its normal metal value, as indicated in Fig. 13.8g. The current voltage characteristic for two identical superconductors is, of course, antisymmetric about the point $V = 0$. By antisymmetric we mean that when V → −V we will have I → −I. Note that the onset of tunneling for S–I–S junctions occurs at $V = \pm 2\Delta/e$, which is twice the value for the N–I–S case.

Figure 13.8 was drawn for the case $T = 0$. For a finite temperature there will be some quasiparticles in each superconductor, so that a small tunneling current will flow for bias voltages below $2\Delta/e$, as shown in Fig. 13.9 in the boson condensation representation (a) and in the semiconducting representation (b). The current–voltage characteristic is given in Fig. 13.9c.

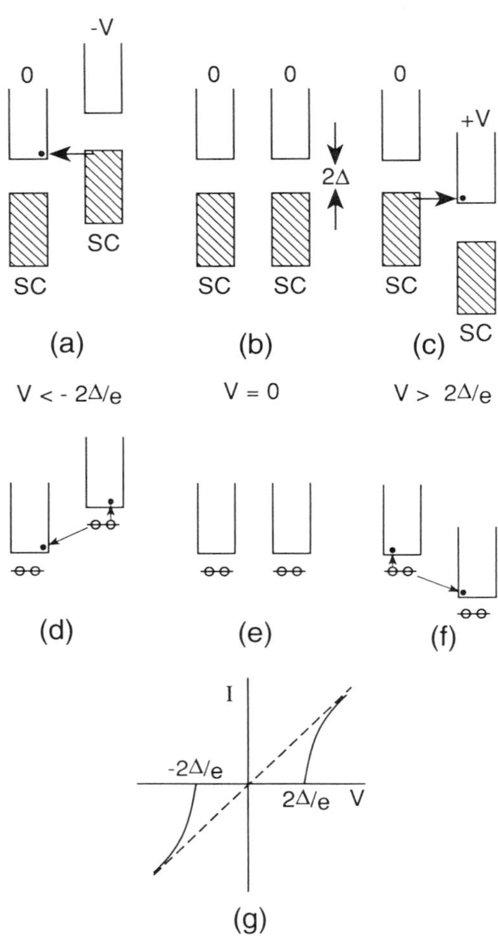

Figure 13.8 Superconductor-to-superconductor tunneling at absolute zero. The semiconductor representation shows (a) super electron tunneling for $V < -2\Delta/e$, (b) zero tunneling current for bias voltages in the range $-2\Delta/e < V < 2\Delta/e$, and (c) opposite-direction super electron tunneling for $2\Delta/e < V$, which are also shown in the boson condensation representation (d), (e), and (f), respectively, where the tunneling arises from the breaking of a Cooper pair. The current–voltage characteristic is given in (g). The arrows show the electron tunneling directions, which are opposite to the current flow directions. The sketches are drawn with the superconductor on the left grounded.

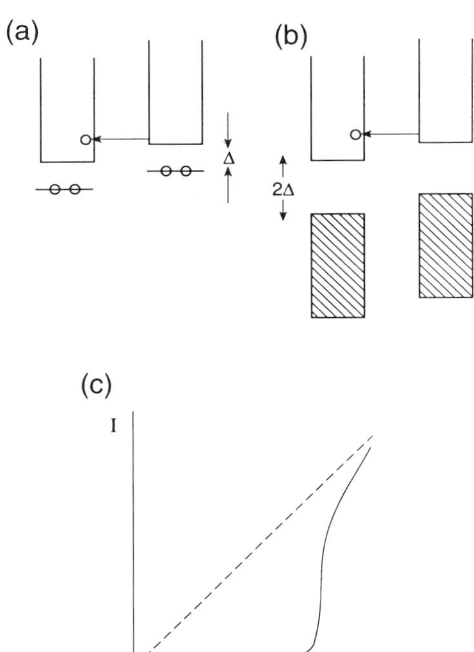

Figure 13.9 Superconductor-to-superconductor tunneling at finite temperatures, $T > 0$. The semiconductor representation (a) and boson condensation representation (b) show finite tunneling between upper quasiparticle levels sparsely populated by thermal excitation. The current–voltage characteristic (c) shows a small current flow for $0 < V < 2\Delta/e$, and the usual larger current flow for $2\Delta/e < V$.

V. QUANTITATIVE TREATMENT OF TUNNELING

The previous section discussed the different tunneling processes in terms of both the boson condensation and the semiconductor representations. The former seems to give a better physical picture of what is happening because it involves the breakup of Cooper pairs, whereas the latter provides a framework for carrying out quantitative calculations of the tunneling current as a function of temperature. We will now apply the Fermi statistics approach of Chapter 1, Section IX, to the semiconductor representation to derive quantitative expressions for the tunneling current.

A. Distribution Function

We explained in Chapter 1, Section IX, that the concentration of conduction electrons as a function of their energy is given by the product of the Fermi–Dirac

(F–D) distribution function, $f(E)$, and the density of states, $D(E)$. We begin by expressing the former in a form that is convenient for treating tunneling problems, and then make use of the latter, which we write $D_n(E)$ for the normal electrons involved in the tunneling. The super electrons have a different density of states, $D_s(E)$, which was derived in Chapter 6, Section V.

In Chapter 1, Section IX, we expressed the energies of a conductor relative to the Fermi energy E_F. In the present discussion it is convenient to select the Fermi level as the zero of energy, i.e., to set $E_F = 0$. With this in mind the F–D distribution of Eq. (1.35) for electrons assumes the form

$$f(E) = \frac{1}{\exp(E/k_BT) + 1}, \quad (13.2)$$

which at absolute zero equals 1 for negative energies and 0 for positive energies. The corresponding distribution function for unoccupied states, sometimes called holes, is $1 - f(E)$. If a bias voltage V is applied, the distribution function for electrons becomes

$$f(E + eV) = \frac{1}{\exp[(E + eV)/k_BT] + 1}, \quad (13.3)$$

and for holes is given by

$$1 - f(E + eV) = \frac{1}{1 + \exp[-(E + eV)/k_BT]}. \quad (13.4)$$

These distributions functions for $T > 0$ are plotted in Fig. 13.10 for zero, positive, and negative biases. Figure 1.4 shows the effect of temperature on the F–D distribution.

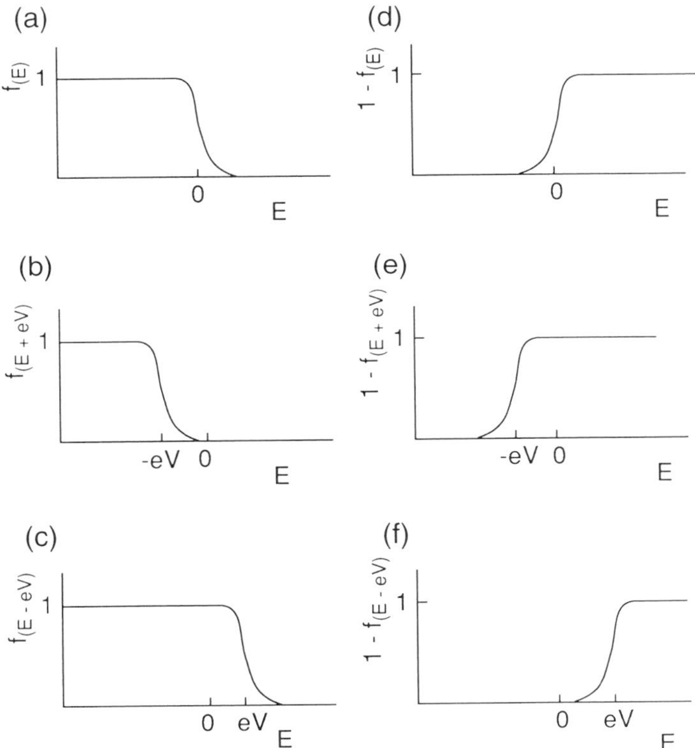

Figure 13.10 The dependence of the Fermi–Dirac distribution function $f(E)$ on the energy for zero bias (a), positive bias (b), and negative bias (c). The dependence of the distribution function in the case of holes, $1 - f(E)$, on the energy for the same bias conditions (d, e, and f).

B. Density of States

Now that we have rewritten the F–D distribution functions relative to the zero of energy set at the Fermi level Eq. (1.41) should similarly be rewritten for the density of states $D_n(E)$ of normal electrons with this same zero of energy,

$$D_n(E) = D_n(0)\left(\frac{E_F + E}{E_F}\right)^{1/2}, \quad (13.5)$$

where $D_n(0)$ is the density of states at the Fermi level ($E = 0$). Plots of $D_n(E)f(E)$ against the energy are shown in Fig. 1.7; we will make use of these plots with the zero of energy set at the Fermi level. Since E in Eq. (13.5) is usually very small compared with the Fermi energy $|E| \ll E_F$, and since the energies of interest are generally limited by the maximum applied bias voltage V_{max}, in tunneling calculations it is usually valid to write

$$D_n(E) \approx D_n(0) \qquad -eV_{max} < E < eV_{max}. \quad (13.6)$$

The density of states in the superconducting state is given by the BCS expression (6.103),

$$D_s(E) = \begin{cases} \dfrac{D_n(0)|E|}{(E^2-\Delta^2)^{1/2}} & E < -\Delta \quad (13.7) \\ 0 & -\Delta < E < \Delta \quad (13.8) \\ \dfrac{D_n(0)E}{(E^2-\Delta^2)^{1/2}} & \Delta < E, \quad (13.9) \end{cases}$$

which is plotted in Figs. 6.4 and 6.5 over narrow and broad energy ranges, respectively.

Another property of the density of states that has important implications for superconductivity is the conservation of states in k-space that was mentioned in Chapter 1, Section X. This is reflected in the conservation of energy levels at the onset of superconductivity. When a material becomes superconducting, an energy gap forms, with some energy states shifting upward above the gap and some falling below it, with the total number of states remaining the same, in the manner illustrated in Fig. 1.9. Comparing Eqs. (13.7) and (13.9) shows that the level spacing is the same just above and just below the gap. The area under the curve for $D(E)$ versus E, which is numerically equal to the total number of energy levels, is unchanged during the passage through T_c,

$$\int_{T>T_c} D_n(E)dE = \int_{T<T_c} D_s(E)dE. \quad (13.10)$$

We will use these expressions for $f(E)$ and $D(E)$ to write down an analytic expression for the tunneling current.

C. Tunneling Current

We begin by deducing a general expression for the tunneling current and then discuss the cases N–I–N, N–I–S, and S–I–S. Note, first, that in our qualitative discussion we were assuming that tunneling takes place in one direction only. A more careful analysis shows that it actually occurs in both directions, and that the net tunneling current is the difference between forward and backward tunneling processes.

Consider the case of a bias voltage V,

$$V = V_1 - V_2, \quad (13.11)$$

applied to a metal–insulator–metal junction, M_1–I–M_2, as shown in Fig. 13.11, where each M can be a normal metal or a superconductor. To simplify the mathematics we treat the special case of metal 1 at the potential $V_1 = V$ and metal 2 at zero potential, $V_2 = 0$. Electrons can tunnel to the right with the current density J_{12} and to the left with the current density J_{21}, as shown. The tunneling current is, therefore,

$$I = (J_{12} - J_{21})A, \quad (13.12)$$

where A is the area of the junction. Most electrons that impinge on the barrier from the left are reflected, but a few penetrate it, as illustrated in Fig. 13.12, and con-

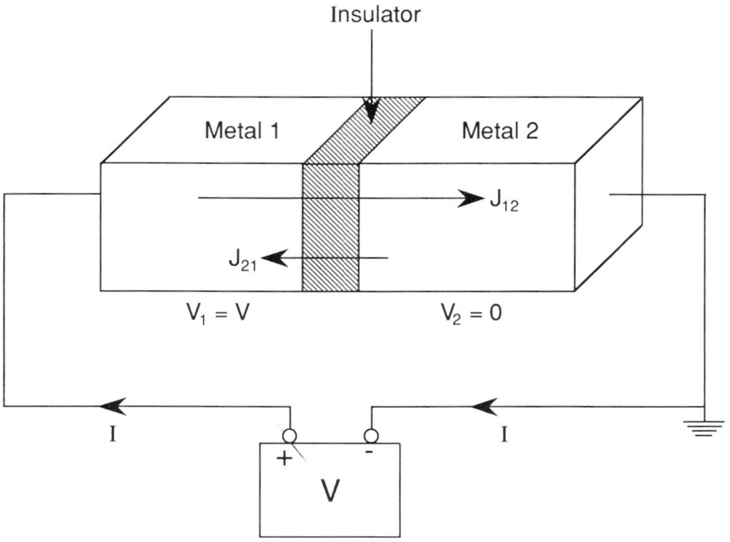

Figure 13.11 Metal-to-metal tunnel junction in the form of a thin uniform insulating layer between two metals M_1 and M_2. Tunneling current densities J_{12} and J_{21} in both directions are shown, with $J_{12} > J_{21}$ for the bias indicated.

tribute to the tunneling current. We recall from quantum mechanics that Fermi's Golden Rule from time-dependent perturbation theory provides the probability per unit time W that an electron will undergo a transition from state 1 to state 2 in the energy range from E to $E + \Delta E$,

$$W_{1\to 2} = (2\pi/\hbar)|\langle 2|H_{\text{pert}}|1\rangle|^2\, \delta_2(E), \tag{13.13}$$

where

$$\delta_2(E) = D_2(E)[1 - f(E)] \tag{13.14}$$

is the "target" density of empty states in the energy range into which the electron tunnels. We assume that the tunneling matrix element H_T,

$$H_T = H_{12} = H_{21}^* = \langle 2|H_{\text{pert}}|1\rangle \tag{13.15}$$

of the perturbation Hamiltonian H_{pert} responsible for the penetration at the barrier can be evaluated. The tunneling current from metal 1 to metal 2 is related to the transition probability through the expression

$$\begin{aligned} J_{12} &= e\int W_{1\to 2} D_1(E - eV) f(E - eV)\,dE \\ &= \frac{2\pi e}{\hbar}\int |H_T|^2 D_1(E - eV) f(E - eV) \\ &\quad \times D_2(E)[1 - f(E)]\,dE, \end{aligned} \tag{13.16a}$$

where $W_{1\to 2}$ is given by Eq. (13.13). We note that the integrand is proportional to the overlap between the two densities of states. By the same reasoning, the tunneling current in the reverse direction, J_{21}, is

$$\begin{aligned} J_{21} &= \frac{2\pi e}{\hbar}\int |H_T|^2 D_2(E) f(E) \\ &\quad \times D_1(E - eV)[1 - f(E - eV)]\,dE, \end{aligned} \tag{13.16b}$$

where again the integrand contains the density of states overlap. Inserting Eqs. (13.16a) and (13.16b) in Eq. (13.12) gives

$$\begin{aligned} I &= \frac{2\pi eA}{\hbar}|H_T|^2 \int D_1(E - eV) D_2(E) \\ &\quad \times [f(E - eV) - f(E)]\,dE \end{aligned} \tag{13.17}$$

for the total tunneling current I in the direction from metal 1 to metal 2, where it

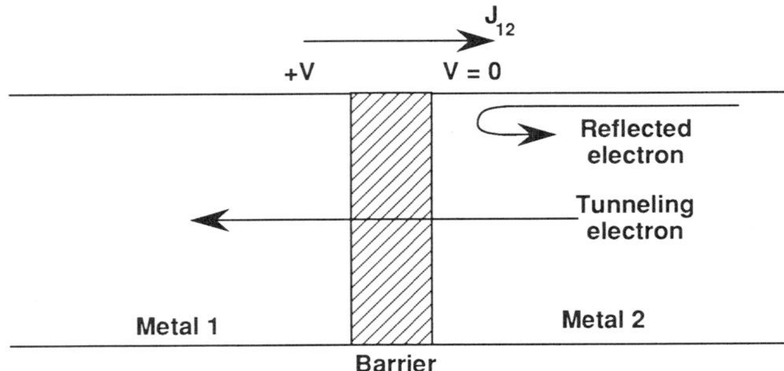

Figure 13.12 One electron reflected from, and another tunneling through, the insulating barrier of a tunnel junction for the bias of Fig. 13.11. The tunneling of the electron is in the direction from the negative to the positive side of the junction, but the corresponding current flow J_{12} is from + to − because it is based on the convention of positive-charge carriers.

is assumed that the tunneling matrix element H_T is independent of the energy near $E = 0$.

In this expression the density-of-states functions $D_1(E - eV)$ and $D_2(E)$ depend on the nature of the source and target states, whether they are normal or superconducting. The distribution function difference $[f(E - eV) - f(E)]$, on the other hand, depends only on the potential V, being close to 1 for energies between 0 and eV and approaching zero rapidly outside this range. Therefore, strong tunneling can occur only where $D_1(E - eV)$ and $D_2(E)$ are both appreciable in magnitude in this energy range. Weak tunneling could occur in the tails of the function

$$[f(E - eV) - f(E)]$$

just beyond this range. These characteristics will be illustrated in the next three sections for N–I–N, N–I–S, and S–I–S tunneling, respectively.

D. N–I–N Tunneling Current

Normal metal-to-normal metal tunneling depends mainly on the difference in the distribution functions

$$[f(E - eV) - f(E)],$$

since there is very little difference in the two normal metal densities of states. The analysis of this case is left as an exercise (see Problem 1).

E. N–I–S Tunneling Current

For tunneling between a normal metal and a superconductor, the normal-metal density of states $D_n(E - eV)$ can be approximated by $D_n(0)$ and factored out of the integral (13.17). N–I–S tunelling will then occur when the superconducting density of states $D_s(E)$ overlaps with

$$[f(E - eV) - f(E)].$$

At absolute zero $[f(E - eV) - f(E)]$ is 1 in the range $0 \leq E \leq eV$ and zero outside this range, so that for a positive bias, $V > 0$, the integrand of (13.17) becomes

$$D_n(E - eV)D_s(E)[f(E - eV) - f(E)] = \begin{cases} D_n(0)D_s(E) & \Delta < E < eV \\ 0 & \text{otherwise.} \end{cases} \tag{13.18}$$

A similar reasoning shows that for a negative bias,

$$D_n(E - eV)D_s(E)[f(E - eV) - f(E)] = \begin{cases} -D_n(0)D_s(E) & -eV < E < -\Delta \\ 0 & \text{otherwise.} \end{cases} \tag{13.19}$$

With the aid of Eqs. (13.7)–(13.9) we see that Eq. (13.17) can be integrated in closed form, as shown in Problem 4, to give

$$I_{ns} = \begin{cases} G_n\left[V^2 - \left(\frac{\Delta}{e}\right)^2\right]^{1/2} & \frac{\Delta}{e} < V \\ 0 & \frac{-\Delta}{e} < V < \frac{\Delta}{e} \\ -G_n\left[V^2 - \left(\frac{\Delta}{e}\right)^2\right]^{1/2} & V < -\frac{\Delta}{e} \end{cases} \quad (13.20)$$

where G_n, which has the value

$$G_n = \left(\frac{2\pi A}{\hbar}\right)|H_T|^2 D_n(0), \quad (13.21)$$

is the normal metal electron tunneling conductance defined by

$$G_n = \frac{I_n}{V} \quad (13.22)$$

and plotted as on asymptotic (dashed) straight line in Figs. 13.7g, 13.8g, and 13.9c. The N–I–S tunneling current (13.20) is shown plotted in Fig. 13.7g. We see that as the voltage increases, the current approaches the normal conductor value. Experimentally determined plots of this type can be used to evaluate the energy gap Δ from N–I–S tunneling measurements.

The expression (13.20) is valid for absolute zero. For $T > 0$ the tails of the distribution function difference,

$$[f(E - eV) - f(E)],$$

produce weak tunneling for potentials $|V|$ in the gap close to the value Δ/e.

The significance of the overlap conditions in producing N–I–S tunneling is illustrated in Fig. 13.13. Figure 13.13a shows the lack of overlap when $|V| \ll \Delta/e$, so that no tunneling occurs. Figure 13.13b indicates the small overlap when V is in the gap near the edges and there is weak tunneling. Finally, Fig. 13.13c shows the strong overlap for $V > \Delta/e$ which pro-

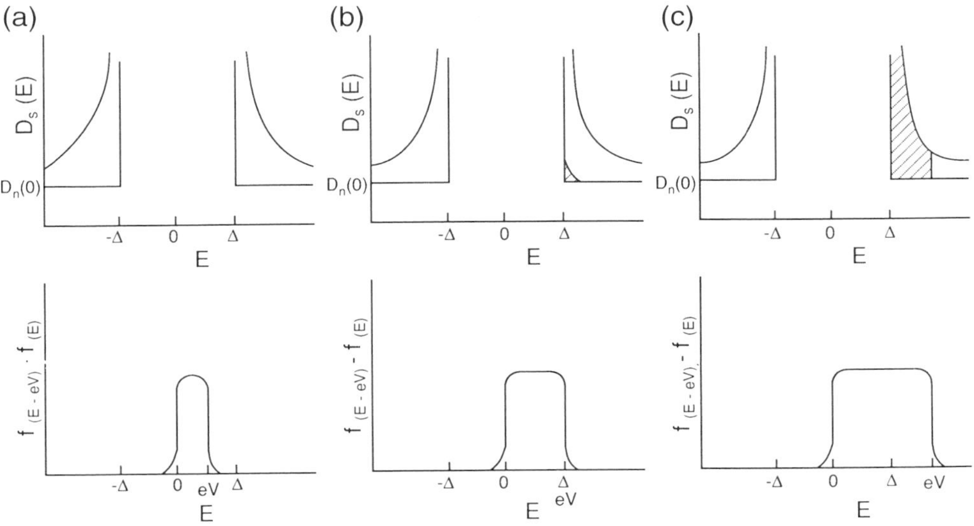

Figure 13.13 Contribution of the occupancy (shaded) of the superconductor density of states $D_s(E)$ to N–I–S tunneling for (a) small positive bias, $0 < V \ll \Delta/e$, and no tunneling current, (b) small bias, $V = \Delta/e$, for which the tail of the distribution function difference $[f(E - eV) - f(E)]$ overlaps $D_s(E)$, the occupancy of $D_s(E)$ is small, and a weak tunneling current flows, and (c) more positive bias, $V > \Delta/e$, producing a strong overlap so that the occupancy of $D_s(E)$ is large near the gap and a strong tunneling current flows.

duces strong tunneling. These figures should be compared with the more qualitative representations sketched in Fig. 13.7.

F. S–I–S Tunneling Current

Superconductor–superconductor tunneling is treated in a manner similar to the treatment we have just used for the N–I–S case. Unfortunately, in the S–I–S case the tunneling current equation (13.17) cannot be integrated in closed form, and instead we present the more qualitative treatment that is outlined in Fig. 13.14. We select $D_1(E - eV)$ as a superconductor with a small gap Δ_1 and $D_2(E)$ as a superconductor with a larger gap $\Delta_2 > \Delta_1$. As the bias voltage V is increased, the lower bands of D_1 and D_2 are made to coincide at a bias $V = (\Delta_2 - \Delta_1)/e$, as indicated in Fig. 13.14b, and also in the semiconductor and boson condensation representation plots of Fig. 13.15. This coincidence and the overlap of the bands results in weak tunneling because of the very low concentration of electrons in each level, and a small peak appears in the current versus voltage plot of Fig. 13.16b. The current is less on either side of the peak because the amount of overlap of the bands is less. The decrease of I with increasing V beyond the peak at finite temperatures constitutes a

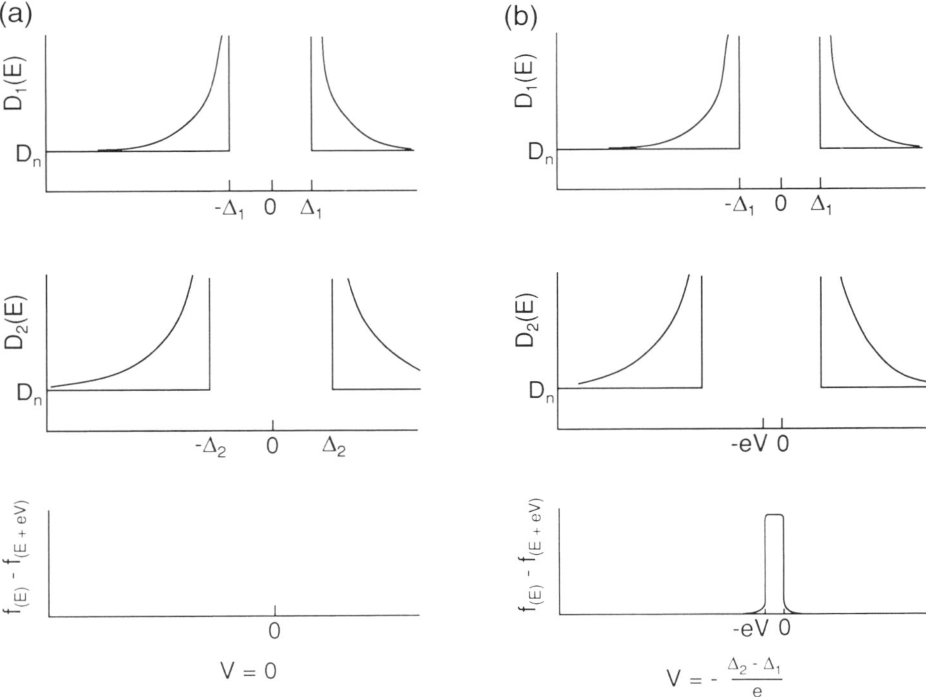

Figure 13.14 Densities of states for superconductor 1 (top) with the gap Δ_1, for superconductor 2 (middle) with the gap Δ_2, and distribution function difference (bottom) for (a) zero bias in which no tunneling current flows, (b) bias $V = (\Delta_2 - \Delta_1)/e$ producing weak tunneling current due to the overlap of the two quasiparticle bands, (c) bias $V = (\Delta_2 + \Delta_1)/e$ for onset of strong tunneling, and (d) bias $V \gg (\Delta_2 + \Delta_1)/e$ producing strong tunneling current due to the large overlap between the occupied superconductor band of the first superconductor and the empty quasiparticle band of the second superconductor. Quasiparticle tunneling (b) arises from the tail of the distribution function difference $f(E) - f(E + eV)$, hence is very weak and vanishes at absolute zero. Figure 13.13 presents energy-level diagrams for these four cases. (*Continues*)

negative resistance region of the I-versus-V characteristic of Fig. 13.16b. At absolute zero this current vanishes, as indicated in Fig. 13.16a, because the quasiparticle levels are all empty.

We see from Figs. 13.14c and 13.15b that when the magnitude of the bias reaches the value $V = (\Delta_1 + \Delta_2)/e$, the densities of states $D_1(E)$ and $D_2(E)$ begin to overlap at their infinity points, and there is a large jump in the tunneling current, as indicated in Figs. 13.16a for $T = 0$ and in Fig. 13.16b for $T > 0$. The tunneling current is now large because it flows from a nearly full level to a nearly empty level. An evaluation of the integral (13.17) at $T = 0$ and $V = (\Delta_1 + \Delta_2)/e$ gives for the jump in current as this bias

$$\Delta I_s = \frac{\pi G_n (\Delta_1 \Delta_2)^{1/2}}{2e}, \qquad (13.23)$$

where G_n is the normal tunneling conductance defined by Eq. (13.22). In Problem 5 we show that the ratio of the jump in current ΔI_s to the normal tunneling current I_n at the bias $V = 2\Delta/e$ is given by

$$\Delta I_s / I_n = \frac{\pi}{4}, \qquad (13.24)$$

which represents a jump of around 80%. Van Duzer and Turner (1981, p. 87) have shown that, for a finite temperature, this jump has the magnitude

$$\Delta I_s = \frac{\pi G_n \sqrt{\Delta_1 \Delta_2}}{4e} \times \frac{\sinh\left(\dfrac{\Delta_1 + \Delta_2}{2k_B T}\right)}{\left(\cosh \dfrac{\Delta_1}{2k_B T}\right) \cdot \left(\cosh \dfrac{\Delta_2}{2k_B T}\right)}, \qquad (13.25)$$

which reduces to Eq. (13.23) for $T = 0$.

If the gaps are the same for the two superconductors, $\Delta_1 = \Delta_2 = \Delta$, there will be no maximum in the weak quasiparticle

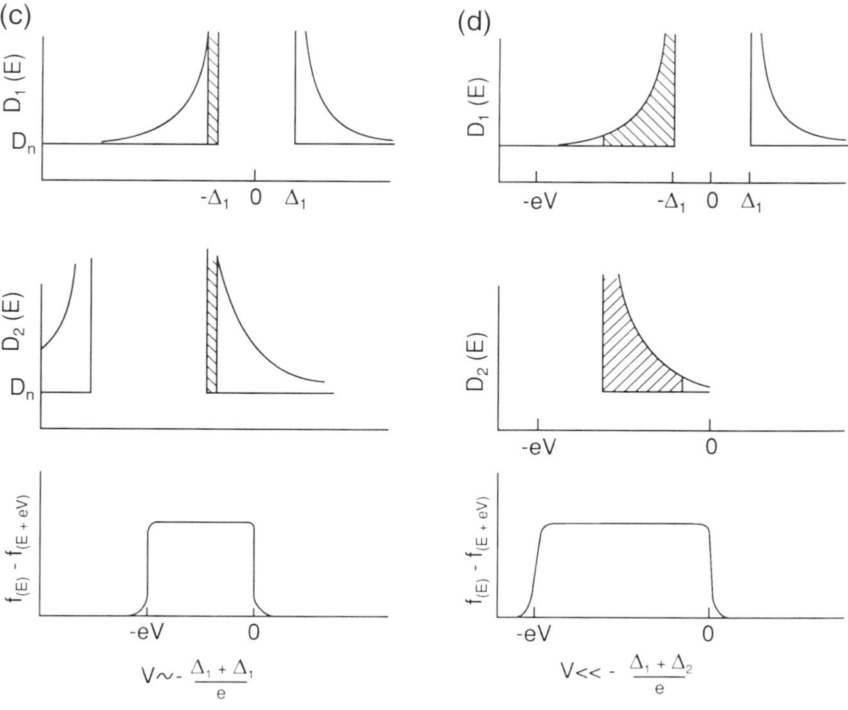

Figure 13.14 (*Continued*)

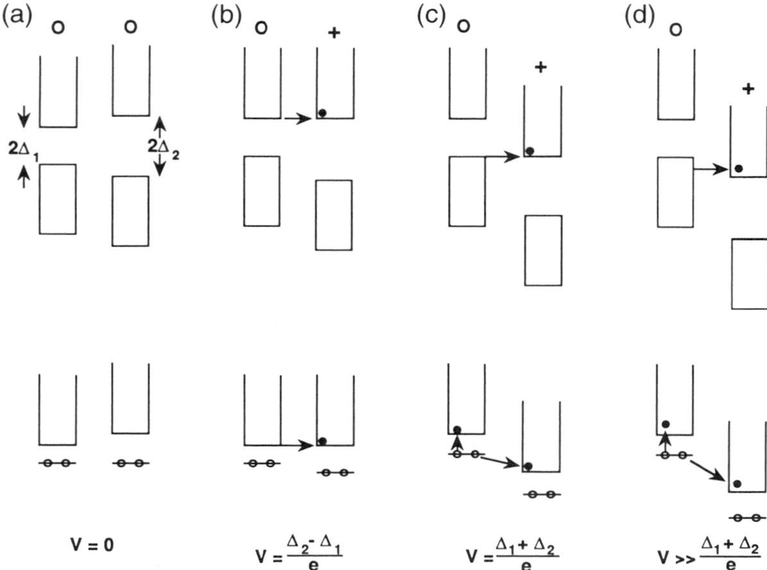

Figure 13.15 Semiconductor (top) and boson condensation (bottom) representations of the S–I–S tunneling cases of Fig. 13.14 for (a) zero bias, (b) quasiparticle band alignment and weak tunneling, (c) onset of Cooper pair tunneling, and (d) strong tunneling of Cooper pair electrons.

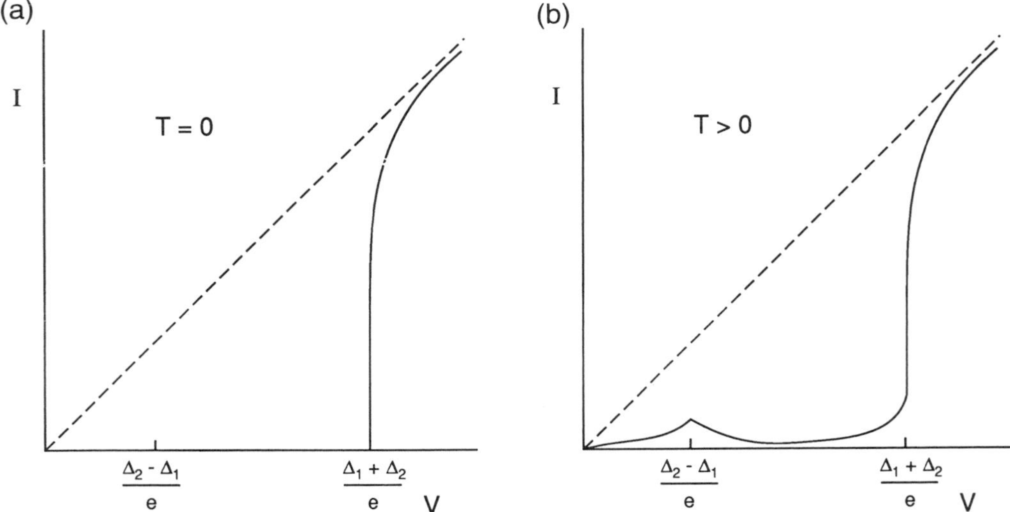

Figure 13.16 Tunneling current versus bias voltage for S–I–S tunneling involving two superconductors with energy gaps $\Delta_2 > \Delta_1$ for (a) $T = 0$, with no tunneling occurring until the bias $V = (\Delta_2 + \Delta_1)/e$ is reached, and (b) $T > 0$, with weak tunneling at the bias $V = (\Delta_2 - \Delta_1)/e$ and strong tunneling for $V > (\Delta_1 + \Delta_2)/e$.

tunneling current, but such a weak current does flow for $V < 2\Delta/e$, as shown in Fig. 13.9. For $T \ll \Delta/k_B$, this current is given approximately by (Van Duzer and Turner, 1981, p. 87)

$$T_s = \frac{2G_n}{e}(\Delta + eV) \times \sqrt{\frac{2\Delta}{2\Delta + eV}} \sinh\left(\frac{eV}{2k_B T}\right) K_0 \times \left(\frac{eV}{2k_B T}\right) e^{-\Delta/k_B T}, \qquad (13.26)$$

where K_0 is the zeroth-order modified Bessel function (cf. Chapter 9, Section III.B).

In Eq. (13.22) we defined the normal metal electron tunneling conductance G_n as the asymptotic slope of the I-versus-V characteristic curve for very large V. We can also define the differential conductance,

$$G_d = \frac{dI}{dV}, \qquad (13.27)$$

which is the slope at any point of the I-versus-V curve. Many workers report their tunneling measurements as plots of G_d versus V. This has the advantage of providing greater resolution, since structural features tend to be better resolved in plots of differential conductance than in plots of I versus V (see example in Section VI.C, especially Fig. 13.24).

G. Nonequilibrium Quasiparticle Tunneling

So far we have assumed that the super electrons in the ground energy band and the quasiparticles in the excited band are in thermal equilibrium both between the bands and within each individual band. The tunneling process, of course, disturbs this equilibrium, but this effect is negligible.

We now wish to treat so-called branch imbalance, in which the number of quasiparticles n_+ with momentum in one direction, $\mathbf{p}_i$, is greater than the number of quasiparticles n_- with momentum in the opposite direction, $-\mathbf{p}_i$, in accordance with Fig. 13.17b. The imbalance $[n_+ - n_-]$ can be brought about by injecting quasiparticles across an N–I–S junction (Clarke, 1972).

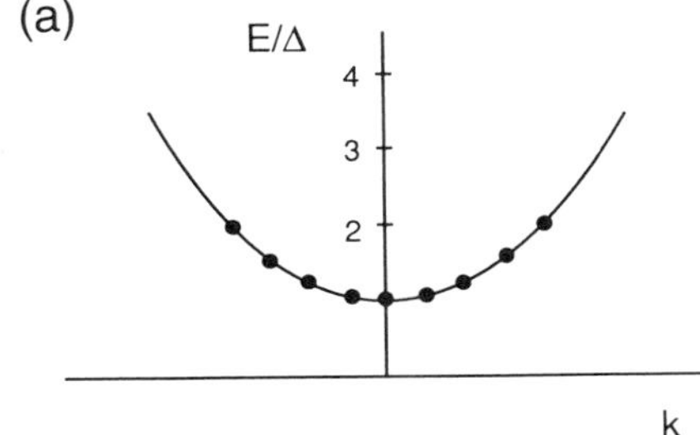

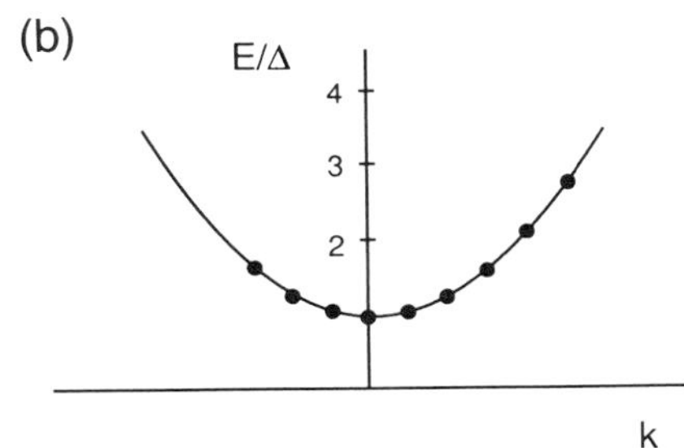

Figure 13.17 Branch imbalance illustrated using a one-electron energy parabola (a) for the usual case of no imbalance where the number of electrons with positive momentum is equal to the number of electrons with negative momentum, and (b) for the branch-imbalance state (number of electrons with positive momentum greater than numbers of electrons with negative momentum).

When a quasiparticle imbalance exists in the neighborhood of a tunnel junction, a current flows,

$$I = e\frac{d}{dt}[n_+ - n_-], \qquad (13.28)$$

to reestablish balance between the positive and negative momentum states in the quasiparticle band. Equilibrium is restored in a time τ_Q, called the *branch imbalance relaxation time*, and we can write

$$[n_+ - n_-] = \frac{I\tau_Q}{e}. \qquad (13.29)$$

For temperatures near T_c the relaxation time is predicted, assuming a spacially uniform case, to have the temperature dependence (Schmid, 1968)

$$\tau_Q(T) \approx \tau_Q(T_c)\left(1 - \frac{T}{T_c}\right)^{-1/2}. \qquad (13.30)$$

This relation has also been found experimentally (Clarke and Patterson, 1974).

A more extensive discussion of quasiparticle imbalance may be found in the texts by Tinkham and Clarke (1972) and by Tilley and Tilley (1986).

VI. TUNNELING MEASUREMENTS

Let us now say a few words about the experimental arrangements used in carrying out tunneling experiments, followed by a discussion of representative experimental data that have appeared in the literature.

Tunneling, like photoemission, represents a surface-probe sampling of a region of dimensions determined by the coherence length (Cucolo *et al.*, 1991; J.-X. Liu *et al.*, 1991; Pierson and Valls, 1992), which, for high-temperature superconductors, can be only one or two unit cell dimensions in magnitude. Lanping *et al.* (1989) constructed a histogram of the distribution of gap parameters Δ determined from tunneling measurements made at 600 different surface locations on the same $YBa_2Cu_3O_{7-\delta}$ sample, obtaining values ranging from 15 to 50 meV, with several data points outside this range.

A. Weak Links

If a superconducting rod is cut at some point and then joined through an intervening insulating section, either (1) the insulating region will turn out to be so thick that the two separated superconducting sections lose contact and have no interaction, or (2) the insulator layer will consist of a monolayer of foreign atoms, so that strong contact is maintained across it, or (3) the section will be intermediate in thickness, so that the superconductors are weakly coupled and electrons can tunnel. The third case, called a *weak link*, is the one that is most commonly used for tunneling studies and Josephson effect measurements (Furusaki and Tsukada, 1991).

B. Experimental Arrangements for Measuring Tunneling

The overall structure at the interface between two superconductors or between a normal metal and a superconductor is called a *microbridge*. The barrier region is the crucial part of the microbridge. A typical barrier thickness is a coherence length or less in magnitude, so that barriers must be much thinner for high-temperature superconductors ($\xi \approx 2$ nm) than for an element like lead ($\xi = 80$ nm). Beenakker and van Houten (1991) discussed weak links that are quantum point contacts.

The original Zeller and Giaever technique (1969; Fulton *et al.*, 1989; Giaever and Zeller, 1968) for making an $Al{-}Al_2O_3{-}Al$ sandwich-type tunneling junction was to embed Sn particles in the aluminum oxide since Al oxidizes much faster than Sn, making it easier to form a thin insulating oxide layer on the tin. The preparation method shown in Fig. 13.18 consists in evaporation of a strip of aluminum film on a glass substrate, oxidizing the strip, evaporating tin on the film, and oxidizing once again. Then a second strip of aluminum at right angles to the first strip is evaporated on the substrate, as shown in the figure. An arrangement of the Sn particles for four different samples is shown in the electron micrographs of Fig. 13.19. The final asymmetric feed configuration arrangement is sketched in Fig. 13.20, with the details of the junction area

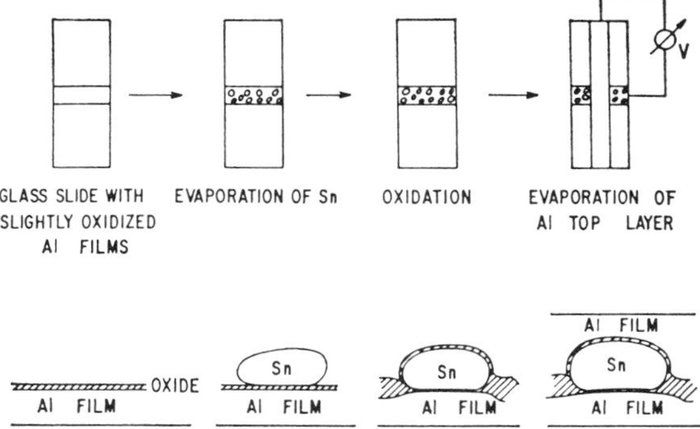

Figure 13.18 Preparation of a tunnel junction containing tin particles. Aluminum oxidizes faster than tin, so that the oxide layer is thicker between the particles than on their surface, as indicated (Zeller and Giaever, 1969).

indicated in Fig. 13.21 (Florjanczyk and Jaworski, 1989; Monaco, 1990a, b). To make a tunneling measurement, a bias voltage is applied across the junction using two of the leads, with the current monitored at the other two leads, as shown. Sandwich-tunnel junctions of this type have been made for many S–I–S and N–I–S cases using various combinations of superconducting and normal metals, but the use of embedded particles, such as Sn, to control the film thickness is not generally employed.

Another common technique for tunneling measurements employs a scanning electron microscope (SEM). A probe ground to a point with a very small tip radius makes contact with the superconductor surface, as indicated in Fig. 13.22a (typical probe materials are Au, Nb, Pt–Rh, Pt–Ir, and W.) The tip touches the surface, or comes very close to it. If contact is made, tunneling probably takes place through a layer of inhomogeneous insulating or semiconducting material, such as the oxide coating the surface. Tunneling can also occur at a constriction, as shown in Fig. 13.22b.

Mooreland's group developed what they call a break junction technique for tunneling measurements (Mooreland *et al.*, 1987a, b). A small piece of bulk material is

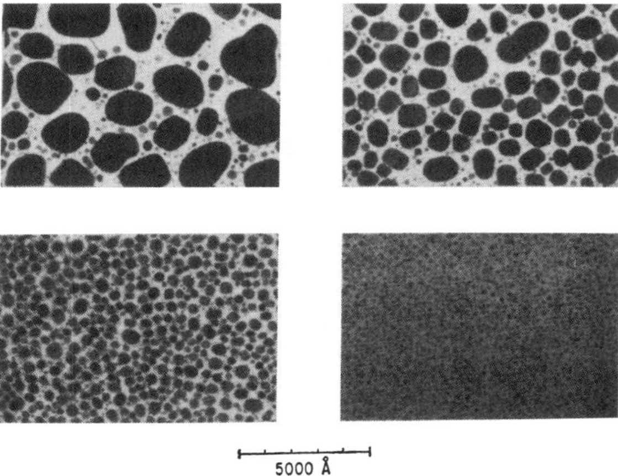

Figure 13.19 Electron micrograph of tin particles on an oxidized aluminum film for four particle sizes (Zeller and Giaever, 1969).

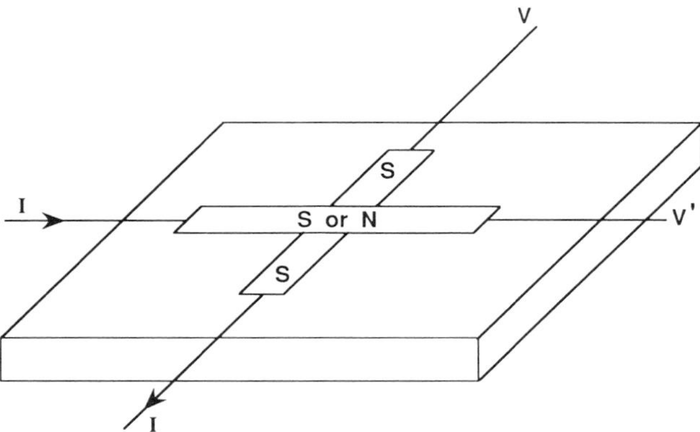

Figure 13.20 Sketch of tunnel junction on a glass substrate, showing current (I) and applied bias voltage (V, V') leads.

electromechanically broken under liquid helium and the freshly fractured interfaces joined to form a tunneling barrier with the liquid helium acting as insulator.

We will now present some typical experimental tunneling measurements that were made using these techniques.

C. N–I–S Tunneling Measurements

Gallagher *et al.* (1988) made an N–I–S tunneling study of $YBa_2Cu_3O_7$ using a scanning tunneling microscope operating in liquid helium. A coarse-adjust screw and fine-adjust piezoelectric transducer provided the desired tip-to-sample contact, where the tip is embedded in an insulating surface layer with a typical 1 MΩ resistance, which causes the junction to end up in the tunneling regime. Figure 13.23 shows the I-versus-V characteristic made with an Nb tip and Fig. 13.24 the dI/dV-versus-V characteristic for a similar sample made using a W tip. Note the increased resolution of the differential curve.

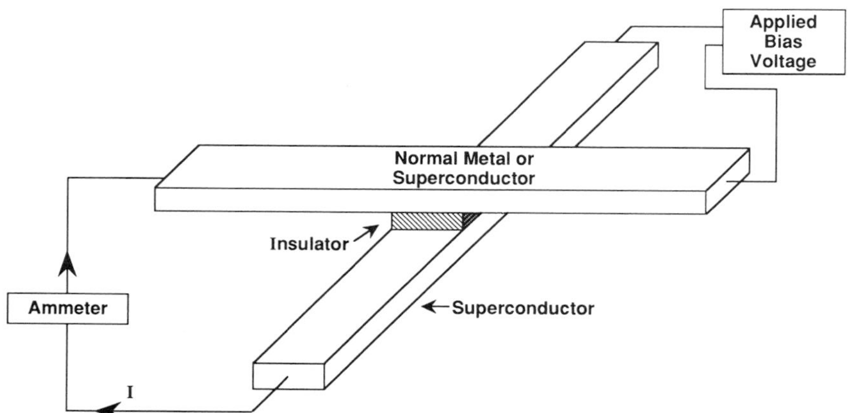

Figure 13.21 Details of the tunnel junction sketched in Fig. 13.20, showing the insulating layer between the metal strips. The ammeter measures the tunneling current produced by the bias voltage.

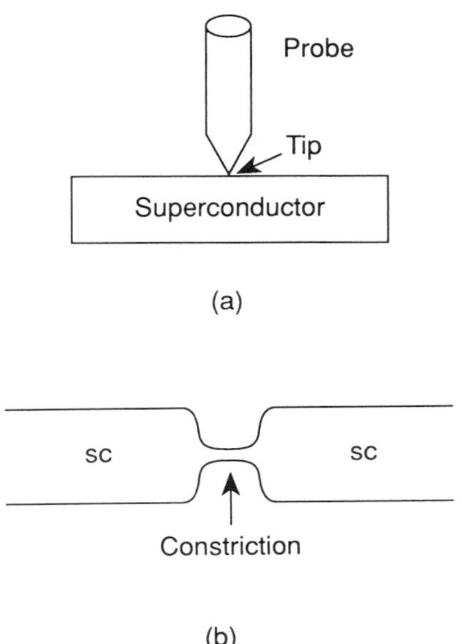

Figure 13.22 Tunnel junction formed from (a) a probe tip in contact with, or almost in contact with, the superconducting surface, and (b) a constricted region in a superconductor.

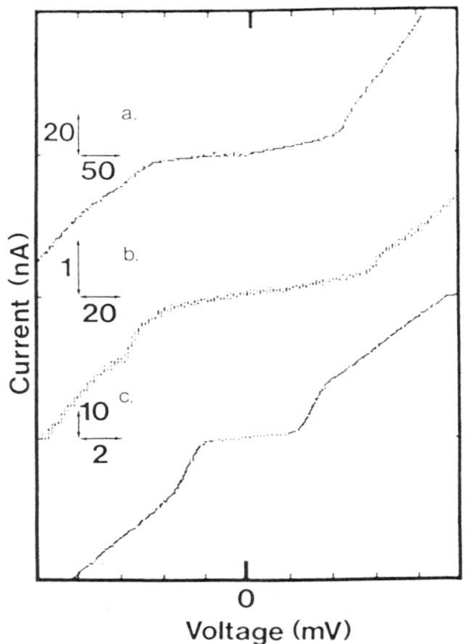

Figure 13.23 Current–voltage characteristics at different positions on the surface of aluminum-doped $YBa_2Cu_3O_{6.5+x}$ determined by scanning tunneling microscopy using an Nb tip at 4.2 K. A jump in the current is observed at 95, 30, and 2.5 mV, respectively. Note the changes in scale for each curve (Gallagher *et al.*, 1988).

Ekino and Akimitsu (1989a, b) reported point-contact electron tunneling studies of BiSrCaCuO and BiSrCuO bulk, monocrystal, and sputtered film samples. Figure 13.25 presents the I versus V differential conductance plot for the 2:2:2:3 sample, and we again see that the differential data exhibit much more structure. Many tunneling recordings similar to those presented in Figs. 13.23–13.25 have appeared in the literature (e.g., Escudero *et al.*, 1990a, b; Kirtley, 1990a, b; Lesueur *et al.*, 1992; Tao *et al.*, 1992; Valles *et al.*, 1991; Wittlin *et al.*, 1988).

D. S–I–S Tunneling Measurements

Figure 13.26 shows some experimental data on tunneling across the $Al{-}Al_2O_3{-}Al$ junction formed from an oxide layer between two aluminum samples with $T_c = 1.25$ K. We see from the figure that quasiparticle tunneling is negligible at $T = 0$, becoming dominant just below $T = T_c$. The jump in current at $V = 2\Delta/e$ and $T = 0$ appears to be less than the expected 80%.

E. Energy Gap

We saw in Sections IV.C and IV.D, respectively, that N–I–S tunneling occurs

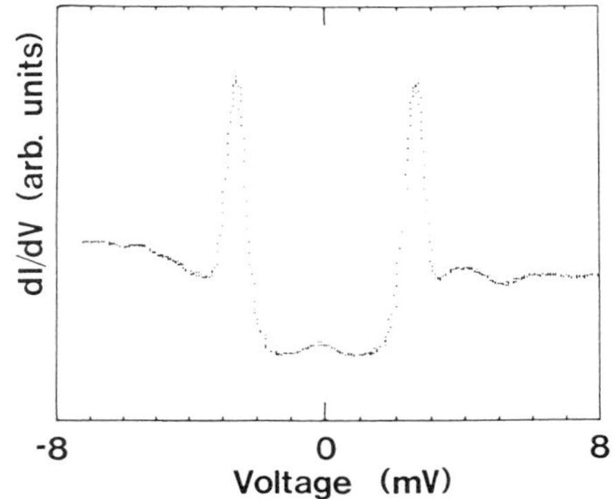

Figure 13.24 Recording of dI/dV obtained for $YBa_2Cu_3O_{6.5+x}$ at 4.2 K using a tungsten tip (Gallagher *et al.*, 1988).

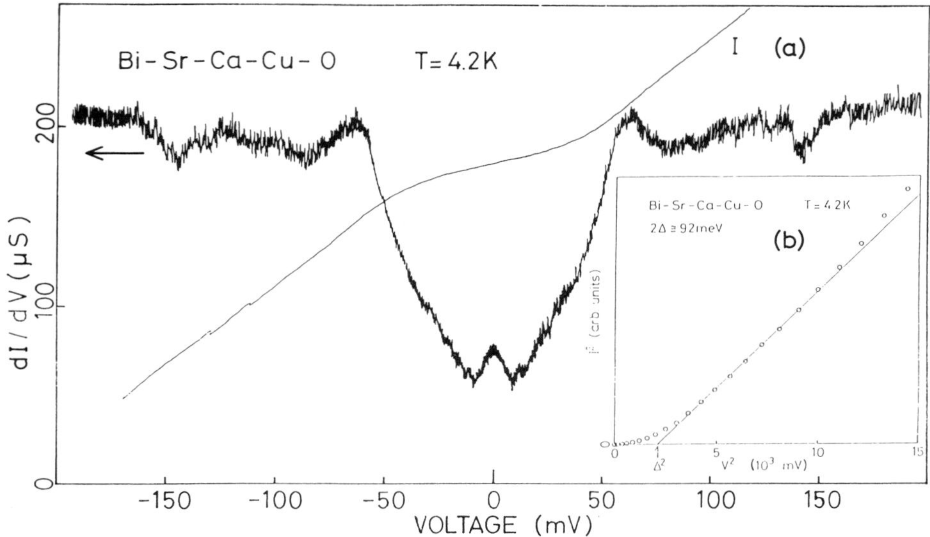

Figure 13.25 Current–voltage and dI/dT characteristic curves of $Bi_2Sr_2Ca_2Cu_3O_{10}$ determined by point contact tunneling. The inset shows a plot of I^2 versus V^2 made from the I-versus-V curve (Ekino and Akimitsu, 1989a).

for biases with magnitudes greater than Δ/e, and that S–I–S tunneling occurs for biases exceeding $2\Delta/e$, as indicated in Figs. 13.7 and 13.8. The abrupt rise in current at these biases gives us the superconducting energy gap Δ. When the two superconductors that form an S–I–S junction have different gaps Δ_1 and Δ_2, a finite temperature tunneling measurement can give us the values of both gaps, as pointed out in Section V.F and indicated in Fig. 13.16. Thus a tunneling experiment provides a

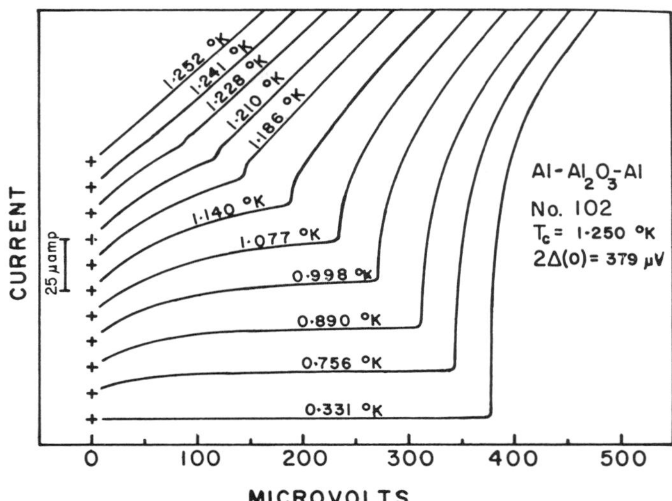

Figure 13.26 Current–voltage measurements on an $Al–Al_2O_3–Al$ tunnel junction. Zero-current positions for each curve are staggered for clarity (Blackford and March, 1968).

convenient way of measuring the energy gap.

As an example of a gap determination, note that the peaks on the derivative N–I–S tunneling curve of Fig. 13.24 are separated by 5 meV, which gives $2\Delta \approx 5$ meV. The inset of Fig. 13.25 shows the I^2 versus V^2 plot of Eq. (13.22), giving us the value of the energy gap Δ from the intercept at zero current. The temperature dependence of the energy gap, $\Delta(T)$, obtained by fitting the tunneling data to a broadened BCS density-of-states function,

$$D(E) = \mathrm{Re}\left\{ \frac{E - i\Gamma}{[(E - i\Gamma)^2 - \Delta^2]^{1/2}} \right\}, \tag{13.31}$$

where Γ is the gap broadening parameter, provided a good fit to the experimental data for two Bi superconductors, as shown in Fig. 13.27. However, the ratio

$$2\Delta/k_B T_c \approx 10.5$$

is about three times the BCS value of 3.53. Plots similar to Fig. 13.27 have been reported elsewhere (e.g., Ekino and Akimitsu, 1990; Escudero *et al.*, 1989, 1990a; Flensberg and Hansen, 1989).

Figure 13.26 shows S–I–S tunneling with the sharp rises occurring at the values 2Δ(T). We see from the figure that the gap 2Δ(T) decreases with temperature, as expected. The ratio $2\Delta/k_B T_c = 3.52$ is almost precisely the BCS value.

The ratio $2\Delta(0)/k_B T_c$ has been reported in the range 2–10 for high-temperature superconductors (Mattis and Molina, 1991); the older superconductors usually had valves near the range 3–5 (Schlesinger *et al.*, 1990a), much closer to the BCS value of 3.54. Figure 13.28 shows the dependence of Δ(0) on T_c for several high-temperature superconductors, all with reported ratios $2\Delta(0)/k_B T_c \approx 6$ (Takeuchi *et al.*, 1989).

The energy gaps of high-temperature superconductors are anisotropic (Bulaevskii and Zyskin, 1990; Mahan, 1989; Spalek and Gopalan, 1989), being much larger in the a, b-plane than in the c direction. Some reported values are $2\Delta_{ab} \approx 6.2 k_B T_c$ and $2\Delta_c \approx 2 k_B T_c$ for $(La_{1-x}Sr_x)_2CuO_4$ (Kirtley, 1990a, b), $2\Delta_{ab} \approx 8 k_B T_c$ and $2\Delta_c \approx 2.5 k_B T_c$ for $YBa_2Cu_3O_{7-\delta}$ (Collins

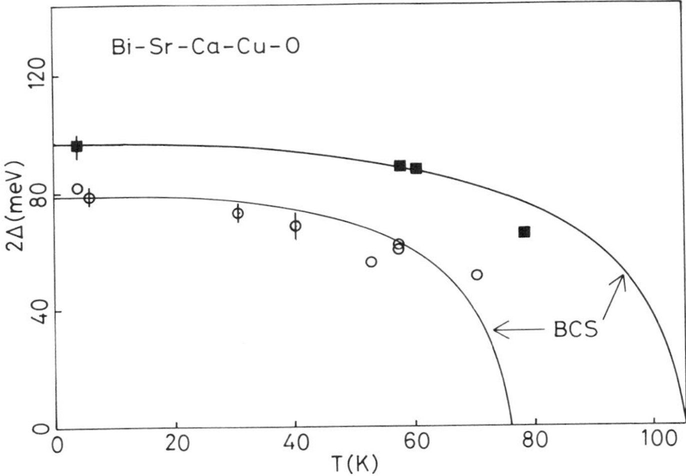

Figure 13.27 Temperature dependence of the energy gap 2Δ of $Bi_2Sr_2CaCu_2O_8$ (○) and $Bi_2Sr_2Ca_2Cu_3O_{10}$ (■) obtained by point-contact tunneling. Some of the data points have vertical error bars. The solid lines are fits to a broadened BCS density of states (Ekino and Akimitsu, 1989a).

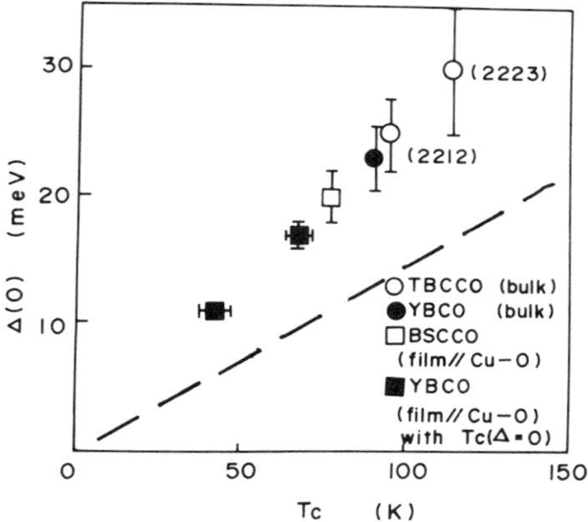

Figure 13.28 Correlation of the energy gap $\Delta(0)$ with the transition temperature T_c for bulk $YBa_2Cu_3O_7$ (YBCO), two $YBa_2Cu_3O_7$ films with tunneling in the Cu–O plane direction, $Bi_2Sr_2CaCu_2O_8$ (BSCCO) film with tunneling in the Cu–O plane direction, bulk $Tl_2Ba_2CaCu_2O_8$ (TBCCO, 2212), and bulk $Tl_2Ba_2Ca_2Cu_3O_{10}$ (2223). The dashed line is drawn for the BCS slope $2\Delta(0)/k_BT_c = 3.53$ (Takeuchi *et al.*, 1989).

et al., 1989a), $2\Delta_{ab} \approx 8k_BT_c$ for $Bi_2Sr_2CaCu_2O_{8-\delta}$, and $2\Delta_{ab} \approx 8k_BT_c$ and $2\Delta_c \approx 4k_BT_c$ for $Ba_{0.6}K_{0.4}BiO_3$ (Schlesinger *et al.*, 1990a; see also Kussmanl *et al.*, 1990; Takada *et al.*, 1989). The existence of these high anisotropies could account for much of the scatter in the reported gaps for the high-temperature superconductors.

F. Proximity Effect

We have been discussing the effect of an insulating layer between a normal metal and a superconductor or between two superconductors. If no intervening layer is present, another effect, called the *proximity effect*, comes into play. The direct contact at the junction and the overlap of wave functions causes the density n_s of electron pairs to differ in the neighborhood of the surface from its value in the bulk. At an N–S interface some electron pairs leak into the normal metal while some quasiparticles leak into the superconductor, thereby reducing the transition temperature of the superconductor. The proximity effect can cause two superconductors with different T_c that are in contact with each other to exhibit the same intermediate T_c. Theoretical treatments of this effect, such as the tunneling approach of McMillan (DiChiara *et al.*, 1991, 1993; Kadin, 1990; McMillan, 1968; Noce and Maritato, 1989; Stephen and Carbotte, 1991) have been published. (Ashida *et al.*, 1989; Broussard, 1991; Chen, 1990a, b; de Gennes, 1964; Hara *et al.*, 1993; Tanaka and Tsukada, 1990; Totsujii, 1991).

To elucidate this T_c reduction, an experimental study of composite films (Werthamer, 1963) was undertaken, each film consisting of a superconductor of thickness d_s and a normal metal of thickness d_n. Figure 13.29 shows a plot of the critical temperature T_c' of layered PbCu composite relative to $T_c = 7.2$ K of bulk Pb versus the Cu layer thickness d_n for vari-

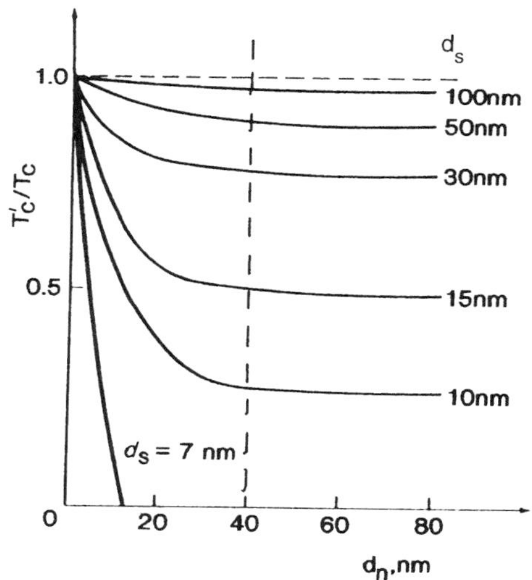

Figure 13.29 Proximity effect for a PbCu composite illustrating how the critical temperature T_c' of superconducting Pb in a copper–lead composite relative to T_c of bulk lead depends on the Cu film thickness d_n for several Pb film thicknesses d_s from 7 to 100 nm. The vertical dashed line indicates the characteristic thickness L_n (adapted from Werthamer, 1963).

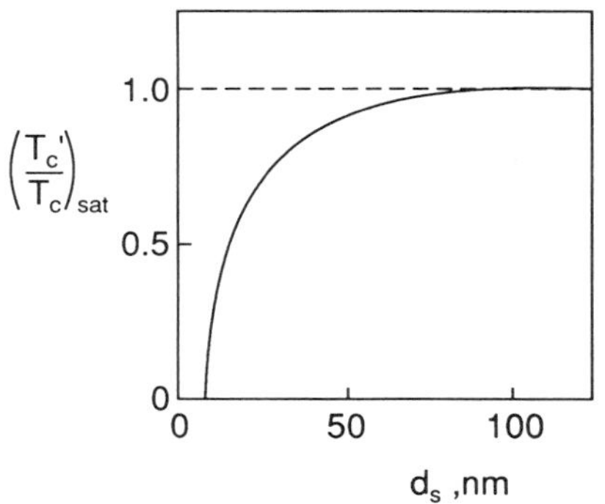

Figure 13.30 Dependence of the limiting value of the relative transition temperature $(T_c'/T_c)_{sat}$ of the PbCu composite of Fig. 13.29 on the thickness d_s of the superconducting component Pb (data from Fig. 13.29).

ous thickness d_s of Pb. The reduction of the critical temperature is small for superconductor layer thicknesses greater than the coherence length $\xi = 80$ nm irrespective of the normal layer thickness. The data show that there is a characteristic thickness $L_n \approx 40$ nm of the normal layer beyond which $(d_n > L_n)$ there is no additional reduction of the transition temperature. Van Duzer and Turner (1981, p. 301) associate this effect with the diffusion constant D of the normal metal through the expression

$$L_n = (\hbar D/2\pi k_B T)^{1/2}. \quad (13.32)$$

The thinnest sample studied, $d_s = 7$ nm, went normal before this characteristic could be attained. Figure 13.30 shows how the limiting value of T_c' obtained from Fig. 13.29 for the condition $d_n > L_n$ depends on the superconducting layer thickness.

Similar experiments have been carried out with layers of the superconductor $YBa_2Cu_3O_{7-\delta}$ containing N_y Cu–O layers ($\frac{1}{2}N_y$ unit cells thick) adjacent to N_{Pr} Cu–O layers of the nonsuperconducting material $PrBa_2Cu_3O_{7-\delta}$. The calculations by Wu *et al.* (1991a), which are compared in Fig. 13.31 with the experimental data of Lowndes *et al.* (1990), provide results comparable with those presented in Fig. 13.29 (see also Aarts *et al.*, 1990; Ashida *et al.*, 1992; Brunner *et al.*, 1991; Jakob *et al.*, 1991; Norton *et al.*, 1991; Polturak *et al.*, 1991; Terashima *et al.*, 1991; Visani *et al.*, 1990). Layered compounds are useful for studying other properties as well, such as resistivity (Minnhagen and Olson, 1992).

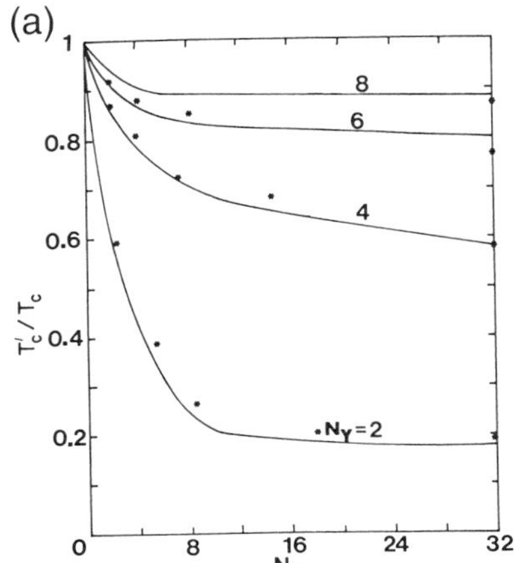

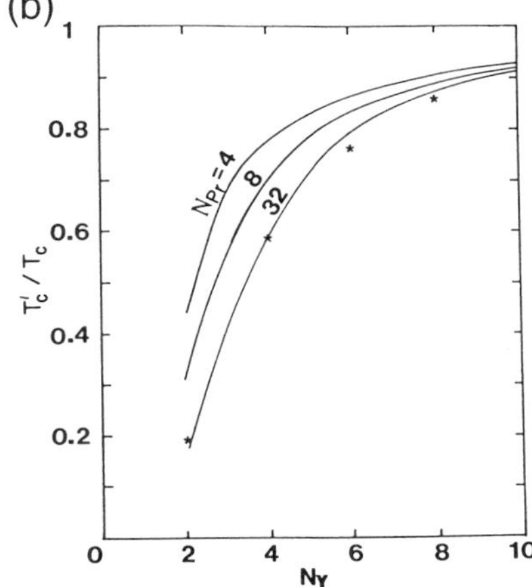

Figure 13.31 Normalized transition temperature T_c'/T_c of $YBa_2Cu_3O_{7-\delta}/PrBa_2Cu_3O_{7-\delta}$ layers plotted (a) versus the number of Cu–O planes N_{Pr} in $PrBa_2Cu_3O_{7-\delta}$, and (b) versus the number of Cu–O planes N_Y in $YBa_2Cu_3O_{7-\delta}$. The calculated curves are drawn to fit the data points (Wu *et al.*, 1991a).

Radousky (1992) reviewed the superconducting and normal state properties of the $Y_{1-x}Pr_xBa_2Cu_3O_{7-\delta}$ system.

Proximity junctions are (S–N–S) Josephson junctions in which the Cooper pair and quasiparticle transfer arises from the proximity effect (Agrait *et al.*, 1992; Braginski, 1991; Claasen *et al.*, 1991; Gijs *et al.*, 1990a; Han *et al.*, 1990a; Harris *et al.*, 1991; Jung *et al.*, 1990; Klein and Aharony, 1992; Maritato *et al.*, 1988; Polturak *et al.*, 1991; J. Yu *et al.*, 1991). Studies have also been carried out on S-Semicond-S or S-Semicond junctions (Furusaki *et al.*, 1991, 1992; Kastalsky *et al.*, 1991; Kleiner *et al.*, 1992; van Wees *et al.*, 1991) and arrays (Hebboul and Garland, 1991; Kwong *et al.*, 1992; Lerch *et al.*, 1990; Sohn *et al.*, 1992).

G. Even–Odd Electron Number Effect

Measurements of single-electron tunneling through a small superconducting island of volume 3×10^6 containing 6×10^8 conduction electrons exhibited a $2e$ periodicity in the tunneling current for $T < 0.2T_c$. Such a parity effect arises from the electron pairing, whereby the free energy of the superconducting island depends on whether there is an even or odd number of electrons in the island (Tuominen *et al.*, 1993).

VII. JOSEPHSON EFFECT

Until now we have been discussing the participation of quasiparticles in tunneling. The S–I–S processes that have concerned us included the strong tunneling current that flows between an occupied super electron band and an empty quasiparticle band ($S \to Q$), as well as the relatively weak tunneling between two quasiparticle bands ($Q \to Q$). There is also a third case—tunneling between two occupied super electron bands at zero bias ($S \to S$). In this process there is transfer of Cooper pairs across the junction through an effect predicted by Josephson in 1962 and observed experimentally shortly thereafter (Anderson and Rowell, 1963). Figure 13.32 compares these processes. In the following treatment we assume that the Josephson junction is of the weak-link type referred to in Section VI.A.

A. Cooper Pair Tunneling

When two superconductors are separated by a thin layer of insulating material, electron pairs are able to tunnel through the insulator from one superconductor to the other. There are four modes of pair tunneling: (1) the dc Josephson effect, or flow of a dc current $J = J_0 \sin \delta$ across the junction in the absence of an applied electric or magnetic field, where δ is a phase factor and J_0 the maximum zero voltage current, (2) the ac Josephson effect, relating to the flow of a sinusoidal current, $J = J_0 \sin[\delta - (4\pi eVt/h)]$, across a junction with an applied voltage V, where $\nu = 2eV/h$ is the frequency of oscillation, (3) the inverse ac Josephson effect, whereby dc voltages are induced across an unbiased junction by incident radiation or an impressed rf current, and (4) macroscopic quantum interference effects, involving a tunneling current J with an oscillatory dependence on the applied magnetic flux $\sin(\pi\Phi/\Phi_0)$, where Φ_0 is the quantum of magnetic flux.

B. dc Josephson Effect

In deriving the basic equations for the dc Josephson effect we follow the classic approach of Feynman (1965). Consider two superconductors, 1 and 2, separated by an insulating barrier, as shown in Fig. 13.11. If the barrier is thick enough so that the superconductors are isolated from each other, the time-dependent Schrödinger

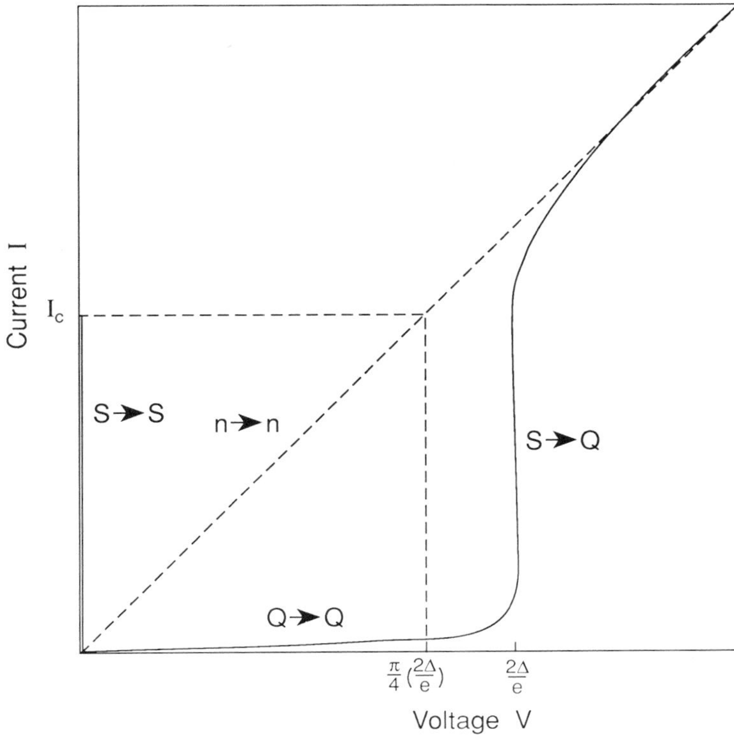

Figure 13.32 Current–voltage characteristic curves for tunneling via the S → S, Q → Q, and S → Q processes. All three processes follow the linear n → n tunneling slope at high voltages, above $2\Delta/e$.

equation for each side is

$$i\hbar\frac{d\Psi_1}{dt} = H_1\Psi_1, \qquad (13.33a)$$

$$i\hbar\frac{d\Psi_2}{dt} = H_2\Psi_2, \qquad (13.33b)$$

where Ψ_i and H_i are the wavefunctions and Hamiltonians on either side of the barrier. We assume that a voltage V is applied between the two superconductors. If the zero of potential is assumed to occur in the middle of the barrier between the two superconductors, superconductor 1 will be at the potential $-\frac{1}{2}V$ with Cooper-pair potential energy +eV, while superconductor 2 will be at the potential $+\frac{1}{2}V$ with Cooper-pair potential energy $-eV$. (The factor of $\frac{1}{2}$ does not appear in the potential energy terms because the charge of each Cooper pair is $2e$.)

The presence of the insulating barrier couples together the two equations,

$$i\hbar\frac{d\Psi_1}{dt} = eV\Psi_1 + K\Psi_2, \qquad (13.34a)$$

$$i\hbar\frac{d\Psi_2}{dt} = -eV\Psi_2 + K\Psi_1, \qquad (13.34b)$$

where K is the coupling constant for the wavefunctions across the barrier. Since the square of each wavefunction is the probability density that super electrons are present, the two wavefunctions can be written in the form

$$\Psi_1 = (n_{s1})^{1/2}e^{i\Theta_1}, \qquad (13.35a)$$

$$\Psi_2 = (n_{s2})^{1/2}e^{i\Theta_2}, \qquad (13.35b)$$

$$\phi = \Theta_2 - \Theta_1, \qquad (13.35c)$$

where n_{s1} and n_{s2} are the densities of super electrons in the two superconductors and ϕ is the phase difference across the barrier. If the two wavefunctions (13.35a) and (13.35b) are substituted in the coupled wave equations (13.34) and the results separated into real and imaginary parts, we obtain equations for the time dependence of the pair densities and the phase difference:

$$\hbar \frac{d}{dt} n_{s1} = 2K(n_{s1}n_{s2})^{1/2} \sin \phi, \qquad (13.36a)$$

$$\hbar \frac{d}{dt} n_{s2} = -2K(n_{s1}n_{s2})^{1/2} \sin \phi, \qquad (13.36b)$$

$$\frac{d}{dt}\phi = \frac{2e}{\hbar} V. \qquad (13.37)$$

We can specify the current density in terms of the difference between Eqs. (13.36a) and (13.36b) times e

$$J = e\frac{d}{dt}(n_{s1} - n_{s2}), \qquad (13.38)$$

which has the value

$$J = J_c \sin \phi, \qquad (13.39)$$

where

$$J_c = \frac{4eK(n_{s1}n_{s2})^{1/2}}{\hbar}, \qquad (13.40)$$

and the coupling constant K is an unknown quantity. Equations (13.37) and (13.39) are called Josephson relations; they are the basic equations for the tunneling behavior of Cooper pairs. Multiplying Eq. (13.39) by the area A of the junction gives the current $I = JA$,

$$I = I_c \sin \phi, \qquad (13.41)$$

where $I_c = J_c A$ is the critical current.

Ambegaokar and Baratoff (1963a, b; cf. Aponte *et al.*, 1989) showed that for Cooper-pair tunneling between two identical superconductors with temperature-dependent gaps, $\Delta(T)$, the critical current is given by

$$I_c(T) = \tfrac{1}{4}\pi G_n[2\Delta(T)/e] \times \tanh[\Delta(T)/2k_B T], \qquad (13.42)$$

where the normal tunneling conductance G_n is given by Eq. (13.22). In the respective limiting cases $T \to 0$ and $T \to T_c$, this is

$$I_c(0) = \tfrac{1}{4}\pi G_n[2\Delta(0)/e] \quad T \approx 0 \quad (13.43a)$$

$$I_c(0) = \tfrac{1}{4}\pi G_n\left[\Delta^2(T)/ek_B T_c\right] \quad T \approx T_c. \quad (13.43b)$$

The voltage $\tfrac{1}{4}\pi(2\Delta/e)$ on the right side of Eq. (13.43) has the physical significance indicated in Fig. 13.32. Thus the maximum Josephson current $I_c(0)$, which occurs for $T = 0$ and $\phi = \pi/2$, is equal to $\tfrac{1}{4}\pi$, or $\approx 80\%$ of the normal-state current at the gap voltage $V = 2\Delta/e$. Figure 13.33 com-

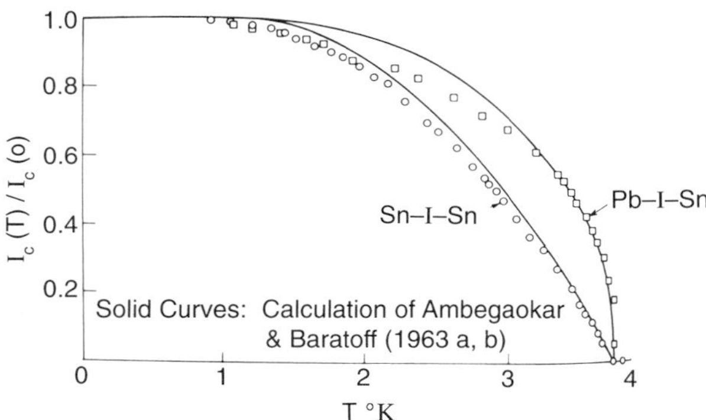

Figure 13.33 Temperature dependence of the maximum zero-voltage current, showing fit of theoretical curves to the experimental tunneling data of Pb–I–Sn (□) and Sn–I–Sn (○) junctions. The normalized tunneling current $I_c(T)/I_c(0)$, is plotted (Fiske, 1964).

pares the temperature dependence of the maximum zero-voltage tunneling currents of an Sn–I–Sn junction measured by Fiske (1964) with the values predicted by Eq. (13.42). The fit of the Pb–I–Sn Josephson junction data to the same theory is also shown.

Copper-oxide superconductors are often granular in texture with Josephson junctions forming at the intergranular boundaries and perhaps at defect centers as well. Current flows through and between the Josephson junctions, and sometimes the intrajunction phases are favorable for the formation of complete circuits, which can produce flux shielding (Jung *et al.*, 1990; vide Doyle and Doyle, 1993). There is a grain-decoupling (or phase-locking) temperature T_g below which the Josephson junction network exhibits coherent properties, as well as a grain depairing (or critical) temperature T_c below which individual grains superconduct, where $T_g < T_c$ (Sergeenkov and Ausloos, 1993).

C. ac Josephson Effect

We have been discussing the dc effect, whereby a phase difference $\phi = \Theta_2 - \Theta_1$ between either side of a superconductor junction causes a dc current to spontaneously flow at zero voltage. Now let us examine what happens when a dc voltage is applied across the junction.

From Eq. (13.37) we know that a rate of change of phase accompanies the presence of a voltage across a Josephson junction. Since the applied voltage is a constant, this equation can be integrated directly to give

$$\phi(t) = \phi_0 + \left(\frac{2e}{\hbar}\right)Vt, \qquad (13.44)$$

which provides a characteristic frequency ν_J known as the Josephson frequency

$$\nu_J = \frac{2eV}{h} = \frac{V}{\Phi_0} = 483.6 \times 10^{12}\ V\ \text{Hz}, \qquad (13.45)$$

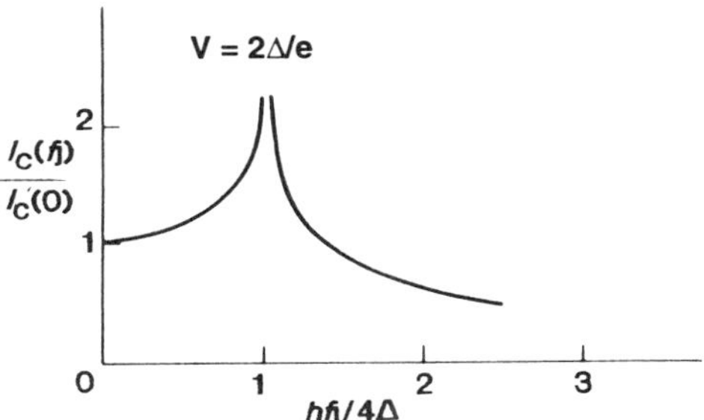

Figure 13.34 Dependence of the critical current on the Josephson frequency, showing the peak at the applied dc voltage $V = 2\Delta/e$ (Van Duzer and Turner, 1981, p. 144).

where Φ_0 is the quantum of flux (Tsai *et al.*, 1983). A more practical expression to remember is

$$\frac{\nu_J}{V} = 483.6\ \text{MHz}/\mu\text{V}. \qquad (13.46)$$

With the aid of these expressions and Eq. (13.39), the critical current density can be written in the form

$$J = J_c \sin(\omega_J t + \phi_0), \qquad (13.47)$$

where $\omega_J = 2\pi\nu_J$.

It can be shown that the critical current density J_c depends on the frequency in terms of the voltage and that it reaches a maximum when the applied voltage is equal to the gap voltage, $V = 2\Delta/e$. This voltage dependence of $J_c(V)$, which was predicted by Reidel (1964) and confirmed experimentally by Hamilton (1972), is sketched in Fig. 13.34.

The ac Josephson effect that we have been describing occurs when current flows across a junction at the frequency given by Eq. (13.45) when a dc voltage V is applied across it. There is also an inverse ac Josephson effect, whereby a dc voltage is induced across the junction when an ac current is caused to flow through it, or when an electromagnetic field is incident on it. This will be discussed in Section VII.E.

D. Driven Junctions

The Josephson relations (13.41) and (13.37),

$$I = I_c \sin \phi, \qquad (13.48)$$

$$\frac{d}{dt}\phi = \frac{2e}{\hbar}V, \qquad (13.49)$$

apply to an idealized case in which all the current is carried by electron pairs. In the more general case there can be other types of current flowing, such as displacement current, quasiparticle tunneling current, and perhaps conduction current, if the barrier is not a perfect insulator. It will be instructive to analyze the junction in terms of the equivalent circuit shown in Fig. 13.35, which contains the current source $I_c \sin \phi$ of the junction, a capacitor to represent the displacement current, and a conductance to account for the quasiparticle tunneling and capacitor leakage currents. We assume that the dc current source $I = I(V)$, shown on the left, drives the junction circuit.

The differential equation for the current flow I in the equivalent circuit is

$$I = I_c \sin \phi + GV + C\frac{dV}{dt}, \qquad (13.50)$$

where G is assumed to be constant, although in a more general analysis it can be taken as voltage dependent. Equation (13.49) can be used to eliminate the voltage and write the circuit equation in terms of the phase $\phi(t)$:

$$I = \frac{\hbar C}{2e}\cdot\frac{d^2\phi}{dt^2} + \frac{\hbar G}{2e}\cdot\frac{d\phi}{dt} + I_c \sin \phi. \qquad (13.51)$$

With the aid of the Josephson angular frequency $\omega_c = (2e/\hbar)V_c$ for the voltage V_c,

$$V_c = I_c/G, \qquad (13.52)$$

obtained from Eq. (13.45), and a new dimensionless variable Θ,

$$\Theta = \omega_c t, \qquad (13.53)$$

the circuit equation assumes a simplified form,

$$\frac{I}{I_c} = \beta_c\frac{d^2\phi}{d\Theta^2} + \frac{d\phi}{d\Theta} + \sin \phi, \qquad (13.54)$$

where β_c is the admittance ratio,

$$\beta_c = \omega_c C/G. \qquad (13.55)$$

The solutions to this second-order differential equation exhibit complex time variations of the current. We will not try to interpret these time dependences, and instead we will find the average value of the voltage, from Eq. (13.49)

$$V = \langle V \rangle = \left\langle \frac{d\phi}{dt} \right\rangle \frac{\hbar}{2e}, \qquad (13.56)$$

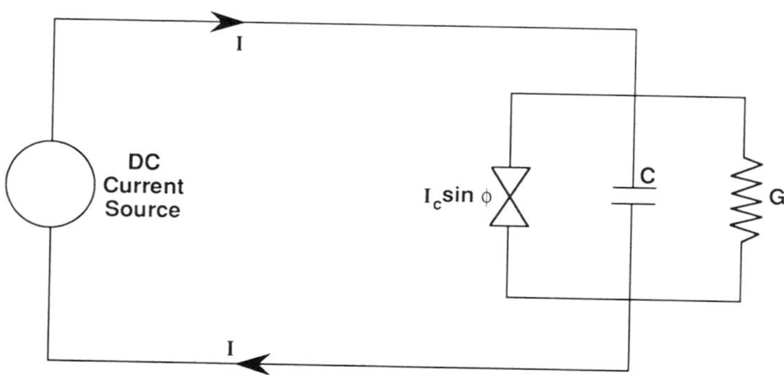

Figure 13.35 Josephson junction represented by the parallel circuit on the right consisting of a junction current source $I_c \sin \phi$, a capacitor C, and a conductance G. The circuit is driven by the dc current source I shown on the left.

for simple cases. We readily see from the form of Eq. (13.54) that when $I \le I_c$ there is a solution corresponding to Eq. (13.48),

$$I = I_c \sin \phi \qquad I \le I_c, \qquad (13.57)$$

with all of the time derivatives equal to zero. In other words, this is the zero-voltage solution. At the other extreme, when $I \gg I_c$, the term $I_c \sin \phi$ becomes negligible, and we can use Eq. (13.49) to obtain the constant-voltage solution

$$I = GV \qquad I \gg I_c, \qquad (13.58)$$

where $dV/dt = 0$. Hence, from Eq. (13.49) we have $d^2\phi/dt^2 = 0$ in this limit. The situation is more complex for driving currents that are near the critical current I_c.

The case in which $C \approx 0$, so that $\beta_c \ll 1$, which corresponds to

$$\frac{I}{I_c} = \frac{d\phi}{d\Theta} + \sin \phi, \qquad (13.59)$$

can be solved analytically (see Problem 9), and has the solution

$$V = 0 \qquad \text{for } I < I_c, \qquad (13.60a)$$

$$V = V_c\left[\left(\frac{I}{I_c}\right)^2 - 1\right]^{1/2} \qquad \text{for } I > I_c, \qquad (13.60b)$$

$$V \approx \frac{I}{G} \qquad \text{for } I \gg I_c. \qquad (13.60c)$$

This is plotted in Fig. 13.36a. Pairs of arrows pointing in opposite directions mean that there is no hysteresis. Figure 13.37 shows how the voltage oscillates with the average values from Eqs. (13.60b) and (13.60c), respectively, indicated by points A and B in Fig. 13.36a (for further details, see Orlando and Delin, 1991, pp. 458ff; Van Duzer and Turner, 1981, pp. 170ff).

When $\beta_c \gg 1$, the two solutions (13.57) and (13.58) apply with $\phi = \pi/2$ so that $\sin \phi = 1$. The I-versus-V characteristic plotted in Fig. 13.36c shows that there is a hysteresis in which V remains pinned at the value $V = 0$ as the current is initially increased from zero until the critical current is reached, at which point the voltage jumps to the value $V = I_c/G$ and thereafter follows the diagonal line upwards. Subsequent reduction of the voltage follows the diagonal line into the origin.

For immediate values of β_c, the I-versus-V characteristic follows the behavior illustrated in Fig. 13.36b for the case $\beta_c = 4$. Again there is hysteresis, with the initial rise of the current to I_c and its return to the value I_{min} at $V = 0$. Figure 13.38 shows how I_{min} depends on the value of β_c.

The solutions that we have been discussing were average values (13.56) of the voltage V involving a time average $\langle d\phi/dt \rangle$ of the derivative of the phase. The voltage itself oscillates in time, and Fig. 13.37 shows two examples of these oscillations.

E. Inverse ac Josephson Effect

We have found that applying a dc voltage across a Josephson junction causes an ac current to flow. In the reverse ac Josephson experiment, dc voltages are induced across an unbiased junction by introducing an rf current into the junction,

To explain this effect we assume that the Josephson junction can be represented by the parallel equivalent circuit of Fig. 13.39. The circuit consists of the usual Josephson current $I_c \sin \phi$ in parallel with a conductance G. In addition, it has as inputs a dc source current I_0 and an rf source current $I_s \cos \omega_s t$, with the total source current I given by

$$I(t) = I_0 + I_s \cos \omega_s t. \qquad (13.61)$$

When $I(t)$ is inserted into Eq. (13.50), a nonlinear differential equation that is difficult to solve results. A numerical solution provides the staircase I versus V characteristic presented in Fig. 13.40. Measurements carried out by Taur *et al.* (1974)

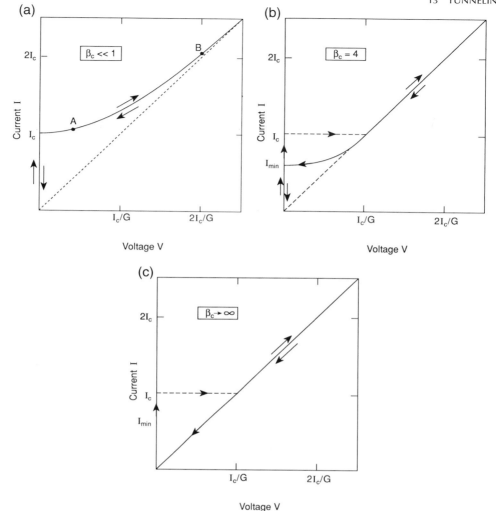

Figure 13.36 Current–voltage characteristics, I, versus V, for the Josephson junction circuit of Fig. 13.35 with: (a) negligible capacitance, $\beta_c \ll 1$, (b) appreciable capacitance, $\beta_c = 4$, and (c) dominating capacitance, $\beta_c \to \infty$, where $\beta_c = \omega_c C/G$.

with a 35-GHz source satisfying the condition $\omega_s = 0.16\omega_c$ compare well with the calculated curves shown in the figure, where the zero rf power curve ($I_s = 0$) is shown for comparison. This staircase pattern, which is referred to as *Shapiro steps* (Eikmans and van Himbergen, 1991; Shapiro, 1963; W. Yu *et al.*, 1992), has been reported by many observers (e.g., Kriza *et al.*, 1991; Kvale and Hebboul, 1991; Larsen *et al.*, 1991; H. C. Lee *et al.*, 1991; Rzchowski *et al.*, 1991; Sohn *et al.*, 1991).

It is mathematically easier to analyze this problem in terms of the circuit of Fig. 13.41 where the source is an applied voltage,

$$V(t) = V_0 + V_s \cos \omega_s t. \quad (13.62)$$

It is shown in Van Duzer and Turner (1981; see also Orlando and Delin, 1991) that the current $I(t)$ through the Joseph-

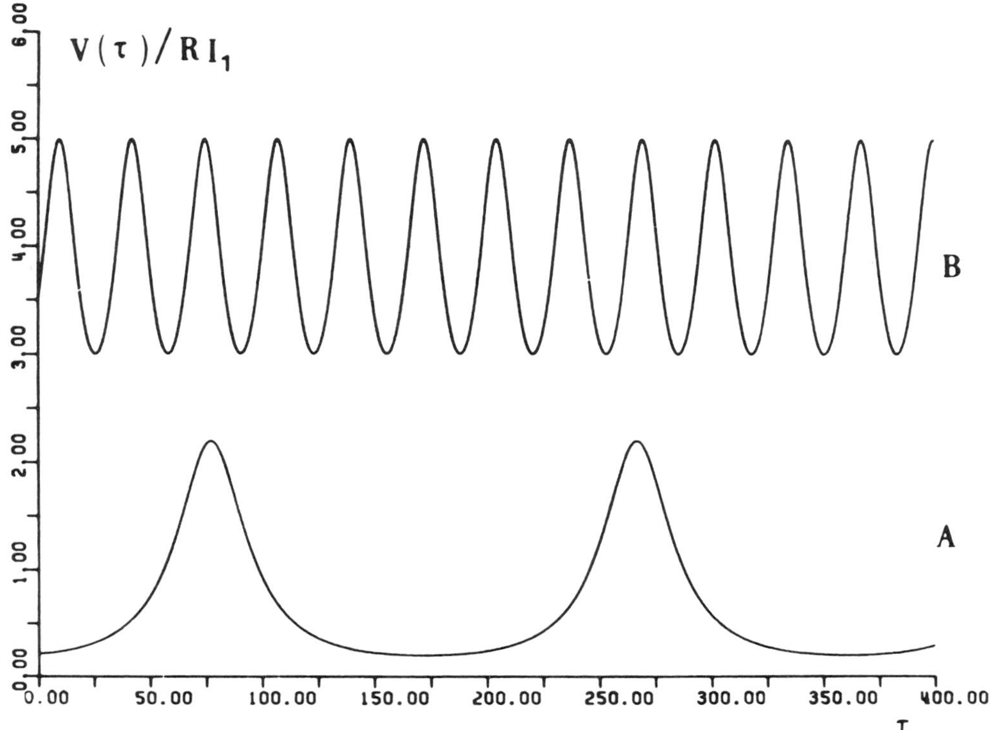

Figure 13.37 Voltage oscillations across the Josephson junction of Fig. 13.35 for the negligible capacitance case $\beta_c \ll 1$, and small and large dc bias voltages as marked at points A and B, respectively, of Fig. 13.36a (Barone and Paterno, 1982, p. 128).

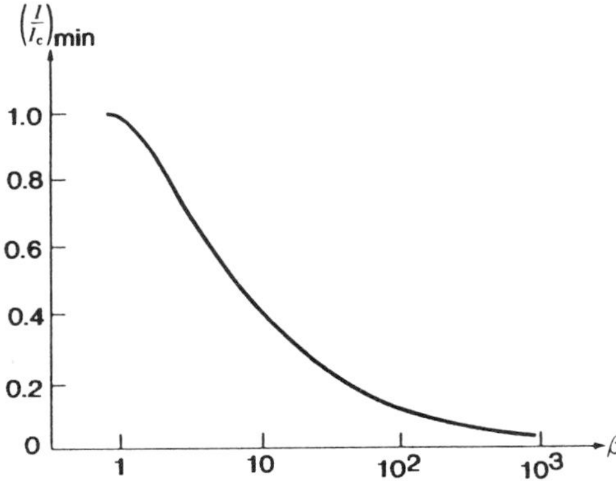

Figure 13.38 Dependence on β_c of the minimum current I_{min} (indicated in Fig. 13.36b) in the circuit of Fig. 13.35 when the current is decreased from values above I_c (Van Duzer and Turner, 1981, p. 173).

son junction can be written as an infinite series of products of Bessel functions J_n and sine waves,

$$I(T) = I_c \sum_n (-1)^n J_n\left(\frac{2eV_s}{\hbar\omega_s}\right) \times \sin[(\omega_J - n\omega_s)t + \phi'], \quad (13.63)$$

where ϕ' is a constant of integration. Since the I versus V characteristic is drawn for the average current, $I \approx \langle I(t)\rangle$, and since the sine term averages to zero unless $\omega_J = n\omega_s$, there are spikes appearing on this characteristic for voltages equal to

$$V = \frac{n\hbar\omega_s}{2e}, \quad (13.64)$$

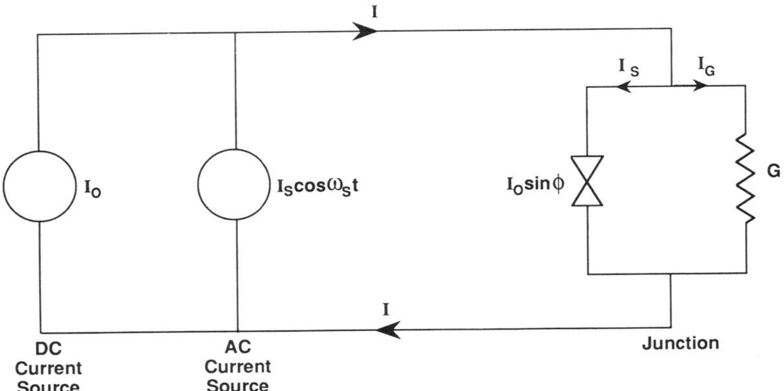

Figure 13.39 Equivalent circuit of a Josephson junction represented by a junction current $I_c \sin\phi$ in parallel with a conductance $G = 1/R$ irradiated with rf power. The junction is shown driven by a dc current source I_0 in parallel with an rf current source $I_s \cos \omega_s t$.

with the maximum amplitude,

$$I_{\max} = I_c J_n\left(\frac{2eV_s}{\hbar\,\omega_s}\right), \qquad (13.65)$$

occurring for the phase $\phi' = \pi/2$. Figure 13.42 shows these spikes at intervals proportional to the source frequency and indicates their maximum amplitude range. The

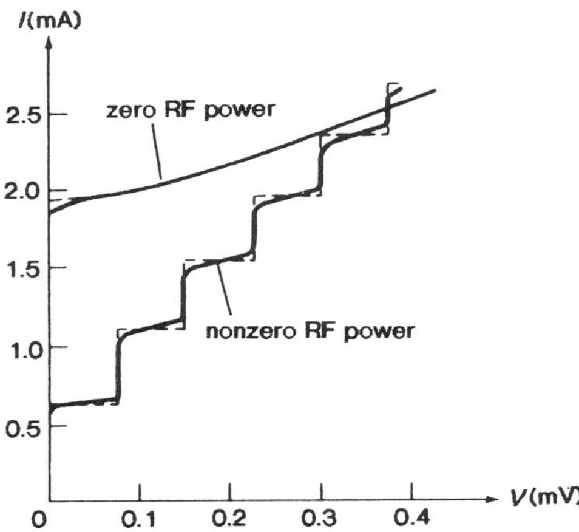

Figure 13.40 Current–voltage characteristics of a point-contact Josephson junction with applied rf power at 35 GHz, for the ac current source of Fig. 13.39. The solid curves calculated for 100 K thermal noise fit the experimental data well. The dashed line is from calculations done without noise (Van Duzer and Turner, 1981, p. 184).

current can be anywhere along a particular spike, depending on the initial phase (see Orlando and Delin, 1991).

Estéve *et al.* (1987) reported that an LaSrCuO sample with the current–voltage characteristic shown in Fig. 13.43a exhibited the I–V characteristic of Fig. 13.43b when irradiated with x-band (9.4 GHz) microwaves. The microwaves produced the spike-step pattern we have already described. The researchers attributed the results to the beating of the oscillating Josephson supercurrent with the microwaves. The separation in voltage between these steps is proportional to the microwave frequency, and their amplitude is Bessel-like. The Josephson junction characteristics were observed even when the point-contact metal tip was itself superconducting, which indicates that the junction was inside the material underneath the tip. There is a rather extensive literature (e.g. Bhagavatula *et al.*, 1992; Chi and Wanneste, 1990; Marcon *et al.*, 1989; Monaco, 1990a, b; and Hübler, 1990).

F. Analogues of Josephson Junctions

Josephson tunneling involves a quantum phenomenon that is difficult to grasp intuitively. This is especially true when we

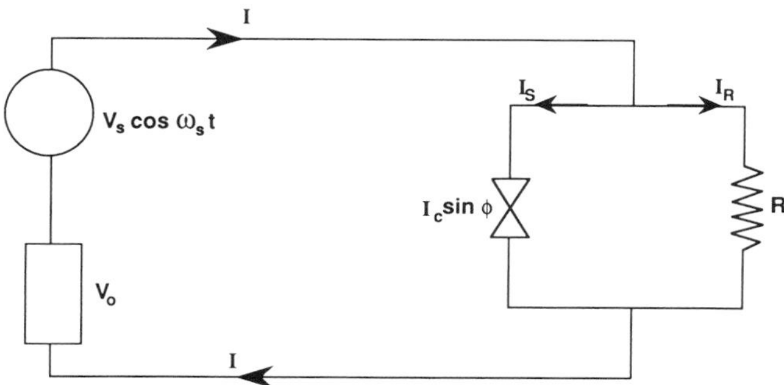

Figure 13.41 Josephson junction of Fig. 13.39 driven by a dc voltage V_0 in series with an rf voltage $V_s \cos \omega_s t$.

try to picture how the total current flowing through a Josephson junction depends on the phase difference of the electron pairs on either side of the junction. The differential equation for this phase difference ϕ happens to be the same as the differential equation for the rotational motion of a driven pendulum. We will describe this motion and then relate it to the Josephson junction.

Consider a simple pendulum consisting of a mass M attached to a pivot by a

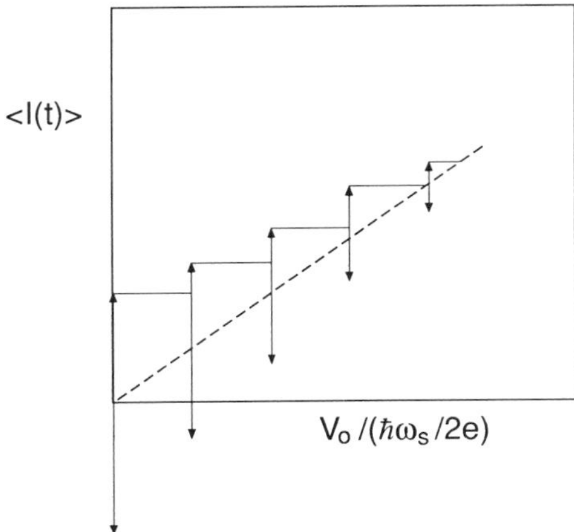

Figure 13.42 I–V characteristic for the dc component of the current versus applied dc voltage V_0 of the equivalent circuit of Fig. 13.41. The value of the current can be anywhere along a particular current spike, depending on the initial phase. The dashed line is for $\langle I(t)\rangle = V_0 G$.

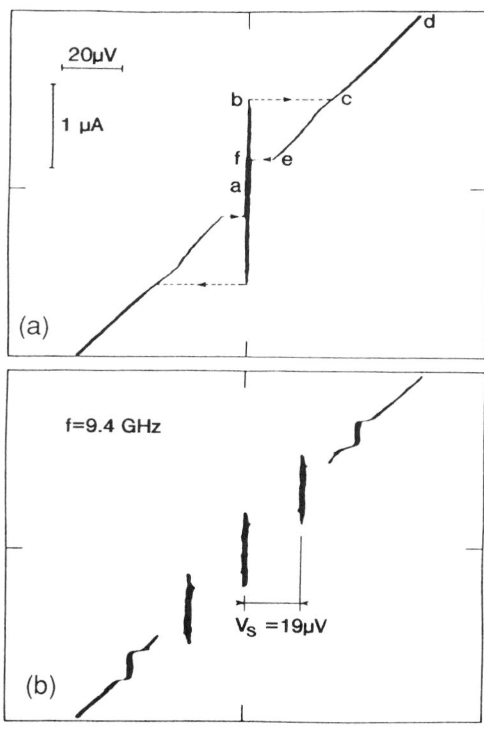

Figure 13.43 Oscilloscope presentation of current-versus-voltage characteristics of a tunnel junction at 4.2 K formed by an Al tip on a $(La_{0.925}Sr_{0.075})_2CuO_4$ sample (Estève *et al.*, 1987). (a) Trace obtained in the absence of rf power, the letters a through f giving the sense of the trace and the dashed lines indicating switching between branches, (b) steps induced by incident microwave radiation at 9.4 GHz (Estève *et al.*, 1987).

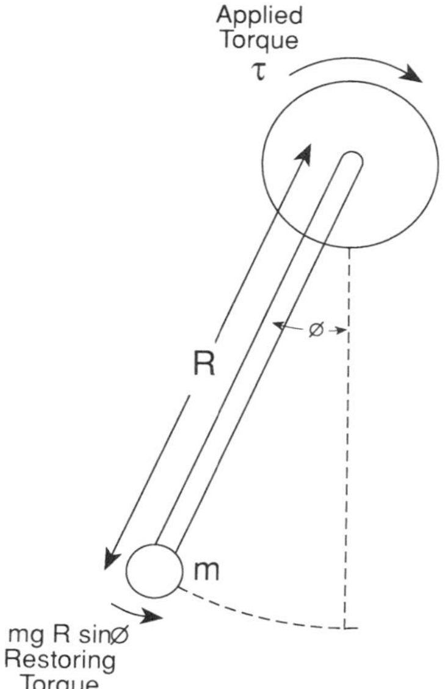

Figure 13.44 Pendulum model of a Josephson junction showing the counterclockwise restoring torque $mgR \sin \phi$ arising from the presence of a clockwise applied torque τ.

massless rod of length R. If a constant torque τ is applied by a motor, it will move the mass through an angle ϕ, as shown in Fig. 13.44. We know from our study of mechanics that the force of gravity acting on the mass m produces a restoring torque $mgR \sin \phi$. For a relatively small applied torque the pendulum assumes an equilibrium position at the angle given by

$$\tau = mgR \sin \phi \qquad \left(\frac{d\phi}{dt} = 0 \right), \quad (13.66)$$

as indicated in Fig. 13.45b. The greater the torque, the larger the angle ϕ. There is a critical torque τ_c indicated in Fig. 13.45c for the angle $\phi = \pi/2$,

$$\tau_c = mgR. \quad (13.67)$$

If the applied torque exceeds this critical value, the pendulum will continue its motion beyond the angle $\phi = \pi/2$ and rotate continuously as long as the applied torque $\tau > \tau_c$ operates. The motion is fast at the bottom and slow at the top, corresponding to a large angular velocity $\omega = d\phi/dt$ at the bottom and a small ω at the top. For a large torque, $\tau \gg mgR$, the average angular velocity of the motion $\langle \omega \rangle$ increases linearly with the torque, reaching a limit determined by retarding drag forces coming from, for example, the viscosity η of the air or mechanical friction. The drag force is assumed to be proportional to the angular velocity ω, and is written as $\eta\omega$.

The dependence of the average angular velocity on the applied torque is shown in Fig. 13.46. We see from the figure that ω remains zero as the torque τ is increased until the critical value $\tau_c = mgR$ of Eq. (13.67) is reached. Beyond this point ω jumps to a finite value and continues to rise in the way we have already described. If the torque is now decreased down from a large magnitude, once it passes the critical value (13.67) the pendulum will have sufficient kinetic energy to keep it rotating

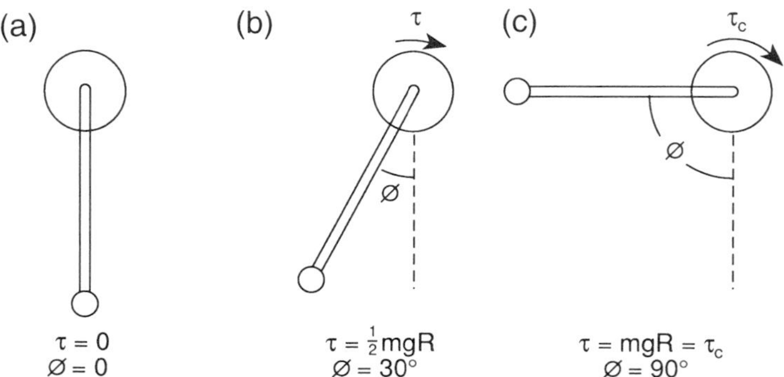

Figure 13.45 Pendulum (a) with no applied torque, $\tau = 0$, (b) with the torque $\tau = \frac{1}{2}mgR$, and (c) with the critical torque applied, $\tau_c = mgR$.

for torques below τ_c, as indicated in the figure. The torque must be reduced much further, down to the value τ_c', before friction begins to dominate and motion stops, as indicated in the figure. Thus we have hysteresis of motion for low applied torques, and no hysteresis for high torques. If we compare Fig. 13.46 with Fig. 13.36b, we see that the torque–angular velocity characteristic curve of the driven pendulum has the same shape as the current–voltage characteristic of the Josephson junction.

The correspondence between the driven pendulum and a Josephson junction can be demonstrated by writing down a differential equation that governs the motion of the pendulum, setting the applied torque τ equal to the rate of change of the angular momentum L,

$$\frac{d}{dt}L = mR^2\frac{d\omega}{dt}, \qquad (13.68)$$

and then adding the restoring and damping torques,

$$\tau = mR^2\frac{d^2\phi}{dt^2} + \eta\frac{d\phi}{dt} + mgR\sin\phi, \qquad (13.69)$$

where mR^2 is the moment of inertia; here we have made use of the expression $\omega = d\phi/dt$. This equation is mathematically equivalent to its Josephson counterpart (13.51), so we can make the following identifications:

applied current	$I \leftrightarrow \tau$	applied torque
average voltage term	$V(2e/\hbar) = d\phi/dt \leftrightarrow$	average angular velocity $\omega = d\phi/dt$
phase difference	$\phi \leftrightarrow \phi$	angular displacement
capacitance term	$\hbar C/2e \leftrightarrow mR^2$	moment of inertia
conductance term	$\hbar G/2e \leftrightarrow \eta$	viscosity
critical current	$I_c \leftrightarrow mgR$	critical torque

This analogue has been found useful in the study of the behavior of Josephson junctions.

Another mechanical device that illustrates Josephson junction-type behavior is the washboard analogue sketched in Fig. 13.47, in which a particle of mass m moves down a sloped sinusoidal path in a viscous fluid, passing through regularly spaced minima and maxima along the way.

Electrical analogues have been proposed (Bak and Pedersen, 1973; Hamilton, 1972; Hu and Tinkham, 1989; cf. Goodrich and Srivastava, 1992; Goodrich *et al.*, 1991) that do not give as much insight into Josephson junction behavior as the mechanical analogues, however, though they are useful for studying the behavior of Josephson junctions when the parameters are varied.

VIII. MAGNETIC FIELD AND SIZE EFFECTS

Until now we have assumed that no magnetic fields are applied, and that the currents circulating in the Josephson junctions produce a negligible amount of magnetic flux. The next few sections examine the effect of applying a magnetic field parallel to the plane of a single Josephson junction as well as perpendicular to a loop containing two such junctions. To determine how the presence of the field affects the phase Θ, we make use of Eq. (5.33):

$$\oint \nabla\Theta \cdot d\mathbf{l} = \frac{2\pi}{\Phi_0}\oint \mathbf{A}\cdot d\mathbf{l}. \qquad (13.70)$$

This shows that $\nabla\Theta$ and the vector potential $\mathbf{A}$ play similar roles in determining the phase. In writing out this expression we have assumed that the line integration is performed over regions of the supercon-

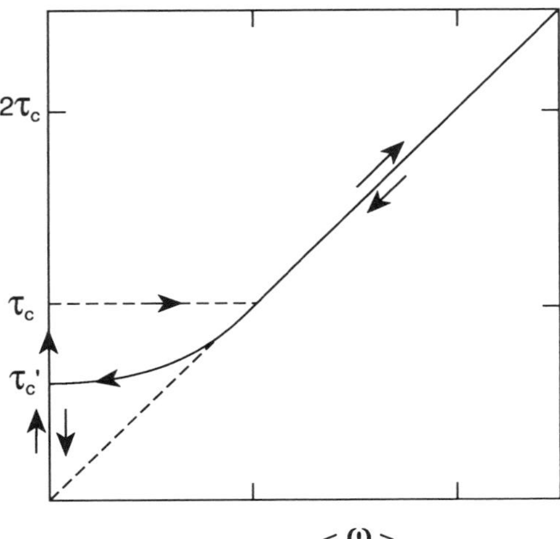

Figure 13.46 Relationship between the average angular velocity of the pendulum $\langle \omega \rangle$ and the applied torque τ. For low applied torques the pendulum oscillates and the average velocity is zero, whereas at high torques, $\tau > \tau_c$, motion is continuous with $\langle \omega \rangle$ proportional to τ. Note the hysteresis for increasing and decreasing torques.

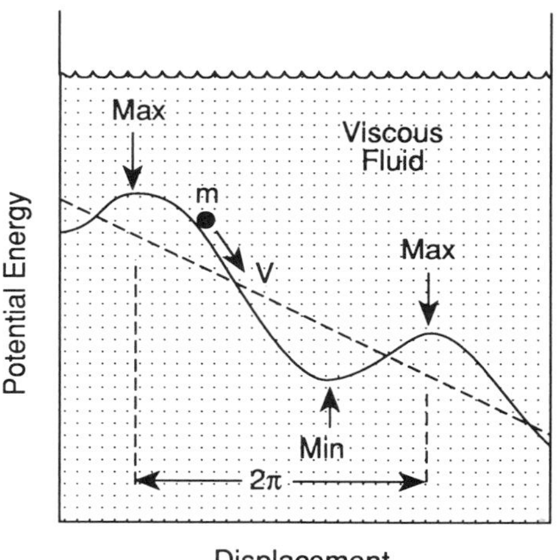

Figure 13.47 Washboard analogue of the Josephson junction showing a particle of mass m descending along a sloped wavy path in a viscous fluid.

ductor where the current density is either zero or makes no contribution to the integral, so that the term $\mathbf{J} \cdot d\mathbf{l}$ of Eq. (5.33) is omitted. This expression will be applied to several cases.

We begin with a discussion of a short Josephson junction in which the magnetic fields produced by the currents are negligible compared with the externally applied field, and then will treat long junctions where this is not so. We first examine two-junction loops and arrays of many junctions, followed by ultra-small junctions in which single-electron tunneling is observable. We will conclude with a brief section on superconducting quantum interference devices (SQUIDS).

A. Short Josephson Junction

Consider a weak-link tunnel junction of the type sketched in Fig. 13.11 with a magnetic field $B_0\hat{\mathbf{k}}$ applied along the vertical z direction, as shown in Fig. 13.48. The junction is of thickness d normal to the y-axis with cross-sectional dimensions a and c along x and z, respectively. It is small enough so that the applied magnetic field is larger than the field produced by the currents. One superconductor SC_1 is to the right of the insulating barrier and the other SC_2 to the left, as indicated in the figure.

Because of symmetry, the magnetic field $B_z(y)$ has no x or z dependence, but does vary with distance along y into the superconductors,

$$\mathbf{B} = B_z(y)\hat{\mathbf{k}}. \qquad (13.71)$$

This applied field is derived from the vector potential $\mathbf{B} = \nabla \times \mathbf{A}$,

$$\mathbf{A} = A_x(y)\hat{\mathbf{i}}, \qquad (13.72)$$

which has the value (cf. Eq. (5.41)),

$$\mathbf{A} = -yB_0\hat{\mathbf{i}} \qquad |y| \le \tfrac{1}{2}d, \qquad (13.73)$$

in the barrier layer where the material is normal and $B_z = B_0$, as sketched in Fig. 13.49. We assume that the magnetic field decays exponentially into the superconductors on either side of the barrier, as indicated in Fig. 13.49b. If we proceed far enough inside where B_z drops to zero $\mathbf{A}$ will become constant, as seen from the

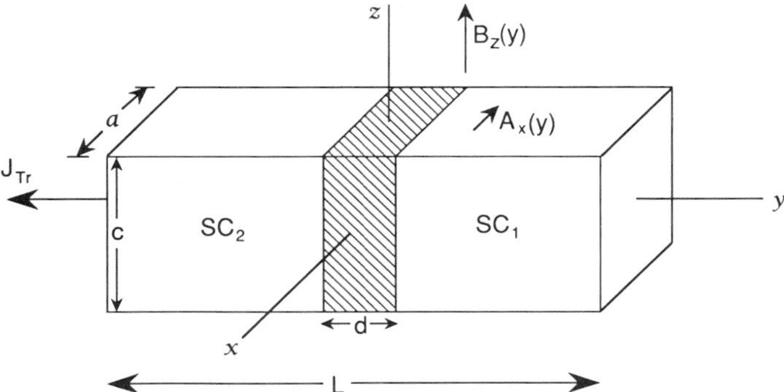

Figure 13.48 Application of a magnetic field $B_0(y)$ transverse to the Josephson junction of Fig. 13.11 with a transport current of density J_{Tr} flowing to the left. The vector potential $A_x(y)$ of the applied field is indicated.

expression $\mathbf{B} = \nabla \times \mathbf{A}$. We assign it the value $A_{1\infty}$ far inside SC_1 and the value $-A_{2\infty}$ far inside SC_2, as indicated in Fig. 13.49a.

We start by calculating the value for the phase $\Theta_1(x)$ at an arbitrary point x along the interface between the barrier and the superconductor SC_1, as shown in Fig. 13.50. This phase will be found relative to the phase Θ_{10} of $\Theta_1(x)$ at a reference point x_0 on the interface, as indicated. The phase difference $\Theta_1(x) - \Theta_{10}$ may be determined by integrating $\nabla\Theta \cdot d\mathbf{l}$ along the path A → B → C → D in Fig. 13.50, but it is easier to make use of Eq. (13.70) and carry out the equivalent integration of $\mathbf{A} \cdot d\mathbf{l}$ along this same path. Since $\mathbf{A}$ is a vector in the x direction, it is perpendicular to the vertical paths A → B and C → D, so that the line integral vanishes for these two segments of the path. We already mentioned that the vector potential has the constant value $A_{1\infty}$ along the path B → C, so that integration gives

$$\Theta_1(x) = \Theta_{10} + (2\pi/\Phi_0)A_{1\infty}(x - x_0). \quad (13.74)$$

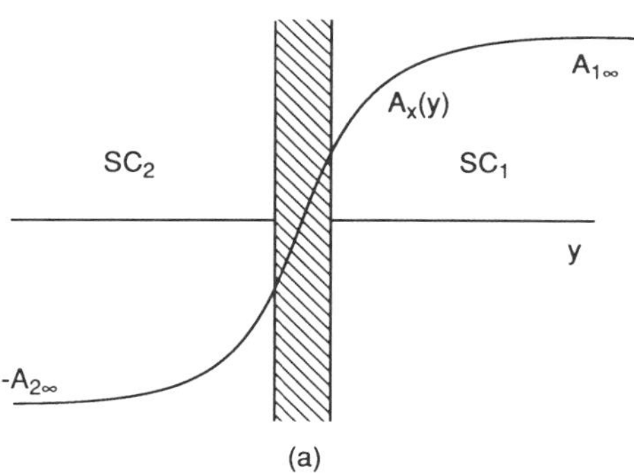

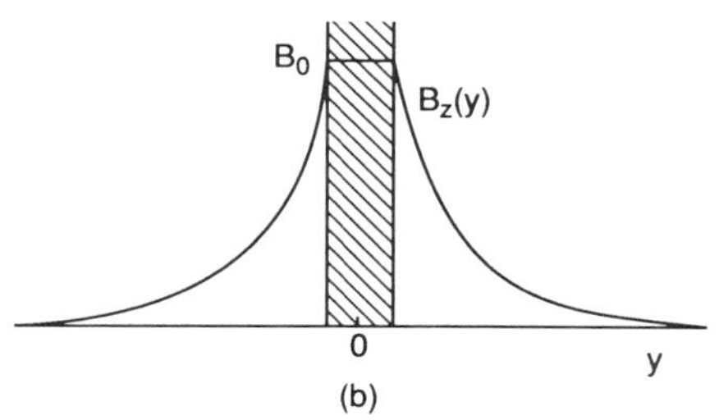

Figure 13.49 Variation of (a) vector potential $A_x(y)$ and (b) magnetic field $B_z(y)$ in the neighborhood of the junction of Fig. 13.48.

An analogous expression can be obtained for the other superconductor SC_2 using the path A′ → B′ → C′ → D′. Hence we can write for the phase difference $\phi(x)$

$$\phi(x) = \Theta_2(x) - \Theta_1(x) \quad (13.75)$$

$$= \phi_0 + \frac{2\pi}{\Phi_0}[A_{1\infty} + A_{2\infty}]x, \quad (13.76)$$

where $\phi_0 = \Theta_2(x_0) - \Theta_1(x_0) = \Theta_{20} - \Theta_{10}$ at the reference point x_0. The quantity $[A_{1\infty} + A_{2\infty}]$ is evaluated by integrating along a similar closed path that has been

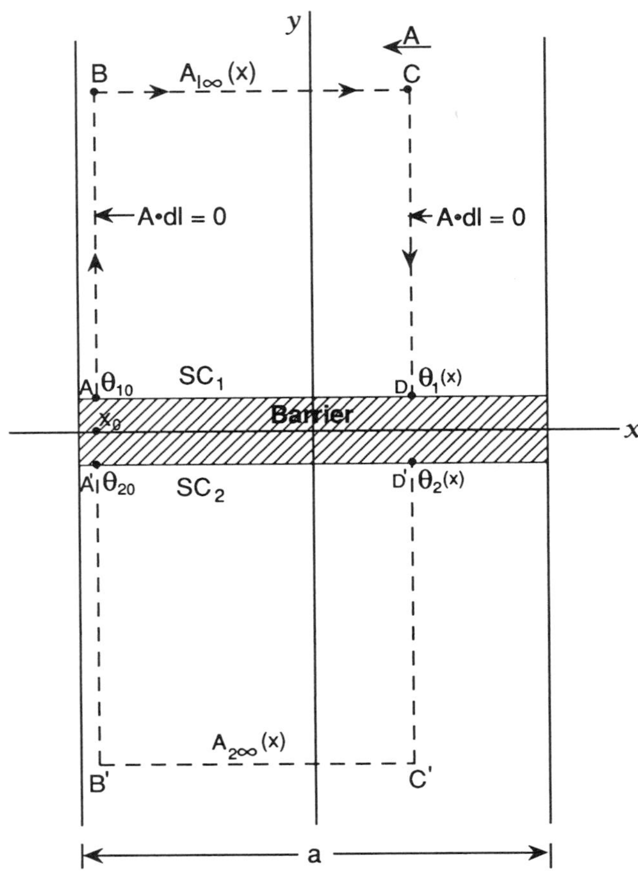

Figure 13.50 Path of integration around the junction of Fig. 13.48 for determining the phase difference $\phi(x) = \Theta_2(x) - \Theta_1(x)$ across the junction at a position x relative to the phase difference $\phi_0 = \Theta_{20} - \Theta_{10}$ at the position x_0 on the left.

enlarged to enclose the entire barrier and extend deep into both superconductors. The resulting line integral equals the total flux through the barrier,

$$\Phi = \int \mathbf{A} \cdot d\mathbf{s}. \qquad (13.77)$$

Again, the integrand vanishes for the two vertical paths along which the vectors $\mathbf{A}$ and $d\mathbf{s}$ are antiparallel. The contributions from the top and bottom paths far inside the two superconductors add to give for the total enclosed flux

$$\Phi = a[A_{1\infty} + A_{2\infty}]. \qquad (13.78)$$

This total flux is approximately equal to the applied magnetic field strength B_0 times the effective area of the junction,

$$\Phi = a(d + 2\lambda)B_0. \qquad (13.79)$$

The quantity $d + 2\lambda$ constitutes the effective thickness of the junction,

$$d_{\text{eff}} = d + 2\lambda. \qquad (13.80)$$

Inserting Eq. (13.78) in (13.76) gives

$$\phi(x) = \phi_0 + \frac{2\pi\Phi}{\Phi_0} \cdot \frac{x}{a}. \qquad (13.81)$$

If this is substituted in Eq. (13.39) and integrated over the area $A = ac$ of the junction (see Problem 12),

$$I = J_c \int \sin[\phi(x)] dx\, dz, \quad (13.82)$$

we obtain

$$I = I_c \sin \phi_0 \frac{\sin(\pi \Phi / \Phi_0)}{\pi \Phi / \Phi_0}, \quad (13.83)$$

where $I_c = AJ_c$ is the critical current. This has a maximum for the phase difference $\phi_0 = \pi/2$,

$$I_{max} = I_c \frac{\sin(\pi \Phi / \Phi_0)}{\pi \Phi / \Phi_0}. \quad (13.84)$$

We call this the *Josephson junction diffraction equation.*

The plot of Eq. (13.84) sketched in Fig. 13.51, which has been seen for many samples (e.g., Rosenthal *et al.*, 1991; Seidel *et al.*, 1991), illustrates how the tunneling current varies with increasing magnetic flux Φ through the junction. Figure 13.52 presents four special cases. When there is no flux, $\Phi = 0$, the current in the junction is uniform, as shown in Fig. 13.52a, and has the critical value I_c. When half a flux quantum is present, $\Phi = \frac{1}{2}\Phi_0$, as in Fig. 13.52b, the average value of the current is the average of a sine wave over a half-cycle, namely $(2/\pi)I_c$. For the next maximum, $\Phi = 3\Phi_0/2$, two of the half-cycles cancel to give the current $I = (2/3\pi)I_c$, which is one-third of the half-cycle case. By induction, the nth maximum of the current $I_c/[\pi(n + \frac{1}{2})]$ occurs at the flux value $\Phi = (n + \frac{1}{2})\Phi_0$. We also deduce from Fig. 13.52c that the current cancels for even cycles, where $\Phi = n\Phi_0$.

For the case of Fig. 13.52c, in which the total phase change across the length of the junction is 2π, one flux quantum fits in it. We see from the directions of the arrows on the figure that the super current flows down across the junction on the left and up on the right. To complete the circuit it flows horizontally within a penetration depth λ inside the superconductor to form closed loops, as illustrated in Fig. 13.53a. These current loops encircle flux, and the resulting configuration is known as a Josephson vortex (Miller *et al.*, 1985). There is no core because none is needed; the super current density is already zero in the center.

When the phase change across the junction is $2\pi n$, where n is an integer, there will be n Josephson vortices side by side in the junction, each containing one flux quantum and each having a horizontal

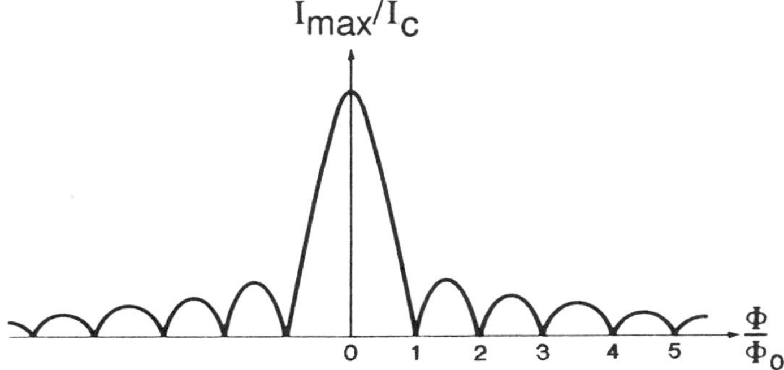

Figure 13.51 Josephson Fraunhofer diffraction pattern showing the maximum normalized zero-voltage current I_{max}/I_c versus Φ/Φ_0 through the parallel junction of Fig. 13.48 when the current density is uniform across the x, z-plane of the junction. The values of I_{max}/I_c at the peaks of the curve, from the center outwards, are 1, $2/3\pi$, $2/5\pi$, $2/7\pi, \ldots$ (from Van Duzer and Turner, 1981, p. 155).

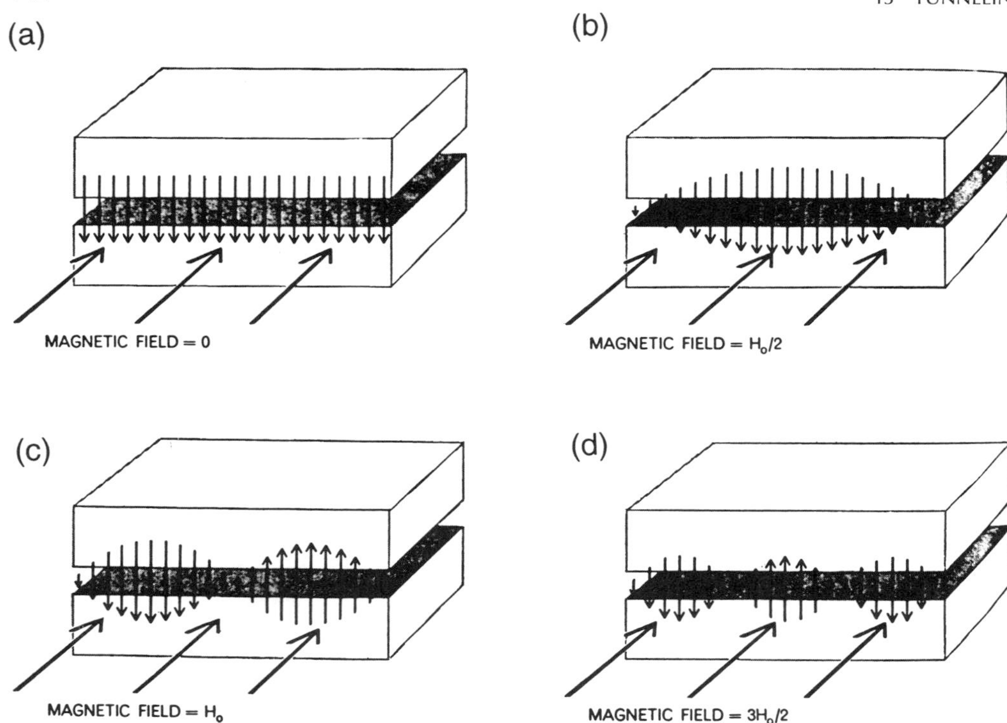

Figure 13.52 Effect of an applied magnetic field on the tunneling-current oscillations across a uniform Josephson junction, where the field $\mu_0 H_0$ corresponds to one flux quantum Φ_0 in the junction (Langenberg *et al.*, 1966).

length $1/n$ times the length for the single-flux quantum case. Figure 13.53b sketches two fluxons in the gap for $n = 2$, with total phase change of 4π.

Equation (13.83) is mathematically equivalent to the well-known expression for single-slit Fraunhofer diffraction in optics. Peterson and Ekin (1989, 1990; cf. Barone and Paterno, 1982) suggest that an Airy diffraction pattern, in which the quantity $\sin(\pi\Phi/\Phi_0)$ in Eq. (13.83) is replaced by twice a first-order Bessel function, $2J_1(\pi\Phi/\Phi_0)$, more properly characterizes superconductors with grain-boundary barriers in bulk materials. They also give a figure in which the Airy and Fraunhofer diffraction patterns are compared.

B. Long Josephson Junction

We will now examine what are called long Josephson junctions, leaving discussion of the criterion for longness until the next section, VIII.C. We begin by taking the derivative of Eq. (13.81),

$$\frac{d\phi}{dx} = \frac{2\pi\Phi}{\Phi_0 a} \tag{13.85}$$

$$= \frac{2\pi B}{\Phi_0} d. \tag{13.86}$$

Although we will not prove it, this expression is more general than (13.81), which may be obtained from it by assuming that Φ arises from a constant magnetic field and then integrating. If we make use of the Maxwell relation $\nabla \times \mathbf{B} = \mu_0 \mathbf{J}$, which for the present case is given explicitly by

$$\frac{dB_z(x)}{dx} = \mu_0 J_y(x), \tag{13.87}$$

we obtain

$$\frac{d^2\phi}{dx^2} = \frac{2\pi\mu_0 J_y(x) d}{\Phi_0}. \tag{13.88}$$

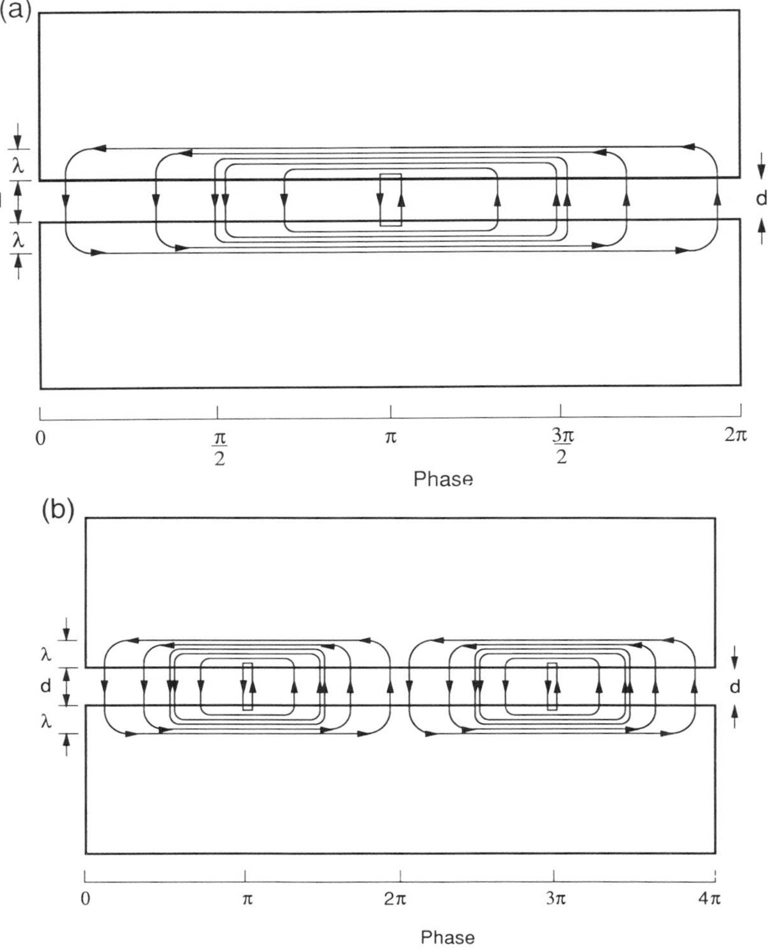

Figure 13.53 Current distribution around Josephson vortices in the junction of Fig. 13.52 for (a) single-vortex case of Fig. 13.52c when the magnetic flux in the junction is Φ_0, and (b) the double-vortex case when the magnetic flux is $2\Phi_0$.

Using Eq. (13.39) this becomes the pendulum equation (which might also be called the stationary sine Gordon equation) (Fehrenbacher *et al.*, 1992).

$$\frac{d^2\phi}{dx^2} = \frac{\sin \phi(x)}{\lambda_J^2}, \qquad (13.89)$$

where $\lambda_J = (\phi_0/2\pi\mu_0 J_c d)^{1/2}$ called the Josephson penetration depth, is the natural length scale for the junction. A long junction is one whose length a is greater than λ_J, while for a short junction $a \ll \lambda_J$.

If the time dependence is taken into account, it can be shown that the sine Gordon equation is obtained (Orlando and Delin, 1991, p. 437),

$$\frac{d^2\phi}{dx^2} - \frac{1}{u_p^2} \cdot \frac{d^2\phi}{dt^2} = \frac{\sin \phi(x)}{\lambda_J^2}, \qquad (13.90)$$

where u_p, given by

$$u_p = \frac{1}{(\mu_0 \epsilon)^{1/2}} \cdot \frac{d}{d + 2\lambda}, \qquad (13.91)$$

is the velocity of a transverse electromagnetic (TEM) mode wave in the junction region. This equation has two types of solitary wave (soliton) solutions. The first, called kinks or topological solitons, are able to propagate and have the property that $\phi(x)$ increases monotonically from 0 to 2π as x increases from $-\infty$ to ∞. There are also propagating antikink solutions for which $\phi(x)$ decreases monotonically from 2π to 0 as x increases from $-\infty$ to ∞. Kinks represent magnetic flux quanta Φ_0 in superconductivity (Holst *et al.*, 1990; Kivshar and Soboleva, 1990), and domain walls in the theory of two-dimensional magnetism. The second type of solution, called a *breather*, is a nontopological variety of soliton which is stationary, i.e., does not travel (Dodd *et al.*, 1982; Drazin and Johnson, 1989; Kivshar *et al.*, 1991).

Sometimes, perturbation terms for dissipation and energy (current) input are added to the sine Gordon equation (Grønbech-Jensen *et al.*, 1991; Holst *et al.*, 1990; Malomed, 1989, 1990; Malomed and Nepomnyashchy, 1992; Olsen and Samuelsen, 1991; Pagano *et al.*, 1991; Petras and Nordman, 1989; Ustinov *et al.*, 1992). Phase locking can also occur, in which the fluxon motion in the long junction follows the frequency of the external field, or two such junctions can be phase-locked to each other (Fernandez *et al.*, 1990; Grønbech-Jensen, 1992; Grønbech-Jensen *et al.*, 1990; Pedersen and Davidson, 1990). Frequency locking to the external field produces an ordered state and can lead to the appearance of Shapiro steps. The absence of phase locking can produce a disordered state and a condition of chaos (Chi and Vanneste, 1990). Other sine Gordon theory applications have appeared (e.g., Weber and Minnhagen, 1988).

C. Josephson Penetration Depth

Equation (13.89) was obtained from Eq. (13.88) by defining the Josephson penetration depth λ_J,

$$\lambda_J = (\Phi_0/2\pi\mu_0 J_c d_{\mathrm{eff}})^{1/2}, \quad (13.92)$$

which is the length criterion that distinguishes short from long junctions. To obtain a physical significance for this characteristic length, let us compare the energies associated with the stored fields and with the current flow through the junction which is sketched in Fig. 13.48. For a constant magnetic field, the stored magnetic energy U_B is

$$U_B = \int \frac{B^2}{2\mu_0}\, dx\, dy\, dz \quad (13.93)$$

$$= (B_0^2/2\mu_0)acd_{\mathrm{eff}}$$

$$= \frac{\Phi_0^2}{2\mu_0} \cdot \frac{c}{ad_{\mathrm{eff}}}, \quad (13.94)$$

where we have assumed one Josephson vortex present in the junction, as in Figs. 13.52c and 13.53a, with $B_0 = \Phi_0/ad$. The energy U_J associated with the current flow is

$$U_J = \int JV\, dx\, dy\, dt, \quad (13.95)$$

and, using Eqs. (13.39) and (13.49), this becomes

$$U_J \approx \frac{\Phi_0 J_c ac}{2\pi} \int \sin\phi\, d\phi. \quad (13.96)$$

If we equate the magnetic and current energies, $U_B = U_J$, we obtain

$$a = \lambda_J \left[2\pi^2 / \int \sin\phi\, d\phi \right]^{1/2}, \quad (13.97)$$

where the factor in the square brackets is close to but larger than unity. Thus the two energies become comparable when the junction length a approaches the Josephson penetration depth λ_J.

A short junction is one for which $a \ll \lambda_J$, $U_J \ll U_B$, the magnetic fields arising from the current flow are much less than the applied field, and the field B is effectively constant over the junction region. A long junction is one for which $a > \lambda_J$, $U_J > U_B$, etc.

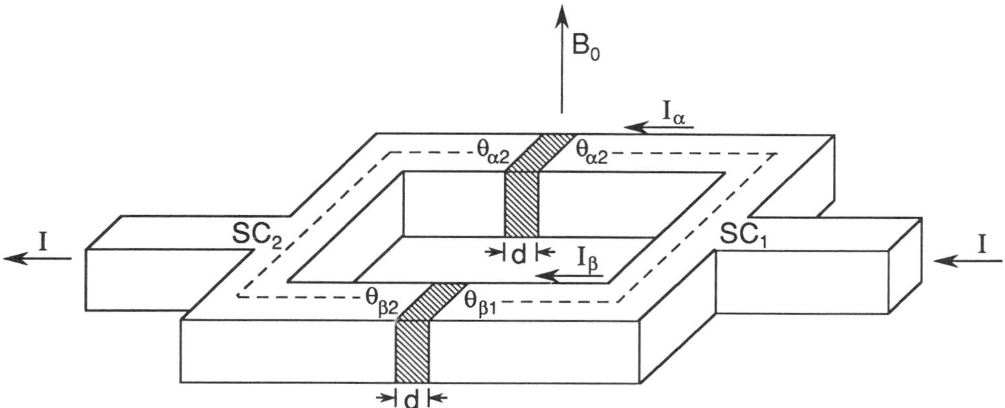

Figure 13.54 Superconducting loop containing two weak links α and β of thickness d showing the phases Θ_{ij} at the junctions and the direction of current flow I. The dashed line indicates the path of integration around the loop.

D. Two-Junction Loop

In Section A we derived the diffraction equation (13.84) for a short Josephson junction in the presence of an applied magnetic field. In a typical case applied fields in the millitesla range (see Problem 13) are used to see the current pattern. We will now consider the case of a superconducting loop containing two weak links (short junctions) in parallel, as shown in Fig. 13.54. For this arrangement flux quantization occurs in the area of the loop, which we are assuming to be considerably larger than the area of either junction, so that the system is sensitive to much smaller changes in applied flux.

In analyzing the two-junction loop we are assuming that the individual junction areas are small enough so as to be negligible. Integration of Eq. (13.70) around the dashed path shown in Fig. 13.54 gives

$$(\Theta_{\alpha 2} - \Theta_{\alpha 1}) - (\Theta_{\beta 2} - \Theta_{\beta 1}) = \left(\frac{2\pi\Phi}{\Phi_0} \right), \tag{13.98}$$

where again we are neglecting current flow effects on the phases. Using the phase difference notation of Eq. (13.75), this becomes

$$\phi_\alpha = \phi_\beta + 2\pi\Phi/\Phi_0. \tag{13.99}$$

The total current I flowing through this parallel arrangement of weak links is the sum of the individual currents I_α and I_β in the two arms,

$$I = I_\alpha + I_\beta, \tag{13.100}$$

and each current satisfies its own individual Josephson equation (13.40), to give

$$I = I_{c\alpha} \sin \phi_\alpha + I_{c\beta} \sin \phi_\beta \tag{13.101}$$

$$= I_{c\alpha} \sin \phi_\alpha + I_{c\beta} \sin\left[\phi_\alpha - 2\pi \left(\frac{\Phi}{\Phi_0} \right) \right] \tag{13.102}$$

where we have used Eq. (13.99). For equal individual currents,

$$I_{c\alpha} = I_{c\beta} = I_c, \tag{13.103}$$

the total current I is maximized by the choice of phase

$$\phi_\alpha = \tfrac{1}{2}\pi + \pi\Phi/\Phi_0, \tag{13.104a}$$

$$\phi_\beta = \tfrac{1}{2}\pi - \pi\Phi/\Phi_0, \tag{13.104b}$$

to give for the magnitude of the maximum current

$$I_{max} = 2I_c|\cos(\pi\Phi/\Phi_0)|, \tag{13.105}$$

an expression that we call the *Josephson loop interference equation*. It has its optical analogue in Young's experiment for detecting the interference of light from two identical slits. The phase ϕ_α is an unknown function of the flux in the ring, and adjusts itself to maximize the current. The dependence of I_{max} on the applied flux Φ given by Eq. (13.105) for this equal-current case is plotted in Fig. 13.55.

When the two currents are not the same (Saito and Oshiyama, 1991), it is more complicated to calculate the phase which maximizes the total current (13.101) subject to the condition (13.99). Van Duzer and Turner (1981) give

$$I_{max} = \left[(I_{c\alpha} - I_{c\beta})^2 + 4 I_{c\alpha} I_{c\beta} \cos^2\left(\frac{\pi \Phi}{\Phi_0}\right) \right]^{1/2}. \tag{13.106}$$

This has the minimum and maximum values

$$\text{minimum} = I_{c\alpha} - I_{c\beta} \qquad \Phi = (n + \tfrac{1}{2})\Phi_0, \tag{13.107a}$$

$$\text{maximum} = I_{c\alpha} + I_{c\beta} \qquad \Phi = n\Phi_0, \tag{13.107b}$$

where we have assumed that $I_{c\alpha} > I_{c\beta}$. The dependence of I_{max} on the applied flux given by Eq. (13.106) for the case $I_{c\alpha} = 2I_{c\beta}$ is plotted in Fig. 13.56 to the same scale as in Fig. 13.55 with the ordinate scale labeled with the limits of Eq. (13.107). Equation (13.106) reduces to Eq. (13.105) for the equal-current case $I_{c\alpha} = I_{c\beta}$.

Equation (13.106) corresponds to the analogue of light interference from two nonidentical slits, but this is rarely studied in optics because it is so easy to make matching slits. Identical Josephson junctions are not so easy to fabricate, so the case of Eq. (13.106) is of interest in superconductivity.

E. Self-Induced Flux

In the previous two sections we tacitly assumed that the flux Φ in the circuits is the applied flux, which meant neglecting the contribution of the currents. We noted in Chapter 2, Section X that the current I_{circ} circulating in a loop can contribute the amount $I_{circ}L$ to the flux, where L is the inductance of the loop. The currents I_α and I_β in the two arms of the loop flow in the same direction, as indicated in Figs. 13.54 and 13.57, producing magnetic fields (cf. Fig. 2.35) pointed in opposite direc-

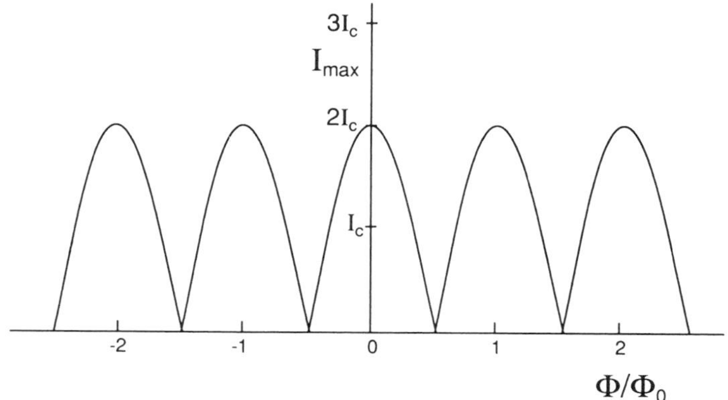

Figure 13.55 Dependence of the current maximum I_{max} of Eq. 13.105 of a balanced Josephson junction loop ($I_{c\alpha} = I_{c\beta}$) on the applied flux Φ normalized relative to the quantum of flux Φ_0.

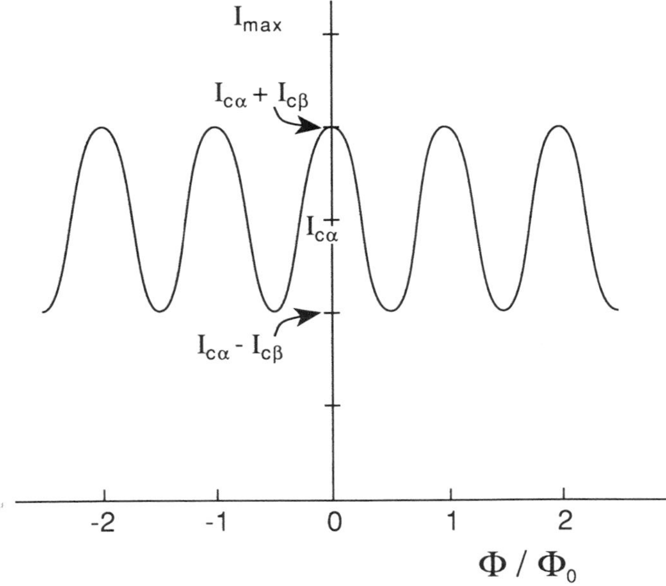

Figure 13.56 Dependence of the current maximum I_{max} of an unbalanced Josephson junction loop ($I_{c\alpha} \neq I_{c\beta}$) on the applied flux Φ normalized relative to the quantum of flux Φ_0. The plot is made for the case of setting $I_{c\alpha} = 2I_{c\beta}$ in Eq. (13.106).

tions through the loop; for $I_\alpha = I_\beta$ these fields cancel each other. However, when the two currents are not equal, we can decompose them into a symmetrical component $\frac{1}{2}(I_\alpha + I_\beta)$, which flows in the same direction in each arm of the loop and does not contribute to the flux, and an antisymmetrical circulating component,

$$I_{circ} = \tfrac{1}{2}(I_\alpha - I_\beta), \qquad (13.108)$$

as indicated in Fig. 13.57, which contributes to the flux. The total flux Φ is

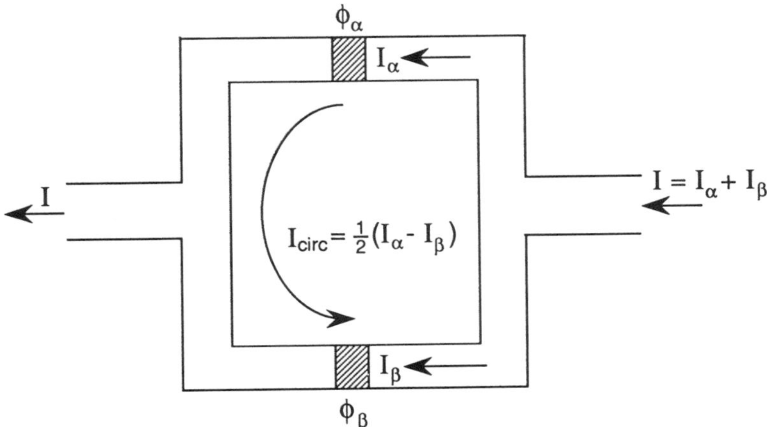

Figure 13.57 Circulating current $I_{circ} = \frac{1}{2}(I_\alpha - I_\beta)$ in an unbalanced Josephson junction loop, $I_\alpha \neq I_\beta$, in the presence of an applied current $I = I_\alpha + I_\beta$.

then the sum of the applied flux Φ_{app} and the self-induced flux arising from the circulating current,

$$\Phi = \Phi_{app} + \tfrac{1}{2}L(I_\alpha - I_\beta). \quad (13.109)$$

This self-induced flux should be taken into account for a proper treatment of Josephson junctions.

F. Junction Loop of Finite Size

As a final example of tunneling we examine a loop in which the two identical junctions are large enough in area so as to contribute to the observed oscillatory current pattern. For this case we can combine the diffraction equation (13.84) for the junction and the interference equation (13.105) for the loop,

$$I = 2I_c \left| \cos(\pi\Phi_L/\Phi_0) \cdot \frac{\sin(\pi\Phi_J/\Phi_0)}{\pi\Phi_J/\Phi_0} \right|, \quad (13.110)$$

where $\Phi_J = B_{app}A_J$ and $\Phi_L = B_{app}A_L$ are the amounts of flux in the junctions of area A_J and in the loop of area A_L, respectively, for a particular applied field B_{app}. We can define the critical applied fields B_J and B_L for which one flux quantum Φ_0 is present in the junction and in the loop in terms of their respective areas:

$$B_J = \Phi_0/A_J, \quad (13.111a)$$

$$B_L = \Phi_0/A_L. \quad (13.111b)$$

These expressions permit us to write Eq. (13.110) in terms of the applied field,

$$I = 2I_c \left| \cos(\pi B_{app}/B_L) \frac{\sin(\pi B_{app}/B_J)}{\pi B_{app}/B_J} \right|. \quad (13.112)$$

We call this the *Josephson loop diffraction equation*. Since $A_L \gg A_J$, we have $B_L \ll B_J$, and the expected current pattern is sketched in Fig. 13.58 for the case $B_J = 3B_L$. We see from the figure that the slower individual junction variations constitute an envelope for the more rapid loop oscillations. Figure 13.59a shows some experimental results for a small loop in which the self-induced flux is negligible, so that a pattern similar to the pattern in the center of Fig. 13.58, but with more oscillations, is obtained. Figure 13.59b shows a large loop result in which the self-induced flux is appreciable, so that the minima in the oscillations do not reach zero, as in the pattern of Fig. 13.56. Because this second loop is larger, it has more rapid oscillations, as

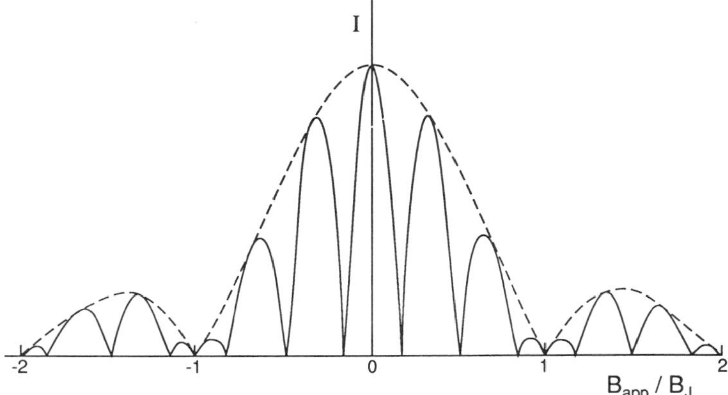

Figure 13.58 Josephson loop diffraction pattern showing the dependence of the super current I flowing through a two-element junction loop of finite size on the magnetic field B_{app}, producing the flux Φ passing through the plane of the loop. The figure is drawn for the case of setting $B_J = 3B_L$ in Eq. (13.112).

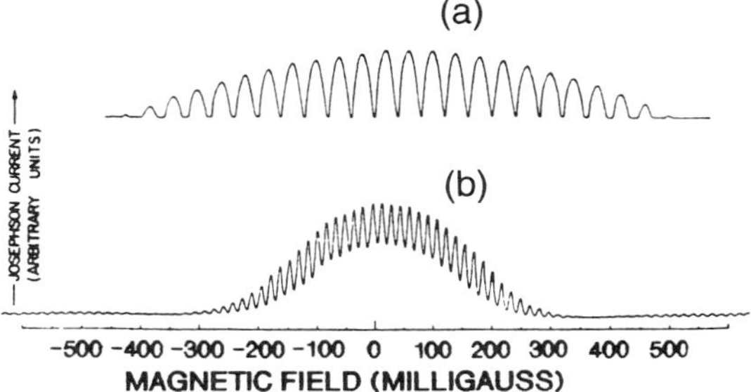

Figure 13.59 Experimentally measured dependence of the Josephson current in a two-element loop on the applied magnetic field. (a) is for the case $B_L \approx 4$ mT and $B_J \approx 50$ mT with negligible self-induced flux ($I_\alpha \approx I_\beta$), so that the oscillations all return to the baseline, as in Fig. 13.58. (b) is for $B_L \approx 1.5$ mT and $B_J \approx 35$ mT, so that the self-induced flux is appreciable ($I_\alpha \not\approx I_\beta$) and the rapid oscillations do not return to the baseline (Jaklevic *et al.*, 1965).

shown. On each side of Fig. 13.59b we can see traces of the next set of oscillations arising from the second cycles of Eq. (13.112) for the field range $B_J < B_{app} < 2B_J$, and these are also shown in the pattern of Fig. 13.58.

Equation (13.112) corresponds to the optics analogue of Fraunhofer diffraction from two identical wide slits.

G. Ultrasmall Josephson Junction

When a Josephson junction becomes much smaller than a typical weak link or short junction, new phenomena can appear. As an example, consider an ultrasmall junction or nanobridge with area A of 0.01 μm^2, thickness d of 0.1 nm, and capacitance estimated from the expression $C = \epsilon_0 A/d$ of about 10^{-15} F (Kuzmin and Haviland, 1991; Kuzmin *et al.*, 1991). The change in voltage ΔV brought about by the tunneling of one electron across the junction barrier is given by $\Delta V = e/C = 0.16$ mV, which is an appreciable fraction of a typical junction voltage. This can be enough to impede the tunneling of the next electron. Blocking of current flow has been termed Coulomb blockade (Furusaki and Veda, 1992; Tagliacozzo *et al.*, 1989). Note that it is only in recent years that techniques such as electron-beam lithography have developed to the point where nanobridges with capacitances in the range 10^{-15}–10^{-16} F can be fabricated (Ralls *et al.*, 1989).

Single-electron tunneling manifests itself by the appearance of fluctuations, called a Coulomb staircase, on an I versus V or dI/dt versus V characteristic, as illustrated in Fig. 13.60 (McGreer *et al.*, 1989).

The Coulomb blockade is a quantum effect that represents the quantization of charge q transferred across a junction. The Hamiltonian, considered as a function of the two conjugate variables q and the phase ϕ, may be written as the sum of a capacitor charging energy, a current bias term, and a Josephson coupling energy.

$$H(q,\phi) = q^2/2C - (\hbar/2e)I\phi - (\pi\hbar/4e^2)(\Delta/R_n)\cos\phi, \tag{13.113}$$

where I is the bias current, Δ is the energy gap, C is the junction capacitance, and R_n is the normal resistance (Iansiti *et al.*, 1989; Shimshoni and Ben-Jacobs, 1991). For conventional junctions the coupling energy, $(h/8e^2)(\Delta/R_n)\cos\phi$, is dominant

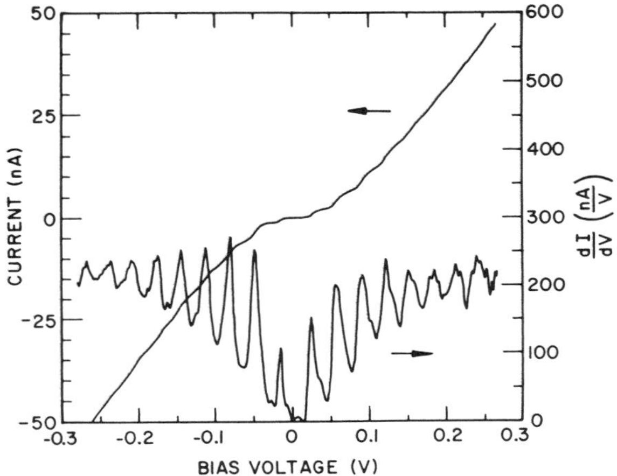

Figure 13.60 Coulomb staircase structure on plots of I versus V and dI/dT versus V of tunneling between a granular lead film and the tip of a scanning tunneling microscope (McGreer *et al.*, 1989).

and for ultrasmall junctions the charging energy $q^2/2C$ predominates.

There is an uncertainty relationship between the two conjugate variables

$$\Delta\phi\,\Delta q \geq e$$

(Graham *et al.*, 1991; Kuzmin *et al.*, 1991). In observing the Coulomb blockade, $q = e$ and the phase (or voltage) fluctuations are large. In the usual dc Josephson effect, which we described in Section VII.B, the phase is well defined and the fluctuations occur in the charge.

In an ultra-small Josephson junction biased by a dc current, correlated tunneling of Cooper pairs can lead to what are called *Bloch oscillations* at the frequency

$$\nu_B = I/2e. \tag{13.114}$$

One-dimensional N-junction arrays have been used to observe single electron tunneling (Delsing *et al.*, 1989; Kuzmin *et al.*, 1989). A current of 0.01 $\mu A = 10^{-8}$ C/s corresponds to a frequency $\approx 3 \times 10^{10}$ Hz, which is in the microwave region. Singularities of the observed microwave resistance at the current values corresponding to Eq. (13.114) have provided evidence for the presence of Bloch oscillations (Furusaki and Veda, 1992; Geerligs *et al.*, 1989; Hu and O'Connell, 1993; Kuzmin and Haviland, 1991; Shimshoni *et al.*, 1989). Another type of quantum oscillation is resonant tunneling in the tilted cosine potential illustrated in Fig. 13.61. The figure shows a microwave photon emission accompanying resonant tunneling between levels aligned in adjacent wells (Hatakenaka *et al.*, 1990; Schmidt *et al.*, 1991).

In an ultra-small junction the effect of a bias current is taken into account by the term $-\hbar I\phi/2e$ in the Hamiltonian (13.113). When the bias current energy $\hbar I_c/2e$ becomes comparable with the

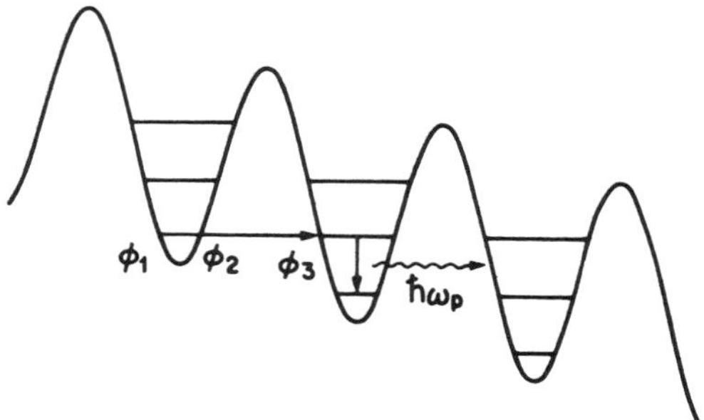

Figure 13.61 Macroscopic resonant tunneling of phase in a tilted cosine potential showing the phases ϕ_i at the turning points and the microwave photon emission $\hbar\omega$ accompanying the tunneling (Hatakenaka *et al.*, 1990).

charging energy $e^2/2C$, we have $I_cC \approx 3.9 \times 10^{-23}$ AF, where I_c is the ideal critical current (Kautz and Martinis, 1990). This relation is satisfied by representative current and capacitance values of some ultra-small junctions.

H. Arrays and Models for Granular Superconductors

The discussion until now has concerned single Josephson junctions, although we have also examined a pair of junctions. There is an extensive literature on chains, arrays, and layers of Josephson junctions that are coupled together in various ways, and sometimes coupled to an applied current, magnetic field, or radiation (e.g., Eckern and Sonin, 1993; van der Zant *et al.*, 1993). High-temperature superconductors can be modeled by arrays of superconducting grains coupled together by Josephson junctions at their interfaces (Babic *et al.*, 1991; Cai and Welch, 1992; Deutscher and Chaudhari, 1991; Fishman, 1988, 1989; Majhofer *et al.*, 1990; Saslow, 1989; Sugano *et al.*, 1992). The junctions in the arrays are often phase locked to each other (cf. Section VIII.B).

I. Superconducting Quantum Interference Device

Flux changes in a loop with two weak lines were shown to produce oscillatory variations in the supercurrent through the loop. A Superconducting Quantum Interference Device (SQUID), is a practical circuit that measures these current variations to quantitatively determine the strength of the applied field. It is more acurate to call such a device a dc SQUID since it measures a slowly changing applied field. A dc SQUID can detect much smaller changes in field than is possible by non-superconducting technology.

One example of a dc SQUID is described in Chapter 10, Section VIII (see Fig. 10.12), and another version is sketched in Fig. 13.62. In the latter arrangement the current change through the weak links is detected as a voltage change across the pair of weak links and then amplified by a step-up transformer followed by further amplification and measurement.

Another type of SQUID, called an rf SQUID, is shown in Fig. 13.63. It consists of a loop with one weak link W coupled to an LC-tuned circuit driven by an rf current source. A change in flux in the loop pro-

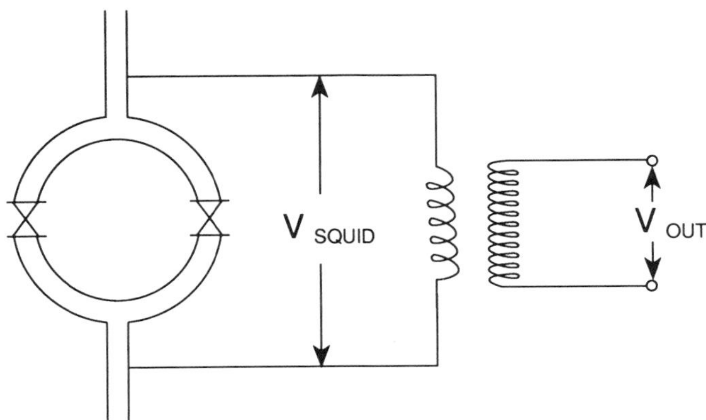

Figure 13.62 Diagram of a Superconducting Quantum Interference Device consisting of two weak links, on the left, showing the voltage V_{SQUID} across them coupled via the transformer to produce the output voltage V_{out}.

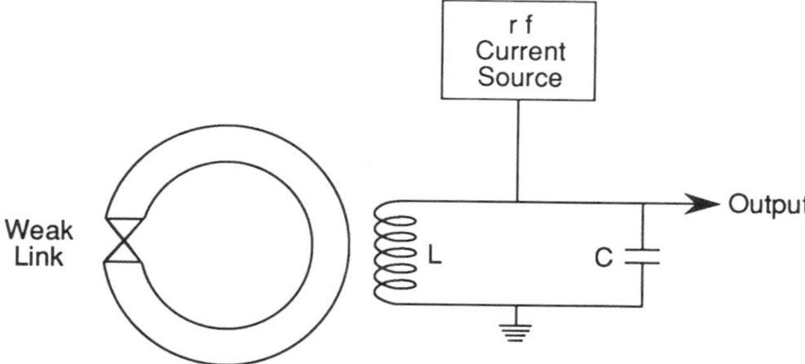

Figure 13.63 An rf SQUID consisting of one weak link coupled to an LC-tuned circuit driven by an rf current source. The rf output voltage is a measure of the change in loading of the tuned circuit produced by a change in flux through the loop.

duces a change in the loading of the tuned circuit, and this is detected by measuring the change in rf voltage across the circuit.

More information on SQUIDs may be obtained from the texts by Van Duzer and Turner (1981) and by Orlando and Delin (1991). SQUIDs have been fabricated from high-temperature superconductors (Gross *et al.*, 1990b; Siegel *et al.*, 1991; Vasiliev, 1991). Miller *et al.* (1991) discussed a Superconducting Quantum Interference Grating (SQUIG), an interferometer consisting of several Josephson junctions in parallel.

Since a SQUID easily detects a change in one quantum of flux in an area with dimensions in the centimeter range, it is said to measure a macroscopic quantum phenomenon. Orlando and Delin (1991) build on Fritz London's observation that superconductivity is inherently a quantum mechanical phenomenon with macroscopic manifestations in their utilization of a macroscopic quantum model to describe superconductivity.

FURTHER READING

Several superconductivity texts discuss tunneling and the Josephson effect, for example (Kresin and Wolf, 1990; Rose Innes and Rhoderick, 1994; Tilley and Tilley, 1986; Tinkham, 1985). Van Duzer and Turner (1981), and Orlando and Delin (1991) have good discussions of the Josephson effect.

Tunneling is also discussed in several monographs, including Huebener, 1979; Phillips, 1989a, Poole *et al.*, 1988; Van Duzer and Turner, 1981; Wilson, 1983. Tallon (1994) reviewed tunneling measurements.

Likharev (1986) wrote a monograph on Josephson junctions and circuits. The Josephson effect has been reviewed by Ferrell (1990). A number of workshops and conferences have been held on critical currents, such as the 7th International Workshop in Tyrol, Austria in January, 1994 (Proc. by World Publ.). Kiss and Svedlindh (1995) surveyed conductance noise in high temperature superconductors. Several recent works (Chrisey and Hubler, 1994; Phillips, 1994; Salama *et al.*, 1994, and Scheel, 1994) are devoted to materials processing to obtain high critical currents.

Mück (1994) and Weinstock (1991) reviewed SQUID magnetometry.

PROBLEMS

1. Describe N–I–N tunneling in the manner that N–I–S and S–I–S are described in Sections V.E and V.F, respectively. Calculate D_1 and D_2.
2. Derive and justify Eqs. (13.15) and (13.16).
3. Show how to derive the expressions (13.18) and (13.19) for the integrand of Eq. (13.17).
4. Show how to obtain the N–I–S expression (13.20) for I_{ns} from the general tunneling current equation (13.17).
5. Show that in S–I–S tunneling the ratio of the jump in current ΔI_s to the

normal tunneling current I_n at the bias $V = 2\Delta/e$ is given by

$$\Delta I_s/I_n = \pi/4,$$

which is about an 80% jump.

6. Show how Eq. (13.25) for ΔI_s reduces to the simple expression (13.23) in the limit $T \to 0$.
7. Plot $\Delta(T)$ versus T using the data of Fig. 13.26, and compare the plot with Fig. 13.27. How well are the data fit by Eq. (2.65)?
8. Derive Eq. (13.96) from Eq. (13.95).
9. Show how to solve Eq. (13.59):

$$\frac{I}{I_c} = \frac{d\phi}{d\Theta} + \sin\phi.$$

10. Explain how the washboard analogue sketched in Fig. 13.47 mimics the behavior of a Josephson junction.
11. Justify and derive Eq. (13.70). What is the physical significance of each term?
12. Carry out the integration (13.81) to obtain

$$I = I_c \sin\phi_0 \frac{\sin(\pi\Phi/\Phi_0)}{\pi\Phi/\Phi_0}. \quad (13.83)$$

13. Consider a barrier junction that is 35 μm long and has a 45-nm oxide layer. What applied magnetic field will put one flux quantum in the junction if the superconductors are Nb on one side of the oxide layer and Sn on the other? At what value of the applied field will the first maximum following the principal center maximum appear in the "diffraction pattern"?
14. Find the values of ϕ_α and ϕ_β that maximize the current (13.88), subject to the condition (13.85).

14

Transport Properties

I. INTRODUCTION

In the previous chapter we discussed electron and Cooper pair tunneling, phenomena that constitute important mechanisms of charge transport in superconductors. There are other processes, occurring both above and below T_c, which provide additional information on transport in superconductors, and we will proceed to discuss a number of them. Most of these processes are summarized in Table 14.1.

The chapter begins with a model for ac current flow in superconductors. This is followed by a discussion of the influence of electric and magnetic fields, as well as heat and light, on electrical conductivity. Discussion of the spectroscopic aspects of the interaction with light will be postponed to the next chapter.

II. INDUCTIVE SUPERCONDUCTING CIRCUITS

In earlier chapters, when we remarked that the phenomenon of superconductivity is characterized by a zero-resistance flow of electrical current, we were referring to dc current. There is no heat dissipation when current flows without resistance. We also pointed out, in Chapter 9, Section V.C, that unpinned vortices set into motion by a transport current experience a viscous drag force, and that both ac and dc currents can produce this flux flow dissipation. We will now discuss another process that leads to heat loss.

A. Parallel Inductances

In Chapter 2, Section X, we examined a perfect conductor in terms of a simple circuit resistance in parallel with an inductance. Here we wish to consider two possible zero-resistance paths, or channels, through a superconducting grain that can be taken by a super current. Figure 14.1a shows a sketch of the circuit. (It is assumed that the current loops have self and mutual inductance, as in Fig. 14.1b.) The following equations for the two current paths can be deduced from simple circuit

Table 14.1 Thermoelectric and Thermomagnetic Effects[a]

Effect	Electric field (E or ∇V)	Electric current (I)	Temp. gradient (∇T)	Heat current (dQ/dt)	Magnetic field B_{app}	Figure
Resistivity	Meas. y	Appl. y	0	—	0	2.9
Magnetores. longitud.	Meas. y	Appl. y	0	—	Appl. y	14.16
Magnetores. transv.	Meas. y	Appl. y	0	—	Appl. z	14.16
Thermal cond.	—	0	Meas. y	Appl. y	—	—
Hall	Meas. x	Appl. y	0	—	Appl. z	1.18
Righi–Leduc	—	0	Meas. x	Appl. y	Appl. z	14.47
Seebeck	Meas. y	0	Appl. y	—	0	14.34
Magneto-Seebeck	Meas. y	0	Appl. y	—	Appl. x, y, or z	14.39
Nernst	Meas. x	0	Appl. y	—	Appl. z	14.40
Peltier	—	Appl. y	0	Meas. y	—	14.44
Ettinghausen	Meas. y	Appl. y	Meas. x	—	Appl. z	14.45

[a] The applied quantity is in the y direction, an effect measured longitudinally is also in the y direction, an effect measured transversely is in the x direction, and in most cases an applied magnetic field is in the z direction.

analysis by observing that the voltage drop between points A and B of Fig. 14.1b must be the same for the two paths,

$$L_1 \frac{dI_1}{dt} + M \frac{dI_2}{dt} = L_2 \frac{dI_2}{dt} + M \frac{dI_1}{dt}. \quad (14.1)$$

These can be integrated to give for steady-state current flow

$$(L_1 - M)I_1 = (L_2 - M)I_2. \quad (14.2)$$

In practice, the mutual inductances are negligible, and we have

$$L_1 I_1 = L_2 I_2, \quad (14.3)$$

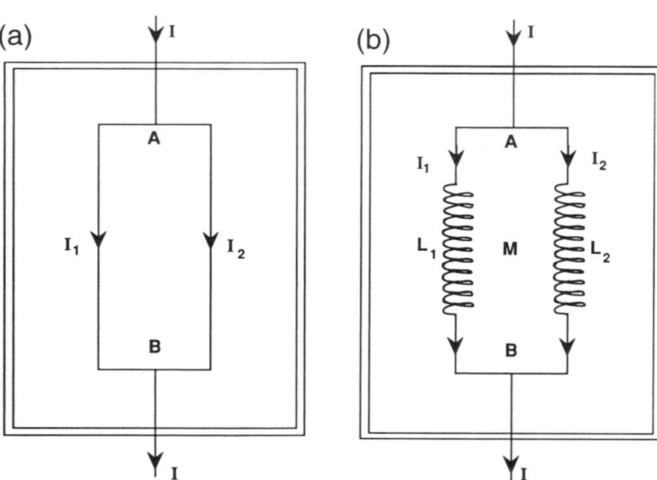

Figure 14.1 A super current I split into two parallel currents I_1 and I_2, perhaps passing through different grains of a superconductor, is shown on the left (a), and the equivalent circuit involving parallel inductances is shown on the right (b).

so that the total current I,

$$I = I_1 + I_2, \tag{14.4}$$

splits between the two paths in the inverse ratio of their inductances $I_1/I_2 = L_2/L_1$.

B. Inductors

To gain some perspective on the magnitudes of the inductances that are involved in superconducting loops around, for example, the grains of superconductors, it will be helpful to recall some of the expressions that apply to simple inductor geometries with dimensions $\approx 20\ \mu$m. A closely wound N-turn coil of radius r and length d has inductance

$$L \approx \frac{\pi\mu_0 r^2 N^2}{d + 0.9r}, \tag{14.5}$$

where the permeability μ_0 is assumed to be that of free space. A long straight wire of length d has the much smaller inductance

$$L \approx \frac{\mu_0 d}{8\pi}. \tag{14.6}$$

A loop of wire with the wire radius a and loop radius r has the inductance given by Eq. (2.14), which may be written in the form

$$L \approx \mu_0 r\left(0.0794 + \ln\left(\frac{r}{a}\right)\right), \tag{14.7}$$

which is small for current paths through grains where r would typically be 20 μm or less, as is shown in Problem 2.

C. Alternating Current Impedance

When superconducting and normal-charge carriers are present during dc current flow, the resistive circuits of the normal carriers will be short-circuited by the zero-resistance circuits of the super current. When the current is alternating, it is assumed that the super current flows in paths with inductance L, and hence with the reactance $i\omega L$, and that the normal current flows in paths with the resistance R, as shown in Fig. 14.2. The super current will lag behind the normal current in phase. The voltage drop between points A and B in the figure due to the super current will be the same as that due to the normal current, which corresponds to $I_n R = iI_s L\omega$. This gives us for the ratio between the magnitudes of the two currents

$$I_n/I_s \approx L\omega/R. \tag{14.8}$$

The circular loop of Eq. (14.7) has the resistance $R = 2\pi r\rho/\pi a^2$, which gives, inserting the numerical value of μ_0,

$$L/R = 2\pi \times 10^{-7}(a^2/\rho) \times [0.0794 + \ln(r/a)]. \tag{14.9}$$

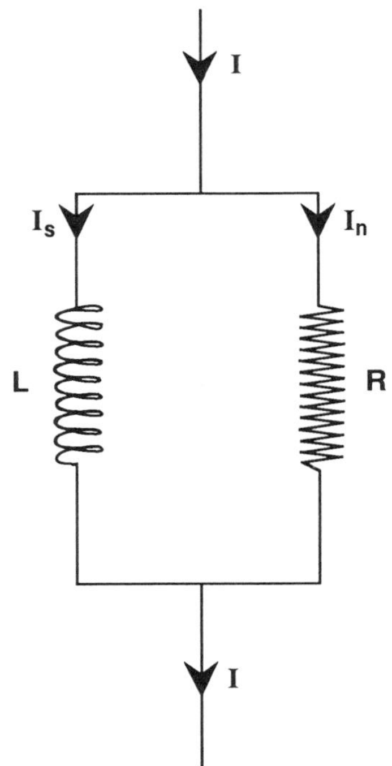

Figure 14.2 Current split similar to that of Fig. 14.1 in which one parallel current, I_s, flows through superconducting material represented by an inductance L, while the other, I_n, flows through normal material represented by a resistance R.

We estimate the dimensions $r \approx 30\ \mu\text{m}$ and $a \approx 3\ \mu\text{m}$ for typical grain sizes. Using the normal state low-temperature resistivity $\rho \approx 500\ \mu\Omega$ cm of YBaCuO from Table 2.2, we have

$$L/R \approx 2.7 \times 10^{-12}\ H/\Omega. \quad (14.10)$$

Equation (14.8) provides an estimate of the dimensionless ratio

$$I_n/I_s \approx 1.7 \times 10^{-11}\nu, \quad (14.11)$$

where ν is in Hz, so that very little normal current will flow at low frequencies, and we will have $I \approx I_s$. The current ratio I_n/I_s will be negligible until the frequency exceeds about 10^{11} Hz, at which point ωL becomes comparable to R. At optical frequencies, $\nu > 10^{14}$ Hz, the current flow is mostly normal, $I \approx I_n$. Figure 14.3 shows these frequency dependences of the currents.

The total dissipated power is the power loss due to the normal current in the resistance,

$$P = I_n^2 R. \quad (14.12)$$

From this discussion and Eq. (14.8) we find limiting expressions for the power dissipation at low and high frequencies, respectively:

$$P \approx I^2 R \left(\frac{L\omega}{R} \right)^2 \qquad \omega \ll R/L, \quad (14.13a)$$

$$P \approx I^2 R \qquad \omega \gg R/L. \quad (14.13b)$$

Thus the dissipation is negligible at low frequencies, $L\omega \ll R$, while at high frequencies the effective resistance is similar to the normal state value, as in the optical range.

The impedance Z of the parallel L–R circuit shown in Fig. 14.2 has the magnitude and phase angle

$$Z = R\omega L/(R^2 + L^2\omega^2)^{1/2} \quad (14.14)$$

$$\phi = \arctan(R/\omega L), \quad (14.15)$$

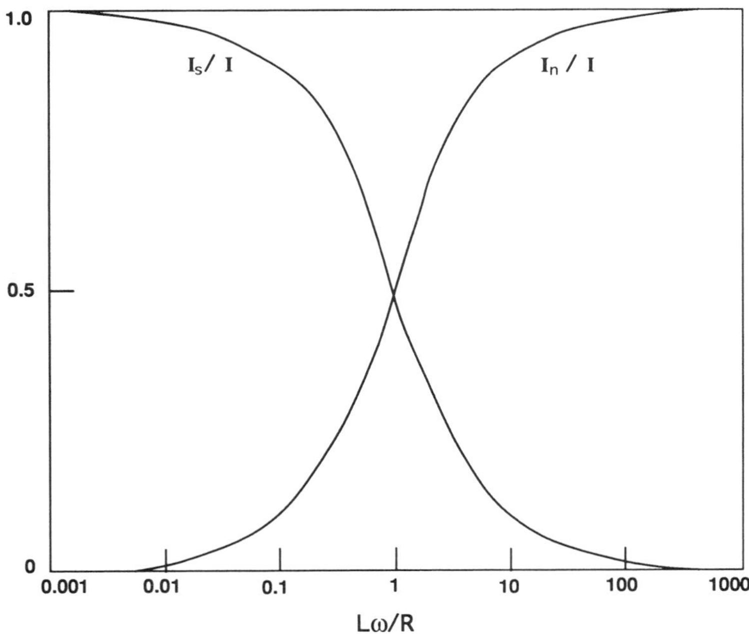

Figure 14.3 Dependence of the superconducting I_s and normal I_n components of the current $I = I_s + I_n$ on the impedance ratio $L\omega/R$ for the loop shown in Fig. 14.2 representing an admixture of superconducting (L) and normal (R) material.

with the following limiting values for small and large ω, respectively:

$$Z \approx L\omega, \quad \phi \approx \frac{\pi}{2} \quad \omega \ll \frac{R}{L}, \quad (14.16a)$$

$$Z \approx R \quad \phi \approx 0 \quad \omega \gg \frac{R}{L}. \quad (14.16b)$$

The result $Z \approx R$ at high frequencies confirms the change to normal state (resistive) behavior that we have already noted.

III. CURRENT DENSITY EQUILIBRATION

Ordinarily, we are interested in equilibrium current flow in which the radial dependence of the current density remains constant along the wire. In this section we will examine what happens when a discontinuity disturbs this regularity. We will find that the disturbance persists for a transition distance along the wire, beyond which spatial equilibrium is restored. For simplicity, we will assume that the undisturbed super current flows with uniform density throughout the entire cross section.

Consider the situation depicted in Fig. 14.4, in which current is flowing in a Type II superconducting wire of radius a. The current enters and leaves the wire radially at the ends, where the radius has been increased to a much larger value c, as indicated in Fig. 14.5. This figure shows how the current-flow contours change gradually from radially directed to longitudinally directed as the current proceeds into the wire. When it enters the narrower part of the wire, it has a greater density toward the surface than in the center. As it proceeds along the wire it gradually becomes uniformly distributed.

To treat this situation quantitatively, we divide the wire into an inner cylinder of radius b and an outer cylindrical shell with inner and outer radii b and a, respectively, as shown in Fig. 14.4. The upper graph in the figure presents a plot of the average

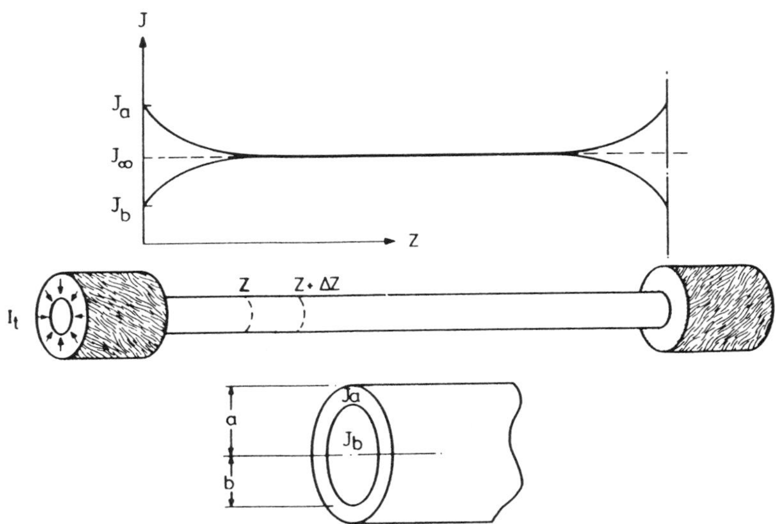

Figure 14.4 Average current densities in the outer (J_a) and inner (J_b) concentric regions of a superconducting wire as a function of the distance z from the junctions at the ends where the current enters and leaves. Within the characteristic distance Λ from the ends $J_a > J_b$ (Wilson, 1983, p. 239).

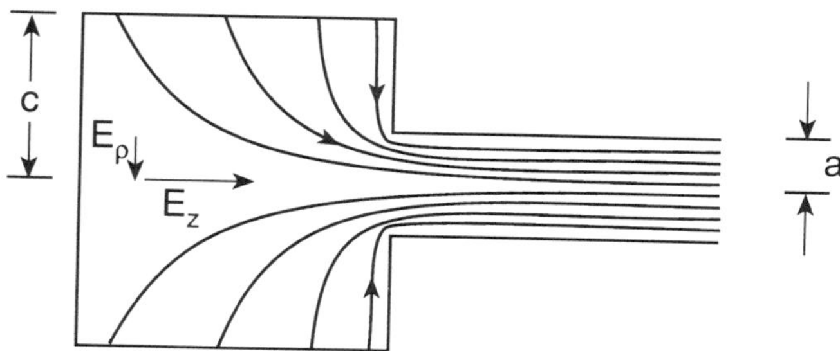

Figure 14.5 Cross section of the current flow contours at one of the junctions of Fig. 14.4. The radial (E_ρ) and axial (E_z) electric fields associated with the current flow in the junction region are indicated.

current densities J_b and J_a in these two regions, respectively, as a function of the distance z along the wire. The figure shows that uniform-density flow is established after a distance Λ, called the characteristic length. That is, if we wish to measure the critical-current density by gradually increasing the total current until the appearance of a voltage drop ΔV between two electrodes placed along the wire, as shown in Fig. 14.6, care must be taken to locate the electrodes a distance from the ends greater than the characteristic distance Λ.

Near the ends of the wire, where $J_a > J_b$, equilibrium is brought about by the presence of a voltage drop between the inner and outer parts of the wire. The radial electric field component E_ρ associated with this voltage drop, shown in Fig. 14.5, causes an inward current flow, and E_ρ decreases with the approach to uniform density as we proceed along the wire. This situation is treated theoretically by Dresner (1978). Plots made from Dresner's equations, which show how the voltage drop in the wire associated with this radial electric field varies with the distance from the ends, are presented in Fig. 14.7. We see from the figure that as z increases from 0.01Λ to Λ, the radial electric field decreases by over four orders of magnitude to a negligibly small value, confirming that the current density has now become uniform throughout the wire. The three curves in the figure are for the three values of the parameter n occurring in Dresner's model. The results for $z > \Lambda$ are insensitive to n, since the residual radial field approaches zero for all three choices of n.

We have been assuming that equilibrium current flow occurs with uniform density. There are also cases of a pronounced radial distribution, as in the Bean model, when the penetration depth λ limits the current to a surface layer. When this is the case, λ is the distance that must be traveled to reach the equilibrium state.

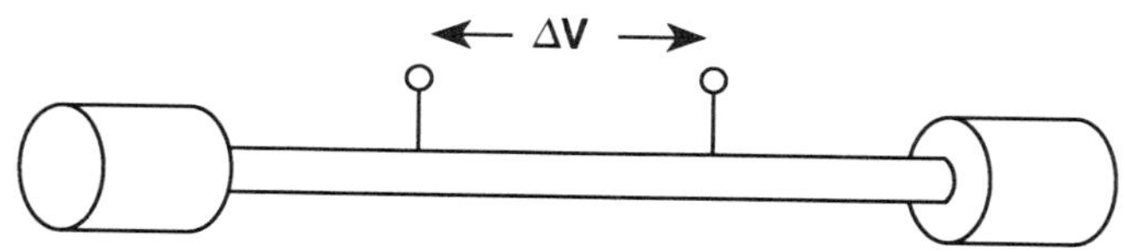

Figure 14.6 Location of the voltage probes on the wire of Fig. 14.4 at positions a distance more than one characteristic length Λ from the junctions.

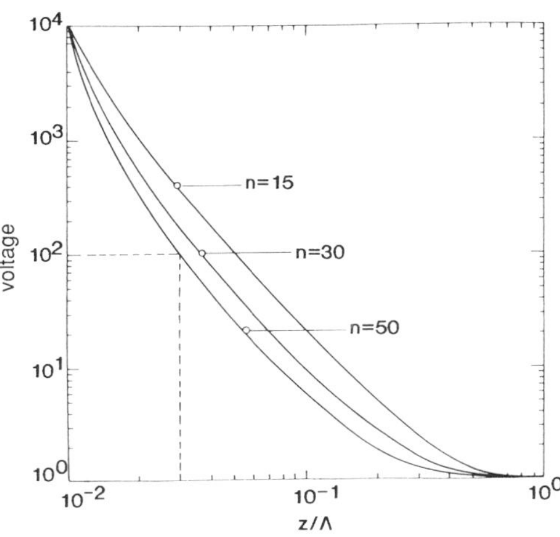

Figure 14.7 Voltage V associated with the radial electric field inside a filamentary wire as a function of the distance z from the junction. Figure 14.5 shows the radial field in the junction and also plots E_ρ in the wire. The constant in Dresner's (1978) equations was selected to make $V = 10^4$ at $z/\Lambda = 10^{-2}$. We see that for all values of Dresner's parameter n, the radial voltage becomes negligibly small within a distance equal to one characteristic length from the junction (Wilson, 1983, p. 241).

IV. CRITICAL CURRENT

The most important transport property of a superconductor is its ability to carry a super current without any dissipation. Chapter 2 discussed transport current and its characteristics of zero resistance and persistence. Some additional understanding of super current flow was provided by the Bean model treatment of Chapter 12. Shielding current was treated at length in Chapters 9, 10, and 12, while the previous chapter discussed current flow through tunnel junctions. The current-induced intermediate state was described in Chapter 11. We now wish to examine the anisotropy of current flow, and its dependence on the magnitude and direction of an applied magnetic field.

A. Anisotropy

Super current flows more easily in the Cu–O planes of high-temperature superconductors than perpendicular to these planes (Gross *et al.*, 1990a). The data in Table 12.3 demonstrate that the critical transport current for flow in the a, b-planes is much greater than for flow perpendicular to these planes, i.e., parallel to the c direction, $(J_c)_{ab} \gg (J_c)_c$. Because of this high anisotropy, almost all critical current measurements on single crystals or epitaxial films are for flow in the a, b-planes.

A good way of showing that J_c in the Cu–O plane is much greater than J_c perpendicular to this plane is to use the magnetization current method of determining J_c, as explained in Chapter 12, Section VI.E. Figure 14.8 shows the temperature dependence of the in-plane magnetization critical current determined by this method for a $Bi_2Sr_2CaCu_2O_8$ monocrystal. The crystal was in the shape of a platelet with the c-axis along the short direction, and the magnetic field was applied along c for this measurement. When the field was applied parallel to the plane, as indicated in the inset to Fig. 14.9, the current exhibited a similar decrease with increasing temperature. The plot was constructed with the aid of the Bean model, taking into account the difference in the field penetration along the narrow (t) as opposed to along the broad (w) faces, as indicated in the inset. The researchers found that w/t ratios from 8 to 23 gave the same value of J_c. Of course, for a continuous current flow path, the larger current density, J_{cy} in the figure, is associated with a smaller effective penetration depth, as indicated. Farrell *et al.* (1989a) reported that the magnetization current anisotropy of yttrium cuprate is much larger than those of the lanthanum and thallium cuprates.

B. Magnetic Field Dependence

We will describe the effects of applied magnetic fields on the flow of transport current in the a, b-plane of cuprates

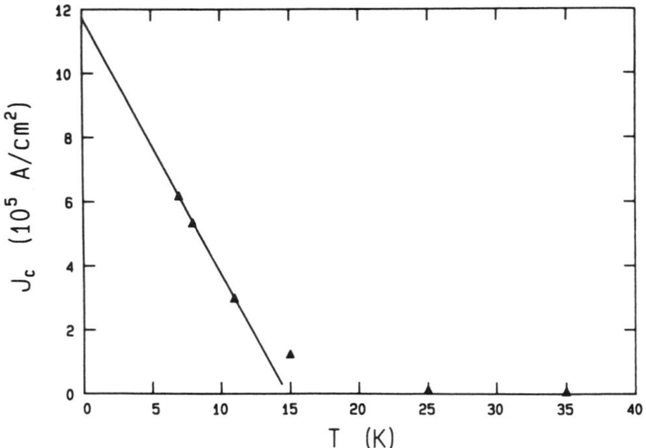

Figure 14.8 Temperature dependence of the critical current J_c determined by magnetization method for a magnetic field applied perpendicular to the Cu–O planes (Biggs *et al.*, 1989).

(Satchell *et al.*, 1988). We see from Fig. 14.10 that for an epitaxial film of $Bi_2Sr_2CaCu_2O_8$ ($T_c \approx 80$ K), the critical current is smaller when the field is applied along c, perpendicular to the plane of the film, than when it is applied in the plane. We also see that the drop-off in J_c with increasing field is especially pronounced at 60 K, and much less so at 4.2 K.

Ekin *et al.* (1990, 1991; cf. Lan *et al.*, 1991) made a more comprehensive study of the magnetic field dependence of transport current in grain-aligned

$$YBa_2Cu_3O_{7-\delta}.$$

The grains were platelets, $\approx 5\ \mu$m in diameter, with the c-axis perpendicular to

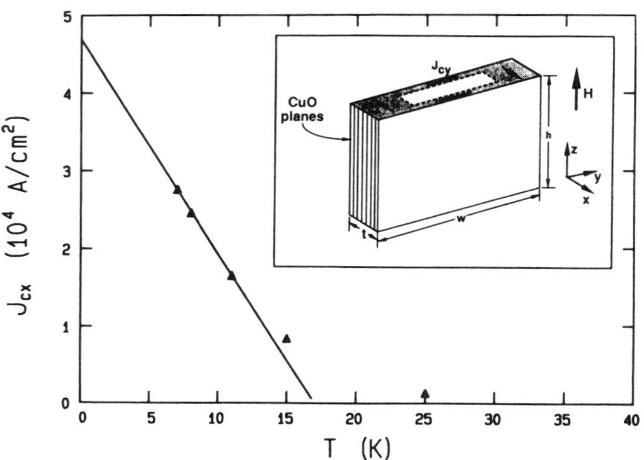

Figure 14.9 Temperature dependence of the critical current J_c determined by the magnetization method for a magnetic field applied parallel to the Cu–O planes. Shaded areas of the inset represent flux penetration (Biggs *et al.*, 1989).

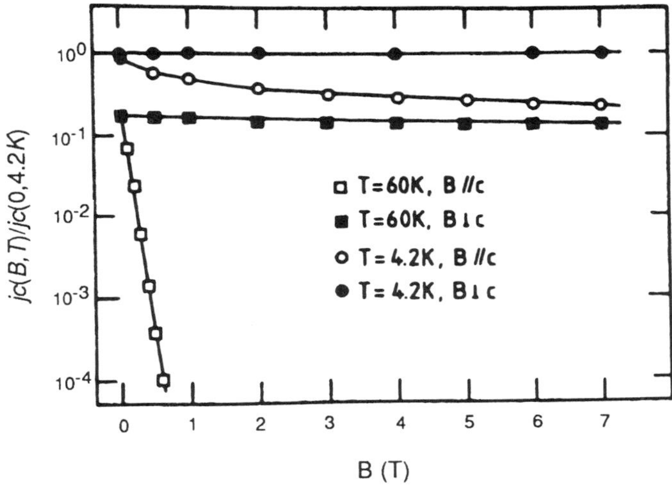

Figure 14.10 Magnetic field dependence of the transport critical current of a $Bi_2Sr_2CaCu_2O_8$ epitaxial film determined at 4.2 K and 60 K for magnetic fields applied parallel to the Cu–O planes, i.e. $\mathbf{B} \perp c$, and perpendicular to the Cu–O planes, i.e. $\mathbf{B} \| c$, (Schmitt *et al.*, 1991).

the plane, forming blocks approximately 1 mm × 1 mm in cross section and 15 mm long. Their results, summarized in Figs. 14.11 and 14.12, show that applying the field along the c direction ($B \perp ab$) causes a greater decrease in the current than applying the field in the plane ($B \| ab$). The current decreased much less with increasing field at 4 K than it did at 76 K. In all of these measurements, the applied field and

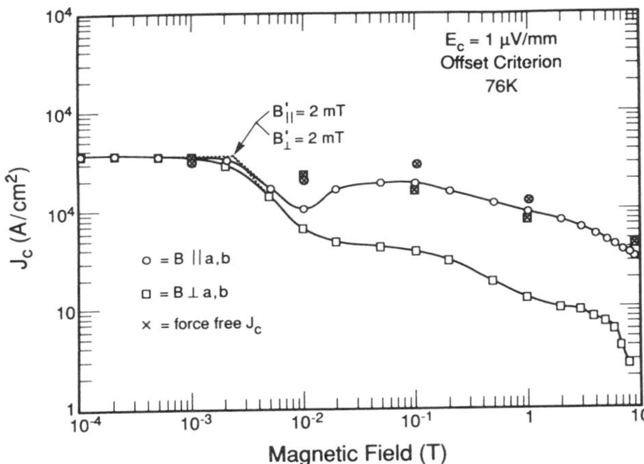

Figure 14.11 Magnetic field dependence of the transport critical current of grain-aligned $YBa_2Cu_3O_{7-\delta}$ determined at 76 K for magnetic fields applied parallel ($\mathbf{B} \| a, b$) and perpendicular ($\mathbf{B} \perp a, b$) to the Cu–O planes. The darkened symbols (filled with ×) are measurements made with the applied field rotated so as to be directed along the current direction (Ekin *et al.*, 1990).

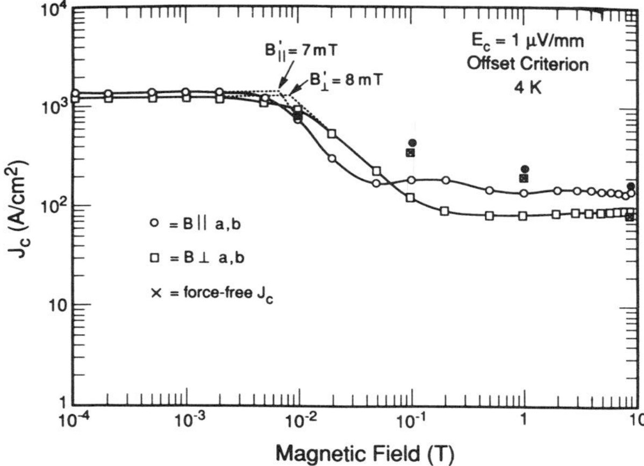

Figure 14.12 Same geometry and physical situation as in Fig. 14.11, at liquid helium temperature (Ekin *et al.*, 1990).

current flow were perpendicular. Force-free values of J_c were determined by rotating the applied field along J_c, which increased the critical current, as indicated in the figures.

The dependence of the critical current on the angle which the applied field makes with the c direction, shown in Fig. 14.13, suggests that J_c depends on $\cos\Theta$ (Maley *et al.*, 1992; Schmitt *et al.*, 1991; Ekin *et al.*, 1991; Fukami *et al.*, 1991b; Miu, 1992). Maley *et al.* (1992) found that the applied field degraded the magnetization currents

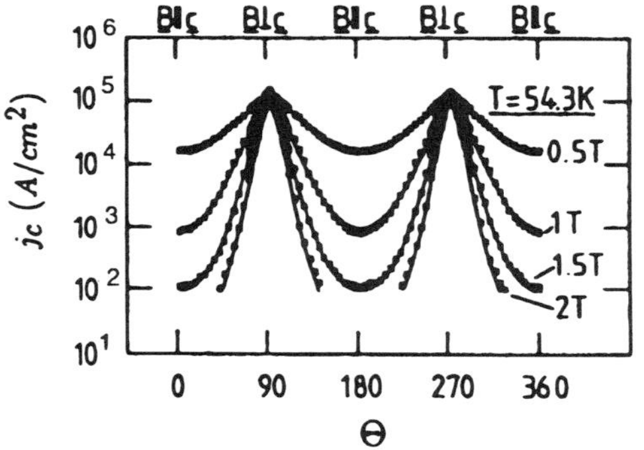

Figure 14.13 Dependence of the critical-current density of a $Bi_2Sr_2CaCu_2O_{8+x}$ epitaxial film on the angle Θ between the applied magnetic field and the c-axis. For all angles the field was perpendicular to the current direction (Schmitt *et al.*, 1991).

to a greater extent than did the transport currents, as indicated in Fig. 14.14.

V. MAGNETORESISTANCE

In this section we will discuss the resistivity of a wire in the presence of a magnetic field, which ordinarily is applied transverse to the current direction. This resistivity, called the *magnetoresistivity* ρ_m, is the same as the ordinary zero-field resistivity ρ for some metals, though it has a different value for others. First we will treat the case of a superconductor above and in the neighborhood of its transition temperature, and then we will show that below T_c the resistance can arise from flux flow.

A. Applied Fields above T_c

Consider the current flow situation illustrated in Fig. 1.16 in the absence of a magnetic field. The flowing current produces the potential difference $V_2 - V_1$ between the ends of the wire. The resistance R as given by Ohm's law,

$$R = (V_2 - V_1)/I, \qquad (14.17)$$

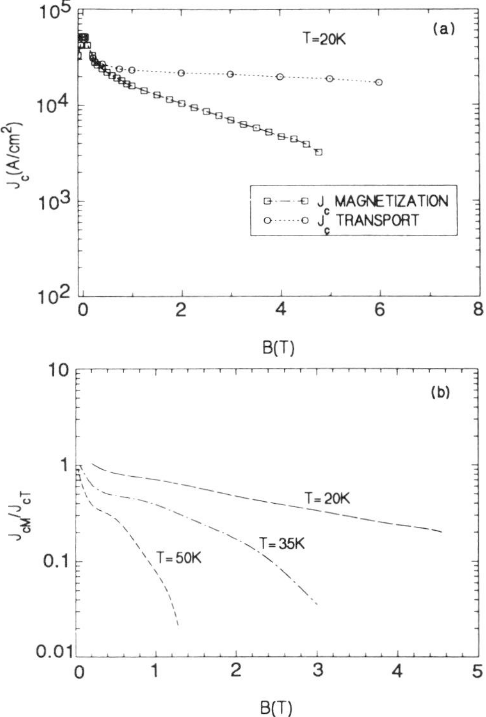

Figure 14.14 Magnetic field dependence of critical-current densities obtained from magnetization and transport measurements of a Pb-doped BiSrCaCuO/Ag superconducting tape showing (a) individual critical current densities at 20 K, and (b) ratio of magnetization to transport critical-current densities at 20 K, 35 K, and 50 K (Maley *et al.*, 1992).

may be written in terms of the current density $J = I/ad$ and the longitudinal electric field $E_y = (V_2 - V_1)/L$ to give for the resistivity (1.97)

$$\rho = E_y/J, \tag{14.18}$$

which is equivalent to Eq. (1.21).

When a transverse magnetic field is applied, as shown in the figure, a transverse Hall effect field $\mathbf{E}_x = \mathbf{v} \times \mathbf{B}_{\text{app}}$ (see Eq. (1.90)) is induced that separates the charge on either side of the wire, as explained in Chapter 1, Section XVI. A resistance measurement provides the magnetoresistivity ρ_m,

$$\rho_m = E_y/J, \tag{14.19}$$

which is more precisely called the transverse magnetoresistivity. The longitudinal magnetoresistivity is defined for a magnetic field aligned along the direction of current flow.

For ordinary (normal-state) conductors the applied field does not affect the longitudinal current flow, so that the resistance of a wire is field independent, and ρ_m from (14.19) equals the ordinary resistivity from (14.18). However, at very high magnetic fields the trajectories of the electrons deflected by the field can be open, i.e., extending from one Brillouin zone into the next, or they can close on themselves in k-space, making the situation complicated. The magnetoresistivity often tends to increase with increasing magnetic field strength, but in some cases it saturates, that is, approaches a field-independent value at the highest fields.

The magnetoresistance of the cuprates in the normal state is not very much affected by the application of small or moderate magnetic fields. This can be seen from Fig. 14.15 for fields up to 7 T applied to $Bi_2Sr_2CaCu_2O_8$ several degrees above T_c (Briceño *et al.*, 1991). A study of $YBa_2Cu_3O_7$ showed very little change in the transverse and longitudinal magnetoresistance for fields up to 10 T and temperatures up to 200 K. For higher fields at 200 K, the longitudinal magnetoresistance was found to increase and its transverse counterpart was found to decrease slightly as the applied field was raised to 43 T (Oussena *et al.*, 1987).

B. Applied Fields below T_c

In the superconducting state the presence of an applied magnetic field shifts the transition temperature downward and broadens the transition in the manner illustrated in Figs. 14.15 and 14.16 for two bismuth cuprates (Ando *et al.*, 1991a, b; Briceño *et al.*, 1991; Fiory *et al.*, 1990; Palstra *et al.*, 1988). Such a downward shift is also to be expected from Fig. 2.46. Similar results have been reported for

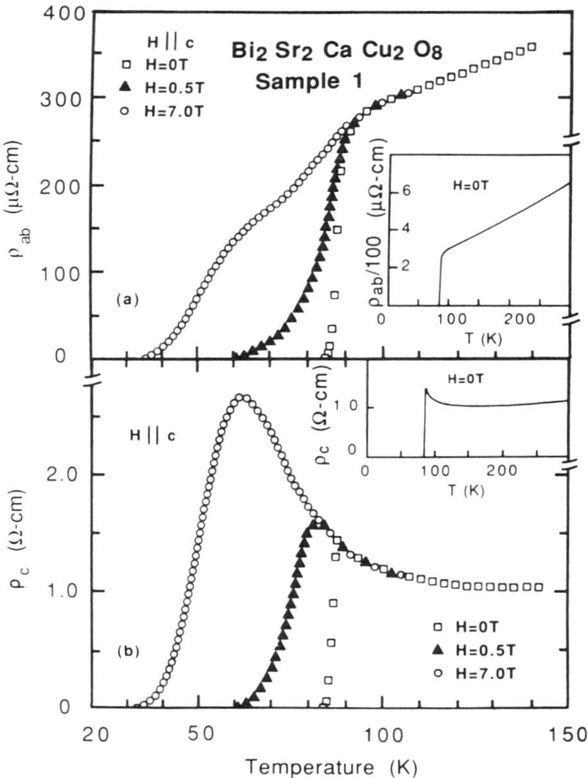

Figure 14.15 Temperature dependence of the magnetoresistance in the a, b-plane (top) and along the c direction (bottom) of a $Bi_2Sr_2CaCu_2O_8$ monocrystal for magnetic fields of 0, 0.5, and 7.0 T applied along the c direction (Briceño *et al.*, 1991).

$(La_{1-\delta}Sr_\delta)_2CuO_4$ (Preyer *et al.*, 1991; Suzuki and Hikita, 1991), $Nd_{1.85}Se_{0.15}CuO_4$ (Suzuzi and Hikita, 1990), $YBa_2Cu_3O_7$ (Blackstead, 1992, 1993; Hikita and Suzuki, 1989; Kwok *et al.*, 1990a), and $Tl_2Ba_2CaCu_2O_8$ (Kim and Riseborough, 1990; Poddar *et al.*, 1989). The zero-field plots of Fig. 14.15 are similar to the $YBa_2Cu_3O_7$ resistivity plots of Fig. 2.7 for ρ_{ab} and ρ_c, with the c-direction resistivity exhibiting a rise slightly above T_c for both compounds. The figures provide ratios $\rho_c/\rho_{ab} \approx 150$ for $YBa_2Cu_3O_7$ and

$$\rho_c/\rho_{ab} \approx 5600$$

for $Bi_2Sr_2CaCu_2O_8$, which demonstrate that the bismuth compound is much more anisotropic. This also constitutes one of the principal differences between the two superconductors (Raffy *et al.*, 1991). The n-type superconductor $Nd_{1.85}Se_{0.15}CuO_4$ has $\rho_c/\rho_{ab} \approx 310$, which is closer to the value for the yttrium compound (Crusellas *et al.*, 1991). Figure 14.16 shows that the shift and broadening of the in-plane resistivity (ρ_{ab}) plots are greater for applied fields along the c-axis than for applied fields in the a, b-plane. (The lower part of Fig. 14.16 is magnified by a factor of 100 to emphasize the rapid exponential drop of the resistivity down to zero for all applied field magnitudes and directions.)

We see from Figs. 14.15 and 14.16 that the in-plane magnetoresistivity ρ_{ab} has a bulge halfway down the curve. The expansion of the sharp zero-field resistivity tran-

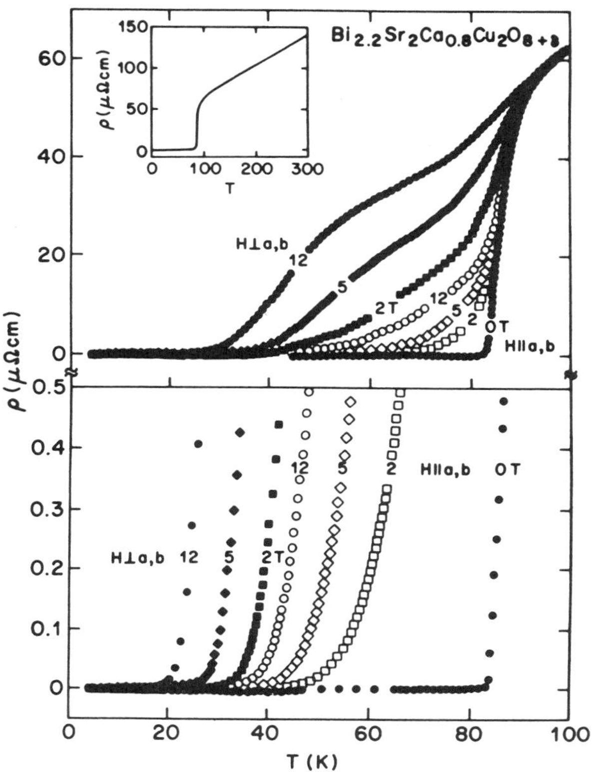

Figure 14.16 Temperature dependence of the magnetoresistance in the *a*, *b*-plane of a $Bi_{2.2}Sr_2Ca_{0.8}Cu_2O_{8+\delta}$ monocrystal with magnetic fields of 0, 2, 5, and 12 T applied parallel to and perpendicular to the *a*, *b*-plane. The lower part shows the ordinate scale magnified by a factor of about 100 to emphasize the exponential behavior (Palstra *et al.*, 1988).

sition of a $YBa_2Cu_3O_7$ monocrystal, shown in Fig. 14.17, reveals that there are actually two very close transition temperatures, $T_{c1} = 90.71$ K and $T_{c2} = 90.83$ K, which separate in applied magnetic fields and are responsible for the observed bulge.

C. Fluctuation Conductivity

The cuprate superconductors exhibit strong temperature and magnetic field dependencies just above T_c that are responsible for the rounding of the resistivity plots of Figs. 14.15, 14.16, and 2.21 at the knee just above T_c. Aronov *et al.* (1989) assumed that the field-dependent part of the electrical conductivity $\Delta\sigma$,

$$\Delta\sigma = \sigma(B) - \sigma(0), \qquad (14.20)$$

obtained from the resistivity measurements, where $\sigma = 1/\rho$, is due entirely to the superconductivity fluctuations. Another approach (Bieri and Maki, 1990; D. H. Kim *et al.*, 1991a; Semba *et al.*, 1991; Suzuki and Hikita, 1989) for explaining fluctuation conductivity made use of the Aslamazov–Larkin (Aslamazov and Larkin, 1968) term due to the excess current carried by Cooper pairs and the Maki–Thompson mechanism (Maki, 1968; Thompson, 1970) of forward scattering on quasiparticles due to Cooper pairs. Several researchers have provided plots of $\Delta\sigma$

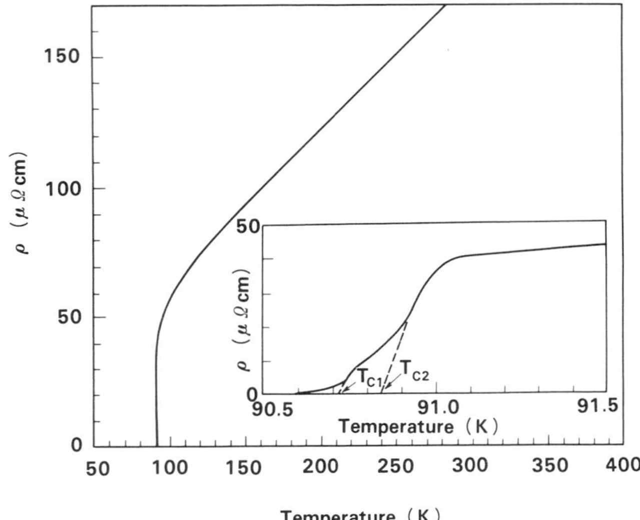

Figure 14.17 Expansion of the region near T_c of a ρ_{ab}-versus-T curve of a $YBa_2Cu_3O_7$ monocrystal showing the two closely spaced critical temperatures T_{c1} and T_{c2} (Hikita and Suzuki, 1989).

versus the field or temperature for $La_{2-x}Sr_xCuO_4$ (Suzuki and Hikita, 1989), $YBa_2Cu_3O_7$ (Bieri and Maki, 1990; Matsuda *et al.*, 1989; Osofsky *et al.*, 1991; Semba *et al.*, 1991), $Nd_{1.85}Se_{0.15}CuO_4$ (Kussmaul *et al.*, 1991), and

$$Tl_2Ba_2CaCu_2O_8$$

(Kim *et al.*, 1991a). The Hikami–Larkin approach (Hikami and Larkin, 1988) has been used to obtain values of the coherence length $\xi_{ab} = 15.6$ Å and $\xi_c = 3.6$ Å for $YBa_2Cu_3O_7$ (Andersson and Rapp, 1991).

D. Flux Flow Effects

When transport current flows in the presence of an applied magnetic field, the vortices arising from the field interact with the current, as was shown in Chapter 9, Section VI.G. This interaction can lead to vortex motion and heat dissipation, and the result is a resistive term called flux-flow resistance. It is a type of magnetoresistance, and limits the achievable critical current in many samples.

We showed in Chapter 9, Section V.C, that when the Lorentz force $\mathbf{J} \times \mathbf{\Phi}_0$ exceeds the pinning force F_P,

$$|\mathbf{J} \times \mathbf{\Phi}_0| > F_P, \qquad (14.21)$$

where $\mathbf{\Phi}_0$ is the quantum of flux, the vortices move with the velocity $\mathbf{v}_\phi$ in accordance with the equation of motion Eq. (9.75). The vortex velocity is limited by the frictional drag force $\beta \mathbf{v}_\phi$, while the Magnus force $\alpha n_s e(\mathbf{v} \times \mathbf{\Phi}_0)$ shifts the direction of this motion through an angle Θ_ϕ away from the direction perpendicular to $\mathbf{J}$, as shown in Fig. 14.18.

By Faraday's law, the motion of the vortices transverse to the current density induces a time-averaged macroscopic electric field $\mathbf{E}$, which is given by

$$\mathbf{E} = -\mathbf{v}_\phi \times \mathbf{B}_{in} \qquad (14.22)$$

as indicated in Fig. 14.18, where $\mathbf{B}_{in}$ is the average internal field due to the presence of the vortices. The component of this electric field E_y along the current-flow direction,

$$E_y = E \cos \Theta_\phi, \qquad (14.23)$$

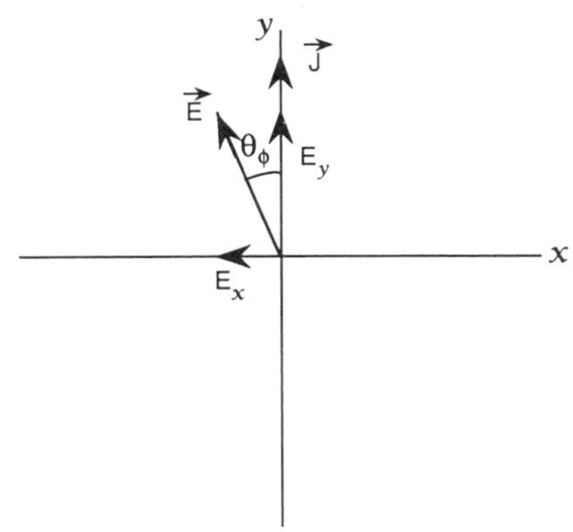

Figure 14.18 Electric field E induced by the motion of a vortex Φ_0 moving at a velocity v_ϕ through an applied magnetic field B_{app} directed upward from the page. The vectors $\mathbf{E}$, $\mathbf{v}_\phi$, and $\mathbf{B}_{app}$ are mutually perpendicular. The vectors for the current density $\mathbf{J}$ and the Lorentz force $\mathbf{J} \times \mathbf{\Phi}_0$ are also indicated.

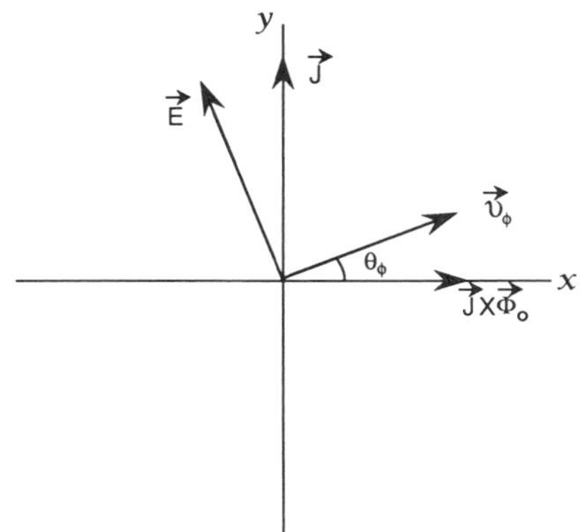

Figure 14.19 Resolution of the induced electric field of Fig. 14.18 into components transverse (E_x) and longitudinal (E_y) to the current density direction (**J**).

shown in Fig. 14.19, produces a voltage drop along this direction. The other component of the induced electric field, $E_x = E \sin \Theta_\phi$, produces a Hall effect, as we will show in the following section.

Figure 14.20 shows how the longitudinal voltage drop along the wire depends on the applied current for two $Nb_{1/2}Ta_{1/2}$ samples with different concentrations of pinning centers (Strnad *et al.*, 1964; see also Tilley and Tilley, 1986, p. 229). Beyond the initial curvature, the V versus I curves of Fig. 14.20 may be represented at low voltage by the equation

$$V = R_{ff}(I - I_c), \qquad (14.24)$$

where the slopes of the lines provide the flux-flow resistance R_{ff}. The flux-flow resistivity is given by

$$\rho_{ff} = \frac{R_{ff}ad}{L}, \qquad (14.25)$$

where the sample dimensions are shown in Fig. 1.16. In Fig. 14.20 we see that the slopes of the straight lines for the two samples are the same, while the intercepts differ due to the variation in pinning force. Measurements made with different magnetic field strengths, shown in Fig. 14.21, exhibit slopes that increase with the magnetic field (Huebener *et al.*, 1970; see also Huebener, 1979, p. 126). Figure 14.22 shows the magnetic field dependence of the flux-flow resistance for three temperatures.

To a first approximation, the critical-current density J_c is obtained by setting the Lorentz force $\mathbf{J} \times \mathbf{\Phi}_0$ equal to the pinning force $\mathbf{F}_P$,

$$\mathbf{J}_c \times \mathbf{\Phi}_0 = \mathbf{F}_P. \qquad (14.26)$$

After the onset of flux flow, increasing $\mathbf{J}$ increases the fluxon velocity $\mathbf{v}_\phi$, which may be calculated using the models introduced in Chapter 9, Section V.F. If the Magnus force is neglected, then, as we show in Problem 7, the flux-flow resistivity is

$$\rho_{ff} = \Phi_0 B_0/\beta. \qquad (14.27)$$

Strnad *et al.* (1964) found that ρ_{ff} is given by the following empirical relation:

$$\rho_{ff} = \rho_n(B_{in}/B_{c2}). \qquad (14.28)$$

The ratio B_{in}/B_{c2} is approximately proportional to the fraction of the material that is "occupied" by the "normal" vortex cores. Thus the resistivity can be imagined as arising from electric current flowing through the normal material that constitutes the vortex cores.

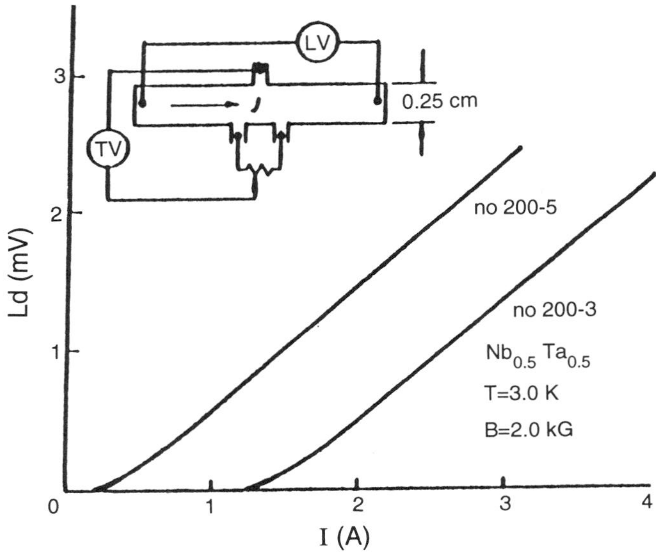

Figure 14.20 Voltage–current characteristics of an $Nb_{1/2}Ta_{1/2}$ superconductor in a magnetic field of 0.2 T at the temperature 3.0 K. The two curves, for samples containing different concentrations of defect pinning centers, have almost the same slope and hence the same flux–flow resistance, but differ in their critical currents I_c. The inset shows the experimental arrangement for the measurements (Strnad *et al.*, 1964).

VI. HALL EFFECT

The Hall effect provides information on the sign, concentration, and mobility of charge carriers in the normal state, with a positive sign for the Hall coefficient $R_H = E_x/JB_0 = \pm 1/ne$ of Eqs. (1.91) and (1.92) indicating that the majority carriers are holes. In the superconducting state, the Hall voltage arises from the electric field induced by flux motion. Chapter 1, Section XVI, describes a Hall effect measurement made with the experimental arrangement of Fig. 1.16. Hall effect probes have been used to measure the local field at the surface of a superconductor in an applied field B_{app}(Brawner *et al.*, 1993).

A. Hall Effect above T_c

Perhaps the most important result that has been obtained from Hall effect measurements above T_c is that the charge carriers in the copper-oxide planes of most of the high-temperature superconductors are holes. Included in this group are the lanthanum, yttrium, bismuth, thallium, and mercury classes of compounds. The major exception is compounds with the Nd_2CuO_4 T′ structure described in Chapter 7, Section VIII.E; their charge carriers are electron-like.

It is easy to argue on the basis of chemical considerations as to why the lanthanum and yttrium compounds are hole-like. Replacing a La^{3+} by a Sr^{2+} without changing the oxygen content can convert a Cu^{2+} to Cu^{3+} on one of the CuO_2 planes, which is the same thing as introducing a hole in a plane. The stoichiometric $YBa_2Cu_3O_7$ compound has an average Cu charge of 2.33, corresponding to one Cu^{3+} and two Cu^{2+} ions, so there is already one trivalent copper ion to contribute a hole. It has also been suggested that the hole might exist on oxygen, corresponding to the ion

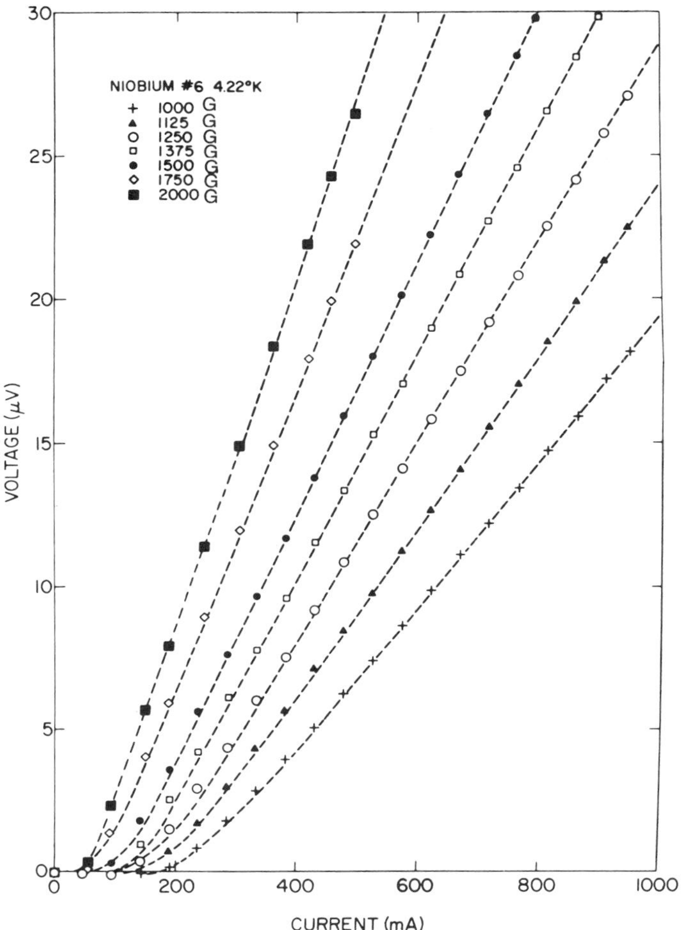

Figure 14.21 Voltage–current characteristics of a niobium foil at 4.22 K in magnetic fields ranging from 0.1 to 0.2 T (1000 to 2000 G), as indicated. The flux-flow resistances evaluated from the linear portions of the plots range from 25 $\mu\Omega$ for the lowest (0.1 T) curve to 64 $\mu\Omega$ for the highest (0.2 T) curve (Huebener *et al.*, 1970).

O^-. From a band structure viewpoint we can say that the hole is in an oxygen $2p$ band.

In contrast, an electron superconductor can be created by doping with a cation having a higher charge, such as substituting Ce^{4+} for Nd^{3+} in $(Nd_{1-x}Ce_x)_2CuO_4$, or substituting a trivalent rare earth such as $R = Gd^{3+}$ for Ca^{2+} in the compound $TlCa_{1-x}R_xSr_2Cu_2O_7$, perhaps to convert Cu^{2+} to Cu^+, or to add an electron to the conduction band. In addition, it has been found that the Hall effect is negative for the applied field aligned in the a, b-plane of $YBa_2Cu_3O_7$ (Penney *et al.*, 1988; Tozer *et al.*, 1987). Table X-3 of our earlier work (Poole *et al.*, 1988) summarizes some Hall effect results.

Several research groups have found that the Hall number V_0/R_He of

$$YBa_2Cu_3O_7$$

has a temperature dependence of the form

$$V_0/R_He = A + BT, \qquad (14.29)$$

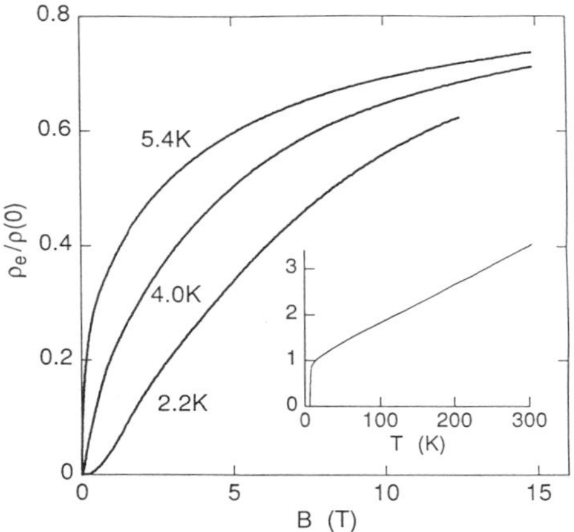

Figure 14.22 Magnetic field dependence of the flux-flow resistivity of $Bi_{2+x}Sr_{2-y}CuO_{6\pm\delta}$ for transport in the a, b-plane at three temperatures below the 7 K transition temperature. The inset shows the temperature dependence of the resistivity at zero field. The abscissae are normalized to the value $\rho(0) = 90\ \mu\Omega$ cm (Fiory *et al.*, 1990).

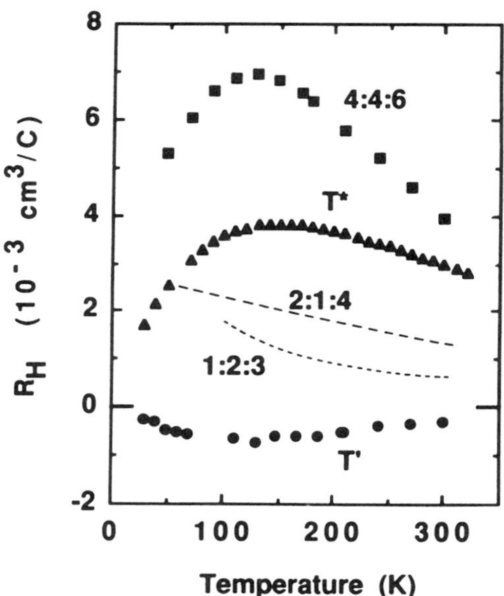

Figure 14.24 Temperature dependence of the Hall coefficient R_H for $(Nb_{0.925}Ce_{0.075})_2CuO_{4-\delta}$ ($T_c = 18$ K, e type, T′ phase), $(Nb_{0.7}Ce_{0.1}Sr_{0.2})_2CuO_{4-\delta}$ (T_c = 23 K, T* phase), $(Nb_{2/3}Ce_{1/3})_4(Nd_{1/3}Ba_{5/12}Sr_{1/4})_4Cu_6O_y$ ($T_c = 38$ K, 4:4:6 compound), $(La_{0.925}Sr_{0.075})_2CuO_{4-\delta}$ ($T_c = 38$ K, T phase, 2:1:4 compound), and $GdBa_2Cu_3O_{7-\delta}$ (1:2:3 compound) (Cheong *et al.*, 1987; Ikegawa *et al.*, 1990).

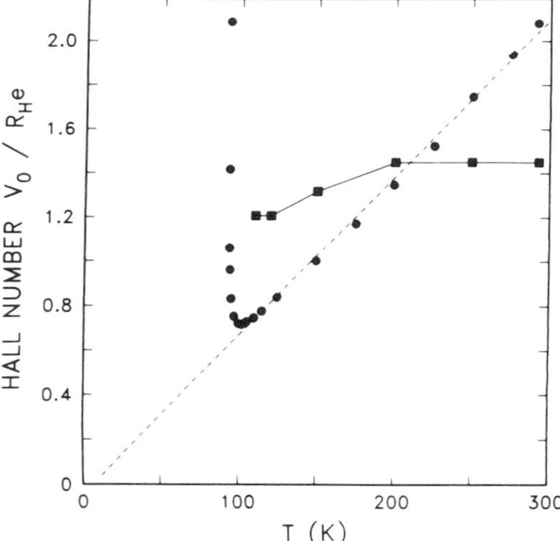

Figure 14.23 Temperature dependence of the Hall number V_0/R_He of $YBa_2Cu_3O_{7-\delta}$. The squares show the nearly temperature independent n-type Hall number of one sample for the magnetic field in the a,b-plane, while the circles show the p-type Hall number for another sample in which the applied field is perpendicular to the a, b-plane. The dashed curve is a linear fit to the data above T_c; the solid curve is provided as a visual aid. Near T_c the Hall number diverges, so that the Hall voltage tends to zero (Penney *et al.*, 1988).

as shown in Fig. 14.23, where $V_0 = 174$ Å and $A \approx 0$ (Penney *et al.*, 1988; a, b-plane data from Tozer *et al.*, 1987). The divergence at T_c shown in the figure arises from the Hall voltage (i.e., E_x) going to zero at the transition. Many superconductors do not exhibit the temperature dependence of Eq. (14.29), as the data plotted in Fig. 14.24 demonstrate (Ikegawa *et al.*, 1990; Gd compound data from Cheong *et al.*, 1987). The data for the electron superconductor $(Nd_{0.925}Ce_{0.075})_2CuO_{4-\delta}$ that are plotted in this figure show that R_H is negative, as expected. In Problem 8 it is necessary to show that the other four compounds in this figure are hole-like. The Hall coefficient is also strongly affected by the oxygen content, as shown by the plots in Fig. 14.25.

Mandal *et al.* (1989) compared the Hall numbers per Cu ion for the various high-temperature superconductors. Their results are plotted in Fig. 14.26. We see from

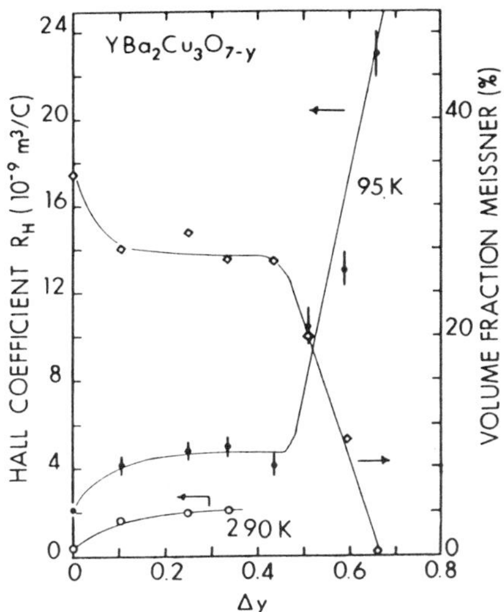

Figure 14.25 Dependence of the Hall coefficient R_H on the oxygen content δ at 77 K (filled circles) and 290 K (open circles). The percentage of the sample exhibiting the Meissner effect is also plotted, with the scale on the right. The solid lines are provided as visual aids (Z. Z. Wang *et al.*, 1987).

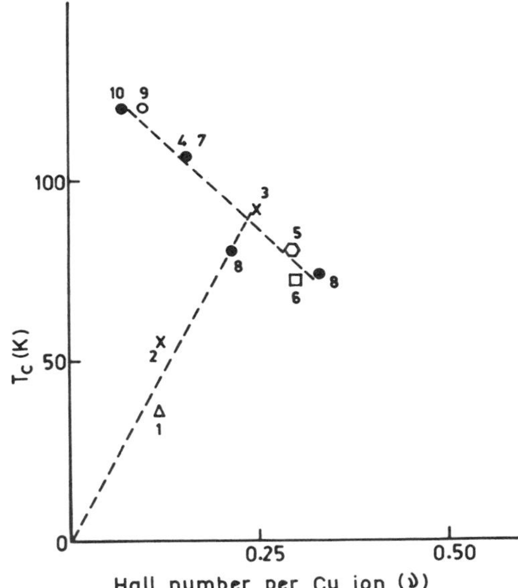

Figure 14.26 Plot of the superconducting transition temperature T_c versus the Hall number per Cu ion, for (1) $(La_{0.925}Sr_{0.075})_2CuO_4$ (Ong *et al.*, 1987), (2) $YBa_2Cu_3O_{6.7}$ (Z. Z. Wang *et al.*, 1987), (3) $YBa_2Cu_3O_7$ (Z. Z. Wang *et al.*, 1987), (4) BiSrCaCuO (Clayhold *et al.*, 1988), (5) $Bi_2(Sr, Ca)_3Cu_2O_{8.35}$ (Takagi *et al.*, 1988), (6) $BiSrCaCu_2O_x$ (Skumryev *et al.*, 1988), (7) $BiSrCaCu_2O_x$ (mixed phase, Mandal *et al.*, 1989), (8) $BiSrCaCu_2O_x$ (85 K phase, Mandal *et al.*, 1989), (9) $Tl_2Ca_2Ba_2Cu_3O_x$ (Clayhold *et al.*, 1988), and (10) $TlCa_3BaCu_3O_x$ (Mandal *et al.*, 1989). The dashed lines are provided as visual aids (figure from Mandal *et al.*, 1989).

the figure that these Hall numbers lie along two straight lines.

Hall effect measurements, like other transport measurements, are affected by sample quality, and hence strongly dependent on factors such as sample preparation, defects, and grain boundaries. This can be deduced from the scatter in some of the data listed in Table X-3 of our earlier work (Poole *et al.*, 1988).

B. Hall Effect below T_c

We have shown that flux flow arising from a transport current in a superconductor below T_c induces an electric field $\mathbf{E}$ given by Eq. (14.22). The component of this electric field perpendicular to the direction of the current, $E_x = E \sin \Theta_\phi$, shown in Fig. 14.19, produces a Hall-effect voltage. The Hall resistivity ρ_{xy}, defined by Eq. (1.99),

$$\rho_{xy} = E_x/J_y, \tag{14.30}$$

is close to zero for low applied fields in the mixed state below T_c and negative for higher fields. Thereafter, it becomes positive and increases linearly with further increases in field, as shown in Fig. 14.27 for $YBa_2Cu_3O_7$ (Hagen *et al.*, 1990a; Rice *et al.*, 1992). The inset of this figure shows the Hall resistance of a niobium film versus the applied field at $T = 9.16$ K, which is slightly below T_c.

We see from Fig. 14.28 that in the mixed state below T_c the Hall mobility $\mu_H = R_H/\rho$ of $Bi_{2+x}Sr_{2-y}CuO_{6\pm\delta}$ increases as the magnetic field is increased and also as the temperature is increased.

VII. THERMAL CONDUCTIVITY

In Chapter 1, Section VIII, we saw that the heat currents carried by conduc-

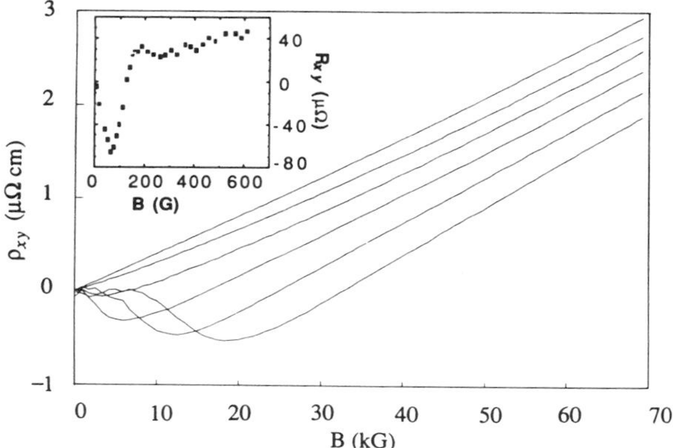

Figure 14.27 Dependence of the Hall resistivity p_{xy} of a $YBa_2Cu_3O_7$ film on the magnetic field from 0 to 7 T for six temperatures near $T_c \approx 90$ K, illustrating the linearity at high field. The temperatures from bottom to top are 88.4, 89.1, 89.8, 90.5, 91.5, and 93.0 K. The inset shows the Hall resistance of a niobium film versus the field from 0 to 60 mT at a temperature of 9.6 K (Hagen *et al.*, 1990a).

tion electrons are closely related to electrical currents. An additional complication in the heat transport case is that the carriers of heat can be either charge carriers like electrons or electrically neutral phonons, whereas electrical current arises only from charge carrier transport. The transformation to the superconducting state changes the nature of the carriers of the electric current, so it is to be expected that the transport of heat will be strongly affected. In this section we will examine how this comes about. Some theoretical treatments are available (e.g., Oguri and Maekawa, 1990; Peacor *et al.*, 1991a; Tewordt and Wölkhausen, 1989; Wermter and Tewordt, 1991a, b).

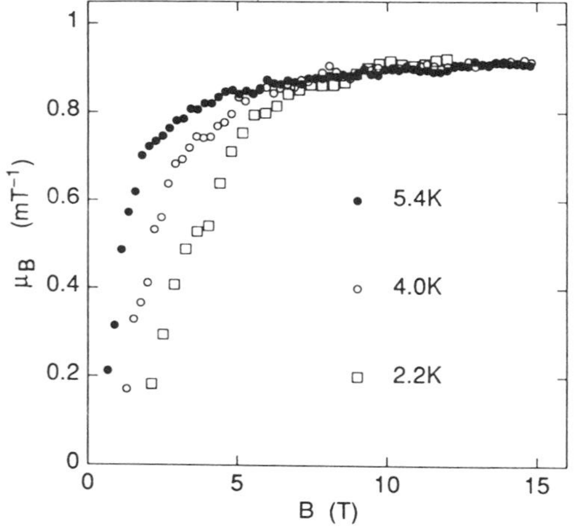

Figure 14.28 Magnetic field dependence of the Hall mobility μ_H in the mixed state of $Bi_{2+x}Sr_{2-y}CuO_{6\pm\delta}$ at three temperatures (Fiory *et al.*, 1990).

A. Heat and Entropy Transport

The thermal current density **U** is the thermal energy per unit time crossing a unit area aligned perpendicular to the direction of heat flow. It is a vector representing the transport of entropy density S_ϕ at the velocity **v**,

$$\mathbf{U} = TS_\phi \mathbf{v}, \tag{14.31}$$

from the hotter to the cooler regions of the material (Maki, 1991). It proportional to the gradient of the temperature ∇T through Fourier's law,

$$\mathbf{U} = -K\nabla T, \tag{14.32}$$

where K is the coefficient of thermal conductivity.

In the normal state, electrical conductors are good conductors of heat in accordance with the law of Wiedermann and Franz (1.33). In the superconducting state, in contrast, the heat conductivity can be much lower because, as Uher (1990) points out, Cooper pairs carry no entropy and do not scatter phonons.

B. Thermal Conductivity in the Normal State

The principal carriers of thermal energy through metals in the normal state are conduction electrons and phonons. Heat conduction via each of these two channels acts independently, so that the two channels constitute parallel paths for the passage of heat. A simple model for the conduction of heat between two points A and B in the sample is to represent the two channels by parallel resistors with conductivities K_e and K_{ph} for the electronic and phonon paths, respectively, as shown in Fig. 14.29a. The conductivities add directly, as in the electrical analogue of parallel resistors, to give the total thermal conductivity K,

$$K = K_e + K_{ph}. \tag{14.33}$$

The electronic path has an electron–lattice contribution K_{e-L}, which is always present, and an impurity term K_{e-I}, which becomes dominant at high defect concentrations. In like manner, the phonon path has a phonon–electron contribution K_{ph-e} plus an additional contribution k_{ph-I} from impurities. Since each pair of terms involves the same carriers of heat, they act in series and add as reciprocals, as in the electrical analogue case of Mattheissen's rule (1.30), where the resistivities (reciprocals of conductivities) add directly. The result is

$$\frac{1}{K_e} = \frac{1}{K_{e-L}} + \frac{1}{K_{e-I}}, \tag{14.34}$$

$$\frac{1}{K_{ph}} = \frac{1}{K_{ph-e}} + \frac{1}{K_{ph-I}}, \tag{14.35}$$

which corresponds to Fig. 14.29b.

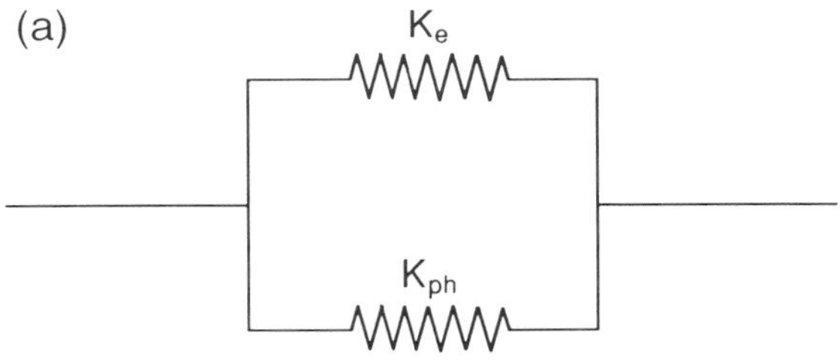

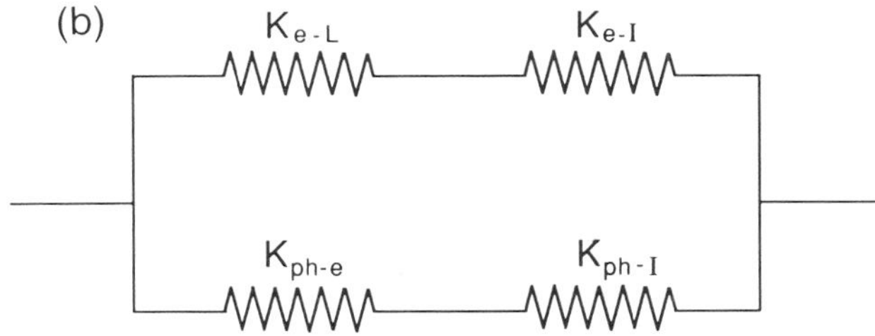

Figure 14.29 Representation of the electron and phonon heat-conduction paths between two points A and B (a) by parallel resistors with respective conductivities K_e and K_{ph}, and (b) representation of the interaction mechanisms with the lattice (L) and impurities (I) operative along each of these two paths by a pair of series resistors.

It is shown in standard solid-state physics texts that the electronic contribution to the thermal conductivity has the form

$$K_{e-L} = \tfrac{1}{3} v_F l C_e \qquad (14.36)$$

$$= \tfrac{1}{3} \gamma v_F^2 \tau T, \qquad (14.37)$$

where we have used Eqs. (1.5), the electron mean free path $l = v_F \tau$, and, from (1.51), $C_e = \gamma T$. If we recall from Eq. (1.23) that $\tau \approx T^{-3}$ at low temperatures and $\tau \approx T^{-1}$ at high temperatures, applying the law of Wiedermann and Franz (1.33) gives us

$$K_{e-L} \approx \begin{cases} \dfrac{[\text{const}]}{T^2} & T \ll \Theta_D \\ [\text{const}] & T \gg \Theta_D \end{cases} \qquad (14.38)$$

for temperatures that are low and high, respectively, relative to the Debye temperature Θ_D. In Chapter 1, Section VII, we saw that at the lowest temperatures the electrical conductivity $\sigma(T)$ approaches a limiting value, $\sigma(T) \to \sigma_0$, arising from the impurity contribution. For this term the law of Wiedermann and Franz gives

$$K_{e-I} \to [\text{const}]T \qquad T \to 0. \quad (14.39)$$

The temperature dependence of the thermal conductivity of copper, shown in Fig. 14.30, seems to follow this behavior. There is an initial linear region corresponding to Eq. (14.39), a maximum in the curve due to the $1/T^2$ term, which should dominate in the intermediate temperature region, perhaps near $T/\Theta_D \approx 0.05–0.5$, and a final asymptotic term at high temperatures.

The lattice contribution to the thermal conductivity has a form which is the phonon analogue of Eq. (14.36),

$$K_{ph-L} = \tfrac{1}{3} v_{ph} l_{ph} C_{ph}, \qquad (14.40)$$

where Eqs. (1.62a) and (1.62b) give the low- and high-temperature limits of C_{ph}, respectively. The temperature dependence is, however, more complicated than that predicted by the specific heat term, since C_{ph} increases with T, whereas the phonon

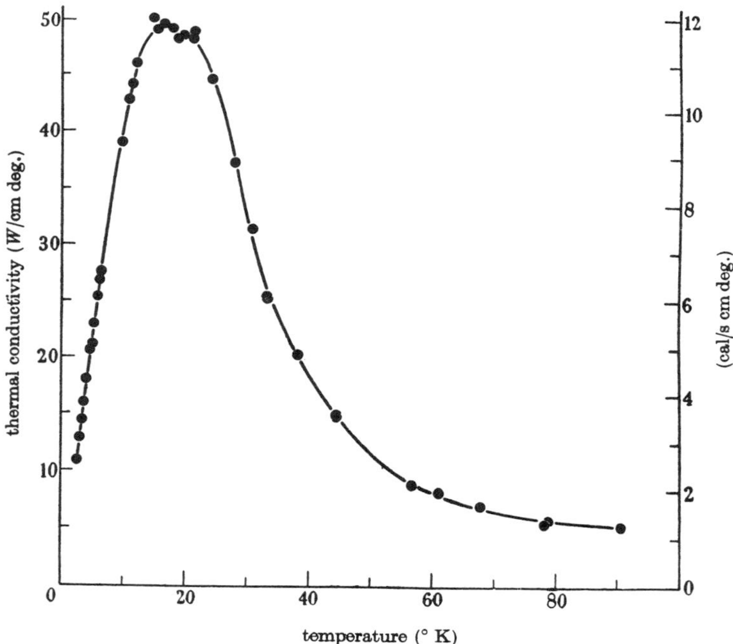

Figure 14.30 Temperature dependence of the thermal conductivity of copper (Berman and MacDonald, 1952).

mean free path l_{ph} decreases with increasing temperature, which not only compensates for C_{ph}, but also tends to cause K_{ph-L} to drop.

In pure metals the electronic contribution to the thermal conductivity tends to dominate at all temperatures, as in the Cu case of Fig. 14.30. When many defects are present, as in disorganized alloys, they affect K_{ph} more than K_e, and the phonon contribution can approach or exceed that of the conduction electrons.

C. Thermal Conductivity below T_c

Thermal conductivity involves the transport of entropy S_ϕ; super electrons, however, do not carry entropy nor do they scatter phonons. We also know from Eq. (4.47) (cf. Fig. 4.8) that below T_c the entropy of a superconductor drops continuously to zero, so that the thermal conductivity can be expected to decrease toward zero also. Figure 14.31 shows this behavior for aluminum (Burns, 1985, p. 657; Scatterthwaite, 1962). The figure plots the ratio of the superconducting state to normal state thermal conductivities K_s/K_n as a function of temperature, where K_n was measured in the presence of a magnetic field $B > B_c$ that extinguished the superconductivity. We see from the figure that the data fit the BCS theory very well. The behavior shown in Fig. 14.31 is typical of elemental superconductors. A material of this type could be employed as a heat switch by using a magnetic field to change its thermal conductivity by more than a factor of 100.

In high-temperature superconductors the phonon contribution to the thermal conductivity is predominant above T_c. The onset of superconductivity can have the effect of first increasing the conductivity until it reaches a maximum, beyond which it decreases at lower temperatures, as shown in Fig. 14.32 (Cohn *et al.*, 1992a, c; Heremans *et al.*, 1988; Pillai, 1991; Terzijska *et al.*, 1992; Uher and Huang, 1989; R. C. Yu *et al.*, 1992; cf. Marshall *et*

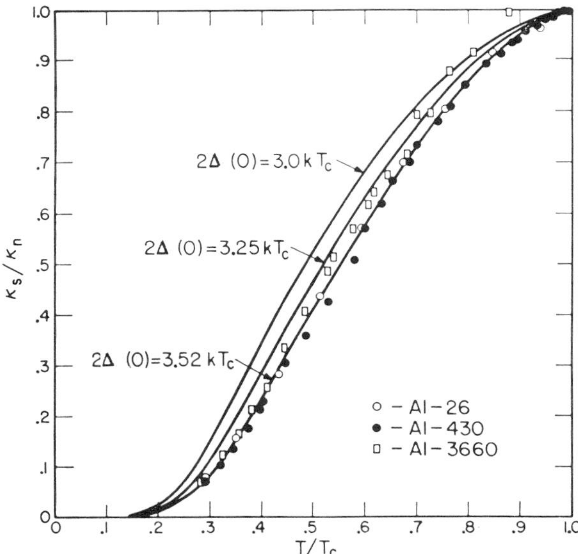

Figure 14.31 Dependence on temperature of the ratio between the electronic thermal conductivity of Al in the superconducting state and its normal-state value. The normal-state data were obtained in a magnetic field that extinguished the superconductivity. The data were fitted by the curve calculated from the BCS theory for $2\Delta/k_BT = 3.52$ (Satterthwaite, 1962).

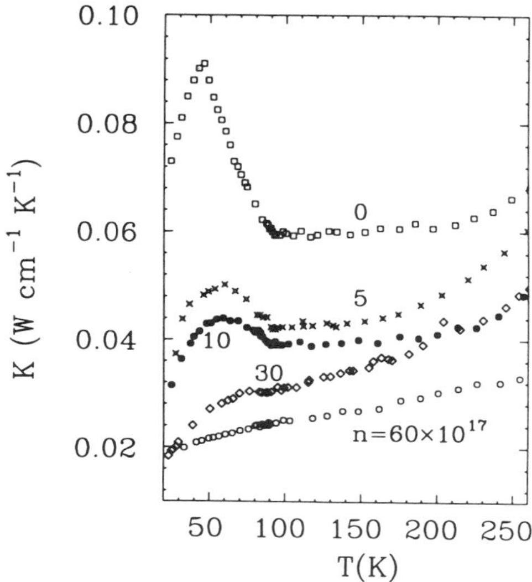

Figure 14.32 Decrease in the thermal conductivity of $YBa_2Cu_3O_{7-\delta}$ brought about by irradiation with fast neutrons at given levels of irradiation, with fluences from 0 up to 6×10^{18} neutrons/cm^2, as indicated (Uher and Huang, 1989).

al., 1992; Szasz *et al.*, 1990). This increase can occur when the thermal conductivity arises mainly from the phonon–electron contribution to K_{ph}. The onset of the superconducting state causes normal electrons to condense into Cooper pairs. These no longer undergo collisions with the phonons and hence do not participate in the phonon–electron interaction. The result is a longer mean free path l_{ph} in Eq. (14.40) and a larger conductivity, as shown in Fig. 14.32 for the unirradiated sample below T_c (Uher, 1990). Irradiating the sample produces defects that limit the mean free paths of the phonons and charge carriers, and leads to a decrease in the thermal conductivity and a suppression of the peak of Fig. 14.32. At lower temperatures, freezing out of the lattice vibrations is reflected in the $C_{ph} \approx AT^3$ term (see Eq. 1.62a) of Eq. (14.40), which becomes negligible relative to the impurity term (14.39). In turn, the latter becomes enhanced by irradiation, and the result is a decrease in K.

Thermal conductivity measurements have been reported, inter alia, on the lanthanum (Bartkowski *et al.*, 1987), bismuth (Peacor and Uher, 1989; Zhu *et al.*, 1989), and $Ba_{1-x}K_xBi$ superconductors (Peacor *et al.*, 1990).

D. Magnetic Field Effects

In a Type II superconductor the thermal conductivity begins to decrease at the lower-critical field B_{c1}, passing through a minimum and then increasing with increasing field until it reaches its normal-state value at the upper critical field B_{c2} (Zhu *et al.*, 1990), as shown in Fig. 14.33 for the superconductor Bi doped In (Dubeck *et al.*, 1964). This behavior is explained by Uher (1990) as due to the presence of vortices acting as additional scattering centers for phonons, which dominate transport far below T_c, and electronic excitations, which are more important near T_c, where quasiparticles are still plentiful. At the lower critical field, vortices enter the superconductor, degrading the thermal conductivity by inducing increased scattering. As the upper critical field is approached, the electronic excitations associated with the normal core of the vortices begin to enhance the thermal conductivity. The fact that the conductivity is independent of the magnetic field in the Meissner state, where $B_{app} < B_{c1}$, as shown in Fig. 14.33, provides further support for this explanation.

The magnetic field dependence of the thermal conductivity of superconductors has been studied (Peacor *et al.*, 1991b; Regueiro *et al.*, 1991; Richardson *et al.*, 1991); entropy transport due to vortex motion (Palstra *et al.*, 1990) and magnetocaloric cooling (Rey and Testardi, 1991) have been observed in $YBa_2Cu_3O_7$.

E. Anisotropy

The planar structure of high-temperature superconductors makes the thermal conductivity anisotropic. This type of

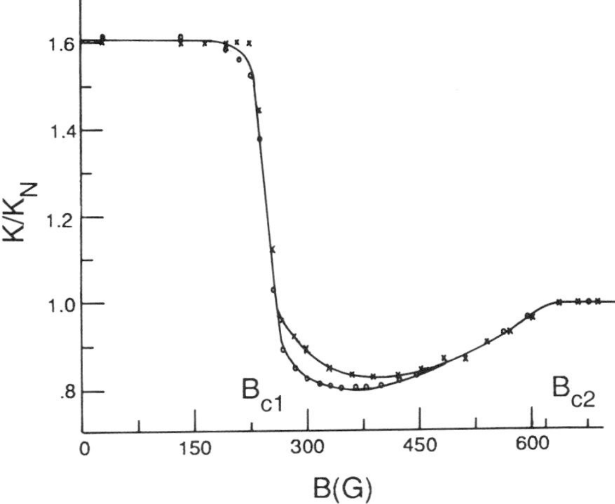

Figure 14.33 Dependence of the thermal conductivity of $In_{0.96}Bi_{0.04}$ normalized to its normal-state value on applied magnetic fields up to 70 mT (700 G). Changes in conductivity occur at the lower- (B_{c1}) and upper- (B_{c2}) critical fields, as indicated (Dubeck *et al.*, 1964).

anisotropy has been observed in polycrystalline samples of $YBa_2Cu_3O_{7-\delta}$ with the microcrystallites aligned along the compression axis (Kirk *et al.*, 1989). For a single crystal, the ratio of the in-plane K_{ab} to the out-of-plane (c-axis) K_c thermal conductivity has been observed to have the value

$$K_{ab}/K_c \approx 17$$

for $YBa_2Cu_3O_7$ (Shao-Chun *et al.*, 1991), $K_{ab}/K_c \approx 6$ for $BiSr_2CaCu_2O_8$ (Crommie and Zettl, 1991), and $K_{ab}/K_c \approx 9$ for $Tl_2Ba_2CaCu_2O_8$ (Shao-Chun *et al.*, 1991). The electrical-conductivity anisotropy is much greater, amounting to $\sigma_{ab}/\sigma_c \approx 10^4$ for the bismuth crystal (Crommie and Zettl, 1991).

VIII. THERMOELECTRIC AND THERMOMAGNETIC EFFECTS

A conductor which is open circuited, as shown in Fig. 14.34, and possesses a temperature gradient can develop an electric field along the gradient direction, called the *thermopower* or *Seebeck effect*, and an electric field perpendicular to this gradient, called the *Nernst effect*. When an isothermal electric current flows, a thermal current can appear flowing parallel to the electric current direction. This is called the *Peltier effect*; an electric current flowing perpendicular to the electric current direction is called the *Ettinghausen effect*. In the two longitudinal effects, Seebeck and Peltier, the central role is played by normal state charge carriers, or quasiparticles. The two transverse effects, Nernst and Ettinghausen, require the presence of an applied magnetic field. For superconductors the central role is played by the motion of vortices (Huebener *et al.*, 1990; Palstra *et al.*, 1990; Ullah and Dorsey, 1990). This means that the Seebeck and Peltier effects are useful for studying superconductors above T_c, whereas the Nernst and Ettinghausen effects provide important information below T_c. Finally, there is a fifth effect, called the *Righi–Leduc effect*, which is the thermal analogue of the Hall effect. This effect can be observed in superconductors below T_c. The characteristics of these five effects are summarized in Table 14.1.

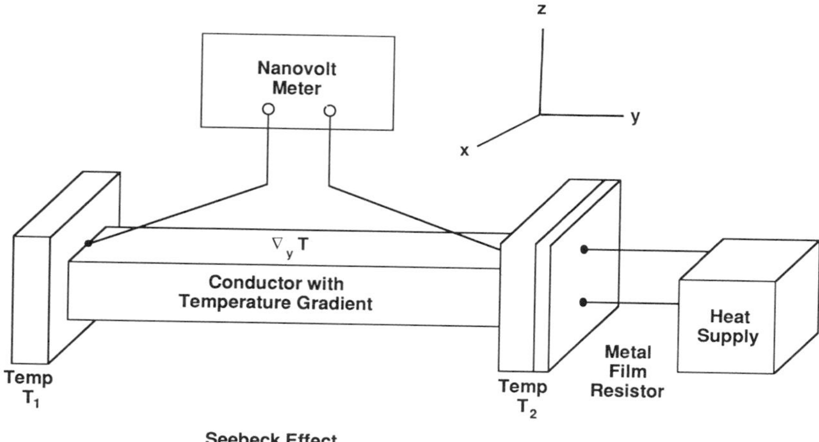

Figure 14.34 Experimental arrangement for measuring the Seebeck effect (thermopower) voltage across a conductor mounted between two temperature reservoirs T_1 and T_2. The heat dissipation in the metal film raises the temperature T_2 above T_1.

The thermoelectric effects are generally described in solid-state physics texts in terms of bimetallic circuits. A bimetallic circuit can be in the form of a superconducting rod in series with an ordinary conducting wire, such as one made of copper. Since the transverse effects (Nernst, Ettinghausen, Righi–Leduc) occur in the presence of a magnetic field, they are also referred to as thermomagnetic effects. Some articles report more than one transport measurement, such as magnetoresistivity, Hall effect, thermopower, etc., on the same sample (e.g., Burns *et al.*, 1989; Freimuth *et al.*, 1991; Fujishita *et al.*, 1991; Ikegawa *et al.*, 1992; Kaiser and Uher, 1988; Ohtani, 1989; Sugiyama *et al.*, 1992; Z. H. Wang *et al.*, 1993). The polar Kerr effect (Spielman *et al.*, 1992) and the magneto optical Faraday effect (Forkl *et al.*, 1990) of high-temperature superconductors have also been reported.

A. Thermal Flux of Vortices

In Chapter 9, Section VI, we discussed the motion of vortices in the presence of applied magnetic fields and currents; some of the thermomagnetic effects can be explained in terms of this motion. Vortex motion can also be induced by the presence of a temperature gradient. We will say a few words about the origin of this motion before proceeding to describe the effects themselves.

Consider a Type II superconductor in a uniform applied magnetic field. The density of vortices, which is equal to B_{in}/Φ_0, where Φ_0 is the quantum of flux, is independent of the temperature. If a temperature gradient is established perpendicular to the magnetic field direction, the vortices in the high-temperature regions will have larger radii than those in the low-temperature regions because the effective radius, which is equal to the penetration depth, increases with the temperature, as indicated in Fig. 2.42. We saw in Chapter 9, Section V.A, that the range of the repulsive force between two vortices increases with the penetration length. This means that the vortices at the hot end of a sample will exert a force on their neighbors that pushes them toward the cooler end of the sample. Thus the uniformity of the magnetic field tends to preserve a constant flux density while the thermally induced forces tend to produce flux motion. The result is a continual flux flow, with vortices entering the sample at the hot end and leaving at

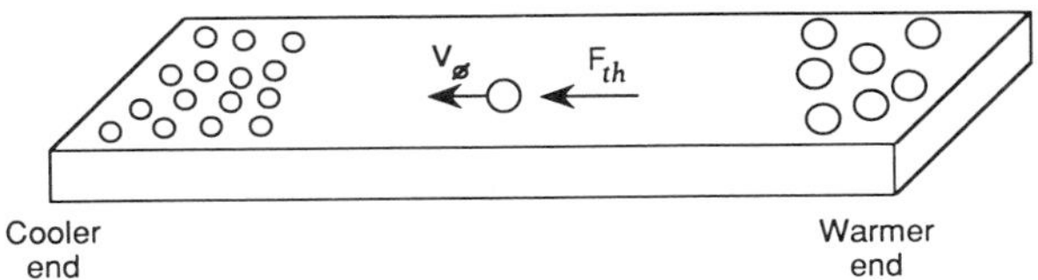

Figure 14.35 Vortex flow from the warmer end of a superconducting slab, where the vortex density is low, to the cooler end, where the density is high, under the action of the thermal force $\mathbf{F}_{\text{th}} = -S_\phi \nabla T$. Note that the vortices have larger radii at the warmer end.

the cold end. This is illustrated in Fig. 14.35.

From a thermodynamic viewpoint, the force producing flux motion can be looked upon as a thermal force $-S_\phi \nabla T$ equal to the product of what is called the transport entropy S_ϕ (which is measured per vortex unit length) (de Lange and Gridin, 1992; Samoilov *et al.*, 1992) and the temperature gradient ∇T. The motion of vortices subject to this force can be described by a vortex equation of motion similar to Eq. (9.75), with the Lorentz force term replaced by the thermal force.

The entropy density within a vortex core, where the material is in the normal state, is higher than it is in the surrounding superconducting medium, causing the vortex to move toward the lower temperature region, where the medium will have a lower entropy density. This pecularity of the entropy density represents one explanation for the flux flow in the presence of a temperature gradient. Increasing the vortex density where the temperature is lower tends to equalize throughout space the average entropy density arising from the superconducting medium plus the vortices with their normal-state cores.

An additional effect can arise from the electrons and holes in the vortex cores, where the material is in the normal state. These will have different thermal distributions in the cores at the hot and cold ends, and the drift velocity of the cores will cause these charge carriers to experience a Lorentz force with the applied magnetic field. This effect can also influence the vortex motion (Wang and Ting, 1992a).

Now that we have seen the ways in which thermal effects can cause vortex motion we are better prepared to understand the thermomagnetic effects that are based on this notion, such as the Seebeck effect arising from thermal diffusion of quasiparticles under the influence of a temperature gradient, and the Nernst effect, which is due to the thermal diffusion of magnetic flux lines (Ri *et al.*, 1993).

B. Seebeck Effect

A conductor which has no electric current flowing through it, but which has a temperature gradient along its length, can develop a steady-state electric field in the gradient direction,

$$\mathbf{E} = S \nabla T. \tag{14.41}$$

This gives rise to an electrostatic potential difference $V_2 - V_1$ between the ends,

$$V_2 - V_1 = S(T_2 - T_1), \tag{14.42}$$

where S is called the thermopower, thermoelectric power, or Seebeck coefficient. We should be careful not to confuse this symbol with the symbol S_ϕ for the transport entropy. A typical experimental arrangement for determining the thermopower, shown in Fig. 14.34, consists of a conducting rod connecting two copper blocks that serve as temperature reservoirs. One block is heated with a metal film resistor that raises its temperature above that of the other block, so that the thermal conductivity of the two copper blocks exceeds that of the conducting rod. A nanovolt meter is employed to measure the thermoelectric voltage between the ends of the rod arising from the longitudinal temperature gradient along the rod.

In the free-electron approximation, the thermopower has the value (Ashcroft and Mermin, 1976, p. 52; MacDonald, 1962)

$$S = \frac{\pi^2}{2} \cdot \frac{k_B}{e} \cdot \frac{T}{T_{\mathrm{F}}} \tag{14.43}$$

$$= 142 \left(\frac{T}{T_{\mathrm{F}}} \right) \mu\mathrm{V/K}, \tag{14.44}$$

where the Fermi temperature can be, typically, 10^4 to 10^5 K. Devaux *et al.* (1990) added a temperature-dependent electron diffusion term to this expression. Thermopower results have been explained using percolation (Rajput and Kumar, 1990) and Hubbard (Oguri and Maekawa, 1990) models. Sergeenkov and Ausloos (1993) discussed thermopower of granular superconductors in terms of the superconductive glass model of Ebner and Stroud (1985). They define a phase locking or grain decoupling temperature below which the grains form a coherent Josephson junction network and above which (but still below T_c) the grains react independently (Mocaër *et al.*, 1991).

A number of workers have reported thermopower measurements on polycrystalline samples of high-temperature superconductors including the recent mercury compounds (Ren *et al.*, 1993; Subramanian *et al.*, 1994; Xiong *et al.*, 1994), but we will confine our attention to the single crystal and mixed state results. Figures 14.36, 14.37, and 14.38 compare the temperature dependence of the thermopower measured in the a, b-plane (S_{ab}) and perpendicular to this plane (S_c) for single crystals of $YBa_2Cu_3O_{7-\delta}$, $Bi_2Sr_2CaCu_2O_8$, and TlBaCaCuO. In all these cases the thermopower was zero below T_c and showed a sharp rise in magnitude at the transition

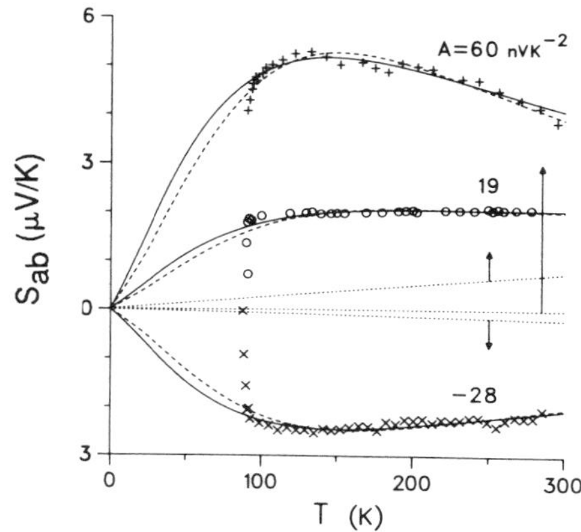

Figure 14.36 In-plane thermopower measurements S_{ab} of $YBa_2Cu_3O_{7-\delta}$ monocrystals. The data are from Lin *et al.* (1989) (+), Sera *et al.* (1988) (○), and Yu *et al.* (1988) (×); fits to the data are explained in Kaiser and Mountjoy (1991); the figure is from Kaiser and Mountjoy (1991).

temperature. Its subsequent behavior above T_c was not predictable.

Kaiser and Mountjoy (1991) explained these results in terms of the metallic-diffusion approach, an approach that has

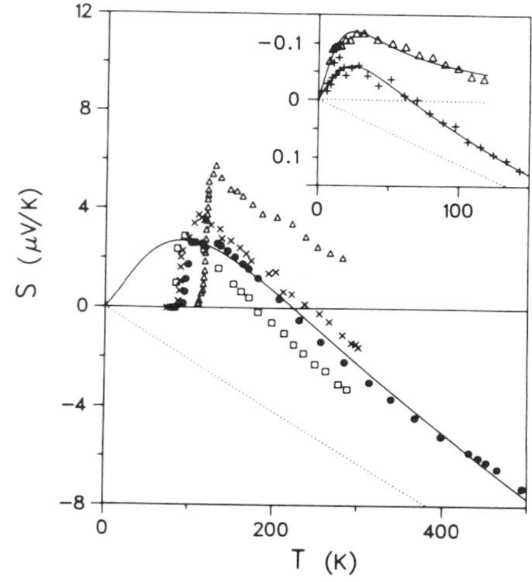

Figure 14.37 In-plane thermopower measurements S_{ab} of sintered $Bi_2Sr_2CaCu_2O_{8+\delta}$ [Pekala *et al.* (1989) (●), Crommie *et al.* (1989) (□), Chen *et al.* (1989) (×), and TlCaBaCuO (Bhatnagar *et al.* (1990) (△)]. Fits to the data are from Pekala *et al.* (1989); the inset data for $Sn_{0.8}Ag_{0.2}$ (△) and $Sn_{0.6}Ag_{0.4}$ (+) are from Compans and Baumann (1987); the figure is from Kaiser and Mountjoy (1991).

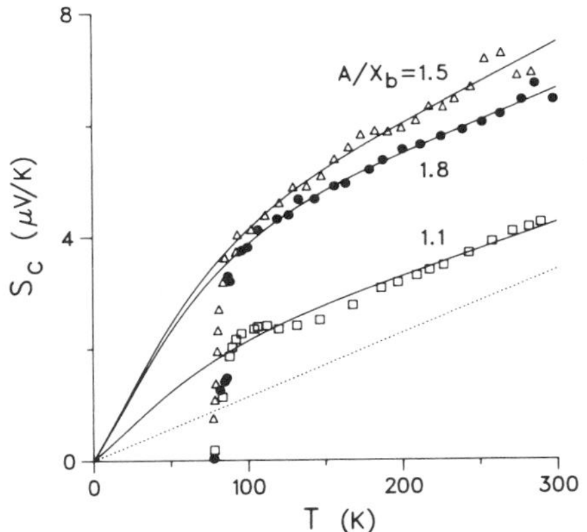

Figure 14.38 Out-of-plane thermopower measurements S_c of monocrystals fitted to a diffusion model (——). The data for $YBa_2Cu_3O_{7-\delta}$ are from Crommie *et al.* (1988) (△), the data for $Bi_2Sr_2CaCu_2O_{8+\delta}$ are from Crommie *et al.* (1989) (□) and Chen *et al.* (1989) (●); the figure is from Kaiser and Mountjoy (1991).

been successfully applied to other metal systems in which phonon drag is suppressed. They found that the bare thermopower term, which is linear in the temperature, like the simpler free-electron expression (14.43), is strongly enhanced in high-temperature superconductors by the electron–phonon interaction. Their fits to the data shown in Figs. 14.36, 14.37, and 14.38 are quite good. The figures also show the bare thermopower contribution, which is especially small for the in-plane yttrium data. In the free-electron approximation (14.43), the bare phonon lines of Figs. 14.37 and 14.38 correspond to Fermi temperatures of about 20,000 K and 40,000 K, respectively; actual Fermi temperatures are expected to be smaller than these values. Doyle *et al.* (1992) estimated the Fermi energy of $Bi_{1.6}Pb_{0.4}Sr_2Ca_2Cu_2O_y$ from thermopower measurements.

Single-crystal thermopower results have been reported for the lanthanum (Cheong *et al.*, 1989a; Nakamura and Uchida, 1993), yttrium (J. L. Cohn *et al.*, 1991, 1992b; Lengfellner *et al.*, 1992; Lowe *et al.*, 1991), bismuth (Obertelli *et al.*, 1992; Song *et al.*, 1990), and thallium (Obertelli *et al.*, 1992; Shu Yuan *et al.*, 1993) compounds and monocrystals of K- and Rb-doped C_{60} (Inabe *et al.*, 1992). The thermoelectric power of the Nd–Ce and Nd–Pr electron superconductors was found to be similar in sign (positive) and in terms of temperature dependence to those of hole-type cuprates (Lim *et al.*, 1989; Xu *et al.*, 1992). The thermopower of the organic superconductor

$$\text{K–(BEDT–TTF)}_2\text{Cu[N(CN)}_2\text{]Br}$$

was positive along the *a* direction and negative along *c*, suggesting that the carriers in the *a* direction are hole-like, whereas those along *c* are electron-like (J. Yu *et al.*, 1991). Electron–phonon enhancement of the thermopower was found in the Chevrel compounds $Cu_{1.8}Mo_6S_{8-x}M_x$, where $M = $ Se or Te (Kaiser, 1987, 1988).

We have been discussing the Seebeck effect in the normal state. Several workers have applied a magnetic field for thermopower measurements to study the mixed state. Figure 14.39 shows the results obtained for $YBa_2Cu_3O_7$ (Hohn *et al.*, 1991) with the applied field along the *x*, *y*, and *z* directions, respectively, of Fig. 14.34. Gridin *et al.* (1989) obtained results similar to those shown in Fig. 14.39c for the compound $Bi_2Sr_2CaCu_2O_8$ with the applied field along the *z* direction. It was found that the area between the curve for the thermopower in a field **B** and the curve for zero field (**B** = 0) (cf. Fig. 14.39c) was proportional to the applied field **B**. The Seebeck effect in the mixed state has been attributed to counter flow of quasiparticles (normal current) and super current in the presence of a temperature gradient (Huebener *et al.*, 1990; Ri *et al.*, 1991).

While resistivity and Hall effect experiments determine the density and mobility of charge carriers, thermopower experiments are intended to measure their energy distribution. From Eq. (14.32), we see that the electric field (14.41) associated with the thermopower is proportional to the thermal-energy flux **U** of the charge

(a)

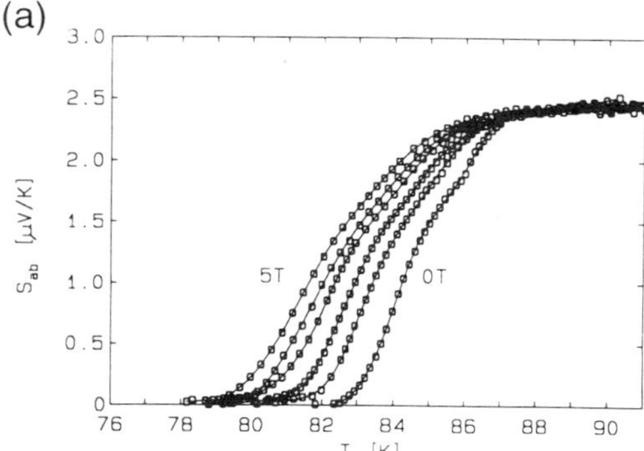

(b)

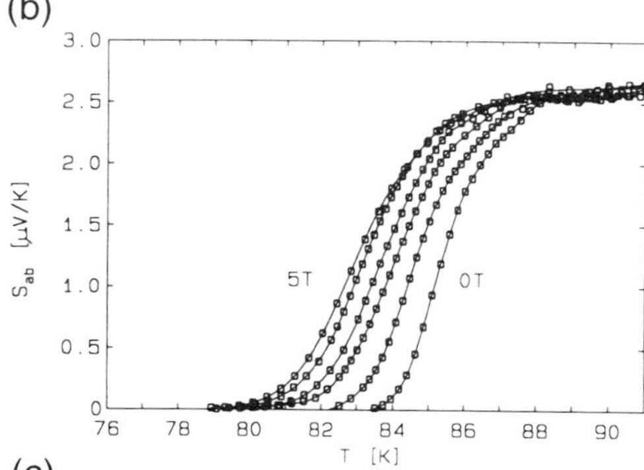

(c)

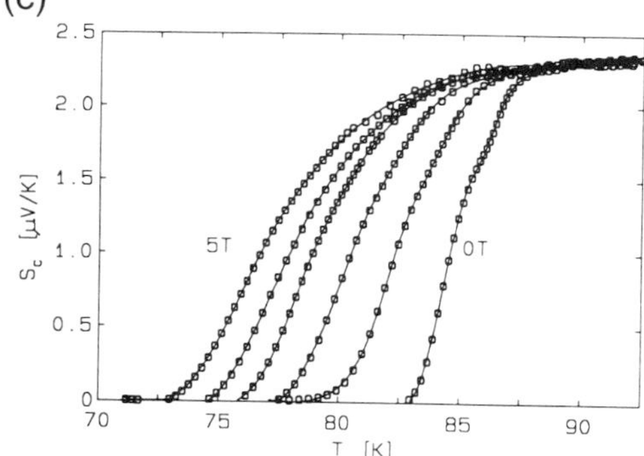

Figure 14.39 Temperature dependence of the magnetothermopower S for the experimental arrangement of Fig. 14.34 in magnetic fields of 1, 2, 3, 4, and 5 T. The c-axis oriented film is in the x, y-plane of Fig. 14.34 so that the temperature gradient $\nabla_y T$ and the measured electric field E_y are both along the y direction. Results are shown for the applied magnetic field oriented (a) in the a, b-plane along the y direction parallel to ∇T, (b) in the a, b-plane, along the x direction perpendicular to ∇T, and (c) perpendicular to the a, b-plane, along the z direction (Hohn *et al.*, 1991).

carriers, which, in turn, from Eq. (14.31), is proportional to the entropy flow $S_\phi \mathbf{v}$. Thermopower measurements have been looked upon as measuring the entropy S_ϕ per carrier (Burns *et al.*, 1989).

C. Nernst Effect

In the presence of an applied magnetic field, a conductor with a temperature gradient and no electric-current flow can develop a steady-state electric field transverse to the gradient direction, a phenomenon that is called the Nernst effect. The effect is very small in normal conductors, but can be appreciable in superconductors if flux flow occurs.

To explain the origin of the transverse electric field, Huebener (1979, pp. 155ff.) made use of the vortex equation of motion (9.76) with the thermal force $-S_\phi \boldsymbol{\nabla} T$ introduced in Section VII.A replacing the Lorentz force $\mathbf{J} \times \mathbf{\Phi}_0$ as the driving force,

$$S_\phi \boldsymbol{\nabla} T + \beta \mathbf{v}_\phi + \mathbf{F}_\mathrm{p} = \mathbf{0}, \qquad (14.45)$$

where $\beta \mathbf{v}_\phi$ is the drag force, $\mathbf{F}_\mathrm{p}$ the pinning force, and the Magnus force $\alpha n_\mathrm{S} e(\mathbf{v}_\phi \times \mathbf{\Phi}_0)$ is neglected (Zeh *et al.*, 1990). The thermal force does not induce flux flow until it exceeds the pinning force F_p, and this occurs at the critical gradient $(\boldsymbol{\nabla} T)_c$,

$$S_\phi (\boldsymbol{\nabla} T)_\mathrm{c} = -\mathbf{F}_\mathrm{p}, \qquad (14.46)$$

to give

$$S_\phi [\boldsymbol{\nabla} T - (\boldsymbol{\nabla} T)_\mathrm{c}] + \beta \mathbf{v}_\phi = \mathbf{0}. \quad (14.47)$$

Therefore, a thermal gradient that exceeds the critical gradient causes vortices to move from the high-temperature end of the material to the low-temperature end, in accordance with Fig. 14.35.

Consider a magnetic field B applied perpendicular to the thermal gradient, as shown in Fig. 14.40. The vortex moving at the velocity v_ϕ in the gradient direction entrains its encircling screening currents, as described in Chapter 2, Section VIII. Let v_e be the velocity of an electron encircling the vortex when the vortex is stationary. Once the vortex starts to move, the velocity of this electron on one side is $v_\mathrm{e} + v_\phi$, and on the other side $v_\mathrm{e} - v_\phi$, as shown in Fig. 14.41. As a result, a Lorentz force $e\mathbf{v} \times \mathbf{B}$ that is stronger on one side of the vortex than on the other comes into

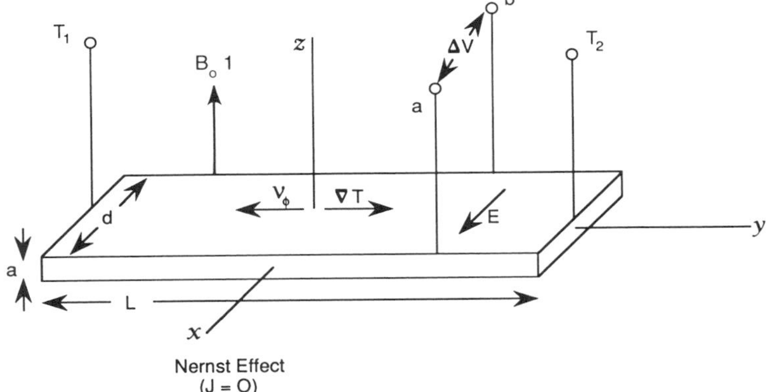

Figure 14.40 Experimental arrangement for measuring the Nernst effect of a superconducting slab in a transverse magnetic field B. The vortices flow to the left with the velocity v_ϕ under the action of the temperature gradient ∇T directed to the right. The temperature (T_1, T_2) and voltage (a, b) measuring leads are indicated.

play, and the vortices move in a direction perpendicular to both **v** and **B**. This causes a flux gradient to be established along the x direction of Fig. 14.40, producing an electric field **E**,

$$\mathbf{E} = -\mathbf{v}_\phi \times \mathbf{B}, \tag{14.48}$$

which cancels the effect of the magnetic field and causes the flux motion along ∇T to proceed undeflected. An analogous transverse electric field (1.90) arises in the Hall effect. As a result, a voltage difference ΔV, called the Nernst voltage, is established between the terminals a and b on the two sides of the superconductor.

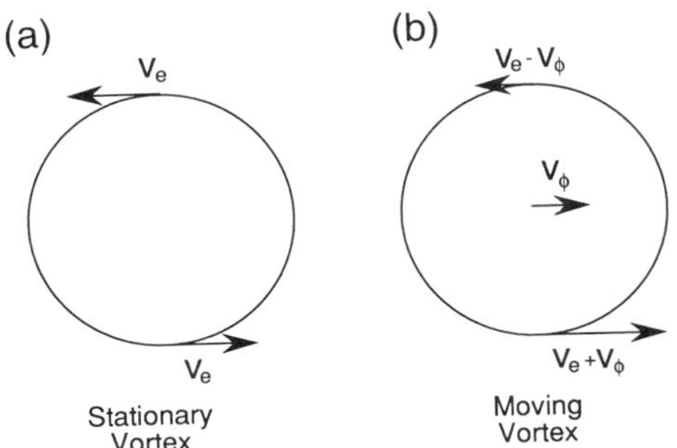

Figure 14.41 Electron circulation around a vortex which is (a) stationary, and (b) moving to the right with the velocity v_ϕ.

To obtain the equation for the Nernst effect we can substitute v_ϕ from Eq. (14.47) into Eq. (14.48) and write, in scalar notation,

$$\frac{dV}{dx} = -S_\phi(B/\beta)[\nabla T - (\nabla T)_c]. \tag{14.49}$$

We define the Nernst coefficient Q as the ratio between the transport entropy S_ϕ and the vortex friction coefficient β,

$$Q = S_\phi/\beta,$$

and note that the electric field E is the gradient of the potential V in the x direction, and this gives for the Nernst voltage ΔV across a sample of width w

$$\Delta V = -QBw[\nabla T - (\nabla T)_c]. \tag{14.50}$$

Figure 14.42 shows plots of the Nernst voltage measured across thin films of Sn at 2 K ($T_c = 3.7$ K for $B = 0$) for several magnetic field strengths. We see from the figure that the critical thermal gradient $(\nabla T)_c$ is less for higher fields, which is to be expected because the Lorentz force adds to the thermal force in Eq. (14.45) to overcome the pinning. The slopes of the lines $\Delta V/\Delta T$ increase with the applied field, but this increase does not have the proportionality to B predicted by Eq. (14.50).

A measurement of the temperature and field dependence of the Nernst coefficient of $YBa_2Cu_3O_7$ below T_c showed that the vortex entropy per unit length S_ϕ increases with decreasing temperature, with very little field dependence (Hagen *et al.*, 1990b), as shown in Fig. 14.43. The measured magnetoresistance ρ_{xx} together with the expression (Kim and Stephan, 1969)

$$\rho_{xx} = E_x/J_y = \phi_0 B/\beta \qquad (14.51)$$

has been used to evaluate β and to determine S_ϕ. K. Kober *et al.* (1991) found a pronounced dependence of S_ϕ on B, with the entropy tending to decrease in higher fields. The Nernst effect below T_c has been reported for thallium superconductors (Koshelev *et al.*, 1991; Lengfellner *et al.*, 1990).

The existence of the Nernst effect in the superconducting state indicates the presence of flux flow or vortex motion; flux depinning activation energies can be deduced (Lengfellner and Schnellbögl, 1991) and entropy is transported by a moving flux line. These factors distinguish the effect from the thermoelectric voltage which, since it is produced in the absence of an

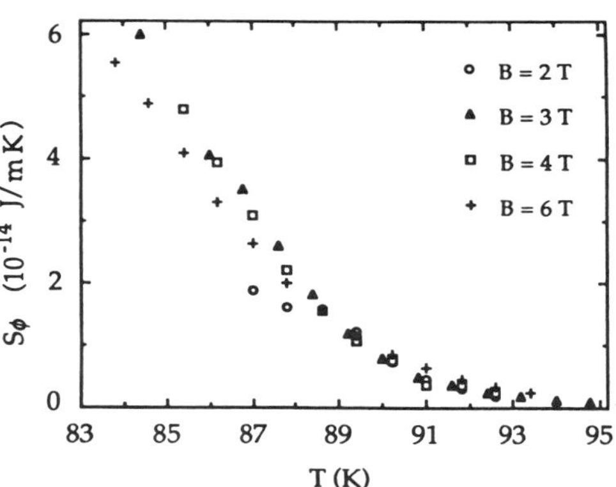

Figure 14.43 Dependence of the vortex entropy per unit length S_ϕ of epitaxial $YBa_2Cu_3O_7$ on the temperature for applied magnetic fields of 2, 3, 4, and 6 T. The entropy was determined from Nernst effect measurements (Hagen *et al.*, 1990b).

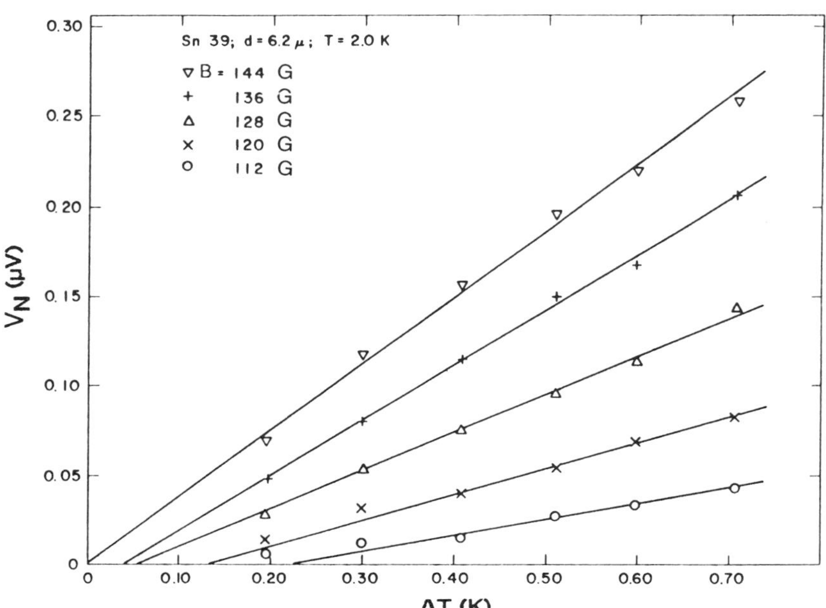

Figure 14.42 Dependence of the Nernst voltage at 2.0 K of a 6.2-μm thick Sn thin film on the longitudinal temperature difference $\Delta T = T_2 - T_1$ for the range of applied magnetic fields from 11.2 to 14.4 mT (110 to 144 G) (Rowe and Huebener, 1969).

applied magnetic field, is due to dissipation processes other than flux motion (Hagen *et al.*, 1990b; Hohn *et al.*, 1991; Lengfellner *et al.*, 1991a).

D. Peltier Effect

When a conductor is maintained at a constant temperature with a uniform electric current flowing through it, the electric current flow is accompanied by a thermal current, a phenomenon called the Peltier effect. (The thermal current serves to carry away the Joule heat generated by the electric current.) The electric current density J and thermal current density U are related by the Peltier coefficient π_P

$$U = \pi_P J. \tag{14.52}$$

This effect is demonstrated experimentally by driving a current through a bimetallic circuit maintained at a constant temperature and measuring the heat absorbed at one junction and released at the other, as shown in Fig. 14.44. Lord Kelvin deduced the relation (Thomson relation)

$$\pi_P = ST \tag{14.53}$$

between the Peltier coefficient π_P and the thermopower S. The Peltier effect is strong in a normal metal but has yet to be observed in a superconductor, perhaps because super current does not carry entropy. In a superconductor the effect could arise from dissipative electric current associated with flux motion carrying Peltier heat across the sample in the mixed state, with the Thomson relation satisfied (Huebener, 1990; Logvenov *et al.*, 1991). The calculations of Maki (1991) provide an extra Peltier effect due to fluctuations.

E. Ettinghausen Effect

When a conductor in an applied magnetic field is maintained at a constant temperature with a uniform electric current flowing through it, heat energy (i.e., a thermal current) can travel in the transverse direction to establish a transverse temperature gradient, a phenomenon called the Ettinghausen effect. This is the transverse analogue of the longitudinal Peltier effect. It is very small in a normal metal, but can be large in a superconductor in the presence of an applied magnetic field because a heat current will be generated by he motion of the vortices in the field. We will

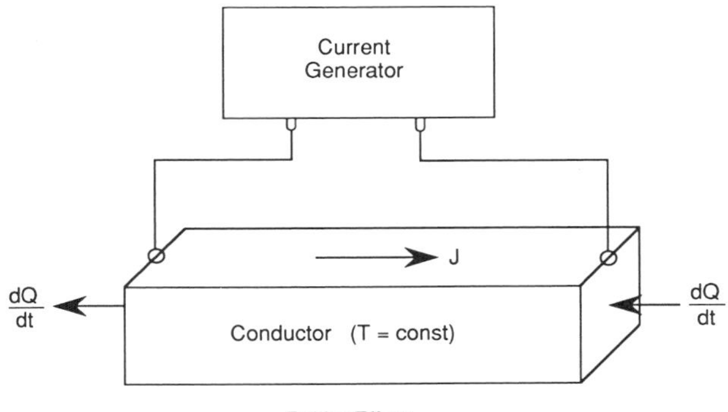

Figure 14.44 Experimental arrangement for the Peltier effect. An electric current is passed through a metal maintained at a constant temperature T, and the heat current dQ/dt that enters at the right and leaves at the left is measured.

analyze this situation for the experimental arrangement of Fig. 14.45, which shows the magnetic field $\mathbf{B}_0$, current density $\mathbf{J}$, and flux flow direction $\mathbf{v}_\phi$, as well as the temperature change ΔT that is developed across the sample.

The electric current flowing through the wire exerts a Lorentz force on the flux structures, which are vortices with quantized flux in the mixed state of Type II superconductors and nonquantized domains in the intermediate state of Type I superconductors, as explained in Chapter 11, Section III. This causes the structures to move with the velocity $\mathbf{v}_\phi$; in addition, their motion is dissipative, as explained in Chapter 9, Section V.C, so that it is accompanied by a flow of heat. The heat-current density $\mathbf{U}_\phi = nTS_\phi \mathbf{v}_\phi$ may be equated to the heat flux $K\nabla T$ through Fourier's law (14.31, 14.32),

$$nTS_\phi \mathbf{v}_\phi = -K\nabla T, \qquad (14.54)$$

where n is the number of flux structures per unit area and S_ϕ is the entropy transported per unit length of such a structure. If we substitute v_ϕ from Eq. (14.48) in Eq. (14.54), where $v_\phi = E/B$ (since $\mathbf{v}_\phi$ and $\mathbf{B}$ are mutually perpendicular), and express $\mathbf{E}$ as the gradient of a potential ∇V, we obtain the scalar expression

$$\left|\frac{dT}{dx}\right| = \left(\frac{TS_\phi}{K\Phi}\right)\left|\frac{dV}{dy}\right|, \qquad (14.55)$$

which is the fundamental equation for the Ettinghausen effect in the superconducting state. Figure 14.45 clarifies the directions of these gradients, and shows the terminals a and b across which the temperature change is measured. It should be emphasized that the potential gradient ∇V, which is in the $\mathbf{J}$ direction, arises from the motion of the vortices, since the super current flow itself is not accompanied by any potential gradient. The potential difference, $\Delta V = V_2 - V_1$, measured between the two ends of the sample, as shown in Fig. 14.46, provides the magnitude of the gradient $\nabla V = \Delta V/L$.

Figure 14.46 shows some experimental data obtained with the Type II alloy $In_{0.6}Pb_{0.4}$ in several applied magnetic fields. We see from the figure that ΔT is almost linear in $\Delta V = V_2 - V_1$, especially for small applied fields. The slopes of the lines, however, do not exhibit the inverse dependence on the applied field which is

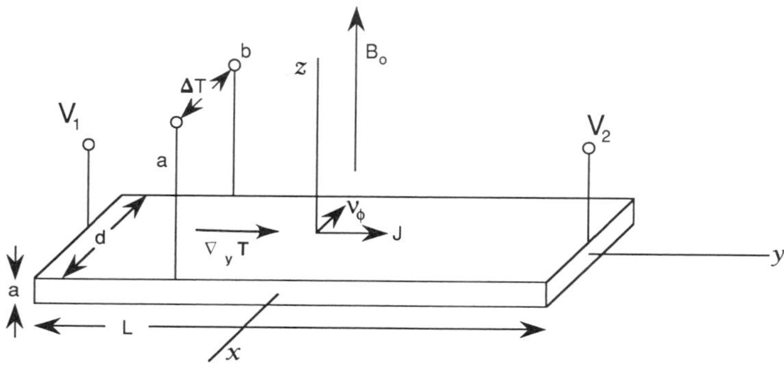

Figure 14.45 Experimental arrangement for measuring the Ettinghausen effect of a superconducting slab carrying a transport current density J in a transverse magnetic field B_0. To establish the transverse temperature difference ΔT, the vortices are caused to flow towards the side of the slab with the velocity v_ϕ. The temperature (a, b) and voltage (V_1, V_2) measuring leads are indicated.

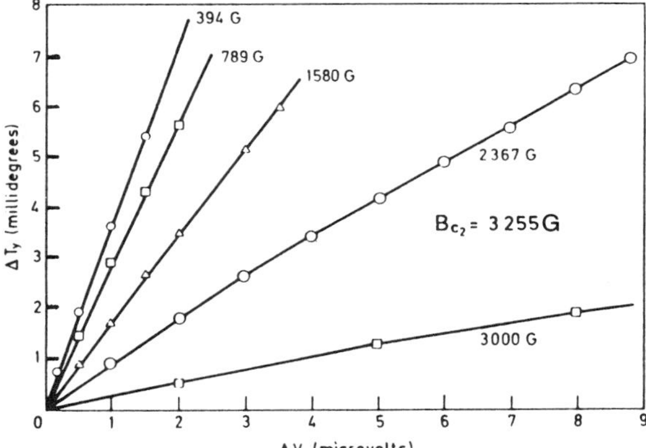

Figure 14.46 Transverse temperature difference ΔT_x arising from the Ettinghausen effect in the Type II superconducting alloy $In_{0.6}Pb_{0.4}$ plotted versus the longitudinal flux-flow voltage ΔV_y for the range of applied magnetic fields 39.4–300 mT (394 to 3000 G) (Solomon and Otter, 1967).

expected from Eq. (14.55). Part of this discrepancy is explained by the fact that Φ in Eq. (14.55) is the internal flux, which is not proportional to the applied field (cf. Fig. 9.5). The extra Peltier effect arising from fluctuations (Maki, 1991) gives rise to the Ettinghausen effect in the presence of an applied magnetic field. Others have reported Ettinghausen effects in the superconducting state (Freimuth *et al.*, 1991; Palstra *et al.*, 1990; Ullah *et al.*, 1990).

F. Righi–Leduc Effect

The Righi–Leduc effect is the thermal analogue of the Hall effect. One end of the sample is heated and the resulting temperature gradient $\nabla_y T$ in the y direction produces a thermal current of density U_y that flows from the hot end to the cold end, as shown in Fig. 14.47. The application of a magnetic field B_0 along z produces a temperature gradient $\nabla_x T$ along x

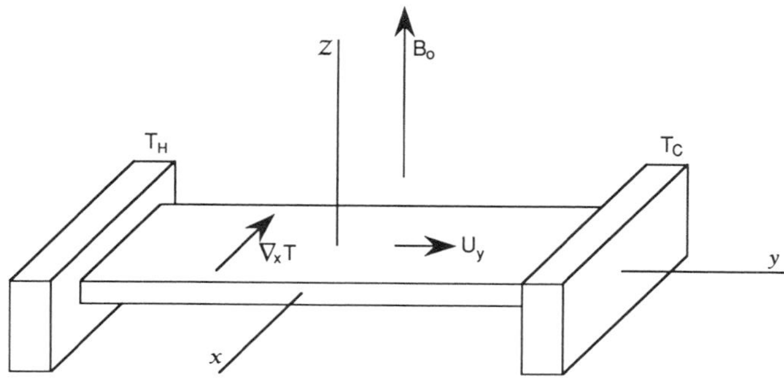

Figure 14.47 Experimental arrangement for measuring the Righi–Leduc effect of a superconducting slab mounted between a cold (T_c) and a hot (T_h) temperature reservoir in a transverse magnetic field B_0. The transverse temperature gradient $\nabla_x T$ and the longitudinal thermal current flow U_y are indicated.

given by

$$\nabla_x T = R_L B_0 U_y, \qquad (14.56)$$

where R_L is the Righi–Leduc coefficient. For metals in which the law of Wiedermann and Franz (1.33) is valid, this coefficient is related to the Hall coefficient R_H (14.28) by the expression

$$R_H = R_L L_0 T, \qquad (14.57)$$

where $L_0 = \frac{3}{2}(k_B/e)^2$ is the Lorentz number that appears in Eq. (1.33). The thermal Hall angle Θ_{th} can be defined by analogy to its ordinary Hall effect counterpart (1.93):

$$\tan \Theta_{th} = \nabla_x T / \nabla_y T. \qquad (14.58)$$

Figure 14.48 shows that the y direction temperature gradient behaves differently in the Meissner state ($B_{app} < B_{c1}$), the mixed state ($B_{c1} < B_{app} < B_{c2}$), and the normal state ($B_{c2} < B_{app}$).

IX. PHOTOCONDUCTIVITY

Photoconductivity is the increase in conductivity produced by shining light on a material. A related effect, called the *photovoltaic effect* is the inducing of voltages by light. This latter phenomenon is particularly pronounced in semiconductors when the band gap is small and light is able to excite electrons from the full valence band into the empty conduction band.

Figure 14.49 shows the time dependence of the voltage responses of

$$YBa_2Cu_3O_7$$

to a high-power laser pulse of energy density 2 mJ/cm^2 at 99 K in the normal state. This same laser pulse produced no photoresponse in the superconducting state since more energy was needed. The figure shows the delayed and weaker photoresponse at 57 K obtained with the pulse of higher energy density 4.5 mJ/cm^2 (C. L. Chang *et al.*, 1990).

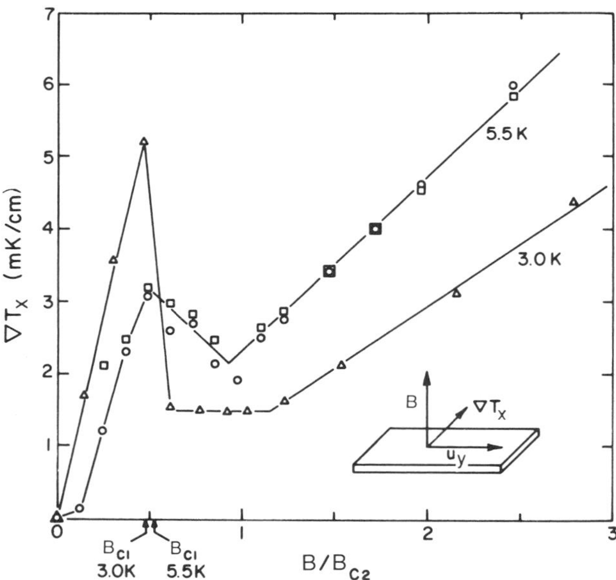

Figure 14.48 Righi–Leduc effect determination of the magnetic field dependence of the transverse temperature gradient for two temperatures. The behavior changes at the lower-critical field B_{c1}, as shown (Stephan and Maxfield, 1973).

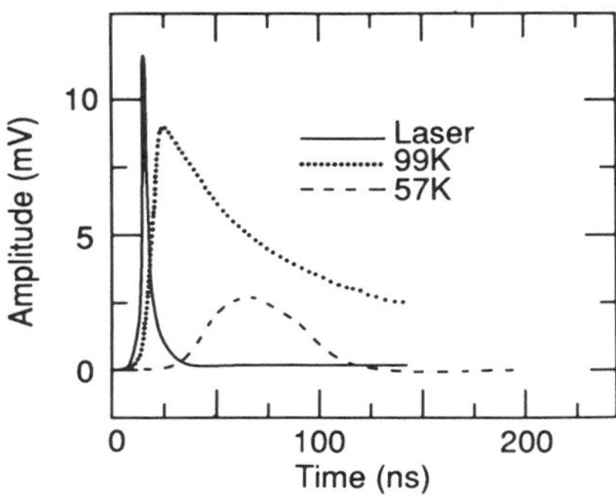

Figure 14.49 Laser intensity in arbitrary units (——) and photoresponse voltages from $YBa_2Cu_3O_7$ above T_c at 99 K (·······) and with higher laser intensity below T_c at 57 K (– – –) (C. L. Chang *et al.*, 1990).

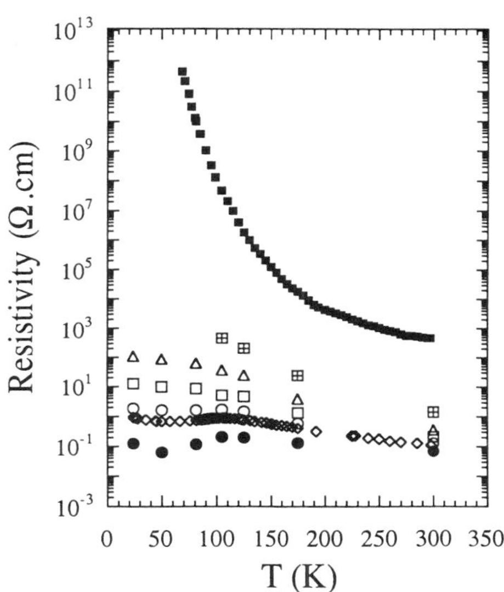

Figure 14.50 Temperature dependence of the photoresistivity in $YBa_2Cu_3O_{6.3}$ for 0.18-eV incident photons with the intensity 10^{13} (⊞), 10^{14} (△), 10^{15} (□), 7×10^{15} (○), 1.1×10^{16} (◇), and 2.3×10^{16} photons/cm^2 (●). The dark resistivity, with no incident light (■), is also shown (G. Yu *et al.*, 1992).

The compound $YBa_2Cu_3O_{7-\delta}$ is a conductor for δ below about 0.4 and a semiconductor for δ between 0.4 and 1; the quantity δ correlates with the number of charge carriers n_c. Figure 14.50 shows the dramatic lowering of the resistivity of the semiconductor $YBa_2Cu_3O_{6.3}$ from its dark value to its value for several light intensities, with higher intensities producing a greater reduction. Figure 14.51 shows how the resistivity of the related superconductor $DyBa_2Cu_3O_{7-\delta}$ decreases as δ decreases, i.e., as the oxygen content increases. Comparing these two figures shows that irradiating the sample has an effect similar to that of increasing the oxygen content, since both processes have the effect of increasing the number of carriers n_c (G. Yu *et al.*, 1990, 1992).

Bluzer (1991) studied transient photoresponse relaxation in $YBa_2Cu_3O_{7-\delta}$ using very short pulses, 0.3 ns in length, and obtained sharply rising signals followed by signals that decreased more slowly, as shown in Fig. 14.52b for the case $T < T_c$. In this temperature regime the response signal is proportional to the derivative of the quasiparticle concentration n_c. According to the n_c-versus-time curve reconstructed from this response and shown in Fig. 14.52a, a laser pulse arrives at time

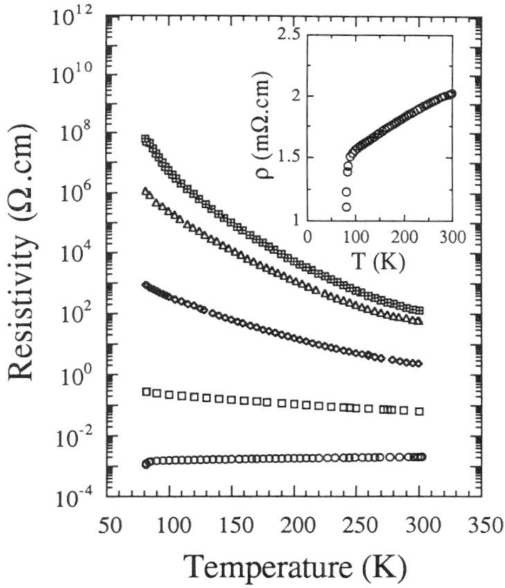

Figure 14.51 Temperature dependence of the dark resistivity of several $DyBa_2Cu_3O_{7-\delta}$ thin films with their oxygen contents increasing over the range from the insulating antiferromagnetic phase at the top to the metallic and superconducting phase at the bottom. The inset shows the resistivity of a $\delta = 0$ superconducting sample on a linear scale (G. Yu *et al.*, 1992).

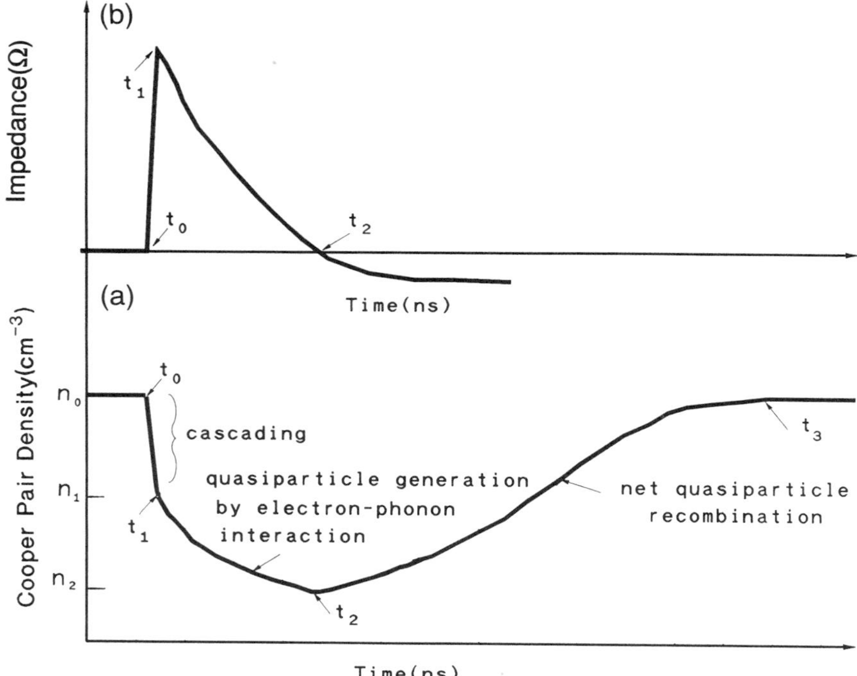

Figure 14.52 (a) Time dependence of the Cooper pair density with the quasiparticle formation and recombination processes indicated, and (b) measured photoresponse produced by these processes. The response signal is proportional to the derivative of the Cooper pair density (Bluzer, 1991).

t_0, with each photoabsorbed photon, of energy $E_0 = 2$ eV, splitting a Cooper pair to form two quasiparticles. Energetic quasiparticles break up additional Cooper pairs via an avalanche or cascade process that eventually produces, on average, 32 quasiparticles per adsorbed photon by the time t_1 (Han *et al.*, 1990b). Following this cascading process, the lower-energy quasiparticles thermalize between t_1 and t_2 by emitting 40–50 meV phonons, which break up additional Cooper pairs. The subsequent quasiparticle recombination process is detected as a negative photoresponse until the Cooper pair density n_c returns to its initial equilibrium value, as shown in the figure.

Such a cascade process produces quasiparticles at much higher temperatures than the phonons, which are at the lattice temperature. The thermalization process involves electron–phonon interaction; this interaction has been measured by determining the relaxation rate via femtosecond spectroscopy (Brorson *et al.*, 1990; Chekalin *et al.*, 1991; Rice *et al.*, 1993).

We have discussed what might be called transient photoconductivity. Persistent photoconductivity has also been observed, in which the photoinduced conductivity change persists for a long time following excitation (Ayache *et al.*, 1992; Kreins and Kudinov, 1992).

Photoconductivity has also been discussed by Boyn *et al.* (1991), Culbertson *et al.* (1989, 1991), Eesley *et al.* (1990), Glass and Rogovin (1989), D. H. Kim *et al.* (1991a), Kleinhammes *et al.* (1991), Kwok (1991), and Otis and Dreyfus (1991).

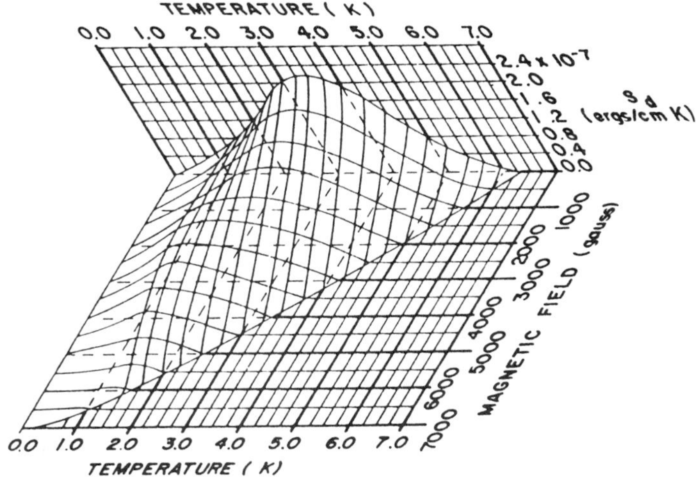

Figure 14.53 Three-dimensional plot showing the dependence of the transport entropy S_ϕ of a vortex per unit length on the magnetic field and the temperature for the Type II alloy $In_{0.6}Pb_{0.4}$ with $T_c = 6.3$ K and $B_{c2} = 6500$ G (0.65 T) (adapted from Huebener, 1979, p. 161).

X. TRANSPORT ENTROPY

The principal quantity obtained from thermomagnetic measurements is the transport entropy S_ϕ. Experimentally, flux-flow resistance measurements determine $\Phi_0 B/\beta$ with the aid of Eq. (14.27), the Nernst effect then gives $S_\phi B/\beta$ from Eq. (14.50), and the Ettinghausen effect provides S_ϕ/Φ from Eq. (14.55). These results provide the entropy per unit length of a vortex in a Type II superconductor.

Figure 14.53 shows how the transport entropy S_ϕ of the Type II alloy $In_{0.6}Pb_{0.4}$ varies with the temperature and magnetic field in the superconducting state. We see from the figure that, for a fixed magnetic field, the entropy increases from zero at $T = 0$, passes through a maximum (at 3.4 K for $B = 0$), and then decreases to zero at $T = T_c(B)$.

FURTHER READING

Several superconductivity texts discuss transport properties (e.g., Kresin and Wolf, 1990; Orlando and Delin, 1991; Rose Innes and Rhoderick, 1994; Tilley and Tilley, 1986; Tinkham, 1985).

Transport is also discussed in several monographs (Phillips, 1989a; Poole *et al.*, 1988; Van Duzer and Turner, 1981; Wilson, 1983) and review articles (Allen *et al.*, 1989; Iye, 1992; Lemberger, 1992). Huebener (1979), in particular, covers many of the transport effects. Callen (1985) provides a good discussion of these effects in the normal state from the thermodynamic point of view.

The Hall effect of high-temperature superconductors is discussed by Allen, Fisk, and Migliori (1989) and Crow and Ong (1990), and reviewed by Ong (1990).

Huebener *et al.* (1990) compare and discuss the Ettinghausen, Nernst, Peltier, and Seebeck effects in the mixed state of high-temperature superconductors.

Transport measurements on alkali metal doped buckminsterfullerenes are reviewed by Palstra and Haddon (1995).

Thermal conductivity is reviewed by Allen *et al.* (1989), Kaiser and Uher (1990), and Uher (1990).

PROBLEMS

1. Find the impedance Z of the circuit of Fig. 14.2, and find the ratios I_n/I and I_s/I of the two currents to the total current $I = I_n + I_s$.
2. For the case of relatively small grains, compare the inductances of (a) a 5-turn coil of radius 5 μm and length 10 μm,

(b) a straight wire 10 μm long, and (c) a loop of radius 10 μm and wire radius 1 μm.

3. For the case of relatively large grains, compare the inductances of (a) a 10-turn coil of radius 40 μm and length 80 μm, (b) a straight wire 80 μm long, and (c) a loop of radius 80 μm and wire radius 5 μm.
4. For the case of typical electrical circuits, compare the inductances of (a) the coil of Chapter 2, Section IV.B, (b) a straight wire 10 cm long, and (c) a loop of radius 10 cm with wire radius 0.3 mm.
5. Determine the flux-flow resistance determined for each of the V-versus-R curves shown in Fig. 14.21. Make a plot of R versus B.
6. Show that if the Magnus force is neglected, the electric field induced by flux flow is aligned along the transport current direction with the magnitude

$$E \approx 0 \qquad J\Phi_0 < F_P$$

$$E \approx \frac{B_0(J\Phi_0 - F_P)}{\beta} \qquad J\Phi_0 > F_P.$$

7. Show that if the Magnus force is neglected, the differential flux-flow resistivity is given by

$$\rho_{ff} = \frac{\Phi_0 B_0}{\beta}.$$

8. Show that four of the compounds with Hall effect data in Fig. 14.24 are hole-like and that the fifth is electron-like.
9. Derive the Ettinghausen equation (14.55). Deduce the polarity of the potential drop along y (i.e., determine which end is + and which is −), and determine which of the two terminals, a or b, is at the higher temperature in Fig. 14.45.
10. Thermopower and Peltier experiments were carried out using the same superconducting rod. In the former experiment the temperature difference ΔT across the sample produced the voltage difference ΔV, and in the latter experiment the input electrical current I produced the thermal current dQ/dt. Show that

$$I\Delta V = \Delta T \frac{dS_\phi}{dt},$$

where the temperature and its gradient are assumed independent of time.

11. Why does the absence of a Peltier effect in a superconductor show that super current does not transport entropy?
12. Derive the Righi–Leduc expression (14.57),

$$R_H = R_L L_0 T.$$

What do you assume about Θ_{th} and Θ_H?

13. Describe the details of the quasiparticle production and recombination processes outlined on Fig. 14.52 (consult the original reference).

15

Spectroscopic Properties

I. INTRODUCTION

Several standard spectroscopic techniques have been widely used for the study of superconductors. We will start by describing the principles of each of these techniques and what can be learned from it, and then present some of the results that have been reported for superconductors.

Most branches of spectroscopy are concerned with the absorption by the sample of an incoming photon of radiation $h\nu$, where h is Planck's constant and ν is the frequency. The photon transfers its energy to the sample by inducing a transition from a ground state E_0 into an excited state E_e. The difference in energy between the two states is equal to the energy of the photon,

$$E_e - E_0 = h\nu, \tag{15.1}$$

as indicated in Fig. 15.1. The intensity of light I_0 incident on the sample is partly transmitted, I_t, and partly reflected, I_r, so that the amount absorbed is given by

$$I_a = I_0 - I_r - I_t, \tag{15.2}$$

as shown in Fig. 15.2. Transmission spectrometers measure I_t, generally when I_r is small, while reflectance spectrometers measure I_r, generally when I_t is small. Either way, the spectrometer provides the frequency dependence of the ratio I_a/I_0, a maximum in I_a indicating the center of an absorption line. In a single-beam measurement, I_a itself is determined, while in a double-beam technique the absorption of a sample is measured relative to that of a reference material. Superconductors tend to be opaque at infrared and visible frequencies, so that reflectance techniques apply; higher frequencies in the x-ray region can penetrate, and transmission is

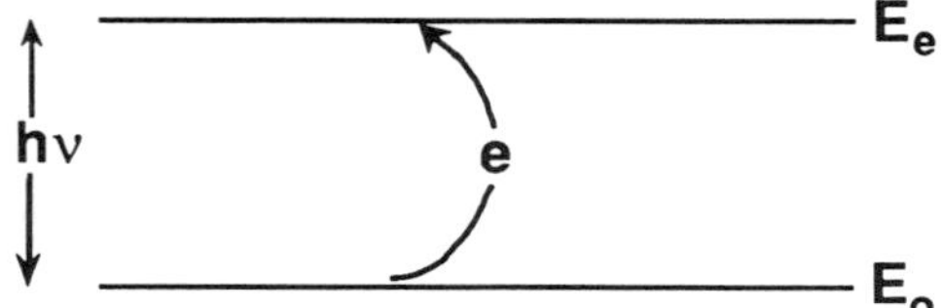

Figure 15.1 Incoming photon $h\nu$ inducing an electron to jump from a ground state energy level E_0 into an excited state level E_e.

often employed here. In the next section we will say a few words about reflection before proceeding to the various individual spectroscopies.

A number of acronyms, such as ACAR, ARPES, BIS, EELS, EPR, ESR, EXAFS, IPS, IR, μSR, NMR, NQR, PAS, PES, UPS, UV, XAFS, XANES, and XPS, in common use in the field will be defined in the appropriate sections. In addition, spectroscopists report their results in terms of different energy units. If we had standardized this chapter, by for example, converting all energies to joules, it would have been difficult to compare the results we wish to present with those found in the literature. The appropriate conversion factors are

$$\begin{aligned} 1000\ \text{cm}^{-1} &\equiv 0.124\ \text{eV} \\ &\equiv 30\ \text{THz} \equiv 10^5\ \text{Å}. \end{aligned} \quad (15.3)$$

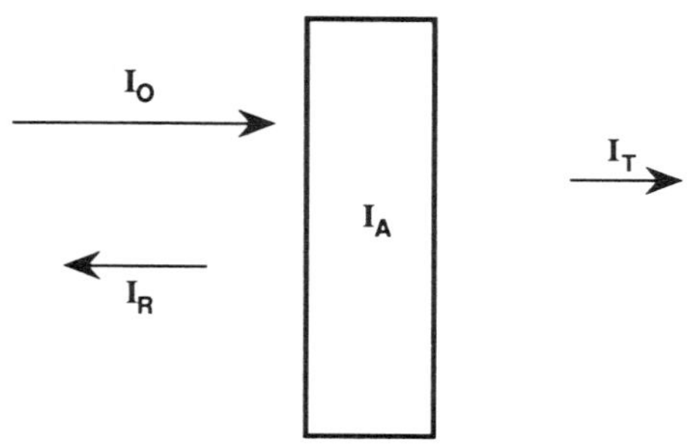

Figure 15.2 Absorbed (I_A), transmitted (I_T), and reflected (T_R) components of an incident light beam of intensity I_0.

II. VIBRATIONAL SPECTROSCOPY

Vibrational spectroscopy involves photons that induce transitions between vibrational states in molecules or solids. These transitions generally fall within the frequency band of infrared (IR) spectrometers, typically from 2×10^{13} to 12×10^{13} Hz, but sometimes over a wider range. It is customary for workers in the field to use the unit of reciprocal centimeters, which corresponds to 650 to 4000 cm^{-1} for the above range. The conversion factor between the two is the velocity of light, 2.9979×10^{10} cm/s.

The energy gaps of high-temperature superconductors are in the infrared region, so that a change in the absorption can occur when the vibrational frequency equals the energy gap,

$$hv = E_g \quad (15.4)$$

which gives us a value of $E_g = 2\Delta$, (see Chapter 6, Section VI).

A. Vibrational Transitions

In infrared spectroscopy, an IR photon $h\nu$ is absorbed directly to induce the vibrational transition (15.1), while in the case of Raman spectroscopy an incident optical photon of frequency $h\nu_{inc}$ is absorbed and a second optical photon $h\nu_{emit}$ is emitted, with the transition induced by the difference frequency $h\nu$,

$$hv = |\hbar\nu_{inc} - \hbar\nu_{emit}|, \quad (15.5)$$

where $\nu_{inc} > \nu_{emit}$ for what is called a Stokes line and $\nu_{inc} < \nu_{emit}$ for an anti-Stokes line. The fundamental vibrational energy levels E_n have the energies

$$E_n = (n_v + \tfrac{1}{2})h\nu_0, \quad (15.6)$$

where the vibrational quantum number $n_v = 0, 1, 2, 3, \ldots$ is a positive integer and ν_0 is the characteristic frequency for a particular vibrational mode. Transitions occur for the condition

$$\nu = (n'_v - n_v)\nu_0. \quad (15.7)$$

The lowest frequency transition with $n'_v - n_v = 1$ is called a fundamental band.

B. Normal Modes

Molecular vibrations occur in what are called normal modes. These involve the coherent oscillations of atoms in the unit cell relative to each other at a characteristic frequency. The oscillations occur in such a manner that the center of gravity is preserved. The normal modes of $Tl_2Ba_2CaCu_2O_8$ (also isomorphous $Bi_2Sr_2CaCu_2O_8$; cf. Kulkarni *et al.*, 1989, 1990; R. Liu *et al.*, 1992b; Prade *et al.*, 1989) are sketched in Fig. 15.3, with the arrows indicating the motion of the various atoms during a normal mode oscillation. Analogous mode diagrams have been published for the compounds La_2CuO_4 (Mostoller *et al.*, 1990; Pintschovius *et al.*, 1989), $(Nd, Ce)_2CuO_4$ (Zhang *et al.*, 1991a), and $YBa_2Cu_3O_6$, and $YBa_2Cu_3O_7$ (Bates, 1989). Isomorphous compounds have the same normal modes, but different frequencies of oscillation because of the differences in the masses and bonding strengths of the atoms.

Spectroscopists have developed a notation for these modes based on characteristics of the oscillations. The one-dimensional A and B modes refer to atom motions parallel to the c-axis, i.e., in the vertical (z) direction. The A mode is symmetrical for a 90° rotation about z, which means that all of the arrows on the atoms of Fig. 15.3 are coincident under this operation. The B mode is antisymmetrical under this 90° rotation, so that the arrows reverse direction. The subscript 1 is for a mode which is symmetrical for a 180° rotation about x or y, while the subscript 2 is for a mode which is antisymmetrical for this rotation. We see from the figure that there is a center of inversion, so that atoms interchange positions under the inversion operation $x \to -x$, $y \to -y$, and $z \to -z$. The even, or gerade (g), vibrations, which preserve this center of symmetry, are said to be symmetric with respect to inversion, and the odd, or ungerade (u), vibrations are antisymmetric with respect to inversion.

There are also two-dimensional modes E_g and E_u involving atom motions in the a, b-plane, but these are more difficult to characterize.

C. Soft Modes

A phase transition in which the low- and high-temperature crystal structures differ by only small lattice displacements is often accompanied by what are called soft vibrational modes (Burns, 1985, Section 14-3). Most vibrational modes increase in frequency as the temperature is lowered, but the soft modes decrease in frequency as the transition temperature is approached from above, reaching very low frequencies near the transition. Further cooling below the transition temperature causes the modes to increase in frequency again, and sometimes a split into two modes occurs. Phase transitions associated with high-temperature superconductors often involve orthorhombic-to-tetragonal changes in crystal structure, in which individual atoms undergo very small shifts in position, so that soft modes are to be expected.

D. Infrared and Raman Active Modes

Two important characteristics of a vibrating system are its electric dipole moment and its polarizability. The electric dipole moment μ_D arises from the separation of charge. For point charges $-Q$ and $+Q$ separated by the distance d, it is

$$\mu_D = dQ. \qquad (15.8)$$

The polarizability P causes the electric vector of an incident light wave to induce a dipole moment μ_{ind}. It is defined as the ratio of the induced moment to the applied field,

$$P = \mu_{ind}/E. \qquad (15.9)$$

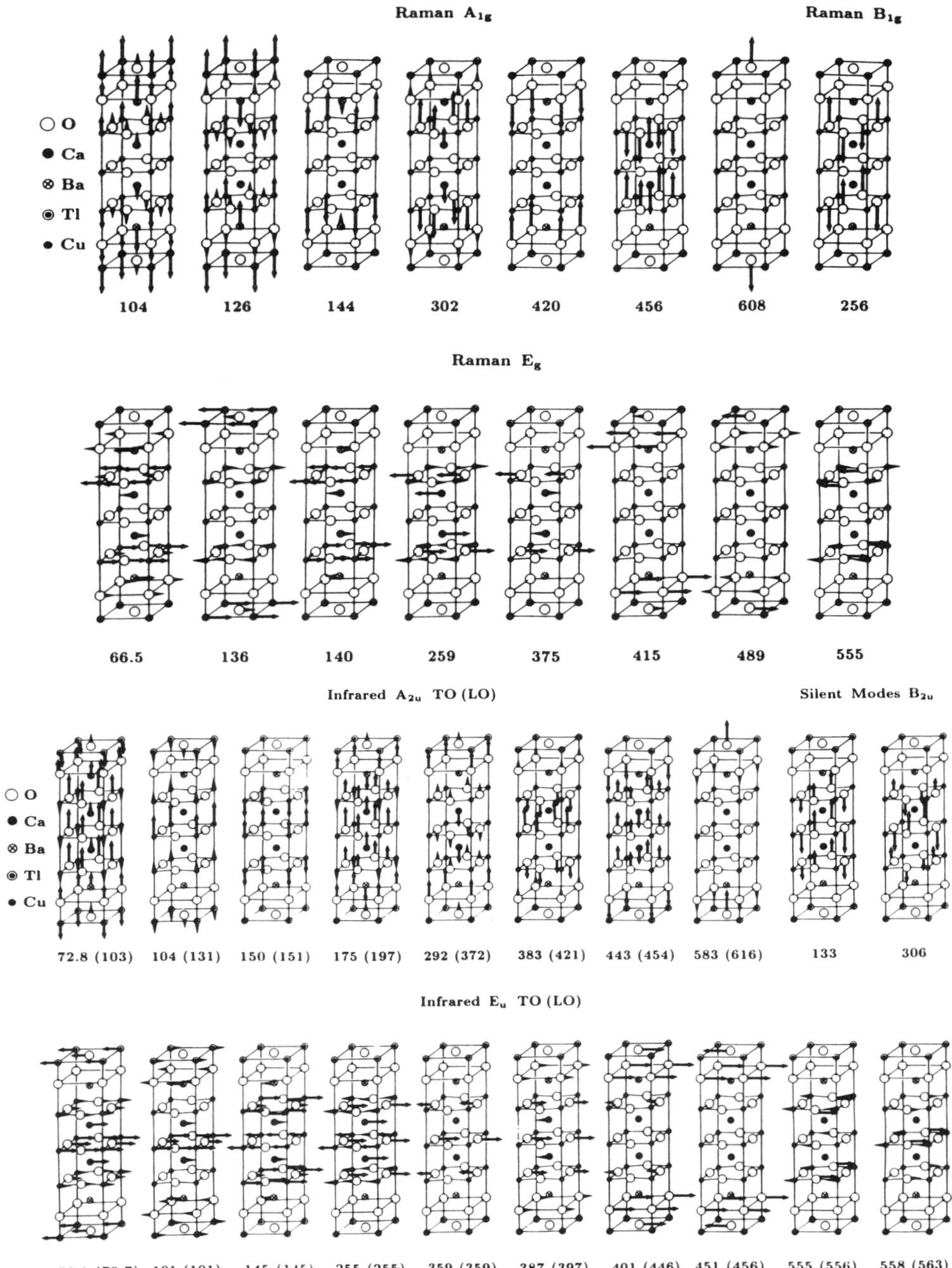

Figure 15.3 Raman-active (top panels) and infrared-active (bottom panels) normal vibrational modes of $Tl_2Ba_2CaCu_2O_8$, a body-centered tetragonal compound. For the infrared-active modes, the transverse optical (TO) frequencies, in cm^{-1}, are given first, followed by the corresponding longitudinal optical (LO) values in parentheses. The lengths of the arrows, although not drawn to scale, are indicative of the relative vibrational amplitudes (Kulkarni *et al.*, 1989, 1990).

Therefore, the polarizability is a measure of the deformability of the electron cloud of the molecule in the presence of an electric field. Infrared spectral lines are due to a change in the electric dipole moment of the molecule, while Raman lines appear when there is a change in the polarizability. These two spectroscopies are complementary to each other because some vibrational transitions are IR active while others are Raman active. Infrared active modes are of odd (u) type, where the oscillating atoms produce a dipole moment, while Raman active modes are of even (g) type, having no moment themselves, though a moment is induced by the electric field of the incident radiation. Thomsen and Cardona (1989) give lists of infrared and Raman active modes for some high-temperature superconductors. Kulkarni *et al.* (1989) obtained good agreement between calculated vibrational frequencies of $Tl_2Ba_2CaCu_2O_8$ and experimental values from infrared and Raman studies (Kostadinov *et al.*, 1988; McCarty *et al.*, 1988, 1989).

E. Kramers–Kronig Analysis

Infrared and optical reflectance measurements of superconductors can provide information on the conductivity. In this section we will explain how the conductivity is obtained from these data.

The reflectance (or reflectivity) R represents the fraction of reflected light,

$$R = \frac{I_r}{I_0}. \tag{15.10}$$

For normal incidence it is related to the dielectric constant ϵ through the expression

$$R = \frac{|\sqrt{\epsilon} - 1|}{|\sqrt{\epsilon} + 1|}, \tag{15.11}$$

where ϵ has real and imaginary parts ϵ' and ϵ'',

$$\epsilon = \epsilon' + i\epsilon'', \tag{15.12}$$

corresponding to dispersion and absorption, respectively. The limiting dielectric constant for large ω, ϵ_∞, is obtained from a fit to the data, so that it is a limiting value for the range of frequencies under investigation, rather than the ultimate limit $\epsilon_\infty = 1$ of free space. Equation (15.11) is more complicated for oblique incidence. A Kramers–Kronig analysis (Wooten, 1972) can be performed to extract the frequency dependence of $\epsilon'(\omega)$ and $\epsilon''(\omega)$, and the data can sometimes be fitted to an expression containing Drude-like terms (Lupi *et al.*, 1992; Terasaki *et al.*, 1990a; Wang, 1990; Wooten, 1972; Zhang *et al.*, 1991a; cf. Arfi, 1992; Calvani *et al.*, 1991; Monien and Zawadowski, 1989; Peiponen and Vartiainen, 1991; Watanabe *et al.*, 1989), such as

$$\epsilon(\omega) = \epsilon_\infty - \frac{\omega_p^2}{\omega^2 + i\omega/\tau} + \sum_i \frac{f_i \omega_i^2}{(\omega_i^2 - \omega^2) - i\omega/\tau_i}, \tag{15.13}$$

where f_i is the oscillator strength and the relaxation times τ and τ_i provide the broadening of the resonances. The summation terms are Lorentz oscillator types that account for features arising, for example, from vibrational absorption lines. The second term corresponds to Eq. (1.27), with the damping factor $i\omega/\tau$ added, where ω_p is the plasma frequency (1.28),

$$\omega_p = (ne^2/\epsilon_0 m)^{1/2}, \tag{15.14}$$

which was introduced in Chapter 1, Section V.

Some experimentalists report their data as plots of $\epsilon'' = Im[\epsilon(\omega)]$ versus the frequency. Others present plots of the high-frequency conductivity σ_1,

$$\sigma_1(\omega) = (\omega\epsilon''/4\pi). \tag{15.15}$$

We see from a comparison of Figs. 15.4a and 15.4b that the ϵ'' (or σ_1) plots are superior to reflectance plots for determining the positions and widths of individual absorption lines arising from the summa-

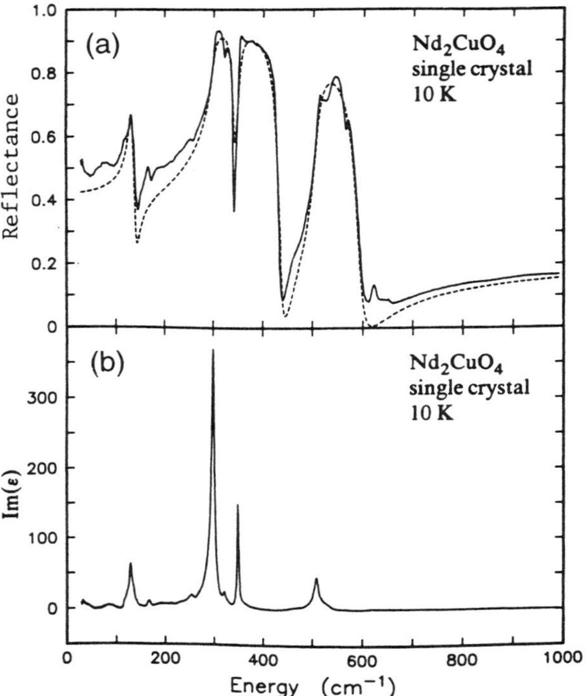

Figure 15.4 Infrared spectrum of an Nd_2CuO_4 single crystal at 10 K showing (a) the reflectance, and (b) the imaginary part of the dielectric constant ϵ'' determined by a Kramers–Kronig analysis using the value $\epsilon_\infty = 6.8$ (Crawford *et al.*, 1990a).

tion terms of Eq. (15.13). To see why this is so, consider the real and imaginary parts of one of the terms in the summation of Eq. (15.13),

$$f_i \omega_i^2 \left[\frac{\omega_i^2 - \omega^2}{(\omega_i^2 - \omega^2)^2 + (\omega/\tau_i)^2} + i \frac{\omega/\tau_i}{(\omega_i^2 - \omega^2)^2 + (\omega/\tau_i)^2} \right], \quad (15.16)$$

which, in the usual limit of narrow lines, $\omega_i \tau_i \gg 1$, can be written

$$\frac{2(\omega_i - \omega)\tau_i}{4(\omega_i - \omega)^2 \tau_i^2 + 1} + i \frac{1}{4(\omega_i - \omega)^2 \tau_i^2 + 1}, \quad (15.17)$$

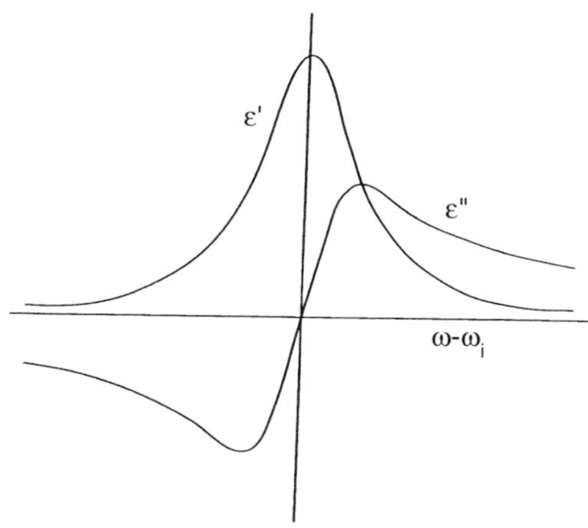

Figure 15.5 Normalized line shape of the dielectric constant $\epsilon = \epsilon' + i\epsilon''$ showing the real part ϵ', called the dispersion, and the imaginary part ϵ'', called the absorption.

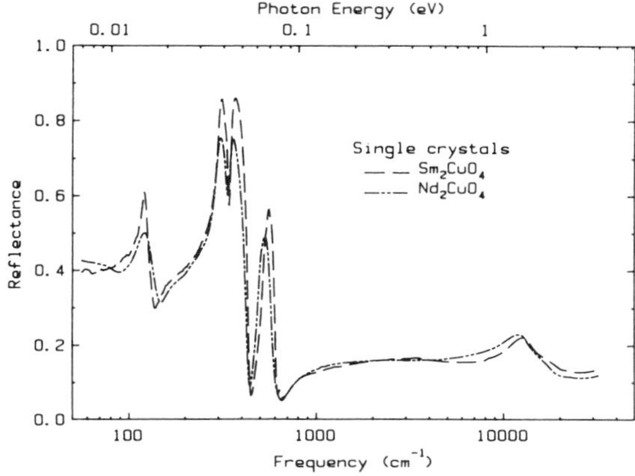

Figure 15.6 Frequency dependence of the reflectance of Sm_2CuO_4 (– – –) and Nd_2CuO_4 (–···–···–) in the infrared and visible regions (Herr *et al.*, 1991).

where the factor $f_i \omega_i \tau_i$ has been omitted. This corresponds to a Lorentzian line shape. The sketches of this function in Fig. 15.5 show that the real (dispersion) and imaginary (absorption) parts produce resonant lines centered at ω_i, where $1/\tau_i$ is the full linewidth. The reflectance plotted in Fig. 15.4a is a mixture of absorption and dispersion, hence it cannot provide the resonant frequencies ω_i with any precision.

F. Infrared Spectra Results

Figure 15.4 shows an example of an infrared spectrum of Nd_2CuO_4 with the T′ structure in the far-infrared region where the fundamental band vibrations are found. Figure 15.6 shows a much broader scan for this same compound, from 50 to 32,000 cm^{-1} (4 eV), and Fig. 15.7 presents the conductivity of three R_2CuO_4 compounds, where $R =$ Nd, Sm, or Gd, calculated

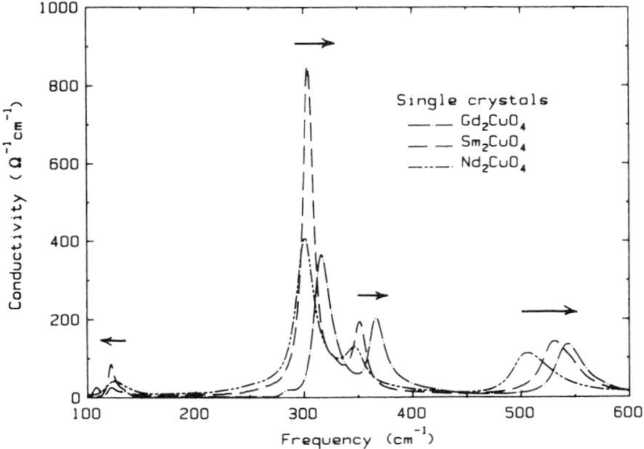

Figure 15.7 Frequency dependence of the conductivity of three monocrystals determined from a Kramers–Kronig analysis. The arrows indicate the shift direction with increasing mass Nd → Sm → Gd (Herr *et al.*, 1991).

from their infrared reflectances. The mid-infrared spectrum is devoid of features that are typical of an insulating compound. What is referred to as a charge-transfer transition appears at 12,000 cm^{-1} (1.5 eV).

The far-infrared reflectance and conductivity spectra of $YBa_2Cu_3O_{7-\delta}$ in the normal (at 110 K) and superconducting (at 2 K) states are compared for ceramic samples in Figs. 15.8 and 15.9 (Bonn *et al.*, 1988). The ranges of reflectance and conductivity values are much higher than in the Nd_2CuO_4 case of Figs. 15.6 and 15.7, and data for single crystals and oriented films have even higher reflectances. The low-frequency conductivity 2100 $(\Omega\ cm)^{-1}$ of Fig. 15.9 approaches the measured dc value of 3300 $(\Omega\ cm)^{-1}$. The plasma frequency ω_p is 6,000 cm^{-1} (0.75 eV), and $1/\tau = 300\ cm^{-1}$.

Isotopic substitutions have been employed to identify modes. For example, it was observed that enriching $YBa_2Cu_3O_{7-\delta}$ with the heavy isotope ^{65}Cu causes the 148.6 cm^{-1} line, which involves Cu vibrations, to shift downward in energy by 1.8 cm^{-1}, whereas the 112.5 cm^{-1} line, which does not involve Cu motion, remained at the same frequency. A similar result occurs with ^{18}O enrichment of

$$(Pr_{1-x}Ce_x)CuO_4,$$

where three modes involving oxygen vibrations were observed to shift downward by 3–4%, whereas a fourth mode, which involves Pr vibrations, did not change. These downward shifts occur because, classically, the vibrational frequency depends on the mass, in accordance with the expression

$$\nu_0 = (1/2\pi)(k/m)^{1/2} \quad (15.18)$$

where k is the spring constant, so that higher masses produce lower frequencies, assuming that the substitution does not change k. Table 15.1 lists spring constants for various atom pairs in

$$(La_{0.925}Sr_{0.75})_2CuO_4$$

and $YBa_2Cu_3O_7$ that were deduced from measured vibrational frequencies (Bates, 1989; Brun *et al.*, 1987).

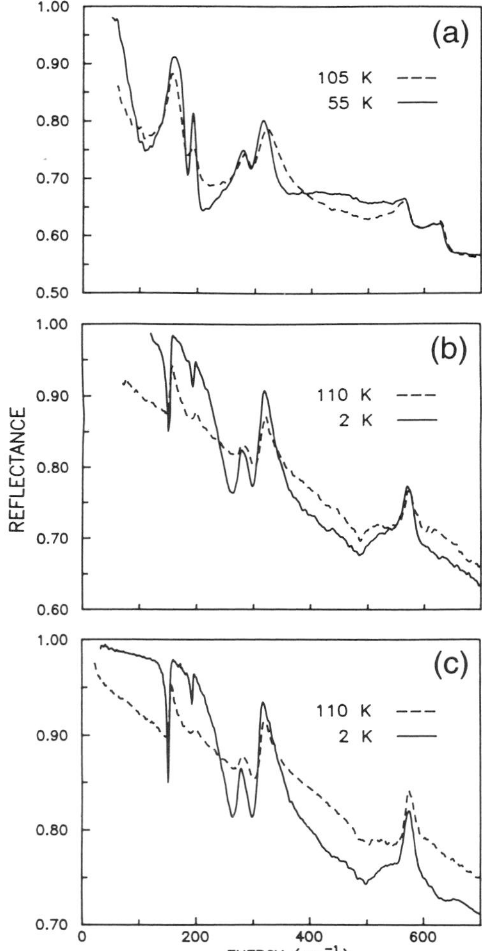

Figure 15.8 Optical reflectance of $YBa_2Cu_3O_{7-\delta}$ in the superconducting state (dashed curves) and in the normal state (solid curves) for (a) polished sample, (b) unpolished sample following several days exposure to the air, and (c) unpolished sample immediately after annealing in oxygen (Bonn *et al.*, 1988).

We see from Fig. 15.7 and Table 15.2 how the low-frequency infrared line shifts downward in frequency from 125 to 115 cm^{-1} as R of the compound R_2CuO_4 changes in the order Pr–Nd–Sm–Gd of increasing mass. This is expected behavior for a mass change effect. At higher field, the other three lines shift in the opposite direction, which may be attributed to the decrease in bond length with a consequent

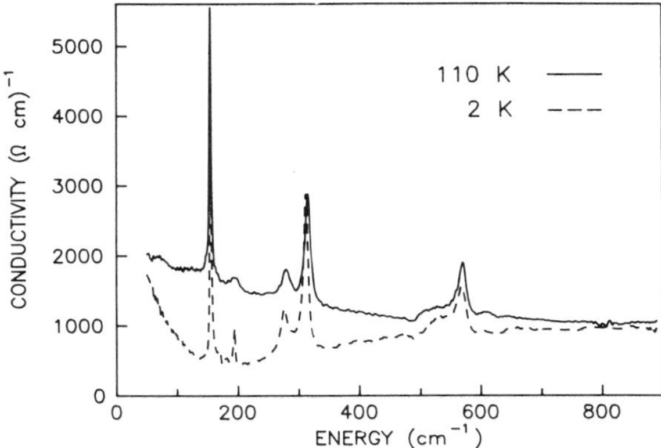

Figure 15.9 Real part of the conductivity of $YBa_2Cu_3O_{7-\delta}$ determined from a Kramers–Kronig analysis of the reflectance of the unpolished sample of Fig. 15.8 immediately after annealing in oxygen, shown in the superconducting state (dashed curve) and in the normal state (solid curve) (Bonn *et al.*, 1988).

increase in the spring constant in the order Pr–Nd–Sm–Gd, with the spring constant effect dominating in Eq. (15.18).

Recent high-temperature superconductor (HTSC) IR spectroscopy articles on single crystals and thin films (f) are, for the lanthanum (Collins *et al.*, 1989b; Tamasaku *et al.*, 1992), Nd–Ce (Lupi *et al.*, 1992; Zhang *et al.*, 1991a), yttrium (Budhani *et al.*, 1991 (f); Collins *et al.*, 1989a, (f); Cooper *et al.*, 1989; Crawford *et al.*, 1989; F. Gao *et al.*, 1991 (f), Knoll *et al.*, 1990; Pham *et al.*, 1991; Rotter *et al.*, 1991; Williams *et al.*, 1990 (f)), and bismuth

Table 15.1 Bond Lengths and Effective Spring Constants (k_{eff}) of Atom Pairs in Lanthanum and Yttrium Compounds

$(La_{0.925}Sr_{0.075})_2CuO_4$			$YBa_2Cu_3O_{7-\delta}$		
Bond	Length[a] Å	k_{eff}[b] N/m	Bond	Length[c] Å	k_{eff}[c] N/m
Cu–O(2)	1.89	85	Cu(1)–O(2)	1.83	176
Cu–O(2)	2.40	20	Cu(1)–O(1)	1.94	152
La–O(1)	2.64	160	Cu(2)–O(3)	1.93	155
La–O(2)	2.39	105	Cu(2)–O(4)	1.96	149
La–O(2)	2.73	50	Cu(2)–O(2)	2.33	103
La–La		30	Ba–O(2)	2.75	58
La–Cu		10	Ba–O(1)	2.91	55
O(1)–O(1)	2.67	20	Ba–O(3)	2.94	54
O(2)–O(2)	3.77	7	Ba–O(4)	2.94	54
O(1)–O(2)	3.05	4	Y–O(4)	2.38	79
			Y–O(3)	2.42	77

[a] From Collin and Comes (1987).
[b] From Brun *et al.* (1987).
[c] From Bates (1989).

Table 15.2 Shift of Infrared Frequency ω_i of the Series of Tetragonal R_2CuO_4 Compounds (R Changing in the Order Pr, Nd, Sm, and Gd of Increasing Mass Number)

Atom	Mass number	Lattice constant, Å		Infrared frequency, cm^{-1}			
		a	c	ω_1	ω_2	ω_3	ω_4
Pr[a]	140.9	3.95	12.17	126	299	336	495
Nd[b]	144.2	3.94	12.15	127	301	346	510
Sm[b]	150.4	3.91	11.93	123	304	351	534
Gd[b]	157.3	3.89	11.85	121	318	368	545

[a] From Crawford *et al.* (1990b).
[b] Lattice constants from Wyckoff (1965); IR frequencies from Burns (1989).

(Calvani *et al.*, 1991; Romero *et al.*, 1991; Y. Watanabe *et al.*, 1991) compounds and K_xC_{60} (FitzGerald *et al.*, 1992; M. C. Martin *et al.*, 1993).

G. Light-Beam Polarization

In conventional Raman spectroscopy, an incident unpolarized light beam simultaneously excites many of the A_g, B_g, and E_g Raman active modes. Polarized light enhances some of these modes and diminishes or eliminates others. A variety of directions and polarizations of the incident and scattered light beams can be employed to sort out and identify the modes.

To label the polarized spectra we will use the notation $k_i(E_i, E_s)k_s$ to denote the orientations of the incident (i) and scattered (s) light propagation directions **k** and electric vector **E** polarizations. Sometimes, the polarization will be along x', y'-axes that are oriented at 45° with respect to the x, y-axes, as shown in the inset of Fig. 15.10. A horizontal bar will be printed over the coordinate (e.g., $\bar{x}$) to denote the nega-

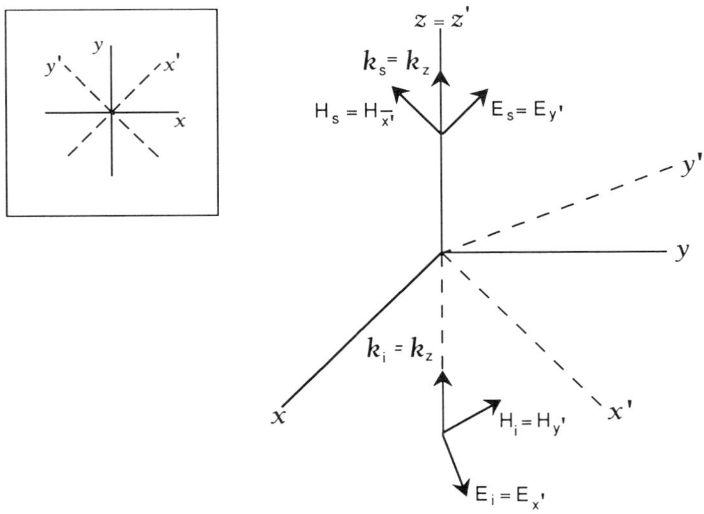

Figure 15.10 Experimental conditions for a $z(x', y')z$ polarization measurement. (The abbreviated notation $x'y'$ is sometimes employed.) The inset shows the orientation of the x'-y'-axes relative to x-y-axes.

tive (e.g., $-x$) direction. Figure 15.10 illustrates the $z(x', y')z$ case, in which an incident light beam travels along z and is polarized along x', while a scattered beam departs along z and is polarized along y'. Some authors (e.g., Weber *et al.*, 1988) use a shorthand notation, specifying only the polarization directions, writing $x'y'$ for the case of Fig. 15.10.

H. Raman Spectra Results

In the previous section, we discussed that the Raman active modes can be sorted by using polarized light sources and detectors. For example, $YBa_2Cu_3O_7$ has five observed A_g modes, at 116, 149, 335, 435, and 495 cm^{-1}, plus some weaker B_{2g} and B_{3g} modes. Figures 15.11a, 15.11b, and 15.11c show how to distinguish between these modes by changing the polarization conditions. For example, the $z(y', x')\bar{z}$ spectrum contains only the 335 cm^{-1} line, while $y(zz)\bar{y}$ exhibits only the other four A_g types. These spectra were obtained with twin-free monocrystals. B_{2g} and B_{3g} are essentially the same modes with atomic vibrations along a and b, respectively, and are detectable using the respective polarizations $y(z, x)\bar{y}$ and $x(z, y)\bar{x}$. These two modes differ because of the chains running along the b direction.

Weber and Ford (1989) published a Raman study of undoped La_2CuO_4 in which they demonstrated the superiority of single crystal samples by means of the spectra presented in Fig. 15.12. This figure compares a powder sample with micrometer-sized particles with the freshly broken surface of a ceramic sample composed of 1–10 μm grains, and an optically polished single crystal. Figure 15.13 shows a soft mode at 104 cm^{-1} observed below the transition temperature 573 K from the high-temperature tetragonal phase to the low-temperature *ortho*-rhombic phase. Figure 15.14 shows the pronounced decrease in frequency of this soft mode as the transition temperature is approached from below. We see from Fig. 15.15, which compares spectra of the superconductor $(La_{1.85}Sr_{0.15})_2CuO_4$ at room temperature and at 8 K below T_c, that there is no sign of a phonon mode associated with the superconducting transition.

Table 15.3 compares frequencies of the Raman active modes of several of the high-temperature superconductors. Each mode in the table is labeled with the atom that dominates the particular vibration. Figure 15.16 shows the

$$Bi_2Sr_2Ca_nCu_{n+1}O_{2n+6}$$

Raman spectra for $n = 0$ and $n = 1$ (M. J. Burns *et al.*, 1989), the frequencies of which are presented in the table. Some recent HTSC Raman spectroscopy articles on single crystals and thin films (f) are, for LaSrCuO (C. Y. Chen *et al.*, 1991; Ohana *et al.*, 1989), ReCeCuO (Heyen *et al.*, 1990c; Tomeno *et al.* 1991), YBaCuO (Bist *et al.*, 1991 (f); Guha *et al.*, 1991; Heyen *et al.*, 1990b; Kirillov *et al.*, 1991 (f); Liu *et al.*, 1990; Poberaj *et al.*, 1990; Slakey *et al.*, 1989, 1991; Wake *et al.*, 1991), (Y, Pr)BaCuO (I. S. Yang *et al.*, 1990), (Bi, Pb)SrCaCuO (Boekhoet *et al.*, 1991; Burns *et al.*, 1988; Hangyo *et al.*, 1993; Sapril *et al.*, 1991; Staufer *et al.*, 1992), TlBaCuO (Nemetschek *et al.*, 1993), BaKBiO at high pressure (Bonner *et al.*, 1990), K_3C_{60} (Pichler *et al.*, 1992 (f), Zhou *et al.*, 1992), Rb_xC_{60} (Mitch *et al.*, 1992), and Cs_xC_{60} (K. A. Wang *et al.*, 1992 (f)). Raman studies have been made of polycrystalline $HgBa_2CuO_{4+\delta}$ (Hul *et al.*, 1993; Ren *et al.*, 1994). Some theory articles are Monien and Zawadowski (1989, 1990), Shastry and Shraiman (1991), Tüttö and Zawadowski (1992), Virosztek and Ruvalds (1992), and Weber *et al.* (1989).

I. Energy Gap

Tunneling and vibrational spectroscopy are complementary ways of determining the energy gap of a superconductor (see Chapter 13, Section VI.E, for a discussion of tunneling spectroscopy and energy gaps). In the present section we will

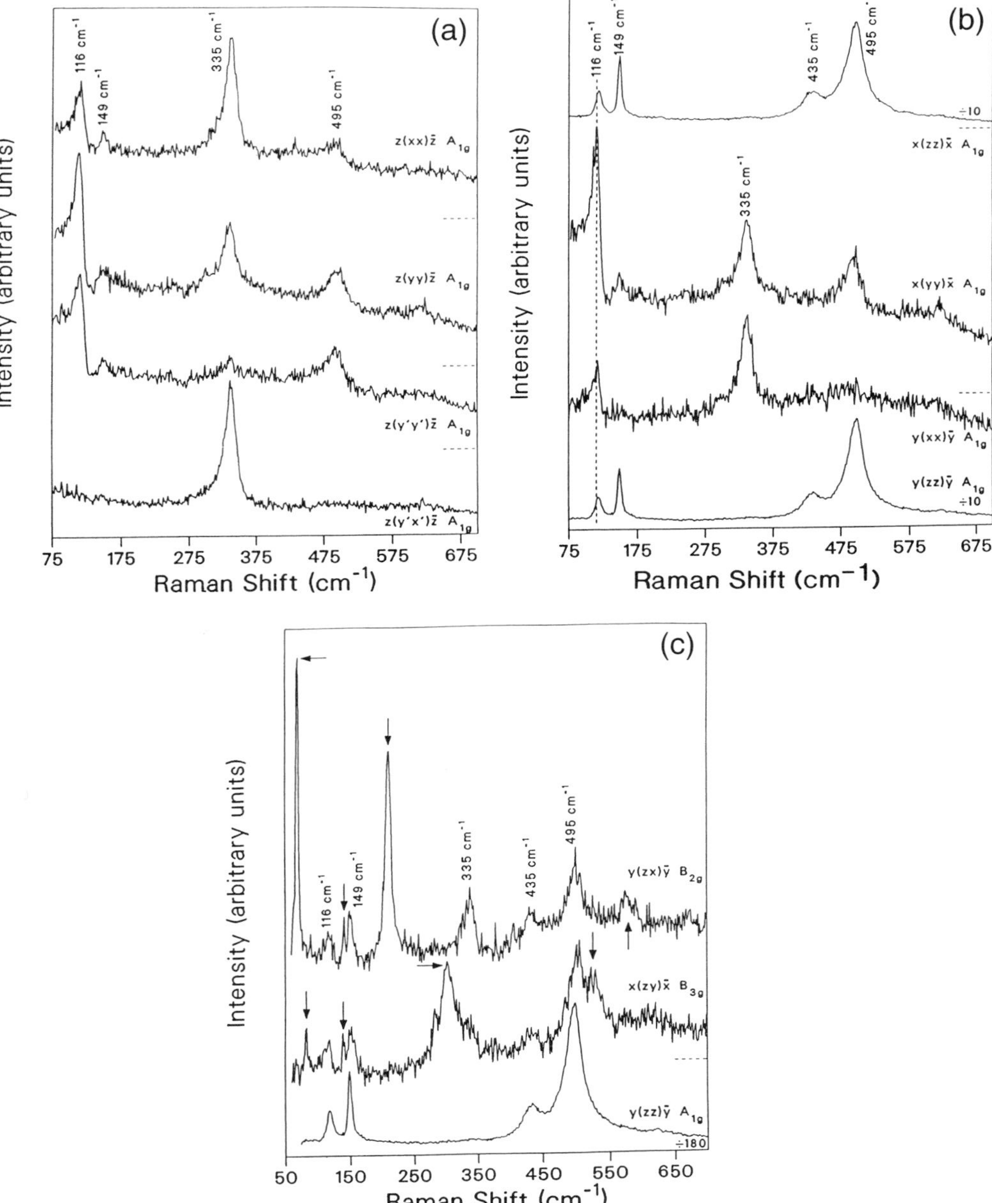

Figure 15.11 (a) Raman spectra of twin-free $YBa_2Cu_3O_7$ recorded with the laser beam directed along the *c*-axis and the indicated polarizations. The *x*-axis is the base line for the power spectrum, while the dotted lines indicate the base lines of the three upper spectra (McCarty *et al.*, 1990a, b). (b) Raman spectra of twin-free $YBa_2Cu_3O_7$ recorded with the laser beam propagating in the *x*, *y*-plane, using the notation of Fig. 15.11a. Note the scale factor change for the two middle spectra (McCarty *et al.*, 1990a, b). (c) Raman spectra of twin-free $YBa_2Cu_3O_7$ recorded with the laser beam directed along the *c*-axis, and the indicated polarizations selected to enhance the B_{2g} (top spectrum) and B_{3g} (middle spectrum) modes. Note the scale factor change for the lower A_g mode spectrum. The five A_g modes, with their frequencies labeled, appear on the upper spectrum due to polarization leakage (McCarty *et al.*, 1990a, b).

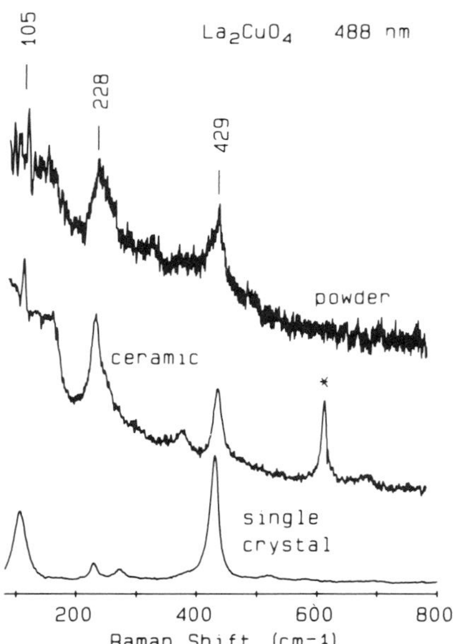

Figure 15.12 Raman spectra of La_2CuO_4. The laser powers and exposure times were 15 mW and 50 min for the powder, 15 mW and 10 hr for the ceramic (1–10 μm grains), and 50 mW and 50 min for the single crystal. Different polarization conditions were used, and the spectra have a common baseline (Weber *et al.*, 1989).

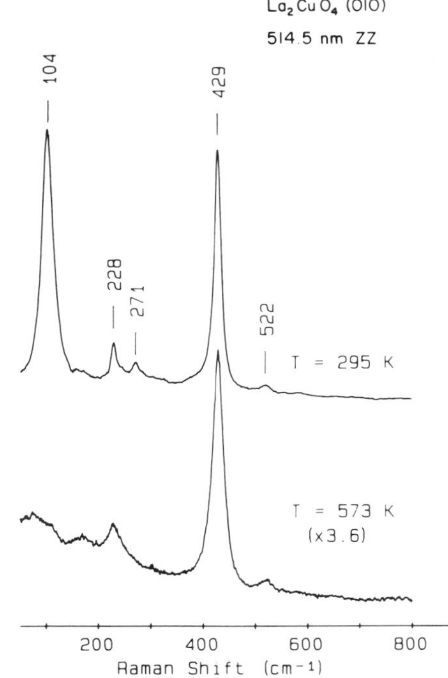

Figure 15.13 Low-frequency zz spectra from the [010] surface of orthorhombic La_2CuO_4 at 295 K (top), and tetragonal La_2CuO_4 at 573 K (bottom). A 50 mW, 514.5 nm laser was employed (Weber *et al.*, 1989).

say a few words concerning the spectroscopic determination of gaps.

For a superconductor at absolute zero, we expect light with frequencies lower than $2\Delta/h$ to be transmitted and light with frequencies $\nu > 2\Delta/h$ to be reflected, as in the case of a normal metal. Above absolute zero these latter frequencies can excite quasiparticles and induce a photoconductive response. Figure 15.17 shows low-temperature experimental data for transmission of infrared radiation at frequencies below the gap value $2\Delta \sim 70$ cm^{-1}, with a drop to zero transmittance for frequencies above this value for the superconductor $Ba_{0.6}K_{0.4}BiO_3$ (Schlesinger *et al.*, 1989). The figure also shows the drop in reflectivity when the temperature is increased, and also when a magnetic field is applied. Figure 15.18 presents infrared reflectivity (reflectance) spectra for two single-domain (untwinned) $YBa_2Cu_3O_{7-\delta}$ crystals arising from the Cu–O planes when the electric field is polarized parallel to the *a*-axis, and with possible contributions from the chains as well for polarization parallel to *b* (Schlesinger *et al.*, 1990b; vide also Friedl *et al.*, 1990; McCarty *et al.*, 1991). In both cases, the superconducting-to-normal state resistivity ratios R_s/R_n, obtained at the temperatures 35 K and 100 K, respectively, peak near ≈ 500 cm^{-1}, indicative of an energy gap. Brunel *et al.* (1991) measured the sharp infrared reflectivity discontinuity at the gap for the superconductor $Bi_2Sr_2CaCu_2O_8$.

III. OPTICAL SPECTROSCOPY

Visible (13,000 to 25,000 cm^{-1}, or 1.6 to 2.5 eV) and ultraviolet (UV) (3.1 to ≈ 40 eV) spectroscopy, both often referred to as optical spectroscopy, have been em-

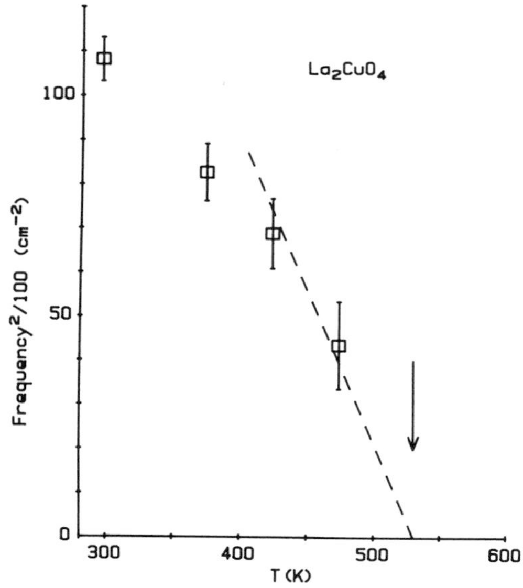

Figure 15.14 Plot of frequency squared versus T for the soft phonon observed in La_2CuO_4 for zz scattering under the same conditions as Fig. 15.13 (Weber *et al.*, 1989).

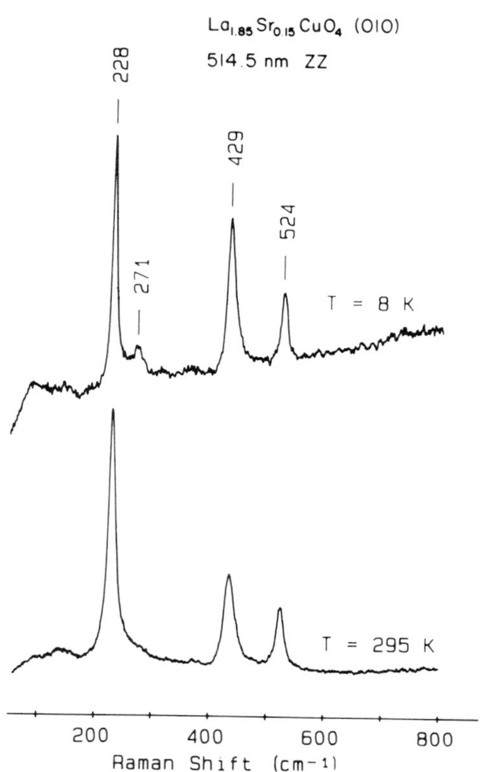

Figure 15.15 Raman spectra from the orthorhombic form of $(La_{0.925}Sr_{0.075})_2CuO_4$ at 8 K (top), and from the tetragonal form at 295 K, for the same conditions as Fig. 15.13 (Weber *et al.*, 1989).

ployed to detect crystal field-split electronic energy levels in insulating solids containing transition ions and to determine energy gaps in semiconductors as well as the locations of impurity levels within these gaps. The response of metals to incident optical radiation depends on the plasma frequency ω_p (15.14) which, as noted in Section II, lies in the near-infrared region for the high-temperature superconductors.

A study was made of the optical reflectance (reflectivity) of the series of $La_{2-x}Sr_xCuO_4$ compounds prepared for the range of compositions indicated in Fig. 15.19. The broad spectral scan, up to 37 eV, which is shown in Fig. 15.20 exhibits three reflectivity edges. The highest frequency edge, near 30 eV, falls off as $1/\omega^4$, which was attributed to excitations involving some of the valence electrons (see discussion in Chapter 8, Section XIII.A). The midfrequency band, from 3 to 12 eV, was assigned to interband excitations from O $2p$ valence bands to La $5d/4f$ orbitals, with the semiconductor La_2CuO_4 having an optical energy gap of ≈ 2 eV (Uchida *et al.*, 1991). The low-frequency edge is absent in the $x = 0$ insulating compound and present in the two doped conductors. Figure 15.21 presents a set of Bi compound spectra in the range 0.1–3 eV. The superconductor $Bi_2Sr_2CaCu_2O_8$ and the metal $Bi_2Sr_2(Ca,Nd)Cu_2O_8$ both exhibit the absorption edge near 1.1 eV, whereas the other two compounds, which are semiconductors, do not.

A Kramers–Kronig analysis carried out for the reflectance spectra of Fig. 15.20 provided the conductivity spectra presented in Fig. 15.22 for the low-energy region. At the low-frequency limit $\sigma(\omega)$ increases continuously with the level x of doping, being low for the insulators ($x = 0$, 0.2, 0.6), high for the superconductors

Table 15.3 Measured Raman Frequencies in cm^{-1} of the A_{1g} and B_{1g} Modes of High-Temperature Superconductors[a]

$(La_{1-x}Sr_x)_2CuO_4$[b]	$Bi_2Sr_2Ca_nCu_{n+1}O_{6+2n}$		$Tl_2Ba_2Ca_nCu_{n+1}O_{6+2n}$		$YBa_2Cu_3O_7$[c]	$TlBa_2CaCu_2O_7$[c,d]
$x = 0.075$	$n = 0$	$n = 1$	$n = 1$	$n = 2$		
				92		
	196 Bi	164 Bi	134 Tl	129 Tl	108 Ba	120 Ba
			157 Cu_p		152 Cu_p	148 Cu_p
226 La	309 Sr	292 Sr				
		282			340	278
			409 O_p		440 O_p	
430 O_z	455 O_z	464 O_z	493 O_z	498 O_z	504 O_z	525 O_z
	625 O_0	625 O_0	599 O_0	599 O_0		

[a] Most of the modes are labeled with their dominant vibrating atom; Cu_p and O_p denote copper and oxygen atoms in the planes; O_0, oxygens centered in an axially distorted octahedron of six heavy-atom nearest neighbors; and O_z, oxygens on the *c*-axis above and below the Cu atoms. From Burns *et al.* (1989).

[b] An Sr atom replacing La in the compound $(La_{1-x}Sr_x)_2CuO_4$ is expected to have its frequency raised from 226 to 284 cm^{-1}.

[c] Isostructural compound.

[d] $T_c = 60$ K.

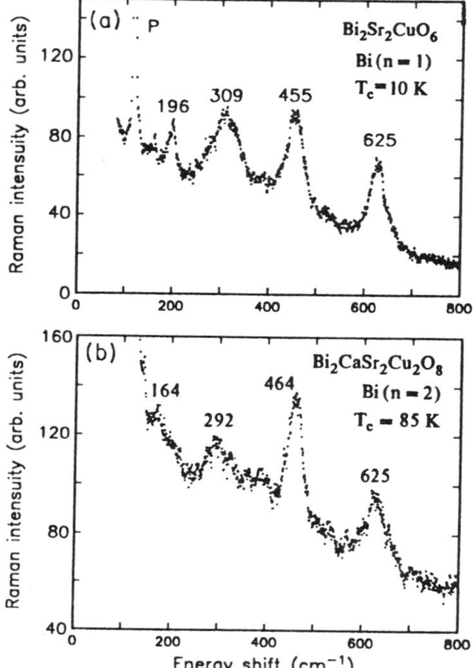

Figure 15.16 Polarized Raman spectra obtained at room temperature from single crystal $Bi_2Sr_2CuO_6$ (2201, a) and $Bi_2Sr_2CaCu_2O_8$ (2212, b) (Burns *et al.*, 1989).

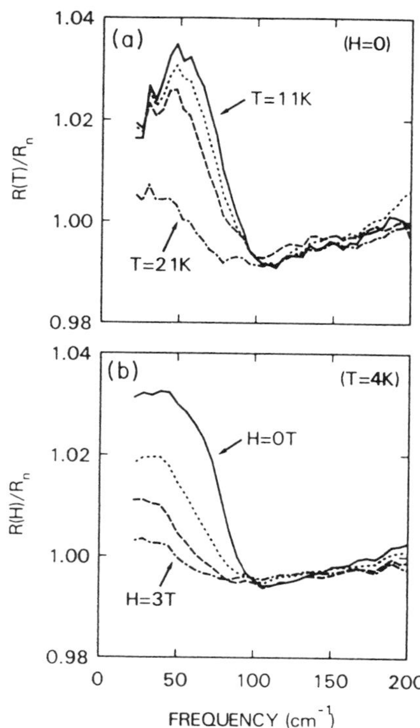

Figure 15.17 The frequency dependence of the reflectivity R in the superconducting state of $Ba_{0.6}K_{0.4}BiO_3$ normalized relative to its normal state value R_n showing the low frequency enhancement associated with the superconducting energy gap. The suppression of the low frequency enhancement by (a) a change in temperature ($T = 11, 14, 17, 21$ K) in zero field $B_{app} = 0$, and (b) the effect of applying a field ($B_{app} = 0, 1, 2, 3$ T) at the temperature 4 K are shown.

($x = 0.1$, 0.15, 0.2), and highest for the nonsuperconducting metal ($x = 0.34$). (Recall that $La_{2-x}Sr_xCuO_4$ is a hole superconductor.) A similar set of spectra obtained for the electron superconductor $Nd_{2-x}Ce_xCuO_{4-y}$ exhibited the same dependence of the low-frequency conductivity on x as in the hole case.

The rare-earth ions have crystal-field energy-level splittings in the optical region, and transitions between them can be observed. As an example, the energy levels of six erbium compounds are given in Fig. 15.23, and the optical transitions in the green region of the visible spectrum are shown in Fig. 15.24 for three of them (Jones *et al.*, 1990). This technique could be employed for checking the purity of a sample.

Representative optical spectroscopy studies include articles on single crystals of $(La_{1-x}Sr_x)_2CuO_4$ (S. Tajima *et al.*, 1988) and $YBa_2Cu_3O_{7-\delta}$ (Božovic *et al.*, 1988; Cooper *et al.*, 1992, 1993; Kircher *et al.*, 1991, Orenstein *et al.*, 1990; Y. Wang *et al.*, 1992; Z. D. Wang and Ting, 1992a (with Br)), and on films of K_3C_{60} (Iwasa *et al.*, 1992), $Ba_{1-x}K_xBiO_3$ (Božovic *et al.*, 1992), and $Tl_2Ba_2Ca_2Cu_3O_{10}$ (Božovic *et al.*, 1991).

IV. PHOTOEMISSION

Photoemission spectroscopy (PES) measures the energy distribution of the electrons emitted by ions in various charge and energy states. These electrons have energies characteristic of particular atoms

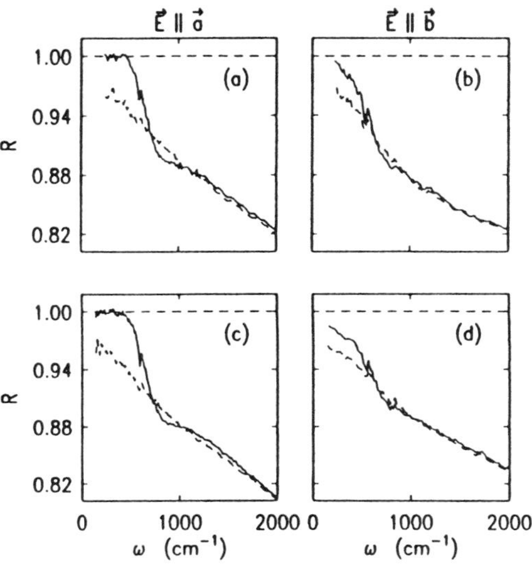

Figure 15.18 Polarized infrared reflectance spectra for two untwinned $YBa_2Cu_3O_{7-\delta}$ samples in the normal state ($T = 100$ K, dashed curves) and in the superconducting state ($T = 35$ K, solid curves). Polarization parallel to the a-axis is on the left, while polarization parallel to the b-axis is on the right. Two samples were used, spectra (a) and (b) from one of them and spectra (c) and (d) from the other (Schlesinger *et al.*, 1990b).

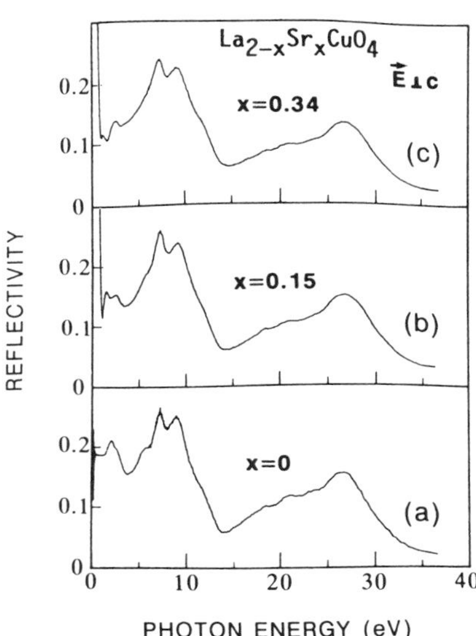

Figure 15.20 Optical reflectivity (reflectance) spectra with the **E** vector polarized in the a, b-plane for $La_{2-x}Sr_xCuO_4$ single crystals with three of the compositions x indicated in Fig. 15.19 (Uchida *et al.*, 1991).

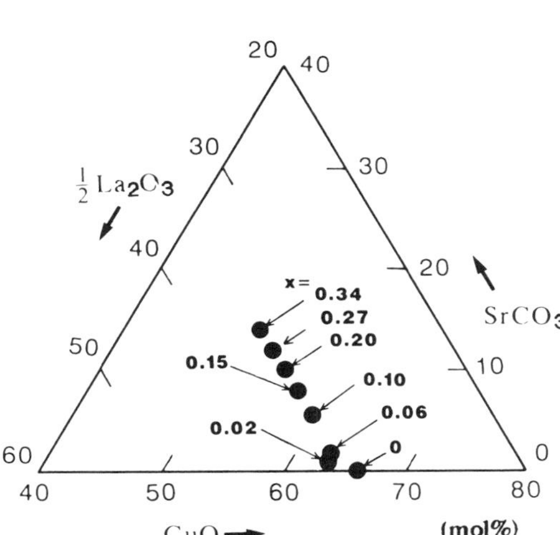

Figure 15.19 Compositions of the starting materials La_2O_3, $SrCO_3$, and CuO used to grow single crystals of $La_{2-x}Sr_xCuO_4$ with the indicated x values (Uchida *et al.*, 1991).

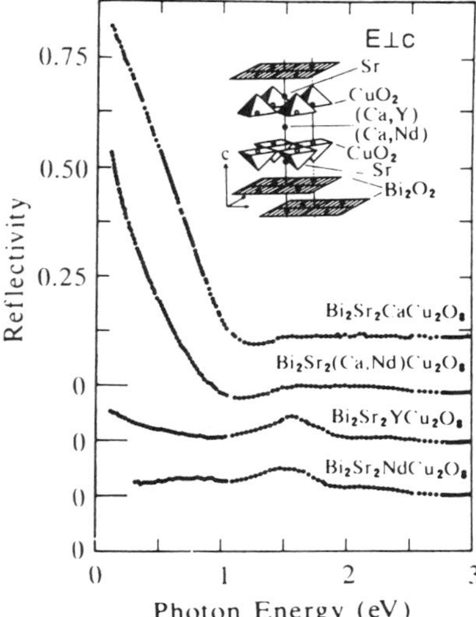

Figure 15.21 Room-temperature optical reflectivity (reflectance) spectra for four Bi-cuprates with the electric field E of the incident light polarized in the a, b-plane (Terasaki *et al.*, 1990a, b).

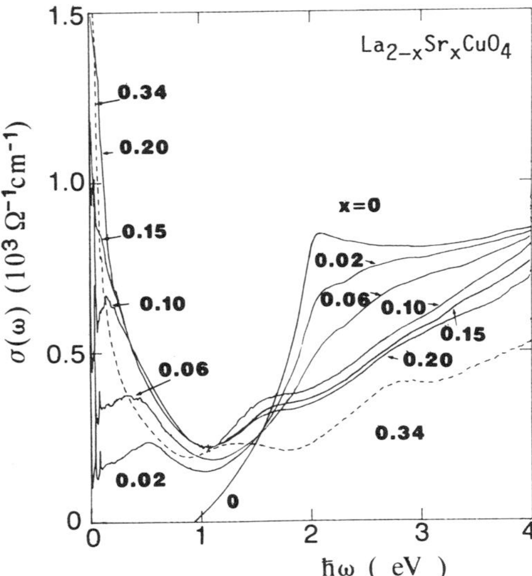

Figure 15.22 Frequency dependence of the optical conductivity $\sigma(\omega)$ of $La_{2-x}Sr_xCuO_4$ obtained from a Kramers–Kronig analysis of reflectance spectra for the **E** vector polarized in the a, b-plane. Results for several compositions x from Fig. 15.19 are shown (Uchida *et al.*, 1991).

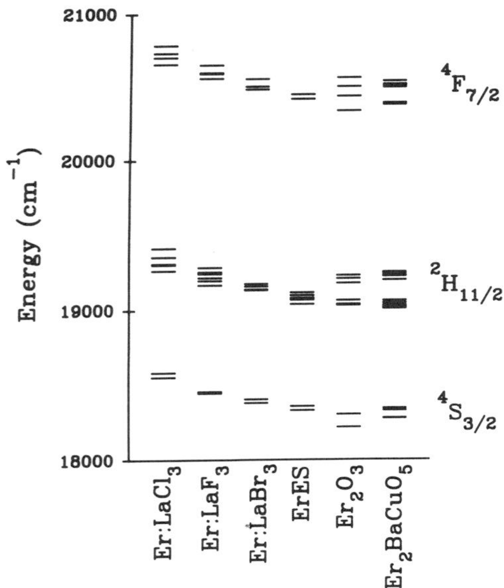

Figure 15.23 Crystal-field energy levels of Er^{3+} in several compounds, including the erbium green-phase Er_2BaCuO_5, which has levels close to those of the oxide Er_2O_3. ErES denotes erbium ethyl sulphate (Jones *et al.*, 1990).

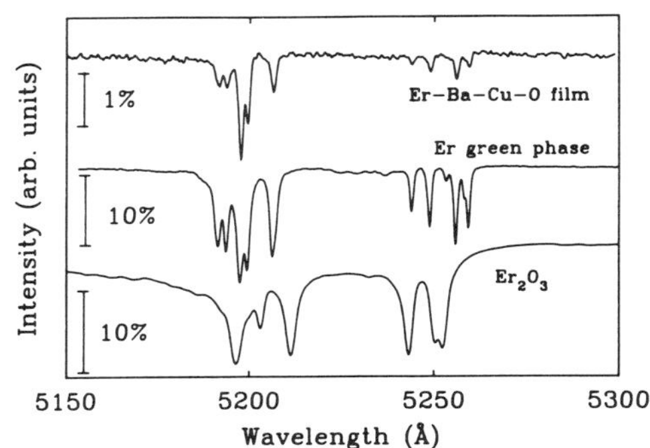

Figure 15.24 Optical spectra for the $^4I_{15/2} \rightarrow {}^2H_{11/2}$ transition in $ErBa_2Cu_3O_{7-\delta}$ (top), Er_2BaCuO_5 (middle), and Er_2O_3 (bottom) (Jones *et al.*, 1990).

in particular valence states. We will describe the technique, say something about the energy states that are probed, and describe what the technique tells us about superconductors. A number of pertinent theory articles have recently appeared (Folkerts and Haas, 1990; Gröbke, 1990; Kim and Riseborough, 1990; Sá de Melo and Doniach, 1990).

A. Measurement Technique

To carry out this experiment, the material is irradiated with ultraviolet light or x-rays, and the incoming photons cause electrons to be ejected from the atomic energy levels. The emitted electrons, called *photoelectrons*, have a kinetic energy KE which is equal to the difference between the photon energy $h\nu_{ph}$ and the ionization energy E_{ion} required to remove an electron from the atom, as follows:

$$KE = h\nu_{ph} - E_{ion}. \qquad (15.19)$$

The detector measures the kinetic energy of the emitted electrons, and since $h\nu_{ph}$ is known, the ionization energy is determined from Eq. (15.19). Each atomic energy state of each of the ions has a characteristic ionization energy, so that the measured kinetic energies provide information about the valence states of the atoms. In addition, many ionization energies are perturbed by the surrounding lattice environment, so that this environment is also probed by the measurement.

In ultraviolet photoemission spectroscopy (UPS), the excitation energy comes from a high-intensity UV source, such as the 21.2-eV resonance line (He–I) or the higher-frequency 40.8-eV line (He–II) of a helium-gas discharge tube. In the x-ray analogue (XPS), the radiation used to excite the photoelectrons is obtained from an Mg–K_α (1253.6), Al–K_α, (1486.7 eV) or other convenient x-ray source.

It is also possible to carry out the reverse experiment, called inverse photoelectron spectroscopy (IPS), in which the sample is irradiated with a beam of electrons and the energies of the emitted photons are measured. When UV photons are detected, the method is sometimes called bremsstrahlung isochromat spectroscopy (BIS). A related experiment is electron energy-loss spectroscopy (EELS) in which the decrease in energy of the incident electron beam is measured. Another technique, called Auger electron spectroscopy is based on a radiationless transition, whereby an x-ray photon generated within an atom does not leave the atom as radiation, but instead ejects an electron from a higher atomic level.

B. Energy Levels

We know from the quantum theory of atoms that, to first order, the frequency ν of a transition from the energy level with principal quantum number n_1 to the level n_2 is

$$\nu = \frac{me^4Z^2}{8\epsilon_0^2h^3}\left(\frac{1}{n_1^2} - \frac{1}{n_2^2}\right), \qquad (15.20)$$

where Z is the atomic number and the other symbols have their usual meaning. Figure 15.25 gives the energy level scheme for molybdenum, with additional fine-structure splittings not included in Eq. (15.20). For the atomic number (Z) dependence of the K_α line, which represents the innermost x-ray transition from $n_1 = 1$ to $n_2 = 2$, Eq. (15.20) gives Moseley's law

$$\sqrt{\nu} = a_K(Z - 1). \qquad (15.21)$$

The factor ($Z - 1$) in Eq. (15.21) in place of Z takes into account shielding of the nucleus by the remaining $n_1 = 1$ electron, whose apparent charge falls to ($Z-1$). A similar expression applies to the next highest frequency L_α line, which has $n_1 = 2$ and $n_2 = 3$.

Figure 15.26 presents a plot of $\sqrt{\nu}$ versus the atomic number Z for the experimentally measured K_α and L_α lines of the elements in the periodic table from $Z = 15$ to $Z = 60$, showing that Moseley's law is obeyed. These inner-level transitions are very little disturbed when the atom is bound in a solid because of shielding by the outer electrons, so that the regularity of Moseley's law applies to bound as well as free atoms, permitting atoms to be unambiguously identified. This law only holds for the innermost atomic electrons, however.

The ionization energies of the outer electrons of atoms are more dependent on the number of electrons outside the closed shells than on the atomic number, as shown by the data in Fig. 15.27. The ionization energies are in the visible or near-ultraviolet region. When the atom is bound in a solid, its valence electrons form ionic or covalent bonds, drastically modifying their upper energy level schemes and ionization energies. Atomic electrons below the valence electrons but not in the deepest levels undergo shifts in energy that are intermediate between the two extreme cases of the outermost and innermost electrons.

C. Core-Level Spectra

The four parts of Fig. 15.28b presents core-level XPS spectra arising from the atoms Ba, Cu, O, and Y of $YBa_2Cu_3O_7$

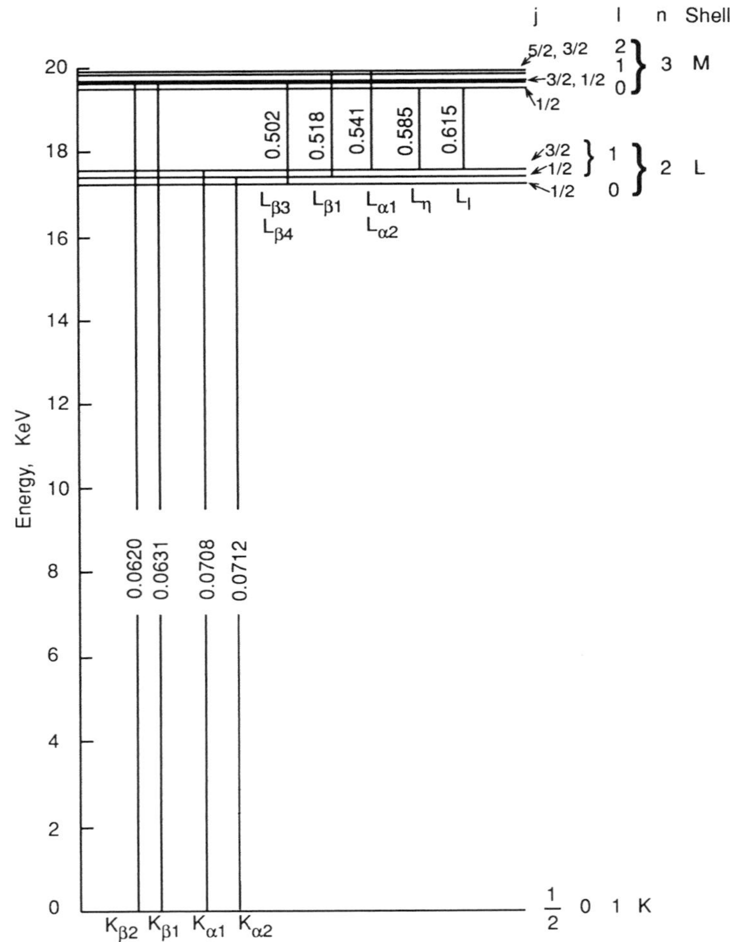

Figure 15.25 Energy-level diagram of Mo showing the wavelengths, in nanometers, of the *K* and *L* series lines, each of which is labeled using the Siegbahn notation. The *j*, *l*, and *n* quantum number is given for each energy level.

(Steiner *et al.*, 1987). Figure 15.29 gives corresponding spectra from Bi, O, and Sr of $Bi_2(Sr, Ca)_3Cu_2O_{8+\delta}$ (Fujimori *et al.*, 1989). The latter figure shows the decomposition of each line into components. Figure 15.30 shows how the lines in the $Cu(2p_{3/2})$ spectral region with binding energy from 934 to 937 eV vary in position and intensity for the four compounds $LaCuO_3$, La_2CuO_4, CuO, and Cu_2O. The $Cu(2p_{1/2})$ transition is near 954 eV. (Allan *et al.* (1990), and Yeh *et al.* (1990) show similar Cu(2*p*) spectra for $YBa_2Cu_3O_{7-\delta}$ at three temperatures in the superconducting region.) The lines, near 944 and 963 eV in the spectra of Fig. 15.30, are satellites of the two main lines. Several researchers have studied the photoemission of the oxides Cu_2O, CuO, and $NaCuO_2$, which have monovalent, divalent, and trivalent Cu, respectively, for comparison with cuprate spectra (Brandow, 1990; Ghijsen *et al.*, 1990; Karlsson *et al.*, 1992; Sacher and Klemberg-Sapieha, 1989; Shen *et al.*, 1990).

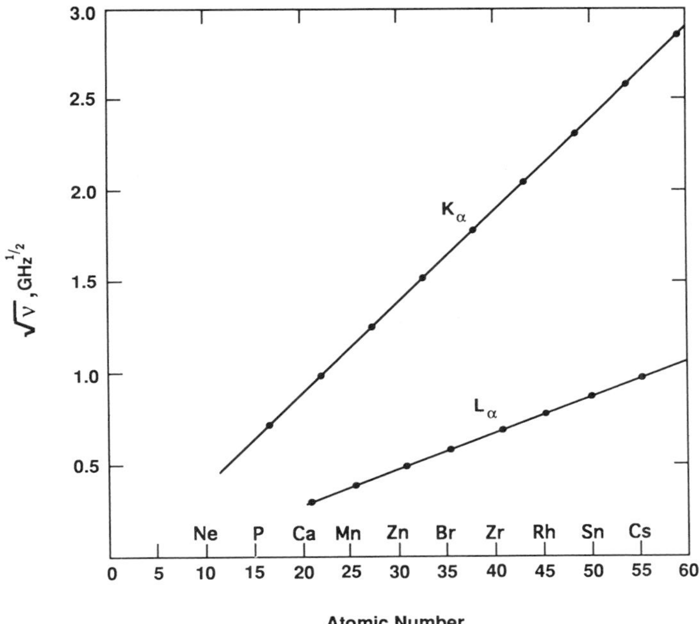

Figure 15.26 Moseley plot of the K_α and L_α characteristic x-ray lines of a number of elements in the atomic number range from 17 to 59.

The shapes of photoemission core spectra provide information on various sample characteristics.

1. The spectra from the six atoms in the compound $Bi_2Sr_2Ca_{1-x}Y_xCu_2O_y$ are presented in Fig. 15.31 for $x = 0$, 0.5, 0.8, and 1.0, with the $x = 0$ scan omitted for Y (Itti *et al.*, 1991). The decline in the intensity of the Ca line for these four x values is evident. Figure 15.32 clarifies how the various line positions shift toward higher values with the increase in x.

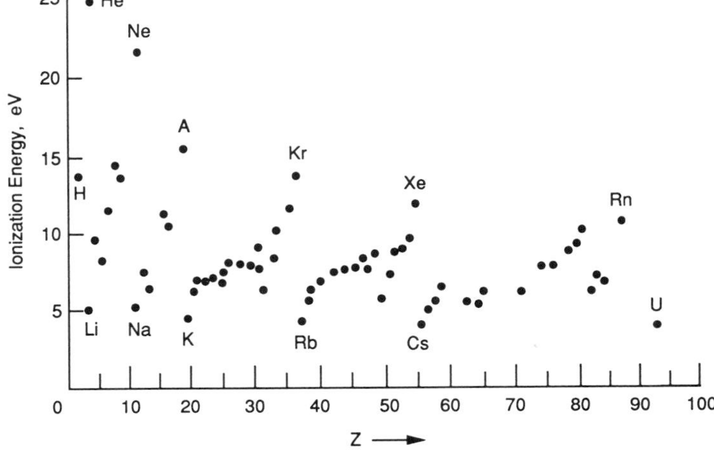

Figure 15.27 Experimentally determined ionization energy of the outer electron in various elements (Eisberg and Resnick, 1974, p. 364). Copyright © 1974. Reprinted by permission of John Wiley & Sons, Inc.

2. The core spectra from the four atoms of $YBa_2Cu_3O_{7-\delta}$ are compared in Fig. 15.33 for no pretreatment, following two high-temperature heat treatments in an ultra-high vacuum (UHV), and following heating and annealing in oxygen (Frank *et al.*, 1991). The decomposition into component lines arising from the surface and from the bulk is shown for three of the spectra. As the treatment proceeds, the bulk fraction increases relative to the surface fraction, as shown.
3. A combined photoelectron microscopy and spectroscopy experiment compared the Bi spin-orbit split $d_{5/2}$, $d_{3/2}$ doublet obtained from different regions, ≈ 20 μm in diameter, on the surface of cleaved monocrystals of $Bi_2Sr_{2-x}Ca_{1+x}Cu_2O_{8+\delta}$. The spectra are given in Fig. 15.34 (Komeda *et al.*, 1991).

We see that some spectra exhibit a doublet from a highly oxidized form of Bi shifted by about 2 eV to higher binding energies. The change in the Bi oxidation state at the crystal edges could degrade the superconducting properties.

Recent core level spectra of monocrystals and films (f) have been published by: BaPbBiO (Matsuda *et al.*, 1989), NdCeCuO (Allen *et al.*, 1990), YBaCuO (Rao *et al.*, 1992 (f); Tressand *et al.*, 1990; Yeh, 1992; Yeh *et al.*, 1990 (f)), BiSrCaCuO (Kohiki *et al.*, 1988; Lindberg *et al.*, 1989). Photoemission has been reported for K_3C_{60} (Merkel *et al.*, 1993; Poirier and Weaver, 1993; Takahashi *et al.*, 1992).

D. Valence Band Spectra

The spectra of the outer, or valence, electrons occur at lower energies, 0 to 16 eV, as shown on the panel of Fig. 15.28a. The overlapping of the O2*p* and Cu3*d* bands depends on the conditions under which they are obtained, and these conditions can be varied to enhance certain features relative to others. For example,

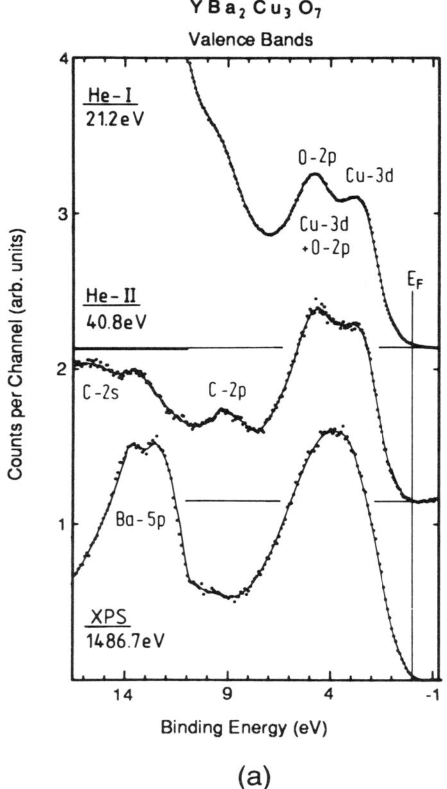

Figure 15.28 Photoemission spectra of $YBa_2Cu_3O_7$, showing (a) valence band spectra at low energies, and (b) core-level x-ray spectra (XPS) (Steiner *et al.*, 1987). (*Continues*)

Fig. 15.35 shows angle-resolved photoemission spectra (ARPES) obtained from $(Bi_{0.8}Pb_{0.2})_2Sr_2CaCu_2O_8$ single crystals cleaved *in vacuo* for electron-emission angles in the range from 30° to 61.5° (Böttner *et al.*, 1990), while Fig. 15.36 (Arko *et al.*, 1989) presents spectra of $YBa_2Cu_3O_{6.9}$ for different incident-photon energies between 14 and 70 eV. The peaks B to F in Fig. 15.35 are associated with flat regions of the energy bands. The A and D peaks of Fig. 15.36, which vary in the extent of their resolution, are assigned to the O2*p* and Cu3*d* states, respectively. From Fig. 15.36 it is clear that the discontinuity in intensity at the absorption edge itself, the zero of energy, is small compared with the atomic absorptions that start near 1 eV. This edge

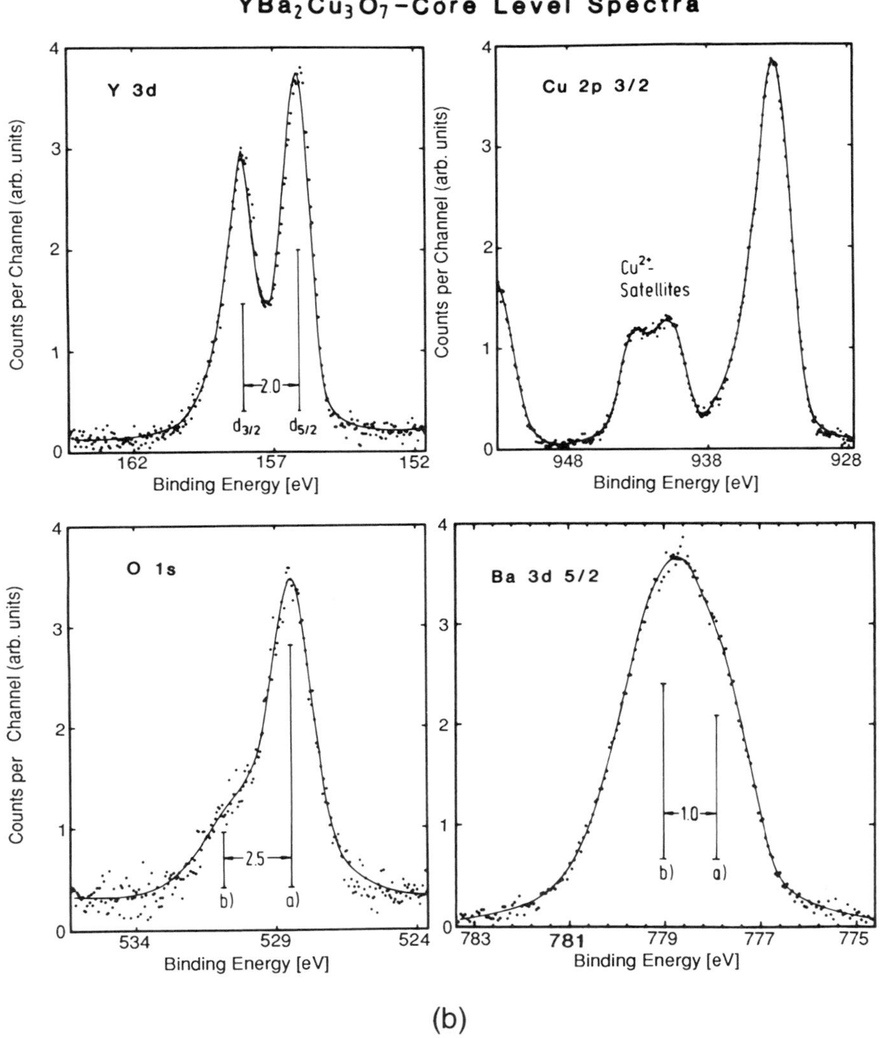

(b)

Figure 15.28 (*Continued*)

has been resolved using UPS with the 21.7-eV exciting line (Imer *et al.*, 1989). Figure 15.37 illustrates how excitation with the photon energies of one element, O1*s* in this case, enhances the spectral features from another element, the Cu3*d* lines at 13 eV (Sarma *et al.*, 1989). The figure also shows an Auger signal (Bar-Deroma *et al.*, 1992; Cota *et al.*, 1988).

Angle resolved photoemission spectra have been analyzed by Fermi liquid theory (Kim and Riseborough, 1990). Some typical articles on valence bands are (Brookes *et al.* 1989; Dessau *et al.*, 1991; Matsuyama *et al.*, 1989; Mehl *et al.* 1990; Wells *et al.*, 1990).

E. Energy Bands and Density of States

Various investigators have employed photoemission to obtain information on, for example, energy bands (Dessau *et al.*, 1992; Liu *et al.*, 1992a; Takahashi *et al.*,

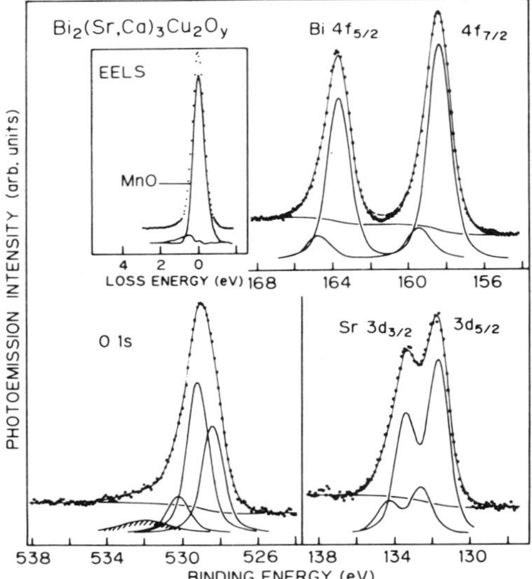

Figure 15.29 Core-level XPS spectra of $Bi_2(Sr, Ca)_3Cu_2O_y$ shown fit with calculated line shapes. The shaded part of the O–1s spectrum is due to contamination. The inset shows the elastic peak of the electron energy loss spectrum (EELS, dots, $E_0 \approx 2$ kV) decomposed into a dominant, purely elastic part characteristic of the wide-gap insulator MnO and a weak residual signal (Fujimori *et al.*, 1989).

1989), the Fermi surface (Campuzano *et al.*, 1991; Mazin *et al.*, 1992; Tobin *et al.*, 1992), and the Eliashberg function $\alpha^2 D_{\mathrm{ph}}(W)$ (Arnold *et al.*, 1991; Bulaevskii *et al.*, 1988) of high-temperature superconductors.

V. X-RAY ABSORPTION EDGES

A. X-ray Absorption

An energetic photon is capable of removing electrons from all occupied atomic energy levels with ionization energies less than the photon energy. When the photon energy drops below the highest ionization energy, which corresponds to the K-level, the $n = 1$ electron can no longer be removed and the x-ray absorption coefficient abruptly drops. It does not, however, drop to zero, because the x-ray photon is still energetic enough to knock out electrons in the L ($n = 2$), M ($n = 3$), etc., levels, as is clear from Fig. 15.25. The abrupt drop in the absorption coefficient is referred to as

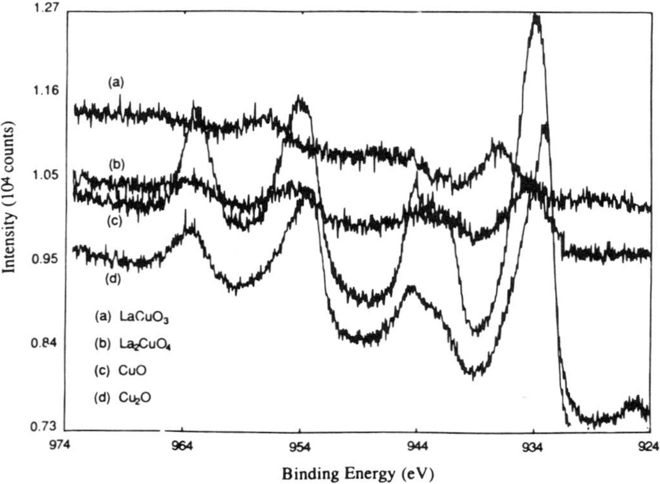

Figure 15.30 XPS spectra of the Cu $2p_{3/2}$ (≈ 935 eV) and Cu $2p_{1/2}$ (≈ 954 eV) regions of four copper oxides: CuO, Cu_2O, La_2CuO_4, and $LaCuO_3$. The lines near ≈ 944 eV and ≈ 963 eV are satellites (Allan *et al.*, 1990).

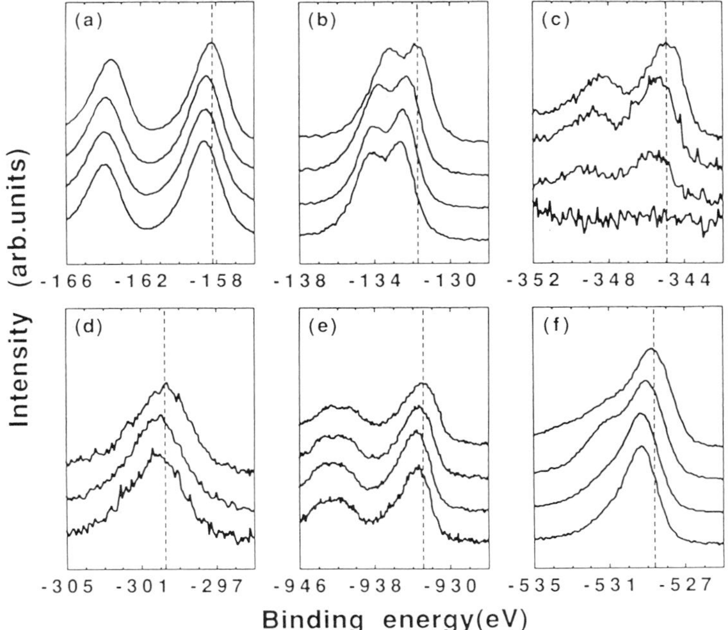

Figure 15.31 XPS core spectra of $Bi_2Sr_2Ca_{1-x}Y_xCu_2O_y$, with $x = 0$, 0.5, 0.8, and 1.0 from top to bottom, for the atoms: (a) Bi $4f$, (b) Sr $3d$, (c) Ca $2p$, (d) Y $3p$, (e) Cu $2p_{3/2}$, and (f) O $1s$. There is, of course, no $x = 0$ spectrum for Y (Itti *et al.*, 1991).

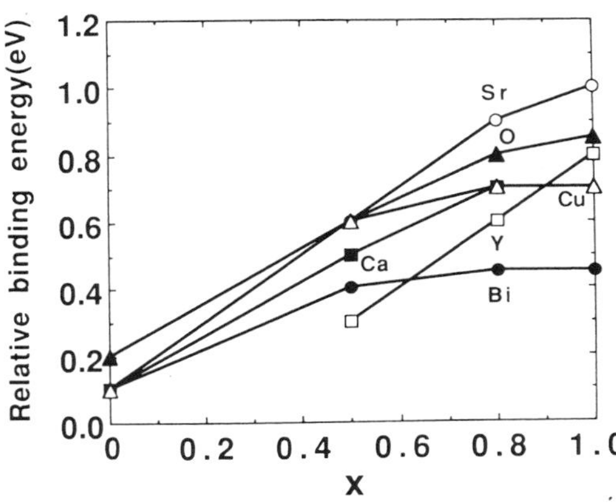

Figure 15.32 Shift of the binding energy of each core level with the Y content x obtained from the spectra of Fig. 15.31. The vertical axis indicates the shift in binding energy relative to the offset values of 158.1 (Bi $4f_{7/2}$), 131.6 (Sr $3d_{5/2}$), 344.8 (Ca $2p_{3/2}$), 299.4 (Y $2p_{3/2}$), 932.7 (Cu $2p_{3/2}$), and 528.4 eV (O $1s$) (Itti *et al.*, 1991).

an absorption edge; in this case, it is a K-absorption edge.

A photon with energy slightly below the ionization energy can raise the $n = 1$ electron to a higher unoccupied level, such as a $3d$ or $4p$ level. Transitions of this type provide what is called fine structure on the absorption edge, furnishing information on the bonding states of the atom in question. The resolution of individual fine-structure transitions can be improved with the use of polarized x-ray beams (Abbate *et al.*, 1990). Among the specialized x-ray absorption spectroscopy (XAS) techniques that have been used we may note x-ray absorption near-edge structure (XANES), x-ray absorption fine-structure (XAFS), and extended x-ray absorption fine-structure (EXAFS) spectroscopy.

Figure 15.38 shows the O $1s$ x-ray absorption edges obtained with twin-free monocrystals of

$$YBa_2Cu_3O_7 \quad \text{and} \quad YBa_2Cu_4O_8$$

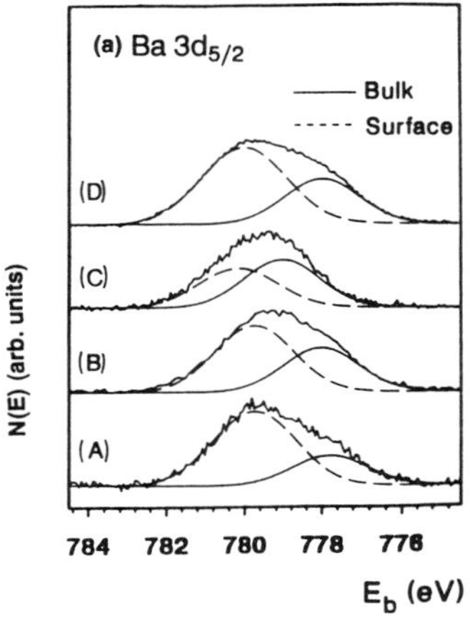

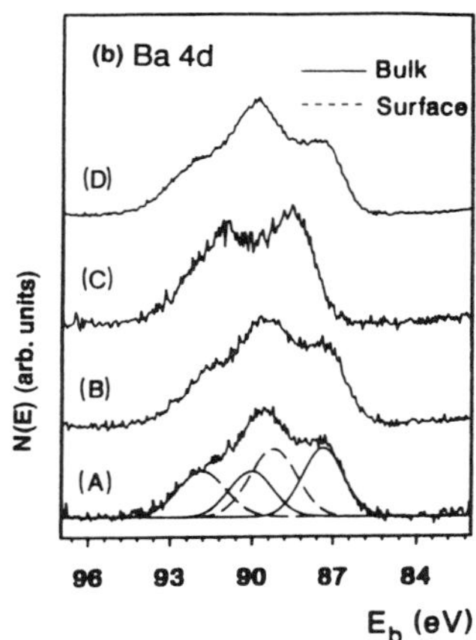

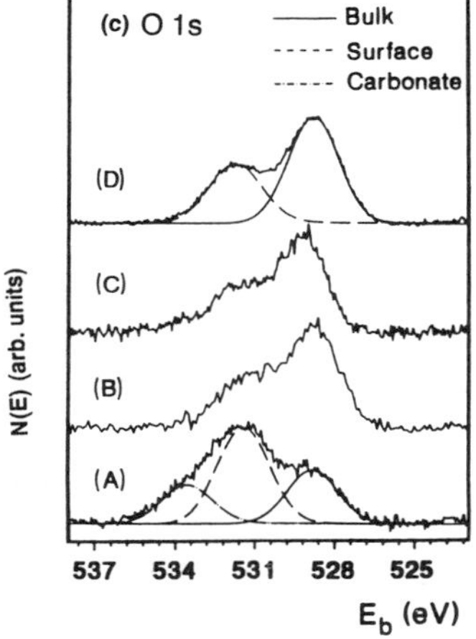

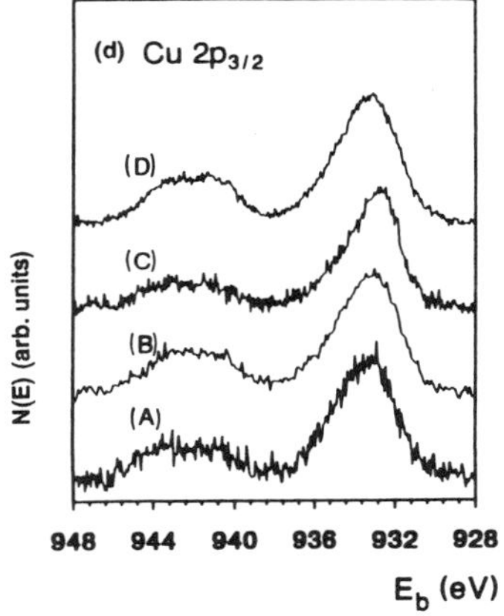

Figure 15.33 XPS spectra and line shape decomposition of three atoms in $YBa_2Cu_3O_{7-\delta}$: (a) Ba $3d_{5/2}$, (b) Ba $4d$, (c) O $1s$, and (d) Cu $2p_{3/2}$. Spectra are presented for samples before pretreatment (A), after heating *in vacuo* to 520 K (B), after heating *in vacuo* to 650 K (C), and after annealing in pure oxygen at 700 K (D). The spectra were recorded at room temperature *in vacuo*, and both the S-shaped background and the $K_{\alpha 3,4}$ satellite contributions have been subtracted out (Frank *et al.*, 1991).

for the case of polarization parallel to the a and b directions (Krol *et al.*, 1992). The difference spectrum is also shown. The XAS spectrum for $E \| a$ is due to the O(2) atoms in the CuO_2 planes, while that for $E \| b$ arises from the O(3) atoms in the planes and the O(1) atoms in the chains. For $YBa_2Cu_3O_7$, the O(1) and apex oxygen O(4) binding energies determined from the absorption edge were found to be 0.4 and 0.7 eV, respectively, both of which is lower than the binding energies of the oxygens O(2, 3).

Figure 15.39 shows how varying the angle between the incident beam and the c-axis of $YBa_2Cu_3O_{6.9}$ monocrystals resolves oxygen–hole structure into a small, lower-energy peak (A) at 526.4 eV at-

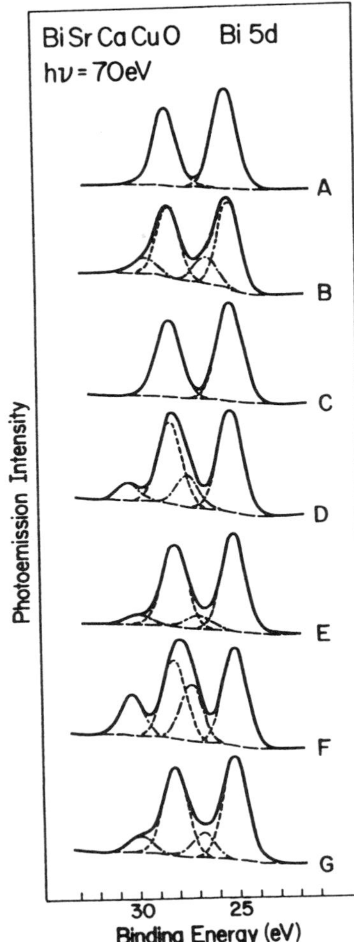

Figure 15.34 Core-level photoemission energy distribution curves for Bi $5d_{3/2}$ and $5d_{5/2}$ lines in $Bi_2Sr_{2-x}Ca_{1+x}Cu_2O_{8+\delta}$. Spectra were obtained from seven $\approx 20\ \mu m$ diameter regions located at different places on an electron micrograph of the surface (not shown). Regions A and C, which are representative of the clean surface, exhibit a single Bi doublet. Regions B and D–G, which are situated near the border between the monocrystal stacks, show a second-position dependent component suggestive of highly oxidized Bi (Komeda *et al.*, 1991).

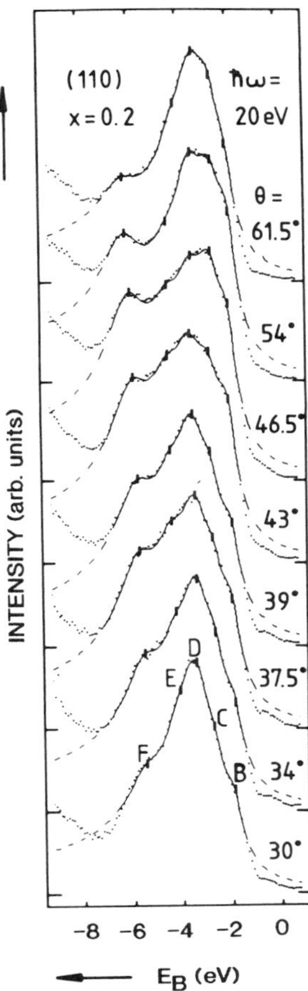

Figure 15.35 Angle-resolved photoemission spectra (ARPES) of $(Bi_{0.8}Pb_{0.2})_2Sr_2CaCu_2O_8$ single crystals for emission angles between 30° and 61.5°. Calculated curves fit to the spectra are shown as solid lines inside the fit range and as dashed lines outside. Calculated peak positions are shown as tick marks labeled B, C, D, E, and F (Böttner *et al.*, 1990).

tributed to holes on O(2) and O(3) in the CuO_2 planes and a more prominent (B) peak at 529.2 eV assigned to holes on O(4) along the chains (Alp *et al.*, 1989b).

The XANES spectra presented in Fig. 15.40 show the effect of doping the hole superconductor LaSrCuO and the electron superconductor NdCeCuO by comparing the absorption with that of the respective undoped compounds. The results indicate that substitution has less effect on the Cu bonding in La_2CuO_4 than in Nd_2CuO_4, and suggest that electron doping occurs mainly at the Cu atom of CuO_2 in the Nd compound, mainly at the O atom in the La compound.

Substitution of first-transition ions for Cu in $YBa_2Cu_3O_{7-\delta}$ produces the changes in the K-absorption edge that are shown in Fig. 15.41. These changes provide evidence

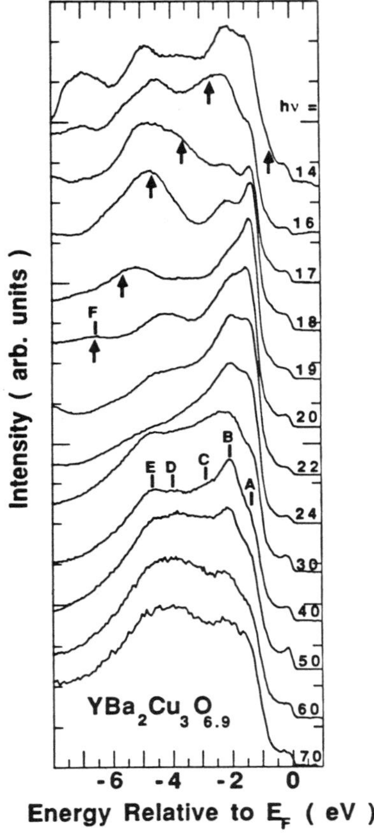

Figure 15.36 Valence-band photoemission spectra of $YBa_2Cu_3O_{6.9}$ for a series of incident photon energies from 14 to 70 eV, normalized to equal maximum intensities. The symbols A, B, C, D, E, and F indicate identifiable peaks. Peak F, which shifts with $h\nu$ to apparent higher binding energies, is labeled by arrows. The O $2p$ intensity is strongly concentrated in peaks A and B, while the Cu $3d$ intensity is partly centered on peak D, and partly distributed throughout the valence bands (Arko *et al.*, 1989).

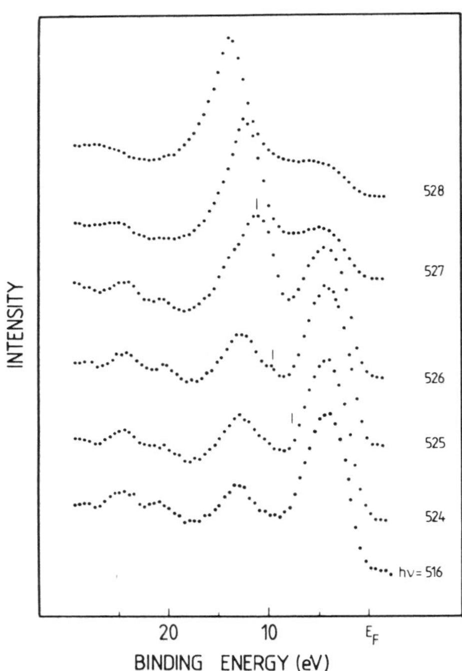

Figure 15.37 Valence-band photoemission spectra of $YBa_2(Cu_{0.9}Fe_{0.1})_3O_{6.9}$. The oxygen Auger line is indicated by a vertical tick on three of the spectra (Sarma *et al.*, 1989).

that Fe and Co substitute for Cu(1) in the linear chain site, Zn occupies only the in-plane position at Cu(2) and Ni resides in both.

X-ray absorption studies have been reported on single crystals and films (f), including theoretical analyses (t); for LaCuO (Pompa *et al.*, 1991a; Tolentino *et al.*, 1992 (t)), for LaSrCuO (C. T. Chen *et al.*, 1992; Pompa *et al.*, 1991b; Wu *et al.*, 1992 (t)), for YBaCuO (Curtiss and Tam, 1990 (t), Guo *et al.*, 1990 (t), Krol *et al.*, 1992; Werfel *et al.*, 1992), for BiSrCaCuO (Abbate *et al.*, 1990 (with Pb), Parmigiani *et al.*, 1991; Suzuki *et al.*, 1991), and for TlBaCaCuO (DiMarzio *et al.*, 1990 (f)). Superconductors have also been studied by related x-ray techniques, such as Rutherford backscattering (Sharma *et al.*, 1991).

B. Electron-Energy Loss

Another technique for obtaining absorption edges, called electron-energy loss spectroscopy (EELS), involves irradiating a thin film with a beam of monoenergetic electrons with energies of, for example, 170 keV. As the electrons pass through the film, they exchange momentum with the lattice and lose energy by exciting or ionizing the atoms. An electron-energy analyzer is then used to determine the energy E_{abs} that is absorbed. This energy corresponds

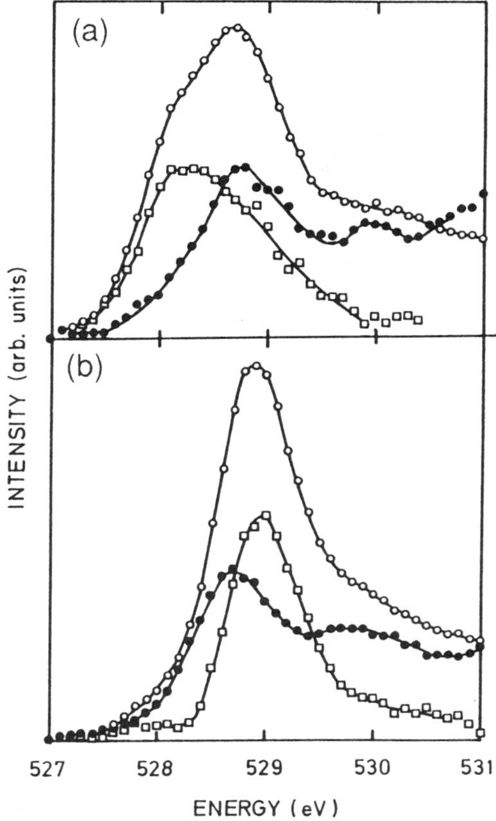

Figure 15.38 X-ray absorption spectrum of the O 1*s* line of (a) $YBa_2Cu_3O_7$, and (b) $YBa_2Cu_4O_8$ for E$\|a$ (●), E$\|b$ (○), and difference spectrum E$\|b$ − E$\|a$ (□) (Krol *et al.*, 1992).

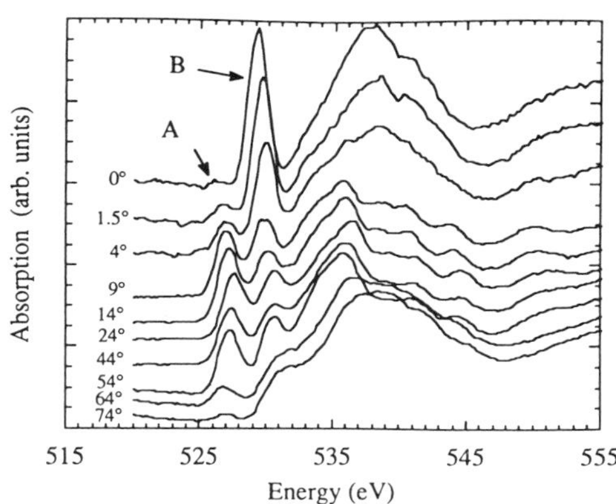

Figure 15.39 Energy dependence of the x-ray absorption of $YBa_2Cu_3O_{6.9}$ by O 1*s* electrons for a series of angles between the electric field and the *c*-axis. Peak A arises from the oxygens O(2) and O(3) in the CuO_2 planes, while peak B is from O(4) along the chains (Alp *et al.*, 1989b).

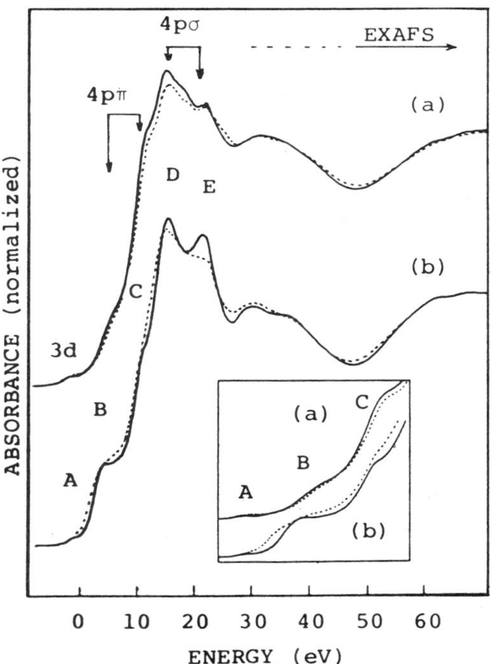

Figure 15.40 X-ray absorption Cu*K* near-edge spectra (XANES) for (a) La_2CuO_4 (solid line) and $(La_{0.925}Sr_{0.075})_2CuO_4$ (dashed line), and (b) Nd_2CuO_4 (solid line) and $(Nd_{0.925}Ce_{0.075})_2CuO_4$ (dashed line). The inset presents the 1*s* 3*d* and 1*s* 4*p*π regions on a magnified scale. The excitation energy was measured relative to the first inflection point of the Cu foil *K*-edge (Kosugi *et al.*, 1990).

to a transition of the type shown in the energy level diagram of Fig. 15.25, and equals the difference between the kinetic energy KE_0 of the incident electrons and the kinetic energy KE_{sc} of the scattered electrons

$$E_{abs} = KE_0 - KE_{sc}. \qquad (15.22)$$

In a plot of the intensity of the scattered electrons as a function of the absorbed energy, peaks will be found at the binding energies of the various electrons in the sample.

An analogue of optical and x-ray polarization experiments can be obtained from EELS by varying the direction of the momentum transfer **q** between the incom-

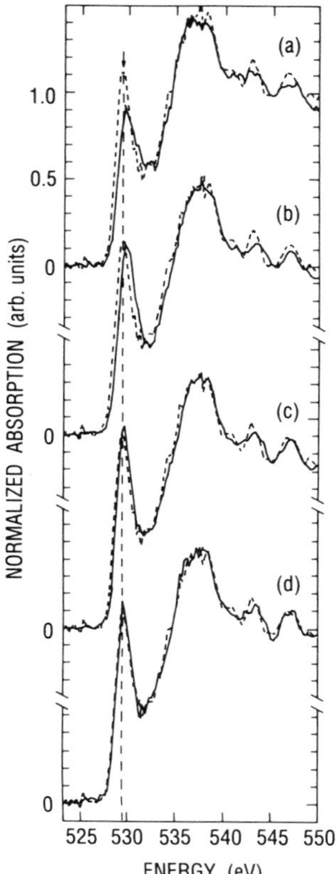

Figure 15.41 Comparison of the x-ray absorption oxygen K near-edge absorption spectra of $YBa_2(Cu_{0.96}M_{0.04})_3O_{7-\delta}$ for the metal substitutions M given by (a) Fe, (b) Co, (c) Ni, and (d) Zn. The 4%-doped (solid curves) and undoped (dashed curves) spectra are compared for each case (C. Y. Yang *et al.*, 1990).

ing electron and the lattice relative to the c-axis of the crystal. The vector **q** plays the role of the electric polarization vector **E** in photon spectroscopy.

Typical EELS studies include articles on LaSrCuO (Nücker *et al.*, 1988), NdCeCuO (Alexander *et al.*, 1991), YBaCuO (Batson *et al.*, 1991; Nücker *et al.*, 1989; Vaishnava *et al.*, 1990 (with Fe)), BiSrCaCuO (Nücker *et al.*, 1989; Wang *et al.*, 1990), and (TlPb)(CaY)SrCuO (Yuan *et al.*, 1991).

VI. INELASTIC NEUTRON SCATTERING

A neutron is a particle with almost the same mass as a proton, but, unlike the proton, it is electrically neutral. Despite this lack of charge, it has a magnetic moment, which enables it to interact with local magnetic moments as it passes through matter. When it scatters elastically, it has the same kinetic energy after the scattering event as it had beforehand. In nonmagnetic materials, neutrons scatter elastically off atomic nuclei; coherent scattering experiments, called neutron diffraction, are similar to their x-ray diffraction counterparts, likewise helping to determine crystal structures. Neutrons interact strongly with the magnetic moments of any transition ions that are present, and the resulting diffraction pattern provides the spin directions, as illustrated in Fig. 10.23, for antiferromagnetic alignment.

When neutrons scatter inelastically in matter, their kinetic energy changes through the creation (+) or absorption (−) of a phonon with energy $\hbar\,\omega_{ph}$,

$$\tfrac{1}{2}mv^2 = \tfrac{1}{2}mv'^2 \pm \hbar\,\omega_{ph}, \quad (15.23)$$

so that energy is exchanged with the lattice vibrations. A measurement of the angular distribution of neutrons scattered at various energies provides detailed information about the phonon spectrum, such as the dispersion curves and the phonon density of states $D_{ph}(\omega)$. The latter determines the dimensionless electron–phonon coupling constant λ through the Eliashberg relation (6.121),

$$\lambda = 2\int \frac{\alpha^2(\omega)D_{ph}(\omega)d\omega}{\omega}, \quad (15.24)$$

where $\alpha(\omega)$ is the electron–phonon coupling strength; $\alpha^2(\omega)D_{ph}(\omega)$ is called the Eliashberg function. Inelastic scattering has also resolved spin waves in La_2CuO_4 (Aeppli *et al.*, 1989). Thus inelastic neutron scattering measurements can provide us with important information about su-

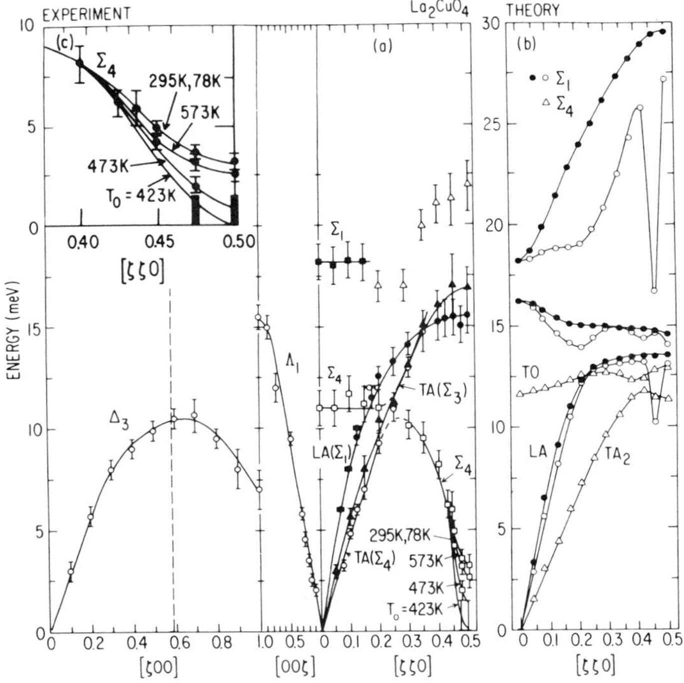

Figure 15.42 Low-lying phonon branches in $(La_{1-x}Sr_x)_2CuO_4$ with the modes labeled according to Weber (1987), showing (a) experimental results, (b) calculated dispersion curves, and (c) inset, measured temperature dependence near the X point. The dispersion curves are only weakly temperature dependent, except for the TO phonon near the X point shown in inset (c). The filled symbols show unrenormalized bare phonons, and the open circles indicate Σ_1 symmetry phonons renormalized by interactions with conduction electrons (Böni *et al.*, 1988).

perconductors, including, for example, K_3C_{60} (Prassides *et al.*, 1991) and monocrystals of LaSrCuO (Birgeneau *et al.*, 1988; Chou *et al.*, 1990; Hedegård and Pedersen, 1990; Shirane *et al.*, 1989) and YBaCuO (Shirane *et al.*, 1990; Tranquada *et al.*, 1989, 1992). We will give some representative results obtained using this experimental tool.

Figure 15.42 presents the dispersion curves, determined by inelastic neutron scattering, of the low-lying phonon branches for the superconductor

$$(La_{1-x}Sr_x)_2CuO_4.$$

Phonon dispersion curves were determined over a much broader energy range for the isomorphous nonsuperconducting compound La_2NiO_4. A soft mode (cf. Section II.C) exists in La_2NiO_4 at the point X (point $(\frac{1}{2}, 0, \frac{1}{2})$) in the Brillouin zone sketched in Fig. 8.37. Figure 15.43 shows that this soft mode decreases in frequency by 15% when the temperature is reduced from 300 K to 12 K. Phase transitions in crystals often involve soft modes, as was mentioned in Section II.C.

The experimental phonon density of states $D_{exp}(\omega)$ corresponding to the phonon dispersion curves of Li_2NiO_4 is plotted in Fig. 15.44. The calculated values, also shown in the figure, are in moderate agreement with experiment. The corrected density of states $D_{ph}(\omega)$ is obtained

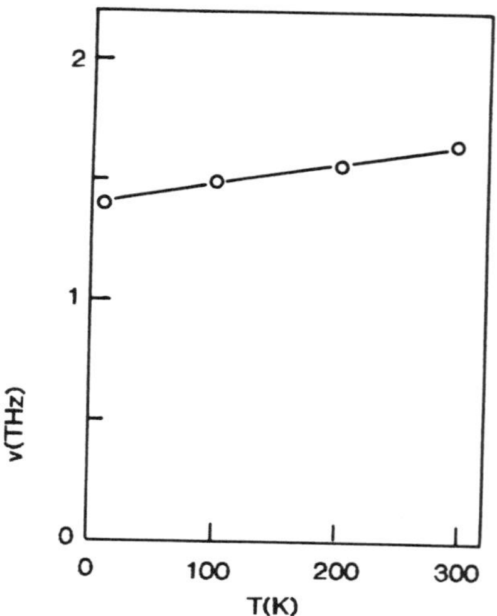

Figure 15.43 Temperature dependence of the frequency of the soft mode of La_2NiO_4 at $q = (\frac{1}{2}, 0, \frac{1}{2})$ (Pintschovius *et al.*, 1989).

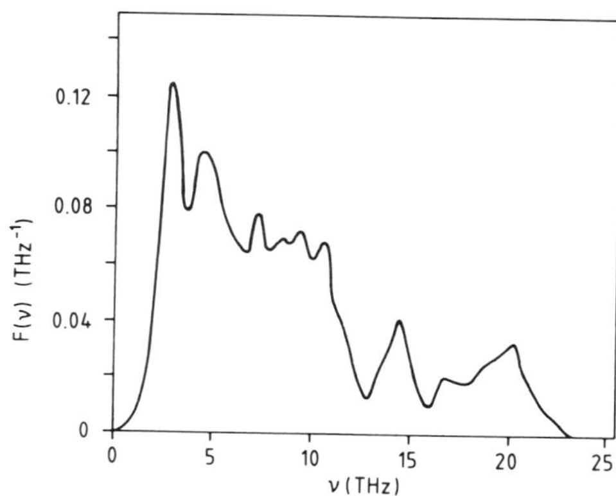

Figure 15.45 Corrected phonon density of states of La_2NiO_4 obtained by applying corrections to the experimentally determined curve of Fig. 15.44 (Pintschovius *et al.*, 1989).

from the experimental DOS by weighting the vibrations of the ith atom with the ratio σ_i/M_i, where σ_i is the neutron-scattering cross section and M_i is the mass of the ith atom. The result is plotted in Fig. 15.45.

The phonon DOS for the cubic superconductor $Ba_{0.6}K_{0.4}BiO_3$, is presented in Fig. 15.46 together with its counterpart, which was calculated by molecular dynamics simulation (Loong *et al.*, 1989, 1991, 1992). The random nature of the substitution of K on the Ba sites of this compound causes the experimental spectrum to be broader and less well resolved than the calculated spectrum. The partial DOS calculated for the atoms Ba, Bi, and K, shown in Fig. 15.47, are responsible for the peaks

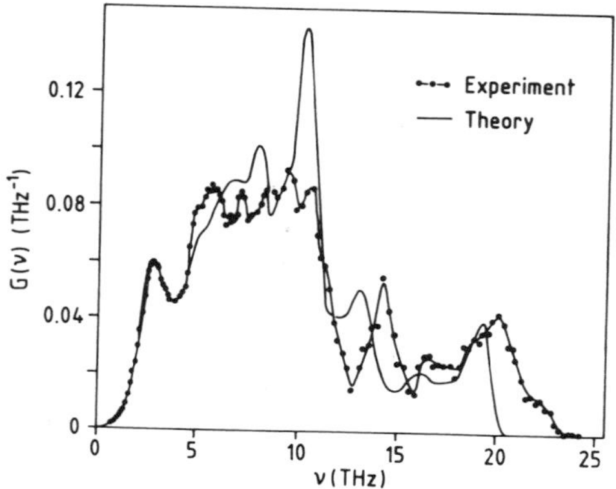

Figure 15.44 Comparison of the phonon density of states experimentally determined by inelastic neutron scattering (····) and calculated (——) (Pintschovius *et al.*, 1989).

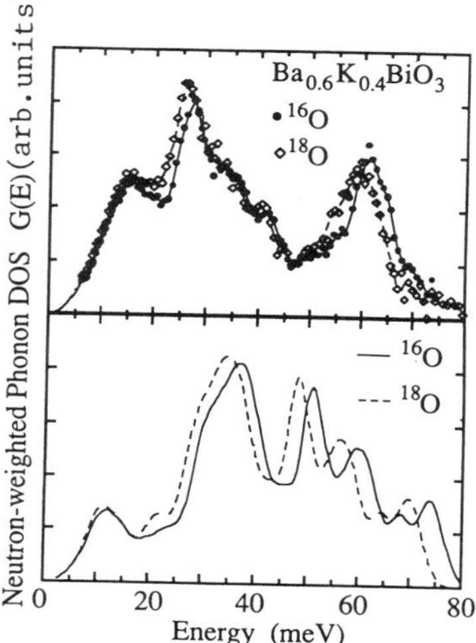

Figure 15.46 Comparison of the phonon density of states of $Ba_{0.6}K_{0.4}BiO_3$ (a) determined experimentally by inelastic neutron scattering, and (b) calculated by molecular dynamics simulations (Loong *et al.*, 1991, 1992).

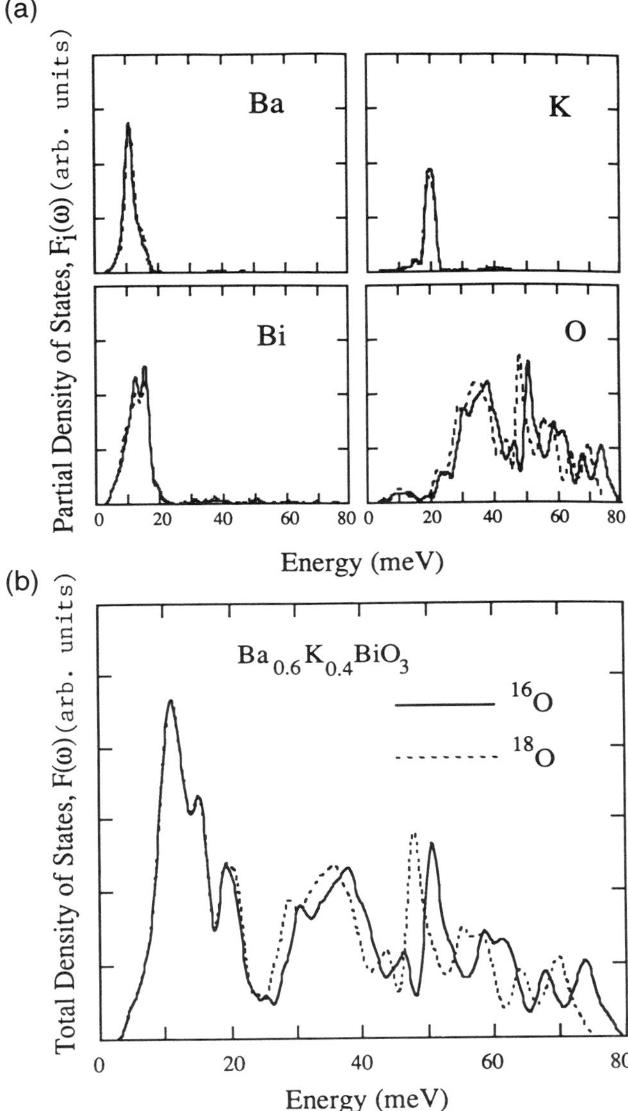

Figure 15.47 $Ba_{0.6}K_{0.4}BiO_3$ partial phonon density of states calculated for the atoms Ba, K, Bi, and O (upper panels), and the total density of states (lower panel). Isotopic substitution of ^{18}O for ^{16}O shifts the oxygen partial DOS and the total DOS in the oxygen region, but does not affect the Ba, K, or Bi partial DOS curves (Loong *et al.*, 1991, 1992).

seen in the total DOS at around 11 and 15 meV, while the more spread-out region beyond 20 meV arises from the oxygen atoms. The phonon density of states reported here is analogous to the more familiar electron density of states discussed at length in Chapter 8.

We see from these figures that the replacement of ^{16}O by ^{18}O shifts the phonon DOS frequencies to lower values. This shift gives an isotope effect exponent of $\alpha = 0.42$, which is close to the two values of $\alpha = 0.35$ and $\alpha = 0.41$ obtained from the variation of T_c (Hinks *et al.*, 1988b).

Phonons $\hbar\omega$ are also capable of probing the phonon spectrum by inelastic scattering. This is monitored by measuring the frequency shifts,

$$\hbar\omega = \hbar\omega' \pm \hbar\omega_{ph}, \qquad (15.25)$$

and scattering angles. When the emitted or absorbed phonon $\hbar\omega_{ph}$ is acoustic, the process is called *Brillouin scattering*, and when it is optical, it is referred to as *Raman scattering*.

VII. POSITRON ANNIHILATION

In positron annihilation spectroscopy (PAS), a sample is irradiated by a radioactive source, such as ^{22}NaCl, which emits high-energy (545-keV) electrons with positive charges e^+, called *positrons* (Benedek and Schüttler, 1990, Chakraborty, 1991). When the positron enters the solid, it rapidly loses most of its kinetic energy and approaches thermal energy, $\approx \frac{3}{2}k_BT \approx 0.04$ eV, in 0.001 to 0.01 ns. Following thermalization, the positron diffuses like a free particle, although its motion is correlated with nearby conduction electrons, until it encounters an electron e^- and annihilates in about 0.1 ns, producing two 0.51-MeV gamma (γ) rays in the process

$$e^+ + e^- \rightarrow \gamma + \gamma. \qquad (15.26)$$

The electron has much more momentum than the positron, and momentum balance causes the two gamma rays, to make a slight angle with respect to each other as they depart in opposite directions. The Angular Correlation of this Annihilation Radiation (ACAR) is one of the important parameters which is measured in this technique. The positron lifetime, τ, is determined by the time delay between the 1.28-MeV gamma ray emitted by the radioactive ^{22}Na simultaneously with the positron, and the pair of 0.51-MeV gamma rays produced by the annihilation event. The emitted gamma rays have a spread in energy due to Doppler broadening. The positrons can become trapped in vacancies before annihilation, with oxygen vacancies the likely trapping sites in high-temperature superconductors. A positron is sensitive to the details of the local electronic environment, which are reflected in its mean lifetime τ, its angular correlation, and its Doppler broadening parameters S and W. In $YBa_2Cu_3O_{7-\delta}$ there is a lifetime $\tau_1 \approx 0.2$ ns due to a shoft-lived component, perhaps from annihilation in the grains, and a lifetime $\tau_2 \approx 0.7$ ns of a long-lived component, perhaps from annihilation at grain surfaces. These parameters exhibit discontinuities at the transition temperature (Barbiellini *et al.*, 1991; Huang *et al.*, 1988; McMullen, 1990; Tang *et al.*, 1990; Wang *et al.*, 1988). Figure 15.48 shows four of these discontinuities for the superconductor $YBaCu_3O_{7-\delta}$ with a midpoint $T_c = 85.7$ K. One theoretical study has suggested that BCS pairing could be responsible for the measured shifts in positron properties near T_c (Benedek and Schüttler, 1990).

The positron annihilation characteristics are determined by the overlap of the positron and electron densities (Bharathi *et al.*, 1990; Sundar *et al.*, 1990b). Figure 15.49 shows the positron densities in the [020] vertical plane of the three $Tl_2Ba_2Ca_nCu_{n+1}O_{2n+6}$ superconductors 2201, 2212, and 2223. In the 2201 compound, the positron density is quite generally spread out, while in the other two compounds it is concentrated within the sets of copper-oxide layers, especially between the layers where the calcium atoms are located. The lack of concentration in the CuO_2 layers in the former case is consistent with the electron density plot for the 2201 Tl compound that is presented in Fig. 7.27. In contrast to the situation in the hole-type thallium superconductors, the positron density is found to be fairly generally distributed throughout the unit cell of the electron superconductor $(Nd_{0.925}Ce_{0.075})_2CuO_{3.98}$ (Sundar *et al.*, 1990a). The BiSrCaCuO superconductor (Sundar *et al.*, 1991), as well as K_3C_{60} (Lou *et al.*, 1992), has also been studied.

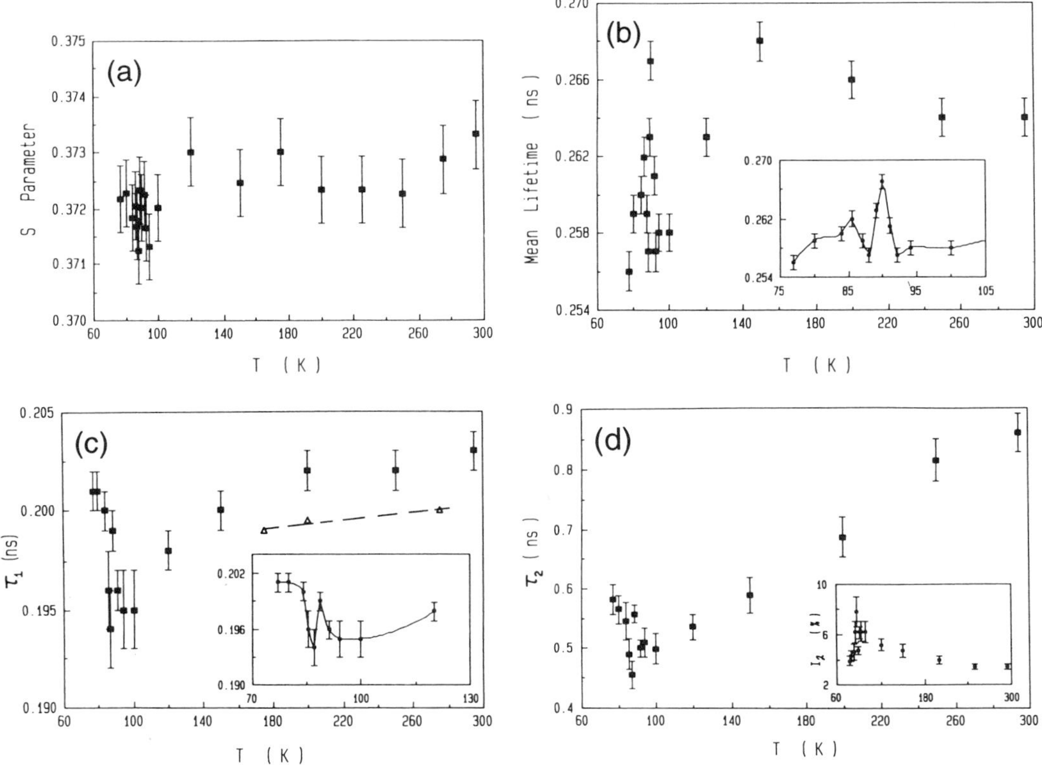

Figure 15.48 Temperature dependence of positron annihilation results obtained with a $YBa_2Cu_3O_{7-\delta}$ sample showing (a) Doppler broadening line shape parameter S, (b) mean lifetime τ, (c) lifetime τ_1 of short-lived component, and (d) lifetime τ_2 of long-lived component. The insets of (b) and (c) show data on an expanded scale. The inset of (d) shows the temperature dependence of the relative intensity of the long-lived component. The dashed curve of (c) represents "delayed" data taken 40 hours later. The curves are drawn as visual aids (Wang *et al.*, 1988).

The upper half of each positron density plot of Fig. 15.49 shows the Ba of the Ba–O and the O of Tl–O; the lower halves show the Tl of the Tl–O and the O of Ba–O. A comparison with the unit cells of Figs. 7.25 and 7.26 shows that this is in accord with the atom positions.

A two-dimensional angular correlation technique, called 2D-ACAR, is designed to sample the anisotropy of the conduction electron motion, thus providing information on the topology of the Fermi surface (Barbiellini *et al.*, 1991; Rozing *et al.*, 1991). For example, Bansil *et al.* (1991) published plots of Fermi surface sheets of $YBa_2Cu_3O_7$ similar to some of those presented in Fig. 8.34, and Tanigawa *et al.* (1988) provided three-dimensional sketches of the first Brillouin zone of $La_2CuO_{4-\delta}$, a zone that exhibits electron regions at the point Γ similar to the regions in the upper part of Fig. 8.40. 2D-ACAR studies have been reported for single crystals of $YBa_2Cu_3O_{6.9}$ (Smedskjaer *et al.*, 1992).

VIII. MAGNETIC RESONANCE

Another branch of spectroscopy that has provided valuable information on superconductors is magnetic resonance, the study of microwave and radio frequency transitions. We will comment on several types of magnetic resonance, including nu-

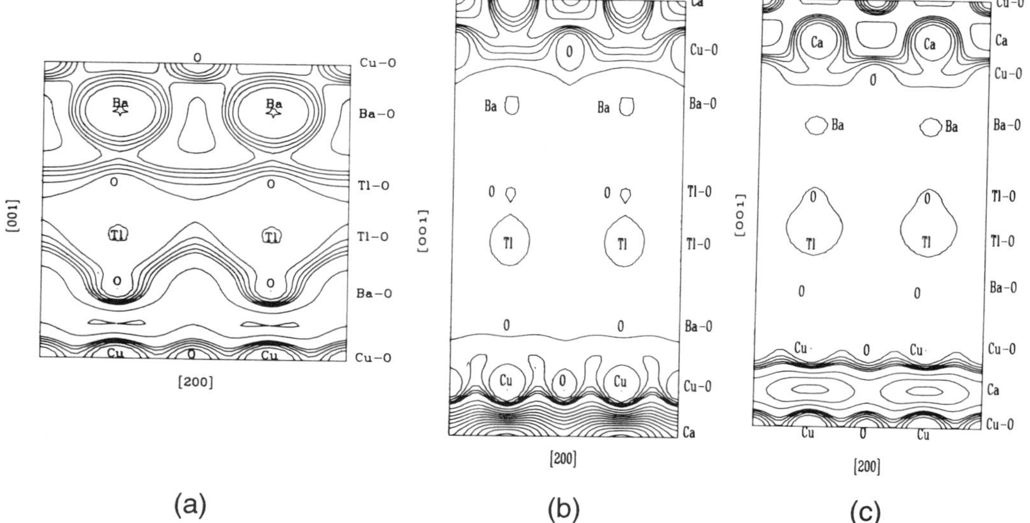

Figure 15.49 Contour plots of the positron density distribution in the [020] vertical plane of (a) $Tl_2Ba_2CuO_6$, (b) $Tl_2Ba_2CaCu_2O_8$, and (c) $Tl_2Ba_2Ca_2Cu_3O_{10}$ crystals. See Fig. 7.27 for corresponding charge density plots of $Tl_2Ba_2CuO_6$ (Sundar *et al.*, 1990b).

clear magnetic resonance (NMR), nuclear quadrupole resonance (NQR), electron-spin resonance (ESR or EPR), microwave absorption, muon spin resonance (μSR), and Mössbauer resonance, all of which have been used to study superconductors, and we will discuss some of the results that have been obtained.

Magnetic resonance measurements are made in fairly strong magnetic fields, typically ≈ 0.33 T for ESR and ≈ 10 T for NMR, which are considerably above the lower-critical field B_{c1} of a high-temperature superconductor. At these fields most of the external magnetic flux penetrates into the sample, so that the average value of B inside is not very different from the value of B outside.

A. Nuclear Magnetic Resonance

Nuclear magnetic resonance involves the interaction of a nucleus possessing a nonzero nuclear spin I with an applied magnetic field B_{app}, giving the energy level splitting into $2I + 1$ lines with energies

$$E_m = \hbar \gamma B_{app} m, \qquad (15.27)$$

where γ is the gyromagnetic ratio, sometimes called the magnetogyric ratio, characteristic of the nucleus and m assumes integer or half-integer values in the range $-I < m < I$, depending on whether I is an integer or a half-integer (Poole and Farach, 1987). Figure 15.50 shows the energy levels and the NMR transition for the case $I = \frac{1}{2}$, $m = \pm \frac{1}{2}$. Typical NMR frequencies range from about 60 to 400 MHz. Several nuclei common to superconductors are listed in Table 15.4 together with their spins, natural abundances and other characteristics. The isotopes of Tl and Y are particularly favorable for NMR because they have nuclear spin $I = 1/2$, so that they lack a quadrupole moment and their lines are not broadened by noncubic crystalline electric fields. The dominant isotope of oxygen, ^{16}O, which is 99.76% abundant, has $I = 0$, so that it does not exhibit NMR. Zero-spin nuclei are not listed in the table.

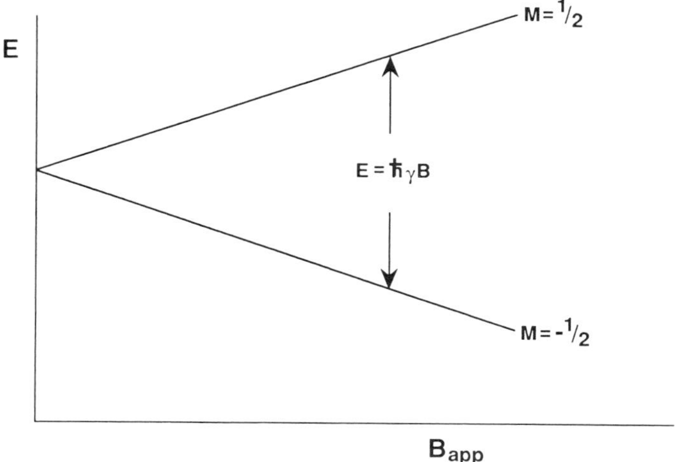

Figure 15.50 Zeeman splitting of a spin-$\frac{1}{2}$ energy state in a magnetic field.

The importance of NMR arises from the fact that the value of γ is sensitive to the local chemical environment of the nucleus. It is customary to report the chemical shift δ,

$$\delta = \frac{\gamma - \gamma_R}{\gamma_R}, \qquad (15.28)$$

which is the extent to which γ deviates from γ_R, the value of a reference sample, where, for photon reference samples, $\gamma_R/2\pi$ is close to 42.576 MHz/T. Chemical shifts are small, and are usually reported in parts per million (ppm). In addition, spin–spin interactions with neighboring nuclei can split the line into a multiplet, providing further information on the coordination to surrounding atoms. Relaxation-time measurements determine the efficiency of spin-energy transfer to the lattice (Poole and Farach, 1971).

Pulsed NMR of ^{89}Y nuclei has been observed in $YBa_2Cu_3O_{7-\delta}$ at 12.2 MHz and 5.9 T in the temperature range from 59 to 295 K (Mali *et al.*, 1987; Markert *et al.*, 1987). The value of $T_c = 86$ K at 5.9 T was determined by the onset of line broadening from a width of 0.31 mT above T_c to 0.71 mT ten degrees below T_c. This broadening arises from the spatial variation in the internal field, as sketched at the top of Fig. 9.10, which causes each ^{89}Y nucleus to experience a slightly different local field. The fraction of ^{89}Y detected decreased from 100% above T_c to about 80% at 59 K due to incomplete rf penetration in the mixed state. The spin-lattice relaxation time T_1 increased below T_c. Preparation conditions influence the Y site, since different ^{89}Y chemical shifts have been observed under different conditions (slowly cooled, rapidly cooled, or water-exposed $YBa_2Cu_3O_{7-\delta}$).

Most NMR studies are carried out with the isotope ^{63}Cu (nuclear spin $I = 3/2$) since it is 69% abundant. Figure 15.51 presents the ^{63}Cu NMR spectra obtained at 100 K for the applied field parallel to c and in the a, b-plane. The resonances attributed to the four-coordinated chain Cu(1) sites and to the five-coordinated plane Cu(2) sites are indicated. Nuclei in metals have their frequency ν_m shifted in position relative to its value ν_i in a diamagnetic insulator by their nuclear spin interaction with the spin paramagnetism of the conduction electrons and the relative frequency shift $K = (\nu_m - \nu_i)/\nu_i$ is called

Table 15.4 NMR Data on Nuclei Commonly Found in High-temperature Superconductors[a]

Z	A	Elem	*I*	% Abund	Mag Mon[b]	MHz / T[c]	Sensit / B[d]	Sensit / f[e]	eqQ[f]
1	1	H	1/2	99.985	2.79268	42.5759	1.000	1.000	0
1	2	D	1	0.015	0.85739	6.5357	0.00965	0.409	0.0029
6	13	C	1/2	1.108	1.216	10.705	0.016	2.51	0
8	17	O	5/2	0.037	−1.8930	−5.7719	0.0291	1.58	−0.26
19	39	K	3/2	93.08	0.39094	1.987	0.0005	0.233	0.049
20	43	Ca	7/2	0.145	−1.3153	−2.8646	0.0640	1.41	−0.065
29	63	Cu	3/2	69.09	2.2206	11.285	0.0931	1.33	−0.209
29	65	Cu	3/2	30.91	2.3790	12.090	1.14	1.42	−0.195
38	87	Sr	9/2	7.02	−1.0893	1.845	0.00269	1.43	0.15
39	89	Y	1/2	100.0	−0.13682	2.086	0.000118	0.0005	0
41	93	Nb	9/2	100.0	6.1435	10.407	0.482	8.07	−0.36
56	135	Ba	3/2	6.59	0.83229	4.230	0.0049	0.497	0.18
56	137	Ba	3/2	11.32	0.93107	4.732	0.00686	0.556	0.28
57	139	La	7/2	99.911	2.7615	6.014	0.0592	2.97	0.22
60	143	Nd	7/2	12.20	−1.25	2.72	0.00549	1.34	−0.48
60	145	Nd	7/2	8.30	−0.78	1.7	0.00133	0.838	−0.25
80	199	Hg	1/2	16.9	0.498	7.60	0.0057	0.178	0
81	203	Tl	1/2	29.5	1.5960	24.332	0.187	0.571	0
81	205	Tl	1/2	70.5	1.6115	24.570	0.192	0.577	0
82	207	Pb	1/2	22.1	0.5837	8.899	0.00913	0.209	0
83	209	Bi	9/2	100.0	4.0389	6.842	0.137	5.30	−0.46

[a] The nucleus ^{16}O (99.8%) has no nuclear spin ($I = 0$) and thus cannot be observed. Data from Harris (1981); see also Emsley, Feeney, and Sutcliffe (1965), and Poole and Farach (1994).

[b] Magnetic moment in units of nuclear magneton.

[c] Resonant frequency for a field of 1 T in units of MHz.

[d] Relative sensitivity at constant field.

[e] Relative sensitivity of constant frequency.

[f] Quadrupole moment eqQ in units of 10^{-24} cm^2. Data from Landolt–Börnstein, New Series III/20a, 1988.

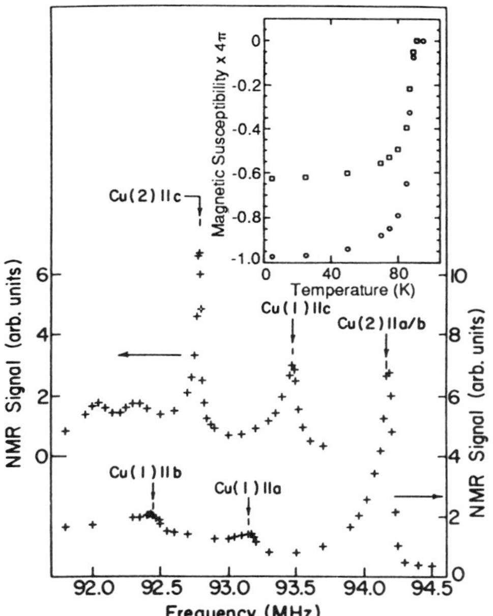

Figure 15.51 NMR spectrum of ^{63}Cu in $YBa_2Cu_3O_{7-\delta}$ at 100 K with the applied magnetic field parallel to the *c*-axis (above) and with the applied field in the *a*, *b*-plane (below). The resonances attributed to Cu(1) in the chains and to Cu(2) in the planes are indicated. The inset shows the magnetization for zero field cooling (open circles) and field cooling in 1.6 mT (open squares), with the sharp superconducting transition evident (Barrett *et al.*, 1990).

the Knight shift (Lane, 1962; Pennington and Slichter, 1990). When normal conduction electrons convert to super electrons as the temperature is lowered in the range below T_c the Knight shift K is expected to decrease. Figure 15.52 shows this decrease for Cu(1) and Cu(2) nuclei in the temperature range from 0 to 120 K. These shifts were found to obey BCS expressions for a strong coupling-spin singlet state (Barrett *et al.*, 1990). NMR of ^{63}Cu provided the energy-gap ratio $E_g/k_B T_C = 1.3$ in

$$(La_{0.915}Sr_{0.085})_2CuO_{4-\delta}$$

(Lee *et al.*, 1987).

Several high-temperature superconductors enriched with the rare isotope ^{17}O, which has nuclear spin $I = 5/2$, have been studied by NMR. The broad-scan room-temperature spectrum of $YBa_2Cu_3O_{7-\delta}$ presented in Fig. 15.53 exhibits 20 lines from the various oxygens and these are identified in the caption. The use of aligned grains considerably increased the resolution of this spectrum, indicating a considerable amount of anisotropy. The narrower scans of Fig. 15.54 show that the

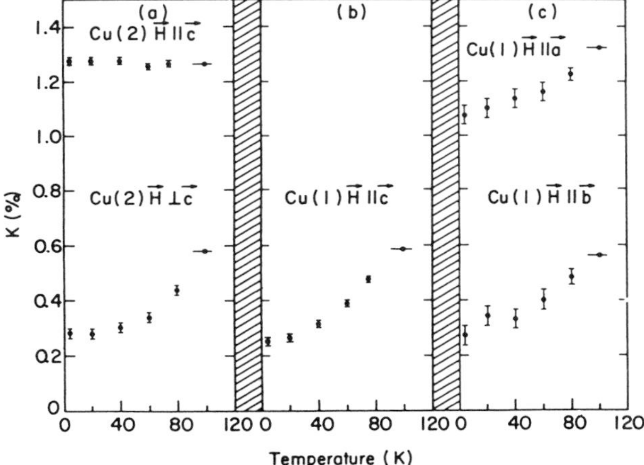

Figure 15.52 Temperature dependence below T_c of the five NMR signals of Fig. 15.51 arising from ^{63}Cu in the planes and chains with the applied field along the *a*-, *b*-, and *c*-axes, as indicated (Barrett *et al.*, 1990).

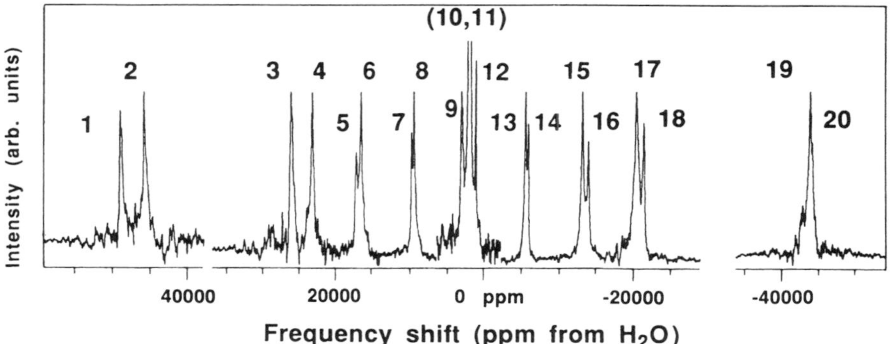

Figure 15.53 Room-temperature ^{17}O NMR spectra at 48.8 MHz (8.45 T) of $YBa_2Cu_3O_{7-\delta}$ magnetically aligned in a field parallel to the *c*-axis. The measured relative intensities for central and satellite transitions have the expected 9:8:5 ratio, but here the peak intensities have been equalized for clarity. All but one of the 20 expected transitions, five lines from each of the four oxygens, are shown: peaks 2, 4, 12, 18, and 19 from O(1), peaks 5–8, (10, 11), 13–16 from O(2, 3), and peaks 1, 3, 9, 17, and 20 from O(4) (Oldfield *et al.*, 1989).

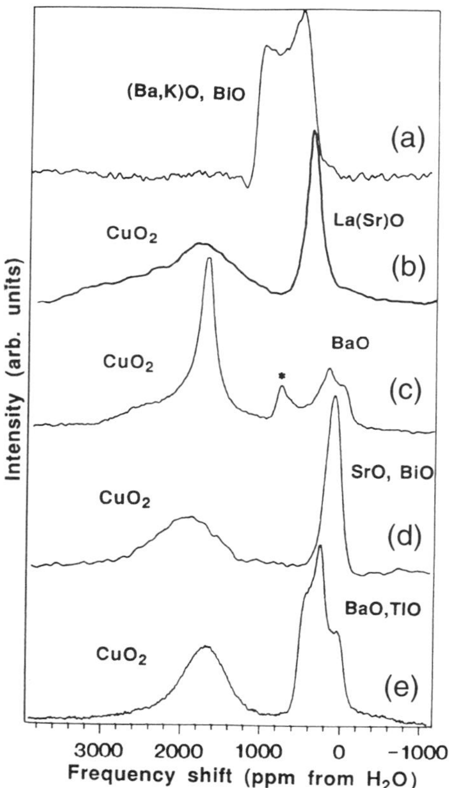

Figure 15.54 Room-temperature ^{17}O NMR spectra at 67.8 MHz (11.7 T) of (a) $(Ba_{0.6}K_{0.4})BiO_3$, (b) $(La_{0.925}Sr_{0.075})_2CuO_4$, (c) $YBa_2Cu_3O_{7-\delta}$, (d) $Bi_2Sr_2CaCu_2O_{8+\delta}$, and (e) $Tl_2Ba_2CaCu_2O_{8+\delta}$. The * line in (c) arises from O(1) sites in a small population of aligned crystallites, which also contribute to the absorption at 18 ppm (Oldfield *et al.*, 1989).

compounds

$$(La_{0.925}Sr_{0.075})_2CuO_4,$$

$$Bi_2Sr_2CaCu_2O_{8+\delta},$$

and $Tl_2Ba_2CaCu_2O_{8+\delta}$, all of which have similar structures (cf. Chapter 7), exhibit similar spectra. These spectra differ from those of the compounds $(Ba_{0.6}K_{0.4})BiO_3$ and $YBa_2Cu_3O_{7-\delta}$, which have different structures. This result is to be expected, since NMR probes the local environment of the nucleus.

NMR spectroscopy has been instrumental in confirming the structures of the fullerenes, such as C_{60} and C_{70}. The room-temperature ^{13}C NMR spectrum of C_{60}, shown at the top of Fig. 15.55, is a single narrow line with a chemical shift of 143 ppm relative to the standard compound tetramethylsilane (TMS), confirming the equivalence of all of the carbons as well as demonstrating that the molecule is rapidly and isotropically reorientating. We see from the figure that when the molecule is cooled, the NMR line broadens. At 77 K its spectrum is a typical asymmetric chemical shift pattern with the principal values 220, 186, and 25 ppm, which are typical of aromatic hydrocarbons. This suggests that the molecules are now stationary and randomly oriented in the solid. The chemical shift tensor is expected to have one princi-

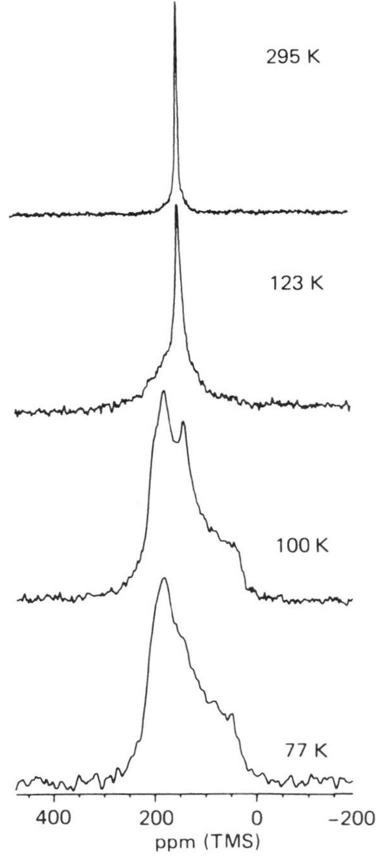

Figure 15.55 Temperature dependence of the 15-MHz ^{13}C NMR spectrum of C_{60}. The single narrow line at room temperature shows that all of the carbons are equivalent. The sequence of spectra suggests rapid reorientation at room temperature and the lack of rotational motion at liquid nitrogen temperature on the NMR timescale of $\tau \approx 0.1$ ms (R. D. Johnson *et al.*, 1992).

pal value in the direction perpendicular to the approximate plane of the sp^3 hybrid CC_3 group. Within this plane the three C–C bonds are not equivalent, since two of them connect a five-membered and a six-membered ring, whereas the third connects two six-membered rings, thereby explaining the lack of axial symmetry in the chemical-shift powder pattern.

The fullerene C_{70} has the five inequivalent carbons labeled a, b, c, d, and e on the left side of Fig. 15.56, giving rise to five lines in the ^{13}C NMR spectrum displayed at the top of Fig. 15.56. These lines have the respective intensity ratios 10 : 10 : 20 : 20 : 10, corresponding to the numbers of their respective carbon atoms in the C_{70} molecule. The two-dimensional spectrum given in the figure provides the measured spin–spin coupling constants between the carbons. The C–C bond lengths of C_{60} and C_{70} determined by NMR agreed with those deduced from crystallographic studies.

The ^{13}C NMR of alkali metal-doped fullerenes, such as K_xC_{60}, which are both conducting and superconducting, exhibit a second narrow ^{13}C resonance at 186 ppm in addition to the usual resonance at 143 ppm. This resonance appears for $0 < x < 3$ and arises from K_3C_{60} molecules with the K^+ ions at interstitial sites adjacent to the C_{60}^{3-} ions. The C_{60}^{3-} ions rotate rapidly at room temperature to average out the chemical shift anisotropy. Thus K_xC_{60} constitutes a two-phase system. The chemical shift is identified with a Knight shift arising from hyperfine coupling between the ^{13}C nuclei and the conduction electrons (Tycko *et al.*, 1991, 1992).

Some relevant articles on NMR are: ^1H (DeSoto *et al.*, 1993; Le Dang *et al.*, 1989; Maniwa *et al.*, 1991b), $^9Be(3/2)$ (Tien and Jiang, 1989), ^{13}C (Antropov *et al.*, 1993 (t)), $^{17}O(5/2)$ (Asayama *et al.*, 1991; Coretsopoulos *et al.*, 1989; Howes *et al.*, 1991; Reveu *et al.*, 1991; Trokiner *et al.*, 1990, 1991), $^{63,65}Cu(3/2)$ (Horvatic *et al.*, 1993; Millis and Monien, 1992; Millis *et al.*, 1990 (t), Walstedt *et al.*, 1990, 1992), ^{89}Y (Alloul *et al.*, 1993; Barrett *et al.*, 1990; Carretta and Corti, 1992; Carretta *et al.*, 1992; Millis and Monien, 1992; Millis *et al.*, 1990 (t)), $^{139}La(7/2)$ (Hammel *et al.*, 1990), $^{203,205}Tl$ (Fujiwara *et al.*, 1991; Kitaoka *et al.*, 1991; Song *et al.*, 1991a). Articles on NMR relaxation include $^{17}O(5/2)$ (Barrett *et al.*, 1991; Hammel *et al.*, 1989; Takigawa *et al.*, 1991a), $^{63,65}Cu(3/2)$ (Anikenok *et al.*, 1991; Borsa *et al.*, 1992; Martindale *et al.*, 1992; Mila and Rice, 1989 (t); Pennington *et al.*, 1989; Reyes *et al.*, 1991; Taki-

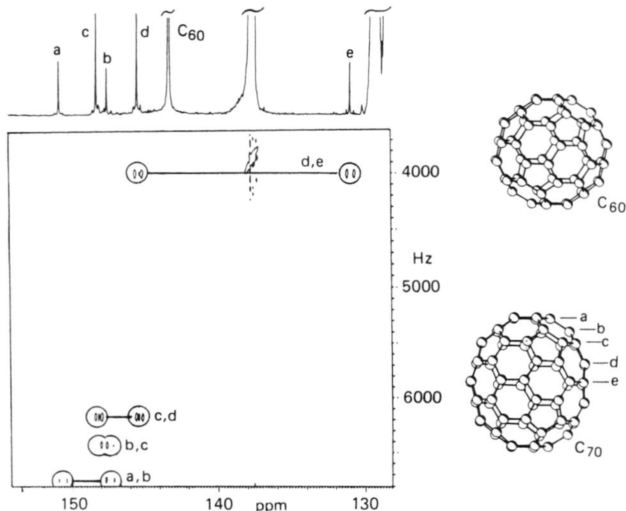

Figure 15.56 The upper trace is the 125.7-MHz ^{13}C NMR spectrum of a ^{13}C enriched mixture of C_{60} and C_{70}. The C_{60} line and the five C_{70} lines labeled a, b, c, d, and e with the respective relative intensities 10 : 10 : 20 : 20 : 10 are indicated. The two-dimensional spectrum presented on the lower left shows doublets arising from the various bonded carbon pairs. Reprinted by permission from R. D. Johnson *et al.*, 1992. Copyright (1992) by the American Chemical Society.

gawa *et al.*, 1991b; Walstedt *et al.*, 1991), ^{89}Y (Adrian, 1988, 1989; Alloul *et al.*, 1989; Z. P. Han *et al.*, 1991, 1992), $^{141}Pr(5/2)$ (Teplov *et al.*, 1991), ^{169}Tm (Bakharev *et al.*, 1991; Teplov *et al.*, 1991), ^{195}Pt (Vithayathil *et al.*, 1991), and $^{203,205}Tl$ (Lee *et al.*, 1989; Nishihara *et al.*, 1991; Song *et al.*, 1993). (Theory and calculation articles are indicated by (t); the nuclear spin is given when it is not $\frac{1}{2}$.)

B. Quadrupole Resonance

A nucleus with spin $I > \frac{1}{2}$ has an electric quadrupole moment. Several such nuclei are listed in Table 15.4. The crystalline electric fields at an atomic site with symmetry less than cubic split the nuclear-spin levels in a manner that depends on the site symmetry, and the spacings between the levels are measured experimentally by nuclear quadrupole resonance (NQR). The frequencies used for making these measurements are similar to those employed for NMR. Table VI-14 of our earlier work (Poole *et al.*, 1988) lists the point symmetries for the occupied atomic sites in some of the high-temperature superconductors. Babu and Remakrishna (1992) reviewed the NQR of superconductors.

The ^{139}La NQR spectrum of the prototype compound La_2CuO_4 in zero magnetic field at 1.3 K is shown in Fig. 15.57. It has five main lines from 2.4 to 19.3 MHz, arising from the five $m \to m'$ transitions $-\frac{1}{2} \to \frac{1}{2}$, $+\frac{1}{2} \to \pm\frac{3}{2}$, $-\frac{1}{2} \to \pm\frac{3}{2}$, $\pm\frac{3}{2} \to \pm\frac{5}{2}$, and $\pm\frac{5}{2} \to \pm\frac{7}{2}$ of the $I = \frac{7}{2}$ ^{139}La nucleus. Additional doublet splittings are caused by internal magnetic fields that arise from the magnetic ordering of the copper ions occurring below 240 K (Kitaoka *et al.*, 1987a). The doublet splittings are not resolved in the barium- and strontium-substituted compounds, as shown in Fig. 15.58, suggesting that the internal magnetic fields decrease with al-

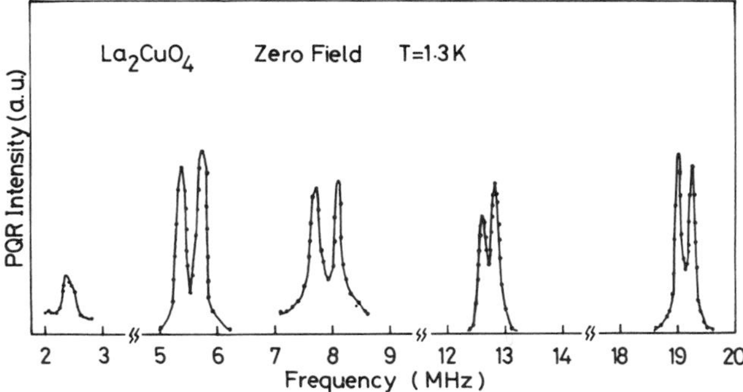

Figure 15.57 Nuclear quadrupole resonance spectrum of ^{139}La in La_2CuO_4 in zero field at 1.3 K. Reprinted by permission from Kitaoka *et al.*, 1987a. Copyright (1987) American Chemical Society.

kaline earth doping. The internal field parallel to c is about 35 mT for low barium contents ($\approx 1\%$) in the superconducting region (Kitaoka *et al.*, 1987b). The electric field gradient at the La site also changes on passing from the normal to the superconducting state (Watanabe *et al.*, 1989). Cho *et al.* (1992) used ^{139}La NQR relaxation to study magnetic ordering in $(La_{1-x}Sr_x)_2CuO_4$.

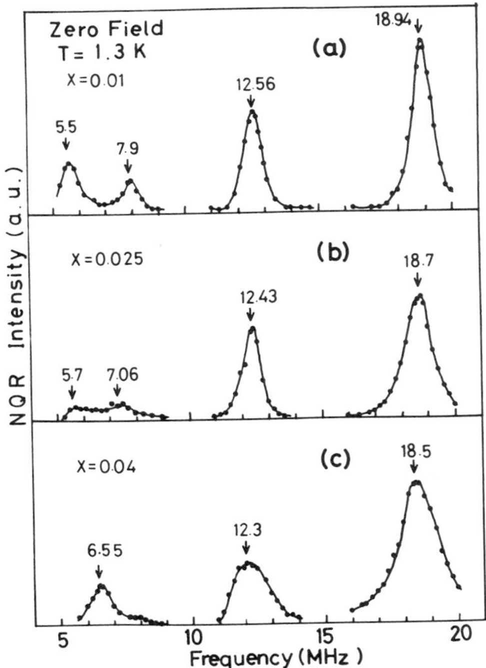

Figure 15.58 Nuclear quadrupole resonance spectrum at 1.3 K of ^{139}La in $(La_{1-x}Ba_x)_2CuO_4$ for (a) $x = 0.01$, (b) $x = 0.025$, and (c) $x = 0.04$. The calculated resonant frequencies are indicated by arrows (Kitaoka *et al.*, 1987a).

The room-temperature ^{63}Cu NQR spectrum of $YBa_2Cu_3O_x$ presented in Fig. 15.59 consists of one line at 22.1 MHz arising from Cu(1) in the chains and another at 31.2 MHz arising from Cu(2) in the basal plane (Vega *et al.*, 1989a). The ^{65}Cu isotope produces NQR lines shifted 6.7% lower in frequency; these are not shown. The symmetry was found to be close to axial for Cu(2), deviating considerably from axial for Cu(1), as would be expected from an examination of the structural drawings in Figs. 7.8, 7.10, and 7.11. We see from Fig. 15.59 that the linewidth strongly depends on the oxygen content. The sharpest line occurs in the stoichiometric compound $YBa_2Cu_3O_7$. Removal of oxygen lowers the symmetries of the two sites, broadening the lines and shifting them toward each other. This means that oxygen is being removed adjacent to both sites.

When the temperature of the sample is gradually lowered from room temperature to 20 K, the ^{63}Cu(1) resonance decreases in frequency by 0.5% while the ^{63}Cu(2) line increases in frequency by 1.1% (Mali *et al.*, 1987), as shown in Fig. 15.60.

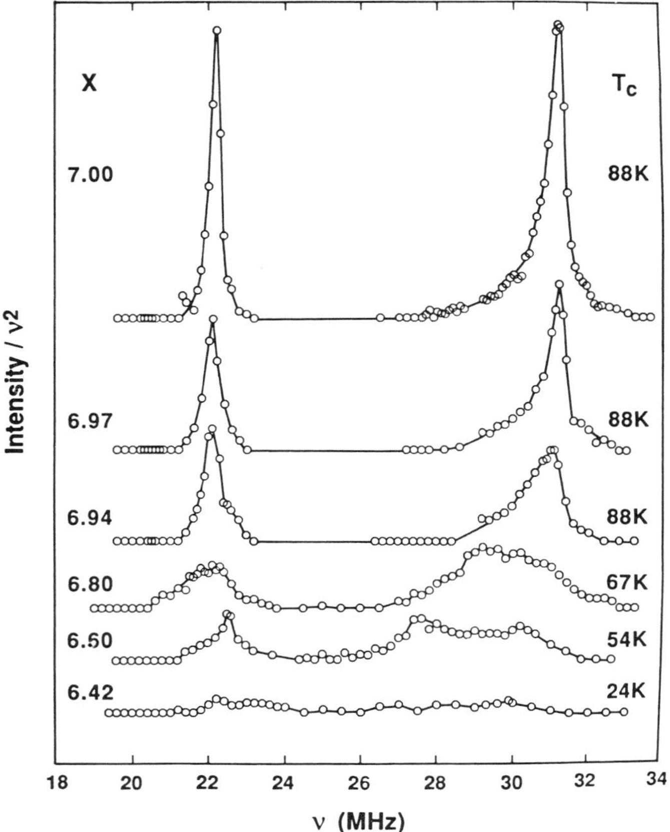

Figure 15.59 Short T_1 components of the NQR spectrum of ^{63}Cu in high-temperature quenched $YBa_2Cu_3O_x$ with the indicated x and T_c values. The 22.1 MHz line arises from Cu(1) in the chains, while the 31.2 MHz line is from Cu(2) in the planes (Vega *et al.*, 1989a).

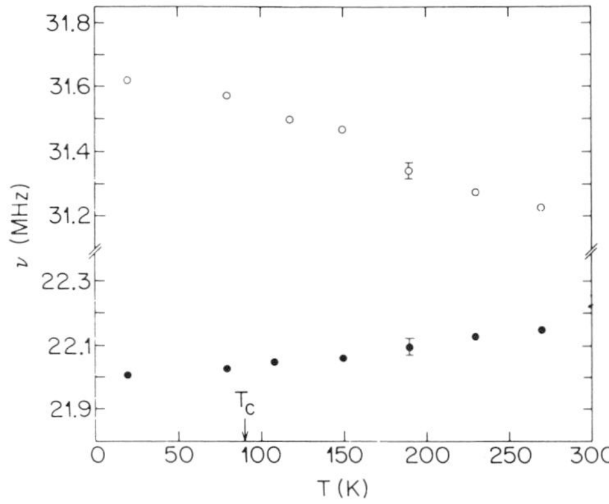

Figure 15.60 Temperature dependence of the ^{63}Cu nuclear quadrupole frequencies arising from Cu(1) (●) in the chain sites and Cu(2) (○) in the planar sites of $YBa_2Cu_3O_{7-\delta}$ (Mali *et al.*, 1987).

The variation in the electric field gradients at the two Cu sites can be accounted for by lattice compression. There is no discontinuity at the transition temperature.

Relevant NQR articles for several nuclei are ^{17}O (Sahoo *et al.*, 1990), $^{63,65}Cu$ (Carretta *et al.*, 1992; Fujiwara *et al.*, 1991; Ishida *et al.*, 1991; Kitaoka *et al.*, 1991; Pennington *et al.*, 1988, 1990; Pieper, 1992; Reyes *et al.*, 1990; Saul and Weissmann, 1990; Song *et al.*, 1991b; Sulaiman *et al.*, 1991; Vega *et al.*, 1989b), $^{69,71}Ga$ (Pieper, 1992), ^{135}Ba (Sulaiman *et al.*, 1992), ^{139}La (Song and Gaines, 1991; Sulaiman *et al.*, 1992), ^{141}Pr (Erickson, 1991).

C. Electron-Spin Resonance

Electron-spin resonance (ESR) detects unpaired electrons in transition ions, especially those with odd numbers of electrons, such as Cu^{2+} ($3d^9$) and Gd^{3+} ($4f^7$). Free radicals, like those associated with defects or radiation damage, can also be detected. The Zeeman energy level diagram of Fig. 15.50 also applies to ESR, except that the energies or resonant frequencies are three orders of magnitude higher for the same magnetic field. A different notation is employed for the energy,

$$E_m = g\mu_B B_{app} m, \qquad (15.29)$$

where μ_B is the Bohr magneton and g is the dimensionless g-factor; g has the value 2.0023 for a free electron. Equations (15.27) and (15.29) are related through the expression $g\mu_B = \hbar\gamma$ (Poole, 1983; Poole and Farach, 1987).

Some oxide superconductors exhibit an ESR signal, with g in the range from ≈ 2.05 to ≈ 2.27, arising from the divalent copper ions. This signal does not appear in high-purity samples, so that its appearance indicates the presence of a nonsuperconducting fraction, such as the green-phase Y_2BaCuO_5 admixed with $YBa_2Cu_3O_{7-\delta}$. We say that the high-temperature superconductors are ESR silent so far as the Cu^{2+} signal is concerned (McKinnon *et al.*, 1987, 1988; Simon *et al.*, 1993).

The magnetic field inside a superconducting sample was probed by placing one free radical marker on the face of a sample normal to the magnetic field direction and another free radical marker on the face of the sample parallel to the external magnetic field (Bontemps *et al.*, 1991; Davidov *et al.*, 1992; Farach *et al.*, 1990; Frait *et al.*, 1988a, b; Koshta *et al.*, 1993; Maniwa *et al.*, 1990; Poole *et al.*, 1988; Rakvin *et al.*, 1989; Shvachko *et al.*, 1991). In the superconducting state the two markers experience different local magnetic fields, so that the resonant positions of the lines shift in the manner shown in Fig. 15.61. The observed shift occurs because the free radicals respond to the surface field, which differs from the applied field in accordance with Eq. (10.35). Thus the observed shift in line position is a measure of the magnitude of the internal field B_{in} within the sample. With this result we are able to determine the temperature dependence of the susceptibility, with the results presented in Fig. 15.62. A related NMR method of probing the surface measures proton signals in a silicone oil coating (Maniwa *et al.*, 1991a).

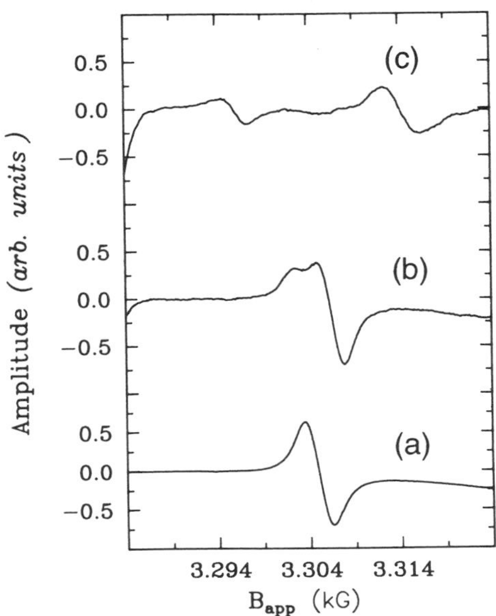

Figure 15.61 Shift of the ESR signals of paramagnetic markers located on the side and end of a $YBa_2Cu_3O_{7-\delta}$ sample from their superposed position (a) above T_c to different field positions (b, c) below T_c. The separation of the lines is proportional to the susceptibility (Farach *et al.*, 1990).

The ESR spectrum of the compound LaC_{82}, illustrated in Fig. 15.63, consists of an unresolved hyperfine octet. This is well resolved by dissolving the LaC_{82} in degassed 1,1,2,2-tetrachloroethane, as shown. The spectrum is interpreted as arising from an unpaired electron delocalized in the π-electron system of the triply negative fullerene anion $(C_{82})^{3-}$ and interacting with the La^{3+} inside (i.e. endohedral). The 99.9%-abundant ^{139}La nucleus has spin

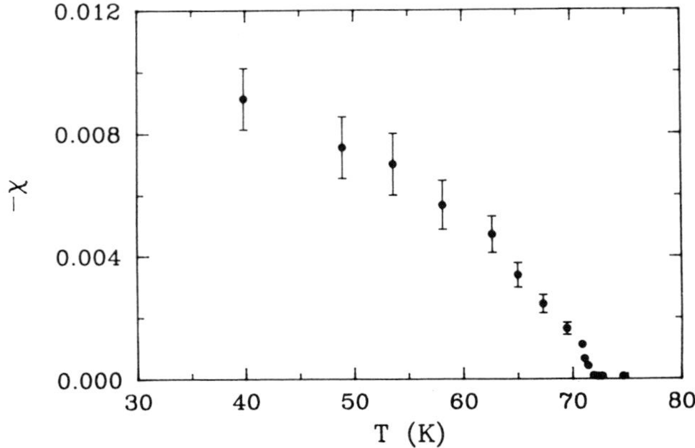

Figure 15.62 Temperature dependence of the susceptibility of $YBa_2Cu_3O_{7-\delta}$ determined by the ESR method of Fig. 15.61 (Farach *et al.*, 1990).

$I = 7/2$, which gives the $2I + 1 = 8$ observed hyperfine multiplet. The hyperfine coupling is only 0.125 mT, indicating that the interaction with the La nucleus is very weak.

Acrivos *et al.* (1994) used ESR dynamic measurements to compare the paramagnetic and antiferromagnetic properties of various superconducting oxides.

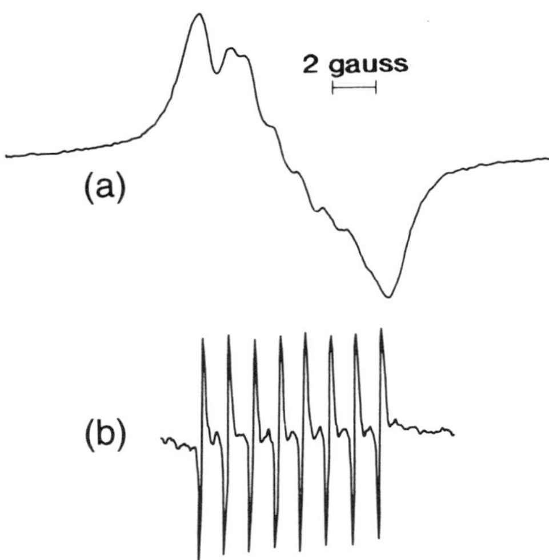

Figure 15.63 Electron-spin resonance spectrum of LaC_{82} mixed with C_{60} and C_{70} with the La inside the C_{82} fullerene cage. The poorly resolved octet (a) in the solid-state spectrum becomes well resolved (b) after the compound has been dissolved in degassed 1,1,2,2-tetrachloroethane solution. The linewidth is 12.5 μT and $g = 2.0010$ in the latter case (R. D. Johnson *et al.*, 1992).

D. Nonresonant Microwave Absorption

Below the transition temperature a superconductor has a microwave absorption signal that increases in amplitude as the temperature is lowered. There are often superimposed fluctuations that exhibit regularities, as shown in Fig. 15.64. These closely spaced oscillations have been attributed to Josephson junctions in the sample. Irradiating a Josephson junction with microwaves induces an oscillating voltage that depends on the microwave power and frequency. This phenomenon, called the inverse Josephson effect, was explained in Chapter 13, Section VII.E. If the magnetic field is scanned through zero to negative fields, the absorption exhibits a hysteresis, as shown in Fig. 15.65. The absorption is called nonresonant because it does not involve transitions between the Zeeman energy levels such as those which are characteristic of NMR and ESR absorption lines.

Xia and Stroud (1989) suggested that the absorption takes place in superconducting grains whose dimensions are small compared with the penetration depth and

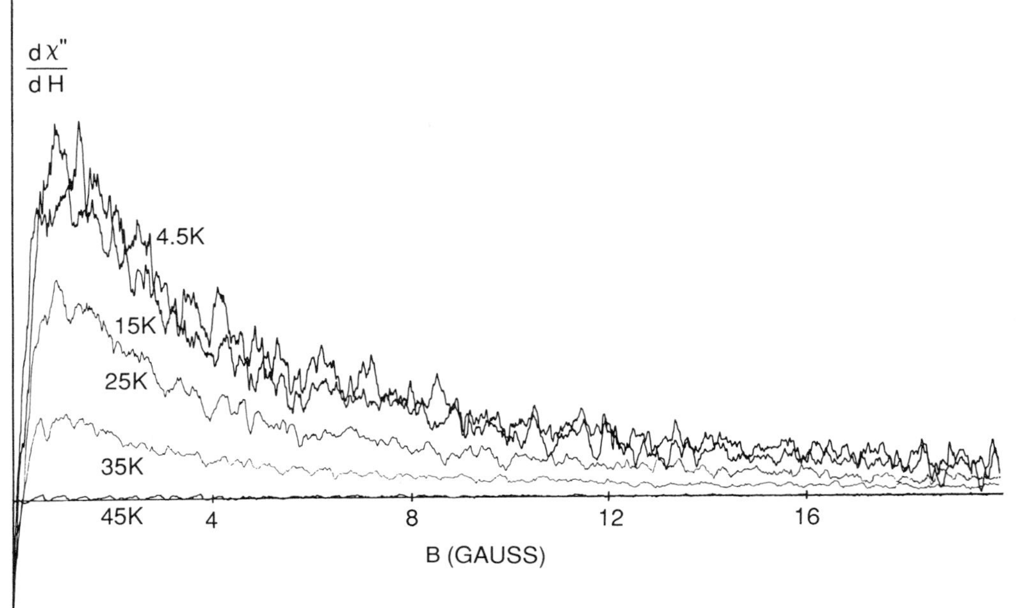

Figure 15.64 Low-field microwave absorption of $La_{2.8}Sr_{0.2}Cu_2O_7$ after field cooling at several temperatures in the range 4.5–45 K. Reprinted with permission from Blazey *et al.*, 1987. Copyright (1987) American Chemical Society.

which are coupled together in closed loops. Imperfect monocrystalline samples could also contain weakly linked loops. These loops support screening currents in response to an external magnetic field. The presence of a dc field perpendicular to the plane of the loop and an incident microwave field can cause phase slips via jumps from one energy state into another as the flux through the loop changes with time. The phase slip generates a voltage difference between neighboring grains, and hence leads to energy absorption.

Blazey *et al.* (1987) identified the field B_{max} in which the low-field absorption reaches a maximum as the field where flux slippage starts to occur. This phenomenon was used to estimate the average radius r_L of the superconducting loops,

$$\pi r_L^2 = \frac{\phi_0}{2B_{max}}. \qquad (15.30)$$

Various samples gave loop radii in the range 0.6–2.5 μm. Zero field cooled samples exhibit a minimum absorption at zero field; stored flux shifts this minimum in field cooled samples (Mzoughi *et al.*, 1992).

Microwave absorption measurements have been reported for most recent superconductors, such as

$$\kappa\text{-(BEDT-TTF)}_2\text{Cu(NCS)}_2$$

(Achkir *et al.*, 1993), YBaCuO (Dulčic *et al.*, 1989; Glarum *et al.*, 1990), BiSrCaCuO (Owens, 1992; Rakvin *et al.*, 1990), TlBaCaCuO (Owens, 1991; Pöppl *et al.*, 1992), and Rb- and K-doped C_{60} (Rosseinsky *et al.*, 1991; Zakhidov *et al.*, 1991).

E. Microwave Energy Gap

Energy gaps of superconductors with low transition temperatures, $T_c < 1$ K, occur in the microwave region. Since a temperature of 1 K is equivalent to 20.8 GHz, we can estimate from the BCS result $E_g/k_BT_c = 3.53$ that 1 K corresponds to an energy gap of ≈ 74 GHz, which is in the

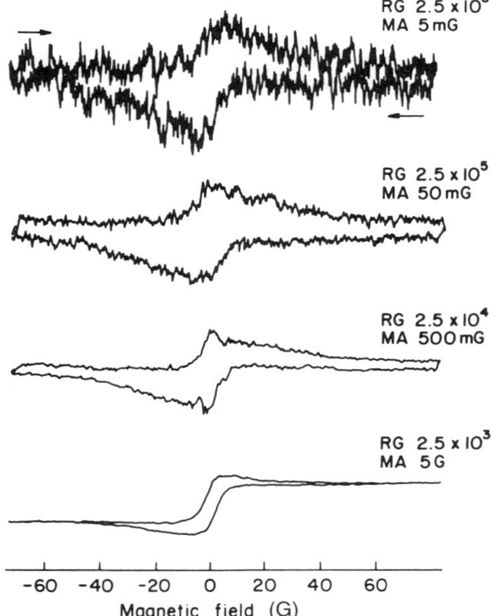

Figure 15.65 Hysteresis loops for microwave absorption of $YBa_2Cu_3O_{7-\delta}$ single crystals at 56 K cycled through zero field for four different modulation amplitudes (MA). Higher receiver gain (RG) settings were needed for the lower modulation amplitudes due to the decrease in sensitivity (Dulčic *et al.*, 1989).

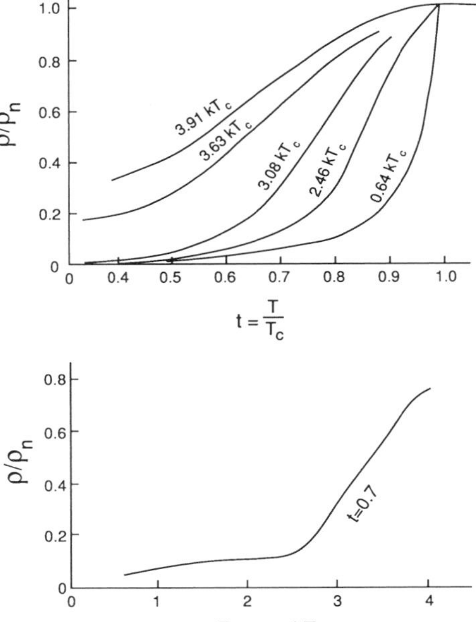

Figure 15.66 Temperature dependence of the normalized microwave resistivity ρ/ρ_n of aluminum (top) for a range of microwave frequencies, where ρ_n is the normal-state resistivity. Each curve is labeled with its equivalent k_BT_c value. The plot of the normalized resistivity versus k_BT_c (bottom) exhibits a break in the curve at the temperature $T = 0.7T_c$ corresponding to the energy gap $E_g \approx 2.6\ k_BT_c$ (Biondi and Garfunkel, 1959).

upper range of readily available microwave frequencies. Most microwave absorption studies of the type described in the previous section were carried out at ≈ 9 GHz, which is almost three orders of magnitude below the energy gap frequency of a high-temperature superconductor.

A study made of the temperature dependence of the normalized microwave resistivity of aluminum ($T_c = 1.2$ K) for a range of microwave frequencies of 12–80 GHz, shown in Fig. 15.66a, illustrates how the gap can be estimated (Biondi and Garfunkel, 1959). Each curve is labeled with its equivalent k_BT_c value. The curves for photon energies less than $3k_BT_c$ extrapolate to zero, which indicates that super electrons are not excited above the gap for microwave energies less than $3k_BT_c$. Above this energy the curves extrapolate to a finite resistivity, indicative of the presence of excited quasiparticles. In carrying out this experiment the lowest temperatures, ≈ 0.3 K, were reached with the aid of a He^3 refrigerator.

To determine the gap energy a plot was made of the microwave resistivity of each frequency at the temperature $T = 0.7T_c$ versus the energy, as shown in Fig. 15.66b. We see from the plot that the resistivity has a small slope up to the energy $2.6k_BT_c$ and a larger slope beyond this point, indicating that the gap energy is $E_g \approx 2.6k_BT_c$. The more rapid rise in resistivity beyond this point arises from super electrons that have become excited to the quasiparticle state.

F. Muon-Spin Relaxation

The negative muon μ^- acts in all respects like an electron and the positive muon μ^+ like a positron except for each having a mass 206.77 times larger (Poole, 1983). In this experiment positive polarized muons are implanted into a sample that had been placed in a magnetic field. The precession of the muons at $\gamma_\mu/2\pi = 135.5$ MHz/T provides a microscopic probe of the distribution of the local magnetic fields (Budnick *et al.*, 1987). In particular, the width of the muon spin relaxation (μSR) signal from a superconductor provides an estimate of this field distribution and of the penetration depth λ (Ansaldo *et al.*, 1991a; Pümpin *et al.*, 1990). The measurements are carried out in an external field that is significantly larger than the lower-critical field, so that the separation between the vortices is smaller than λ, and the μSR signal represents a simple average over the internal field in different parts of the sample.

As an example of a penetration depth determination we present in Fig. 15.67 the temperature dependence measured using a single crystal of $YBa_2Cu_3O_{7-\delta}$ with an 11-T applied magnetic field aligned parallel to the c-axis ($\Theta = 0$). The distribution of the internal magnetic field B_{in} depends on the anisotropy in the fall-off of the magnetic field in various directions around a vortex. The fall-off is, in turn, governed by the corresponding penetration depth in the plane perpendicular to the field direction. Figure 15.67 compares the temperature dependence of the measured values λ_{ab} with the dependence expected from Eq. (2.57),

$$\lambda = \lambda(0)\left[1 - \left(\frac{T}{T_c}\right)^4\right]^{-1/2} \qquad (15.31)$$

with $\lambda(0) = 141.5$ nm. We see that the fit to the data is good.

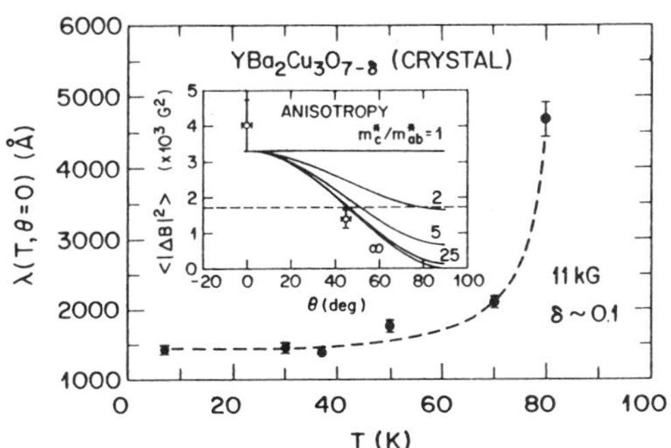

Figure 15.67 Temperature dependence of the penetration depth λ in a single crystal of $YBa_2Cu_3O_{7-\delta}$ for an 11-T magnetic field aligned along the c-axis, showing measured data points and fits to the data (dashed curves). The inset shows the average field squared $\langle|\Delta B|^2\rangle$ for two data points and several calculated mass anisotropy curves m_c^*/m_{ab}^* as a function of the angle Θ of the magnetic field relative to the c-axis. The data were obtained from muon spin relaxation (Harshman *et al.*, 1989).

We saw in Chapter 9, Section IV.A, that for a high-temperature superconductor $m_{ab}^* < m_c^*$, and hence that $\lambda_{ab} < \lambda_c$. We can conclude from a comparison of Eqs. (9.59) and (9.61) that the area enclosed by a vortex within a distance from the origin that satisfies Eq. (9.49) and makes the modified Bessel function assume the value $K_0(1)$ is larger when the magnetic field is aligned in the a, b-plane than when B_{app} is along c (see Fig. 9.23). This means that the vortices overlap more when the applied field is in the a, b-plane than when it is along the c direction; the variation in space of the internal field ΔB about its average value, shown plotted in Fig. 9.19, is also less for the former case. An intermediate amount of overlap, and hence of ΔB, will occur for intermediate angular orientations.

To check the anisotropy, Harshman *et al.* (1989) oriented the applied magnetic field at an angle $\pi/4$ relative to the c direction and found that the measured variation in the average field squared $\langle|\Delta B|^2\rangle$ had decreased. The result is compared in the inset of Fig. 15.67 with representative curves calculated for various ratios m_c^*/m_{ab}^* that show the existence of strong anisotropy, since $m_c^*/m_{ab}^* > 25$. This means that $\lambda_c/\lambda_{ab} > 5$; in other words, $\lambda_c > 700$ nm. The effective mass m_c^* relative to the electron rest mass m_0 was also determined and it was found that $m_c^* \approx 10m_0$. For further discussion, see Amato *et al.* (1992), Ansaldo *et al.* (1991a, b) (C_{60}/C_{70}), Barsov *et al.* (1987), Budnick *et al.* (1988), Heffner *et al.* (1990), Kiefl *et al.* (1992) (C_{60}), Le *et al.* (1992), Lichti *et al.* (1991), Pümpin *et al.* (1988), Rammer (1988), and Schenk *et al.* (1990).

G. Mössbauer Resonance

Mössbauer resonance measures gamma rays emitted by a recoilless nucleus when it undergoes a transition from a nuclear ground state to a nuclear excited state. For ^{57}Fe the emitted gamma ray has an energy of 14.4 KeV and a linewidth typically of 5×10^{-9} eV. The gamma ray can shift in energy, called an isomer shift, or its spectrum can split into a multiplet by hyperfine interaction from the nuclear spin, by crystal field effects, or by the quadrupole interaction. Line broadening and relaxation provide additional information. These factors are sensitive to the chemical environment of the nucleus in the lattice. Mössbauer workers frequently quote energy shifts in velocity units, mm/s.

In a typical experiment, one of the atoms of a superconductor such as Cu, Y, or Tl, is partially replaced by a small concentration of a nucleus, such as ^{57}Co, ^{57}Fe, ^{151}Eu, or ^{119}Sn, any one of which is favorable for Mössbauer studies. Sometimes, the replacement is 100%, as in the compound $EuBa_2Cu_3O_{7-\delta}$. The partial substitution can have the effect of lowering the transition temperature, particularly when Cu is being replaced, as shown in Fig. 15.68. The spectra provide information on the valence state of the nucleus (e.g., Fe^{2+} or Fe^{3+}), for example, whether it is high spin (e.g., $S = 5/2$) or low spin (e.g., $S = 1/2$), which is the dominant substitutional site (e.g., Cu(1) or Cu(2)), etc. Perhaps of greater interest is the information that Mössbauer gives us about the magnetic changes that occur.

Mössbauer data from $YBa_2Cu_3O_{7-\delta}$ with ^{57}Fe substituted for 10% of the Cu are shown in Fig. 15.69. The Fe is magnetically ordered at low temperature, with the

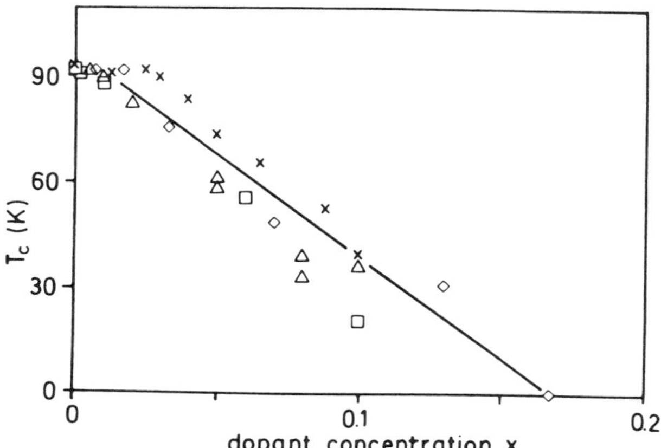

Figure 15.68 Dependence of the zero-resistance midpoint transition temperature T_c on the concentration x of the transition ion dopant M in $YBa_2(Cu_{1-x}M_x)_3O_{7-\delta}$ for: (a) Fe (□) (Bottán *et al.*, 1988), (b) Fe (△) (Oda *et al.*, 1987), (c) Fe (◇) (Tarascon *et al.*, 1988a), and (d) Co (×) (Langen *et al.*, 1988) (figure from Bottayán *et al.*, 1988).

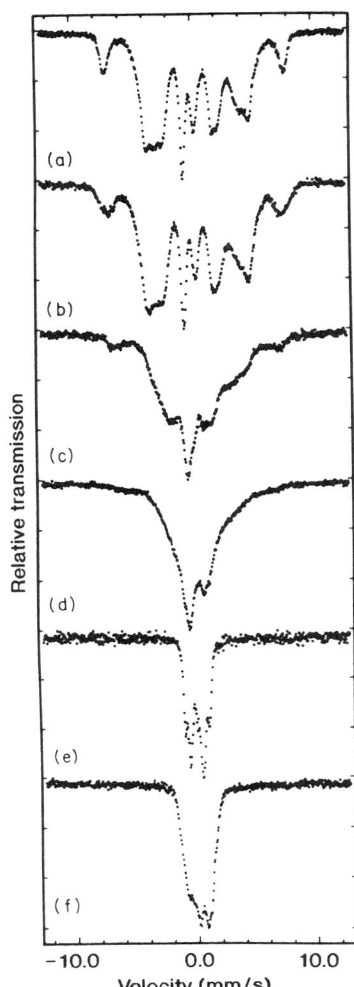

Figure 15.69 Comparison of Mössbauer absorption spectra of $YBa_2(Cu_{0.9}Fe_{0.1})_3O_7$ in zero field below T_c at (a) 4.2 K, (c) 15 K, and (d) 19 K, and above T_c at (e) 100 K. Spectra are also shown in a 5-T magnetic field at (b) 4.2 K, and (f) 100 K (Bottyán *et al.*, 1988).

ordering identified as antiferromagnetic since turning on a magnetic field of 5 T has the effect of broadening and producing a small inside shift of the outer lines of the spectrum, as shown in the figure. Increasing the temperature produces a decrease in the magnetic splitting accompanied by relaxation-time broadening. The onset of magnetic splitting occurs at 50 K; it appears in the wings in Figs. 15.69c and 15.69d, and is resolved in Figs. 15.69a and 15.69b. Below T_c, which from Fig. 15.69 is about 25 K for 10% Fe, the spectra of Fig. 15.69 appear more spread out. Bottyan *et al.* (1988) conclude that there are four Fe species that appear as the oxygen content (δ) and the Fe/Cu ratio of Fe varies, with three high-spin Fe^{4+} and one high-spin Fe^{3+}, with a preference for the Cu(1) sites. Pissas *et al.* (1992) found Fe equally distributed between the chain and plane Cu sites, being high-spin ($S = 5/2$) at the latter site.

Shinjo and Nasu (1989) reviewed magnetic order at very low temperatures in the superconductors

$$YbBa_2Cu_3O_7 \quad \text{and} \quad GdBa_2Cu_3O_7.$$

The isomer shift values indicate that the conduction-electron densities in the rare-earth ions are close to zero (Smit *et al.*, 1987), and suggest that they do not contribute to the electrical conductivity. The pronounced change in the spectrum with the temperature, shown in Figs. 15.70 and 15.71 for the two compounds, indicates the change from a low-temperature ordered state into a high-temperature paramagnetic-type state with the respective Néel temperatures T_N of 0.35 K and 2.5 K, both of which are far below the superconducting transition temperature $T_c \approx 90$ K. The authors suggest that the rare earth sheets sandwiched by superconducting layers may be an ideal two-dimensional magnetic lattice.

Relevant Mössbauer articles on several isotopes are, for ^{119}Sn (Kuzmann *et al.*, 1989; Matsumoto *et al.*, 1991; Nishida *et al.*, 1990a, b; Shiujo *et al.*, 1989; Shinjo and Nasu, 1989; Smith *et al.*, 1992), for ^{121}Sb (Smith *et al.*, 1992), for ^{151}Eu (Kuzmann *et al.*, 1989; Malik *et al.*, 1988; Shinjo and Nasu, 1989; Stadnik *et al.*, 1989, 1991; Yoshimoto *et al.*, 1991), for ^{155}Gd (Bornemann *et al.*, 1991; Shinjo and Nasu, 1989), and for ^{170}Yb (Shinjo and Nasu, 1989). The literature on ^{57}Fe studies is

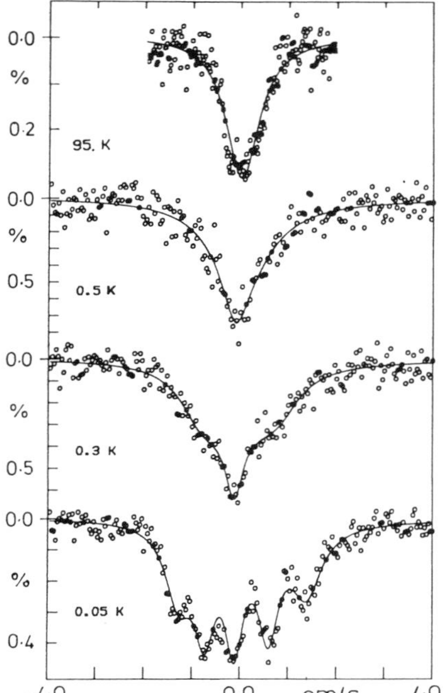

Figure 15.70 Mössbauer spectra of ^{170}Yb in $YbBa_2Cu_3O_7$ at four temperatures showing resolution of structure in the millidegree region (Hodges *et al.*, 1987).

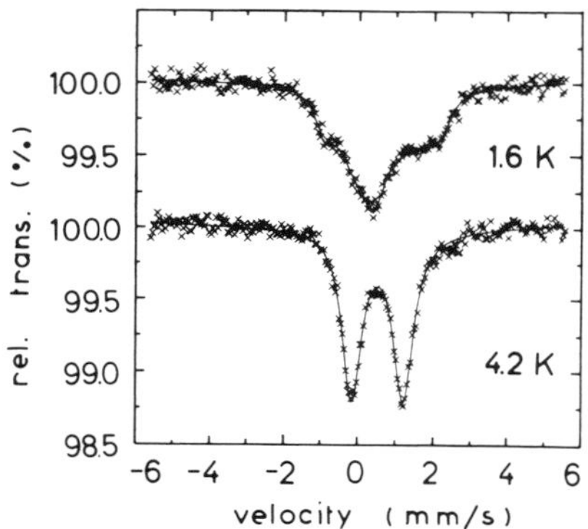

Figure 15.71 Mössbauer spectra of ^{155}Gd in $GdBa_2Cu_3O_7$ at 1.6 K and 4.2 K, showing resolved structure at the lower temperature (van den Berg *et al.*, 1987; Smit *et al.*, 1987).

extensive; two grain-aligned studies are Boolchand *et al.* (1988, 1992).

FURTHER READING

Other textbooks in superconductivity do not generally discuss the topics in this chapter, but many monographs do (Ashkenazi *et al.*, 1991; Bednorz and Müller, 1990; Lynn *et al.*, 1990; Vonsovsky *et al.*, 1982).

Reviews have appeared on infrared (Litvinchuk *et al.*, 1994; Timusk and Tanner, 1989) and Raman (Cardona, 1991; Faulques and Russo, 1991; Thomsen and Cardona, 1989) spectroscopies. Pintschovius and Reichardt (1994) surveyed inelastic neutron scattering. Femtosecond optoelectronic response of $YBa_2Cu_3O_7$ is reviewed by Sobolewski *et al.* (1994). Absorption and dispersion are discussed by Poole (1983) and Poole and Farach (1971). Portis (1993) discusses the electrodynamics of high temperature superconductors.

Photoemission and XPS are surveyed by Shen *et al.* (1995), Meyer and Weaver (1990), Kurmaev and Finkelstein (1989, 1991), and Kurmaev *et al.* (1988). Electron energy loss spectroscopy (EELS) was recently reviewed by Nücker *et al.* (1992).

Nuclear magnetic resonance is reviewed by Brinkmann (1994) and by Pennington and Slichter (1990), and the NMR of $YBa_2Cu_3O_{7-\delta}$ by Walstedt and Warren (1990). Nuclear quadrupole resonance is reviewed by Babu and Ramakrishna (1992) and Brinkmann (1994).

The status of buckminsterfullerene research is summarized by the articles in the special issue of *Accounts of Chemical Research* (Vol. 25, No. 3, March 1992) devoted to the topic and provided by Dresselhaus *et al.* (1994). A conference on spectroscopies of high temperature, fullerence, and heavy fermion superconductors was held in March 1995 (Bansil, 1995). The status of superconductor microwave circuits is provided by Shen (1994). The isotope effect in high temperature superconductors has been reviewed by Franck (1994).

PROBLEMS

1. What are the real and imaginary parts of $\sqrt{\epsilon}$ if $\epsilon = 2 + 3j$?
2. A metal is opaque for incident radiation below 2 eV and transparent for higher incident energies. Find the plasma frequency and the density of conduction electrons.

3. An ionically bonded molecule A^+B^- with the bond length 0.17 nm has polarizability 2×10^{-16} cm^2/V. It is irradiated with light with the power density 3 W/m^2. Find the permanent and induced dipole moments.
4. Calculate the frequency and the energy of the $n = 1$ to $n = 4$ transition of Cu. Find the Moseley law constant for this transition.
5. A neutron moving through a lattice at the velocity $v = 2 \times 10^5$ m/sec creates a phonon of frequency 5×10^{11} Hz. Find its new velocity v'.
6. Show that Eq. (15.16) reduces to Eq. (15.17) in the limit $\omega_i \tau_i \gg 1$. Show that $1/\tau_i$ is the linewidth for the real and imaginary parts of the expression.
7. A molecule with the vibrational frequency $\nu_0 = 10^{12}$ Hz is irradiated with visible light of wavelength 600 μm. What are the first five Stokes line frequencies in the Raman spectrum?
8. What is the gyromagnetic ratio γ for a Cu^{2+} ion with $g = 2.17$? What will be its ESR frequency in a magnetic field of 0.3 T?
9. Derive an expression for the dependence of the shift in the resonant lines of Fig. 15.62 on the applied field for the sample geometry of Fig. 10.17, taking into account the demagnetization factor.
10. A material has a characteristic vibrational frequency $\nu_0 = 3 \times 10^{12}$ Hz and an index of refraction $n = 2$. Find the energies E of the three lowest vibrational transitions, the frequency ν of the fundamental vibrational band, the spring constant k and the reflection coefficient R for normal incidence.
11. Sketch the energy level diagram and indicate all of the transitions appearing on the ^{139}La nuclear quadrupole resonance spectrum of Fig. 15.57.

References

J. Aarts, J. Meiresonne, H. Sprey, W. Maj, and P. Zagwijn, Phys. Rev. B *41*, 4739 (1990).

M. Abbate, M. Sacchi, J. J. Wnuk, L. W. M. Schreurs, Y. S. Wang, R. Lof, and J. C. Fuggle, Phys. Rev. B *42*, 7914 (1990).

M. Abramowitz and I. A. Stegun, Eds., "Handbook of Mathematical Functions," NBS, U.S. Gov. Printing Off., Washington, DC, 1964.

A. A. Abrikosov, Zh. Eksp. Teor. Fiz. *35*, 1442 (1957); Sov. Phys. JETP *5*, 1174 (1957).

D. Achkir, D. Poirier, C. Bourbonnais, G. Quirion, C. Lenoir, P. Batail, and D. Jérome, Phys. Rev. B *47*, 11595 (1993).

J. V. Acrivos, L. Chen, C. M. Burch, P. Metcalf, J. M. Honig, R. S. Liu, and K. K. Singh, Phys. Rev. B, in press.

S. Adachi, A. Tokiwa-Yamamoto, M. Itoh, K. Isawa, and H. Yamauchi, Physica C *214*, 313 (1993).

F. J. Adrian, Phys. Rev. Lett. *61*, 2148 (1988); see also *63*, 688, 690 (1989).

F. J. Adrian, Chem. Eng. News *Dec. 21*, 24 (1992).

G. Aeppli, S. M. Hayden, H. A. Mook, Z. Fisk, S.-W. Cheong, D. Rytz, J. P. Remeika, G. P. Espinosa, and A. S. Cooper, Phys. Rev. Lett. *62*, 2052 (1989).

N. Agraït, J. G. Rodrigo, and S. Vieira, Phys. Rev. B *46*, 5814 (1992).

N. Agraït, J. G. Rodrigo, and S. Vieira, Phys. Rev. B *47*, 12345 (1993).

I. J. R. Aitchison and N. E. Mavromatos, Phys. Rev. B *39*, 6544 (1989).

M. Akera and T. Andu, *in* "Proc. 8th Int. Conf. on Elect. Prop. of 2-Dimensional Systems," Grenoble, France, 1989.

H. Akera, A. H. MacDonald, S. M. Girvin, and M. R. Norman, Phys. Rev. Lett. *67*, 2375 (1991).

J. Akimitsu, S. Suzuki, M. Wantanabe, and H. Sawa, Jpn. J. Appl. Phys. *27*, L1857 (1988).

S. Aktas, "A Numerical Study of Magnetic Vortices in High Kappa Superconductors," Ph.D. thesis, University of South Carolina, 1993.

S. Aktas, C. P. Poole, Jr., and H. A. Farach, J. Phys. Condens. Matter *6*, 7373 (1994).

N. E. Alekseevskii, N. M. Dobrovolskii, D. Ekkert, and V. I. Tsebro, JETP *72*, 1145 (1977).

M. Alexander, H. Romberg, N. Nücker, P. Adelmann, J. Fink, J. T. Markert, M. B. Maple, S. Uchida, H. Takagi, Y. Tokura, A. C. W. P. James, and D. W. Murphy, Phys. Rev. B *43*, 333 (1991).

A. A. Aligia, Phys. Rev. B *39*, 6700 (1989).

K. Allan, A. Campion, J. Zhou, and J. B. Goodenough, Phys. Rev. B *41*, 11572 (1990).

P. B. Allen, Z. Fisk, and A. Migliori, in "Physical Properties of High Temperature Superconductors" (D. M. Ginsberg, Ed.), Vol. 1, Chap. 5, World Scientific, Singapore, 1989.

P. B. Allen, *in* "High-Temperature Superconductivity" (J. W. Lynn, Ed.), Chap. 9, Springer-Verlag, Berlin, 1990.

J. W. Allen, C. G. Olson, M. B. Maple, J.-S. Kang, L. Z. Liu, J.-H. Park, R. O. Anderson, W. P. Ellis, J. T. Market, Y. Dalichaouch, and R. Liu, Phys. Rev. Lett. *64*, 595 (1990).

H. Alloul, T. Ohno, and P. Mendels, Phys. Rev. Lett. *63*, 1700 (1989).

H. Alloul, A. Mahajan, H. Casalta, and O. Klein, Phys. Rev. Lett. *70*, 1171 (1993).

C. C. Almasan, J. Estrada, C. P. Poole, Jr., T. Datta, H. A. Farach, D. U. Gubser, S. A. Wolf, and L. E. Toth, Mater. Res. Soc. Symp. Proc. *99*, 451 (1987).

C. C. Almasan and M. B. Maple, *in* "Chemistry of High Temperature Superconductors" (C. N. R. Rao, Ed.), World Scientific, Singapore, 1991.

C. C. Almasan, S. H. Han, E. A. Early, B. W. Lee, C. L. Seaman, and M. B. Maple, Phys. Rev. B *45*, 1056 (1992).

E. E. Alp, J. C. Campuzano, G. Jennings, J. Guo, D. E. Ellis, L. Beaulaigue, S. Mini, M. Faiz, Y. Zhou, B. W. Veal, and J. Z. Liu, Phys. Rev. B *40*, 9385 (1989).

A. Amato, C. Geibel, F. N. Gygax, R. H. Heffner, E. Knetsch, D. E. MacLaughlin, C. Schank, A. Schenck, F. Steglich, and M. Weber, Z. Phys. B *86*, 159 (1992).

V. Ambegaokar and A. Baratoff, Phys. Rev. Lett. *10*, 468 (1963a); *11*, 104 (1963b).

V. Ambegaokar and U. Eckern, Phys. Rev. B *44*, 10358 (1991).

J. R. Anderson, D. A. Papaconstantopoulis, J. W. McCaffrey, and J. E. Schirber, Phys. Rev. B *7*, 5115 (1973).

P. W. Anderson, Phys. Rev. *112*, 1900 (1958).

P. W. Anderson, Phys. Rev. B *115*, 2 (1959).

P. W. Anderson, Phys. Rev. Lett. *9*, 309 (1962).

P. W. Anderson and J. M. Rowell, Phys. Rev. Lett. *10*, 230 (1963).

P. W. Anderson and Y. B. Kim, Rev. Mod. Phys. *36*, 39 (1964).

P. W. Anderson, Science *235*, 1196 (1987a); Phys. Rev. Lett. *59*, 2497 (1987b).

P. W. Anderson, *in* "Frontiers and Borderlines of Many Particle Physics," Int. School of Physics "Enrico Fermi," North-Holland, Varenna, 1987.

P. W. Anderson and Z. Zou, Phys. Rev. Lett. *60*, 132 (1988). (Reprinted in Halley, 1988).

P. W. Anderson, Phys. Rev. Lett. *64*, 1839 (1990a); *65*, 2306 (1990b).

P. W. Anderson, Physica C *185–189*, 11 (1991).

P. W. Anderson and R. Schrieffer, Physics Today, June 1991, p. 54.

P. W. Anderson, Science *256*, 1526 (1992).

P. W. Anderson, "A Career in Theoretical Physics," World Scientific, Singapore, 1994a.

P. W. Anderson, Amer. Scientist *82*, 379 (1994b).

M. Andersson and Ö. Rapp, Phys. Rev. B *44*, 7722 (1991).

Y. Ando, N. Motohira, K. Kitazawa, J. Takeya, and S. Akita, Jpn. J. Appl. Phys. *30*, L1635 (1991a); Phys. Rev. Lett. *67*, 2737 (1991b).

B. Andraka, J. S. Kim, G. R. Stewart, K. D. Carlson, H. H. Wang, and J. M. Williams, Phys. Rev. B *40*, 11345 (1989).

O. A. Anikeenok, M. V. Eremin, R. Sh. Zhdanov, V. V. Naletov, M. P. Rodionova, and M. A. Teplov, JETP Lett. *54*, 149 (1991).

V. I. Anisimov, M. A. Korotin, J. Zaanen, and P. L. Andersen, Phys. Rev. Lett. *68*, 345 (1992).

S. M. Anlage, M. Pambianchi, A. T. Findikoglu, C. Doughty, D.-H. Wu, J. Mao, S.-N. Mao, X. X. Xi, T. Venkatesan, J. L. Peng, and R. L. Greene, Proc. SPIE Conf. on Oxide Superconductivity, Vol. 2158 (D. Pavuna, Ed.), in press, 1994.

J. F. Annett, N. Goldenfeld, and S. R. Renn, *in* "Physical Properties of High Temperature Superconductors" (D. M. Ginsberg, Ed.), Vol. 2, Chap. 9, World Scientific, Singapore, 1990.

E. J. Ansaldo, C. Niedermayer, H. Glückler, C. E. Stronach, T. M. Riseman, R. S. Cary, D. R. Noakes, X. Obradors, A. Fuetes, J. M. Navarro, P. Gomez, N. Casañ, B. Martinez, F. Perez, J. Rodriguez-Carvajal, and K. Chow, Physica C *185–189*, 1213 (1991a).

E. J. Ansaldo, C. Niedermayer, and C. E. Stronach, Nature *353*, 121 (1991b).

V. P. Antropov, I. I. Mazin, O. K. Andersen, A. I. Liechtenstein, and O. Jepsen, Phys. Rev. B *47*, 12373 (1993).

J. Aponte, H. C. Abache, A. Sa-Neto, and M. Octavio, Phys. Rev. B *39*, 2233 (1989).

B. Arfi, Phys. Rev. B *45*, 2352 (1992).

G. Arfken, "Mathematical Methods for Physicists," 3rd ed., Wiley, New York, 1985.

T. A. Arias and J. D. Joannopoulos, Phys. Rev. B *39*, 4071 (1989).

D. N. Aristov, S. V. Maleyev, and A. G. Yashenkin, Phys. Rev. B *48*, 3527 (1993).

A. J. Arko, R. S. List, R. J. Bartlett, S.-W. Cheong, Z. Fisk, J. D. Thompson, C. G. Olson, A.-B. Yang, R. Liu, C. Gu, B. W. Veal, J. Z. Liu, A. P. Plaulikas, K. Vandervoort, H. Claus, J. C. Campuzano, J. E. Schirber, and N. D. Shinn, Phys. Rev. B *40*, 2268 (1989).

G. B. Arnold, F. M. Mueller, and J. C. Swihart, Phys. Rev. Lett. *67*, 2569 (1991).

A. G. Aronov, S. Hikami, and A. I. Larkin, Phys. Rev. Lett. *62*, 965, 2236(E), (1989).

D. Arovas, J. R. Schrieffer, and F. Wilczek, Phys. Rev. Lett. *53*, 722 (1984).

E. Arrigoni, G. Strinati, and C. Castellani, Phys. Rev. B *41*, 4838 (1990).

E. Arrigoni and G. C. Strinati, Phys. Rev. B *44*, 7455 (1991).

K. Asayama, G.-Q. Zheng, Y. Kitaoka, K. Ishida, and K. Fujiwara, Physica C *178*, 281 (1991).

N. W. Ashcroft and N. D. Mermin, "Solid State Physics," Saunders, Philadelphia, 1976.

M. Ashida, S. Aoyama, J. Hara, and K. Nagai, Phys. Rev. B *40*, 8673 (1989).

M. Ashida, J. Hara, and K. Nagai, Phys. Rev. B *45*, 828 (1992).

J. Ashkenazi, S. E. Barnes, F. Zuo, G. C. Vezzoli, and B. M. Klein, Eds., "High Temperature Superconductivity," Plenum, New York, 1991.

T. R. Askew, R. B. Flippen, K. J. Leary, M. N. Kunchur, J. Mater. Res. *6*, 1135 (1991).

L. G. Aslamazov and A. I. Larkin, Fiz. Tverd. Tela *10*, 1104 (1968) [Sov. Phys. Solid State *10*, 875 (1968)].

W. Assmus, M. Herrmann, U. Rauchschwalbe, S. Riegel, W. Lieke, H. Spille, S. Horn, G. Weber, F. Steglich, and G. Cordier, Phys. Rev. Lett. *52*, 469 (1984).

K. S. Athreya, O. B. Hyun, J. E. Ostenson, J. R. Clem, and D. K. Finnemore, Phys. Rev. B *38*, 11846 (1988).

A. Auerbach, "Interacting Electrons and Quantum Magnetism," Springer-Verlag, Berlin, 1994.

B. Aurivillus, Ark. Kemi *1*, 463, 499 (1950).

B. Aurivillus, Ark. Kemi *2*, 519 (1951).

B. Aurivillus, Ark. Kemi *5*, 39 (1952).

C. Ayache, I. L. Chaplygin, A. I. Kirilyuk, N. M. Kreines, and V. I. Kudinov, Solid State Commun. *81*, 41 (1992).

J. Azoulay, Phys. Rev. B *44*, 7018 (1991).

E. Babic, M. Prester, D. Drobac, Ž. Marohnic, and N. Biškup, Phys. Rev. B *43*, 1162 (1991).

E. Babic, M. Prester, D. Drobac, Ž. Marohnic, P. Nozar, P. Stastny, F. C. Matacotta, and S. Bernik, Phys. Rev. B *45*, 913 (1992).

P. K. Babu and J. Ramakrishna, Supercond. Rev. *1*, 75 (1992).

D. C. Baird and B. K. Mukherjee, Phys. Rev. Lett. *21*, 996 (1968).

D. C. Baird and B. K. Mukherjee, Phys. Rev. *3*, 1043 (1971).

C. K. Bak and N. F. Pedersen, Appl. Phys. Lett. *22*, 149 (1973).

O. N. Bakharev, A. V. Dooglav, A. V. Egorov, V. V. Naletov, M. P. Rodionova, M. S. Tagirov, and M. A. Teplov, Appl. Magn. Reson. *2*, 559 (1991).

J. Bala and A. M. Oleś, Phys. Rev. B *47*, 515 (1993).

C. J. Ballhausen, "Introduction to Ligand Field Theory," McGraw-Hill, New York, 1962.

M. Ban, T. Ichiguchi, and T. Onogi, Phys. Rev. B *40*, 4419 (1989).

A. Bansil, P. E. Mijnarends, and L. C. Smedskjaer; Phys. Rev. B *43*, 3667 (1991).

A. Bansil and S. Kaprzyk, Phys. Rev. B *43*, 10335 (1991).

A. Bansil (Ed.), "Proc. Conf. Spectroscopies in Novel Superconductors," J. Phys. Chem. Solids, in press.

A. Baratoff and G. Binnig, Physica B *188*, 1335 (1981).

B. Barbiellini, P. Genoud, J. Y. Henry, L. Hoffmann, T. Jarlborg, A. A. Manuel, S. Massidda, M. Peter, W. Sadowski, H. J. Scheel, A. Shukla, A. K. Singh, and E. Walker, Phys. Rev. B *43*, 7810 (1991).

J. Bardeen, L. N. Cooper, and J. R. Schrieffer, Phys. Rev. *108*, 1175 (1957).

J. Bardeen and M. P. Stephen, Phys. Rev. A *140*, 1197 (1965).

R. Bar-Deroma, J. Felsteiner, R. Brener, J. Ashkenazi, and D. van der Marel, Phys. Rev. B *45*, 2361 (1992).

C. Barlingay, V. Garcia-Vázquez, C. M. Falco, S. Mazumdar, and S. H. Risbud, Phys. Rev. B *41*, 4797 (1990).

A. Barone and G. Paterno, "Physics and Applications of the Josephson Effect," Wiley, New York, 1982.

S. E. Barrett, D. J. Durand, C. H. Pennington, C. P. Slichter, T. A. Friedmann, J. P. Rice, and D. M. Ginsberg, Phys. Rev. B *41*, 6283 (1990).

S. E. Barrett, J. A. Martindale, D. J. Durand, C. H. Pennington, C. P. Slichter, T. A. Friedmann, J. P. Rice, and D. M. Ginsberg, Phys. Rev. Lett. *66*, 108 (1991).

J. C. Barry, Z. Iqbal, B. L. Ramakrishna, H. Eckhardt, F. Reidinger, and R. Sharma, Appl. Phys. Lett. No. 71.70-b (1989).

S. G. Barsov, A. L. Getalov, V. P. Koptev, L. A. Kuz'min, S. M. Mikirtych'yants, N. A. Tarasov, G. V. Shcherbakov, N. M. Kotov, A. S. Nigmatulin, Ya. M. Mukovskii, V. G. Grebinnik, V. N. Duginov, V. A. Zhukov, A. B. Lazarev, V. G. Ol'shevskii, S. N. Shilov, and E. P. Krasnoperov, JETP Lett. *46*, 499 (1987).

K. Bartkowski, R. Horyn, A. J. Zaleski, Z. Bukowski, M. Horobiowski, C. Marucha, J. Rafalowicz, K. Rogacki, A. Stepien-Damm, C. Sulkowski, E. Trojnar, and J. Klamut, Phys. Status Solidi *103*, K37 (1987).

F. E. Bates, Phys. Rev. B *39*, 322 (1989).

B. Batlogg, A. P. Ramirez, R. J. Cava, R. B. van Dover, and E. A. Reitman, Phys. Rev. B *35*, 5340 (1987).

B. Batlogg, R. J. Cava, L. W. Rupp, Jr., A. M. Mujsce, J. J. Krajewski, J. P. Remeika, W. F. Peck, Jr., A. S. Cooper, and G. P. Espinosa, Phys. Rev. Lett. *61*, 1670 (1988).

P. E. Batson, T. M. Shaw, D. Dimos, and P. R. Duncombe, Phys. Rev. B *43*, 6236 (1991).

B. W. Batterman and C. S. Barrett, Phys. Rev. Lett. *13*, 390 (1964).

G. Baym and C. Pethick, "Landau Fermi-Liquid Theory," Wiley, New York, 1991.

C. P. Bean, Phys. Rev. Lett. *8*, 250 (1962).

C. P. Bean, Rev. Mod. Phys. *36*, 31 (1964).

M. R. Beasley, R. Labusch, and W. W. Webb, Phys. Rev. *181*, 682 (1969).

J. G. Bednorz and K. A. Müller, Z. Phys. B *64*, 189 (1986).

J. G. Bednorz and K. A. Müller (Eds.), "Earlier and Recent Aspects of Superconductivity," Springer-Verlag, Berlin, 1990.

C. W. J. Beenakker and H. van Houten, Phys. Rev. Lett. *66*, 3056 (1991).

V. P. Belash, E. Z. Kurmaev, and S. A. Nemnonov, Fiz. Met. Metalloved *37*, 659 (1974).

D. Belitz, *in* "High Temperature Superconductivity" (J. W. Lynn, Ed.), Chap. 2, Springer-Verlag, Berlin, 1990.

R. Benedek and H.-B. Schüttler, Phys. Rev. B *41*, 1789 (1990).

L. H. Bennett, Y. Flom, and G. C. Vezzoli, *in* "Proceedings High T_c Superconductors: Magnetic Interactions, Gaithersburg, MD," World Scientific, Singapore, 1989.

M. A. Beno, L. Soderholm, D. W. Capone II, D. G. Hinks, J. D. Jorgensen, J. D. Grace, I. K. Schuller, C. U. Segre, and K. Zhang, Appl. Phys. Lett. *51*, 57 (1987).

V. L. Berezinskii, Sov. Phys. JETP *34*, 610 (1972).

D. D. Berkley, E. F. Skelton, N. E. Moulton, M. S. Osofsky, W. T. Lechter, V. M. Browning, and D. H. Liebenberg, Phys. Rev. B *47*, 5524 (1993).

A. J. Berlinsky, C. Kallin, G. Rose, and A.-C. Shi, Phys. Rev. B *48*, 4074 (1993).

R. Berman and D. K. C. MacDonald, Proc. R. Soc. London Ser. A *211*, 122 (1952).

R. Beyers and T. M. Shaw, Solid State Phys. *42*, 135 (1989).

R. Beyers and B. T. Ahn, Annu. Rev. Mater. Sci. *21*, 335 (1991).

A. Bezinge, K. Yvon, and J. Muller, Solid State Commun. *63*, 141 (1987).

R. Bhagavatula, C. Ebner, and C. Jayaprakash, Phys. Rev. B *45*, 4774 (1992).

K. V. Bhagwat and P. Chaddah, Physica C *166*, 1 (1990).

K. V. Bhagwat and P. Chaddah, Physica C *190C*, 444 (1992).

A. Bharathi, C. S. Sundar, W. Y. Ching, Y. C. Jean, P. H. Hor, Y. Y. Xue, and C. W. Chu, Phys. Rev. B *42*, 10199 (1990).

A. K. Bhatnagar, R. Pan, D. G. Naugle, P. J. Sqattrito, A. Clearfield, Z. Z. Sheng, Q. A. Shams, and A. M. Hermann, Solid State Commun. *73*, 53 (1990).

R. N. Bhatt, Phys. Rev. B *16*, 1915 (1977).

R. N. Bhatt, Phys. Rev. B *17*, 2947 (1978).

A. Bhattacharya and C. S. Wang, Phys. Rev. B. *45*, 10826 (1992).

J. B. Bieri and K. Maki, Phys. Rev. B *42*, 4854 (1990).

B. D. Biggs, M. N. Kunchur, J. J. Lin, S. J. Poon, T. R. Askew, R. B. Flippen, M. A. Subramanian, J. Gopalakrishnan, and A. W. Sleight, Phys. Rev. B *39*, 7309 (1989).

S. J. L. Billinge, G. H. Kwei, and J. D. Thompson, *in* "Strongly Correlated Electronic Materials" (K. S. Bedell, Ed.) Addison–Wesley, New York, 1994.

W. E. Billups and M. A. Ciufolini (Eds.), "Buckminster Fullerenes," VCH, New York, 1993.

K. Binder and A. P. Young, Rev. Mod. Phys. *58*, 801 (1986).

B. Binnig, A. C. Castellano, M. De Santis, P. Rudolf, P. Lagarde, A. M. Frank, and A. Marcelli, Solid State Commun. *63*, 1009 (1980).

M. A. Biondi and M. P. Garfunkel, Phys. Rev. *116*, 853 (1959).

R. J. Birgeneau, C. Y. Chen, D. R. Gabbe, H. P. Jenssen, M. A. Kastner, C. J. Peters, P. J. Picone, T. Thio, T. R. Thurston, H. L. Tuller, A. D. Axe, P. Böni, and G. Shirane, Phys. Rev. Lett. *59*, 1329 (1987).

R. J. Birgeneau, D. R. Gabbe, H. P. Jenssen, M. A. Kastner, P. J. Picone, T. R. Thurston, G. Shirane, Y. Endoh, M. Sato, K. Yamada, Y. Hidaka, M. Oda, Y. Enomoto, M. Suzuki, and T. Murakami, Phys. Rev. B *38*, 6614 (1988).

R. J. Birgeneau and G. Shirane, *in* "Physical Properties of High Temperature Superconductors" (D. M. Ginsberg, Ed.), Vol. 1, Chap. 4, World Scientific, Singapore, 1989.

P. Birrer, F. N. Gygax, B. Hitti, E. Lippelt, A. Schenck, M. Weber, D. Cattani, J. Cors, M. Decroux, and Ø. Fischer, Phys. Rev. B *48*, 15689 (1993).

D. Bishop, C. M. Varma, B. Batlogg, E. Bucher, Z. Fisk, and J. L. Smith, Phys. Rev. Lett. *53*, 1009 (1984).

H. D. Bist, T. Datta, T. S. Little, J. C. Thigpen, Jr., J. R. Durig, P. R. Broussard, and D. D. Berkley, J. Raman Spectrosc. *22*, 639 (1991).

F. Bitter, Phys. Rev. *38*, 1903 (1931).

B. L. Blackford and R. H. March, Canad. J. Phys. *46*, 141 (1968).

H. A. Blackstead, J. Supercond. *5*, 67 (1992).

H. A. Blackstead, Phys. Rev. B *47*, 11411 (1993).

E. J. Blagoeva, G. Busiello, L. De Cesare, Y. T. Millev, I. Rabuffo, and D. I. Uzunov, Phys. Rev. B *40*, 7357 (1989).

E. J. Blagoeva, G. Busiello, L. De Cesare, Y. T. Millev, I. Rabuffo, and D. I. Uzunov, Phys. Rev. B *42*, 6124 (1990).

G. Blatter, J. Rhyner, and V. M. Vinokur, Phys. Rev. B *43*, 7826 (1991a).

G. Blatter, V. B. Geshkenbein, and V. M. Vinokur, Phys. Rev. Lett. *66*, 3297 (1991b).

G. Blatter, B. I. Ivlev, and J. Rhyner, Phys. Rev. Lett. *66*, 2392 (1991c).

G. Blatter, V. B. Geshkenbein, and A. I. Larkin, Phys. Rev. Lett. *68*, 875 (1992).

G. Blatter and B. Ivlev, Phys. Rev. Lett. *70*, 2621 (1993).

G. Blatter, M. V. Feigel'man, V. B. Geshkenbein, A. I. Larkin, and V. M. Vinokur, Rev. Mod. Phys. *66*, 1125 (1994).

K. W. Blazey, K. A. Müller, J. G. Bednorz, W. Berlinger, G. Amoretti, E. Buluggiu, A. Vera, and F. C. Matacotta, Phys. Rev. B *36*, 7241 (1987).

J. E. Blendell, C. K. Chiang, D. C. Cranmer, S. W. Freiman, E. R. Fuller, Jr., E. Drescher-Krasicka, W. L. Johnson, H. M. Ledbetter, L. H. Bennett, L. J. Swartzendruber, R. B. Marinenko, R. L. Mykleburst, D. S. Bright, and D. E. Newbury, ACS Symp. Ser. *351*, 240 (1987).

B. Blok and H. Monien, Phys. Rev. B *47*, 3454 (1993).

N. Bluzer, Phys. Rev. B *44*, 10222 (1991).

G. S. Boebinger, T. T. M. Palstra, A. Passner, M. J. Rosseinsky, D. W. Murphy, and I. I. Mazin, Phys. Rev. B *46*, 5876 (1992).

M. Boekholt, M. Hoffmann, and G. Güntherodt, Physica C *175C*, 127 (1991).

C. A. Bolle, P. I. Gammel, D. G. Grier, C. A. Murray, D. J. Bishop, D. B. Mitzi, and A. Kapitulnik, Phys. Rev. Lett. *66*, 112 (1991).

J. E. Bonevich, K. Harada, T. Matsuda, H. Kasai, T. Yoshida, G. Pozzi, and A. Tonomura, Phys. Rev. Lett. *70*, 2952 (1993).

P. Böni, J. D. Axe, G. Shirane, R. J. Birgeneau, D. R. Gabbe, H. P. Jenssen, M. A. Kastner, C. J. Peters, P. J. Picone, and T. R. Thurston, Phys. Rev. B *38*, 185 (1988).

E. Bonjour, R. Calemczuk, J. Y. Henry, and A. F. Khoder, Phys. Rev. B *43*, 106 (1991).

D. A. Bonn, A. H. O'Reilly, J. E. Greedan, C. V. Stager, T. Timusk, K. Kamarás, and D. B. Tanner, Phys. Rev. B *37*, 1574 (1988).

B. P. Bonner, R. Reichlin, S. Martin, H. B. Radousky, T. J. Folkerts, and R. N. Shelton, Phys. Rev. B *41*, 11576 (1990).

N. Bontemps, D. Davidov, P. Monod, and R. Even, Phys. Rev. B *43*, 11512 (1991).

P. Boolchand, C. Blue, K. Elgaid, I. Zitkovsky, D. McDaniel, W. Huff, B. Goodman, G. Lemon, D. E. Farrell, and B. S. Chandrasekhar, Phys. Rev. B *38*, 11313 (1988).

P. Boolchand, S. Pradhan, Y. Wu, M. Abdelgadir, W. Huff, D. Farrell, R. Coussement, and D. McDaniel, Phys. Rev. B *45*, 921 (1992).

P. Bordet, C. Chaillout, J. Chenavas, J. L. Hodeau, M. Marezio, J. Karpinski, and E. Kaldis, Nature (London) *334*, 596 (1988).

H. J. Bornemann, D. E. Morris, C. Steinleitner, and G. Czjzek, Phys. Rev. B *44*, 12567 (1991).

F. Borsa, A. Rigamonti, M. Corti, J. Ziolo, O. Hyun, and D. R. Torgeson, Phys. Rev. Lett. *68*, 698 (1992).

I. Bose, Phys. Rev. B *43*, 13602 (1991).

R. Böttner, N. Schroeder, E. Dietz, U. Gerhardt, W. Assmus, and J. Kowalewski, Phys. Rev. B *41*, 8679 (1990).

L. Bottyán, B. Molnár, D. L. Nagy, I. S. Szücs, J. Tóth, J. Dengler, G. Ritter, and J. Schober, Phys. Rev. B *38*, 11373 (1988).

R. Boyn, K. Löbe, H.-U. Habermeier, and N. Pruß, Physica C *181C*, 75 (1991).

I. Božovic, K. Char, S. J. B. Yoo, A. Kapitulnik, M. R. Beasley, T. H. Geballe, Z. Z. Wang, S. Hagen, N. P. Ong, D. E. Aspnes, and M. K. Kelly, Phys. Rev. B *38*, 5077 (1988).

I. Božovic, J. H. Kim, J. S. Harris, Jr., and W. Y. Lee, Phys. Rev. B *43*, 1169 (1991).

I. Božovic, J. Supercond. *4*, 193 (1991).

I. Božovic, J. H. Kim, J. S. Harris, Jr., E. S. Hellman, E. H. Hartford, and P. K. Chan, Phys. Rev. B *46*, 1182 (1992).

A. I. Braginski, Physica C *180*, 642 (1991).

B. H. Brandow, J. Solid State Chem. *88*, 28 (1990).

E. H. Brandt, Physica C *162C–164C*, 1167 (1989).

E. H. Brandt, Physica B *165–166*, 1129 (1990); Int. Conf. Low Temperature Phys., Brighton, U.K., August 1990.

E. H. Brandt and A. Sudbø, Physica C *180C*, 426 (1991).

E. H. Brandt, Phys. Rev. Lett. *69*, 1105 (1992).

A. Brass, H. J. Jensen, and A. J. Berlinsky, Phys. Rev. B *39*, 102 (1989).

A. Brass and H. J. Jensen, Phys. Rev. B *39*, 9587 (1989).

D. A. Brawner, A. Schilling, H. R. Ott, R. J. Haug, K. Ploog, and K. von Klitzing, Phys. Rev. Lett. *71*, 785 (1993).

Y. J. M. Brechet, B. Douçot, H. J. Jensen, and A.-C. Shi, Phys. Rev. B *42*, 2116 (1990).

E. Brézin, A. Fujita, and S. Hikami, Phys. Rev. Lett. *65*, 1949 (1990).

G. Briceño, M. F. Crommie, and A. Zettl, Phys. Rev. Lett. *66*, 2164 (1991).

D. Brinkmann and M. Mali, *in* "NMR Basic Principles and Progress," Vol. 31, Springer-Verlag, Berlin, 1994.

M. B. Brodsky, R. C. Dynes, K. Kitazawa, and H. L. Tuller (Eds.), "High Temperature Superconductors," Vol. 99, Materials Research Society, Pittsburgh, 1988.

C. Broholm, G. Aeppli, R. N. Kleiman, D. R. Harshman, D. J. Bishop, E. Bucher, D. Ll. Williams, E. J. Ansaldo, and R. H. Heffner, Phys. Rev. Lett. *65*, 2062 (1990).

N. B. Brookes, A. J. Viescas, P. D. Johnson, J. P. Remeika, A. S. Cooper, and N. V. Smith, Phys. Rev. B *39*, 2736 (1989).

J. S. Brooks, C. C. Agosta, S. J. Klepper, M. Tokumoto, N. Kinoshita, H. Anzai, S. Uji, H. Aoki, A. S. Perel, G. J. Athas, and D. A. Howe, Phys. Rev. Lett. *69*, 156 (1992).

S. D. Brorson, A. Kazeroonian, J. S. Moodera, D. W. Face, T. K. Cheng, E. P. Ippen, M. S. Dresselhaus, and G. Dresselhaus, Phys. Rev. Lett. *64*, 2172 (1990).

P. R. Broussard, Phys. Rev. B *43*, 2783 (1991).

P. Brüll, D. Kirchgässner, and P. Liederer, Physica C *182C*, 339 (1991).

T. Brun, M. Grimsditch, K. E. Gray, R. Bhadra, V. Maroni, and C. K. Loong, Phys. Rev. B *35*, 8837 (1987).

L. C. Brunel, S. G. Louie, G. Martinez, S. Labdi, and H. Raffy, Phys. Rev. Lett. *66*, 1346 (1991).

O. Brunner, L. Antognazza, J.-M. Triscone, L. Miéville, and Ø. Fischer, Phys. Rev. Lett. *67*, 1354 (1991).

R. Brusetti, A. J. Dianoux, P. Gougeon, M. Potel, E. Bonjour, and R. Calemczuk, Phys. Rev. B *41*, 6315 (1990).

W. Buckel, "Superconductivity, Fundamentals and Applications," VCH, Weinheim, Germany, 1991.

R. C. Budhani, L. Lesyna, D. DiMarzio, H. Wiesmann, and G. P. Williams, Phys. Rev. B *44*, 7087 (1991).

J. I. Budnick, A. Golnik, Ch. Niedermayer, E. Recknagel, M. Rossmanith, A. Weidinger, B. Chamberland, M. Filipkowski, and D. P. Yang, Phys. Lett. A *124*, 103 (1987).

J. I. Budnick, B. Chamberland, D. P. Yang, Ch. Niedermayer, A. Golnik, E. Recknagel, M. Rossmanith, and A. Weidinger, Europhys. Lett. *5*, 651 (1988).

L. N. Bulaevskii, Zh. Eksp. Teor. Fiz. *64*, 2241 (1973); Sov. Phys. JETP (Engl. Trans.) *37*, 1133 (1988).

L. N. Bulaevskii, O. V. Dolgov, and M. O. Ptitsyn, Phys. Rev. B *38*, 11290 (1988).

L. N. Bulaevskii and M. V. Zyskin, Phys. Rev. B *42*, 10230 (1990).

L. N. Bulaevskii and I. D. Vagner, Phys. Rev. B *43*, 8694 (1991).

L. N. Bulaevskii, Phys. Rev. B *44*, 910 (1991).

L. N. Bulaevskii, M. Ledvij, and V. G. Kogan, Phys. Rev. B *46*, 366, 11807 (1992).

N. Bulut and D. J. Scalapino, Phys. Rev. B *45*, 2371 (1992).

G. Burns, "Solid State Physics," Academic Press, Orlando, FL 1985.

G. Burns, B. V. Chandrashekhar, F. H. Dacol, M. W. Shafer, and P. Strobel, Solid State Commun. *67*, 603 (1988).

G. Burns, P. Strobel, G. V. Chandrashekhar, F. H. Dacol, F. Holtzberg, and M. W. Shafer, Phys. Rev. B *39*, 2245 (1989).

G. Burns and A. M. Glazer, "Space Groups for Solid State Scientists," Academic Press, San Diego, 1990.

G. Burns, "High Temperature Superconductivity: An Introduction," Academic Press, Boston, 1992.

M. J. Burns, Phys. Rev. B *40*, 5473 (1989).

R. Busch, G. Ries, H. Werthner, G. Kreiselmeyer, and G. Saemann-Ischenko, Phys. Rev. Lett. *69*, 522 (1992).

G. Busiello and D. I. Uzunov, Phys. Rev. B *42*, 1018 (1990).

G. Busiello, L. De Cesare, Y. T. Millev, I. Rabuffo, and D. I. Uzunov, Phys. Rev. B *43*, 1150 (1991).

A. Bussmann-Holder and A. R. Bishop, Phys. Rev. B *44*, 2853 (1991).

A. I. Buzdin, Phys. Rev. B *47*, 11416 (1993).

B. Cabrera, C. E. Cunningham, and D. Saroff, Phys. Rev. Lett. *62*, 2040 (1989).

Z.-X. Cai and D. O. Welch, Phys. Rev. B *45*, 2385 (1992).

J. Callaway, "Energy Band Theory," Academic Press, New York, 1964.

H. B. Callen, "Thermodynamics and an Introduction to Thermostatics," Wiley, New York, 1985.

P. Calvani, M. Capizzi, S. Lupi, P. Maselli, and E. Agostinelli, Physica C *180C*, 116 (1991).

I. A. Campbell, L. Fruchter, and R. Cabanel, Phys. Rev. Lett. *64*, 1561 (1990).

J. C. Campuzano, G. Jennings, M. Faiz, L. Beaulaigue, B. W. Veal, J. Z. Liu, A. P. Paulikas, K. Vandervoort, H. Claus, R. S. List, A. J. Arko, and R. L. Bartlett, Phys. Rev. Lett. *64*, 2308 (1990).

J. C. Campuzano, L. C. Smedskjaer, R. S. Benedek, G. Jennings, and A. J. Bansil, Phys. Rev. B *43*, 2788 (1991).

J. J. Capponi, C. Chaillout, A. W. Hewat, P. LeJay, M. Marezio, N. Nguyen, B. Raveau, J. L. Soubeyroux, J. L. Tholence, and R. Tournier, Europhys. Lett. *3*, 1301 (1987).

J. P. Carbotte, Rev. Mod. Phys. *62*, 1027 (1990).

J. R. Carbotte and C. Jiang, Phys. Rev. B *48*, 4231 (1993).

M. Cardona, Physica C *185C–189C*, 65 (1991).

G. Carneiro, Phys. Rev. B *45*, 2391 (1992).

P. Carretta and M. Corti, Phys. Rev. Lett. *68*, 1236 (1992).

P. Carretta and M. Corti, Phys. Rev. Lett. *68*, 1236 (1992).

P. Carretta, M. Corti, A. Rigamonti, R. De Renzi, F. Licci, C. Paris, L. Bonoldi, M. Sparpaglione, and L. Zini, Physica C *191C*, 97 (1992).

C. Castellani, M. Grilli, and G. Kotliar, Phys. Rev. B *43*, 8000 (1991).

R. J. Cava, A. Santoro, D. W. Johnson, Jr., and W. W. Rhodes, Phys. Rev. B *35*, 6716 (1987).

R. J. Cava, B. Batlogg, J. J. Krajewski, R. Farrow, L. W. Rupp, Jr., A. E. White, K. Short, W. F. Pick, and T. Kometani, Nature *332*, 814 (1988).

M.-C. Cha, M. P. A. Fisher, S. M. Girvin, M. Wallin, and A. P. Young, Phys. Rev. B *44*, 6883 (1991).

P. Chaddah, K. V. Bhagwat, and G. Raulkumaer, Physica C *159C*, 570 (1989).

P. Chaddah and K. Bhagwat, *in* "High Temperature Superconductivity" (S. K. Malik and S. S. Shah, Eds.), Nova Science, New York, 1992.

C. Chaillout, J. P. Remeika, A. Santoro, and M. Marezio, Solid State. Commun. *56*, 829 (1985).

T. K. Chaki and M. Rubinstein, Phys. Rev. B *36*, 7259 (1987).

B. Chakraborty, Phys. Rev. B *43*, 378 (1991).

S. Chakravarty, B. I. Halperin, and D. R. Nelson, Phys. Rev. Lett. *60*, 1057 (1988).

S. Chakravarty, B. I. Ivlev, and Y. N. Ovchinnikov, Phys. Rev. B *42*, 2143 (1990).

S. Chakravarty, A. Sudbø, P. W. Anderson, and S. Strong, Science *261*, 337 (1993).

L. P. Chan, D. R. Harshman, K. G. Lynn, S. Massidda, and B. D. Mitzi, Phys. Rev. Lett. *67*, 1350 (1991).

B. S. Chandrasekhar, Appl. Phys. Lett. *1*, 7 (1962).

B. S. Chandrasekhar, *in* "Superconductivity" (R. D. Parks, Ed.), Vol. 1, Chap. 1, Dekker, New York, 1969.

N. Chandrasekhar, O. T. Valls, and A. M. Goldman, Mod. Phys. Letters *8*, 1863 (1994).

C. L. Chang, A. Kleinhammes, W. G. Moulton, and L. R. Testardi, Phys. Rev. B *41*, 11564 (1990).

M. Charalambous, J. Chaussy, and P. Lejay, Phys. Rev. B *45*, 5091 (1992).

T. Chattopadhyay, P. J. Brown, B. C. Sales, L. A. Boatner, H. A. Mook, and H. Maletta, Phys. Rev. B *40*, 2624 (1989).

P. Chaudhari, R. T. Collins, P. Freitas, R. J. Gambino, J. R. Kirtley, R. H. Koch, R. B. Laibowitz, F. K. LeGoues, T. R. McGuire, T. Penney, Z. Schlesinger, A. P. Segmüller, S. Foner, and E. J. McNiff, Jr., Phys. Rev. B *36*, 8903 (1987).

S. V. Chekalin, V. M. Farztdinov, V. V. Golovlyov, V. S. Letokhov, Yu. E. Lozovlk, Yu. A. Matveets, and A. G. Stepanov, Phys. Rev. Lett. *67*, 3860 (1991).

J. Chela-Flores, M. P. Das, and A. G. Saif, Solid State Commun. *65*, 77 (1988).

D.-X. Chen, R. B. Goldfarb, J. Nogués, and K. V. Rao, J. Appl. Phys. *63*, 980 (1988).

D.-X. Chen and R. B. Goldfarb, J. Appl. Phys. *66*, 2489 (1989).

G. H. Chen, J. H. Wang, D. N. Zheng, Y. F. Yan, S. L. Jia, Q. S. Yang, Y. M. Ni, and Z. X. Zhao, Mod. Phys. Lett. B *3*, 295 (1989).

H. Chen and J. Callaway, Phys. Rev. B *40*, 8800 (1989).

C. H. Chen, *in* "Physical Properties of High Temperature Superconductors" (D. M. Ginsberg, Ed.), Vol. 2, Chap. 4, World Scientific, Singapore, 1990.

J. H. Chen, Solid State Commun. *75*, 557, 563, 567, 573 (1990a); Phys. Rev. B *42*, 3952, 3957 (1990b).

D.-X. Chen, A. Sanchez, J. Nogués, and J. S. Muñoz, Phys. Rev. B *41*, 9510 (1990a).

D.-X. Chen, A. Sanchez, and J. Muñoz, J. Appl. Phys. *67*, 3430 (1990b).

D.-X. Chen, A. Sanchez, T. Puig, L. M. Martinez, and J. S. Muñoz, Physica C *168C*, 652 (1990c).

C. Y. Chen, R. J. Birgeneau, M. A. Kastner, N. W. Preyer, and T. Thio, Phys. Rev. B *43*, 392 (1991).

D.-X. Chen, J. A. Brug, and R. B. Goldfarb, IEEE Trans. Magn. *27*, 3601 (1991).

D.-X. Chen and A. Sanchez, J. Appl. Phys. *70*, 5463 (1991).

C.-C. Chen, S. P. Kelty, and C. M. Lieber, Science *253*, 886 (1991).

B. Chen and J. Dong, Phys. Rev. B *44*, 10206 (1991).

C. T. Chen, L. H. Tjeng, J. Kwo, H. L. Kao, P. Rudolf, F. Sette, and R. M. Fleming, Phys. Rev. Lett. *68*, 2543 (1992).

T.-P. Chen, Z. X. Zhao, H. D. Yang, E. L. Wolf, R. N. Shelton, and P. Klavins, Phys. Rev. *45*, 7945 (1992).

Q. Y. Chen, *in* "Magnetic Susceptibility of Superconductors and other Spin Systems" (R. A. Hein, T. L. Francavilla, and D. H. Liebenberg, Eds.), Plenum, New York, 1992.

L. Chengren and D. C. Larbalestier, Cryogenics *27*, 171 (1987).

S.-W. Cheong, S. E. Brown, Z. Fisk, R. S. Kwok, J. D. Thompson, E. Zirngiebl, G. Gruner, D. E. Peterson, G. L. Wells, R. B. Schwarz, and J. R. Cooper, Phys. Rev. B *36*, 3913 (1987).

S.-W. Cheong, M. F. Hundley, J. D. Thompson, and Z. Fisk, Phys. Rev. B *39*, 6567 (1989a).

S.-W. Cheong, Z. Fisk, J. D. Thompson, and R. B. Schwarz, Physica C *159C*, 407 (1989b).

H.-F. Cheung, Y. Gefen, E. K. Riedel, and W.-H. Shih, Phys. Rev. B *37*, 6050 (1988).

X.-F. Chen, M. J. Marone, G. X. Tessema, M. J. Skove, M. V. Nevitt, D. J. Miller, and B. W. Veal, Phys. Rev. B *48*, 1254 (1993).

C. C. Chi and C. Vanneste, Phys. Rev. B *42*, 9875 (1990).

T. R. Chien, T. W. Jing, N. P. Ong, and Z. Z. Wang, Phys. Rev. Lett. *66*, 3075 (1991).

J. H. Cho, F. Borsa, D. C. Johnston, and D. R. Torgeson, Phys. Rev. B *46*, 3179 (1992).

C. H. Choi and P. Muzikar, Phys. Rev. B *39*, 11296 (1989); Phys. Rev. B *40*, 5144 (1989).

J. Choi and J. V. José, Phys. Rev. Lett. *62*, 320 (1989).

M. Y. Choi and S. Kim, Phys. Rev. B *44*, 10411 (1991).

M. Y. Choi, C. Lee, and J. Lee, Phys. Rev. B *46*, 1489 (1992).

H. Chou, K. Yamada, J. D. Axe, S. M. Shapiro, G. Shirane, I. Tanaka, K. Yamane, and H. Kojima, Phys. Rev. B *42*, 4272 (1990).

D. B. Chrisey and G. K. Hubler (Eds.), "Pulsed Laser Deposition of Thin Films," Wiley, New York, 1994.

C. W. Chu, P. H. Hor, R. L. Meng, L. Gao, and Z. J. Huang, Science *235*, 567 (1987).

C. W. Chu, P. H. Hor, R. L. Meng, L. Gao, Z. J. Huang, and Y. Q. Wang, Phys. Rev. Lett. *58*, 405 (1987).

C. W. Chu, P. H. Hor, R. L. Meng, L. Gao, Z. J. Huang, J. Bechtold, M. K. Wu, and C. Y. Huang, Mater. Res. Soc. Symp. Proc. *99*, 15 (1987).

C. W. Chu, J. Bechtold, L. Gao, P. H. Hor, Z. J. Huang, R. L. Meng, Y. Y. Sun, Y. Q. Wang, and Y. Y. Xue, Phys. Rev. Lett. *60*, 941 (1988).

C. W. Chu, L. Gao, F. Chen, Z. J. Huang, R. L. Meng, and Y. Y. Xue, Nature *365*, 323 (1993b).

C. W. Chu, J. Superconductivity *7*, 1 (1994).

C. W. Chu, "Unusual High Temperature Superconductors," Proc. Symp. Quantum Theory of Real Materials, Berkeley, California, Aug. 1994, in press.

E. M. Chudnovsky, Phys. Rev. B *40*, 11355 (1989).

E. M. Chudnovsky, Phys. Rev. Lett. *65*, 3060 (1990).

E. M. Chudnovsky, Phys. Rev. B *43*, 7831 (1991).

F. Chung and S. Sternberg, Amer. Scientist *81*, 56 (1993).

L. Civale, A. D. Marwick, M. W. McElfresh, T. K. Worthington, A. P. Malozemoff, F. H. Holtzberg, J. R. Thompson, and M. A. Kirk, Phys. Rev. Lett. *65*, 1164 (1990).

L. Civale, T. K. Worthington, and A. Gupta, Phys. Rev. B *43*, 5425 (1991).

L. Civale, A. D. Marwick, T. K. Worthington, M. A. Kirk, J. R. Thompson, L. Krusin-Elbaum, Y. Sun, J. R. Clem, and F. Holtzberg, Phys. Rev. Lett. *67*, 648 (1991).

J. H. Claassen, J. F. Evetts, R. E. Somekh, and Z. H. Barber, Phys. Rev. B *44*, 9605 (1991).

J. Clarke, Phys. Rev. Lett. *28*, 1363 (1972).

J. Clarke and J. L. Paterson, J. Low Temp. Phys. *15*, 491 (1974).

J. Clayhold, N. P. Ong, P. H. Hor and C. W. Chu, Phys. Rev. B *38*, 7016 (1988).

J. R. Clem, Physica C *162–164*, 1137 (1989).

J. R. Clem and M. W. Coffey, Phys. Rev. B *42*, 6209 (1990).

J. R. Clem, Phys. Rev. B *43*, 7837 (1991).

J. Clem, "A. C. Losses in Type-II Superconductors, Chap. in Magnetic Susceptibility of Superconductors and Other Spin Systems," (R. A. Hein, T. L. Francavilla, and D. H. Liebenberg, Eds.), Plenum, New York, 1992.

J. R. Clem and Z. Hao, Phys. Rev. B *48*, 13774 (1993).

A. M. Clogston and J. Jaccarino, Phys. Rev. *121*, 1357 (1961).

A. M. Clogston, Phys. Rev. Lett. *9*, 266 (1962).

M. W. Coffey and J. R. Clem, Phys. Rev. B *44*, 6903 (1991).

M. W. Coffey, Phys. Rev. B *46*, 567 (1992).

M. W. Coffey, Phys. Rev. B *47*, 12284 (1993).

M. W. Coffey, Phys. Rev. B *49*, 9774 (1994).

R. W. Cohen, G. D. Cody, and L. J. Vieland, *in* Electronic Density of States, NBS Spec. Publ. (U.S.) *323*, 767 (1971).

M. L. Cohen, *in* "Novel Superconductivity" (S. A. Wolf and V. Z. Kresin, Eds.), p. 1095. Plenum, New York, 1987.

M. L. Cohen and D. R. Penn, Phys. Rev. B *42*, 8702 (1990).

R. E. Cohen, Computers in Phys., in press, 1994.

J. L. Cohn, S. A. Wolf, V. Selvamanickam, and K. Salama, Phys. Rev. Lett. *66*, 1098 (1991).

M. B. Cohn, M. S. Rzchowski, S. P. Benz, and C. J. Lobb, Phys. Rev. B *43*, 12823 (1991).

J. L. Cohn, S. A. Wolf, and T. A. Vanderah, Phys. Rev. B *45*, 511 (1992).

J. L. Cohn, E. F. Skelton, S. A. Wolf, J. Z. Liu, and R. N. Shelton, Phys. Rev. B *45*, 13144 (1992).

J. L. Cohn, E. F. Skelton, S. A. Wolf, J. Z. Liu, and R. N. Shelton, Phys. Rev. B *45*, 13140, 13144 (1992).

B. R. Coles, Cont. Phys. *28*, 143 (1987).

G. Collin and R. Comes, C. R. Acad. Sci. Paris *304*, 1159 (1987).

R. T. Collins, Z. Schlesinger, F. Holtzberg, and C. Feild, Phys. Rev. Lett. *63*, 422 (1989).

R. T. Collins, Z. Schlesinger, G. V. Chandrasehekhar, and M. W. Shafer, Phys. Rev. B *39*, 2251 (1989).

R. T. Collins, Z. Schlesinger, F. Holtzberg, P. Chaudhari, and C. Feild, Phys. Rev. B *39*, 6571 (1989).

R. T. Collins, Z. Schlesinger, F. Holtzberg, C. Feild, U. Welp, G. W. Crabtree, J. Z. Liu, and Y. Fang, Phys. Rev. B, *43*, 8701 (1991).

S. J. Collocott, R. Driver, and E. R. Vance, Phys. Rev. B *41*, 6329 (1990a).

S. J. Collocott, N. Savvides, and E. R. Vance, Phys. Rev. B *42*, 4794 (1990b).

S. Collocott, R. Driver, and C. Andrikidis, Phys. Rev. B *45*, 945 (1992).

R. Combescot, Phys. Rev. Lett. *67*, 148 (1991a); Phys. Rev. B *42*, 7810 (1991b).

E. Compans and F. Baumann, Jpn. J. Appl. Phys. *26* Suppl. 3, 805 (1987).

L. D. Cooley, G. Stejic, and D. C. Labalestier, Phys. Rev. B *46*, 2964 (1992).

L. N. Cooper, Phys. Rev. *104*, 1189 (1956).

S. L. Cooper, G. A. Thomas, J. Orenstein, D. H. Rapkine, M. Capizzi, T. Timusk, A. J. Millis, L. F. Schneemeyer, and J. V. Waszczak, Phys. Rev. B *40*, 11358 (1989).

S. L. Cooper, A. L. Kotz, M. A. Karlow, M. V. Klein, W. C. Lee, J. Giapintzakis, and D. M. Ginzberg, Phys. Rev. B *45*, 2549 (1992).

S. L. Cooper, D. Reznik, A. Kotz, M. A. Karlow, R. Liu, M. V. Klein, W. C. Lee, J. Giapintzakis, D. M. Ginsberg, B. W. Veal, and A. P. Paulikas, Phys. Rev. B *47*, 8233 (1993).

S. L. Cooper and K. E. Gray *in* "Physical Properties of High Temperature Superconductors" (D. M. Ginsberg, Ed.), Vol. 4, Chap. 3, World Scientific, Singapore, 1994.

C. Coretsopoulos, H. C. Lee, E. Ramli, L. Raven, T. B. Rauchfuss, and E. Oldfield, Phys. Rev. B *39*, 781 (1989).

D. L. Cox and M. B. Maple, Phys. Today, Feb. 1995, p. 32.

L. A. Curtiss and S. W. Tam, J. Mater. Res. *3*, 1269 (1988).

J. Costa-Quintana, F. López-Aguilar, S. Balle, and R. Salvador, Phys. Rev. B *39*, 9675 (1989).

L. Cota, L. Morales de la Garza, G. Hirata, L. Martínez, E. Orozco, E. Carrillo, A. Mendoza, J. L. Albarrán, J. Fuentes-Maya, J. L. Boldú, J. G. Pérez-Ramírez, R. Pérez, J. Reyes Gasga, M. Avalos, and M. José-Yacamán, J. Mater. Res. *3*, 417 (1988).

R. Côte and A. Griffin, Phys. Rev. B *48*, 10404 (1993).

F. A. Cotton, "Chemical Applications of Group Theory." Wiley, New York, 1963.

D. E. Cox and A. W. Sleight, Solid State Commun. *19*, 969 (1976).

D. E. Cox and A. W. Sleight, Acta Cryst. B *35*, 1 (1979).

D. L. Cox and M. B. Maple, *Phys. Today*, February 1995, p. 32.

G. W. Crabtree, J. Z. Liu, A. Umezawa, W. K. Kwok, C. H. Sowers, S. K. Malik, B. W. Veal, D. J. Lam, M. B. Brodsky, and J. W. Downey, Phys. Rev. B *36*, 4021 (1987).

M. K. Crawford, G. Burns, and F. Holtzberg, Solid State Commun. *70*, 557 (1989).

M. K. Crawford, G. Burns, G. V. Chandrashekhar, F. H. Dacol, W. E. Farneth, E. M. McCarron III, and R. J. Smalley, Phys. Rev. B *41*, 8933 (1990a).

M. K. Crawford, G. Burns, G. V. Chandrashekhar, F. H. Dacol, W. E. Farneth, E. M. McCarron III, and R. J. Smalley, Solid State Commun. *73*, 507 (1990b).

R. J. Creswick, H. A. Farach, C. P. Poole, Jr., "Introduction to Renormalization Group Methods in Physics," Wiley, New York, 1992.

M. Crisan, Phys. Lett. A *124*, 195 (1987).

M. F. Crommie, A. Zettl, T. W. Barbee, III, and M. L. Cohen, Phys. Rev. B *37*, 9734 (1988).

M. F. Crommie, G. Briceño, and A. Zettl, Physica C *162C–164C*, 1397 (1989).

M. F. Crommie and A. Zettl, Phys. Rev. B *43*, 408 (1991).

R. W. Cross and R. B. Goldfab, Appl. Phys. Lett. *58*, 415 (1991).

J. E. Crow and N.-P. Ong, *in* "High Temperature Superconductivity" (J. W. Lynn, Ed.), Chap. 7, Springer-Verlag, Berlin, 1990.

M. A. Crusellas, J. Fontcuberta, S. Piñol, T. Grenet, and J. Beille, Physica C *180*, 313 (1991).

A. M. Cucolo, R. Di Leo, P. Romano, L. F. Schneemeyer, and J. V. Waszczak, Phys. Rev. B *44*, 2857 (1991).

S.-M. Cui and C.-H. Tsai, Phys. Rev. B *44*, 12500 (1991).

J. C. Culbertson, U. Strom, S. A. Wolf, P. Skeath, E. J. West, and W. K. Burns, Phys. Rev. B *39*, 12359 (1989).

J. C. Culbertson, U. Strom, S. A. Wolf, and W. W. Fuller, Phys. Rev. B *44*, 9609 (1991).

L. A. Curtiss and S. W. Tam, J. Mater. Res. *3*, 1269 (1988).

L. A. Curtiss and S. W. Tam, Phys. Rev. B *41*, 1824 (1990).

L. L. Daemen and J. E. Gubernatis, Phys. Rev. *43*, 413, 2625 (1991).

L. L. Daeman, L. J. Campbell, and V. G. Kogan, Phys. Rev. B *46*, 3631 (1992).

L. L. Daemen, L. J. Campbell, A. Yu Simonov, and V. G. Kogan, Phys. Rev. Lett. *70*, 2948 (1993).

L. L. Daemen and A. W. Overhauser, Phys. Rev. B *40*, 10778 (1989).

L. L. Daemen and J. E. Gubernatis, Phys. Rev. B *45*, 314 (1992b).

E. Dagotto, A. Moreo, R. Joynt, S. Bacci, and E. Gagliano, Phys. Rev. B *41*, 2585, (1990).

E. Dagotto, Rev. Mod. Phys. *66*, 763 (1994).

E. Dagotto, A. Moreo, F. Ortolani, D. Poilblanc, and J. Riera, Phys. Rev. B *45*, 10741 (1992).

P. Dai, Y. Zhang, and M. P. Sarachik, Phys. Rev. Lett. *67*, 136 (1991).

Y. Dalichaouch, M. B. Maple, J. Y. Chen, T. Kohara, C. Rossel, M. S. Torikachvili, and A. L. Giorgi, Phys. Rev. *41*, 1829 (1990a).

Y. Dalichaouch, B. W. Lee, C. L. Seaman, J. T. Markert, and M. B. Maple, Phys. Rev. Lett. *64*, 599 (1990b).

Y. Dalichaouch, B. W. Lee, S. E. Lambert, M. P. Maple, J. L. Smith, and Z. Fisk, Phys. Rev. B *43*, 299 (1991).

M. Das, *in* "Studies in High Temperature Superconductors" (A. V. Narlikar, Ed.), Vol. 3, Nova Sci., New York, 1989.

C. Dasgupta and T. V. Ramakrishnan, Physica C (Amsterdam) *183C*, 62 (1991).

T. Datta, C. P. Poole, Jr., H. A. Farach, C. Almasan, J. Estrada, D. U. Gubser, and S. A. Wolf, Phys. Rev. B *37*, 7843 (1988).

M. Däumling and G. V. Chandrashekhar, Phys. Rev. B *46*, 6422 (1992).

D. Davidov, P. Monod, and N. Bontemps, Phys. Rev. B *45*, 8036 (1992).

M. C. de Andrade, C. C. Almasan, Y. Dalichaouch, and M. B. Maple, Physica C *184C*, 378 (1991).

L. De Cesare, Phys. Rev. B *43*, 10555 (1991).

P. G. de Gennes, Rev. Mod. Phys. *36*, 225 (1964).

P. G. de Gennes, "Superconductivity of Metals and Alloys," Benjamin, New York, 1966.

L. Degiorgi, P. Wachter, G. Grüner, S.-M. Huang, J. Wiley, and R. B. Kaner, Phys. Rev. Lett. *69*, 2987 (1992).

L. J. de Jongh, Europ. J. Solid State Inorg. Chem., in press.

C. Dekker, W. Eidelloth, and R. H. Koch, Phys. Rev. Lett. *68*, 3347 (1992).

O. L. de Lange and V. V. Gridin, Phys. Rev. B *46*, 5735 (1992).

P. Delsing, K. K. Likharev, L. S. Kuzmin, and T. Claeson, Phys. Rev. Lett. *63*, 1180, 1861 (1989).

H. Dersch and G. Blatter, Phys. Rev. B *38*, 11391 (1988).

S. M. DeSoto, C. P. Slichter, H. H. Wang, U. Geiser, and J. M. Williams, Phys. Rev. Lett. *70*, 2956 (1993).

D. S. Dessau, B. O. Wells, Z.-X. Shen, W. E. Spicer, A. J. Arko, R. S. List, D. B. Mitzi, and A. Kapitulnik, Phys. Rev. Lett. *66*, 2160 (1991).

D. S. Dessau, Z.-X. Shen, B. O. Wells, D. M. King, W. E. Spicer, A. J. Arko, L. W. Lombardo, D. B. Mitzi, and A. Kapitulnik, Phys. Rev. B *45*, 5095 (1992).

G. Deutscher and P. Chaudhari, Phys. Rev. B *44*, 4664 (1991).

F. Devaux, A. Manthiram, and J. B. Goodenough, Phys. Rev. B *41*, 8723 (1990).

M. J. DeWeert, D. A. Papaconstantopoulos, and W. E. Pickett, Phys. Rev. B *39*, 4235 (1989).

F. W. de Wette, A. D. Kulkarni, J. Prade, U. Schröder, and W. Kress, Phys. Rev. B *42*, 6707 (1990).

A. DiChiara, F. Fontana, G. Peluso, and F. Tafuri, Phys. Rev. B *44*, 12026 (1991).

A. DiChiara, F. Fontana, G. Peluso, and F. Tafuri, Phys. Rev. B *48*, 6695 (1993).

F. Diederich and R. L. Whetten, Acc. Chem. Res. *25*, 119 (1992).

D. DiMarzio, H. Wiesmann, D. H. Chen, and S. M. Heald, Phys. Rev. B *42*, 294 (1990).

H.-Q. Ding, Phys. Rev. Lett. *68*, 1927 (1992).

T. R. Dinger, T. K. Worthington, W. J. Gallagher, and R. L. Sandstrom, Phys. Rev. Lett. *58*, 2687 (1987).

M. I. Dobroliubov and S. Yu. Khlebnikov, Phys. Rev. Lett. *67*, 2084 (1991).

R. K. Dodd, J. C. Eilbeck, J. D. Gibbon, and H. C. Morris, "Solitons and Nonlinear Wave Equations." Academic Press, New York, 1982.

I. Doi, K. Sano, and K. Takano, Phys. Rev. B *45*, 274 (1992).

G. J. Dolan, G. V. Chandrashekhar, T. R. Dinger, C. Feild, and F. Holtzberg, Phys. Rev. Lett. *62*, 827 (1989a).

G. J. Dolan, F. Holtzberg, C. Feild, and T. R. Dinger, Phys. Rev. Lett. *62*, 2184 (1989b).

S. Doniach, *in* "Proceedings Los Alamos Symposium, High Temp. Supercond.," p. 406. Addison–Wesley, New York, 1989.

S. Doniach and M. Inui, Phys. Rev. B *41*, 6668 (1990).

G. Dopf, A. Muramatsu, and W. Hanke, Phys. Rev. Lett. *68*, 353 (1992).

M. M. Doria, J. E. Gubernatis, and D. Rainer, Phys. Rev. B *41*, 6335 (1990).

A. T. Dorsey, M. Huang, and M. P. A. Fisher, Phys. Rev. B *45*, 523 (1992).

S. X. Dou, H. K. Liu, A. J. Bourdillon, M. Kviz, N. X. Tan, and C. C. Sorrell, Phys. Rev. B *40*, 5266 (1989).

R. A. Doyle, O. L. deLange, and V. V. Gridin, Phys. Rev. B *45*, 12580 (1992).

T. B. Doyle and R. A. Doyle, Phys. Rev. B *47*, 8111 (1993).

P. G. Drazin and R. S. Johnson, "Solitons: An Introduction." Cambridge Univ. Press Cambridge, U.K., 1989.

L. Dresner, Cryogenics, May, p. 285, 1978.

M. S. Dresselhaus, G. Dresselhaus, and R. Saito, *in* "Physical Properties of High Temperature Superconductors" (D. M. Ginsberg, Ed.), Vol. 4, Chap. 7, World Scientific, Singapore, 1994.

H. Drulis, Z. G. Xu, J. W. Brill, L. E. De Long, and J.-C. Hou, Phys. Rev. Lett. *44*, 4731 (1991).

Q. Du, M. D. Gunzberger, and J. S. Peterson, Phys. Rev. B *46*, 9027 (1992).

L. Dubeck, P. Lindenfeld, E. A. Lynten, and H. Rohrer, Rev. Mod. Phys. *36*, 110 (1964).

A. Dulčic, R. H. Crepeau, and J. H. Freed, Phys. Rev. B *39*, 4249 (1989).

B. D. Dunlap, M. Slaski, Z. Sungaila, D. G. Hinks, K. Zhang, C. Segre, S. K. Malik, and E. E. Alp, Phys. Rev. B *37*, 592 (1988).

C. Duran, J. Yazyi, F. de la Cruz, D. J. Bishop, D. B. Mitzi, and A. Kapitulnik, Phys. Rev. B *44*, 7737 (1991).

C. H. Eab and I. M. Tang, Phys. Rev. B *40*, 4427 (1989).

D. E. Eastman, Solid State Commun. *7*, 1697 (1969).

C. Ebner and D. Stroud, Phys. Rev. B *31*, 165 (1985).

C. Ebner and D. Stroud, Phys. Rev. B *39*, 789 (1989).

U. Eckern and E. B. Sonin, Phys. Rev. B *47*, 505 (1993).

G. L. Eesley, J. Heremans, M. S. Meyer, G. L. Doll, and S. H. Liou, Phys. Rev. Lett. *65*, 3445 (1990).

H. Eikmans and J. E. van Himbergen, Phys. Rev. *44*, 6937 (1991).

R. Eisberg and R. Resnick, "Quantum Physics," Wiley, New York, 1974.

J. W. Ekin, H. R. Hart, and A. R. Gaddipati, J. Appl. Phys. *68*, 2285 (1990).

J. W. Ekin, K. Salama, and V. Selvamanickam, Appl. Phys. Lett. *59*, 360 (1991).

T. Ekino and J. Akimitsu, Phys. Rev. B *40*, 6902, 7364 (1989a).

T. Ekino and J. Akimitsu, J. Phys. Soc. Jpn. *58*, 2135 (1989b).

T. Ekino and J. Akimitsu, Phys. Rev. B *42*, 8049 (1990).

G. M. Eliashberg, Zh. Eksp. Teor. Fiz. *38*, 966 (1960a).

G. M. Eliashberg, Zh. Eksp. Teor. Fiz. *39*, 1437 (1960b).

B. Ellman, J. Yang, T. F. Rosenbaum, and E. Bucher, Phys. Rev. Lett. *64*, 1569 (1990).

V. J. Emery, Phys. Rev. Lett. *58*, 2794 (1987) (reprinted in Halley, 1988, p. 227).

V. J. Emery and G. Reiter, Phys. Rev. B *38*, 4547 (1988).

J. W. Emsley, J. Feeney, and L. H. Sutcliffe, "High Resolution Nuclear Magnetic Resonance Spectroscopy, Pergamon, New York, 1965, Vol. 1.

D. Emin, Phys. Rev. B *49*, 9157 (1994).

P. Entel and J. Zielinski, Phys. Rev. B *42*, 307 (1990).

O. Entin-Wohlman and Y. Imry, Phys. Rev. B *40*, 6731 (1989).

L. E. Erickson, Phys. Rev. B *43*, 12723 (1991).

R. Escudero, E. Guarner, and F. Morales, Physica C *162C–164C*, 1059 (1989).

R. Escudero, F. Morales, and E. Guarner, Physica C *166C*, 15 (1990).

D. Estève, J. M. Martinis, C. Urbina, M. H. Devoret, G. Collin, P. Monod, M. Ribault, and A. Revcolevschi, Europhys. Lett. *3*, 1237 (1987).

J. P. Estrera and G. B. Arnold, Phys. Rev. B *39*, 2094 (1989).

L. M. Falicov and C. R. Proetto, Phys. Rev. B *47*, 14407 (1993).

M. M. Fang, J. E. Ostenson, D. K. Finnemore, D. E. Farrell, and N. P. Bansal, Phys. Rev. B *39*, 222 (1989).

H. A. Farach, E. Quagliata, T. Mzoughi, M. A. Mesa, C. P. Poole, Jr., and R. Creswick, Phys. Rev. B *41*, 2046 (1990).

D. E. Farrell, B. S. Chandrasekhar, M. R. DeGuire, M. M. Fang, V. G. Kogan, J. R. Clem, and D. K. Finnemore, Phys. Rev. B *36*, 4025 (1987).

D. E. Farrell, C. M. Williams, S. A. Wolf, N. P. Bansal, and V. G. Kogan, Phys. Rev. Lett. *61*, 2805 (1988).

D. E. Farrell, M. M. Fang, and N. P. Bansal, Phys. Rev. B *39*, 718 (1989a).

D. E. Farrell, S. Bonham, J. Foster, Y. C. Chang, P. Z. Jiang, K. G. Vandervoort, D. J. Lam, and V. G. Kogan, Phys. Rev. Lett. *63*, 782 (1989b).

D. E. Farrell, R. G. Beck, M. F. Booth, C. J. Allen, E. D. Bukowski, and D. M. Ginsberg, Phys. Rev. B *42*, 6758 (1990a).

D. E. Farrell, C. J. Allen, R. C. Haddon, and S. V. Chichester, Phys. Rev. B *42*, 8694 (1990b).

D. E. Farrell, J. P. Rice, D. M. Ginsberg, and J. Z. Liu, Phys. Rev. Lett. *64*, 1573 (1990c).

D. E. Farrell, J. P. Rice, and D. M. Ginsberg, Phys. Rev. Letts. *67*, 1165 (1991).

D. E. Farrell, *in* "Physical Properties of High Temperature Superconductors," (D. M. Ginsberg, Ed.), Vol. 4, Chap. 2, World Scientific, Singapore, 1994.

E. Faulques and R. E. Russo, *in* "Applications of Analytical Techniques to the Characterization of Materials" (D. L. Perry, Ed.), p. 59. Plenum, New York, 1991.

R. Fazio and G. Schön, Phys. Rev. B *43*, 5307 (1991).

J. F. Federici, B. I. Greene, H. Hartford, and E. S. Hellman, Phys. Rev. B *42*, 923 (1990).

R. Feenstra, D. K. Christen, C. Klabunde, and J. D. Budai, Phys. Rev. B *45*, 7555 (1992).

R. Fehrenbacher, V. B. Geshkenbein, and G. Blatter, Phys. Rev. B *45*, 5450 (1992).

D. Feinberg and C. Villard, Phys. Rev. Lett. *65*, 919 (1990).

L. F. Feiner, M. Grilli, and C. DiCastro, Phys. Rev. B *45*, 10647 (1992).

L. F. Feiner, Phys. Rev. B *48*, 16857 (1993).

I. Felner, U. Yaron, Y. Yeshurun, G. V. Chandrashekhar, and F. Holtzberg, Phys. Rev. B *40*, 5329 (1989).

J. C. Fernandez, R. Grauer, K. Pinnow, and G. Reinisch, Phys. Rev. B *42*, 9987 (1990).

M. J. Ferrari, M. Johnson, F. C. Wellstood, J. Clarke, D. Mitzi, P. A. Rosenthal, C. B. Eom, T. H. Geballe, A. Kapitulnik, and M. R. Beasley, Phys. Rev. Lett. *64*, 72 (1989).

M. J. Ferrari, F. C. Wellstood, J. J. Kingston, and J. Clarke, Phys. Rev. Lett. *67*, 1346 (1991).

R. A. Ferrell, *in* "High Temperature Superconductivity" (J. W. Lynn, Ed.), Chap. 3, Springer-Verlag, Berlin, 1990.

K. Fesser, U. Sum, and H. Büttner, Phys. Rev. B *44*, 421 (1991).

A. L. Fetter and J. D. Walecka, "Quantum Theory of Many Particle Systems." McGraw–Hill, New York, 1971.

R. P. Feynman, "Lectures on Physics," Vol. 3, Chap. 21. Addison–Wesley, New York, 1965.

W. A. Fietz, M. R. Beasley, J. Silcox, and W. W. Webb, Phys. Rev. *136*, A335 (1964).

A. T. Fiory, M. Gurvitch, R. J. Cava, and G. P. Espinosa, Phys. Rev. B *36*, 7262 (1987).

A. T. Fiory, G. P. Espinosa, R. M. Fleming, G. S. Grader, M. Gurvitch, A. F. Hebard, R. E. Howard, J. R. Kwo, A. F. J. Levi, P. M. Mankiewich, S. Martin, C. E. Rice, L. F. Schneemeyer, and A. E. White, International Conference on Electronic Materials, Tokyo, 1988.

A. T. Fiory, S. Martin, R. M. Fleming, L. F. Schneemeyer, J. V. Waszczak, A. F. Hebard, and S. A. Sunshine, Physica C *162C–164C*, 1195 (1989).

A. T. Fiory, M. A. Paalanen, R. R. Ruel, L. F. Schneemeyer, and J. V. Waszczak, Phys. Rev. B *41*, 4805 (1990).

O. Fischer, Appl. Phys. *16*, 1 (1978).

O. Fischer, *in* "Earlier and Recent Aspects of Superconductivity" (J. G. Bednorz and K. A. Müller, Eds.), p. 96, Springer, Berlin, 1990.

P. Fischer, K. Kakurai, M. Steiner, K. N. Clausen, B. Lebech, F. Hulliger, H. R. Ott, P. Brüesch, and P. Unternährer, Physica C *152C*, 145 (1988).

P. Fischer, H. W. Neumüller, B. Roas, H. F. Braun, and G. Saemann-Ischenko, Solid State Commun. *72*, 871 (1989).

J. E. Fischer, P. A. Heiney, A. R. McGhie, W. J. Romanow, A. M. Denenstein, J. P. McCauley, Jr., and A. B. Smith, III, Science *252*, 1288 (1991).

J. E. Fischer, P. A. Heiney, and A. B. Smith, III, Acc. Chem. Res. *25*, 97 (1992).

K. H. Fischer, Physica C *178C*, 161 (1991).

D. S. Fisher and D. A. Huse, Phys. Rev. B *38*, 373, 386 (1988).

M. P. A. Fisher, Phys. Rev. Lett. *62*, 1415 (1989).

R. A. Fisher, S. Kim, B. F. Woodfield, N. E. Phillips, L. Taillefer, K. Hasselbach, J. Flouquet, A. L. Giorgi, and J. L. Smith, Phys. Rev. Lett. *62*, 1411 (1989).

M. P. A. Fisher, Phys. Rev. Lett. *65*, 923 (1990).

D. S. Fisher, M. P. A. Fisher, and D. A. Huse, Phys. Rev. B *43*, 130 (1991).

R. S. Fishman, Phys. Rev. B *38*, 11996 (1988).

R. S. Fishman, Phys. Rev. Lett. *63*, 89 (1989).

Z. Fisk, P. C. Canfield, W. P. Beyermann, J. D. Thompson, M. F. Hundley, H. R. Ott, E. Felder, M. B. Maple, M. A. Lopez de la Torre, P. Visani, and C. L. Seaman, Phys. Rev. Lett. *67*, 3310 (1991).

M. D. Fiske, Rev. Mod. Phys. *36*, 221 (1964).

S. A. FitzGerald, S. G. Kaplan, A. Rosenberg, A. J. Sievers, and R. A. S. McMordie, Phys. Rev. B *45*, 10165 (1992).

R. L. Fleisher, H. R. Hart, Jr., K. W. Lay, and F. E. Luborsky, Phys. Rev. B *40*, 2163 (1989).

R. B. Flippen, Phys. Rev. B *44*, 7708 (1991).

M. Florjanczyk and M. Jaworski, Phys. Rev. B *40*, 2128 (1989).

R. Flükiger and W. Klose, "Landolt–Börnstein, Group III Solid State Physics," Vol. 21, Superconductors. Springer-Verlag, Berlin/New York, 1993.

M. Foldeaki, M. E. McHenry, and R. C. O'Handley, Phys. Rev. B *39*, 2883 (1989).

W. Folkerts and C. Haas, Phys. Rev. B *41*, 6341 (1990).

S. Foner, E. J. McNiff, Jr., D. Heiman, S.-M. Huang, and R. B. Kaner, Phys. Rev. B *46*, 14936 (1992).

J. Fonteuberta and L. Fàbrega, *in* "Studies in High Temperature Superconductors" (A. V. Narlikar, Ed.), Vol. 16, Nova Sci., New York, in press.

A. Forkl, T. Dragon, and H. Kronmüller, J. Appl. Phys. *67*, 3047 (1990).

A. Forkl, H. U. Habermeier, B. Liebold, T. Dragon, and H. Kronmüller, Physica C *180C*, 155 (1991).

M. Forsthuber and G. Hilscher, Phys. Rev. B *45*, 7996 (1992).

N. A. Fortune, K. Murata, K. Ikeda, and T. Takahashi, Phys. Rev. Lett. *68*, 2933 (1992).

C. M. Foster, K. F. Voss, T. W. Hagler, D. Mihailovic, A. S. Heeger, M. M. Eddy, W. L. Olsen, and E. J. Smith, Solid State Commun. *76*, 651 (1990).

H. Frahm, S. Ullah, and A. T. Dorsey, Phys. Rev. *66*, 3067 (1991).

Z. Frait, D. Fraitová, and L. Pust, J. Phys. Colloque C *8*, 2235 (1988a); Z. Frait, D. Fraitová, E. Pollert, and L. Pust, Phys. Status Solidi *146*, K119 (1988b).

J. P. Franck, *in* "Physical Properties of High Temperature Superconductors," (D. M. Ginsberg, Ed.), Vol. 4, Chap. 4, World Scientific, Singapore (1994).

G. Frank, Ch. Ziegler and W. Göpel, Phys. Rev. B *43*, 2828 (1991).

A. Freimuth, C. Hohn, and M. Galffy, Phys. Rev. B *44*, 10396 (1991).

T. Freltoft, G. Shirane, S. Mitsuda, J. P. Remeika, and A. S. Cooper, Phys. Rev. B *37*, 137 (1988).

T. Freltoft, H. J. Jensen, and P. Minnhagen, Solid State Commun. *78*, 635 (1991).

T. A. Friedmann, J. P. Rice, J. Giapintzakis, and D. M. Ginsberg, Phys. Rev. B *39*, 4258 (1989).

B. Friedl, C. Thomsen, and M. Cardona, Phys. Rev. Lett. *65*, 915 (1990).

T. A. Friedmann, M. W. Rabin, J. Giapintzakis, J. P. Rice, and D. M. Ginsberg, Phys. Rev. B *42*, 6217 (1990).

H. Fröhlich, Phys. Rev. *79*, 845 (1950).

L. Fruchter and I. A. Campbell, Phys. Rev. B *40*, 5158 (1989).

A. Fujimori, S. Takekawa, E. Takayama-Muromachi, Y. Uchida, A. Ono, T. Takahashi, Y. Okabe and H. Katayama-Yoshida, Phys. Rev. B *39*, 2255 (1989).

H. Fujishita, M. Sera, and M. Sato, Physica C *175*, 165 (1991).

K. Fujiwara, Y. Kitaoka, K. Ishida, K. Asayama, Y. Shimakawa, T. Manako, and Y. Kubo, Physica C *184*, 207 (1991).

T. Fukami, T. Kamura, A. A. Youssef, Y. Hori, and S. Mase, Physica C *159*, 427 (1989).

T. Fukami, K. Hayashi, T. Yamamoto, T. Nishizaki, Y. Horie, F. Ichikawa, T. Aomine, V. Soares, and L. Rinderer, Physica C *184*, 65 (1991a).

T. Fukami, T. Yamamoto, K. Hayashi, T. Nishizaki, Y. Hori, F. Ichikawa, and T. Aomine, Physica C *185*, 2255 (1991b).

T. A. Fulton, P. L. Gammei, D. J. Bishop, L. N. Dunkleberger, and G. J. Dolan, Phys. Rev. Lett. *63*, 1307 (1989).

P. Fumagalli and J. Schoenes, Phys. Rev. B *44*, 2246 (1991).

A. Furusaki, H. Takayanagi, and M. Tsukada, Phys. Rev. Lett. *67*, 132 (1991).

A. Furusaki and M. Tsukada, Phys. Rev. B *43*, 10164 (1991).

A. Furusaki and M. Ueda, Phys. Rev. B *45*, 10576 (1992).

A. Furusaki, H. Takayanagi, and M. Tsukada, Phys. Rev. B *45*, 10563 (1992).

M. Furuyama, N. Kobayashi, and Y. Muto, Phys. Rev. B *40*, 4344 (1989).

E. Gagliano and S. Bacci, Phys. Rev. B *42*, 8772 (1990).

P. L. Gai and J. M. Thomas, Supercond. Rev. *1*, 1 (1992).

M. C. Gallagher, J. G. Adler, J. Jung, and J. P. Franck, Phys. Rev. B *37*, 7846 (1988).

W. J. Gallagher, J. Appl. Phys. *63*, 4216 (1988).

C. F. Gallo, L. R. Whitney, and P. J. Walsh, *in* "Novel Superconductivity" (S. A. Wolf and V. Z. Kresin, Eds.), p. 385, Plenum, New York, 1987.

C. F. Gallo, L. R. Whitney, and P. J. Walsh, Mater. Res. Soc. Symp. Proc. *99*, 165 (1988).

P. L. Gammel, D. J. Bishop, G. J. Dolan, J. R. Kwo, C. A. Murray, L. F. Schneemeyer, and J. V. Waszczak, Phys. Rev. Lett. *59*, 2592 (1987).

P. L. Gammel, L. F. Schneemeyer, J. K. Waszczak, and A. J. Bishop, Phys. Rev. Lett. *61*, 1666 (1988).

P. L. Gammel, A. Hebard, and D. J. Bishop, Phys. Rev. B *40*, 7354 (1989).

J. T. Gammel, R. J. Donohoe, A. R. Bishop, and B. I. Swanson, Phys. Rev. B *42*, 10566 (1990).

P. L. Gammel, L. F. Schneemeyer, and D. J. Bishop, Phys. Rev. Lett. *66*, 953 (1991).

P. L. Gammel, D. J. Bishop, T. P. Rice, and D. M. Ginsberg, Phys. Rev. Lett. *68*, 3343 (1992).

F. Gao, G. L. Carr, C. D. Porter, D. B. Tanner, S. Etemad, T. Venkatesan, A. Inam, B. Dutta, X. D. Wu, G. P. Williams, and C. J. Hirschmugl, Phys. Rev. B *43*, 10383 (1991).

L. Gao, R. L. Meng, Y. Y. Xue, P. H. Hor and C. W. Chu, Appl. Phys. Lett. *58*, 92 (1991).

L. Gao, Z. J. Huang, R. L. Menag, J. G. Lin, F. Chen, L. Beauvais, Y. Y. Sun, Y. Y. Xue, and C. W. Chu, Physica C *213*, 261 (1993).

L. Gao, Y. Y. Xue, F. Chen, Q. Xiong, R. L. Meng, D. Ramirez, C. W. Chu, J. H. Eggert, and H. K. Mao, submitted for publication.

M. M. Garland, J. Mater. Res. *3*, 830 (1988).

P. Garoche, R. Brusett, D. Jérome, and K. Bechgaard, J. Physique Lett. *43*, L147 (1982).

L. J. Geerligs, M. Peters, L. E. M. de Groot, A. Verbruggen, and J. E. Mooij, Phys. Rev. Lett. *63*, 326 (1989).

I. I. Geguzin, I. Ya. Nikiforov, and G. I. Alperovitch, Fiz. Tverd. Tela. *15*, 931 (1973).

C. Geibel, S. Thies, D. Kaczorowski, A. Mehner, A. Grauel, B. Seidel, U. Ahlheim, R. Helfrich, K. Petersen, C. D. Bredl, and F. Steglich, Z. Phys. B—Cond. Matt. *83*, 305 (1991a).

C. Geibel, C. Schank, S. Thies, H. Kitzawa, C. D. Bredl, A. Böhm, M. Rau, A. Grauel, R. Caspary, R. Helfrich, U. Anlheim, G. Weber, and F. Steglich, Z. Phys. B—Cond. Matt. *84*, 1 (1991b).

C. Geibel, U. Ahlheim, C. D. Bredl, J. Diehl, A. Grauel, R. Helfrich, H. Kitazawa, R. Köhler, R. Modler, M. Lang, C. Schank, S. Thies, F. Steglich, N. Sato, and T. Komatsubara, Physica C *185*, 2651 (1991c).

A. K. Geim, I. V. Grigorieva, and S. V. Dubonos, Phys. Rev. B *46*, 324 (1992).

B. Y. Gelfand and B. I. Halperin, Phys. Rev. B *45*, 5517 (1992).

A. Gerber, Th. Grenet, M. Cyrot, and J. Beille, Phys. Rev. B *45*, 5099 (1992).

W. Gerhäuser, G. Ries, H. W. Neumüller, W. Schmidt, O. Eibl, G. Saemann-Ischenko, and S. Klaumünzer, Phys. Rev. Lett. *68*, 879 (1992).

V. B. Geshkenbein, V. M. Vinokur, and R. Fehrenbacher, Phys. Rev. B *43*, 3748 (1991).

S. K. Ghatak, A. Mitra, and D. Sen, Phys. Rev. B *45*, 951 (1992).

J. Ghijsen, L. H. Tjeng, H. Eskes, G. A. Sawatzky, and R. L. Johnson, Phys. Rev. B *42*, 2268 (1990).

I. Giaever and H. R. Zeller, Phys. Rev. Lett. *20*, 1504 (1968).

J. Giapintzakis, W. C. Lee, J. P. Rice, D. M. Ginsberg, I. M. Robertson, R. Wheeler, M. Kirk, and M.-Q. Ruault, Phys. Rev. B *45*, 10677 (1992).

M. A. M. Gijs, D. Scholten, Th. van Rooy, and A. M. Gerrits, Phys. Rev. B *41*, 11627 (1990a).

M. A. M. Gijs, A. M. Gerrits, and C. W. J. Beenakker, Phys. Rev. B *42*, 10789 (1990b).

L. R. Gilbert, R. Messier, and R. Roy, Thin Solid Films *54*, 129 (1978).

D. B. Gingold and C. J. Lobb, Phys. Rev. B *42*, 8220 (1990).

M. J. P. Gingras, Phys. Rev. B *45*, 7547 (1992).

D. M. Ginsberg (Ed.), "Physical Properties of High Temperature Superconductors," Vol. 1, World Scientific, Singapore, 1989.

D. M. Ginsberg (Ed.), "Physical Properties of High Temperature Superconductors," Vol. 2, World Scientific, Singapore, 1990.

D. M. Ginsberg (Ed.), "Physical Properties of High Temperature Superconductors," Vol. 3, World Scientific, Singapore, 1992.

D. M. Ginsberg (Ed.), "Physical Properties of High Temperature Superconductors," Vol. 4, World Scientific, Singapore, 1994.

V. L. Ginzburg and L. Landau, Zh. Eksp. Teor. Fiz. *20*, 1064 (1950).

V. L. Ginzburg and D. A. Kirzhnits, "High Temperature Superconductivity," Nauka, Moscow, 1977 [Engl. Transl. Consultants Bureau, New York, 1982].

S. L. Ginzburg, V. P. Khavronin, G. Yu. Logvinova, I. D. Luzyanin, J. Herrmann, B. Lippold, H. Börner and H. Schmiedel, Physica C *174*, 109 (1991).

J. I. Gittleman and B. Rosenblum, J. Appl. Phys. *39*, 2617 (1968).

S. H. Glarum, L. F. Schneemeyer, and J. V. Waszczak, Phys. Rev. B *41*, 1837 (1990).

N. E. Glass and D. Rogovin, Phys. Rev. B *39*, 11327 (1989).

D. Glatzer, A. Forkl, H. Theuss, H. U. Habermeier, and Kronmüller, Phys. Status Solidi *170*, 549 (1992).

L. I. Glazman and A. E. Koshelev, Phys. Rev. B *43*, 2835 (1991a); Physica C *173*, 180 (1991b).

H. R. Glyde, L. K. Moleko, and P. Findeisen, Phys. Rev. B *45*, 2409 (1992).

A. Gold and A. Ghazali, Phys. Rev. B *43*, 12952 (1991).

R. B. Goldfarb, A. F. Clark, A. I. Braginski, and A. J. Panson, Cryogenics *27*, 475 (1987a); R. B. Goldfarb, A. F. Clark, A. I. Braginski, and A. J. Panson, *in* "High Temperature Superconductors" (D. U. Grubser and M. Schluter, Eds.), p. 261, Mater. Res. Soc., Pittsburgh (1987b).

D. Goldschmidt, Phys. Rev. B *39*, 2372 (1989).

M. J. Goldstein and W. G. Moulton, Phys. Rev. B *40*, 8714 (1989).

M. Golosovsky, D. Davidov, E. Farber, T. Tsach, and M. Schieber, Phys. Rev. B *43*, 10390 (1991).

M. Golosovsky, Y. Naveh, and D. Davidov, Phys. Rev. B *45*, 7495 (1992).

J. B. Goodenough and J.-S. Zhou, Phys. Rev. B *42*, 4276 (1990).

J. B. Goodenough, J.-S. Zhou, and J. Chan, Phys. Rev. B *47*, 5275 (1993).

L. F. Goodrich, A. N. Srivastava, and T. C. Stauffer, J. Res. NIST *96*, 703 (1991).

L. F. Goodrich and A. N. Srivastava, Supercond. Industry Spring, 28 (1992).

L. P. Gor'kov, Zh. Eksp. Teor. Fiz. *36*, 1918. [Sov. Phys. JETP *36*, 1364] (1959).

L. P. Gor'kov, Sov. Phys. JETP *38*, 830 (1973).

L. P. Gor'kov, Sov. Phys. JETP Lett. *20*, 260 (1974).

U. Gottwick, R. Held, G. Sparn, F. Steglich, H. Rietschel, D. Ewert, B. Renker, W. Bauhofer, S. von Molnar, M. Wilhelm, and H. E. Hoenig, Europhys. Lett. *4*, 1183 (1987).

C. E. Gough, M. S. Colclough, E. M. Forgan, R. G. Jordan, M. Keene, C. M. Muirhead, A. I. M. Rae, N. Thomas, J. S. Abell, and S. Sutton, Nature *326*, 855 (1987).

M. E. Gouvêa, G. M. Wysin, A. R. Bishop, and F. G. Mertens, Phys. Rev. B *39*, 11840 (1989).

G. S. Grader, P. K. Gallagher, and E. M. Gyorgy, Appl. Phys. Lett. *51*, 1115 (1987).

J. E. Gaebner, R. C. Haddon, S. V. Chichester, and S. M. Glarum, Phys. Rev. B *41*, 4808 (1990).

R. Graham, M. Schlautmann, and D. L. Shepelyansky, Phys. Rev. Lett. *67*, 255 (1991).

K. E. Gray, R. T. Kampwirth, and D. E. Farrell, Phys. Rev. B *41*, 819 (1990).

K. E. Gray, D. H. Kim, B. W. Veal, G. T. Seidler, T. F. Rosenbaum, and D. E. Farrell, Phys. Rev. B *45*, 10071 (1992).

K. E. Gray, Appl. Supercond., (D. Shi, Ed.), in press.

L. H. Green and B. G. Bagley, *in* "Physical Properties of High Temperature Superconductors" (D. M. Ginsberg, Ed.), Vol. 2, Chap. 8, World Scientific, Singapore, 1990.

R. L. Greene, *in* "Organic Superconductivity" (V. Z. Kresin and W. A. Little, Eds.), p. 7. Plenum, New York, 1990.

E. Gregory, T. S. Kreilick, J. Wong, A. K. Ghosh, and W. B. Sampson, Cryogenics *27*, 178 (1987).

V. V. Gridin, P. Pernambuco-Wise, C. G. Trendall, W. R. Datars, and J. D. Garrett, Phys. Rev. B *40*, 8814 (1989).

D. G. Grier, C. A. Murray, C. A. Bolle, P. L. Gammel, D. J. Bishop, D. B. Mitzi, and A. Kapitulnik, Phys. Rev. Lett. *66*, 2770 (1991).

R. Griessen, Phys. Rev. Lett. *64*, 1674 (1990).

M. Grilli, R. Raimondi, C. Castellani, C. DiCastro, and G. Kotliar, Phys. Rev. Lett. *67*, 259 (1991).

H. H. Gröbke, Phys. Rev. B *41*, 11047 (1990).

N. Grønbech-Jensen, Phys. Rev. B *45*, 7315 (1992).

N. Grønbech-Jensen, N. F. Pedersen, A. Davidson, and R. D. Parmentier, Phys. Rev. B *42*, 6035 (1990).

N. Grønbech-Jensen, S. A. Hattel, and M. R. Samuelsen, Phys. Rev. *45*, 12457 (1991).

R. Gross, P. Chaudhari, M. Kawasaki, and A. Gupta, Phys. Rev. B *42*, 10735 (1990a).

R. Gross, P. Chaudhari, M. Kawasaki, M. B. Ketchen, and A. Gupta, Appl. Phys. Lett. *57*, 727 (1990b).

R. Gross and D. Koelle, Rept. Prog. Phys., *57*, 651 (1994).

D. Y. Gubser and M. Schluter, Eds., "High Temperature Superconductors," Proc. Symp., Spring Meet., Anaheim, CA, Apr. 1987, Mater. Res. Soc., Pittsburgh, 1987.

S. Guha, D. Peebles, and T. J. Wieting, Phys. Rev. B *43*, 13092 (1991).

F. Guinea and G. Zimanyi, Phys. Rev. B *47*, 501 (1993).

B. Gumhalter and V. Zlatic, Phys. Rev. B *42*, 6446 (1990).

J. Guo, D. E. Ellis, E. E. Alp, and G. L. Goodman, Phys. Rev. B *42*, 251 (1990).

A. Gupta, P. Esquinazi, and H. F. Braun, Physica C *184*, 393 (1991).

R. P. Gupta and M. Gupta, Phys. Rev. B *47*, 11635 (1993a); Phys. Rev. B *48*, 16068 (1993b).

A. Gurevich and H. Küpfer, Phys. Rev. B *48*, 6477 (1993).

M. Gurvitch and A. T. Fiory, Phys. Rev. Lett. *59*, 1337 (1987a); Appl. Phys. Lett. *51*, 1027 (1987b); *in* "Novel Superconductivity" (S. A. Wolf and V. Z. Kresin, Eds.), p. 663, Plenum, New York, 1987c.

M. Gurvitch, A. T. Fiory, L. F. Schneemeyer, R. J. Cava, G. P. Espinosa, and J. V. Waszczak, Physica C. *153–155*, 1369 (1988).

H. Gutfreund and W. A. Little, *in* "Highly Conducting One Dimensional Solids" (J. T. Devreese, R. P. Evrard, and V. E. van Doren, Eds.), Chap. 7, Plenum, New York, 1979.

F. Gygi and M. Schlüter, Phys. Rev. Lett. *65*, 1820 (1990a); Phys. Rev. B *41*, 822 (1990b); *43*, 7609 (1991).

R. C. Haddon, Acc. Chem. Res. *25*, 127 (1992).

S. J. Hagen, T. W. Jing, Z. Z. Wang, J. Horvath, and N. P. Ong, Phys. Rev. B *37*, 7928 (1988).

S. J. Hagen, C. J. Lobb, R. L. Greene, M. G. Forrester and J. H. Kang, Phys. Rev. B *41*, 11630 (1990a).

S. J. Hagen, C. J. Lobb, R. L. Greene, M. G. Forrester and J. Talvacchio, Phys. Rev. B *42*, 6777 (1990b).

R. R. Hake, Phys. Rev. *166*, 471 (1968).

K. Hallberg, A. G. Rojo, and C. A. Balseiro, Phys. Rev. B *43*, 8005 (1991).

J. W. Halley, (Ed.), "Theories of High Temperature Superconductivity," Addison Wesley, Reading, MA, 1988.

B. I. Halperin, Phys. Rev. Lett. *52*, 1583, 2390 (1984).

N. Hamada, S. Massidda, A. J. Freeman, and J. Redinger, Phys. Rev. B *40*, 4442 (1989).

D. R. Hamann and L. F. Mattheiss, Phys. Rev. B *38*, 5138 (1988).

P. D. Hambourger and F. J. Di Salvo, Physica B *99*, 173 (1980).

C. A. Hamilton, Phys. Rev. B *5*, 912 (1972).

P. C. Hammel, M. Takigawa, R. H. Heffner, Z. Fisk, and K. C. Ott, Phys. Rev. Lett. *63*, 1992 (1989).

P. C. Hammel, A. P. Reyes, Z. Fisk, M. Takigawa, J. D. Thompson, R. H. Heffner, W.-W. Cheong, and J. E. Schirber, Phys. Rev. B *42*, 6781 (1990).

D. P. Hampshire, X. Cai, J. Seuntjens, and D. C. Larbalestler, Supercond. Sci. Technol. *1*, 12 (1988).

S. Han, L. F. Cohen, and E. L. Wolf, Phys. Rev. B *42*, 8682 (1990a).

S. G. Han, Z. V. Vardeny, K. S. Wong, and O. G. Symko, Phys. Rev. Lett. *65*, 2708 (1990b).

Z. P. Han, R. Dupree, D. McK. Paul, A. P. Howes, and L. W. J. Caves, Physica C *181*, 355 (1991).

Z. P. Han, R. Dupree, A. Gencten, R. S. Liu, and P. P. Edwards, Phys. Rev. Lett. *69*, 1256 (1992).

S. H. Han, C. C. Almasan, M. C. de Andrade, Y. Dalichaouch, and M. B. Maple, Phys. Rev. B *46*, 14290 (1992).

T. Hanaguri, T. Fukase, I. Tanaka, and H. Kojima, Phys. Rev. B *48*, 9772 (1993).

M. Hangyo, N. Nagasaki, and S. Nakashima, Phys. Rev. B *47*, 14595 (1993).

Z. Hao and J. R. Clem, Phys. Rev. Lett. *67*, 2371 (1991).

Z. Hao, J. R. Clem, M. W. McElfresh, L. Civale, A. P. Malozemoff, and F. Holtzberg, Phys. Rev. B *43*, 2844 (1991).

Z. Hao and J. R. Clem, Phys. Rev. B *46*, 5853 (1992).

J. Hara, M. Ashida, and K. Nagai, Phys. Rev. B *47*, 11263 (1993).

A. B. Harris and R. V. Lange, Phys. Rev. *157*, 295 (1967).

R. K. Harris, "Nuclear Magnetic Resonance Spectroscopy," Halsted, 1986.

D. C. Harris, S. T. Herbert, D. Stroud, and J. C. Garland, Phys. Rev. Lett. *67*, 3606 (1991).

D. R. Harshman, L. F. Schneemeyer, J. V. Waszczak, G. Aeppli, R. J. Cava, B. Batlogg, L. W. Rupp, E. J. Ansaldo, and D. Ll. Williams, Phys. Rev. B *39*, 851 (1989).

D. R. Harshman, R. N. Kleiman, R. C. Haddon, S. V. Chichester-Hicks, M. L. Kaplan, L. W. Rupp, Jr., T. Pfiz, D. Ll. Williams, D. B. Mitzi, Phys. Rev. Lett. *64*, 1293 (1990).

D. R. Harshman and A. P. Mills, Jr., Phys. Rev. B *45*, 10684 (1992).

M. Hase, I. Terasaki, A. Maeda, K. Uchinokura, T. Kimura, K. Kishio, I. Tanaka, and H. Kojima, Physica C *185–189*, 1855 (1991).

T. Hasegawa, H. Ikuta, and K. Kitazawa, *in* "Physical Properties of High Temperature Superconductors" (D. M. Ginsberg, Ed.), Vol. 3, Chap. 7, World Scientific, Singapore, 1992.

S. Hasegawa, T. Matsuda, J. Endo, N. Osakabe, M. Igarashi, T. Kobayashi, M. Naito, A. Tonomura, and R. Aoki, Phys. Rev. B *43*, 7631 (1991).

K. C. Hass, Solid State Phys. *42*, 213 (1989).

N. Hatakenaka, S. Kurihara, and H. Takayanagi, Phys. Rev. B *42*, 3987 (1990).

J. Hauck, S. Denker, H. Hindriks, S. Ipta, and K. Mika, Z. Phys. B *84*, 31 (1991).

D. B. Haviland, Y. Liu, and A. M. Goldman, Phys. Rev. Lett. *62*, 2180 (1989).

R. M. Hazen, L. W. Finger, R. J. Angel, C. T. Prewitt, N. L. Ross, H. K. Mao, C. G. Hadidiacos, P. H. Hor, A. L. Meng, and C. W. Chu, Phys. Rev. B *35*, 7238 (1987).

R. M. Hazen, L. W. Finger, R. J. Angel, C. T. Prewitt, N. L. Ross, C. G. Hadidiacos, P. J. Heaney, D. R. Veblen, Z. Z. Sheng, A. El Ali, and A. M. Hermann, Phys. Rev. Lett. *60*, 1657 (1988).

R. M. Hazen, *in* "Physical Properties of High Temperature Superconductors" (D. M. Ginsberg, Ed.), Vol. 2, Chap. 3, World Scientific, Singapore, 1990.

A. F. Hebard, P. L. Gammel, C. E. Rice, and A. F. J. Levi, Phys. Rev. B *40*, 5243 (1989).

A. F. Hebard and M. A. Paalanen, Phys. Rev. Lett. *65*, 927 (1990).

A. F. Hebard, M. J. Rosseinsky, R. C. Haddon, D. W. Murphy, S. H. Glarum, T. T. M. Palstra, A. P. Ramirez, and A. R. Kortan, Nature, *350*, 600 (1991).

A. F. Hebard, *in* "Proc. R. L. Orbach Symp. Random Magnetism and High T_c Supercond.," World Scientific, Singapore, 1994.

A. F. Hebard, *in* "Strongly Correlated Electronic Materials" (K. S. Bedell, Ed.), Addison-Wesley, New York, 1994.

S. E. Hebboul and J. C. Garland, Phys. Rev. B *43*, 13703 (1991).

P. Hedegård and M. B. Pedersen, Phys. Rev. B *42*, 10035 (1990).

R. H. Heffner, J. L. Smith, J. O. Willis, P. Birrer, C. Baines, F. N. Gygax, B. Hitti, E. Lippelt, H. R. Ott, A. Schenck, E. A. Knetsch, J. A. Mydosh, and D. E. MacLaughlin, Phys. Rev. Lett. *65*, 2816 (1990).

R. Heid, Phys. Rev. B *45*, 5052 (1992).

R. A. Hein, T. L. Francavilla, and D. H. Liebenberg (Eds.), "Magnetic Susceptibility of Superconductors and Other Spin Systems," Plenum, New York, 1991.

C. S. Hellberg and E. J. Mele, Phys. Rev. B *48*, 646 (1993).

E. S. Hellman, B. Miller, J. M. Rosamilia, E. H. Hartford, and K. W. Baldwin, Phys. Rev. B *44*, 9719 (1991).

F. Hellman and T. H. Geballe, Phys. Rev. B *36*, 107 (1987).

N. F. M. Henry and K. Lonsdale, "International Tables for X-Ray Crystallography," Kynboh, Birmingham, England, 1965.

J. Heremans, D. T. Morelli, G. W. Smith, and S. C. Strite III, Phys. Rev. B *37*, 1604 (1988).

F. Herman, R. V. Kasowski, and W. Y. Hsu, Phys. Rev. B *36*, 6904 (1987).

A. M. Hermann and J. V. Yakhmi, "Thallium Based High Temperature Superconductors," Dekker, Basel, 1993.

S. L. Herr, K. Kamarás, D. B. Tanner, S.-W. Cheong, G. R. Stewart, and Z. Fisk, Phys. Rev. B *43*, 7847 (1991).

D. W. Hess, T. A. Tokuyasu, and J. A. Sauls, Phys. Condens. Matt. *1*, 8135 (1989).

H. F. Hess, R. B. Robinson, R. C. Dynes, J. M. Valles, Jr., and J. V. Waszczak, Phys. Rev. Lett. *62*, 214 (1989).

H. F. Hess, R. B. Robinson, and J. V. Waszczak, Phys. Rev. Lett. *64*, 2711 (1990).

H. F. Hess, R. B. Robinson, and J. V. Waszczak, Physica B *169*, 422 (1991).

J. D. Hettinger and D. G. Steel, *in* "High Temperature Superconducting Science" (D. Shi, Ed.), Pergamon, New York, 1994.

J. M. Hettinger, A. G. Swanson, W. J. Skocpol, J. S. Brooks, J. M. Graybeal, P. M. Mankiewich, R. E. Howard, B. L. Straughn, and E. G. Burkhardt, Phys. Rev. Lett. *62*, 2044 (1989).

R. E. Hetzel, A. Sudbø, and D. Huse, Phys. Rev. Lett. *69*, 518 (1992).

E. T. Heyen, R. Liu, C. Thomsen, R. Kremer, M. Cardona, J. Karpinski, E. Kaldis, and S. Rusiecki, Phys. Rev. B *41*, 11058 (1990a).

E. T. Heyen, S. N. Rashkeev, I. I. Mazin, O. K. Andersen, R. Liu, M. Cardona, and O. Jepsen, Phys. Rev. Lett. *65*, 3048 (1990b).

E. T. Heyen, G. Kliche, W. Kress, W. König, M. Cardona, E. Rampf, J. Prade, U. Schröder, A. D. Kulkarni, F. W. de Wette, S. Piñol, D. McK. Paul, E. Morán, and M. A. Alario-Franco, Solid State Commun. *74*, 1299 (1990c).

E. Heyen, M. Cardona, J. Karpinski, E. Kaldis, and S. Rusiecki, Phys. Rev. B *43*, 12958 (1991).

Y. Hidaka, Y. Enomoto, M. Suzuki, M. Oda, and T. Murakami, Jpn. J. Appl. Phys. *26*, L377 (1987).

S. Hikami and A. I. Larkin, Mod. Phys. Lett. B *2*, 693 (1988).

S. Hikami and A. Fujita, Prog. Theor. Phys. *83*, 443 (1990a); Phys. Rev. B *41*, 6379 (1990b).

S. Hikami, A. Fujita, and A. Larkin, Phys. Rev. B *44*, 10400 (1991).

M. Hikita, Y. Tajima, A. Katsui, Y. Hidaka, T. Iwata, and S. Tsurumi, Phys. Rev. B *36*, 7199 (1987).

M. Hikita and M. Suzuki, Phys. Rev. B *39*, 4756 (1989).

G. Hilscher, H. Michor, N. M. Hong, T. Holubar, W. Perthold, M. Vybornov, and P. Rogl, *in* Int. Conf. Strongly Correlated Electron Systems, Amsterdam, Netherlands, Aug. 1994.

D. G. Hinks, D. R. Richards, B. Dabrowski, D. T. Marx, and A. W. Mitchell, Nature *335*, 419 (1988).

D. G. Hinks, B. Dabrowski, D. R. Richards, J. D. Jorgensen, S. Pei, and J. F. Zasadzinski, Mat. Res. Soc. Symp. Proc. *156*, 357 (1989).

J. E. Hirsch, Phys. Rev. B *31*, 4403 (1985a); Phys. Rev. Lett. *54*, 1317 (1985b).

J. E. Hirsch, Phys. Rev. Lett. *59*, 228 (1987).

T. Hocquet, P. Mathieu, and Y. Simon, Phys. Rev. B *46*, 1061 (1992).

J. A. Hodges, P. Imbert, and G. Jéhanno, Solid State Commun. *64*, 1209 (1987).

U. Hofmann and J. Keller, Z. Phys. B. Cond. Matter *74*, 499 (1989).

C. Hohn, M. Galffy, A. Dascoulidou, A. Freimuth, H. Soltner, and U. Poppe, Z. Phys. B *85*, 161 (1991).

K. Holczer, O. Klein, G. Grüner, J. D. Thompson, F. Deiderich, and R. L. Whetten, Phys. Rev. Lett. *67*, 271 (1991).

T. Holst, J. B. Hansen, N. Grønbech-Jensen, and J. A. Blackburn, Phys. Rev. B *42*, 127 (1990).

T. Holst and J. B. Hansen, Phys. Rev. B *44*, 2238 (1991).

X. Q. Hong and J. E. Hirsch, Phys. Rev. B *46*, 14702 (1992).

T. Honma, K. Yamaya, F. Minami, and S. Takekawa, Physica C *176*, 209 (1991).

B. Hopfengärtner, B. Hensel, and G. Saemann-Ischenko, Phys. Rev. B *44*, 741 (1991).

M. L. Horbach, F. L. J. Vos, and W. van Saarloos, Phys. Rev. B *48*, 4061 (1993).

M. L. Horbach, F. L. J. Vos, and W. van Saarloos, Phys. Rev. B *49*, 3539 (1994).

M. Horvatic, T. Auler, C. Berthier, Y. Berthier, P. Butaud, W. G. Clark, J. A. Gillet, P. Ségransan, and J. Y. Henry, Phys. Rev. B *47*, 3461 (1993).

A. Houghton, R. A. Pelcovits, and A. Sudbø, Phys. Rev. B *40*, 6763 (1989); *41*, 4785(E) (1990).

A. P. Howes, R. Dupree, D. McK. Paul, and S. Male, Physica C *185–189*, 1137 (1991).

T. C. Hsu and P. W. Anderson, Physica C *162–164*, 1445 (1989).

Q. Hu and M. Tinkham, Phys. Rev. B *39*, 11358 (1989).

Q. Hu, C. A. Mears, P. L. Richards, and F. L. Lloyd, Phys. Rev. Lett. *64*, 2945 (1990).

G. Y. Hu and R. F. O'Connell, Phys. Rev. B *47*, 8823 (1993).

W. F. Huang, P. J. Ouseph, K. Fang, and Z. J. Xu, Solid State Commun. *66*, 283 (1988).

Z. J. Huang, Y. Y. Xue, P. H. Hor, and C. W. Chu, Physica C *176*, 195 (1991a).

Z. J. Huang, H. H. Fang, Y. Y. Xue, P. H. Hor, C. W. Chu, M. L. Norton, and H. Y. Tang, Physica C *180*, 331 (1991b).

M.-Z. Huang, Y.-N. Xu, and W. Y. Ching, Phys. Rev. B *46*, 6572 (1992).

Z. J. Huang, Y. Y. Xue, R. L. Meng, and C. W. Chu, Phys. Rev. B *49*, 4218 (1994).

J. Hubbard, Proc. Royal Soc. London A *276*, 238 (1963).

J. Hubbard, Proc. Royal Soc. London A *281*, 401 (1964).

R. P. Huebener, R. T. Kampwirth, and A. Seher, J. Low Temp. Phys. *2*, 113 (1970).

R. P. Huebener, "Magnetic Flux Structures in Superconductors," Springer Verlag, Berlin, 1979.

R. P. Huebener, A. V. Ustinov, and V. K. Kaplunenko, Phys. Rev. B *42*, 4831 (1990).

R. P. Huebener, Physica C *168*, 605 (1990).

H. J. Hug, A. Moser, I. Parashikov, O. Fritz, B. Stiefel, H.-J. Güntherodt, and H. Thomas, Physica, to be published.

N. H. Hur, H.-G. Lee, J.-H. Park, H.-S. Shin, and I.-S. Yang, Physica C *218*, 365 (1993).

N. H. Hur, N. H. Kim, S. H. Kim, Y. K. Park, and J. C. Park, Physica C, *231*, 227 (1994).

M. S. Hybertsen, E. B. Stechel, W. M. C. Foulkes, and M. Schlüter, Phys. Rev. B *45*, 10032 (1992).

T. L. Hylton and M. R. Beasley, Phys. Rev. B *41*, 11669 (1990).

O. B. Hyun, D. K. Finnemore, L. Schwartzkopf, and J. R. Clem, Phys. Rev. Lett. *58*, 599 (1987).

O. B. Hyun, J. R. Clem, and D. K. Finnemore, Phys. Rev. B *40*, 175 (1989).

M. Iansiti, M. Tinkham, A. T. Johnson, W. F. Smith, and C. J. Lobb, Phys. Rev. B *39*, 6465 (1989).

H. Ihara, R. Sugise, K. Hayashi, N. Terada, M. Jo, M. Hirabayashi, A. Negishi, N. Atoda, H. Oyanagi, T. Shimomura, and S. Ohashi, Phys. Rev. B *38*, 11952 (1988).

H. Ihara, M. Hirabayashi, H. Tanino, K. Tokiwa, H. Ozawa, Y. Akahama, and H. Kawamura, Jpn. J. Appl. Phys. *32*, L1732 (1993).

J. Ihm and B. D Yu, Phys. Rev. B *39*, 4760 (1989).

S. Ikegawa, T. Wada, A. Ichinose, T. Yamashita, T. Sakurai, Y. Yaegashi, T. Kaneko, M. Kosuge, H. Yamauchi, and S. Tanaka, Phys. Rev. B *41*, 11673 (1990).

S. Ikegawa, T. Wada, T. Yamashita, H. Yamauchi, and S. Tanaka, Phys. Rev. B *45*, 5659 (1992).

J.-M. Imer, F. Patthey, B. Dardel, W. D. Schneider, Y. Baer, Y. Petroff, and A. Zettl, Phys. Rev. Lett. *62*, 336 (1989).

T. Inabe, H. Ogata, Y. Maruyama, Y. Achiba, S. Suzuki, K. Kikuchi, and I. Ikemoto, Phys. Rev. Lett. *69*, 3797 (1992).

S. E. Inderhees, M. B. Salamon, J. P. Rice, and D. M. Ginsberg, Phys. Rev. Lett. *66*, 232 (1991).

Y. Inoue, Y. Shichi, F. Munakata, and M. Yamanaka, Phys. Rev. B *40*, 7307 (1989).

M. Inui, P. B. Littlewood, and S. N. Coppersmith, Phys. Rev. Lett. *63*, 2421 (1989).

L. Ioffe and G. Kotliar, Phys. Rev. B *42*, 10348 (1990).

I. Ioffe and V. Kalmeyer, Phys. Rev. B *44*, 750 (1991).

Z. Iqbal, J. C. Barry, and B. L. Ramakrishna, *in* "Studies in High Temperature Superconductors" (A. V. Narlikar, Ed.), Nova Sci., New York, 1989.

Z. Iqbal, G. H. Kwei, B. L. Ramakrishna, and E. W. Ong, Physica C *167*, 369 (1990).

Z. Iqbal, R. H. Baughman, B. L. Ramakrishna, S. Khare, N. S. Murthy, H. J. Bornemann, and D. E. Morris, Science *254*, 826 (1991).

Z. Iqbal, Supercond. Rev. *1*, 49 (1992).

Z. Iqbal, T. Datta, D. Kirven, A. Longu, J. C. Barry, F. J. Owens, A. G. Rinzler, D. Yang, and F. Reidinger, Phys. Rev. B *49*, 12322 (1994).

F. Irie and K. Yamafuji, J. Phys. Soc. Jpn. *23*, 255 (1976).

E. D. Isaacs, D. B. McWhan, R. N. Kleiman, D. J. Bishop, G. E. Ice, P. Zschack, B. D. Gaulin, T. E. Mason, J. D. Garrett, and W. J. L. Buyers, Phys. Rev. Lett. *65*, 3185 (1990).

K. Isawa, A. Tokiwa-Yamamoto, M. Itoh, S. Adachi, and H. Yamauchi, Physica C *217*, 11 (1993).

K. Isawa, A. Tokiwa-Yamamoto, M. Itoh, S. Adachi, and H. Yamauchi, Physica C *222*, 33 (1994a).

K. Isawa, T. Higuchi, T. Machi, A. Tokiwa-Yamamoto, S. Adachi, M. Murakami, and H. Yamauchi, Appl. Phys. Lett. *64*, 1301 (1994b).

T. Ishida and R. B. Goldfarb, Phys. Rev. B *41*, 8937 (1990).

T. Ishida, R. B. Goldfarb, S. Okayasu, and Y. Kazumata, Physica C *185–189*, 2515 (1991).

T. Ishiguro and K. Yamaji, "Organic Superconductors," Springer-Verlag, Berlin, 1990.

A. Isihara, "Statistical Physics," Academic Press, New York, 1971.

Y. Ishii and J. Ruvalds, Phys. Rev. B *48*, 3455 (1993).

T. Itoh and H. Uchikawa, Phys. Rev. B *39*, 4690 (1989).

M. Itoh, A. Tokiwa-Yamamoto, S. Adachi, and H. Yayauchi, Physica C *212*, 271 (1993).

R. Itti, F. Munakata, K. Ikeda, H. Yamauchi, N. Koshizuka, and S. Tanaka, Phys. Rev. B *43*, 6249 (1991).

Yu. M. Ivanchenko, Phys. Rev. B *48*, 15966 (1993).

B. I. Ivlev and N. B. Kopnin, Phys. Rev. Lett. *64*, 1828 (1990).

B. I. Ivlev, N. B. Kopnin, and M. M. Salomaa, Phys. Rev. B *43*, 2896 (1991a).

B. I. Ivlev, Yu. N. Ovchinnikov, and R. S. Thompson, Phys. Rev. B *44*, 7023 (1991b).

B. I. Ivlev and R. S. Thompson, Phys. Rev. B *45*, 875 (1992).

Y. Iwasa. K. Tanaka, T. Yasuda, T. Koda, and S. Koda, Phys. Rev. Lett. *69*, 2284 (1992).

Y. Iye, T. Tamegai, T. Sakakibara, T. Goto, and N. Miura, Physica C *153–155*, 26 (1988).

Y. Iye, S. Nakamura, T. Tamegai, T. Terashima, K. Yamamoto, and Y. Bundo, "High-Temperature Superconductors: Fundamental Properties and Novel Materials Processing" (D. Christen, J. Narayan, and L. Schneemeyer, Eds.), MRS Symposia Proceedings, No. 169, p. 871. Material Research Soc., Pittsburgh, 1990.

Y. Iye, *in* "Physical Properties of High Temperature Superconductors" (D. M. Ginsberg, Ed.), Vol. 3, Chap. 4, World Scientific, Singapore, 1992.

J. D. Jackson, "Classical Electrodynamics," Wiley, New York, 1975.

H. M. Jaeger, D. B. Haviland, B. G. Orr, and A. M. Goldman, Phys. Rev. *40*, 182 (1989).

K. P. Jain and D. K. Ray, Phys. Rev. B *39*, 4339 (1989).

R. C. Jaklevic, J. Lambe, J. E. Mercereau, and A. H. Silver, Phys. Rev. A *140*, 1628 (1965).

G. Jakob, P. Przyslupski, C. Stölzel, C. Tomé-Rose, A. Walkenhorst, M. Schmitt, and H. Adrian, Appl. Phys. Lett. *59*(13), 1626 (1991).

G. M. Japiassú, M. A. Continentino, and A. Troper, Phys. Rev. B *45*, 2986 (1992).

M. Jarrell, H. R. Krishnamurthy, and D. L. Cox, Phys. Rev. B *38*, 4584 (1988).

B. Jeanneret, Ph. Flückiger, J. L. Gavilano, Ch. Leemann, and P. Martinoli, Phys. Rev. B *40*, 11374 (1989).

C. S. Jee, B. Andraka, J. S. Kim, H. Li, M. W. Meisel, and G. R. Stewart, Phys. Rev. B *42*, 8630 (1990).

J. H. Jefferson, H. Eskes, and L. F. Feiner, Phys. Rev. B *45*, 7959 (1992).

C. D. Jeffries, Q. H. Lam, Y. Kim, C. M. Kim, A. Zettl and M. P. Klein, Phys. Rev. B *39*, 11526 (1989).

H. J. Jensen, A. Brass, An-C. Shi, and A. J. Berlinsky, Phys. Rev. B *41*, 6394 (1990).

H. J. Jensen and P. Minnhagen, Phys. Rev. Lett. *66*, 1630 (1991).

Y. Jeon, G. Liang, J. Chen, M. Croft, M. W. Ruckman, D. Di Marizo, and M. S. Hegde, Phys. Rev. B *41*, 4066 (1990).

L. Ji, R. H. Sohn, G. C. Spalding, C. J. Lobb, and M. Tinkham, Phys. Rev. B *40*, 10936 (1989).

Y. X. Jia, J. Z. Liu, M. D. Lan, P. Klavins, R. N. Shelton, and H. B. Radousky, Phys. Rev. B *45*, 10609 (1992).

H. Jiang, Y. Huang, H. How, S. Zhang, C. Vittoria, A. Widom, D. B. Chrisey, J. S. Horwitz, and R. Lee, Phys. Rev. Lett. *66*, 1785 (1991).

C. Jiang and J. P. Carbotte, Phys. Rev. B *45*, 10670 (1992a).

C. Jiang and J. P. Carbotte, Phys. Rev. B *45*, 7368 (1992b).

W. Jin, C. K. Loong, D. G. Hinks, P. Vashishta, R. K. Kalia, M. H. Degani, D. L. Price, J. D. Jorgensen, and B. Dabrowski, Mat. Res. Soc. Symp. Proc. *209*, 895 (1991).

S. Jin, G. W. Kammlott, S. Nakahara, T. H. Tiefel, and J. Graebner, Science *253*, 427 (1991).

S. Jin, T. H. Tiefel, R. C. Sherwood, M. E. Davis, R. B. van Dover, G. W. Kammlott, R. A. Fastnacht, and H. D. Keith, Appl. Phys. Lett. *52*, 2074 (1988).

W. Jin, M. H. Dagani, R. K. Kalia, and P. Vashishta, Phys. Rev. B *45*, 5535 (1992).

T. W. Jing and N. P. Ong, Phys. Rev. B *42*, 10781 (1990).

R. Job and M. Rosenberg, Supercond. Sci. Technol. *5*, 7 (1992).

K. H. Johnson, Phys. Rev. B *42*, 4783 (1990).

M. W. Johnson, D. H. Douglass, and M. F. Bocko, Phys. Rev. B *44*, 7726 (1991).

R. D. Johnson, D.S. Bethune, and C. S. Yannoni, Acc. Chem. Res. *25*, 169 (1992).

C. E. Johnson, H. W. Jiang, K. Holczer, R. B. Kaner, R. L. Whetten, and F. Diederich, Phys. Rev. B *46*, 5880 (1992).

D. C. Johnston, H. Prakash, W. H. Zachariasen, and R. Viswanathan, Mat. Res. Bull. *8*, 777 (1973).

D. C. Johnston and J. H. Cho, Phys. Rev. B *42*, 8710 (1990).

Th. Jolicoeur and J. C. LeGuillou, Phys. Rev. B *44*, 2403 (1991).

M. L. Jones, D. W. Shortt, and A. L. Schawlow, Phys. Rev. B *42*, 132 (1990).

J. D. Jorgensen, M. A. Beno, D. G. Hinks, L. Soderholm, K. J. Volin, R. L. Hitterman, J. D. Grace, I. K. Schuller, C. U. Segre, K. Zhang, and M. S. Kleefisch, Phys. Rev. B *36*, 3608 (1987a); see also Schuller *et al.* (1987).

J. D. Jorgensen, B. W. Veal, W. K. Kwok, G. W. Crabtree, A. Umezawa, L. J. Nowicki, and A. P. Paulikas, Phys. Rev. B *36*, 5731 (1987b).

J. D. Jorgensen, B. W. Veal, A. P. Paulikas, L. J. Nowicki, G. W. Crabtree, H. Claus, and W. K. Kwok, Phys. Rev. B *41*, 1863 (1990).

B. D. Josephson, Phys. Lett. *1*, 251 (1962).

J. Jung, M. A.-K. Mohamed, S. C. Cheng, and J. P. Franck, Phys. Rev. B *42*, 6181 (1990).

A. Junod, A. Bezinge, and J. Muller, Physica C *152*, 50 (1988).

A. Junod, *in* "Physical Properties of High Temperature Superconductors" (D. M. Ginsberg, Ed.), Vol. 2, Chap. 2, World Scientific, Singapore, 1990.

A. Junod, D. Sanchez, J.-Y. Genoud, T. Graf, G. Triscone, and J. Muller, Physica C *185–189*, 1399 (1991).

V. V. Kabanov and O. Yu. Mashtakov, Phys. Rev. B *47*, 6060 (1993).

K. K. Kadish and R. S. Ruoff, (Eds.), "Recent Advances in the Chemistry and Physics of Fullerenes and Related Materials," Electrochemical Society, Pennington, N. J., 1994.

K. Kadowski, Y. Songliu, and K. Kitazawa, "Lorentz Force Independent Dissipation in HTSC," submitted for publication.

A. Kadin, Phys. Rev. B *41*, 4072 (1990).

A. Kahan, Phys. Rev. B *43*, 2678 (1991).

A. B. Kaiser, Phys. Rev. B *35*, 4677 (1987).

A. B. Kaiser and C. Uher, Aust. J. Phys. *41*, 597 (1988).

A. B. Kaiser, Phys. Rev. B *37*, 5924 (1988).

A. B. Kaiser and C. Uher, *in* "Studies in High Temperature Superconductors" (A. V. Narlikar, Ed.), Vol. 7, Nova Sci., New York, 1990.

A. B. Kaiser and G. Mountjoy, Phys. Rev. B *43*, 6266 (1991).

E. Kaldis, P. Fischer, A. W. Hewat, E. A. Hewat, J. Karpinski, and S. Rusiecki, Physica C *159*, 668 (1989).

C. Kallin, A. J. Berlinsky, and W.-K. Wu, Phys. Rev. B *39*, 4267 (1989).

V. Kalmeyer and R. B. Laughlin, Phys. Rev. Lett. *59*, 2095 (1987).

A. Kampf and J. R. Schrieffer, Phys. Rev. B *41*, 6399 (1990).

K. Kanoda, H. Mazaki, T. Mizutani, H. Hosoito, and T. Shinjo, Phys. Rev. B *40*, 4321 (1989).

K. Kanoda, K. Akiba, K. Suzuki, T. Takahashi, and G. Saito, Phys. Rev. Lett. *65*, 1271 (1990).

A. Kapitulnik and M. R. Beasley, C. Castellani, and D. DiCastro, Phys. Rev. B *37*, 537 (1988).

S. G. Kaplan, T. W. Noh, A. J. Sievers, S.-W. Cheong and Z. Fisk, Phys. Rev. B *40*, 5190 (1989).

V. R. Karasik, N. G. Vasil'ev, and V. G. Ershov, Zh. Eksp. Teor. Fiz. *59*, 790 (1970); Sov Phys—JETP *32*, 433 (1971).

K. Karlsson, O. Gunnarsson, and O. Jepsen, Phys. Rev. B *45*, 7559 (1992).

K. Karraï, E. J. Choi, F. Dunmore, S. Liu, H. D. Drew, Q. Li, D. B. Fenner, Y. D. Zhu, and F.-C. Zhang, Phys. Rev. Lett. *69*, 152 (1992).

H. Kasatani, H. Terauchi, Y. Hamanaka, and S. Nakashima, Phys. Rev. B *47*, 4022 (1993).

A. Kastalsky, A. W. Kleinsasser, L. H. Greene, R. Bhat, F. P. Milliken, and J. P. Harbison, Phys. Rev. Lett. *67*, 3026 (1991).

R. Kato, Y. Enomoto, and S. Maekawa, Phys. Rev. B *44*, 6916 (1991).

R. Kato, Y. Enomoto, and S. Mackawa, Phys. Rev. B *47*, 8016 (1993).

K. Katti and S. H. Risbud, Phys. Rev. B *45*, 10155 (1992).

R. L. Kautz and J. M. Martinis, Phys. Rev. B *42*, 9903 (1990).

M. Kaveh and N. H. Mott, Phys. Rev. Lett. *68*, 1904 (1992).

Z. A. Kazei and I. B. Krynetskii, "Landolt–Börnstein, Group III," Solid State Physics, Vol. 27, Subvol. f2, Springer, Heidelberg, 1992.

A. Kebede, C. S. Jee, J. Schwegler, J. E. Crow, T. Mihalisin, G. H. Myer, R. E. Salomon, P. Schlottmann, M. V. Kuric, S. H. Bloom, and R. P. Guertin, Phys. Rev. B *40*, 4453 (1989).

F. J. Kedves, S. Mészáros, K. Vad, G. Halász, B. Keszei, and L. Mihály, Solid State Commun. *63*, 991 (1987).

O. Keller, Phys. Rev. B *43*, 10293 (1991).

P. H. Kes, C. J. van der Beek, M. P. Maley, M. E. McHenry, D. A. Huse, M. J. V. Menken, and A. A. Menovsky, Phys. Rev. Lett. *67*, 2383 (1991).

I. B. Khalfin and B. Ya. Shapiro, Phys. Rev. B *46*, 5593 (1992).

A. F. Khoder, M. Couach, and J. L. Jorda, Phys. Rev. B *42*, 8714 (1990).

A. F. Khoder and M. Couach, *in* "Magnetic Susceptibility of Superconductors and Other Spin Systems" (R. A. Hein, T. L. Francavilla, and D. H. Liebenberg, Eds.), Plenum, New York, 1992.

A. Khurana, Phys. Rev. B *40*, 4316 (1989).

R. F. Kiefl, J. W. Schneider, A. MacFarlane, K. Chow, T. L. Duty, T. L. Estle, B. Hitti, R. L. Lichti, E. J. Ansaldo, C. Schwab, P. W. Percival, G. Wei, S. Wlodek, K. Kojima, W. J. Ronamov, J. P. McCauley, Jr., N. Coustel, J. E. Fischer, and A. B. Smith III, Phys. Rev. Lett. *68*, 1347 (1992).

Y. B. Kim, C. F. Hempstead, and A. R. Strnad, Phys. Rev. Lett. *9*, 306 (1962).

Y. B. Kim, C. F. Hempstead, and A. R. Strand, Phys. Rev. *129*, 528 (1963).

Y. B. Kim and M. J. Stephan, *in* "Superconductivity" (R. D. Parks, Ed.), Vol. 2, p. 1107, Dekker, New York, 1969.

H. K. Kim and P. S. Riseborough, Phys. Rev. B *42*, 7975 (1990).

D. H. Kim, K. E. Gray, R. T. Kampwirth, K. C. Woo, D. M. McKay, and J. Stein, Phys. Rev. B *41*, 11642 (1990).

D. H. Kim, K. E. Gray, R. T. Kampwirth, and D. M. McKay, Phys. Rev. B *42*, 6249 (1990); *43*, 2910 (1991a).

D. H. Kim, D. J. Miller, J. C. Smith, R. A. Holoboff, J. H. Kang, and J. Talvacchio, Phys. Rev. B *44*, 7607 (1991b).

J.-J. Kim, H.-K. Lee, J. Chung, H. J. Shin, H. J. Lee, and J. K. Ku, Phys. Rev. B *43*, 2962 (1991).

D. M. King, Z.-X. Shen, D. S. Dessau, B. O. Wells, W. E. Spicer, A. J. Arko, D. S. Marshall, J. DiCarlo, A. G. Loeser, C. H. Park, E. R. Ratner, J. L. Peng, Z. Y. Li, and R. L. Greene, Phys. Rev. Lett. *70*, 3159 (1993).

K. Kinoshita, F. Izumi, T. Yamada, and H. Asano, Phys. Rev. B *45*, 5558 (1992).

J. Kircher, M. K. Kelly, S. Rashkeev, M. Alouani, D. Fuchs, and M. Cardona, Phys. Rev. B *44*, 217 (1991).

D. Kirillov, C. B. Eom, and T. H. Geballe, Phys. Rev. B *43*, 3752 (1991).

W. P. Kirk, P. S. Kobiela, R. N. Tsumura, and R. K. Pandey, Ferroelectrics *92*, 151 (1989).

T. R. Kirkpatrick and D. Belitz, Phys. Rev. Lett. *68*, 3232 (1992).

J. R. Kirtley, R. T. Collins, Z. Schlesinger, W. J. Gallagher, R. L. Sandstrom, T. R. Dinger and D. A. Chance, Phys. Rev. B *35*, 8846 (1987).

J. R. Kirtley, Phys. Rev. *41*, 7201 (1990a); Int. J. Mod. Phys. B *4*, 201 (1990b).

L. B. Kiss and P. Svedlindh, IEEE Trans. Electronic Devices, *41*, 2112 (1994).

T. J. Kistenmacher, Phys. Rev. B *39*, 12279 (1989).

Y. Kitaoka, S. Hiramatsu, T. Kohara, K. Asayama, K. Oh-ishi, M. Kikuchi, and N. Kobayashi, Jpn. J. Appl. Phys. *26*, L397 (1987a).

Y. Kitaoka, S. Hiramatsu, K. Ishida, T. Kohara, and K. Asayama, J. Phys. Soc. Jpn. *56*, 3024 (1987b).

Y. Kitaoka, K. Fujiwara, K. Ishida, K. Asayama, Y. Shimakawa, T. Manako, and Y. Kubo, Physica C *179*, 107 (1991).

K. Kitazawa and S. Tajima, *in* "Some Aspects of Superconductivity," (L. C. Gupta, Ed.), Nova Sci., New York, 1990.

C. Kittel, "Introduction to Solid State Physics," Wiley, New York, 1976.

S. Kivelson, Phys. Rev. B *39*, 259 (1989).

Y. S. Kivshar and T. K. Soboleva, Phys. Rev. B *42*, 2655 (1990).

Y. S. Kivshar, B. A. Malomed, Z. Fei, and L. Vázquez, Phys. Rev. B *43*, 1098 (1991).

A. K. Klehe, A. K. Gangopadhyay, J. Diederichs, and J. S. Schilling, Physica C *213*, 266 (1992).

A. K. Klehe, J. S. Schilling, J. L. Wagner, and D. G. Hinks, Physica C *223*, 313 (1994).

B. M. Klein, L. L. Boyer, D. A. Papaconsantopoulos, and L. F. Mattheiss, Phys. Rev. B *18*, 6411 (1978).

B. M. Klein, L. L. Boyer, and D. A. Papaconstantopoulos, Phys. Rev. Lett. *42*, 530 (1979).

U. Klein, Phys. Rev. B *40*, 6601 (1989); *41*, 4819 (1990).

D. J. Klein, T. G. Schmalz, M. A. García-Bach, R. Valenti, and T. P. Zivkovic, Phys. Rev. B *43*, 719 (1991).

U. Klein and B. Pöttinger, Phys. Rev. B *44*, 7704 (1991).

L. Kleion and A. Aharony, Phys. Rev. B *45*, 9926 (1992).

R. Kleiner, F. Steinmeyer, G. Kunkel, and P. Müller, Phys. Rev. Lett. *68*, 2394 (1992).

A. Kleinhammes, C. L. Chang, W. G. Moulton, and L. R. Testardi, Phys. Rev. B *44*, 2313 (1991).

A. W. Kleinsasser and T. N. Jackson, Phys. Rev. B *42*, 8716 (1990).

R. A. Klemm and S. H. Liu, Phys. Rev. B *44*, 7526 (1991).

R. A. Klemm, Phys. Rev. B *47*, 14630 (1993).

R. A. Klemm, "Layered Superconductors," Oxford, NY, in press.

P. Knoll, C. Thomsen, M. Cardona, and P. Murugaraj, Phys. Rev. B *42*, 4842 (1990).

F. Kober, H.-C. Ri, R. Gross, D. Koelle, R. P. Huebener, and A. Gupta, Phys. Rev. B *44*, 11951 (1991).

J. Kober, A. Gupta, P. Esquinazi, H. F. Braun, E. H. Brandt, Phys. Rev. Lett. *66*, 2507 (1991).

R. H. Koch, V. Foglietti, W. J. Gallagher, G. Koren, A. Gupta, and M. P. A. Fisher, Phys. Rev. Lett. *63*, 1511 (1989).

B. N. Kodess, Ph.D. thesis, Perm State University, Perm. Cited as Ref. 6.130 of Vonsovsky *et al.* (1982).

V. G. Kogan, M. M. Fang, and S. Mitra, Phys. Rev. B *38*, 11958 (1988).

V. G. Kogan, Phys. Rev. B *38*, 7049 (1988).

V. G. Kogan and L. J. Campbell, Phys. Rev. Lett. *62*, 1552 (1989).

V. G. Kogan, N. Nakagawa, and S. L. Thiemann, Phys. Rev. B *42*, 2631 (1990).

S. Kohiki, T. Wada, S. Kawashima, H. Takagi, S. Uchida, and S. Tanaka, Phys. Rev. B *38*, 7051, 8868 (1988).

S. Kohiki, S.-I. Hatta, K. Setsune, K. Wasa, Y. Higashi, S. Fukushima, and Y. Gohshi, Appl. Phys. Lett. *56*, 298 (1990).

S. Koka and K. Shrivastava, Physica B *165–166*, 1097 (1990).

S. Kolesnik, T. Skoskiewicz, J. Igalson, and Z. Korczak, Phys. Rev. B *45*, 10158 (1992).

T. Komeda, G. D. Waddill, P. J. Benning, and J. H. Weaver, Phys. Rev. B *43*, 8713 (1991).

M. Konczykowski, F. Rullier-Albenque, E. R. Yacoby, A. Shaulov, Y. Yeshurun, and P. Lejay, Phys. Rev. B *44*, 7167 (1991).

J. Konior, Phys. Rev. B *47*, 14425 (1993).

J. Konior, "Some Properties of Narrow Band Systems Coupled to Phonons," submitted for publication.

R. Konno and K. Ueda, Phys. Rev. B *40*, 4329 (1989).

P. Koorevaar, J. Aarts, P. Berghuis, and P. H. Kes, Phys. Rev. B *42*, 1004 (1990).

Y. Kopelevich, A. Gupta, P. Esquinazi, C.-P. Heidmann, and H. Müller, Physica C *183*, 345 (1991).

P. Kopietz, Phys. Rev. Lett. *70*, 3123 (1993).

A. E. Koshelev, G. Yu. Logvenov, V. A. Larkin, V. V. Ryazanov, and K. Ya. Soifer, Physica C *177*, 129 (1991).

A. A. Koshta, Yu. N. Shvachko, A. A. Romanyukha, and V. V. Ustinov, Zh. Eksp. Teor. Fiz. *103*, 629 (1993); Transl. Sov. Phys. JETP *76*, 314 (1993).

I. Z. Kostadinov, V. G. Hadjiev, J. Tihov, M. Mateev, M. Mikhov, O. Petrov, V. Popov, E. Dinolova, Ts. Zheleva, G. Tyuliev, and V. Kojouharov, Physica C *156*, 427 (1988).

J. M. Kosterlitz and D. Thouless, J. Phys. C *5*, L124 (1972); *6*, 1181 (1973).

N. Kosugi, Y. Tokura, H. Takagi and S. Uchida, Phys. Rev. B *41*, 131 (1990).

V. Kovachev, "Energy Dissipation in Superconducting Materials," Clarendon, Oxford (1991).

Y. Koyama and M. Ishimaru, Phys. Rev. B *45*, 9966 (1992).

H. Krakauer, W. E. Pickett, D. A. Papaconstantopoulos, and L. L. Boyer, Jpn. J. Appl. Phys. *26*, Suppl. 26-3. (1987)

H. Krakauer, W. E. Pickett and R. E. Cohen, J. Supercond. *1*, 111 (1988).

H. Krakauer and W. E. Pickett, Phys. Rev. Lett. *60*, 1665 (1988).

V. M. Krasnov, V. A. Larkin, and V. V. Ryazanov, Physica C *174*, 440 (1991).

V. M. Krasnov, Physica C *190*, 357 (1992).

N. M. Kreines and V. I. Kudinov, Mod. Phys. Lett. B *6*, 6 (1992).

V. Z. Kresin and S. A. Wolf, *in* "Novel Superconductivity" (S. A. Wolf and V. Z. Kresin, Eds.), p. 287, Plenum, New York, 1987.

V. Z. Kresin, and S. A. Wolf, "Fundamentals of Superconductivity," Plenum, New York, 1990.

V. Z. Kresin and W. A. Little (Eds.), "Organic Superconductivity," Plenum, New York, 1990.

V. Z. Kresin, H. Morawitz, and S. A. Wolf, "Mechanisms of Conventional and High T_c Superconductivity," Oxford Univ. Press, Oxford, 1993.

A. Krimmel, P. Fischer, B. Roessli, H. Maletta, C. Geibel, C. Schank, A. Grauel, A. Loidl, and F. Steglich, Z. Phys. B *86*, 161 (1992).

G. Kriza, G. Quirion, O. Traetteberg, W. Kang, and D. Jérome, Phys. Rev. Lett. *66*, 1922 (1991).

A. Krol, Z. H. Ming, Y. H. Kao, N. Nücker, G. Roth, J. Fink, G. C. Smith, K. T. Park, J. Yu, A. J. Freeman, A. Erband, G. Müller-Vogt, J. Karpinski, E. Kaldis, and K. Schönmann, Phys. Rev. B *45*, 2581 (1992).

H. W. Kroto and D. R. M. Walton, *in* "The Fullerenes: New Horizons for the Chemistry, Physics, and Astrophysics of Carbon," Cambridge Univ. Press, New York, 1993.

E. Krüger, Phys. Stat. Sol. B *156*, 345 (1989).

L. Krusin-Elbaum, A. P. Malozemoff, Y. Yeshurun, D. C. Cronemeyer and F. Holtzberg, Phys. Rev. B *39*, 2936 (1989).

H. C. Ku, H. D. Yang, R. W. McCallum, M. A. Noack, P. Klavins, R. N. Shelton, and A. R. Moodenbaugh, *in* "High Temperature Superconductors" (U. Gubser and M. Schluter, Eds.), p. 177, Mater. Res. Soc., Pittsburgh, 1987.

R. Kuentzler, C. Hornick, Y. Dossmann, S. Wegner, R. El Farsi, and M. Drillon, Physica C *184*, 316 (1991).

M. L. Kulic and R. Zeyher, Phys. Rev. B *49*, 4395 (1994).

J. Kulik, Y. Y. Xue, Y. Y. Sun, and M. Bonvalot, J. Mater. Res. *5*, 1625 (1990).

A. D. Kulkarni, J. Prade, F. W. de Wette, W. Kress, and U. Schröder, Phys. Rev. B *40*, 2642 (1989).

A. D. Kulkarni, F. W. de Wette, J. Prade, U. Schröder, and W. Kress, Phys. Rev. B *41*, 6409 (1990).

A. D. Kulkarni, F. W. de Wette, J. Prade, U. Schröder, and W. Kress, Phys. Rev. B *43*, 5451 (1991).

H. Kumakura, M. Uehara, and K. Togano, Appl. Phys. Lett. *51*, 1557 (1987).

G. R. Kumar and P. Chaddah, Phys. Rev. B *39*, 4704 (1989).

N. Kumar and A. M. Jayannavar, Phys. Rev. B *45*, 5001 (1992).

M. N. Kunchur and S. J. Poon, Phys. Rev. B *43*, 2916 (1991).

P. J. Kung, M. P. Maley, M. E. McHenry, J. O. Willis, J. Y. Coulter, M. Murakami, and S. Tanaka, Phys. Rev. B *46*, 6427 (1992).

S. Kurihara, Phys. Rev. B *39*, 6600 (1989).

E. Z. Kurmaev, V. P. Belash, S. A. Nemnonov, and A. S. Shulakov, Phys. Stat. Solid B *61*, 365 (1974).

E. Z. Kurmaev, V. I. Nefedov, and L. D. Finkelstein, Int. J. Mod. Phys. B *2*, 393 (1988).

E. Z. Kurmaev and L. D. Finkelstein, Int. J. Mod. Phys. B *3*, 973 (1989).

E. Z. Kurmaev and L. D. Finkelstein, Int. J. Mod. Phys. *5*, 1097 (1991).

H. Kuroda, K. Yakushi, H. Tasima, A. Ugawa, Y. Okawa, A. Kobayashi, R. Kato, H. Kobayashi, and G. Saito, Synth. Metals A *27*, 491 (1988).

A. Kussmaul, J. S. Moodera, G. M. Roesler, Jr., and P. M. Tedrow, Phys. Rev. B *41*, 842 (1990).

A. Kussmaul, J. S. Moodera, P. M. Tedrow, and A. Gupta, Physica C *177*, 415 (1991).

A. L. Kuzemsky, *in* Int. Conf. Supercond. and Strongly Correlated Electron Systems, Amalfi, Italy, 1993.

E. Kuzmann, Z. Homonnay, A. Vértes, M. Gál, K. Torkos, B. Csákvári, G. K. Solymos, G. Horváth, J. Bánkuti, I. Kirschner, and L. Korecz, Phys. Rev. B *39*, 328 (1989).

L. S. Kuzmin, P. Delsing, T. Claeson, and K. Likharev, Phys. Rev. Lett. *62*, 2539 (1989).

L. S. Kuzmin, Yu. V. Nazarov, D. B. Haviland, P. Delsing, and T. Claeson, Phys. Rev. Lett. *67*, 1161 (1991).

L. S. Kuzmin and D. Haviland, Phys. Rev. Lett. *67*, 2890 (1991).

M. Kvale and S. E. Hebboul, Phys. Rev. B *43*, 3720 (1991).

G. H. Kwei, J. A. Goldstone, A. C. Lawson, Jr., J. D. Thompson, and A. Williams, Phys. Rev. B *39*, 7378 (1989).

G. H. Kwei, R. B. Von Dreele, S.-W. Cheong, Z. Fisk, and J. D. Thompson, Phys. B *41*, 1889 (1990).

W. K. Kwok, U. Welp, G. W. Crabtree, K. G. Vandervoort, R. Hulscher, Y. Zheng, B. Dabroski, and D. G. Hinks, Phys. Rev. B *40*, 9400 (1989).

W. K. Kwok, U. Welp, G. W. Crabtree, K. G. Vandervoort, R. Hulscher, and J. Z. Liu, Phys. Rev. Lett. *64*, 966 (1990a).

W. K. Kwok, U. Welp, K. D. Carlson, G. W. Crabtree, K. G. Vandervoort, H. H. Wang, A. M. Kini, J. M. Williams, D. L. Stupka, L. K. Montgomery, and J. E. Thompson, Phys. Rev. B *42*, 8686 (1990b).

H. S. Kwok, J. P. Zheng, and S. Y. Dong, Phys. Rev. B *43*, 6270 (1991).

Y. K. Kwong, K. Lin, M. Park, M. S. Isaacson, and J. M. Parpia, Phys. Rev. B *45*, 9850 (1992).

J. Labbé, Phys. Rev. *158*, 647, 655 (1967a).

J. Labbé, S. Barisic, and J. Friedel, Phys. Rev. Lett. *19*, 1039 (1967b).

B. M. Lairson, S. K. Streiffer, and J. C. Bravman, Phys. Rev. B *42*, 10067 (1990a).

B. M. Lairson, J. Z. Sun, J. C. Bravman, and T. H. Geballe, Phys. Rev. B *42*, 1008 (1990b).

B. M. Lairson, J. Z. Sun, T. H. Geballe, M. R. Beasley, and J. C. Bravman, Phys. Rev. B *43*, 10405 (1991).

R. Lal and S. K. Joshi, Phys. Rev. B *45*, 361 (1992).

Q. H. Lam, Y. Kim, and C. D. Jeffries, Phys. Rev. B *42*, 4846 (1990).

M. D. Lan, J. Z. Liu, and R. N. Shelton, Phys. Rev. B *44*, 233 (1991).

L. D. Landau, Sov. Phys. JETP *3*, 920 (1957a); *5*, 101 (1957b).

C. T. Lane, "Superfluid Physics," Chap. 9, McGraw–Hill, New York, 1962.

M. Lang, N. Toyota, T. Sasaki, and H. Sato, Phys. Rev. Lett. *69*, 1443 (1992a); Phys. Rev. B *46*, 5822 (1992b).

J. Langen, M. Veit, M. Galffy, H. D. Jostarndt, A. Erle, S. Blumenröder, H. Schmidt, and E. Zirngiebl, Solid State Commun. *65*, 973 (1988).

D. N. Langenberg, D. J. Scalapino, and B. N. Taylor, Sci. Amer. *214*, 30 (May 1966).

W. Lanping, H. Jian, and W. Guowen, Phys. Rev. B *40*, 10954 (1989).

D. C. Larbalestier, M. Daeumling, X. Cai, J. Suentjens, J. McKinnell, D. Hampshire, P. Lee, C. Meingast, T. Willis, H. Muller, R. D. Ray, R. G. Dillenburg, E. E. Hellstrom, and R. Joynt, J. Appl. Phys. *62*, 3308 (1987a).

D. C. Larbalestier, M. Daeumling, P. J. Lee, T. F. Kelly, J. Seuntjens, C. Meingast, X. Cai, J. McKinnell, R. D. Ray, R. G. Dillenburg, and E. E. Hellstrom, Cryogenics *27*, 411 (1987b).

A. I. Larkin and Yu. N. Ovchinnikov, Sov. Phys. JETP *38*, 854 (1974).

A. Larsen, H. D. Jensen, and J. Mygind, Phys. Rev. B *43*, 10179 (1991).

R. B. Laughlin, Phys. Rev. Lett. *60*, 2677 (1988a).

R. B. Laughlin, Science *242*, 525 (1988b).

W. E. Lawrence and S. Doniach, *in* "Proc. 12th Int. Conf. Low Temp. Phys. Kyoto, 1970" (E. Kanda, Ed.), p. 361. Keigaku, Tokyo, 1971.

L. P. Le, G. M. Luke, B. J. Sternlieb, W. D. Wu, Y. J. Uemura, J. H. Brewer, T. M. Riseman, C. E. Stronach, G. Saito, H. Yamochi, H. H. Wang, A. M. Kini, K. D. Carlson, and J. M. Williams, Phys. Rev. Lett. *68*, 1923 (1992).

M. A. R. LeBlanc, D. LeBlanc, A. Golebiowski, and G. Fillion, Phys. Rev. Lett. *66*, 3309 (1991).

D. LeBlanc and M. A. R. LeBlanc, Phys. Rev. B *45*, 5443 (1992).

K. Le Dang, J. P. Renard, P. Veillet, E. Vélu, J. P. Burger, J. N. Daou, and Y. Loreaux, Phys. Rev. B *40*, 11291 (1989).

E. Lederman, L. Wu, M. L. denBoer, P. A. van Aken, W. F. Müller, and S. Horn, Phys. Rev. B *44*, 2320 (1991).

T.-K. Lee, J. L. Birman, and S. J. Williamson, Phys. Rev. Lett. *39*, 839 (1977a); Phys. Lett. A *64*, 89 (1977b).

T.-K. Lee and J. L. Birman, Phys. Rev. B *17*, 4931 (1978).

M. Lee, M. Yudkowsky, W. P. Halperin, J. Thiel, S.-J. Hwu, and K. R. Poeppelmeier, Phys. Rev. B *36*, 2378 (1987).

M. Lee, Y.-Q. Song, W. P. Halperin, L. M. Tonge, T. J. Marks, H. O. Marcy, and C. R. Kannewurf, Phys. Rev. B *40*, 817 (1989).

S. J. Lee and J. B. Ketterson, Phys. Rev. Lett. *64*, 3078 (1990).

S.-I. Lee, Y. H. Jeong, K. H. Han, Z. S. Lim, Y. S. Song, and Y. W. Park, Phys. Rev. B *41*, 2623 (1990).

W. C. Lee and D. C. Johnston, Phys. Rev. B *41*, 1904 (1990).

H. C. Lee, R. S. Newrock, D. B. Mast, S. E. Hebboul, J. C. Garland, and C. J. Lobb, Phys. Rev. *44*, 921 (1991).

W. C. Lee and D. M. Ginsberg, Phys. Rev. B *44*, 2815 (1991).

W. C. Lee, J. H. Cho, and D. C. Johnston, Phys. Rev. B *43*, 457 (1991).

T. R. Lemberger, *in* "Physical Properties of High Temperature Superconductors" (D. M. Ginsberg, Ed.), Vol. 3, Chap. 6, World Scientific, Singapore, 1992.

S. Lenck, S. Wermbter, and L. Tewordt, J. Low Temp. Phys. *80*, 269 (1990).

S. Lenck and J. P. Carbotte, Phys. Rev. B *49*, 4176 (1994).

H. Lengfellner, A. Schnellbögl, J. Betz, W. Prettl, and K. F. Renk, Phys. Rev. B *42*, 6264 (1990).

H. Lengfellner, A. Schnelbögl, J. Betz, K. Renk, and W. Prettl, Appl. Phys. Lett. *60*, 1991 (1991a).

H. Lengfellner and A. Schnellbögl, Physica C *174*, 373 (1991).

H. Lengfellner, G. Kremb, A. Schnellbögl, J. Betz, K. F. Renk, and W. Prettl, Appl. Phys. Lett. *60*, 501 (1992).

Y. Le Page, T. Siegrist, S. A. Sunshine, L. F. Schneemeyer, D. W. Murphy, S. M. Zahurak, J. V. Waszczak, W. R. McKinnon, J. M. Tarascon, G. W. Hull, and L. H. Greene, Phys. Rev. B *36*, 3617 (1987).

F. Lera, R. Navarro, C. Rillo, L. A. Angurel, A. Badia, and J. Bartolome, J. Magn. Mag. Mater. *104–107*, 615 (1992).

Ph. Lerch, Ch. Leemann, R. Theron, and P. Martinoli, Phys. Rev. B *41*, 11579 (1990).

J. Lesueur, L. H. Greene, W. Feldmann, and A. Inam, Physica C *191*, 325 (1992).

B. G. Levi, Phys. Today 19 (May 1988), p. 19.

G. Levin, Phys. Rev. B *47*, 14634 (1993).

L. Levitov, Phys. Rev. Lett. *66*, 224 (1991).

J. A. Lewis, C. E. Platt, M. Wegmann, M. Teepe, J. L. Wagner, and D. G. Hinks, Phys. Rev. B *48*, 7739 (1993).

J. Q. Li, C. Chen, D. Y. Yang, F. H. Li, Y. S. Yao, Z. Y. Ran, W. K. Wang, and Z. X. Zhao, Z. Phys. B. *74*, 165 (1989).

J. Q. Li, X. X. Xi, X. D. Wu, A. Inam, S. Vadlamannati, W. L. McLean, T. Venkatesan, R. Ramesh, D. M. Hwang, J. A. Martinez, and L. Nazar, Phys. Rev. Lett. *64*, 3086 (1990).

Z.-Z. Li and Y. Qiu, Phys. Rev. B *43*, 12906 (1991).

Y.-H. Li and S. Teitel, Phys. Rev. Lett. *66*, 3301 (1991).

C. Li, M. Pompa, S. D. Longa, and A. Bianconi, Physica C *178*, 421 (1991).

Y.-H. Li and S. Teitel, Phys. Rev. B *45*, 5718 (1992).

Q. Li, M. Suenaga, T. Hikata, and K. Sato, Phys. Rev. B *46*, 5857 (1992).

Q. Li, M. Suenaga, T. Kimura, and K. Kishio, Phys. Rev. *47*, 11384 (1993).

R. L. Lichti, D. W. Cooke, and C. Boekema, Phys. Rev. B *43*, 1154 (1991).

A. I. Liechtenstein, I. I. Mazin, C. O. Rodriguez, O. Jepsen, O. K. Andersen, and M. Methfessel, Phys. Rev. B *44*, 5388 (1991).

K. K. Likharev, "Dynamics of Josephson Junctions and Circuits," Gordon & Breach, New York, 1986.

L. Lilly, A. Muramatsu, and W. Hanke, Phys. Rev. Lett. *65*, 1379 (1990).

Z. S. Lim, K. H. Han, S.-I. Lee, Y. H. Jeong, S. H. Salk, Y. S. Song, and Y. W. Park, Phys. Rev. B *40*, 7310 (1989).

S.-Y. Lin, L. Lu, H.-M. Duan, B.-H. Ma, and D.-L. Zhang, Int. J. Mod. Phys. B *3*, 409 (1989).

J. J. Lin, Phys. Rev. B *44*, 789 (1991).

P. A. P. Lindberg, I. Lindau, and W. E. Spicer, Phys. Rev. B *40*, 6822 (1989).

F. Lindemann, Phys. Z. *11*, 609 (1910).

A. P. Litvinchuk, C. Thomsen, and M. Cardona, *in* "Physical Properties of High Temperature Superconductors," Vol. 4, Chap. 6, World Scientific, Singapore, 1994.

W. A. Little and R. D. Parks, Phys. Rev. Lett. *9*, 9 (1962).

Y. Liu, J. Y. Lee, M. J. Sumner, R. Sooryakumar, and T. R. Lemberger, Phys. Rev. B *42*, 10090 (1990).

J.-X. Liu, J.-C. Wan, A. M. Goldman, Y. C. Chang, and P. Z. Jiang, Phys. Rev. Lett. *67*, 2195 (1991).

J.-Z. Liu, Y. X. Jia, R. N. Shelton, and M. J. Fluss, Phys. Rev. Lett. *66*, 1354 (1991).

J. Z. Liu, L. Zhang, M. D. Lan, R. N. Shelton, and M. J. Fluss, Phys. Rev. B *46*, 9123 (1992).

L. Liu, J. S. Kouvel, and T. O. Brun, Phys. Rev. B *43*, 7859 (1991).

R. Liu, B. W. Veal, A. P. Paulikas, J. W. Downey, H. Shi, C. G. Olson, C. Gu, A. J. Arko, and J. J. Joyce, Phys. Rev. B *45*, 5614 (1992).

R. Liu, M. V. Klein, P. D. Han, and D. A. Payne, Phys. Rev. B *45*, 7392 (1992).

L. Liu, J. S. Kouvel, and T. O. Brun, Phys. Rev. B *45*, 3054 (1992).

C. J. Lobb, Phys. Rev. *36*, 3930 (1987).

G. Yu. Logvenov, V. V. Ryazanov, A. V. Ustinov, and R. P. Huebener, Physica C *175*, 179 (1991).

L. W. Lombardo, D. B. Mitzi, A. Kapitulnik, and A. Leone, Phys. Rev. B *46*, 5615 (1992).

F. London and H. London, Proc. Roy. Soc. (London) A *141*, 71 (1935).

P. London, "Une Conception Nouvelle de la Superconductibilité," Hermann, Paris, 1937.

F. London, "Superfluids," Wiley, New York, Vol. 1, 1950; Vol. 2, 1954, Dover, New York, 1961.

J. M. Longo and P. M. Raccah, J. Solid State Chem. *6*, 526 (1973).

C.-K. Loong, P. Vashishta, R. K. Kalia, M. H. Degani, D. L. Price, D. J. Jorgensen, D. G. Hinks, B. Dabrowski, A. W. Mitchell, D. R. Richards, and Y. Zheng, Phys. Rev. Lett. *62*, 2628 (1989).

C.-K. Loong, D. G. Hinks, P. Vashishta, W. Jin, R. K. Kalia, M. H. Degani, D. L. Price, J. D. Jorgensen, B. Dabrowski, A. W. Mitchell, D. R. Richards, and Y. Zheng, Phys. Rev. Lett. *66*, 3217 (1991).

C.-K. Loong, P. Vashishta, R. K. Kalia, W. Jin, M. H. Degani, D. G. Hinks, D. L. Price, J. D. Jorgensen, B. Dabrowski, A. W. Mitchell, D. R. Richards, and Y. Zheng, Phys. Rev. B *45*, 8052 (1992).

Y. Lou, X. Lu, G. H. Dai, W. Y. Ching, Y.-N. Xu, M.-Z. Huang, P. K. Tseng, Y. C. Jean, R. L. Meng, P. H. Hor, and C. W. Chu, Phys. Rev. B *46*, 2644 (1992).

A. J. Lowe, S. Regan, and M. A. Howson, Phys. Rev. B *44*, 9757 (1991).

D. H. Lowndes, D. P. Norton, and J. D. Budai, Phys. Rev. Lett. *65*, 1160 (1990).

J. P. Lu, K. Arya, and J. L. Birman, Phys. Rev. B *40*, 7372 (1989).

J.-T. Lue and J. S. Sheng, Phys. Rev. B *47*, 5469 (1993).

G. M. Luke, L. P. Le, B. J. Sternlieb, Y. J. Uemura, J. H. Brewer, R. Kadono, R. F. Kiefi, S. R. Kreitzman, T. M. Riseman, C. E. Stronach, M. R. Davis, S. Uchida, H. Takagi, Y. Tokura, Y. Hidaka, T. Murakami, J. Gopalakrishnan, A. W. Sleight, M. A. Subramanian, E. A. Early, J. T. Markert, M. B. Maple, and C. L. Seaman, Phys. Rev. B *42*, 7981 (1990).

S. Lupi, P. Calvani, M. Capizzi, P. Maselli, W. Sadowski, and E. Walker, Phys. Rev. B *45*, 12470 (1992).

J. Luzuriaga, M.-O. Andrè, and W. Benoit, Phys. Rev. B *45*, 12492 (1992).

J. W. Lynn, T. W. Clinton, W.-H. Li, R. W. Erwin, J. Z. Liu, K. Vandervoort, and R. N. Shelton, Phys. Rev. Lett. *63*, 2606 (1989).

J. W. Lynn, (Ed.), "High Temperature Superconductivity," Springer-Verlag, Berlin, 1990a.

J. W. Lynn, "High Temperature Superconductivity," Chap. 8, Springer-Verlag, Berlin, 1990b.

J. W. Lynn, I. W. Sumarlin, S. Skanthakumar, W.-H. Li, R. N. Shelton, J. L. Peng, Z. Fisk, and S.-W. Cheong, Phys. Rev. B *41*, 2569 (1990).

J. W. Lynn, J. Alloys Compounds *181*, 419 (1992).

E. A. Lynton, "Superconductivity," Methuen, London, 1962.

D. K. C. MacDonald, "Thermoelectricity, An Introduction To the Principles," Wiley, New York, 1962.

H. Maeda, Y. Tanaka, M. Fukutomi, and T. Asano, Jpn. J. Appl. Phys. Lett. *27*, 209 (1988).

A. Maeda, T. Shibauchi, N. Kondo, K. Uchinokura, and M. Kobayashi, Phys. Rev. B *46*, 14234 (1992).

Y. Maeno, T. Tomita, M. Kyogoku, S. Awaji, Y. Aoki, K. Hoshino, A. Minami, and T. Fujita, Nature *328*, 512 (1987).

G. D. Mahan, Phys. Rev. B *40*, 11317 (1989).

G. D. Mahan, Phys. Rev. B *48*, 16557 (1993).

F. Mahini, F. S. Razavi, and Z. Altounian, Phys. Rev. B *39*, 4677 (1989).

R. Mailfert, R. W. Batterman, and J. J. Hanak, Phys. Lett. A *24*, 315 (1967).

A. Majhofer, L. Mankiewicz, and J. Skalski, Phys. Rev. B *42*, 1022 (1990).

K. Maki, Prog. Theoret. Phys. *39*, 897 (1968).

K. Maki, Phys. Rev. B *43*, 1252 (1991); erratum, *43*, 13685 (1991).

H. Maletta, A. P. Malozemoff, D. C. Cronemeyer, C. C. Tsuei, R. L. Greene, J. G. Bednorz, and K. A. Müller, Solid State Commun. *62*, 323 (1987).

M. P. Maley, J. Appl. Phys. *70*, 6189 (1991).

M. P. Maley, P. J. Kung, J. Y. Coulter, W. L. Carter, G. N. Riley, and M. E. McHenry, Phys. Rev. B *45*, 7566 (1992).

M. Mali, D. Brinkmann, L. Pauli, J. Roos, H. Zimmermann, and J. Hulliger, Phys. Lett. A *124*, 112 (1987).

S. K. Malik, C. V. Tomy, D. T. Adroja, R. Nagarajan, R. Prasad, and N. C. Soni, Solid State Commun. *66*(10), 1097 (1988).

B. A. Malomed, Phys. Rev. *39*, 8018 (1989).

B. A. Malomed, Phys. Rev. B *41*, 2616 (1990).

B. A. Malomed and A. Weber, Phys. Rev. B *44*, 875 (1991).

B. A. Malomed and A. A. Nepomnyashchy, Phys. Rev. B *45*, 12435 (1992).

A. P. Malozemoff, *in* "Physical Properties of High Temperature Superconductors" (D. M. Ginsberg, Ed.), Vol. 1, Chap. 3, World Scientific, New York, 1989.

P. Mandal, A. Poddar, A. N. Das, B. Ghosh and P. Choudhury, Phys. Rev. B *40*, 730 (1989).

V. Manivannan, J. Gopalakrishnan, and C. N. R. Rao, Phys. Rev. B *43*, 8686 (1991).

Y. Maniwa, H. Sato, K. Mizoguchi, and K. Kune, Jpn. J. Appl. Phys. *29*, 268 (1990).

Y. Maniwa, T. Mituhashi, K. Mizoguchi, and K. Kume, Physica C *175*, 401 (1991a).

Y. Maniwa, S. Sato, T. Mituhaski, K. Mizoguchi, and K. Kume, Physica C *185–189*, 1761 (1991b).

M. B. Maple, J. W. Chen, S. E. Lambert, Z. Fisk, J. L. Smith, and H. R. Ott, cited in Stewart (1984).

M. C. Marchetti and D. R. Nelson, Phys. Rev. B *41*, 1910 (1990).

M. C. Marchetti, Phys. Rev. B *43*, 8012 (1991).

R. Marcon, R. Fastampa, M. Giura, C. Matacotta, Phys. Rev. B *39*, 2796 (1989).

R. Marcon, E. Silva, R. Fastampa, and M. Giura, Phys. Rev. B *46*, 3612 (1992).

J. Marcus, C. Escribe-Filippini, C. Schlenker, R. Buder, J. Devenyi, and P. L. Reydet, Solid State Commun. *63*, 129 (1987).

M. Marder, N. Papanicolaou, and G. C. Psaltakis, Phys. Rev. B *41*, 6920 (1990).

L. Maritato, A. M. Cucolo, R. Vaglio, C. Noce, J. L. Makous, and C. M. Falco, Phys. Rev. B *38*, 12917 (1988).

J. T. Markert, T. W. Noh, S. E. Russek, and R. M. Cotts, Solid State Commun. *63*, 847 (1987).

J. T. Markert, Y. Dalichaouch, and M. B. Maple, *in* "Physical Properties of High Temperature Super Conductors" (D. M. Ginsberg, Ed.), Vol. 1, Chap. 6, World Scientific, Singapore, 1989.

R. S. Markiewicz, Physica C *177*, 171 (1991).

R. S. Markiewicz, Int. J. Mod. Phys. B *5*, 2037 (1991).

P. Marsh, R. M. Fleming, M. L. Mandich, A. M. DeSantolo, J. Kwo, M. Hong, and L. J. Martinez-Miranda, Nature *334*, 141 (1988).

C. D. Marshall, I. M. Fishman, R. C. Dorfman, C. B. Eom, and M. D. Fayer, Phys. Rev. B *45*, 10009 (1992).

F. Marsiglio and J. E. Hirsch, Phys. Rev. B *44*, 11960 (1991).

F. Marsiglio, Phys. Rev. B *44*, 5373 (1991).

F. Marsiglio, Phys. Rev. B *45*, 956 (1992).

F. Marsiglio and J. E. Hirsch, Phys. Rev. B *49*, 1366 (1994).

S. Martin, A. T. Fiory, R. M. Fleming, L. F. Schneemeyer, and J. V. Waszczak, Phys. Rev. Lett. *60*, 2194 (1988).

S. Martin, A. T. Fiory, R. M. Fleming, G. P. Espinosa, and A. S. Copper, Phys. Rev. Lett. *62*, 677, (1989); see *63*, 582 (1989) for comment by P. C. E. Stamp and a reply by the authors.

S. Martin, A. T. Fiory, R. M. Fleming, L. F. Schneemeyer, and J. V. Waszczak, Phys. Rev. B *41*, 846 (1990).

S. Martin and A. F. Hebard, Phys. Rev. B *43*, 6253 (1991).

M. Martin, C. Kendziora, L. Mihaly, and R. Lefferts, Phys. Rev. B *46*, 5760 (1992).

M. C. Martin, D. Koller, and L. Mihaly, Phys. Rev. B *47*, 14607 (1993).

T. P. Martin, U. Näher, H. Schaber, and U. Zimmermann, Phys. Rev. Lett. *70*, 3079 (1993).

C. Martin, M. Hervieu, M. Huvé, C. Michel, A. Maignan, G. van Tendeloo, and B. Raveau, Physica C *222*, 19 (1994).

J. A. Martindale, S. E. Barrett, C. A. Klug, K. E. O'Hara, S. M. DeSoto, C. P. Slichter, T. A. Friedmann, and D. M. Ginsberg, Phys. Rev. Lett. *68*, 702 (1992).

J. L. Martins and N. Troullier, Phys. Rev. B *46*, 1766 (1992).

A. Masaki, H. Sato, S.-I. Uchida, K. Kitazawa, S. Tanaka, and K. Inoue, Jpn. J. Appl. Phys. *26*, 405 (1987).

H. Mathias, W. Moulton, H. K. Ng, S. J. Pan, K. K. Pan, L. H. Peirce, L. R. Testardi, and R. J. Kennedy, Phys. Rev. B *36*, 2411 (1987).

P. Mathieu and Y. Simon, Europhys. Lett. *5*, 67 (1988).

I. Matsubara, H. Tanigawa, T. Ogura, H. Yamashita, M. Kinoshita, and T. Kawai, Phys. Rev. B *45*, 7414 (1992).

Y. Matsuda, N. P. Ong, Y. F. Yan, J. M. Harris, and J. B. Peterson, Phys. Rev. B *49*, 4380 (1994).

Y. Matsuda, T. Hirai, S. Komiyama, T. Terashima, Y. Bando, K. Iijima, K. Yamamoto, and K. Hirata, Phys. Rev. B *40*, 5176 (1989).

Y. Matsumoto, M. Katada, and T. Nishida, Physica C *185*, 1229 (1991).

T. Matsuura and K. Miyake, Jpn. J. Appl. Phys. *26*, L407 (1987).

H. Matsuyama, T. Takahashi, H. Katayama-Yoshida, Y. Okabe, H. Takagi, and S. Uchida, Phys. Rev. B *40*, 2658 (1989).

L. F. Mattheiss, Phys. Rev. B *1*, 373 (1970).

L. F. Mattheiss and D. R. Hamann, Phys. Rev. B *28*, 4227 (1983).

L. F. Mattheiss, Jpn. J. Appl. Phys. 24(2), 6 (1985).

L. F. Mattheiss and D. R. Hamann, Solid State Commun. *63*, 395 (1987).

L. F. Mattheiss, Phys. Rev. Lett. *58*, 1028 (1987).

L. F. Mattheiss, E. M. Gyrogy, and D. W. Johnson, Jr., Phys. Rev. B *37*, 3745 (1988).

L. F. Mattheiss and D. R. Hamann, Phys. Rev. Lett. *60*, 2681 (1988).

L. F. Mattheiss and D. R. Hamann, Phys. Rev. B *39*, 4780 (1989).

L. F. Mattheiss, Phys. Rev. B *42*, 359 (1990).

B. Matthias, Phys. Rev. *92*, 874 (1953).

B. Matthias, Phys. Rev. *97*, 74 (1955).

D. C. Mattis and M. Molina, Phys. Rev. B *44*, 12565 (1991).

E. Maxwell, Phys. Rev. *78*, 477 (1950).

I. I. Mazin, O. Jepsen, O. K. Andersen, A. I. Liechtenstein, S. N. Rashkeev, and Y. A. Uspenskii, Phys. Rev. B *45*, 5103 (1992).

K. F. McCarty, D. S. Ginley, D. R. Boehme, R. J. Baughman, and B. Morosin, Solid State Commun. *68*, 77 (1988).

K. F. McCarty, B. Morosin, D. S. Ginley, and D. R. Boehme, Physica C *157*, 135 (1989).

K. F. McCarty, J. Z. Liu, R. N. Shelton, and H. B. Radousky, Phys. Rev. B *41*, 8792 (1990a); *42*, 9973 (1990b).

K. F. McCarty, H. B. Radousky, J. Z. Liu, and R. N. Shelton, Phys. Rev. B *43*, 13751 (1991).

K. A. McGreer, J.-C. Wan, N. Anand, and A. M. Goldman, Phys. Rev. B *39*, 12260 (1989).

M. E. McHenry, S. Simizu, H. Lessure, M. P. Maley, J. Y. Coulter, I. Tanaka, and H. Kojima, Phys. Rev. B *44*, 7614 (1991).

W. R. McKinnon, J. R. Morton, K. F. Preston, and L. S. Selwyn, Solid State Commun. *65*, 855 (1988).

W. L. McMillan, Phys. Rev. *167*, 331 (1968).

T. McMullen, Phys. Rev. B *41*, 877 (1990).

G. A. Medina and M. D. N. Regueiro, Phys. Rev. B *42*, 8073 (1990).

N. I. Medvedeva, S. A. Turzhevsky, V. A. Gubanov, and A. J. Freeman, Phys. Rev. B *48*, 16061 (1993).

D. Mehl, A. R. Köymen, K. O. Jensen, F. Gotwald, and A. Weiss, Phys. Rev. B *41*, 799 (1990).

W. Meissner and R. Ochsenfeld, Naturwissenschaft *21*, 787 (1933).

K. Mendelssohn, "Cryophysics," Chap. 6, Interscience, New York, 1960.

R. L. Meng, Y. Y. Sun, J. Kulik, Z. J. Huang, F. Chen, Y. Y. Xue, and C. W. Chu, Physica C *214*, 307 (1993a).

R. L. Meng, L. Beauvais, X. N. Zhang, Z. J. Huang, Y. Y. Sun, Y. Y. Zue, and C. W. Chu, Physica C *216*, 21 (1993b).

M. Merkel, M. Knupfer, M. S. Golden, J. Fink, R. Seemann, and R. L. Johnson, Phys. Rev. B *47*, 11470 (1993).

R. Meservey and B. B. Schwartz, *in* "Superconductivity" (R. D. Parks, Ed.), Vol. 1, Chap. 3, Dekker, New York, 1969.

J. Metzger, T. Weber, W. H. Fietz, K. Grube, H. A. Ludwig, T. Wolf, and H. Wühl, Physica C, *214*, 371 (1993).

H. M. Meyer III, D. M. Hill, T. J. Wagener, Y. Gao, J. H. Weaver, D. W. Capone II, and K. C. Goretta, Phys. Rev. B *38*, 6500 (1988).

H. M. Meyer III and J. H. Weaver, *in* "Physical Properties of High Temperature Superconductors" (D. M. Ginsberg, Ed.), Vol. 2, Chap. 6, World Scientific, Singapore, 1990.

P. F. Miceli, J. M. Tarascon, L. H. Greene, P. Barboux, M. Giroud, D. A. Neumann, J. J. Rhyne, L. F. Schneemeyer, and J. V. Waszczak, Phys. Rev. B *38*, 9209 (1988).

C. Michel and B. Raveau, Rev. Chim. Miner. *21*, 407 (1984).

C. Michel, M. Hervieu, M. M. Borel, A. Grandin, F. Deslandes, J. Provost, and B. Raveau, Z. Phys. B. Cond. Matt. *68*, 421 (1987).

R. Micnas, J. Ranninger, and S. Robaszkiewicz, Phys. Rev. B *36*, 4051 (1987).

R. Micnas, J. Ranninger, and S. Robaszkiewicz, Rev. Mod. Phys. *62*, 113 (1990).

A. R. Miedema, J. Phys. (Paris) *F3*, 1803 (1973).

A. R. Miedema, J. Phys. (Paris) *F4*, 120 (1974).

F. Mila, Europhys. Lett. *8*, 555 (1989).

F. Mila and T. M. Rice, Physica C *157*, 561 (1989).

S. L. Miller, K. R. Biagi, J. R. Clem, and D. K. Finnemore, Phys. Rev. B *31*, 2684 (1985).

J. H. Miller, Jr., G. H. Gunaratne, J. Huang, and T. D. Golding, Appl. Phys. Lett. *59*, 3330 (1991).

A. J. Millis, H. Monien, and D. Pines, Phys. Rev. B *42*, 167 (1990).

A. J. Millis and S. N. Coppersmith, Phys. Rev. B *43*, 13770 (1991).

A. J. Millis and H. Monien, Phys. Rev. B *45*, 3059 (1992).

L. Mingzhu, T. Weihua, M. Xianren, L. Zhenjin, H. Wei, T. Qingyun, R. Yanru, and L. Zhenxing, Phys. Rev. B *41*, 2517 (1990).

P. Minnhagen and P. Olsson, Phys. Rev. Lett. *67*, 1039 (1991).

P. Minnhagen and P. Olsson, Phys. Rev. B *45*, 5722 (1992).

N. Missert and M. R. Beasley, Phys. Rev. Lett. *63*, 672 (1989).

M. G. Mitch, S. J. Chase, and J. S. Lannin, Phys. Rev. Lett. *68*, 883 (1992).

L. Miu, A. Crisan, S. Popa, V. Sandu, and L. Nistor, J. Supercond. *3*, 391 (1990).

L. Miu, Phys. Rev. B *45*, 8142 (1992).

M. Miyazaki, J. Inoue, and S. Maekawa, Phys. Rev. B *40*, 6611 (1989).

P. Mocaër, L. Tessler, M. Laguës, F. Laher-Lacour, C. Lacour, U. Dai, N. Hess, and G. Deutscher, Physica C *185–189*, 2505 (1991).

M. A. K. Mohamed, J. Jung, and J. P. Franck, Phys. Rev. B *39*, 9614 (1989).

M. A. K. Mohamed, J. Jung, and J. P. Franck, Phys. Rev. B *41*, 4286, 6466 (1990).

M. A.-K. Mohamed and J. Jung, Phys. Rev. B *44*, 4512 (1991).

R. Monaco, Int. J. Infrared. Millimeter Waves II, 533 (1990a); J. Appl. Phys. *68*, 679 (1990b).

H. Monien and A. Zawadowski, Phys. Rev. Lett. *63*, 911 (1989).

H. Monien and A. Zawadowski, Phys. Rev. B *41*, 8798 (1990).

H. C. Montgomery, J. Appl. Phys. *42*, 2971 (1971).

P. Monthoux and D. Pines, Phys. Rev. B *49*, 4261 (1994).

J. S. Moodera, R. Meservey, J. E. Tkaczyk, C. X. Hao, G. A. Gibson, and P. M. Tedrow, Phys. Rev. B *37*, 619 (1988).

H. A. Mook, D. McK. Paul, B. C. Sales, L. A. Boatner, and L. Cussen, Phys. Rev. B *38*, 12008 (1988).

F. C. Moon, "Superconducting Levitation," Wiley, New York, 1994.

J. Moreland, A. F. Clark, H. C. Ku, and R. N. Shelton, Cryogenics *27*, 227 (1987).

J. Moreland, J. W. Ekin, L. F. Goodrich, T. E. Capobianco, A. F. Clark, J. Kwo, M. Hong, and S. H. Liou, Phys. Rev. B *35*, 8856 (1987).

H. Mori, Phys. Rev. B *43*, 5474 (1991).

D. E. Morris, J. H. Nickel, J. Y. T. Wei, N. G. Asmar, J. S. Scott, U. M. Scheven, C. T. Hultgren, A. G. Markelz, J. E. Post, P. J. Heaney, D. R. Veblen, and R. M. Hazen, Phys. Rev. B *39*, 7347 (1988).

D. E. Morris, N. G. Asmar, J. Y. T. Wei, J. H. Nickel, R. L. Sid, J. S. Scott, and J. E. Post, Phys. Rev. B *40*, 11406 (1989).

D. C. Morse and T. C. Lubensky, Phys. Rev. B *43*, 10436 (1991).

A. Moser, H. J. Hug, O. Fritz, I. Parashikov, H.-J. Güntherodt, and Th. Wolf, J. Vacuum Sci. and Technol., submitted for publication.

A. Moser, H. J. Hug, I. Parashikov, B. Stiefel, O. Fritz, H. Thomas, A. Baratoff, H.-J. Güntherodt, and P. Chaudhari, Phys. Rev. Lett. *74*, 1847 (1995).

M. Mostoller, J. Zhang, A. M. Rao, and P. C. Eklund, Phys. Rev. B *41*, 6488 (1990).

M. Mück, SPIE Symp. Adv. Electronic and Optoelectronic Matter, Los Angeles, California, 1994, submitted for publication.

H. Mukaida, K. Kawaguchi, M. Nakao, H. Kumakura, D. Dietderich, and K. Togano, Phys. Rev. B *42*, 2659 (1990).

K. A. Müller, M. Takashige, and J. G. Bednorz, Phys. Rev. Lett. *58*, 1143 (1987).

K.-H. Müller, Physica C *159*, 717 (1989).

K.-H. Müller and A. J. Pauza, Physica C *161*, 319 (1989).

K.-H. Müller, IEE Trans. Magn. *March* (1991).

K.-H. Müller, M. Nikolo, and R. Driver, Phys. Rev. B *43*, 7976 (1991).

H. Muller, M. Suenaga, and Y. Yokoyama, J. Appl. Phys. *70*, 4409 (1991).

M. Murakami, H. Fujimoto, S. Gotoh, K. Yamaguchi, N. Koshizuka, and S. Tanaka, Physica C *185–189*, 321 (1991).

M. Murakami, *in* "Studies of High Temperature Superconductors" (A. V. Narlikar, Ed.), Vol. 9, Nova Sci., New York, 1991.

D. W. Murphy, S. Sunshine, R. B. van Dover, R. J. Cava, B. Batlogg, S. M. Zahurak, and L. F. Schneemeyer, Phys. Rev. Lett. *58*, 1888 (1987).

P. Muzikar, D. Rainer, and J. A. Sauls, Proc. NATO Adv. Study Inst. Vortices in Superfluids, Cargèse, Corsica (N. Bontemps, Ed.), Kluwer, Dordrecht, 1994.

J. A. Mydosh, Phys. Scripta *T19*, 260 (1987).

T. Mzoughi, H. A. Farach, E. Quagliata, M. A. Mesa, C. P. Poole, Jr., and R. Creswick, Phys. Rev. B *46*, 1130 (1992).

N. Nagaosa and P. Lee, Phys. Rev. B *43*, 1233 (1991).

M. Nagoshi, Y. Fukuda, T. Suzuki, K. Ueki, A. Tokiwa, M. Kikuchi, Y. Syono, and M. Tachiki, Physica C *185*, 1051 (1991).

M. Naito, A. Matsuda, K. Kitazawa, S. Kambe, I. Tanaka, and H. Kojima, Phys. Rev. B *41*, 4823 (1990).

Y. Nakamura and S. Uchida, Phys. Rev. B *47*, 8369 (1993).

K. Nakao, N. Miura, K. Tatsuhara, H. Takeya, and H. Takei, Phys. Rev. Lett. *63*, 97 (1993).

A. V. Narlikar, Ed., "Studies of High Temperature Superconductors," Nova Sci., New York, 1989.

K. Nasu, Phys. Rev. B *42*, 6076 (1990).

B. Nathanson, O. Entin-Wohlman, and B. Mühlschlegel, Phys. Rev. B *45*, 3499 (1992).

R. Navarro and L. J. Campbell, Phys. Rev. B *44*, 10146 (1991).

D. R. Nelson, *in* "Fundamental Problems in Structural Mechanics V" (E. G. D. Cohen, Ed.), North-Holland, Amsterdam, 1980.

D. L. Nelson, M. S. Whittingham, and T. F. George, Eds., "Chemistry of High-Temperature Superconductors," ACS Symposium Series No. 351, American Chemical Society, Washington, DC, 1987.

D. R. Nelson and H. S. Seung, Phys. Rev. B *39*, 9153 (1989).

D. R. Nelson and P. Le Doussal, Phys. Rev. B *42*, 10113 (1990).

D. R. Nelson and V. M. Vinokur, Phys. Rev. Lett. *68*, 2398 (1992).

E. Nembach, K. Tachikawa, and S. Takano, Philos. Mag. *21*, 869 (1970).

R. Nemetschek, O. V. Misochko, B. Stadlober, and R. Hackl, Phys. Rev. B *47*, 3450 (1993).

S. A. Nemnonov, E. Z. Kurmaev, and V. I. Minin, IMF Akad. Nauk. USSR (Kiev) *1*, 87 (1969).

S. J. Nettel and R. K. MacCrone, Phys. Rev. B *47*, 11360 (1993).

M. V. Nevitt, G. W. Crabtree, and T. E. Klippert, Phys. Rev. B *36*, 2398 (1987).

V. L. Newhouse, *in* "Superconductivity" (R. D. Parks, Ed.), Vol. 2, p. 1283, Dekker, New York, 1969.

E. J. Nicol and J. P. Carbotte, Phys. Rev. B *43*, 10210 (1991).

E. J. Nicol and J. P. Carbotte, Phys. Rev. B *47*, 8205 (1993).

Ch. Niedermayer, H. Glückler, A. Golnik, U. Binninger, M. Rauer, E. Recknagel, J. I. Budnick, and A. Weidinger, Phys. Rev. B *47*, 3427 (1993).

L. Niel and J. E. Evetts, Supercond. Sci. Technol. *5*, S347 (1992).

G. Nieva, E. N. Martinez, F. de la Cruz, D. A. Esparza, and C. A. D'Ovidio, Phys. Rev. B *36*, 8780 (1987).

M. Nikolo and R. B. Goldfarb, Phys. Rev. B *39*, 6615 (1989).

M. Nikolo, W. Kiehl, H. M. Duan, and A. M. Hermann, Phys. Rev. B *45*, 5641 (1992).

H. Ning, H. Duan, P. D. Kirven, A. M. Hermann, and T. Datta, J. Supercond. *5*, 503 (1992).

T. Nishida, M. Katada, and Y. Matsumoto, Physica B *165–167*, 1327 (1990a); Jpn. J. Appl. Phys. *29*, 259 (1990b).

H. Nishihara, T. Ohtani, Y. Sano, and Y. Nakamura, Physica C *185–189*, 2733 (1991).

T. Nitta, K. Nagase, S. Hayakawa, and Y. Iida, J. Am. Ceram. Soc. *48*, 642 (1965).

R. K. Nkum and W. R. Datars, Physica C *192*, 215 (1992).

C. Noce and L. Maritato, Phys. Rev. B *40*, 734 (1989).

H. Noel, P. Gougeon, J. Padiou, J. C. Levet, M. Potel, O. Laborde, and P. Monceau, Solid State Commun. *63*, 915 (1987).

T. Nojima and T. Fujita, Physica C *178*, 140 (1991).

F. Nori, E. Abrahams, and G. T. Zimanyi, Phys. Rev. B *41*, 7277 (1990).

M. R. Norman, Phys. Rev. B *42*, 6762 (1990).

D. P. Norton, D. H. Lowndes, S. J. Pennycook, and J. D. Budai, Phys. Rev. Lett. *67*, 1358 (1991).

D. L. Novikov, V. A. Gubanov, and A. J. Freeman, Physica C *191*, 399 (1992).

P. Nozières and W. F. Vinen, Philos. Mag. *14*, 667 (1966).

N. Nücker, J. Fink, J. C. Fuggle, P. J. Durham, and W. M. Temmerman, Phys. Rev. B *37*, 5158 (1988).

N. Nücker, H. Romberg, X. X. Xi, J. Fink, B. Gegenheimer, and Z. X. Zhao, Phys. Rev. B *39*, 6619 (1989).

N. Nücker, H. Romberg, M. Alexander, and J. Fink, *in* "Studies of High Temperature Superconductors" (A. V. Narlika, Ed.), Nova Sci., New York, 1992.

B.-H. O and J. T. Markert, Phys. Rev. B *47*, 8373 (1993).

S. D. Obertelli, J. R. Cooper, and J. L. Tallon, Phys. Rev. B *46*, 14928 (1992).

B. Obst, Phys. Status Solidi B *45*, 467 (1971).

S. P. Obukhov and M. Rubinstein, Phys. Rev. Lett. *65*, 1279 (1990).

Y. Oda, H. Fujita, H. Toyoda, T. Kaneko, T. Kohara, I. Nakada, and K. Asayama, Jpn. J. Appl. Phys. *26*, L1660 (1987).

T. Oguchi, Jpn. J. Appl. Phys. *26*, L417 (1987).

A. Oguri and S. Maekawa, Phys. Rev. B *41*, 6977 (1990).

I. Ohana, A. Kazeroonian, D. Heiman, M. Dresselhaus, and P. J. Picone, Phys. Rev. B *40*, 2255, 2562 (1989).

K. Ohbayashi, N. Ogita, M. Udagawa, Y. Aoki, Y. Maeno, and T. Fujita, Jpn. J. Appl. Phys. *26*, L423 (1987).

F. J. Ohkawa, Phys. Rev. B *42*, 4163 (1990).

Y. Ohta and S. Maekawa, Phys. Rev. B *41*, 6524 (1990).

T. Ohtani, Mater. Res. Bull. *24*, 343 (1989).

N. Okazaki, T. Hasegawa, K. Kishio, K. Kitazawa, A. Kishi, Y. Ikeda, M. Takano, K. Oda, H. Kitaguchi, J. Takada, and Y. Miura, Phys. Rev. B *41*, 4296 (1990).

E. Oldfield, C. Coretsopoulos, S. Yang, L. Reven, H. C. Lee, J. Shore, O. H. Han, E. Ramli, and D. Hicks, Phys. Rev. B *40*, 6832 (1989).

O. H. Olsen and M. R. Samuelsen, Phys. Rev. B *43*, 10273 (1991).

N. P. Ong, Z. Z. Wang, J. Clayhold, J. M. Tarascon, L. H. Greene, and W. R. McKinnon, Phys. Rev. B *35*, 8807 (1987).

N. P. Ong, *in* "Physical Properties of High Temperature Superconductors" (D. M. Ginsberg, Ed.), Vol. 2, Chap. 7, World Scientific, Singapore, 1990.

N. P. Ong, Phys. Rev. *43*, 193 (1991).

H. Kamerlingh Onnes, Leiden Commun., 120b, 122b, 124c (1911).

M. Onoda, S. Shamoto, M. Sato, and S. Hosoya, Jap. J. Appl. Phys. *26*, L363 (1987).

J. Orenstein, G. A. Thomas, A. J. Millis, S. L. Cooper, D. H. Rapkine, T. Timusk, L. F. Schneemeyer, and J. V. Waszczak, Phys. Rev. B *42*, 6342 (1990).

T. P. Orlando and K. A. Delin, "Foundations of Applied Superconductivity," Addison–Wesley, Reading, MA, 1991.

K. Osamura, Ed., "Composite Superconductors," Dekker, New York, 1993.

J. A. Osborn, Phys. Rev. *67*, 351 (1945).

S. B. Oseroff, D. C. Vier, J. F. Smyth, C. T. Salling, S. Schultz, Y. Dalichaouch, B. W. Lee, M. B. Maple, Z. Fisk, J. D. Thompson, J. L. Smith, and E. Zirngiebl, *in* "Novel Superconductivity" (S. A. Wolf and V. Z. Kresin, Eds.), p. 679, Plenum, New York, 1987.

M. S. Osofsky, H. Rakoto, J. C. Ousset, J. P. Ulmet, J. Leotin, S. Askenazy, D. B. Crisey, J. S. Horwitz, E. F. Skelton, and S. A. Wolf, Physica C *182*, 257 (1991).

J. G. Ossandon, J. R. Thompson, D. K. Christen, B. C. Sales, Y. Sun, and K. W. Lay, Phys. Rev. B *46*, 3050 (1992).

J. G. Ossandon, J. R. Thompson, D. K. Christen, B. C. Sales, H. R. Kerchner, J. O. Thomson, Y. R. Sun, K. W. Lay, and J. E. Tkaczyk, Phys. Rev. B *45*, 12534 (1992).

S. B. Ota, Phys. Rev. B *35*, 8730 (1987).

S. B. Ota, V. S. Sastry, E. Gmelin, P. Murugaraj, and J. Maier, Phys. Rev. B *43*, 6147 (1991).

C. E. Otis and R. W. Dreyfus, Phys. Rev. Lett. *67*, 2102 (1991).

H. R. Ott, H. Rudigier, Z. Fisk, and J. L. Smith, Phys. Rev. Lett. *50*, 1595 (1983).

H. R. Ott, *in* "Novel Superconductivity" (S. A. Wolf and V. Z. Kresin, Eds.), p. 187, Plenum, New York.

H. R. Ott, "Ten Years of Superconductivity: 1980–1990," Kluwer, 1993.

M. Oussena, S. Senoussi, G. Collin, J. M. Broto, H. Rakoto, S. Askenazy, and J. C. Ousset, Phys. Rev. B *36*, 4014 (1987).

Yu. N. Ovchinnikov and B. I. Ivlev, Phys. Rev. B *43*, 8024 (1991).

A. W. Overhauser, Phys. Rev. Lett. *4*, 462 (1960).

A. W. Overhauser, Phys. Rev. *128*, 1437 (1962).

A. W. Overhauser and L. L. Daemen, Phys. Rev. Lett. *62*, 1691 (1989).

F. J. Owens, Physica C *178*, 456 (1991).

F. J. Owens, Physica C *195*, 225 (1992).

M.-A. Ozaki and K. Machida, Phys. Rev. B *39*, 4145 (1989).

S. Pagano, B. Ruggiero, and E. Sarnelli, Phys. Rev. B *43*, 5364 (1991).

E. J. Pakulis, Phys. Rev. B *42*, 10746 (1990).

T. T. M. Palstra, A. A. Menovsky, and J. A. Mydosh; Coles (1987); Phys. Rev. B *33*, 6527 (1988).

T. T. M. Palstra, B. Batlogg, L. F. Schneemeyer, and J. V. Waszczak, Phys. Rev. Lett. *61*, 1662 (1988).

T. T. M. Palstra, B. Batlogg, R. B. Van Dover, L. F. Schneemeyer, and J. V. Waszczak, Appl. Phys. Lett. *54*, 763 (1989).

T. T. M. Palstra, B. Batlogg, L. F. Schneemeyer, J. V. Waszczak, Phys. Rev. Lett. *64*, 3090 (1990).

T. T. M. Palstra, R. C. Haddon, A. F. Hebard, and J. Zaanen, Phys. Rev. Lett. *68*, 1054 (1992).

T. T. M. Palstra and R. C. Haddon, Solid State Commun., *92*, 71 (1994).

M. Palumbo, P. Muzikar, and J. A. Sauls, Phys. Rev. B *42*, 2681 (1990a).

M. Palumbo, C. H. Choi, and P. Muzikar, Physica B *165–166*, 1095 (1990b).

W. Pan and S. Doniach, Phys. Rev. B *49*, 1192 (1994).

D. A. Papaconstantopoulos, A. Pasturel, J. P. Julien, and F. Cyrot-Lackmann, Phys. Rev. B *40*, 8844 (1989).

M. Paranthaman, J. R. Thompson, Y. R. Sun, and J. Brynestad, Physica C *213*, 271 (1993).

M. Paranthaman, Physica C *222*, 7 (1994).

G. S. Park, C. E. Cunningham, B. Cabrera, and M. E. Huber, Phys. Rev. Lett. *68*, 1920 (1992).

R. D. Parks and W. A. Little, Phys. Rev. *133*, A97 (1964).

R. D. Parks, Ed. "Superconductivity," Vols. 1 and 2, Dekker, New York, 1969.

Y. S. Parmar and J. K. Bhattacharjee, Phys. Rev. B *45*, 814 (1992).

F. Parmigiani, Z. X. Shen, D. B. Mitzi, I. Lindau, W. E. Spicer, and A. Kapitulnik, Phys. Rev. B *43*, 3085 (1991).

P. C. Pattnaik, C. L. Kane, D. M. Newns, and C. C. Tsuei, Phys. Rev. B *45*, 5714 (1992).

D. McK. Paul, H. A. Mook, A. W. Hewat, B. C. Sales, L. A. Boatner, J. R. Thompson, and M. Mostoller, Phys. Rev. B *37*, 2341 (1988).

D. McK. Paul, H. A. Mook, L. A. Boatner, B. C. Sales, J. O. Ramey, and L. Cussen, Phys. Rev. B *39*, 4291 (1989).

L. Pauling and E. B. Wilson, "Introduction to Quantum Mechanics," McGraw–Hill, New York, 1935.

L. M. Paulius, C. C. Almasan, and M. B. Maple, Phys. Rev. B *47*, 11627 (1993).

S. D. Peacor and C. Uher, Phys. Rev. B *39*, 11559 (1989).

S. D. Peacor, R. Richardson, J. Burm, C. Uher, and A. Kaiser, Phys. Rev. B *42*, 2684 (1990).

S. D. Peacor, J. L. Cohn, and C. Uher, Phys. Rev. B *43*, 8721 (1991).

S. D. Peacor, R. A. Richardson, F. Nori, and C. Uher, Phys. Rev. B *44*, 9508 (1991).

W. B. Pearson, "Handbook of Lattice Spacings and Structures of Metals," p. 79, Pergamon, New York, 1958.

M. J. Pechan and J. A. Horvath, Am. J. Phys. *58*, 642 (1990).

N. F. Pedersen and A. Davidson, Phys. Rev. B *41*, 178 (1990).

S. Pei, J. D. Jorgensen, B. Dabrowski, D. G. Hinks, D. R. Richards, A. W. Mitchell, J. M. Newsam, S. K. Sinha, D. Vaknin, and A. J. Jacobson, Phys. Rev. B *41*, 4126 (1990).

K. E. Peiponen and E. Vartiainen, Phys. Rev. B *44*, 8301 (1991).

M. Pekala, K. Pekala, and A. Pajaczkowska, Phys. Status Solidi B *152*, K1 (1989).

M. T. Pencarinha, C. P. Poole, Jr., H. A. Farach, and O. A. Lopez, J. Phys. Chem. Solids *56*, 301 (1995).

D. R. Penn and M. L. Cohen, Phys. Rev. B *46*, 5466 (1992).

T. Penney, S. von Molnár, D. Kaiser, F. Holtzberg, and A. W. Kleinsasser, Phys. Rev. B *38*, 2918 (1988).

C. H. Pennington, D. J. Durand, D. B. Zax, C. P. Slichter, J. P. Rice, and D. M. Ginsberg, Phys. Rev. B *37*, 7944 (1988).

C. H. Pennington, D. J. Durand, C. P. Slichter, J. P. Rice, E. D. Bukowski, and D. M. Ginsberg, Phys. Rev. B *39*, 274, 2902 (1989).

C. H. Pennington and C. P. Slichter, *in* "Physical Properties of High Temperature Superconductors" (D. M. Ginsberg, Ed.), Chap. 5, World Scientific, Singapore, 1990.

C. H. Pennington and C. P. Slichter, Phys. Rev. Lett. *66*, 381 (1991).

S. J. Pennycook, M. F. Chisholm, D. E. Jesson, D. P. Norton, D. H. Lowndes, R. Feenstra, H. R. Kerchner, and J. O. Thomson, Phys. Rev. Lett. *67*, 765 (1991).

F. Perez, X. Obradors, J. Fontcuberta, M. Vallet, and J. Gonzalez-Calbet, Physica C *185–189*, 1843 (1991).

A. Pérez-González and J. P. Carbotte, Phys. Rev. B *45*, 9894 (1992).

A. Pérez-González, E. J. Nicol, and J. P. Carbotte, Phys. Rev. B *45*, 5055 (1992).

R. L. Peterson and J. W. Ekin, Physica C *157*, 325 (1989).

R. L. Peterson and J. W. Ekin, Phys. Rev. B *42*, 8014 (1990).

M. F. Petras and J. E. Nordman, Phys. Rev. B *39*, 6492 (1989).

B. W. Pfalzgraf and H. Spreckels, J. Phys. C *27*, 4359 (1987).

T. Pham, M. W. Lee, H. D. Drew, U. Welp, and Y. Fang, Phys. Rev. B *44*, 5377 (1991).

J. C. Phillips, Phys. Rev. B *36*, 861 (1987).

J. C. Phillips, "Physics of High-T_c Superconductors," Academic Press, New York, 1989a.

J. C. Phillips, Phys. Rev. B *40*, 7348, 8774 (1989b).

J. C. Phillips, Mater. Lett. *18*, 106 (1993).

J. M. Phillips, *in* "High Temperature Superconducting Thin Films" (D. Shi, Ed.), Pergamon, New York, 1994.

N. E. Phillips, Phys. Rev. *114*, 676 (1959).

N. E. Phillips, R. A. Fisher, and J. E. Gordon, Prog. Low Temp. Phys. *19* (1991).

T. Pichler, M. Matus, J. Kürti, and H. Kuzmany, Phys. Rev. B *45*, 13841 (1992).

W. E. Pickett, H. Krakauer, D. A. Papaconstantopoulos, and L. L. Boyer, Phys. Rev. B *35*, 7252 (1987).

W. E. Pickett, Rev. Mod. Phys. *61*, 433 (1989).

W. E. Pickett, R. E. Cohen, and H. Krakauer, Phys. Rev. B *42*, 8764 (1990).

W. E. Pickett, H. Krakauer, R. E. Cohen, and D. J. Singh, Science *255*, 46 (1992).

M. W. Pieper, Physica C *190*, 261 (1992).

S. W. Pierson and O. T. Valls, Phys. Rev. B *45*, 2458 (1992).

C. G. S. Pillai, Solid State Commun. *80*, 277 (1991).

W. Pint and E. Schachinger, Phys. Rev. B *43*, 7664 (1991).

L. Pintschovius, J. M. Bassat, P. Odier, F. Gervais, G. Chevrier, W. Reichardt, and F. Gompf, Phys. Rev. B *40*, 2229 (1989).

L. Pintschovius and W. Reichardt, *in* "Physical Properties of High Temperature Superconductors" (D. M. Ginsberg, Ed.), Vol. 4, Chap. 5, World Scientific, Singapore, 1994.

A. B. Pippard, Proc. R. Soc. London A *216*, 547 (1953).

M. Pissas, G. Kallias, A. Simopoulos, D. Niarchos, and A. Kostikas, Phys. Rev. B *46*, 14119 (1992).

F. Pistolesi and G. C. Strinati, Phys. Rev. B *49*, 6356 (1994).

B. Plaçais and Y. Simon, Phys. Rev. B *39*, 2151 (1989).

B. B. Plapp and A. W. Hübler, Phys. Rev. Lett. *65*, 2302 (1990).

I. Poberaj, D. Mihailovic, and S. Bernik, Phys. Rev. B *42*, 393 (1990).

A. Poddar, P. Mandal, K. G. Ray, A. N. Das, B. Ghosh, P. Choudhury, and S. Lahiri, Physica C *159*, 226 (1989).

D. Poilblanc and E. Dagotto, Phys. Rev. B *42*, 4861 (1990).

D. M. Poirier and J. H. Weaver, Phys. Rev. B *47*, 10959 (1993).

C. Politis, V. Buntar, W. Krauss, and A. Gurevich, Europhys. Lett. *17*, 175 (1992).

E. Polturak, G. Koren, D. Cohen, E. Aharoni, and G. Deutscher, Phys. Rev. Lett. *67*, 3038 (1991).

A. Pomar, A. Díaz, M. V. Ramallo, C. Torrón, J. A. Veira, and F. Vidal, Physica C *218*, 257 (1993).

M. Pompa, C. Li, A. Bianconi, A. C. Castellano, S. Dalla Longa, A. M. Flank, P. Lagarde, and D. Udron, Physica C *184*, 51 (1991a).

M. Pompa, P. Castrucci, C. Li, D. Udron, A. M. Flank, P. Lagarde, H. Katayamaa-Yoshida, S. Dalla Longa, and A. Bianconi, Physica C *184*, 102 (1991b).

C. P. Poole, Jr., and H. A. Farach, "Relaxation in Magnetic Resonance," Academic Press, New York, 1971.

C. P. Poole, Jr., "Electron Spin Resonance," 2nd ed., Wiley, New York, 1983.

C. P. Poole, Jr., and H. A. Farach, "Theory of Magnetic Resonance," 2nd ed., Wiley, New York, 1987.

C. P. Poole, Jr., T. Datta, and H. A. Farach, "Copper Oxide Superconductors," Wiley, New York, 1988.

C. P. Poole, Jr. and H. A. Farach, Magn. Reson. Relat. Phenom., Proc. 24th Ampere Congr., Poznan, p. 601 (1988).

C. P. Poole, Jr., T. Datta, and H. A. Farach, J. Supercond. *2*, 369 (1989).

C. P. Poole, Jr., and H. A. Farach, Eds., "Handbook of Electron Spin Resonance," Amer. Inst. Phys., New York, 1994.

A. Pöppl, L. Kevan, H. Kimura, and R. N. Schwartz, Phys. Rev. B *46*, 8559 (1992).

A. M. Portis, K. W. Blazely, and F. Waldner, Physica C *153–155*, 308 (1988).

A. M. Portis, "Electrodynamics of High Temperature Superconductors," World Scientific, Singapore, 1993.

J. Prade, A. D. Kulkarni, and F. W. de Wette, U. Schröder, and W. Kress, Phys. Rev. B *39*, 2771 (1989).

A. K. Pradhan, S. J. Hazell, J. W. Hodby, C. Chen, Y. Hu, and B. M. Wanklyn, Phys. Rev. B *47*, 11374 (1993).

R. Prange and S. Girvin, Eds., "The Quantum Hall Effect," Springer-Verlag, Heidelberg, 1987.

K. Prassides, J. Tomkinson, C. Christides, M. J. Rosseinsky, D. W. Murphy, and R. C. Haddon, Nature *354*, 462 (1991).

K. Prassides, M. J. Rosseinsky, A. J. Dianoux, and P. Day, J. Phys. Condens. Matter *4*, 965 (1992).

N. W. Preyer, M. A. Kastner, C. Y. Chen, R. J. Birgeneau, and Y. Hidaka, Phys. Rev. B *44*, 407 (1991).

D. Prost, L. Fruchter, I. A. Campbell, N. Motohira, and M. Konczykowski, Phys. Rev. B *47*, 3457 (1993).

T. Puig, L. M. Martinez, M. T. Aurell, A. Sanchez, D.-X. Chen, and J. S. Muñoz, *in* "Physics and Materials Science of High-Temperature Superconductivity" (R. Kossowsky, S. Methfessel, and D. Wohlbeben, Eds.), p. 467, Kluwer Academic, Dordrecht, 1990.

B. Pümpin, H. Keller, W. Kündig, W. Odermatt, B. D. Patterson, J. W. Schneider, H. Simmler, S. Connell, K. A. Müller, J. G. Bednorz, K. W. Blazey, I. Morgenstern, C. Rossel, and I. M. Savic, Z. Phys. B *72*, 175 (1988).

B. Pümpin, H. Keller, W. Kündig, W. Odermatt, I. M. Savic, J. W. Schneider, H. Simmler, P. Zimmermann, E. Kaldis, S. Rusiecki, Y. Maeno, and C. Rossel, Phys. Rev. B *42*, 8019 (1990).

P. Pureur and J. Schaf, Solid State Commun. *78*, 723 (1991).

S. N. Putilin, I. Bryntse, and E. V. Antipov, Mater. Res. Bull. *26*, 1299 (1991).

S. N. Putilin, E. V. Antipov, E. V. Chmaissem, and M. Marezio, Nature *362*, 266 (1993).

D. S. Pyun and T. R. Lemberger, Phys. Rev. B *44*, 7555 (1991).

K. F. Quader and E. Abrahams, Phys. Rev. B *38*, 11977 (1988).

R. M. Quick, C. Esebbag, and M. de Llano, Phys. Rev. B *47*, 11512 (1993).

M. Rabinowitz and T. McMullen, Chem. Phys. Lett. *218*, 437 (1994).

H. B. Radousky, J. Mater. Res. *7*, 1917 (1992).

R. J. Radtke, K. Levin, H.-B. Shüttler, and M. R. Norman, Phys. Rev. B *48*, 653 (1993).

H. Raffy, S. Labdi, O. Laborde, and P. Monceau, Phys. Rev. Lett. *66*, 2515 (1991).

D. Rainer and J. A. Sauls, *in* "Proc. 1992 Spring School on Cond. Mat. Phys., Trieste, Italy," World Scientific, Singapore, 1994, to be published.

D. Rainer and J. A. Sauls, "Proc. 1992 Spring School on Cond. Matter Phys., Trieste, Italy," World Scientific, Singapore, 1994.

A. K. Rajagopal and S. D. Mahanti, Phys. Rev. B *44*, 10210 (1991).

P. F. Rajam, C. K. Subramaniam, S. Kasiviswanathan, and R. Srinivasan, Solid State Commun. *71*, 475 (1989).

R. Rajput and D. Kumar, Phys. Rev. B *42*, 8634 (1990).

B. Rakvin, M. Pozek, and A. Dulcic, Solid State Commun. *72*, 199 (1989).

B. Rakvin, T. A. Mahl, A. S. Bhalla, Z. Z. Sheng, and N. S. Dalal, Phys. Rev. B *41*, 769 (1990).

K. S. Ralls, D. C. Ralph, and R. A. Buhrman, Phys. Rev. B *40*, 11561 (1989).

S. Ramakrishnan, R. Kumar, P. L. Paulose, A. K. Grover, and P. Chaddah, Phys. Rev. B *44*, 9514 (1991).

R. Ramakumar, R. Kumar, K. P. Jain, and C. C. Chancey, Phys. Rev. B *48*, 6509 (1993).

S. Ramasesha and C. N. R. Rao, Phys. Rev. B *44*, 7046 (1991).

A. P. Ramirez, T. Siegrist, T. T. M. Palstra, J. D. Garrett, E. Bruck, A. A. Menovsky, and J. A. Mydosh, Phys. Rev.B *44*, 5392 (1991).

A. P. Ramirez, A. R. Kortan, M. J. Rosseinsky, S. J. Duclos, A. M. Mujsce, R. C. Haddon, D. W. Murphy, A. V. Makhija, S. M. Zahurak, and K. B. Lyons, Phys. Rev. Lett. *68*, 1058 (1992a).

A. P. Ramirez, M. J. Rosseinsky, D. W. Murphy, and R. C. Haddan, Phys. Rev. Lett. *69*, 1687 (1992b).

J. Rammer, Phys. Rev. B *36*, 5665 (1987).

J. Rammer, Europhys. Lett. *5*, 77 (1988).

J. Rammer, Phys. Rev. B *43*, 2983 (1991).

C. N. R. Rao, P. Ganguly, A. K. Raychaudhuri, R. A. Mohan Ram, and K. Sreedhar, Nature *326*, 856 (1987).

K. V. Rao, D.-X. Chen, J. Nogues, C. Politis, C. Gallo, and J. A. Gerber, *in* "High Temperature Superconductors" (D. U. Gubser and M. Schluter, Eds.), p. 133, Mater. Res. Soc., Pittsburgh, 1987.

C. N. R. Rao, Philos. Trans. R. Soc. London Ser. A *336*, 595 (1991).

C. N. R. Rao, A. K. Santra, and D. D. Sarma, Phys. Rev. B *45*, 10814 (1992).

U. Rauchschwalbe, F. Steglich, G. R. Stewart, A. L. Giorgi, P. Fulde, and K. Maki, Europhys. Lett. *3*, 751 (1987).

J. Redinger, A. J. Freeman, J. Yu, and S. Massidda, Phys. Lett. A *124*, 469 (1987).

M. Reedyk, T. Timusk, J. S. Xue, and J. Greedan, Phys. Rev. B *45*, 7406 (1992a).

M. Reedyk, C. V. Stager, T. Timusk, J. S. Xue, and J. E. Greedan, Phys. Rev. B *45*, 10057 (1992b).

M. E. Reeves, S. E. Stupp, T. Friedmann, F. Slakey, D. M. Ginsberg, and M. V. Klein, Phys. Rev. B *40*, 4573 (1989).

J. D. Reger, T. A. Tokuyasu, A. P. Young, and M. P. A. Fisher, Phys. Rev. B *44*, 7147 (1991).

M. N. Regueiro, B. Salce, R. Calemczuk, C. Marin, and J. Y. Henry, Phys. Rev. B *44*, 9727 (1991).

W. Rehwald, M. Rayl, R. W. Cohen, and G. D. Cody, Phys. Rev. B *6*, 363 (1972).

F. Reif, "Fundamentals of Statistical and Thermal Physics," McGraw–Hill, New York, 1965.

Y. T. Ren, J. Clayhold, F. Chen, Z. J. Huang, X. D. Qiu, Y. Y. Sun, R. L. Meng, Y. Y. Xue, and C. W. Chu, Physica C *217*, 6 (1993).

Y. T. Ren, H. Chang, Q. Xiong, Y. Y. Xue, and C. W. Chu, Physica C, *226*, 209 (1994).

B. Renker, F. Gompf, E. Gering, D. Ewert, H. Rietschel, and A. Dianoux, Z. Phys. B *73*, 309 (1988).

Ch. Renner, A. D. Kent, Ph. Niedermann, Ø. Fischer, and F. Lévy, Phys. Rev. Lett. *67*, 1650 (1991).

L. Reven, J. Shore, S. Yang, T. Duncan, D. Schwartz, J. Chung, and E. Oldfield, Phys. Rev. B *43*, 10466 (1991).

C. M. Rey and L. R. Testardi, Phys. Rev. B *44*, 765 (1991).

A. P. Reyes, D. E. MacLaughlin, M. Takigawa, P. C. Hammel, R. H. Heffner, J. D. Thompson, J. E. Crow, A. Kebede, T. Mihalisin, and J. Schwegler, Phys. Rev. B *42*, 2688 (1990).

A. P. Reyes, D.E. MacLaughlin, M. Takigawa, P. C. Hammel, R. H. Heffner, J. D. Thompson, and J. E. Crow, Phys. Rev. B *43*, 2989 (1991).

C. A. Reynolds, B. Serin, W. H. Wright, and L. B. Nesbitt, Phys. Rev. *78*, 487 (1950).

H.-C. Ri, F. Kober, R. Gross, R. P. Huebener, and A. Gupta, Phys. Rev. B *43*, 13739 (1991).

H.-C. Ri, J. Kober, A. Beck, L. Alff, R. Gross, and R. P. Huebener, Phys. Rev. B *47*, 12312 (1993).

J. K. Rice, S. W. McCauley, A. P. Baronavski, J. S. Horwitz, and D. B. Chrisey, Phys. Rev. B *47*, 6086 (1993).

J. P. Rice, N. Rigakis, D. M. Ginsberg, and J. M. Mochel, Phys. Rev. B *46*, 11050 (1992).

P. L. Richards and M. Tinkham, Phys. Rev. *119*, 575 (1960).

R. A. Richardson, S. D. Peacor, F. Nori, and C. Uher, Phys. Rev. Lett. *67*, 3856 (1991).

C. T. Riecke, Th. Wölkhausen, D. Fay, and L. Tewordt, Phys. Rev. B *39*, 278 (1989).

E. Riedel, Z. Naturforsch A. *190*, 1634 (1964).

E. K. Riedel, H.-F. Cheung, and Y. Gefen, Phys. Scr. T *25*, 357 (1989).

P. S. Riseborough, Phys. Rev. B *45*, 13984 (1992).

B. Roas, L. Schultz, and G. Saemann-Ischenko, Phys. Rev. Lett. *64*, 479 (1990).

B. W. Roberts, J. Phys. Chem. Ref. Data *5*, 581 (1976).

G. I. Rochlin, Am. J. Phys. *43*, 335 (1975).

C. O. Rodriguez, Phys. Rev. B *49*, 1200 (1994).

E. Rodriguez, J. Luzuriaga, C. D'Ovidio, and D. A. Esparza, Phys. Rev. B *42*, 10796 (1990).

J. P. Rodriguez, Phys. Rev. B *36*, 168 (1987).

J. P. Rodriguez and B. Douçot, Phys. Rev. B *42*, 8724 (1990).

J. P. Rodriguez and B. Douçot, Phys. Rev. B *45*, 971 (1992).

C. T. Rogers, K. E. Myers, J. N. Eckstein, and I. Bozovic, Phys. Rev. Lett. *69*, 160 (1992).

D. S. Rokhsar, Phys. Rev. Lett. *65*, 1506 (1990).

D. B. Romero, G. L. Carr, D. B. Tanner, L. Forro, D. Mandrus, L. Mihaly, and G. P. Williams, Phys. Rev. B *44*, 2818 (1991).

M. Rona, Phys. Rev. *42*, 4183 (1990).

A. C. Rose-Innes and E. H. Rhoderick, "Introduction to Superconductivity," Pergamon, Oxford, 1994.

P. A. Rosenthal, M. R. Beasley, K. Char, M. S. Colclough, and G. Zaharchuk, Appl. Phys. Lett. *59*, 3482 (1991).

J. Rossat-Mignod, P. Burlet, M. J. G. M. Jurgens, J. Y. Henry, and C. Vettier, Physica C *152*, 19 (1988).

M. J. Rosseinsky, A. P. Ramirez, S. H. Glarum, D. W. Murphy, R. C. Haddon, A. F. Hebard, T. T. M. Palstra, A. R. Kortan, S. M. Zahurak, and A. V. Makhija, Phys. Rev. Lett. *66*, 2830 (1991).

C. Rossel, Y. Maeno, and I. Morgenstein, Phys. Rev. Lett. *62*, 681 (1989a); C. Rossel, Y. Maeno, and F. H. Holtzberg, IBM J. Res. Dev. *33*, 328 (1989b).

C. Rossel, O. Peña, H. Schmitt, and M. Sergent, Physica C *181*, 363 (1991).

S. J. Rothman, J. L. Routbort, U. Welp, and J. E. Baker, Phys. Rev. B *44*, 2326 (1991).

L. D. Rotter, Z. Schlesinger, R. T. Collins, F. Holtzberg, C. Field, U. W. Welp, G. W. Crabtree, J. Z. Liu, Y. Fang, K. G. Vandervoort, and S. Fleshler, Phys. Rev. Lett. *67*, 2741 (1991).

J. L. Routbort and S. J. Rothman, Appl. Phys. Rev., submitted for publication.

V. A. Rowe and R. P. Huebener, Phys. Rev. *185*, 666 (1969).

J. M. Rowell and R. C. Dynes, *in* "Phonons" (M. A. Nusimovici, Ed.), Flammarion, Sciences, Paris, 1972.

G. J. Rozing, P. E. Mijnarends, A. A. Menovsky, and P. F. de Châtel, Phys. Rev. B *43*, 9523 (1991).

A. E. Ruckenstein, P. J. Hirschfeld, and J. Appel, Phys. Rev. B *36*, 857 (1987); reprinted *in* "Theories of High Temperature Superconductivity" (J. W. Halley, Ed.), p. 137. Addison–Wesley, Reading, MA, 1988.

S. Ryu, S. Doniach, G. Deutscher, and A. Kapitulnik, Phys. Rev. Lett. *68*, 710 (1992).

M. S. Rzchowski, L. L. Sohn, and M. Tinkham, Phys. Rev. B *43*, 8682 (1991).

S. Sachdev and Z. Wang, Phys. Rev. B *43*, 10229 (1991).

S. Sachdev, Phys. Rev. B *45*, 389 (1992).

E. Sacher and J. E. Klemberg-Sapieha, Phys. Rev. B *39*, 1461 (1989).

C. A. R. Sá de Melo and S. Doniach, Phys. Rev. B *41*, 6633 (1990).

C. A. R. Sá de Melo, Z. Wang, and S. Doniach, Phys. Rev. Lett. *68*, 2078 (1992).

H. Safar, C. Durán, J. Guimpel, L. Civale, J. Luzuriaga, E. Rodriguez, F. de la Cruz, C. Fainstein, L. F. Schneemeyer, and J. V. Waszczak, Phys. Rev. B *40*, 7380 (1989).

H. Safar, H. Pastoriza, F. de la Cruz, D. J. Bishop, L. F. Schneemeyer, and J. Waszczak, Phys. Rev. B *43*, 13610 (1991).

H. Safar, P. L. Gammel, D. J. Bishop, D. B. Mitzi, and A. Kapitulnik, Phys. Rev. Lett. *68*, 2672 (1992).

H. Safer, P. L. Gammel, D. A. Huse, D. J. Bishop, W. C. Lee, J. Giapintzakis, and D. M. Ginsberg, Phys. Rev. Lett. *70*, 3800 (1993).

L. Sagdahl, S. Gjølmesli, T. Laegreid, K. Fossheim, and W. Assmus, Phys. Rev. B *42*, 6797 (1990).

L. Sagdahl, T. Laegreid, K. Fossheim, M. Murkami, H. Fujimoto, S. Gotoh, K. Yamaguchi, H. Yamauchi, N. Koshizuka, and S. Tanaka, Physica C *172*, 495 (1991).

N. Sahoo, S. Markert, T. P. Das, and K. Nagamine, Phys. Rev. B *41*, 220 (1990).

D. Saint-James and P. D. de Gennes, Phys. Lett. *7*, 306 (1963).

D. Saint-James, E. J. Thomas, and G. Sarma, "Type II Superconductivity," Pergamon, Oxford, 1969.

K. Saitoh and T. Nishino, Phys. Rev. B *44*, 7070 (1991).

S. Saito and A. Oshiyama, Phys. Rev. Lett. *66*, 2637 (1991).

K. Salama, Y. H. Zhang, M. Mironova, and D. F. Lee, "Proc. Sixth US–Japan Joint Workshop on High T_c Superconductivity," Houston, Texas, December 1993.

M. B. Salamon and J. Bardeen, Phys. Rev. Lett. *59*, 2615 (1987).

M. B. Salamon, *in* "Physical Properties of High Temperature Superconductors," (D. M. Ginsberg, Ed.), Vol. 1, Chap. 2, World Scientific, Singapore, 1989.

S. Salem-Sugui, Jr., E. E. Alp, S. M. Mini, M. Ramanathan, J. C. Campuzano, G. Jennings, M. Faiz, S. Pei, B. Dabrowski, Y. Zheng, D. R. Richards, and D. G. Hinks, Phys. Rev. B *43*, 5511 (1991).

A. V. Samoilov, A. A. Yurgens, and N. V. Zavaritsky, Phys. Rev. B *46*, 6643 (1992).

B. A. Sanborn, P. B. Allen, and D. A. Papaconstantopoulos, Phys. Rev. B *40*, 6037 (1989).

A. Sanchez and D.-X. Chen, *in* "Susceptibility of Superconductors and Other Spin Systems" (T. Francavilla, R. A. Hein, and D. Leiberger, Eds.), Plenum, New York, 1991.

A. Sanchez, D.-X. Chen, J. Muñoz, and Y.-Z. Li, Physica C *175*, 33 (1991).

P. Santhanam and C. C. Chi, Phys. Rev. B *38*, 11843 (1988).

A. Santoro, *in* "High Temperature Superconductivity" (J. W. Lynn, Ed.), Chap. 4, Springer-Verlag, Berlin, 1990.

J. Sapriel, J. Schneck, J. F. Scott, J. C. Tolédano, L. Pierre, J. Chavignon, C. Daguet, J. P. Chaminade, and H. Boyer, Phys. Rev. B *43*, 6259 (1991).

E. Sardella, Phys. Rev. B *45*, 3141 (1992).

D. D. Sarma, P. Sen, C. Carbone, R. Cimino, and W. Gudat, Phys. Rev. B *39* 12387 (1989).

W. M. Saslow, Phys. Rev. B *39*, 2710 (1989).

J. S. Satchell, R. G. Humphreys, N. G. Chew, J. A. Edwards, and M. J. Kane, Nature *334*, 331 (1988).

N. Sato, T. Sakon, N. Takeda, T. Komatsubara, C. Geibei, and F. Steglich, J. Phys. Soc. Jpn. *61*, 32 (1992).

S. Satpathy and R. M. Martin, Phys. Rev. B *36*, 7269 (1987).

C. B. Satterthwaite, Phys. Rev. *125*, 873 (1962).

A. Saul and M. Weissmann, Phys. Rev. B *42*, 4196 (1990).

J. A. Sauls, APS Fall Mtg. Symp. Kondo Lattices and Heavy Fermions, to be published.

D. J. Scalapino, Phys. Reports, to be published.

J. Schaf, P. Pureur, and J. V. Kunzler, Phys. Rev. B *40*, 6948 (1989).

H. J. Scheel *in* "Adv. in Supercond.," Springer-Verlag, Berlin, New York, to be published.

S. Scheidl and G. Hackenbroich, Phys. Rev. B *46*, 14010 (1992).

A. Schenck, P. Birrer, F. N. Gygax, B. Hitti, E. Lippelt, M. Weber, P. Böni, P. Fischer, H. R. Ott, and Z. Fisk, Phys. Rev. Lett. *65*, 2454 (1990).

A. Schenstrom, M-F. Xu, Y. Hong, D. Bein, M. Levy, B. K. Sarma, S. Adenwalla, Z. Zhao, T. Tokuyasu, D. W. Hess, J. B. Ketterson, J. A. Sauls, and D. G. Hinks, Phys. Rev. Lett. *62*, 332 (1989).

A. Schilling, M. Cantoni, J. D. Guo, and H. R. Ott, Nature *363*, 56 (1993).

A. Schilling, M. Catoni, O. Jeandupeux, J. D. Guo, and H. R. Ott, *in* "Advances in Superconductivity" (T. Fujita and Y. Shiohara, Eds.), Vol. 6, Springer-Verlag, Berlin, 1994a.

A. Schilling, O. Jeandupeux, S. Büchi, H. R. Ott, and C. Rossel, Physica C *235*, in press (1994b).

J. E. Schirber, D. L. Overmyer, K. D. Carlsan, J. M. Williams, A. M. Kini, H. H. Wang, H. A. Charlier, B. J. Love, D. M. Watkins, and G. A. Yaconi, Phys. Rev. B *44*, 4666 (1991).

Z. Schlesinger, R. L. Greene, J. G. Bednorz, and K. A. Müller, Phys. Rev. B *35*, 5334 (1987).

Z. Schlesinger, R. T. Collins, J. A. Calise, D. G. Hinks, A. W. Mitchell, Y. Zheng, B. Dabrowski, N. E. Bickers, and D. J. Scalapino, Phys. Rev. B *40*, 6862 (1989).

Z. Schlesinger, R. T. Collins, F. Holtzberg, C. Feild, G. Koren, and A. Gupta, Phys. Rev. B *41*, 11237 (1990a).

Z. Schlesinger, R. T. Collins, F. Holtzberg, C. Feild, S. H. Blanton, U. Welp, G. W. Crabtree, Y. Fang, and J. Z. Liu, Phys. Rev. Lett. *65*, 801 (1990b).

D. Schmeltzer, Phys. Rev. B *49*, 6944 (1994).

A. Schmid, Phys. Kondens. Mat. *8*, 129 (1968).

J. M. Schmidt, A. N. Cleland and J. Clarke, Phys. Rev. B *43*, 229 (1991).

P. Schmitt, P. Kummeth, L. Schultz, and G. Saemann-Ischenko, Phys. Rev. Lett. *67*, 267 (1991).

H. Schnack and R. Griessen, Phys. Rev. Lett. *68*, 2706 (1992).

L. F. Schneemeyer, J. K. Thomas, T. Siegrist, B. Batlogg, L. W. Rupp, R. L. Opila, R. J. Cava, and D. W. Murphy, Nature *335*, 421 (1988).

T. Schneider, Z. Phys. B *85*, 187 (1991).

T. Schneider, Z. Gedik, and S. Ciraci, Z. Phys. B *83*, 313 (1991).

T. Schneider, Physica C *195*, 82 (1992).

T. Schneider and H. Keller, Phys. Rev. Lett. *69*, 3374 (1992).

A. J. Schofield and Wheatley, Phys. Rev. B *47*, 11607 (1993).

K. Schönhammer, Phys. Rev. B *42*, 2591 (1990).

J. R. Schrieffer, "Theory of Superconductivity," Addison–Wesley, New York, 1964.

J. R. Schrieffer, X.-G. Wen, and S.-C. Zhang, Phys. Rev. Lett. *60*, 944 (1988).

E. A. Schuberth, B. Strickler, and K. Andres, Phys. Rev. Lett. *68*, 117 (1992).

I. K. Schuller, D. G. Hinks, M. A. Beno, S. W. Capone II, L. Soderholm, J. P. Locquet, Y. Bruynseraede, C. U. Segre, and K. Zhang, Solid State Commun. *63*, 385 (1987).

L. Schultz, E. Hellstern, and A. Thomä, Europhys. Lett. *3*, 921 (1987).

H. J. Schulz, Phys. Rev. Lett. *64*, 2831 (1990).

J. Schwartz, S. Nakamae, G. W. Raban, Jr., J. K. Heuer, S. Wu, J. L. Wagner, and D. G. Hinks, Phys. Rev. B *48*, 9932 (1994).

H. Schwenk, F. Gross, C. P. Heidmann, K. Andres, D. Schweitzer, and H. Keller, Mol. Cryst. Liq. Cryst. *119*, 329 (1985); Phys. Rev. B *31*, 3138 (1985).

P. Seidel, E. Heinz, M. Siegel, F. Schmidl, K. J. Zach, and H.-J. Köhler, *in* "Proc. 4th Int. Conf. on Superconducting and Quantum Effect Devices and Their Applications," Berlin, June 1991.

G. T. Seidler, T. F. Rosenbaum, D. L. Heinz, J. W. Downey, A. P. Paulikas, and B. W. Veal, Physica C *183*, 333 (1991).

G. T. Seidler, T. F. Rosenbaum, and B. W. Veal, Phys. Rev. B *45*, 10162 (1992).

K. Semba, T. Ishii, and A. Matsuda, Phys. Rev. Lett. *67*, 769 (1991).

S. Sengupta, C. Dasgupta, H. R. Krishnamurthy, G. I. Menon, and T. V. Ramakrishnan, Phys. Rev. Lett. *67*, 3444 (1991).

S. Sengupta and D. Shi, *in* "High Temperature Superconducting Materials Science and Engineering," to be published.

S. Senoussi, M. Oussena, and S. Hadjoudi, J. Appl. Phys. *63*, 4176 (1988).

A. Sequeira, H. Rajagopal, P. V. P. S. S. Sastry, J. V. Yakhmi, R. M. Iyer, and B. A. Dasannacharya, Physica B *180–181*, 429 (1992).

M. Sera, S. Shamoto, and M. Sato, Solid State Commun. *68*, 649 (1988).

S. Sergeenkov and M. Ausloos, Phys. Rev. B *47*, 14476 (1993).

R. D. Shannon and P. E. Bierstedt, J. Am. Ceram. Soc. *58*, 635 (1970).

C. Shao-Chun, Z. Dong-Ming, Z. Dian-Lin, H. M. Duan, and A. M. Hermann, Phys. Rev. B *44*, 12571 (1991).

B. Ya. Shapiro, Phys. Rev. B *48*, 16722 (1993).

S. Shapiro, Phys. Rev. Lett. *11*, 80 (1963).

R. P. Sharma, L. E. Rehn, and P. M. Baldo, Phys. Rev. B *43*, 13711 (1991).

B. S. Shastry and B. I. Shraiman, Int. J. Mod. Phys. *5*, 365 (1991).

T. P. Sheahen, "Introduction to High T_c Superconductivity," Plenum, New York, 1994.

V. Sh. Shekhtman, Ed., The Real Structure of High-T_c Superconductors, Springer-Verlag, Berlin, 1993.

S.-Q. Shen and W. Lu, Phys. Rev. B *48*, 1105 (1993).

Z.-X. Shen, P. A. P. Lindberg, B. O. Wells, D. S. Dessau, A. Borg, I. Lindau, W. E. Spicer, W. P. Ellis, G. H. Kwei, K. C. Ott, J.-S. Kang, and J. W. Allen, Phys. Rev. B *40*, 6912 (1989).

Z.-X. Shen, R. S. List, D. S. Dessau, F. Parmigiani, A. J. Arko, R. Bartlett, B. O. Wells, I. Lindau, and W. E. Spicer, Phys. Rev. B *42*, 8081 (1990).

Z.-X. Shen, W. E. Spicer, D.M. King, D. S. Dessau, and B. O. Wells, Science *267*, 343 (1995).

Z.-Y. Shen, "High Temperature Superconducting Microwave Circuits," Artech House, Norwood, Massachusetts, 1994.

Z. Z. Sheng, A. M. Hermann, A. El Ali, C. Almasan, J. Estrada, T. Datta, and R. J. Matson, Phys. Rev. Lett. *60*, 937 (1988).

Z. Z. Sheng and A. M. Hermann, Nature *332*, 55 (1988).

D.-N. Sheng, Z.-B. Su, and L. Yu, Phys. Rev. B *42*, 8732 (1990).

D. Shi, M. S. Boley, U. Welp, J. G. Chen, and Y. Liao, Phys. Rev.B *40*, 5255 (1989).

D. Shi, M. Xu, M. M. Fang, J. G. Chen, A. L. Cornelius, and S. G. Lanan, Phys. Rev. B *41*, 8833 (1990a).

D. Shi, M. Xu, A. Umezawa, and R. F. Fox, Phys. Rev. B *42*, 2062 (1990b).

D. Shi, X. S. Ling, M. Xu, M. M. Fang, S. Luo, J. I. Budnick, B. Dabrowski, D. G. Hinks, D. R. Richards, and Y. Zheng, Phys. Rev. B *43*, 3684 (1991).

D. Shi and M. Xu, Phys. Rev. B *44*, 4548 (1991).

D. Shi, Ed., "High Temperature Superconducting Materials Science and Engineering," Elsevier, Oxford, 1994.

J. S. Shier and D. M. Ginsberg, Phys. Rev. *147*, 384 (1966).

E. Shimizu and D. Ito, Phys. Rev. B *39*, 2921 (1989).

E. Shimshoni, Y. Gefen, and S. Levit, Phys. Rev. *40*, 2147 (1989).

E. Shimshoni and E. Ben-Jacob, Phys. Rev. B *43*, 2705 (1991).

S. L. Shindé, J. Morrill, D. Goland, D. A. Chance, and T. McGuire, Phys. Rev. B *41*, 8838 (1990).

T. Shinjo, T. Mizutani, N. Hosoito, T. Kusuda, T. Takabatake, K. Matsukuma, and H. Fujii, Physica C *159*, 869 (1989).

T. Shinjo and S. Nasu, *in* "Mechanisms of High Temperature Superconductivity" (H. Kamimura and A. Oshiyama, Eds.), p. 166, Springer Series in Material Science, Springer-Verlag, Heidelberg, 1989.

G. Shirane, R. J. Birgeneau, Y. Endoh, P. Gehring, M. A. Kastner, K. Kitazawa, H. Kojima, I. Tanaka, T. R. Thurston, and K. Yamada, Phys. Rev. Lett. *63*, 330 (1989).

G. Shirane, J. Als-Nielsen, M. Nielsen, J. M. Tranquada, H. Chou, S. Shamoto, and M. Sato, Phys. Rev. B *41*, 6547 (1990).

K. N. Shrivastava and K. P. Sinha, "Magnetic Superconductors: Model Theories and Experimental Properties of Rare-Earth Compounds," North-Holland, Amsterdam, 1984.

K. N. Shrivastava, Phys. Rev. B *41*, 11168 (1990).

J. D. Shore, M. Huang, A. T. Dorsey, and J. P. Sethna, Phys. Rev. Lett. *62*, 3089 (1989).

L. Shu-yuan, L. Li, and Z. Dian-lin, H. M. Duan, W. Kiel, and A. M. Hermann, Phys. Rev. B *47*, 8324 (1993).

Yu. N. Shvachko, A. A. Koshta, A. A. Romanyukha, V. V. Ustinov, and A. I. Akimov, Physica C *174*, 447 (1991).

Q. Si and G. Kotliar, Phys. Rev. Lett. *70*, 3143 (1993).

M. Siegel, F. Schmdl, K. Zach, E. Heinz, J. Borck, W. Michalke, and P. Seidel, Physica C *180*, 288 (1991).

T. Siegrist, S. Sunshine, D. W. Murphy, R. J. Cava, and S. M. Zahurak, Phys. Rev. B *35*, 7137 (1987).

T. Siegrist, S. M. Zahurak, D. W. Murphy, and R. S. Roth, Nature *334*, 231 (1988).

M. Sigrist, T. M. Rice, and K. Ueda, Phys. Rev. Lett. *63*, 1727 (1989).

P. Simon, J. M. Bassat, S. B. Oseroff, Z. Fisk, S.-W. Cheong, A. Wattiaux, and S. Schultz, Phys. Rev. Lett. *48*, 4216 (1993).

R. R. P. Singh, P. A. Fleury, K. B. Lyons, and P. E. Sulewski, Phys. Rev. Lett. *62*, 2736 (1989).

D. Singh, W. E. Pickett, E. C. von Stetten, and S. Berko, Phys. Rev. B *42*, 2696 (1990).

D. J. Singh, Physica C *212*, 228 (1993a).

D. J. Singh, Phys. Rev. B *48*, 3571 (1993b).

D. J. Singh and W. E. Pickett, Phys. Rev. Lett., *73*, 476 (1994).

K. Sinha, Ind. J. Phys. *66A*, 1 (1992) *in* K. P. Sinha, "Magnetic Superconductors; Recent Developments," Nova, New York, 1989.

S. Skanthakumar, H. Zhang, T. W. Clinton, W.-H. Li, J. W. Lynn, Z. Fisk, and S.-W. Cheong, Physica C *160*, 124 (1989).

H. L. Shriver and I. Mertig, Phys. Rev. B *41*, 6553 (1990).

V. Skumryev, R. Puzniak, N. Karpe, H. Zheng-he, M. Pout, H. Medelius, D.-X. Chen, and K. V. Rao, Physica C *152*, 315 (1988).

V. Skumryev, M. R. Koblischka, and H. Kronmüller, Physica C *184*, 332 (1991).

F. Slakey, S. L. Cooper, M. V. Klein, J. P. Rice, and D. M. Ginsberg, Phys. Rev. B *39*, 2781 (1989).

F. Slakey, M. V. Klein, J. P. Rice, and D. M. Ginsburg, Phys. Rev. B *43*, 3764 (1991).

A. W. Sleight, J. L. Gillson, and P. E. Bierstedt, Solid State Commun. *17*, 27 (1975).

A. W. Sleight, Am. Chem. Soc. Symp. Ser. *351*, 2 (1987).

L. C. Smedskjaer, A. Bansil, U. Welp, Y. Fang, and K. G. Bailey, Phys. Rev. B *46*, 5868 (1992).

H. H. A. Smit, M. W. Dirken, R. C. Thiel, and L. J. de Jongh, Solid State Commun. *64*, 695 (1987).

M. G. Smith, A. Manthiram, J. Zhou, J. B. Goodenough, and J. T. Markert, Nature *351*, 549 (1991).

M. G. Smith, J. B. Goodenough, A. Manthiram, R. Taylor, W. Peng, C. Kimball, J. Solid State Chem. *98*, 181 (1992).

R. Sobolewski, L. Shi, T. Gong, W. Xiong, X. Weng, Y. Kostoulas, and P. M. Fauchet, "Proc. High Temperature Supercond. Detectors," *2159*, to be published.

J. O. Sofo, C. A. Balseiro, and H. E. Castillo, Phys. Rev. B *45*, 9860 (1992).

L. L. Sohn, M. S. Rzchowski, J. U. Free, S. P. Benz, M. Tinkham, and C. J. Lobb, Phys. Rev. B *44*, 925 (1991).

L. L. Sohn, M. S. Rzchowski, J. U. Free, M. Tinkham, and C. J. Lobb, Phys. Rev. *45*, 3003 (1992).

P. R. Solomon and F. A. Otter, Phys. Rev. *164*, 608 (1967).

S. N. Song, Q. Robinson, S.-J. Hwu, D. L. Johnson, K. R. Poeppelmeier, and J. B. Ketterson, Appl. Phys. Lett. *51*, 1376 (1987).

Y. S. Song, H. Park, Y. S. Choi, Y. W. Park, M. S. Jang, H. C. Lee, and S. I. Lee, J. Korean Phys. Soc. *23*, 492 (1990).

Y. Song and J. R. Gaines, J. Phys. Condens. Matter. 3, 7161 (1991).

Y.-Q. Song, M. Lee, W. P. Halperin, L. M. Tonge, and T. J. Marks, Phys. Rev. B *44*, 914 (1991a).

Y.-Q. Song, M. A. Kennard, M. Lee, K. R. Poeppelmeier, and W. P. Halperin, Phys. Rev. B *44*, 7159 (1991b).

Y. Song, A. Misra, P. P. Crooker, and J. R. Gaines, Phys. Rev. B *45*, 7574 (1992).

Y.-Q. Song, W. P. Halperin, L. Tonge, T. J. Marks, M. Ledvij, V. G. Kogan, and L. N. Bulaevskii, Phys. Rev. Lett. *70*, 3127 (1993).

J. Spalek and P. Gopalan, J. Phys. France *50*, 2869 (1989).

J. Spalek and J. M. Honig, *in* "Studies in High Temperature Superconductors" (A. V. Narlikar, Ed.), Vol. 8, Chap. 1. Nova Sci., New York (1991).

J. Spalek and W. Wojcik, Phys. Rev. B *45*, 3799 (1992).

G. Sparn, J. D. Thompson, R. L. Whetten, S.-M. Huang, R. B. Kaner, F. Diederich, G. Grüner, and K. Holczer, Phys. Rev. Lett. *68*, 1228 (1992).

P. N. Spathis, M. P. Soerensen, and N. Lazarides, Phys. Rev. B *45*, 7360 (1992).

S. Spielman, J. S. Dodge, L. W. Lombardo, C. B. Eom, M. M. Fejer, T. H. Geballe, and A. Kapitulnik, Phys. Rev. *68*, 3472 (1992).

Z. M. Stadnik, G. Stroink and R. A. Dunlap, Phys. Rev. B *39*, 9108 (1989).

Z. M. Stadnik, G. Stroink, and T. Arakawa, Phys. Rev. B *44*, 12552 (1991).

B. W. Statt and A. Griffin, Phys. Rev. B *48*, 619 (1993).

T. Staufer, R. Nemetschek, R. Hackl, P. Müller, and H. Veith, Phys. Rev. Lett. *68*, 1069 (1992).

D. G. Steel and J. M. Graybeal, Phys. Rev. B *45*, 12643 (1992).

F. Steglich, J. Aarts, C. D. Bredl, W. Lieke, D. Meschede, W. Franz, and H. Schäfer, Phys. Rev. Lett. *43*, 1892 (1979).

P. Steiner, V. Kinsinger, I. Sander, B. Siegwart, S. Hüfner, and C. Politis, Z. Phys. B Cond. Mat. *67*, 19 (1987).

C. H. Stephan and B. W. Maxfield, J. Low Temp. Phys. *10*, 185 (1973).

W. Stephan and J. P. Carbotte, Phys. Rev. B *43*, 10236 (1991).

G. R. Stewart, Rev. Mod. Phys. *56*, 755 (1984).

G. R. Stewart, Z. Fisk, J. O. Willis, and T. J. Smith, Phys. Rev. Lett. B *52*, 679 (1984).

G. R. Stewart, J. O'Rourke, G. W. Crabtree, K. D. Carlson, H. H. Wang, J. M. Williams, F. Gross, and K. Andres, Phys. Rev. B *33*, 2046 (1986).

S. T. Stoddart, H. I. Mutlu, A. K. Geim, and S. J. Bending, Phys. Rev. B *47*, 5146 (1993).

E. C. Stoner, Phil. Mag. *36*, 803 (1945).

H. T. C. Stoof, Phys. Rev. B *47*, 7979 (1993).

J. A. Stratton, "Electromagnetic Theory," McGraw Hill, New York, 1941.

S. K. Streiffer, B. M. Lairson, C. B. Eom, B. M. Clemens, J. C. Bravman, and T. H. Geballe, Phys. Rev. B *43*, 13007 (1991).

A. R. Strnad, C. F. Hempstead, and Y. B. Kim, Phys. Rev. Lett. *13*, 794 (1964).

S. E. Stupp, M. E. Reeves, D. M. Ginsberg, D. G. Hinks, B. Dabrowski, and K. G. Vandervoort, Phys. Rev. B *40*, 10878 (1989).

S. E. Stupp, T. A. Friedmann, J. P. Rice, R. A. Schweinfurth, D. J. Van Harlingen, and D. M. Ginsberg, Phys. Rev. B *43*, 13073 (1991).

C. K. Subramanian, M. Paranthaman, and A. B. Kaiser, Physica C *222*, 47 (1994).

M. A. Subramanian, C. C. Torardi, J. C. Calabrese, J. Gopalakrishnan, K. J. Morrissey, T. R. Askew, R. B. Flippen, U. Chowdhry, and A. W. Sleight, Science *239*, 1015 (1988a).

M. A. Subramanian, J. C. Calabrese, C. C. Torardi, J. Gopalakrishnan, T. R. Askew, R. B. Flippen, K. J. Morrissey, U. Chowdhry, and A. W. Sleight, Nature *332*, 420 (1988b).

A. Sudbø and E. H. Brandt, Phys. Rev. B *43*, 10482 (1991a); Phys. Rev. Lett. *67*, 3176 (1991b).

M. Suenaga, A. K. Ghosh, Y. Xu, and D. O. Welch, Phys. Rev. Lett. *66*, 1777 (1991).

R. Sugano, T. Onogi, and Y. Murayama, Phys. Rev. B *45*, 10789 (1992).

J. Sugiyama, S. Tokuono, S.-I. Koriyama, H. Yamauchi, and S. Tanaka, Phys. Rev. B *43*, 10489 (1991).

J. Sugiyama, K. Matsuura, M. Kosuge, H. Yamauchi, and S. Tanaka, Phys. Rev. B *45*, 9951 (1992).

S. B. Sulaiman, N. Sahoo, T. P. Das, O. Donzelli, E. Torikai, and K. Nagamine, Phys. Rev. B *44*, 7028 (1991).

S. B. Sulaiman, N. Sahoo, T. P. Das and O. Donzelli, Phys. Rev. B *45*, 7383 (1992).

P. E. Sulewski, A. J. Sievers, R. A. Buhrman, J. M. Tarascon, L. H. Greene, and W. A. Curtin, Phys. Rev. B *35*, 8829 (1987).

P. E. Sulewski, P. A. Fleury, K. B. Lyons, S.-W. Cheong, and Z. Fisk, Phys. Rev. B *41*, 225 (1990).

J. Z. Sun, D. J. Webb, M. Naito, K. Char, M. R. Hahn, J. W. P. Hsu, A. D. Kent, D. B. Mitzi, B. Oh, M. R. Beasley, T. H. Geballe, R. H. Hammond, and A. Kapitulnik, Phys. Rev. Lett. *58*, 1574 (1987).

K. Sun, J. H. Cho, F. C. Chou, W. C. Lee, L. L. Miller, D. C. Johnston, Y. Hidaka and T. Murakami, Phys. Rev. B *43*, 239 (1991).

C. S. Sundar, A. Bharathi, Y. C. Jean, P. H. Hor, R. L. Meng, Z. J. Huang, and C. W. Chu, Phys. Rev. B *42*, 426 (1990a).

C. S. Sundar, A. Bharathi, W. Y. Ching, Y. C. Jean, P. H. Hor, R. L. Meng, Z. J. Huang and C. W. Chu, Phys. Rev. B *42*, 2193 (1990b).

C. S. Sundar, A. Bharathi, W. Y. Ching, Y. C. Jean, P. H. Hor, R. L. Meng, Z. J. Huang, and C. W. Chu, Phys. Rev. B *43*, 13019 (1991).

C. Sürgers, H. v. Löhneysen, and L. Schultz, Phys. Rev. B *40*, 8787 (1989).

Y. Suwa, Y. Tanaka, and M. Tsukada, Phys. Rev. B *39*, 9113 (1989).

M. Suzuki, Y. Enemoto, T. Murakami, and T. Inamura, "Proc. 3rd Meeting Ferroelectric Materials and Their Applications," Kyoto, 1981a.

M. Suzuki, Y. Enemoto, T. Murakami, and T. Inamura, Jpn. J. Appl. Phys. *20*, Suppl. 20–24, 13 (1981b).

M. Suzuki and M. Hikita, Jpn. J. Appl. Phys. *28*, L1368 (1989).

M. Suzuki and M. Hikita, Phys. Rev. B *41*, 9566 (1990).

M. Suzuki and M. Hikita, Phys. Rev. B *44*, 249 (1991).

S. Suzuki, T. Takahashi, T. Kusunoki, T. Morikawa, S. Sato, H. Katayama-Yoshida, A. Yamanaka, F. Minami, and S. Takekawa, Phys. Rev. B *44*, 5381 (1991).

H. Svensmark and L. M. Falicov, Phys. Rev. B *42*, 9957 (1990).

A. Szasz, J. Hajdu, J. Kojnok, Z. Dankhazi, W. Krasser, T. Trager, and J. Bankuti, J. Supercond. *3*, 425 (1990).

B. Szpunar and V. Smith, Jr., Phys. Rev. B *45*, 10616 (1992).

M. Tachiki and S. Takahashi, Solid State Commun. *70*, 291 (1989).

A. Tagliacozzo, F. Ventriglia, and P. Apell, Phys. Rev. B *40*, 10901 (1989).

L. Taillefer, *in* "Hyperfine Interactions," to be published.

S. Tajima, S. Uchida, A. Masaki, H. Takaki, K. Kitazawa, S. Tanaka, and A. Katsui, Phys. Rev. B *32*, 6302 (1985).

S. Tajima, S. Uchida, H. Ishii, H. Takagi, S. Tanaka, U. Kawabe, H. Hasegawa, T. Aita, and T. Ishiba, Mod. Phys. Lett. B *1*, 353 (1988).

Y. Tajima, M. Hikita, T. Ishii, H. Fuke, K. Sugiyama, M. Date, A. Yamagishi, A. Katsui, Y. Hidaka, T. Iwata, and S. Tsurumi, Phys. Rev. B *37*, 7956 (1988).

S. Tajima and K. Kitazawa, *in* "Some Aspects of Superconductivity" (L. C. Gupta, Ed.) Nova Scientific Publ., New York, 1990.

J. Takada, T. Terashima, Y. Bando, H. Mazaki, K. Iijima, K. Yamamoto, and K. Hirata, Phys. Rev. B *40*, 4478 (1989).

Y. Takada, Phys. Rev. B *39*, 11575 (1989).

H. Takagi, H. Eisaki, S. Uchida, A. Maeda, S. Tajima, K. Uchinokura, and S. Tanaka, Nature *332*, 236 (1988).

T. Takahashi, H. Matsuyama, H. Katamaya-Yoshida, Y. Okabe, S. Hosoya, K. Seki, H. Fujimoto, M. Sato, and H. Inokuchi, Phys. Rev. B *39*, 6636 (1989).

T. Takahashi, S. Sukuki, T. Morikawa, H. Katayama-Yoshida, S. Hasegawa, H. Inokuchi, K. Seki, K. Kikuchi, S. Suzuki, K. Ikemoto, and Y. Achiba, Phys. Rev. Lett. *68*, 1232 (1992).

I. Takeuchi, J. S. Tsai, Y. Shimakawa, T. Manako, and Y. Kubo, Physica C *158*, 83 (1989).

M. Takigawa, A. P. Reyes, P. C. Hammel, J. D. Thompson, R. H. Heffner, Z. Fisk, and K. C. Ott, Phys. Rev. B *43*, 247 (1991).

M. Takigawà, J. L. Smith, and W. L. Hults, Phys. Rev. B *44*, 7764 (1991).

J. L. Tallon, Proc. 7th Intl. Workshop Critical Currents *in* "Superconductors, Tyrol, Austria," World Scientific, Singapore, 1994.

K. Tamasaku, Y. Nakamura, and S. Uchida, Phys. Rev. Lett. *69*, 1455 (1992).

T. Tamegai, K. Koga, K. Suzuki, M. Ichihara, F. Sadai, and Y. Iye, Jpn. J. Appl. Phys. Lt. 2, *28*, L112 (1989).

Z. Tan, J. I. Budnick, W. Q. Chen, D. L. Brewe, S.-W. Cheong, A. S. Cooper, and L. W. Rupp, Jr., Phys. Rev. B *42*, 4808 (1990).

Z. Tan, J. I. Budnick, S. Luo, W. Q. Chen, S.-W. Cheong, A. S. Cooper, P. C. Canfield, and Z. Fisk, Phys. Rev. B *44*, 7008 (1991).

Y. Tanaka and M. Tsukada, Phys. Rev. B *40*, 4482 (1989a); Solid State Commun. *69*, 195, 491 (1989b).

Y. Tanaka and M. Tsukada, Phys. Rev. B *42*, 2066 (1990).

Y. Tanaka and M. Tsukada, Phys. Rev. B *44*, 7578 (1991).

S. Tanda, M. Honma, and T. Nakayama, Phys. Rev. B *43*, 8725 (1991).

C. Q. Tang, B. R. Li, and A. Chen, Phys. Rev. B *42*, 8078 (1990).

X. X. Tang, D. E. Morris, and A. P. B. Sinha, Phys. Rev. B *43*, 7936 (1991).

T. Tani, T. Itoh and S. Tanaka, J. Phys. Soc. Jpn. Suppl. A *49*, 309 (1980).

S. Tanigawa, Y. Mizuhara, Y. Hidaka, M. Oda, M. Suzuki, and T. Murakami, Mater. Res. Soc. Symp. Proc. *99*, 57 (1988).

D. B. Tanner and T. Timusk, *in* "Physical Properties of High Temperature Superconductors" (D. M. Ginsberg, Ed.), Vol. 3, Chap. 5, World Scientific, Singapore, 1992.

H. J. Tao, A. Chang, F. Lu, and E. L. Wolf, Phys. Rev. B *45*, 10622 (1992).

J. M. Tarascon, L. H. Greene, W. R. McKinnon, and G. W. Hull, Phys. Rev. B *35*, 7115 (1987a).

J. M. Tarascon, W. R. McKinnon, L. H. Greene, G. W. Hull, and E. M. Vogel, Phys. Rev. B *36*, 226 (1987b).

J. M. Tarascon, L. H. Greene, W. R. McKinnon, G. W. Hull, and T. H. Geballe, Science *235*, 1373 (1987c).

J. M. Tarascon, P. Barboux, P. F. Miceli, L. H. Greene, D. W. Hull, M. Eibschutz, and S. A. Sunshine, Phys. Rev. B *37*, 7458 (1988a).

J. M. Tarascon, Y. LePage, P. Barboux, B. G. Bagley, L. H. Greene, W. R. McKinnon, G. W. Hull, M. Giroud, and D. M. Hwang, Phys. Rev. *37*, 9382 (1988b).

J. M. Tarascon, E. Wang, L. H. Greene, B. G. Bagley, G. W. Hull, S. M. D'Egidio, P. F. Miceli, Z. Z. Wang, T. W. Jing, J. Clayhold, D. Brawner, and N. P. Ong, Phys. Rev. B *40*, 4494 (1989a).

J. M. Tarascon, Y. LePage, W. R. McKinnon, E. Tselepis, P. Barboux, B. G. Bagley, and R. Ramesh, *in* "Proc. Mater. Res. Soc. Symp.," San Diego, Apr. 23–28, 1989b.

V. V. Tatarskii, M. Paranthaman, and A. M. Hermann, Phys. Rev. B *47*, 14489 (1993).

Y. Taur, P. L. Richards, and T. Auracher, Low Temp. Phys. *3*, 276 (1974).

W. M. Temmerman, G. M. Stocks, P. J. Durham, and P. A. Sterne, J. Phys. F *17*, L135 (1987).

M. A. Teplov, O. N. Bakharev, A. V. Dooglav, A. V. Egorov, M. V. Eremin, M. S. Tagirov, A. G. Volodin, and R. Sh. Zhdanov, Physica C *185–189*, 1107 (1991).

I. Terasaki, T. Nakahashi, S. Takebayashi, A. Maeda and K. Uchinokura, Physica C *165*, 152 (1990a).

I. Terasaki, S. Tajima, H. Eisaki, H. Takigi, K. Uchinokura, and S. Uchida, Phys. Rev. B *41*, 865 (1990b).

T. Terashima, K. Shimura, Y. Bando, Y. Matsuda, A. Fujiyama, and S. Komiyama, Phys. Rev. Lett. *67*, 1362 (1991).

B. M. Terzijska, R. Wawryk, D. A. Dimitrov, Cz. Marucha, V. T. Kovachev, and J. Rafalowicz, Cryogenics *32*, 53 (1992).

Z. Tesanovic and M. Rasolt, Phys. Rev. B *39*, 2718 (1989).

Z. Tesanovic, M. Rasolt, and L. Xing, Phys. Rev. B *43*, 288 (1991).

Z. Tesanovic, Phys. Rev. B *44*, 12635 (1991).

L. R. Tessler, J. Provost, and A. Maignan, Appl. Phys. Lett. *58*, 528 (1991).

L. Tewordt, S. Wermbter, and Th. Wölkhausen, Phys. Rev. B *40*, 6878 (1989).

L. Tewordt and Th. Wölkhausen, Solid State Commun. *70*, 839 (1989).

T. D. Thanh, A. Koma, and S. Tanaka, Appl. Phys. *22*, 205 (1980).

S. Theodorakis and Z. Tesanovic, Phys. Rev. B *40*, 6659 (1989).

S. L. Thiemann, Z. Radovic, and V. G. Kogan, Phys. Rev. B *39*, 11406 (1989).

T. Thio, T. R. Thurston, N. W. Preyer, P. J. Picone, M. A. Kastner, H. P. Jenssen, D. R. Gabbe, C. Y. Chen, R. J. Birgeneau, and A. Aharony, Phys. Rev. B *38*, 905 (1988).

G. Thomas, M. Capizzi, J. Orenstein, D. Rapkine, A. Millis, P. Gammel, L. Gammel, L. F. Schneemeyer, and J. Waszczak, *in* "Proceedings of the International Symposium on the Electronic Structure of High T_c Superconductors," p. 67, (A. Bianconi, Ed.) Pergamon Press, Oxford, 1988.

R. S. Thompson, Phys. Rev. B *1*, 327 (1970).

J. R. Thompson, J. Brynestad, D. M. Kroeger, Y. C. Kim, S. T. Sekula, D. K. Christen, and E. Specht, Phys. Rev. B *39*, 6652 (1989).

J. R. Thompson, D. K. Christen, H. A. Deeds, Y. C. Kim, J. Brynestad, S. T. Sekula, and J. Budai, Phys. Rev. B *41*, 7293 (1990).

J. R. Thompson, J. G. Ossandon, D. K. Christen, B. C. Chakoumakos, Y. R. Sun, M. Paranthaman, and J. Brynestad, Phys. Rev. B *48*, 14031 (1993).

C. Thomsen and M. Cardona, *in* "Physical Properties of High Temperature Superconductors" (D. M. Ginsberg, Ed.), Vol. 1, Chap. 8, World Scientific, Singapore, 1989.

R. J. Thorn, ACS Symp. Ser. *351*, Chap. 3, (1987).

C. Tien and I. M. Jiang, Phys. Rev. B *40*, 229 (1989).

T. S. Tighe, M. T. Tuominen, J. M. Hergenrother, and M. Tinkham, Phys. Rev. B *47*, 1145 (1993).

D. R. Tilley and J. Tilley, "Superfluidity and Superconductivity," Hilger, Boston (1986).

T. Timusk and D. B. Tanner, *in* "Physical Properties of High Temperature Superconductors" (D. M. Ginsberg, Ed.), Vol. 1, Chap. 7, World Scientific, New York (1989)

M. Tinkham, Phys. Rev. Lett. *13*, 804 (1964).

M. Tinkham and J. Clarke, Phys. Rev. Lett. *28*, 1366 (1972).

M. Tinkham, "Introduction to Superconductivity," Krieger, FL (1985).

M. Tinkham and C. J. Lobb, Solid State Phys. *42*, 91 (1989).

J. E. Tkaczyk, R. H. Arendt, M. F. Garbauskas, H. R. Hart, K. W. Lay, and F. E. Luborsky, Phys. Rev. B *45*, 12506 (1992).

J. G. Tobin, C. G. Olson, C. Gu, J. Z. Liu, F. R. Solal, M. J. Fluss, R. H. Howell, J. C. O'Brien, H. B. Radousky, and P. A. Sterne, Phys. Rev. B *45*, 5563 (1992).

B. H. Toby, T. Egami, J. D. Jorgensen, and M. A. Subramanian, Phys. Rev. Lett. *64*, 2414 (1990).

K. Togano, H. Kumakura, K. Fukutomi, and K. Tachikawa, Appl. Phys. Lett. *51*, 136 (1987).

A. Tokiwa, M. Nagoshi, and Y. Syono, Physica C *170*, 437 (1990).

A. Tokiwa, Y. Syono, T. Oku, and M. Nagoshi, Physica C *185–189*, 619 (1991).

A. Tokiwa-Yamamoto, K. Isawa, M. Itoh, S. Adachi, and H. Yamauchi, Physica C *216*, 250 (1993).

A. Tokumitu, K. Miyake, and K. Yamada, Phys. Rev. B *47*, 11988 (1993).

M. Tokumoto, H. Bando, H. Anzai, G. Saito, K. Murata, K. Kajimura, and T. Ishiguro, J. Phys. Soc. Jpn. *54*, 869 (1985).

T. A. Tokuyasu, D. W. Hess, and J. A. Sauls, Phys. Rev. B *41*, 8891 (1990); T. A. Tokuyasu and J. A. Sauls, Physica B *165–166*, 347 (1990).

J. C. Tolédano, A. Litzler, J. Primot, J. Schneck, L. Pierre, D. Morin, and C. Daguet, Phys. Rev. B *42*, 436 (1990).

H. Tolentino, M. Medarde, A. Fontaine, F. Baudelet, E. Dartyge, D. Guay, and G. Tourillon, Phys. Rev. B *45*, 8091 (1992).

I. Tomeno, M. Yoshida, K. Ikeda, K. Tai, K. Takamuku, N. Koshizuka, S. Tanaka, K. Oka, and H. Unoki, Phys. Rev. B *43*, 3009 (1991).

J. Toner, Phys. Rev. Lett. *66*, 2523 (1991a).

J. Toner, Phys. Rev. Lett. *67*, 2537 (1991b); see comment by T. Nattermann and I. Lyuksyutov, and reply *68*, 3366 (1992).

C. C. Torardi, M. A. Subramanian, J. C. Calabrese, J. Gopalakrishnan, K. J. Morrissey, T. R. Askew, R. B. Flippen, U. Chowdhry, and A. W. Sleight, Science *240*, 631 (1988a).

C. C. Torardi, M. A. Subramanian, J. C. Calabrese, J. Gopalakrishnan, E. M. McCarron, K. J. Morrissey, T. R. Askew, R. B. Flippen, U. Chowdhry, and A. W. Sleight, Phys. Rev. B *38*, 225 (1988b).

J. B. Torrance, J. Solid State Chem. *96*, 59 (1992).

J. B. Torrance, P. Lacorre, A. I. Nazzal, E. J. Ansaldo, and Ch. Niedermayer, Phys. Rev. B *45*, 8209 (1992).

H. Totsuji, Phys. Rev. B *43*, 5287 (1991).

M. Touminen, A. M. Goldman, Y. Z. Chang, and P. Z. Jiang, Phys. Rev. B *42*, 412 (1990).

S. W. Tozer, A. W. Kleinsasser, T. Penney, D. Kaiser, and F. Holtzberg, Phys. Rev. Lett. *59*, 1768 (1987).

J. M. Tranquada, G. Shirane, B. Keimer, S. Shamoto, and M. Sato, Phys. Rev. B *40*, 4503 (1989).

J. M. Tranquada, *in* "Early and Recent Aspects of Superconductivity" (J. G. Bednorz and K. A. Müller, Eds.), p. 422, Springer-Verlag, Berlin, 1990.

J. M. Tranquada, P. M. Gehring, G. Shirane, S. Shamoto, and M. Sato, Phys. Rev. B *46*, 5561 (1992).

A. Tressaud, K. Amine, J. P. Chaminade, J. Etourneau, T. M. Duc, and A. Sartre, J. Appl. Phys. *68*, 248 (1990).

J.-M. Triscone, Ø. Fischer, O. Brunner, L. Antognazza, A. D. Kent, and M. G. Karkut, Phys. Rev. Lett. *64*, 804 (1990).

V. N. Trofimov, A. V. Kuznetsov, P. V. Lepeschkin, K. A. Bolschinskov, A. A. Ivanov, and A. A. Mikhailov, Physica C *183*, 135 (1991).

A. Trokiner, R. Mellet, A-M. Pougnet, D. Morin, Y. M. Gao, J. Primot, and J. Schneck, Phys. Rev. B *41*, 9570 (1990).

A. Trokiner, L. LeNoc, J. Schneck, A. M. Pougnet, R. Mellet, J. Primot, H. Savary, Y. M. Gao, and S. Aubry, Phys. Rev. B *44*, 2426 (1991).

N. Troullier and J. L. Martins, Phys. Rev. B *46*, 1754 (1992).

S. A. Trugman, Phys. Rev. Lett. *65*, 500 (1990).

J.-S. Tsai, A. K. Jain, and J. E. Lukens, Phys. Rev. Lett. *51*, 316 (1983).

S.-F. Tsay, S.-Y. Wang, and T. J. W. Yang, Phys. Rev. B *43*, 13080 (1991).

C. C. Tsuei, A. Gupta, and G. Koren, Physica C *161*, 415 (1989).

M. Touminen, A. M. Goldman, Y. Z. Chang, and P. Z. Jiang, Phys. Rev. B *42*, 412 (1990).

M. T. Tuominen, J. M. Hergenrother, T. S. Tighe, and M. Tinkham, Phys. Rev. B *47*, 11599 (1993).

I. Tüttö, L. M. Kahn and J. Ruvalds, Phys. Rev. B *20*, 952 (1979).

I. Tüttö and A. Zawadowski, Phys. Rev. B *45*, 4842 (1992).

R. Tycko, G. Dabbagh, M. J. Rosseinsky, D. W. Murphy, R. M. Fleming, A. P. Ramirez, and J. C. Tully, Science *253*, 884 (1991).

R. Tycko, G. Dabbagh, M. J. Rosseinsky, D. W. Murphy, A. P. Ramirez, and R. M. Fleming, Phys. Rev. Lett. *68*, 1912 (1992).

S. Uchida, H. Takagi, K. Kitazawa and S. Tanaka, Jpn. J. Appl. Phys. *26*, L1 (1987).

S. Uchida, T. Ido, H. Takagi, T. Arima, Y. Tokura and S. Tajima, Phys. Rev. B *43*, 7942 (1991).

Y. J. Uemura, L. P. Le, G. M. Luke, B. J. Sternlieb, W. D. Wu, J. H. Brewer, T. M. Riseman, C. L. Seaman, M. B. Maple, M. Ishikawa, D. G. Hinks, J. D. Jorgensen, G. Saito and H. Yamochi, Phys. Rev. Lett. *66*, 2665 (1991).

A. Ugawa, K. Iwasaki, A. Kawamoto, K. Yakushi, Y. Yamashita, and T. Suzuki, Phys. Rev. B *43*, 14718 (1991).

C. Uher and W.-N. Huang, Phys. Rev. B *40*, 2694 (1989).

C. Uher, J. Supercond. *3*, 337 (1990).

C. Uher, *in* "Physical Properties of High Temperature Superconductors" (D. M. Ginsberg, Ed.), Vol. 3, Chap. 3, World Scientific, Singapore, 1992.

S. Ullah, A. T. Dorsey, and L. J. Buchholtz, Phys. Rev. B *42*, 9950 (1990).

S. Ullah and A. T. Dorsey, Phys. Rev. Lett. *65*, 2066 (1990).

S. Ullah and A. T. Dorsey, Phys. Rev. B *44*, 262 (1991).

J. S. Urbach, D. B. Mitzi, A. Kapitulnik, J. Y. T. Wei, and D. E. Morris, Phys. Rev. B *39*, 12391 (1989).

A. V. Ustinov, T. Doderer, R. P. Huebener, N. F. Pederson, B. Mayer, and V. A. Oboznov, Phys. Rev. Lett. *69*, 1815 (1992).

P. P. Vaishnava, C. A. Taylor, II, and C. L. Foiles, Phys. Rev. B *41*, 4195 (1990).

J. M. Valles, Jr., R. C. Dynes, and J. P. Garno, Phys. Rev. B *40*, 6680 (1989).

J. M. Valles, Jr., R. C. Dynes, A. M. Cucolo, M. Gurvitch, L. F. Schneemeyer, J. P. Garno, and J. V. Waszczak, Phys. Rev. B *44*, 11986 (1991).

J. van den Berg, C. J. van der Beek, P. H. Kest, J. A. Mydosh, M. J. V. Menken, and A. A. Menovsky, Supercond. Sci. Tech. *1*, 249 (1989).

J. van den Berg, C. J. van der Beek, P. H. Kes, J. A. Mydosh, G. J. Nieuwenhuys, and L. J. de Jongh, Solid State Commun. *64*, 699 (1987).

A. M. van den Brink, G. Schön, and L. Geerligs, Phys. Rev. Lett. *67*, 3030 (1991).

D. Van Der Marel, Physica C *165*, 35 (1990).

H. P. van der Meulen, A. de Visser, J. J. M. Franse, T. T. J. M. Berendschot, J. A. A. J. Perenboom, H. van Kempen, A. Lacerda, P. Lejay, and J. Fouquet, Phys. Rev. B *44*, 814 (1991).

H. S. J. van der Zant, F. C. Fritschy, T. P. Orlando, and J. E. Mooij, Phys. Rev. Lett. *66*, 2531 (1991).

K. G. Vandervoort, U. Welp, J. E. Kessler, H. Claus, G. W. Crabtree, W. K. Kwok, A. Umezawa, B. W. Veal, J. W. Downey, A. P. Paulikas, and J. Z. Liu, Phys. Rev. B *43*, 13042 (1991).

H. S. J. van der Zant, F. C. Fritschy, T. P. Orlando, and J. E. Mooij, Phys. Rev. B *47*, 295 (1993).

R. B. VanDover, E. M. Gyorgy, A. E. White, L. F. Schneemeyer, R. J. Felder, and J. V. Waszczak, Appl. Phys. Lett. *56*, 2681 (1990).

T. Van Duzer and C. W. Turner, "Principles of Superconductive Devices and Circuits," Elsevier, New York (1981).

A. G. Van Vijfeijkenand and A. K. Niessen, Philips Res. Rep. *20*, 505 (1965a); Phys. Lett. *16*, 23 (1965b).

B. J. van Wees, K.-M. H. Lenssen, and C. J. P. M. Harmans, Phys. Rev. B *44*, 470 (1991).

C. M. Varma, S. Schmitt-Rink, and E. Abrahams, *in* "Theories of High Temperature Superconductivity" (J. W. Halley, Ed.), p. 211, Addison Wesley, Reading, MA, 1988.

C. M. Varma, P. B. Littlewood, S. Schmitt-Rink, E. Abrahams, and A. E. Ruckenstein, Phys. Rev. Lett. *63*, 1996 (1989).

B. V. Vasiliev, J. Supercond. *4*, 271 (1991).

D. R. Veblen, P. J. Heaney, R. J. Angel, L. W. Finger, R. M. Hazen, C. T. Prewitt, N. L. Ross, C. W. Chu, P. H. Hor, and R. L. Meng, Nature *332*, 334 (1988).

A. J. Vega, W. E. Farneth, E. M. McCarron, and R. K. Bordia, Phys. Rev. B *39*, 2322 (1989a).

A. J. Vega, M. K. Crawford, E. M. McCarron, and W. E. Farneth, Phys. Rev. B *40*, 8878 (1989b).

E. L. Venturini, D. S. Ginley, J. F. Kwak, R. J. Baughman, J. E. Schirber, and B. Morosin, *in* "High Temperature Superconductors" (D. U. Gubser and M. Schluter, Eds.), p. 97, Mater. Res. Soc., Pittsburgh, 1987.

R. Vijayaraghavan, A. K. Ganguli, N. Y. Vasanthacharya, M. K. Rajumon, G. U. Kulkarni, G. Sankar, D. D. Sarma, A. K. Sood, N. Chandrabhas, and C. N. R. Rao, Supercond. Sci. Technol. *2*, 195 (1989).

P. Villars and J. C. Phillips, Phys. Rev. B *37*, 2345 (1988).

L. Ya. Vinnikov and I. V. Grigor'eva, JETP Lett. *47*, 106 (1988).

V. M. Vinokur, M. V. Feigel'man, V. B. Geshkenbein, and A. I. Larkin, Phys. Rev. Lett. *65*, 259 (1990).

V. M. Vinokur, M. V. Feigel'man, and V. B. Geshkenbein, Phys. Rev. Lett. *67*, 915 (1991).

A. Virosztek and J. Ruvalds, Phys. Rev. Lett. *67*, 1657 (1991).

V. M. Virosztek and J. Ruvalds, Phys. Rev. B *45*, 347 (1992).

P. Visani, A. C. Mota, and A. Pollini, Phys. Rev. Lett. *65*, 1514 (1990).

J. P. Vithayathil, D. E. MacLaughlin, E. Koster, D. Ll. Williams, and E. Bucher, Phys. Rev. B *44*, 4705 (1991).

B. M. Vlcek, M. C. Frischherz, S. Fleshler, U. Welp, J. Z. Liu, J. Downey, K. G. Vandervoort, G. W. Crabtree, M. A. Kirk, J. Giapintzakis, and J. Farmer, Phys. Rev. Lett. *46*, 6441 (1992).

N. V. Volkenshteyn *et al.*, Fiz. Met. Metalloved *45*, 1187 (1978).

A. R. von Hippel, "Dielectrics and Waves," p. 255, MIT Press, Cambridge, MA, 1954.

S. von Molnár, A. Torresson, D. Kaiser, F. Holtzberg, and T. Penney, Phys. Rev. B *37*, 3762 (1988).

F. von Oppen and E. K. Riedel, Phys. Rev. Lett. *66*, 84 (1991).

S. V. Vonsovsky, Yu. A. Izyumov, and E. Z. Kurmaev, "Superconductivity in Transition Metals," Springer, New York (1982).

A. Wadas, O. Fritz, H. J. Hug, and H.-J. Güntherodt, Z. Phys. B *88*, 317 (1992).

S. F. Wahid and N. K. Jaggi, Physica C *184*, 88 (1991).

D. R. Wake, F. Slakey, M. V. Klein, J. P. Rice, and D. M. Ginsberg, Phys. Rev. Lett. *67*, 3728 (1991).

M. Wallin, Phys. Rev. B *41*, 6575 (1990).

R. E. Walstedt and W. W. Warren, Jr., Science *248*, 1082 (1990).

R. E. Walstedt, W. W. Warren, Jr., R. F. Bell, R. J. Cava, G. P. Espinosa, L. F. Schneemeyer, and J. V. Waszczak, Phys. Rev. B *41*, 9574 (1990).

R. E. Walstedt, R. F. Bell, and D. B. Mitzi, Phys. Rev. B *44*, 7760 (1991).

R. E. Walstedt, R. F. Bell, L. F. Schneemeyer, J. V. Waszczak, and G. P. Espinosa, Phys. Rev. B *45*, 8074 (1992).

B. L. Walton, B. Rosenblum, and F. Bridges, Phys. Rev. Lett. *32*, 1047 (1974).

Z. Wang, N. Zou, J. Pang, and C. Gong, Solid State Commun. *64*, 531 (1987).

Z. Z. Wang, J. Clayhold, N. P. Ong, J. M. Tarascon, L. H. Greene, W. R. McKinnon, and G. W. Hull, Phys. Rev. *36*, 7222 (1987).

S. J. Wang, S. V. Naidu, S. C. Sharma, D. K. De, D. Y. Jeong, T. D. Black, S. Krichene, J. R. Reynolds, and J. M. Owens, Phys. Rev. B *37*, 603 (1988).

Y. R. Wang, Phys. Rev. B *40*, 2698 (1989).

C-P. S. Wang, *in* "High Temperature Superconductivity" (J. W. Lynn, Ed.), Chap. 5, Springer-Verlag, Berlin, 1990.

Y.-Y. Wang, G. Feng, and A. L. Ritter, Phys. Rev. B *42*, 420 (1990).

T. Wang, K. M. Beauchamp, D. D. Berkley, B. R. Johnson, J.-X. Liu, J. Zhang, and A. M. Goldman, Phys. Rev. B, *43*, 8623 (1991).

Z. Wang, Y. Bang, and G. Kotliar, Phys. Rev. Lett. *67*, 2733 (1991).

Y. Wang, A. M. Rao, J.-G. Zhang, X.-X. Bi, P. C. Eklund, M. S. Dresselhaus, P. P. Nguyen, J. S. Moodera, G. Dresselhaus, H. B. Radousky, R. S. Glass, M. J. Fluss, and J. Z. Liu, Phys. Rev. B *45*, 2523 (1992).

K-A. Wang, Y. Wang, P. Zhou, J. M. Holden, S.-L. Ren, G. T. Hager, H. F. Ni, P. C. Eklund, G. Dresselhaus, and M. S. Dresselhaus, Phys. Rev. B *45*, 1955 (1992).

Z. D. Wang and C. S. Ting, Phys. Rev. Lett. *69*, 1435 (1992a).

Z. D. Wang and C. S. Ting, Phys. Rev. *46*, 284 (1992b).

Z. D. Wang and C.-R. Hu, Phys. Rev. B *44*, 11918 (1991).

C. A. Wang, R. L. Wang, H. C. Li, H. R. Yi, C. G. Cui, S. L. Li, X. N. Jing, J. Li, P. Xu, and L. Li, Physica C *191*, 52 (1992).

Z. H. Wang, A. W. P. Fung, G. Dresselhaus, M. S. Dresselhaus, K. A. Wang, P. Zhou, and P. C. Eklund, Phys. Rev. B *47*, 15354 (1993).

N. L. Wang, Y. Chong, C. Y. Wang, D. J. Huang, Z. Q. Mao, L. Z. Cao, and Z. J. Chen, Phys. Rev. B *47*, 3347 (1993).

Y. Watanabe, Z. Z. Wang, S. A. Lyon, D. C. Tsui, N. P. Ong, J. M. Tarascon, and P. Barboux, Phys. Rev. B *40*, 6884 (1989).

Y. Watanabe, D. C. Tsui, J. T. Birmingham, N. P. Ong, and J. M. Tarascon, Phys. Rev. B *43*, 3026 (1991).

K. Watanabe, S. Awaji, N. Kobayashi, H. Yamane, T. Hirai, and Y. Muto, J. Appl. Phys. *69*, 1543 (1991).

J. H. P. Watson, J. Appl. Phys. *39*, 3406 (1968).

H. L. Watson and R. P. Huebener, Phys. Rev. B *10*, 4577 (1974).

C. H. Watson, D. A. Browne, J.-C. Xu, and R. G. Goodrich, Phys. Rev. B *40*, 8885 (1989).

W. J. Wattamaniuk, J. P. Tidman, and R. F. Frindt, Phys. Rev. Lett. *35*, 62 (1975).

B. D. Weaver, J. M. Pond, D. B. Chrisey, J. S. Horwitz, H. S. Newman, and G. P. Summers, Appl. Phys. Lett. *58*, 1563 (1991).

W. Weber, Phys. Rev. Lett. *58*, 1371 (1987).

W. Weber, Z. Phys. B *70*, 323 (1988).

H. Weber and P. Minnhagen, Phys. Rev. B *38*, 8730 (1988).

W. H. Weber, C. R. Peters, B. M. Wanklyn, C. Chen, and B. E. Watts, Phys. Rev. *38*, 917 (1988).

W. H. Weber, C. R. Peters, and E. M. Logothetis, J. Opt. Soc. Am. B *6*, 455 (1989).

W. H. Weber and G. W. Ford, Phys. Rev. B *40*, 6890 (1989).

H. Weber and H. J. Jensen, Phys. Rev. B *44*, 454 (1991).

M. Weger, Rev. Mod. Phys. *36*, 175 (1964).

H. Weinstock, IEEE Trans. Magn. *27*, 3231 (1991).

B. O. Wells, Z.-X. Shen, D. S. Dessau, W. E. Spicer, C. G. Olson, D. B. Mitzi, A. Kapitulnik, R. S. List, and A. Arko, Phys. Rev. Lett. *65*, 3056 (1990).

U. Welp, W. K. Kwok, G. W. Crabtree, K. G. Vandervoort, and J. Z. Liu, Phys. Rev. Lett. *62*, 1908 (1989); Phys. Rev. B *40*, 5263 (1989).

U. Welp, S. Fleshler, W. K. Kwok, J. Downey, Y. Fang, G. W. Crabtree, and J. Z. Liu, Phys. Rev. B *42*, 10189 (1990).

Z. Y. Weng, C. S. Ting, and T. K. Lee, Phys. Rev. B *41*, 1990 (1990).

F. Wenger and S. Östlund, Phys. Rev. B *47*, 5977 (1993).

F. Werfel, G. Dräger, J. A. Leiro, and K. Fischer, Phys. Rev. B *45*, 4957 (1992).

S. Wermbter and L. Tewordt, Phys. Rev. B *44*, 9524 (1991); Physica C *183*, 365 (1991).

N. R. Werthamer, Phys. Rev. *132*, 2440 (1963).

M.-H. Whangbo and C. C. Torardi, Acct. Chem. Res. *24*, 127 (1991).

J. M. Wheatley, T. C. Hsu, and P. W. Anderson, Phys. Rev. B *37*, 5897 (1988).

A. Widom, Y. N. Srivastava, C. Vittoria, H. How, R. Karim, and H. Jiang, Phys. Rev. B *46*, 1102 (1992).

R. J. Wijngaarden, E. N. van Eenige, J. J. Scholtz, and R. Griessen, High Pressure Res. *3*, 105 (1990).

F. Wilczek, Phys. Rev. Lett. *48*, 1144 (1982a).

F. Wilczek, Phys. Rev. Lett. *49*, 957 (1982b).

N. K. Wilkin and M. A. Moore, Phys. Rev. B *48*, 3464 (1993).

G. P. Williams, R. C. Budhani, C. J. Hirschmugl, G. L. Carr, S. Perkowitz, B. Lou, and T. R. Yang, Phys. Rev. B *41*, 4752 (1990).

P. J. Williams and J. P. Carbotte, Phys. Rev. B *43*, 7960 (1991).

M. N. Wilson, "Superconducting Magnets," Clarendon Press, Oxford, 1983.

A. Wittlin, L. Genzel, M. Cardona, M. Bauer, W. König, E. Garcia, M. Barahona, and M. V. Cabañas, Phys. Rev. B *37*, 652 (1988).

T. Wittmann and J. Stolze, Phys. Rev. B *48*, 3479 (1993).

S. A. Wolf and V. Z. Kresin, Eds., "Novel Superconductivity," Plenum, New York, 1987.

Y. Wolfus, Y. Yeshurun, I. Felner, and H. Sompolinsky, Phys. Rev. B *40*, 2701; see B *39*, 11690 (1989).

H. Won and K. Maki, Phys. Rev. B *49*, 1397 (1994).

R. F. Wood and J. F. Cooke, Phys. Rev. B *45*, 5585 (1992).

F. Wooten, "Optical Properties of Solids," p. 244, Academic Press, New York, 1972.

R. Wördenweber, Phys. Rev. B *46*, 3076 (1992).

A. H. Worsham, N. G. Ugras, D. Winkler, D. E. Prober, N. R. Erickson, and P. F. Goldsmith, Phys. Rev. Lett. *67*, 3034 (1991).

T. K. Worthington, W. J. Gallagher, and T. R. Dinger, Phys. Rev. Lett. *59*, 1160 (1987).

W. H. Wright, D. J. Holmgren, T. A. Friedmann, M. P. Maher, B. G. Pazol, and D. M. Ginsberg, J. Low Temp. Phys. *68*, 109 (1987).

M. K. Wu, J. R. Ashburn, C. J. Torng, P. H. Hor, R. L. Meng, L. Gao, Z. J. Huang, Y. Q. Wang, and C. W. Chu, Phys. Rev. Lett. *58*, 908 (1987).

J. Z. Wu, C. S. Ting, and D. Y. Xing, Phys. Rev. B *40*, 9296 (1989).

D.-H. Wu and S. Sridhar, Phys. Rev. Lett. *65*, 2074 (1990).

J. Z. Wu, C. S. Ting, W. K. Chu, and X. X. Yao, Phys. Rev. B *44*, 411 (1991a).

J. Z. Wu, P. Y. Hsieh, A. V. McGuire, D. L. Schmidt, L. T. Wood, Y. Shen, and W. K. Chu, Phys. Rev. B *44*, 12643 (1991b).

Z. Y. Wu, M. Benfatto, and C. R. Natoli, Phys. Rev. B *45*, 531 (1992).

D. H. Wu, J. Mao, S. N. Mao, J. L. Peng, X. X. Xi, T. Venkatesan, R. L. Greene, and S. M. Anlage, Phys. Rev., Lett. *70*, 85 (1993).

R. W. G. Wyckoff, "Crystal Structures," Vol. 1, Wiley, New York, 1963.

R. W. G. Wyckoff, "Crystal Structures," Vol. 2, Wiley, New York, 1964.

R. W. G. Wyckoff, "Crystal Structures," Vol. 3, Wiley, New York, 1965.

R. W. G. Wyckoff, "Crystal Structures," Vol. 4, Wiley, New York, 1968.

D. G. Xenikos and T. R. Lemberger, Phys. Rev. B *41*, 869 (1990).

T.-K. Xia and D. Stroud, Phys. Rev. B *39*, 4792 (1989).

W. Xia and P. L. Leath, Phys. Rev. Lett. *63*, 1428 (1989).

G. Xiao, F. H. Streitz, A. Gavrin, Y. W. Du, and C. L. Chien, Phys. Rev. B *35*, 8782 (1987a).

G. Xiao, F. H. Streitz, A. Gavrin, M. Z. Cieplak, J. Childress, M. Lu, A. Zwicker, and C. L. Chien, Phys. Rev. B *36*, 2382 (1987b).

G. Xiao, M. Z. Cieplak, and C. L. Chien, Phys. Rev. B *40*, 4538 (1989).

D. Y. Xing and M. Liu, Phys. Rev. B *43*, 3744 (1991).

Q. Xiong, Y. Y. Xue, Y. Cao, F. Chen, Y. Y. Sun, J. Gibson, L. M. Liu, A. Jacobson, and C. W. Chu, Phys. Rev. Lett., in press (1994).

Y. Xu, M. Suenaga, A. R. Moodenbaugh, and D. O. Welch, Phys. Rev. B *40*, 10882 (1989).

Y. Xu, M. Suenaga, Y. Gao, J. E. Crow, and N. D. Spencer, Phys. Rev. B *42*, 8756 (1990).

M. Xu, D. Shi, and R. F. Fox, Phys. Rev. B *42*, 10773 (1990).

Y. Xu and M. Suenaga, Phys. Rev. B *43*, 5516 (1991).

Y.-N. Xu, M.-Z. Huang, and W. Y. Ching, Phys. Rev. B *44*, 13171 (1991).

X-Q. Xu, S. J. Hagen, W. Jiang, J. L. Peng, Z. Y. Li, and R. Greene, Phys. Rev. B *45*, 7356 (1992).

J. V. Yakhmi and R. M. Iyer, *in* "High Temperature Superconductors" (S. K. Malik and S. S. Shah, Eds.), Nova Sci., New York, 1992.

J. V. Yakhmi, "Chemistry and Physics of Fullerenes," *in* "Atomic and Molecular Physics," (S. A. Ahmad, Ed.) Narosa Publishing, New Delhi, India, 1994.

K. Yamada, K. Kakurai, Y. Endoh, T. R. Thurston, M. A. Kaster, R. J. Birgeneau, G. Shirane, Y. Hidaka, and T. Murakami, Phys. Rev. B *40*, 4557 (1989).

K. Yamamoto, H. Mazaki, H. Yasuoka, S. Katsuyama, and K. Kosuge, Phys. Rev. *46*, 1122 (1992).

K. Yamamoto, H. Mazaki, and H. Yasuoka, Phys. Rev. B *47*, 915 (1993).

H. Yamasaki, K. Endo, S. Kosaka, M. Umeda, S. Yoshida, and K. Kajimura, Phys. Rev. Lett. *70*, 3331 (1993).

Y. Yan and M. G. Blanchin, Phys. Rev. B *43*, 13717 (1991).

K. N. Yang, J. M. Ferreira, B. W. Lee, M. B. Maple, W.-H. Li, J. W. Lynn, and R. W. Erwin, Phys. Rev. B *40*, 10963 (1989).

C. Y. Yang, A. R. Moodenbaugh, Y. L. Wang, Y. Xu, S. M. Heald, D. O. Welch, D. A. Fischer, and J. E. Penner-Hahn, Phys. Rev. B *42*, 2231 (1990).

I.-S. Yang, G. Burns, F. H. Dacol, and C. C. Tsuei, Phys. Rev. B *42*, 4240 (1990).

S. Yarlagadda and S. Kurihara, Phys. Rev. B *48*, 10567 (1993).

Z. Ye and S. Sachdev, Phys. Rev. B *44*, 10173 (1991).

Z. Ye, H. Umezawa, and R. Teshima, Phys. Rev. B *44*, 351 (1991).

W.-J. Yeh, L. Chen, F. Xu, B. Bi, and P. Yang, Phys. Rev. B *36*, 2414 (1987).

N.-C. Yeh and C. C. Tsuei, Phys. Rev. B *39*, 9708 (1989); N.-C. Yeh, Phys. Rev. B *40*, 4566 (1989).

J.-J. Yeh, L. Lindau, J.-Z. Sun, K. Char, M. Missert, A. Kapitulnik, T. H. Geballe, and M. R. Beasley, Phys. Rev. B *42*, 8044 (1990).

N.-C. Yeh, Phys. Rev. B *42*, 4850 (1990).

N.-C. Yeh, Phys. Rev. B *43*, 523 (1991).

J.-J. Yeh, Phys. Rev. B *45*, 10816 (1992).

N.-C. Yeh, W. Jiang, D. S. Reed, A. Gupta, F. Holtzberg, and A. Kussmaul, Phys. Rev. B *45*, 5710 (1992b).

N.-C. Yeh, D. S. Reed, W. Jinag, U. Kriplani, F. Holtzberg, A. Gupta, B. D. Hunt, R. P. Vasquez, M. C. Foote, and L. Bajuk, Phys. Rev. B *45*, 5654 (1992a).

Y. Yeshurun, A. P. Malozemoff, F. Holtzberg, and T. R. Dinger, Phys. Rev. B *38*, 11828 (1988).

S.-K. Yip, O. F. DeA. Bonfim, and P. Kumar, Phys. Rev. B *41*, 11214 (1990).

F. Yndurain and G. Martínez, Phys. Rev. B *43*, 3691 (1991).

M. Yoshimoto, H. Koinuma, T. Hashimoto, J. Tanaka, S. Tanabe, and N. Soga, Physica C *181*, 284 (1991).

M. Yoshimura, H. Shigekawa, H. Nejoh, G. Saito, Y. Saito, and A. Kawazu, Phys. Rev. B *43*, 13590 (1991).

J. Yu, A. J. Freeman, and S. Massidda, *in* "Novel Superconductivity" (S. A. Wolf and V. Z. Kresin, Eds.), p. 367, Plenum, New York, 1987.

R. C. Yu, M. J. Naughton, X. Yan, P. M. Chaikin, F. Holtzberg, R. L. Greene, J. Stuart, and P. Davies, Phys. Rev. B *37*, 7963 (1988).

G. Yu, C. H. Lee, A. J. Heegar, N. Herron, and E. M. McCarron, Phys. Rev. Lett. *67*, 2581 (1990).

J. Yu, A. J. Freeman, R. Podloucky, P. Herzig, and P. Weinberger, Phys. Rev. B *43*, 532 (1991).

X.-J. Yu and M. Sayer, Phys. Rev. B *44*, 2348 (1991).

R. C. Yu, J. M. Williams, H. H. Wang, J. E. Thompson, A. M. Kini, K. D. Carlson, J. Ren, M.-H. Whangbo, and P. M. Chaikin, Phys. Rev. B *44*, 6932 (1991).

W. Yu, E. B. Harris, S. E. Hebboul, J. C. Garland, and D. Stroud, Phys. Rev. B *45*, 12624 (1992).

B. D. Yu, H. Kim, and J. Ihm, Phys. Rev. B *45*, 8007 (1992).

G. Yu, C. H. Lee, A. J. Heeger, N. Herron, E. M. McCarron, L. Cong, G. C. Spalding, C. A. Nordman, and A. M. Goldman, Phys. Rev. B *45*, 4964 (1992).

R. C. Yu, M. B. Salamon, J. P. Lu, and W. C. Lee, Phys. Rev. Lett. *69*, 1431 (1992).

J. Yuan, L. M. Brown, W. Y. Liang, R. S. Liu, and P. P. Edwards, Phys. Rev. B *43*, 8030 (1991).

B. J. Yuan and J. P. Whitehead, Phys. Rev. B *44*, 6943 (1991).

K. Yvon and M. François, Z. Phys. B *76*, 413 (1989).

A. A. Zakhidov, A. Ugawa, K. Imaeda, K. Yakushi, H. Inokuchi, K. Kikuchi, I. Ikemoto, S. Suzuki, and Y. Achiba, Solid State Commun. *79*, 939 (1991).

M. Zeh, H.-C. Ri, F. Kober, R. P. Huebener, J. Fischer, R. Gross, H. Müller, T. Sermet, A. V. Ustinov, and H.-G. Wener, Physica C *167*, 6 (1990).

E. Zeldov, N. M. Amer, G. Koren, A. Gupta, R. J. Gambino, and M. W. McElfresh, Phys. Rev. Lett. *62*, 3093 (1989).

H. R. Zeller and I. Giaever, Phys. Rev. *181*, 789 (1969).

X. C. Zeng, D. Stroud, and J. S. Chung, Phys. Rev. B *43*, 3042 (1991).

R. Zeyher, Phys. Rev. B *44*, 10404 (1991).

F. C. Zhang and T. M. Rice, Phys. Rev. B *37*, 3759 (1988).

L. Zhang, M. Ma, and F. C. Zhang, Phys. Rev. B *42*, 7894 (1990).

J.-G. Zhang, X.-X. Bi, E. McRae, P. C. Eklund, B. C. Sales, and M. Mostoller, Phys. Rev. B *43*, 5389 (1991a).

Z. Zhang, C.-C. Chen, and C. M. Lieber, Science *254*, 1619 (1991).

H. Zhang, J. W. Lynn and D. E. Morris, Phys. Rev. B *45*, 10022 (1992).

L. Zhang, J. Z. Liu, and R. N. Shelton, Phys. Rev. B *45*, 4978 (1992).

L. Zhang, J. K. Jain, and V. J. Emery, Phys. Rev. B *47*, 3368 (1993).

Z. Zhang and C. M. Lieber, Phys. Rev. B *47*, 3423 (1993).

H. Zhang and H. Sato, Phys. Rev. Lett. *70*, 1697 (1993).

Z. Zhao, L. Chen, Q. Yang, Y. Huang, G. Chen, R. Tang, G. Liu, C. Cui, L. Chen, L. Wang, S. Guo, S. Li, and J. Bi, *in* "Cooper Oxide Superconductors" (C. P. Poole, Jr., T. Datta, and H. A. Farach, Eds.), p. 274, Wiley, New York, 1987.

Z. Zhao, F. Behrooze, S. Adenwalla, Y. Guan, J. B. Ketterson, B. K. Sarma, and D. G. Hinks, Phys. Rev. B *43*, 13720 (1991).

B.-R. Zhao, S.-I. Kuroumaru, Y. Horie, E. Yanada, T. Aomine, X.-G. Qiu, Y.-Z. Zhang, Y.-Y. Zhao, P. Xu, L. Li, H. Ohkubo, and S. Mase, Physica C *179*, 138 (1991).

G. L. Zhao and J. Callaway, Phys. Rev. B *49*, 6424 (1994).

C. Zhaojia, Z. Yong, Y. Hongshun, C. Zuyao, Z. Donquin, Q. Yitai, W. Baimei, and Z. Qirui, Solid State Commun. *64*, 685 (1987).

H. Zheng, M. Avignon, and K. H. Bennemann, Phys. Rev. B *49*, 9763 (1994).

X. Zhengping, J. Chunlin, and Z. Lian, J. Supercond. *3*, 421 (1990).

H. Zhou, J. Rammer, P. Schleger, W. N. Hardy, and J. F. Carolan, Phys. Rev. B *43*, 7968 (1991).

P. Zhou, K.-A. Wang, A. M. Rao, P. C. Eklund, G. Dresselhaus, and M. S. Dresselhaus, Phys. Rev. B *45*, 10838 (1992).

J. Zhou and S. G. Chen, Phys. Rev. B *47*, 8301 (1993).

D.-M. Zhu, A. C. Anderson, E. D. Bukowski, and D. M. Ginsberg, Phys. Rev. B *40*, 841 (1989).

D.-M. Zhu, A. C. Anderson, T. A. Friedmann, and D. M. Ginzberg, Phys. Rev. B *41*, 6605 (1990).

S. Zhu, D. K. Christen, C. E. Klabunde, J. R. Thompson, E. C. Jones, R. Feenstra, D. H. Lowndes, and D. P. Norton, Phys. Rev. B *46*, 5576 (1992).

Y. Zhu *in* "High Temperature Superconducting Materials Science and Engineering," (D. Shi, Ed.), Pergamon, New York, 1994.

G. T. Zimanyi and K. S. Bedell, Phys. Rev. B *48*, 6575 (1993).

P. Zolliker, D. E. Cox, J. B. Parise, E. M. McCarron III, and W. E. Farneth, Phys. Rev. B *42*, 6332 (1990).

X. Zotos, P. Prelovšek, and I. Sega, Phys. Rev. B *42*, 8445 (1990).

Z. Zou and P. W. Anderson, Phys. Rev. B *37*, 627 (1988); (reprinted in Halley, 1988, p. 163).

V. E. Zubkus, E. E. Tornau, S. Lapinskas, and P. J. Kundrotas, Phys. Rev. B *43*, 13112 (1991).

F. Zuo, M. B. Salamon, T. Datta, K. Ghiron, H. Duan, and A. M. Hermann, Physica C *176*, 541 (1991).

W. Zwerger, Phys. Rev. B *42*, 2566 (1990).

Appendix

Harshman and Mills (1992) published an excellent survey of experimental properties of superconductors, including the cuprates, other recently discovered varieties, and several classical types. Their results are presented in the following series of four tables: Table A.1, Measured Electronic Parameters, A.2, Measured Structural Parameters, A.3, Relevant Derived Quantities, and A.4, Dimensionless Ratios using Standard Notations. The article contains plots of the dependence of several tabulated parameters on the transition temperature and the interplanar spacing, and should be consulted for the many references.

The transition temperatures quoted are the maximum known values for the particular structure and symmetry. The penetration depths $\lambda(0) = \lambda_{ab}(0)$ were extracted from bulk μ^+SR measurements on phase-pure samples, when available. Upper critical fields $B_{c2}(0)$ were obtained from the expression

$$B_{c2}(0) \approx \frac{T_c(B=0)}{dT_c/dB}, \qquad (A.1)$$

where the derivative is determined at T_c. The characteristic critical field B_0 listed in Table A.3 is given by

$$B_0 = \frac{\phi_0}{\pi}\left(\frac{k_B T_c}{\hbar v_F}\right)^2 = 0.3725\left(\frac{m^*}{m_e}\right)\frac{T_c^2}{T_F}, \qquad (A.2)$$

and on the basis of the Ginzburg–Landau theory, it has the approximate value

$$B_0 = \frac{B_{c2}}{15}. \qquad (A.3)$$

Table A.1 Measured Electronic Parameters for Various Materials[abc]

No.	Compound	T_c (K)	ΔT_c (K)	α	$\rho(0)$; $\rho(T_c)$ ($\mu\Omega$ cm)	Coherence effects	$\Delta C/T_c$ (mJ/mol K^2)	$B_{1/2}$ (Tesla)	$\lambda(0)$ (Å)	$B_{c2}(0)$ (Tesla)
1.	$La_{1.9}Sr_{0.1}CuO_4$	33	6.5	0.4 ± 0.05	$\alpha T + c$	No	3.85*[b]	x	3200 ± 160	x
	$La_{1.075}Sr_{0.125}CuO_4$	36	4.5	x	$\alpha T + c$	x	5.3*	x	2700 ± 135	x
	$La_{1.85}Sr_{0.15}CuO_4$	39	2	0.140 ± 0.008	0; 33 ± 6	No	17.5 ± 2	x	2185 ± 100	45 ± 10
	$La_{1.775}Sr_{0.225}CuO_4$	29	7.5	0.1 ± 0.03	$\alpha T + c$	x	5.3*	x	x	x
	$La_{1.725}Sr_{0.275}CuO_4$	22	10	0.12 ± 0.05	$\alpha T + c$	x	1.75*	x	x	x
2.	$La_{1.6}Sr_{0.4}CaCu_2O_6$	60	~ 10	x	$\alpha T + c$	x	x	x	x	x
3.	$YBa_2Cu_3O_{6.67}$	60	1	x	x; 63 ± 8	No	11 ± 4	x	2550 ± 125	87
4.	$YBa_2Cu_3O_7$	92	1	0.018 ± 0.004	0; 40 ± 5	No	50 ± 10	3 ± 1	1415 ± 30	140 ± 30
	$HoBa_2Cu_3O_7$	92	1	x	x; x	x	55 ± 10	x	x	x
	$PrBa_2Cu_3O_7$	0	—	—	—	—	—	—	—	—
	$(Y_{0.95}Pr_{0.05})Ba_2Cu_3O_7$	90	1.8	x	50; 100	x	x	x	1500 ± 75	x
	$(Y_{0.9}Pr_{0.1})Ba_2Cu_3O_7$	86	2.2	0.04 ± 0.03	50; 100	x	x	x	1730 ± 85	x
	$(Y_{0.8}Pr_{0.2})Ba_2Cu_3O_7$	78	3.2	0.07 ± 0.03	100; 140	x	x	x	1980 ± 100	x
	$(Y_{0.7}Pr_{0.3})Ba_2Cu_3O_7$	60	7.5	0.22 ± 0.03	250; 280	x	x	x	2150 ± 100	x
	$(Y_{0.6}Pr_{0.4})Ba_2Cu_3O_7$	40	13.2	0.34 ± 0.03	300; 335	x	x	x	2900 ± 150	x
	$YBa_2(Cu_{0.99}Fe_{0.01})_3O_7$	91	1.3	x	x; x	x	46	x	x	x
	$YBa_2(Cu_{0.975}Fe_{0.025})_3O_7$	85	3	x	x; x	x	24	x	x	x
	$YBa_2(Cu_{0.95}Fe_{0.05})_3O_7$	75	5	x	x	x	< 10	x	x	x
	$YBa_2(Cu_{0.99}Zn_{0.01})_3O_7$	78	1	x	αT	x	32 ± 4	x	x	x
	$YBa_2(Cu_{0.975}Zn_{0.025})_3O_7$	62	1	x	150; 200	x	13	x	x	x
	$YBa_2(Cu_{0.95}Zn_{0.05})_3O_7$	36	1.5	x	240; 300	x	7.2	x	x	x
5.	$YBa_2Cu_4O_8$ (ambient press.)	80	2	x	$\alpha T + c$*	No	16 ± 2	x	1980 ± 100	x
	$YBa_2Cu_4O_8$ (10 GPa)	107	x	x	x	x	x	x	x	x
	$HoBa_2Cu_4O_8$ (ambient press.)	80	5	x	100; 400	x	~ 5*	x	1610 ± 80	x

6.	$Bi_2Sr_2CuO_{6\pm\delta}$	8.5	1	x	100; 109	x	x	x	x	x
	$Bi_2(Sr_{1.6}La_{0.4})CuO_{6+8}$	23.5	x	x	320; 360	x	x	x	x	x
	$(Bi, Pb)_2(Sr_{1.75}La_{0.25})CuO_6$	24	x	x	x; x	x	x	x	x	x
	$(Bi, Pb)_2(Sr_{1.8}Pr_{0.2})CuO_6$	15	x	x	x; x	x	x	x	x	x
	$(Bi, Pb)_2(Sr_{1.75}Nd_{0.25})CuO_6$	17	x	x	x; x	x	x	x	x	x
7.	$Bi_2Sr_2CaCu_2O_8$ (no-anneal)	89	1	x	0; 38 ± 2	No	32 ± 10*	x	2500 ± 500*	107
	$Bi_2Sr_2CaCu_2O_{8+\delta}$($O_2$ annealed)	75	1	0.037 ± 0.004	αT	No	x	x	x	x
8.	$Bi_2Sr_2Ca_2Cu_3O_{10}$	107	~ 10	x	0; < 60*	x	18 ± 5	x	x	x
	$(Bi_{1.6}Pb_{0.4})Sr_2Ca_2Cu_3O_{10}$	107	1	0.026 ± 0.002	0; x	No	32 ± 8	x	2525 ± 300	184
9.	$Tl_2Ba_2CuO_6$	85	x	x	40; 100*	No	x	x	1700 ± 100*	x
10.	$Tl_2Ba_2CaCu_2O_8$	99	x	x	0 ± 20; 300 ± 80	No	35 ± 10*	x	2210 ± 100	99
11.	$Tl_2Ca_2Ba_2Cu_3O_{10}$	125	x	x	20; 83	No	24 ± 8	x	1960 ± 100	75
12.	$(Tl_{0.7}Cd_{0.3})BaLaCuO_3$	48	x	x	x; x	x	x	x	x	x
13.	$(Tl_{0.5}Pb_{0.5})Sr_2CaCu_2O_7$	80	x	x	x; x	x	13 ± 2*	x	1816 ± 90*	x
	$(Tl_{0.5}Pb_{0.5})Sr_2(Ca_{0.8}Y_{0.2})Cu_2O_7$	107	x	x	0; 250	x	22 ± 2*	x	x	x
14.	$(Tl_{0.5}Pb_{0.5})Sr_2Ca_2Cu_3O_9$	122	x	x	$\alpha T + c$	x	x	x	1580 ± 80	x
15.	$Pb_2(Y_{1-x}Ca_x)Sr_2Cu_3O_8$	80	x	x	50; 250	No	x	x	x	x
16.	$(Nd_{2-x}Ce_x)CuO_4$	24	2	≤ 0.05	250; 250*	No	~ 1.7	x	x	x
17.	κ–$[BEDT–TTF]_2Cu[NCS]_2$	10.5	x	x	10 ± 150*; 330 ± 50	No	48 ± 5	0.2 ± 0.1	7500 ± 1000	~ 10*
18.	$[TMTSF]_2ClO_4$	1.2	0.1	x	nonlinear	No	15 ± 2	0.02 ± 0.005	12000 ± 2000	0.10 ± 0.02*
19.	$A_3C_{60}(K_3C_{60}, \ldots, Rb_2CsC_{60})$	31.3	1	0.30 ± 0.06*	~ const.*	No	x	x	< 5100*	~ 30*
20.	KC_8	0.55	0.05	x	x	x	x; x	x	x	x
21.	$TaS_2(Py)_{1/2}$	3.4	x	x	const.	x	36 ± 8	x	2350 ± 250	0.214 ± 0.025
22.	$Ba_{0.6}K_{0.4}BiO_3$	32	8	0.42 ± 0.05	2000; 2000	x	2.2 ± 0.4	~ 6	3450 ± 200	17
23.	$BaPb_{0.75}Bi_{0.25}O_3$	11	x	0.22 ± 0.03	340; 340	x	2.3 ± 0.3	x	10000 ± 500	7 ± 1.5
24.	$PbMo_6S_8$ (Chevrel)	12	0.5	0.27 ± 0.04*	50; 90	x	10 ± 2	9 ± 1	2700 ± 400	55
25.	UPt_3	0.53	x	x	0.2; 0.5	No*	300 ± 50	~ 0.6	6960 ± 37	2.1
26.	Nb_3Sn	17.9	0.2	0.08 ± 0.2*	8.8; 8.8	x	25 ± 3	8 ± 1	640 ± 30*	37

[a] From Harshman and Mills (1992). [b] See original article for a discussion of the items denoted by an asterisk. [c] The columns give the transition temperature T_c, the transition width (10%–90%) ΔT_c, the isotope effect coefficient α, the resistivity $\rho(0)$ extrapolated to 0 K and at the transition temperature $\rho(T_c)$, the presence of coherence effects (the BCS theory predicts coherence effects having to do with the electron–lattice interaction), the specific heat anomaly $\Delta C/T_c$, the field $B_{1/2}$ at which $\Delta C/T_c$ decreases to half of its zero field value, the magnetic penetration depth $\lambda(0) = \lambda_{ab}(0)$ and the upper critical field $B_{c2}(0)$. Compounds 1–16 are high temperature superconductors, and those 17–26 are miscellaneous other types.

Table A.2 Measured Unit Cell Dimensions, Volume V_m Per Formula Unit and Density, and Assumed Values for the Average Plane Spacing δ and Interlayer Spacings d. (Other Possible Values of δ and d Are Given in Square Brackets)[a]

No.	Compound	a (Å)	b (Å)	c (Å)	δ (Å)	d (Å)	Space group	V_m (Å^3/f.u.)	Density ($g\ cm^{-3}$)
1.	$La_{1.9}Sr_{0.1}CuO_4$	3.7839(8)	3.7839(8)	13.211(4)	6.61	6.61	I4/mmm	94.6	7.02
	$La_{1.875}Sr_{0.125}CuO_4$	3.7784(8)	3.7784(8)	13.216(4)	6.61	6.61	I4/mmm	94.4	7.02
	$La_{1.85}Sr_{0.15}CuO_4$	3.7793(1)	3.7793(1)	13.2260(3)	6.62	6.62	I4/mmm	95.0	7.00
	$La_{1.775}Sr_{0.225}CuO_4$	3.7708(8)	3.7708(8)	13.247(3)	6.62	6.62	I4/mmm	94.2	6.94
	$La_{1.775}Sr_{0.225}CuO_4$	3.7666(8)	3.7666(8)	13.225(3)	6.61	6.61	I4/mmm	93.8	6.92
2.	$La_{1.6}Sr_{0.4}CaCu_2O_6$	3.8208(1)	3.8208(1)	19.5993(7)	4.9	3.39	I4/mmm	143.1	6.04
3.	$YBa_2Cu_3O_{6.67}$	3.831(2)	3.889(2)	11.736(6)	5.87	11.74[5.9?]	Pmmm	174.9	6.28
4.	$YBa_2Cu_3O_7$	3.8198(1)	3.8849(1)	11.6762(3)	5.84	3.36	Pmmm	173.3	6.39
	$HoBa_2Cu_3O_7$	3.846(1)	3.881(1)	11.640(2)	5.82	3.31	Pmmm	173.7	7.09
	$PrBa_2Cu_3O_7$	3.905(2)	3.905(2)	11.660(10)	5.83	3.47	Pmmm	177.8	6.71
	$(Y_{0.25}Pr_{0.05})Ba_2Cu_3O_7$	x	x	x	x	≈ 3.36	Pmmm	x	x
	$(Y_{0.9}Pr_{0.1})Ba_2Cu_3O_7$	x	x	x	x	≈ 3.36	Pmmm	x	x
	$(Y_{0.8}Pr_{0.2})Ba_2Cu_3O_7$	x	x	x	x	≈ 3.36	Pmmm	x	x
	$(Y_{0.7}Pr_{0.3})Ba_2Cu_3O_7$	x	x	x	x	≈ 3.36	Pmmm	x	x
	$(Y_{0.6}Pr_{0.4})Ba_2Cu_3O_7$	x	x	x	x	≈ 3.36	Pmmm	x	x
	$YBa_2(Cu_{0.99}Fe_{0.01})_3O_7$	x	x	x	x	≈ 3.36	Pmmm	x	x
	$YBa_2(Cu_{0.975}Fe_{0.025})_3O_7$	x	x	x	x	≈ 3.36	Pmmm	x	x
	$YBa_2(Cu_{0.95}Fe_{0.05})_3O_7$	x	x	x	x	≈ 3.36	Pmmm	x	x
	$YBa_2(Cu_{0.99}Zn_{0.01})_3O_7$	x	x	x	x	≈ 3.36	Pmmm	x	x
	$YBa_2(Cu_{0.975}Zn_{0.025})_3O_7$	3.820(2)	3.890(2)	11.673(6)	5.84	≈ 3.36	Pmmm	173.5	6.39
	$YBa_2(Cu_{0.95}Zn_{0.05})_3O_7$	3.820(2)	3.885(2)	11.671(6)	5.84	≈ 3.36	Pmmm	173.2	6.39
5.	$YBa_2Cu_4O_8$ (ambient press.)	3.86(1)	3.86(1)	27.24(6)	6.81	3.38	Ammm	202.9	6.10
	$YBa_2Cu_4O_8$ (10 GPa)	3.79	3.79	26.75	6.69	3.32	Ammm	192.1	6.45
	$HoBa_2Cu_4O_8$ (ambient press.)	3.855(1)	3.874(1)	27.295(11)	6.82	~ 3.3	Ammm	203.8	6.70

6.	$Bi_2Sr_2CuO_{6\pm\delta}$	5.361(2)	5.370(1)	24.369(6)	12.15	12.15	Cmmm	175.4	7.13
	$Bi_2(Sr_{1.6}La_{0.4})CuO_{6+\delta}$	5.370(5)	5.400(5)	24.50(2)	12.25	12.25	Cmmm	177.6	7.23
	$(Bi, Pb)_2(Sr_{1.25}La_{0.25})CuO_6$	5.282(5)	5.410(5)	24.62(1)	12.31	12.31	Cmmm	175.9	7.23
	$(Bi, Pb)_2(Sr_{1.8}Pr_{0.2})CuO_6$	5.264(5)	5.412(5)	24.27(1)	12.14	12.14	Cmmm	172.9	7.33
	$(Bi, Pb)_2(Sr_{1.75}Nd_{0.25})CuO_6$	5.249(5)	5.419(5)	24.24(1)	12.12	12.12	Cmmm	172.4	7.39
7.	$Bi_2Sr_2CaCu_2O_8$ (no-anneal)	5.413(2)	5.411(2)	30.91(1)	7.73	3.35	Fmmm	226.4	6.52
	$Bi_2Sr_2CaCu_2O_{8+\delta}$ (O_2 annealed)	5.408(2)	5.413(2)	30.81(1)	7.70	3.35	Fmmm	225.5	6.54
8.	$Bi_2Sr_2Cu_3O_{10}$	5.39	5.39	37.1	9.27	6.70 [3.35?]	Fmmm	269.5	6.26
	$(Bi_{1.6}Pb_{0.4})Sr_2Ca_2Cu_3O_{10}$	5.413(3)	5.413(3)	37.100(12)	9.27	6.70 [3.35?]	Fmmm	271.8	6.31
9.	$Tl_2Ba_2CuO_6$	3.866(1)	3.866(1)	23.239(6)	11.60 [5.8?]	11.60 [5.8?]	I4/mmm	173.7	8.06
10.	$Tl_2Ba_2CaCu_2O_8$	3.8550(6)	3.8550(6)	29.318(4)	7.33	3.2	I4/mmm	217.8	7.46
11.	$Tl_2Ca_2Ba_2Cu_3O_{10}$	3.8503(6)	3.8503(6)	35.88(3)	8.97	6.42 [3.21?]	I4/mmm	266.6	6.96
12.	$(Tl_{0.7}Cd_{0.3})BaLaCuO_5$	3.844(2)	3.844(2)	9.16(1)	9.16[4.6?]	9.16 [4.6?]	P4/mmm	135.4	7.32
13.	$(Tl_{0.5}Pb_{0.5})Sr_2CaCu_2O_7$	3.8023(3)	3.8023(3)	12.107(1)	6.05	3.24	P4/mmm	174.0	6.30
	$(Tl_{0.5}Pb_{0.5})Sr_2(Ca_{0.8}Y_{0.2})Cu_2O_7$	3.8075(3)	3.8075(3)	12.014(2)	6.05	3.24	P4/mmm	174.2	6.30
14.	$(Tl_{0.5}Pb_{0.5})Sr_2Ca_2Cu_3O_8$	3.8206(2)	3.8206(2)	15.294(1)	7.65	6.46[3.23?]	P4/mmm	220.9	5.98
15.	$Pb_2(Y_{1-x}Ca_x)Sr_2Cu_3O_4$	5.3933(2)	5.4311(2)	15.7334(6)	7.87	3.45	Cmmm	230.4	7.20
16.	$(Nd_{2-x}Ce_x)CuO_4$	3.9469(2)	3.9469(2)	12.0776(5)	6.04	6.04	I4/mmm	188.2	6.72
17.	κ–$[BEDT–TTF]_2Cu[NCS]_2$	16.248	8.440	13.124	15.24	15.24	$P2_1$	844.0	1.87
18.	$[TMTSF]_2ClO_4$	7.266(1)	7.678(1)	13.275(2)	13.275	13.275	$P\bar{1}$	694.3	2.38
19.	$A_3C_{60}(K_3C_{60}, \ldots, Rb_2 2CsC_{60})$	14.436(2)	14.436(2)	14.436(2)	(14.44)	(14.44)	$Fm\bar{3}m$	752.1	2.11
20.	KC_8	4.961	8.592	23.76	5.94	5.94	Fdd2	126.6	1.77
21.	$TaS_2(Py)_{1/2}$	3.326	3.326	12.02	12.02	12.02	$P\bar{3}m1$	133.0	3.63
22.	$Ba_{0.6}K_{0.4}BiO_3$	4.287(6)	4.287(6)	4.287(6)	(4.29)	(4.29)	$Pm\bar{3}m$	79.0	7.47
23.	$BaPb_{0.75}Bi_{0.25}O_3$	6.0496(1)	6.0696(1)	8.6210(2)	(8.62)	(8.62)	I4/mcm	79.2	8.24
24.	$PbMo_6S_8$ (Chevrel)	6.5759(9)	6.5383(9)	6.4948(8)	(6.49)	(6.49)	$R\bar{3}/P\bar{1}$	279.2	6.18
25.	UPt_3	5.754	5.754	4.890	(4.89)	(4.89)	$P6_3/mmc$	81.0	16.88
26.	Nb_3Sn	5.289(2)	5.289(2)	5.289(2)	(5.29)	(5.29)	$Pm\bar{3}m$	74.0	8.92

[a] See discussion in the original article by Harshman and Mills (1992).

Table A.3 Relevant Derived Quantities Calculated from the Information in Tables A.1 and A.2[a]

No.	Compound	n_{3D} (10^{21} cm^{-3})	n_{2D} (10^{14} cm^{-2})	m^*/m_0	E_F^{3D}/k_B (K)	E_F^{2D}/k_B (K)	$R(0)$; $R(T_c)$ (Ω)	B_0 (T)
1.	$La_{1.9}Sr_{0.1}CuO_4$	~ 0.5	~ 0.4	~ 2.0	—	510 ± 50	x	x
	$La_{1.875}Sr_{0.125}CuO_4$	~ 1.0	~ 0.6	~ 2.6	—	710 ± 70	x	x
	$La_{1.85}Sr_{0.15}CuO_4$	5.2 ± 0.8	3.4 ± 0.5	8.6 ± 1.0	—	1090 ± 100	0; 500 ± 90	4.6 ± 0.6
	$La_{1.775}Sr_{0.225}CuO_4$	x	x	~ 2.6	—	x	x	x
	$La_{1.725}Sr_{0.175}CuO_4$	x	x	~ 0.9	—	x	x	x
2.	$La_{1.6}Sr_{0.4}CaCu_2O_6$	x	x	x	—	x	x	x
3.	$YBa_2Cu_3O_{6.67}$	1.1 ± 0.4	1.3 ± 0.5	2.6 ± 1.0	—	710 ± 70	x; 535 ± 70	2.5 ± 1.0
4.	$YBa_2Cu_3O_7$	16.9 ± 3.4	9.9 ± 2.0	12.0 ± 2.4	—	2290 ± 100	0; 685 ± 85	16.5 ± 3.4
	$HoBa_2Cu_3O_7$	x	x	x	—	x	x; x	x
	$PrBa_2Cu_3O_7$	x	x	x	—	x	x; x	x
	$(Y_{0.95}Pr_{0.05})Ba_2Cu_3O_7$	x	x	x	—	2040 ± 200	855; 1700	x
	$(Y_{0.9}Pr_{0.1})Ba_2Cu_3O_7$	x	x	x	—	1530 ± 150	855; 1700	x
	$(Y_{0.8}Pr_{0.2}Ba_2Cu_3O_7$	x	x	x	—	1170 ± 120	1700; 2400	x
	$(Y_{0.7}Pr_{0.3})Ba_2Cu_3O_7$	x	x	x	—	990 ± 90	4300; 4800	x
	$(Y_{0.6}Pr_{0.4})Ba_2Cu_3O_7$	x	x	x	—	550 ± 60	5100; 5700	x
	$YBa_2(Cu_{0.99}Fe_{0.01})_3O_7$	x	x	11.0	—	x	x; x	x
	$YBa_2(Cu_{0.975}Fe_{0.025})_3O_7$	x	x	5.7	—	x	x; x	x
	$YBa_2(Cu_{0.95}Fe_{0.05})_3O_7$	x	x	2.4	—	x	x; x	x
	$YBa_2(Cu_{0.99}Zn_{0.01})_3O_7$	x	x	7.7 ± 1.0	—	x	x; x	x
	$YBa_2(Cu_{0.975}Zn_{0.025})_3O_7$	x	x	3.1	—	x	2570; 3420	x
	$YBa_2(Cu_{0.95}Zn_{0.05})_3O_7$	x	x	x	—	x	4110; 5140	x
5.	$YBa_2Cu_4O_8$ (ambient press.)	2.8 ± 0.44	1.9 ± 0.3	3.8 ± 0.5	—	1360 ± 140	x; x	6.7 ± 1.1
	$YBa_2Cu_4O_8$ (10 GPa)	x	x	x	—	x	x; x	x
	$HoBa_2Cu_4O_8$ (ambient press.)	~ 1.3	~ 0.9	~ 1.2	—	2070 ± 200	1470; 5870	~ 1.4

6.	$Bi_2Sr_2CuO_{6\pm\delta}$	x	x	x	—	x	825; 900	x
	$Bi_2(Sr_{1.6}La_{0.4})CuO_{6+\delta}$	1.17 ± 0.33	1.44 ± 0.41	x	—	x	2610; 2940	x
	$(Bi, Pb)_2(Sr_{1.75}La_{0.25})CuO_6$	0.69 ± 0.14	0.84 ± 0.17	x	—	x	x	x
	$(Bi, Pb)_2(Sr_{1.8}Pr_{0.2})CuO_6$	0.57 ± 0.11	0.70 ± 0.14	x	—	x	x	x
	$(Bi, Pb)_2(Sr_{1.75}Nd_{0.25})CuO_6$	0.75 ± 0.15	0.92 ± 0.18	x	—	x	x	x
7.	$Bi_2Sr_2CaCu_2O_8$ (no-anneal)	3.5 ± 1.8	2.7 ± 1.4	7.8 ± 2.4	—	970 ± 390	0; 490 ± 26	24 ± 12
	$Bi_2Sr_2CaCu_2O_{8+\delta}(O_2$ annealed)	x	x	x	—	x	x; x	x
8.	$Bi_2Sr_2Ca_2Cu_3O_{10}$	x	x	4.4 ± 1.2	—	x	0; < 650	x
	$(Bi_{1.6}Pb_{0.4})Sr_2Ca_2Cu_3O_{10}$	3.4 ± 1.2	3.2 ± 1.1	7.8 ± 1.9	—	1140 ± 270	0; x	29 ± 10
9.	$Tl_2Ba2CuO_6$	x	x	x	—	3150 ± 370	345; 860	x
10.	$Tl_2Ba_2CaCu_2O_8$	4.9 ± 1.5	3.6 ± 1.1	8.4 ± 2.4	—	1180 ± 110	0 ± 20; 4100 ± 110	26 ± 8
11.	$Tl_2Ca_2Ba_2Cu_3O_{10}$	4.2 ± 1.5	3.8 ± 1.3	5.7 ± 1.9	—	1830 ± 190	220; 925	18 ± 6
12.	$(Tl_{0.7}Cd_{0.3})BaLaCuO_5$	x	x	x	—	x	x	x
13.	$(Tl_{0.5}Pb_{0.5})Sr_2CaCu_2O_7$	2.8 ± 0.5	1.7 ± 0.3	3.2 ± 0.5	—	1440 ± 140	x; x	5.3 ± 1.0
	$(Tl_{0.5}Pb_{0.5})Sr_2(Ca_{0.8}Y_{0.2})Cu_2O_7$	x	x	5.5 ± 0.5	—	x	0; 4130	x
14.	$(Tl_{0.5}Pb_{0.5})Sr_2Ca_2Cu_3O_9$	x	x	x	—	2410 ± 240	x; x	x
15.	$(Pb_2(Y_{1-x}Ca_x)Sr_2Cu_3O_8$	x	x	x	—	x	635; 3180	x
16.	$(Nd_{2-x}Ce_x)CuO_4$	x	x	~ 0.4*	—	x	4140; 4140	x
17.	κ–[BEDT–TTF]$_2$Cu[NCS]$_2$	0.31 ± 0.09	0.47 ± 0.13	6.2 ± 0.6	—	213 ± 57	~ 0; 2160 ± 330	1.2 ± 0.3
18.	[TMTSF]$_2$ClO$_4$	0.38 ± 0.17	0.5 ± 0.22	20.0 ± 5.0	—	72 ± 24	x; x	0.15 ± 0.07
19.	$A_3C_{60}(K_3C_{60}, \ldots, Rb_2CsC_{60})$	x	x	x	x	—	x; x	x
20.	KC_8	x	x	x	—	x	x; x	x
21.	$TaS_2(Py)_{1/2}$	12 ± 3.6	14.2 ± 4.4	23.0 ± 5.0	—	1710 ± 360	x; x	0.058 ± 0.018
22.	$Ba_{0.6}K_{0.4}BiO_3$	0.57 ± 0.09	(0.24 ± 0.04)	2.4 ± 0.3	1210 ± 120	—	(47k; 47k)	0.76 ± 0.16
23.	$BaPb_{0.75}Bi_{0.25}O_3$	0.12 ± 0.015	(0.10 ± 0.01)	4.2 ± 0.4	242 ± 20	—	(3940; 3940)	0.79 ± 0.13
24.	$PbMo_6S_8$ (Chevrel)	1.0 ± 0.27	(0.65 ± 0.18)	2.6 ± 0.4	1640 ± 370	—	(770; 1390)	0.085 ± 0.030
25.	UPt_3	7.8 ± 1.0	(3.83 ± 0.49)	135.0 ± 17.0	124 ± 5	—	(4.1; 10)	0.11 ± 0.02
26.	Nb_3Sn	47 ± 5	(24.7 ± 2.8)	6.8 ± 0.6	8100 ± 620	—	(170; 170)	0.10 ± 0.02

[a] $YBa_2Cu_3O_{6.67}$ (point no. 3), n_{3D}, n_{2D} and the sheet resistances are calculated for double layers, while the value for E_F^{2D} is calculated for a single sheet. [b] The table lists the superfluid (single particle), three dimensional carrier density n_{3D}, the two-dimensional carrier density, $n_{2D} = n_{3D}\delta$, the ratio of the average two-dimensional effective mass $m^* = m^*_{ab}$, in the plane of the conducting sheets to the free electron mass m_e, the 3D and 2D Fermi temperatures $T_F = E_F/k_B$, the sheet resistance $R(T) = \rho(T)\delta$, and a characteristic magnetic field B_0 given by Eq. A.2.

Table A.4 Dimensionless Quantities Derived from Tables A.2 and A.3[ab]

No.	Compound	$n_{2D}d^2$	$m^*d/m_e a_0$	$E_F/k_B T_c$	$B_0/B_{1/2}$	$R(T_c)/R_K$	$m^*d\,R_k/m_e a_0 R(T_c)$
1.	$La_{1.9}Sr_{0.1}CuO_4$	x	x	15.5 ± 1.5	x	x	x
	$La_{1.875}SR_{0.125}CuO_4$	x	x	19.7 ± 2.0	x	x	x
	$La_{1.85}Sr_{0.15}CuO_4$	1.48 ± 0.22	108 ± 12	27.9 ± 2.6	x	0.019 ± 0.04	5680 ± 1360
	$La_{1.775}Sr_{0.225}CuO_4$	x	x	x	x	x	x
	$La_{1.725}Sr_{0.275}Cu)_4$	x	x	x	x	x	x
2.	$La_{1.6}Sr_{0.4}CaCu_2O_6$	x	x	x	x	x	x
3.	$YBa_2Cu_3O_{6.67}$	1.83 ± 0.69	58 ± 22	11.8 ± 1.2	x	0.021 ± 0.003	2760 ± 1120
4.	$YBa_2Cu_3O_7$	1.11 ± 0.23	76 ± 15	24.9 ± 1.1	5.5 ± 2.2	0.027 ± 0.003	2815 ± 640
	$HoBa_2Cu_3O_7$	x	x	x	x	x	x
	$PrBa_2Cu_3O_7$	x	x	x	x	x	x
	$(Y_{0.95}Pr_{0.05})Ba_2Cu_3O_7$	x	x	22.1 ± 2.2	x	x	x
	$(Y_{0.9}Pr_{0.1})Ba_2Cu_3O_7$	x	x	17.8 ± 1.7	x	x	x
	$(Y_{0.8}Pr_{0.2})Ba_2Cu_3O_7$	x	x	15.0 ± 1.5	x	x	x
	$(Y_{0.7}Pr_{0.3})Ba_2Cu_3O_7$	x	x	16.5 ± 1.5	x	x	x
	$(Y_{0.6}Pr_{0.4})Ba_2Cu_3O_7$	x	x	13.8 ± 1.5	x	x	x
	$YBa_2(Cu_{0.99}Fe_{0.01})_3O_7$	x	70	x	x	x	x
	$YBa_2(Cu_{0.975}Fe_{0.025})_3O_7$	x	36	x	x	x	x
	$YBa_2(Cu_{0.95}Fe_{0.05})_3O_7$	x	15	x	x	x	x
	$YBa_2(Cu_{0.99}Zn_{0.01})_3O_7$	x	49 ± 6	x	x	x	x
	$YBa_2(Cu_{0.975}Zn_{0.025})_3O_7$	x	20	x	x	x	x
	$YBa_2(Cu_{0.95}Zn_{0.05})_3O_7$	x	11	x	x	x	x
5.	$YBa_2Cu_4O_8$ (ambient press.)	0.21 ± 0.03	24 ± 3	17.0 ± 1.8	x	x	x
	$YBa_2Cu_4O_8$ (10 GPa)	x	x	x	x	x	x
	$HoBa_2Cu_4O_8$ (ambient press.)	~ 0.1	~ 7	25.8 ± 2.5	x	~ 0.23	~ 30

No.	Compound						
6.	$Bi_2Sr_2CuO_{6\pm\delta}$	x	x	x	x	x	x
	$Bi_2(Sr_{1.6}La_{0.4})CuO_{6+\delta}$	2.16 ± 0.61	x	x	x	x	x
	$(Bi, Pb)_2(Sr_{1.75}La_{0.25})CuO_6$	1.27 ± 0.25	x	x	x	x	x
	$(Bi, Pb)_2(Sr_{1.8}Pr_{0.2})CuO_6$	1.03 ± 0.20	x	x	x	x	x
	$(Bi, PB)_2(Sr_{1.75}Nd_{0.25})CuO_6$	1.35 ± 0.26	x	x	x	x	x
7.	$Bi_2Sr_2CaCu_2O_8$ (no-anneal)	0.27 ± 0.14	49 ± 15	10.9 ± 4.4	x	0.019 ± 0.001	2580 ± 800
	$Bi_2Sr_2CaCu_2O_{8+\delta}$ (O_2 annealed)	x	x	x	x	x	x
8.	$Bi_2Sr_2CaCu_3O_{10}$	x	55 ± 15	x	x	< 0.025	< 2200
	$(Bi_{1.6}Pb_{0.4})Sr_2Ca_2Cu_3O_{10}$	1.44 ± 0.50	98 ± 25	10.7 ± 2.5	x	x	x
9.	$Tl_2Ba_2CuO_6$	x	x	37.1 ± 4.4	x	0.033	x
10.	$Tl_2Ba_2CaCu_2O_8$	0.36 ± 0.11	51 ± 15	11.9 ± 1.1	x	0.16 ± 0.04	320 ± 120
11.	$Tl_2Ca_2Ba_2Cu_3O_{10}$	1.56 ± 0.54	70 ± 23	14.6 ± 1.5	x	0.036	1940
12.	$(Tl_{0.7}Cd_{0.3})BaLaCuO_5$	x	x	x	x	x	x
13.	$(Tl_{0.5}Pb_{0.5})Sr_2CaCu_2O_7$	0.17 ± 0.03	20 ± 3	18.0 ± 1.8	x	x	x
	$(Tl_{0.5}Pb_{0.5})Sr_2(Ca_{0.8}Y_{0.2})Cu_2O_7$	x	34 ± 3	x	x	0.16	x
14.	$(Tl_{0.5}Pb_{0.5})Sr_2Ca_2Cu_3O_9$	x	x	19.7 ± 2.0	x	x	x
15.	$Pb_2(Y_{1-x}Ca\ x)Sr_2Cu_3O_8$	x	x	x	x	0.12	x
16.	$(Nd_{2-x}Ce\ x)CuO_4$	x	~ 4	x	x	0.16	~ 25
17.	κ–$[BEDT–TTF]_2Cu[NCS]_2$	1.10 ± 0.31	177 ± 118	20.2 ± 5.4	6.0 ± 3.5	0.084 ± 0.013	2110 ± 390
18.	$[TMTSF]_2ClO_4$	0.88 ± 0.39	500 ± 125	60 ± 20	7.5 ± 3.8	x	x
19.	$A_3C_{60}(K_3C_{60}, \ldots, Rb_2CsC_{60})$	x	x	x	x	x	x
20.	KC_8	x	x	x	x	x	x
21.	$TaS_2(Py)_{1/2}$	20.5 ± 6.3	525 ± 117	500 ± 105	x	x	x
22.	$Ba_{0.6}K_{0.4}BiO_3$	(0.044 ± 0.007)	(19.6 ± 2.9)	37.8 ± 3.7	~ 0.13	(1.8)	(10.9)
23.	$BaPb_{0.75}Bi_{0.25}O_3$	(0.074 ± 0.01)	(69.3 ± 7.7)	22 ± 1.8	x	(0.15)	(460)
24.	$PbMo_6S_8$ (Chevrel)	(0.27 ± 0.08)	(31.7 ± 6.6)	137 ± 31	0.01 ± 0.003	(0.056)	(587)
25.	UPt_3	(1.87 ± 0.24)	(1243 ± 155)	234 ± 10	~ 0.18	(0.0004)	(3108k)
26.	Nb_3Sn	(13.1 ± 1.5)	(67.7 ± 6.9)	453 ± 35	0.013 ± 0.002	(0.007)	(9670)

[a] $YBa_2Cu_3O_{6.67}$, $n_{2D}d^2$ is calculated for a double layer, while E_F/k_BT_c is derived for a single conducting sheet. [b] The Bohr radius $a_0 = 0.52918$ Å and the Klitzing quantum Hall-effective resistance $R_k = h/e^2 = 25812.8\ \Omega$.

Index

Main topics are indicated by page numbers in boldface. Many chemical formulae are abbreviated, such as YBaCuO.

D

E

I

J

K

L

P

S

T